INDUSTRIAL INORGANIC CHEMICALS AND PRODUCTS

AN ULLMANN'S ENCYCLOPEDIA

VOLUME **2**

WILEY-VCH

Weinheim New York Chichester Brisbane Singapore Toronto

AN ULLMANN'S ENCYCLOPEDIA

INDUSTRIAL INORGANIC CHEMICALS AND PRODUCTS

VOLUME 1
Aluminum Compounds, Inorganic to **Carbon Monoxide**

VOLUME 2
Cement and Concrete to **Cyano Compounds, Inorganic**

VOLUME 3
Fertilizers to **Hydrogen**

VOLUME 4
Hydrogen Peroxide to **Phosgene**

VOLUME 5
Phosphate Fertilizers to **Sodium Carbonates**

VOLUME 6
Sodium Chloride to **Zirconium Compounds**

Index

AN ULLMANN'S ENCYCLOPEDIA

INDUSTRIAL INORGANIC CHEMICALS AND PRODUCTS

VOLUME 2

**Cement and Concrete
to Cyano Compounds, Inorganic**

WILEY-VCH

Weinheim New York Chichester Brisbane Singapore Toronto

> This book was carefully produced. Nevertheless, authors and publisher do not warrant the information contained therein to be free of errors. Readers are advised to keep in mind that statements, data, illustrations, procedural details or other items may inadvertently be inaccurate.

Library of Congress Card No.: Applied for.
British Library Cataloguing-in-Publication Data: A catalogue record for this book is available from the British Library.

Die Deutsche Bibliothek – CIP-Einheitsaufnahme
Industrial Inorganic Chemicals and Products. – Weinheim ; New York ; Chichester ; Brisbane ; Singapore ; Toronto : Wiley-VCH
  ISBN 3-527-29567-4
Vol. 2. Cement and Concrete to Cyano Compounds, Inorganic. – 1998.

© WILEY-VCH Verlag GmbH, D-69469 Weinheim (Federal Republic of Germany), 1999
Printed on acid-free and chlorine-free paper.
All rights reserved (including those of translation in other languages). No part of this book may be reproduced in any form – by photoprinting, microfilm, or any other means – nor transmitted or translated into machine language without written permission from the publishers. Registered names, trademarks, etc. used in this book, even when not specifically marked as such, are not to be considered unprotected by law.

Composition, Printing and Bookbinding: Rombach GmbH, Druck- und Verlagshaus, D-79115 Freiburg.
Cover design: mmad, Michel Meyer, D-69469 Weinheim.
Printed in the Federal Republic of Germany.

Contents

1 Cement and Concrete

1. Cement 852
2. Concrete 894
3. References 924

2 Cements, Chemically Resistant

1. Introduction 931
2. Types of Cement 932
3. Specifications and Testing 938
4. Storage and Transportation 941
5. Toxicology 941
6. References 941

3 Ceramics, General Survey

1. Traditional and Advanced Ceramics 944
2. Raw Materials for Traditional Ceramics 956
3. Raw Materials for Advanced Ceramics 968
4. Processing Ceramic Ware 970
5. Glazes and Glazing 990
6. Glass 994
7. Refractories 996
8. Abrasives 996
9. Cement 997
10. Properties of Ceramic Materials and Products 997
11. Testing Ceramic Raw Materials and Products 998
12. Economic Aspects 1003
13. References 1005

4 Ceramics, Advanced Structural Products

1. Introduction 1011
2. Mechanical Properties 1013
3. Uses 1018
4. Specific Materials 1020
5. References 1026

5 Ceramics, Ceramic – Metal Systems

1. Introduction 1029
2. Fundamentals of Bonding 1030
3. Glass-to-Metal Joining 1035
4. Ceramic-to-Metal Joining 1044
5. Thin Films and Coatings 1049
6. Ceramic – Metal Composites 1060
7. References 1066

6 Ceramic Colorants

1. Introduction 1069
2. Classification 1070
3. Pigment Structures 1070
4. Colorant Systems 1072
5. Industrial Production of Ceramic Pigments 1082
6. Use of Ceramic Pigments in Glazes . 1083
7. Ceramic Glazes 1083
8. Application Media 1084
9. Quality Control 1084
10. References 1085

V

7 Ceramics, Electronic

1. Introduction 1087
2. Linear Dielectrics 1087
3. Nonlinear Dielectrics 1091
4. Semiconducting Ceramics 1102
5. Sensors 1105
6. References 1106

8 Cerium Compounds

1. Production 1111
2. Uses 1112
3. Analysis 1114
4. Economic Aspects 1114
5. Toxicology and Occupational Health 1115
6. References 1115

9 Cesium Compounds

1. Individual Compounds 1117
2. Resources 1117
3. Production 1118
4. Chemical Analysis 1119
5. Storage and Transportation 1119
6. Uses 1119
7. Economic Aspects 1120
8. Toxicology and Occupational Health 1120
9. References 1121

10 Chlorine

1. Introduction 1124
2. Physical Properties 1126
3. Chemical Properties 1130
4. Chlor-Alkali Process 1134
5. Mercury Cell Process 1141
6. Diaphragm Process 1162
7. Membrane Process 1185
8. Electrodes 1205
9. Comparison of the Processes ... 1214
10. Other Production Processes ... 1220
11. Chlorine Purification and Liquefaction 1225
12. Chlorine Handling 1232
13. Quality Specifications and Analytical Methods 1241
14. Uses 1242
15. Economic Aspects 1244
16. Toxicology 1247
17. References 1248

11 Chlorine Oxides and Chlorine Oxygen Acids

1. Introduction 1258
2. Hypochlorous Acid 1262
3. Solid Hypochlorites 1263
4. Hypochlorite Solutions 1265
5. Chlorine Dioxide 1278
6. Sodium Chlorite 1285
7. Chloric Acid and Chlorates ... 1287
8. Perchloric Acid and Perchlorates .. 1308
9. Toxicology and Occupational Health 1318
10. References 1319

12 Chlorosulfuric Acid

1. Introduction 1329
2. Physical Properties 1329
3. Chemical Properties 1331
4. Production 1332
5. Safety Precautions, Transportation . 1333
6. Uses and Economic Aspects 1334
7. Toxicology and Occupational Hygiene 1335
8. References 1335

13 Chromium Compounds

1. Introduction 1338
2. Chromium Ores 1339
3. Production of Sodium Dichromate . 1345
4. Chromium Oxides 1350
5. Chromium(III) Salts 1357
6. Chromic Acids and Chromates(VI) . 1364
7. Other Chromium Compounds 1371
8. Analysis 1372
9. Transportation, Storage, and Handling 1373
10. Environmental Protection 1373
11. Ecotoxicology 1374
12. Nutrition 1376
13. Toxicology and Occupational Health 1376
14. Economic Aspects 1380
15. References 1383

14 Clays

1. Introduction 1389
2. Structure and Composition of Clay Minerals 1390
3. Geology and Occurrence of Major Clay Deposits 1394
4. Mining and Processing 1407
5. Properties and Uses 1417
6. Environmental Aspects 1429
7. Production and Consumption 1430
8. References 1431

15 Coal

1. Coal Petrology 1436
2. Coalification 1441
3. Occurrence 1443
4. Classification 1445
5. Chemical Structure of Coal 1449
6. Coal Mining 1456
7. Hard Coal Preparation 1471
8. Coal Conversion (Uses) 1478
9. Agglomeration 1487
10. Transportation 1488
11. Coal Storage 1489
12. Quality and Quality Testing 1490
13. Economic Aspects 1493
14. References 1501

16 Cobalt Compounds

1. Occurrence 1503
2. Individual Compounds 1506
3. Cobalt Powders 1523
4. Physiology and Toxicology 1524
5. References 1525

17 Composite Materials

1. Definition 1527
2. Classification 1530
3. Principles of Fiber Reinforcement .. 1533
4. Materials Used in Advanced Composites 1538
5. Processing and Fabrication Technology 1559
6. Properties of Advanced Composite Materials 1568
7. Applications 1580
8. References 1588

18 Construction Ceramics

1. Bricks and Structural Tiles 1591
2. Stoneware 1613
3. References 1646

19 Copper Compounds

1. Introduction 1649
2. The Copper Ions 1650
3. Basic Copper Compounds 1652
4. Salts and Basic Salts 1659
5. Compounds and Complexes of Minor Importance 1674
6. Copper Reclamation 1681
7. Copper and the Environment 1682
8. Economic Aspects 1684
9. Toxicology and Occupational Health 1686
10. References 1687

20 Cyanates, Inorganic Salts

1. Constitution 1693
2. Properties 1693
3. Production and Uses 1695
4. References 1696

21 Cyano Compounds, Inorganic

1. Hydrogen Cyanide 1698
2. Metal Cyanides 1706
3. Detoxification of Cyanide-Containing Wastes 1727
4. Cyanogen Halides 1730
5. Cyanogen 1734
6. Toxicology and Occupational Health 1736
7. References 1739

Cement and Concrete

FRIEDRICH W. LOCHER, Forschungsinstitut der Zementindustrie, Düsseldorf, Federal Republic of Germany (Chap. 1)

JÖRG KROPP, Institut für Massivbau und Baustofftechnologie, Universität Karlsruhe, Karlsruhe, Federal Republic of Germany (Chap. 2)

1.	Cement	852
1.1.	Cement Components	853
1.2.	Standard Cements, Types of Cement	853
1.3.	History	855
1.4.	Cement Clinker	857
1.4.1.	Composition	857
1.4.2.	Raw Materials	860
1.4.2.1.	Type of Raw Materials	860
1.4.2.2.	Mining of Raw Materials	860
1.4.2.3.	Grinding, Mixing, and Homogenization of Raw Materials	861
1.4.3.	Burning and Cooling of Cement Clinker	862
1.4.3.1.	Reactions Occurring on Burning and Cooling	862
1.4.3.2.	Burning and Cooling Technology	863
1.4.3.3.	Thermochemistry of the Burning Process	867
1.4.4.	Characterization of Cement Clinker	868
1.5.	Other Components of Cement	871
1.5.1.	Glassy Blast-Furnace Slag	871
1.5.1.1.	Properties	871
1.5.1.2.	Processing	871
1.5.2.	Pozzolana	872
1.5.3.	Fly Ash	873
1.5.4.	Burnt Oil Shale	874
1.5.5.	Filler	874
1.6.	Storage, Grinding, and Transportation of Cement	875
1.7.	Hardening of Cement	876
1.7.1.	Phases of Hydration	876
1.7.2.	Course of Hydration	877
1.7.2.1.	Mixing Water	877
1.7.2.2.	Setting	877
1.7.2.3.	Hardening	878
1.7.3.	Heat of Hydration	881
1.8.	Formation and Properties of Hardened Cement Paste	881
1.8.1.	Binding of Water	881
1.8.2.	Structure	883
1.8.3.	Strength	884
1.8.4.	Deformation	885
1.8.4.1.	Modulus of Elasticity	885
1.8.4.2.	Shrinkage on Drying and Swelling	885
1.8.4.3.	Creep	886
1.8.4.4.	Thermal Expansion	886
1.8.5.	Permeability to Water	887
1.8.6.	Resistance to Chemical Attack	887
1.9.	Testing of Cement	888
1.9.1.	Setting Time	888
1.9.2.	Strength	889
1.9.3.	Constancy of Volume, Soundness	890
1.9.4.	Fineness	890
1.9.5.	Heat of Hydration	891
1.10.	Environmental Protection	891
1.10.1.	Particulate Emissions (Dust and Smoke)	891
1.10.2.	Gaseous Emissions	892
1.10.3.	Noise and Vibrations	893
1.11.	Toxicology	893
1.12.	Economic Aspects	893
2.	Concrete	894
2.1.	Introduction	894
2.2.	Materials	896
2.2.1.	Cement	896
2.2.2.	Aggregates	896

2.2.3.	Water	897	2.5.3.3.	Creep	913
2.2.4.	Additives	898	2.5.3.4.	Shrinkage and Swelling	914
2.2.4.1.	Setting and Hardening Additives	898	**2.6.**	**Physical Properties**	914
2.2.4.2.	Workability Additives	898	2.6.1.	Density	915
2.2.4.3.	Porosity Additives	898	2.6.2.	Porosity	915
2.2.4.4.	Other Additives	899	2.6.3.	Thermal Conductivity	915
2.2.4.5.	Fine-grained Material and Polymers	899	2.6.4.	Electrical Conductivity	915
2.3.	**Production**	900	2.6.5.	Permeability	916
2.3.1.	Definition of Concrete Properties	900	2.6.6.	Thermal Expansion	917
2.3.2.	Proportioning of Materials	901	2.6.7.	Shielding Properties	917
2.3.3.	Production Process	903	**2.7.**	**Durability**	918
2.4.	**Fresh Concrete**	904	2.7.1.	Chemical Attack	918
2.4.1.	Workability	904	2.7.2.	Physical Attack	919
2.4.2.	Placement, Consolidation, and Finish	906	2.7.2.1.	Frost Action	919
2.4.3.	Curing	906	2.7.2.2.	Elevated Temperature	920
2.4.4.	Special Methods	907	2.7.2.3.	Radiation	920
2.5.	**Mechanical Properties**	908	2.7.2.4.	Abrasion and Wear	920
2.5.1.	Crack Development	908	2.7.3.	Corrosion of the Reinforcement	920
2.5.2.	Strength	909	**2.8.**	**Special Concretes**	921
2.5.2.1.	Compressive Strength	909	2.8.1.	Lightweight Concrete	921
2.5.2.2.	Tensile Strength	910	2.8.2.	Heavyweight Concrete	922
2.5.2.3.	Fracture Mechanics	911	2.8.3.	Massive Concrete	922
2.5.3.	Deformation Characteristics	911	2.8.4.	Fiber-reinforced Concrete	923
2.5.3.1.	Relationship of Stress to Strain	912	2.8.5.	Polymer Concrete	923
2.5.3.2.	Modulus of Elasticity and Poisson's Ratio	913	**2.9.**	**Mortar**	923
			3.	**References**	924

1. Cement

Cement is a hydraulic binder. It is a finely ground, nonmetallic, inorganic material. When mixed with water it forms a paste that sets and hardens by hydration. After hardening, it retains its strength and stability even under water. The hydraulic hardening is primarily due to the formation of calcium silicate hydrates. Accordingly, cements consist of those substances which react with the mixing water to form calcium silicate hydrates. Aluminous cements harden hydraulically by forming calcium aluminate hydrates.

1.1. Cement Components

The most important component of silicate cements is *portland cement clinker*, also known as cement clinker or just clinker. Crystallized calcium silicates that are rich in calcium oxide and can rapidly react with the mixing water make up at least two-thirds of portland cement clinker.

Another hydraulically reactive material is predominantly glassy *blast-furnace slag* which is obtained by rapid cooling and granulation. At least two-thirds of it consist of calcium oxide, magnesium oxide, and silicon dioxide. It can also react with water to form calcium silicate hydrates, but this reaction proceeds very slowly. If the slag is mixed with cement clinker or calcium hydroxide, the reaction proceeds relatively rapidly. The glassy blast-furnace slag is known as a latent hydraulic material.

There are *other natural or synthetic substances* which are also capable of hydraulically hardening. They have a high content of reactive silicon dioxide and can react with calcium hydroxide in the presence of water to form calcium silicate hydrates. Materials having these properties are, e.g., certain types of volcanic ash and the coal fly ash from power plants. They are called pozzolanic materials after the town Pozzuoli near Naples, where the occurrence of volcanic ash was first registered by the Romans. Ash that accumulates during the burning of oil shale can also be regarded as a component of cement. Depending on its composition, it can either hydraulically harden on its own or form compounds, which are capable of hardening, with the components of cement clinker.

Cements may also contain small amounts of *fillers*. These substances are inert or have weakly expressed hydraulic, latent hydraulic, or pozzolanic properties. As a result of their grain size distribution, they are capable of improving the physical properties of cement, especially its workability. An auxiliary component of cements, used to control setting, is calcium sulfate in the form of gypsum, $CaSO_4 \cdot 2\,H_2O$, or anhydrite, $CaSO_4$.

1.2. Standard Cements, Types of Cement

Differing economic and industrial developments, sources of raw materials, and climatic conditions have led to different developments in building methods and construction materials and, therefore, to different types of cement. Thus, there are substantial differences in the cement standards of composition, designation, and minimum requirements in different countries.

The term *portland cement* generally refers to a cement which completely or predominantly consists of cement clinker. Calcium sulfate, which is present in practically all cements, is ignored when defining the composition. *White portland cement* is made from a cement clinker that does not contain any colored components, e.g., calcium aluminoferrite.

A cement that consists of at least 60 wt% cement clinker and a ground additive is classified as *portland slag cement, portland pozzolana cement,* or *portland fly-ash cement*.

Portland slag cement is called iron portland cement in the Federal Republic of Germany. *Pozzolanic cement* is of substantial economic importance, especially in Italy. It generally contains up to 40 wt% pozzolana as a ground additive. This amount is not limited, but the cement must pass the pozzolanicity test, which tests the capability of the pozzolana to bind lime. The probability of passing this test increases with the reactivity of the pozzolana and with the pozzolana content of the cement. The *trass cement* in the Federal Republic of Germany corresponds to the portland pozzolana cement and to pozzolanic cement.

Blast-furnace cement refers to cement which contains about 35–80 wt% glassy blast-furnace slag. Other substances with lower hydraulic reactivity may only be added in quantities of up to about 40 wt%.

Cements having special properties are produced for specific purposes. Cements with a *high sulfate resistance* are used for concrete constructions exposed to sulfate solutions. Portland cement with a low tricalcium aluminate content and blast-furnace cement with a high blast-furnace slag content belong to this category. Indeed, the cement standards in different countries specify definite limits for the components of these cements. The content of tricalcium aluminate in highly sulfate-resistant portland cement is limited to 3–5 wt%, and the slag content in blast-furnace cement with a high sulfate resistance must be at least 65 wt%.

Cements with a *low heat of hydration*, used in the production of massive concrete constructions, are either portland cements containing small amounts of tricalcium aluminate and tricalcium silicate or blast-furnace cements containing a large amount of slag. According to various national cement standards the heat of hydration developed in the first 7 or 28 days should not exceed 250–290 J/g as measured by the heat-of-solution method. It may not exceed 290 J/g after 28 days.

Low-alkali cements are used in the production of concrete that contains large amounts of aggregates sensitive to alkali. Harmful alkali reactions would otherwise be expected in the presence of adequate moisture from the surroundings. Low-alkali cements are portland cements with a maximum permissible alkali content of 0.60 wt% total alkali (calculated as Na_2O). In the Federal Republic of Germany, low-active-alkali cements include blast-furnace cements containing at least 65 wt% blast-furnace slag and not more than 2.00 wt% total alkali and blast-furnace cements containing at least 50 wt% blast-furnace slag and not more than 1.10 wt% total alkali.

Supersulfated cement consists of blast-furnace slag, at least 5 wt% anhydrous calcium sulfate, and a maximum of 5 wt% cement clinker.

The use of *water-repellent cement* is advantageous in certain types of construction, e.g., soil compaction. Water-repellent cement is portland cement to which small amounts of water-repellent substances have been added during production.

Oil-well cements are used for cementing the steel casing of gas and oil wells to the walls of the bore-hole and to seal porous formations. They are portland and pozzolanic cements which set normally even at high temperatures because of their composition.

Regulated-set cement, known in Japan as jet cement and as Schnellzement in the Federal Republic of Germany, sets quickly and hardens very rapidly because of its high content of aluminate ($11\ CaO \cdot 7\ Al_2O_3 \cdot CaF_2$).

Expanding cement generally contains large amounts of aluminates and sulfates and expands on hardening.

Masonry cements are binders for brickwork and plasterwork mortar with an adequate strength development. Masonry cement is usually a mixture of cement clinker and crushed rocks to which air-entraining agents are added to improve the workability, water retention, and frost resistance.

In contrast to the silicate cements, *high-alumina cement* consists mainly of calcium aluminates. Hardening is based on the formation of calcium aluminate hydrate. Under certain conditions this compound can be converted into compounds with a higher density. This increases the porosity of the concrete and simultaneously decreases its strength. The steel reinforcement is then no longer protected adequately against corrosion. For this reason, the use of aluminous cement for concrete, reinforced concrete, and prestressed concrete constructions is not permitted in some countries.

1.3. History

The name cement dates back to the Romans, who called a concrete-like brickwork made from stones and a burnt lime binder "opus caementitium". Later, additives such as ground bricks and volcanic tuff, which were mixed with burnt lime to obtain a hydraulic binder, were given the names cementum, cimentum, caement, and cement.

The Englishman JOHN SMEATON (1724–1792) discovered that clay plays a very important part in the hardening properties of hydraulic lime, which is produced from a natural mixture of limestone and clay. He was looking for a binder for water-resisting mortar to build the Eddystone lighthouse near Plymouth. His countryman JAMES PARKER in 1796 called the Roman lime, which he obtained by burning London marl clay, "Roman cement." The Frenchman LOUIS-JOSEPH VICAT (1786–1861) and the German JOHANN FRIEDRICH JOHN (1782–1847) independently discovered that mixtures of limestone with 25–30% clay were most suitable for the production of hydraulic lime. Even the binder produced in 1824 by JOSEPH ASPDIN by burning an artificial mixture of limestone and clay corresponded, at first, to Roman lime in its composition and properties because it had not been burnt to sintering. He called this binder portland cement. The artificial stones made from portland cement were similar to portland stone, an oolitic limestone that is quarried on the channel coast of the Portland peninsula in Dorsetshire. WILLIAM ASPDIN, the son of JOSEPH ASPDIN, began producing portland cement in 1843 in his newly established plant in Rotherhithe, near London. His cement proved to be far superior to "Roman cement," as seen during the construction of the parliament building in London. The reason for this difference was that a considerable part of the mixture had been sintered during burning.

The importance of the sintering process was apparently pointed out for the first time by ISAAC CHARLES JOHNSON (1811–1911) in 1844. The first German portland cement, based on the English version, was produced in Buxtehude in 1850. The foundations of the production of portland cement in Germany were laid by HERMANN BLEIBTREU (1824–1881), who built two cement plants, in Züllchow near Szczecin (1855) and in Oberkassel near Bonn (1858). The production of portland cement in France began in 1850. It was found that the sintered residue obtained when burnt lime was hydrated gave a slow-setting binder when ground between grindstones.

In the United States, DAVID SAYLOR first produced sintered cement clinker in 1870. He homogenized the raw material by grinding it and then formed the resulting powder into bricks which were burnt.

WILHELM MICHAELIS (1840–1911) played an important part in further developments. His book "The Hydraulic Mortar," published in 1868, for the first time gave details on the optimal composition of the raw material mixture. Valuable information on the highest possible content of calcium oxide in the raw material mixture, i.e., the amount that can bind to silicon dioxide, aluminum oxide, and iron(III) oxide on burning, and on reactions during burning and cooling of cement clinker resulted from the studies of S. B. and W. B. NEWBERRY [1], E. WETZEL [2], E. SPOHN [3], F. M. LEA and T. W. PARKER [4], and H. KÜHL [5].

At first, only the simple, discontinuously run shaft kiln was available, but later the annular brick kiln was also employed to burn cement clinker. The term *cement clinker* originates from this time because the material to be burnt in the annular kiln was formed into bricks, which were then burnt like ordinary bricks. The first rotary cement kilns were operated in the United States in 1895 and in Germany in 1896. Indeed, the first grate preheater kiln was introduced in Germany in 1929 and the first cyclone or suspension preheater kiln in 1950.

The latent hydraulic properties of granulated, i.e., rapidly cooled and largely glassy, blast-furnace slag were discovered by EMIL LANGEN in 1862. He showed that a mixture of granulated blast-furnace slag and burnt lime can attain considerable strength. In 1882 GODHARD PRÜSSING (1828–1903) was the first to add granulated blast-furnace slag to portland cement. Cement with a lower slag content was designated *iron portland cement* in Germany in 1901, and the cement with a higher content of slag has been called *blast-furnace cement* since 1907. In 1908, H. KÜHL discovered the sulfate activation of granulated blast-furnace slag, which was the basis of the production of *supersulfated cement*. The first cement to have a higher *early strength* was produced in an Austrian cement plant in Lorüns (Vorarlberg) in 1912. This is a very finely ground portland cement from clinker obtained by burning a carefully prepared mixture of raw materials at a higher temperature.

The first portland cement with a high sulfate resistance was the *"Erzzement"* patented by Krupp-Grusonwerk in Magdeburg in 1901 and produced in the cement plant in Hemmoor. It contained a lower amount of aluminum oxide and a high amount of iron oxide and its iron modulus (see Section 1.4.4) was only 0.3. Portland cement with a high sulfate resistance is similar to the so-called *Ferrari cement*, a portland cement with an iron modulus of 0.64, which was first produced in Italy in 1919. The importance of a high slag content for the sulfate resistance of blast-furnace cements was established in the 1920s. White portland cement was produced in small amounts in the 1880s in the portland cement plant in Heidelberg. Later, other cement plants also started producing this cement.

The development of *oil-well cement* began ca. 1930, when deep wells became necessary for the recovery of petroleum. Types of cement that set and harden very gradually, even at high temperatures and pressures, were required to line the oil wells. The first description of *expanding cement* appeared in 1920 [6]. The expanding cements produced today are based on the work of V. V. MIKHAILOV and A. KLEIN [7]. The Portland Cement Association of the United States developed *regulated-set cement* [8]. A similar cement is the *jet cement* produced in Japan [9]. *High-alumina cement* was first produced in France during World War I, based on a patent received by the French chemist J. BIED in 1908. He discovered that crystallized melts with the composition of monocalcium aluminate can harden hydraulically and attain high strength.

The first cement standards were introduced in 1878 in Germany. In France, the use of cement for the construction of government buildings was permitted in 1885. The cement standards in the United States and Great Britain were established in 1904.

1.4. Cement Clinker

The production of cement includes (1) isolation and preparation of the raw materials, (2) burning of the raw material mixture to give cement clinker, (3) preparation of the other components of the cement, and (4) grinding of the cement components with calcium sulfate to regulate setting. These continuous processes demand the monitoring of large material flows with respect to amount and required composition. This procedure is necessary to compensate for variations in the raw materials and in the production procedure during the preparation of cement clinker and the grinding of cement.

1.4.1. Composition

Cement clinker consists principally of tricalcium silicate, dicalcium silicate, tricalcium aluminate, and calcium aluminoferrite. It is made from a mixture of raw materials which mainly contains calcium oxide, silicon dioxide, aluminum oxide, and iron(III) oxide in definite proportions. When this mixture is heated to the sintering temperature, new compounds form, the *clinker phases*. Table 1 shows the amounts of these compounds formed. The terms *alite* and *belite* were coined by A. E. TÖRNEBOHM who examined clinker microscopically in 1897 [10]. He used the first letters of the alphabet to represent the main components because he did not know their composition. These names are still used to distinguish between clinker silicates, which always contain small amounts of aluminum, iron, and magnesium oxides in addition to alkali compounds, and pure silicates.

Table 1. Composition of cement clinker

Constituent	Chemical formula	Content, wt%		
		minimum	average	maximum
Tricalcium silicate	$3\,CaO \cdot SiO_2$	45	62	75
Dicalcium silicate	$2\,CaO \cdot SiO_2$	5	15	35
Calcium aluminoferrite	$2\,CaO \cdot (Al_2O_3, Fe_2O_3)$	4	8	15
Tricalcium aluminate	$3\,CaO \cdot Al_2O_3$	4	11	15
Free calcium oxide	CaO	0.1	1	4
Free magnesium oxide	MgO	0.5	1.5	4.5

Tricalcium silicate is the compound responsible for most of the properties of cement. It hardens quickly and attains a very high strength when it is finely ground and mixed with water to give a paste. It is formed by the chemical reaction between calcium oxide and silicon dioxide and can be produced from, e.g., limestone and quartz sand. The starting materials must be very finely ground and burnt at extremely high temperatures. The reaction proceeds rapidly in the presence of a melt consisting of calcium oxide, aluminum oxide, and iron(III) oxide at ca. 1450 °C. For this reason, raw materials used in the production of portland cement clinker contain predominantly calcium oxide and silicon dioxide with small amounts of aluminum oxide and iron(III) oxide.

Tricalcium silicate melts incongruently at 2070 °C, releasing calcium oxide [11]. Tricalcium silicate is unstable at temperatures below 1250 °C and decomposes to calcium oxide and dicalcium silicate. However, decomposition occurs only with very slow cooling or tempering. Foreign ions, especially Fe^{2+}, accelerate the process [12]. Therefore, cement clinker that contains iron oxide must always be burnt and cooled under oxidizing conditions. In the metastable temperature region below 1250 °C, six polymorphic forms of tricalcium silicate have been identified [13]. These are stabilized by solid solution of such ions as magnesium and aluminum. However, the crystalline structures and hydraulic properties of all of these forms are very similar.

Dicalcium silicate occurs when cement clinker is not completely saturated with calcium oxide. It hardens hydraulically the same as tricalcium silicate, but at a much slower rate, and after a long period of time, it attains the same, if not a higher, strength. Dicalcium silicate melts congruently at 2130 °C [11]. It crystallizes into four polymorphic forms which are known as α, α', β, and γ [13]. The β-form is metastable at all temperatures. It is converted to the γ-form, which is 10% less dense, when it is cooled to about 500 °C. This conversion is the cause of the disintegration (dusting) which is occasionally experienced with cement clinker rich in dicalcium silicate and crystallized blast-furnace slag. It is undesirable because γ-dicalcium silicate hydraulically hardens extremely slowly. Dicalcium silicate exists in its β-form in cement clinker. Polished sections of clinker seen under the microscope often reveal different forms of dicalcium silicate. Therefore, it can be expected that industrial clinker also contains the α'-form and, perhaps, the α-form. The α-, α'-, and β-forms seem to have similar hydraulic properties.

Foreign ions, such as aluminum, iron(III), magnesium, and alkali ions, stabilize the different dicalcium silicate forms in cement clinker. Pure β-dicalcium silicate can be prepared by adding 0.5 wt% boric oxide. Tricalcium phosphate in varying quantities can be used to stabilize the α- and α'-forms.

Calcium aluminoferrite contains all of the iron and a part of the aluminum present in cement clinker. It is a phase in the limited solid solution series $2\,CaO \cdot (Al_2O_3, Fe_2O_3)$ with $2\,CaO \cdot Fe_2O_3$ (dicalcium ferrite) and $2\,CaO \cdot (0.69\,Al_2O_3, 0.31\,Fe_2O_3)$ at its ends. In normal cement clinker, the composition of calcium aluminoferrite roughly corresponds to the formula $4\,CaO \cdot Al_2O_3 \cdot Fe_2O_3$. It can accept up to 2 wt% of magnesium oxide in its crystal lattice. This causes a change in color from brown to gray – the color of portland cement [14]. Calcium aluminoferrite contributes little to the hydraulic hardening process.

Tricalcium aluminate contains the aluminum oxide that is not combined in calcium aluminoferrite. It melts incongruently, releasing calcium oxide. Its crystal lattice can accept ca. 0.5 wt% calcium oxide in excess of the stoichiometric amount and up to 2.5 wt% magnesium oxide and alkali compounds [13]. Tricalcium aluminate reacts rapidly with water, but its hydraulic properties are not very pronounced. However, it improves the initial strength of cement when combined with silicates.

Minor components of cement clinker are free calcium oxide (free lime) and free magnesium oxide (periclase). They react with water to form calcium hydroxide and magnesium hydroxide, which occupy more space than the original oxides. Indeed, free calcium oxide and free magnesium oxide can cause expansion when they are present in larger amounts in a coarse-crystalline formation. The reaction with water would then proceed very slowly, and the cement would have started hardening before the reaction went to completion. Hence, all cement standards contain appropriate specifications to ensure that expansion does not occur.

Alkali compounds, such as alkali sulfates and alkali-containing calcium aluminate, are also minor components of cement clinker. Clinker contains up to 2.0 wt% alkali oxides ($Na_2O + K_2O$) and up to 2.0 wt% sulfate (SO_4^{2-}). Molten alkali sulfate forms at the sintering temperatures and does not mix with the aluminoferrite melt in the clinker. Solid solutions of alkali sulfate crystallize on cooling with a composition roughly corresponding to the formula $3\,K_2SO_4 \cdot Na_2SO_4$. If an excess of alkali is present, alkali-containing calcium aluminate forms with a composition and crystalline structure similar to that of tricalcium aluminate [15]. It can be easily recognized by microscopically examining clinker.

Table 2. Essential components of the raw materials mixture for making cement clinker

Raw material	Chemical formula	Content, wt%		
		minimum	average	maximum
Calcium oxide	CaO	63	66	68
Silicon dioxide	SiO_2	20	21.5	23
Aluminum oxide	Al_2O_3	4.5	5.5	6.5
Iron(III) oxide	Fe_2O_3	1	3	5

1.4.2. Raw Materials

1.4.2.1. Type of Raw Materials

A mixture of raw materials with the chemical composition given in Table 2 is required for the production of cement clinker with the composition described in Table 1. The chemical components are found in limestone, chalk, or clay, and in naturally occurring mixtures, i.e., argillaceous limestone. Limestone and chalk provide the necessary calcium oxide for clinker, and clay provides the silicon, aluminum, and iron oxides. In many cases it is necessary to raise the content of silicon dioxide and iron oxide by adding quartz sand and iron ore.

The composition of the raw material mixture is usually adjusted in such a manner that its calcium carbonate content is 75–79 wt%. A constant calcium carbonate content is necessary to prevent an increase or decrease in the tricalcium silicate content. This, in turn, influences the hardening properties of the cement.

In the combined production of portland cement and sulfuric acid (Müller-Kühne process), the mixture of raw materials consists of anhydrite, $CaSO_4$, or gypsum, $CaSO_4 \cdot 2 H_2O$, clay, and coke. The sulfur dioxide in the exhaust gas is converted into sulfuric acid by the usual processes (→ Sulfuric Acid and Sulfur Trioxide).

1.4.2.2. Mining of Raw Materials

Limestone and argillaceous limestone are usually quarried by blasting. The large rocks are loaded onto heavy-duty trucks by high-power excavators or bucket loaders and are transported to a hammer crusher. They are then crushed to a particle size of less than 30 mm. Mobile crushers on full-track vehicles, which can follow the excavators along the working front, are also employed. Loose rocks are torn away without blasting by caterpillars with bucket teeth and pushed toward the excavators. Chalk and clay are directly removed from the quarry site by using bucket or rotary bucket excavators or a dragline.

The main prerequisite for the quality and uniformity of cement is that the composition of the raw materials before entering the kiln is exact and constant. If the chemical composition of the deposit itself varies greatly, then the broken raw material is often prehomogenized. To achieve this, two or more rectangular or circular blending beds are piled in layers. The blending beds contain, as a rule, a one-week supply of raw material, i.e., 10–50 kt. The material is subsequently reclaimed in the transverse direction of the pile. Variations in the deposits can be compensated largely in this manner. At the same time, samples can be constantly taken during the forming of the stockpile, and the average chemical composition can be regulated by adding raw material from a particular part of the quarry or from a previously prepared material depot.

1.4.2.3. Grinding, Mixing, and Homogenization of Raw Materials

The raw materials are further processed to form a dry powder or a slurry.

Dry processing entails the fine grinding of the raw material components, separated according to their lime and clay contents. The ratio of the raw materials is regulated with dosing equipment. If necessary, components such as quartz sand and iron ore can be added to obtain the desired chemical composition of the raw meal. During the grinding process, the material to be ground is dried with hot gas, which is drawn through the mill. The heat of the exhaust gases from the kiln is generally used for this purpose. Depending on the construction of the mill, raw material containing 8–12 wt% moisture can be dried in this manner. Supplementary firing is required if the raw material contains more moisture. Extremely moist raw materials are dried in a rotary dryer before they are ground.

Tube mills or roller mills are employed to grind the raw materials. In tube mills, the material is ground by impact and friction of steel balls, and in roller mills, it is comminuted on a rotating pan by rollers. In both cases, separators connected at the outlet separate the coarse part of the material from the finely divided part, and feed it back into the mill. Roller mills require less energy and have a larger drying capacity than tube mills, but they are of limited use for very abrasive materials. Tube mills for raw meal have a diameter of up to 6 m and a throughput capacity of up to 400 t/h. Their specific energy requirements are ca. 12–16 kWh/t. Roller mills are built with throughput capacities of up to 400 t/h and specific energy requirements of ca. 10–14 kWh/t.

Wet processing is especially effective when the raw materials contain more than 20% water. The mixture of raw materials is then ground in a tube mill with more water to form a slurry. Soft raw materials are converted to an aqueous slurry by washmills or washdrums. Before the mixing and homogenization of dry raw powders became well-known, wet processing was also applied to materials which were less moist but had a variable chemical composition.

The finished raw meal or slurry is continually analyzed. Changes in the chemical composition are compensated by adjusting the proportions of the components. Any remaining variations are evened out in homogenizing silos, which can hold an 8–10 h mill throughput. Homogenization is carried out by fluidizing the raw powder with air and circulating it in a fluidized bed. Alternatively the raw meal can be mechanically circulated through several silos. It can also be homogenized in a blending-chamber silo by alternately draining different parts of the silo. Raw slurry is homogenized by agitating both mechanically or using compressed air.

1.4.3. Burning and Cooling of Cement Clinker

1.4.3.1. Reactions Occurring on Burning and Cooling

The raw material mixture is burnt in a cement kiln to give cement clinker. It is heated slowly to the sintering temperature of 1450 °C, which takes 40 min to 5 h, depending on the type of kiln. It remains for ca. 10–20 min at this temperature and is cooled as quickly as possible after leaving the kiln.

The mixture being burnt first loses liquid water at ca. 100 °C. At higher temperatures up to ca. 600 °C adsorbed water and water that is chemically bound in the clay minerals is released. The decomposition of calcium carbonate to calcium oxide starts at temperatures as low as 550–600 °C in the presence of silicon dioxide, aluminum oxide, and iron(III) oxide and proceeds very rapidly at temperatures above 900 °C. The material being burnt loses more than a third of its dry weight during this process. Dicalcium silicate and various transient phases containing aluminum oxide and iron(III) oxide are obtained. These transient compounds in turn disintegrate when melt starts to form at a temperature of about 1280 °C.

The melting process plays a very important part in the burnring of clinker. The amount of free calcium oxide present at the start of melting is high because all of the silicon dioxide has been used in the formation of dicalcium silicate. In the subsequent reaction, dicalcium silicate and calcium oxide form tricalcium silicate. This occurs in the presence of the melt, which facilitates the diffusion of the reactants.

The melt also dissolves coarse particles of quartz and limestone. Therefore, the proportion of melt present, which is ca. 25 wt% at the sintering temperature, must increase with a greater particle size of the material being burnt.

Equilibrium is rapidly reached in the clinker in the presence of the melt at the sintering temperature. This means that the phase composition of the clinker at the sintering temperature corresponds to the equilibrium composition at melting. The main correlations can be deduced from the quaternary system $CaO-Al_2O_3-Fe_2O_3-SiO_2$ [16]. It follows that the clinker melt, which contains practically all of the aluminum and iron oxides originally present in the raw materials, is deficient in calcium oxide to form calcium aluminoferrite and tricalcium aluminate on cooling. If equilibrium is maintained during cooling, that is, on very slow cooling of the clinker, a part of the solid tricalcium silicate is resorbed. Some of the tricalcium silicate reacts with the melt to form dicalcium silicate and release calcium oxide, which is used to make tricalcium aluminate. Slow cooling would, thus, yield less tricalcium silicate, which is very important for the hardening properties of the cement. Therefore, clinker must be cooled sufficiently rapidly to prevent the resorption of tricalcium silicate and cause the melt to crystallize independently of the solid phases already present. However, because the calcium oxide content of the melt is then insufficient for the formation of calcium aluminoferrite and tricalcium aluminate, a corresponding amount of calcium aluminate with a lower calcium oxide content is formed or the remaining melt solidifies forming a glass.

1.4.3.2. Burning and Cooling Technology

In Europe, cement clinker is predominantly burnt in rotary kilns; shaft kilns are seldom used.

Rotary kilns are refractory-lined tubes with a diameter up to about 6 m. They are inclined at an angle of 3–4° and rotate at 1.2–2 times/min. As a result of the inclination and rotation of the tube, the material to be burnt, fed into the top of the kiln, moves down the tube toward the coal dust, oil, or gas flame burning at the bottom of the tube. Near the flame, in the sintering or clinkering zone of the rotary kiln with a gas temperature of 1800–2000 °C, the temperature of the material being burnt reaches 1350–1500 °C, which is necessary for the formation of clinker.

The clinker leaves the rotary kiln after being sintered and enters a *grate cooler, tube cooler, or planetary cooler,* in which it is cooled by an air stream. In the grate cooler, the clinker is transported on a moving grate passed by a flow of air. In the tube cooler and the planetary or satellite cooler (which is made of 10 or 11 tube coolers attached to the circumference of the rotary kiln), the clinker is cooled by a counterflow air stream.

The air used to cool clinker in the clinker cooler reaches a temperature of 800–900 °C. It flows into the rotary tube as a secondary air stream, thus serving as combustion air. In this manner, the largest part of the energy contained in the clinker is led back to the kiln. The function of the primary air stream is to blow the fuel into the kiln.

Grate coolers require more cooling air than is necessary for combustion. The excess air is employed either for drying or passed through a dust filter and into the atmosphere.

The energy released on fuel combustion in the rotary kiln is transferred to the material being burnt. This material forms a coherent stream, which fills only a small part of the cross-section of the kiln, and because the kiln gas only flows over the material, the area which takes part in heat transmission is small. On the other hand, this coherent stream of material facilitates the formation of cement clinker and tricalcium silicate. The latter is formed rapidly in the compact cement agglomerate at the sintering temperature of 1450 °C and in the presence of the clinker melt.

To ensure effective heat transfer in the temperature range below clinkering temperature, the rotary kiln must be appropriately long or it must be connected to a preheater.

The length of a long rotary kiln is 32 – 35 times its diameter. The largest kilns have lengths of more than 200 m. Long kilns, with clinker capacities of more than 3 kt/d, are used in both the wet and dry process. They are known for their simplicity and their operating reliability. The upper part of the long kilns is equipped with chain curtains and fixed installations to improve heat transfer. The energy requirements of a long kiln per ton of clinker are 5.0 – 6.0 GJ for the wet process and 3.4 – 5.0 GJ for the dry process.

Rotary kilns equipped with preheaters are only 10 – 17 times longer than they are wide. They consume 3.1 – 4.2 GJ/t, which is less energy than long kilns. Two types of preheaters exist, grate and suspension preheaters. *Grate preheaters* are fed with pellets or briquettes. Pellets are made from raw meal and water in a pelletizer (semidry process). The rope-shaped briquettes are prepared from the raw material slurry, which is desiccated in a mechanical filter press and subsequently extruded in a screen compactor. In the grate preheater (Fig. 1), pellets or briquettes of the material being burnt are placed on a grate which travels through a closed tunnel. The tunnel is separated into a hot gas chamber and a drying chamber by a partition with an opening for the grate. The first induced draught fan draws the exhaust gas from the rotary kiln into the top of the preheater, through the pellet layer in the hot gas chamber, and then through the cyclones of the intermediate dust collector. In these cyclones, large dust particles, which would otherwise wear out the fan, are removed. The second induced draught fan then draws the gas into the top of the drying chamber, through the moist layer of pellets, and finally pushes it out into the dust collector.

The exhaust gas leaves the rotary kiln with a temperature of 1000 – 1100 °C. As it flows through the layer of material in the hot gas chamber, the exhaust gas cools down to 250 – 300 °C, and it leaves the drying chamber at 90 – 150 °C. As a result of its intensive contact with the exhaust gas, the material being burnt reaches a temperature of ca. 150 °C in the drying chamber and 700 – 800 °C in the heating chamber. The largest grate preheater kilns have a clinker capacity of more than 3 kt/d.

Different types of *suspension preheaters* are built. The first system of this kind consists of four cyclone stages, which are arranged one above the other in a tower 50 – 70 m high (Fig. 2). The uppermost stage comprises two parallel cyclones for better dust separation. The exhaust gases from the rotary kiln flow through the cyclone stages from the bottom upward. The dry, powdery raw material mixture is added to the exhaust gas before the uppermost cyclone stage. It is separated from the gas in the cyclones and rejoins it before the next cyclone stage. This procedure repeats itself three times before the material is discharged from the last stage into the rotary kiln. This alternate mixing, separation, and remixing at higher temperature is necessary for optimal heat transfer. Indeed, the transfer of heat between gas and solid occurs quickly, and an almost complete temperature balance is rapidly reached in the short pipes between the individual cyclone stages.

Figure 1. Schematic of grate preheater

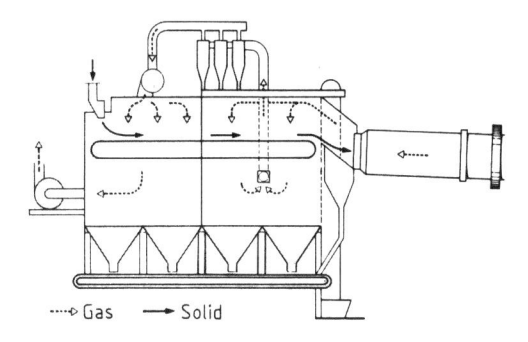

Figure 2. Schematic of suspension preheater

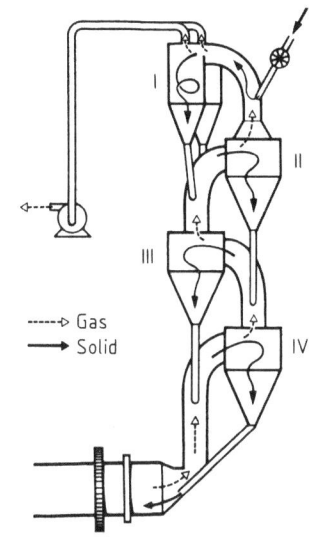

The raw meal is preheated to 800 °C. The temperature of the exhaust gas leaving the uppermost cyclone stage is as high as 380 °C, and it can be used to dry raw material and coal. Indeed, it contains sufficient energy to dry raw material with up to 12% moisture. The energy of the exhaust gas from a grate preheater cannot be utilized. Its temperature is lower because the material to be calcined is not a dry powder but rather pellets or briquettes containing at least 12% of moisture. The largest conventional suspension preheater kilns have a clinker capacity of more than 4 kt/d.

The *precalcination* technique is a new development in cement clinker burning; it is available to the cement industry since about 1970. In this procedure, the energy supply is split up into two burnings. The secondary burning, between rotary kiln and preheater, supplies so much energy that calcium carbonate in the material to be burnt is 80–90% dissociated when it enters the rotary kiln. Figure 3 shows this procedure

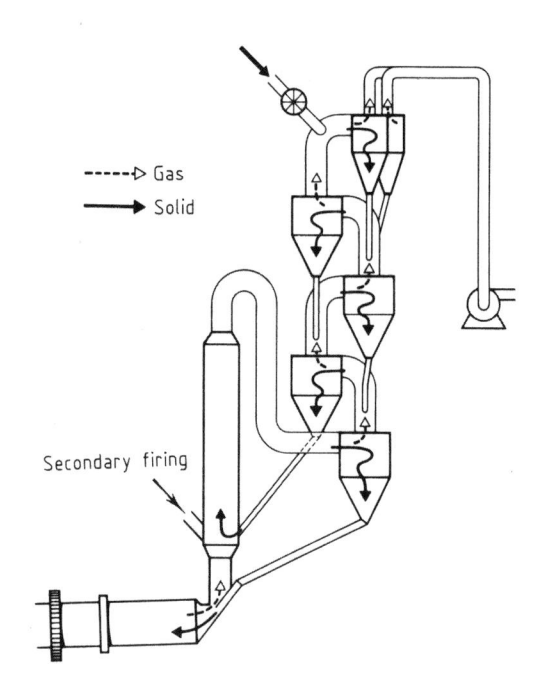

Figure 3. Schematic of suspension preheater with precalcination [17]

applied to a kiln with a suspension preheater. In principle, secondary burning can also be employed in a kiln with a grate preheater.

In conventional burning methods, the energy required to calcine calcium carbonate, which is about two-thirds of the total energy consumed in the burning of cement clinker (see Section 1.4.3.3., Table 3), is released at high temperatures in the sintering zone at the lower end of the rotary kiln. The hot kiln gas then transfers most of this energy to the calcination zone of the kiln and to the preheater. In this case, calcium carbonate is 40–50% calcined when it enters the rotary kiln. The main advantage of the precalcination process is that the energy required to calcine calcium carbonate is released during the secondary burning (by flameless combustion at 900 °C) and can be used immediately. For this reason, low-grade fuels, such as low-grade brown and hard coal and combustible waste, can be employed for the secondary burning. For a given rotary kiln, precalcining increases the clinker capacity. The largest suspension preheater kilns with precalcination have a capacity of more than 8 kt/d.

Shaft kilns are also approved and are widely used. They consist of a refractory-lined, vertical cylinder 2–3 m in diameter and 8–10 m high. They are fed from the top with raw meal pellets and fine-grained coal or coke. The material being burnt travels through a short sintering zone in the upper, slightly enlarged part of the kiln. It is then cooled by the combustion air blown in from the bottom and leaves the lower end of the kiln on a discharge grate in the form of clinker.

Shaft kilns produce less than 300 t/d of clinker and require 3.1–4.2 GJ/t. They are only economical for small plants, and for this reason, their number has been diminishing.

1.4.3.3. Thermochemistry of the Burning Process

Several chemical reactions occur during the heating of raw materials to produce cement clinker. Carbon dioxide and water vapor are thereby added to the gas stream and affect its composition.

Liquid water and adsorbed water first escape from the material as it is heated, followed by water chemically bound in the clay minerals. After the calcination of calcium carbonate, the material is sintered to give cement clinker, which is subsequently cooled. There are five distinguishable zones in a long rotary kiln. In the *drying zone*, the material is heated to 100 °C, and in the *preheating zone* it attains a temperature of 550 °C. The *calcining zone*, with material temperatures of 550–1200 °C, is followed by the *sintering* or *clinkering zone*, where the material reaches a temperature between 1200 and 1450 °C. The cooling of the clinker commences in the *precooling zone* of the kiln. Subsequently the clinker leaves the kiln for the clinker cooler at a temperature of at least 1150 °C. This temperature can be as high as 1350 °C, if the sintering zone is near the kiln exit. In preheater kilns drying, preheating, and some calcination take place in the preheater. In precalcining kilns calcination is almost complete before the material enters the rotary kiln.

The total energy required to burn cement clinker comprises several parts. The theoretical energy for clinker formation is the difference in enthalpy between the products (clinker) and the reactants (raw materials) [18]. To this must be added the energy needed for water evaporation, the energy lost with exhaust gas, the wall loss of preheater, kiln, and cooler, and the heat loss with hot clinker.

The dessication and decomposition of the clay minerals and the calcination of calcium carbonate and magnesium carbonate are all endothermic reactions. The formation of clinker phases from calcium oxide and the decomposition products of the clay components involves exothermic reactions. Hence, the theoretical energy requirements depend on the chemical composition of the raw material mixture and on the mineral composition of the clay components and amount to 1.6–1.9 GJ/t.

The energy required for the evaporation of water depends on the water content of the material. Raw powder has less than 1 wt% water as it enters the kiln. On the other hand, pellets fed into a kiln with a grate preheater generally contain 11–13 wt% water, and raw slurry contains 32–42 wt% water.

Table 3 shows the energy requirements of modern cement kilns. It contains also the specific electrical energy required for the induced draught fans for primary air, kiln, and cooler, and the electrical energy for the kiln drive. In addition, the specific volume and the temperature of exhaust gas are indicated.

Table 3. Energy requirements and exhaust gas characteristics for the production of cement clinker in kilns of various types

Process stage or characteristic	Dry-process kiln		Long wet-process kiln	Shaft kiln
	Suspension preheater	Grate preheater		
Throughput capacity, t/d	300–4000	300–3300	300–3600	120–300
Energy consumption, MJ/t				
Clinker formation	1590–1840	1590–1840	1590–1840	1590–1840
Water vaporization	8–38	420–590	1670–2720	420–590
Exhaust gas (waste)	600–1200	250–380	500–1050	80–840
Exhaust air from cooler (waste)	0–500	0–500	0–300	
Residual heat in clinker (waste)	40–210	40–170	40–210	80–250
Radiation and other wall losses (waste)	250–750	330–670	330–700	20–80
Electrical energy for motors	36–72	43–72	36–72	43–58
Exhaust gas, m^3/kg	2.1–2.5	1.8–2.2	3.2–4.2	2.0–2.8
Temperature of exhaust gas, °C	330–380	90–150	130–180	45–125

1.4.4. Characterization of Cement Clinker

Microscopic examination of clinker discloses the type, development, and distribution of the compounds present in clinker. The type of compounds produced depends primarily on the chemical composition of the material being burnt, but the structure, i.e., the development, distribution, and intergrowth of the compounds, depends on the preparation of the raw material mixture and on the conditions of clinker burning and cooling.

Examination under a normal microscope is carried out on a polished, etched section of clinker in reflected light. In general, distilled water, potassium hydroxide, alcoholic nitric acid, or dimethyl ammonium citrate is used for etching [19]. The individual constituents of the clinker can be identified by their light reflexion and by their behavior on etching. Studies on thin sections and on powder preparations using transmitted light are often difficult because of the small particle size of the clinker phases. Figure 4 shows a microphotograph of a polished clinker section that has been etched with water. The heavily etched and, hence, dark idiomorphic alite (a) and the weakly etched spherical belite (b) can be distinguished. The matrix surrounding these particles consists of solidified clinker melt; it shows dark tricalcium aluminate (c) and light calcium aluminoferrite (d).

For scanning electron microscopic studies a freshly fractured surface of a clinker piece is used. Metal atoms, generally gold, are electrically sputtered onto this surface under vacuum. Figure 5 is an electron micrograph showing alite, belite, and the matrix. It is impossible to differentiate here between tricalcium aluminate and calcium aluminoferrite as components of the matrix.

The composition of the clinker phases can be deduced from a chemical analysis [20]. However, the prerequisites are, first, that the composition of the clinker

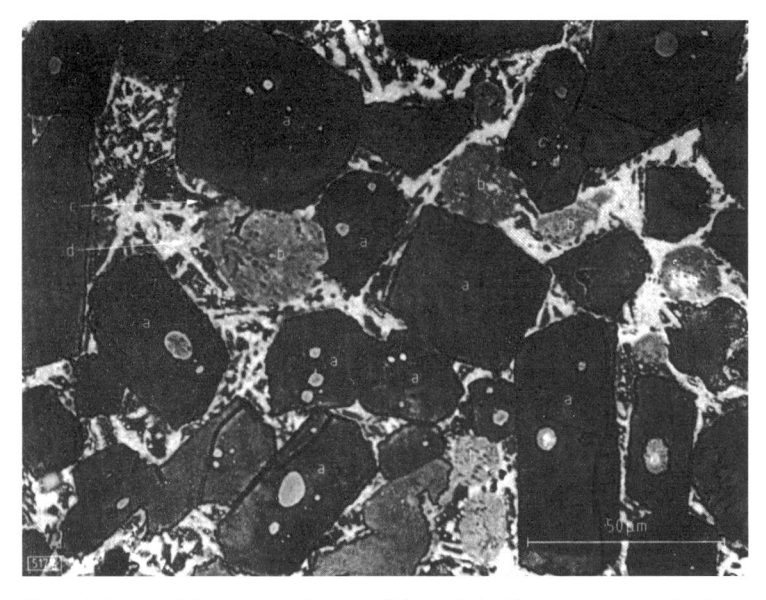

Figure 4. Cement clinker: section of cement clinker etched with water as seen under the microscope

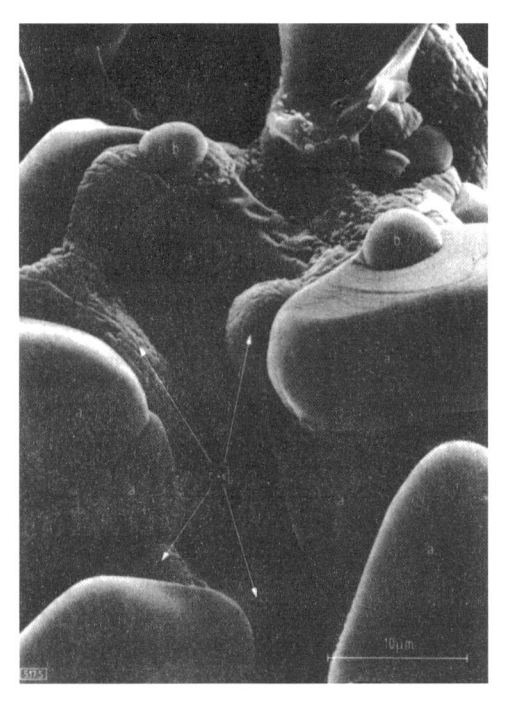

Figure 5. Cement clinker: picture of a fractured surface as seen under the scanning electron microscope

phases corresponds to their formulas and, second, that the clinker melt is in continuous equilibrium with the solid phases, not only at the sintering temperature, but also during crystallization when it is cooled. This means that tricalcium silicate is resorbed in an amount corresponding to the tricalcium aluminate content. In practice, these requirements are not met. Consequently, the calculated clinker composition values are too low for tricalcium silicate and too high for dicalcium silicate. Calcium aluminoferrite is a solid solution showing relatively large variations in the ratio of aluminum oxide to ferric oxide. The lime content of the melt is too low to form tricalcium aluminate in addition to calcium aluminoferrite. All the clinker phases also contain foreign substances in solid solution. For these reasons, a calculation gives only a rough estimate of the real clinker composition [21].

In practice, the raw material and clinker compositions are usually defined by their *lime saturation factor* (LSF), their *silica modulus,* and their *iron modulus.* The *lime saturation factor* is a measure of the amount of calcium oxide actually present in the raw material mixture or clinker, based on the maximum amount of calcium oxide which could be bound to silicon dioxide, aluminum oxide, and iron(III) oxide under industrial burning and cooling conditions. The maximum calcium oxide value is the sum of the amount of calcium oxide present in solid tricalcium silicate at the sintering temperature and in the corresponding equilibrium melt. This maximum calcium oxide content, $w_{CaO_{max.}}$, at which no free calcium oxide is present at equilibrium, can be directly deduced from the quaternary system of calcium oxide, aluminum oxide, iron(III) oxide, and silicon dioxide [4].

$$w_{CaO_{max.}} = 2.80 w_{SiO_2} + 1.18 w_{Al_2O_3} + 0.65 w_{Fe_2O_3}$$

The lime saturation factor is then

$$LSF = w_{CaO}/w_{CaO_{max}}$$

The silica modulus is the ratio of the amount of silicon dioxide to the sum of the amounts of aluminum oxide and iron oxide [22]. At the sintering temperature, silicon is primarily bound in the solid phases tricalcium silicate and dicalcium silicate; aluminum and iron are contained in the melt. For this reason, the silica modulus defines the solid:liquid ratio in the sintering zone of the cement kiln. The silica modulus is generally 1.8–3.0, but the most frequent and favorable value is 2.3–2.8.

The iron modulus is the ratio of the amount of aluminum oxide to the amount of iron oxide present [22]. It gives information about the ratio of calcium aluminate to calcium aluminoferrite and, therefore, about the clinker melt. Clinker with a normal composition has an iron modulus of 1.5–4.0. If the iron modulus is 0.638, then arithmetically the entire aluminum oxide contained in the clinker is bound as calcium aluminoferrite, $4\,CaO \cdot Al_2O_3 \cdot Fe_2O_3$.

The amount of free lime present, determined by X-ray diffraction or analytically after extraction with certain organic liquids, indicates the degree of burning of cement clinker.

1.5. Other Components of Cement

1.5.1. Glassy Blast-Furnace Slag

1.5.1.1. Properties

Glassy blast-furnace slag is known as a latent hydraulic material because it needs to be activated to harden hydraulically in a commercially useful period of time. This is typical for rapidly cooled (in general, with water), granulated, glassy blast-furnace slag, but not for the slowly cooled, crystalline lump slag. The activator in the case of portland slag cement and blast-furnace cement is cement clinker (clinker activation). In the case of supersulfated cement, it is anhydrite or gypsum (sulfate activation).

The hydraulic properties of glassy blast-furnace slag depend on the microscopically determined glass content and on its chemical composition [23]. Table 4 gives the chemical composition of blast-furnace slags used in Europe in the production of cement. In other countries the chemical composition can differ to a certain extent. For example, Indian slag usually has more aluminum oxide, and South African slag contains more magnesium oxide.

In general, the ability of glassy blast-furnace slag to harden hydraulically is directly proportional to the amounts of calcium oxide and magnesium oxide that it contains. A higher content of aluminum oxide especially increases the initial strength. The sulfide content promotes glass formation because the larger sulfide ions are incorporated into the glass structure rather than the smaller oxide ions.

Simple formulas, based on the chemical composition, are used to evaluate the hydraulic properties of glassy slag. However, these formulas can only be applied to slag of the same origin. According to the European cement standard, the contents of calcium oxide and magnesium oxide must together exceed that of silicon dioxide. A reliable measure of the hydraulic properties of glassy blast-furnace slag is the strength achieved by mortar test specimens made from a slag-containing binder. The strength of mixtures of portland cement with blast-furnace slag is compared with similar mixtures containing an inert material of the same fineness, e.g., finely ground quartz [24], [25].

The strength of a test specimen obtained from a mixture of finely ground slag and calcium oxide is another property which can be used to evaluate the hydraulic value of a slag [26].

1.5.1.2. Processing

Blast-furnace slag is formed from iron ore, coke ash, and limestone additives during the manufacture of pig iron in a blast furnace. It is removed from the furnace separately as running slag or together with molten pig iron as tapping slag. Slag has a temperature of ca. 1350–1600 °C when it leaves the blast furnace; its solidification temperature is 1250 °C. Glassy blast-furnace slag is produced by rapidly cooling the

Table 4. Concentrations of some important chemical components in granulated blast-furnace slag, natural pozzolana, hard-coal fly ash, and burnt oil shale

Com-ponent*	Concentration, wt%			
	Granulated blast-furnace slag	Natural pozzolana	Fly ash	Burnt oil shale
CaO	34–46	1–15	1–10	32
SiO_2	30–43	45–75	34–61	34
Al_2O_3	9–18	10–20	17–30	10
TiO_2	0.3–1.4	0.5–1.0	0.7–1.5	
P_2O_5	0.1–0.4		0.1–0.8	
Fe_2O_3	0.1–1.0	1–12	5–16	
Mn_2O_3	0.2–3.0		0.1–0.4	
MgO	2–13	0.2–4	1.5–4	
Na_2O	0.3–1.2	1–7	0.2–1.3	
K_2O	0.4–1.3	1–10	0.7–4.7	
SO_3	0.0–0.2	0.0–0.8	0.2–1.8	9
S^{2-}	0.5–1.8			

* The components are usually given as simple oxides.

molten slag with water and finely dispersing the material (granulation). The glass content, which is decisive for the hydraulic properties of slag, must be kept as high as possible.

Granulation is carried out by allowing the molten slag to flow into a trough with a tub-shaped inlet, the transfer pot. In this pot, pressurized water jets disrupt the slag stream, and the resulting granulated material is washed out of the trough and into a tank. The granulated slag is removed from the tank by a gripping device or a bucket conveyor with perforated buckets. In some blast-furnace plants, the slag is desiccated in a box-car, or in an overhead hopper with a perforated bottom, or on a sloping concrete surface. The granulated slag retains at least 7–20% residual moisture. For this reason it is dried in rotating drying drums or in rapid dryers before grinding. In some cases, it is directly dried in the mill using the heat produced during grinding and hot gas. The grinding of slag, cement clinker, and calcium sulfate to give slag cement requires more energy than the grinding of portland cement, because the components are harder to grind and because the product must be finer to achieve a definite strength.

1.5.2. Pozzolana

Pozzolana is a naturally occurring or synthetic fine-grained material which at normal temperatures can only harden in the presence of calcium hydroxide solutions. A major component of pozzolana is silicon dioxide which reacts with calcium hydroxide in the presence of water. Products of this reaction are calcium silicate hydrates as strength-giving constituents. Even calcium aluminate hydrates, obtained from the reaction of calcium hydroxide solution with aluminum oxide, which is also present in pozzolanas, can play a part in hydraulic hardening.

These properties are primarily exhibited by volcanic tuff, which originates in volcanic ash deposits. The material is named after the tuff found in Pozzuoli near Naples (Italy) and has been used since ancient times. Even today, the volcanic tuff found in Campagna and Latium is of considerable importance to the Italian cement industry. The Rhenish trass found in the Neuwied basin near Koblenz (Germany) is also volcanic tuff, but the Suevit or Bavarian trass occurring in the Nördlinger Ries was originally sedimentary rock, transformed by the impact of a meteorite.

Sedimentary rocks containing a large amount of reactive silicon dioxide are also natural pozzolanic materials, and need only to be dried to be used as secondary constituent in the production of cement. Rocks with a high content of silicon dioxide of organic origin, e.g., kieselguhr found in northern Germany, Danish moler, and French gaize, also belong to this category. These rocks are often heated to 400–800 °C to increase their reactivity and are then considered to be synthetic pozzolana. This also applies to phonolite, an alkali-rich volcanic rock found in southwestern Germany. The zeolite present in phonolite is converted to reactive silicon dioxide when it is heated. Clay burnt at temperatures of 500–800 °C and fly ash from power plant boilers are also synthetic pozzolanas. Table 4 presents the chemical compositions of various pozzolanas.

The reactivity of pozzolanic materials can be determined, as in the case of blast-furnace slag, by their contribution to the strength of mortar test specimens [24]–[26]. The future European cement standard will require a special test of pozzolanic cement, a test which has been a part of the Italian cement standard for a long time. This test shows that the pozzolanic portion of cement has the ability to bind a certain minimum amount of the calcium hydroxide released during hydration of the clinker portion.

Hard rocks, generally including tuff, are quarried by blasting and are then crushed to the size of gravel. Loose rocks are directly removed from the quarry front. The rocks are dried in drying drums or in rapid dryers and, if necessary, heated in a kiln. They are then ground with cement clinker and calcium sulfate to give pozzolanic cement.

1.5.3. Fly Ash

Fly ash is a fine-grained residue that is obtained from the combustion of coal dust. It accumulates in the electrical or mechanical dust-collecting equipment that is connected to the steam generators of power plants. The particle size, defined as the specific surface [27], is ca. 2100–6200 cm^2/g. Fly ash consists mainly of fine-grained, glassy particles which are usually spherical but can also have an irregular shape. It also contains crystalline particles of quartz, magnetite, and mullite; unburnt and partially burnt coal; and small amounts of such alkali compounds as the chlorides and sulfates. These compounds are volatile at high temperature. When the combustion gases are cooled, they condense onto the solid particles. Table 4 shows the chemical composition of fly ash. The properties of fly ash, especially the amount of unburnt and partially burnt coal, quartz, and other crystalline components are chiefly determined by the ash content of the coal but also by the firing and combustion conditions; the quality of the fly ash increases with higher load of the steam generator and with combustion temperature.

Fly ash rich in lime, e.g., from the Rhenish brown coal, generally contains a larger amount of anhydrite. As it also usually contains free calcium oxide, it cannot be used for the production of fly-ash cement. Fly ash obtained from brown coal from other regions, however, can be deficient in calcium oxide and have a composition similar to that of normal hard-coal fly ash.

Both fly ash deficient in lime and fly ash rich in lime are used in the production of cement. The Austrian standard for fly ash used in cement production even stipulates a minimum calcium oxide content; on the other hand there exists also an upper limit for the content of free lime and sulfate.

Fly ash obtained from hard coal is a synthetic pozzolana. The glass it contains can chemically react with dissolved calcium hydroxide, e.g., from cement clinker, at normal temperatures to form compounds capable of hardening. A higher proportion of unburnt or partially burnt coal can increase the water requirement of concrete and reduce its frost resistance.

Hard-coal fly ash, cement clinker, and calcium sulfate are ground together to produce fly-ash cement. In this process, only the coarse ash particles are ground, the other components of fly-ash cement are mixed to homogeneity.

1.5.4. Burnt Oil Shale

Another secondary constituent for the manufacture of cement is produced from oil shale found on the northwestern slopes of the Swabian Mountains (Germany). Oil shale is a lime-containing bituminous shale with ca. 11 wt% organic substances, 41 wt% calcium carbonate, 27 wt% clay minerals, and 12 wt% quartz [28]. It is burnt at 800 °C in a fluidized bed furnace. The energy that is released when oil shale is burned is employed to generate electricity and the ash produced is ground with cement clinker to give oil-shale cement.

The chemical composition of burnt oil shale is characterized by the values shown in Table 4. The burnt oil shale itself is able to harden hydraulically. This is due to its content of dicalcium silicate, different calcium aluminates, and reactive silicon dioxide.

1.5.5. Filler

A filler is a cement component which does not participate, or only slightly participates, in the hardening reactions of the hydraulically active components. However, as a result of its physical properties, especially its particle size distribution, it improves the properties of fresh and hardened cement paste, mortar, or concrete. It is effective because it can be incorporated into the hollow spaces between the cement particles. In this way, the space which must first be filled by the mixing water and during hardening by the hydration products of cement can be reduced. Thus, the filler, despite its extreme fineness, lowers the water demand, improves the workability, and helps to form a denser structure. Therefore, the filler can increase the strength to a limited extent without taking part in the hardening reactions [29]. The most important filler is limestone. Because it is easier to grind, it has a broader particle size distribution after grinding than other cement components. In addition, it accumulates in the fines. In this way, the grain size distribution of the cement is optimized by the filler with respect to water demand, workability, and structure of the hardened cement paste. The amount of filler should not exceed 15 wt%, and the filler must not influence other properties; e.g., it may not reduce the frost resistance of concrete made from the cement.

1.6. Storage, Grinding, and Transportation of Cement

Clinker and other cement components are stored in silos or in closed sheds. Larger stocks can be stored in the open if the necessary precautions against dust formation are taken.

Cement is produced by finely grinding either clinker alone or clinker with blast-furnace slag, pozzolana, fly ash, or filler. Gypsum or anhydrite is added to the material being ground to control setting.

In general, cement is ground in tube mills; but roller mills have also proved to be suitable. Large tube mills are constructed with a diameter of 4–6 m and a length of 14–16.5 m for an hourly throughput of 300 t. Most mills work in a closed circuit, that is, they have adjustable air separators which can separate cement with the required fineness from the material being ground and can return coarse material to the mill.

Mills usually take up 3–6 MW. For modern cement mills the specific energy requirements for different cement fineness can be characterized by the following standard values:

Specific surface area, cm^2/g	3000	4000	5000
Specific energy requirement, kW h/t	25–50	40–60	60–90

The fineness and composition of cement are constantly monitored, and the mass flow of the components into the cement mill is regulated.

Finished cement is also stored in silos, tested in detail, and then filled into bags by using a bagging machine, or it is loaded as such onto trucks for transportation.

1.7. Hardening of Cement

Setting and hardening of cement is based on the formation of water-containing compounds, the hydrate phases, which are produced by the reaction of cement components with mixing water. Hydration of cement occurs in a paste that contains relatively little water; the reactions take place in the very thin film of water surrounding each cement particle.

The plastic mixture of cement and water is called *cement paste,* and the hardened mixture is known as *hardened cement paste* or *cement stone.* Mixing cement with sand whose grain size is less than 4 mm gives mortar. If coarser aggregates are also added, concrete is obtained. The properties of cement stone, mortar, and concrete depend on the structure developed by the hydration products of cement, i.e., on their formation, spatial arrangement, and packing density.

1.7.1. Phases of Hydration

The calcium silicate hydrates, m $CaO \cdot SiO_2 \cdot n$ H_2O, are the most important water-containing compounds formed on hydration by all cements, except aluminous cement. They are responsible for the strength of the cement. The molar ratio of calcium oxide to silicon dioxide, m, is 1.5, if the compound is formed in the presence of excess water. If less water is added, as is the case with concrete, the value of m increases to 2.0.

The degree of crystallization of the calcium silicate hydrates is low. Electron microscopic studies show that calcium silicate hydrates occur in the hollow spaces in hardened cement in the form of needles or prisms. A large part of the calcium silicate hydrates is composed of very fine, irregular particles. The results of some studies indicate that calcium silicate hydrates are formed in layers, with water molecules between the layers.

Cement aluminates rich in calcium oxide form tabular tetracalcium aluminate hydrate, $4 CaO \cdot Al_2O_3 \cdot 19 H_2O$, on hydration. This irreversibly converts to the compound $4 CaO \cdot Al_2O_3 \cdot 13 H_2O$ when it is air-dried. Calcium aluminoferrite yields these same compounds on hydration except that aluminum oxide is partially replaced by iron(III) oxide. In the presence of sulfate, cement aluminates are preferably converted to two calcium aluminate sulfate hydrates: the needle-shaped trisulfate, $3 CaO \cdot Al_2O_3 \cdot 3 CaSO_4 \cdot 32 H_2O$ (ettringite), is formed in sulfate-rich solutions; whereas in solutions deficient in sulfate but rich in lime, the tabular monosulfate, $3 CaO \cdot Al_2O_3 \cdot CaSO_4 \cdot 12 H_2O$, is formed. Iron(III) oxide can replace the aluminum oxide in these compounds. Carbon dioxide in the air can convert the monosulfate to the corresponding monocarbonate, $3 CaO \cdot Al_2O_3 \cdot CaCO_3 \cdot 11 H_2O$, and chloride solutions can give rise to the formation of monochloride, $3 CaO \cdot Al_2O_3 \cdot CaCl_2 \cdot 10 H_2O$ (Friedel salt), which has a low solubility in water.

1.7.2. Course of Hydration

1.7.2.1. Mixing Water

The water:cement ratio is the mass of the mixing water divided by that of the cement. Mortar and concrete generally have a water:cement ratio of 0.35–0.80. The water:cement ratio determines the consistency of the cement paste. The consistency also depends to a small extent on the cement, especially on the degree of fineness to which it has been ground. In general, the water:cement ratio needed to attain a definite consistency increases with increasing fineness. However, this effect is negligible at 2500–4000 cm^2/g (according to Blaine). It becomes noticeable only when the specific surface area exceeds 4000 cm^2/g.

The cement in cement paste tends to settle because it has a density three times that of water. Thus, a clear water layer is formed on the surface of cement paste. This process is called bleeding. The tendency to segregate increases with an increasing water:cement ratio; as the cement is coarser this tendency increases. In concrete the water secretion is not as pronounced as in the case of pure cement paste because the fine-grained aggregate particles also consume a part of the mixing water for wetting.

Chemical reactions occur immediately after the mixing of cement paste with water. The alkali sulfate contained in clinker, the calcium sulfate added to control setting, and small parts of the tricalcium aluminate and tricalcium silicate dissolve very rapidly. Thus, a saturated solution of calcium hydroxide containing alkali hydroxide forms quickly. The pH of the solution is at least 12.5. Alkali sulfate reacts with dissolved aluminate to form alkali hydroxide. As the content of alkali hydroxide raises, the content of calcium hydroxide decreases and the pH of the solution raises.

1.7.2.2. Setting

The chemical reactions taking place between the cement components and the mixing water, which start immediately after mixing, always result in a stiffening of the cement paste that increases with time. The setting commences when the stiffening has proceeded to a certain extent. This state is called *initial set*, and the further solidification of the cement paste is called *setting*. This, in turn, is followed by *hardening*. The reaction between tricalcium aluminate and calcium sulfate is decisive for setting, and the hydration of calcium silicate is decisive for hardening [30].

Immediately after mixing, a small part of the tricalcium aluminate dissolves and reacts with dissolved calcium sulfate, forming trisulfate. This reaction stops after a few minutes and starts again after the *dormant* or *induction period*, which lasts several hours and is practically free of chemical reactions. Trisulfate forms a thin layer of very fine, prismoidal crystals on the surface of the cement particle. The crystals are too small to bridge the space between the cement particles and to form a solid structure (Fig. 6). Thus, the cement particles are movable with respect to each other; i.e., the viscosity of the cement paste is still low. Setting of cement paste according to cement standard starts 1–3 h after mixing, during the dormant period. This is probably due to the recrystallization of trisulfate; the crystals that were already larger at the beginning of the reaction grow and the smaller ones dissolve and disappear. The larger trisulfate crystals can then bridge the space between the cement particles and thus cause the beginning of solidification (Fig. 7). Later, the trisulfate reacts with calcium hydroxide and aluminate in the presence of moisture to give monosulfate.

If finely ground cement clinker is mixed with water without added gypsum, the mixture sets immediately as a rule. In the sulfate-free solution, large, flat crystals of tetracalcium aluminate hydrate are formed from the small amount of tricalcium aluminate that dissolved after mixing. These hydrate crystals grow together to form a structure like a house of cards in the space between the cement particles filled with mixing water. Bridges between the cement particles are formed, and in this way rapid setting occurs.

The calcium sulfate added to control setting should be completely used up within 24 h. Otherwise, trisulfate and monosulfate would be formed in the hardened cement stone, and the pressure exerted by their crystallization would impair strength development. The amount of added sulfate must therefore be limited.

1.7.2.3. Hardening

The hardening of *portland cement* is primarily due to the hydration of fast-reacting tricalcium silicate and of slower reacting dicalcium silicate. Both give fibrous calcium silicate hydrate which is poorer in lime than the unhydrated starting compounds. As a result, calcium hydroxide is released, forming larger crystals in the hardened cement. Any tricalcium aluminate that has not reacted with sulfate forms tetracalcium aluminate hydrate. The additional calcium hydroxide that is required is released primarily by the hydration of tricalcium silicate. The slower reacting calcium aluminoferrite also forms tetracalcium aluminate hydrate, with iron(III) oxide replacing a part of the aluminum oxide.

These hydration products, especially at the start of hydration, are not formed simultaneously, but one after the other. X-ray diffraction and electron microscopic studies show three hydration stages, which are presented schematically in Figure 8. Only small amounts of calcium hydroxide and calcium trisulfate are formed in the first stage, the plastic cement paste. The fine-grained trisulfate crystals on the surface of the cement particles lead to slight stiffening in the first stage. The recrystallization, which gives longer trisulfate needles, causes setting. Although the formation of calcium silicate hydrate is detectable after one hour, substantial amounts are produced only in the second stage, which commences after ca. four hours. In this stage, the hydration of tricalcium silicate and dicalcium silicate proceeds rapidly. The initial long fibers of calcium silicate hydrate bridge the water-filled spaces between the cement particles and give the structure a definite strength. In the third stage, the remaining pores are filled

Figure 6. Picture of plastic cement paste after 45 min of hydration as seen under the scanning electron microscope [30]

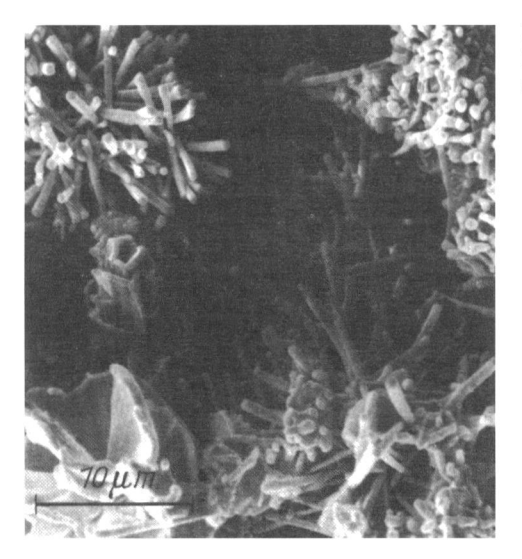

Figure 7. Picture of set cement paste after 6 h of hydration as seen under the scanning electron microscope [30]

with fine crystals of calcium silicate hydrate and calcium aluminate hydrate and the pores are reduced in size. In this way, the basic structure is densified and the strength is increased. The simultaneously produced calcium hydroxide, in the form of large, hexagonal crystals incorporated into the structure, does not contribute to the strength.

If the hardening process is delayed, e.g., by low temperature or by added foreign substances, the second hydration stage is prolonged, more of the long, fibrous calcium silicate hydrate is formed, and higher ultimate strength is attained than with normal

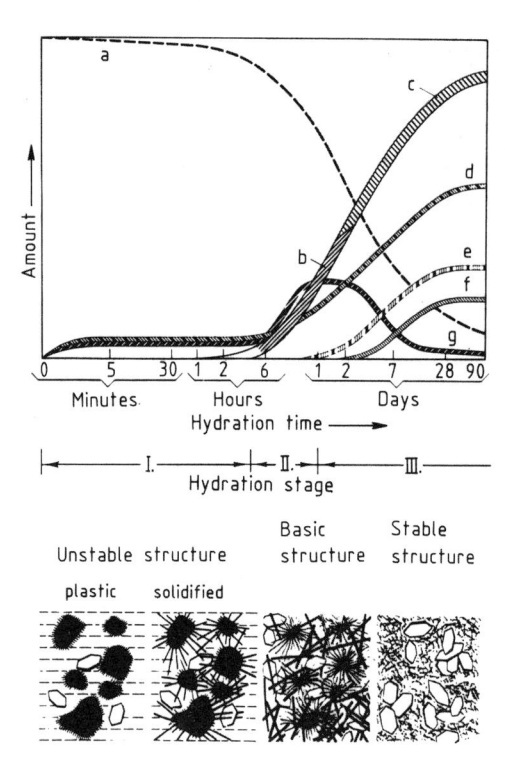

Figure 8. Schematic of the formation of hydrate phases and of the structural development during hydration [31]
a) Porosity; b) Calcium silicate hydrate, long fibers; c) Calcium silicate hydrate, shorr fibers; d) Calcium hydroxide; e) Calcium aluminate hydrate containing iron(III) oxide; f) Monosulfate; g) Trisulfate

hardening. If the reaction is accelerated, the initial strength is higher, but the final strength is usually lower because in the shortened second stage, less of the long, fibrous calcium silicate hydrate is formed.

Slag cements harden in a similar manner. The hydration products are practically the same as those obtained from portland cement. Slag cement, however, hardens slower and releases less calcium hydroxide because blast-furnace slag contains less lime than portland cement clinker and reacts slower with water.

In the hydration of *supersulfated cement*, calcium sulfate, usually added as anhydrite, reacts with the aluminum oxide in blast-furnace slag to form trisulfate. Calcium silicate hydrate is the strength-determining compound in the hardening process.

In *pozzolanic cements*, the natural or synthetic pozzolana reacts with calcium hydroxide, released during hydration of the clinker, to produce calcium silicate hydrate and tetracalcium aluminate hydrate. Hardened pozzolanic cement, therefore, contains much less calcium hydroxide than the corresponding portland cement. *Fly-ash cement* hardens in a similar manner. As pozzolana usually contains less than 5 wt% CaO, it reacts and hardens even slower than blast-furnace slag and needs extended moist curing.

1.7.3. Heat of Hydration

The hydration of cement is an exothermic process. The heat released increases the temperature of the concrete, a fact which must especially be considered in the production of bulk concrete. The heat developed during complete hydration under isothermal conditions depends on the composition of the cement. Table 5 shows the isothermal heat of complete hydration for different types of cement. Table 6 shows the contributions of the main clinker phases to the total heat of hydration of portland cement, calculated from studies with cements of different composition [32].

The relationship between the heat released and its contribution to strength varies with the cement component. Tricalcium aluminate and tricalcium silicate contribute more heat per unit of strength than dicalcium silicate and blast-furnace slag. The more reactive a cement is, the faster it hardens and the quicker it releases heat of hydration. The hardening rate increases with the fineness of the cement and, in the case of portland cement, with the contents of tricalcium silicate and tricalcium aluminate; it also increases with a decreasing amount of blast-furnace slag in slag cement. Table 7 shows the cumulative heat of hydration of cement after varying periods of hydration.

1.8. Formation and Properties of Hardened Cement Paste

1.8.1. Binding of Water

The individual cement particles in fresh cement paste are surrounded by a water film and are movable with respect to each other. The average thickness of the water film between two neighboring particles is a few micrometers. It increases slightly less than proportionally with the water:cement ratio; e.g., a doubling of the water:cement ratio from 0.4 to 0.8 causes a 1.7-fold increase in the thickness of the water film.

Initially, the hydration products form a very loose structure in the water-filled space between the cement particles; this structure slowly becomes increasingly dense as the fresh cement paste converts to cement stone. A part of the mixing water is chemically bound in the hydration products of cement. This water becomes a component of cement stone, which can also contain water adsorbed onto the surface of the hydration products and liquid water which partly or completely fills the pores.

A clear distinction between the chemically bound water and that which is adsorbed or contained in pores cannot be made, but "evaporable" and "nonevaporable" water can be determined [33].

The evaporable water escapes when the hydrated cement is dried at 20 °C and at the partial water vapor pressure of ice at −79 °C (temperature of a cooling trap operated with dry ice and acetone) [34]. Nonevaporable water escapes completely only at

Table 5. Isothermal heat of complete hydration for different types of cement

Type of cement	Heat of hydration, J/g
Portland	375–525
Portland blast-furnace, blast-furnace	355–440
Portland pozzolana, pozzolanic, fly-ash	315–420
High alumina	545–585

Table 6. Heats of hydration for various clinker phases of portland cement

Clinker phases	Heat of hydration, J/g
Tricalcium silicate	500
Dicalcium silicate	250
Tricalcium aluminate	1340
Calcium aluminoferrite	420
Free calcium oxide	1150
Free magnesium oxide	840

Table 7. Cumulative heat of hydration for cements with different rates of hardening

Period of hydration	Cumulative heat of hydration, J/g		
	Slow hardening	Average hardening	Fast hardening
1 day	60–175	125–200	200–275
3 days	125–250	200–335	300–350
1 week	150–300	275–375	325–375
4 weeks	200–375	300–425	375–425

1000 °C. It is, in any case, chemically bound and, therefore, a part of cement stone. After complete hydration, nonevaporable water accounts for ca. 25 wt% of the original portland cement. This value is nearly independent of the composition of the cement. In addition, a part of the evaporable water can also be considered as chemically bound because of its alignment in the hydration phases. This is the case with the water of crystallization in calcium aluminate hydrates, and possibly also with a part of the water bound in calcium silicate hydrate. The proportion of this water, which can be regarded as part of the solid material, is ca. 5–15 wt%.

The volume of the hydration products is less than the volume originally occupied by the cement and by the chemically bound water before hydration because the water bound in the hydration products occupies less space than liquid water. Accordingly, a decrease of 6 cm^3 per 100 g of original cement is observed when hydration is complete. The outer dimensions of cement stone remain practically unchanged, but cement stone stored under water absorbs a corresponding amount of water. The volume decrease caused by the incorporation of water into the hydration products is called chemical shrinkage or intrinsic shrinkage because the outer dimensions remain unchanged.

1.8.2. Structure

The structure of cement stone depends on the grain size, the shape of the hydration products, and the volume and size distribution of the pores. The specific surface area, measured by the BET method, is an indication of the particle size of the hydrate phases. A value of ca. 250 m^2/g is measured for completely hydrated portland cement using water vapor, and a value of about 150 m^2/g using nitrogen or argon. The higher value obtained with water vapor can be explained in two ways: (1) it is possible that cement stone possesses very small pores that are accessible to water molecules but not to nitrogen or argon or (2) part of the chemically bound water, driven out during the vigorous drying before the BET measurement, is taken up again during the measurement, which results in an unrealistically high specific surface value.

The average particle size of the hydration products can be calculated from the specific surface. Assuming that the particles are spherical, an average diameter of ca. 15 nm is obtained, based on a surface area of 150 m^2/g (nitrogen or argon measurement). Because of their extreme fineness the hydration products of cement are also known as cement gel.

The hydration products, which constitute the cement gel, are 1000 times smaller than cement particles. They cannot completely fill the space, even when densely packed. The remaining space is known as the *gel pores*. The average diameter of a gel pore corresponds to the particle size of the gel.

The hydrate phases have a lower density and occupy more space than the unhydrated cement. Cement gel including the gel pores occupies more than twice as much space as the cement from which it originates.

A thin layer of cement gel first surrounds the cement particle. On further hydration, water diffuses through the gel pores of this layer to reach the core of the cement particle which is still unhydrated. Water dissolves more of the cement particle, and a part of the dissolved substances immediately precipitate as cement gel in the inner free space. The rest of the dissolved substances diffuse through the gel layer already present and precipitate at the boundary between gel layer and water. Thus, the hydration products grow into the water space surrounding the individual cement particles. After complete hydration, which can take days or months depending on the size of the cement particle, the gel occupies more than twice the space occupied by the original cement particle. The space available to the cement gel is dependent on the original volume of cement and mixing water. The amount of gel formed at complete hydration totally fills the space available when the amount of mixing water corresponds to a water:cement ratio of slightly less than 0.4. At a lower water:cement ratio, the spacing is filled with cement gel even before the cement has been completely hydrated. The cement stone then contains cement particles with a non-hydrated core. If the water:cement ratio is greater than 0.4, the cement gel cannot fill the space between the cement particles even after complete hydration of the cement. Water-filled hollow spaces remain, the *capillary pores*, which are 1000 times larger than the gel pores.

Accordingly, the water:cement ratio mainly determines the capillary porosity of the cement stone. Another important factor is the degree of hydration. Capillary porosity to a large extent decreases with proceeding hydration, which requires water. In addition, the components of cement gel can be transported only by diffusion in a solution and, hence, can only occupy the pore space that is filled with water. It is therefore necessary to keep the cement stone moist for a sufficiently long period to produce cement stone with as low a capillary porosity as possible.

The capillary porosity strongly determines the technical properties of cement stone. Its permeability decreases with decreasing capillary porosity, and its strength, resistance against chemical attack, and frost resistance increase. The same is true for mortar and concrete.

1.8.3. Strength

Cement gel is the basis of strength. It forms a continuous framework of very fine calcium silicate hydrate and calcium aluminate hydrate particles, in which coarser calcium hydroxide crystals and unhydrated clinker remain embedded and which can be interspersed with pores of different sizes and shapes. The phenomenon of strength is not completely understood. It may be due to the cross-linking of the needle-shaped crystals of the hydrate phase or to adhesion of the fine-grained hydration products under strong surface forces. Recent studies indicate that adhesion is enhanced by meshing of the rough surfaces of calcium silicate hydrates, in particular.

The strength of the gel is generally higher with a higher proportion of long, fibrous calcium silicate hydrate and a lower calcium hydroxide content. When cement hardens slowly at low temperatures or under the influence of retarding additives the proportion of long, fibrous calcium silicate hydrate increases. The amount of calcium hydroxide present in the completely hardened cement stone is lower with lower alite contents and higher belite contents in portland cement and with higher blast-furnace slag contents in slag cement. However, these cements harden very slowly, so that the high final strength becomes apparent only after long moist curing.

After hardening, dry cement stone attains a higher strength than wet cement stone. Soaking with liquids other than water, e.g., mineral oil, also reduces strength. The porosity has the greatest influence on the strength of cement stone. In fact, the strength decreases even faster than the porosity increases. Although the effect varies with the type and size of the pores, all pores participate in the reduction of strength.

The influence on the strength of cement stone exerted by the porosity can be described by the following formula:

$$D = D_0(1-P)^n$$

D is the compressive strength of cement stone, D_0 is the compressive strength of pore-free cement stone, P is the part of the volume occupied by the pores, and n is a constant.

Air voids and capillary pores contribute to the porosity and affect strength. The total pore space is usually recorded because this value is easily obtained from the amount of evaporable water. A part of the evaporable water is chemically bound in calcium aluminate hydrates and in calcium silicate hydrates; thus, hollow spaces inside the solids are also determined in this manner. These hollow spaces, however, do not influence the strength to the same extent as the much larger capillary pores and air voids.

These differences affect the constant D_0, i.e., the compressive strength of pore-free cement stone, and the exponent n. If the pore space is calculated from the total amount of evaporable water, then a compressive strength of ca. 300 MPa is obtained for D_0, and a value of 2.5–3.0 for n. If only the capillary pores and air voids are considered, then D_0 is about 200 MPa and n about 4.5. The flexural and tensile strength of cement stone with a very low water:cement ratio is about one tenth the compressive strength. As the water:cement ratio increases and with it the porosity, the flexural strength diminishes more rapidly than the compressive strength. At high water:cement ratios, the flexural strength diminishes less rapidly than the compressive strength.

1.8.4. Deformation

1.8.4.1. Modulus of Elasticity

Cement stone is a viscoelastic material; i.e., deformation caused by stress is partly elastic and reversible and partly viscous and irreversible. The modulus of elasticity E is a measure of the relationship between stress and the elastic part of deformation. Elastic deformation increases more than proportionally with stress. Hence, Hooke's law does not apply to cement stone. For reasons of comparison, the elastic deformation used to calculate E is one that reaches one-third of the fracture stress after repeated loading and unloading.

E is influenced by the same factors as the compressive strength. It decreases with increasing porosity. The interrelation can be expressed by the following formula:

$$E = E_0(1-P)^n$$

When the total evaporable water is used to determine the porosity, E_0 is ca. 70 000 – 80 000 MPa and n ca. 2.5 – 3.0.

1.8.4.2. Shrinkage on Drying and Swelling

Shrinkage and swelling are volume changes that occur when cement stone is dried or wetted. This behavior is due either to volume changes of the calcium silicate hydrate layers, caused by the reversible diffusion of water into and out of interlayer sites [35], or to a change of the force on solid particles as a result of the surface tension of water [36].

The amount of shrinkage after the first drying of cement stone can be as much as 1%. The subsequent changes in length occurring on redrying and rewetting are reversible and are in the range of 0.3–0.4% [37].

Shrinkage, caused by the first drying, increases with increasing capillary porosity, i.e., with an increasing water:cement ratio. Shrinkage is a property of cement gel; thus, more shrinkage occurs when larger amounts of gel are present in cement stone. Finely ground cements tend to shrink more than coarser cements because they need more mixing water, but the composition of the cement is of little importance. Expansion of cement stone stored constantly under water is much less than shrinkage, only about 0.1%.

1.8.4.3. Creep

Creep is a deformation occurring under constant load in addition to elastic deformation and drying shrinkage. Creep proceeds rapidly at first and slows down after 1–2 years. After the load has been removed creep partially reverses. Creep is roughly proportional to the load up to 50% of the breaking load. Creep is generally lower with higher strength.

Moisture also affects creep. When the cement stone is loaded and dried simultaneously, creep is larger than in the case of drying prior to loading. If the change in moisture content occurred before the load is applied, creep decreases with the content of evaporable water. Because drying takes months, even in the case of concrete, substantial creep must always be expected even when the load is applied late after the manufacture of the concrete construction unit.

The changes in the structure of cement stone that cause creep are not yet fully understood. Creep may be caused either by the diffusion of adsorbed water under a stress gradient [36] or by the relative displacement of adjacent particles [35]. Creep is of the same order of magnitude as elastic deformation, but it can increase to a multiple of that under unfavorable conditions.

1.8.4.4. Thermal Expansion

The thermal expansion of cement stone depends largely on its moisture content. It is the lowest in the case of completely dried and water-saturated cement stone. Indeed, the coefficient of thermal expansion is then 11×10^{-6} K^{-1}. The expansion is twice as high when the moisture content of the surrounding air corresponds to a relative humidity of 70% [38].

1.8.5. Permeability to Water

The permeability of cement stone to water depends exclusively on the number of capillary pores. Cement gel is practically impermeable. The permeability to water is very small, almost zero, when the proportion of capillary pores is less than 20 vol% [39]. This capillary porosity is obtained at a water:cement ratio of ca. 0.5 and on complete hydration.

Cement stone with a water:cement ratio of 0.7 has so many capillary pores that it remains permeable to water even after complete hydration. Cement stone with a water:cement ratio of 0.5 shows approximately the same permeability to water when the cement is only 60% hydrated. However, the permeability to water decreases rapidly with an increasing degree of hydration so that at complete hydration the cement stone is practically impermeable.

1.8.6. Resistance to Chemical Attack

Hardened cement can react with certain gases and liquids. This process can reduce the strength and ultimately the structure of the cement stone can disintegrate. Acids and dissolved salts with exchangeable cations, e.g., ammonium or magnesium salts, can dissolve the cement stone from the surface because the reaction products are water-soluble. Sulfate can also penetrate into cement stone and react with calcium aluminate hydrates to form trisulfate (ettringite). Gypsum can also form, if the solution contains high concentrations of sulfate. The pressure due to crystal growth of the newly formed sulfate compounds can cause sulfate expansion.

Resistance to chemical attack depends primarily on the tightness of the cement stone, mortar, or concrete. Results of different studies show that cement stone from slag-rich blast-furnace cements is more resistant to the diffusion of various ions, in particular, sulfate and chloride ions, than cement stone from portland cement or from slag cements containing less slag. Blast-furnace cements rich in slag are generally more resistant to weak acids and solutions containing salts with exchangeable ions than other cements are. However, this difference is relatively small when compared with the effect of other parameters, and all cements are considered to be equally resistant to dissolution.

Larger differences exist with respect to sulfate resistance. Cement with a high sulfate resistance must be used as binder in concrete that is exposed to water and soil containing large amounts of sulfate. According to various national cement standards, sulfate-resisting cements are portland cement with a low tricalcium aluminate content and blast-furnace cement with a minimum content of blast-furnace slag of 65 wt%.

Gases can also penetrate the cement stone and react with its components in the presence of moisture. Hydrogen sulfide and sulfur dioxide are oxidized to sulfuric acid, which dissolves cement stone. Carbon dioxide reacts with calcium hydroxide in cement stone to form calcium carbonate and with the calcium silicate hydrates and calcium aluminate hydrates. The strength of the cement stone is generally increased in this process.

Carbonation, however, can affect the corrosion protection of the steel reinforcement in concrete. In non-carbonated concrete the pore solution of cement stone protects the reinforcement against corrosion. As a result of its high alkalinity (pH > 12), the surface of the steel is passivated. In carbonated concrete, the pH of the pore solution can fall to such an extent that the steel is no longer passive. Access of sufficient moisture and oxygen can then cause corrosion of the reinforcement. Penetrated chloride ions do not attack cement stone, but they can destroy the passivity of the reinforcement, even at high pH values, and thus promote (or produce) corrosion.

1.9. Testing of Cement

The requirements for the composition and properties of cements and the corresponding testing methods are stipulated in the various national cement standards [40]. The main properties of cement are strength, setting, dimensional stability (soundness), and fineness of grinding. Depending on the type of cement, limits for the contents of blast-furnace slag, pozzolana, fly ash, and filler are established. In addition, the cement standards specify certain characteristics, such as limits for the residue that is insoluble in dilute hydrochloric acid, the loss on ignition, and the contents of carbon dioxide, magnesium oxide, sulfate, and chloride. Cements with special properties must meet additional requirements concerning composition or heat of hydration.

In many countries compliance with the standard requirements for composition and properties of cement must be checked by internal and external quality control. Internal control must be carried out by the producer and external control by an authorized laboratory, which also supervises the internal control.

All test procedures are specified in the national cement standards. The composition of the cements must be determined using standard analytical methods combined with microscopic methods. The standard procedures specified to determine setting time, strength, soundness, fineness, and heat of hydration are described in the following sections.

1.9.1. Setting Time

Setting of mortar and concrete should allow enough time to work with the plastic cement paste, mortar, or concrete. In practice, the final set is not important. According to the various national cement standards the initial setting time must not be less than 0.75 – 2 h and the final setting time not more than 6 – 12 h, depending on the type of cement.

The Vicat needle is universally used for the determination of the setting time. It measures the resistance of a pure cement paste to the penetration of a cylindrical needle with a cross section of 1 mm². The sample is prepared by mixing cement with water; after mixing the paste must have a standard consistence, which is also measured with the Vicat instrument but with a plunger replacing the needle. Depending on the fineness and the composition of the cement, 23–35 wt% water must be added.

The initial set is the moment when the needle under a load of 300 g stops penetrating the 40-mm high cement paste at 5 ± 1 mm from the bottom. The final set is the instant at which the needle penetrates no more than 1 mm into the sample.

Quick set, changed setting, and false set interfere with normal setting. When setting starts within the first hour it is called *quick set* or in extreme cases *flash set*. *Changed setting* also refers to a quick set and occurs when cement with an originally normal set has been stored for a long time. Cement paste occasionally stiffens during mixing and becomes plastic again when it is further mixed; this phenomenon is called *false set.*

1.9.2. Strength

The strength of cement is tested on mortar samples. The dimensions of the specimen, the composition of the mortar, the type of curing, the duration of hardening and the test procedures are defined in various cement standards. Many countries have adopted the ISO-RILEM-CEM procedure, which will also be incorporated into the European cement standard. The method was worked out in the 1950s by the European cement institutes which are joined in Cembureau. (Cembureau is the union of the western European cement industries, ISO is the International Standard Organization, and RILEM is the Réunion Internationale des Laboratoires d'Essais et de Recherches sur les Matériaux et les Constructions.)

In this method, strength is tested on mortar prisms, $4 \times 4 \times 16$ cm, consisting of 1 part by weight of cement, 3 parts by weight of standard sand, and 0.5 part by weight of water. The standard sand has an even particle size distribution, which consists of equal amounts of fine- (0.08–0.5 mm), medium- (0.5–1.0 mm), and coarse- (1.0–2.0 mm) grained sand. The mortar is mechanically mixed under defined conditions, filled into molds, and compacted by using a shock table or vibration. The mortar prisms are kept in the molds surrounded by moist air for one day and subsequently under water at a constant temperature of 20 °C. The compressive strength of the prisms is tested after 2 or 7 and 28 days, as established in the cement standards. Definite minimum values must be attained. The bending strength, which is usually tested simultaneously, is of less importance. The cement standards of several countries also specify maximum values for the compressive strength at 28 days. The objective of this additional limitation is to obtain similar strength and comparable properties of cements of different origin. This is desirable because the 28-day standard strength is the basis for the mixing proportions of concrete.

1.9.3. Constancy of Volume, Soundness

Cement must be dimensionally stable, which means that hardened cement must not undergo any appreciable change of volume and thereby loosen or even destroy the solid cement stone structure. Expansion is observed in cements that contain large amounts of free calcium oxide or free magnesium oxide because reaction with water forms hydroxides that occupy more space than the oxides. Sulfate expansion can occur if excess gypsum is included and free gypsum is still present when hardening commences. The trisulfate then forms in the hardened cement stone and its pressure of crystallization loosens the structure.

The Le Chatelier test is used to determine the constancy of volume. Cylindrical test specimens are made from cement and water, stored moist for 24 h, and then boiled for 3 h. The mold is a flexible strip of brass with needles 150 mm long at each end. It is bent into the form of a cylinder 30 mm in diameter. The distance between the needles is a criterion for the expansion. Cement is dimensionally stable if expansion does not exceed a specified limit.

This method only discovers expansion caused by free lime. Expansion caused by magnesia or sulfate is prevented by limiting the contents of magnesium oxide in clinker and of sulfur in cement, respectively. The magnesium oxide content may be 3–6 wt% and the sulfate content 2.5–4.5 wt%, depending on the type of cement.

Expansion caused by free lime and free magnesium oxide can be determined using the autoclave method specified in the United States by ASTM standard. In this method, prisms with the dimensions $2.5 \times 2.5 \times 25$ cm are exposed to the action of steam under pressure at 216 °C in an autoclave. Cement is considered to be dimensionally stable if the expansion observed does not exceed 0.8%.

1.9.4. Fineness

The specific surface according to BLAINE [27] is a measure of the fineness of cement. It is calculated from the air permeability of a cement bed, its porosity, the density of the cement, and the viscosity of the air. The measure of air permeability is the time taken for a defined quantity of air to pass through the cement bed unter specified conditions. The specific surface of cement determined in this way is 2000–6000 cm^2/g.

A sieve test to measure the content of particles coarser than 0.09–0.2 mm is sometimes required. Laser granulometry is used increasingly to monitor the fineness of cement in the cement works.

1.9.5. Heat of Hydration

The testing of cement with a low heat of hydration is generally conducted by the heat-of-solution method. The heat of hydration is the difference between the amounts of heat given off when original cement and hydrated cement are dissolved in a mixture of nitric acid and hydrofluoric acid. The hydrated cement must have been subjected to hydration in a sealed tube at a constant temperature for 7 or 28 d.

1.10. Environmental Protection

The production of cement can result in finely divided particulate and gaseous air pollutants, noise, and vibrations due to the operation of machines and blasting at the quarry. Measures must be taken to protect the environment, measures which can amount to 20% of the total production cost. The reclamation of abandoned quarries and their integration into the countryside is also part of environmental preservation.

1.10.1. Particulate Emissions (Dust and Smoke)

In the course of cement production, 2.6–2.8 t of raw materials, cement clinker, gypsum, and, if necessary, blast-furnace slag, pozzolana, and coal are ground to make a ton of cement [41]. Drying, grinding, transport, burning, and cooling stir up 5–10% of this amount as finely divided particles of dust or smoke. The dust is sucked out at different points. The exhaust gas and exhaust air streams, amounting to 6000–12 000 m^3/t and depending on the plant, are freed of dust, and the recovered dust is usually recycled to the production process. The main dust producers in the cement plant are the kilns, along with their coolers, dryers, grinding and transport equipment, silos, loading devices, and the open-air storage areas and traffic on the plant site.

Electric filters and cloth fabric filters are predominantly used to remove dust. Modern, continuously operating dust collectors can reduce the dust concentration to less than 50 mg/m^3. The installation of modern dust collectors has reduced dust emission substantially. In 1950 dust emission amounted to more than 3.5% of cement production, today this value is less than 0.06%.

In general, dust from cement kilns is a mixture of the components of the raw material and alkali salts [42], in particular, alkali sulfate and chloride, that have been concentrated by internal or external recycling [43], [44]. An internal cycle is formed when volatile compounds evaporate in hot zones and condense in cooler zones of the kiln, in the preheater, or in the connected drying plants, then reach again hot kiln zones and re-evaporate. An external cycle is formed when kiln dust together with the condensed volatile compounds is precipitated in the dust collectors and is returned to the raw mix.

Dust from cement kilns also contains heavy-metal compounds, e.g., arsenic, lead, cadmium, chromium, cobalt, nickel, and thallium. The concentrations of these compounds depend on the corresponding concentrations in the raw materials and in the fuel, as well as on the recycling conditions of the kiln system.

The volatilities of the heavy-metal compounds vary considerably at the high temperature and in the alkaline and oxidizing atmosphere of the cement kiln. The *rate of retention* in the clinker is defined as the ratio of the amount removed from the kiln with the clinker to the total amount introduced into the kiln. Elements forming highly volatile compounds give a low rate of retention, and high rates of retention characterize elements with low volatility.

Thallium, in particular, gives low rates of retention. Therefore, thallium compounds concentrate in the kiln dust. This concentration is decreased, however, when the dust is no longer totally returned from the collectors to the kiln, i.e., when the cycle is interrupted. The other heavy metals form less volatile compounds under the conditions in the kiln. Consequently these elements are largely bound to the cement clinker and do not occur in the off-gas. In these cases, the rate of retention is raised when an increased portion of the separated dust is returned to the kiln.

1.10.2. Gaseous Emissions

The exhaust gas from the cement kiln consists primarily of nitrogen, carbon dioxide, oxygen, and water vapor. In addition, it can contain small amounts of sulfur dioxide, nitrogen oxides, carbon monoxide, and organic carbon compounds [41].

Under oxidizing burning conditions, the sulfur introduced into the cement kiln with the raw materials and fuel is converted into gaseous sulfur dioxide. Most of the sulfur dioxide reacts with oxygen and the alkali compounds vaporized at sintering temperatures to form alkali sulfate [45], [46], which is bound in cement clinker and in kiln dust. In the presence of excess alkali, the sulfur dioxide emission is very low, even when sulfur-rich fuel is employed.

The high temperature and the oxidizing conditions required to burn cement clinker lead to the formation of nitrogen oxides. Nitrogen monoxide and nitrogen dioxide are produced in a volume ratio of 9:1. The total emission of nitrogen oxides is between 500 and more than 2000 mg/m^3, calculated as NO_2.

Fluorine, contained in all cement raw materials to an extent of 0.05 wt%, is released at high temperature. It reacts with the material being burnt and becomes part of the cement clinker melt and kiln dust. Gaseous fluorine compounds are not emitted [47].

Chlorine, contained in the raw material mixture to an extent of 0.01–0.1 wt%, vaporizes at the sintering temperature and reacts with vaporized alkali compounds. The alkali chlorides formed then condense at lower temperature and are removed from the kiln with the clinker and dust. Chlorine gas is not present in the kiln gas [46].

The exhaust gas generally contains very little carbon monoxide because cement kilns are operated with excess air. The concentrations of organic compounds in exhaust gas are very low and depend on the types of raw material and fuel.

Grinding aids, e.g., ethylene glycol, can be added during the grinding of cement. At least 85% of it attaches to the cement; the rest is emitted as gas [48].

1.10.3. Noise and Vibrations

Noisy parts of a cement plant, e.g., tube mills with the revolving charge of grinding balls, are either accommodated in sound-absorbing buildings, covered by a soundproof capsule, or enclosed within soundproof walls. Disturbing noises produced by fans at the intake and exhaust outlets are eliminated by the installation of silencers. In addition, the working area of the operating personnel is soundproofed. Vibrations produced by blasting can be significantly reduced by using appropriate blasting methods, e.g., a series of boreholes can be blasted, not simultaneously, but sequentially at millisecond intervals.

1.11. Toxicology

Cement forms an alkaline aqueous solution, saturated with calcium hydroxide and containing alkali hydroxides. The pH of this solution is generally more than 12.5. Fresh mortar and concrete can cause burns, if allowed to remain on the skin for a long time.

Bricklayer's eczema is occasionally observed in people working regularly with cement. This is a skin disease caused in individuals sensitive to the low amounts of chromate present in cement. The effect of the chromate is probably aggravated by the alkalinity of the cement. The total chromium content of cement is 20–100 mg/kg. Chromium originates exclusively in the raw materials, not the equipment, and only 1–20 mg/kg of the total chromium is dissolved in the mixing water as chromate. Chromate is firmly bound in the hydration products of cement during the hardening process [49]. The chromate dissolved in the mixing water is responsible for bricklayer's eczema. It can be greatly reduced by the addition of iron(II) sulfate to cement.

1.12. Economic Aspects

Table 8 presents a review of the 12 countries with the largest cement production in 1985 [50]. The data for their cement production in 1975 and 1980 are also given for reasons of comparison. The amounts of the different types of cement produced in the various countries differ [51]. For example, Finland, Great Britain, Ireland, Portugal,

Table 8. The production of major cement-producing countries [50]

Country	Annual production of cement, Mt		
	1975	1980	1985
China (People's Republic)	29.0	73.5	135.0
Soviet Union	122.1	124.8	129.0
Japan	65.2	87.4	81.7
United States	59.2	68.2	70.3
Italy	34.6	41.9	37.3
India	16.2	17.8	31.0
Spain	24.4	29.6	24.2
France	30.7	30.6	23.5
Germany (Federal Republic)	33.0	33.1	22.9
Korea (Republic)	10.1	15.6	22.0
Mexico	11.6	16.3	21.3
Brazil	16.7	27.2	19.0

Sweden, and Switzerland produce more than 90% portland cement. Belgium, Denmark, and the Federal Republic of Germany produce ca. 60–80% portland cement and France, Italy, Luxembourg, the Netherlands, and Norway not more than 50%. Slag-containing cements make up approximately 10–50% of the total cement production in Austria, the Benelux countries, France, and the Federal Republic of Germany. Pozzolanic cement is produced in Italy to an extent of 45% and fly-ash cement in Denmark and France to an extent of 15–30%.

In the future, more waste materials, especially blast-furnace slag and power-plant fly ash, will be used in the production of cement. The 5% natural gypsum and anhydrite, added during grinding to control setting, will probably be at least partially replaced by calcium sulfate, which is a waste product of the chemical industry and the desulfuring of power-plant exhaust gases.

The production of cement clinker involves a relatively large expenditure of energy. The burning of cement clinker by using the dry method consumes an average fuel energy of 3.2 GJ/t and electricity of 110 kW h/t (0.396 GJ/t). Approximately 15% of the energy is used for grinding the raw materials, and 40% is used for grinding cement. A modern cement plant needs a working time of less than 0.5 h/t.

2. Concrete

2.1. Introduction

Concrete is an artificial stone in which natural stones of well-graded sizes are bound together by a cement matrix, thus providing strength. Although this principle of manufacturing building materials can be traced back at least to the Roman Empire, it

Table 9. Total energy content of some building materials

Material	Mean strength, MPa	Total energy content, MJ/m^3	Total energy content/strength
Steel	370 *	237000	640
Aluminum	320 *	730000	2280
Clay brick	25 **	4200	168
Sand lime brick	25 **	1220	49
Polyethylene		92800	
Concrete	35 **	1242	35.5

* Tensile strength.
** Compressive strength.

was later neglected for centuries. Concrete regained importance in the 19th century, when methods for large-scale production of cement were developed [52], [53].

Concrete is a brittle material with steel bars (reinforced concrete) or with prestressed tendons (prestressed concrete) has greatly expanded its applicability. It is now one of the most important building materials because of its structural, physical, and chemical properties and its economics. Fresh concrete can easily be shaped and most concrete is poured in place on the construction site. However, there is increasing use of concrete elements or concrete products that have been prefabricated in plants.

In 1983, ca. 920×10^6 t of cement was produced worldwide to manufacture ca. 3×10^9 m^3 of concrete for the construction industry. The seven countries – The People's Republic of China, the Soviet Union, Japan, the United States, Italy, Spain, and the Federal Republic of Germany – consumed more than half of the world production of cement and concrete [54].

Concrete is one of the less energy-intensive building materials. Concrete based on portland cement has a total energy content of only ca. 1250 MJ/m^3 [55], and even less when waste materials or industrial byproducts such as fly ash, silica fume, or granulated slags are incorporated. Table 9 gives a comparison of total energy content and total energy content per unit of strength of different building materials [55] – [57].

The supply of natural concrete aggregates of good quality is not guaranteed worldwide, and because of continuous consumption, a shortage of suitable material is observed in some locations. Aside from more sophisticated methods of recovery or treatment, the use of low-quality aggregates for special applications as well as recycling of old concrete, other building materials, or waste products must be considered in the future [58].

The properties of a structural concrete depend to a great extent on its age and curing history. Physical and chemical properties must be tested by standardized procedures both before and during the construction process. National standards, as well as international standards or recommendations given by the International Organization for Standardization (ISO), Comité Européen de Normalisation (CEN), Comité Européen du Béton (CEB), or Réunion Internationale des Laboratoires d'Essais et de Recherche sur les Matériaux et les Constructions (RILEM), cover quality control of concrete raw materials, as well as that of the concrete itself. Additional guidelines may be specified by the purchaser.

2.2. Materials

Basic materials for the production of concrete are cement, aggregates, and water. Up to certain limits, additional compounds – concrete additives – may be used to control some properties of the fresh or hardened concrete. These materials must meet specifications set forth in standards or in approval documents. The compatibility of all of the compounds used must also be considered.

2.2.1. Cement

The hydraulic binder cement reacts with water to form a highly dispersed cement paste matrix in which the aggregates are embedded. This cement matrix normally constitutes ca. 25% of the volume of concrete. Different types of cement can be distinguished according to cement standard compressive strength, chemical composition, kinetics and heat of hydration, or resistance to chemical or physical attack [59]–[62]. Economic considerations may influence the choice of the type of cement, depending on the local supply of raw materials for cement production or concrete aggregates.

2.2.2. Aggregates

The remaining 75% of the volume of concrete is aggregates, which act as a filler and which give concrete the properties needed in a structural building material (e.g., volume stability, limited crack development, modulus of elasticity, wear resistance, ductility). The mechanical and physical properties of the aggregates may differ considerably from those of the cement paste. They are reflected in the properties of the hardened concrete but not necessarily in proportion to the concentrations of the aggregates. Concrete aggregates can be categorized according to their origin as nautral or artificial aggregates; according to their density as light, normal, or heavyweight concrete aggregates; or according to their composition as mineral, metallic, or organic materials. The last two types are used only for special concretes.

Grading of Aggregates. Aggregates used in producing high-quality concrete have a particle-size distribution in the range 0.01–100 mm. The finer fraction is classified as sand, which comprises particles up to 4–5 mm in diameter; larger particles are called coarse aggregates. As a general rule, the size distribution of aggregates is selected to maximize their concentration in the concrete while maintaining sufficient workability of fresh concrete with a low water content. Therefore, the total surface area of the aggregates should be kept as small as possible [63].

The particle-size distribution of aggregates is determined by sieve analysis, which subdivides a sample into fractions. Within each fraction, particles of sizes between a lower and an upper limit are present, depending on the mesh sizes of two consecutive sieves. For two consecutive sieves, these mesh openings are normally increased by a factor of 2. Specifications for sieves, as well as procedures for sieve analyses, are given in various national and international standards or recommendations [64] – [66].

After subdivision of the aggregates into fractions, grading is based on the specific surface area, water requirement factor, average particle size, fineness modulus, etc. The grading curve is a graphic presentation of the cumulative percentage retained on or passing through successive sieves.

In general, specifications of limiting grading curves for the aggregates as a whole, or for the sand only, are commonly used [67], [68]. The maximum aggregate size is usually specified. Concrete aggregates must not contain compounds that inhibit the hardening of the cement paste or that are vulnerable to chemical or physical degradation. Therefore, the permissible contents of organic or expansive compounds, soluble sulfates, silt, and clay are limited. Aggregates to be used in reinforced or prestressed concrete must not contain soluble compounds, such as chlorides, that attack steel [64], [68]. Aggregates that contain reactive silica, e.g., opal, demand special, low-alkali cements to prevent expansive alkali–silica reactions in the hardened concrete [69], [70].

Other characteristics of aggregates, such as modulus of elasticity, hardness, density, porosity, thermal conductivity, thermal expansion, and resistance to wear and frost action, must be considered in many applications.

2.2.3. Water

The mixing water added in the preparation of fresh concrete serves to hydrate the cement and assures the workability of the fresh concrete. Although only 20 – 25 % of the mass of cement must be water for the chemical reaction with cement, additional water is needed to achieve complete hydration, as well as good workability.

Impurities in the mixing water, such as sugar, organic compounds, oil, carbonic acid, sulfates, or other salts, may affect the hardening of the concrete or reduce the concrete strength. The concentration of dissolved chlorides is restricted for reinforced and prestressed concrete. Doubtful cases should be subjected to chemical analysis, but normally hints of excessive amounts of impurities can be derived from the color, odor, taste, turbidity, or formation of bubbles or foam. Guidelines for permissible impurities can be found in [63], [71].

As a general rule, potable water is suitable as mixing water. In case of doubt, a trial mix with the water should be prepared and the development of concrete strength compared to that of a control mix. A trial mix is also important because the effect of impurities on concrete properties may depend on the type and amount of cement used.

2.2.4. Additives

Compounds may be added to control specific properties of the fresh or hardened concrete. These additives are either water-soluble (mainly organic) compounds in low concentrations or finely divided mineral material in higher concentrations. The classification of additives with respect to their effect is arbitrary, because in most cases several concrete properties are influenced simultaneously [67], [72], [73]. The majority of additives function to control setting and hardening, workability, or concrete porosity.

2.2.4.1. Setting and Hardening Additives

In some instances, the setting and hardening development of concrete must be accelerated or retarded to meet the needs of a construction job. Set-retarding agents are used to prolong the period of workability during hot weather, to avoid discontinuities in subsequent placements of concrete, or to reduce crack formation due to deformations in an early stage. Accelerators are used when concrete is placed under cold weather conditions or when there is need to strip the formwork earlier than normal. Use of high concentrations of accelerators in shotcrete (see Section 2.4.4) leads to rapid development of strength only seconds after mixing.

2.2.4.2. Workability Additives

Many types of additives exert a significant influence on concrete workability while also affecting other properties such as the setting time. The main objective of using such additives is to improve concrete workability. These plasticizers or superplasticizers reduce the amount of water needed and are the most commonly used additives, because they facilitate the placing and compacting of the concrete. Because they permit the use of lower water: cement ratios, the strength and durability of the concrete is improved [74].

2.2.4.3. Porosity Additives

The resistance of concrete to frost and deicing salt can be improved substantially by the use of air-entraining additives that form spherical pores with diameters considerably less than 0.5 mm [53]. To prevent frost damage, these pores must be closely spaced and uniformly distributed in the cement paste matrix of the concrete. The necessary volume of entrained air depends on the mix proportion of the concrete and is normally 2–6% by volume of the concrete [67], [75]. Foam-forming additives are used to make concrete with fine aggregates and a cellular structure. Such special lightweight concretes are primarily used for thermal insulation.

Air-removing additives are also available, and are often added to plasticizers.

2.2.4.4. Other Additives

Other additives are used for special applications, for example, expansion-producing agents; corrosion inhibitors; fungicidal, insecticidal, or germicidal agents; damp-proofing and permeability-reducing additives; bonding agents; or pigments. Their total consumption, however, is low [63], [73].

Much experience and careful control of concrete properties is required when additives are used because their effects may depend on many parameters, such as concrete composition, type of cement, and temperature. The simultaneous use of more than one additive may lead to problems because of interferences and to undesired concrete properties. In such situations a trial batch is mandatory [67].

2.2.4.5. Fine-grained Material and Polymers

Fine-grained Material. The main groups of fine-grained mineral solids used as additives are powdered unreactive rock, such as limestone or quartz or finely divided pozzolanas, which are either natural or industrial byproducts. Unreactive rock material is an inert filler that can improve the workability of the fresh concrete, especially when the available aggregates lack fines. Pozzolanas may increase the strength and reduce the permeability of the hardened concrete because of chemical reactions with hydration products of the cement. These reactions consume the calcium hydroxide formed during the hydration of portland cement. Accordingly, excessive amounts of pozzolanas may lower the alkaline reserve of the concrete.

Reactive Minerals. Natural pozzolanas include volcanic ashes, opaline shales, and cherts that contain reactive compounds of silicon or aluminum. Specifications for natural pozzolanas are given in [76], [77]. Among the synthetic pozzolanas, pulverized fuel ash (fly ash) is the most important additive and is used in large quantities. Its requirements as a concrete additive are specified in standards [76] or approval documents. Aside from improving certain concrete properties, the use of fly ash can reduce the amount of energy-expensive cement and save natural resources by recycling waste material [78].

Silica Fume. New developments in concrete technology have shown that silica fume – a pozzolanic waste material in silicon and ferrosilicon industries — opens new areas of application for advanced cement-based materials. Owing to its small particle sizes (ca. 0.1 μm) silica fume acts as a microfiller in the cement paste, thus reducing the total porosity considerably. Compressive strengths exceeding 250 MPa have been reported for concretes containing silica fume. For special applications, the new material can be substituted for cast iron, other metals, ceramics, or polymers [79].

Metal Oxide Coloring Agents. Coloring additives are generally alkali-resistant oxides of iron, titanium, or chromium. A uniform coloring effect is best achieved when such additives are used in combination with white cement.

Polymers. A considerable increase in concrete strength and durability under severe conditions is achieved by the addition of certain polymers to concrete (polymer cement concrete, PCC). However, the properties of polymer-modified concrete may deteriorate on continuous exposure to high humidity [80].

2.3. Production

2.3.1. Definition of Concrete Properties

Advances in concrete technology have established the basic rules for making concrete with desired properties in both the fresh and hardened states. Although the desired properties of the fresh concrete must be deduced from the working conditions and equipment at a particular construction site, the requirements for hardened concrete are determined by the expected service conditions. Strength, physical properties, and durability must be considered. A careful analysis of all required concrete properties is necessary before a selection of materials and their proportions is made.

In many instances, the need for concrete durability may lead to more stringent requirements for materials selection and proportions than those for the desired strength. Expected chemical attack of the hardened concrete, e.g., by sulfates, may require the selection of special cements. Anticipated exposure to severe weather and deicing salts requires the preparation of air-entrained concrete. Alkali-sensitive aggregates should be used only in combination with low-alkali cements [67], [81]. In all cases where external attack is expected, a concrete with a dense pore structure and low permeability should be used. In reinforced concrete, corrosion protection of the steel reinforcement must be assured by a low water:cement ratio. Therefore, for moderate service conditions, the water:cement ratio is limited to a maximum of 0.65, whereas severe conditions may require water:cement ratios not above 0.4 [67], [81], [82]. Other uses may require similar restrictions for the water:cement ratio, type of cement, or cement content.

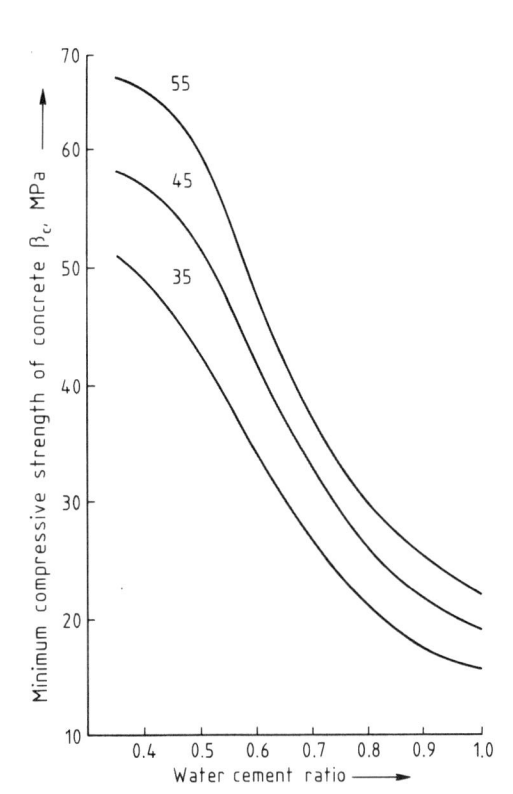

Figure 9. Minimum compressive strength β_c of concrete cubes for cements with different standard compressive strengths as a function of the water:cement ratio [89]

2.3.2. Proportioning of Materials

Basic concrete materials and additives must be proportioned in such a manner that all desired properties of the concrete are achieved.

There is no general approach, but rather empirical methods for determining the proper mix proportions, based on either volume or density. The volume or mass of the individual concrete constituents is determined in consecutive steps.

For low-grade concrete with a characteristic strength $f_{ck} < 25$ MPa, suitable mix proportions are presented in tables given in [67]. High-grade concrete or concrete containing additives requires a thoroughly detailed mix design for the first trial batch. The mix design may be changed after the properties of the trial mix have been tested.

One of several possible approaches to obtain a first estimate of the mix proportions is presented below. The concrete quality is specified by its characteristic compressive strength, which normally is determined after 28 days on cubes or cylinders [83]–[86].

Because of variations in the strength of concrete, the mean strength f_{cm} of a set of test samples must exceed the required characteristic strength f_{ck} by a certain amount, which can be derived by statistical methods [87], [88].

Table 10. Relationship between water:cement ratio and compressive strength of concrete cylinders [82]

Compressive strength, MPa	Water : cement ratio	
	Non-air-entrained	Air-entrained
45	0.38	
40	0.43	
35	0.48	0.40
30	0.55	0.46
25	0.62	0.53
20	0.70	0.61
15	0.80	0.71

Table 11. Approximate water content of fresh concrete [82]

Slump, mm	Water content, kg/m³ Maximum diameter of aggregate, mm				
	10	12.5	20	25	40
30–50	205	200	185	180	160
80–100	225	215	200	195	175
150–180	240	230	210	205	185

For normal aggregates and a given type of cement, the strength of completely compacted concrete is a function of the water:cement ratio. Approximate values of this ratio for achieving different strength levels are given in Table 10 for non-air-entrained and air-entrained concretes [82]; these values can be obtained from diagrams for different types of cement [89], such as Figure 9 for non-air-entrained concrete.

The content of mixing water required to achieve a certain consistency of the fresh concrete depends mainly on the maximum size and the grading of the aggregates. As a general rule, more sand or fine particles and smaller maximum aggregate size require more mixing water. For angular coarse aggregates, Table 11 gives maximum quantities of mixing water as a function of aggregate size and consistency as described by the slump (see Section 2.4.1.) of the fresh concrete.

The amount of water required for a concrete of given consistency may be estimated from the parameter k, which is defined as the sum of the percentages of aggregate retained on 9 standard sieves (0.25; 0.5; 1.0; 2.0; 4.0; 8.0; 16.0; 31.5; and 63 mm mesh), divided by 100:

$$k = 1/100 \sum R_i$$

R_i is the percentage of material retained on each of nine sieves, each of which is charged with the complete range of aggregate sizes.

In Figure 10 the water content of concrete with rounded aggregates is shown as a function of k for various ranges of consistency as described by the compaction factor v.

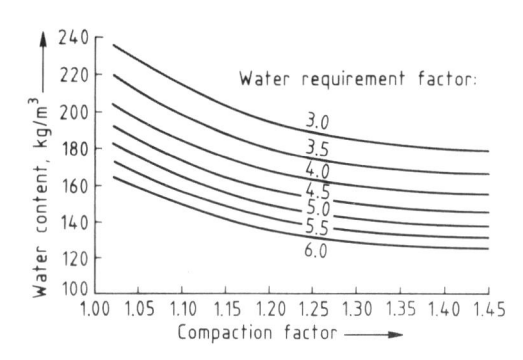

Figure 10. Water content of fresh concrete as a function of the compaction factor for different water requirement factors [90]

The amount of water added to the mix may need to be corrected for the moisture content of the aggregates, particularly sand. After the water content is estimated, the cement content of the mix is determined from the water:cement ratio. The amount of aggregates can be calculated by considering the absolute volume of the various concrete constituents contained in 1 m³ of compacted fresh concrete.

In this example, no restrictions on the water:cement ratio or cement content are considered. Furthermore, the results obtained by any method of mix design will give only a first estimate of the concrete composition. Necessary corrections to the individual parameters are determined by trial batches. Trial batches are mandatory whenever additives are used because the effects of the additives cannot be reliably predicted and may depend on the temperature, intensity of mixing, and composition of the concrete.

2.3.3. Production Process

For the production of uniform concrete of high quality, adequate storage facilities for the individual components must be provided. The cement should be protected from moisture to prevent partial prehydration. Fine and coarse aggregates stored in separate fractions must be handled carefully to avoid segregation. Sand, if not stored in protected bins, should be drained to avoid excessive water. Liquid additives must be protected from freezing and thoroughly mixed before use.

All solid materials are measured by mass; liquids can be measured by volume. All measuring devices should be calibrated regularly to operate accurately within 1–3% [67], [91]. On large construction sites or in ready-mixed concrete plants, measurement of the components may be automated at the bin discharge. They are carried to the mixing station by conveyor belts. Automatic sensing of the surface moisture of the aggregates allows for the correction of the amount of water to be added to the mix. Solid additives are added to the mix with the cement, whereas liquid additives are mixed with the water. Superplasticizers are effective for only a limited period of time.

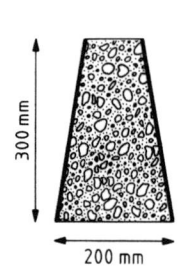

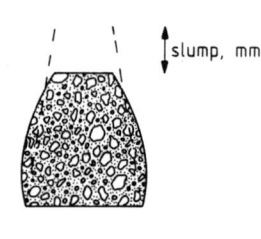

Figure 11. Test setup for the determination of the slump of fresh concrete [93]

Therefore, they are added to the fresh concrete during a second mixing immediately before placing the concrete.

To achieve a thoroughly mixed and homogeneous fresh concrete, the materials are charged into mixers which generally consist of revolving drums equipped with blades, stationary or revolving pans in which blades rotate on vertical shafts, or horizontal drums in which spiral blades operate on a rotating horizontal shaft. The necessary mixing time depends on the intensity of mixing, as well as on the batch size, and normally ranges from 30 to 120 s. Truck-mixed concrete may require up to 100 revolutions of the truck-mounted drum rotating at mixing speed [53], [67], [90], [91].

2.4. Fresh Concrete

2.4.1. Workability

The term workability describes several characteristics of fresh concrete which are of particular importance for the placement, consolidation, and finish of it. These characteristics include flowability, resistance to plastic deformation, compactibility, cohesiveness, and tendency to bleed or segregate. There is no single test method for evaluating all aspects of the workability of fresh concrete. Accordingly, test methods commonly determine either the flowability, expressed in terms of consistency, or the compactibility, which may be described by different compaction factors.

A widely used method for evaluating the consistency of fresh concrete is the slump test, which is specified (with minor differences) in various standards [92]–[94]. In principle, in this test a frustum of a cone which serves as a mold is filled with concrete in a standard manner. After the mold is removed, the subsidence of the concrete cone is measured to determine the slump (Fig.11).

In [83], the spread of a cone of fresh concrete is determined on a flow table, which is dropped in a standard manner (15 times within 15 seconds one side is lifted 40 mm and dropped) to agitate the fresh concrete. In the Vebe test, a cone of fresh concrete is remolded to a cylindrical shape by standard vibration. The time required to achieve

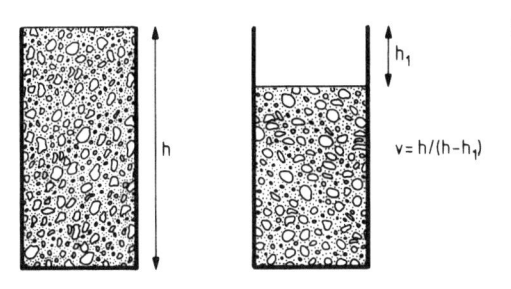

Figure 12. Test setup for the determination of the compaction factor v of fresh concrete [83]

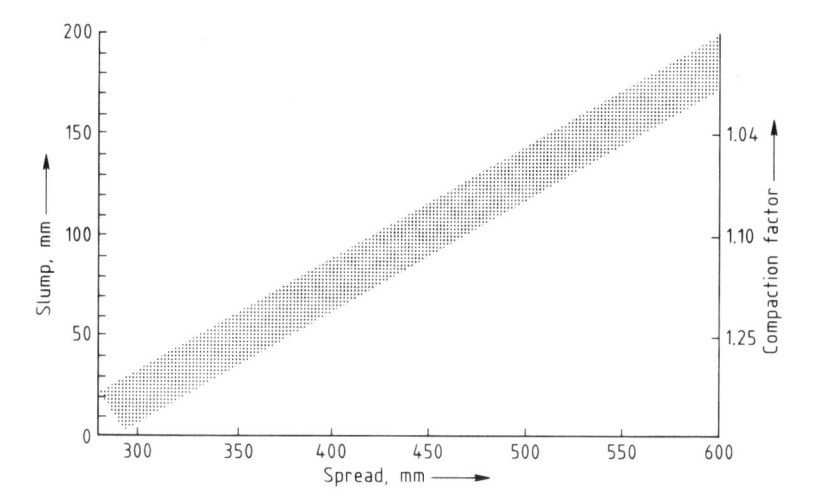

$v = h/(h-h_1)$

Figure 13. Approximate correlation of the slump, spread, and compaction factor of fresh concrete [90]

complete remolding describes the workability of the fresh concrete in terms of Vebe seconds [95], [96].

Tests for the compactibility of fresh concrete are specified in [83], [97]. In general, a mold is filled with concrete in such a manner that any uncontrolled compaction is avoided. After the mold is filled, the fresh concrete is consolidated by vibration. Compaction factors are then derived from density measurements or a comparison of the original volume (height) of the fresh concrete to the volume (height) after compaction. The test setups are shown schematically in Figure 12.

There is no strict correlation of the results obtained from the different methods for evaluating the workability of fresh concrete, and no single test method is universally applicable. Although plastic concrete mixes can be characterized by the slump or spread, the compacting tests are more appropriate for stiff and dry mixes. The different methods of testing workability are correlated in Figure 13 [90].

2.4.2. Placement, Consolidation, and Finish

After mixing, the concrete must be placed in its final position, consolidated, and eventually surface-finished within 1–2 h and before the initial set. The method of placing concrete into forms depends on its consistency and the equipment available at the site. Separate placement of single batches by buckets or buggies, as well as continuous placement by concrete pumps and hoses or conveyor belts, is successful if care is taken to avoid separation of the concrete constituents. To prevent cold joints, subsequent layers must be placed while the first layer is still plastic and while knitting of the two layers can be achieved by vibration.

Consolidation of the fresh concrete after casting removes entrapped air. Allowing for a monolithic and densely packed concrete structure, the reinforcement must be embedded thoroughly to provide a good bond with the concrete and to be protected from corrosion.

Fresh concrete is consolidated by vibration, which is usually provided by vibrators inserted into the concrete. Vibrators mounted on vertical formworks of thin concrete cross-sections or surface vibrators are also used. The vibration energy necessary for complete compaction decreases as the workability of the fresh concrete increases. Plasticizing additives may substantially reduce the consolidation work, and concrete containing superplasticizers may not require any consolidation. Excess vibration can cause segregation [90], [98].

Concrete may benefit from a second vibration after the initial setting to repair cracks that have developed due to settlement, early shrinkage, or deformations of the formwork.

The top surface of fresh concrete can be troweled or screeded. The durability of horizontal concrete surfaces, such as pavements and parking garage decks, is increased by vacuum processing of the fresh concrete surface. Partial withdrawal of the mixing water results in lower porosity and higher strength of the surface region [53], [90], [98].

2.4.3. Curing

The development of concrete strength and impermeability requires favorable conditions for hydration of the cement. In addition, crack formation should be avoided. Curing of concrete, therefore, comprises all measures undertaken to protect the hardening concrete from drying and to control the temperature across a section [99].

Early loss of water from the young concrete may prevent further hydration of the cement, especially in the surface region, resulting in insufficient strength development and high permeability. Furthermore, premature drying causes extensive shrinkage, which will lead to random cracks in the surface layer. Protection against moisture loss is normally achieved by covering the concrete surface with water-absorbent materials such as burlap or straw, which are kept wet with damp-proof plastic sheets, or by the application of membrane-forming curing compounds. The last-named compounds contain wax or resins and are sprayed or brushed on the concrete surface; however, they do not provide sufficient protection for hot weather concrete curing [100], [101].

Control of concrete temperature may be necessary when concrete is placed at low ambient temperature. Because the hydration of cement proceeds only slowly or even ceases at temperatures $< 5\,°C$, thermal insulation of the concrete surface is used to prevent rapid loss of the heat of hydration. Thermal insulation also prevents excessive temperature gradients in a concrete cross-section and protects the young concrete from freezing [102]. Thermal insulation is particularly important for massive concrete sections.

The required curing period depends primarily on the ambient climatic conditions and on the hydration kinetics of the cement used; it may vary from two days for rapidly hardening cement under moderate climatic conditions up to 14 days for massive concrete structures. Guidelines can be found in [99], [101].

Interior sections of a concrete member are less affected by curing than the surface regions. Adequate curing is, therefore, essential for achieving the required durability of a concrete structure.

The accelerating effect of elevated temperature on the hardening of concrete is used in the prefabrication of concrete elements; however, curing temperatures exceeding ca. 60 °C may reduce the final strength of the concrete. Steam curing at atmospheric or high pressure is applied in the preparation of certain concrete products such as pavement units, masonry units, and precast elements [103], [104].

2.4.4. Special Methods

A variety of special methods have been developed to simplify or increase the efficiency of preparing, transporting, placing, consolidating, and curing concrete.

For repair work, for strengthening existing structures, or for special applications, shotcrete may be superior to conventional concrete placing practice. In shotcrete, concrete is transferred pneumatically in a hose and projected through a nozzle onto a rigid surface at high velocity; consolidation is readily achieved by impact [105] – [109].

In construction of large concrete elements such as dams or for concrete placed under water, grout is injected into the voids of preplaced coarse aggregates to produce concrete [91].

Special procedures also exist for the production of concrete products such as masonry units, concrete pavements, or concrete tubes [103], [104].

2.5. Mechanical Properties

The most important design characteristics of concrete needed for structural analysis are the compressive strength and the modulus of elasticity of the hardened concrete; in many instances the tensile strength is not a design parameter. For concrete sections subjected to tensile stresses, a fracture mechanics approach is recommended because the tensile strength of concrete is controlled by such material defects as voids, pores, or microcracks. Structural design also requires information on dimensional changes due to creep, moisture migration, or temperature variation.

2.5.1. Crack Development

The mechanical behavior of concrete is controlled by defects or microcracks, which are caused by the nature of the hydrated cement paste or which are generated during the hardening of concrete. The cement hydrates continuously, thereby dissicating the matrix and causing it to shrink. Additional shrinkage is due to evaporation of water from the concrete surface. Because the stiff aggregates restrain the shrinkage of the cement paste, microscopic cracks are formed in the interfaces between mortar and coarse aggregates. Crack propagation continues from these defects when the concrete is subjected to stress. Tensile stresses and compressive stresses cause different types of crack propagation [110].

When concrete is loaded in compression, microcracks start to propagate when the stress exceeds ca. 40% of the maximum stress. With an increasing stress level, crack growth proceeds mainly along the interfaces between coarse aggregates and the matrix. At ca. 80% of the maximum stress, the cracks start to extend into the mortar matrix. Frequently cracks are arrested at aggregates, which results in branch cracking. As a consequence, the nonlinearity of the stress–strain diagram increases substantially. The predominant orientation of the cracks is then parallel to the axis of loading. At high stress, microcracks join to form longer cracks. Thus, the cracks reach a critical length, and unstable crack growth can only be avoided by reduction of the stress. As long as crack propagation remains stable, the strains continue to increase and the stress–strain relationship shows a descending part (stress σ = force/area, strain ε = change in length/original length).

If the load is kept constant at a strain greater than the strain at maximum stress, a continuous fracture plane is formed, causing the complete collapse of the concrete [111].

Concrete loaded in tension behaves like a brittle material. Initiation of crack growth takes place at preexisting microcracks, but multiple crack growth and branch cracking are much less pronounced than when under compression. Crack propagation is usually limited to one section. Nevertheless, a *process zone* is formed in front of a propagating crack, thus increasing the energy dissipation. At tensile failure stress, unstable crack growth forms a fracture surface perpendicular to the axis of applied tensile stress.

Sophisticated testing procedures can detect some stable crack propagation, which is more pronounced for larger amounts and sizes of aggregates [112].

2.5.2. Strength

2.5.2.1. Compressive Strength

The compressive strength of normal concrete is in the range 10–100 MPa (MN/m²), depending on the composition and preparation procedures. The compressive strength of most concrete placed, however, is in the range 25–50 MPa (MN/m²). The most important parameter controlling concrete strength is the water:cement ratio [113]. Other factors are the type of cement, curing conditions, moisture content, degree of hydration of the cement, and the grading and shape of aggregates. If sufficient water is provided, the hydration of cement will continue over a period of several years, and concrete strength may increase continuously during this time [52], [53]. It is customary to classify concrete according to the compressive strength attained after a standardized curing of 28 days.

In certain cases, however, the compressive strength after shorter or longer curing periods may be specified.

Because of inevitable differences in the properties of concrete components, preparation procedures, handling, consolidation, and curing, the strength of any particular type of concrete varies within a range. For the classification of concrete strength, therefore, a characteristic strength f_{ck} is defined, below which only a certain percentage, usually 5%, of the entire population of data falls. This value of f_{ck} defines a 5% defective concrete [83], [88]. Assuming a normal frequency distribution of the individual strength values, the mean strength, f_{cm} required to reach a certain 5% defective for a given standard deviation s can be estimated.

$$f_{cm} = f_{ck} + 1.64\,s$$

When good quality control is maintained, the standard deviation s is ca. 5 MPa. If a sufficient number of individual strength data are available, the standard deviation can be computed from the observed frequency distribution of individual strength results [83], [87], [114].

In general, the strength of concrete is determined by testing separately cast companion specimens, which are either cured under standard laboratory conditions or are exposed to the same curing conditions as the ultimate structure. Because the test results depend on the size and slenderness (i.e., height/thickness) of the concrete specimens tested, standard test specimens must be used. The most common specimens are cubes with a side length of 150 or 200 mm [83], [85] or cylinders with a diameter of 150 mm and 300 mm high [84], [85]. Specimen preparation and testing procedures are specified in the corresponding standards [83], [115].

The use of drilled cores for destructive laboratory testing is the most reliable method for testing the strength of an existing concrete structure. Nondestructive methods such

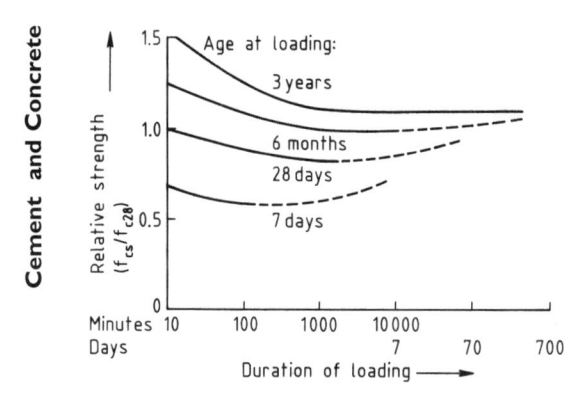

Figure 14. Relative sustained load strength of concrete f_{cs}/f_{c28} for different durations of loading and ages at the time of loading

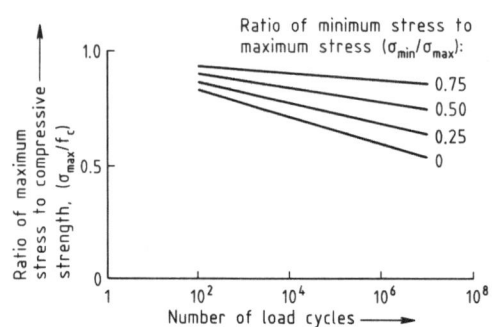

Figure 15. Fatigue strength of concrete for different stress levels

as ultrasonic pulse velocity, rebound, or penetration resistance require calibration [83], [116]–[118].

The sustained load strength of concrete is controlled by two counteracting mechanisms. While a continuous increase of strain under a sustained load may weaken the integrity of the concrete, a strengthening effect is exerted by the continuing hydration of the cement, which is more pronounced for younger concrete than for older concrete. The sustained load strength of concrete loaded at an age of 28 days is ca. 80 % of its short-time strength, which is less influenced by creep (Fig. 14) [119].

The fatigue strength of concrete decreases with an increasing number of stress cycles and an increasing stress range. The relations between the maximum stress and the average number of cycles leading to failure for various ratios of minimum to maximum stress are given in Figure 15. Concrete does not show a fatigue limit up to 10^7 load repetitions [120].

2.5.2.2. Tensile Strength

The tensile strength of normal concrete is approximately one-tenth the compressive strength. Detailed analyses by CEB [121] predict that the tensile strength should not increase in proportion to the compressive strength. In general, the tensile strength of concrete is determined by testing beams subjected to flexure or cylinders that are subjected to diametrically opposed axial line loads, giving the so-called tensile splitting strength [83], [122]–[126]. The direct tensile strength can be determined by axial tension according to [127]. Different test methods yield different results because of differences in stress state, stress distribution, and probability of failure.

2.5.2.3. Fracture Mechanics

The application of linear elastic fracture mechanics to concrete is limited because of the heterogeneity of the material and because of the development of a process zone, i.e., a region of microcracks in front of a major crack.

Accordingly, specimens with minimum dimensions at least 20 times the maximum aggregate size should be used when the conventional fracture mechanics characteristics of a concrete are determined.

The fracture toughness K_{IC} of cement paste, mortar, and concrete are as follows:

hydrated cement paste	$K_{IC} = 5-15 \text{ N/mm}^{3/2}$
mortar	$K_{IC} = 10-20 \text{ N/mm}^{3/2}$
concrete	$K_{IC} = 15-50 \text{ N/mm}^{3/2}$

K_{IC} increases with increasing degree of hydration and decreasing water:cement ratio. Because aggregates act as crack arresters and cause multiple crack growth, K_{IC} increases with increasing content and size of aggregates.

Fracture mechanics concepts can be applied successfully in concrete technology and design, e.g., for the description of fracture processes, for estimating notch sensitivity, and for the design of large unreinforced concrete members [128].

More recently, the fracture energy G_F has been introduced as a fracture parameter. It corresponds to the total energy required to separate a concrete section loaded in tension. Since G_F is less dependent on size than K_{IC}, it can be used to analyze smaller concrete members [129].

2.5.3. Deformation Characteristics

The deformations of concrete under load can be separated into elastic and time-dependent deformations. In structural analysis, time-dependent strains are of considerable importance in estimating the serviceability of constrained members or prestressed structures in which a loss of prestress may occur. All time-dependent deformations are markedly influenced by the concrete composition and by the prevailing climatic conditions. Thermal expansion is discussed in Section 2.6.6.

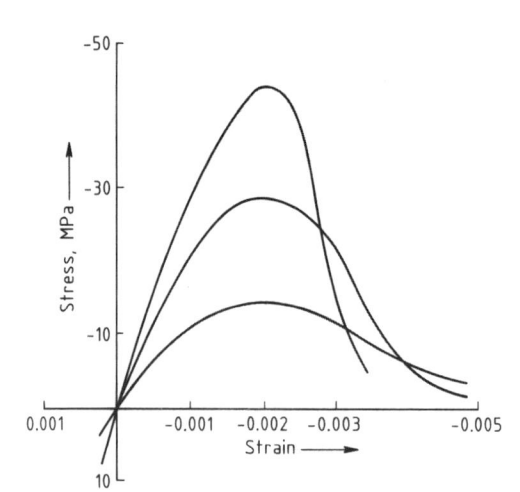

Figure 16. Stress–strain relationship for concretes of different strengths

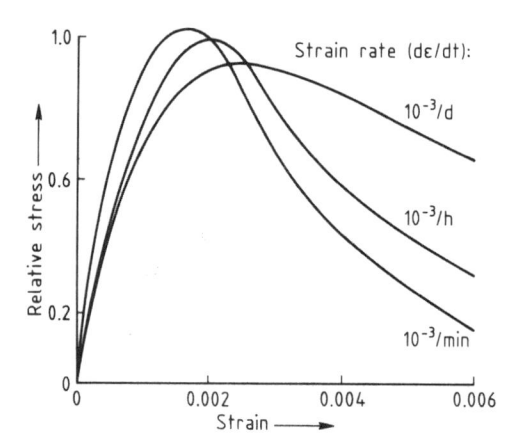

Figure 17. Influence of strain rate on the stress–strain relationship of concrete

2.5.3.1. Relationship of Stress to Strain

Stress–strain diagrams of normal concretes with different strengths are presented for compression and tension in Figure 16. The diagrams were recorded at a constant strain rate. Under compression, these diagrams are nonlinear even at low stress levels. With increasing stress, the nonlinearity increases. The strain at maximum compressive stress is almost independent of the strength of the concrete and amounts to ca. 2×10^{-3}. The stress–strain relationships are also influenced by the strain rate, as shown in Figure 17. A reduction in strain rate causes a slight reduction in strength and a pronounced increase in the ductility of the concrete.

The stress–strain relationship for concrete loaded in tension is almost linear up to failure, which is normally brittle. Failure strains range from ca. 0.1×10^{-3} for a concrete loaded in axial tension to 0.3×10^{-3} for concrete subjected to flexure [53], [110].

2.5.3.2. Modulus of Elasticity and Poisson's Ratio

The data in Figure 16 show that there is no constant modulus of elasticity for concrete in any part of the stress–strain curve. It is, therefore, customary to define an initial tangent modulus given by the slope of the stress–strain curve at the origin, a tangent modulus at a given stress, or a secant modulus over a certain range of the diagram at ca. $0 < \sigma \leq 0.5\, f_c$. The lower and upper limits of this range vary in different standards and recommendations [83], [130], [131]. The modulus of elasticity E_c for normal concrete is in the range 20 000 – 40 000 MPa. A higher compressive strength of the concrete leads to a higher modulus of elasticity. This correlation is expressed empirically as follows:

$$E_c = 9500 f_{cm}^{1/3}$$

where f_{cm} represents the mean compressive strength of a concrete cylinder [121].

The modulus of elasticity can also be deduced from the natural frequency of concrete specimens subjected to longitudinal, transverse, or torsional vibration as the dynamic modulus of elasticity [53]. These nondestructive tests yield higher values of E_c.

Poisson's ratio ν of concrete is about 0.2 as long as the stress level σ does not exceed $0.8\, f_c$. At higher stress levels, values of Poisson's ratio exceeding 0.5 are observed due to multiple crack formation [110].

2.5.3.3. Creep

Creep is defined as the time-dependent increase of load-dependent strain under constant stress. It is often expressed in terms of the creep coefficient, φ_c, which is the ratio of creep strain, ε_{cc}, to initial strain, ε_{ce}. For the range of working loads, the creep coefficient of concrete ranges from ca. 1 to 4. Creep is partially reversible. The part of creep that can be recovered on unloading is referred to as *delayed elastic strain*, whereas the irreversible part is generally described as *flow*.

The water content of concrete exerts an important influence on creep deformation. *Basic creep* is defined as the creep of concrete in a state of moisture equilibrium. Basic creep is smaller as the water content of the concrete prior to load application decreases. *Drying creep* is defined as the difference between basic creep and the creep of concrete that is allowed to dry under load. Drying creep is thus proportional to moisture loss. Creep deformation increases because of microcracking, particularly at higher stress levels [132].

Creep of concrete increases with an increase in the amount of cement paste, of the water content of the concrete prior to loading, and of the water:cement ratio. Creep

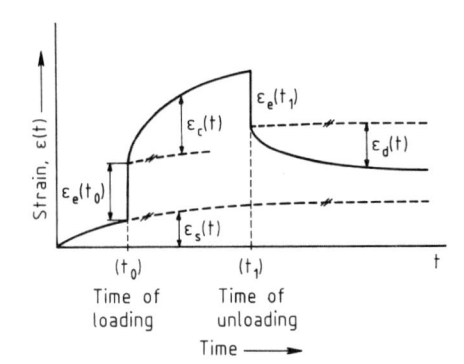

Figure 18. Total deformation of concrete loaded at t_0 and unloaded at t_1 as a function of time.
ε_e is elastic strain, ε_s is shrinkage strain, ε_c is total creep strain, ε_d is delayed elastic strain

increases at low relative humidity and with decreasing member size. A higher degree of hydration at the time of loading reduces creep.

There are several mathematical methods for estimating the time development of creep. In these formulas, which may incorporate viscoelastic models, terms or functions represent characteristics of the concrete, geometry of the member, and nature of the environment [132], [133].

2.5.3.4. Shrinkage and Swelling

Shrinkage and swelling are load- and temperature-independent volume changes essentially caused by changes in the moisture content of the concrete. As a first approximation, shrinkage is proportional to the moisture loss. When the concrete is rewetted, swelling occurs, although some irreversible shrinkage may remain.

The total shrinkage of hydrated cement paste may amount to several percent. Aggregates restrain shrinkage so that shrinkage of concrete is considerably less and is ca. 0.1×10^{-3} to 1×10^{-3}. Rapid drying of concrete generates shrinkage-induced tensile stresses in the surface zones that may lead to cracks [132], [133].

The total deformation of concrete is shown as a function of time in Figure 18.

2.6. Physical Properties

Most of the physical properties of concrete are controlled by its composition, mainly by the type and amount of aggregates used and the water content of the fresh or hardened concrete.

2.6.1. Density

Concrete is classified according to its density:

lightweight concrete	$\varrho \leq 1850$ kg/m^3 [134]
	$\varrho \leq 2000$ kg/m^3 [135]
normal-weight concrete	$\varrho \leq 2800$ kg/m^3
heavyweight concrete	$\varrho > 2800$ kg/m^3, [67], [82]

Structural lightweight concrete can be reinforced or prestressed like normal concrete. Heavyweight concrete is used mainly as a radiation-shielding material.

2.6.2. Porosity

The total porosity of concrete comprises gel pores and capillary pores in the hydrated cement paste, the porosity of aggregates and aggregate interfaces, and voids due to incomplete consolidation. Additionally, a protective pore system may be introduced into the matrix by air-entraining additives or air-entraining cements. The size of pores ranges from small gel pores with diameters of ca. 10^{-6} mm up to several millimeters for the voids [136]. In normal concrete with dense aggregates, the total porosity is 10–15% by volume. Air-entrained concrete contains an additional 2–6% pores. Lightweight concrete with porous aggregates, lightweight concrete with a honeycomb structure, or cellular lightweight concrete may have a total porosity of 60–90% by volume. Such concretes have low strength and are used only for thermal insulation [53].

2.6.3. Thermal Conductivity

The thermal conductivity of concrete increases with decreasing porosity and increasing moisture content. Crystalline aggregates give higher conductivity than amorphous aggregates. For normal concrete the thermal conductivity is 1.5–3.7 W m^{-1} K^{-1}. The thermal conductivity is shown as a function of the density of dry concrete in Figure 19 [53].

2.6.4. Electrical Conductivity

Dissolved ions present in the water in the pores of hardened concrete are primarily responsible for the electrical conductivity. For dry concrete, specific ohmic resistance is 10^{11} $\Omega \cdot$ cm, whereas for concrete saturated with water it is ca. 10^4 $\Omega \cdot$ cm. Concrete contaminated with soluble salts has even lower resistance. Because the water in the pores is an electrolyte, conductivity increases with increasing water:cement ratio [53].

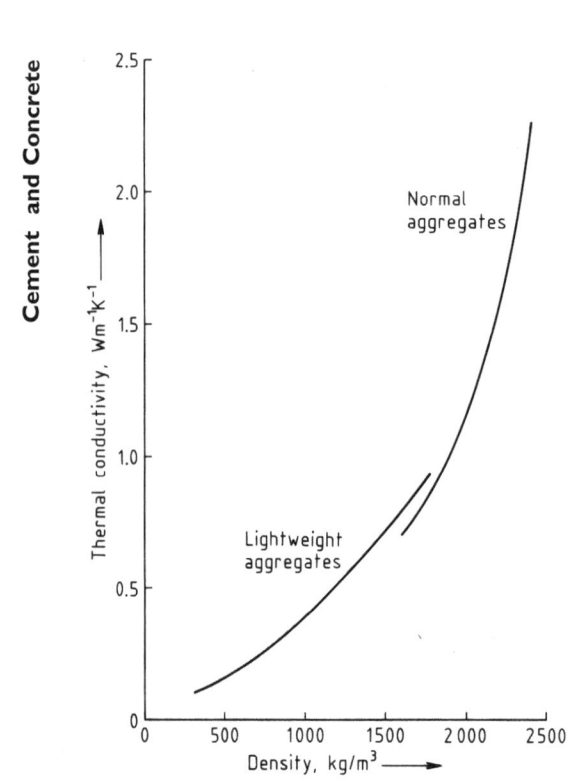

Figure 19. Thermal conductivity of concrete as a function of concrete density

2.6.5. Permeability

An important parameter for concrete durability is its permeability, which describes the ability of gases or liquids to penetrate into and through concrete sections. The movement of gases or liquids in concrete primarily occurs in the capillary pores of the cement paste matrix. Because the volume of capillary pores increases with an increasing water:cement ratio, permeability increases correspondingly. High values of permeability can be expected when the capillaries form an interconnected continuous network caused by a low degree of hydration. For water:cement ratios exceeding ca. 0.7, continuity of capillaries must be assumed even after complete hydration [53].

For normal concrete mixes with dense aggregates, the *coefficient of water permeability* is 10^{-11} to 10^{-13} m/s. Little information is available on *gas permeability*. Coefficients of gas permeability of 10^{-15} to 10^{-17} m^2/s have been reported [137].

Moisture movement is important in the drying of concrete. The diffusion of water molecules in the vapor phase or in condensed films depends strongly on the local moisture concentration and can be 10^{-8} to 10^{-11} m^2/s for concretes of normal compositions at relative humidity of 0–100 % [137].

2.6.6. Thermal Expansion

The thermal expansion of concrete is controlled by the thermal expansion of the cement paste and the aggregates. Therefore, different aggregates lead to different coefficients of thermal expansion of the concrete. Restraint of the matrix by coarse aggregates or the formation of microcracks, however, causes a total deformation of concrete that is not equal to the sum of the individual deformations of matrix and aggregates weighted by their volume concentrations. The coefficient of thermal expansion for concrete made with quartzitic aggregates is ca. 12×10^{-6} K^{-1}. For concrete made with calcareous aggregates, it is ca. 6×10^{-6} K^{-1} [53], [138].

Changing temperatures always cause a moisture movement in the matrix, which may lead to shrinkage or swelling. These deformations are added to the thermal dimensional changes.

2.6.7. Shielding Properties

Concrete is an economical material for shielding against high-energy X-rays, gamma rays, or neutrons because it combines structural properties with good shielding characteristics. Because of its heterogeneity, concrete contains a variety of different elements for attenuation or absorption of the different types of radiation. Although these elements exist in normal concrete, special designs can improve the shielding properties.

Neutrons are classified according to their energy as fast, intermediate, and slow (thermal) neutrons. The attenuation of fast and intermediate neutrons by elastic scattering by light elements, such as hydrogen or boron, and inelastic scattering by heavy elements lead to slow neutrons that are finally absorbed by heavy elements. Gamma rays are emitted in the steps of neutron attenuation [53], [139].

Because the attenuation of X-rays and gamma rays is proportional to the density of the material penetrated, heavy elements are required for effective shielding.

Special aggregates for shielding concrete can be natural heavy-element aggregates like baryte, iron ores, or granulated iron. Light elements can be added by using aggregates that contain large amounts of crystal water such as serpentine or by the introduction of insoluble boron compounds [140].

In the structural design of a biological shield, one should keep in mind that the attenuation of radiation generates heat that causes stress. Furthermore, high-energy radiation causes crystal defects in the aggregates that may reduce the strength of the concrete [141].

2.7. Durability

Structural concrete is subjected to a variety of physical and physicochemical influences that result from environmental conditions or other types of attack. In most instances, concrete needs no particular protective measures. In reinforced or prestressed concrete structures, the concrete must prevent corrosion of the reinforcement by providing a high alkalinity to passivate the steel surface.

Concrete aggregates are inert to most types of external attack. However, the hydrated cement paste may interact with the environment because of its interconnected *pore structure*. Water, ionic solutions, or gases may penetrate into the concrete through the capillary pores or microcracks to react chemically with constituents of the cement paste, thus causing deterioration. A durable concrete, therefore, should have a dense structure with a low total porosity. Low water: cement ratios and sufficient curing of the concrete are mandatory.

Another important factor in durability is the *moisture content* of the concrete. Because most damage mechanisms require a certain amount of water, dry concrete exhibits a high durability. Continuous wetting of concrete may increase the depth of penetration of dissolved compounds, and repeated drying–wetting cycles can accelerate the deterioration [142].

2.7.1. Chemical Attack

Chemical attack on hydrated cement paste involves either partial or complete dissolution of certain hydration products or formation of new compounds, accompanied by an increase in solid volume or in expansive gels.

Water that does not contain excessive amounts of ions or industrial waste pollutants does not dissolve cement paste compounds to a significant degree. However, continuing exposure to distilled water may leach the calcium hydroxide from the paste. Natural waters containing considerable amounts of *carbonic acid* will attack the cement paste and calcareous aggregates, forming soluble calcium hydrogen carbonate. Efflorescence of $CaCO_3$ is often observed.

The compounds of hydrated cement paste are dissolved by *organic and inorganic acids*, and the rate of dissolution increases as the pH of the environment decreases. Moderate attack can be seen when the pH of the surrounding solution drops below 6.5–6.8.

The formation of expansive phases in interior concrete sections can be detected by the formation of a network of surface cracks, through which white efflorescence or penetration of a gellike material can occur.

Sulfates penetrating into the matrix will react with either calcium hydroxide or calcium aluminates to form gypsum ($CaSO_4 \cdot 2 H_2O$) or ettringite ($3 CaO \cdot Al_2O_3 \cdot 3 CaSO_4 \cdot 32 H_2O$). The disruptive effect of ettringite formation is more pronounced because it involves a considerable increase in solid volume [143].

Magnesium compounds, either contained in the aggregates or penetrating from outside, can be decomposed to form an expansive gel of magnesium hydroxide.

Alkali–silica reactions occur between the alkali materials of the cement and amorphous silica of some siliceous aggregates. When moisture is available, an expansive gel is formed that causes a continuous deterioration of the concrete. Questionable aggregates can be tested for the presence of reactive silica [70]. Their use as concrete aggregates may require special cements with low effective alkali contents.

The activity of natural waters and soils is evaluated in standards, where recommendations are given for the composition of concrete to be used in contact with them. Concrete that is subject to strong chemical attack may require protective coatings of organic materials on the surface [67], [143]–[145]. A comprehensive list of chemical compounds and their effect on concrete is given in [144].

2.7.2. Physical Attack

2.7.2.1. Frost Action

The mechanism of freezing of water in concrete is rather complex and not entirely clear. Thorough discussions of this subject may be found in [136]. The following description is highly simplified.

Water contained in the larger capillary pores of concrete starts to freeze at temperatures below ca. − 8 °C. If the concrete is critically saturated, i.e., more than ca. 90 % of its pores are water-filled, the expansion of ice may have a disruptive effect if free expansion of ice or migration of the remaining liquid water into free spaces cannot take place. Thus, important parameters for controlling frost damage are the amount of freezable water in the concrete and the rate of ice formation.

The application of deicing salts to a frozen concrete surface may cause further damage. While the ice on the concrete surface will melt, a sudden drop in the temperature of the underlying concrete layer occurs. Furthermore, deicing salts can increase the degree of saturation and the number of freeze–thaw cycles. Additional damaging effects due to osmotic pressure or pressure from salt crystallization have not been clarified [136].

Preparing concrete with high resistance to freezing and deicing agents requires aggregates with a high resistance to freezing and a dense pore structure of the cement paste that is provided by low water : cement ratios and sufficient curing. Furthermore, when critical saturation is likely to occur, a protective system of small, spherical, closely spaced pores must be introduced into the mortar matrix by air-entraining additives or air-entraining cements. Because these pores do not fill with water under normal conditions, they provide space for the expansion of ice or water.

2.7.2.2. Elevated Temperature

The resistance of concrete to elevated temperatures depends on the mineral composition of the aggregates and the water content of the concrete. The mortar matrix in a concrete that is allowed to dry while hot will exhibit large shrinkage deformations, whereas the aggregates expand as the temperature increases. These opposing deformations lead to the formation of cracks, thus reducing concrete strength [146]. The cement paste decomposes at temperatures > 250 °C [147].

Concrete exposed to temperatures > 100 °C in a moist state undergoes phase transformations in the hydrated cement paste. In the absence of finely divided siliceous material, these phase transformations increase the total porosity of the paste and cause a significant reduction in the concrete strength. If fine siliceous aggregates or siliceous additives are present, new compounds are formed that can compensate for the loss of strength [146].

2.7.2.3. Radiation

Concrete subjected to high-energy radiation may lose strength. While there is no damaging effect of radiation on the hydrated cement paste, lattice defects can be formed in some types of crystalline rocks used as aggregates. Heterogeneous rock material with crystalline compounds such as granites undergoes a substantial loss in strength [141].

2.7.2.4. Abrasion and Wear

The surface regions of concrete structures can be subjected to abrasion and wear due to traffic loads, sliding, or scraping of hard objects, or repeated impact. Damage is often encountered on pavements, walls of silos or bunkers, and in pipes and channels for water that carries solids (erosion). Water flowing at a high velocity can cause cavitation. Concrete with a high resistance to wear requires a low water:cement ratio and careful curing. Vacuum processing of the fresh concrete and the use of hard aggregates or additives can substantially improve the resistance to wear.

2.7.3. Corrosion of the Reinforcement

The surface zones of reinforced concrete structures can be damaged by corrosion of the embedded reinforcement. The formation of voluminous iron hydroxides generates tensile stresses that cause spalling of the concrete cover. The most common causes of corrosion of the reinforcement are carbonation of the concrete or intrusion of reactive ions such as chlorides.

Carbon dioxide in the atmosphere penetrates the surface zones of concrete and reacts with the alkaline hydration products of the cement. In this diffusion-controlled reaction, insoluble calcium carbonates are formed and the pH of the pore solution of the matrix drops from more than 12 to ca. 8.7. In this environment, an embedded steel reinforcement is no longer protected by passivation because the passive layers become unstable, and anodic dissolution of the iron occurs when oxygen and moisture is available [148].

The passive layer of steel in concrete can also be destroyed by chloride ions. The major sources of chloride contamination of concrete are deicing agents and seawater.

Although some chloride can be immobilized chemically or physically by the hydration products of the cement, additional chloride ion greater than a threshold concentration will destroy the passive layers of the steel reinforcement locally, and corrosion occurs [81], [137]. Structures exposed to seawater may, therefore, require an increased concrete cover of the reinforcement [53].

Carbonation and intrusion of active ions are controlled by diffusion. A dense pore structure of the cement paste matrix, especially in the surface zones, reduces the penetration rate and thus improves durability. Under very severe conditions, additional protective measures such as surface coating of the concrete or the steel reinforcement may be required.

2.8. Special Concretes

The basic principles of concrete technology for normal concrete are generally valid for other types of concrete, which may differ from normal concrete with respect to composition or properties tailored for special applications. Preparation, handling, and testing of such concretes are often specified in separate standards or guidelines. The most important types of concrete that differ from normal concrete are lightweight and heavyweight concrete, mass concrete, fiber-reinforced concrete, and polymer concrete.

2.8.1. Lightweight Concrete

Lightweight concrete has a lower bulk density than normal concrete, which can be achieved in several ways. Structural lightweight concrete with a dense structure is prepared with lightweight aggregates, such as tuff, expanded shales, or calcined waste products. Structural lightweight concrete may be reinforced or prestressed like normal concrete.

Lightweight concrete with a honeycomb structure is prepared with only coarse aggregates, which are enveloped by a thin shell of mortar, thus providing a large volume of interstitial voids. Such concretes can be used for masonry units or drainage purposes. Cellular lightweight concrete can be prepared by introducing foam-forming additives into a mortar mix. Because of its low compressive strength, cellular lightweight concrete is used only for making masonry units or for thermal insulation.

The basic relations between the water:cement ratio and concrete strength are valid for structural lightweight concrete, although with limitations because the strength of the matrix may be higher than the strength of the aggregates. Furthermore, the definition of a water:cement ratio is uncertain because the porous aggregates can absorb various amounts of the water added to the mix. Preparing lightweight concrete, therefore, always requires trial mixes. In general, compressive strength increases with increasing bulk density. For densities of $1.0-2.0$ kg/dm^3, the compressive strength is $10-60$ MPa. Compared with normal concrete, lightweight concrete has a lower compressive strength, lower modulus of elasticity, lower thermal conductivity, and lower thermal expansion, but higher creep and shrinkage deformations [53].

2.8.2. Heavyweight Concrete

The major applications of heavyweight concrete are as shielding against nuclear radiation, as ballast, and as anchor bodies. Heavyweight aggregates like baryte, ores with densities of $3.6-4.6$ kg/dm^3, slags, or iron can be used to achieve bulk densities of the concrete exceeding 2.8 kg/dm^3. The mix design for heavyweight concrete follows the principles for normal concrete except that additional care must be taken to ensure the specified concrete density. Segregation must be avoided in handling heavyweight concrete. Therefore, comparatively dry mixes are preferred [82], [90].

2.8.3. Massive Concrete

The term massive concrete refers not to a particular composition but rather to a concrete structure of such size that the heat of hydration of the cement must be considered. In interior parts of large concrete cross-sections, the heat of hydration may generate temperatures > 80 °C, causing thermal stresses that can lead to crack formation in the concrete during early hydration. The temperature increase in the concrete member can be kept low by using cements with a low heat of hydration, addition of finely divided pozzolanas, and reduction of the cement content of the mix. The size of the aggregates may be as great as $125-150$ mm. The curing of massive concrete may require thermal insulation of the surfaces to avoid steep temperature gradients over the cross-section. Otherwise cracking of the concrete is likely to occur [149], [150].

2.8.4. Fiber-reinforced Concrete

The mortar matrix of concrete can be reinforced by short fibers randomly oriented and homogeneously distributed. The diameter of these fibers may vary from several micrometers to ca. 1 mm and the length from a few millimeters to 50 or 60 mm. High-strength steel fibers are used in fiber-reinforced shotcrete, and alkali-resistant glass fibers are sometimes used in thin shells or cladding elements. New types of organic fibers are being tested as substitutes for asbestos.

Fiber-reinforced concrete has good ductility in both compression and tension [151]. Problems in manufacturing fiber-reinforced concrete are the limited workability of the fresh concrete and the diminished effectiveness of the fibers because of insufficient bonding between fiber and matrix.

2.8.5. Polymer Concrete

In polymer concrete, organic resins form an essential part of the binding material. Polymer concrete has high strength in tension and compression, and exhibits a superior resistance to chemical attack, freezing, and thawing as compared to normal concrete. Disadvantages may result from the high cost of the resins and reduced thermal stability compared to that of normal concrete. The polymers may be added to the fresh mix as the only binding material, e.g., epoxy concrete; they can be combined with water and cement, as in polymer cement concrete (PCC); or an already hardened concrete can be impregnated with monomers for subsequent polymerization as in (polymer impregnated concrete (PIC)) [80].

2.9. Mortar

Mortar is distinguished from concrete by the maximum size of its aggregates. In general, mortar contains only sand with a maximum grain size of 2–4 mm. Other binding materials such as hydraulic lime can be used in addition to cement.

In concrete construction, cement mortars are used for connecting precast elements, as injection mortars in ducts of posttensioned structures with bonded tendons, and for the repair of old concrete surfaces with areas of deterioration.

Because of the small maximum size of aggregates, mortars contain higher water:cement ratios than are normally encountered in concrete technology. Therefore, they exhibit lower compressive strength and volume stability. These disadvantages can be offset by the use of water-reducing and expansion-producing additives [53], [90], [110].

The composition and strength requirements for masonry mortars are specified in national standards [152], [153]. Lightweight aggregate mortars with reduced thermal conductivity have been developed to improve the thermal insulation properties of a masonry structure. Polymer-modified cement-based mortars with fine sand as the aggregate make possible bed joints only ca. 3 mm thick (thin-layer mortar). Centrally mixed mortars may contain retarding additives to allow placement of the mortar up to 48 h after mixing.

A variety of polymer-modified mortars designed for the repair of concrete structures with deteriorated surface zones are available. Basic requirements of mortar for such applications are the ability to form strong bonds with old concrete and a low shrinkage strain.

3. References

[1] S. B. Newberry, W. B. Newberry: "The constitution of hydraulic cements," *J. Soc. Chem. Ind. London* **16** (1897) 887–893.

[2] E. Wetzel: "Bericht über den Stand der auf Antrag des Vereins Deutscher Portland-Cement-Fabrikanten im Kgl. Materialprüfungsamt ausgeführten Arbeiten über die Konstitution des Portlandzements," *Protokolle des Vereins deutscher Portland-Zement-Fabrikanten* (1911) 281–306; *Protokolle des Vereins deutscher Portland-Zement-Fabrikanten* (1912) 217–249; *Protokolle des Vereins deutscher Portland-Zement-Fabrikanten* (1913) 347–358; *Protokolle des Vereins deutscher Portland-Zement-Fabrikanten* (1914) 145–160358.

[3] E. Spohn: "Die Kalkgrenze des Portlandzementes und die technischen Eigenschaften seiner Klinkermineralien," *Zement* **21** (1932) no. 702–706, 717–723, 731–736.

[4] F. M. Lea, T. W. Parker: "The quaternary system $CaO-Al_2O_3-SiO_2-Fe_2O_3$ in relation to cement technology," *Build. Res. Techn. Paper* nr. 16, published by His Majesty's Stationery Office, London (1935).

[5] H. Kühl: "Gelöste und ungelöste Aufgaben der Zementforschung," *Protokolle des Vereins deutscher Portland-Zement-Fabrikanten* (1936) 196–216.

[6] A. Guttmann: "Der Einfluß von Gips- und Chlorkalziumzusätzen zum Zement auf sein Schwinden," *Zement* **9** (1920) 310–313, 429–432. DE 330 784, 1920 (A. Guttmann).

[7] *Klein Symposium on Expansive Cement Concretes*, Publication SP-38, Am. Concrete Assoc., 1973.

[8] Portland Cement Assoc., DE-OS 1 929 684, 1969.

[9] Onoda Cement Co., DE-OS 2 165 434, 1971.

[10] A. E. Törnebohm: "Die Petrographie des Portlandzements," *Tonind. Ztg.* **21** (1897) no. 1148–1151, 1157–1159.

[11] J. H. Welch, W. Gutt: "Tricalcium silicate and its stability within the system $CaO-SiO_2$," *J. Am. Ceram. Soc.* **42** (1959) 11–15.

[12] E. Woermann: "Decomposition of alite in technical portland cement clinker," *Proc. Int. Sympos. Chem. Cem.*, 4th (1960) vol. I, 119–128.

[13] A. Guinier, M. Regourd: "Structure of portland cement minerals," *Proc. Int. Sympos. Chem. Cem.*, 5th (1968) vol. I, 1–32.

[14] H. E. Schwiete, H. zur Strassen: "Über den Einfluß des Magnesiagehalts im Portlandzementklinker auf das Tetracalciumaluminatferrit," *Zement* **23** (1934) 511–514.

[15] M. Regourd, A. Guinier: "The crystal chemistry of the constituents of portland cement clinker," *Proc. Int. Sympos. Chem. Cem.* 6th (1974) Principal Paper I–2.

[16] M. A. Swayze: "A report on studies of 1. the ternary system $CaO-C_5A_3-C_2F$, 2. the quaternary system $CaO-C_5A_3-C_2F-C_2S$, 3. the quaternary system as modified by 5% magnesia," *Am. J. Sci.* **244** (1946) no. 1–30, 65–94.

[17] H. Rechmeier: "Der fünfstufige Wärmetauscherofen zum Brennen von Klinker aus Kalkstein und Ölschiefer," *Zement-Kalk-Gips* **23** (1970) 249–253.

[18] H. zur Strassen: "Der theoretische Wärmebedarf des Zementbrandes," *Zement-Kalk-Gips* **10** (1957) 1–12.

[19] Verein Deutscher Zementwerke e.V.: *Mikroskopie des Zementklinkers*, Beton-Verlag, Düsseldorf 1965.

[20] R. H. Bogue: "Calculation of the compounds in portland cement", *Ind. Eng. Chem., Analyt. Ed.* **1** (1929) 192–197.

[21] F. W. Locher: "Berechnung der Klinkerphasen," *Zement-Kalk-Gips* **14** (1961) 573–580; *Schriftenr. Zementind.* **29** (1962) 7–19.

[22] H. Kühl: *Zement-Chemie*, 3rd. ed.,vol. **2, 3** VEB Verlag Technik, Berlin 1961.

[23] H. G. Smolczyk: "Zum Einfluß der Chemie des Hüttensands auf die Festigkeiten von Hochofenzementen," *Zement-Kalk-Gips* **31** (1978) 294–296.

[24] F. Keil: "Zur Bewertung der Zementschlacken," *Zement* **33** (1944) 90–93.

[25] G. Haegermann: "Über die Bewertung der Erhärtungsfähigkeit hydraulischer Zusatzstoffe in Mischungen mit Portlandzement," *Zement* **33** (1944) 93–97.

[26] W. Wittekindt: "Zur Qualitätsbeurteilung von Hochofenschlacken und Puzzolanen," *Zement-Kalk-Gips* **16** (1963) 314–320.

[27] R. L. Blaine: "A simplified air permeability fineness apparatus," *ASTM Bull.* **123** (1943) 51–55.

[28] R. Rohrbach: "Herstellung von Ölschieferzement und Gewinnung elektrischer Energie aus Ölschiefer nach dem Rohrbach-Lurgi-Verfahren," *Zement-Kalk-Gips* **22** (1969) 293–296.

[29] G. Wischers, W. Richartz: *Einfluß der Bestandteile und der Granulometrie des Zements auf das Gefüge des Zementsteins*, Beton-Verlag, Düsseldorf 1982.

[30] F. W. Locher, W. Richartz, S. Sprung: "Erstarren von Zement," Part I: "Reaktion und Gefügeentwicklung," *Zement-Kalk-Gips* **29** (1976) 435–442; Part II: "Einfluß des Calciumsulfatzusatzes," *Zement-Kalk-Gips* **33** (1980) 271–277.

[31] W. Richartz: "Einfluß von Zusätzen auf das Erstarrungsverhalten von Zement," *Beton* **33** (1983) no. 425–429, 465–471.

[32] L. E. Copeland, D. L. Kantro, G. Verbeck: "Chemistry of hydration of portland cement," *Proc. Int. Sympos. Chem. Cem. 4th* (1960) vol. **1**, 429–465.

[33] T. C.Powers, T. L. Brownyard: "Studies of the physical properties of hardened portland cement paste," Part 2: "Studies of water fixation," *Proc. Am. Concr. Inst.* **43** (1946) 249–336; *Res. Lab. Portl. Cem. Assoc., Bull.* 22 (1947).

[34] T. C. Powers: "The nonevaporable water content of hardened portland cement paste – Its significance for concrete research and its method of determination," *ASTM Bull.* **158** (1949) 68–76; *Res. Lab. Portl. Cem. Assoc., Bull.* 29 (1949).

[35] R. F. Feldman, P. Sereda: "A model for hydrated portland cement paste as deduced from sorption-length change and mechanical poperties," *Matér. Constr.* **1** (1968) 509–520.

[36] T. C. Powers: "The thermodynamics of volume change and creep," *Matér. Constr.* **1** (1968) 487–507.

[37] R. A. Helmuth, D. H. Turk: "The reversible and irreversible drying shrinkage of hardened portland cement and tricalcium silicate pastes," *J. Res. Portl. Cem. Assoc. Res. Dev. Lab.* **9** (1967) no. 2, 8–21; *Res. Dev. Lab. Portl. Cem. Assoc. Bull.* 215 (1967).

[38] G. Wischers: "Physikalische Eigenschaften des Zementsteins," *Beton* **11** (1961) 481–486.
[39] T. C. Powers: "Structure and physical properties of hardened portland cement paste," *J. Am. Ceram. Soc.* **41** (1958) 1–6.
[40] G. Wischers: "Zur Normung von Zement," *Beton* **21** (1971) no. 147–151, 193–197, 241–245.
[41] Verein Deutscher Ingenieure: VDI-Richtlinie 2094 *"Emissionsminderung Zementwerke;"* Sept. 1985.
[42] S. Sprung: "Die chemische und mineralogische Zusammensetzung von Zementofenstaub," *Tonind. Ztg.* **90** (1966) 441–449.
[43] F. W. Locher: "Stoffkreisläufe und Emissionen beim Brennen von Zementklinker," *Fortschr. Mineral.* **60** (1982) 215–234.
[44] S. Sprung: "Technologische Probleme beim Brennen des Zementklinkers, Ursache und Lösung," *Schriftenr. Zementind.* no. 43 (1982).
[45] S. Sprung: "Das Verhalten des Schwefels beim Brennen von Zementklinker," *Schriftenr. Zementind.* no. 31 (1964).
[46] F. W. Locher, S. Sprung, D. Opitz: "Reaktionen im Bereich der Ofengase," *Zement-Kalk-Gips* **25** (1972) 1–12.
[47] S. Sprung, H.-M. v. Seebach: "Fluorhaushalt und Fluoremission von Zementöfen," *Zement-Kalk-Gips* **21** (1968) 1–8.
[48] W. Rechenberg: Über das Verhalten von Mahlhilfen bei der Zementmahlung, in preparation.
[49] H. Pisters: "Chrom im Zement und Chromatekzem," *Zement-Kalk-Gips* **19** (1966) 467–472.
[50] Cembureau: *Statistical Review* no. 34, 1977; *World Statistical Review* no. 3, 1981; no. 7, Paris 1985. Zement-Nachrichten no. 5–6 (1986) des Bundesverb. Dt. Zementind.
[51] Cembureau: *European Annual Review* no. 7, Paris 1985.
[52] S. Mindess, J. F. Young: *Concrete*, Prentice-Hall, Inc., Englewood Cliffs 1981.
[53] A. M. Neville: *Properties of Concrete*, Pitman Publ., London 1973.
[54] *Zahlen aus der Zementindustrie*, Bundesverband der Deutschen Zementindustrie e. V., Köln 1984.
[55] W. Marmé, J. Seeberger: "Der Primärenergieinhalt von Baustoffen," *Bauphysik* **4** (1982) no. no. 5, 155–165; **4** (1982) no. 6, 208–214.
[56] W. A. Gutteridge, C. D. Pommery: "Cement in its Conventional Uses: Problems and Possibilities," *Technology in the 1990s: Developments in Hydraulic Cement*, The Royal Society, London 1983, pp. 7–15.
[57] J. D. Birchall: "Cement in the context of new materials for an energy-expensive future," *Technology in the 1990s: Developments in Hydraulic Cements*, The Royal Society, London 1983, pp. 31–42.
[58] H. K. Hilsdorf, J. Kropp: "Entwicklungstendenzen der Baustoffe," *Beratende Ingenieure* **1** (1982) 18–23.
[59] DIN 1164: Portland-, Eisenportland-, Hochofen-, und Trasszement (1978).
[60] ASTM C 150-84: Specification for Portland Cement.
[61] ASTM C 595-83: Specification for Blended Hydraulic Cements.
[62] EN 197: Definitions and Specifications for Cements (1984).
[63] S. Popovics: *Concrete-Making Materials*, Hemisphere Publ. Corp., Washington 1979.
[64] DIN 4226: Zuschlag für Beton (1983).
[65] ASTM C 136-84: Method for Sieve Analysis of Fine and Coarse Aggregates.
[66] ISO 6274: Concrete – Sieve Analysis of Aggregates.

[67] DIN 1045: Beton und Stahlbeton – Bemessung und Ausführung (1978).
[68] ASTM C 33-84: Standard Specification for Concrete Aggregates.
[69] G. M. Idorn, S. Rostam: "Alkalis in Concrete – Research and Practice," *Proc. 6th Int. Conf.*, Copenhagen, 22–25 June 1983.
[70] ASTM C 227-81: Potential Alkali Reactivity of Cement-Aggregate Combinations.
[71] "Zugabewasser für Beton. Merkblatt für die Vorabprüfung und die Beurteilung vor Baubeginn sowie die Prüfungswiederholung während der Bauausführung, Jan. 1982," *Beton Stahlbetonbau* **77** (1982) no. 5, 137–140.
[72] ASTM C 494-82: Specification for Chemical Admixtures for Concrete.
[73] M. R. Rixom, *Chemical Admixtures for Concrete*, J. Wiley & Sons, Inc., New York 1978.
[74] *Superplasticizers in Concrete*, American Concrete Institute, Detroit 1979, Publication SP–62.
[75] ASTM C 457-82 a: Standard Practice for Microscopical Determination of Air-Void Content and Parameters of the Air-Void System in Hardened Concrete.
[76] ASTM C 618-84: Standard Specification for Fly Ash or Calcined Pozzolan for Use as a Mineral Admixture in Portland Cement Concrete.
[77] DIN 51043: Trass, Anforderungen, Prüfung (1979).
[78] *Fly Ash, Silica Fume, Slag and Other Mineral By-Products in Concrete*, vol. **I**, II, American Concrete Institute, Detroit 1983, Publication SP 79.
[79] L. Hjorth: "Development and Application of High-Density Cement-Based Materials," *Technology in the 1990s: Developments in Hydraulic Cements*, The Royal Society, London 1983, pp. 167–174.
[80] *Polymers in Concrete*, American Concrete Society, Detroit 1978, Spec. Publ. SP 58.
[81] "Guide to Durable Concrete," ACI Committee 201, *ACI Manual of Concrete Practice*, Part 1, American Concrete Institute, Detroit 1979.
[82] "Recommended Practice for Selecting Proportions for Normal and Heavyweight Concrete," ACI Committee 211, *ACI Manual of Concrete Practice*, Part 1, American Concrete Institute, Detroit 1979.
[83] DIN 1048: Prüfverfahren für Beton (1978).
[84] ASTM C 39-83 b: Test Method for Compressive Strength of Cylindrical Concrete Specimens.
[85] ISO 4012: Concrete – Determination of Compressive Strength of Test Specimens.
[86] RILEM Recommendation CPC 4: *Compression Test*.
[87] DIN 1084: Überwachung im Beton- und Stahlbetonbau (1978).
[88] ISO 3893: Concrete – Classification by Compressive Strength.
[89] K. Walz: *Herstellung von Beton nach DIN 1045*, Bauverlag GmbH, Düsseldorf 1972.
[90] Deutscher Betonverein e. V.: *Betonhandbuch*, Bauverlag, Wiesbaden 1984.
[91] "Recommended Practice for Measuring, Mixing Transporting and Placing Concrete," ACI Committee 304, *ACI Manual of Concrete Practice*, Part I, American Concrete Institute, Detroit 1979.
[92] ASTM C 143-78: Test Method for Slump of Portland Cement Concrete.
[93] RILEM Recommendation CPC 2.1: *Slump Test*.
[94] ISO 4109: Fresh Concrete – Determination of the Consistency – Slump Test.
[95] ISO 4110: Fresh Concrete – Determination of the Consistency – Vebe Test.
[96] RILEM Recommendation CPC 2.2: Vebe Test.
[97] ISO 4111: Fresh Concrete – Determination of the Consistency – Degree of Compactibility.
[98] ACI Standard 309-72:*Recommended Practice for Consolidation of Concrete*.
[99] "Recommended Practice for Curing Concrete," ACI Committee 308, *ACI Manual of Concrete Practice*, Part I, American Concrete Institute, Detroit 1979.

[100] "Hot Weather Concreting," ACI Committee 305, *ACI Manual of Concrete Practice*, Part I, American Concrete Institute, Detroit 1979.
[101] *Empfehlungen zur Nachbehandlung von Beton*, ed. 1983, Deutscher Ausschuß für Stahlbeton, Berlin 1984.
[102] ACI Standard 306: *Recommended Practice for Cold Weather Concreting*.
[103] "High Pressure Steam Curing: Modern Practice, and Properties of Autoclaved Products," ACI Committee 516, *ACI Manual of Concrete Practice*, Part III, American Concrete Institute, Detroit 1978.
[104] ACI Standard 517: *Recommended Practice for Atmospheric Pressure Steam Curing of Concrete*.
[105] ACI Standard 506: *Recommended practice for shotcreting*.
[106] ACI Standard 506.2: Specification for Materials, Proportioning, and Application of Shotcrete.
[107] DIN 18551: Spritzbeton; Herstellung und Prüfung.
[108] *Richtlinie für die Ausbesserung und Verstärkung von Betonbauteilen mit Spritzbeton*, Deutscher Ausschuß für Stahlbeton, Berlin 1976.
[109] "Stahlfaserspritzbeton," *Beton Stahlbeton* **79** (1984) no. 5, 134–136.
[110] K. Wesche: *Baustoffe für tragende Bauteile*, vol. **2**, Bauverlag GmbH, Wiesbaden 1981.
[111] J. Eibl, G. Ivanyi: *Studie zum Trag- und Verformungsverhalten von Stahlbeton*, no. 260, Deutscher Ausschuß für Stahlbeton, Verlag Ernst & Sohn, Berlin 1976.
[112] P.-E. Petersson: "Crack Development and Fracture Zones in Plain Concrete and Similar Materials," *Report TVMB – 1006*, Lund, Sweden, 1981.
[113] T. C. Powers, T. L. Brownyard: "Studies of the Physical Properties of Hardened Cement Paste," *Res. Dev. Lab. Portland Cem. Assoc. Res. Dep. Bull.* 1948, no. 22.
[114] ACI Standard 214:Recommended Practice for Evaluation of Strength Test Results of Concrete.
[115] ASTM C 192-81: Method of Making and Curing Concrete Test Specimens in the Laboratory.
[116] ASTM C 597-83: Test Method for Pulse Velocity Through Concrete.
[117] ASTM C 803-82: Test Method for Penetration Resistance of Hardened Concrete.
[118] ASTM C 805-79: Test Method for Rebound Number of Hardened Concrete.
[119] H. Rüsch, R. Sell, C. Rasch, E. Grasser, A. Hummel, K. Wesche, H. Flatten: *Festigkeit und Verformung von unbewehrtem Beton unter konstanter Dauerlast*, no. 198, Deutscher Ausschuß für Stahlbeton, Verlag Ernst & Sohn, Berlin 1968.
[120] RILEM Report 36 RDL: *Long Term Random Dynamic Loading of Concrete Structures*, Feb.1980.
[121] *Model Code for Concrete Structures*, CEB-FIP International Recommendations, 3rd ed., Comité Euro-International du Béton (CEB), 1978.
[122] ASTM C 78-84: Test Method for Flexural Strength of Concrete (Third Point Loading).
[123] ASTM C 293-79: Test Method for Flexural Strength of Concrete (Center Point Loading).
[124] ASTM C 496-71: Test Method for Splitting Tensile Strength of Cylindrical Concrete Specimens.
[125] RILEM Recommendation CPC 5: *Flexural Test*.
[126] RILEM Recommendation CPC 6: *Tension by Splitting*.
[127] RILEM Recommendation CPC 7: *Direct Tension*.
[128] F. H. Wittman (ed.): *Fracture mechanics of concrete*, Elsevier Science Publ., Amsterdam 1983.
[129] H. K. Hilsdorf, W. Brameshuber: *Size Effects in the Experimental Determination of Fracture Mechanics Parameters*, Nato Adv. Res. Workshop, Evanston, Illinois, USA, 1984.
[130] ASTM C 469-83: Test Method for Static Modulus of Elasticity and Poisson's Ratio of Concrete in Compression.
[131] RILEM Recommendation CPC 8: *Modulus of Elasticity of Concrete in Compression* .

[132] H. Rüsch: *Stahlbeton-Spannbeton*, vol. **1**: "Werkstoffeigenschaften und Bemessungsverfahren," Werner-Verlag, Düsseldorf 1972.

[133] H. Rüsch, D. Jungwirth, H. K. Hilsdorf: *Creep and Shrinkage*, Springer Verlag, New York 1983.

[134] ACI Standard 211.2: *Recommended Practice for Selecting Proportions for Structural Lightweight Concrete*.

[135] DIN 4219: Leichtbeton und Stahlleichtbeton mit geschlossenem Gefüge (1979).

[136] M. J. Setzer: *Einfluß des Wassergehaltes auf die Eigenschaften des erhärteten Betons*, no. 280, Deutscher Ausschuß für Stahlbeton, Verlag Ernst & Sohn, Berlin 1977.

[137] "Durability of Concrete Structures Under Normal Outdoor Exposure," *Proceedings of the RILEM Seminar*, Hannover, 26th – 29th March 1984.

[138] S. Ziegeldorf, K. Kleister, H. K. Hilsdorf: *Vorherbestimmung und Kontrolle des thermischen Ausdehnungskoeffizienten von Beton*, no. 305, Deutscher Ausschuß für Stahlbeton, Verlag Ernst & Sohn, Berlin 1979.

[139] T. Jaeger, *Grundzüge der Strahlenschutztechnik*, Springer Verlag, Berlin 1960.

[140] R. G. Jaeger: *Engineering Compendium on Radiation Shielding*, vol. **II**: "Shielding materials," Springer Verlag, Berlin 1975.

[141] J. Seeberger, H. K. Hilsdorf: *Einfluß von radioaktiver Strahlung auf die Festigkeit und Struktur von Beton*, Institut für Massivbau und Baustofftechnologie, Universität Karlsruhe 1982.

[142] S. Rostam: "Durability of concrete structures," *CEB-RILEM International Workshop*, Copenhagen, 18 – 20 May 1983.

[143] I. Biczok: *Betonkorrosion, Betonschutz*, Bauverlag GmbH, Wiesbaden 1968.

[144] DIN 4030: Beurteilung betonangreifender Wässer, Böden und Gase (1969).

[145] "Guide for the Protection of Concrete Against Chemical Attack by Means of Coatings and Other Corrosion-Resistant Materials," ACI Committee 515, *ACI Manual of Concrete Practice*, Part 3, American Concrete Institute, Detroit 1978.

[146] J. Seeberger, J. Kropp, H. K. Hilsdorf: *Festigkeitsverhalten und Strukturänderungen von Beton bei Temperaturbeanspruchung bis 250 °C*, no. 360, Deutscher Ausschuß für Stahlbeton, Verlag für Architektur und technische Wissenschaften, Berlin 1985.

[147] U. Schneider: *Verhalten von Beton bei hohen Temperaturen – Behavior of Concrete at High Temperatures*, no. 337, Deutscher Ausschuß für Stahlbeton, Verlag Ernst & Sohn, Berlin 1982.

[148] K. Wesche: *Baustoffe für tragende Bauteile*, vol. 3, Bauverlag GmbH, Wiesbaden 1985.

[149] "Mass Concrete for Dams and Other Massive Structures," ACI Committee 207, *ACI Manual of Concrete Practice*, Part 1, American Concrete Institute, Detroit 1979.

[150] *Sachstandbericht Massenbeton*, Deutscher Ausschuß für Stahlbeton, no. 329, Verlag Ernst & Sohn, Berlin 1982.

[151] A. Neville: *Fibre Reinforced Cement and Concrete*, The Construction Press, Hornby, Lancaster, 1975.

[152] ASTM C 270-82: Specification for Mortar for Unit Masonry.

[153] DIN 1053: Mauerwerk – Berechnung und Ausführung (1974).

Cements, Chemically Resistant

Jürgen Fenner, Keramchemie GmbH, Siershahn/Westerwald, Federal Republic of Germany

1.	Introduction	931	2.2.2. Furan Resin Cements	934
2.	Types of Cement	932	2.2.3. Epoxy Resin Cements	935
2.1.	Cements Containing Inorganic Binders	933	2.2.4. Unsaturated Polyester Resin Cements	936
2.1.1.	Silicate Cements	933	2.2.5. Vinylester Resin Cements	937
2.1.2.	Sulfur Cements	933	2.2.6. Bituminous Material Cements	938
2.1.3.	Hydraulic Cement Mortars	933	3. Specifications and Testing	938
2.2.	Cements Containing Organic Binders	934	4. Storage and Transportation	941
2.2.1.	Phenol–Formaldehyde Resin Cements	934	5. Toxicology	941
			6. References	941

1. Introduction

Chemically resistant cements are materials used in chemically resistant tiling or brick lining to join together such chemically resistant, nonmetallic units as tiles, bricks, stones, blocks, etc., and to the membrane (Figs. 1 and 2) [1], [2].

Chemically resistant units are, for example, ceramic tiles and bricks, refractory tiles and bricks, carbon bricks, parts made from porcelain, cast basalt, graphite, silicon carbide, granite, etc., alone or in combination with other materials. Chemically resistant bricklining is used, for instance, to protect the floors of production and storage buildings, pits, trenches, emergency containments, and production and storage tanks against chemicals that are produced or stored in these places.

Chemically resistant brick lining can be self-supporting, e.g., in chimneys, or it can serve as a mechanical, thermal, and, to some extent, a chemical barrier on top of a liquid-tight membrane [3]–[5].

Acid-proof, or rather chemically resistant, cement or mortar always consists of an inorganic or organic binder and inorganic fillers. Most of these materials are similar to mortar at normal temperatures, and they harden when the binder undergoes a chemical reaction. Some cements, however, can also be worked at higher temperatures; these harden as a result of physical setting or crystallize when cooled.

The cement of choice depends primarily on the desired chemical resistance and physical properties. Further factors to be considered are the compatibility with other components of the masonry, working time, setting time, stability during storage, transportation requirements, and price [6].

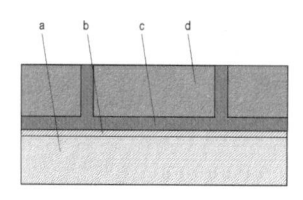

Figure 1. Fully bedded and jointed bricks
a) Substrate (concrete or steel); b) Membrane; c) Cement; d) Tiles or bricks

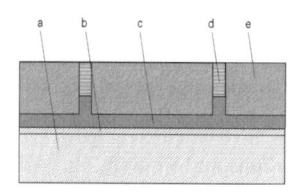

Figure 2. Brick lining with open joints
a) Substrate (concrete or steel); b) Membrane; c) Bedding cement; d) Jointing cement; e) Tiles or bricks

Stress caused by shrinkage and by changes in volume that occur during the hardening process must also be considered [7].

The oldest chemically resistant cements are the silicate cements, which date from about 1920. Even bitumen-bound materials gained early importance and were used especially for flooring. In the 1930s, cements based on phenol–formaldehyde resins were developed. Furan resins were introduced in the next decade, followed in the 1950s by unsaturated polyester and epoxy resins [4] and vinylester resins.

2. Types of Cement

Chemically resistant mortar is classified according to the binder it contains, and is further classified according to the type of filler it contains. Even if the binder and filler are the same, the properties of cements from different producers can differ with respect to mixing and application.

The following specifications, outlined according to the respective ASTM standards [8], are characteristic of each binder, but do not consider properties typical of any particular cement manufacturer.

2.1. Cements Containing Inorganic Binders

2.1.1. Silicate Cements

Soluble silicate cements contain potassium silicate or sodium silicate as the binder. The filler is quartz or another inert material. Sodium fluorosilicate, potassium fluorosilicate, aluminum phosphate, organic acid esters, amines, or formamide may be used as the hardener, which is usually mixed with the filler. Soluble silicate cements are delivered as one liquid component and one powder. They are used according to the instructions given in ASTM C397. The mixture is placed at ambient temperature and hardens without heating as a result of the coagulation of silicon dioxide caused by a change in pH. The hardened cement can withstand temperatures up to 850 °C.

Silicate cements are used to produce such self-supporting, chemically resistant constructions as the linings of chimneys, foundations, and other containers and equipment subject to high temperatures or exposed to acids. Because they are sensitive to water, they are not recommended for use in flooring.

Trade name: Keranol WG 300 (Keramchemie, FRG), Hoechst Acidproof Cement HB (Permatex/Hoechst, FRG), Acidproof Cement S 50 HF (Steuler Industriewerke, FRG).

2.1.2. Sulfur Cements

Quartz or carbon is the filler used in cements with a sulfur binder. The powdered cement is used in accordance with ASTM C386. It is heated until the sulfur melts and becomes free flowing (138 – 149 °C). The molten cement is poured into the open joints between spaced blocks, where it hardens upon cooling. The hardened cement can resist temperatures up to 88 °C.

Sulfur cements are used, primarily in the United States, for the linings of containers used in the pickling of steel with mixtures of nitric acid and hydrofluoric acid. Their use is limited because they are difficult to handle.

Trade name: Corobond (Ceilcote, USA), Basolit no. 600, Basolit no. 610 Sauereisen, USA).

2.1.3. Hydraulic Cement Mortars

Hydraulic cement mortars contain blast-furnace cement or portland cement which binds a quartz sand filler. They are delivered in the form of a cement powder accompanied by dry or moist sand and are used in accordance with ASTM C398. The mortar is worked using routine construction industry methods. When hydrated, the mortar hardens to a crystalline structure that resists temperatures up to 400 °C.

Hydraulic cement mortars are used for ceramic flooring that is laid with open joints which are subsequently filled with synthetic resin cements. Other uses include ceramic container linings (especially those used in the cellulose industry) and self-supporting constructions.

2.2. Cements Containing Organic Binders

2.2.1. Phenol–Formaldehyde Resin Cements

Phenol–formaldehyde resins or modified phenol-formaldehyde resins are used as the binder in some cements. The latter often contain furfural [98-01-1] to increase the chemical resistance. Today it is possible to produce furfural-free cements with adequate chemical stabilities. The filler is quarz, another inert material, or carbon and is usually mixed with a hardener such as *p*-toluenesulfonic acid, naphthalenesulfonic acid, or sulfonyl chloride. The cement is delivered as a liquid accompanied by a powder and is used in accordance with ASTM C399. The mixture is worked at ambient temperature and hardens (without heating) by the polycondensation of phenol–formaldehyde resin. The resulting bond is resistant to temperatures up to 180 °C.

Steel and concrete should not come into direct contact with phenolic resin-based mortars because of corrosive active hardeners.

Phenol–formaldehyde resin cements are used in laying and jointing of flooring made from nonmetallic inorganic tiles or bricks. They are used to produce chemically resistant masonry. Containers and equipment subject to high pressure and high temperature can be lined with compressive prestressed bricklining by using a mortar made with a modified phenol–formaldehyde resin cement [3], [5].

Trade name: Asplit CN (Permatex/Hoechst, FRG), Asplit CN 916 (Permatex/Hoechst, FRG).

2.2.2. Furan Resin Cements

Furan resin cements employ the polycondensation products of furfuryl alcohol [98-00-0] as the binding agent. Furfural [98-01-1] is usually included as a reactive thinner. The filler is quartz, another inert material, or carbon and is usually mixed with the hardener. The hardener can be an acid, such as amidosulfuric acid [5329-14-6] or *p*-toluenesulfonic acid [104-15-4], an acid chloride, a urea salt, or an amine salt. The cement is delivered as a liquid accompanied by a powder and is used in accordance with ASTM C399. The mixture is worked at ambient temperature and hardens (without heating) due to polycondensation of the furan resin to form a bond capable of withstanding temperatures up to 200 °C. With special formulations temperatures even

Figure 3. Application of ceramic tiles in furan resin cement (applied with open joints)

up to 250 °C are possible. Steel and concrete should not come into direct contact with furan resin-based mortars because of the corrosive active hardeners (Fig. 3.)

Furan resin cements containing quartz or another inert material are widely used because of their resistance to a broad spectrum of chemicals and their excellent storage properties. They are used to lay and joint flooring and to line trenches, pits, and collecting tanks. The cements that contain a carbon filler are used to solve special problems.

Trade name: Keranol FU 310 (Keramchemie, FRG), Keranol FU 315 (furan-free; Keramchemie, FRG), Asplit FQ (Permatex/Hoechst, FRG), Furadur Mortar (Steuler Industriewerke, FRG).

2.2.3. Epoxy Resin Cements

Epoxy resin cements use the product of condensation of bisphenol A [*80-05-7*] with epichlorohydrin as their binder, if necessary with a reactive thinner. The filler is quartz, another inert material, or (infrequently) carbon, and the hardener is a polyamine, modified polyamine, or polyamidoamine. The cement is delivered either as two liquids and a powder or as a liquid and a paste; it is used in accordance with ASTM C399. The

Figure 4. Floor tiling with chemical resistent tiles bedded and jointed with epoxy resin cement

mixture is worked at ambient temperature and hardens (without heating) due to polyaddition of the epoxy resin to the polyamine to form a bond capable of resisting temperatures up to 100 °C (Fig. 4).

Epoxy resin cements are widely used to bond light-colored ceramic surfaces, such as those used in the food and beverage industry, the production of drinking water, and the treatment of cellulose. They are also used to lay and joint flooring.

Trade names: Keranol EP 310 (Keramchemie, FRG), Keranol EP 110 (Keramchemie, FRG) Asplit ET (Permatex/Hoechst, FRG), Alkadur K 75 (Steuler Industriewerke, FRG).

2.2.4. Unsaturated Polyester Resin Cements

The products of the condensation of unsaturated polycarboxylic acids with polyalcohols dissolved in styrene serve to bind unsaturated polyester resin cements. Products formed by the reaction of unsaturated polyester resins with polyisocyanate or modified polyisocyanate dissolved in styrene are also added occasionally. The filler is quartz, another inert material, or carbon. The hardener is an organic peroxide, e.g., benzoyl peroxide; it is mixed with the filler unless the filler is carbon. Unsaturated polyester resin cements that do not contain carbon as the filler are delivered as one or two liquids along with a powder. Those with a carbon filler are delivered as two liquids, one powder, and one paste. These cements are used in accordance with ASTM C397. The mixture is worked at ambient temperature and hardens (without heating) by

Figure 5. Application of brick lining with vinylester resin cement at the bottom of a rubber lined tank

the polymerization of the unsaturated polyester resin with styrene to form a bond that resists temperatures up to 100 °C.

Unsaturated polyester resin cements are used primarily where oxidation reactions are performed, e.g., bleaching in the cellulose industry. They are also employed for light-colored jointing in the food and beverage industry, in flooring made from nonmetallic inorganic tiles, and in lining foundations. Cements containing carbon are used to prepare the brick linings of containers used for pickling steel with nitric acid and hydrofluoric acid. Their relatively large shrinkage upon hardening should be kept in mind.

Trade names: Asplit OC (Permatex/Hoechst, FRG), Keranol UP 311 (Keramchemie, FRG), Keranol UP 320 (Keramchemie, FRG), Oxydur A Mortar (Steuler Industriewerke, FRG).

2.2.5. Vinylester Resin Cements

The vinylester resin based on bisphenol-A-acrylate and novolak–acrylate contains styrene as reactive thinner. The filler is quarz, another inert material or carbon. The hardener is a peroxide. The mixture is worked at ambient temperature and hardeners (without heating) by polymerization of the vinylester resin with styrene to form a bond that resists temperatures up to 140 °C (Fig. 5).

Vinylester resin are widely used because of their resistance to a broad spectrum of chemicals, especially against oxidizing chemicals like nitric acid, chromic acid and sodium hypochlorite. They are used to lay and joint flooring made from acidproof ceramic tiles and cements containing carbon are used to prepare the brick linings of containers used for pickling steel with nitric acid and hydrofluoric acid. Vinylester resin cements are especially in textile industry, bleaching in the cellulose industry and for electroplating plants.

Trade names: Asplit VE-series (Permatex/Hoechst, FRG), Keranol VE 310 (Keramchemie, FRG), Keranol VE 311 (Keramchemie, FRG), Oxydur VE Mortar (Steuler Industriewerke, FRG).

2.2.6. Bituminous Material Cements

Blown bitumen is the binder in some cements. Quartz, another inert material, or carbon is used as the filler. The cement is delivered as a powder or in blocks and is heated to 200–220 °C before it is applied hot by means of a trowel. The material hardens as it cools to form a bond resistant to temperatures up to 80 °C.

Cements based on bituminous materials are used to produce chemically resistant flooring; to line trenches, pits, and emergency containments; and to repair damaged containers. Bituminous materials are inexpensive and easy to apply. They are resistant to a wide spectrum of chemicals. As a result of high labor costs, however, the use of these materials is declining in Europe and is almost unknown in the United States.

3. Specifications and Testing

The American Society for Testing and Materials (1916 Race Street, Philadelphia, PA 19103, USA) has published a series of standards for testing the properties and physical characteristics of cements (applied in accordance with ASTM C308 and C414 [8]). Standard values are also available in the DECHEMA Richtlinien [9] and in DIN 28062 [10] (Table 1).

Table 2 presents the materials specifications for chemically resistant cements with quartz or carbon fillers. These specifications are always greatly surpassed by all major products.

Chemical Resistance. Tests of chemical resistance are carried out according to such standards as ASTM C267 and DIN ISO 175.

Table 3 provides information on the resistance of different types of cement to a variety of groups of chemicals [3], [6], [11].

Table 2. Properties of chemically resistent cements

	Silicate cements	Sulfur cements	Hyraulic cement mortars	Bituminous material cements	Phenolic resin cements	Furan resin cements	Epoxy resin cements	Unsaturated polyester resin cements	Vinylester resin cements
Binder	Potassium or sodium silicates	Sulfur	Blast-furnace or portland cement	Blown bitumen	Phenol–formaldehyde resins	Furan resins	Epoxy resins	Unsaturated polyester resins	Vinylester resins
Filler	Quartz or other inert material	Quartz or carbon	Quartz	Quartz, carbon, kaolin, baryte	Quartz, carbon, baryte	Quartz, carbon, baryte	Quartz, carbon, baryte	Quartz, carbon, baryte	Quartz, carbon, baryte
Hardener	Neutralization agent		Water		Organic acid	Organic acid	Polyamine	Organic peroxide	Organic peroxide
Processing aids								Organic accelerator	Organic accelerator
Hardening reaction	Coagulation	Solidification	Hydration	Solidification	Polycondensation	Polycondensation	Polyaddition	Polymerization	Polymerization
Pot life (at 20 °C)	≥ 0.5 h ≤ 2h		0.5 to several hour		≥ 0.5 h ≤ 1 h	≥ 0.5 h ≤ 1 h	≥ 0.5 h ≤ 1 h	0.5 h	0.5 h
Fit for exposure to stress	Several days	After cooling	Several days	After cooling	2 to 7 days	2 to 7 days	2 to 7 days	1 to 7 days	1 to 7 days
Adhesion to:									
Carbon	not to apply	O	not to apply	+	++	++	++	+	+
Ceramic	+	O	+	+	++	++	++	+	+
Steel	+[a]	O	+	+	(+)[b]	(+)[b]	++	+	+
Concrete	–	O	–/+[a]	+	(+)[b]	(+)[b]	++	+	+
Rubber	–/+[a]	O	–/+[a]	+	+	+	++	+[a]	+[a]
Thermoplasts	–/+[a]	O	–/+[a]		– to +	– to +	– to +	– to +	– to +
Resin coatings					O to ++	O to ++	O to ++	O to ++	O to ++
Used for:									
Flooring and walls	O	–		+	++	++	++	++	++
Tanks and apparatus	++	++	++	+	++	++	+	++	++
Chemical loads	++ (pH < 5)	+	O (pH > 4)	+	++	++	+	+	+
Mechanical loads	+	O	+	O	++	++	++	++	++
Thermal loads	++	O	++	–	+	++	O	+	+

[a] Primed and sanded off; [b] In combination with membrane; ++ Very good and/or very suitable; + Good and/or suitable; O Conditional suitable; – Insufficient and/or unsuitable

Cements, Chemically Resistant

Table 3. Resistance of cements to various chemicals

Chemical	Silicate cement	Sulfur cement	Hydraulic cement mortar	Bituminous material cement	Phenolic resin cement	Furan resin cement	Epoxy resin cement	Unsaturated polyester resin cement	Vinylester resin cement
Non-oxidizing acids	+	+	−	+	+	+	O	+	+
Oxidizing acids	+	+	−	O	O	O	O	+	+
Silica-dissolving acids	−	+	−	+	+	+	+	O	+
Bases	−	+	O	+	O	+	+	+	+
Oxidizing bases	−	−	O	O	−	O	O	+	+
Salts	+	+/−[a]	O	+	+	+	+	+	+
Water	−	+	+	+	+	+	+	+	+
Organic acids	+	O	−	O	+	+	O	+	+
Aliphatic compounds	+	+	O	−	+	+	+	+	+
Aromatic compounds	+	+	O	−	+	+	O	−	+
Alcohols	+	+	+	O	+	+	O	+	+
Ketones, esters	+	−	−	−	+	+	O	−	O
Aliphatic chlorinated hydrocarbons	+	−	O	−	+	+	−	−	O
Aromatic chlorinated hydrocarbons	+	−	O	−	+	+	−	−	O
Aldehydes	+	−	O	O	+	+	O	O	O
Aliphatic amines	+	−	+	−	+	+	−	O	+
Aromatic amines	+	−	+	−	+	+	−	−	O
Phenoles	O	−	−	−	+	+	−	−	O
Fats and oils	+	−	O	−	+	+	O	+	+

[a] resistant with pH ≤ 7; + resistant; − not resistant; O resistant under special conditions

Table 1. Tests and Specifications of Physical Properties

Property	US Standard	DIN Standard
Tensile strength	ASTM C307	DIN 53455
Flexural strength	ASTM C580	DIN 53452
Compressive strength	ASTM C579	DIN 51067
Bond strength	ASTM C321	
Thermal expansion	ASTM C531	
Shrinkage	ASTM C531	
Absorption	ASTM C413	DIN 51056

4. Storage and Transportation

Beside phenol–formaldehyde, unsaturated polyester and vinylester resins, the individual components, sealed in their original packing, can generally be stored for 12 months or longer at room temperature and under dry conditions.

Special care must be taken in the transportation of the binders because of their reactivity and toxicity.

5. Toxicology

The national guidelines for the handling of the binder must be carefully observed. Indeed, the binding agents can be toxic, irritating, and detrimental to health. Some powders can also be physiologically active because of the hardener they contain.

6. References

[1] DIN 28052-5, Chemische Apparate; Oberflächenschutz mit nichtmetallischen Werkstoffen für Bauteile aus Beton in verfahrenstechnischen Anlagen; Kombinierte Beläge, 1997.
[2] AGI Arbeitsblatt S10 Part 3, Schutz von Baukonstruktionen mit Plattenbelägen gegen chemische Angriffe—Plattenlagen, part 3, Vincentz Verlag, Hannover.
[3] F. K. Falcke: *Kleines Handbuch des Säureschutzbaues,* Verlag Chemie, Weinheim 1966.
[4] W. L. Sheppard: *Handbook of Chemically Resistant Masonry,* C.C.R.M. Inc., Havertown, Pennsylvania, 1977.
[5] F. K. Falcke, G. Lorentz (eds.): *Handbook of Acid-Proof Construction,* VCH Verlagsgesellschaft, Weinheim 1985.
[6] E. Schacht: "Ausmauerungen in chemisch beanspruchten Behältern und Apparaten," *Z. Werkstofftech.* **5** (1974) 297–307.
[7] J. Dück: "Schwund- und Quellverhalten reaktionshärtender Kunstharzkitte für den Säureschutzbau (Shrinking and Swelling Properties of Chemically Cured Resin Mortars for Chemical Resistant Linings)," *Z. Werkstofftech.* **12** (1981) 73–83.

[8] *ASTM-Standards* 1977, Annual Book of ASTM Standards, vol. 04.05.1997. Chemical Resistent Materials; Vitrified Clay; Fiber-Cement Products; Mortars; Masonry, American Society for Testing and Materials, 1916 Race St., Philadelphia, Pa. 19103, USA.

[9] *DECHEMA Richtlinie: Bestimmung physikalischer, insbesondere mechanischer Kennwerte von Kitten für den Säureschutzbau*, Dechema, Frankfurt.

[10] DIN 28062, Chemische Apparate; Bau-und Werkstoffe für Ausmauerungen; Einteilung – Eigenschaften – Prüfung; 1978.

[11] W. A. Kuenning: "Guide for the Protection of Concrete against Chemical Attack by Means of Coatings and other Corrosion-Resistant Materials," *Proc. Amer. Concr. Inst.* **63** (1966) 1305–1391.

Ceramics, General Survey

GIRARD W. PHELPS, Department of Ceramics, Rutgers – The State University, Piscataway, New Jersey 08854, United States

JOHN B. WACHTMAN, JR., Center for Ceramics Research, Rutgers – The State University, Piscataway, New Jersey 08854, United States

1.	Traditional and Advanced Ceramics	944	4.5.	Firing Ceramic Products 985
1.1.	Traditional Ceramics	945	4.5.1.	Firing Traditional Ceramics 985
1.2.	Advanced Ceramics	947	4.5.2.	Densification of Advanced Ceramic Products 986
1.2.1.	Advanced Structural Ceramics	948	4.6.	Kilns and Firing Conditions .. 988
1.2.2.	Electronic Ceramics	951	4.6.1.	Modern Periodic Kilns 988
1.2.3.	Other Advanced Ceramics	953	4.6.2.	Tunnel Kilns 989
1.3.	Characterization of Ceramic Materials	953	4.6.3.	Advanced Ceramics Furnaces ... 989
			4.6.4.	Kiln Atmosphere 989
2.	Raw Materials for Traditional Ceramics	956	4.6.5.	Fired Ware Finishing 990
2.1.	The Structure of Clays and Nonplastics	957	5.	Glazes and Glazing 990
			5.1.	The Nature of Glazes 990
2.2.	Clay – Water System	958	5.2.	Preparation of Glazes 994
2.3.	Commercial Ceramic Clays	960	5.3.	Glaze Application 994
2.4.	Commercial Nonplastics for Ceramics	964	6.	Glass 994
			7.	Refractories 996
3.	Raw Materials for Advanced Ceramics	968	8.	Abrasives 996
3.1.	Metal Oxides and Carbonates .	969	9.	Cement 997
3.2.	Borides, Carbides, and Nitrides	970	10.	Properties of Ceramic Materials and Products 997
4.	Processing Ceramic Ware	970	11.	Testing Ceramic Raw Materials and Products 998
4.1.	Preparation of Clay-based Forming Systems	970	11.1.	Raw Material and Product Tests 998
4.2.	Preparation of Advanced Ceramic Systems	975	11.2.	Simplified Testing of Clay Body Materials 1000
4.3.	Forming Ceramic Articles	978	11.3.	Quality Control of Advanced Ceramics 1003
4.4.	Drying and Finishing	981	12.	Economic Aspects 1003
			13.	References 1005

1. Traditional and Advanced Ceramics

This general survey covers the fields of traditional ceramics and advanced (or high-technology) ceramics, touching on the materials employed, processing and forming, firing and finishing, and the use of products. Advantages and disadvantages of various types of ceramic ware are discussed.

The word ceramic is a "general term applied to the art or technique of producing articles by a ceramic process, or to articles so produced" [1]. In general, it applies to any of a class of inorganic, nonmetallic products subjected to high temperature during manufacture or use. "High temperature" means any temperature above red heat, ca. 540 °C [2].

Typically, although not exclusively, a ceramic item is a metal oxide, boride, carbide, or nitride, or a compound of such materials. Thus, a *ceramic article* is "a glazed or unglazed object of crystalline or partly crystalline structure (or of glass), produced from essentially inorganic, nonmetallic substances; such objects are made from either a molten mass which solidifies upon cooling or which is formed and matured simultaneously or subsequently by action of heat" [3, p. 197].

The noun ceramic is derived from the Greek *keramos* meaning "burned earth." *Traditional ceramics* refers to ware prepared from an unrefined clay or to combinations of one or more refined clays in combination with one or more powdered or granulated nonplastic minerals or prereacted ceramic compositions. Traditional ceramics also refers to ware or products made from compositions or naturally occurring materials in which clay mineral substance exceeds 20 %. Traditional ceramics and clay ceramics are synonymous expressions.

The past 50 years have seen an increasing interest in ceramic items made from highly refined natural or synthetic compositions designed to provide special properties [4, pp. 150–153]. These objects are termed *advanced*, new, or (in Japan) fine *ceramic products*, and find use as key components in such high-technology fields as electronics, computers, optical communication, cutting tools, metal forming dies, wear-resistant parts, high-temperature reactors, high-temperature engine parts, medical implants, and many other special purpose applications. Advanced ceramics must be considered as an enabling technology — one essential to competitive or functional performance of larger systems.

Advanced new roles for ceramics depend on properties inherent in basic structure and composition. Recognition of special capabilities of ceramics is largely due to progress over the past 30 years in relating physical to compositional and structural features [5]. Two recent developments are responsible for the exponential growth in applications for advanced ceramics: first, advances in systems that require special, highly developed ceramics; and second, advances in ceramic processing that permit production of usable ceramic parts.

Historical Aspects. The qualities of plasticity, dried strength, and fired hardness of clays were discovered and used possibly as long ago as 10 000 B.C. [6], and certainly by 5500 B.C. [7].

The earliest societies that give reasonable evidence of a ceramic industry seem to have been in the Near East, where a pottery tradition dates back ca. 7000–8000 years [8]. In the Far East, Neolithic villagers at Banpo in Shaanxi Province of China were making fine red, gray, black, and painted pottery at least as early as 4500 B.C. [9, pp. 135 –162].

Urban planners at Mohenjo-Daro in the Indus Valley were using fired clay bricks and tile for public building, water-supply conduits, and an advanced sewer system 4000 years ago [10]. A cuneiform tablet of the 17th century B.C. describes the making of a copper–lead glaze [11].

By the Shang Dynasty (1500–1025 B.C.) the Chinese had changed Neolithic earthenware to a fine-grained white stoneware [9, pp. 135–162]. A primitive feldspathic glaze appeared during the Zhou Dynasty (1000–771 B.C.), followed by a soft green to brown lead silicate glaze in the Han Period (206 B.C.–220 A.D.). In the Six Dynasties time period (265–907 A.D.), marked developments occurred in art pottery. Ceramics of the Song Dynasties (960–1127 and 1127–1279 A.D.) and into the Ming Dynasty (1368 –1644 A.D.) showed increasing use of hard-paste porcelain formulas. An egg-shaped kiln with a special stack designed for high draft was developed for firing Ming porcelains; the fuel was pine wood [12].

Beginning in the sixth century B.C., Attic vases of ancient Greece represented a ceramic art milestone. Use was made of a local illitic clay [13] to prepare a levigated glaze that was black under reducing conditions and red under oxidizing conditions. Rome seems to have had no ceramic tradition, but drew on ware and workmen from various parts of its empire. The famous Arretine ware was made near what is now Florence by Oriental Greeks, and terra sigillata, sometimes thought to have been invented by Romans, originated on the island of Samos [14].

Chinese ceramic technology is thought to have been transferred to the Near and Middle East by cultural contacts and by Chinese prisoners following defeat of a Chinese army by Persian Abbasids in 751 A.D. Islamic influence was felt late in Italy and France [15], and later still in Germany, the Netherlands, and England [16].

Salt-glazed stoneware was developed in High Germany near the end of the fourteenth century [17], and soft-paste porcelain was being manufactured in Meissen as early as 1730 and in Sèvres, France, by 1751–1754. In England, COOKWORTHY developed and patented a hard-fire porcelain (1768), and sometime before 1750, calcined bone was being used in making chinaware [18].

Stoneware and hard porcelain served as prototypes for development in the United States, England, and Europe of mechanically strong, vitreous bodies for use in manufacture of sanitary ware, high-tension electrical porcelain, and impact-resistant dinnerware [19, p. 4]. The advent of steam power in the eighteenth century permitted WEDGWOOD and others in England to mechanize preparation and forming operations. Although acceptance of new ideas in the way of processing has been slow over the centuries, within the past two generations "felt wants" in industry have stimulated research in the areas of advanced ceramics.

1.1. Traditional Ceramics

Clay is the oldest ceramic material. The earliest ceramic ware was most likely made from natural clay, selected by the potter for its forming properties. However, at very early times, it was customary to add some other nonclay materials. A sticky, high-shrinkage clay might be modified by addition of crushed stone, sand, or crushed shell to reduce shrinkage and cracking. Currently, the major nonclay materials used in making clay-based ceramic items are silica powder and certain alkali-containing minerals added as fluxes. Traditional ceramics can be regarded as ware made from formulations in which clay provides the plastic and dry bonding properties required for shaping and handling. Analyses of natural clay bodies show that the actual clay mineral content is 25–40%.

Pottery is sometimes used as a generic term for all fired ceramic wares that contain clay in their compositions, except technical, structural, and refractory products [3, p. 201].

The term *whiteware* was originally applied to white tableware and artware [19, p. 4], but has been broadened to include ware that is ivory colored or has a light gray appearance in the fired state. Fine ceramic whitewares are conveniently divided into two classes: (1) formulas consisting primarily of clay minerals, feldspathics, and quartz; and (2) nontriaxial bodies made entirely or predominantly of other materials. For purposes of this discussion, ceramic whiteware is placed into five categories, namely, (1) earthenware, (2) stoneware, (3) chinaware, (4) porcelain, and (5) technical ceramics.

Earthenware is defined as glazed or unglazed nonvitreous (porous) clay-based ceramic ware. NORTON subdivides earthenware into four categories: (1) natural clay body, (2) refined clay body, (3) talc body, and (4) semivitreous triaxial body [19, p. 4]. Fired absorptions may range from 4–5% for semivitreous ware to 20% for the high-talc formulas. Fired color may range from red for high iron oxide bodies to white for the talc and triaxial formulas.

Stoneware is a vitreous or semivitreous ceramic ware of fine texture, made primarily from nonrefractory fireclay or some combination of clays, fluxes, and silica that matches the forming and fired properties of a natural stoneware. Thus, stoneware may be made either from a clay or may be a synthesized stoneware. Synthesized stoneware can range from highly refined, zero-absorption chemical stoneware to less demanding dinnerware and artware formulas.

Chinaware is vitreous ware of zero or low-fired absorption used for nontechnical applications. It can be either glazed or unglazed. The expression soft-paste porcelain has the same meaning [19, p. 4]. Formulas can be simple clay–flux–silica triaxial bodies or bodies containing significant percentages of alumina, bone ash, frit, or low-expansion cordierite or lithium mineral powders. Fired absorptions range from 0 to 5% for ovenware.

Porcelain is defined as glazed or unglazed vitreous ceramic ware used primarily for technical purposes. Formulations are generally of the triaxial type although some or all of the silica can be replaced by calcined alumina to increase mechanical strength. Firing of ware may be bisque (unglazed) at low temperature with glazing at high temperature or by single-firing at high temperature.

Technical ceramics include vitreous (i.e., nonporous) ceramic whiteware used for such products as electrical insulation, chemical ware, mechanical and structural items, and thermal ware.

The clays used for making *common brick* are usually of low grade and in most cases red-burning. The main requirements are that they are easy to form and fire hard at as low a temperature as possible, with a minimum loss from cracking and warping. An average of analyses of a number of brick clays from sources in New Jersey [20] showed approximately 67% SiO_2, 18% Al_2O_3, 3% Fe_2O_3, 2% alkaline-earth oxides, and 4% alkalies, with an ignition loss of about 4%.

Bodies [19, p. 2] can be classified as being either *fine* (having particles not larger than ca. 0.2 mm) or *coarse* (having the largest particle ca. 8 mm). These can, in turn, be subdivided into bodies fired to a *porous* state and those with a fired absorption [3, p. 197] not exceeding 5%, i.e., a *dense* state. The classes of fine clay ceramics and product uses are arranged in Table 1 to show the percent absorptions and body colors. The classes of coarse clay ceramics and their fired porosities are given in Table 2.

1.2. Advanced Ceramics

Advanced ceramics are generally used as components in processing equipment by virtue of such ceramic properties as special electromagnetic qualities, relative chemical inertness, hardness and strength, and temperature capabilities, sometimes in combination.

A systematic classification of advanced ceramics based on function is presented in Table 3, and examples of materials and uses are shown. A broader system classifies all applications into structural, electronic, and other. *Structural* applications are mechanical, but do include chemical aspects where these are required to carry out the mechanical function. The *electronic* category covers electric, magnetic, and optical functions plus chemical functions that involve direct use of electronic properties. The *other* classification includes strictly chemical functions, for example, catalysis, as well as biological functions.

Of course, any classification is likely to be inexact because many applications involve simultaneous use of several functions. However, a functional classification system does point to the fact that, in contrast to metals, ceramics can be made to embody a wide variety of electronic functions while also having desirable chemical and mechanical properties.

Ceramics are already widely used in process industries, especially where corrosion, wear, and heat resistance are important. Excellent examples are found in metallurgical refractories [22], an area already feeling the effect of new developments in ceramics and the new demands of advanced metallurgical processing.

Table 1. Fine ceramic products

Type of product	Earthenware		Stoneware		Chinaware		Porcelain	
	% Absorption	Color	% Absorption	Color	% Absorption	Color	% Absorption	Color
Artware	10–20	red-white	0–5	red-white	0–1	white	0–0.5	white
Ballmill balls							0–0.2	white
Ballmill liners							0–0.2	white
Chemical ware			0–0.2	gray-white			0–0.2	white
Cookware			0–5	gray-white				
Drainpipe			0–5	gray				
Insulators							0–0.2	white
Kitchenware	10–15	white	0–5	gray				
Ovenware	10–20	white			1–5	tan		
Sanitary ware					0.1–0.3	tan-white		
Tableware	5–20	white	0–5	white	0–1	white	0–0.5	white
Tile	10–20	white	0–5	red-white				

Table 2. Coarse ceramic products

Porous (> 5% (+ 5%) absorption)		Dense (< 5% (– 5%) absorption)	
Building materials	Refractory	Chemical	Structural
bricks	flue linings	acid-resistant bricks	quarry tile
terra cotta	fireclay bricks		sewer pipe
roofing tile	insulating bricks		fireclay sanitary ware
drain tile			

1.2.1. Advanced Structural Ceramics (→ Ceramics, Advanced Structural Products)

The prominent families of advanced structural ceramics and structural materials involving ceramics include

 alumina
 silicon carbide
 silicon nitride
 partially stabilized zirconia
 transformation-toughened alumina
 lithium aluminosilicates
 ceramic–ceramic composites
 ceramic-coated materials

These materials are widely used in diesel, turbocharger, and gas-turbine engines; in high-temperature furnaces; and in the machines and equipment needed for manufacturing.

Table 3. Classification of high-technology ceramics by function [21]

Function	Material	Uses	
Electric functions	insulation materials (Al_2O_3, BeO, MgO)	IC circuit substrate, package, wiring substrate, resistor substrate, electronics interconnection substrate	
	ferroelectric material ($BaTiO_3$, $SrTiO_3$)	ceramic capacitor	
	piezoelectric materials (PZT)	vibrator, oscillator, filter transducer, ultrasonic humidifier, piezoelectric spark generator	
	semiconductor materials ($BaTiO_3$, SiC, $ZnO-Bi_2O_3$, V_2O_5, and other transition-metal oxides)	NTC thermistor:	temperature sensor, temperature compensation
		PTC thermistor:	heater element, switch, temperature compensation
		CTR thermistor:	heat sensor element
		thick-film thermistor:	infrared sensor
		varistor:	noise elimination, surge current absorber, lighting arrestor
		sintered CdS material:	solar cell
		SiC heater:	electric furnace heater, miniature heater
	ion-conducting materials (β-Al_2O_3, ZrO_2)	solid electrolyte for sodium battery	
		ZrO_2 ceramics:	oxygen sensor, pH meter, fuel cells
Magnetic functions	soft ferrite	magnetic recording head, temperature sensor	
	hard ferrite	ferrite magnet, fractional-horsepower motor	
Optical functions	translucent alumina	high-pressure sodium vapor lamp	
	translucent magnesia, mullite	lighting tube, special purpose lamp, infrared transmission window	
	translucent Y_2O_3–ThO_2 ceramics	laser material	
	PLZT ceramics	light memory element, video display and storage system, light modulation element, light shutter, light valve	
Chemical functions	gas sensor (ZnO, Fe_2O_3, SnO_2)	gas leakage alarm, automatic ventilation fan, hydrocarbon detector, fluorocarbon detector	
	humidity sensor ($MgCr_2O_4$–TiO_2)	cooking control element in microwave oven	
	catalyst carrier (cordierite)	catalyst carrier for emission control	
	organic catalyst	enzyme carrier, zeolite	
	electrodes (titanates, sulfides, borides)	electrowinning aluminum, photochemical processes, chlorine production	
Thermal functions	ZrO_2, TiO_2 ceramics	infrared radiator	
Mechanical functions	cutting tools (Al_2O_3, TiC, TiN)	ceramic tool, sintered SBN, cermet tool, artificial diamond, nitride tool	
	wear-resistant materials (Al_2O_3, ZrO_2)	mechanical seal, ceramic liner, bearings, thread guide, pressure sensor	
	heat-resistant materials (SiC, Al_2O_3, SiN_4)	ceramic engine, turbine blade, heat exchangers, welding-burner nozzle, high-frequency combustion crucible	

Table 3. (continued)

Function	Material	Uses
Biological functions	alumina ceramics implantation hydroxyapatite bioglass	artificial tooth root, bone, and joint
Nuclear functions	nuclear fuels (UO_2, $UO_2 - PuO_2$)	
	cladding material (C, SiC, B_4C)	
	shielding material (SiC, Al_2O_3, C, B_4C)	

(Courtesy of the American Ceramic Society)

Although *alumina* [1344-28-1] denotes pure Al_2O_3, the term is commonly applied to any ceramic whose major constituent is alumina, even if the ceramic contains other components. Commercial alumina microelectronic substrates with strengths above 350 MPa are obtained by conventional sintering. Hot-pressing techniques result in strengths of ca. 750 MPa, although parts are expensive with limited size and geometries. A recent development [23] involving a variation on conventional sintering produces a glass-bonded alumina with strengths of ca. 700 MPa. Although the glassy phase limits applications to moderate temperatures, this new alumina ceramic should compete with other more expensive, advanced ceramic items.

Fibrous alumina is employed as a reinforcing agent in metal matrix composites and offers promise for filtration of hot gases and as high-temperature insulation. Alumina is used with SiO_2 in making such fibers [24]. Pure Al_2O_3 fibers are made by a variety of solution processes to produce fibers with strength of 1400 MPa.

Silicon carbide [409-21-2], a synthetic product, has good wear and erosion resistance and can be produced in either cubic or hexagonal crystal structure. Unfortunately, SiC is inherently unstable in oxygen so that long life under oxidizing conditions requires a surface coating of protective oxide.

Silicon nitride [12033-89-5], Si_3N_4, is likewise a synthetic product, existing in two phases, alpha and beta, each having hexagonal crystal structures. Silicon nitride ceramics include hot-pressed, reaction-bonded, and sintered products. The SiAlON family is a solid solution of Al_2O_3 and/or other metal oxides in the β-Si_3N_4 structure [25]. Reaction-bonded Si_3N_4 is made by nitriding cast or cold-pressed shapes of silicon powder, whereas hot-pressed Si_3N_4 is made from silicon nitride powder as a sintered Si_3N_4 powder product. Reaction-bonded Si_3N_4 retains its strength at high temperature if it is protected from oxidation [26]. Hot-pressed Si_3N_4 has high short-term strength and better oxidation resistance, but needs additives to facilitate compaction [27].

The advanced cutting tool industry is dominated by cemented carbides [28]. Ceramic vapor-deposited coatings have extended tool life. Efforts are under way to increase tool use by basing tools on Si_3N_4 and SiAlON to reduce dependence on strategic W, Ta, and Co [29].

Silicon nitride possesses many interesting properties that suggest use in bearings [30]. Tests showed an estimated life for Si_3N_4 bearings of 8 times that of steel bearings. The economics of machining and finishing is the biggest obstacle to widespread use of Si_3N_4 bearings.

Zirconia [1314-23-4], ZrO_2, finds widespread use in a stabilized cubic form as an oxygen sensor in process industries and the automobile industries [31]. The destructive transformation of ZrO_2 at 1100 °C from monoclinic to cubic form has been overcome by keeping unstabilized particle size of ZrO_2 grains below 1 μm diameter. Then an alumina matrix toughens the Al_2O_3 ceramic [32]. Hot pressing was initially used, but slip-cast forming and sintering has been found to be feasible [33], [34].

Cordierite [12182-53-5], $2\,MgO \cdot 2\,Al_2O_3 \cdot 5\,SiO_2$, has a thermal expansion of $(8-12) \times 10^{-7}$ over the range 20–1000 °C and is widely used as a catalyst support for automobile emission control units. Similar materials are used as heat exchangers in automotive gas-turbine prototypes and can be considered candidates for other heat-exchanger applications where good thermal shock resistance and moderate crushing strength are required [35]. Silicon carbide and silicon nitride also find application in heat exchangers [36].

Ceramic–ceramic composites and *ceramic–metal composites* (→ Ceramics, Ceramic–Metal Systems) are receiving increasing attention. Silicon carbide fibers in glass–ceramic matrices have shown toughness values up to 24 MPa m$^{0.5}$ at 1000 °C with cross-plied and unidirectional strengths of 500 and 900 MPa [37]. The reinforcing action of 60 % alumina fibers in aluminum gave a tensile strength of 690 MPa up to 316 °C [38]. Use of as little as 3 % of pure Al_2O_3 particles in aluminum increased strength and wear resistance [39].

A thickness of 10–15 mils (25–38 mm) of plasma-sprayed porous *ceramic coating* such as ZrO_2 can reduce the temperature of the metal surface under the coating by 160 °C [40]. Such coatings are used on aircraft burners and aircraft afterburners, but not in critical parts of aircraft gas turbines. Pore-free coatings applied by chemical vapor deposition, sputtering, or reactive evaporation are 70–80 times as resistant to wear and erosion as porous coatings. High-temperature lubrication may make use of solid ceramic lubricants.

1.2.2. Electronic Ceramics

Ceramics are involved in electronics as discrete units; however, as component sizes become progressively smaller, they are increasingly integrated into overall electronic assemblies. FISHER [41] has classified discrete ceramic parts into three categories: insulators, magnetic ceramics, and transducers.

Insulators represent a complex category including integrated circuit packages, insulating substrates, and a variety of special tube circuits. Electrical insulation materials are, in a sense, descended from traditional electrical porcelains, but property requirements plus the complex nature of integrated circuits make them a new family.

Aluminum oxide is the dominant advanced ceramic insulator [19, pp. 426–429]. Tape-cast alumina ceramics dominate in uses requiring high heat dissipation and hermeticity. Alumina ceramics also compete with polymers and coated metals as supports for electronic chips. As excellent as alumina is for this purpose, alternative materials are being studied in an effort to lower the dielectric constant, permit higher frequency operation, and provide a closer match to silicon thermal expansion. Multiphase ceramics in the Al_2O_3–SiO_2–MgO family may be the second generation of ceramics, with Si_3N_4 as the third generation.

Several trends are apparent in the development of later-generation ceramic substrates. One line of development seeks to use low-firing, glass-bonded aluminas that can be cofired with copper, silver, or gold electrodes. A second line of development seeks to exploit the high thermal conductivity of AlN [24304-00-5]. Another candidate is BeO-doped SiC. A third line of development is concerned with finding lower-loss materials for microwave applications.

Ferroelectric ceramics, primarily high dielectric constant $BaTiO_3$ [12047-27-7] and related materials, find use in capacitors, which are indispensable in electronics. The use of cheaper metals as electrodes may lower unit costs [19, pp. 415–417].

Piezoelectrics are crystals whose charge centers are offset: a mechanical stress alters the polarization of the crystal just as an electrical field would. Piezoelectric crystals are widely used for voltage–pressure *transducers*. Piezoelectric ceramics, such as lead zirconate titanate [12626-81-2], are used in a wide variety of devices to convert motion into electrical signals and vice versa. Vibrators, oscillators, filters, loudspeakers, all using piezoelectric devices, are essential parts of many industrial and consumer products [4, pp. 55, 287].

Certain ceramics are termed semiconductors, electrical conduction occurring only if external energy is applied to fill energy gaps between filled and empty electron bands. An increase in temperature can also provide the required energy. Ceramic semiconductor materials include titanates, SiC, ZnO, NiO, and Fe_2O_3. In some instances, they are used as thermistors for temperature control. They may be used as voltage-sensitive resistors (varistors) to protect against voltage surges, as chemical sensors, or as miniheaters [4, pp. 47–50].

Ion-conducting ceramics, such as β-alumina and stabilized ZrO_2, are employed as oxygen sensors in automobiles and as electrolytes in fuel cells [42], [43].

Ceramic materials having magnetic properties are commonly termed ferrites. *Magnetic ceramics,* such as ferrites of Fe_2O_3 in combination with one or more of the oxides of Ba, Pb, Sr, Mn, Ni, and Zn, can be made into either hard or soft magnets. These are widely used in loudspeakers, motors, transformers, recording heads, and the like [19, pp. 417–421].

The optical properties of a material include absorption, transparency, refractive index, color, and phosphorescence. Optical transparency is often important. Glass and various ionic ceramics are transparent to visible light, and there are many applications for windows, lenses, prisms, and the like. Fiber optics offer enormous potential for communication; small fiber bundles transmitting coherent laser light can carry many times the information carried by wire cables. Magnesium oxide, Al_2O_3, and fused SiO_2 are transparent in the ultraviolet and a portion of the infrared and radar wavelengths. Magnesium fluoride, ZnS, ZnSe, and CdTe are transparent to infrared and radar wavelengths [4, p. 59].

Special pore-free Al_2O_3 is widely used as the inner envelope of high-pressure sodium vapor lights. Lead zirconate titanate ceramics are finding increasing use in light modulation and displays. Translucent Y_2O_3–ThO_2 ceramics are also useful optical materials.

Ceramic sensors can use bulk grain phenomena (such as piezoelectric effects, oxygen-ion conductivity, or negative temperature coefficient of resistivity), grain boundary phenomena (such as positive temperature coefficient of resistivity, voltage-dependent resistivity, or gas absorption), or controlled pore structure (moisture absorption). Occasionally all three microstructural features come into play, with different levels of importance. A broad class of sensors is based on optical fibers [44]. New types of optical sensors using optical fibers can measure temperature, pressure, sound, rotation, current, and voltage. A blood oxygen meter using optical fibers measures light transmission at eight different wavelengths, thus permitting blood oxygen determination.

1.2.3. Other Advanced Ceramics

One of the oldest uses of ceramics is as a thermal insulator at high temperature, and this role is continued in modern form, e.g., as super insulators such as the silica tile used on the U.S. space shuttle. Modern ceramics such as silicon carbide and silicon nitride are increasingly attractive as heat exchangers, as are low-expansion ceramics such as cordierite.

A potentially important market for new ceramics is as implants to replace teeth, bone, and joints.

Ceramics have long been used in the nuclear field as a fuel, cladding material, and shielding material. They are leading candidates as matrices to contain radioactive wastes for long-term storage.

1.3. Characterization of Ceramic Materials

The technology of ceramic manufacturing rests on measurement of the structural and chemical properties of the raw materials used in ceramic forming systems. The need for adequate test procedures is being met by continuing advances in materials science. Many sophisticated instruments and equally sophisticated techniques are available for evaluation of formula ingredients and of forming systems at various stages of manufacture [45, chap. 1].

Purity of ingredients has a profound influence on high-temperature properties of advanced ceramics, including strength, stress rupture life, and oxidation resistance. The presence of Ca^{2+} is known to sharply decrease the creep resistance of Si_3N_4 hot pressed with MgO sintering aid [46], but seems to have little effect on Si_3N_4 hot pressed with Y_2O_3 densifying aid [47]. Electrical, magnetic, and optical properties must be carefully tailored by additions of a dopant; slight variations in distribution or concentration can alter final properties significantly. Ceramic materials can occur in different geometries. As an example α-Si_3N_4 is preferred over β-Si_3N_4 for hot pressing or ordinary sintering.

In recognition of the importance of consistent properties of raw materials and synthetic powders used for advanced ceramic items, an ad hoc committee appointed by the Materials Advisory Board of the National Research Council (United States) gave the term *characterization* a special, restrictive meaning in the following definition [48]: "Characterization describes those features of the composition and structure (including defects) of a material that are significant for a particular preparation, study of properties, and suffice for the reproduction of the material." True characterization involves a direct correlation between test results and properties. The mere taking of data is not characterization unless the test procedure serves a particular function in predicting properties of the material under test.

Although this definition was designed as an aid in establishing significant features for advanced ceramic products and their constituents, the concept has been successfully applied in the field of traditional ceramics. The many properties encountered in forming and firing are found to be consequences of the interaction of two or more of a limited list of fundamental characterizing features [49]. Table 4 provides a listing of significant, interacting features for traditional clay-based ceramics, with a partial list of the more important consequential properties encountered in forming and firing. An exhaustive survey of pertinent literature, in addition to a continuing review of plant and laboratory results, has shown no exceptions to the list of characterizing features of Table 4.

Characterization, itself rapidly developing as a discipline, has suggested ways whereby selected properties of materials or a body can be used in development and control of clay bodies. Sanitary ware and vitreous chinaware are typical clay-based traditional ceramic products. The chemical, mineral, particulate, and surface data of Table 5 constitute complete characterizing descriptions of examples of formulas used in making these products.

Two terms require definition. The *mole of flux* is the sum of the percentages of CaO, MgO, K_2O, and Na_2O divided by their respective molecular masses. The *MBI* (methylene blue index) is the milliequivalents of methylene blue cation (chlorine salt) absorbed per 100 g of clay and is a measure of surface area [50].

Table 4. Characterizing features and ceramic properties

Characterizing features
chemical composition
mineral composition
particle-size distribution
specific surface
colloid modifiers

Ceramic properties

Unfired/forming	Firing/fired
slip viscosity	vitrification
water of plasticity	shrinkage
workability	pyroplasticity
shrinkage	absorption
strength	strength
slip dispersion	color
casting rheology	thermal behavior
casting rate	microstructure

Table 5. Characterization of two clay-based bodies

Properties	Vitreous sanitary ware	Vitreous china
Chemical, wt%		
SiO_2	65.0 [x]	69.4 [x]
Al_2O_3	23.1 [x]	19.5 [x]
Fe_2O_3	0.44	0.30 [x]
TiO_2	0.28	0.14 [x]
CaO	0.33	1.33 [x]
MgO	0.13	0.11
K_2O	2.68 [x]	1.45 [x]
Na_2O	2.41 [x]	1.14 [x]
Ignition loss	5.67	6.46
Mole of flux	0.0766 [x]	0.0604 [x]
Minerals, wt%		
Smectite	3.7	3.0
Kaolin group	32.7	33.3
Mica	8.8 [x]	5.8 [x]
Free quartz	23.7 [x]	39.6 [x]
Organic	0.46 [x]	0.23 [x]
Auxiliary flux		2.0
Particle size		
% < 20 µm	76	76
% < 5 µm	47	45
% < 2 µm	33	36
% < 1 µm	25 [x]	28 [x]
% < 0.5 µm	19	21
Surface		
MBI, meq/100 g	3.3 [x]	2.7 [x]

[x] Key indicators.

Reproducibility of desired forming, firing, and fired properties is ensured by maintaining these characterizing features within prescribed limits [49].

Experience has shown that when any or all of the ingredients of a clay body must be replaced the 20-odd characterizing values of a full description may be reduced to 8–10 key indicators. A key indicator is a feature that is critical to controlling a particular property. The superscript x's of Table 5 label the key indicators for the two examples.

Because fired body color is much more critical for vitreous chinaware than for sanitary ware, the coloring effect of Fe_2O_3 and TiO_2 must be taken into account when, for example, vitreous chinaware is reformulated [51]. The presence of mica in sanitary ware slip-casting significantly improves the casting rate and the quality of cast [52]. The presence of colloidal organic matter can increase response to deflocculants and result in significant increases in dry bonding power [53]. The rheology of clay-based forming systems can be altered adversely by apparently minor changes in subsieve particle-size distribution [54]: the percentage finer than 1 µm equivalent spherical diameter is an excellent indicator of any change [55]. The methylene blue indices (MBI) correlate with plastic forming properties and dry strength of unfired ware, both of which are functions of specific surface [50].

2. Raw Materials for Traditional Ceramics

Clay-based ceramics are predominant among ceramic products. Clay formulas (or bodies) may consist of a single clay or one or more clays mixed with mineral modifiers such as powdered quartz and feldspar. The special properties of the clay minerals that permit preparation of high-solids fluid systems and plastic forming masses are critical in the shaping of ware.

In developed countries, ceramic manufacturers and raw material suppliers usually work together in establishing standards [56]. The supplier assumes responsibility for continuity of material quality and works closely with the manufacturer in solving material-related plant problems.

However, in less developed countries, manufacturers may need to depend on suppliers who lack facilities and expertise for maintaining material uniformity. An alternative is that the manufacturer may be forced to mine and refine his own materials. In either case, the potter must be prepared to cope with variation in material properties, either by active supervision of supplier mining or through in-plant beneficiation prior to use. The characterization concept (Section 1.3) has permitted development of objective, simple test procedures for use in mining and beneficiation control [57].

2.1. The Structure of Clays and Nonplastics

The atomic structures of the common clay minerals are based on Pauling's generalizations for the structure of the micas and related minerals [58]. Two structural units are involved in most clay mineral lattices. One is the *silica sheet*, formed of tetrahedra consisting of a Si^{4+} surrounded by four oxygen ions. These tetrahedra are arranged to form a hexagonal network repeated to make a sheet of composition $Si_2O_5^{2-}$. The tetrahedral apex oxygens all point in the same direction with pyramid bases in the same plane.

The other structural unit is the *aluminum hydroxide, or gibbsite, sheet*, consisting of octahedra in which an Al^{3+} ion is surrounded by six hydroxyl groups. These octahedra make up a sheet, owing to sharing of edges: two layers of hydroxyls have cations embedded in octahedral coordination, equidistant from six hydroxyls. These octahedral sheets condense with silica sheets to form important clay minerals.

Kaolinite [1318-74-7] is the main mineral of kaolins, with usually tabular particles made up from units resulting from the interaction of gibbsite and silica sheets:

$$Al_2(OH)_6 + (Si_2O_5)^{2-} \longrightarrow Al_2(OH)_4(Si_2O_5) + 2\ OH^-$$

The kaolinite platelets have negative charges on their faces (or basal planes) due to an occasional Al^{3+} ion missing from the octahedral (gibbsite) layer or an Si^{4+} from the tetrahedral (silica) layer.

Disordered kaolinite is a variant of kaolinite in which Fe^{2+} and Mg^{2+} are thought to replace some Al^{3+} in the octahedral layer [59, p. 59]:

$$Al^{3+}_{1.8}Ca^{2+}_{0.1}Fe^{2+}_{0.1}(Si_2O_5)(OH)_4 \cdots M^{2+}_{0.1}$$

The M^{2+}, usually Ca^{2+}, is a balancing exchangeable cation. Hydrogen bonds between gibbsite and the silica layers can be weakened by changes in the octahedral dimensions caused by replacement of the small Al^{3+} (ionic radius of 0.051 nm) by the larger Fe^{2+} (0.074 nm) and Mg^{2+} (0.066 nm) ions. This produces the smaller grain size of disordered kaolinite found in some sedimentary kaolin and ball clay deposits.

Kaolinite crystals consist of a large number of two-layer units held together by hydrogen bonds acting between OH groups of the gibbsite structural layer of one unit and oxygens of adjacent silica structural layers. Unit layers are displaced regularly with respect to one another along the *a* axis. In the case of *halloysite*, the unit layers are stacked along both *a* and *b* axes in random fashion; because of less hydrogen bonding, water can penetrate between successive layers, thereby forming a hydrated variety of kaolinite,

$$Al_2(OH)_4(Si_2O_5) \cdots 2\ H_2O$$

According to KELLER [60], halloysite can exist as spheres, tubular elongates, or polygonal tubes; thus, kaolin occurs in a number of morphologies ranging from worms through stacks, irregular platelets, to euhedral kaolinite crystals. Particle morphology can have significant effects on ceramic forming systems [57].

The montmorillonites result from isomorphous replacements of portions of Al^{3+} or Si^{4+} in the three-layer mineral *pyrophyllite* [12269-78-2], which is formed by fusion of two silica sheets with one gibbsite sheet [61]:

$$Al_2(OH)_6 + 2\,(Si_2O_5)^{2-} \longrightarrow Al_2(OH)_2 \cdot 2\,(Si_2O_5) + 4\,OH^-$$

When Mg^{2+} replaces some of the Al^{3+} in the octahedral layer, the result is *montmorillonite* [1318-93-0] (smectite),

$$Al_{1.67}Mg_{0.33}(OH)_2 \cdot 2\,(Si_2O_5) \cdots M^+_{0.33}$$

M^+ lying between two adjacent three-layer units as an exchangeable cation, offsetting the excess basal-plane negative charge. Because the SiO_2 of adjacent unit layers are held together only by weak van der Waals attraction, montmorillonite particles are thin and small.

If one-quarter of the Si^{4+} ions of the tetrahedral layers of pyrophyllite are replaced by Al^{3+}, a charge of sufficient magnitude is produced to bind univalent cations in regular 12-fold coordination. If the cation is K^+, the result is *muscovite mica* [1318-94-1] [59, p. 23]:

$$KAl_2(OH)_2 \cdot 2\,(Si_{1.5}Al_{0.5}O_5)$$

If the cation is Na^+, the result is *paragonrite mica* [12026-53-8]:

$$NaAl_2(OH)_2 \cdot 2\,(Si_{1.5}Al_{0.5}O_5)$$

Many natural clays contain a micaceous mineral, resembling muscovite but containing less M^+ and more combined water than normal muscovite. This *illite* [12173-60-3] occurs in sedimentary clays sometimes associated with montmorillonite and kaolinite. Analyses of illites from various localities show K_2O contents of 3–7.5%; SiO_2 of 38–53%; and Al_2O_3 of 9–32%. Knowledge of illite is as yet incomplete [59, pp. 24–25].

Table 6 shows the names and chemical compositions of plastic clay minerals and nonplastic layered aluminum and alkaline-earth silicate minerals commonly encountered in ceramic clays.

2.2. Clay–Water System

When a clay is dispersed in water, its balancing *exchangeable cations* retreat to a distance from the clay determined by their size and charge, forming an electrical double layer. If the water contains cations of a different kind and charge, an exchange of solution cations for clay-held cations may occur. Some cations are attracted more

Table 6. Layer lattice minerals

Mineral	Composition
Plastic	
Kaolinite	$Al_2O_3 \cdot 2\ SiO_2 \cdot 2\ H_2O$
Fireclay	$(Al_{1.8} \cdot Fe_{0.1} \cdot Mg_{0.1})O_3 \cdot 2\ SiO_2 \cdot 2\ H_2O \cdots Ca_{0.05}$
Montmorillonite	$(Al_{1.67} \cdot Mg_{0.33})O_3 \cdot 4\ SiO_2 \cdot 2\ H_2O \cdots Ca_{0.165}$
Illite group	muscovite $\longrightarrow$ illites $\longrightarrow$ montmorillonite
Halloysite [12244-16-5]	$Al_2O_3 \cdot 2\ SiO_2 \cdot 4\ H_2O$
Nonplastic	
Muscovite	$K_2O \cdot 3\ Al_2O_3 \cdot 6\ SiO_2 \cdot 4\ H_2O$
Pyrophyllite [12269-78-2]	$Al_2O_3 \cdot 4\ SiO_2 \cdot H_2O$
Talc [14807-96-6]	$3\ MgO \cdot 4\ SiO_2 \cdot H_2O$
Tremolite [14567-73-8]	$5\ MgO \cdot 2\ CaO \cdot 8\ SiO_2 \cdot H_2O$
Chlorite [14998-27-7]	$5\ MgO \cdot Al_2O_3 \cdot 3\ SiO_2 \cdot 4\ H_2O$

strongly to the clay than others. Cations can be arranged in a lyotropic (Hofmeister) series [62, p. 24]; hydrogen is held most strongly and lithium least:

H Al Ba Sr Ca Mg NH$_4$ K Na Li

The capacity of a clay for absorbed cations is termed its cation exchange capacity (c.e.c.) and is a function of clay specific surface [63]. The usual measure of the cation exchange capacity is the MBI (see Section 1.3).

The stability of a suspension of clay particles in water depends on the *degree of deflocculation* of the particles. Deflocculation depends on the character of an electrical double layer made up of the following parts [62, pp. 92–110]:

1) Negative surface charge consisting of the inherent negative planar surface charge plus absorbed OH on normally positively charged edges
2) Absorbed layer of cations at the negative surface, the Stern layer
3) Diffuse cloud of cations that extends to a distance from the charged particle that is determined by the
   a) concentration of ions in the bulk solution away from diffuse cation cloud
   b) size and charge of the cations

The thickness of the electrical double layer is a maximum when the concentration of hydroxides or hydrolyzable salts of the monovalent cations of the Hofmeister series is the minimum needed to fully charge the clay surface. Excess deflocculant reduces the extent of the diffuse layer.

In the absence of a double layer, the bringing together of two clay particles by Brownian motion results in formation of a doublet. Attraction between platelets is either by edge–face attraction or by van der Waals force, or both. Where the normally positive edge has been neutralized or made negative, there is only van der Waals attraction. Particles provided with diffuse, extended counterion clouds cannot approach one another closely enough to allow the inherent van der Waals forces to function fully [64, pp. 183–212], so deflocculation or reduced flocculation is the result.

The very polar *water molecules* are attracted strongly to negative faces or positive edges of clay particles. The adsorbed water molecules, in turn, attract other water molecules, and these, in turn, attract yet other water molecules. Thus, a water structure is built on the surfaces of clay platelets or rods. The extension of the water envelope from the particle surface is thought [65] to depend on the size and valence of the cations present in the water. Exchangeable cations can adsorb water molecules and build up a structure whose extension from the clay surface depends on the amount and kind of cations present. Where large singly charged cations are present, a loose, wide extension occurs; for small multiply charged cations, the counterion cloud is compact and less extended [65]. Water of plasticity and plastic qualities are functions of surface area, particle geometry, and exchangeable cations.

However, if a clay is allowed to absorb *organic colloids,* such as tannic acid or humic acid colloids derived from soil organic matter or lignites, the attraction between clay particles is greatly reduced, water of plasticity drops significantly, response to deflocculants is enhanced, and dry strength rises [53]. Apparently the absorbed organic particles with their absorbed water layers neutralize positive edges and provide a measure of steric hindrance to the close approach of particles.

Deflocculation is, thus, a neutralization reaction between acidic groups of absorbed organic colloids and the monovalent cations and hydroxyl groups provided by the deflocculating compound, rather than a reaction between clay and the deflocculant. Some functional groups are more responsive than others; as a consequence, organic-bearing ball clays vary in their forming properties.

The hydroxyl ion is necessary in the defloccution of clays [53]. The presence of any soluble sulfate or chloride salts in the clay–water system reduces the formation of OH^- and lessens the deflocculating effect of a given quantity of deflocculant.

2.3. Commercial Ceramic Clays

In the United States and the United Kingdom, the major classes of ceramic clays are termed kaolin (or china clay) and ball clay. *Kaolin* may occur at its point of origin in primary deposits or in sedimentary deposits composed of clay particles washed from the point of formation by stream action and laid down in quiet water. Kaolin deposits are widely distributed in the temperate zone. However, in the tropics alteration may be rapid, resulting in bauxite [66].

The term *ball clay* has no technological significance; it is derived from older mining practice in England, whereby cubes of moist, plastic clay were cut from the working face with a special tool, rolled down the clay face, assuming a vaguely spherical shape, and loaded onto wagons by women workers (ball maidens). A general definition of ball clay would be sedimentary clay of fine to very fine grain size, consisting mainly of ordered and disordered kaolinite with varying percentages of illite, mica, montmorillonite, free quartz, and organic matter.

Clays classified as ball clays are widely used in North and South America, England, and to an increasing extent, in Asia. Ball clay is far less used in Europe. The use of ball clays in clay-based forming systems is designed to improve plasticity, reduce water of plasticity, increase unfired strength, improve casting slip properties, and in some cases, improve firing and fired properties. The unfired functions of a ball clay can sometimes be matched by treating fine-grained kaolins with colloidal organic substances [53].

Table 7 characterizes representative china and ball clays from major producing areas in England and the United States. The china and ball clays from Thailand provide examples of ceramic clays available in less-developed nations. The mineral constituents of the clays of Table 7 were calculated from the chemical analyses with a procedure suggested by HOLDRIDGE [67].

The primary kaolins of the china clay deposits of England and Thailand contain more mica than the sedimentary kaolins of Georgia (United States), as demonstrated by their higher K_2O contents [49]. English ball clays are much higher in mica than their U.S. analogues [67]. Mica has favorable effects in slip casting and provides a measure of fluxing.

The flow diagrams of Figure 1 are representative of mining and refining practices in ball clay producing areas of Dorsetshire and Devonshire and china clay deposits of Cornwall in England. The ball clay deposits are very thick with relatively thin overlaying soil. The china clay deposits are kaolinized granite and consist largely of mixtures of kaolinite, muscovite mica, quartz, and small amounts of accessory minerals. Over the past 40 years, the clay producers of England have raised mining and refining of their materials to a very high level of technology. As a consequence of already desirable clay properties, coupled with close control and technical competence, a large export trade has been developed.

The thin overburden and thick deposits of English ball clay permit both open-pit mining and underground mining. Open-pit operations are of two types: (1) excavating of uniform seams with backhoes and (2) selective mining of some clays with a spade-carrying version of the pneumatic jackhammer. Air-spaded clay is lifted from the pit with a boom, placed in a truck, and transported to a processing center. Backhoe-dug clay is placed directly into the truck for transport to a processing center. Underground ball clay mining is done either with air-spading for selective mining or by a rotating cutter that loads the clay directly into the mine car for transport to a processing, storage, and refining center.

English ball clays are stored in accordance with types determined by characterizing feature tests. Clays are sliced (shredded) into thumb-size pieces and often blended with one or more other selections to provide controlled, specified properties. Such blends may be extruded in the form of pellets for bulk shipment or dried and subjected to grinding to a refined powder in an air-elutriation grinding mill. Air-floated clay is usually bagged for shipment.

English china clay is recovered by subjecting the parent ore to "hydraulicking" (high-pressure jet of water). The clay and fine muscovite mica are separated from the ore and transported by the resulting stream to a classifier for removal of the coarser mica and quartz. Further nonclay impurities are removed with a hydrocyclone. The low-solids slip is thickened, characterized, and stored as a 20%

Table 7. Characterizations of typical clay

Properties	Clay*									
	A	B	C	D	E	F	G	H	J	K
Chemical, wt %										
SiO_2	45.7	46.7	50.5	60.4	46.6	47.2	48.5	59.8	49.0	58.5
Al_2O_3	38.3	38.2	28.7	27.0	38.1	37.6	32.3	26.4	34.6	24.4
Fe_2O_3	0.41	0.60	0.91	0.93	0.69	0.50	0.98	1.00	0.71	1.26
TiO_2	1.55	1.42	1.48	1.62	0.07	0.05	1.16	1.39	0.02	0.92
CaO	0.08	0.12	0.40	0.28	0.19	0.20	0.18	0.20	0.35	0.05
MgO	0.06	0.20	0.30	0.26	0.20	0.08	0.21	0.51	0.34	1.05
K_2O	0.06	0.15	0.89	1.70	1.47	1.35	1.89	2.42	2.52	2.36
Na_2O	0.14	0.03	0.18	0.50	0.08	0.07	0.19	0.38	0.48	0.12
Ignition loss	13.65	13.79	16.58	7.59	12.66	12.62	14.78	7.88	10.66	8.64
Minerals, wt %										
Montmorillonite	nil	3	8	7	6	2	6	14	9	29
Kaolin group	96	93	58	44	80	82	60	34	60	23
Mica	2	2	10	21	12	11	18	25	27	22
Free quartz	trace	1	14	26	1	1	9	23	3	19
Organic	trace	trace	8	0.5	trace	trace	5	2	nil	3
Particle size										
% < 20 µm	95	99	99	98	100	88	96	97	77	95
% < 5 µm	69	88	95	79	97	74	91	85	52	87
% < 2 µm	52	72	82	61	64	30	83	81	36	72
% < 1 µm	35	56	69	43	57	23	77	75	19	56
% < 0.5 µm	28	41	51	29	35	15	62	65	16	43
Surface										
MBI, meq/100 g	1.6	10.5	12.1	5.6	5.4	2.4	8.7	12.8	3.4	16.5

* Key to designations:
  A) Coarse kaolin, sedimentary, Washington County, Georgia, United States
  B) Fine kaolin, sedimentary, Wilkinson County, Georgia, United States
  C) Dark fine ball, Graves County, Kentucky, United States
  D) Coarse light ball, Weakley County, Tennessee, United States
  E) Fine china clay, Cornwall, England, United Kingdom
  F) Coarse china clay, Cornwall, England, United Kingdom
  G) Dark ball, Devonshire, England, United Kingdom
  H) Light ball, Dorset, England, United Kingdom
  J) China clay, primary, Thailand
  K) Ball clay, Thailand

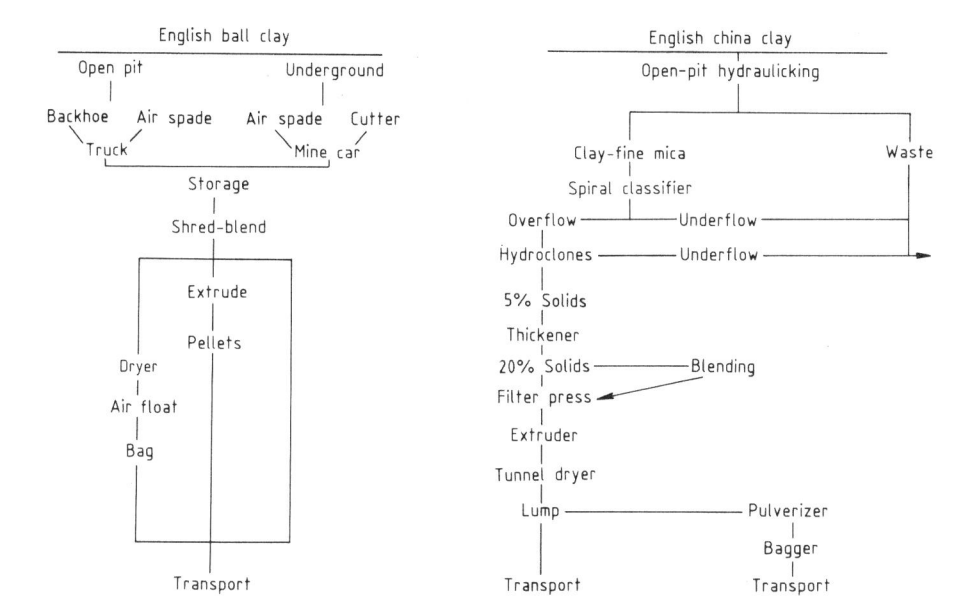

```
English ball clay                                English china clay
Open pit          Underground                    Open-pit hydraulicking
   |                  |                                    |
Backhoe  Air spade  Air spade  Cutter          ┌───────────┴───────────┐
    \   /              \      /              Clay-fine mica          Waste
    Truck              Mine car                   |
              |                              Spiral classifier
           Storage                     Overflow───┴──────Underflow────┐
              |                                   |                   |
         Shred-blend                         Hydroclones──────Underflow┤
     ┌────────┴────────┐                          |                   |
     |              Extrude                   5% Solids
     |                 |                          |
     |              Pellets                   Thickener
   Dryer               |                          |
     |                 |                    20% Solids──────Blending
  Air float            |                          |
     |                 |                    Filter press ◄────┘
    Bag                |                          |
     |                 |                      Extruder
     |                 |                          |
     |                 |                    Tunnel dryer
     |                 |                          |
     |                 |                       Lump ────────── Pulverizer
     |                 |                          |                |
     |                 |                          |              Bagger
     |                 |                          |                |
  Transport       Transport                  Transport          Transport
```

Figure 1. Flow diagrams showing mining and processing methods for the English ball clay deposits of Devonshire and Dorsetshire and the English china clay deposits of Cornwall

solids slurry. Two or more slurried china clay selections may be blended to give desired, controlled properties before filter pressing and drying. The dried clay may be shipped in bulk pellet form or passed through a pulverizer and shipped in bags.

Figure 2 provides flow diagrams representative of mining and refining methods employed in U.S. sedimentary kaolin deposits of Georgia and South Carolina and in ball clay deposits of Tennessee and Kentucky. Overburden is usually no more than 8–10 m thick. Neither ball clay nor kaolin deposits exceed $\approx$ 15 m. All mining is open pit.

Selective mining based on drill hole and working face characterization tests is done with dragline or power shovel. Transport from pit to processing and storage sites is by trucks carrying 5–10 t up to 10 km. Storage is in the form of shredded clay.

Kaolins are blended to specification and either dry-ground for bulk or bagged shipment or subjected to wet processing. High-solids slurries (70%) are prepared for tank-car shipment to ceramic plants using slip-cast manufacturing. Low-solids slurries are subjected to centrifugal fractionation with subsequent thickening, filtration, and drying. The dried filter cake may be shipped in bulk, air-floated and sent in hopper cars as bulk, or pulverized for bagged shipment.

The ball clays are blended to specification and shipped as is, as high-solids slurries, or dried.

The processing of clays for use in ceramics is also described under → Clays.

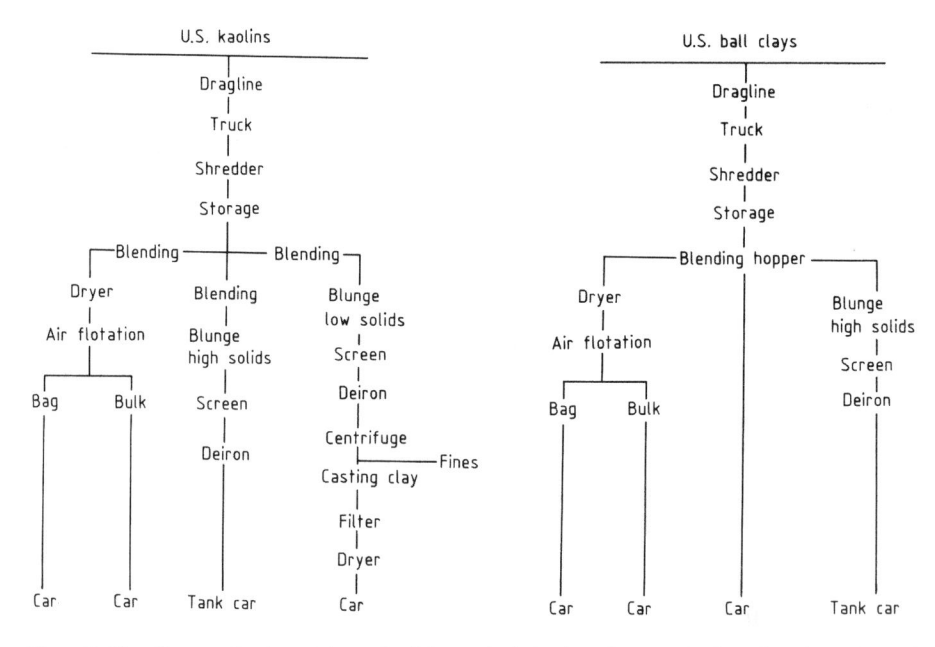

Figure 2. Flow diagrams showing mining and refining methods for the sedimentary kaolins of South Carolina and Georgia and the ball clay deposits of Kentucky and Tennessee

2.4. Commercial Nonplastics for Ceramics

A large proportion of ceramic ware is made from clay-based formulas whose major constituents are clay minerals, powdered silica, and powdered feldspar or a related feldspathoid. Such bodies are termed triaxial [19, pp. 178–183]. The fluxing feldspathoids and silica minerals are termed nonplastics. The term *flint* is properly used only with reference to powdered flint pebbles.

The *feldspar group* of minerals is the most important source of fluxing oxides for clay bodies. All are framework aluminosilicates based on an SiO_2 structure. Replacement of Si^{4+} by Al^{3+} results in charge deficits that are balanced by K^+, Na^+, or Ca^{2+} lying in framework voids. The smaller Na^+ and Ca^{2+} ions confer a different crystal structure than the larger K^+ ion. Albite [12244-10-9] ($NaAlSi_3O_8$) and anorthite [1302-54-1] ($CaAl_2Si_2O_8$) are isomorphous and form the plagioclase solid-solution series. Albite and anorthite are triclinic, whereas microcline [12251-43-3] ($KAlSi_3O_8$) is monoclinic. Nepheline syenite is a type of rock consisting of nepheline [12251-27-3] ($K_2O \cdot 3\,Na_2O \cdot 4\,Al_2O_3 \cdot 9\,SiO_2$) mixed with microcline and albite.

An old saying, attributed to the Chinese [19, p. 92], says in effect that *silica* [7631-86-9] is the skeleton and clay the flesh of a ceramic body. There is a tendency to regard silica as an inert substance in the body. However, this is far from the case: the silica can have profound effects both in forming and firing.

Table 8 provides examples of fluxing feldspathoids and silicas used in clay-based ceramic formulations. The mineral constituents of the feldspars and silicas of Table 8 were calculated from the chemical analyses with a method by KOENIG [68]. Feldspar A is a froth-floated feldspar recovered from North Carolina alaskite granite. Material C is dry-ground, selectively mined nepheline syenite from Ontario, Canada. Material E is wet-ground feldspar from Thailand. All are successfully used in clay-based ceramic formulations.

In addition to the feldspathics and silica, some clay-based bodies contain calcined Al_2O_3 to increase fired strength; ground limestone and/or dolomite as auxiliary flux; talc for special heatshock bodies and wall tile; chlorite to lower the maturing temperature of slip-cast porcelains; or wollastonite, a wall-tile body constituent.

The principal sources of pottery and glass grade feldspar in the United States are deposits in Connecticut, North Carolina, South Carolina, Oklahoma, and California [69]. Nepheline syenite, also widely used in ceramic formulations and in glass batches, is produced from deposits in Methuen Township, Ontario, Canada [70].

Prior to 1940 all feldspar mined in the United States was selectively quarried, crushed, and hand-cobbed on picking belts before being ground. Just after World War II a froth floating procedure began to be applied to mixed-mineral rocks containing feldspar. At the present time over 80% of the feldspar produced in the United States is recovered by froth flotation from a variety of ores, including alaskite granite, pegmatite, graphic granite, beach sand, and weathered granite. The remaining feldspar, mainly high K_2O feldspar, is block mined, hand-cobbed, and processed dry. Nepheline syenite is also selectively mined and subjected to dry processing.

Figure 3 provides a generalized flow diagram for froth flotation recovery of feldspar from coarse granites.

After (normally) thin overburden has been removed from the ore, the granite is blasted and transported to a processing plant. The large pieces are passed through, successively, a jaw crusher and cone crusher to prepare rodmill feed.

From the feed bins the thumb-sized pieces of ore pass through rodmills where they are reduced to millimeter-sized grains. The rodmilled pulp then goes onto rotating screens to remove oversize, which is returned to the rodmill for further grinding. Passage of screened pulp suspended in water through a hydroseparator removes most of the fines that might interfere with the chemistry of flotation processes. The sized, de-slimed pulp is then sent to a chemical conditioner where the mica particles are treated to promote bubble adherence. The underflow (feldspar, quartz, and garnet) is conditioned chemically to allow only the iron-containing garnet to be attracted to bubbles and so removed in the froth overflow. Next comes separation of feldspar from the quartz by adjusting the reagents to cause feldspar particles to adhere to the froth and the quartz to be rejected.

Final steps involve draining, rewashing to remove reagents and draining of the cleaned products, passage of drained material through a dryer and through a magnetic field, and finally storage. Pottery uses require fine grinding; glass grade requires no grinding of the granular feldspar or quartz.

Table 8. Characterizations of typical nonplastics

Properties	Feldspathic/Feldspathoid*					Flint/Quartz*				
	A	B	C	D	E	F	G	H	J	K
Chemical, wt %										
SiO_2	66.8	68.5	60.7	79.5	71.1	96.6	98.5	97.9	99.5	95.7
Al_2O_3	19.6	17.5	23.3	12.0	16.0	0.2	0.9	0.5	0.2	2.1
Fe_2O_3	0.04	0.08	0.07	0.08	0.26	0.10	0.09	0.40	0.06	0.03
TiO_2				0.01	0.34	0.01	0.06			0.09
CaO	1.70	0.30	0.70	0.20	1.54	0.20	0.02			0.20
MgO	trace	trace	0.10	0.09	0.37	0.09	0.03			0.20
K_2O	4.80	10.40	4.60	3.80	0.06	0.39	0.05			0.30
Na_2O	6.90	3.00	9.80	3.90	8.64	0.15	0.04			0.03
Ignition loss	0.20	0.30	0.70	0.45	0.42	1.58	0.17	0.20	0.12	1.60
Mole of flux	0.1728	0.1644	0.2220	0.1092	0.1768	0.0124	0.0028			0.0123
Minerals, wt %										
Feldspars	92	83	75	48	81					
Nepheline			24							
Mica	4	7			trace					2
Quartz	4	9		41	18	96	97	97	99	91
Clay				3		1	1		trace	6
Organic						trace				trace
Other			1	8	1	4	2	3	1	1
Particle size										
% < 20 µm	67	64	68	60	53	56	57	75	58	53
% < 5 µm	26	26	22	23	20	16	18	26	15	12
% < 2 µm	11	12	9	11	10	5	7	9	5	5
% < 1 µm	9	9	3	6	5	1	2	3	1	3
% < 0.5 µm	trace	6	trace	2	2			1		2

* Key to designations:
  A) Flotation feldspar, Mitchell County, North Carolina, United States
  B) Block feldspar, Custer County, South Dakota, United States
  C) Nepheline syenite, Ontario, Canada
  D) Cornish stone, Cornwall, England, United Kingdom
  E) Feldspar, Thailand
  F) Flint, France
  G) Quartzite, Pennsylvania, United States
  H) Quartzite, Venezuela
  J) Silica sand, California, United States
  K) Silica sand, Philippines

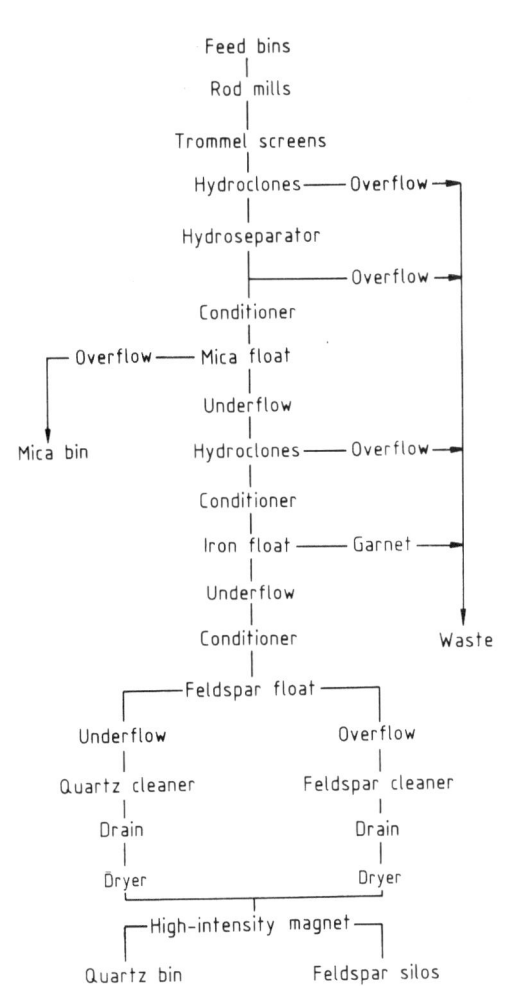

```
                Feed bins
                    |
                Rod mills
                    |
               Trommel screens
                    |
                Hydroclones ─── Overflow ──┐
                    |                       |
                Hydroseparator              |
                    ├─────── Overflow ──────┤
                    |                       |
                Conditioner                 |
                    |                       |
  ┌── Overflow ── Mica float                |
  |                 |                       |
  |             Underflow                   |
  ↓                 |                       |
Mica bin        Hydroclones ─── Overflow ──┤
                    |                       |
                Conditioner                 |
                    |                       |
                Iron float ─── Garnet ─────┤
                    |                       |
                Underflow                   |
                    |                       |
                Conditioner                Waste
                    |
          ┌── Feldspar float ──┐
          |                    |
     Underflow              Overflow
          |                    |
     Quartz cleaner      Feldspar cleaner
          |                    |
        Drain                Drain
          |                    |
        Dryer                Dryer
          |                    |
          ├─ High-intensity magnet ─┐
          |                         |
     Quartz bin              Feldspar silos
```

Figure 3. Flow diagram showing typical froth flotation recovery of muscovite mica, feldspar, and quartz from a coarse granite found in western North Carolina

Where a deposit is sufficiently pure, block feldspar may be processed as shown by the diagram of Figure 4.

The blasted block material is passed through a jaw crusher prior to passage through a rotary dryer into a surge bin. The crushed, dried product passes through a cone crusher onto a 2.4-mm vibrating screen, with any oversize being returned for further crushing.

The minus-2.4-mm product goes into a surge bin that feeds a high-intensity magnetic separator; magnetic particles pass to waste, while nonmagnetics go to mill feed bins. Milling is by pebble mills; ground product goes to air classifiers, with any oversize returned for further milling. Undersize passes to storage silos for bagged or bulk-loaded shipment. Nepheline syenite is processed in much the same manner, with an additional step designed to produce a minus-1.0-mm size for glass batching.

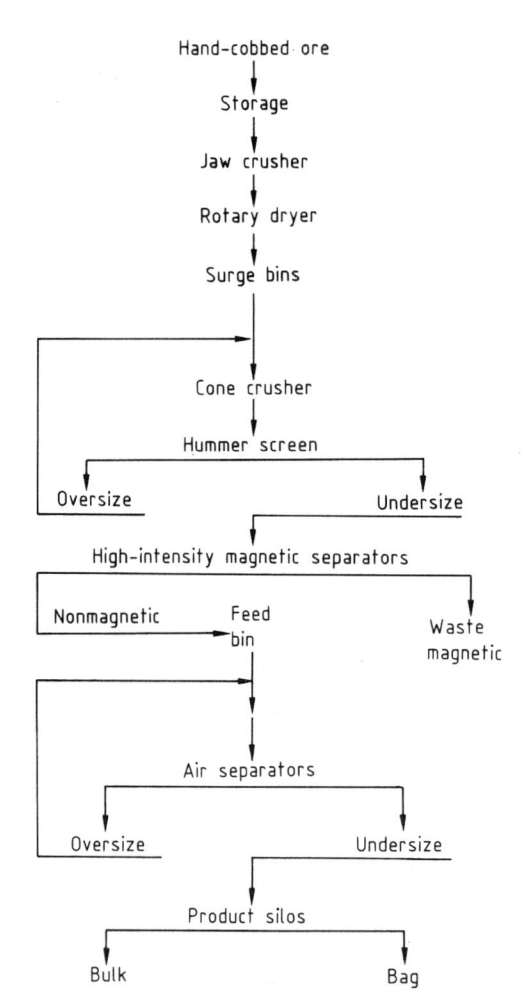

```
Hand-cobbed ore
       ↓
    Storage
       ↓
   Jaw crusher
       ↓
   Rotary dryer
       ↓
   Surge bins
       ↓
   Cone crusher
       ↓
   Hummer screen
   ↓         ↓
Oversize  Undersize
             ↓
High-intensity magnetic separators
   ↓       ↓        ↓
Nonmagnetic Feed  Waste
            bin   magnetic
             ↓
        Air separators
        ↓         ↓
    Oversize   Undersize
                  ↓
             Product silos
             ↓        ↓
           Bulk      Bag
```

Figure 4. Flow diagram showing a typical cobbing (hand selection) method for mining and processing a pegmatite feldspar in southwestern South Dakota

3. Raw Materials for Advanced Ceramics

Although traditional ceramics are composed of natural raw materials that are physically separated and reduced in size, advanced ceramics require chemical conversion of raw materials into intermediate compounds. These intermediates lend themselves to purification and eventual chemical conversion into a final desired form.

Oxides and carbonates available in powder form include those of Al, Sb, Ba, Be, Bi, Co, Mn, Mg, Ni, Si, Th, Ti, and Zr. Also available are carbides of Si, Ti, and W and the nitrides of Al, B, Hf, Si, and Zr. However, needs exist for specialized powders for some advanced ceramics, and a variety of chemical routes can be used to synthesize these powders. Chemical routes, such as sol–gel processing, can bypass the powder stage.

Requirements for high strength and smooth finishes, particularly of small parts, necessitate fine-grained powders. Thus, one line of advanced ceramic research aims at producing very fine, essentially spherical, monosize particle powders. These are typically made by colloidal chemistry for oxides. Nitrides and carbides involve controlled nucleation and growth in gas-phase reactions. However, most high-technology ceramics are still made from powders with broad size distributions in the submicrometer (under 1 µm) range.

3.1. Metal Oxides and Carbonates

Alumina [1344-28-1] is derived from bauxite by selective leaching with NaOH, precipitation of purified Al(OH)$_3$, and thermal conversion of the resulting fine-size precipitate to Al$_2$O$_3$ powder (→ Aluminum Oxide) for use in polycrystalline Al$_2$O$_3$-based ceramics. Antimony [7440-36-0] is derived from Sb$_2$S$_3$ (stibnite) by reduction with iron scrap, and antimony trioxide [1314-60-9] is formed by burning antimony in air.

Barium oxide [1304-28-5] is obtained by decomposition of BaCO$_3$ at high temperature; the carbonate itself is made by reaction of Na$_2$CO$_3$ with BaS. Beryllium oxide [1304-56-9] is prepared by heating Be(NO$_3$)$_2$ or Be(OH)$_2$. Bismuth oxide [1332-64-5] is obtained by heating Bi(NO$_3$)$_3$ in air.

Cobalt compounds are derived from ore concentrates by roasting and leaching with acid or ammonia; the oxide [1307-96-6] is formed by calcination of the carbonate or sulfate. Magnesium oxide [1309-48-4] is readily available as the 99.5 % pure grade powder, but greater purity may require calcining of high-purity salt solutions. Manganese oxide [1344-43-0] can be prepared by calcination of manganous nitrate.

Nickel ores are either sulfidic or oxidic. Sulfides are flotation-separated and roasted to sintered oxide. Oxides are treated by hydrometallurgical leaching with ammonia. Nickel oxide [1313-99-1] is then prepared by gentle heating of Ni(NO$_3$)$_2 \cdot$ 6 H$_2$O. Strontium carbonate is formed by boiling celestite, SrSO$_4$, in a solution of (NH$_4$)$_2$CO$_3$; SrO [1314-11-0] is formed by decomposition of the resulting SrCO$_3$.

Vanadium pentoxide [1314-62-1] is prepared by ignition of alkali solutions from vanadium minerals. Zinc carbonate [3486-35-9] is prepared by action of sodium bicarbonate on a zinc salt, such as zinc chloride. Zirconia [1314-23-4], ZrO$_2$, is derived from Zr(OH)$_4$ or Zr(CO$_3$)$_2$ by heating.

3.2. Borides, Carbides, and Nitrides

Boron and carbon can be made into B_4C [12069-32-8] by heating B_2O_3 and carbon in an electric furnace. Boron nitride [10043-11-5] is made by heating B_2O_3 and tricalcium phosphate in an ammonia atmosphere in an electric furnace (→ Boron Carbides, Boron Nitride, and Metal Borides).

Boron, carbon, and nitrogen can be made into other synthetic compounds with refractory and wear properties. Examples are silicon carbide (SiC), silicon nitride (Si_3N_4), tungsten carbide (WC), titanium carbide (TiC), titanium nitride [25583-20-4] (TiN), tungsten boride (WB_2). A translucent AlN has been developed that is 5 times as thermally conductive as Al_2O_3 ceramics.

4. Processing Ceramic Ware

Traditional and advanced ceramic industries use many techniques for processing their products. The exact process is governed by the nature of the forming system, the size and geometry of the piece, product specification, and practices in various areas of the ceramic industry.

Most ceramic manufacturing processes start with formulas consisting of one or more particulate materials. These formulas are used for shaping products that are further processed by firing and by finishing of the fired items.

In many cases products have complex shapes made by use of one or another of such forming techniques as dry or isostatic pressing, plastic shaping, extrusion, slip casting, injection molding, tape casting, and green finishing. These shaping techniques are based on some old — the potter's wheel, for example — and some new procedures developed from recent research findings.

Forming systems employed in making traditional and advanced ceramic ware are (1) liquid suspensions, (2) plastic masses, or (3) more or less dry granulated or powdered formulations.

4.1. Preparation of Clay-based Forming Systems

The clay bodies of traditional ceramics are normally mixtures of clays and powdered nonclay minerals or else natural mixtures of clay substances and nonclay particulate materials. Most clays occur as aggregates of clay particles. When contacted with water, such aggregates tend to break apart or slake. The development of a water structure on the surfaces of the particles results in plasticity (see Section 2.2). If sufficient water is added to the clay and the mixture is agitated, a dispersion forms. Because the powdered nonplastics, i.e., the nonclays, do not develop any great degree of plasticity when moistened with water, the various ceramic systems of traditional ceramics depend on the plastic component (usually but not always clay) to provide (1) the workability required in plastic forming or dry pressing, (2) the deflocculant response of fluid systems in slip casting, and (3) the green and dry strength of unfired ware.

Figure 5 shows the moisture-content variation and forming-pressure ranges for soft plastic shaping, extrusion, dry pressing, dust pressing, isostatic pressing, and slip casting of clay-based bodies.

Because the ingredients used by any given plant may range from highly purified to as-mined lump materials, the body preparation process must vary with the particular circumstances. However, the main objectives of processing are always (1) to arrive at as intimate a mixture of clay and nonplastic particles as possible, (2) to provide uniformity of shaping properties from lot to lot, and (3) to maintain uniformity of firing and fired properties from lot to lot.

Preparation processes for these forming systems can be divided into two general classes: (1) wet processing and (2) dry processing.

Wet Processing. Wet processing is usually employed whenever one or more of the ingredients needs initial or supplementary beneficiation. General practice in the United States and the United Kingdom subjects dinnerware bodies (Table 5, Vitreous china) to wet processing to ensure adequate dispersion of clay constituents, permit sieving for removal of oversize, and allow magnetic treatment to remove iron particles. Such a process uses relatively unrefined shredded or lump ball clays and filter cake or coarsely pulverized china clay.

Third-world ceramic manufacturers may have access to producer-beneficiated materials but often must depend upon their own mines for at least a portion of their raw materials. In some instances beneficiation of local materials becomes an integral part of body preparation. In the People's Republic of China and Thailand, for example, the silica and fluxing feldspars may be received in block form and ground during the body preparation process.

Because grinding is readily accomplished by dry crushing, followed by wet ball milling, one approach is to wet-grind the nonplastics along with a small, fixed percentage of suspending fine clay. The nonplastic slop (suspension) is then sieved, deironed by magnets, and stored in agitators. Clays are wet-dispersed as suspensions, sieved and deironed, and then blended by formula with nonplastic slop in agitator tanks.

A modification of this method is to simply weigh all formula ingredients as a unit, transfer the batch to a ball mill with the required water, and mill to a specified sieve residue percentage.

For plastic or dry-press forming systems, wet processing is done in the flocculated state. Indeed, flocculation is often enhanced in preparation of high-tension electrical porcelain bodies by addition of $AlCl_3$ or $MgSO_4$. Sufficient water must be used to allow sieving and passage through a magnetic separator.

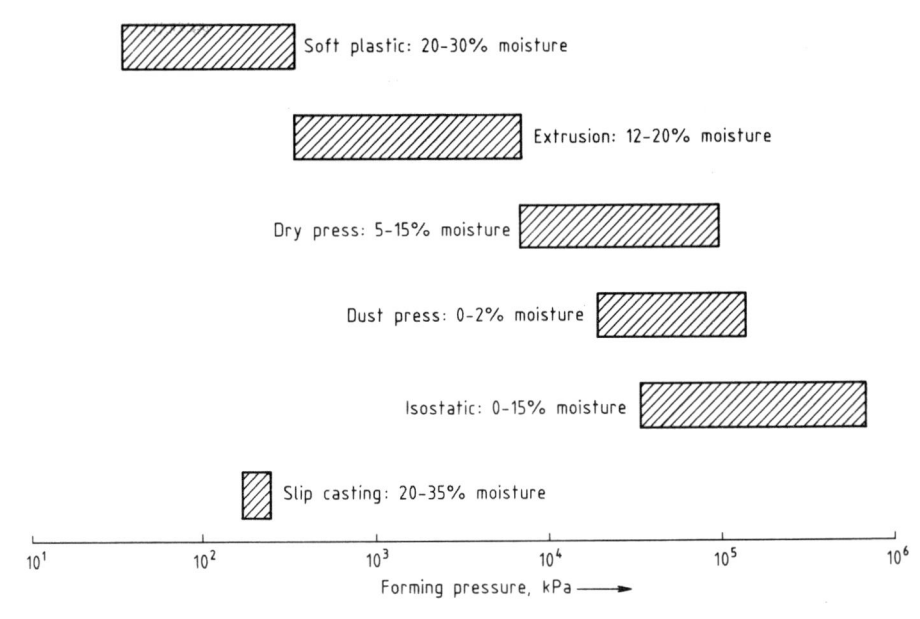

Figure 5. Moisture content and pressure ranges required for shaping clay-based forming systems

Bars shown (moisture % vs forming pressure in kPa):
- Soft plastic: 20–30% moisture
- Extrusion: 12–20% moisture
- Dry press: 5–15% moisture
- Dust press: 0–2% moisture
- Isostatic: 0–15% moisture
- Slip casting: 20–35% moisture

X-axis: Forming pressure, kPa (10^1 to 10^6)

Consistency of plastic masses is controlled by four major factors: (1) specific surface of the body, (2) modifying inorganic ions such as Ca^{2+}, Mg^{2+}, Al^{3+}, SO_4^{2-}, and Cl^-, (3) the amount and kinds of organic colloid present, and (4) the proportion and temperature of water present.

The water content of plastic forming systems is reduced to a working level by filter-pressing and, when necessary, by further drying of the filter cake. Air is removed from the filter cake by passage through a vacuum pug mill.

Pressing dust can be made from filter cake by drying the cake and passing the dried cake through a granulating hammer mill. Otherwise, sieved and deironed slip can be diverted to a spray drier and formed into pressing-size granules.

Dry Processing. The flow diagrams of Figures 1–4 (Chap. 2) indicate that finely ground, deironed clays and nonplastics can be obtained in both North America and the United Kingdom. Such materials make it possible to prepare both plastic forming systems and pressing dust without a slip stage. Dry pressing dusts are prepared by dry blending the ingredients with ribbon blenders or rotating cone mixers and then incorporating the required moisture with a muller mixer. If sufficient water is mulled into the mix and the resulting plastic mass is passed through a vacuum pug mill, the resulting forming system can be used for plastic forming.

Refractories and heavy clay products are usually made from combinations of clay and coarse nonplastics by crushing them in a wet pan (heavy rollers revolving in a pan) and adding water plus other modifiers. By variation of the moisture, the mulled mixes can be made into pressing dusts by granulation or into plastic systems by a deairing operation.

Casting Slip. Although filter-cake clay body is sometimes made into casting slip by addition of deflocculating agents, by far most casting slips are made by direct wet methods.

Clay-based casting slips must be made to cast to a firmly plastic state within a prescribed time range. Casting properties, such as rate, amount of retained water, and plastic quality of casts, are each in some way related to freshly stirred consistency of the slip and its tendency to thicken on standing. Common practice in industry is to control casting properties by maintaining a constant *solids concentration* by measuring slip specific gravity and adjusting *slip rheology* to targeted freshly stirred viscosity and thixotropic gelling. Unfortunately, the mere meeting of a targeted rheology is no guarantee of constant casting performance.

Variation in *slip temperature* can alter slip viscosity and casting rate significantly [71]. Thus, it is possible for two slip batches at different temperatures to have identical viscosities and thixotropies, yet to cast in decidedly different ways. RYAN and WORRALL [72] found that the nature of exchangeable cations in casting slip governs the rate of cast under constant temperature and rheological conditions. The custom in sanitary ware slip control is to buffer the effect of deflocculant-enhancing organic colloid by addition of divalent alkaline-earth-metal carbonates or sulfates to control the rate and structure of the cast [73].

The rheology and casting properties of casting slips are strongly influenced by apparently minor changes in the distribution of *particle sizes* in the subsieve region. BROCINER and BAILEY [74] have shown that the coarse kaolin component of a casting slip can be made variably finer as the input of energy imparted in blunging or ball milling is varied: the mixing or milling operation must be very carefully controlled, and both equipment and time of mixing should be kept constant.

In direct preparation of casting slips, on occasion a standard sequence and timing of additive and raw material introduction into the mixer is not followed. If, for example, a light ball clay is added first with the Na_2CO_3, followed by an organic-bearing ball clay, the amount of adjusting sodium silicate required is significantly greater than if the reverse order were used, and the resulting slip requires a longer aging period. If deflocculation is initiated with sodium silicate and the Na_2CO_3 is added later, the aging time [75] is greatly extended. When slips prepared by using differing sequences of addition are adjusted to the same viscosity and thixotropy, their casting rates and cast structures are also likely to differ significantly.

Equipment. Those plants that grind their own nonplastics use *ball milling,* either continuous dry grinding in an air-swept conical ball mill or batch wet grinding in a cylindrical ball mill. Dry grinding demands that the feed material be dry to avoid packing and to allow air sweeping of fines to a collector. Wet grinding is claimed to require less power than dry grinding, but dry grinding produces less wear on the mill lining and grinding media [76, sect. 6, pp. 20–25].

Ball mills belong to a class of grinding devices termed tumbling mills. The rotating container is a cylinder mounted with its axis horizontal. The grinding action is due to the tumbling of the grinding media, which are cast iron or steel balls, hard rock (e.g., flint pebbles), or some nonmetallic material such as high-alumina porcelain [76, sect. 6, pp. 20–25].

Blunging refers to the agitation or blending of ceramic materials in a mixing tank equipped with an impeller to stir the suspension and baffles to direct the suspension to the impeller. Impellers may be simple paddles or specially designed shapes for increased efficiency of dispersion [77].

Screening (or sieving) of fluid dispersions is termed wet screening. Two general types are employed in the sieving of blunged or ball-milled slips: (1) an inclined rectangular panel of wire mesh having the proper openings and (2) circular screens. The *inclined rectangular panels* are subjected to vibration that agitates and separates the coarser particles during transit of the slip. Vibration can be by shaking or electromagnetic pulse [76, sect. 7, pp. 34–37].

Circular vibratory screens can [78] effectively separate particles as fine as 44 µm in diameter. The basic arrangement consists of a motor plus interchangeable frames that hold screening wire cloth and discharge ports. The frame is held rigidly to a main screen assembly. The motor has a vertical upward and downward extended shaft fitted with eccentric weights. The main screen assembly is mounted on a circular base by springs that permit the assembly to vibrate freely, while preventing vibration of the floor. A number of three-dimensional patterns of the suspension on the screen can be developed by varying the angle between top and bottom weights. This type of screen is widely used in the United States and the United Kingdom.

Screens used for pressing-dust sizing are relatively coarsely meshed (2.0–3.0 mm), whereas those used for plastic body systems and casting slips are much finer (0.20–0.05 mm).

To remove magnetic particles, granular nonclay ball-mill feed can be subjected to a *magnetic separator*, passage either through a magnetic field or over a magnetic pulley prior to the grinding operation. Suspensions of clays or nonclay powders can be passed through the grid of an electromagnetic purifier prior to the dry or pugging operations. High-gradient magnetism is capable of removing such colorants as TiO_2 from kaolin slurries; this can transform the high-TiO_2 Georgia and South Carolina kaolins into very white-firing fine-china constituents [79].

Dewatering of slips for preparation of plastic forming systems or pressing dusts is usually by *filter pressing*. The basic concept of filter presses involves feeding the slurry under pressure into the space between square, round, or rectangular plates. This space is created by frames that alternate with the plates. Plates are hollow and normally covered with filter cloth. As the space fills with suspension under pressure, the liquid is forced through the cloth and drains away as the solids form a cake [80].

Pugging is the process of blending clays and water by manual or mechanical means. A pug mill is an open trough with a lengthwise shaft on which are mounted blades that blend the clay and water to a plastic forming system of the desired consistency. Filter-cake bodies are subjected to a combination of pugging in an auger trough, coupled with passage through a vacuum chamber, followed by extrusion. Vacuum pug milling (or deairing) makes the plastic mass more workable and cohesive by elimination of the air from the system [2, p. 267].

A fluid suspension of particulate material can be dried and formed into pressing dusts or granules by *spray drying*. A spray dryer consists essentially of a drying chamber. A downward flow of heated air is introduced at the top of this chamber. A flow of suspension is transformed into an upward flowing cloud of droplets by a nozzle atomizer. The droplets are dewatered and fall to a product outlet at the bottom of the drying chamber. An attached exhaust removes excessively fine particles to a cyclone collector. Relatively uniform spheres are formed, and moisture content is also uniform [81].

When fully purified clays and nonclay powders are available, *dry blending and tempering* are employed. A shell or ribbon mixer may be used as an initial step, followed by addition of water in a mixing muller. Otherwise, dry blending and tempering may be done in stages in a mixing muller. Mixing mullers normally have heavy wheels, under which the moistened body rotates. First, a smearing action occurs, and second, a rotating plow scrapes the compressed body up and turns it under the mullers for additional mulling [82]. The tempered body then goes to dust mill/sieving operation for lower-moisture-dust processing or dry pressing or to a deairing pug mill for plastic systems.

4.2. Preparation of Advanced Ceramic Systems

Traditional ceramic forming systems are nearly always polydisperse, with particle size ranging more or less continuously from an upper to a lower limit. As a distribution ranging between definite limits approaches linearity on an arithmetic plot, optimum packing results in minimum voids [83]. The more extended the range between upper and lower limiting sizes, the lower the void volume for a given distribution [84]. However, the more extended the distribution is, the more sensitive it is, with respect to void volume, to deficits or excesses of intermediate particle sizes. This finding has been related to differences in calcined alumina slip occasioned by altering particle-size distribution size limits and intermediate size distribution [85].

Although a controlled optimum particle size distribution is needed for maximum, reproducible strength, sometimes a mono-size distribution must be approached to avoid growth of larger particles at the expense of the smaller: very fine particles are much more reactive than larger particles, and quite porous initial compacts can be sintered at high temperature to nearly theoretical density. Transparent polycrystalline Al_2O_3 is an example. The finer the powder, the more rapid the sintering and the lower the densification temperature, thereby reducing grain growth and increasing fired strength.

Sizing of Advanced Ceramic Materials. Because particle size and distribution are so important for controlling properties of advanced ceramic products, the manufacturer must often further refine an already refined as-received material to meet his specifications. A variety of techniques are used for modifying particle size and distribution:

screening
air elutriation
ball milling
attrition milling
vibratory milling
fluid energy milling
liquid elutriation
precipitation
freeze drying
laser
plasma
calcining
sol–gel

Dry *screening* is used for sizing particles down to 44 µm, whereas wet, slurry screening is often employed for subsieve sizes. *Air elutriation* (or classification) is used to separate coarse and fine fractions. Special air classifiers are available for separating minus-20-µm particles, but care must be exercised to avoid contamination. *Liquid elutriation* can be used to separate a single specific material into fractions or to separate materials having different specific gravities.

Ball milling [86, pp. 410–438] consists in placing either a dry or a suspension charge in a closed container with appropriate grinding media and rotating the container to create a cascading action of the media. Media selection is important. Higher density pebbles or cylinders will grind more quickly than lower density media. Wear of media creates contamination that can be controlled by careful selection of wear-resistant mill lining and hard grinding media. Wet ball milling requires removal of water from the powder. Dry ball milling requires additional grinding aids such as a lubricating stearate. A very small amount of moisture has been found to prevent packing of high-alumina prereacted body during dry grinding.

Attrition milling is similar to ball milling, but the container is held in a fixed vertical position and the grinding media agitated by arms attached to a rotating shaft. The attrition mill can be used for dry grinding or wet grinding with vacuum or various controlled atmospheres [86, pp. 439–443].

Vibratory milling uses fixed containers typically lined with polyurethane or rubber. Suitable grinding media are placed in the container with the material to be ground, and a vibration is transmitted through the bottom center. The resulting cascading–mixing action leads to shear and impact breaking of particles between grinding media [86, pp. 410–438].

Fluid energy milling functions by causing particles of the material to be ground to impact one another. They are carried at high velocity in a fluid — air, water, superheated steam. Jet mills are lined with wear-resistant materials [87].

Precipitation of soluble salts and pyrolysis to the oxide has been used to provide controlled particle size and high purity. Calcined alumina has been made by precipitating alumina trihydrate from solution by changing pH and using seed crystals. The very fine, reactive alumina greatly extends the uses for alumina [4, p. 165].

Freeze drying involves forming drops from solutions of metal salts, freezing them rapidly, removing water by sublimation under vacuum, and calcining the crystallized salts [88]. Another method for preparing pressing granules is by dispersing the powder and additives as a slurry and drying by spraying the slurry or solution into a chamber where the drops fall through hot gases. Surface tension holds the drops in spherical form. These drops, when dry, flow readily into a die [89].

Slip-cast advanced ceramic forming systems require a particle distribution [85] that provides maximum packing. Often sizing is accomplished by blending several narrow distributions [83], or the material may be ball-milled with binder, wetting agents, deflocculants, and densification aids. Disk mills [86, pp. 468–488] are especially effective in dispersing agglomerated powder. The liquid phase normally used in mold casting is water, whereas in tape casting the liquid is usually nonaqueous [90]. In each instance, all air bubbles must be removed from slips by vacuum treatment prior to use [91].

A number of glasses have been prepared in the laboratory by hot pressing or sintering gels of single oxides or combinations of two or more oxides, such as SiO_2, Al_2O_3, and TiO_2. Carefully controlled processing makes monolithic objects possible [92]. Commercial uses of *sol–gel* are fibers, powders, bulk shapes, and oxide coatings of films [93]. Of these uses, film or oxide coatings are regarded as very important.

Processing of a sol–gel starts with a metal alkoxide: $Si(OC_2H_5)_4$, $Ti(OC_2H_5)_4$, as well as $Al(OC_2H_5)_3$ are examples. Alcohol and distilled water are hydrating reagents. A wide variety of silicate and aluminosilicate systems have been made with other cations, such as those of Li, Na, K, Rb, Mg, Ca, Sr, Ba, Pb, Ga, Fe, Ln, Ti, Zr, and Th, as well as ternary or quaternary compositions with two or more of these elements [94].

The basic procedure for making SiO_2 and metal oxide gels is to dissolve $Si(OC_2H_5)_4$ in ethyl alcohol and add alcohol or water solutions of the desired metal nitrate. Hydrolysis is effected with an excess of distilled water. At 60 °C the SiO_2 precipitates as a stiff gel.

Preconsolidation of Advanced Ceramics. Preparation of a pressing dust sometimes involves addition of a binder, a lubricant, possibly a sintering aid, and finally, development of a free-flowing powder by granulation. This may be done by blending the fine, low-bulk-density powder with binder solution and lubricant, and then compacting the mass into blocks that are chopped, crushed, or coarsely pulverized. The resulting granules are screened to obtain proper size for die filling.

4.3. Forming Ceramic Articles

Forming systems used to make traditional and advanced ceramic ware include slip casting, soft plastic, stiff plastic, dust pressing, dry casting, and a number of modified or special systems for advanced ceramics (see Fig. 5).

Soft Plastic Forming. The simplest method of forming plastic masses is by hand molding. This requires a soft plastic system. Currently, soft plastic forming systems are used in the production of soft mud bricks; pottery by throwing; jiggered or roller-formed tableware; hot-plunge insulators; and ram process products. In *soft mud brick making,* the selected clays are prepared by wet panning and passed through a pug mill that forces the plastic clay through a die into wooden molds. *Throwing* on the wheel is a soft plastic method for making vases and the like, used in simple cultures and by art potters. The wheel is a disk on top of a shaft turned by a weighed kickwheel or by a motor.

Jiggering was developed from throwing. A measured slug of soft plastic body is placed on a plaster form that revolves on a wheel head. A template tool is brought down onto the moist bat, pressing it down onto the plaster mold and so forming the upper part of the piece. At the same time, the template tool scrapes away excess body from the moist piece with the aid of a spray of water. Automation requires carefully controlled, deaired forming masses [95].

The *roller-head method* for soft plastic forming is an alternative to the jigger, especially for less plastic formulations such as bone china and hard porcelain. Instead of a scraping template blade, a polished (and sometimes heated) contoured metal roller is brought down and rolls out the plastic body onto the plaster form. In this case, the form remains stationary [96].

Hot-plunging or jollying of plastic body articles involves the placing of a measured slug of body in a plaster mold and having a heated, revolving polished metal tool press down and form hollow objects, such as pin insulators or cups. The term *hot pressing* is sometimes applied to the hot-plunging operation, but hot pressing is more generally used for special, advanced ceramics processed by application of high pressure to fine-grained oxides in refractory molds held at high temperature [97].

The *ram process* involves pressing a lump of soft plastic body between two hard plaster molds and squeezing them together to form a plate, ash tray, or similar object. In the pressing stage, water is squeezed out of the piece and a vacuum pulls moisture into the molds. In the removal step, the vacuum is maintained on the upper mold and pressure is applied to the lower mold to release the piece. The upper mold then lifts the piece free, and pressure is applied to free the object from the upper mold. Pressure is also applied to blow moisture from the mold halves before another cycle starts [98].

Stiff Plastic Forming. Stiff plastic systems are extruded through a die, either by auger extrusion or piston extrusion. Auger extrusion is a continuous operation, whereas piston extrusion is necessarily intermittent. *Piston extrusion* is used for extruding fine-grained refractories, cermets, and electronic bodies. A preformed, deaired slug is placed in the cylinder and forced through a die at pressures up to 35 MPa. Pieces as small as 1 mm in diameter with a half-dozen 0.1-mm-diameter holes can be made. Large sewer pipes are piston-extruded with a vertical piston extruder [19, p. 147].

Auger extrusion finds use in extruding bricks and hollow tile on a continuous basis. Short sections are cut off at desired lengths. The auger device consists of a pugging trough that feeds a screw, which in turn pushes the clay through a shredder into a vacuum chamber. The deaired shreds are recompacted with a screw and pushed through the die. High-tension insulator blanks of up to 1 m in diameter [99, pp. 111–112] are extruded with auger deairing pug mills and are used in lathe-turning segments of very large electrical insulators.

Dust Pressing. The term applied to forming of damp, granulated body batches containing 5–15% moisture that are formed at high pressure in a steel die is dust pressing. All wall tile, floor tile, some quarry tile, and most low-tension electrical porcelain is formed by dust pressing. More than 85% of all fireclay brick and nearly all silica brick and basic brick are formed by dust pressing. Hydraulic presses and hydraulic toggle presses are used [19, pp. 149–151].

Dry Pressing. Dry pressing is similar to dust pressing, but the moisture content is < 2.0%, so that a binder and internal lubricant must be employed. Dry pressing is employed for advanced ceramic products in two ways [19, pp. 149–151]. Small shapes are pressed by *uniaxial compaction* [19, pp. 151–152], wherein the pressure is applied (usually) in a downward, vertical direction, thereby producing pressure variations due to wall friction and particle–particle friction. This results in nonuniform density. *Isostatic compaction* involves application of pressure equally to all sides of the charge. An isostatic press consists of a thick-walled pressure vessel. Powder is enclosed in a liquid-proof rubber mold that is immersed in a noncompressible fluid. The fluid is pressurized and transmits pressure equally to all sides of the mold; pressures can range from 35 MPa to as much as 1400 MPa, but usually ca. 210 MPa [100].

Slip Casting. Deflocculated liquid systems are made into ware by slip casting [57, chap. 10]. Formation is accomplished by consolidation of the particles into a semirigid state through removal of a portion of the liquid phase by an absorbent, porous mold. The most common mold material is the hemihydrate of gypsum, $CaSO_4 \cdot 0.5\,H_2O$ [*10034-76-1*], which when mixed with water rehydrates and forms needles of gypsum crystals as an interlocked mass [101], thus forming continuous capillary pores (→ Calcium Sulfate). The size and liquid-carrying capacity of plaster molds is controlled by varying the plaster:water ratio.

WALKER [102] observed that as water rises from the stoichiometric 18.5 kg of water to the region of 60–100 kg (depending on the processing and ore source of the plaster) per 100 kg of $CaSO_4 \cdot 0.5\,H_2O$, the rate of casting increases to a maximum and then decreases with greater water ratio. Both the specific gravity and the compressive strength decrease with further increases in water.

Although the suction pressure of porous plaster decreases as the amount of water increases, a larger pore size allows freer passage of moisture from the developing cast and provides a larger reservoir for liquid as it is removed from the slip [102]. The loss of moisture from the exterior of the mold by evaporation is a significant controlling factor in governing the rate and condition of casts; high external humidity reduces, and low external humidity raises, the rate of cast and time of setup.

Slip casting takes two general forms. In the first, slip is poured into the mold where water is absorbed, leaving a semirigid layer of particles next to the mold wall. After a sufficiently thick layer has developed, the excess slip is poured out. The cast wall continues to pass moisture into the mold, thus reducing the moisture gradient from drain to wall, and allowing the cast to assume the firmly plastic state needed for cast removal. This is *drain casting*, which is used for hollow items. In the second, a slip at a somewhat higher solids concentration (55 vol % against 50 vol % for the drain-cast slip) and a greater thixotropy (reversible thickening) is poured into the mold and allowed to cast solid. This is termed *solid casting*. On occasion, solid casting and drain casting are used on the same piece. The character of the cast and its rate of buildup are controlled by manipulation of the particle size [55] and colloid modifiers [57, chap. 1].

Special Systems for Advanced Ceramics. Advanced ceramics can be consolidated and formed by the following methods:

1) Pressing

   uniaxial pressing
   isostatic pressing
   hot pressing
   hot isostatic pressing

2) Casting

   slip casting
   soluble-mold casting
   thixotropic casting

3) Plastic forming

   extrusion
   injection molding
   transfer molding
   transfer molding
   compression molding

4) Others

   tape forming
   flame spraying
   green machining

To this point discussion has focused on the shaping methods originally used for the less-demanding clay-based formulas, but which have been refined for use in making small, more-demanding advanced ceramics. Certain advanced ceramic products require very thin sheets. A method for making such products makes use of casting or spreading a specially prepared slip or slurry onto a moving carrier surface and controlling its thickness with a doctor blade [103]. In such cases, the system resembles an oil-base paint. The powder is dispersed in a volatile solvent (nonaqueous organic liquid) with unsaturated organic acids of 18–20 carbons, and a polymer binder and plasticizer are added. Drying consists primarily in removal of the volatile solvent, which leaves a thin flexible tape.

An interesting and useful modification of slip casting also involves an adaptation of investment casting. First, a water-soluble wax is injection molded to make a pattern. The pattern is then coated with a water-insoluble wax, and the water-soluble part is dissolved away. The wax mold is fastened to an absorbent plaster block and is filled with slip. Once casting is completed, the water-insoluble wax is dissolved from the cast with an organic solvent, and the cast is dried, machined as needed, and fired to the proper temperature for densification [4, pp. 197–199].

Injection molding makes use of the techniques for molding plastic combs and the like, the difference being that the polymer, either thermosetting or thermoplastic, serves only to disperse the ceramic powder and to provide lubrication [104]. A sized powder is milled dry with organic binders and made plastic by preheating. The plastic mass may require as little as 24 % or as much as 50 % binder by volume, depending on particle size and particle-size distribution. Complex shapes can be made [4, pp. 200–203].

4.4. Drying and Finishing

Drying of ceramic products is one of the more critical processing operations. The moisture must be removed as rapidly as possible without generating stresses great enough to cause cracking or distortion.

A plastic ceramic piece contains liquid in three forms: (1) adsorbed liquid on the colloidal particles; (2) liquid films on particles of noncolloidal dimensions; and (3) free liquid held in pores between the particles. Liquid must leave the system in three distinct stages [105, pp. 82–84]:

1) By evaporation from the surface of the piece, bringing the particles closer together, decreasing the volume of the piece proportionately, and eventually allowing the particles to come into contact, at which time shrinkage ceases
2) By removal of the remaining free moisture
3) By removal of the adsorbed moisture

As moisture leaves the piece, a gradient is established between the surface and interior of the ware. Because of the shrinkage factor, this gradient must not be too great; otherwise, excess shrinkage at the surface will cause cracking.

Moisture Stress. Many mechanisms affect the behavior of clay-based ceramic forming systems during dewatering processes such as slip casting, filtration, and drying. Some of the mechanisms involved are capillarity, adsorption, osmotic pressure, the electrical double-layer, and pore water structure [106]. The moisture changes in unfired ceramic bodies can be studied by measuring the specific energy of the water as the fundamental parameter. This energy concept is termed moisture stress [107], [108].

Moisture stress is defined as the work done per unit mass of water when a small amount of the water is moved from the clay–water system to a free water surface at the same temperature. The SI unit is J/kg. Expressed in simpler terms, moisture stress is a measure of the affinity of a porous system, such as a moist clay body or liquid slip, for its moisture.

A number of methods have been proposed for measuring moisture stress. PACKARD [108] employed a direct suction device in which moist clay was placed on a water-saturated, fritted glass plate that was in contact with water. A tube connected to the water vessel could be raised and lowered to increase or decrease suction, thus altering the moisture content of the clay. COLEMAN and MARSH [109] used a pressure-membrane apparatus for very high moisture stress. PACKARD [108] proposed use of a series of evacuated closed containers where moist clay was suspended over solutions of known relative humidity. SAMUDIO [110] was able to assign moisture stress (pF) values to casts made from slips containing small percentages of various inorganic salts by using a pressure-membrane method.

Moisture stress ranges from nearly 0 for dilute suspensions to ca. 10^6 J/kg for oven-dried clay bodies. Soft plastic bodies have moisture stresses in the region of 40 – 50 J/kg. Leatherhardness occurs at moisture stress of in the region of 5×10^3 J/kg, whereas air-dried clay bodies have moisture stresses of ca. 10^5 J/kg.

Rate of Drying. The moisture stress concept implies that for a body of a given particle-size distribution, the rate of drying depends on a structure imparted by the interaction of many factors. A simple example is the control of permeability and water holding in slip-cast pieces by interaction of organic matter, deflocculants, and flocculating salts [111]. Organic colloids are known to reduce the water of plasticity of fine-grained clays [57, p. 45], yet that same organic matter has a strong affinity for water and can retard moisture loss [112]. Similarly, the presence of fine-grained muscovite mica in plastic clay bodies or moist cast appears to slow drying [52].

If a granular solid is involved, moisture loss proceeds at about the same rate as from a pan of water under the same conditions [19, p. 161]. However, from a plastic mass of fine clay, the rate of moisture loss may be less than that of the coarser system [113]. Ceramic clay bodies would be expected to show loss rates between those for granular masses and clay masses. In any one of these circumstances, the drying rate is constant until particles touch and shrinkage stops. At this point the continuum of the pore water ceases and flow of water from the interior of the piece cannot maintain the surface film needed for rapid evaporation. The rate of drying then falls.

The shrinkage properties of clay bodies are useful in setting up an efficient drying schedule. Natural clays give shrinkage traces that vary with clay fineness and packing characteristics of the noncolloidal particles [113]. Contrary to other opinion [114, p. 554], NORTON stated that the rate of water removal from the surface of a plastic clay piece is approximately one half that from a free water surface under the same conditions. Accordingly, evaporation rate from a moist granular solid is much greater than that from a plastic fine-grained clay [115]. The rate of flow of liquid from the interior can be increased by decreasing the viscosity of the liquid and this is accomplished by raising its temperature.

Speed of drying is also governed by the moisture capacity of the air surrounding the piece, i.e., the relative humidity, and the volume of air passing over the ware. Because the moist piece is a porous system, a balance must be struck between loss of moisture at the surface and movement of moisture through the particles from the interior to the surface. If liquid loss at or near the surface exceeds liquid movement from interior pores, differential shrinkage can result in cracking or warping.

Defects. Drying defects can originate wherever there are discontinuities in the formed piece. *Cracks* can develop at these points during shrinkage associated with drying. Clay particles tend to orient with their long dimension normal to the direction of pressure. Because shrinkage is least in the direction parallel to particle orientation and greatest in the normal direction, solid-cast test bars made in an open mold tend to *warp* on the exposed face away from the oriented mold-face layer. Similarly, frictional forces in extrusion force moisture into the interior and cause orientation of clay particles parallel to the direction of extrusion. Differential shrinkage and excessive moisture gradient are leading causes for cracking and warping.

Even though a moist object has a uniform distribution of moisture, it warps unless it is evenly dried. A wall tile, for example, placed on a plate so that one face is protected from air flow and evaporation warps. During initial drying, an originally uniformly moist piece can develop a moisture gradient through (1) loss of moisture at the surface or (2) uneven heating of the pore water, which lowers its viscosity at the warmer surface but not in the cool interior.

Some operations subsequent to slip casting can cause livering (dilatant consolidation) in one part of the cast and not in adjacent areas, resulting in a lower moisture-release rate and lower shrinkage in the livered area than in the adjoining portion.

Even after defect sources originating in the forming operation have been eliminated, the problem of removing the moisture without rupturing or warping the ware remains. This objective is achieved through the techniques of *humidity drying* that (1) lower the viscosity of the water, (2) uniformly warm the pore water without causing differential shrinkage, and (3) remove the water economically with respect to both time and fuel consumption.

The principle of humidity drying involves (1) heating greenware all the way through in a saturated atmosphere, (2) reducing the humidity as fast as possible without stressing the ware, and (3) once shrinkage ceases, raising the temperature and reducing the humidity to zero relative humidity [105, pp. 82–84].

Drying Methods. Drying methods fall into two classes: (1) convection and (2) radiation [116], [117]. *Convection methods* circulate warm air around the ware being dried, the warm air serving the dual purpose of heating the pieces and removing moisture by convection circulation.

A *simple tunnel dryer* requires passage of cars loaded with ware through the dryer, while heat is supplied by steam coils underneath, hot air from a heater, or waste heat from kilns. Drying is likely to be uneven from top to bottom of the load in tunnel dryers, but cross-circulation of heated air by fans or jets improves uniformity in such dryers.

Controlled humidity can also be attained in a tunnel dryer by introducing the moist ware at one end into hot moist air sent in originally as warm dry air at the exit end, the *counterflow method.* The dry air picks up moisture in its passage over the loaded cars, thus becoming saturated with moisture. Sometimes an auxiliary heating unit is located at the ware entrance end. In this way the moist ware is heated uniformly with no initial loss of moisture or shrinkage. As the ware moves toward the exit end, the surrounding air becomes progressively cooler and drier. The exterior parts of the pieces are reduced in temperature, owing to evaporative cooling, while the interior remains warmer: in a clay mass with a temperature gradient moisture moves toward the cooler part [118].

A refinement of the counterflow method has the *tunnel divided into sections,* each with its own independent heat and humidity controls. The ware enters a hot, saturated zone and is warmed without moisture loss. The car then passes through progressively drier zones, each held at constant but higher temperature.

A *chamber humidity dryer* operates by stages on a stationary load. Hot, saturated air warms the ware. When the chamber is at a uniformly high temperature, drier and slightly cooler air is passed through the load and evaporation occurs. Ware can be dried more quickly if air is directed at right angles than if it is blown parallel to a surface [119]. For small, simple shapes, a method termed *jet drying* is sometimes used: air is blown in a definite pattern at right angles to the surface of the piece, thereby saving fuel, space, and time [19, p. 167].

Radiation methods involve transfer of heat to the moist ware as infrared or high-frequency radio waves. The principle of heat transfer by radiation methods is being employed increasingly in the drying of ceramic ware. Over 40 years ago, infrared lamps were used to dry large slip-cast units in open settings while ordinary air movement was used to remove the moisture. Drying times were reduced from 14 days to 12 h [120]. Hotel china from automatic jiggers can be dried sufficiently for removal from the bat in 10–15 min [117]. The infrared drying is followed by hot-air jet drying [119].

Finishing. Nearly all ceramic ware, however formed, must be subjected to finishing operations. These may be as simple as removal of casting spares, mold seams, and fins. However, the operation may involve the turning of a foot on a leather-hard cup or an elaborate turning of a high-tension insulator from a 1-m-diameter extruded blank. The term *trimming* means the shaving away of seams from a cast piece or cutting off the casting spare. *Fettling* refers to removal of fins, mold seams, and rough edges from dry, or nearly dry, ware.

4.5. Firing Ceramic Products

The terms used to describe the densifying processes that occur during heat treatment of ceramic items can be confusing. The expression *sintering* is used [2, p. 232] to describe a process by which a substance is bonded together, stabilized, or agglomerated by being heated to a point close to, but below, the melting point. *Vitrification* is defined as a progressive reduction of the pores of a ceramic piece as a result of heat treatment [3, p. 202], but says nothing about formation of a liquid, glassy phase. However, the dictionary definitions [121] dealing with the terms vitreous, vitrify, vitrification, and vitrifiable all center on some aspect of glass. BURKE [122] notes that the word sintering is generally used in referring to processes that assume no liquid is formed during heat treatment.

Here the term *densification* [4, chap. 7] is applied to processes where removal of pores from a ceramic product by heat treatment can take place either by formation of a glassy phase or by solid-state material transport, or both. However, traditional ceramics and advanced ceramics are described separately because clay-based products invariably involve development of some glass, whereas advanced ceramic products nearly always involve solid-state reactions.

4.5.1. Firing Traditional Ceramics

The main reactions occurring in the course of heating a clay-based product to maturity are summarized in Table 9 [19, p. 267]. As shown by the expansion curves in Figure 6 [123], in the initial stages of firing clay-based bodies, there is an expansion to a peak at ca. 600 °C with a small inflection (the quartz inversion) at 573 °C, followed by a gradual drop up to ca. 775 °C. After remaining level to ca. 850 °C, a rise in expansion follows to a peak at ca. 900 °C. Shortly thereafter, the bodies begin a decided contraction. Above 1300 °C (not shown in Figure 6) irreversible thermal expansion occurs in all types of clay-based bodies [124], and this expansion must be taken into account in devising firing schedules [123]. The expansion on heatup of body A, which contains pyrophyllitic South American clays, is much greater than that of body B, a U.S. clay body containing no pyrophyllite.

Shrinkage and porosity changes with increasing temperature vary, depending on the body composition and the porosity at maturity. Vitrification/shrinkage curves can be shifted downward in the maturing range while their configuration and the forming properties of the bodies can be maintained by the judicious selection of auxiliary fluxing constituents [125]. An understanding of those factors affecting the nature of the glassy phase permits ready control of the pyroplastic deformation characteristic of large clay body units [126] without altering the unfired forming properties of the body.

4.5.2. Densification of Advanced Ceramic Products

The densification of formed advanced ceramic items is generally referred to as sintering. Sintering is, in simple terms, removal of voids (pores) of the formed piece, accompanied by shrinkage. Criteria to be met before sintering can take place are (1) an available means for material transport and (2) an energy source to initiate and promote material transport. Diffusion and viscous flow are transport mechanisms. Heat is the primary source of energy, functioning with particle–particle contact and surface tension to produce energy gradients [4, pp. 217–223].

Vapor-Pressure Sintering. Difference in vapor pressure as a function of surface curvature provides the driving energy. Material goes from particle surfaces with a positive curvature radius and high vapor pressure to contact regions with negative curvature radius and much lower vapor pressure. The smaller the particles, the greater the driving force. While vapor-phase sintering bonds particles, it does not eliminate pores.

Diffusion Sintering. Diffusion can be the movement of atoms or vacancies along a surface or grain boundary through the body. Only grain boundary or body diffusion results in sintering. The driving force is differential free energy or chemical potential between free surfaces and contact points of adjacent particles. Finer particles sinter more rapidly and at lower temperatures than coarser particles. Uniformity of particle shape, size, and distribution governs the uniformity of the final product.

Table 9. Reactions occurring in firing clay bodies

Temperature, °C	Reactions in the course of firing
≤ 100	free moisture removed from the piece
100–200	loss of adsorbed moisture
200–400	gradual loss of H_2O from halloysite and montmorillonite, pyrophyllite and fine sericitic mica begin decided expansion
400–700	organic matter oxidized, breakup of clay mineral structures, pyrophyllite starts sharp expansion
573	quartz inversion
700–950	pyrophyllite attains maximum expansion, spinel forms in clays
950–1000	muscovite structure destroyed, γ-Al_2O_3 or mullite forms
1000–1100	mullite, $3\,Al_2O_3 \cdot 2\,SiO_2$, forms from clay
1100–1200	feldspars melt, clay and cristobalite dissolve, porosity decreases, shrinkage increases rapidly
> 1300	glass increases, ware expands, absorption increases, strength decreases

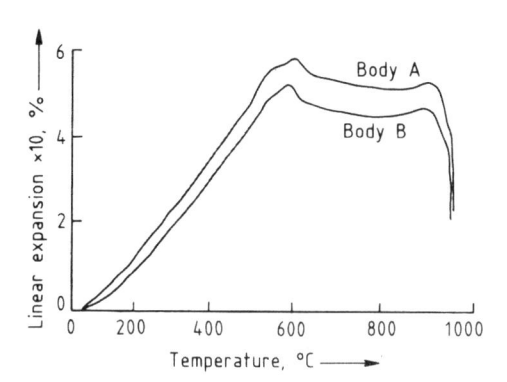

Figure 6. Irreversible heatup thermal-expansion traces for vitreous sanitary ware bodies containing pyrophyllite and sericite (A) and no pyrophyllite and little mica (B)

Liquid-Phase Sintering. The main densification mechanism for most silicate systems is liquid-phase sintering. The sintering occurs best where the liquid phase thoroughly wets the solid grains at sintering heat. Capillary pressure in narrow pores between particles may be ≥ 7 MPa. Because small particles have higher surface energy and form smaller pores, there is more densification driving energy than for compacts of larger particles. Temperature strongly affects sintering; generally, small increases in temperature cause significant increases in the amount of liquid. In some cases, this is desirable, but in others excessive grain growth and fire distortion occur. The amount of liquid at a given temperature can be predicted from phase equilibrium diagrams.

Hot Pressing. Hot pressing resembles sintering except that temperature and pressure are applied at the same time [127]. Pressure speeds densification by increasing particle packing and by stressing points of contact. The densification energy can be increased 20-fold by applying pressure. *Hot isostatic pressing* can be done with special heat-treating equipment [128] and provides results superior to those attained with conventional hot pressing. Hot pressing injection-molded items gives results superior to those obtained with simple isopressing/sintering, slipcasting/sintering, or injection molding/sintering. The Weibull probability plot of Figure 7 shows the superior uniformity and higher strength of hot-pressed, injection-molded products.

4.6. Kilns and Firing Conditions

The furnaces in which ceramic products are heat-treated are usually termed *kilns*. Kilns, depending on the manner of operation, can be termed periodic (intermittent) or tunnel (continuous):

1) Intermittant firing kilns
   stationary periodic
   lifting charge
   lifting kiln
   moving charge
   moving kiln

2) Continuous firing kilns
   chamber
   conveyor belt
   roller slab
   muffled tunnel
   direct-fire tunnel

Periodic kilns are heated and cooled in accordance with prescribed schedules that differ with the kind of product. A *tunnel kiln* has temperature zones held at specific temperatures through which kiln cars (or other supports) are passed to provide the specified time–temperature cycle. Tunnel kilns are adapted to firing one type of body in long runs, whereas periodic kilns can be adapted to a variety of products.

Periodic kilns can be heated by electrical elements or fired with gas or oil. Traditionally, where hot gases are involved, heating is accomplished by having the combustion products pass through the load of ware, either upward (updraft kiln) or downward (downdraft kiln), before going out in a flue. The disadvantages of such kilns is that (1) they must be loaded and unloaded by hand, (2) there is a long cooling period, and (3) the entire kiln must be reheated in the next firing.

4.6.1. Modern Periodic Kilns

Elevator kilns are of three types: (1) ware to be fired is placed on a refractory protected car that is pushed in position under a suspended kiln that is then lowered over the car (a top-hat kiln); (2) the car with its ware is elevated into the kiln that is permanently fixed; and (3) an elevated kiln is moved horizontally over a series of cars and placed over any one of them, as desired.

Shuttle kilns are positioned permanently. One end has a movable door. Ware is loaded onto a car that is then run on rails into the kiln, the door is closed, and the ware is fired. A variant is the envelope kiln, which is rolled over and encloses ware placed on a permanent hearth.

4.6.2. Tunnel Kilns

Tunnel kilns are refractory chambers, sometimes 90–100 m long, through which ware is moved to achieve gradual heating and cooling. The entry section is the preheat zone, the middle section is the firing zone, and the exit portion is the cooling zone.

Cooling air is blown into the cooling zone, is heated, and moves into the firing zone, where it improves combustion and preserves an oxidizing atmosphere. Combustion gases from the firing zone are conveyed into the preheat zone to heat and dry the ware.

Refractory-topped cars riding on insulated rails carry the ware into and through tunnel kilns. Pushing is done on a prescribed schedule, expressed in terms of "cars per 24 hours." Some smaller tunnel kilns have positively rotated refractory (alumina) rollers, on which refractory slabs carry the ware. Other tunnel kilns use sled hearths, which are intermediate between cars and roller slabs, for smaller fast-fired products.

Advances in ceramic-fiber technology [128], [129] have provided alumina, silica, and kaolin fiber products for use in place of high-density castables for kiln cars and for insulating replacements for higher-density brick. This has made possible kiln designs that greatly reduce fuel consumption and permit faster firing of ware [130].

4.6.3. Advanced Ceramics Furnaces

Because advanced ceramic products often have special sintering requirements, the furnaces differ from those of ordinary ceramic kilns and require new furnace technology [131]. A common requirement is the need for total control of kiln atmosphere, as well as control of temperature and time scheduling. Atmospheric control can be achieved by sintering in a vacuum furnace or an autoclave. Initial air can be pumped out so that contaminants are vaporized and evacuated. The atmosphere around the ware can then be controlled with respect to composition and pressure.

Separate furnaces may be needed to eliminate lubricants and volatile binders. However, delubing can be done at low temperature, along with degassing, and the ware can be fired in an inert atmosphere on a controlled schedule to a required temperature.

4.6.4. Kiln Atmosphere

The atmosphere has a profound influence on the fired properties of clay-based ceramics. If there is enough O_2 to permit the piece to absorb some, the atmosphere is regarded as oxidizing. An oxidizing atmosphere helps eliminate carbon and converts salts to oxides. A low-oxygen atmosphere reduces multivalent ions to their lowest positive state, thus causing color and other changes.

Sulfur-bearing fuels provide SO_2, which is harmful to body and especially to glazes of clay-based ceramics. The ware must be protected by saggers (refractory boxes) or by keeping the combustion gases away from the work with a refractory wall (muffles), through which heat is radiated.

Electric kilns of all types [19, pp. 305–306], which avoid contamination from burned fuels, are used widely in Europe. In the United States, such kilns are mainly used for decorating, special ceramic products, and wall tile. The heating elements are nichrome, kanthal, or silicon carbide. Nichrome (a nickel–iron–chromium alloy) elements are used for decorative materials firing, kanthal (iron) for intermediate-temperature kilns, and silicon carbide for high-fire kilns.

4.6.5. Fired Ware Finishing

Postfiring processes fall in the category of finishing. Finishing may include grinding to size and removing kiln marks (grains of kiln dirt). Technical ceramics must be examined for flaws.

5. Glazes and Glazing

Glazes are applied to clay-based ceramic products to provide a shiny, generally smooth surface that seals the body. The surface may be either matt or bright. Glazes resemble glass in structure and texture, but have greater viscosity in the molten state. Glaze adheres strongly and uniformly to the ware. Application of glaze suspension to ware is by spraying or dipping.

Glazes can be made for maturing from ca. 600 °C up to ca. 1500 °C, depending on the items to which they are applied. If necessary, surfaces can be made resistant to various corrosive liquids and gases. Semiconducting glazes can be prepared for electrical porcelains.

5.1. The Nature of Glazes

Glasses and glazes used by ceramists are normally combinations of oxides (Table 10). Oxides that form glasses by themselves are termed *network formers;* by Zachariasen's rules [19, p. 130], SiO_2, B_2O_3, and P_2O_5 should, and do, form glasses. The holes in the network are filled by network *modifiers,* which weaken the bonds. Such modifiers are usually Na_2O, K_2O, CaO, and MgO. Generally speaking, the more modifier present, the lower the glass viscosity and chemical resistance. Oxides such as Al_2O_3, PbO, ZnO, ZrO_2, and CdO can enter the network by replacing some Si^{4+} or B^{3+}. These are the *intermediate glass formers.* The relative single-bond strengths of the oxides correlate with glass formation: network formers have high bond strengths, modifiers have low bond strengths, and intermediates fall in between.

The conventional representation of glaze formulas is the Seger convention, by which the formula is expressed with the fluxing oxides R_2O and RO, including PbO and ZnO, adding up to unity on a molar basis. The Al_2O_3 and SiO_2 plus B_2O_3 are listed as separate items, also on a molar basis, in the following manner:

$$1.0\ (R_2O + RO) \cdot x\ Al_2O_3 \cdot y\ (SiO_2 + B_2O_3)$$

Glazes can be classified as raw glazes and fritted glazes. In the case of *raw glazes*, the oxides are introduced in the form of compounds or minerals, such as feldspar, which melt readily and act as solvents for the other ingredients. A *frit* is a prereacted glass containing ingredients, such as Na_2CO_3, which are soluble in water. Fritting fixes the desired oxide in a relatively insoluble form. Fritted glazes are used for whitewares. Frits may form only a part of the whiteware glaze formula, or the glaze may be composed entirely of frit [19, p. 190].

Raw Glazes. Raw glazes include (1) porcelain glazes, (2) Bristol glazes, (3) raw lead glazes, (4) raw leadless glazes, and (5) slip glazes. According to the Seger convention the basic *porcelain glaze* is expressed as follows [133]:

0.3 K_2O 0.4 Al_2O_3 4.0 SiO_2
0.7 CaO

It matures as a bright glaze in the region of Orton pyrometric cones 8–10 (1236–1285 °C). When Al_2O_3 is raised to 0.5 and SiO_2 to 5.0, a bright glaze is obtained at cone 12 (1306 °C). At SiO_2 7.0 and Al_2O_3 0.5, the maturity of a bright glaze rises to cone 14 (1388 °C), whereas at SiO_2 8.0 and Al_2O_3 1.0, maturity occurs at cone 16 (1455 °C). Increasing the Al_2O_3 and SiO_2 content produces semimatt and matt low-gloss surfaces.

Bristol glazes are raw glazes containing ZnO, which are used on terra-cotta clayware and sometimes on stoneware items [134]. A typical Bristol glaze falls near the following molar composition:

0.40 K_2O 0.5 Al_2O_3 3.4 SiO_2
0.35 CaO
0.25 ZnO

Maturity occurs at cones 4–8 (1168–1236 °C).

Raw lead glazes [135] are used only on artware, never on commercial ware, owing to health hazards from soluble lead. A typical bright glaze maturing at ca. cone 05 (1031 °C) has roughly the following composition:

0.55 PbO 0.23 Al_2O_3 1.55 SiO_2
0.36 CaO
0.09 Na_2O

Table 10. Oxide glasses [132]

Oxide	Oxidation state	Coordination number	Bond strength, relative
Network formers			
SiO_2	4	4	106
B_2O_3	3	3	119
Intermediates			
Al_2O_3	3	4	90
Al_2O_3	3	6	60
ZnO	2	2	72
ZnO	2	4	36
PbO	2	2	73
PbO_2	4	6	39
Modifiers			
Na_2O	1	6	20
CaO	2	8	32
Substitutions for Na_2O			
K_2O	1	9	13
Rb_2O	1	10	12
Cs_2O	1	12	10
Li_2O	1	4	36
Substitutions for CaO			
MgO	2	6	37
BaO	2	8	33
SrO	2	8	32

Lead-free glazes, or leadless glazes, are designed to provide lower maturity than true porcelain glazes without the use of lead oxide. A typical glaze has the following composition [136]:

0.2 K_2O 0.3 Al_2O_3 3.0 SiO_2
0.3 SrO
0.1 CaO
0.4 BaO

Glazes built around this composition have an excellent maturing range as low as cone 03 and up to cone 9 (1086–1260 °C), but lack the covering power and brightness of lead glazes.

Slip glazes [137] are natural clays having the following approximate composition:

0.20 K_2O 0.60 Al_2O_3 4.00 SiO_2
0.45 CaO   0.08 Fe_2O_3
0.35 MgO

Such clays are used for artware glazing and often for high-tension electrical porcelain insulators. Slip clay glazes have a maturing range from ca. 1200 °C to 1300 °C.

Frits. Because B_2O_3 and most borates are soluble in water, the B_2O_3 must be added in a frit. A typical lead-free borate frit has the following composition:

0.69 CaO	0.37 Al_2O_3	2.17 SiO_2
0.19 Na_2O		1.16 B_2O_3
0.12 K_2O		

Although lead bisilicate [*11120-22-2*] ($PbO \cdot 2\, SiO_2$ or $PbSi_2O_5$) is relatively insoluble in water, lead is normally introduced into commercial glazes in fritted form. A more complex frit might have a composition such as the following:

0.50 PbO	0.10 Al_2O_3	2.70 SiO_2
0.30 Na_2O		
0.20 K_2O		

Boric oxide and lead oxide often appear together in low-temperature glaze. A typical composition is the following:

0.50 PbO	0.10 Al_2O_3	2.70 SiO_2
0.10 Na_2O		0.60 B_2O_3
0.20 K_2O		
0.30 CaO		

Special Glazes. Low-expansion glazes are required by zircon, cordierite, and low-expansion lithium silicates [138]. Recommended glazes for zircon porcelain in the range of cones 10–14 (1285–1400 °C) approximate the following composition:

1.00 RO	0.6–0.7 Al_2O_3	9–11 SiO_2

Semiconducting glazes are used to remove charges from surfaces of electrical insulators. This property can be produced with a high concentration of Fe_2O_3 in the glaze, crystals of ZnC_2O_4, or activated SnO_2 [19, p. 197].

Glaze opacity is obtained by mill additions of zirconium-type opacifiers [139].

Lithium oxide additions increase the hardness of commercial glazes [140]. High-compression glazes resist scratching. Crystalline glazes grow large crystals in the firing and cooling of the glaze. NORTON [141] demonstrated that crystals grow when temperatures of nucleation and crystal growth do not overlap. For example, proper control of the heating schedule allowed the growth of large willemite crystals.

Salt glazing is an old method for glazing stoneware. The glaze is formed by throwing damp common salt into the kiln during the sintering stage of firing. The NaCl decomposes to form Na_2O and HCl, the Na_2O combining with the Al_2O_3 and SiO_2 of the body to form complex silicates. BARRINGER [142] found the limits of the $Al_2O_3 : SiO_2$ ratio within which it is commercially possible to produce good salt glaze to be $1:4.6 - 1:12.5$.

5.2. Preparation of Glazes

Prepared frit, clay, and materials not incorporated in the frit are ground generally in ball mills with water. Grinding is followed by sieving and magnetic treatment.

The ball mills are large steel drums lined with quartzite or ceramic blocks. Flint pebbles of various size grades are frequently used. Recently, high-density (usually high-alumina) balls or cylindrical rods have found favor for this operation. The higher-density media reduce grinding times and lessen contamination from pebble wear. The media, batch, and water occupy ca. 60% of mill volume. Media weight is around 3 times that of batch. Water content runs 30–50%, depending on the material being processed. Mill speed varies with mill size: the smaller the mill, the faster the rotation.

The practice of grinding to a stated percentage of residue remaining on a test sieve does not take into account differences in particle-size distribution brought about by changes in media size and size distribution and linear wear. PHELPS [143] has shown that apparently minor differences in glaze and enamel slip particle-size distribution can cause marked differences in slip rheology.

While fine grinding improves glaze brightness, hardness, and chemical stability, grinding too fine can result in crawling (parting of the glaze, which leaves bare spots) or peeling (flaking away) of the glaze. Changes in the soluble materials content of the water can lead to difficulties in consistency and fineness of glazes.

5.3. Glaze Application

The prepared glaze slip is adjusted with additives designed to control consistency and adhesion. Hand dipping of prefired (bisque) ware in glaze slip was the general practice prior to 1920. However, spraying then became common. Automatic spraying on a conveyor line is now used. Tiles are glazed by spraying or by passing them unter a falling sheet of glaze slip.

Aside from grinding-induced differences, the nature of the materials can result in glaze settling. Suspending agents, such as bentonite or organic agents, are then used. Flocculants, such as $MgSO_4$ or $CaCl_2$, can be used.

6. Glass (→ Glass)

Glass has been defined in simple terms as a fusion product of an inorganic material that has been cooled to a reasonably rigid, noncrystalline state [56, p. 42]. Objects made of glass are simply called glass, although specific kinds of glass are qualified by types, such as flint, barium, lead, container, or window glass.

As shown by Table 10 glasses consist of the following types of oxides: (1) network formers (form glasses by themselves); (2) network modifiers (alkali-metal ions and alkaline-earth-metal ions); (3) intermediate glass formers (partial substitutes for network formers).

A fairly high PbO content (10–45%) is characteristic of *flint glasses*. Low-expansion, chemically stable glasses contain significant levels of B_2O_3 (6–12%). Optical glasses contain variable percentages of lead and barium oxides.

Soda–lime–silica glass represents the major proportion of commercial glass [99, chap. 13]. It is made by melting more or less pure silica sand by fluxing with soda (Na_2O) and stabilizing with CaO or CaO · MgO.

Container glass is a typical soda–lime–silica glass containing smaller percentages of materials having special functions. Alumina (from feldspar or nepheline syenite) helps chemical stability, sodium nitrate functions as an oxidizing agent, and arsenic is a fining agent (for elimination of bubbles and undissolved gases). Clear glass requires low Fe_2O_3 content, whereas high Fe_2O_3 materials are used in colored glasses [144]. Manganese blanks out the green color of iron.

Container glass production is usually large scale. Batches are weighed automatically, blended, and conveyed to glass melting tanks (refractory containers for melting). Tanks hold 180–275 t of glass. Batches are fed into tanks as layers 15–20 cm thick and melted by heat from side burners fired with oil or gas. Exhaust heat passes into checker chambers below the burner ports; flow is reversed 3–4 times per hour. Combustion air passes through the heated checkerwork.

Large refractory blocks made of fusion-cast Al_2O_3 or Al_2O_3–ZrO_2–SiO_2 mixtures function as side walls and end walls, while the bottom is $ZrSiO_4$ and the roof is SiO_2. A container glass tank may have melting areas of 90–150 m^2 with glass depths of 1–1.5 m. A campaign (working life before major repair) of a tank may be 4–5 years.

Sheet glass tanks, holding 1200–1500 t, are much larger than container glass tanks. Such a tank can supply 180–275 t every 24 h.

Optical glass tanks are much smaller, with outputs of perhaps 40–200 kg every 24 h.

Forming is by pressing, vacuum, or blowing (using air pressure to transform the gob, i.e., mass of molten glass, into a hollow piece). Otherwise, a sheet is formed by pulling (drawing) a continuous sheet of molten glass from the tank and passing it through a flattener or roller. Float glass involves a process in which the sheet floats on a bath of molten tin with heaters above and in the bath; the glass settles to an even ribbon and is allowed to cool slowly. Sheet glass requires polishing and grinding for use as plate glass, whereas float glass does not.

Fiberglass takes two forms: (a) continuous thread for textiles or (b) discontinuous fiber for insulation, filtering, or reinforced fiberglass. Glass marbles are fed continuously into a melt-ing chamber, and filaments are pulled through platinum spinnerets. Discontinuous fibers are blown by striking a molten stream of glass with a high-velocity steam jet. Such glass generally has a lower viscosity than textile fiberglass. Fiberglass, with its high surface area, also must be chemically stable.

Certain glasses [145] and glass-bonded ceramics [146] have application as refractory substances. Included are vitreous silica, high-silica glasses, aluminosilicate glasses, aluminate glasses, mullite glass, and barium feldspar glass ceramics.

Table 11. Classes and types of refractories

Class	Type	PCE*	MOR**, MPa
Fireclay	Super duty	33	4.14
	High duty	31.5	3.45
	Medium duty	29	3.45
	Low duty	15	4.14
High Al$_2$O$_3$	50% Al$_2$O$_3$	34	
	60% Al$_2$O$_3$	36	
	70% Al$_2$O$_3$	36	
	80% Al$_2$O$_3$	37	
	90% Al$_2$O$_3$		
	99% Al$_2$O$_3$		

* PCE pyrometric cone equivalent.
** MOR modulus of rupture.

7. Refractories

Refractory materials are essential to the manufacture of all forms of ceramic products, including refractories themselves. Table 11 [3, pp. 13–15] gives classes and types of refractory brick. Insulating firebrick are rated in eight progressively more refractory groups where reheat shrinkages are not more than 2% at testing temperatures of 845 °C (group 16), 1065 °C (group 20), 1230 °C (group 23), 1400 °C (group 26), 1510 °C (group 28), 1620 °C (group 30), 1730 °C (group 32), and 1790 °C (group 33) [3, p. 104].

Special refractories include zircon (ZrSiO$_4$), zirconia (ZrO$_2$), silicon carbide (SiC), chromic oxide (Cr$_2$O$_3$), and graphite (C). Refractory specialties include nonformed products such as mortars, castables, plastics, and ramming mixes. Basic refractories include the chrome brick, chrome–magnesite brick, and magnesite brick used in basic oxygen steelmaking [147].

8. Abrasives

Several natural minerals are employed as abrasives for cutting, grinding, and polishing. These include quartz, garnet, corundum, emery, and diamond. Manufactured abrasives include boron carbide, boron nitride, diamond, fused alumina, silicon carbide, titanium carbide, tungsten carbide, and zirconium silicate. Abrasive products include loose grains, wheels, coated abrasives, and grinding pebbles.

9. Cement (→ Cement and Concrete),

Cement is a synthetic mineral mixture (clinker) that when ground to a powder and mixed with water forms a stonelike mass, and is thus a ceramic product [148]. A primary requirement for cement manufacture is a source of CaO; this can be limestone, oyster shell, slag, etc. Also necessary is a source of Al_2O_3 and SiO_2, most commonly clay and, where needed, a quartz rock or sand.

Processing involves (1) grinding of rock material, (2) blending of ground materials to a desired chemical composition or slurry blending of the powders, (3) burning the blended material to form a clinker, (4) blending the clinker and gypsum, and (5) grinding the gypsum–clinker. Grinding can be accomplished by dry grinding — ball milling, rod milling, roller, race, tube mill — with air classification. Oversize is recycled.

The American Society for Testing and Materials has listed specifications for eight types of cement in accordance with chemical composition and physical requirements [149].

10. Properties of Ceramic Materials and Products

Ceramic products have relatively high strength associated with brittle fracture, high thermal stability, and low electrical conductivity. These properties are related to structure and depend on the size and arrangement of multiphase polycrystalline constituents and the glassy phase. Size, type, and distribution of pores must be considered because pores affect strength, ther-mal expansion, heat insulation, corrosion and weathering resistance, and electrical properties.

Table 12 shows mechanical properties of a number of representative ceramic products. In all cases, there is a characteristic direct transition from a small elastic deformation, with no or small plastic deformation, to fracture. Irreversible deformations from above the elastic region up to fracture may be due to viscous processes within the particle structure.

Important ceramic oxides have high melting points (°C):

Al_2O_3	2050	Fe_3O_4	1600
MgO	2800	FeO	1360
CaO	2600	$MgO \cdot Al_2O_3$	2135
SiO_2	1780	$2\,MgO \cdot SiO_2$	1890
ZrO_2	2700	$3\,Al_2O_3 \cdot 2\,SiO_2$	1810
Cr_2O_3	2265	$ZrO_2 \cdot SiO_2$	1775

Table 13 gives thermal expansion coefficients for a number of ceramic product constituents. Because high melting points generally correlate with low thermal expansion, these materials generally have low coefficients of thermal expansion. The anisotropic structure results from polycrystalline mixed phases and varying amounts of glassy phases, which explains the relatively poor thermal-shock resistance.

Most ceramic products have thermal conductivities lower than platinum, for example, but higher than, for example, insulating firebrick or organic polymers. Table 14 shows thermal conductivity coefficients for a number of refractory brick products [114, p. 942], and Table 15 provides thermal conductivity ranges for a number of electrical porcelain types [4, p. 47].

Table 16 gives resistivities for a number of ceramic products that serve as electrical insulators: their resistivities are of the order of $10^{12} - 10^{13}$ Ω cm — several orders of magnitude higher than for metals [4, p. 47].

Table 17 provides a tabulation of mechanical and thermal properties of materials employed in making advanced ceramics articles. The temperature for which these values are valid is room temperature or somewhat above, unless specified otherwise.

Chemical stability of nonporous ceramic products in the presence of acids or alkalies is adequate although it decreases as the temperature is increased. Nonporous ceramics can withstand atmospheric effects up to their melting points.

11. Testing Ceramic Raw Materials and Products

A distinction is made between tests made to determine the suitability of raw materials or ceramic products for particular applications and quality control test procedures. Tests for suitability can be very involved, whereas acceptance tests agreed upon between the raw materials supplier and the user or employed by the supplier in mining and refining and the manufacturer in processing can be simple. The characterization concept is considered the basis for determining the suitability of raw materials and products [56], [57, pp. 195–249], [48].

11.1. Raw Material and Product Tests

Various ceramic manufacturers and suppliers of raw materials for ceramic products have joined together in "the development of standards on characteristics and performance of materials, products, systems, and services; and the promotion of related knowledge" [56, p. iii].

Table 18 lists the volume identifications and subject areas of interest to the several areas of ceramic endeavor [56, pp. vi–vii].

Table 12. Mechanical properties of ceramic materials

Type of material	Compressive strength, MPa	Flexural strength, MPa	Modulus of elasticity, GPa
Solid brick	10–25	5–10	5–20
Roof tile	10–25	8–15	5–20
Steatite	850–1000	140–160	1–3
Silica refractories, 96–97% SiO_2	15–40	30–80	8–14
Fireclay refractories, 10–44% Al_2O_3	10–80	5–15	20–45
Corundum refractories, 75–90% Al_2O_3	40–200	10–150	30–120
Forsterite refractories	20–40	5–10	25–30
Magnesia refractories	40–100	8–200	30–35
Zircon refractories	30–60	80–200	35–40
Whiteware	30–40	20–25	10–20
Stoneware	40–100	20–40	30–70
Electrical porcelain	350–850	90–145	55–100
Capacitor ceramics	300–1000	90–160	

Table 13. Thermal expansion coefficients, 10^{-6} K^{-1}, of components of ceramic materials

Component	Temperature ranges, °C		
	20–300	20–900	20–1400
Silica	36.5	15.5	10.0
Magnesite	10	12.7	14.2
Chrome magnesite	8.3	9.4	10.5
Chromite	8.3	9.1	9.5
Corundum 99	7.3	7.2	*
Corundum 90	4.3	5.2	6.5
Zircon	2.7	3.8	**
Sillimanite	3.3	4.4	4.8
Silicon carbide	1.6	3.5	4.4

* 7.9 over the range 20–1200 °C.
** 4.5 over the range 20–1200 °C.

Table 14. Thermal conductivity of refractory brick

Material	% Porosity	Thermal conductivity, $W\ m^{-1}\ K^{-1}$	
		371 °C	1000 °C
28% Al_2O_3	22	0.84	1.72
42% Al_2O_3	19	1.21	1.42
72% Al_2O_3	22	1.55	1.42
99% Al_2O_3	24	3.77	2.47
Silica	23	1.34	1.76
Mullite	23	0.92	1.76
94% MgO	20	6.86	2.76
Chrome magnesite	22	1.72	1.80
Zircon	17	2.76	2.38

Table 15. Thermal conductivity of electrical ceramics at room temperature

Body type	Thermal conductivity, W m^{-1} K^{-1}
Electrical porcelain	0.8–1.7
Steatite porcelain	2.1–2.5
Cordierite	1.3–2.1
Zircon porcelain	4.6–5.0
Titania porcelain	2.9–4.2
Titanate	3.3–4.2

Table 16. Resistivities of some metals, ceramic insulators, and semiconductors at room temperature

Ceramic material	Resistivity, Ω cm
Insulators	
Low-voltage porcelain	$10^{12} - 10^{14}$
Steatite porcelain	10^{14}
Mullite porcelain	10^{13}
Cordierite porcelain	10^{13}
Zircon porcelain	10^{14}
Alumina porcelain	10^{16}
Silica	10^{19}
Semiconductors	
Silicon carbide	10
Boron carbide	0.5
Ferric oxide	10^{-2}

11.2. Simplified Testing of Clay Body Materials

The tests listed in Table 19 were derived from the characterization concept to serve as control tests by miners and refiners of clay body materials and by manufacturers in acceptance testing and for plant control. Table 19 lists suggested control tests for clays and nonplastics. These tests are designed as surrogate procedures for the methods that require complex, expensive equipment and highly trained personnel. Chemical analysis, particle-size distribution, and mineral constituents govern fired properties and glaze fit of clay-based bodies. A simple test, using two clear glazes (one fitted to a standard kaolin and the other to a standard ball clay) make possible detection of variation in free silica. The presence of specking impurities is shown more strongly by clear glazes than by bisque clay. Deflocculation tests indicate changes in particle size and soluble salts. The solubles test also relates to deflocculation and the rate of cast. Nonplastics should be evaluated in a standard body for deflocculation, rheology, and fired properties. A fusion test, compared against standard specimens [151], is indicative of changes in mineral composition and particle size.

Table 17. Properties of advanced ceramics [150]

Material	Crystal structure	Theoretical density, Mg/m^3	Knoop or Vickers hardness, GPa	Transverse rupture strength, MPa	Fracture toughness K_{IC}, MPa·m$^{0.5}$	Young's modulus, GPa	Thermal expansion, 10^{-6} K^{-1}	Thermal conductivity W m^{-1} K^{-1}		Specific heat, J kg^{-1} K^{-1}
								400 K	1400 K	
Alumina, Al_2O_3	hexagonal	3.97	18–23	276–1034	2.7–4.2	380	7.2–8.6	27.2	5.8	1088
Mullite, 3 Al_2O_3 · SiO_2	orthorhombic	2.8	10–11	185	2.2	145	5.7	5.2	3.3	1046
Partially stabilized zirconia, ZrO_2	cubic, monoclinic, tetragonal	5.70–5.75		600–700	8–9	205	8.9–10.6	1.8–2.2		400
Titanium dioxide, TiO_2	tetragonal (rutile)	4.25	7–11	69–103	2.5	283	9.4	8.8	3.3	799
Silicon carbide, SiC	hexagonal (α) cubic (β)	3.21	20–30	230–825 (hot pressed)	4.8–6.1	207–483	4.3–5.6	63–155	21–33	628–1046
Silicon nitride, Si_3N_4	hexagonal (α) hexagonal (β)	3.18–3.19	8–19	700–1000 (hot pressed)	3.6–6.0	304	3.0	9–30		400–1600
Titanium nitride, TiN	cubic	5.43–5.44	16–20			251	8.0	24	67.8*	628

* At 1773 K.

Table 18. Listing of ASTM book of standards (1985) for ceramic-related test compilations

Volume identification	Subject area
03.05	chemical analysis of metals and metal-bearing ores
03.06	emission spectroscopy: surface analysis
04.06	thermal insulation
10.01	electrical insulation, solids, composites and coatings
10.02	electrical insulation (II), wire and cable, heating and electrical tests
10.04	electronics (I)
10.05	electronics (II)
11.01	water (I)
11.02	water (II)
11.03	atmospheric analysis: occupational health and safety
12.01	nuclear energy (I)
12.02	nuclear energy (II), solar and geothermal
14.01	analytical methods — spectroscopy; chromatography; temperature; computerized
14.02	general test methods — nonmetal; lab apparatus; statistical methods; durability
15.01	refractories, carbon and graphite
15.02	glass; ceramic whitewares, porcelain enamels

Table 19. Suggested control tests

China clay	Ball clay	Feldspar	Silica
residue: 100, 200, 325 mesh	residue: 100, 200, 325 mesh	residue: 100, 200, 325 mesh	residue: 100, 200, 325 mesh
solubles: hardness, Cl^-, SO_4^{2-}	solubles: hardness, Cl^-, SO_4^{2-}	particle-size analysis	particle-size analysis
methylene blue index	methylene blue index	body deflocculation	body deflocculation
particle size analysis	particle size analysis	casting cups and bars	casting cups and bars
fired shrinkage and absorption	fired shrinkage and absorption	firing shrinkage and sag bars	firing shrinkage and sag bars
glaze fit	glaze fit	glaze fit and absorption	glaze fit and absorption
fired color, specking	fired color, specking	fusion	
chemical: TiO_2 and Fe_2O_3	ignition loss		
neat clay deflocculation	clay–flint deflocculation		
casting	casting		

11.3. Quality Control of Advanced Ceramics

The degree of quality control needed in a ceramic manufacturing depends on the critical requirements of the application. In advanced ceramic manufacture, most end uses need a specified manufacturing procedure in writing in addition to certification that this procedure has been followed. The more demanding applications must have proof tests, destructive sample tests, and nondestructive inspections of various kinds [4, chap. 9].

A popular method for characterizing flaw distribution is the use of the Weibull approach [152], based on the weakest link theory. This assumes that a given volume of a ceramic material under uniform stress will fail at the worst flaw. Data is shown as a probability of failure F (a function of stress σ and volume V or area A under stress) plotted against σ. The probability F can be estimated from $Fn/(N+1)$, where n = rank of sample and N = total number of samples. When plotted on a log normal grid, as in Figure 7, the result is a straight line.

The data of Figure 7 show the effect of forming technique on the uniformity of an advanced ceramic material, with respect to stress fracture. A vertical plot at a particular fracture stress indicates absolute uniformity. Slip cast and injection-molded parts are much more uniform than simple isopressed items after sintering. Slip-cast pieces are more uniform than injection-molded parts. Hot isopressing of injection-molded pieces gives a significant increase in fracture stress over sintered injection molding, but at about the same level of uniformity.

12. Economic Aspects

The cost of a ceramic product at its point of manufacture depends on such factors as (1) cost of raw materials; (2) cost of energy for processing, forming, firing, and finishing; (3) capital cost and maintenance; and (4) cost of labor. The impact of the various factors necessarily vary with the product being made.

Raw material cost involves the cost of mining, refining, and transporting a given commodity to a point of ceramic manufacture. Equipment cost depends on the country where it is manufactured. However, countries with highly developed, technologically advanced manufacturing capability, coupled with moderate wage scales, can compete in domestic markets of countries having comparable manufacturing ability but higher wage levels. Countries with skilled workers and low wage scales can export products that require detailed handwork and can undersell domestic manufacturers with high wage costs. For example, countries of the Far East, having highly skilled but low-wage labor, make and deliver high-quality tableware to European and American markets at prices lower than domestic potteries can [153].

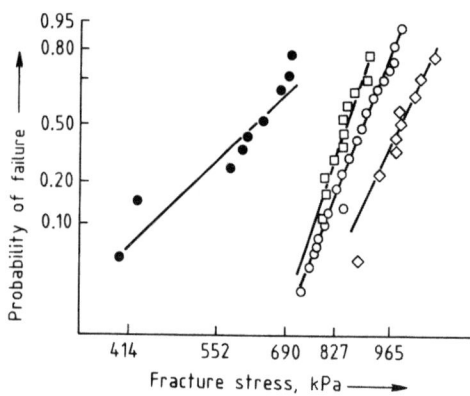

Figure 7. Weibull probability plot showing the effect of forming methods for silicon nitride on the range and level of stress fracture
● Isopressed/sintered; Slipcast/sintered;
○ Injection molded/sintered; ◇ Injection molded/hot isopressed theory

Other high-quality, less labor-intensive products, such as vitreous plumbing ware, are less vulnerable to imports. However, as developing countries gain expertise and improve quality, it is possible that sanitary ware imports may take over some of the sanitary ware markets in developed countries.

In terms of unit weight, the cost of making and delivering a ceramic product depends on the nature of the product and can vary widely. The fine ceramic products of Table 1, in general, cost more to make per unit weight than the coarse ceramic products of Table 2. Furthermore, a vitreous china plumbing fixture costs far less per unit weight than a highly decorated fine china platter. By the same token, a building brick costs far less per unit weight than a high-alumina refractory brick of comparable volume.

Physical and economic geographical factors are likely to limit exports of relatively heavy products from countries where mountains make road or rail transport difficult. The distance from market is a factor in the cost of transport. In the United States, where rail, water, and road transport are reasonably good, the cost of transporting desirable raw materials from sources in the southeast to points on the West Coast can exceed the FOB cost of the raw material. This has encouraged use of local, less desirable raw materials. In glass manufacturing, plants will seek out closest possible sources of silica sand and glass feldspar or nepheline syenite. In smaller countries, transport costs are of less consequence.

Specialty products, such as high-tension electrical porcelain insulators, have stringent mechanical and dielectric strength requirements. Careful processing of controlled compositions is the key to superior quality. Those countries with lower wage scales and good technical capabilities can often meet or exceed these requirements at a lower or equivalent cost than in higher wage North American markets.

Manufacture of structural clay products, such as brick, clay pipe, and tile, has been historically located as close as possible to a given market area, consistent with access to cheap raw material, fuel, and labor. However, structural clay product companies are increasing in size and becoming correspondingly more conscious of the necessity for controlling raw materials and servicing an expanding market [154]. Research is being done on forming methods to reduce losses and improve quality [155]; automated plants are increasingly more common.

The economic health of *traditional, clay-based ceramic* manufacture is closely tied to the state of the economy, especially the level of building [156]–[158]. Traditional ceramics continue to evolve from an art to a science as more use is made of the findings of materials science in better control over raw materials, body preparation, and automated forming. As noted earlier, improvements in kilns and body composition will permit faster firing, lower fuel and refractory costs, and lowered losses.

By way of contrast, *advanced ceramics* is a rapidly developing field of large, although somewhat undefined, potential for growth. Market forecasts for high-technology ceramics vary from one to another with respect to absolute size; although starting bases differ, growth rates are more consistent.

BOWEN [159] estimated sales rising from 1.5×10^9 to 7.0×10^9 dollars in the period 1980–1995 at an annual rate of 11% in the United States. For Japan, the figures are 1.9×10^9 to 9.0×10^9 dollars, also at an annual growth rate of 11%. For the world market, the values are 4.1×10^9 to 17.0×10^9 dollars at an annual growth of 10%. Toshiba [160] forecasts a rise in the Japanese market from 1.26×10^9 to 11.4×10^9 dollars from 1982 to 2000, with an overall growth of 32%. The U.S. Department of Commerce [161] predicts estimated sales for advanced ceramics rising from 0.60×10^9 to 5.9×10^9 dollars over the period 1980–2000 at an annual growth rate of 12–15%.

The Western world in general and the United States in particular depend for economic strength on abundant, readily available natural resources. The countries of Western Europe, the United Kingdom, and North America are either self-sufficient in traditional ceramic materials or have ready access to such materials. This does not appear to be the case for many materials required for advanced ceramics. Of 27 basic industrial minerals or metals listed by the U.S. Department of the Interior, 18 are imported at levels above 50%, including cobalt, manganese, and chromium [162].

13. References

[1] W. W. Perkins (ed.): *Ceramic Glossary – 1984*, American Ceramic Society, Columbus, Ohio, 1984.
[2] L. S. O'Bannon (ed.): *Dictionary of Ceramic Science and Engineering*, Plenum Press, New York 1984.
[3] Annu. Book ASTM Stand. 1972, part 13, Refractories, Glass, and Other Ceramic Materials. Terms Relating to Ceramic Whitewares and Related Products, C242–C272.
[4] D. W. Richerson: *Modern Ceramic Engineering*, Dekker, New York 1982.
[5] J. R. H. Black et al., *Am. Ceram. Soc. Bull.* **64** (1985) no. 39–41, 50.
[6] W. G. Solheim, *Sci. Am.* **226** (1972) 34–41.

[7] T. A. Wertime, *Am. Sci.* **61** (1973) 670–682.
[8] J. G. Ayers, in [163] chap. 2
[9] W. D. Kingery (ed.): *Ceramics and Civilization*, vol. 1, *Ancient Technology to Modern Science*, American Ceramic Society, Columbus, Ohio, 1984.
[10] J. B. Hennessy, in [163] chap. 1
[11] D. J. Hamlin: *The First Cities*, Time-Life Books, New York 1973, chap. 6.
[12] Liu Zhen, Hu Youzhi, Zhen Naizhang, *J. Jingdezhen Ceram. Inst.* **5** (1984) no. 2, 17–36.
[13] M. Farnsworth, *Am. J. Archaeol.* **68** (1964) 221–231.
[14] G. Fehervari, in [163] chap. 3.
[15] A. S. H. Megaw in [163] pp. 100–109.
[16] F. A. Drier in [163] pp. 127–134.
[17] S. Ducret in [163] pp. 216–224.
[18] A. Ray in [163] pp. 246–254.
[19] F. H. Norton: *Fine Ceramics: Technology and Applications*, McGraw-Hill, New York 1970.
[20] H. Ries, H. B. Kummel, G. N. Knapp: *The Clays and Clay Industry of New Jersey*, N.J. State Geological Survey, Trenton, N.J., 1901, chap. 11.
[21] G. B. Kenney, H. K. Bowen, *Am. Ceram. Soc. Bull.* **62** (1983) 591.
[22] E. Ruh: "Metallurgical Refractories," in M. G. Berer (ed.): *Encyclopedia of Materials Science and Engineering*, Pergamon Press, Oxford 1985.
[23] R. A. Haber, V. A. Greenhut, E. J. Smoke, U.S. patent application, 30 April 1984.
[24] J. D. Birchall, *Trans. Br. Ceram. Soc.* **82** (1983) 143–145.
[25] R. A. Katz, *Science (Washington, D.C.)* **208** (1980) 841–847.
[26] M. E. Washburn, H. R. Baumgartner, *Second Annual Army Materials Conference on Ceramics for High Performance Applications*, Hyannis, Mass., 1973.
[27] G. Q. Weaver, J. W. Luckek, *Am. Ceram. Soc. Bull.* **58** (1978) 1131–1135.
[28] J. Friberg, B. Aronsson in S. Somiya (ed.): *Ceramic Science at the Present and in the Future*, Uchido Rokakuyo Publ. Co., Tokyo 1981, pp. 109–130.
[29] B. North, *Materials and Society* **8** (1984) 271–281.
[30] H. R. Baumgartner: "Evaluation of Roller Bearings Containing Hot-Pressed Silicon Nitride Rolling Elements," in *Second Annual Army Materials Conference on Ceramics for High Performance Applications*, Hyannis, Mass., 1973.
[31] J. F. Baumard, B. Cales, A. M. Anthony in S. Somiya (ed.): *Ceramic Science and Technology at the Present and in the Future*, Uchido Rokakuyo Publ. Co., Toyko 1981, pp. 161–191.
[32] P. A. Janeway, *Ceram. Ind. (Chicago)* **122** (1984) 40–45.
[33] P. F. Becher, *J. Am. Ceram. Soc.* **64** (1981) 37–39.
[34] W. R. Cannon, K. Wilfinger, personal communication (1984).
[35] I. M. Lachmann, R. N. McNally, *Ceram. Eng. Sci. Proc.* **2** (1981) 337–351.
[36] D. W. Roy, K. E. Green, *Ceram. Eng. Sci. Proc.* **4** (1983) 510–519.
[37] K. M. Prewo, J. J. Brennan, *J. Mater. Sci.* **15** (1980) 463–468.
[38] A. K. Dhingra, *Philos. Trans. R. Soc. London* **A 294** (1980) 559–564.
[39] M. K. Surappa, P. K. Rohatgi, *J. Mater. Sci.* **16** (1981) 983–993.
[40] W. J. Lackey et al.: "Ceramic Coatings for Heat in Engine Materials — Status and Future Needs," in *Proc. Int. Symp. Ceramic Components for Heat Engines*, Hakone, Japan, 1981.
[41] G. Fisher, *Am. Ceram. Soc. Bull.* **63** (1984) 569–571.
[42] J. T. Kummer, N. Weber, *SAE J.* **76** (1968) 1003–1007.
[43] W. D. Kingery et al., *J. Am. Ceram. Soc.* **42** (1959) 393–398.
[44] J. Hecht, *High Technol.* **3** (1983) no. July/Aug., 49–56.

[45] L. L. Hench, R. W. Gould: *Characterization of Ceramics,* Dekker, New York 1971.
[46] D. W. Richerson, M. E. Washburn, US 3 836 374, 1974.
[47] G. E. Gazza, *Am. Ceram. Soc. Bull.* **54** (1975) 778–781.
[48] *Characterization of Materials,* Materials Adv. Board, Div. Eng., Nat. Research Council, Publ. MAB-229-M, Nat. Acad. Sci.–Nat. Acad. Eng., Washington, D.C., 1967.
[49] G. W. Phelps, *Am. Ceram. Soc. Bull.* **55** (1976) 528–529, 532.
[50] G. W. Phelps, D. L. Harris, *Am. Ceram. Soc. Bull.* **47** (1968) 1146–1150.
[51] W. A. Weyl, N. A. Terhune, *Ceram. Age* **62** (1953) no. 23, 40–41.
[52] D. Arayaphong, M. G. McLaren, G. W. Phelps, *Am. Ceram. Soc. Bull.* **62** (1984) 1181–1185.
[53] G. W. Phelps, *The Role of Naturally Occurring Organic Matter in Clay Casting Slips,* Univ. Microfilms, Ann Arbor, Mich., 1963, pp. 144–180.
[54] G. W. Phelps, M. G. McLaren in [164] pp. 211–225.
[55] G. A. Loomis, *J. Am. Ceram. Soc.* **23** (1940) 159–162.
[56] Annu. Book ASTM Stand. 1985, vol. **15.02,** Glass; Ceramic Whitewares; Porcelain Enamels.
[57] G. W. Phelps et al: *Rheology and Rheometry of Clay-Water Systems,* Cyprus Industrial Minerals, Sandersville, Ga., 1982.
[58] L. Pauling: *The Nature of the Chemical Bond,* 3rd ed., Cornell University Press, Ithaca, N.Y., 1960, p. 544.
[59] W. E. Worrall: *Clays and Ceramic Raw Materials,* 2nd ed., Pergamon, Oxford 1982.
[60] W. D. Keller, *Clays Clay Miner.* **33** (1985) 161–172.
[61] R. E. Grim, *Clay Mineralogy,* 2nd ed., McGraw-Hill, New York 1968, pp. 77–92.
[62] H. van Olphen: *An Introduction to Clay Colloid Chemistry,* 2nd ed., J. Wiley & Sons, New York 1977.
[63] A. L. Johnson, W. G. Lawrence, *J. Am. Ceram. Soc.* **25** (1942) 344–346.
[64] D. J. Shaw: *Introduction to Colloid and Surface Chemistry,* 3rd ed., Butterworth, London 1980.
[65] W. G. Lawrence, R. R. West, *Ceramic Science for the Potter,* 2nd ed., Chilton Book Co., Radnor, Pa., 1982, pp. 45–55.
[66] J. W. Shaffer in [165] pp. 506–508 in vol. 1
[67] D. A. Holdridge, *Trans. Br. Ceram. Soc.* **62** (1963) 857–875.
[68] E. W. Koenig, *J. Am. Ceram. Soc.* **25** (1942) 420–422.
[69] C. P. Rogers, J. P. Neal, K. H. Teague (rev.) in [165] pp. 709–722 in vol. 1
[70] D. G. Minnes et al. in [165] pp. 931–960 in vol. 2
[71] J. van Wunnik, J. S. Dennis, G. W. Phelps, *J. Can. Ceram. Soc.* **30** (1961) 1–7.
[72] W. Ryan, W. E. Worrall, *Trans. Br. Ceram. Soc.* **60** (1961) 540–555.
[73] G. W. Phelps, *Am. Ceram. Soc. Bull.* **38** (1959) 411–414.
[74] R. E. Brociner, R. T. Bailey: "Mechanical Treatment of Ceramic Bodies," in *Trans. Int. Ceram. Congr.* **9** (1966).
[75] V. S. Schory, *J. Am. Ceram. Soc.* **3** (1920) 286–295.
[76] A. F. Taggart (ed.): *Handbook of Mineral Dressing,* J. Wiley & Sons, New York 1945.
[77] N. H. Parker, *Chem. Eng. (N.Y.)* **71** (1964) no. 8 June, 165–220.
[78] L. H. Stone, *Chem. Eng. (N.Y.)* **86** (1979) no. 15 Jan., 125–130.
[79] R. Remirez, *Chem. Eng. (N.Y.)* **85** (1978) 4 Dec., 72, 74.
[80] *Chem. Eng. (N.Y.)* **92** (1985) 5 Aug., 47.
[81] Bowen Engineering, *Spray Dryers for Ceramics,* North Branch, N.J., Bulletin 42–2.
[82] J. T. Jones, M. F. Berard: *Ceramics: Industrial Processing and Testing,* Iowa State University Press, Ames, Iowa, 1972, pp. 30–35.
[83] M. K. Bo et al., *Trans. Inst. Chem. Eng.* **43** (1965) T228–T232.

[84] C. C. Furnas, *Ind. Eng. Chem.* **23** (1931) 1052–1058.
[85] G. W. Phelps et al., *Am. Ceram. Soc. Bull.* **50** (1971) 720–722.
[86] T. C. Patton: *Paint Flow and Pigment Dispersion*, 2nd ed., J. Wiley & Sons, New York 1979.
[87] P. M. Rockwell, A. J. Gitter, *Am. Ceram. Soc. Bull.* **44** (1965) 497–499.
[88] J. G. M. deLau, *Am. Ceram. Soc. Bull.* **49** (1970) 572–574.
[89] D. W. Johnson, F. J. Schettler, *J. Am. Ceram. Soc.* **53** (1970) 440–444.
[90] R. E. Mistler et al. in [164] pp. 411–438.
[91] R. Russell et al., *J. Am. Ceram. Soc.* **32** (1949) 105–113.
[92] L. L. Hench et al., *Ceram. Eng. Sci. Proc.* **3** (1982) 477–483.
[93] L. C. Klein, *Ceram. Eng. Sci. Proc.* **5** (1984) 379–384.
[94] R. Roy, *J. Am. Ceram. Soc.* **52** (1969) 344.
[95] R. E. Gould, R. W. Cline, *Am. Ceram. Soc. Bull.* **29** (1950) 291–292.
[96] R. C. P. Cubbon, *Br. Ceram. Trans. J.* **83** (1984) 121–124.
[97] R. C. Rossi, R. M. Fulrath, *J. Am. Ceram. Soc.* **48** (1965) 558–564.
[98] A. R. Blackburn, *Am. Ceram. Soc. Bull.* **29** (1950) 230–234.
[99] F. H. Norton: *Elements of Ceramics*, 2nd ed., Addison-Wesley, Reading, Mass., 1974.
[100] O. J. Whittemore, Jr. in [164] pp. 343–355.
[101] B. W. Nies, C. M. Lambe, *Am. Ceram. Soc. Bull.* **35** (1956) 319–323.
[102] E. G. Walker, *Trans. Br. Ceram. Soc.* **64** (1965) 233–248.
[103] D. J. Shanefield, R. E. Mistler, *Am. Ceram. Soc. Bull.* **53** (1974) 416–420.
[104] J. A. Mangels, W. Trela in J. A. Mangels, G. L. Messing (eds.): *Forming of Ceramics*, Amer. Ceram. Soc., Columbus, Ohio, 1984, vol. 9, pp. 220–233.
[105] H. Salmang: *Ceramics: Physical and Chemical Fundamentals*, 4th ed., Butterworth, London 1961.
[106] A. Swineford (ed.): "Symposium on the Engineering Aspects of the Physico-Chemical Properties of Clays" in *Proc. 9th Nat. Conf. on Clays and Clay Min.*, Pergamon Press, New York 1962.
[107] R. K. Schofield, *Trans. Int. Congr. Soil Sci. 3rd, 1935*, vol. **2**, pp. 37–48.
[108] R. Q. Packard, *J. Am. Ceram. Soc.* **50** (1967) 223–229.
[109] A. D. Coleman, J. D. Marsh, *J. Soil Sci.* **12** (1961) 342–361.
[110] F. Samudio, unpublished research, Rutgers University, 1981.
[111] G. W. Phelps in [166] pp. 57–65.
[112] G. W. Phelps: *Proceedings of the Materials & Equipment/Whitewares Divisions*, American Ceramic Society, Bedford, Pa., Sept. 1970, pp. 9–14.
[113] F. H. Norton, *J. Am. Ceram. Soc.* **16** (1933) 88–92.
[114] R. W. Grimshaw: *The Chemistry and Physics of Clays and Allied Ceramic Materials*, 4th ed., J. Wiley & Sons, New York 1980.
[115] W. R. Morgan, R. K. Hursh, *J. Am. Ceram. Soc.* **22** (1939) 271–278.
[116] D. Woo et al., *J. Am. Ceram. Soc.* **38** (1955) 383–388.
[117] R. P. Allaire, *Ceram. Ind. (Chicago)* **86** (1966) no. 3, 38.
[118] B. Vassiliou, J. White, *Trans. Br. Ceram. Soc.* **52** (1953) 329–385.
[119] G. W. Bird, A. J. Dale, *Trans. Br. Ceram. Soc.* **51** (1952) 559–573.
[120] L. H. Hepner, *Am. Ceram. Soc. Bull.* **24** (1945) 415–417.
[121] *Webster's Ninth New Collegiate Dictionary*, Merriam-Webster, Springfield, Mass., 1983.
[122] J. E. Burke in W. D. Kingery (ed.): *Ancient Technology to Modern Science*, American Ceramic Society, Columbus, Ohio, 1984, pp. 315–333.
[123] J. G. Weinstein, C. Chanyavanich, unpublished report, Rutgers University, Oct. 1980.

[124] J. R. Schorr, R. Russell, *Am. Ceram. Soc. Bull.* **49** (1970) 1042–1051.
[125] C. R. Moebus et al., *Ceram. Eng. Sci. Proc.* **4** (1983) 935–945.
[126] J. G. Weinstein et al., *Ceram. Eng. Sci. Proc.* **3** (1982) 879–887.
[127] T. Vasilos, R. M. Spriggs, *Proc. Br. Ceram. Soc.* **3** (1967) 195–221.
[128] R. M. Lonero, *Am. Ceram. Soc. Bull.* **62** (1983) no. 1000, 1009.
[129] N. M. Hintz, *Ceram. Eng. Sci. Proc.* **4** (1983) 1014–1022.
[130] C. G. Harmon, Jr., *Ceram. Eng. Sci. Proc.* **2** (1981) 908–916.
[131] S. W. Kennedy, K. W. Doak, *Ceram. Eng. Sci. Proc.* **5** (1984) 1012–1024.
[132] E. C. Bloor, *Trans. Br. Ceram. Soc.* **55** (1956) 631–660.
[133] H. H. Sortwell, *J. Am. Ceram. Soc.* **4** (1921) 718–730.
[134] A. S. Watts, *Trans. Am. Ceram. Soc.* **19** (1917) 301–302.
[135] F. Singer, *Trans. Br. Ceram. Soc.* **53** (1954) 398–421.
[136] C. G. Harmon, H. R. Swift, *J. Am. Ceram. Soc.* **28** (1945) 48–52.
[137] R. P. Isaacs, *Am. Ceram. Soc. Bull.* **45** (1966) 714–715.
[138] C. B. Lutrell, *J. Am. Ceram. Soc.* **32** (1949) 327–332.
[139] C. W. F. Jacobs, *J. Am. Ceram. Soc.* **37** (1954) 216–220.
[140] W. J. Koch et al., *J. Am. Ceram. Soc.* **33** (1950) 1–8.
[141] F. H. Norton, *J. Am. Ceram. Soc.* **20** (1937) 217–224.
[142] L. E. Barringer, *Trans. Am. Ceram. Soc.* **4** (1902) 211–229.
[143] G. W. Phelps, *Proc. Porcelain Enamel Inst. Tech. Forum* **38** (1976) 246–250.
[144] H. N. Mills in [165] pp. 339–347 in vol. 1.
[145] W. H. Dumbaugh, J. W. Malmendier in A. M. Alper (ed.): *High Temperature Oxides*, part 4, Refractory Glasses, Glass-Ceramics, and Ceramics, Academic Press, New York 1971, pp. 1–14.
[146] G. H. Beall in A. M. Alper (ed.): *High Temperature Oxides*, part 4, Refractory Glasses, Glass-Ceramics, and Ceramics, Academic Press, New York 1971, pp. 15–36.
[147] J. A. Crookston, W. D. Fitzpatrick in [165] pp. 373–385 in vol. 1.
[148] J. A. Ames, W. E. Cutliffe in [165] pp. 133–159 in vol. 1.
[149] *Annu. Book ASTM Stand.* 1980, no. part 14, *Concrete and Mineral Aggregates*.
[150] W. J. Lackey, D. P. Stinton, G. A. Cerny, L. L. Fehrenbacher, A. C. Schaffhauser: "Ceramic Coatings for Heat Engine Materials — Status and Future Needs," *Proc. Int. Symp. Ceram. Components for Heat Engines*, 1983 ; ORNL/TM-8959.
[151] H. B. Dubois, *J. Am. Ceram. Soc.* **15** (1932) 144–148.
[152] W. Weibull, *J. Appl. Mech.* **18** (1951) 293–297.
[153] R. J. Beals, *Am. Ceram. Soc. Bull.* **64** (1985) 47–50.
[154] J. H. Belger, *Am. Ceram. Soc. Bull.* **61** (1982) 1285–1286.
[155] J. J. Walsh, *Am. Ceram. Soc. Bull.* **61** (1982) 1284–1285.
[156] "U.S. Industry Trends," *Am. Ceram. Soc. Bull.* **62** (1983) 547.
[157] "U.S. Industry Trends," *Am. Ceram. Soc. Bull.* **63** (1984) 547.
[158] "U.S. Industry Trends," *Am. Ceram. Soc. Bull.* **64** (1985) 11.
[159] H. K. Bowen, personal communication to J. B. Wachtman, Jr., 1985.
[160] J. B. Wachtman, Jr., *Ceram. Ind. (Chicago)* **121** (1983) 24–33.
[161] U.S. Dept. Commerce: *A Competitive Assessment of the U.S. Advanced Ceramics Industry*, NTIS Access. No. PB84-162288 (1984).
[162] P. C. Maxwell, *Am. Ceram. Soc. Bull.* **59** (1980) 1158–1159.
[163] R. J. Charleston (ed.): *World Ceramics*, Paul Hamlyn, London 1968.

[164] G. Y. Onoda, L. L. Hench (eds.): *Ceramic Processing before Firing*, J. Wiley & Sons, New York 1978.
[165] S. J. Lefond (ed.): *Industrial Minerals and Rocks*, 5th ed., 2 vols., Soc. Min. Eng. of AIMME, New York 1983.
[166] H. Palmour et al. (eds.): *Processing of Crystalline Ceramics*, Plenum Press, New York 1978.

Ceramics, Advanced Structural Products

W. Roger Cannon, Department of Ceramics, Rutgers – The State University, Piscataway, New Jersey 08854, United States

1.	Introduction 1011	4.	Specific Materials 1020
2.	Mechanical Properties 1013	4.1.	Alumina 1021
2.1.	Failure of Brittle Materials — Flaws 1013	4.2.	Zirconia 1021
		4.3.	Silicon Carbide 1023
2.2.	The Small Strain to Failure . . 1014	4.4.	Silicon Nitride 1024
2.3.	Effects of Microstructure 1016	4.5.	Ceramic – Ceramic Composites 1025
2.4.	Effects of Temperature 1016	4.6.	Other Materials 1026
2.5.	Wear Resistance and Hardness 1018	5.	References 1026
3.	Uses 1018		

1. Introduction

In recent years interest in use of ceramics in new structural applications has been increasing. One reason is that until recently ceramics have been considered safe for use only in structural applications where stresses were compressive; only small tensile stresses were acceptable. Generally, ceramics support a much higher compressive stress than tensile stress. For instance, a typical supplier of high-strength ceramics tests the tensile strength of a 99.5 % aluminum oxide to be 262 MPa, whereas the compressive strength is 2620 MPa [1]. Recently, however, new ceramics have been developed which are capable of withstanding much higher tensile stresses. Figure 1 presents a historical picture of the development of high-strength ceramics, indicating that the strengths are increasing at an accelerating rate. This trend cannot continue forever, since the ultimate strength is limited by the ionic or covalent bond strength, which is estimated to be 10 – 50 GPa, but progress during the seventies and eighties is impressive.

The trend may be explained as follows. The improvement of high-strength materials from the early to the middle period was achieved primarily by eliminating the glassy material present at the grain boundary and secondly by eliminating porosity in purer materials. The very high strengths of the most recent development, the partially stabilized zirconia and other transformation-toughened ceramics, have been achieved through the development of ceramic – ceramic composites. Even higher strength has been achieved in

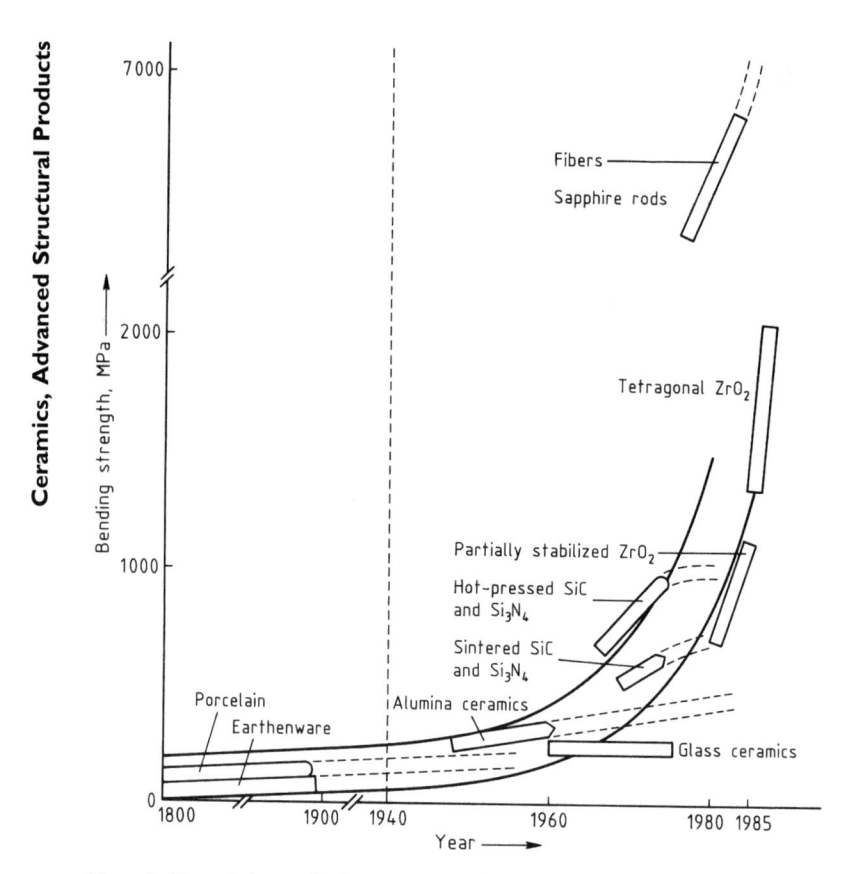

Figure 1. The typical strength of ceramic materials since the mid-1800s

single-crystal aluminum oxide (sapphire) by very careful preparation and protection of the surfaces to almost eliminate flaws. Ceramic fibers are strong for similar reasons.

The field of advanced structural ceramics is rapidly developing, but many applications are yet to be proven. The applications closely related to heat engines will likely have the greatest economic impact on the ceramic industry. Most other applications (see Table 3) are already commercial, with many of these only recently emerging. The trend in development is toward covalent ceramics, carbides, and nitrides that exhibit excellent thermal shock resistance and high-temperature strength and toward composites that have high fracture toughness and high strains to failure.

2. Mechanical Properties

2.1. Failure of Brittle Materials — Flaws

The key to understanding the mechanical properties of ceramics is the Griffith theory for failure of brittle materials [2]. Unlike most metals and polymers, ceramics are extremely brittle and, therefore, extremely sensitive to the presence of any sort of flaw on the surface or in the bulk of the ceramic. Flaws of some size are always present in a material. These may arise during manufacture or in subsequent handling of the material. Since polycrystalline ceramics are usually manufactured by sintering a powder compact at high temperature, the flaws may arise from nonuniformities in the packing of the powder or accidental inclusions in the compact.

The Griffith criterion states that if the flaw is larger than a critical size, then the rate of release of elastic strain energy is greater than the rate of gain of surface energy and the flaw will grow (crack extension). The growth is catastrophic because as the crack grows longer, the driving force becomes greater. The Griffith criterion is a necessary but not sufficient condition for crack growth. In addition, the stress at the tip of the crack must be large enough to break the ionic or covalent bonds. OROWAN estimated this stress and found that for a sharp crack the Griffith stress was sufficient to break the bonds [3]. The *Griffith equation* is

$$\sigma_f = \sqrt{\frac{2E\gamma}{\pi C}}$$

where σ_f is the fracture stress, the applied stress at which the preexisting flaw becomes critical, E is the elastic modulus (Young's modulus), γ is the effective surface energy, and C is the length of a surface flaw or $2C$ is the length of an interior flaw.

One assumption is that only the increasing surface area of the crack acts to resist crack growth. However, other features besides surface energy resist the propagation of the crack: plastic deformation near the tip of the crack, crack branching, and crack deflection. Commonly these contributions are included in γ, which is then renamed the fracture energy, γ_f.

The Griffith equation is sometimes written in terms of the critical stress intensity factor defined as

$$K_C \equiv Y \sigma_f \sqrt{C}$$

where Y is a dimensionless constant that depends on the geometry of loading and the crack configuration. The fracture stress is then written as follows:

$$\sigma_f = \frac{1}{Y}\sqrt{2E\gamma/C} = \frac{K_C}{Y\sqrt{C}} \tag{1}$$

Table 1. Fracture strength vs. flaw size

σ_f, MPa	C, μm
1000	5
500	20
250	80
100	500
50	2025

This second form of the Griffith equation is useful because K_C can be measured by various fracture-mechanics techniques. (K_{IC} is often used instead of K_C, the subscript I indicating the mode of cracking.)

Table 1 shows how the fracture stress and the critical flaw size are related for a promising structural ceramic such as hot-pressed silicon carbide (SiC). The values chosen for this example are a K_C of 4 MPa m$^{0.5}$ and a Young's modulus E of 400 GPa. This table indicates that the flaw size for the high-strength ceramics is very small and that careful manufacturing techniques are necessary to avoid accidentally producing such small flaws in polycrystalline materials. In single crystals, the flaw size is easier to control, hence, the high strength of the sapphire indicated in Figure 1.

Even if great care is taken to manufacture a material so that the largest flaw is on the order of 1 μm, there is no guarantee that new flaws will not be introduced during operation by scratching or particle impact. As a result, most research in developing structural ceramics is now concerned with increasing K_C. Equation (1) shows that acceptable flaw size is proportional to the square of the K_C value, e.g., doubling K_C allows a flaw size 4 times larger.

The K_C value is really an indicator of how brittle or tough the material is. Various methods have been used to increase K_C to as high as 10–15 MPa m$^{0.5}$. These high fracture toughnesses have been achieved with ceramic–ceramic composites.

2.2. The Small Strain to Failure

A second aspect of brittleness in ceramics is the small strain to failure. Ceramics do not undergo plastic deformation at room temperature as do metals. Plastic deformation in both metals and ceramics results from microscopic defects called dislocations. These dislocations move easily in the presence of a stress field in metals, but they do not move easily in ceramics. The stress at which plastic deformation could take place in ceramics is much higher than the fracture stress. The opposite is true for metals. As a consequence, the strain to failure in ceramics is usually on the order of only 0.1%, and ceramics can only be used when the stress is safely below the fracture stress.

The strong sensitivity to flaws and small strains to failure, however, may be compensated by other outstanding properties of ceramics, e.g., wear resistance or corrosion resistance.

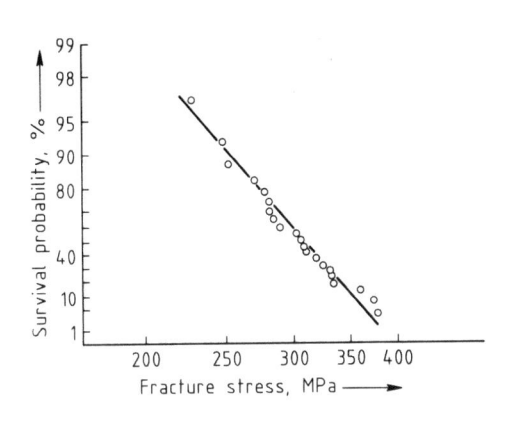

Figure 2. Survival probability vs. fracture stress, a Weibull plot [4]

Safe design of ceramics in tensile or bending load applications requires knowledge of the necessary safety factor. To estimate the safety factor properly, information on the statistics of failure is necessary. The most widely used statistics are Weibull statistics. The *Weibull equation* is as follows:

$$\ln \ln (1/P_s) = \ln V + m \ln (\sigma_f - \sigma_\mu) - m \ln \sigma_0$$

where P_s is the probability of survival of a part subject to the stress, V is the volume of the sample under the tensile stress, σ_μ is the stress of zero probability of failure, usually taken as zero, σ_0 is a normalizing parameter of no physical significance, and m is a constant designated as the Weibull modulus.

This equation indicates that the probability of failure at a certain stress is related to the volume of the material under stress. Such a relationship between volume and stress has been verified experimentally. The equation may be linearized by plotting $\ln \ln (1/P_s)$ vs. $\ln \sigma_f$, allowing the probability of survival at any given stress to be estimated by extrapolating a straight line. This is termed a *Weibull plot* (Fig. 2).

An additional important concern in design is slow crack growth. The crack grows under stress in the presence of atmospheric moisture. The velocity v of crack growth is given by the equation

$$v = \alpha K^N$$

where K is the stress intensity factor, α is a constant, and N is the velocity exponent. Such an equation can be incorporated into Weibull statistics for a more accurate treatment. This subject is reviewed in [4].

2.3. Effects of Microstructure

Microstructure of polycrystalline materials has an important effect on the strength of the ceramic. Generally, strength is improved by achieving a fine grain size. If the flaw size is much larger than the grain size, however, grain size has little effect. For materials having very anisotropic thermal coefficients of expansion and large grain size, *microcracking* develops when the ceramic is cooled to room temperature after sintering. This can be detrimental to strength. In addition, microcracking may occur near second phases having different coefficients of expansion than the continuous phase. Finally, microcracking may occur if the second phase undergoes a phase transformation involving a large change in volume, for instance, the phase transformation of quartz at 573 °C.

2.4. Effects of Temperature

The effect of temperature on the strength of ceramics is illustrated in Figure 3. There is a decrease in strength at high temperatures, where some plastic deformation begins to take place. The temperature at which this decrease in strength begins depends on the material. In materials with a cubic crystal symmetry, such as MgO or UO_2, the temperature where the strength begins to drop is generally low, but in Al_2O_3 the strength drops off little up to 1000 °C. At high temperatures the strength depends on the rate at which the stress is applied.

Above one-half the melting temperature in kelvin, creep (slow time-dependent deformation) becomes appreciable. Since many of the important applications of ceramics are high-temperature applications, creep is important. Creep of ceramic materials has recently been reviewed by CANNON and LANGDON [5]. Most fine-grained polycrystalline ceramics under moderately low applied stresses deform according to one of the following two equations: the Nabarro–Herring equation [6],

$$\dot{\varepsilon} = \frac{10 \Omega \sigma D_v}{k T d^2}$$

or the Coble equation [7],

$$\dot{\varepsilon} = \frac{50 \Omega \sigma w D_b}{k T d^3}$$

where $\dot{\varepsilon}$ is the creep rate, Ω is the volume of a vacancy, σ is the applied stress, D_v is volume self-diffusion coefficient, d is grain diameter, k is Boltzmann's constant, T is temperature (K), w is the effective width of the grain boundary, and D_b is grain boundary diffusion coefficient.

The two equations indicate that the most highly creep-resistant materials are large-grained ceramics. The equations also indicate that the creep resistance is related to the diffusion coefficient. The diffusion coefficient is generally very low in ceramics having a

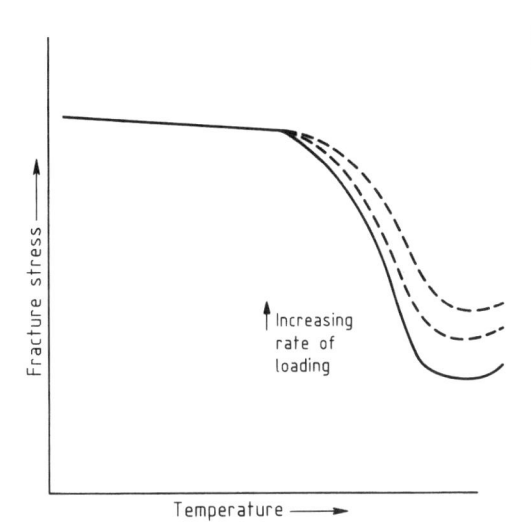

Figure 3. Typical fracture strength vs. temperature curves

high melting temperature, and therefore, ceramics are generally creep-resistant materials at high temperatures (> 1000 °C). If the ceramic contains a glassy grain boundary phase, then creep rates are somewhat higher than those given in these equations. If high creep resistance is desired, the glassy phase should be eliminated.

Ceramics are sometimes limited as a high-temperature material by the inability to withstand *thermal shock*, i.e., rapid drop in temperature. Their lack of thermal shock resistance is primarily due to their inability to accommodate the strain caused by a rapidly cooling surface adjacent to a hot interior. The low thermal conductivity of ceramics contributes to this lack of thermal shock resistance. The critical temperature drop ΔT_c through which a material may be shocked without exceeding the fracture stress at the surface is given by the equation

$$\Delta T_c = \frac{\sigma_f(1-v)}{\psi E \alpha}$$

where α is the thermal coefficient of expansion, v is Poisson's ratio, and ψ is a factor depending on thermal conductivity, size of piece, and heat transfer coefficient.

The equation contains several materials parameters generally available, and, therefore, may be used to estimate the relative thermal shock resistance of various materials. Table 2 compares the thermal shock resistance of various advanced structural ceramic materials for both rapid thermal shock ($R = \sigma_f(1-v)/E\alpha$) and slow thermal shock ($R' = \sigma_f k(1-v)/E\alpha$, where k is the thermal conductivity).

Table 2. Factors affecting the thermal shock behavior

Material	Bending strength, MPa	Modulus of elasticity E, GPa	Poissons ratio ν	Coefficient of thermal expansion α, 273–1273 K, 10^{-6} K^{-1}	Thermal conductivity k at 773 K, W m^{-1} K^{-1}	R, K	R', kW m^{-1}
Hot-pressed Si$_3$N$_4$	850	310	0.27	3.2	17	625	11
Reaction-bonded Si$_3$N$_4$	240	220	0.27	3.2	15	250	3.7
Reaction-bonded SiC	500	410	0.24	4.3	84	215	18
Hot-pressed Al$_2$O$_3$	500	400	0.27	9.0	8	100	0.8

2.5. Wear Resistance and Hardness

Metallic parts that tend to wear out, e.g., in engines, may be replaced by ceramic parts for longer life. Ceramics exhibit good wear resistance and erosion resistance because of their hardness, i.e., they do not plastically deform easily. Those with the greatest resistance to plastic deformation are the covalently bonded materials and the ones of greatest hardness. However, resistance to *crack propagation* also has some influence on the hardness and wear resistance since cracks form under concentrated point loading. These point loadings may be due to dust particles, grit, or other types of concentrated loads. Some of these cracks propagate parallel to the surface at first and then up toward the surface, resulting in small chips dislodged from the surface. After many of these chips dislodge, the surface is rougher and wears even faster. Improving fracture toughness can improve abrasion resistance.

3. Uses

Ceramics have a number of properties not matched by any other potential structural material. Important among these are wear resistance and high-temperature strength. Other outstanding properties are good corrosion resistance, oxidation resistance, and weight-to-strength ratio. Ceramics may someday replace metals for high-technology uses where metals do not perform well enough and the function complies with ceramic capabilities.

A list of applications for advanced structural ceramics is shown in Table 3. At present many of the potential uses of advanced structural ceramics have not been realized. Probably the most important potential application is the use of ceramics in engines. In a *reciprocal engine,* a number of ceramic parts may replace their current

Table 3. Uses of advanced structural ceramics

Heat engine applications			Special refractory applications	Applications in manufacturing	
Diesel	Turbocharger	Gas turbine			
bearings	bearings	blades	furnace tubes	acid bath parts	reachable casting cores
cylinder liners	heat shields	heat exchangers	fibrous insulation	bearings	nozzles
glow plugs	housing	manifolds	kiln furniture	catalyst supports	pump parts
manifolds	turbocharger rotors	rotor shaft		casting dies	protection tubes
piston caps		shrouds		coil forms	rolling jigs
piston rings		stators		cutting tools	textile parts (rollers, guides)
prechamber				drawing dies	
seals				dry-pressing dies	thermal insulators
tappets				electrical insulators	seal faces
valves				extrusion dies	vacuum feed-throughs
valve seats				heaters	wear-resistant parts
				heat exchangers	
				laboratory equipment parts	

metal equivalents. Parts that must be wear resistant, such as the rocker arm chip and the tappet face, may be replaced with ceramics. Ceramics are also being considered for the cylinder lining and the piston caps, both for wear resistance (ceramics may not require lubrication) and heat resistance: the cylinder may be allowed to get much hotter, improving the efficiency of the engine and requiring little or no cooling. These engines are often called *adiabatic* engines, although they are not adiabatic in the strictest sense. In a *turbocharged* engine a ceramic rotor is a good candidate because its moment of inertia is lower than those of metal rotors.

A second type of engine for which ceramics have great potential is the *gas turbine engine*. The primary purpose of ceramics would be to improve the thermodynamic efficiency by allowing the engine to operate at a higher temperature. The expected temperature of the turbine blades is 1200–1350 °C. Not only are high-temperature strength and impact resistance important, thermal shock resistance is also necessary. The candidates for these uses are silicon carbide and silicon nitride.

Other applications for advanced structural materials are heat exchangers, extrusion dies, and prosthetic devices. The use of ceramics in prosthetic devices is favored because of the compatibility of ceramics with the human body. In contrast, current implant materials contain metals, which may be carcinogenic. These new materials also have been used successfully as cutting tools.

Table 4. Some advanced structural ceramics and potential materials with their chemical formulas

Generic name	Chemical formula
Alumina [1344-28-1]	Al_2O_3
Zirconia [1314-23-4]	ZrO_2
Zircon [1490-68-2]	$ZrO_2 \cdot SiO_2$
Spinel [1302-67-6]	$MgO \cdot Al_2O_3$
Mullite [55964-99-3]	$3\,Al_2O_3 \cdot SiO_2$
Cordierite [12182-53-5]	$2\,MgO \cdot 2\,Al_2O_3 \cdot 5\,SiO_2$
Silicon carbide [409-21-2]	SiC
Silicon nitride [12033-89-5]	Si_3N_4
SiAlON	$Si_{6-z}Al_zO_zN_{8-z}$
Boron carbide [12069-32-8]	B_4C
Aluminum nitride [24304-00-5]	AlN
Glass ceramics	a common compositionis Al_2O_3, Li_2O, and SiO_2 with TiO_2 or ZrO_2 nucleating agents

Table 5. Material properties for several oxide and nonoxide ceramics [8], [9]

Material	Density, g/cm^3	Hardness, Knoop, kg/mm^2	Modulus of elasticity, GPa	Fracture toughness K_{IC}, MPa m$^{0.5}$	Coefficient of thermal expansion, 313–1273 K, 10^{-6}K^{-1}	Thermal conductivity, W m^{-1} K^{-1}
Alumina	3.98	1800–2300	300–370	3–4.5	8.1	30
Zirconia (partially stabilized)	≈15.9	1200	205	4–15	10.5	3
Cordierite (porous)	1.5–2.1		130		0.8–1.2	
Silicon carbide (sintered)	3.21	2500–3000	360–460	3–4.5	4.3	58.5
Silicon nitride (dense)	3.2		230	4–7	3.3	30
Silicon nitride (reaction bonded)	2.5–2.7	1100–1500	150	3–4	3.0	20

4. Specific Materials

Table 4 lists a number of ceramics that are considered advanced structural ceramics. The most important advanced structural materials being developed at this time are the oxides zirconia and alumina and the covalent materials silicon carbide and silicon nitride. In addition, cordierite is important because of its use in heat exchangers. Table 5 lists the important mechanical properties of several of these materials. Fracture strength, either in flexure mode or tension, is not included because it depends so strongly on the flaw size, which is not an intrinsic property of the material. The range of values of K_{IC} is included, although these values depend strongly on the method of testing. (For the general production methods and economic aspects of advanced structural products → Ceramics, General Survey.)

4.1. Alumina

Of the advanced structural materials polycrystalline alumina is perhaps the easiest to manufacture, is relatively inexpensive, has high strength, and is the most widely used. It is widely used for crucibles, tubes, and rods for high temperature and for a large number of wear-resistant and corrosion-resistant specialized items. Perhaps the most important single products are spark plugs and the optically translucent polycrystalline alumina lamp hulls for the high-temperature sodium vapor street lamps.

Several grades of sintered alumina are available, the grades being distinguished by the purity. The chief impurity is silica, which forms glass at the grain boundary. This glass acts as a sintering aid, lowering the sintering temperature several hundred kelvin, but as a result these aluminas have lower strength and hardness. The proper glass compositions, however, can improve the low-temperature strength. High-purity alumina is necessary for high-temperature creep resistance, for otherwise the glass in the grain boundary begins to flow, producing grain boundary sliding. Magnesium oxide is used as a sintering aid to sinter high-purity, high-density alumina.

Figure 4 is a summary of bulk diffusion coefficients of various ions in ceramic oxides. Equations (2) and (3) predict that creep rates should be proportional to the diffusion coefficients, and a comparison of diffusion coefficients is a good way of comparing the fundamental resistances to creep. As shown in Figure 4, both the Al ion and the O ion in alumina have two of the lowest diffusion coefficients, accounting for the low creep of high-purity alumina. However, the creep resistance of alumina is exceeded by that of silicon carbide, and thus for many very creep-resistant applications, silicon carbide is being increasingly used.

4.2. Zirconia

Because of its high melting temperature (2764 °C) zirconia can be used for structural applications at higher temperatures than alumina. It is also used widely as a refractory because it does not react with glass. A few percent of MgO, CaO, or Y_2O_3 are added to zirconia to stabilize the high-temperature cubic phase; otherwise, zirconia undergoes a destructive phase transformation from tetragonal to monoclinic when cooled from the sintering temperature. However, *partially stabilized zirconia*, i.e., that containing less additive than necessary to establish 100% cubic, was determined as early as 1947 to be more resistant to thermal shock than the fully stabilized form [11].

The thermal shock resistance of the partially stabilized zirconia may result from a very fine tetragonal precipitate present even at room temperature. The tetragonal precipitate is prevented from transforming to monoclinic by the constraint of the matrix, since the precipitate must necessarily expand 3–5% and shear 8% during the transformation to monoclinic. The toughening effect of the precipitate arises from the transformation that takes place as a crack approaches the precipitate [12]. The stress

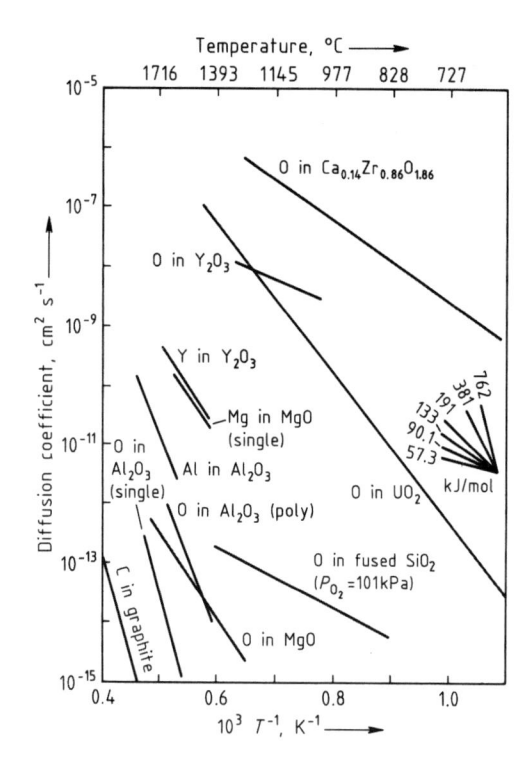

Figure 4. Ionic self-diffusion coefficients in ceramic oxides [10]

field of the crack interacts with the precipitate, allowing the transformation to occur. As the crack passes the precipitate, the transformation of the precipitate then places a compressive stress on the crack, thus reducing its tendency to propagate further. This type of transformation, which takes place in response to stress rather than a change in temperature, is termed a martensitic transformation.

Development has been rapid in improving the strength of partially stabilized zirconia in recent years. Higher and higher strengths and K_{IC} values have been achieved by optimizing the type and percentage of addition and the postsintering annealing, which causes the precipitate to grow.

If less than 4% yttria is added to zirconia and a fine grain structure is maintained, the material can be made 100% tetragonal phase. These materials, called *tetragonal zirconia*, are stronger than partially stabilized zirconia. They can be made even stronger by adding alumina precipitates [13].

Transformation-toughened ceramics [14], as these materials are called, are thus far limited to low-temperature use. The strength drops off linearly as the temperature approaches 1200 °C, where the tetragonal-to-monoclinic transformation normally occurs, as shown in Figure 5. Thermal fatigue is also important. If the materials soak at a high temperature, the microstructure changes slightly, reducing the strength. The high-temperature loss of strength and thermal fatigue characteristics are of concern since one potential use is for the cylinder lining in an "adiabatic" diesel. The high

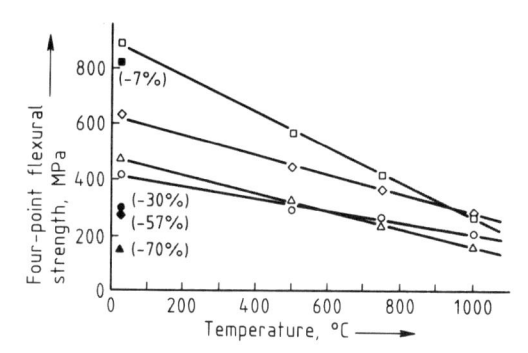

Figure 5. Bending strength of several commercial partially stabilized zirconia materials [15]
The companies and stabilizers: □■ NGK, Y_2O_3; Nilsen, MgO; △▲ Coors, 3% MgO; and ⇆● Feldmühle, MgO. Filled symbols are room-temperature strength after 1000-h exposure at 1000 °C

strength and good insulating ability of transformation-toughened ceramics make them a strong candidate for this use.

4.3. Silicon Carbide

Silicon carbide [16], like most other covalent materials, is not easily sintered from a powdered compact. One process for producing a dense silicon carbide structure is to bond silicon carbide grains, produced by the Acheson process, together with fired clay, glass, silicon nitride, or by other proprietary means. These materials are quite creep resistant and thus can be used for furnace elements and for high-temperature structural use.

The first self-bonding silicon carbide was originally known as Refel silicon carbide, but is now generally known as *reaction-sintered silicon carbide.* It is made by forming a compact of silicon carbide grains with excess carbon and then at a high temperature immersing the compact into a molten silicon bath. The silicon and carbon are then allowed to react at some high soak temperature to form a silicon carbide reaction layer around the original grains. A variation in the process is to introduce the silicon in the vapor phase. Another variation is to introduce the excess carbon as a polymer binder during injection molding. Although properties are usually considerably better than those of the glass-bonded or clay-bonded types, properties are somewhat degraded by the residual unreacted silicon nearly always present in reaction-sintered silicon carbide.

In 1973 PROCHAZKA of the General Electric Co. demonstrated that very fine grained SiC could be *pressureless sintered* with only a small amount of sintering aids, typically 0.5% boron and 1% excess carbon. The sintering temperature is 1900–2100 °C [16]. Later a similar process was patented to sinter α-SiC.

Both reaction-sintered silicon carbide and sintered silicon carbide have structural properties superior to those of the earlier bonded silicon carbides. The sintered silicon carbide, generally, is slightly superior to the reaction-sintered silicon carbide. Even better mechanical properties may be achieved by hot pressing silicon carbide to full density with boron and carbon or alumina sintering aids. Since hot pressing is expensive, this material is not widely produced. The properties listed in Table 5 are for sintered silicon carbide.

4.4. Silicon Nitride

Silicon nitride [17], also a covalent material, is likewise not sinterable from powder without additives. Unlike silicon carbide, however, no sintering aids have been found that allow silicon nitride to be sintered to full density without very high pressures except for intentionally added 5–15% glass phase. The sintering additives commonly used are MgO, Y_2O_3, ZrO_2, Ce_2O_3, and $SiBeN_3$. These sintering additives combine with silica present on the surface of the powder to form a glass that allows the material to densify during the sintering operation.

A sufficient amount of glass of low viscosity is necessary to achieve high densities; however, large amounts of glass of low viscosity degrade the important high-temperature properties of the material. Thus, an optimum is sought between high density, which results in good low-temperature strength, and additives that produce good high-temperature strength. Ideally, the additive after sintering would enter into solid solution with silicon nitride, but thus far only limited success has been achieved in finding such alloys.

A compound related to silicon nitride that has found successful commercialization is *SiAlON*, an aluminum silicon oxide nitride. The most common form has the formula $Si_{6-z}Al_zO_zN_{8-z}$ where z ranges from 0 to 4. Compounds with z near zero have good high-temperature strength, creep resistance, and oxidation resistance, whereas those with z approaching 4 have good low-temperature properties, i.e., strength, toughness, and abrasion resistance.

The properties of silicon nitride are strongly dependent on the additive. Table 6 compares properties of silicon nitride hot pressed with MgO additive, silicon nitride sintered with Y_2O_3 additive, reaction bonded silicon nitride, and SiAlON. Generally higher strengths are achieved both at low and high temperatures by hot pressing; however, this is not an easily commercialized process for mass production. Reaction-bonded silicon nitride, made by exposing a silicon powder compact to nitrogen at high temperatures, has poor low-temperature strength but good high-temperature strength.

Although silicon nitride has high-temperature properties inferior to those of silicon carbide, it has a lower coefficient of thermal expansion and, therefore, is potentially more resistant to thermal shock. Time will tell whether silicon carbide or silicon nitride becomes the more important advanced structural material for engine use. An additional use for silicon nitride only recently introduced is as a cutting tool for rapid machining of cast iron.

Table 6. The effects of additives on the properties of silicon nitride [18]

Material	Bending strength, four point, MPa			Modulus of elasticity, GPa	Coefficient of thermal expansion, $10^{-6} K^{-1}$	Thermal conductivity, $W\ m^{-1}\ K^{-1}$
	RT*	373 K	1648 K			
Hot pressed (MgO additive)	690	620	330	317	3.0	30–15
Sintered (Y_2O_3 additive)	655	585	275	276	3.2	28–12
Reaction bonded (2.45 g/cm^3)	210	345	380	165	2.8	6–3
SiAlON (sintered)	485	485	275	297	3.2	22

* RT = room temperature.

4.5. Ceramic–Ceramic Composites

Both particulate ceramic–ceramic composites and ceramic fiber–ceramic matrix composites are being actively investigated at the current time. The two transformation-toughened ceramics described in Section 4.2, partially stabilized zirconia and tetragonal zirconia, are examples of particulate composites. The transformation-toughened ceramics in which fine zirconia particles are introduced into a ceramic matrix that has a high modulus of elasticity and does not react with zirconia, such as alumina (hence transformation-toughened alumina), are yet another example. The mechanism of toughening is similar to that of the partially stabilized zirconia and tetragonal zirconia.

Other types of particulate composites are possible. Particles deflect cracks if their thermal coefficient of expansion is lower than that of the matrix or if the elastic stiffness is greater; i.e., they increase the K_C value. In contrast to transformation toughening, this type of toughening is rather independent of temperature. A combination of transformation toughening and particle deflection has proved to yield good high-temperature properties.

In particulate composites the value of K_C often parallels the fracture strength, as predicted from Equation (1). The purpose of the particulate composite is to raise the K_C and in turn raise the fracture strength or at least allow the presence of a large flaw for the same fracture strength.

The fiber composite has a different attribute, which is illustrated by Figure 6. In the lower part of the curve is the stress–strain curve for the glass. High-strength fibers added to the glass improve the strength. At the point where the dotted line deviates from the solid line, the matrix, but not the fibers, begins to crack. At higher strain the fibers begin to crack and pull out of the matrix, but the composite still supports some load until all the fibers have fractured. As shown in the figure, the strain can be greater than 1% before complete failure, a contrast to the normal 0.1% strain before failure. This improvement in strain to failure is the most appealing attribute of fiber composites.

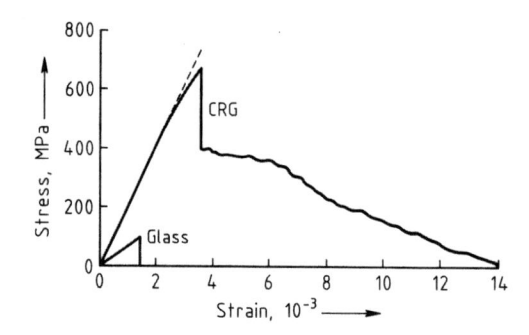

Figure 6. Stress–strain data for glass and carbon fiber reinforced glass (CRG) tested in bending [19]

A second, related benefit is also important. The integrated area under the stress–strain curve, which is large for fiber composites, is related to the fracture energy. The fiber composites, having a large fracture energy, are, therefore, very resistant to thermal shock.

Fiber composites have many potential applications, but are currently expensive to manufacture and thus far have few uses.

4.6. Other Materials

Several other materials listed in Table 4 have a wide range of uses. Glass ceramic has a wide range of compositions and uses. Cordierite has a low coefficient of thermal expansion (see Table 5) and is widely used where good thermal shock resistance is required: cordierite is used for substrates in catalytic converters and in the regenerator core of heat exchangers. Mullite is used much like alumina but at lower temperature. Its thermal shock resistance is better than that of alumina, and mullite of the proper composition and grain structure can be very resistant to creep. Zircon also has an excellent thermal shock resistance. Aluminum nitride, AlN, is a developmental material with some electronic and military applications.

5. References

[1] Coors Porcelain Co.: *AD-995 Alumina.*
[2] A. A. Griffith, *Philos. Trans. R. Soc. London* **A221** (1920) 163.
[3] E. Orowan, *Rep. Prog. Phys.* **12** (1949) 185.
[4] R. W. Davidge: *Mechanical Behavior of Ceramics,* Cambridge University Press, Cambridge 1979.
[5] W. Roger Cannon, T. G. Langdon, *J. Mater. Sci.* **18** (1983) 1–50.
[6] F. R. N. Nabarro: *Report of a Conference on Strength of Solids,* The Physical Society, London 1948, p. 75. C. Herring, *J. Appl. Phys.* **21** (1950) 437.

[7] R. L. Coble, *J. Appl. Phys.* **34** (1963) 1679.
[8] *Engineering Property Data on Selected Ceramics,* vols. I, II, and III, Metals and Ceramics Information Center, Battelle Columbus Laboratories, Columbus, Ohio.
[9] Commercial brochures from Coors Porcelain, Norton Co., Sohio, NGK Automotive Ceramics, and Kyocera.
[10] W. D. Kingery, H. K. Bowen, D. R. Uhlmann: *Introduction to Ceramics,* J. Wiley & Sons, New York 1976, p. 240.
[11] C. E. Curtis, *J. Am. Ceram. Soc.* **30** (1947) 180.
[12] R. C. Garvie, R. H. Hannink, R. T. Pascoe, *Nature (London)* **258** (1975) 703.
[13] K. Tsukuma, K. Ueda, K. Matsushita, M. Shimada, *J. Am. Ceram. Soc.* **68** (1985) C-56.
[14] W. R. Cannon: "Transformation Toughened Ceramics for Structural Uses," in J. B. Wachtman, Jr., (ed.): *Structural Ceramics,* Academic Press, Orlando, Fla. (to be published).
[15] D. C. Larsen, J. W. Adams: "Long-Term Stability and Properties of Partially Stabilized Zirconia," presented at 22nd DOE Contractors Coordination Meeting, Dearborn, Mich., Nov. 1984.
[16] S. Prochazka in J. J. Burke, A. E. Gorum, R. N. Katz (eds.): *Ceramics for High Performance Applications,* Brock Hill Publ. Co., Chestnut Hill, Mass., 1947, pp. 239–252.
[17] F. F. Lange, *Int. Met. Rev.* **25** (1980) 1.
[18] R. N. Katz, *Science (Washington, D.C.)* **208** (1980) 841.
[19] D. H. Bowen, D. C. Phillips, R. A. J. Sambell, A. Briggs, *Mech. Behav. Mater. Proc. Int. Conf., 1st 1971,* 1972 vol. 5, pp. 123–134.

Ceramics, Ceramic – Metal Systems

VICTOR A. GREENHUT, Rutgers – The State University, Piscataway, New Jersey 08854, United States
RICHARD A. HABER, Rutgers – The State University, Piscataway, New Jersey 08854, United States

1.	Introduction ... 1029	4.6.	Vapor-Phase Metallizing ... 1048
2.	Fundamentals of Bonding ... 1030	4.7.	Liquid-Phase Metallizing ... 1048
2.1.	Wetting ... 1030	4.8.	Electroforming ... 1048
2.2.	Bonding and Adherence Mechanisms ... 1032	4.9.	Graded Powder Process ... 1049
		4.10.	Nonmetallic Fusion Process ... 1049
2.3.	Stresses in Ceramic – Metal Systems ... 1033	5.	Thin Films and Coatings ... 1049
3.	Glass-to-Metal Joining ... 1035	5.1.	Uses ... 1050
3.1.	Glass-to-Metal Seals ... 1036	5.2.	Deposition Techniques ... 1051
3.2.	Enamels ... 1037	5.2.1.	Chemical Vapor Deposition (CVD) ... 1052
3.2.1.	The Enameling Process ... 1038	5.2.2.	Evaporation ... 1055
3.2.2.	Theory of Adhesion ... 1041	5.2.3.	Sputtering ... 1056
4.	Ceramic-to-Metal Joining ... 1044	5.2.4.	Plasma Spraying ... 1058
4.1.	Sintered Metal Powder Process (SMPP) ... 1045	5.2.5.	Sol – Gel Processing ... 1058
		5.2.6.	Ion Implantation ... 1059
4.2.	Metal Powder – Glass Frit Method ... 1046	6.	Ceramic – Metal Composites ... 1060
		6.1.	Cermets and Cemented Carbides ... 1061
4.3.	Active Metal Process ... 1047		
4.4.	Gas – Metal Eutectic Process ... 1047	6.2.	Metal Matrix Composites ... 1063
4.5.	Pressed Diffusion Joins ... 1047	7.	References ... 1066

1. Introduction

The use of ceramics and metals in conjunction allows for a combination of properties. Ceramics, here to include glasses, typically show strong covalent/ionic bonding with chemical inertness, high elastic modulus (stiffness) and compressive strength, low electrical and thermal conductivity, low friction and wear behavior, and transparency to electromagnetic radiation. Metallic bonds result in complementary properties, for mechanical ductility is not usually exhibited by ceramic materials. Ceramic – metal systems have been used since prehistoric times in such forms as enamels on metal and metallic decoration on ceramics. The contributions provided by both the ceramic and the metal make such systems important in traditional and advanced technologies.

The ceramic and metal may be coupled as macroscopic pieces or engineering structures. These find use in such components as semiconductor substrates, turbines, electronic equipment, and lamps. Coatings of ceramic or glass can be placed on metal or vice versa. Enamels used as corrosion protection and decoration are an example. A microscopic combination of materials can be produced in the form of a ceramic–metal composite, often termed a cermet when the ceramic is the principal phase. Such materials are used for carbide cutting tools, high-temperature jet orifices, and turbine parts.

In all cases it is necessary to bring the ceramic and metal into intimate contact and create a bond between materials. The distinct differences between ceramic and metal in terms of bonding and properties make it necessary to consider the major factors required to join the two classes of materials and satisfy the physical, chemical, and mechanical properties required of the system.

2. Fundamentals of Bonding

Metals and ceramics may be joined in the solid state or with one component deposited as a liquid or vapor. Intimate contact occurs between the materials by physical means. Bonding may occur by surface or interfacial interaction. Dependent on the chemical nature of the materials, solution and redox reactions may proceed, thus promoting adherence. Quite commonly a liquid phase is present during fabrication of a joint, coating, or composite. In such a case, the wetting behavior of the liquid phase on the solid is quite important. Solution and/or redox reactions are likely, which can affect the wetting behavior and adherence. Wetting is usually required for adherence between ceramic and metal.

A further factor that may affect the bond at a ceramic-to-metal interface is stress between the ceramic and metal. Such stresses may arise in the presence of an applied stress because of differences in elastic constant. As the material cools from fabrication at elevated temperature, differences in thermal expansion can give rise to residual stresses. Changes in temperature during use can have a similar effect. Forces at the ceramic–metal interface can exceed the strength of the bond between materials, preventing a macroscopic join, causing delamination of a coating, or affecting composite properties.

2.1. Wetting

Commonly a bond between metal and ceramic is created by either melting the metal or softening a glassy ceramic material. The success of the bond depends on the wetting

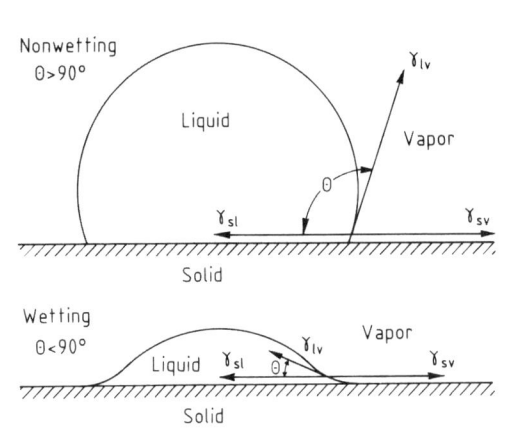

Figure 1. Sessile drop configuration

or capillary action [1], [2]. To join two pieces of material, the liquid must spread across the surface or interface. In a composite material, a phenomenon called sweating occurs in a nonwetting situation such that liquid is exuded at the surface of the material. This may cause nonuniformity or porosity, or it may make the composite impossible to produce.

A number of methods have been developed to determine the wetting response of the liquid phase. These include the geometry of liquid in a capillary or a sessile drop on a flat substrate, the rise of liquid in a capillary, and wicking in a porous material [2]–[4]. Perhaps the simplest and most common approach is the sessile dropmethod, which is shown schematically in Figure 1. A drop of liquid, shown in cross section, is observed at elevated temperature, and the *contact angle* θ is measured. The relationship is as follows:

$$\gamma_{lv} \cos\theta = \theta\gamma_{sv} - \gamma_{sl}$$

where γ_{lv}, γ_{sv}, and γ_{sl} are the interfacial energies between liquid–vapor, solid–vapor, and solid–liquid, respectively. For $\theta > 90°$, a nonwetting situation occurs, while if the angle is $< 90°$, the liquid wets the surface. If $\theta = 0°$, the liquid spreads over the surface as a thin layer. The surface energy can be obtained from this experiment when equilibrium is achieved [3]–[6], providing an estimate of interfacial energy and theoretical bond strength if extrapolated to room temperature.

The wetting angleitself is valuable in evaluating the system. The contact angle may change with small changes in the composition of the atmosphere, liquid, or solid phases. Temperature and time also play an important role. In industrial processes, the liquid is seldom brought to equilibrium as it reacts with the solid oxide films, and contaminants on the surface, as well as the atmosphere of the furnace. While interfacial energy cannot be obtained in this case, the wetting information in itself indicates the degree to which the liquid spreads or infiltrates the solid. Although a strict relationship cannot be drawn between wetting and bonding, usually a good wetting response and a degree of bonding coincide. However, a reasonable bond strength may develop in a nonwetting case, and adequate bond strength may not develop in a wetting case. The wetting response should, therefore, be regarded as a necessary but not a sufficient condition for good ceramic-to-metal bonding.

For ceramic–metal sealing, a transition from metal oxide to metal layers may be used to promote wetting. Such a layer may be applied as a metallizing layer in the form of a moly–manganese paste (see Chap. 4). Such a material may provide a bond to the ceramic because of the oxide ceramic and glass phases present while providing a low interfacial energy with a molten metal because of a metal-rich surface structure [7].

In glass–metal joining, the metal is commonly preoxidized [3], [5], [8]. The oxide layer may be more compatible with the oxide bond of the glass, thereby lowering the interfacial energy between liquid and solid [3], [5] and possibly promoting wetting. Glass-to-metal and ceramic-to-metal bonding often involves a forming gas, which may affect wetting by oxidizing or reducing the metal or ceramic, altering interfacial energies, or diffusing a particular species into solid or liquid. These techniques suggest that some transition between metallic and ionic/covalent bonding enhances the wetting response.

2.2. Bonding and Adherence Mechanisms

The mechanisms of ceramic–metal bonding have been classed in various ways, but perhaps a most general view is that physical, mechanical, and chemical bonding mechanisms can occur. These effects do not occur independently, and all may be involved in a ceramic–metal system.

Physical bonding may be regarded as the bonding effect that occurs when a pair of flat surfaces are brought within atomic-interaction distances. The work of adhesion is given by the specific energy of the metal and ceramic minus that of the metal–ceramic interface. This work of adhesion can yield a theoretical breaking stress near that of the metal or ceramic [4], [9]. Nevertheless, actual interfaces can fail at stresses many orders of magnitude smaller as the result of flaws in the bond.

Mechanical bonding relates to the interlocking structure of rough surfaces, which gives rise to either frictional effects in shear or interlocking of microstructure, which provides mechanical anchors in tension. When ceramic and metal are joined, liquid metal or glass can penetrate pores and cavities, a ductile metal solid can conform to the ceramic surface, or a vapor phase may deposit in surface asperities. A roughened surface, which promotes mechanical bonding, also increases the surface area for physical bonding. However, a mechanically interlocked structure could in itself lead to a bond.

Chemical bonding occurs in a transitional oxide film between the metal and the enamel and is responsible for adherence, the oxide film actually bonding the enamel to the metal. Many complex reactions take place in the transition zone. One accepted model hypothesizes that the adherence is the result of metal-to-metal bonds between the atoms of the base metal and some metallic ions in the enamel. The enamel at the interface must be saturated with the lowest oxidation state oxide of the base metal [5].

Chemical reaction can produce a transition structure that gradually varies from metallic to ceramic. Often such a structure is developed by design. Metals may be preoxidized to provide an oxide that can react with molten glass, saturating the glass near the interface in the metal species. Alternatively, the oxide may adhere to the metal, providing a ceramic surface for the bonding glass. A transition structure may be provided by a mixed oxide and metallic material applied to the ceramic, such as in the sintered powder process, which provides a ceramic-like interface at the ceramic and a sufficiently metallic one at the metal bond [1]–[9].

In practice, complex combinations of these mechanisms may occur. For example, in porcelain enameling, a preoxidized metal surface may undergo galvanic attack, roughening and pitting the metal surface locally, changing the local chemistry of the glass, and redepositing dendritic anchors of metal at the interface. Features of physical, mechanical, and chemical contributions to bonding may all be identified, but their relative contributions may be difficult to assess. Chemical interaction can produce material properties rather different from those of either original material (see also Section 3.2.2).

2.3. Stresses in Ceramic–Metal Systems

Thermally Induced Residual Stresses. The thermal expansion of a metal is generally considerably greater than those of ceramics and glasses, as can be seen in Figure 2. An unconstrained metal may contract significantly more than the ceramic, but if the metal is constrained, the differential expansion can result in residual stresses in the ceramic and the metal. These stresses add to those resulting from applied forces, thermal gradients, phase transformations, and rapid cooling. This residual stress in itself or in combination with external stress may cause failure of the bond.

The greater shrinkage of a metal in a flat sheet causes net compressive stresses in the ceramic and tensile stresses in the metal. Such a situation is favorable because a ceramic typically shows far greater compressive than tensile strength: a residual stress can enhance properties such as fracture strength and wear resistance. However, if the stress is too great, failure may occur in the interface or possibly in either material.

Several approaches may be used to limit the thermally induced residual stresses. The ceramic and metal can be chosen to minimize the mismatch in thermal expansion. For example, molybdenum and tungsten reasonably match several ceramic and glass compositions and can be used where their poor oxidation resistance is not a limitation. For platinum and chromium, several glasses and ceramic compositions show similar expansion, but expense and useful bonding technology, respectively, can limit application. A common alloy in glass sealing, ASTM F15, is a set of iron, 28–29 wt% nickel, and 17–18 wt% cobalt alloys. These alloys show expansion behavior similar to

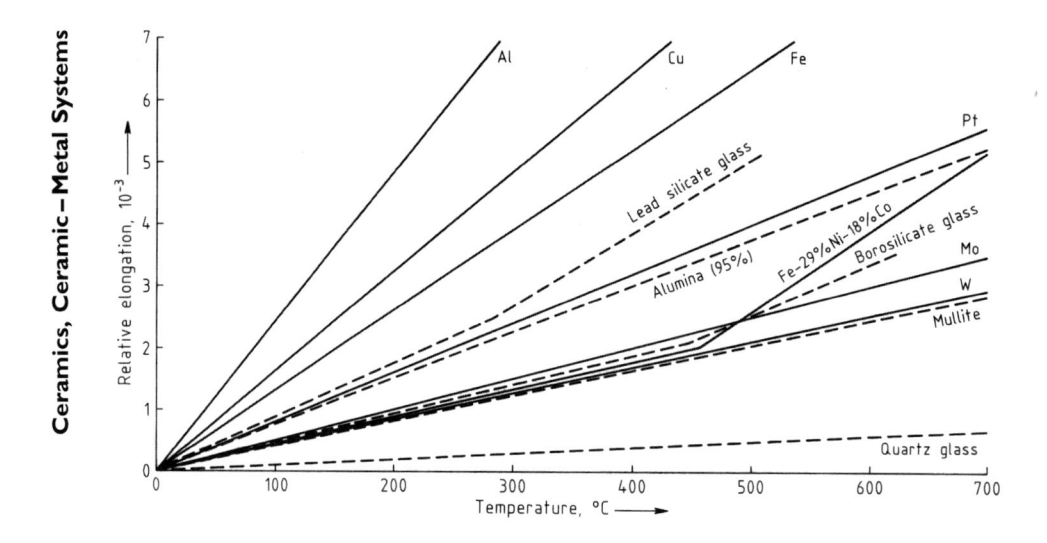

Figure 2. Thermal expansion of typical ceramics (dashed curves) and metals (solid curves)

that of borosilicate glasses, some showing an extraordinary match over both low and high temperature ranges. Such alloys are often referred to by the trade name *Kovar*(Carpenter Technologies). These alloys also show ductility, do not usually embrittle, and can be soldered, brazed, and welded. Some glasses and ceramics have been designed to match the expansions of particular alloys, e.g., lead silicate and other enamel compositions applied to steels.

Another approach to matched expansion is the *Housekeeper seal*, in which a very thin piece of metal is used. The slight size, low yield strength, and ductility of the metal allow it to accommodate the induced stresses [3], [8].

Continuum mechanics can predict the magnitude of residual stresses that result from thermal expansion differences [3], [8], [10]and can be used to design a system configuration or applied loads such that critical stresses are not exceeded. One approach is to design a *compression seal*so that the ceramic is placed in compression and the metal is placed in tension. A compressive seal can be used to obtain a purely mechanical bond between solid metal and ceramic during fabrication. One way is to use the differing thermal expansions to produce a "shrink fit" during cooling. A pair of concentric cylinders with the metal as the outer sleeve is an example, the greater contraction of the metal on cooling resulting in a strong, tight fit. However, different stress components can be in tension or compression for the same location in a material. Caution must be taken not to exceed the fracture stress for any tensile stress component in the ceramic or the interface.

Elastic ModulusEffect. The difference in the elastic modulus or stiffness of the ceramic and the metal components of a system or composite can result in material or

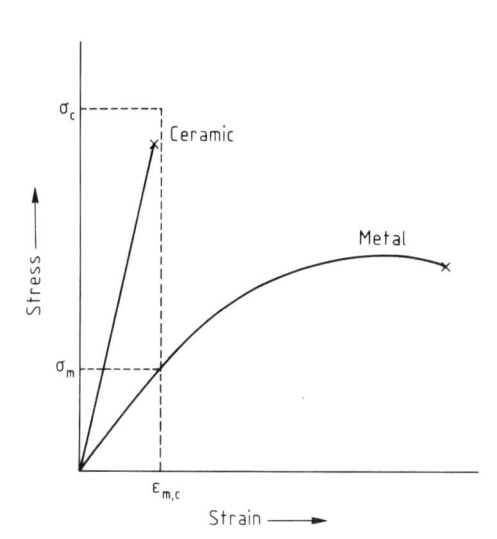

Figure 3. Difference in stresses for a bonded ceramic and metal under applied stress

interfacial failure under applied and residual stresses. This situation is shown in Figure 3. A ceramic is typically stiffer than a metal, as shown by the greater slope of the ceramic curve. For this example the ceramic has been chosen to be stronger in tension than the metal. If the metal is the massive member in the system and is loaded to a stress σ_m at the ceramic–metal interface, it will experience a strain $\varepsilon_{m,c}$. The deformation or strain must be compatible across the interface, thereby imposing a stress σ_c on the ceramic or interface because of the higher modulus of the material. Even when the ceramic is much stronger than the metal, the failure stress may be exceeded, as shown by the x, which indicates failure stress.

This effect can lead to failure of a ceramic–metal joint or coating. For composite material, the situation may be more complex as sharing of load may be desired for properties such as elastic modulus. However, when a failure crack propagates, failure at the ceramic–metal interface may be desirable so that the crack is diverted and fracture energy is expended in the process. Such considerations may be complex, involving the properties of ceramic and metal, chemical interaction zones between materials, and microstructural and geometrical factors. Predictive theoretical treatments for such materials are being developed.

3. Glass-to-Metal Joining

The bonding of glass and metal can be divided into two general categories, glass-to-metal seals and enamels. These two areas are distinguished by use, glass-to-metal seals being used primarily in electronic applications such as hermetic sealing of lamps and devices, isolation of electrical feed throughs, packaging, and vacuum-tube technology.

Enamels are used principally as corrosion protection and decorative coatings on such products as domestic appliances, bathroom fixtures, and cookware.

Enamels and glass-to-metal seals are also distinguished by the base metal. Most enamels are applied to low-carbon steels and cast iron, while glass-to-metal seals are commonly used with nonferrous metals, stainless steels, and the iron–nickel–cobalt Kovar alloys.

A third distinguishing feature is how the carbon and hydrogen are treated. These species may be either in the base metal or on the surface as hydrocarbons. These elements combine with oxygen at elevated temperature and can form gaseous bubbles in the molten glass. In the case of glass-to-metal sealing, the base metal is cleaned of hydrocarbons and then decarburized or the hydrogen baked out. In enameling, higher levels of carbon and hydrogen are permissible, and in many cases these elements are not removed. The formation of bubbles in the enamel is considered undesirable if it weakens the bond with the base metal, results in a large bubble that exposes base metal, or causes aesthetic problems in surface roughness or color. Instead of preventing bubbles from forming, a controlled fine-sized bubble structure is introduced to act as a getter for evolving gases, thus avoiding large, objectionable bubbles in the glass.

3.1. Glass-to-Metal Seals

A good glass-to-metal seal requires a glass and metal with thermal expansions that match adequately over the temperature range of fabrication and application. The base metal is generally cleaned and then decarburized and/or the hydrogen baked out. Other products, such as nitride in the case of titanium alloys, may need to be removed. The metal generally is oxidized to provide a tightly adherent layer of oxide that remains adherent during forming of the glass-to-metal join. The oxide typically is wet by the glass and dissolves in the glass to yield a structure with sufficient chemical bonding [5]. Glass-to-metal sealing has been reviewed [3], [8], [11].

Metals commonly used for glass-to-metal joining include tungsten, molybdenum, titanium, tantalum, nickel–iron alloys, iron–chromium alloys, iron–nickel–cobalt alloys, platinum, copper, gold, and copper-clad nickel or iron–nickel Dumet alloys. A large range of borosilicate and lead silicate glasses are commonly used in glass-to-metal seals; borosilicate glasses are most commonly used. Where copper and copper-clad alloys or iron–chromium alloys such as stainless steels are used, lead silicate glasses are more common.

The glass is usually applied as a frit, bead, or paste. A solid fabricated glass shape may be heated and softened to the required viscosity by electrical, flame, or radio-frequency heating, and then pressed into contact with the preoxidized metal to bond. Frequently a protective or special forming atmosphere is required to promote wetting and bonding. Careful cooling or annealing avoids residual stress because of excessively rapid cooling.

A typical sequence of operations is that of the specialty Kovar alloyASTM F15 (1983) (Fe – 29% Ni – 17% Co) glass-to-metal sealed with borosilicate glass. This system is common because of the excellent thermal expansion match between materials, good wetting response, and strong bond. It is used for microelectronics packages, electrical and high-frequency feed throughs, laser cavity-to-window bonding, and other diverse applications.

The Kovar is first degreased, chemically cleaned, and decarburized in a hydrogen, hydrogen/nitrogen, or dissociated ammonia atmosphere with various water levels at 900 – 1100 °C. The temperature should be higher than final glass-to-metal sealing temperature. A typical cycle is 1100 °C for 10 – 15 min in a hydrogen/nitrogen/water vapor atmosphere, typically 1 vol% water and as little as 10 vol% hydrogen.

The next step is to preoxidize the metal in a controlled fashion. While air and natural gas can be used, the variability of components such as water in these gases makes process control difficult. Pure oxygen gives a thick, flaky coating that does not penetrate metal grain boundaries. An atmosphere of nitrogen/water/hydrogen at 800 – 1050 °C is used: an atmosphere of nitrogen/1 vol% water/0.4 vol% hydrogen at 1000 °C for 10 min produces good oxide structure with optimum thickness and grain boundary penetration.

The final step is the glass sealing. The glass preform or frit is assembled or applied. Firing is performed in a nitrogen-based atmosphere, such as that used for controlled preoxidation, or an exothermic (natural) gas atmosphere is employed. Sealing is done at ca. 1000 °C, the precise time and temperature depending on part geometry, glass viscosity, and wetting response [12].

3.2. Enamels

Enamel is a fused, vitreous, superficial coating on metals or a decorative pattern on a glass surface. Enamels are similar in properties to ceramic glazes.

History. The origin of enameling metals goes back to the Egyptians, centuries before Christ. Enamels were used to provide elaborate designs in jewelry. The materials enameled were typically gold and silver. Later more metals were enameled; bronze and copper jewelry inserts were coated with enamels. By variations in the composition of the enamel the glassy coating could be made to hide defects in the metal.

The application of enamels to iron and steel made enamels available for uses other than jewelry. Enamels were sought to provide decorative coatings on a wide variety of cast-iron shapes. In the earliest operations, cast iron was heated to redness, the enamel was dusted upon the metal as a fine dry powder, and as the powdered glass stuck to the hot metal it softened and adhered to the surface. This operation required a number of successive coating steps to obtain a layer thick enough to provide the desired effect. Many problems were related to the firing: enamels blistered, turned black, lacked the desired color, and exhibited poor adhesion.

Advances corrected or improved these problems. Adherence was related to the surface properties of the metal. Surface pretreatments, such as sand blasting, cleaning, and pickling, evolved. New, cheaper sources of raw materials were found, e.g., borax, allowing lower firing temperatures and producing cheaper enamels. The application of enamels as a suspension, rather than dry powder, was a great improvement. Greater control of the chemistry was found to provide newer, more stable enamels.

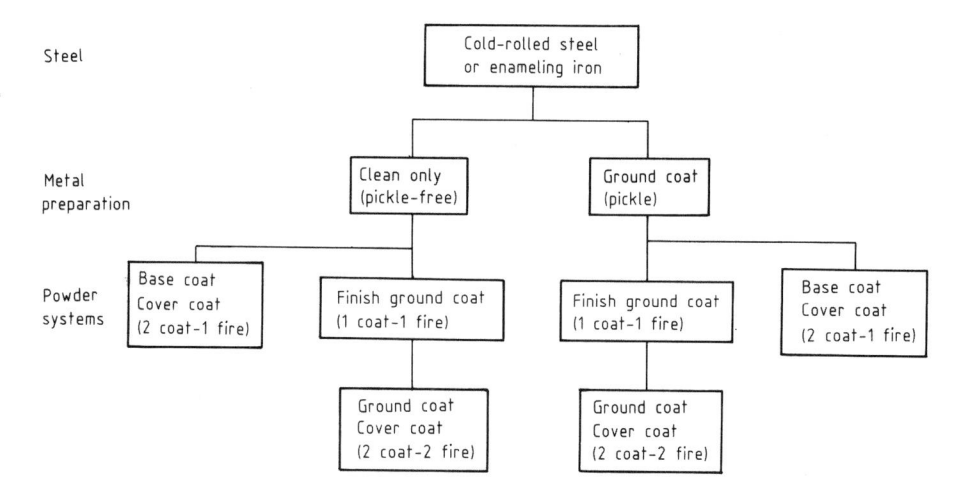

Figure 4. Flow diagram for the enameling of cold-rolled steel or enameling iron

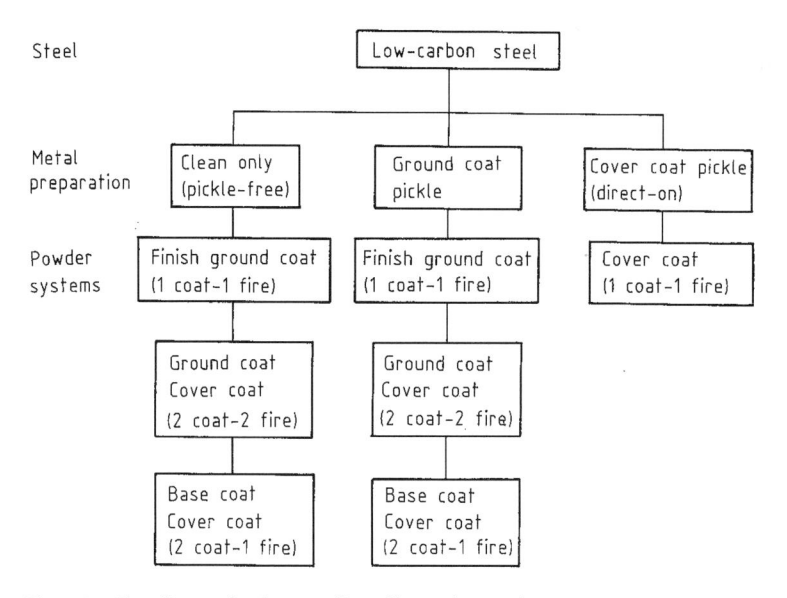

Figure 5. Flow diagram for the enameling of low-carbon steel

3.2.1. The Enameling Process

The enameling process can be simplified into three basic steps: (1) metal preparation, (2) ground-coat application, and (3) cover-coat application. Figures 4 and 5 provide process flow diagrams for three commonly enameled metals — cold-rolled steel, low-carbon steel, and cast (enameling) iron [13].

Metal Preparation. The metal surfaces must be free from grease, machining oils, dirt, oxidation products, and rust. A roughened or suitably etched surface is desirable for most metals. The exact metal preparation depends on the composition of the base metal and the type of enamel to be applied.

No pretreatment is required for gold or silver, provided the metal surfaces are clean. For copper and bronze the cleaning step is often followed with a 2–4% nitric acid pickle and a clean water rinse. Aluminum and its alloys require special surface pretreatment, including cleaning, chromatizing, and preoxidation.

Stainless steel and other metal substrates for high-temperature end use are usually prepared by sand or grit blasting. Otherwise, the metals are processed typically. The pickling acids may be quite reactive; i.e., HF is typically incorporated into the pickle. Cast iron and heavy-gauge sheet steel are prepared by sand or grit blasting and then typical chemical processing.

Metal preparation can be described as the following processes:

1) Annealing. The metal to be enameled is heated to a temperature sufficient to remove organic surface impurities. Such impurities can act as a source of gases that become entrapped in the enamel during the firing.
2) Grit or sand blasting. This acts as part of the cleaning process where heavier rust, scale, and dirt are removed. More importantly, sand blasting creates a rough surface for the enamel to anchor mechanically.
3) Solvent cleaning. Cleaning removes mill oils and the soaps, oils, and rust preservatives added during forming.
4) Acid pickling. Pickling acts to remove metal oxidation products and scale. It also acts to create a rougher surface and increase enamel adherence.
5) Nickel flash. The nickel deposit follows the acid pickle and subsequent acid rinse. Its function is to promote adherence between ground coat and enamel and reduce the occurrence of defects in the ground coat resulting from noncontrolled metal oxidation.
6) Neutralization. This functions to increase the pH of the metal surface.

The type of surface pretreatment is governed not only by the metal, but also by the processing. The three types of process systems are batchtype immersion, continuous or automatic immersion, and continuous spray. For example, the steps used for sheet steel include the following:

1) Physical and chemical cleaning. Heavily oxidized steel is sand blasted prior to chemical cleaning. Alkaline cleaners, typically baths containing caustic soda at 40–60 °C, are used to remove organic soils.
2) Acid pickling. The removal of rust and scale from the metal surface by pickling is carried out in sulfuric acid (6–9 wt% at 65–75 °C) or in hydrochloric (muriatic) acid (10–12 wt% acid at room temperature). Phosphoric acid or an acid salt can also be used. Sulfuric acid is most commonly used because of its speed and lower cost. The sulfuric acid should be pure, and inhibitors must be avoided since they interfere with metal removal and etching, resulting in faulty adherence. After

pickling, the metal should be rinsed in cold running water, usually acidified to pH 3–3.5 to prevent hydrolysis of iron salts.

3) Nickel deposition. Time of treatment, typically 5– 15 min, should be sufficient to provide the desired coating of nickel. The aqueous nickel solutions are maintained at concentrations of 0.25–3 wt% nickel sulfate at 70–80 °C. The pH should be between 2.6 and 3.2. The temperature, pH, and nickel concentration are closely monitored. For conventional enameling, the nickel deposited should range from 0.27 to 1.6 g/m^2.

4) Neutralization. The final step is neutralizing acid residuals on the metal surfaces, and also providing rust protection and minimizing the presence of undesirable surface chemicals. Typically, either alkaline soda ash or borax solutions equivalent to 0.02–0.3 wt% Na_2O are used.

Ground Coats. Porcelain enamel ground coats for sheet steel are basically alkali-metal aluminoborosilicate glasses containing adherence-promoting oxides of such metals as cobalt, nickel, and copper. These coatings serve three purposes. The first is to provide a satisfactory bond or adherence to the base metal by complex reactions with the steel during firing. The second is to provide a protective layer or coating that minimizes surface defects caused by the metal or metal preparation, smoothing the surface for the cover coat. The last is to provide a resistant coating, in some cases decorative, thermally resistant, or chemically resistant. While increased adherence is the primary function of the ground coat, the other functions should not be underemphasized. For some steels satisfactory adherence may be obtained without the use of ground coats; however, special surface treatments must be employed if a ground coat is not used.

Ground coats are most typically composed of *ground-glass powders* or *frits.* In most instances the base ground coat may contain a number of frits, all having specific properties that make them useful. For example, one frit may provide sag resistance to the molten glass when the ground coat is heated. A second frit may fuse at a lower temperature, thus sealing the metal surface and promoting glass–metal bonding at an earlier stage. A third may be used as a decorative or colored base to hide defects prior to the final coat.

Ground coats are applied in the form of a slurry to the treated metal by spraying or dipping, and are then allowed to dry. Typical firing temperatures range from 780 to 870 °C. Table 1 shows typical chemical oxide compositions of ground coats for steel.

Cover Coats. Enamel cover coats are designed to provide specific color and appearance characteristics coupled with resistance to atmospheric and liquid corrosion, abrasion resistance, and thermal shock resistance, to name a few properties. Cover coats are applied either as a suspension, followed by drying and firing, or by spreading a dry powder over the hot metal. Cover coats range in thickness from 0.08 to 0.64 mm, depending on application. The firing temperatures for the cover coat are lower than those for the ground coat to prevent excessive flow in the ground coat.

The corrosion properties of the enameled metal are most often those of the cover coat. For instance, titania-based enamels are used for acid resistance, while zirconia-based cover coats are used for alkali resistance. Table 2 shows typical steel cover coats.

Table 1. Chemical oxide composition of ground coats for steel, parts by weight

Component	I	II	III	IV	V
KNa_2O*	17.8	18.5	21.0	18.0	19.4
B_2O_3	16.0	15.2	11.8	14.1	13.3
Al_2O_3	7.7	8.8	8.7	5.4	8.0
SiO_2	51.1	52.1	44.4	44.1	56.6
CaF_2	5.5	3.8	8.7	6.8	
BaO			2.0		
CoO	0.5	0.4	0.4		
Mn_2O_3	0.9	1.2	0.9		
NiO	0.5		0.2		
CuO			0.7		
CaO			1.2		
Sb_2O_3				4.8	1.4
ZrO_2				3.7	6.2

Ground coats I, II, and III are typical blue ground coats, whereas coats IV and V are white ground coats for use with white enamel cover coats.

* KNa_2O appears often in the enameling literature, indicating that the proportions of K_2O and Na_2O in the oxide raw materials do vary. The formulation K,Na_2O also appears in the literature.

Table 2. Chemical oxide composition of cover coats used for cast iron, parts by weight

Component	I	II	III	IV	V
KNa_2O	37.1	21.8	32.0	37.6	32.4
B_2O_3	11.8	13.4	9.1	11.8	14.9
PbO	10.4	13.4	16.0		
Al_2O_3	6.4	4.6	5.4	6.5	5.3
SiO_2	39.0	2.9	24.0	44.2	14.9
CaF_2	6.2	5.6	7.0	6.2	4.9
CaO	3.9	6.7	2.3	4.0	10.5
ZnO		7.5	7.9	5.0	9.2
Sb_2O_3	3.7		4.2	3.7	
ZrO_2		4.7			5.9
BaO		12.0	8.0		
MgO					5.4

3.2.2. Theory of Adhesion

The theory of adherence can be divided into two main branches: the mechanical theory and the chemical or oxide theory [14]–[16]. Each is characterized by numerous hypotheses that have evolved to explain adhesion. In addition, there is electrolytic theory and dendritic theory. However, no one theory can adequately describe adhesion for all systems.

Mechanical Theory. The mechanical theorystates that adherence is due to mechanical gripping of the roughened base metal. The enamel glass is held to the metal by enclosing projections and by filling the depressions occurring on the metal surface.

The application of porcelain enamels to cast iron was the first commercial application of ceramic coatings. Although the surface of the cast iron was inherently rough, further roughening, usually by sand or grit blasting, improved the adherence. This practice suggested the mechanical theory of adherence. Other metals could be coated with ceramics if their surface was roughened. Treatment included combinations of sandblasting, scouring, electrolytic corosion, pickling, etc. This extension to other metals increased acceptance of the mechanical theory.

To better understand the relationship between the roughness of the ceramic–metal interface and adherence, a model contrasting two measures of roughness was proposed:

1) undercuts/distance in centimeters (or anchor points/centimeter)
2) interface ratio

Both are measuredon microscopic polished sections of the ceramic–metal interface. Figure 6, a schematic section of the enamel–metal interface, shows the two measures of roughness. The undercuts or anchor points (each marked by an x) are counted and expressed as the number per centimeter, in this case, number of x's/AA'. The interface ratio is calculated by measuring the length of the interface line with a map measure and dividing it by the length of the line parallel to the interface, in this case BB'/AA'. These two measures are then correlated to adherences. Adherence is determined by the standard test of deforming the metal and measuring the amount of metal exposed.

If adherence is due to the gripping action of the rough ceramic–metal interface (mechanical theory), then the best correlation should be between adherence and anchor points per centimeter. If adherence is due to chemical bonding (chemical theory), the interface ratio should give a better correlation between adherence and roughness. For enameled cast iron:

1) Positive correlation is found between the adherence of a porcelain enamel ground coat and the roughness of the interface.
2) In general, adherence correlates better with anchor points/centimeter than with the interface ratio.
3) The method of metal preparation has a marked effect on the relationship between roughness and adherence, pickled iron generally producing better adherence than sandblasted iron for the same degree of roughness.
4) Most of the roughness associated with good adherence develops during firing.
5) Roughness of interface is a necessary but not a sufficient condition for the development of good adherence.
6) Factors other than roughness of interface influence adherence.

These conclusions tend to confirm the mechanical theory of adherence. However, some adherence phenomena cannot be explained by the mechanical theory:

1) Excellent adherence can be obtained on extremely smooth surfaces.

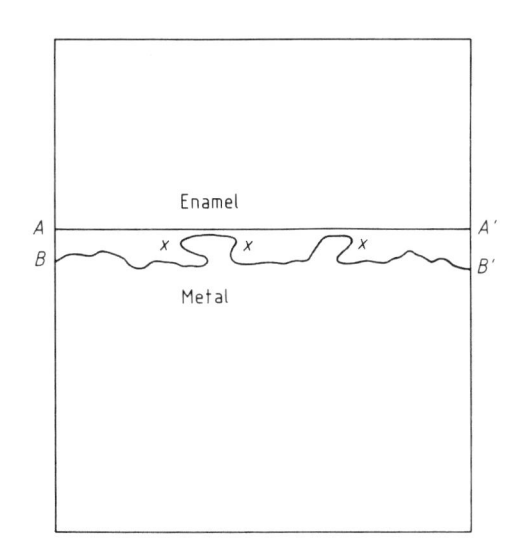

Figure 6. Schematic section of the enamel–metal interface comparing measures of roughness

2) A roughened surface exhibiting many anchor points does not ensure adherence significantly better than chemical adherence alone.
3) The measured values for the strength of the bond necessitate assignment of extremely high values for the strength of iron because the total cross section of the anchor points is always a relatively small portion of the total interface area.

Chemical or Oxide Layer Theory. Chemical reactions include dissolution, redox, interdiffusion, and precipitation, which change the nature of the interface and may result in a transition zone between the metal and the ceramic. The bond in metals and ceramics is fundamentally different, and this may prevent a strong physical bond between bare metal and ceramic. A theory developed for glass-to-metal bonds and verified experimentally may be applied more generally to all ceramic-to-metal bonds. The theory states that chemical bonding occurs across an interface when chemical equilibrium relative to the lowest valence oxide of the metal phase is present [5]. The glass and the metal are saturated in the metal oxide at the interface, and therefore the activity of the lowest valence metal oxide is one. Then there is an idealized molecular monolayer of the metal oxide at the interface, which is part of and compatible with both metal and glass. Bond energies should then balance, and the electronic structure should be continuous across the interface. If oxide activity is less than one, complete continuity is not realized. Reaction of metal and glass with chemical interdiffusion is driven by the lack of chemical equilibrium relative to the lowest valence metal oxide. A chemical bond occurs when equilibrium in terms of chemical potentials or activities is reached at the interface. In glass-to-metal bonding this occurs when the glass near the interface is saturated in the metallic species.

Electrolytic Theory. According to the electrolytic theory, metal oxides dissolved in enamels are reduced during firing to the metallic state by the base metal. The adherence-promoting metal in the enamel is believed to form shorted galvanic cells that strongly corrode the base metal. This concept is illustrated in Figure 7. The current flows from the iron through the melt to the cobalt and back to the iron. These local cells are not exhausted during firing because anodic iron and diffusing oxygen are abundant. Iron goes continuously into solution, the surface becomes roughened, and the glass anchors itself into the depressions produced. Final adhesion is basically mechanical. The electrolytic theory has two main faults:

1) Some metals that exhibit a greater potential difference with iron than does cobalt do not cause greater corrosion.
2) The theory is still based on mechanical theory, which was previously discussed and partially discredited.

Dendritic Theory. Dendrites are the form in which the metal normally crystallizes from the melt, aqueous solutions, or any nonmetallic liquid in which compounds of the metal dissolve. They are branched crystals that occur periodically, giving fernlike, treelike, or toothlike appearance.

The dendrite theory claims that the α-iron dendritic precipitates that occur in the boundary layer between enamel and base metal are responsible for adherence. The dendrites grow from the base metal into the enamel, and after cooling they tend to mechanically attach the enamel to the base metal. This differs from the mechanism of the electrolytic theory, in which the metals and dendrites in the interfacial zone are Co, Ni, or other adherence-promoting metals.

Although a possible adherence mechanism, dendritic theory correlates poorly with adhesion.

4. Ceramic-to-Metal Joining

A ceramic component and a metal component may be joined in various ways, including fasteners, adhesives, and direct joining of ceramic and metal. For durable, integral, hermetic joins capable of extended service, particularly at extreme temperatures, the direct join is used. The successful join usually requires a preparation step to achieve a transition structure, good wetting, and chemical bonding between ceramic and metallic materials. The following are common joining processes:

sintered metal powder (moly–manganese paste)
metal powder–glass frit
active metal
gas–metal eutectic
pressed diffusion joins

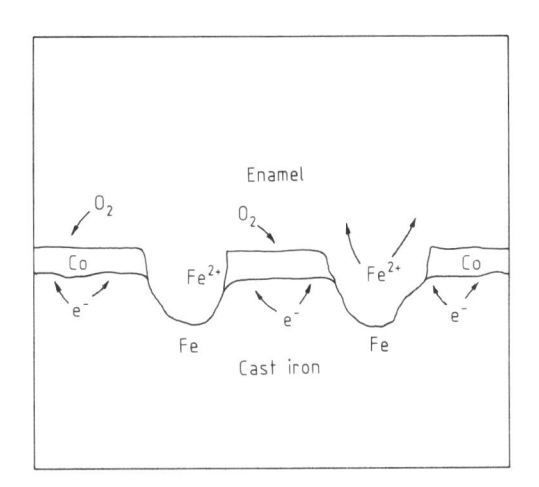

Figure 7. Electrolytic cells between molten enamel and cast iron

vapor-phase metallizing
liquid-phase metallizing
electroforming
graded powder
nonmetallic fusion

These processes are listed in approximate order of frequency of use.

The various joining techniques are often categorized as liquid phase, vapor phase, and solid phase. Ceramic-to-metal joining is discussed comprehensively in [3], [11], [17]–[19].

4.1. Sintered Metal Powder Process (SMPP)

The sintered metal powder process, also referred to as SMPP or moly–manganese paste process, uses a paste composed of molybdenum and manganese metals and their oxides suspended in an organic vehicle. Compositions containing tungsten, tantalum, iron, rhenium, and titanium with their oxides can also be used. Thus, moly–manganese pastes may contain neither molybdenum nor manganese.

A slurry of a composition such as

60 g of molybdenum powder (0.2 µm)
15 g of manganese powder (0.2 µm)
0.2 g of nitrocellulose
30 mL of butyl acetate

is applied by brushing, spraying, roller coating, transfer tape, silk screening, or dipping to a thickness of ≈ 20 µm. For single-phase ceramics such as high alumina classes without a glassy silicate phase, the usual pastes do not bond well. Compositions containing approximately equal amounts of the metal powders and 13 Al_2O_3 – 52 MnO – 35 SiO_2 or 41 Al_2O_3 – 54 CaO – 5 MgO, for example, are used. (The numbers are the weight percents.) After application, the paste is dried in air, often under heat lamps. It is then fired in a hydrogen atmosphere (dew point ≈ 25 °C) at 1200 – 1600 °C for ca. 0.5 h. Manganese or titanium additions lower the firing temperature by 100 – 200 °C.

The process gives rise to a transition coating with a glassy or crystalline ceramic phase that has migrated in the metal – metal oxide powder or that comes from the ceramic added to the paste in the case of single-phase ceramics [1], [2], [7], [20]. The structure of the liquid-phase sintered product consists of a glassy phase with metallic and oxide particles. This structure provides the necessary bridging phases between ceramic and metal. The resulting coating can be directly brazed to metal, but more commonly it is electroplated or electroless plated with a 5 – 8 µm thick nickel layer. Alternatively, copper can be plated, or either metal can be applied as a powder and fired to bond by diffusion in a controlled atmosphere. The resulting nickel surface is suitable for conventional brazing materials and methods.

While SMPP involves several steps, the procedures are relatively straightforward, and the results are reliable, SMPP producing some of the strongest ceramic – metal joins. It is the most common method for producing ceramic-to-metal joins and metallization in industry.

4.2. Metal Powder – Glass Frit Method

For the metallization of hybrid microelectronic circuitry, a finely divided noble metal or alloy (platinum, palladium, gold, silver) mixed in a glass frit and suspended in an organic binder is commonly used. The required pattern is usually screen-printed on an electronic substrate material such as alumina or on a multilayer capacitor material such as barium titanate. These systems permit air firing of the material. The structure of the bridging glassy phase providing a bond between the ceramic and a metal-rich free surface is akin to that produced by SMPP. The metal powder – glass frit method may be regarded as a liquid-phase process.

The incorporation of active metal components such as titanium, zirconium, nickel, or copper can lead to formation of a mixed oxide phase by reaction with the ceramic. These phases, like devitrifying phases, aid in bonding and may be fired in a reducing atmosphere [21].

Copper and nickel have been used as a less expensive alternative to the noble metals. The glass frit is composed of oxides that are stable at low oxygen partial pressures (Al_2O_3, B_2O_3, CaO, MgO, SiO_2, ZnO). Firing is done in low nitrogen with a low partial pressure of oxygen (10 ppm) to achieve slight, but controlled, oxidation of the metal so that it can be wet by the glass. The glass must show sufficiently low viscosity for good flow at firing temperatures [21], [22].

4.3. Active Metal Process

The active metal techniques employ an active constituent such as titanium, zirconium, or hafnium in the braze metal to promote wetting and bonding to the ceramic. The active metal may be alloyed with a metal such as nickel or copper to reduce its reactivity and provide a brazing temperature as low as 800–1200 °C. The braze materials are used in the form of powders, shims, or wires. Brazing is typically performed under vacuum or in an inert atmosphere with careful temperature control. The hydride of the active metal may be applied separately as a layer on the ceramic and fired, a process resembling SMPP [23].

In the liquid braze the active metal is transported to the ceramic surface where it can react [24]. Wetting is enhanced to the point that caution must be taken to prevent braze metal from flowing into unwanted areas. The joints formed by this method cannot withstand elevated temperatures, even in inert gas atmospheres, because of continued reaction of the bonding phase with the ceramic.

4.4. Gas–Metal Eutectic Process

In the gas–metal eutectic technique, a copper sheet is bonded directly to a ceramic substrate by liquid-phase reaction. A carefully controlled oxygen atmosphere and temperature are required. A eutectic copper–copper oxide melt is formed to wet the ceramic. A reaction layer containing $AlCuO_2$ can form and aids in bonding. Strong bonds have been reported for alumina, beryllia, silica, and spinels [25], [26].

4.5. Pressed Diffusion Joins

Joining by solid-phase diffusion and reaction can also be used to create a ceramic-to-metal bond. Such bonds are typically termed pressure diffusion, crunch, or ram joins. A low-yieldstress metal such as copper is pressed to the ceramic, or the ceramic is forced into a cavity in the metal, so that there is good contact between metal and ceramic and there is an applied or residual force at their interface. The parts should be as clean as possible before contact. The parts are held, often with great external force applied to the interface, at as high a temperature as possible above one-half the melting point of the metal in kelvin. A variety of metals and ceramics can be bonded by the resulting diffusion and chemical reaction [27], [28].

4.6. Vapor-Phase Metallizing

A thin film of metal, which may be base, noble, or reactive metal, is applied by various vapor-phase deposition methods, including ion implantation, evaporation, sputtering, reactive sputtering, plasma methods, and chemical vapor deposition (CVD). Often the initial bonding layers are reactive metals such as Ti, Zr, and Cr to promote bonding, followed by base or noble metals such as Cu, Ni, Au, and Pt. Because platinum shows good ductility and its expansion matches that of alumina, it is often directly deposited. Final layers are usually oxidation-resistant metals. Layers up to several micrometers thick can be built up in this manner, although usually the metallization is quite thin.

These layers may be built up by more conventional solder or braze methods for bonding with a metal part. Brazing is usually done in hydrogen, under vacuum, or in cracked ammonia. Strong hermetic seals can be made on various polyphase, single-phase, and single-crystal ceramics. Furthermore, glasses and other ceramics not tolerant of brazing can be soldered at lower temperature.

4.7. Liquid-Phase Metallizing

An alternative metallizing method involves thermal reduction of a salt solution of a precious metal (platinum, gold, silver, or palladium). A thin, conductive coating is obtained and may be electroplated or soldered directly. Molten salts have been used in conjunction with reactive metals (Ti, Zr, Hf, or U) to metallize ceramics [3].

4.8. Electroforming

Electroforming can be used to avoid high-temperature brazing after a ceramic part has been metallized by another method. The metallized ceramic is placed in contact with a metal such as copper under slight pressure. The assembly is then placed in an electroplating bath for several days, and the electrodeposition of metal in and around the gap between metal and metallized ceramic creates the bond. Large, complex parts have been joined in this way. This technique has been coupled with pressure diffusion joining to produce an enhanced bond, but in such a case high temperatures are not avoided.

4.9. Graded Powder Process

The gradedpowder process produces a join by first pressing a powder compact of metal and metal oxide either to the base ceramic or between ceramic and metal. Often this process is similar to SMPP or metal powder–glass frit methods. A gradual change in the powder composition from ceramic-rich to metal-rich is accomplished by producing the preform in layers. This changing composition provided by mechanically mixing powders provides a ceramic-rich layer and a metal-rich layer for the respective materials to be joined, with graded composition and properties in between. The powder is sintered at elevated temperature similar to SMPP. Bond strengths are not especially greater than those of other methods, and the preform is expensive. Only limited commercialization has taken place.

4.10. Nonmetallic Fusion Process

Nonmetallic fusion bonds employ a glass frit to produce a bonding glass between ceramic and metal components. This is a liquid-phase method.

Ceramic glazes and glass frits can be used to join ceramics to each other or to metals if used in layers sufficiently thin to be incorporated into the ceramic and metal on cooling. The incorporation of the glass in the structure may allow use at significantly higher temperatures than the fabrication temperature or softening temperature of the glass suggests.

An oxide composed of constituents such as MnO_2, SiO_2, and Al_2O_3 may be applied to the join interface as a suspension or preform. The assembly is heated in an inert gas atmosphere for up to 15 min at 1200–1500 °C and then cooled relatively rapidly, 100 °C/min, so that the glass does not devitrify. The composition of the frit can be adjusted to match the expansion, wetting, and bonding characteristics of ceramic and metal. The bond may be heated to temperatures substantially higher, ca. 800 °C, than SMPP joins. Success has be obtained with high-purity alumina and sapphire. The high temperatures and cooling rates can lead to defects or interfacial failure.

5. Thin Filmsand Coatings

A coating may be defined as a near-surface region of a material with properties different from those of the bulk. The fabrication of coatings varies widely, depending on application, substrate, coating materials, and economic considerations. Coating techniques may be categorized as follows: (1) those processes that produce a layer of material in the surface region of the bulk material that has properties differing from those of that bulk material and (2) those processes that modify the immediate surface of the bulk material, causing that region to have different properties [29].

Ceramic–metal coatings can be divided into two categories: (1) coatings containing both ceramic and metal phases and (2) ceramic films and coatings on metals. Ceramic–metal coatings are a unique type of coating in that the properties of the coating are strongly controlled by the microstructure of the metallic phase. These composite films exhibit properties unavailable from either the ceramic or metal phases alone. (See Section 6.1.)

Ceramic films and coatings are a special class of ceramics in the sense of their special form and the special preparation techniques required. However, their uses are diverse and exploit the wide range of unique and desirable properties that various bulk ceramics possess. In this sense, ceramic films and coatings should be viewed as a large family of materials with quite diverse compositions and properties [30].

5.1. Uses

Table 3 lists uses of ceramic films and coatings and gives typical examples. The table is illustrative rather than exhaustive, but does show the wide range of ceramic films and coatings, both in terms of uses and in terms of compositions.

Resistance to wear is generally correlated with hardness. Of special interest is the use of ceramics for wear resistance at elevated temperatures. Table 4 gives the microhardness and oxidation temperatures for some hard coat materials. Although some hard-coating materials have good wear resistance, they generally are not low-friction materials. When low friction is required in addition to wear resistance, softer materials may be used. Calcium fluoride and composites including calcium fluoride are among the most promising solid lubricants for service at ca. 650 °C [31].

Corrosion and oxidation resistance at high temperatures constitute a complex field of chemical reactions and protective mechanisms provided by coatings. Often the protective mechanism involves the production of a dense, adherent, diffusion-resistant oxide film by in situ oxidation, e.g., chromia and alumina on high-temperature superalloys and silica on silicon carbide. This requires that the base alloy or ceramic has a composition that is self-passivating. Much work today is directed toward developing pinhole-free coatings with sufficient adherence to use as corrosion protection mechanisms. Another corrosion-related problem involves water attack, especially the stress-assisted water attack involved in slow crack growth in many ceramics. Silicon nitride films are promising barriers for protection of glass, including glass fibers for optical communication [32].

Thermal protection coatings, consisting of a thin layer of ceramic applied over a metallic bond coat, are used in gas turbine components, such as combustion chambers, and are under development for additional uses. Bond coat oxidation and stress development appear to be limiting aspects of present coatings [33].

Table 3. Uses of ceramic films and coatings

Use	Typical ceramic materials
Wear reduction	Al_2O_3, B_4C, Cr_3C_2, CrB_2, $CrSi_2$, Cr_3Si_2, DLC*, Mo_2C, $MoSi_2$, SiC, TiB_2, TiC, TiN, WC
Friction reduction	MoS_2, BN, BaF_2-CaF_2
Corrosion reduction	Cr_2O_3, Al_2O_3, Si_3N_4, SiO_2
Thermal protection	Ca_2Si_4, $MgAl_2O_4$, MgO, ZrO_2 (stabilized with Mg or Ca)
Electrical conductivity	$In_2O_3-SnO_2$
Semiconductors	GaAs, Si
Electrical insulation	SiO_2
Ferroelectricity	$Bi_4Ti_3O_{12}$
Electromechanical	AlN
Selective optical transmission and reflectivity	BaF_2-ZnS, CeO_2, CdS, $CuO-Cu_2O$, Ge-ZnS, SnO_2
Optical wave guides	SiO_2
Optical processing (electrooptic, etc.)	GaAs, InSb
Sensors	SiO_2, SnO_2, ZrO_2

* DLC = diamond-like carbon.

Table 4. Bulk properties of some hard coat materials

Material	Microhardness, kg/mm²	Oxidation temperature,°C *
Al_2O_3	2350	
B_4C	4200	1090
Cr_3C_2	2650	1370
Cr_2O_3	1800	
SiC	2900	1650
Si_3N_4	2000	1400
TiC	3200	540
TiN	1950	540
WC	2050	540

* Temperature of appreciable detrimental oxidation. Passivating oxide films form at lower temperatures.

Thin films play a central role in electronic and optical devices, especially as the dimensions of these devices are reduced. Coating applications include thin-film resistors and photothermal converts for solar energy. Indeed, modern electronics would not be possible without thin-film technology [34].

5.2. Deposition Techniques

Table 5 lists coating techniques that have been used for ceramics. Table 5 includes four subcategories of processes: (1) atomic deposition, (2) particulate deposition, (3) bulk coating, and (4) surface modified coatings. Ceramic coating processes could also

Table 5. Categorization of ceramic coating techniques

Atomic deposition	Particulate deposition	Bulk coatings	Surface modification
chemical vapor environment	fusion coatings	diffusion	chemical (liquid)
chemical vapor deposition	sol–gel	diffusion bonding	oxidation
reduction	thermal spraying	hot isostatic pressing	chemical (vapor)
decomposition	plasma spraying	wetting processes	thermal
plasma enhanced	low-pressure plasma spraying	dipping	plasma
plasma environment	laser-assisted plasma spraying	enameling	ion implantation
sputter deposition	electric arc spraying	spraying	sputtering
diode			
triode			
reactive			
evaporation			
direct			
activated reactive			
vacuum environment			
vacuum evaporation			
ion implantation			

be divided into those that directly apply the ceramic compound or a precursor and those that react metallic species with gases to form the ceramic coating compound [35].

With the exception of CVD-coated cutting tools, traditional ceramic coatings were limited to those materials that could be applied to a surface as a powder and subsequently fused, such as glazes (→ Ceramics, General Survey, Chapter 5) and enamels (see Section 3.2), which were either dipped or sprayed. Recent studies have shown that nontraditional ceramic materials, such as nitrides, carbides, and oxides, could be applied to a variety of materials ranging from polymers to metals to other ceramics. The techniques that are most used include chemical vapor deposition, evaporation, sputtering, plasma spraying, sol–gel processing, and ion implantation [36].

5.2.1. Chemical Vapor Deposition (CVD)

Chemical vapor deposition is the reaction to form the desired compound and the condensation of the compound or compounds from the gas phase onto a substrate where reaction occurs to produce a solid deposit. The compound bearing the deposit material, if not already a gas, is vaporized by either a pressure differential or the action of a carrier gas and is transported to the substrate. Since the vapor condenses on any relatively cool surface that it contacts, all parts of the deposition system must be at least as hot as the vapor source. The reaction portion of the system is generally much hotter than the vapor source but considerably colder than the melting temperature of the deposit.

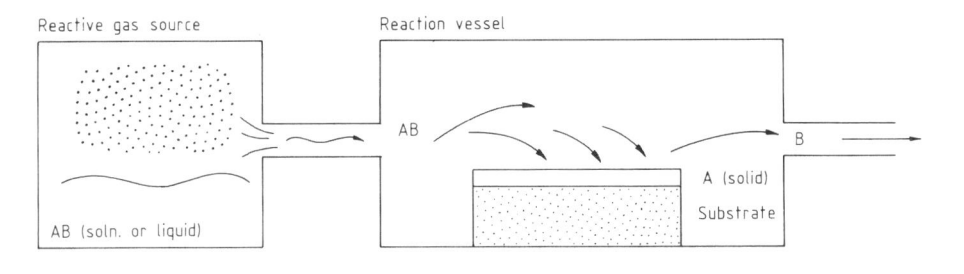

Figure 8. Schematic of CVD reactor

Generally the deposition process is either thermal decomposition (Fig. 8) or chemical reduction. Both thermal decomposition and chemical reduction most frequently involve organometallic compounds, but halides and other simple inorganic compounds can be used. CVD processes are dependent on both the thermodynamics and kinetics of the reaction. Even though the formation of a material is thermodynamically feasible, its growth rate must not be too slow. When reaction occurs on the substrate, the stages required to obtain a deposit are

1) Formation of the gaseous phase
2) Gas transfer to the substrate
3) Adsorption
4) Decomposition of the adsorbed phase
5) Compound deposition and desorption of the other decomposition products
6) Removal of these products

Table 6 lists ceramics that can be deposited by CVD. The CVD process finds its greatest application in the preparation of coatings that are not easily applied by physical vapor deposition (PVD) techniques.

In recent years it has become common to apply thin coatings of carbide and oxide ceramics to cemented carbides and cermets to increase performance. Such coatings are commonly applied by chemical vapor deposition in micrometer thicknesses and can provide abrasion resistance (Al_2O_3, TiC), oxidation resistance (Al_2O_3), friction reduction (TiN), or a thermal or diffusion barrier (Al_2O_3, TiN, TaC). Often many different coatings are used to obtain matched thermal expansions and better adherence properties. In the area of cemented carbide cutting tools, cutting speeds can be increased by three times and tool lifetimes may be two to ten times greater than those of an uncoated tool. In the past decade such coated tools have been introduced and now dominate the marketplace for industrial machining of steel [37].

The advantages of CVD are the following:

1) Various kinds of coatings possible — carbides, nitrides, and oxides
2) Crystal growth controlled by controlling the reactive gas concentrations
3) High deposition rates
4) Possibility to coat complex bulky shapes

Table 6. Chemical vapor deposition of ceramics

Materials	Reactive gas	Temperature, °C
Nitrides		
BN	$BCl_3 + NH_3$	1000–2000
	thermal decomp. $B_3N_3H_3Cl_3$	1000–2000
HfN	$HfCl_x + N_2 + H_2$	950–1300
Si_3N_4	$SiH_4 + NH_3$	950–1050
	$SiCl_4 + NH_3$	1000–1500
TaN	$TaCl_5 + N_2 + H_2$	2100–2300
TiN	$TiCl_4 + N_2 + H_2$	650–1700
VN	$VCl_2 + N_2 + H_2$	1100–1300
ZrN	$ZrCl_4 + N_2 + H_2$	2000–2500
Oxides		
Al_2O_3	$AlCl_3 + CO_2 + H_2$	800–1300
SiO_2	$SiH_4 + O_2$	300– 450
	thermal decomp. $Si(OC_2H_5)_4$	800–1000
Silicon oxide nitride	$SiH_4 + H_2 + CO_2 + NH_3$	900–1000
SnO_2	$SnCl_4 + H_2O$	
TiO_2	$TiCl_4 + O_2$ + hydrocarbon (flame)	
Silicides		
V_3Si	$SiCl_4 + VCl_4 + H_2$	
MoSi	$SiCl_2 + $ Mo (substrate)	800–1100
Borides		
AlB_2	$AlCl_3 + BCl_3$	≈1000
HfB_x	$HfCl_4 + BX_3 (X = Br, Cl)$	1900–2700
SiB_x	$SiCl_4 + BCl_3$	1000–1300
TiB_2	$TiCl_4 + BX_3 (X = Br, Cl)$	1000–1300
VB_2	$VCl_4 + BX_3 (X = Br, Cl)$	1900–2300
ZrB_2	$ZrCl_4 + BBr_3$	1700–2500
Carbides		
B_4C	$BCl_3 + CO + H_2$	1200–1800
	$B_2H_6 + CH_4$	
	thermal decomp. $(CH_3)_3B$	≈ 550
Cr_7C_3	$CrCl_2 + H_2$	≈1000
Cr_3C_2	$Cr(CO)_5 + H_2$	300– 650
HfC	$HfCl_4 + H_2 + C_6H_5CH_3$	2100–2500
	$HfCl_4 + H_2 + CH_4$	1000–1300
Mo_2C	$Mo(CO)_6$	350– 475
	Mo + hydrocarbon	1200–1800
SiC	$SiCl_4 + C_6H_5CH_3$	1500–1800
	$MeSiCl_3 + H_2$	≈1000
TiC	$TiC_4 + H_2 + CH_4$	980–1400
W_2C	thermal decomp. $W(CO)_6$	300– 500
	$WF_6 + C_6H_6 + H_2$	400– 900
VC	$VCl_2 + H_2$	≈1000

Its disadvantages include the following:

1) Rather high temperatures required to produce certain compounds
2) Low pressures restrict the types of substrate that can be coated
3) Uniform heating of the substrate at times difficult
4) Reactor vessel design is critical to coating complex shapes uniformly [38].

5.2.2. Evaporation

In the evaporationtechnique, the coating material is placed in a vacuum chamber opposite the substrate. A relatively high vacuum is used, typically 10^{-6} Torr ($\approx 10^{-4}$ Pa), because the coating material must be evaporated either by heating it to some large fraction of its melting point or by using an electron beam to locally heat it. When it is evaporated, its atoms are ejected in straight lines from the solid and impinge and condense on all exposed surfaces, including the substrate to be coated. Because there is no driving force to accelerate the freed coating atoms, they impinge with ca. 0.2 eV.

Evaporation processes for the deposition of refractory compounds are subdivided into two types:

1) Direct evaporation — the evaporant is the refractory compound itself.
2) Reactive evaporation — a metal or compound in a low valence state is evaporated in the presence of a reactive gas to form a compound deposit.

Direct evaporation of ceramics can occur with or without dissociation of the compound into fragments. However, the observed vapor species show that very few compounds evaporate without dissociation. Examples are MgF_2, B_2O_3, CaF_2, SiO, GeO, and SnO. In the more general cases, ceramic materials are not transformed to the vapor state with the same molecular composition as the solid. Subsequently the fragments must then recombine to reconstitute the compound. Therefore, the stoichiometry of the deposit depends on several factors, which can lead to cation- or anion-deficient structures.

Reactive evaporation broadens the range of ceramics for coatings — the ceramic need not form directly or recombine from a vapor. Rather, a metal or suboxide is evaporated in the presence of a reactive gas to produce a ceramic deposit by reaction.

Figure 9illustrates the evaporation process with an electron-beam source. The entire system is within a vacuum chamber. If reactive gas is allowed to enter the vacuum chamber, evaporation can be reactive rather than direct. However, the reaction must be both thermodynamically and kinetically feasible.

Compounds that have been synthesized by the reactive evaporation process include α-Al_2O_3, γ-Al_2O_3, Y_2O_3, TiO_2, In_2O_3, In_2O_3– SnO_2(ITO), SnO_2, BeO, SiO_2, TiC, ZrC, NbC, Ta_2C, VC, W_2C, HfC, VC–TiC, TiC–Ni, Ti_2N, TiN, MoN, $Cu_xMo_6S_8$, and Cu_xS [39].

5.2.3. Sputtering

Sputtering is a coating process that involves the transport of almost any material from a source, or target, to a substrate of almost any other material. The ejection of the source material is accomplished by the bombardment of the surface of the target with gas ions accelerated by high voltage. Particles of atomic dimensions from the target are ejected as a result of momentum transfer between incident ions and the target. The target-ejected particles traverse the vacuum chamber and are subsequently deposited on a substrate as a thin film. Sputtering can be direct or reactive. It can be accomplished with diode, triode, or magnetron sources. Figures 10 and 11 illustrate diode and triode assemblies.

Diode sputtering uses a simple two-element electrode configuration. The expendable target is attached to the cathode and an anode platen supports the substrates. The system can be modified to allow for reactive sputtering. Diode sputtering may be operated in either the d.c. or RF (radiofrequency) excitation modes, with RF being preferred for ceramics because it permits lower pressures and thus generally purer films. In addition, RF sputtering reduces charge buildup, permitting direct sputtering of ceramics.

In *triode sputtering* a hot electron-emitting filament is added between the anode and cathode and serves to control the plasma. The ion current is significantly increased, and lower gas pressures can be used. These are important advantages for the sputtering rate: production throughput is increased, while potential radiation damage to sensitive electronic devices is minimized.

In *magnetron sputtering* the cathode incorporates a magnetic field aligned perpendicular to the electric excitation field. The magnetic field serves to impart a cycloidal motion to the free electrons being liberated at the cathode surface both by the electric field and by the ionized argon. The great length of the cycloidal path taken by the electrons in response to the presence of the magnetic field greatly enhances the yield of argon ionization for a given current density and voltage. The greater flux of argon ions impinging on the cathode causes rapid disintegration of the cathode and a sputtering rate for a given voltage and current density much higher than that for diode or triode sputtering.

The driving force for sputtering is electrical rather than thermal (as in evaporation). Because of the high kinetic energy transferred, the sputtered atoms impinge on the substrate with ca. 5 eV. Because of this relatively high kinetic energy of impingement, adhesion to the substrate is usually stronger than for evaporation.

Materials having refractive indices ranging between 1.45 and 4.08 have been investigated for optical applications. Indium tin oxide, zinc oxide, and titanium and tantalum oxides have been sputtered for window films, solar cells, and antireflective coatings.

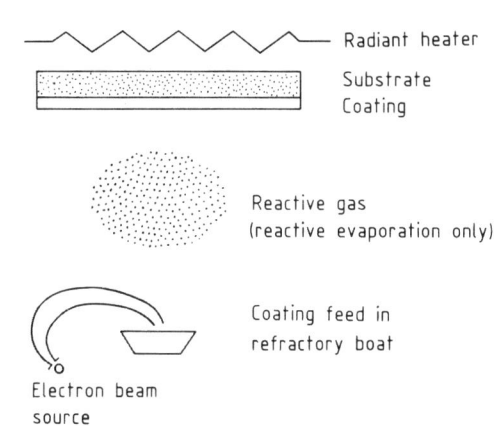

Figure 9. Schematic of evaporation (reactive if reactive gas present)

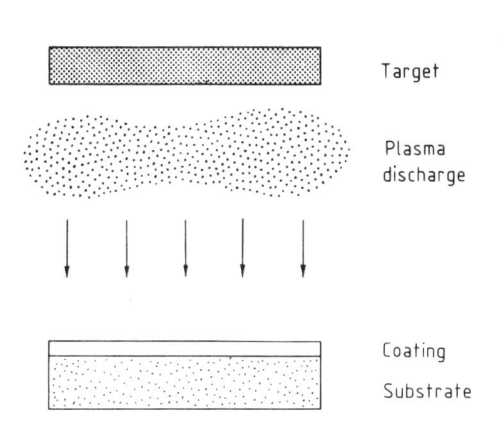

Figure 10. Schematic of a diode sputtering assembly [9]

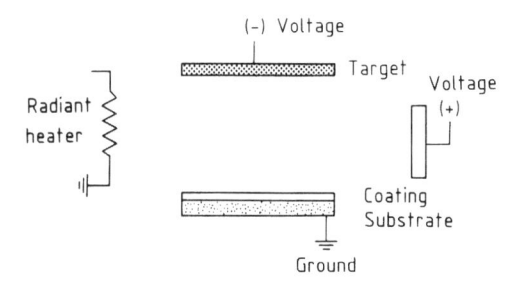

Figure 11. Schematic of a triode sputtering assembly

5.2.4. Plasma Spraying

The plasma spraying process is a technique that combines particle melting, quenching, and consolidation into a single process. The arc plasma spraying process involves injection of powder particles into a plasma jet stream created by heating an inert gas in an electric arc. The particles injected into the plasma jet melt rapidly and at the same time are accelerated toward the workpiece surface. Rapid quenching of the molten particles occurs when the droplets impact the surface. Cooling rates are typically $10^5 - 10^6$ K/s, and the resulting microstructures are fine grained but may retain a substantial amount of porosity.

The torch itself consists of two concentric electrodes that are cooled by water and electrically isolated from each other. Inert or nonoxidizing gases, or a mixture, are passed through the annular space formed by these electrodes, and a d.c. arc is initiated. With proper operating procedure, the arc is continually sustained, forming a high-temperature plasma that issues from the nozzle front. Figure 12 illustrates the configuration of a typical arc plasma torch system.

Plasma-sprayed coatings offer several advantages over other coating processes:

1) High particle velocities result in higher bond strength coupled with higher coating density.
2) The heat source is more efficient because the plasma is in a high-energy state, providing efficient particle heating.
3) The heating source is inert, minimizing oxidation.
4) High plasma temperatures permit the spraying of materials with high melting points.

Disadvantages include the following:

1) Consistency and compositional uniformity of a coating over a large area at times can vary.
2) Spraying tends to produce porous coatings, especially at high velocities, because of gas entrapment.
3) Reactive materials cannot be sprayed in open air; however, low-pressure plasma spraying, LPPS, can be used in air.
4) The deposits contain oxidation products, together with some porosity because of incomplete melting, wetting, and fusing of deposited particles.

In LPPS the reduction of the amount of air present in the torch region is accomplished by maintaining the plasma system under vacuum with only 30–60 Torr (4–8 kPa) of inert gas. This greatly reduces the chances of included porosity; therefore, denser deposits and more uniform deposits are possible.

5.2.5. Sol–Gel Processing

Sol–gel coatings are based on a process of applying sols, stable dispersions of hydrous oxides or hydroxides, to a substrate material. Once these liquid dispersions are applied to a surface, they undergo a catalyzed transition to form a gel. The sol–gel coating is subsequently dried to remove excess water and densified. The specific

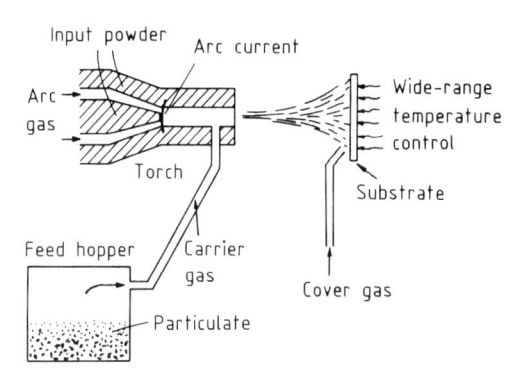

Figure 12. Schematic of an arc plasma spraying assembly

properties of the films, such as pore volume and pore-size distribution, can be controlled during densification. Typical applications of these films include dipping and spin coating [40].

Advantages of sol–gel-derived films include

1) Low densification temperatures
2) High degree of chemical homogeneity
3) Multicomponent films possible

Disadvantages include

1) High shrinkages due to high volume of solvent present
2) Difficulty in making physically stable, thick coatings
3) Coating and substrate densified by heating

5.2.6. Ion Implantation

In ion implantation, ions are formed by collisions with electrons ejected from a heated filament and accelerated in an electric field. In the instances where the element to be ionized is a gaseous species, such as argon, nitrogen, oxygen, or carbon dioxide, which is typical for implanted ceramic coatings, the gas is introduced directly into the ionization chamber, where the high-velocity electrons strip away other electrons, creating a plasma. An electric field then directs the plasma from the chamber to a magnet, which is able to select ions of the desired mass. These ions are then focused in an electric field and accelerated to the final energy to bombard the specimen. The penetration depth depends on the type and energy of the ions and material to be implanted. Figure 13 illustrates an ion implantation system [41].

Sputtered films of carbides, nitrides, and borides exhibited increased adhesion and microhardness when bombarded with heavy inert ions, such as Kr and Xe. The formation of nitride layer on the surface of a Ti–Al–V alloy by bombardment with nitrogen ions has been found to slow the wear rate of the metal by a factor of ten [42].

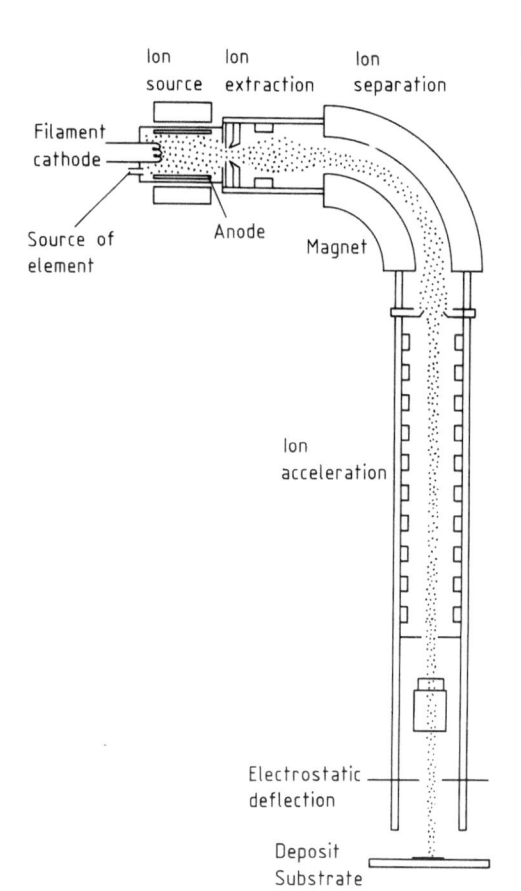

Figure 13. Schematic of an ion implantation system

6. Ceramic–Metal Composites

Ceramic–metal composites may be defined as the heterogeneous combination of one or more ceramic phases with one or more metals or alloys. Systems with very small amounts of metal or ceramic are usually not considered as composite materials. Also not generally viewed as ceramic–metal composites are materials in which metal particles may be in a continuous ceramic — colloidal and other metal particle colorants in glasses, glazes, and enamels, and thin-film materials such as sputtered and CVD resistors and microwave devices for hybrid microelectronics.

The distribution of phases in ceramic–metal composites is typically at a fine structural level, commonly microscopic. The metal and ceramic phases are usually mixed physically, often with a binder, formed, and fired at elevated temperatures. For the usual production times and temperature, there is relatively little or only incomplete solution between phases.

A nomenclature problem exists for the ceramic–metal composite materials. Two classes of materials have evolved: cermets and metal matrix composites (MMC). Neither strict definition nor common technological usage provides a clear distinction. Indeed, some use the two terms interchangeably. Usually, however, a cermet is understood to consist of continuous or partially continuous metal phase with ceramic particles having similar size in all dimensions. The metal matrix composites may be distinguished by the high aspect ratio of the ceramic constituent: the metal surrounds long continuous or chopped fibers or elongated single-crystal whiskers of ceramic composition.

6.1. Cermetsand Cemented Carbides

Withinthe cermet subcategory of ceramic–metal composites, there is a nomenclature difficulty. Tungsten carbide–cobalt is termed a cemented carbide, while other carbide–metal materials, such as titanium carbide–nickel, chromium carbide–nickel, as well as oxide ceramic–metal systems, such as alumina–chromium and thoria–tungsten, are termed cermets. This distinction is largely historical. The cermets were developed during and after World War II as materials with better hardness and high-temperature properties than the cemented carbides. However, the microstructures and properties of the cemented carbides and the cermets, particularly the carbide cermets, are closely related. The structure consists of a continuous metal or alloy cement bonding grains (particles) of carbide and/or oxide ceramic.

Cermets and cemented carbides derive their usefulness from the combination of compressive strength, hardness, and thermal resistance provided by the ceramic with ductility, lubricity, and adhesion provided by the metal. The cermets and cemented carbides are used for their strength, hardness, and abrasion resistance, particularly at elevated temperatures, which may arise from the environment or frictional heating. Uses include metal and rock cutting or grinding tools, friction and glide devices, turbine parts, flame nozzles, magnetron tube cathodes, seals, high-temperature containers and pouring spouts, and the ball for ballpoint pens.

Cermets are generally prepared by mixing carbide and/or oxide powder with metal powder in the presence of a binder. The chemistry of each component is often quite complex. Tungsten carbide typically contains purposeful additions of tantalum carbide and titanium carbide. Molybdenum is added to nickel to promote wetting of titanium carbide. Some cermet is unidirectionally or isostatically pressed and heated to a temperature at which the metal liquifies, coats, and bonds the ceramic. Alternative preparation methods may include slip casting, liquid-metal infiltration, or hot pressing of parts. The factors that dictate the properties of the final part are the following [43]–[45]:

1) Bonding between carbide grains and matrix
2) Composition and structure of the matrix

3) Shape and size of the carbide or oxide grain
4) Dispersion of the carbide grain in the matrix

A good bond between metal binder and ceramic grain is required to transfer load properly and to provide a homogeneous microstructure for uniform load bearing and deformation. Furthermore, since a wetting angle of near zero degrees is usually desired to have uniform spreading of binder, the degree of bonding is usually quite high. The wetting angle is affected by the composition of the metal binder and ceramic grain, as well as by the firing atmosphere [43].

Phase composition, distribution, and stability have major effects on properties. For example, the tungsten carbide–cobalt system is stable only over a narrow carbon content. At 16 wt% cobalt, WC and Co are the sole phases in the compositional range from 6.00 to 6.12 wt% C — an allowable composition range of 0.12%. Graphite forms above 6.12% C and eta phase (η-WC_{1-x}) below 6.00% ($\rightarrow$ Carbides). Both phases embrittle the cemented carbide, causing a loss of one-quarter of the optimum strength for compositions about 0.1% outside the acceptable range [44], [45].

The microstructure of a cemented carbide or cermet has an important influence on its properties. A uniform dispersion of carbide or oxide ceramic grains in a matrix of metal binder gives the best mechanical properties. Rounded grains, which can result either during powder preparation or from partial dissolution by molten metal during firing, improve the fracture strength. The relative amount of binder phase strongly affects the mechanical properties, as can be seen in Figure 14. Fracture strength increases with the amount of binder because the metal ductility provides fracture toughness. Hardness decreases with increased metal binder because the cermet is less resistant to deformation [43], [44]. The size of the ceramic grain also affects hardness: for a 16 wt% Co cemented carbide, the hardness decreases almost linearly from 89 to 85 R_A as the grain size decreases from 1 to 6 µm. Fine submicrometer grain size increases hardness and resistance to deformation. However, for a fixed amount of binder metal, a decrease in the grain size and the corresponding increase in grain surface area may result in inadequate metal to wet the grain, leading to loss of fracture strength.

A useful way of examining the properties of cermets and cemented carbides, as well as other composites, is provided by the concept of mean free path, the average distance between neighbor ceramic grains, i.e., the average thickness of the binder phase. The mean free path depends on the size and shape of the ceramic grain and the amount of metal available. The hardness (as well as a variety of other mechanical properties) depends, within a narrow band, only on this mean free path (see Fig. 15), provided there is adequate metal binder phase to cover the grains uniformly. For the tungsten carbide–cobalt example shown, a mean free path of less than ca. 0.1 µm corresponds to incomplete binder and poorer properties.

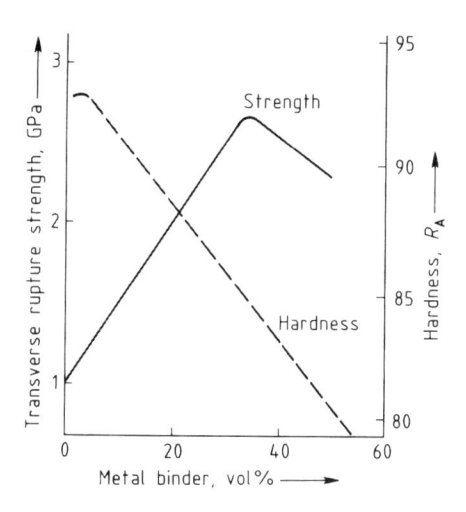

Figure 14. Effect of increased binder on mechanical properties of TiC–Ni–Mo cermet with an average TiC grain size of 1 µm

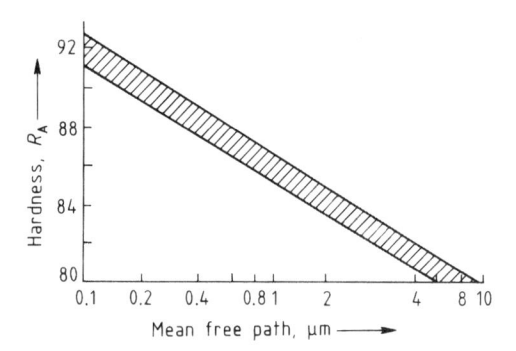

Figure 15. Effect of mean free path on hardness for a cemented carbide

6.2. Metal Matrix Composites

The concept of reinforcing a metal with ceramic is not new. In the 1950s the need for jet engine components that could withstand high temperatures for long periods of time began the search for materials that had properties better than those that could be attained by the metal alone.

The advantages that metal matrix composites offer over conventional metals include (1) taking advantage of the anisotropic character of the composite in the efficient design and fabrication of components, (2) manufacturing an engineered material with a specific set of strength and toughness requirements, and (3) increasing the strength, modulus, creep resistance, and oxidation resistance of common engineering alloys such as aluminum, titanium, and nickel. In addition, a major advantage of metal matrix composites arises from the reduction in the strength:weight ratio. For instance, an

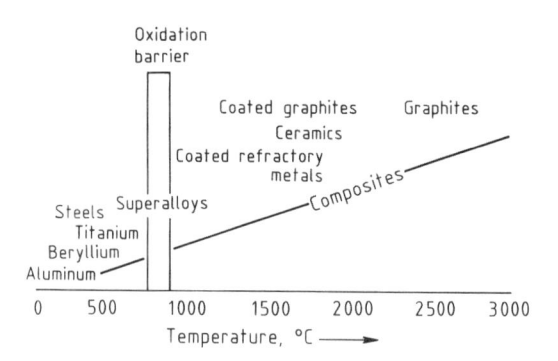

Figure 16. Temperature regimes of usefulness for various materials

aluminum alloy reinforced with graphite exhibits the same strength as steel while exhibiting a 45% increase in the strength:density ratio [46], [47].

In comparison to other composite matrices, the greatest potential advantages of the metal matrix are its resistance to severe environments, toughness, and strength at high temperatures. The temperatures where metal matrix composites exhibit the greatest potential advantages over rival metallic alloys and polymer matrix composites are above 800 °C (Fig. 16). In fiber- or filament-reinforced metals the environmental stability of the matrix at elevated temperatures can be emphasized because the required mechanical strength and stiffness are obtained from the reinforcement. The strength requirements of the matrix are nominal, the matrix only serving to transfer load into the fiber. Up to 540 °C, several potential metal composite systems have been identified. For higher temperatures, at which oxidation presents a major problem, stable reinforcements such as alumina and silicon carbide fibers have considerable potential. Below 320 °C, the potential is less clear, but the advantages of metal matrix composites over polymer matrix composites include improved wear and erosion resistance, greater shear, compressive and transverse strength, increased toughness and hardness, and greater strain prior to failure. The disadvantages include higher density, more difficult fabrication, metal–fiber reactions, and higher cost.

Reinforcement Materials. The mostappropriate reinforcing material depends on the requirements put upon the metal matrix composite. A wide variety of materials of high elastic modulus, high strength, low density, and high-temperature stability are available. Table 7 lists a number of ceramic materials along with typical mechanical properties; in addition, the properties of a few glasses and metals available as fibers are shown for comparison. In many cases the metal or the fabrication technique results in composites with less than theoretical properties. In these instances such things as process defects and matrix–filler incompatibility affect properties. Matrix–filler compatibility can include the following factors: (1) wetting, (2) chemical bonding, (3) filler surface character, i.e., shape, roughness, and porosity, (4) filler volume fraction and size, i.e., particle size or fiber length and diameter, (5) filler orientation, and (6)

Table 7. Filament properties for a variety of reinforcing materials

Material	Tensile strength, (ksi)*	Elastic modulus, GPa	Density, g/cm^3	Strength: density ratio, 10^6
Fibers				
Steel	600	200	7.9	2.1
Tungsten	575	407	19.3	0.83
Molybdenum	320	331	10.2	0.88
S-glass	600	86	2.5	6.52
E-glass	500	72	2.5	5.43
Silica	500	72	2.2	5.43
Boron	450	400	2.6	4.68
Silicon carbide	350	379	3.4	2.84
Alumina	350	483	3.98	2.43
Graphite	285	345	1.66	4.75
Whiskers				
Alumina	2000	483	3.98	13.9
Silicon carbide	1500	483	3.2	13.0

* ksi, a non-SI unit, equals 6.89 MPa.

relative thermal expansion and elastic modulus match. The last can give rise to residual stresses that can affect the mechanical properties.

Fabrication Process. Numeroustechniques have been employed for the consolidation of ceramic fibers, whiskers, and particles into metal matrices. These techniques vary widely, from typical powder-metallurgical techniques such as slip casting to techniques such as diffusion bonding and plasma spray bonding. The most widely used consolidation techniques are the following:

- diffusion bonding
- plasma spray bonding
- electroforming
- liquid-metal infiltration
- high-energy rate forming
- explosive welding
- hot roll bonding
- hot drawing
- cold rolling
- cold drawing
- chemical vapor deposition
- powder slip casting
- hot extrusion

Each technique employed to consolidate a composite requires specific methods for dealing with the anisotropic and fragile reinforcing fillers. The ceramic fiber may be used in oriented or random configurations as individual fiber, bundles, or weaves. Each process must meet to some reasonable degree each of the following objectives:

avoid filament or fiber breakage
consolidate composite with minimal filler degradation
establish and maintain fiber alignment
be able to vary filler loading
achieve a high-density matrix
establish filler–matrix interfacial bond sufficient to transmit applied load from matrix to filler
allow a degree of matrix alloying
allow for postfabrication heat treatment
provide flexibility in filament spacing
allow for cross and angle lay-ups in some instances
minimize product variability
be amenable to scale-up

Compatibility. Filler–matrix compatibility includes the chemical reactions between matrix and filler during consolidation, secondary processing, and use. Some degree of chemical interaction should take place. In a successful composite, the metal completely wets the filler and chemically bonds to it. However, the exact degree of chemical interaction needed to produce optimum properties is not yet known. If the metal and filler are able to bond chemically during consolidation, there is a likelihood that this chemical interaction continues during use, degrading the filler as a result. Knowledge of the thermodynamics for the combination of metal and filler is an important composite design criterion and a suitable starting point for selecting stable systems.

7. References

[1] J. T. Klomp: "Interfacial Reactions between Metals and Oxides during Sealing," *Bull. Am. Ceram. Soc.* **59** (1980) 794–800.
[2] M. E. Twentyman: "High Temperature Metallizing," *J. Mater. Sci.* **10** (1975) 765–766.
[3] W. H. Kohl: *Handbook of Materials and Techniques for Vacuum Devices*, Reinhold Publ. Co., New York 1967.
[4] J. R. Tinklepaugh, W. B. Crandall (eds.): *Cermets*, Reinhold Publ. Co., New York 1960.
[5] J. A. Pask, R. M. Fulrath: "Fundamentals of Glass-to-Metal Bonding: VIII. Nature of Wetting and Adherence," *J. Am. Ceram. Soc.* **45** (1962) 592–596.
[6] F. Bashforth, S. C. Adams: *An Attempt to Test Theories of Capillarity*, Cambridge University Press, London 1883.
[7] A. G. Pincus: "Mechanism of Ceramic-to-Metal Adherence, Adherence of Molybdenum to Alumina Ceramics," *Ceram. Age* **63** (1954) no. 16–20, 30–32.
[8] J. D. Partridge: *Glass-to-Metal Seals*, Society of Glass Technology, Sheffield, England, 1949.
[9] D. Tabor: "Interaction between Surfaces: Adhesion and Friction," in *Surface Physics of Materials*, vol. 11, Academic Press, New York 1975, pp. 475–529.
[10] S. S. Rekhson: "Annealing of Glass-to-Metal and Glass-to-Ceramic Seals," *Glass Technol.* **20** (1979) 27–35.
[11] H. E. Pattee, R. E. Evans, R. E. Monroe: *Joining Ceramics and Graphite to Other Materials*, WASH SP-5052, NASA, Washington, D.C., 1968.

[12] J. J. Schmidt, J. L. Carter: "Using Nitrogen Based Atmosphere for Glass-to-Metal Sealing," *Met. Prog.* **128** (1985) 29–34.

[13] A. I. Andrews: *Enamels: The Preparation, Application and Properties of Vitreous Enamels*, Twin City Printing, Champaign, Ill., 1945.

[14] M. Borom, J. Pask: "Role of Adherence Oxides in the Development of Chemical Bonding at Glass-Metal Interfaces," *J. Am. Ceram. Soc.* **49** (1966) 1–6.

[15] B. W. King, W. H. Duckworth: "Nature of Adherence of Porcelain Enamels to Metals," *J. Am. Ceram. Soc.* **42** (1959) 504–525.

[16] L. S. O'Bannon: *"The Adherence of Enamels, Glass, and Ceramic Coatings to Metals — A Review,"* Batelle Memorial Institute, Columbus, Ohio, 1959.

[17] G. R. Van Houten: "A Survey of Ceramic-to-Metal Bonding," *Ceram. Bull.* **38** (1959) 301–307.

[18] G. P. Chu: "Some Aspects on Glass-Ceramics to Metal Sealing," *Int. Cong. Glass, 10th,* Ceramic Soc. Japan, 1967, pp. 94–102.

[19] C. I. Helgesson: *Ceramic-to-Metal Bonding*, Boston Technical Publ., Cambridge 1968.

[20] S. S. Cole, G. Sommer: "Glass Migration Mechanism of Ceramic to Metal Seal Adherence," *J. Am. Ceram. Soc.* **44** (1961) 265–271.

[21] R. G. Loasby, N. Davey, H. Barlow: "Enhanced Property Thick Film Pastes," *Solid State Technol.* **15** (1972) no. 5, 46–72.

[22] M. Monneraye: "Les Encres Sengraphiables en Microélectronique Hybride — les Matériaux et leur Comportement," *Acta Electron.* **21** (1978) 263–281.

[23] C. S. Pearsall: "New Brazing Methods for Joining Non-Metallic Material to Metals," *Mater. Methods* **30** (1949) 61–62.

[24] J. E. McDonald, A. Eberhart: "Adhesion in Metal Oxide Systems," *Trans. Metall. Soc. AIME* **233** (1964) 512–517.

[25] J. F. Burgess, C. A. Neugebauer, G. Flanagan: "The Direct Bonding of Metals to Ceramics by the Gas–Metal Eutectic Method," *J. Electrochem. Soc. Solid-State Science and Technol.* May, 1975.

[26] J. T. Klomp, P. J. Vrugt: "Interfaces between Metals and Ceramics in Surfaces and Interfaces," *Materials Science Research* **14** (1980) 97–105.

[27] W. Dawihl, E. Klinger: "Mechanical and Thermal Properties of Welded Joints between Alumina and Metals," *Ber. Dtsch. Keram. Ges.* **46** (1969) 12–18.

[28] H. J. de Bruin, A. F. Moodie, C. E. Warble: "Reaction Welding of Ceramics Using Transition Metal Intermediates," *J. Aust. Ceram. Soc.* **7** (1971) no. 2, 57–58.

[29] J. B. Wachtmann, W. H. Rhodes: "Ceramics — Advanced Forms Find Essential Uses as an Enabling Technology," *Science (Washington, D.C. 1979)* **6** (1985) no. May, 6–18.

[30] J. B. Wachtman Jr., R. A. Haber: "Ceramic Films and Coatings," *Proc. Chem. Eng. Prog.* **82** (1986) no. 1, 39–46.

[31] W. A. Brainard: "The Friction and Wear Properties of Sputtered Hard Refractory Compounds," NASA Technical Memorandum TM-78 895, 1978.

[32] E. Lang (ed.): *Coatings for High Temperature Applications*, Applied Science Publishers, New York 1983.

[33] I. Kvernes, P. Fartum: "Use of Corrosion Resistant Plasma Sprayed Coatings in Diesel Engines," *Thin Solid Films* **2** (1978) 119–128.

[34] W. A. Pliskin: "Comparison of Properties of Dielectric Films Deposited by Various Methods," *J. Vac. Sci. Technol.* **14** (1977) 1064–1081.

[35] R. F. Bunshah: *Deposition Technologies of Films and Coatings*, Noyes Press, Park Ridge, N.J., 1982.

[36] D. M. Mattox: "Commercial Applications of Overlay Coating Techniques," *Thin Solid Films* **84** (1981) 361–365.

[37] B. M. Kramer: "Requirements for Wear-Resistant Coatings," *Thin Solid Films* **2** (1980) 117–128.

[38] W. A. Bryant: "Review: The Fundamentals of Chemical Vapor Deposition," *J. Mater. Sci.* **12** (1977) 1285–1306.

[39] R. F. Bunshah, A. C. Raghuram: "Activated Reactive Evaporation Process for High Rate Deposition of Compounds," *J. Vac. Sci. Technol* **9** (1972) 1385–1389.

[40] C. J. Brinker, S. P. Mukherjee: "Comparison of Sol-Gel Derived Thin Films with Monoliths in a Multicomponent Silicate Glass System," *Thin Solid Films* **77** (1981) 141–148.

[41] T. S. Picraux, P. S. Peercey: "Ion Implantation of Surfaces," *Sci. Am.* **252** (1985) no. 3, 102–110.

[42] B. H. Kear, J. W. Mayer, J. M. Poate, P. R. Strutt: "Surface Treatments Using Laser, Electron and Ion Beam Processing Methods" *Metall. Treatises 1981*, 321–342.

[43] J. R. Tinkelpaugh, W. B. Crandall (eds.): *Cermets*, Reinhold Publ. Co., New York 1960.

[44] J. Gurland, P. Bardzil: "Relation of Strength, Composition, and Grain Size of Sintered WC–Co Alloys," *J. Metals* **7** (Trans. Metall. Soc. AIME 203) (1955) 311–315.

[45] J. Gurland, J. T. Norton: "Role of the Binder Phase in Cemented Tungsten Carbide-Cobalt Alloy," *J. Met.* **4** (1952) 1051–1056.

[46] C. T. Lynch, J. P. Kershaw: *Metal Matrix Composites*, CRC Press, Cleveland, Ohio, 1969.

[47] R. W. Rice: "Ceramic Matrix Composites," *Proc. Am. Ceram. Soc.* 1980 no., 661–689.

Ceramic Colorants

Richard A. Eppler, Eppler Associates, Cheshire, Connecticut 06410, United States

1. Introduction 1069
2. Classification 1070
3. Pigment Structures 1070
4. Colorant Systems 1072
5. Industrial Production of Ceramic Pigments 1082
6. Use of Ceramic Pigments in Glazes 1083
7. Ceramic Glazes 1083
8. Application Media 1084
9. Quality Control 1084
10. References 1085

1. Introduction

Color is an important characteristic of many ceramic products. In fact, any product for which aesthetics is a consideration for purchase will be enhanced by proper use of color. Thus, ceramic products which use color include dinnerware, tile, porcelain enamel, sanitary ware, and some glasses and structural clay products.

Ceramic material can be colored in three general ways [1]. First, the ceramic material itself may contain transition metal ions that are colored. This method is rarely used, except for colored glass, because adequate tinting strength and purity of color cannot be obtained this way. A second method to obtain color in ceramics is to induce precipitation of a suitable crystalline phase during processing. For example, certain oxides, such as zirconium oxide and titanium dioxide, dissolve in vitreous material at high temperature. When the temperature is reduced, the solubility is also reduced and precipitation occurs. This method is used for opacification, that is, production of an opaque white color. For oxide colors other than white, this method lacks the control necessary for reproducible results. It is, however, often used for production of nonoxide colorants, such as gold, copper, and cadmium–selenium crystals, dispersed in a vitreous matrix.

The third and most common method of obtaining color in a ceramic material is to disperse within that material a colored crystalline phase that is insoluble in the matrix. This crystalline phase, commonly called a pigment, imparts its color to the matrix.

The materials chosen for this purpose must possess certain properties beyond the tinctorial strength required of any pigment [2]. To be used in a ceramic material, pigments must be resistant to the high temperature and corrosive environment encountered in the firing process. Their rates of dissolution or reaction in the ceramic or molten glaze at high temperature must be low. This problem is aggravated by the fine particle sizes (1–10 μm) that are required for uniform dispersion of the pigment in the ceramic matrix.

2. Classification

Most of the materials used in ceramics as pigments are oxides because of their greater stability in oxygen-containing ceramic systems [1]. The only important exceptions are the cadmium sulfoselenide pigments. These nonoxide-containing materials are used because a bright red color cannot be obtained any other way. However, these pigments are very difficult to use in ceramics. Great care is required to prevent their being oxidized during firing of the ceramic body.

The various oxide pigments have recently been classified according to their crystallographic structure (see Table 1) [3]. These structures vary significantly in the range of concentration of constituent oxides allowed. A wide range of cation substitution may be permitted if the ionic sizes of the substituted ions are similar. Different formulations are used to vary the color obtained. Thus, some of the crystal classes have individual pigments with various primary colors. This effect is particularly evident in the spinel class, wherein are found blue, green, pink, brown, and black pigments.

Color variations within a given structural class are of practical importance. Normally, pigments within a given class have excellent chemical and physical compatibility. Thus, it is possible to mix them to obtain intermediate shades.

3. Pigment Structures

The structures of inorganic solid-state materials are governed by several principles [4]. In the first place, the nearest-neighbor cation–anion distances almost totally determine the lattice energy of an ionic solid material at room temperature. Moreover, the preferred coordination polyhedra of the anions surrounding each cation in ionic solid phases are determined almost solely by the ratio of the ionic size of the cation to that of the anion. Furthermore, the structures that can be built from any combination of cations and anions are subject to the rules of electrostatic neutrality. Thus, in a stable ionic structure, the valence of each anion, with changed sign, is exactly or nearly exactly equal to the sum of the strength of the electrostatic bonds to it from the adjacent cations. Lastly, the coordination of a cation is larger when its charge is less, and is smaller as its field strength increases. The result of these principles is that for any given stoichiometry and ionic size of the materials to be used, only one or two structures exist that will accommodate them.

For example, the spinel class, which contains 19 pigments, is restricted to those materials of A_2BX_4 stoichiometry with:

0.06 nm $< r_A < 0.100$ nm
0.055 nm $< r_B < 0.100$ nm

where r_A and r_B are the ionic radii of the respective cations, and X is oxygen. Similar restrictions apply to the other structures found in Table 1.

Table 1. Classification of mixed-metal oxide inorganic pigments*

Crystal class, name (i.e., category)	CAS registry number	Basic chemical formula	DCMA number
Baddeleyite			
– Zirconium vanadium yellow baddeleyite	[68187-01-9]	$(Zr,V)O_2$	1-01-4
Borate			
– Cobalt magnesium red-blue borate	[68608-93-5]	$(Co,Mg)_2B_2O_5$	2-02-1
Corundum-hematite			
– Chrome alumina pink corundum	[68187-27-9]	$(Al,Cr)_2O_3$	3-03-5
– Manganese alumina pink corundum	[68186-99-2]	$(Al,Mn)_2O_3$	3-04-5
– Chromium green-black hematite	[68909-79-5]	Cr_2O_3	3-05-3
– Iron brown hematite	[68187-35-9]	Fe_2O_3	3-06-7
Garnet			
– Victoria green garnet	[68553-01-5]	$3\ CaO \cdot Cr_2O_3 \cdot 3\ SiO_2$	4-07-3
Olivine			
– Cobalt silicate blue olivine	[68187-40-6]	Co_2SiO_4	5-08-2
– Nickel silicate green olivine	[68515-84-4]	Ni_2SiO_4	5-45-3
Periclase			
– Cobalt nickel gray periclase	[68186-89-0]	$(Co,Ni)O$	6-09-8
Phenacite			
– Cobalt zinc silicate blue phenacite	[68412-74-8]	$(Co,Zn)_2SiO_4$	7-10-2
Phosphate			
– Cobalt violet phosphate	[13455-36-2]	$Co_3(PO_4)_2$	8-11-1
– Cobalt lithium violet phosphate	[68610-13-9]	$CoLiPO_4$	8-12-1
Priderite			
– Nickel barium titanium primrose priderite	[68610-24-2]	$2\ NiO \cdot 3\ BaO \cdot 17\ TiO_2$	9-13-4
Pyrochlore			
– Lead antimonate yellow pyrochlore	[68187-20-2]	$Pb_2Sb_2O_7$	10-14-4
Rutile-cassiterite			
– Nickel antimony titanium yellow rutile	[71077-18-4]	$(Ti,Ni,Sb)O_2$	11-15-4
– Nickel niobium titanium yellow rutile	[68611-43-8]	$(Ti,Ni,Nb)O_2$	11-16-4
– Chrome antimony titanium buff rutile	[68186-90-3]	$(Ti,Cr,Sb)O_2$	11-17-6
– Chrome niobium titanium buff rutile	[68611-42-7]	$(Ti,Cr,Nb)O_2$	11-18-6
– Chrome tungsten titanium buff rutile	[68186-92-5]	$(Ti,Cr,W)O_2$	11-19-6
– Manganese antimony titanium buff rutile	[68412-38-4]	$(Ti,Mn,Sb)O_2$	11-20-6
– Titanium vanadium antimony gray rutile	[68187-00-8]	$(Ti,V,Sb)O_2$	11-21-8
– Tin vanadium yellow cassiterite	[68186-93-6]	$(Sn,V)O_2$	11-22-4
– Chrome tin orchid cassiterite	[68187-53-1]	$(Sn,Cr)O_2$	11-23-5
– Tin antimony gray cassiterite	[68187-54-2]	$(Sn,Sb)O_2$	11-24-8
– Manganese chrome antimony titanium brown	[69991-68-0]	$(Ti,Mn,Cr,Sb)O_2$	11-46-7
– Manganese niobium titanium brown rutile	[70248-09-8]	$(Ti,Mn,Nb)O_2$	11-47-7
Sphene			
– Chrome tin pink sphene	[68187-12-2]	$CaO \cdot SnO_2 \cdot SiO_2 \cdot Cr_2O_3$	12-25-5

Table 1. (continued)

Crystal class, name (i.e., category)	CAS registry number	Basic chemical formula	DCMA number
Spinel			
– Cobalt aluminate blue spinel	[68186-86-7]	$CoAl_2O_4$	13-26-2
– Cobalt tin blue-gray spinel	[68187-05-3]	Co_2SnO_2	13-27-2
– Cobalt zinc aluminate blue spinel	[68186-87-8]	$(Co,Zn)Al_2O_4$	13-28-2
– Cobalt chromite blue-green spinel	[68187-11-1]	$Co(Al,Cr)O_4$	13-29-2
– Cobalt chromite green spinel	[68187-49-5]	$CoCr_2O_4$	13-30-3
– Cobalt titanate green spinel	[68186-85-6]	Co_2TiO_4	13-31-3
– Chrome alumina pink spinel	[68201-65-0]	$Zn(Al,Cr)_2O_4$	13-32-5
– Iron chromite brown spinel	[68187-09-7]	$Fe(Fe,Cr)_2O_4$	13-33-7
– Iron titanium brown spinel	[68187-02-0]	Fe_2TiO_4	13-34-7
– Nickel ferrite brown spinel	[68187-10-0]	$NiFe_2O_4$	13-35-7
– Zinc ferrite brown spinel	[68187-51-9]	$(Zn,Fe)Fe_2O_4$	13-36-7
– Zinc iron chromite brown spinel	[68186-88-9]	$(Zn,Fe)(Fe,Cr)_2O_4$	13-37-7
– Copper chromite black spinel	[68186-91-4]	$CuCr_2O_4$	13-38-9
– Iron cobalt black spinel	[68187-50-8]	$(Fe,Co)Fe_2O_4$	13-39-9
– Iron cobalt chromite black spinel	[68186-97-0]	$(Co,Fe)(Fe,Cr)_2O_4$	13-40-9
– Manganese ferrite black spinel	[68186-94-7]	$(Fe,Mn)(Fe,Mn)_2O_4$	13-41-9
– Chrome iron manganese brown spinel	[68555-06-6]	$(Fe,Mn)(Fe,Cr,Mn)_2O_4$	13-48-7
– Cobalt tin alumina blue spinel	[71750-83-9]	$CoAl_2O_4/Co_2SnO_4$	13-49-2
– Chromium iron nickel black spinel	[71631-15-7]	$(Ni,Fe)(Cr,Fe)_2O_4$	13-50-7
– Chromium manganese zinc brown spinel	[71750-83-9]	$(Zn,Mn)Cr_2O_4$	13-51-7
Zircon			
– Zirconium vanadium blue zircon	[68186-95-8]	$(Zr,V)SiO_4$	14-42-2
– Zirconium praseodymium yellow zircon	[68187-15-5]	$(Zr,Pr)SiO_4$	14-43-4
– Zirconium iron pink zircon	[68187-13-3]	$(Zr,Fe)SiO_4$	14-44-5

* Reprinted with permission [3].

4. Colorant Systems

The pigments given in Table 1 are the alternatives available for coloring ceramics. Their specific applications are described in this chapter.

Opacifiers. Whiteness or opacity is introduced into transparent ceramic materials, such as glazes, by the addition of substances that will disperse in those materials as discrete particles. These particles scatter and reflect some of the incident light. To do this, the dispersed substance must have a *refractive index* that differs appreciably from that of the clear ceramic material. The refractive index (n_D^{20}) of most ceramic materials is 1.5–1.6 and, therefore, the refractive indexes of opacifiers must be either greater or less than this. In practice, opacifiers of higher refractive index are used. Some examples of opacifiers are tin(II) oxide ($n_D^{20} = 2.04$), zirconia ($n_D^{20} = 2.40$), zircon ($n_D^{20} = 1.85$), and titania ($n_D^{20} = 2.5$ for anatase or 2.7 for rutile).

In glazes and other ceramic coatings fired at temperatures in excess of 1000 °C, zircon [1490-68-2] is the opacifier of choice [5]. Its solubility in many ceramic glazes is about 5% at high temperature and 2–3% at room temperature. A customary mill addition would be 8–10% zircon. Consequently, most opacified glazes contain both zircon that was placed in the mill and went through the firing process unchanged and zircon which dissolved in the molten glaze during firing, but which recrystallized on cooling.

The effectiveness of a zircon opacifier is a function of particle size. The finer the particle size, the greater is the opacity of the pigment. Because this greater fineness is achieved by milling, the finer zircons are also the most expensive. On the other hand, zircon for smelting into frit is best when of an intermediate size. Therefore, the effectiveness of zircon opacifiers can be improved in partially or fully fritted glazes by smelting some of the zircon into the frit.

Zirconia is rarely used as an opacifier because in the vast majority of ceramic glazes it reacts with the silica in the glaze to produce zircon [6]. Therefore, because zircon is less expensive than zirconia, zircon is preferred. Tin oxide is a more effective opacifying agent than any of the other possibilities because its solubility is lower. However, the price of tin oxide is so high that it is no longer an economic solution. Its use is restricted to those special cases, such as chrome–tin pinks, where it also enhances the effectiveness of the coloring pigment.

In porcelain enamels and in glazes where the firing temperature is less than 1000 °C, *titania* [13463-67-7] in the anatase crystal phase is the opacifying agent normally used [7], [8]. It is the most effective opacifying agent because it has the highest index of refraction. However, at ca. 850 °C, anatase inverts to rutile in ceramic systems. Once inverted to rutile, titania crystals are able to grow rapidly to sizes that are no longer effective for opacification. Moreover, as the rutile particles grow, their absorption band extends into visible wavelengths, leading to a pronounced cream color. Thus, while titania is an effective opacifier at lower temperatures, it cannot be used above 1000 °C.

The solubility of titania in molten silicates is ca. 8–10% in most cases, and at room temperature is reduced to ca. 5%. Thus, titanium dioxide is customarily used at concentrations of ca. 15%.

When a *pastel color* is required, an opacifier plus a pigment is added to the coating. If this is done, the overall coating–pigment–opacifier system must be considered in selecting materials. The opacifier and the pigment must be compatible. For example, zircon opacifiers should be used with zircon or zirconia pigments. Chrome–tin pinks are stronger if some tin oxide opacifier is used. Titanium-based pigments are used in enamels when titania is the opacifier.

Black Pigments. Black ceramic pigments are formed by calcination of several oxides to develop the *spinel structure* [9], [10]. The formulation of black pigments is an excellent illustration of the wide flexibility of this structure for incorporating various chemical elements. Table 1 lists five different black spinel pigments: copper chromite black spinel, iron cobalt black spinel, iron cobalt chromite black spinel, manganese ferrite black spinel, and chromium iron nickel black spinel.

Selection of a particular black pigment depends somewhat on the specific material with which the pigment is to be used. If care is not taken, the pigment may show a green, blue, or brown tint after firing. A particularly important problem is the tendency of some ceramics to attack the pigment and release any cobalt that may be present. Thus, in some cases, using a cobalt-free pigment is desirable. The relatively high price of cobalt oxide also encourages use of cobalt-free pigments.

One black pigment that is not a spinel is *chromium black hematite.* This pigment is limited to zinc-free systems, as it reacts with zinc oxide to yield a brown spinel.

The basic black pigment is *iron cobalt chromite black spinel.* In some systems, however, it has a slightly greenish tint. In zinc-containing bases, therefore, iron cobalt black spinel is recommended. For a black with a slightly bluish tint, iron cobalt chromite black spinel with some manganese and a higher concentration of cobalt is used, and for a black with a brownish tint, manganese ferrite black spinel is the pigment of choice. In those instances where a cobalt-free system is desirable, copper chromite black spinel can be considered for use in systems fired at < 1000 °C, such as glass colors or porcelain enamels. Chromium black hematite can be used in zinc-free systems. This is the least expensive black on the market for use in materials fired at > 1000 °C. However, the presence of zinc oxide in the material results in a chemical reaction that alters the color. For systems containing zinc and fired at > 1000 °C, chrome iron nickel black spinel can be considered. This pigment can be used with most glaze systems and at all firing temperatures from cone 06 up to sanitary ware firing temperatures of cone 11 or 12 (for cone temperature equivalents, see Table 2).

Gray Pigments. The simplest way to obtain a gray pigment is to dilute a black pigment with a white opacifier. This *dilution* must be done with great care to provide an even color, without specking. Therefore, in most cases, use of a compound that has been formulated to give a gray color is preferred [10].

More uniform results are obtained when a *calcined pigment,* such as cobalt nickel gray periclase, is used. For certain special effects in underglaze decorations, a beautiful deep gray can be prepared by using tin antimony gray cassiterite. The limitation on the use of this material is the high cost of the tin oxide base material.

An important point to note with gray is the many subtle shade variations that are possible. With appropriate blending of three or four carefully chosen pigments, many different shades are possible. On the other hand, uniformity of color in this area requires careful quality control.

Blue Pigments. The traditional way to obtain blue in a ceramic material is with *cobalt,* which has been used as a solution color since antiquity [11]. Today, cobalt is reacted with aluminum oxide to produce the spinel $CoAl_2O_4$ or with silica to produce the olivine Co_2SiO_4. Some formulations are mixtures of these two materials. Cobalt silicate involves the use of a higher percentage of cobalt oxide than does aluminate spinel. However, the color is only modestly more intense.

In the spinel system, the shade can be adjusted toward turquoise or green by additions of chromium oxide replacing alumina and zinc oxide replacing cobalt (see section on Green Pigments).

Table 2. Cone temperature equivalents

Cone number	Orton standard pyrometric cones*, °C		Seger cones, °C (used in Europe)
	Large cones, 150 °C**	Small cones, 300 °C**	
010	894	919	900
09	923	955	920
08	955	983	940
07	984	1008	960
06	999	1023	980
05	1046	1062	1000
04	1060	1098	1020
03	1101	1131	1040
02	1120	1148	1060
01	1137	1178	1080
1	1154	1179	1100
2	1162	1179	1120
3	1168	1196	1140
4	1186	1209	1160
5	1196	1221	1180
6	1222	1255	1200
7	1240	1264	1230
8	1263	1300	1250
9	1280	1317	1280
10	1305	1330	1300
11	1315	1336	1320
12	1326	1335	1350

* From the Edward Orton, Jr., Ceramic Foundation, Columbus, Ohio.
** Temperature rise per hour.

At the lower temperatures encountered in porcelain enamels and glass colors, pigments based on cobalt continue to be fully satisfactory both for stability and for tinting strength, which is quite high. At the higher temperatures encountered with ceramic glazes, however, difficulties arise from partial dissolution of the pigment. The cobalt oxide diffuses into the glaze, giving a defect commonly called *cobalt bleeding*. Thus, in glazes, cobalt pigments have been largely replaced by pigments based on vanadium-doped zircon [2], [12], [13]. These pigments are less intense than cobalt pigments and tend toward turquoise. Therefore, they are not applicable in all cases. However, when they are applicable, they give vastly improved stability.

The *zircon–vanadium blue pigment* is made by calcining a mixture (in the stoichiometry of zircon) of zirconia, silica, and ammonium metavanadate in the presence of a mineralizer [2]. The latter materials, which are selected from various halides and silicohalides, facilitate transport of silica during the reaction forming the pigment. Although there is extensive literature on this subject making many claims with respect to composition, the fact is that for development of a strong blue color, the stoichiometry of zircon must be retained and such mineralizers used as will simultaneously optimize the various transport processes and incorporate the optimum amount of vanadium into the zircon structure when it is formed.

With these pigments, use of zircon for opacification is generally desirable. In addition, at least some zirconium oxide in the glaze is preferred to stabilize the pigment.

Green Pigments. Five of the more important methods to obtain green pigmentation in a ceramic material are discussed in this section [14].

Historically, the basis of most green pigmentation was the chromium ion. Although *chromium oxide* itself may be used to produce a green color, this procedure has a number of limitations. First, pure chromium oxide has some tendency to fume or volatilize during the firing of the ceramic coating, which leads to absorption into the refractory of the furnace used. Second, if tin-containing white pigments or pastel colors containing tin are also in the furnace, the chrome will react with the tin to form a pink coloration. Finally, the ceramic material into which chromium oxide is placed must meet particular requirements. It must not contain zinc oxide, which produces an undesirable dirty brown color. As already mentioned, no tin oxide may be used as opacifier or as a constituent of the glaze.

More satisfactory results are obtained if chromium oxide is used as a constituent in a calcined ceramic pigment. One such system is *cobalt chromite blue-green spinel*. In these spinels, varying amounts of cobalt and zinc appear in tetrahedral sites and varying amounts of alumina and chromium oxide appear in octahedral sites. Greener pigments are obtained by using a higher concentration of chromium oxide and a lower concentration of cobalt oxide. Conversely, shades from blue-green to blue result from lowering the amount of chromium oxide and raising the amount of cobalt oxide. These pigments should not be used in low concentration because they give an undesirable dirty gray color.

The final type of chromium oxide containing green is *Victoria green garnet*. This material is prepared by calcining silica and a dichromate (sodium or potassium) with calcium carbonate to form the garnet $3 \, CaO \cdot Cr_2O_3 \cdot 3 \, SiO_2$. This pigment gives a beautiful bright green color but is transparent. When the color is applied thinly, it has a tendency to blacken. Victoria green garnet is not satisfactory for opaque glazes or pastel shades because the tone always has a gray cast and lacks brilliance. It can be used only in zinc-free coatings with high calcium content. In the presence of zinc, the stability of the garnet structure is inadequate. In addition, because this is a difficult pigment to manufacture correctly, the price is high, reflecting the care required.

Because of all of the difficulties mentioned in the use of chromium-containing pigments, and also because there is a definite limitation on the brilliance of green pigments made with chromium, most ceramic glazes use pigments in the *zircon system* [2]. Originally, pigments in the zirconia–vanadia–silica system were recommended. However, because these pigments are, in fact, in-place mixtures of a zircon–vanadium blue pigment and a zirconia–vanadium yellow pigment, superior products can be obtained by preparing the pigments separately, using the optimum preparative conditions for each one. Moreover, because the zircon–praseodymium yellows are the strongest in their color family, their use as the yellow constituent of a green blend gives even better results. Therefore, the cleanest, brightest, most stable greens are obtained currently by the use of blends of a zircon–vanadium blue and a zircon–praseodymium yellow. The bright green shades are obtained from a mixture of ca. two parts of the yellow pigment to each part of the blue pigment.

Finally, *copper compounds* are used in certain low-temperature firing applications [14]. The use of copper is of little interest to the majority of industrial manufacturers, but the colors obtained from it are of great interest to art potters because of the many subtle shades that can be obtained. This variety arises because the pH of the glaze affects the color obtained from copper. If the glaze is alkaline, a turquoise blue color results; if the glaze is acidic, a beautiful green color develops. Copper oxide dissolves in the glaze composition, and is, therefore, a transparent color. Because copper oxide volatilizes quite readily, it should not be used above 1000 °C.

Another limitation on the use of copper colors is the fact that copper oxide renders many lead-containing glazes unsafe for contact with food or drink. Therefore, copper pigments should never be used on such articles.

Yellow Pigments. Although a number of systems exist for preparing yellow ceramic colors, there are technical and economic reasons for the use of a particular yellow pigment. The pigments of greatest tinting strength, the lead antimonate yellows and the chrome–titania maples, do not have adequate resistance to molten ceramic coatings. Therefore, other systems must be used if the firing temperature exceeds ca. 1000 °C. Three of these higher temperature systems are considered in this section.

Zirconia–vanadium yellows are prepared by calcining zirconium oxide with small amounts of ammonium metavanadate [13], [15]. Titanium dioxide or iron oxide may be used to alter the shade. In the absence of these latter materials, lemon-yellow is obtained; in their presence, orange-yellow results. In ceramic coatings, zirconia–vanadium yellows are usually weaker than tin–vanadium yellows and muddier than praseodymium–zircon yellows. However, they are economical stains for use with either zinc-containing or zinc-free coatings. They are stronger and brighter in low-lead, low-boron glazes. Zirconium silicate is the preferred opacifier.

Tin–vanadium yellows are prepared by introducing small amounts of a vanadium oxide into the cassiterite structure of tin oxide [16]. The shade may be varied by addition of titanium dioxide or iron oxide. In the absence of these materials, a lemon-yellow shade is obtained. A stronger yellow may be made by adding titanium dioxide to the color batch, and the increased strength of this modified yellow is accompanied by an increase in the apparent redness of the pigment when iron oxide is added.

Tin–vanadium yellow pigments develop a yellow color in all ceramic materials, although the actual shade may be influenced by the nature of the substrate material. These are opaque pigments, which need minimum amounts of opacifier. However, these pigments are sensitive to reducing conditions. Moreover, any blends with chrome-bearing pigments should be avoided. The reason is that tin oxide and chrome oxide combine easily to form a compound with a color similar to that of chrome–tin pink, which shows up in the ceramic material as a brown discoloration. Finally, grinding the tin–vanadium yellow pigment should be minimized because it tends to weaken the pigment.

The primary deterrent to the use of tin–vanadium yellows, however, is not any technical deficiency. Rather, it is the high cost of the tin oxide that is the major component. The result of this high cost, together with the quality of the praseodymium–zircon pigments, has been a decline in the use of tin–vanadium yellows.

Praseodymium–zircon pigments are formed by calcination of ca. 5% praseodymium oxide with a stoichiometric mixture of zirconium oxide and silica in the presence of mineralizers to yield a bright yellow pigment [2], [16], [17]. This pigment is quite analogous to zircon–vanadium blue pigments in that the crystal structure is that of zircon. Praseodymium–zircon pigments have excellent tinting strength in high-temperature coatings. They can be used in almost any ceramic coating, although preferably with zircon opacifiers. They blend well with other pigments, particularly with other zircon and zirconia pigments. These pigments are being increasingly used for all applications in which the firing temperature exceeds 1000 °C.

For lower temperature applications, the tinting strength of the *lead antimonate pigments* is unsurpassed, except by cadmium sulfoselenides [18]. Lead antimonate pigments, which have traditionally been called Naples yellow, are exceptionally clean and bright and have good covering power, requiring little or no opacifier.

The primary limitation is their instability in ceramic coatings above ca. 1000 °C, which leads to volatilization of the antimony oxide. Substitutions of cerium oxide, alumina, or tin oxide are sometimes made for a portion of the antimony oxide to improve its stability. Thus, although these materials have limited usage in ceramic glazes, they are the pigment of choice in porcelain enamel.

For the brightest, low-temperature applications, *cadmium sulfoselenide yellow* can be considered [19]. The pure cadmium sulfoselenide colors are produced in a range from primrose yellow through yellow to orange and red. Cadmium sulfide itself is yellow to orange, depending on details of its manufacture and the ratio of the alpha to beta forms of the crystal. The primrose yellow and light yellow shades are made by precipitating small amounts of zinc sulfide along with the cadmium sulfide.

One final orange-yellow pigment needs to be considered. This is the pigment formed when chromium oxide is added with antimony oxide to titanium dioxide to form a *doped rutile* [18]. This material gives an orange-yellow or maple shade, and is useful in lower melting ceramic coatings. Like the lead antimonate yellows, it begins to decompose at ca. 1000 °C. Although it is of limited use in high-firing ceramics, it is one of the largest volume pigments used in porcelain enamel, where it forms the basis for some of the high-volume appliance colors.

Brown Pigments. By far the most important brown pigment used in ceramics is *zinc iron chromite brown spinel* [20]. This family produces a wide palette of tan and brown shades and can be controlled with reasonable care to produce uniformity within the production variables existing in commercial plants. Within the spinel structure, the zinc oxide is found on the tetrahedral sites and the chromium oxide on the octahedral sites. The iron oxide is distributed in such a way as to fulfill the requirements of the structure. Consequently, adjustment of the formula does result in alteration of the shade. For example, a substantial increase in chrome and decrease in zinc results in greener shades in zinc-free coatings and yellow to gray shades in zinc-containing coatings. Minor addition of manganese to this system results in yellowish and grayish shades, whereas addition of minor amounts of nickel oxide results in a much darker brown. Because they are comparatively inexpensive, these pigments are the brown selected for most applications. However, two systems closely related to the zinc iron chromite brown spinel have been developed to improve the firing range and stability of brown pigments.

The first of these is the *addition of alumina to the zinc iron chromite brown spinel*. This creates a pigment that is a hybrid of the zinc iron chromite brown spinel and the chrome alumina pink spinel. It produces warm, orange-brown shades with improved firing stability. This pigment is used in coatings that are high in zinc and alumina and low in calcium oxide. The alumina–to–zinc ratio is kept as high as practical to improve the brightness and cleanliness of the pigment.

Another related pigment is a *tin-containing iron chromite brown spinel*, which is sometimes called a tin tan. As produced, this material is a mixture of chromium oxide, tin oxide, and iron aluminate. It is always used in a zinc-containing coating to obtain optimum brown shades. Most likely, this is because the pigment reacts with zinc from the coating during the firing process to produce a zinc iron chromite brown spinel pigment. In coatings that are free of zinc, this pigment produces shades of gray to dark mahogany. The pigment has excellent stability at low concentration. Therefore, it makes an excellent toner for some tan and beige shades in blends with various pink pigments.

The final brown pigment to be considered is *chrome iron manganese brown spinel*. Manganese is well-known as the colorant in amethyst-stained glass and, with iron oxide, it has been responsible for the deep brown glazes associated with electrical porcelain insulators, artware, and bean pots. It is used, therefore, where a deep brown shade is needed. However, in producing medium to light shades of brown, the presence of manganese often causes poor surface and unstable color with tendencies to volatilization. Therefore, the use of this pigment is rather restricted.

Pink and Purple Pigments. Only a short step in the color spectrum separates brown and pink. This is reflected in the *chrome alumina pink spinels*, which are similar in crystal structure and behavior to the zinc iron chromite brown pigments except for the absence of iron oxide [21]. Chrome alumina pinks are combinations of zinc oxide, aluminum oxide, and chromium oxide. Depending on the concentration of zinc, the crystal structure may be either spinel (zinc aluminate–chromite) or corundum (solid solution of chromium in aluminum oxide). The latter is analogous to the composition of a ruby.

In general, a ceramic coating formulated for chrome alumina pink spinels should be free of calcium oxide, with low concentrations of lead oxide and boric oxide, and with a surplus of zinc oxide and alumina. Using an improper glaze results in a brown pigment in place of the desired pink. Sufficient zinc oxide must be in the coating to prevent the glaze from attacking the pigment and removing zinc from it. A surplus of alumina prevents the molten coating from dissolving the pigment.

A related, but somewhat stronger, pink pigment is *manganese alumina pink corundum*. This pigment is formulated by addition of magnesium oxide and phosphate to aluminum oxide. A pure, clean pigment is obtained. The use of a proper formulation, however, is important. A zinc-free system with a high concentration of alumina is required. Unfortunately, the manufacture of this pigment involves serious pollution problems. As a result, several companies have stopped manufacturing it and there is question as to its continued availability.

The most stable pink pigment is the *iron–zircon* system [2], [22], [23]. It is made by calcining a mixture of zirconium oxide, silica, and iron oxide, using a stoichiometry that will produce zircon. This pigment is sensitive to minor variations in the production process, so that one manufacturer's pigment may not duplicate another's [24]. Shades extend from coral to pink. The pigment is stable in all coating formulations, but those without zinc oxide are bluer in shade.

The final pink system, and the only one to produce purple and maroon shades as well as pinks, is *chrome–tin pink*. These are pigments produced by calcining mixtures of small amounts of chromium oxide with substantial amounts of tin oxide. In addition, most such materials have large quantities of silica and calcium oxide in the formulation. The chemistry of these materials is complex and only recently has their chemical composition been determined [25]. Mixing ca. 90% tin oxide with small amounts of chromium oxide and either calcium oxide or cerium oxide, together with boric oxide as a mineralizer, gives chrome tin orchid cassiterite. This material is a solid solution of chromium oxide in tin oxide. Although this is not the crystal structure of most chrome–tin pinks, residual amounts are present in almost all cases. It is this residual amount of chromium-doped tin oxide that gives most chrome–tin pinks a somewhat gray or purple overtone.

For most chrome–tin pinks, addition of substantial amounts of calcium oxide and silicon oxide is required to make chrome tin pink sphene. Only in the presence of these materials can pink, red, or maroon shades be obtained. In this case the crystal structure is tin sphene ($CaO \cdot SnO_2 \cdot SiO_2$) in which chromium oxide is dissolved as an impurity.

The color of this pigment depends to a great extent on the ratio of the concentration of chromium oxide to that of tin oxide. Generally speaking, when this ratio is 1:5, the resulting color is green; 1:15, purple; 1:17–20, red or maroon; and 1:25, pink.

These pigments are calcined at 1260–1320 °C and, under the right conditions, are stable at these temperatures. They can be used in coating materials that are low in zinc and high in calcium oxide. Either tin oxide or zirconium silicate opacifiers can be used, but tin oxide as a mill addition improves the strength and stability of the pigment.

Gold purple, commonly called Purple of Cassius, is an old pigment consisting of tin oxide gel colored by finely divided gold. This pigment can be used in low-temperature materials, such as porcelain enamels, where it has good coverage and brilliance. It is, however, an expensive pigment. This is due not only to the high price of gold, but also to the difficult methods of preparing the pigment.

Red Pigments. There are no oxide systems which can be used to produce a true red pigment that is stable in ceramic systems. Therefore, orange, red, and dark red pigments are obtained by the use of *cadmium sulfoselenide pigments* [19], [26]. The specific shade results from varying the ratio of the concentration of cadmium sulfide to that of cadmium selenide. An orange pigment is obtained at a ratio of ca. 4:1, a red pigment at 1.7:1, and a deeper red at 1.3:1.

These pigments are prepared by one of several chemical processes involving wet precipitation of suitable raw materials, such as cadmium carbonate and elemental sulfur or selenium, followed by calcination at 500–600 °C under an inert atmosphere.

Cadmium sulfoselenide pigments require the use of a glaze specially designed for this purpose. This glaze contains only small amounts of lead oxide because high-lead flux materials react with selenium in a cadmium sulfoselenide pigment to form lead selenide, which is black. The glaze is a low-alkaline borosilicate type. It contains a few percent of cadmium oxide, which reduces the potential for dissolution of the pigment in the glaze during firing. It is free of vigorous oxidizing agents, such as nitrates, which oxidize the pigment, completely destroying the color.

These pigments are temperature sensitive. Therefore, although they can be used in glass colors, in porcelain enamels, and in low-temperature glazes fired up to ca. 1000 °C, they cannot be used in higher temperature applications.

In order to extend the range of these colors, an inclusion pigment system has recently been developed [2], [27]. In this system, cadmium sulfoselenide is incorporated in a clear zircon lattice during manufacture. In this way, the superior stability of zircon is imparted to the pigment. However, these pigments are difficult to make and not all shades have been made successfully. The color palette extends from yellow through orange to red. Dark reds are not yet available.

Precious Metal Compositions. Although bright gold, burnished gold, and the corresponding silver and platinum preparations are not ceramic pigments in the strict sense, they still play a considerable role in the decoration of ceramic materials [28]. The production of these materials consists essentially of the reaction of pinene with sulfur or hydrogen sulfide to give a pinene thiol, which then reacts with tetrachlorauric acid to yield an auric sulforesinate. This gold resinate is dissolved in an organic solvent and then reacted with various additives, in the form of metal organic compounds, to affect the color tone and to achieve the necessary adherence. These materials are applied to the outside surface of the ceramic object and fired at 500–850 °C.

Lusters. The pigments that are designed as lusters are closely related to the precious metal compositions because they are also usually prepared and applied as organic compounds of metals. After firing, they precipitate as a thin, often irridescent layer on the substrate. They are not a coherent metallic layer as with precious metal coatings, but rather are an oxide layer. Colored lusters result when compounds of transition metals are included in the formulation. Iron lusters give light brown-red to golden coatings; cobalt lusters in high dilution, a chocolate brown; copper lusters, a reddish brown; nickel lusters, a light brown; and manganese lusters, a gray brown.

5. Industrial Production of Ceramic Pigments

Ceramic pigments are customarily prepared by solid-state reactions. For that reason, rapid, uniform, and reproducible conversion requires intimate mixing of the raw materials, which must be of optimum particle size for the given reaction. In most pigment preparations, additives called mineralizers are included to increase the rate of reaction and make the mixture more uniform. For zircon pigments, these mineralizers are usually alkali and alkaline-earth halides [13], [15]. Boric acid is often used as a mineralizer for spinel pigments. Much of the art of ceramic pigment manufacture is concerned with selection of particular mineralizers for a given reaction.

After careful mixing, the pigments are *calcined* in either batch kilns or continuous calciners. In this operation, careful attention must be paid to the control of the temperature of the kiln. The advantages of batch calcination procedures lie in greater production flexibility and the ability to prepare smaller quantities when required. The continuous tunnel kiln provides greater product quality and greater consistency in calcining conditions. However, its use requires a minimum production level.

After calcination, hard clumps of calcined, sintered products are broken in *jaw or roll crushers* and then ground to the necessary fineness in *mills*. Depending on the particular pigment and the particle size required, either wet or dry ball mills may be used. Wet ball milling yields a finer product, but it is considerably more expensive than dry ball milling.

Some pigments must be *washed* to remove soluble constituents that would otherwise cause difficulty in the final application. Washing is particularly required for pigments containing vanadium oxide. The ground, suspended particles are washed in filter presses or decanters. The water is removed mechanically and the remaining pigment slip or filter cake dried.

The final production step involves careful control of color tone by adjustment with toners. *Toners* are formulations at various extremes of the color spectrum covered by a given pigment family and are used to adjust the color of products to specifications.

6. Use of Ceramic Pigments in Glazes

There are five ways in which ceramic colors can be applied to a glazable ceramic article: as a body stain, as an engobe, as an underglaze color, as a colored glaze, and as an overglaze or glass color. The use of a *body stain* refers to a pigment added to the body formulation itself. The technique of using *engobes* may be described as that of applying a ceramic pigment to a raw body. *Underglaze* decorating is the application of color to a bisque body. In *colored glazes,* the pigment is dispersed in the glaze itself. Finally, *overglazes or glass colors* are applied to the already formed and fired glaze as an overcoat.

The *selection of a technique* depends on the requirements of the particular application. For example, if an engobe or body stain is used, it must be stable to the bisque as well as to the glost fires. An underglaze color or a colored glaze need be stable only to the glost fire. On the other hand, an overglaze or glass color need not be stable to either of these firings.

Moreover, the *range of colors and effects* which can be obtained is directly related to the stability. Some colors, such as bright red, can only be obtained in overglaze decoration or in specially formulated glazes. On the other hand, the *durability* of a decoration or color in service depends largely on the distance between the outer surface of the ware and the pigment. Therefore, overglaze decorations are distinctly inferior to other techniques of application with respect to durability in service.

7. Ceramic Glazes

The ceramic pigment and application method must be compatible with the ceramic glaze that is to be used. Glazes having a wide range of firing temperatures are available, from hobby glazes firing at cone 010 to sanitary ware glazes from cone 7 to cone 12 [29]–[31].

Low-firing glazes, which mature at cone 010 to 01, are used primarily by hobbyists and artware potters. The reason is that it is only at these temperatures that a full palette of colors can be used in-glaze as well as overglaze. In particular, the cadmium sulfoselenide red pigments can be used only in this firing range and only with a specially designed glaze formulation. In addition, a wide range of special effects can be used at these firing temperatures. These effects include crackle glazes, high-calcium–alumina matte glazes, and crystalline glazes containing materials such as rutile.

The glazes that fire out at cone 01 to cone 6 consist of most of the dinnerware and tile glazes and a minority of high-temperature artware. The color palette at these temperatures is somewhat reduced, but still quite extensive. Spinel blacks, blues, blue-greens, and browns are suitable, as are all the pigments based on zirconia and zircon: zircon–vanadium blues, zircon–praseodymium yellows, zirconia–vanadium yellows, zircon–iron pinks, and zirconia grays. Tin–vanadium yellows are stable as are chrome–tin pinks. Chrome–alumina pigments are stable with proper glaze formulation.

The colors used in glazes up to cone 12 are much the same as those used at cone 01 to cone 6. Tin–vanadium and zirconia–vanadium yellows are quite stable. Chrome–tin pigments are acceptably stable if a suitable formulation is used. The same is true of chrome–alumina pinks. Chrome–iron–zinc browns are satisfactory with suitable formulations. The zircon pigments are all stable to cone 12, although zircon–praseodymium yellow begins to lose strength at higher temperature. Colors containing chromium oxide are stable to cone 12 in a zinc-free glaze.

8. Application Media

For application methods other than in-glaze, the pigment is mixed with an organic medium, which serves only to apply the decorative color to the ceramic substrate. This medium must then burn or evaporate completely in the firing of the ceramic. The materials that are used are divided into two types: hydrophilic media and lyophilic media.

Hydrophilic media are always needed if the decorating color is to be applied to a wet substrate, or if aqueous suspensions are to be applied over the decoration in a subsequent operating step. A typical example would be underglaze colors applied to bisque-fired pieces that are subsequently to be glazed. Hydrophilic preparations can be based on glycerol, ethylene glycol, poly-(ethylene oxide), or polyglycols as binding agents, and water or other alcohols as solvents.

Lyophilic media are used primarily in overglaze decoration. In one application, they are a component in the decalcomania that are extensively used in overglaze decoration. In terms of volume, however, the largest amount of lyophilic media is used in silk screening. These materials generally consist of a methacrylate-based binding agent, which will depolymerize on firing of the ceramic decoration, and hence, completely evaporate. Pine oil or turpentine usually serves as the solvent.

9. Quality Control

Quality control in the production of ceramic pigments is primarily a matter of controlling the color of the pigment after application [32]. The human eye is an extremely sensitive measuring device for color, particularly on a comparative basis [33]. Thus, control of color solely by electronic procedures is difficult [34]. In many cases, electronic quality control techniques must be backed up with preparation of trial glazes for visual comparison. Moreover, it is usually necessary to adjust the color of each lot of pigment to ensure adequate reproducibility from lot to lot. In addition, for some applications, the particle size of the pigment is critical. If vanadium oxide is used in the formulation, residual free vanadium must be controlled.

10. References

[1] A. Burgyan, R. A. Eppler, *Am. Ceram. Soc., Bull.* **62** (1983) 1001–1003.
[2] R. A. Eppler, *Am. Ceram. Soc., Bull.* **56** (1977) 213–216.
[3] *DCMA Classification and Chemical Description of the Mixed-Metal Oxide Inorganic Colored Pigments,* 2nd ed., Dry Color Manufacturers' Assoc., Alexandria, Va., 1982.
[4] R. A. Eppler, *J. Am. Ceram. Soc.* **66** (1983) 794–801.
[5] F. T. Booth, G. N. Peel, *Trans. Brit. Ceram. Soc.* **58** (1959) 532–564.
[6] C. W. F. Jacobs, *J. Am. Ceram. Soc.* **37** (1954) 216–220.
[7] R. D. Shannon, A. L. Friedberg, *Univ. Ill. Eng. Exp. Sta. Bull.* **456** (1960) 1–49.
[8] R. A. Eppler, *J. Am. Ceram. Soc.* **52** (1969) 89–99.
[9] R. A. Eppler, *Am. Ceram. Soc., Bull.* **60** (1981) 562–565.
[10] W. F. Votava, *Am. Ceram. Soc., Bull.* **40** (1961) 17–18.
[11] R. K. Mason, *Am. Ceram. Soc., Bull.* **40** (1961) 5–6.
[12] Harshaw Chemical Co., US 2 441 447, 1948; US 3 025 178, 1962 (C. A. Seabright).
[13] C. A. Seabright, H. C. Draker, *Am. Ceram. Soc., Bull.* **40** (1961) 1–4.
[14] P. Henry, *Am. Ceram. Soc., Bull.* **40** (1961) 9–10.
[15] F. T. Booth, G. N. Peel, *Trans. Brit. Ceram. Soc.* **61** (1962) 359–400.
[16] E. H. Ray, T. D. Carnahan, R. M. Sullivan, *Am. Ceram. Soc., Bull.* **40** (1961) 13–16.
[17] R. A. Eppler, *Ind. Eng. Chem. Prod. Res. Dev.* **10** (1971) 352–355.
[18] *The Colour Index,* 3rd ed., Soc. Dyers & Colourists, Bradford-London 1971.
[19] R. A. Eppler, D. S. Carr: *Proc. 3rd Int. Cadmium Conference,* International Lead Zinc Research Organization (ILZRO), New York 1982.
[20] J. E. Marquis, R. E. Carpenter, *Am. Ceram. Soc., Bull.* **40** (1961) 19–24.
[21] R. L. Hawks, *Am. Ceram. Soc., Bull.* **40** (1961) 7–8.
[22] Glidden Co., US 3 189 475, 1965 (J. E. Marquis, R. E. Carpenter).
[23] Harshaw Chemical Co., US 3 166 430, 1965 (C. A. Seabright).
[24] R. A. Eppler, *J. Am. Ceram. Soc.* **53** (1970) 457–462.
[25] R. A. Eppler, *J. Am. Ceram. Soc.* **59** (1976) 455.
[26] Glidden Co., US 2 643 196, 1953; US 2 777 778, 1957 (B. W. Allan, F. O. Rummery). Fabriques de Produits Chimiques de Thann et de Mulhouse, US 3 528 834, US 3 528 835, 1970 (J. Gascon).
[27] Fabriques de Produits Chimiques de Thann et de Mulhouse, US 3 445 199, 1969 (B. H. P. Fehr, J. Gascon). H. D. DeAhna, *Ceram. Eng. Sci. Proc.* **1** (1980) 860–862.
[28] Du Pont, US 2 924 540, 1960 (J. B. D'Andrea).
[29] R. A. Eppler in D. R. Uhlmann, N. J. Kreidl (eds.): *Glass Science and Technology,* vol. **1**, Academic Press, New York 1983, pp. 301–338.
[30] C. W. Parmalee, C. G. Harman: *Ceramic Glazes,* 3rd ed., Cahners Publ., Boston, Mass., 1973.

[31] F. Singer, W. L. German: *Ceramic Glazes,* Borax Consolidated, London 1964.
[32] K. Shaw: *Ceramic Colors and Pottery Decoration,* MacLaren & Sons, London 1962.
[33] D. B. Judd, G. Wyszecki: *Color in Business, Science and Industry,* Wiley-Interscience, New York 1963.
[34] D. A. Klimas, A. Canonico, *Am. Ceram. Soc., Bull.* **63** (1984) 445.

Ceramics, Electronic

JOHN B. BLUM, Norton Company, High Performance Ceramics Division, Northboro, Massachusetts 01532, United States

1. Introduction 1087
2. Linear Dielectrics 1087
3. Nonlinear Dielectrics 1091
4. Semiconducting Ceramics 1102
5. Sensors 1105
6. References 1106

1. Introduction

The use of ceramics for electronic applications is widespread and expanding rapidly. Ceramic materials are used for electrical circuit elements, e.g., capacitors, as substrates for circuits, as sensors and transducers, and in a wide variety of magnetic applications. Table 1 is the most recent classification of electronic ceramics, compiled by WAYNE YOUNG of IBM for the Electronics Division of the American Ceramic Society.

While there is no single recent, comprehensive discussion of the field, there are older publications that are still useful. A brief history of the development of electronic ceramics prior to 1945 was given by THURNAUER [1]. In 1964 KOENIG presented an overview [2]; and in 1969 HENRY wrote an introductory book in the field of electronic ceramics [3].

The field of electronic ceramics is very broad, as can be seen from Table 1. In this article, important representative materials, technologies, uses, and background are discussed. Substrates are discussed under the linear dielectrics, while some of the semiconducting ceramics are discussed as sensor materials. Magnetic ceramics are not discussed.

2. Linear Dielectrics

Linear dielectrics are materials in which the variation of electric displacement D is linear with respect to electric field E. An important class of these materials is used as electrical insulators. In such applications it is important to isolate portions of an electric circuit from one another.

Al_2O_3. Aluminum oxide [1344-28-1], or alumina, is a versatile ceramic material ($\rightarrow$ Aluminum Oxide). WEFERS and BELL have reviewed the properties of the oxides

Table 1. Classification of electronic ceramics*

	Dielectric	Magnetic	Semiconductor	Substrates/composites	Optical
Types	linear/nonlinear low/high loss para/ferro/antiferroelectrics piezoelectrics	ferri/ferro/antiferro- magnetics magnetostrictives	luminescents thermoelectrics thermoionics photovoltics conductive ceramics	IC packages films/coatings laminates fiber composites bonded/impregnated composites	lasers windows electro-optical materials fiber optics nonlinear optical materials
Pertinent properties	density resistivity dielectric constant dielectric losses dissipation factor electric strength hysteresis loop properties frequency dependence thermal expansion temperature coefficients Curie temperature piezoelectric constants and coupling	density resistivity permeability coercive force losses residual flux density saturation magnetization resonant frequency hysteresis loop properties frequency dependence thermal expansion coupling coefficient mechanical strength temperature coefficients thermal conductivity Curie point	density resistivity carrier (n, p) properties conductivity resistance vs. temperature Hall coefficient Seebeck coefficient thermal conductivity photoconductivity photovoltage/current electroluminescence thermoelectricity mechanical strength	density resistivity dielectric properties hardness mechanical strength elastic moduli thermal expansion thermal conductivity hermeticity green properties compressibility screenability platability cosinterability single-crystal properties vacuum deposition properties	density optical transmission/ transparency dispersion refractive index ferroelectricity bond levels/transitions mechanical strength melting point Curie temperature dielectric constant piezoelectric modulus fabrication process crystal structure single-crystal properties polarization

Table 1. (continued)

	Dielectric	Magnetic	Semiconductor	Substrates/composites	Optical
Uses	insulators capacitors transducers nonlinear elements attenuators	magnets cores transducers read/write heads microwave devices coils yokes flyback transformers	phosphors sensors infrared detectors thermoelectric devices resistors nonlinear thermistors varistors magnetoresistors lumistors cathodes oxide batteries	electronic packages hermetic seals electrodes thick-film resistors printed circuit materials dielectric films bubble memory devices IC passivation multilayer ceramics	holographic memories optical switches lasers radomes Kerr cells optical doublers windows fiber optics lamp envelopes
Materials	steatites titanates glasses bonded micas aluminas	ferrites — hard/soft garnets magnetoplumbites	elements oxides chalcogenides carbides silicides borides β-aluminas phosphides arsenides	oxides glass – ceramics glass – metals porcelain – metal oxide films garnets cermets	oxides glasses halides gallium arsenide zinc selenide silica gadolinium molybdate

* Prepared by WAYNE YOUNG (IBM).

Linear Dielectrics

Table 2. Properties of commercial alumina substrates

Product no.	Alumina content, wt%	Volume resistivity, Ω cm	Dielectric constant at 1 MHz	Linear expansion[d], 25–800 °C, 10^{-6} K^{-1}	Thermal conductivity, W m^{-1}K^{-1}	Bending strength, MPa
ADO-90[a]	92	> 10^{14}	22	8.1	12.6	365
HA-92[b]	92	> 10^{14}	8.5	7.5	16.8	314
AD-94[a]	94	> 10^{14}	8.9	7.6	18.0	352
AD-96[a]	96	> 10^{14}	9.0	8.0	24.7	358
HA-96[b]	96	> 10^{15}	9.3	7.7	20.1	324
AT-396[c]	96	> 10^{14}	9.6	7.7	33.6	314
HA-995[b]	99.5	> 10^{15}	9.7	8.1	29.4	392
AT-3995[c]	99.5	> 10^{14}	10.6	7.6	37.8	314
AT-3997[c]	99.7	> 10^{14}	10.5	7.8	37.8	392

[a] Product of Coors Ceramics, United States.
[b] Product of NTK Technical Ceramics, Japan.
[c] Product of Narumi Technical Ceramics, Japan.
[d] For comparison, the linear expansion of silicon is $\approx 4 \times 10^{-6}$ K^{-1}.

and hydroxides of aluminum [4]. In this section only aspects relevant to the electronic use of alumina are discussed.

An important use of alumina is in packaging microelectronic circuitry [5]. This use requires a material with a high electrical resistivity, low dielectric constant, good mechanical strength, good thermal conductivity, and a thermal expansion that matches that of the integrated circuit material, which is usually silicon. Alumina is a good compromise.

The material used is polycrystalline corundum with purity generally ranging from 92 to 99 wt % Al_2O_3. The other 1–8 wt % is predominantly MgO and SiO_2. These materials are added to affect the firing behavior. As would be expected, different compositions have different properties, as shown in Table 2. The properties also depend on the manufacturer (Table 2), because the various suppliers use different additives in the same nominal alumina composition (e.g., 96 %). The alumina is processed with careful considerations to purity and is generally formed into substrates by the tape casting, or doctor blading, process (see p. 1096) [6].

A recent development is the use of *multilayer alumina substrates* [7]. In this technology, thinner alumina layers are used than those used as single-layer substrates. The layers are laminated together. In between the individual ceramic layers are metallic lines, as well as holes, or vias, filled with metal. These metal lines and vias act as a three-dimensional conducting path interconnecting individual parts of a circuit. As of 1985, IBM has the most advanced such design in their Thermal Conduction Module (TCM) [7], [8]. In this module is a 90 × 90 × 15 mm ceramic substrate made up of 33 layers of 92% alumina. The substrate supports and, through the three-dimensional conducting network, interconnects, supplies power to, and transmits data to and from up to 118 integrated circuit chips mounted on the substrate surface.

Table 3. Properties of newer electronic substrate materials

Material	Electrical resistivity, Ω cm	Dielectric constant	Thermal conductivity, $W m^{-1} K^{-1}$	Linear expansion, $10^{-6} K^{-1}$
SiC[a]	$> 10^{13}$	42	267	
AlN[b]	$> 10^{13}$	8.8	80	4.5
BeO[c]	10^{17}	6.7	260	6.5
Glass–ceramic[d]	$> 10^{14}$	7.5	4.2	4.2

[a] Hitachiceram, product of Hitachi Corp., Japan [9].
[b] Toshiba Corp., Japan [10].
[c] 99.5 wt% BeO, product of National Beryllia Corp., United States [11].
[d] Lead borosilicate–alumina, NEC Corp., Japan [12].

New Substrate Materials. There are several properties desirable in a substrate for integrated circuitry, and alumina is a good compromise. However, for the increasingly demanding requirements of new integrated circuit technology, materials with properties better than those of alumina are needed. For very high speed integrated circuits (VHSIC), a low-dielectric-constant material is needed to minimize delays in signal propagation. For very large scale integrated circuits (VLSI), the density of circuit elements is so great that the substrate material must have a high thermal conductivity to dissipate the heat. For circuits based on silicon [7440-21-3], a substrate with a close match in thermal expansion to that of Si ($4 \times 10^{-6} K^{-1}$) is preferred. No single material has yet been found that incorporates all of these properties, as can be seen in Table 3, which lists some newer materials being evaluated for use as microelectronic substrates.

3. Nonlinear Dielectrics

Nonlinear dielectrics are materials in which the variation of electric displacement D is a nonlinear function of the electric field E. This class of materials includes ferroelectrics and piezoelectrics. Ceramics displaying these effects are useful in capacitors and transducers. There are books that deal with ferroelectric ceramics [13] and piezoelectric ceramics [14].

Piezoelectricity is defined as electric polarization produced by mechanical strain, the polarization being proportional to the strain and changing sign with it [15]. Ferroelectricity is the spontaneous alignment of electric dipoles by their mutual interaction. The conspicuous property of ferroelectrics is the reversibility of their permanent polarization by an electric field [16]. For a polycrystalline body to exhibit piezoelectricity, the material must also be a ferroelectric. Thus, for ceramics ferroelectricity and piezoelectricity are interrelated.

Ferroelectricity. Ferroelectricity is the dielectric analog of ferromagnetism. The term ferroelectricity is derived from this analogy, but does not imply that iron is important in the phenomenon. Ferroelectricity in a material is a consequence of permanent electric dipoles that result from its crystal structure. These dipoles are aligned because of their mutual interaction. In a ferroelectric material, the dipoles are aligned in parallel. Within a single crystal, all of the dipoles do not point in the same direction; rather, there are small areas within each single crystalline region in which all of these permanent dipoles are aligned in parallel for a net nonzero electric polarization. These areas are called *domains*. In a related class of materials, *antiferroelectrics*, the permanent electric dipoles are aligned antiparallel, which cancels out the individual dipole moments for a net polarization of zero.

If the crystal structure of a material does not yield permanent dipoles, the material does not exhibit ferroelectricity. Therefore, as any temperature or composition variations that lead to phase changes occur in ferroelectric materials, ferroelectric behavior is observed or not, depending on the crystallographic phase present. The temperature below which ferroelectricity is observed is called the *Curie point* of the material. Above the Curie point, the material is nonferroelectric or paraelectric.

The domains in a ferroelectric possess nonzero polarization; however, in the as-formed material the domains are randomly oriented, which leads to an overall polarization of zero in the sample, as seen in Figure 1. The domain structure (i.e., number, size, and direction) in a given sample is the result of several competing factors and, thus, depends on the specific conditions for the sample. The dipole moments in the various domains can be realigned by an externally applied electric field, yielding a nonzero polarization in the sample. The process by which domains change orientation is nucleation and growth; i.e., domains of the new orientation nucleate within the domain and grow preferentially because of the external electric field. The resultant polarization depends on electrical history: if the electrical field is increased and then decreased back to its original value, the final polarization differs. In Figure 2 this *hysteresis* is shown. The hysteresis loop is a consequence of the domains that make up the ferroelectric.

Consider the following experiment on a ferroelectric material:

1) Increase the electric field to the value E_1
2) Decrease the field past zero to the same value but of the opposite direction, $-E_1$
3) Cyclically, repeat steps 1 and 2

As shown in Figure 1, for the starting material the random orientation of the domains results in a net polarization of zero. When not initially subjected to an electric field, the material has no polarization, represented at the origin O of Figure 2. As the electric field is increased, the resulting polarization for low fields is reversible and nearly linear with E. At high fields the domains begin to orient under the influence of the applied electrical field, so the polarization increases more rapidly. At some high value of E, all of the domains that could reorient with the electrical field have done so: no further increase in polarization can occur. This is known as polarization saturation, and extrapolation

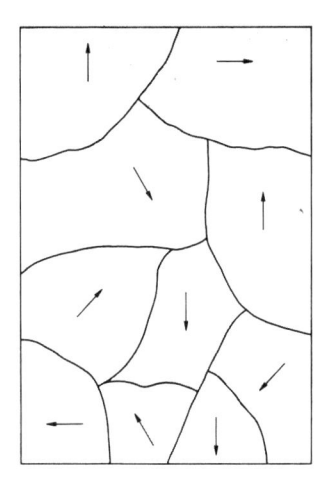

Figure 1. Domain pattern in an unpolarized ferroelectric (schematic), the arrows representing the polarization within each domain

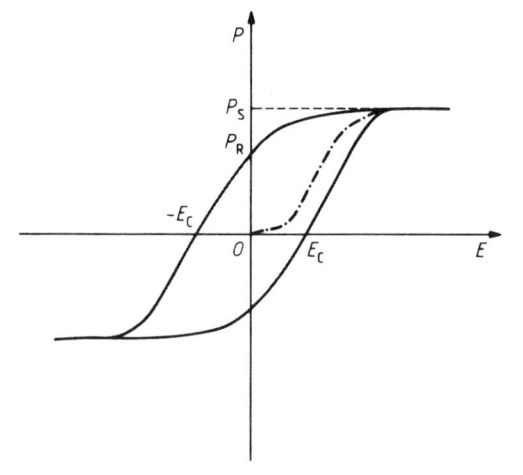

Figure 2. Hysteresis loop for a ferroelectric (schematic)

of this portion of the curve back to the ordinate gives the saturation polarization P_S. The ferroelectric is then said to be *polarized*.

Consider what happens when the field is decreased to 0. The inability of oriented domains to return to their original random ($P = 0$) positions, because of the irreversibility of the nucleation and growth process, results in a nonzero polarization, the remnant polarization P_R. The domains must be forced into a configuration that again results in a net polarization of zero. The magnitude of the field needed to obtain $P = 0$ is called the coercive field E_C. Further decrease in $E < -E_C$ causes domain reorientation, but in the direction opposite to that discussed above. The resulting polarization behavior is symmetrical to that discussed above.

Once the ferroelectric has been polarized, only the outer portion of Figure 2, the hysteresis loop, is followed. The curve in Figure 2 from the origin to saturation is followed only by the unpolarized material. However, a sample can be depolarized by heating it above its Curie point. When the sample is cooled, the domains in the sample again orient randomly.

Piezoelectricity. Piezoelectric materials convert mechanical energy to electrical energy, and vice versa. The conversion of an applied mechanical stress (force) to an electrical polarization is called the direct piezoelectric effect. Conversion of an applied electrical voltage to a mechanical strain (deformation) is called the converse piezoelectric effect. Any material that displays one effect displays the other. An important parameter of a piezoelectric is its electromechanical coupling factor k, a measure of its conversion efficiency.

As with ferroelectricity, piezoelectricity is a consequence of the crystal structure. The piezoelectric effect requires a noncentrosymmetric crystal structure. In materials possessing these structures, the application of a stress creates electrical dipoles. In different grains of a polycrystalline piezoelectric material, these dipoles point in different directions, thus canceling out. For this reason, until the 1940s polycrystalline piezoelectrics were thought to be impossibilities. However, if a ferroelectric material is subjected to a high electric field, aligning the domains, a polycrystalline piezoelectric can be obtained. This process is called *poling*.

The first commercial piezoelectric ceramic was $BaTiO_3$. Although it is a versatile piezoelectric and many uses were found for it, an important advance was the discovery of the piezoelectric properties of the $PbTiO_3$–$PbZrO_3$ solid-solution series. This solid-solution series, often designated PZT for lead zirconate titanate, has better piezoelectric properties than $BaTiO_3$ and can be used at a higher temperature, 300 °C vs. 120 °C, than $BaTiO_3$ (see p. 1097). However, $BaTiO_3$ is still a useful ferroelectric material in capacitors.

$BaTiO_3$. In the 1930s and 1940s, work at the Titanium Alloy Manufacturing Co. (United States) led to the discovery of high dielectric constant titanate compositions, among them barium titanate [*12047-27-7*], $BaTiO_3$ [17]. Independently, researchers in Russia and elsewhere had been conducting similar research [18]. In a series of articles, MCQUARRIE has given an overview of the work and theory of ferroelectric $BaTiO_3$ up to 1955 [19].

In an early report on the dielectric properties of $BaTiO_3$, peaks in the dielectric constant were observed at ca. -70, $+10$, and 120 °C [20]. These correlate to crystallographic phase transformations. $BaTiO_3$ has a Curie temperature of ca. 125 °C. Above this temperature, the crystal structure is cubic (perovskite), while at room temperature there is ca. 1% distortion of one of the cubic axes, yielding a tetragonal structure. At room temperature the lattice constants of $BaTiO_3$ are $a = 0.39860$ nm, $c = 0.40263$ nm, for a c/a ratio of 1.0101 [21]. Below 0 °C the symmetry is orthorhombic; and below -80 °C the symmetry is rhombohedral [22].

$BaTiO_3$ is usually prepared from chemically treated barium and titanium precursors. In an early report, the reaction of $BaCO_3$ and TiO_2 in various atmospheres was studied [23]. More recently, the process patented by PECHINI [24] was used for the study of controlled stoichiometry materials [25]. In this procedure tetraisopropyl orthotitanate and barium carbonate are reacted. Barium titanium oxalate [26] and other barium and titanium compounds have also been used to prepare high-purity $BaTiO_3$. Small, uniform powders have been fabricated in an effort to produce more reactive powders [27]–[29].

There is a large increase in the dielectric constant of $BaTiO_3$ as the temperature approaches the Curie point. To moderate this effect and to alter the dielectric properties, *additives* are usually incorporated into commercial $BaTiO_3$ formulations. If the amount of SiO_2 [7631-86-9] is increased from 0.5 to 2 wt%, there is a progressive lowering of the room-temperature dielectric constant and a suppression of the maximum dielectric constant at the Curie point. Al_2O_3 additions have the same effects but to a lesser degree; i.e., a 0.5% SiO_2 addition has the same effect as 2% Al_2O_3. Variations in the Ba:Ti ratio only slightly affect the dielectric constant and the Curie temperature, but dramatically affect the dielectric loss (dissipation factor). Additions of $PbTiO_3$ increase the Curie point, while additions of $CaSnO_3$ lower it [30]. In the $CaTiO_3$–$BaTiO_3$ series, increasing the percentage of $CaTiO_3$ does not significantly affect the Curie point, but it does lower the dielectric constant at room temperature. Addition of $MgZrO_3$ broadens the sharp temperature dependence near the Curie point [31].

Additions of more than 0.25% La_2O_3 decrease both the dielectric constant and the Curie temperature of coarse-grained materials [32]. Additions of up to 0.5% CeO_2 increase the dielectric constant, but further additions lower it [33]. Many other additives also affect the dielectric properties of $BaTiO_3$ formulations.

The crystallite, or grain, size in polycrystalline $BaTiO_3$ also affects the dielectric properties. The dielectric constant increases to a maximum as grain size decreases to ca. 1 μm [34], [35], phenomenologically explained as being the effect of anomalous variations of permittivity on internal stresses in fine-grained material due to the absence of twinning [36].

In addition to affecting the dielectric properties, additives are also incorporated into $BaTiO_3$ formulations to affect the densification, or sintering behavior. The firing temperature and atmosphere [37], as well as microstructure, affect the dielectric properties of $BaTiO_3$ formulations. To obtain a given set of properties, a particular firing condition or microstructure may be necessary, therefore requiring sintering additives.

Additions of $PbTiO_3$ and of CdO, ZnO, and CuO in amounts above their solid-solubility limits at 1200 °C or additions of more than 3% of NiO lead to the formation of a reactive liquid phase, which leads to rapid densification. Additions of Bi_2O_5, MgO, $CaTiO_3$, and NiO above their solubility limits at 1200 °C lead to precipitates at the grain boundaries that also enhance densification. Additions of Cr_2O_3 and SiO_2 do not enhance densification [38].

A 0.74 Bi_2O_3–0.26 B_2O_3 glass is another liquid-phase former that enhances densification [39]. Lithium fluoride, LiF, is also an effective liquid-phase sintering aid [40]–[42], as is $BaLiF_3$ [43].

To increase the capacitance in a given volume or to decrease the volume for a given capacitance, a multilayer capacitor design is often used [44]. This design takes advantage of the increase in capacitance with decreased dielectric thickness and parallel connection of individual capacitor layers. In this design, thin layers of $BaTiO_3$-based material are used.

Tape Casting Process. The tape casting process is used to achieve thin, stress-free sheets of dielectric. This process is also known as doctor blading, sheet casting, and knife coating [45]. The tape casting process reliably and efficiently can produce layers 25–1250 μm (1–50 mils) thick [46]. Tape casting has its origins in the paint industry [47] and was first applied to ceramics by HOWATT [48], [49] during World War II to form thin titanate sheets to replace mica in capacitors. Since that time, it has been used extensively for a wide variety of ceramic materials [6], [50]–[55].

A slurry that contains the ceramic powder, solvents, and organic compounds is formulated. The organic compounds serve several purposes. To adjust the viscosity of the slurry prior to casting, organics are used to disperse the ceramic particles and to plasticize the mix. Organics are also used to bind the ceramic particles together after casting to provide suitable strength of the cast tape for handling. Table 4 lists one representative formulation [56].

Most commonly a moving carrier film passes under a reservoir containing the slurry. Steel and Mylar are among the materials used as a carrier film. As the film moves beyond the reservoir, a thin layer of the slurry is carried on the film. To maintain a consistent, controllable thickness of the layer, a "doctor blade" is positioned at a predetermined height above the carrier film. This doctor blade trims off excess slurry so that the carrier film beyond the doctor blade has a constant thickness. Variations in design have incorporated the reservoir and doctor blade into an entity that moves over a fixed surface.

$PbTiO_3$. In 1950, indirect evidence suggested that lead titanate [*12060-00-3*] was ferroelectric [57]. Not until 1970 was ferroelectricity in $PbTiO_3$ actually verified [58]. Lead titanate has the cubic perovskite crystal structure above its Curie point, ca. 490 °C. Below the Curie point the structure is tetragonal. At room temperature, the lattice constants are $a = 0.3895$ nm and $c = 0.4146$ nm, for a c/a ratio of 1.064 [59]. This relatively large tetragonal distortion of the cubic perovskite structure leads to a large amount of strain in cooling through the cubic–tetragonal transition temperature. This spontaneous strain and a large thermal expansion anisotropy is thought to break up polycrystalline bodies of pure $PbTiO_3$ during cooling [60]. For these reasons polycrystalline $PbTiO_3$ must be formed in the presence of additives [61].

$PbTiO_3$ is usually formed by the solid-state reaction of PbO and TiO_2. The compound can be formed at temperatures as low as 375 °C [62]. Recently, however, a new technique for forming $PbTiO_3$ has been reported [63]. This technique involves preparation of a porous $PbTiO_3$ body from organometallic precursors at 70 °C and, because it lowers the densification temperature so dramatically, may lead to fully dense, pure $PbTiO_3$.

Because of the difficulty in preparing pure polycrystalline lead titanate, there are few reports on the properties of the material. BANNO, however, has found that pure $PbTiO_3$ is useful in piezoelectric–polymer composites [64]. There are reports on the

Table 4. Batch formulation for tape casting slurry [56]

Ingredient	Function	Parts, by weight
BaTiO$_3$[a]	ceramic	100
Methyl ethyl ketone/ethanol (50/50 mixture)	solvent	20
Menhaden fish oil[b]	dispersant	1
Santicizer 160[c] (butyl benzyl phthalate)	plasticizer	4
Carbowax 400 (poly(ethylene glycol))	plasticizer	4
Cyclohexanone	homogenizer	0.7
Acryloid B-7[d] (methyl ethyl ketone solution, 30%)	binder	13.32

[a] TAM Ceramic Co., Niagara Falls, New York.
[b] Defloc Z-3, Spencer Kellogg Inc., Buffalo, New York.
[c] Monsanto Inc., St. Louis, Missouri.
[d] Rohm and Haas Co., Philadelphia, Pennsylvania.

properties of modified PbTiO$_3$ [60], [65]. However, the commercial application of these materials is still quite small.

PbZrO$_3$. Lead zirconate [12060-01-4] was shown to have unusual dielectric properties in 1950 [66]. Because of its high dielectric constant and temperature dependence, the material was designated a ferroelectric. Actually PbZrO$_3$ is antiferroelectric [67]. Therefore, by itself lead zirconate cannot form a polycrystalline piezoelectric. Below ca. 230 °C, PbZrO$_3$ is orthorhombic with lattice constants $a = 0.587$ nm, $b = 1.174$ nm, and $c = 0.820$ nm. Above 230 °C the material is cubic.

In the original work PbZrO$_3$ was prepared by mixing ZrO$_2$ and PbO and then calcining in a platinum crucible for 1 h at 1050 °C. As with the forming and densification of many lead-containing compounds, the volatility of PbO must be controlled. This is usually done by firing in an atmosphere that contains PbO (see p. 1111).

The properties of PbZrO$_3$ can be altered by modifying the composition. Some of the additions can change the crystallographic phases present as well as alter the properties [68].

Pb(Zr,Ti)O$_3$. Lead zirconate and lead titanate form a complete series of solid solutions [69]. These materials are referred to generically as Pb(Zr,Ti)O$_3$ [12626-81-2], or *PZT*. It is important to realize that PZT refers to a wide range of compositions rather than an individual chemical compound.

The discovery of the piezoelectric properties of PZT ceramics was a result of research around the world. In 1952, SHIRANE and SUZUKI, in Japan, investigated the crystal structure of the solid solution PbZrO$_3$–PbTiO$_3$ [70]. This work was followed in 1953 by SAWAGUCHI's comprehensive report on his investigation of ferroelectricity in the PZT solid solution [71]. In 1954, JAFFE et al., in the United States, published the first report on the piezoelectric properties of the PbZrO$_3$–PbTiO$_3$ solid solution [72]. The

Table 5. Typical properties for piezoelectric ceramics

Property	Barium titanate based	PZT based
Specific gravity	5.60	7.70
Dielectric constant, 1 MHz, 25 °C	1200	1300
Dielectric strength, MV/m	3.5	3.9
Modulus of elasticity, GPa	117	82
Piezoelectric constants		
d_{31}, C/N	-60×10^{-12}	-120×10^{-12}
d_{33}, C/N	150×10^{-12}	290×10^{-12}
g_{31}, V/N	-5×10^{-3}	-10×10^{-3}
g_{33}, V/N	14×10^{-3}	25×10^{-3}
Coupling coefficients		
K_{31}	0.2	0.3
K_{33}	0.5	0.7
K_p	0.3	0.6
Mechanical Q	400	500
Curie temperature, °C	120	345

same authors summarized the work on PZT and expanded it in a more encompassing report in 1955 [73]. The piezoelectric properties of PZT were found to be better than those of the most popular ceramic piezoelectric at that time, $BaTiO_3$, as shown in Table 5.

As Zr replaces Ti in $PbTiO_3$, the amount of tetragonal distortion (c/a ratio) decreases. At ca. 55 mol% $PbZrO_3$, an abrupt change in the crystal structure of the solid solution is observed. This sort of abrupt change in structure with compositional change is known as a morphotropic phase transformation. The *morphotropic phase boundary* (MPB) in PZT is nearly independent of temperature. Compositions near the MPB yield materials with desirable piezoelectric properties over a wide temperature range. Therefore, the most useful commercial PZT compositions are near the MPB. A commonly used composition has a Zr : Ti molar ratio of 53 : 47 [73].

Solid solutions higher in $PbZrO_3$ content than 55 mol% have rhombohedral crystal structures. Above 90 mol% $PbZrO_3$, the crystal structure is orthorhombic. The tetragonal and rhombohedral phases are ferroelectric, while the orthorhombic phase is antiferroelectric.

PZT is commonly formed from the solid-state reaction of PbO, TiO_2, and ZrO_2. There have been many investigations into the calcination reactions. When the three oxide powders are mixed and heated, $PbTiO_3$ first forms exothermically between 450 and 600 °C, and then PZT forms above 700 °C. No measurable amount of $PbZrO_3$ was observed in the mixture. A multistep sequence was proposed to explain the reaction kinetics [74].

As with most technical ceramics, the processing of PZT articles is exacting and critical to their performance. The problems involved in balancing densification with grain growth are compounded in PZT by the *volatility of PbO.* Densification, grain growth, and evaporation rates all increase with temperature. To fully densify a PZT piece while minimizing grain growth and eliminating lead loss is not possible solely by varying the temperature during firing. To minimize lead loss, pieces are usually fired in a closed crucible containing a PbO-rich atmosphere, usually created by packing the PZT pieces in a powder containing lead oxide. This packing powder can be of the same composition as the pieces, or it can be PbO, $PbZrO_3$, etc. The packing powder affects the microstructure and properties of PZT ceramics [75]: the ideal arrangement seems to be to use a closed system with a packing powder of $PbZrO_3$ saturated with PbO.

A major effort through the years has been devoted to preparing PZT ceramics with a good microstructure and good electrical properties. Hot pressing has been used to obtain pieces with a dense, uniform microstructure. However, high-density PZT can be processed without hot pressing by careful control of composition, impurities, calcining, forming, and sintering [76]. An excess of PbO in the PZT composition acts as a liquid-phase densification aid, and is commonly used. However, while a PbO-rich liquid phase increases the initial densification rate, it lowers final density [77]. Additions of V_2O_5 [78] and of LiF [42] also lower the sintering temperature.

Soon after the discovery of the useful piezoelectric properties of PZT, work was directed at improving these properties via *chemical modifications.* Strontium is especially effective [79]. When Sr is substituted for Pb in PZT, the dielectric constant and the electromechanical coupling factor are greater near the MPB. Strontium substitution was also found to lower the Curie point and to shift the MPB to a composition slightly more rich in Zr.

Near the MPB, additions of some +3 or +5 ions (La, Nd, Nb, and Ta) increase the dielectric constant and give the PZT better aging and stability characteristics [80]. Modifying PZT with tungsten and thorium gives results similar to those for PZT containing Nb or La [81]. Additions of Fe_2O_3 to PZT compositions containing 45–60 mol% $PbZrO_3$ (i.e., both sides of the MPB) increase the mechanical and dielectric stiffnesses, while the mechanical and dielectric losses decrease [82]. The effects were greater for the rhombohedral $PbTiO_3$-rich compositions than for the tetragonal compositions.

Solid solutions of PZT with other perovskite ferroelectrics have also been studied. Several of these involved a material $Pb(Nb_{2/3}X_{1/3})O_3$, with X being Mg [83], Co [84], or Ni [85]. These systems contain pseudocubic phases as well as tetragonal and rhombohedral phases. The materials that were produced had high dielectric constants and electromechanical coupling coefficients and good temperature stability.

Uses. There are many possible uses for devices based on piezoelectric materials to convert mechanical to electrical energy, and vice versa. The PZT family of materials is the most commercially important ceramic piezoelectric. By modification of the composition and control of the microstructure through processing variations, materials can be tailored to a specific use. Table 6 lists some uses. A monograph describing the uses of PZT ceramics, giving operating characteristics and a variety of specific applications, is entitled *Piezoelectric Ceramics* [86].

Table 6. Uses of PZT ceramics

Use	Devices
High-voltage generators	igniter for gas appliances
	cigarette lighter
	smallgasoline-motorspark generator
	static remover
Flexural element	buzzer
	liquid-level sensor
	fine-motion controller
	strain gage
	phonograph pickup
Resonators	electrical filters
	electrical resonators
Sound/ultrasound generators	remote control of appliances
	intruder alarm
	tone generator
	echo sounder
	hydrophone
	ultrasonic cleaners
Electro-optics	displays
	flash shield

Tungsten-modified PZT is suitable for piezoelectric ignition of gases, W–Mn-modified PZT is useful as ceramic wave filters, and Nb–Mn-modified PZT is suitable for high-power ultrasonic transducers [87].

PZT can be used to generate high voltages. The voltages produced are high enough to cause sparking (5–20 kV). The spark can ignite gas in an appliance such as a stove or clothes dryer. A piezoelectric igniter makes pilot flames unnecessary, reducing the amount of gas used by the appliance. The mechanical force can be applied to the PZT piece by a spring-loaded hammer or by a lever system. In a related device, air in the vicinity of the tip of a wire connected to a PZT block is ionized to remove static from phonograph records.

Flexural elements are also commonly made from PZT ceramics. The amount of motion generated by a single PZT element is often too small, e.g., 4 µm for an applied 10 kV [86]. However, a laminated cantilever structure, a *bimorph*, can be used. A bimorph consists of two thin layers of PZT laminated together. Often a metallic layer is used between the PZT layers to increase the durability of the bimorph and the electromechanical coupling. The way the bimorph is mounted tailors the device to the application.

Mechanical resonances possible in PZT have been exploited in electrical resonators and filters. The dimensions of the PZT disk and the mode of oscillation (e.g., radial or thickness) determine the frequency.

PZT transducers can be used to generate sound and ultrasound in air. The same materials can be used as receivers. Remote control of slide projectors and televisions has been developed on the basis of these systems. Other uses are intruder alarms, telephone microphones, and tone generators. As with flexural elements, laminated structures are commonly used for ultrasound generators or receivers.

Boats commonly have echo-sounding equipment to determine water depth, and in some cases to locate fish. The systems in small boats usually require a PZT transducer with resonant frequency in the range 150–200 kHz. This transducer is also capable of receiving the signal. A related device is a hydrophone, which is used for underwater detection by passively listening rather than generating and receiving a signal. Composites made from PZT and epoxy offer some unique properties that can be exploited in hydrophone as well as other applications [88].

When high-intensity ultrasonic signals must be generated, such as in ultrasonic cleaning, a sandwich or composite transducer is used. The intensity of the signal necessitates a large-volume transducer.

When doped with rare-earth or a few other additives, PZT was found to act as an electrooptic material [89]; the optical properties could be electrically controlled (see the section on PLZT). An early composition contained 2 mol % Bi and a Zr : Ti molar ratio of 65 : 35 [90].

PLZT. Doped PZT was found to be an electro-optic material. Lanthanum doping produces PZT ceramics with good electro-optical properties and a high optical transparency [91]. This family of materials is designated *PLZT* for lead lanthanum zirconate titanate. A useful electro-optic composition has a Zr : Ti molar ratio of 65 : 35 and 8–9 mol % La. The specific compositions are often designated by a three-term ratio of "mole percents" in the form La/Zr/Ti, e.g., 9/65/35 PLZT.

The phase relations in the $X/65/35$ PLZT system, $0 \leq X \leq 15$, was investigated [92]. At room temperature there are two ferroelectric rhombohedral phases, with the transition occurring near $X = 1$. For $9 \leq X \leq 12$, there is a pseudocubic phase, designated antiferroelectric on the basis of its dielectric properties. For $X > 12$, there is a cubic phase.

In the initial work on PLZT by HAERTLING and LAND [91], the samples were *hot-pressed* to reach optical transparency. The calcined powder was initially cold-pressed at 24 GPa in an alumina hot-press mold [93]. The mold assembly was then placed into the hot-pressing furnace, and pressure was applied from both ends. Typical hot-pressing conditions were 1100 °C for 16 h at 13.8 GPa [91]. Above 1100 °C, ZrO_2 powder acts as a setter to prevent reaction between the alumina mold and the PLZT. Hot pressing is still used to fabricate PLZT pieces.

Transparent PLZT ceramics can be fabricated without hot pressing [94]: cold-pressed 9/65/35 PLZT samples are first sintered in O_2 in platinum crucibles for

45 min at 1180 °C and then heat treated in air for 60 h at 1200 °C in Al_2O_3 crucibles containing $PbZrO_3$ powder.

Uses. Displays made from PLZT can image, store, and display information. PLZT displays have several advantages over other types of electronic displays [95]. They can be selectively erased and do not need to be refreshed: part or all of the information can be erased, and once information is input, no power is required to maintain it. The memory is *nonvolatile*.

The thin PLZT plate used in displays is transparent, initially optically isotropic. Poling reduces the symmetry to uniaxial. A uniaxial material is birefringent; that is, the index of refraction is different for different polarizations of light. By selective reorientation of the domains in the ceramic, the birefringence can be electrically controlled and an image stored.

Electro-optic PLZT has also been used for flash-protection goggles: when a light sensor is triggered by a high-intensity flash, a signal is sent to the transparent PLZT, which reorients the domains and darkens the material, blocking the transmission of the light flash.

$PbNb_2O_6$. Lead metaniobate [*12034-88-7*], $PbNb_2O_6$, was the first nonperovskite oxide ferroelectric discovered. After this discovery by GOODMAN [96], other lead-based nonperovskite ferroelectrics, not discussed here, were identified [97].

The Curie point of $PbNb_2O_6$ is 570 °C [96]. The paraelectric phase has a tetragonal crystal structure similar to that of some alkali-metal tungsten bronzes, such as $K_{0.57}WO_3$ [98], [99]. Below the Curie point, two phases have been observed, a rhombohedral nonferroelectric phase and an orthorhombic ferroelectric phase that is metastable at room temperature [100]. The lattice constants of the ferroelectric form are $a = 1.765$ nm, $b = 1.791$ nm, $c = 0.7736$ nm [98].

In the original work, lead metaniobate was prepared in a two-step process. The first step is a calcination of Nb_2O_5 and $PbSO_4$ in the form of pressed pellets at 1275 °C for 1 h in air, to form $PbNb_2O_6$. The calcined pellets are crushed and repressed. The final firing is carried out in platinum vessels at 1250 °C. Lead volatility is less a problem than for PZT.

Modified lead niobate has been used in transducers. Because of its high Curie point, $PbNb_2O_6$ can be used at higher temperatures than PZT, but the relatively high electrical conductivity of lead metaniobate limits its use at higher temperatures [101].

4. Semiconducting Ceramics

The broad definition that ceramics are nonmetallic inorganic solids classifies the common semiconductors (silicon, germanium, gallium arsenide, etc.) as ceramic materials. However, they are not normally considered ceramics and, therefore, are not discussed here. Electronically, ceramic materials can be regarded as wide-band-gap semiconductors and as such can have useful electronic properties. Semiconducting ceramic materials are commonly used as phosphors, resistors, and sensors, as was shown in Table 1, and can be used as active semiconductors in some applications.

SiC. Silicon carbide [409-21-2] is a semiconducting material that could be useful in high-frequency and high-operating-temperature uses. In addition, SiC-based devices could be useful as blue-light-emitting diodes and high-power–high-frequency devices [102], [103]. Among the devices fabricated from SiC are bipolar transistors [104] and MOS diodes [105].

Silicon carbide exists in a number of crystallographic polytypes, The two polytypes used most commonly for devices are $3C$, which is cubic, and $6H$, which is hexagonal. In the following discussion, the two polytypes will be differentiated where possible.

For high-frequency uses there are different figures of merit in the literature [106], [107]. Silicon is predicted to outperform gallium arsenide, or vice versa, depending upon which figure of merit is used. However, silicon carbide is predicted to outperform both materials in high-frequency applications. This potential advantage is described in the SiC literature [102], [108], and even in the gallium arsenide literature [109].

In choosing a material to be used as a semiconductor at high temperature, several factors must be considered. A wide electronic band gap is necessary so that the thermally generated intrinsic carriers (e.g., electrons excited across the band gap) do not outnumber the extrinsic carriers added intentionally via doping. The carrier mobility should be high at these elevated temperatures so that the carriers can conduct. Any operating solid-state device will generate heat; thus, the material should have good thermal conductivity. In addition, the material must be stable at the operational temperature.

These considerations have been discussed [110] in a comparison of the properties of silicon, gallium arsenide, gallium phosphide, silicon carbide, and diamond. Of these, gallium phosphide [12063-98-8], GaP, and silicon carbide have been actively investigated. While GaP and cubic SiC have similar band gaps and mobilities, the inferior physical stability of GaP at elevated temperature makes it a less useful material than SiC.

To produce SiC-based devices, several technologies must be fairly well developed:

single-crystal growth
doping methods
junction fabrication
contacting

Crystal Growth. Many of the simpler crystal growth techniques developed for silicon and compounds like GaP are inapplicable to SiC because the material sublimes rather than melts. Thus, Czochralski growth and zone refining cannot be used with SiC. In addition, the many polytypes make producing crystals of a desired polytype difficult.

A technique developed at NASA-Lewis (United States) to produce thick single-crystalline layers of cubic SiC grows a buffer layer on single-crystal Si [111]. The SiC layer is then grown on this buffer layer via chemical vapor deposition (CVD).

Research at Siemens in Germany has led to a process for growing relatively large crystals of $6H$ SiC [112]. This process uses a hollow growth tube in which polycrystalline SiC is sublimed at one end and deposited on a seed crystal at the other end. Single crystals up to 24 mm long and 20 mm in diameter have been grown.

Doping and Device Fabrication. Solid-state device technology depends on being able to control precisely the local impurity levels within the host semiconductor. The semiconductor should start with as few impurities as possible, and the desired dopants can then be incorporated where desired with little damage to the rest of the circuit. Nitrogen is an electrically active *n*-type dopant in SiC and is usually present in the atmosphere when crystals are grown. For these reasons, most as-grown SiC crystals start off *n*-type, although overcompensation for the nitrogen impurities by adding a sufficient number of *p*-type dopants to the crystal during the growth process is possible. Therefore, both *n*- and *p*-type starting crystals are available. In silicon-based device technology, two of the more common doping methods are diffusion and ion implantation.

Diffusion of impurities into SiC requires temperatures above 1800 °C. At these high temperatures SiC decomposes; therefore, a protective atmosphere must be created. Light-emitting diodes can be fabricated by diffusing boron and aluminum into $6H$ SiC crystals surrounded by a protective atmosphere [113]. The diffusion is accomplished at 2150–2250 °C for 2–5 h. More recently, the diffusion of boron into *p*-type SiC [114] was studied to determine the effect of temperature (1500–2550 °C) and acceptor dopant concentrations on the diffusivity of boron.

Ion implantation has been accomplished in SiC since the late 1960s. Implantation of nitrogen, boron, and phosphorus was studied [115]. For example, 1-MeV N^+ ions can be implanted to a depth of 1 μm.

To fabricate a device, $p-n$ junctions must be fabricated. This can be done by starting with an *n*- or *p*-type material and selectively counterdoping by diffusion or implantation. A more useful technique has proven to be epitaxial growth of a junction [110]. In this process the dopant source is changed from *n*- to *p*-type (or vice versa) during crystal growth. Thus, a $p-n$ junction is grown directly. A problem is that little postgrowth control is possible.

Device definition is possible by using oxidation and etching processes [102]: an oxide layer of 400 nm thickness was grown at 1070 °C in wet oxygen. Conventional photolithographic techniques were used, and the SiC was selectively etched in an $Ar-Cl_2-O_2$ mixture at 1050 °C.

A solid-state device must have ohmic (nonrectifying) contacts to the *n*- and *p*-type regions. A thin nickel layer for ohmic contact to *n*-type material and an Al–Si eutectic contact for the *p*-type material have been used in $6H$ SiC [102]. On $3C$ SiC, Ni–SiC mixed layers may provide ohmic contact [116].

5. Sensors

Semiconducting ceramics are useful in a wide variety of sensors. The electrical properties of these materials are affected by changes in ambient temperature, atmosphere, and electrical parameters, e.g., voltage fluctuations. Through careful control of the chemical composition and microstructure of the device, inexpensive sensors can be made. The field of ceramic sensors is expanding rapidly and has been the subject of several recent, excellent reviews [117], [118] and conferences.

ZrO_2. Zirconium dioxide [*1314-23-4*] is used for oxygen sensing [119]. An electrochemical cell is made in which there is a known fugacity of oxygen on one side of the sensor and the unknown fugacity on the other side. This sets up an electrochemical potential drop, generating a voltage across the device that is related to the fugacity.

$BaTiO_3$. Semiconducting $BaTiO_3$ has been used as a positive temperature coefficient (PTC) resistor [120]. In most semiconductors, as temperature is increased, the resistivity decreases, which corresponds to a negative temperature coefficient (NTC). However, in appropriately doped $BaTiO_3$, the resistivity increases six orders of magnitude near the Curie point. At first this effect was observed only in donor-doped polycrystalline $BaTiO_3$ that had been heat treated in an oxidizing atmosphere [120]. For such materials the PTC effect was observed across a single-grain boundary but not within a single grain [121], showing that this phenomenon is associated with grain boundaries. The PTC effect has also been observed in porous $BaTiO_3$ [122]. There have been several explanations of this effect [123]–[125].

ZnO. Devices based on ZnO [*1314-13-2*] are useful as varistors. The resistance is voltage dependent, i.e., nonohmic. Additives are used to affect the microstructure, nonohmic behavior, and stability. The microstructure is important to the nonohmic behavior. A typical composition is 96.5 wt% ZnO, 0.5% Bi_2O_3, 1% CoO, 0.5% MnO, 1% Sb_2O_3, and 0.5% Cr_2O_3 [126].

WONG has found that the nonohmic behavior is due to semiconducting ZnO grains and a thin insulating grain-boundary phase [127]. However, CLARKE found no discrete grain-boundary phase. Instead, he observed a bismuth-rich region along the grain boundaries in a commercial ZnO varistor [128]. Several explanations for these effects have been presented [129]–[131].

6. References

[1] H. Thurnauer, *Am. Ceram. Soc. Bull.* **56** (1977) no. 219–220, 224.
[2] J. H. Koenig: *1964 Edgar Marburg Lecture*, Am. Soc. Test. Mater., Philadelphia 1965.
[3] E. C. Henry: *Electronic Ceramics*, Anchor Books, Doubleday & Co., New York 1969.
[4] K. Wefers, G. M. Bell: *Tech. Paper No. 19*, Alcoa Research Laboratories, Pittsburgh 1972.
[5] B. Schwartz, *Am. Ceram. Soc. Bull.* **63** (1984) 577–581.
[6] D. J. Shanefield, R. E. Mistler, *Am. Ceram. Soc. Bull.* **53** (1974) 416–420.
[7] B. Schwartz, *J. Phys. Chem. Solids* **45** (1984) 1051–1068.
[8] A. J. Blodgett, D. R. Barbour, *IBM J. Res. Dev.* **26** (1982) no. 1, 30–36.
[9] Y. Takeda et al., *Adv. Ceram.* **7** (1983) no. 253–259, 260–268.
[10] N. Iwase, A. Tsuge, Y. Sugiura, *Int. J. Hybrid Microelectron.* **7** (1984) no. 4, 49–53.
[11] P. L. Fleischner, *Solid State Technol.* **20** (1977) no. 1, 25–30.
[12] Y. Shimada, K. Utsumi, M. Suzuki, H. Takamizawa et al., *NEC Res. Dev.* **75** (1984) 8–15.
[13] M. Deri: *Ferroelectric Ceramics*, Gordon and Breach Co., New York 1969.
[14] B. Jaffe, W. R. Cook, Jr., H. Jaffe: *Piezoelectric Ceramics*, Academic Press, New York 1971.
[15] W. G. Cady: *Piezoelectricity*, McGraw-Hill, New York 1946, p. 4.
[16] W. Kanzig: *Ferroelectrics and Antiferroelectrics*, Academic Press, New York 1957, p. 5.
[17] E. Wannier, *Trans. Electrochem. Soc.* **89** (1946) 47–71.
[18] H. Hausner, *Ceram. Age* **50** (1947) no. 162–164, 190.
[19] M. McQuarrie, *Am. Ceram. Soc. Bull.* **34** (1955) no. 169–172, 225–230, 256–260, 295–298, 328–331.
[20] A. de Bretteville Jr., *J. Am. Ceram. Soc.* **29** (1946) 303–307.
[21] H. Megaw, *Nature (London)* **155** (1945) 484–485.
[22] E. Sawaguchi, M. L. Charters, *Phys. Rev.* **117** (1960) 465–469.
[23] L. K. Templeton, J. A. Pask, *J. Am. Ceram. Soc.* **42** (1959) 212–216.
[24] M. Pechini, US 3 330 697, 1967.
[25] R. K. Sharma, N.-H. Chan, D. M. Smyth, *J. Am. Ceram. Soc.* **64** (1981) 448–451.
[26] P. K. Gallagher, F. Schrey, *J. Am. Ceram. Soc.* **46** (1963) 567–573.
[27] K. Kiss, J. Magder, M. S. Vukasovich, R. J. Lockhart, *J. Am. Ceram. Soc.* **49** (1966) 291–295.
[28] K. S. Mazdiyasni, R. T. Dolloff, J. S. Smith II, *J. Am. Ceram. Soc.* **52** (1969) 523–526.
[29] Y. Enomoto, A. Yamaji, *Am. Ceram. Soc. Bull.* **60** (1981) 566–570.
[30] E. G. Graf, *Ceram. Age* **58** (1951) no. 6, 16–19.
[31] E.G. Graf, *Am. Ceram. Soc. Bull.* **31** (1952) 279–282.
[32] J. B. MacChesney, P. K. Gallagher, F. V. DiMarcello, *J. Am. Ceram. Soc.* **46** (1963) 197–202.
[33] M. A. A. Issa, N. M. Molokhia, Z. H. Dughaish, *J. Phys. D* **16** (1983) 1109–1114.
[34] L. Egerton, S. E. Koonce, *J. Am. Ceram. Soc.* **38** (1955) 412–418.
[35] K. Kinoshita, A. Yamaji, *J. Appl. Phys.* **47** (1976) 371–373.
[36] W. R. Buessem, L. E. Cross, A. K. Goswami, *J. Am. Ceram. Soc.* **49** (1966) 33–39.
[37] K. Traub, C. A. Best, W. J. Baldwin, *Ceram. Age* **65** (1955) no. 1, 9–14.
[38] M. N. Swilam, A. M. Gadalla, *Trans. J. Br. Ceram. Soc.* **74** (1975) no. 5, 165–169.
[39] K. Ramesh Chowdary, E. C. Subbarao, *Ferroelectrics* **37** (1981) 689–692.
[40] B. E. Walker, Jr., R. W. Rice, R. C. Pohanka, J. R. Spann, *Am. Ceram. Soc. Bull.* **55** (1976) no. 274–276, 284–285.
[41] J. M. Haussonne, G. Desgardin, P. Bajolet, B. Raveau, *J. Am. Ceram. Soc.* **66** (1983) 801–807.
[42] S. L. Fu, C. C. Wei, S. Y. Cheng, T. P. Yeh, *Int. J. Hybrid. Microelectron.* **8** (1985) no. 1, 1–5.

[43] G. Desgardin, I. Mey, B. Raveau, J. M. Haussonne, *Am. Ceram. Soc. Bull.* **64** (1985) 564–570.
[44] D. W. Hamer, *Ceram. Ind. (Chicago)* **93** (1969) no. 1, 49–56, 68–70.
[45] J. C. Williams: "Ceramic Fabrication Processes," in F. F. Y. Wang (ed.): *Treatise on Materials Science and Engineering*, vol. **9**, Academic Press, New York 1979, pp. 173–198.
[46] R. E. Mistler, R. B. Runk, D. J. Shanefield in G. Y. Onoda, Jr., L. L. Hench (eds.): *Ceramic Fabrication Processing Before Firing*, John Wiley & Sons, New York 1978, pp. 411–448.
[47] H. A. Gardner: *Physical and Chemical Examination of Paints, Varnishes, and Colors*, Institute of Paint and Varnish Research, Washington, D.C., 1925.
[48] G. N. Howatt, US 2 582 993, 1952.
[49] G. N. Howatt, R. G. Breckenridge, J. M. Brownlow, *J. Am. Ceram. Soc.* **30** (1947) 237–242.
[50] J. J. Thompson, *Am. Ceram. Soc. Bull.* **42** (1963) 480–481.
[51] C. Wentworth, G. W. Taylor, *Am. Ceram. Soc. Bull.* **46** (1967) 1186–1193.
[52] G. J. Asher, *Ceram. Age* **87** (1971) no. 9, 28–31.
[53] D. J. Shanefield, R. E. Mistler, *West. Electr. Eng.* **15** (1971) no. 2, 26–31.
[54] J. V. Biggers, T. R. Shrout, W. A. Schulze, *Am. Ceram. Soc. Bull.* **58** (1979) no. 516–518, 521.
[55] J. B. Blum, W. R. Cannon, *Mater. Res. Soc. Symp. Proc.* **40** (1985) 77–82.
[56] R. J. MacKinnon, J. B. Blum, *Adv. Ceram.* **9** (1984) 150–157.
[57] G. Shirane, S. Hoshino, K. Suzuki, *J. Phys. Soc. Jpn.* **5** (1950) 453–455.
[58] J. P. Remeika, A. M. Glass, *Mater. Res. Bull.* **5** (1970) 37–45.
[59] T. Y. Tien, E. C. Subbarao, J. Hrizo, *J. Am. Ceram. Soc.* **45** (1962) 572–575.
[60] T. Y. Tien, W. G. Carlson, *J. Am. Ceram. Soc.* **45** (1962) 567–571.
[61] E. C. Subbarao, *J. Am. Ceram. Soc.* **43** (1960) 119–122.
[62] S. S. Cole, H. Espenschied, *J. Phys. Chem.* **41** (1937) 445–451.
[63] S. Gurkovich, J. B. Blum in L. L. Hench, D. R. Ulrich (eds.): *Ultrastructure Processing of Ceramics, Glasses and Composites*, John Wiley & Sons, New York 1984, pp. 152–160.
[64] H. Banno, *Ferroelectrics* **50** (1983) 3–12.
[65] T. Yamamoto, H. Igarashi, K. Okazaki, *J. Am. Ceram. Soc.* **66** (1983) 363–366.
[66] S. R. Roberts, *J. Am. Ceram. Soc.* **33** (1950) 63–66.
[67] E. Sawaguchi, H. Maniwa, S. Hoshino, *Phys. Rev.* **83** (1951) 1078.
[68] Jaffe, Cook, Jaffe [14] pp. 123–131.
[69] S. Fushimi, T. Ikeda, *J. Am. Ceram. Soc.* **50** (1967) 129–132.
[70] G. Shirane, K. Suzuki, *J. Phys. Soc. Jpn.* **7** (1952) 333.
[71] E. Sawaguchi, *J. Phys. Soc. Jpn.* **8** (1953) 615–629.
[72] B. Jaffe, R. S. Roth, S. Marzullo, *J. Appl. Phys.* **25** (1954) 809–810.
[73] B. Jaffe, R. S. Roth, S. Marzullo, *J. Res. Natl. Bur. Stand. U.S.* **55** (1955) 239–254.
[74] S. S. Chandratreya, R. M. Fulrath, J. A. Pask, *J. Am. Ceram. Soc.* **64** (1981) 422–425.
[75] S. S. Chiang, M. Nishioka, R. M. Fulrath, J. A. Pask, *Am. Ceram. Soc. Bull.* **60** (1981) 484–489.
[76] Y. S. Kim, R. J. Hart in H. Palmour III, R. F. Davis, T. M. Hare (eds.): *Processing of Crystalline Ceramics*, Plenum Publ., New York 1978, pp. 323–333.
[77] A. I. Kingon, J. B. Clark, *J. Am. Ceram. Soc.* **66** (1983) 256–260.
[78] D. E. Wittmer, R. C. Buchanan, *J. Am. Ceram. Soc.* **64** (1981) 485–490.
[79] F. Kulcsar, *J. Am. Ceram. Soc.* **42** (1959) 49–51.
[80] T. B. Weston, A. H. Webster, V. M. McNamara, *J. Am. Ceram. Soc.* **52** (1969) 253–257.
[81] F. Kulcsar, *J. Am. Ceram. Soc.* **48** (1965) 54.
[82] F. Kulcsar, *J. Am. Ceram. Soc.* **42** (1959) 343–349.
[83] H. Ouchi, K. Nagano, S. Hayakawa, *J. Am. Ceram. Soc.* **48** (1965) 630–635.

[84] T. Kudo, T. Yazaki, F. Naito, S. Sugaya, *J. Am. Ceram. Soc.* **53** (1970) 326–328.

[85] H. Banno, T. Tsunooka, I. Shimano, *Proc. Meet. on Ferroelectric Mater. and Their Applications 1975, 1st,* 339–344.

[86] J. van Randeraat, R. E. Setterington (eds.): *Piezoelectric Ceramics,* Mullard, London 1974.

[87] H. Banno, paper no. 2-FS-80P, presented at the Pacific Coast Regional Meeting – Am. Ceram. Soc., San Francisco, October 27, 1980; abstract, *Am. Ceram. Soc. Bull.* **59** (1980) 824.

[88] R. E. Newnham, L. J. Bowen, K. A. Klicker, L. E. Cross, *Mater. Eng. (Reigate, U.K.)* **2** (1980) 93–106.

[89] A. H. Meitzler, J. R. Maldonado, D. B. Fraser, *Bell Syst. Tech. J.* **49** (1970) 953–967.

[90] C. E. Land, P. D. Thatcher, *Proc. IEEE* **57** (1969) 751–768.

[91] G. H. Haertling, C. E. Land, *J. Am. Ceram. Soc.* **54** (1971) 1–11.

[92] H. M. O'Bryan, Jr., *J. Am. Ceram. Soc.* **56** (1973) 385–388.

[93] G. Haertling, *J. Am. Ceram. Soc.* **49** (1966) 113–118.

[94] G. S. Snow, *J. Am. Ceram. Soc.* **56** (1973) no. 91–96, 479–480.

[95] A. H. Meitzler, J. R. Maldonado, *Electronics* 1971, Feb. 1, 34–39.

[96] G. Goodman, *J. Am. Ceram. Soc.* **36** (1953) 368–372.

[97] Jaffe, Cook, Jaffe [14] pp. 217–222.

[98] M. H. Francombe, B. Lewis, *Acta Crystallogr.* **11** (1958) 696–703.

[99] R. S. Roth, *Acta Crystallogr.* **10** (1957) 437.

[100] M. H. Francombe, *Acta Crystallogr.* **9** (1956) 683.

[101] Jaffe, Cook, Jaffe [14] pp. 215–216.

[102] E. Pettenpaul, W. von Munch, G. Ziegler, *Conf. Ser. Inst. Phys.* **53** (1980) 21–35.

[103] J. Feitknecht, *Springer Tracts Mod. Phys.* **58** (1971) 48–118.

[104] W. von Munch, P. Hoeck, *Solid-State Electron.* **21** (1978) 479–480.

[105] A. Suzuki, K. Mameno, N. Furui, H. Matasunami, *Appl. Phys. Lett.* **39** (1981) 89–90.

[106] E. O. Johnson, *RCA Rev.* **26** (1965) 163–177.

[107] R. W. Keyes, *Proc. IEEE* **60** (1972) 225.

[108] R. W. Keyes, *Silicon Carbide Proc. Int. Conf. 3rd 1973,* 1974, 534–541.

[109] M. Nowogrodzki (ed.): *Advanced III–V Semiconductor Materials Technology Assessment,* Noyes Publications, Park Ridge, NJ, 1984, p. 178.

[110] J. A. Powell, *NASA Tech. Memo.* **83514** (1983) 1–5.

[111] S. Nishino, J. A. Powell, H. A. Will, *Appl. Phys. Lett.* **42** (1983) 460–462.

[112] G. Ziegler, P. Lanig, D. Theis, C. Weyrich, *IEEE Trans. Electron Devices* **ED-30** (1983) 277–281.

[113] J. M. Blank, *Silicon Carbide Proc. Int. Conf. 2nd 1968,* 1969, 179–186.

[114] E. N. Mokhov, E. E. Goncharov, G. G. Ryabora, *Sov. Phys. Semincond. (Engl. Transl.)* **18** (1984) 27–30.

[115] D. E. Davies, J. J. Comer, *Silicon Carbide Proc. Int. Conf. 3rd 1973,* 1974, 640–644.

[116] D. Fathy, O. Narayan, O. W. Holland, B. R. Appleton et al., *Mater. Lett.* **2** (1984) 324–327.

[117] P. McGeehin, D. E. Williams in H. Krockel, M. Merz, O. van der Biest (eds.): *Ceramics in Advanced Energy Technologies,* D. Reidel, Boston 1984, pp. 422–468.

[118] B. M. Kulwicki, *J. Phys. Chem. Solids* **45** (1984) 1015–1031.

[119] E. M. Logothetis, *Adv. Ceram.* **3** (1981) 388–405.

[120] B. M. Kulwicki, *Adv. Ceram.* **1** (1981) 138–154.

[121] H. Nemoto, I. Oda, *J. Am. Ceram. Soc.* **63** (1980) 398–401.

[122] M. Kuwabara, *J. Am. Ceram. Soc.* **64** (1981) 639–644.

[123] W. Heywang, *J. Am. Ceram. Soc.* **47** (1964) 484–490.

[124] J. Daniels, K. H. Hardtl, R. Wernicke, *Philips Tech. Rev.* **38** (1978/1979) no. 3, 73–82.
[125] M. Kuwabara, *Adv. Ceram.* **7** (1984) 128–136.
[126] M. Matsuoka, *Adv. Ceram.* **1** (1981) 290–308.
[127] J. Wong, *J. Am. Ceram. Soc.* **57** (1974) 357–359.
[128] D. R. Clarke, *J. Appl. Phys.* **50** (1979) 6829–6832.
[129] G. D. Mahan, L. M. Levinson, H. R. Philipp, *J. Appl. Phys.* **50** (1979) 2799–2812.
[130] F. A. Selim, T. K. Gupta, P. L. Hower, W. G. Carlson, *J. Appl. Phys.* **51** (1980) 765–768.
[131] P. Williams, D. L. Krivanek, G. Thomas, M. Yodogawa, *J. Appl. Phys.* **51** (1980) 3930–3934.

Cerium Compounds

KLAUS REINHARDT, Th. Goldschmidt AG, Essen, Federal Republic of Germany

HERWIG WINKLER, Treibacher Chemische Werke AG, Treibach, Austria

1.	Production 1111	4.	Economic Aspects 1114
2.	Uses 1112	5.	Toxicology and Occupational . 1115
3.	Analysis 1114	6.	References 1115

The authors thank I. S. Hirschhorn of Lanthanide Research Corporation, West Orange, New Jersey, United States, previously with Ronson Metals Corporation, for reading the English manuscript and for his helpful suggestions.

Cerium is the most abundant rare-earth element. Nearly 50% of the available rare-earth raw material consists of cerium oxide. Cerium is also the only rare-earth element that can be easily separated from a mixture of rare-earth elements by simple chemical methods, for cerium has a tetravalent state.

1. Production

Cerium is separated from other rare-earth elements by oxidation of solutions resulting from attack of bastnaesite or monazite. The oxidizing agent can be H_2O_2, hypochlorite, or atmospheric oxygen, or anodic oxidation can be used. Cerium precipitates as cerium(IV) oxide hydrate [63394-44-5], [67285-52-3].

For example, cerium(III) rare-earth oxide hydrate is dried and oxidized by air to cerium(IV) rare-earth oxide hydrate. The oxidized hydrate is dissolved in nitric acid, and the solutions are neutralized slowly. Cerium(IV) oxide hydrate is collected by filtration. Initially, of the total rare-earth oxides 45–50 wt% is CeO_2. This is increased to 95 wt% in the product.

In a process of Molycorp [1], ground bastnaesite ore concentrates are roasted, and cerium is oxidized with atmospheric oxygen. Roasted material is treated with hydrochloric acid to dissolve lanthanum, neodymium, praseodymium, etc. The pH is adjusted to 4, and a cerium-rich residue ($\approx$ 90 wt% CeO_2) is collected by filtration. This residue has high fluoride content and poor solubility. To get soluble cerium salts out of this material a subsequent digestion with sulfuric acid or caustic soda is necessary.

Cerium salts are produced today by liquid–liquid extraction from rare-earth–cerium-containing solutions. Cerium can be extracted out of cerium nitrate–nitric acid solutions in a few steps in the form of a cerium(IV) nitrate complex in tributyl phosphate and therefore separated from the accompanying trivalent rare-earth elements, which form less stable nitrate complexes. Purities of 99.99% and better can be achieved easily.

2. Uses

The principal uses for cerium compounds are as polishing agents and as a component in glass.

Cerium Oxide Polishing Compounds. Cerium-(IV) oxide [1306-38-3], CeO_2, has now replaced other polishing oxides like iron oxide (red rouge), silica (white rouge), and zirconium dioxide almost completely. The special merits of cerium oxide are its 100% faster polishing speed and its cleanliness.

Cerium oxide is used for polishing glass mirrors, plate glass, television tubes, ophthalmic lenses, and precision optics. However, the advent of the Pilkington float process in the early 1970s significantly reduced the use of cerium oxide in the manufacture of plate glass.

Cerium oxide polishing powder is produced by calcining oxidic cerium mineral concentrates in rotary kilns or other furnaces at temperatures of ca. 1000 °C. Calcined concentrates are milled and cleaned by sieving or sifting afterwards. The impurities (CaO, SrO, non-rare-earth elements) must not exceed a certain level in the raw material; otherwise, cerium oxide will agglomerate during calcining and lose polishing capacity.

Polishing powders production can also start from cerium salt solutions. Cerium carbonates or hydroxides are precipitated and then calcined in rotary kilns or muffle furnaces to oxide at temperatures of ca. 1000 °C. The calcined oxides are milled and screened or sifted to get the desired grain size distribution and to remove scratching impurities.

Polishing powders made from mineral concentrates polish more slowly and less cleanly than precipitated products. The former are cheaper but are not used for high-performance polishing processes (precision optics).

Additives improve suspension properties of polishing powders in aqueous solutions and increase polishing rate, for instance, preventing foaming (antifoaming) and settling of cerium oxide (anticaking) in tanks and pipes.

Pure cerium oxide is yellowish white. Small amounts of Pr_6O_{11} in combination with other rare-earth oxides give a brown oxide. The color does not appear to affect the polishing properties for fixed CeO_2 content. Currently, standard concentrations are 50, 70, 90, and 100%. The polishing rate increases with CeO_2 content. The average grain size of polishing powders is normally 0.5–5 µm. A narrow grain size distribution is advantageous.

There are many theories about the mechanism of the polishing process [2]. According to these theories, both a chemical and a physical component, among others, are effective during polishing [3]. The glass surface is hydrolyzed by reaction with water (chemical theory) and the silica gel layer formed is removed mechanically by the polishing compound. Another theory states that glass is removed mechanically (wear theory) and to some extent by chemical reaction [4].

Cerium Oxide as a Glass Constituent. Cerium oxide can be used to decolorize soda–lime glasses for bottles, jars, etc. [5]. The Ce^{4+} oxidizes Fe^{2+} impurities, which are always present in the raw materials and therefore in glasses, to Fe^{3+}. A change from the blue-green color of Fe^{2+} to the 10 times weaker yellow Fe^{3+} takes place. Arsenic, zinc selenite, or manganese are used for the same purpose. The combination of these materials with cerium reduces costs. Presently, cerium oxide has been replaced by less expensive decolorizers.

Cerium protects glass against solarization and browning, a discoloration caused by irradiation. It is therefore a constituent of glass for the faceplates of television screens, which are under constant bombardment by an electron beam [6]. Radiation-shielding windows for nuclear and radiochemical uses consist of lead glass stabilized by CeO_2. Since maximum transmission is necessary for these thick windows, only 99.99% cerium oxide (15 ppm Fe max.) can be used.

For phototropic glasses, as in phototropic eye glasses, windshields, and window glasses that darken in the sunlight and lighten in the shade, cerium is a sensitizer.

Other Uses. Cerium rare-earth fluorides improve the brightness of carbon arcs. Cerium fluoride oxide mixtures are mixed with carbon for electrodes. These electrodes are used on movie sets and as military searchlights to increase brightness as much as tenfold [7]. Cerium oxide (99.9%) is used as white pigment in enamels for tiles.

Basic oil-soluble cerium salts of organic acids, such as cerium alkylsulfonates, alkyl sulfates, and alkyl phosphates, as well as cerium octoates, serve as driers in paints and varnishes. Cerium compounds that are soluble in organic liquids find use as combustion additives in fuels [8]. Particle emission in exhaust gases is reduced by 60%. Diesel oil savings in the range 2–3% are achieved. Cerium compounds added to silicones increase the thermal stability.

Cerium oxide is used in self-cleaning ovens as a catalyst [9].

In zeolitic cracking catalysts, cerium is a thermal and hydrothermal stabilizer to extend the life and increase the activity of the catalysts [10]. The following uses of cerium in noncracking catalysts are listed by PETERS and KIM [11]:

ammonia synthesis
hydrogenation
dehydrogenation
polymerization
isomerization
oxidation
automobile emissions control

3. Analysis

Cerium oxide can be analyzed for total rare-earth content (TRO) and for individual rare-earth elements [12], [13].

The determination of the *total rare-earth content* is carried out by dissolving the rare-earth-containing material with acids or by alkaline fusion. The rare-earth elements are precipitated as hydroxides by adding NH_3, dissolved in HCl or HNO_3, and reprecipitated as oxalates, which are ignited to the oxides, R_2O_3, with the exception of cerium, which forms CeO_2.

The *individual rare-earth elements* are determined by X-ray fluorescence (La), atomic absorption spectrometry (Y), and spectrophotometry (Nd, Pr, Sm). Analysis by means of optical plasma emission spectroscopy is a new, efficient method. Cerium can be determined by titration because it can be oxidized to the tetrapositive state.

4. Economic Aspects

Total rare-earth production in the Western world is 20 000 – 25 000 t of rare-earth oxides per year, of which 45 – 50 % is cerium. Raw material prices have stabilized for bastnaesite at $ 1 per pound TRO FAS Los Angeles, for monazite at A$ 450 per tonne FOB Australia, and rare-earth chlorides ($RCl_3 \cdot 6\,H_2O$) at ca. $ 0.70 per pound delivered.

The growing production of rare-earth raw material and products from the People's Republic of China should only stabilize the market and not sensationally affect prices.

Rhône-Poulenc (France, compounds) and Molycorp-Union Oil (United States, bastnaesite and compounds) are the leading producers; other producers are Treibacher (Austria, metals, alloys, and compounds), Santoku (Japan, metals, alloys, and compounds), Ronson and Reactive Metals (United States, metals and alloys), Indian Rare Earths (India, compounds), and Corona and Fluminense (Brazil, metals and alloys). In Germany, Goldschmidt specializes in magnet alloys. Further magnet alloy producers are Hitachi and Research Chemicals (United States) and Sumitomo and Shin-Etsu (Japan).

Important producers of lighter flints in the Western world are Treibacher, Electro Centre (France), Santoku, and Ronson. Total world market for lighter flints is estimated at ca. 700 t/a. The share of flints for disposable lighters is more than 60 %.

Modern steel technology has decreased demand for mischmetal substantially. The metallurgical uses of rare-earth metals consume less than 15 % of the total rare-earth production. The production capacity for mischmetal is 4000 – 5000 t/a, but less than 50 % is in operation. Mischmetal is sold at $ 4 – 5 per pound, depending on quantity, shape, and size.

Total world demand for cerium-based polishing powders, the biggest consumer of cerium oxide at present, is estimated to be 3000 – 4000 t/a. The price for cerium polishing compounds depends on the cerium content and ranges between $ 2 per pound for 50% CeO_2 and $ 5 per pound for 99.9% CeO_2.

The market for more sophisticated uses — electronics, magnets, etc. — is expected to grow.

The production of rare earths is profitable only if the market for all the rare-earth elements, which are produced in the relative amounts occurring in their minerals, is reasonably balanced.

5. Toxicology and Occupational Health

The rare-earth elements and compounds are considered to be only slightly toxic. No toxic effects during production or use have been reported in the rare-earth industry.

Progressive lung retention was observed after inhalation of dust containing rare-earth oxides and fluorides from carbon arc electrodes. The damage to the lung that was assessed by X ray did not seem to be attributable to the rare earths, but to thorium and its disintegration products. Continual progress in rare-earth processing has reduced the radioactive impurities in rare-earth products substantially, so they are practically free of radioactivity today. Intravenous injections of rare-earth salts damage the liver; oral administration has no pathological effect on animals [14, B 2, pp. 282–283], [15].

6. References

[1] H. W. Harrah: *Deco Trefoil*, Denver Equipment Co., Denver, Colorado, 1967, p. 9.
[2] R. V. Horrigan: "Rare Earth Polishing Compounds," [16] pp. 95–100.
[3] A. Kaller, *Silikattechnik* **31** (1980) 208–214.
[4] T. Izumitani, S. Harada, *Wiss. Z. Friedrich-SchillerUniv., Jena, Math.-Naturwiss. Reihe* **28** (1979) no. 2–3, 389–413.
[5] A. P. Herring, R. W. Dean, J. L. Drobnick: "Use of Cerium Concentrate for Decolorizing Soda-Lime Glasses," *Glass Ind.* **51** (1970) no. 316–322, 350–356, 394–399.
[6] L. W. Riker: "The Use of Rare Earths in Glass Compositions," [16] pp. 81–94.
[7] Molycorp-Union Oil: "Cerium," *Overview of Application Information and Possibilities*, no. 58.
[8] Gulf Research & Development Co., US 4 264 335, 1981 (C. Bello, R. J. Hartle, G. M. Singerman).
[9] E. I. Du Pont de Nemours & Co., US 3 266 477, 1966; US 3 271 322, 1966 (A. B. Stiles).
[10] D. N. Wallace: "The Use of Rare Earth Elements in Zeolite Cracking Catalysts," [16] pp. 101–116.
[11] A. W. Peters, G. Kim: "Rare Earths in Noncracking Catalysts," [16] pp. 117–131.

[12] F. Trombe, J. Loriers, P. Gaume-Mahn, C. Henry La Blanchetais: "Scandium-Yttrium-Éléments des Terres Rares-Actinium," in P. Pascal (ed.): *Nouveau Traité de Chimie Minérale*, vol. **VII**, Fascicule 1 + 2, Masson, Paris 1959.

[13] A. Brusdeylins, *Chem. Ztg.* **97** (1973) 343–347. O. B. Michelsen (ed.): *Analysis and Application of Rare Earth Materials*, Universitetsforlaget, Oslo 1973. Cheng Jai-Kai, *Inorg. Chim. Acta* **94** (1984) 249–258. P. Melard: "Quality Control on an Industrial Scale at the LaRochelle Rare Earth Plant," *Rare Earths Mod. Sci. Technol.* **2** (1980) 517–526.

[14] *Gmelin*, system no. 39, Seltene Erden, A 1 (1938); A 2–A 5, A 7, A 8, B 1–B 7, C 1–C 9, D 1–D 6 (1973 to date).

[15] T. J. Healey, [17] vol. 4, pp. 553–585. P. Arvela: "Toxicity of Rare Earths," *Prog. Pharmacol.* **2** (1979) no. 3, 69–114.

Cesium Compounds

MANFRED BICK, Chemetall GmbH, Frankfurt/Main, Federal Republic of Germany

1. Individual Compounds 1117
2. Resources 1117
3. Production 1118
4. Chemical Analysis 1119
5. Storage and Transportation .. 1119
6. Uses 1119
7. Economic Aspects 1120
8. Toxicology and Occupational Health 1120
9. References 1121

1. Individual Compounds

Cesium carbonate [534-17-8], Cs_2CO_3, M_r 325.82, 81.58% Cs, is a colorless, hygroscopic powder, decomp. at 610 °C, ϱ 4.07 g/cm^3, solubility of 261.5 g in 100 g of water, basic solution.

Cesium hydrogen carbonate [15519-28-5], $CsHCO_3$, M_r193.92, 68.54% Cs, colorless crystalline powder, decomp. at 175 °C, solubility of 209 g in 100 g of water, basic solution.

Cesium chloride [7647-17-8], CsCl, M_r168.36, 78.9% Cs, colorless, crystalline, hygroscopic powder, mp 642 °C, ϱ 3.983 g/cm^3, solubility of 186 g in 100 g of water.

Cesium hydroxide [21351-79-1], CsOH, M_r149.91, 88.66% Cs, anhydrous, colorless, lumpy solid, mp 272 °C, ϱ 3.68 g/cm^3, solubility of ca. 400 g in 100 g of water. The solution is a strong base and very caustic.

Cesium hydroxide monohydrate [35103-79-8], $CsOH \cdot H_2O$, M_r167.93, 79.14% Cs, colorless, crystalline, hygroscopic powder, mp 205–208 °C, ϱ 3.5 g/cm^3, solubility of ca. 860 g in 100 g of water. The solution is a strong base and very caustic.

Cesium iodide [7789-17-5], CsI, M_r259.81, 51.2% Cs, colorless, slightly hygroscopic powder, mp 621 °C, ϱ 4.51 g/cm^3, solubility of 74 g in 100 g of water.

Cesium nitrate [7789-18-6], $CsNO_3$, M_r194.91, 68.19% Cs, colorless crystalline powder, mp 414 °C, ϱ 3.69 g/cm^3, oxidant.

Cesium sulfate [10294-54-9], Cs_2SO_4, M_r361.87, 73.46% Cs, colorless hygroscopic powder, mp 1010 °C, ϱ 4.243 g/cm^3.

2. Resources

Cesium is the 40th most abundant among the elements, occurring about as frequently as germanium. Resources can be categorized into two groups. One, of no commercial importance, consists of the diffuse occurrence of the few grams of cesium per ton contained in potassium salt deposits, sedimentary rocks, and seawater [1].

The only sources of cesium of commercial importance originated during the solidification of residual melts of silicate magmas. After the initial formation of huge granite masses, the remaining melt rich in rare elements like lithium, rubidium, cesium, tantalum, niobium, beryllium, and tin then crystallized to form a type of ore body well-known under the name *pegmatite*. Under favorable conditions, which seem to have existed only in relatively few cases, these pegmatites differentiated into separate bodies. Sodium, potassium, and rubidium formed feldspar-type minerals of considerable mutual solubility, while lithium and cesium, because of the considerable difference in ionic radii from the other alkali metals, formed separate minerals. In the case of cesium, this seems to have occurred in far fewer cases than for lithium, yielding the only commercial cesium mineral, *pollucite* [*1308-53-8*] [2], [3].

Pollucite is a cesium aluminum silicate, which typically contains 18–26% cesium oxide. The theoretical content is 42%, but usually pollucite contains considerable quantities of quartz. Its appearance is also very similar to that of quartz. For this reason and because the demand for cesium has not been great enough to result in any systematic exploration for cesium minerals, possibly further deposits may be discovered. Well known are the large deposits at Bernic Lake in Canada, at Bikita in Zimbabwe, and in SouthWest Africa. In the Soviet Union a new mineral, cestibtantite, a mixed cesium–antimony–tantalum oxide, was reported recently [4]. Small concentrations of cesium, less than one percent, are found in lepidolite, a lithium mineral.

3. Production

For any of the production methods described, pollucite must be powdered first. Production processes can be categorized as (1) decomposition with bases and (2) acid digestion. The second category includes the group of processes used industrially.

Alkaline decomposition can be carried out either by mixing ore with lime and calcium chloride and heating to 800–900 °C followed by leaching of the residue [5] or by heating pollucite with sodium chloride and soda ash to 600–800 °C followed by leaching [6]. In both cases, solutions of impure cesium chloride result.

Acid digestion can be carried out with hydrochloric acid, sulfuric acid, or hydrobromic acid. Treatment of pollucite with hydrochloric acid at elevated temperature produces a solution of chlorides of cesium, aluminum, and alkali metals, which is separated from the silica residue. The cesium is precipitated as mixed chloride with lead, antimony, or tin. Hydrolysis precipitates the auxiliary metal [7]. Alternatives are precipitation with hydrogen sulfide or recovery of cesium by solvent extraction from the leach liquor [8] or ion exchange from cesium chloride solution.

Treatment of pollucite with sulfuric acid [5, p. 5] yields sparingly soluble cesium alum, cesium aluminum sulfate [7784-17-0], which is roasted with carbon to convert the aluminum to alumina and the sulfate sulfur to SO_2. The residue is leached to obtain a cesium sulfate solution, which can be converted to the desired salts by ion exchange, treatment with ammonia or lime (to precipitate aluminum), or solvent extraction [9]. Hydrobromic acid converts pollucite into a solution of bromides of cesium, aluminum, and impurity metals, from which cesium can be precipitated by addition of alcohol. Leaching the precipitate with bromine selectively extracts cesium as $CsBr_3$ [10].

Cesium metal can be produced as an amalgam by electrolyzing concentrated cesium salt solutions on a mercury cathode [11], but reduction of solid cesium salts, especially the halides, with calcium or barium at elevated temperature and removal of cesium by vacuum distillation is the usual method [12].

4. Chemical Analysis

Assays and purities of commercial products are derived by subtracting the sum of analyzed impurity levels from unity. Alkali-metal impurities are analyzed by emission spectroscopy, whereas alkaline-earth metals are determined by atomic absorption. Other metals and anions, such as phosphate and sulfate, can be determined by photometric methods; chloride is established argentometrically.

5. Storage and Transportation

Many cesium salts, especially halides, are hygroscopic and must be stored dry. Transportation regulations, where they exist, are governed by the anion, i.e., the hydroxide being caustic, the nitrate being an oxidant, because nothing inherent in the cesium cation calls for special precautions.

6. Uses

The halides (especially the chloride), the trifluoroacetate, and the sulfate are used in ultracentrifuges, where aqueous solutions of high-purity grades are a medium for separation and purification of nucleic acids (DNA and RNA) for biochemical research. At high rates of rotation, these solutions form a density gradient that separates nucleic acids according to their densities [13].

Various catalysts can be doped with cesium salts as activators, much like the corresponding potassium salts. High-purity cesium halides are transparent to infrared radiation; therefore, they are used for cuvettes, prisms, and windows for spectroscopic equipment. Cesium iodide can be doped to make it a scintillator [14]; single crystals are used in scintillation counting equipment. Cesium fluoride is used for fluorination in organic chemistry.

Open-cycle magnetohydrodynamic generators could offer a considerable potential for cesium compounds as a plasma seeding agent. These devices are under development, especially in North America and the Soviet Union, with the hope that they can boost the overall efficiency of power plants that depend on fossil fuels from ca. 35 to ca. 45%. Hot combustion gas is seeded with potassium or cesium carbonate to make a highly conductive plasma, which is passed through a magnetic field. At right angles to both the field and the direction of plasma flow, there is a voltage difference [15]. Nevertheless, the higher seeding efficiency of cesium compounds must compete with the lower price of potassium compounds.

7. Economic Aspects

The producers do not publish production or consumption figures. The U.S. Bureau of Mines estimated world consumption in 1978 at about 20 t of cesium, as metal and in compounds. At that time prices for cesium salts were $ 64–81 per kg for technical grades and $ 147–170 per kg for high-purity products. The price levels in 1984 are $ 40–75 per kg for technical and pure grades and $ 100–150 per kg for high-purity salts.

8. Toxicology and Occupational Health

The *cesium ion* itself is only very slightly toxic, more toxic than the sodium ion, but less toxic than the potassium ion. Typical LD_{50} values of cesium salts are 1400 mg/kg (rat, intraperitoneal) and 1000 mg/kg (rat, oral) [16]. Exceptions are caused by the toxicity of the particular anion. Cesium hydroxide is strongly caustic, cesium nitrate is an oxidant, and cesium fluoride exhibits the typical toxicity of fluoride.

Special precautions are necessary when handling the metal because exposure to this substance results in severe caustic burns.

9. References

[1] C. A. Hampel: *Rare Metals Handbook*, 2nd ed., Reinhold Publ. Co., New York 1969.
[2] V. M. Goldschmidt: *Geochemistry*, Clarendon Press, Oxford 1954.
[3] V. V. Gordiyenko, *Int. Geol. Rev.* **13** (1970) no. 2, 134.
[4] P. Cerny: Mineralogy of Rubidium and Cesium, *Short Course Handb. Mineral. Assoc. Can.* **8** 1982, 149–161.
[5] K. C. Dean, P. H. Johnson, I. L. Nichols, *Rep. Invest. U.S. Bur. Mines* **6387** (1964).
[6] W. D. Arnold, D. J. Crouse, K. B. Brown, *Ind. Eng. Chem.* **4** (1965) 249.
[7] J. C. Bailor, Jr.: *Inorganic Syntheses*, vol. **4,** McGraw-Hill, New York 1965.
[8] *Chem. Eng. News* **41** (1963) no. 51 (Dec. 23), 35.
[9] H. W. Parsons, J. A. Vezina, R. Simard, H. W. Smith, *Can. Dept. Mines, Mineral Branch, Tech. Bull.* **50** (1963), reprint of *Can. Metall. Q.* **2** (1963) 1–13; *Chem. Abstr.* **58** (1963) 12 199c.
[10] V. A. Stenger, US 2 481 455, 1949.
[11] R. E. Davis: "Electrowinning of Rubidium and Cesium," *Encyclopedia of Electrochemistry*, Reinhold Publ. Co., New York 1964.
[12] L. Hackspill, *Helv. Chim. Acta* **11** (1928) 1003.
[13] J. Vinograd, J. E. Hearst: *Equilibrium Sedimentation of Macromolecules and Viruses in a Density Gradient*, Springer-Verlag, Wien 1962.
[14] P. Brinckmann, US 3 446 745, 1966.
[15] J. Melcher, *Min. Eng. (Littleton, Colo.)* **29** (1977) no. 12, 34.
[16] K. W. Cochran et al., *Arch. Ind. Hyg. Occup. Med.* **1** (1950) 637.

Chlorine

PETER SCHMITTINGER, Dynamit Nobel AG, Werk Lülsdorf, Niederkassel, Federal Republic of Germany (Chaps. 2–5, 9 (in part), 10, 11, 13–15)

L. CALVERT CURLIN, OxyTech Systems, Chardon, Ohio 44024, United States (Chaps. 6, 9 (in part))

TATSURO ASAWA, Asahi Glass Company, Tokyo, Japan (Chaps. 7, 9 (in part))

STEPHAN KOTOWSKI, Heraeus Elektroden GmbH, Hanau, Federal Republic of Germany (Chap. 8.1)

HENRI BERNHARD BEER, Scientific Research Society, Essen, Belgium (Chap. 8.1)

ARTHUR M. GREENBERG, ELTECH Systems, Chardon, Ohio 44024, United States (Chap. 8.2)

ERICH ZELFEL, Hoechst Aktiengesellschaft, Werk Knapsack, Hürth-Knapsack, Federal Republic of Germany (Chap. 12)

ROLF BREITSTADT, Hoechst Aktiengesellschaft, Frankfurt/Main, Federal Republic of Germany (Chap. 16)

1.	Introduction	1124
2.	Physical Properties	1126
3.	Chemical Properties	1130
4.	Chlor-Alkali Process	1134
4.1.	Brine Supply	1137
4.2.	Electricity Supply	1140
5.	Mercury Cell Process	1141
5.1.	Principles	1142
5.2.	Mercury Cells	1149
5.2.1.	Uhde Cell	1149
5.2.2.	De Nora Cell	1150
5.2.3.	Krebskosmo Cell	1152
5.2.4.	Olin–Mathieson Cell	1152
5.2.5.	Solvay Cell	1152
5.3.	Operation	1152
5.3.1.	Brine System	1153
5.3.2.	Cell Room	1153
5.3.3.	Treatment of the Products	1154
5.3.4.	Measurement	1158
5.3.5.	Mercury Emissions	1159
6.	Diaphragm Process	1162
6.1.	Principles	1162
6.2.	Diaphragm Cells	1167
6.2.1.	Dow Cell	1168
6.2.2.	Glanor Electrolyzer	1170
6.2.3.	OxyTech Hooker Cells	1170
6.2.4.	HU Monopolar Cells	1173
6.2.5.	OxyTech MDC Cells	1173
6.3.	Operation	1175
6.3.1.	Brine System	1175
6.3.2.	Cell Room	1177
6.3.3.	Diaphragm Aging	1179
6.3.4.	Treatment of the Products	1181
6.3.5.	Measurement	1183
6.3.6.	Replacement of Asbestos	1184
7.	Membrane Process	1185
7.1.	Principles	1185
7.2.	Membranes	1190
7.3.	Membrane Cells	1191
7.3.1.	ACI (Asahi Chemical Industry) Electrolyzer	1194
7.3.2.	PPG BIZEC	1194
7.3.3.	Tokuyama Soda TSE-270	1194
7.3.4.	Hoechst–Uhde Bipolar Cell	1194
7.3.5.	Krebskosmo MZB	1195
7.3.6.	Asahi Glass AZEC	1197
7.3.7.	De Nora K40	1198
7.3.8.	OxyTech MGC Electrolyzer	1199
7.3.9.	ICI FM-21	1199
7.3.10.	Chlorine Engineers MBC	1201
7.4.	Operation	1202
7.4.1.	Brine Purification	1202

7.4.2.	Cell Room	1203
7.4.3.	Treatment of Products	1205
8.	**Electrodes**	1205
8.1.	**Anodes**	1205
8.1.1.	General Properties of the Anodes	1206
8.1.1.1.	Coating Properties and Preparation	1206
8.1.1.2.	Titanium Substrate	1208
8.1.1.3.	Anode and Coating Poisons	1208
8.1.2.	Anodes for Mercury Cells	1208
8.1.3.	Anodes for Diaphragm Cells	1210
8.1.4.	Anodes for Membrane Cells	1211
8.1.5.	Graphite Anodes	1212
8.2.	**Activated Cathode Coatings**	1213
9.	**Comparison of the Processes**	1214
9.1.	**Product Qualities**	1214
9.2.	**Economics**	1216
9.2.1.	Capital Investment	1216
9.2.2.	Operating Costs	1217
9.2.3.	Summary	1219
10.	**Other Production Processes**	1220
10.1.	**Electrolysis of Hydrochloric Acid**	1220
10.2.	**Chemical Processes**	1222
10.2.1.	Catalytic Oxidation of Hydrogen Chloride by Oxygen	1222
10.2.2.	Oxidation of Hydrogen Chloride by Nitric Acid	1225
10.2.3.	Production of Chlorine from Chlorides	1225
11.	**Chlorine Purification and Liquefaction**	1225
11.1.	**Cooling**	1225
11.2.	**Chlorine Purification**	1227
11.3.	**Drying**	1227
11.4.	**Transfer and Compression**	1228
11.5.	**Liquefaction**	1229
11.6.	**Chlorine Recovery**	1231
12.	**Chlorine Handling**	1232
12.1.	**Storage Systems**	1232
12.2.	**Transport**	1233
12.3.	**Chlorine Discharge Systems**	1236
12.4.	**Chlorine Vaporization**	1237
12.5.	**Treatment of Gaseous Effluents**	1237
12.6.	**Materials**	1239
12.7.	**Safety**	1240
13.	**Quality Specifications and Analytical Methods**	1241
13.1.	**Quality Specifications**	1241
13.2.	**Analytical Methods**	1241
14.	**Uses**	1242
15.	**Economic Aspects**	1244
16.	**Toxicology**	1247
17.	**References**	1248

1. Introduction

Although C. W. SCHEELE reported the formation of chlorine gas from the reaction of manganese dioxide with hydrochloric acid in 1774, he did not recognize the gas as an element [1]. H. DAVY is usually accepted as the discoverer (1808), and he named the gas *chlorine* from the Greek κλῶρоσ (chloros), meaning greenish yellow. Chlorine for bleaching textiles was first produced from manganese dioxide and hydrochloric acid by a process developed by WELDON, the yield of chlorine being 35% of the theoretical value. In 1866, DEACON developed a process based on the oxidation of hydrogen chloride gas by atmospheric oxygen in the presence of a copper salt, $CuCl_2$, as the catalyst and obtained yields up to 65% of the theoretical value.

In 1800, CRUICKSHANK was the first to prepare chlorine electrochemically [2]; however, the process was of little significance until the development of a suitable generator by SIEMENS and of synthetic graphite for anodes by ACHESON and CASTNER in 1892. These two developments made possible the electrolytic production of chlorine, the *chlor-alkali process,* on an industrial scale. About the same time, both the *diaphragm cell* process (1885) and the *mercury cell* process (1892) were introduced. The *membrane cell* process was developed much more recently (1970). Currently, 95% of the world chlorine production is obtained by the chlor-alkali process. Since 1970 graphite anodes have been largely superseded by *activated titanium anodes* in the diaphragm and mercury cell processes. The newer membrane cell process uses only activated titanium anodes.

Other electrochemical processes in which chlorine is produced include the electrolysis of hydrochloric acid and the electrolysis of molten alkali-metal and alkaline-earth-metal chlorides, in which the chlorine is a byproduct. Purely chemical methods of chlorine production are currently insignificant.

Since 1975, the membrane cell process has been developed to a high degree of sophistication. It has ecological advantages over the two older processes and has become the most economically advantageous process in recent years [3]. Despite these advantages, the membrane cell process has not become widely accepted because existing chlorine plants have been operating below full capacity.

Today, world capacity for chlorine exceeds 40×10^6 t/a, and production in 1983 was only slightly greater than 30×10^6 t. With an annual energy consumption of about 10^{11} kW h, the chlor-alkali process is one of the largest consumers of electrical energy. The chlorine production of a country is an indicator of the state of development of its chemical industry.

Occurrence and Formation. Chlorine is the 11th most abundant element in the lithosphere. Because it is highly reactive, it is rarely found in the free state and then only in volcanic gases. It exists mainly in the form of chlorides, as in sea water, which contains an average of 2.9 wt% sodium chloride and 0.3 wt% magnesium chloride. In salt deposits formed by evaporation of seas, there are large quantities of rock salt (NaCl) and sylvite (KCl), together with bischofite ($MgCl_2 \cdot 6 H_2O$), carnallite ($KCl \cdot MgCl_2 \cdot 6 H_2O$), tachhydrite ($CaCl_2 \cdot 2 MgCl_2 \cdot 12 H_2O$), kainite ($KCl \cdot MgSO_4 \cdot 3 H_2O$), and others. Occasionally there are also heavy metal chlorides, usually in the form of double salts, such as atacamite ($CuCl_2 \cdot 3 Cu(OH)_2$), and compounds of lead, iron, manganese, mercury, or silver. Chlorates and perchlorates occur to a small extent in Chile saltpeter. Free hydrochloric acid is occasionally found in gases and springs of volcanic origin. Plants and animals always contain chlorine in the form of chlorides or free hydrochloric acid.

Chlorine is formed by oxidation of hydrochloric acid or chlorides by such compounds as manganese dioxide, permanganates, dichromates, chlorates, bleaching powder, nitric acid, or nitrogen oxides. Oxygen, including atmospheric oxygen, acts as an oxidizing agent in the presence of catalysts. Some metallic chlorides produce chlorine when heated, for example, gold(III) chloride or platinum chloride.

2. Physical Properties

Chlorine [7782-50-5] exists in all three physical states. At STP it is a greenish-yellow pungent, poisonous gas, which liquefies to a mobile yellow liquid. Solid chlorine forms pale yellow rhombic crystals. The principal properties are given below; more details, including thermodynamic values are given in [4] and in "New Property Tables of Chlorine in SI Units" [5].

Atomic number Z	17
Relative atomic mass A_r	35.453
Stable isotopes (abundance)	35 (75.53%)
	37 (24.47%)
Electronic configuration in the ground state	[Ne] $3s^2 3p^5$
Term symbol in the ground state	$^2P_{3/2}$
Melting point mp	172.17 K (− 100.98 °C)
Boiling point bp	239.10 K (− 34.05 °C)
Critical density ϱ_{crit}	565.00 kg/m^3
Critical temperature T_{crit} (t_{crit})	417.15 K (144.0 °C)
Critical pressure p_{crit}	7.71083 MPa
Density of gas (0 °C, 101.3 kPa) ϱ	3.213 kg/m^3
Density relative to air d	2.48
Enthalpy of fusion ΔH_f	90.33 kJ/kg
Enthalpy of vaporization ΔH_v	287.1 kJ/kg
Standard electrode potential $E°$	1.359 V
Enthalpy of dissociation ΔH_{diss}	239.44 kJ/mol (2.481 eV)
Electron affinity A	364.25 kJ/mol (3.77 eV)
Enthalpy of hydration ΔH_{hyd} of Cl$^-$	405.7 kJ/mol
Ionization energies ΔE_i	13.01, 23.80, 39.9, 53.3,
	67.8, 96.6, 114.2 eV
EC No.	017-001-00-7

The density of chlorine gas at 101.3 kPa is a function of temperature:

t, °C	0	50	100	150
ϱ, kg/m^3	3.213	2.700	2.330	2.051

The density up to 300 °C is higher than that of an ideal gas because of the existence of more complex molecules, for example, Cl$_4$. In the range 400–1450 °C, the density approximates that of an ideal gas, and above 1450 °C thermal dissociation takes place, reaching 50% at 2250 °C. The density of chlorine gas as a function of temperature and pressure is shown in Figure 1. The gas state can be described by the van der Waals equation with

$a = 6.580$ L^2 bar mol^{-2}, $b = 0.05622$ L/mol

The density of liquid chlorine is given in Figure 2. The compressibility of liquid chlorine is the greatest of all the elements. The volume coefficient per MPa at 20 °C over the range 0–10 MPa is 0.12%. The coefficient increases rapidly with temperature: 0.023% at 35 °C, 0.037% at 64 °C, and 0.064% at 91 °C. One liter of liquid chlorine at 0 °C produces 456.8 dm^3 of chlorine gas at STP; 1 kg of liquid produces 311 dm^3 of gas.

The vapor-pressure curve for chlorine is shown in Figure 3.

The vapor pressure can be calculated over the temperature range 172–417 K from the Martin–Shin–Kapoor equation [5]:

$$\ln P = A + \frac{B}{T} + C \ln T + DT + \frac{E(F-T)\ln(F-T)}{FT}$$

$A = 62.402508$
$B = -4343.5240$
$C = -7.8661534$
$D = 1.0666308 \times 10^{-2}$
$E = 95.248723$
$F = 424.90$

Thermodynamic information is given in Table 1, from which the data required for working with gaseous and liquid chlorine can be obtained [6]. The Joule–Thomson coefficient is 0.0308 K/kPa at STP.

At STP the specific heats of chlorine are

$c_p = 0.481$ kJ kg^{-1} K^{-1}
$c_v = 0.357$ kJ kg^{-1} K^{-1}
$\kappa = c_p/c_v = 1.347$

The molar heat capacity at constant volume c_v increases with temperature [7]:

t, °C	0	100	200	500	1000
c_v, J/mol	24.9	26.4	28.1	28.9	29.7

The heat capacity of liquid chlorine decreases over the temperature range −90 °C to 0 °C:

t, °C	−90	−70	−50	−30	0
c, J kg^{-1}K^{-1}	0.9454	0.9404	0.9341	0.9270	0.9169
c, J mol^{-1}K^{-1}	67.03	66.70	66.23	65.73	65.02

The thermal conductivities of chlorine gas and liquid are almost linear functions of temperature from −50 °C to 150 °C:

t, °C	−50	−25	0	25	50	75	100
λ_g, W m^{-1} K^{-1} × 10^2	6.08	7.06	7.95	8.82	9.75	10.63	11.50
λ_l, W m^{-1} K^{-1}	0.17	0.16	0.15	0.135	0.12	0.11	0.09

The viscosities of chlorine gas and liquid are shown in Figure 4 over the same temperature range. The surface tension at the liquid–gas interface falls rapidly with temperature:

t, °C	−50	−25	0	25	50
σ, mJ/m^2	29.4	25.2	20.9	16.9	13.4

Chlorine

Figure 1. Density of chlorine gas as a function of temperature and pressure

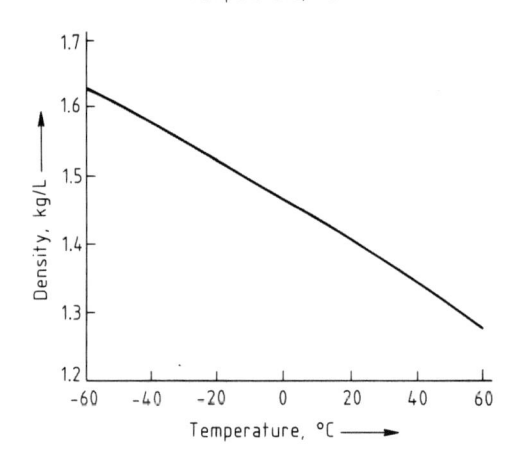

Figure 2. Density of liquid chlorine

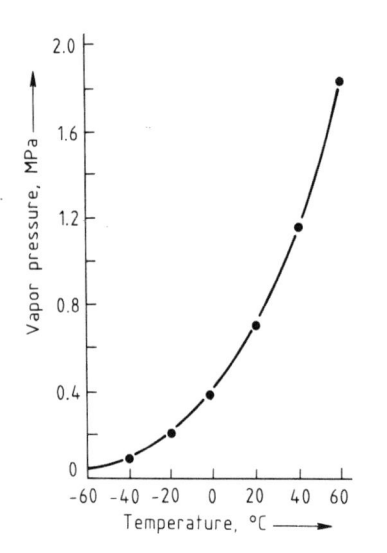

Figure 3. Vapor pressure of liquid chlorine

Table 1. Properties of liquid and gaseous chlorine [5]. Lower values are quoted in more recent literature [2], [3], especially in the region of the critical points.

Temperature t, °C	Pressure, p, bar	Specific volumes, dm³/kg		Specific enthalpies, kJ/kg*			Specific entropies, kJ kg⁻¹ K⁻¹	
		liquid	vapor	liquid	vaporization	vapor	liquid	vaporization	vapor
−70	0.1513	0.6042	1563	351.11	306.89	658.00	3.9021	1.5106	5.4127
−60	0.2768	0.6135	894.4	360.69	301.58	662.27	3.9481	1.4147	5.3629
−50	0.4762	0.6233	541.8	370.15	296.29	666.41	3.9917	1.3276	5.3193
−40	0.7772	0.6336	344.9	379.70	290.73	670.43	4.0336	1.2468	5.2804
−30	1.212	0.6445	229.0	389.37	284.95	674.33	4.0737	1.1719	5.2456
−20	1.816	0.6560	157.7	399.21	278.84	678.05	4.1131	1.1015	5.2147
−10	2.628	0.6682	112.1	408.88	272.73	681.61	4.1508	1.0362	5.1870
0	3.689	0.6812	81.89	418.68***	266.28	684.96	4.1868**	0.9747	5.1615
10	5.043	0.6951	61.26	428.43	259.67	688.10	4.2215	0.9169	5.1385
20	6.731	0.7100	46.77	438.19	252.80	690.99	4.2546	0.8625	5.1171
30	8.800	0.7261	36.35	447.90	245.72	693.63	4.2873	0.8106	5.0978
40	11.30	0.7435	28.66	457.66	238.31	695.97	4.3183	0.7612	5.0790
50	14.27	0.7627	22.88	467.45	230.53	697.98	4.3480	0.7134	5.0614
60	17.76	0.7837	18.44	477.50	222.07	699.57	4.3781	0.6665	5.0447
70	21.84	0.8073	14.97	487.76	212.90	700.66	4.4074	0.6205	5.0279
80	26.55	0.8339	12.20	498.56	202.60	701.16	4.4376	0.5736	5.0112
90	31.95	0.8646	9.944	510.25	190.79	701.04	4.4665	0.5254	4.9919
100	38.14	0.9010	8.082	523.35	176.85	700.20	4.5004	0.4739	4.9743
110	45.18	0.9456	6.508	537.88	160.14	698.02	4.5372	0.4178	4.9551
120	53.18	1.0039	5.169	554.62	139.59	694.21	4.5787	0.3550	4.9337
130	62.24	1.0890	4.001	575.10	113.30	688.39	4.6277	0.2809	4.9086
140	72.50	1.2624	2.842	603.74	71.18	674.91	4.6934	0.1725	4.8659
144	77.01	1.7631	1.763	642.30	0	642.30	4.7825	0	4.7825

* These values have been calculated in S.I. units according to DIN 1345.
** The enthalpy of liquid chlorine at 0 °C was taken to be
H_0 = 418.66 kJ/kg; the entropy of liquid chlorine at 0 °C was taken to be ϱ_0 = 4.1868 kJ kg⁻¹ K⁻¹.

The specific magnetic susceptibility at 20 °C is -7.4×10^{-9} m^3/kg.

Liquid chlorine has a very low electrical conductivity, the value at -70 °C being 10^{-16} Ω^{-1} cm^{-1}. The dielectric constant of the liquid for wavelengths greater than 10 m is 2.15 at -60 °C, 2.03 at -20 °C, 1.97 at 0 °C, and 1.54 at 142 °C, near the critical temperature.

Chlorine gas can be absorbed in considerable quantities onto activated charcoal and silica gel, and this property can be used to concentrate chlorine from gas mixtures containing it.

Chlorine is soluble in cold water, usually less so in aqueous solutions. In salt solutions, the solubility decreases with salt concentration and temperature. In hydrochloric acid, chlorine is more soluble than in water, and the solubility increases with acid concentration (Fig. 5 and Fig. 6). In aqueous solutions, chlorine is partially hydrolyzed, and the solubility depends on the pH of the solution. Below 10 °C chlorine forms hydrates, which can be separated as greenish-yellow crystals. Chlorine hydrate is a clathrate, and there is no definite chlorine:water ratio. The chlorine–water system has a quadruple point at 28.7 °C; the phase diagram has been worked out by KETELAAR [8].

Chlorine is readily soluble in sulfur–chlorine compounds, which can be used as industrial solvents for chlorine. Disulfur dichloride [*10025-67-9*], S$_2$Cl$_2$, is converted to sulfur dichloride (SCl$_2$) and sulfur tetrachloride (SCl$_4$). Some metallic chlorides and oxide chlorides, such as vanadium oxide chloride, chromyl chloride, titanium tetrachloride, and tin(IV) chloride, are good solvents for chlorine. Many other chlorine-containing compounds dissolve chlorine readily. Examples are phosphoryl chloride, carbon tetrachloride (Fig. 7), tetrachloroethane, pentachloroethane, hexachlorobutadiene (Fig. 7), and chlorobenzene. Chlorine also dissolves in glacial acetic acid, dimethylformamide, and nitrobenzene. The solubility of chlorine in a number of these solvents is given in Table 2.

3. Chemical Properties

Inorganic Compounds. Chlorine, fluorine, bromine, and iodine constitute the halogen group, which has marked nonmetallic properties. The valence of chlorine is determined by the seven electrons in the outer shell. By gaining one electron, the negatively charged chloride ion is formed; the chloride ion has a single negative charge and a complete shell of electrons (the argon structure). By sharing one to seven electrons from the outer shell with other elements, the various chlorine oxidation states can be formed, for example, in the oxides of chlorine, hypochlorites (+ 1), chlorates (+ 5), and perchlorates (+ 7).

The bonds between chlorine and the other halogens are mainly covalent. In the chlorine–fluorine compounds ClF and ClF$_3$, there is some ionic character to the bond, with chlorine the anion, and in the chlorine–iodine compounds ICl$_3$ and ICl, there is

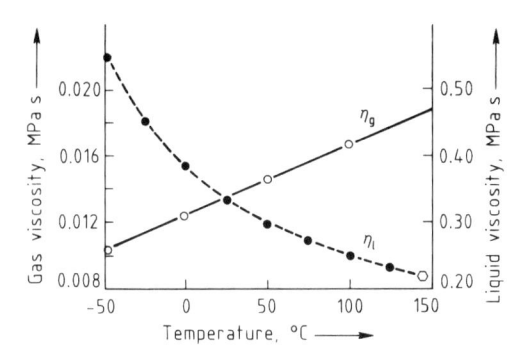

Figure 4. Viscosities of chlorine gas and liquid

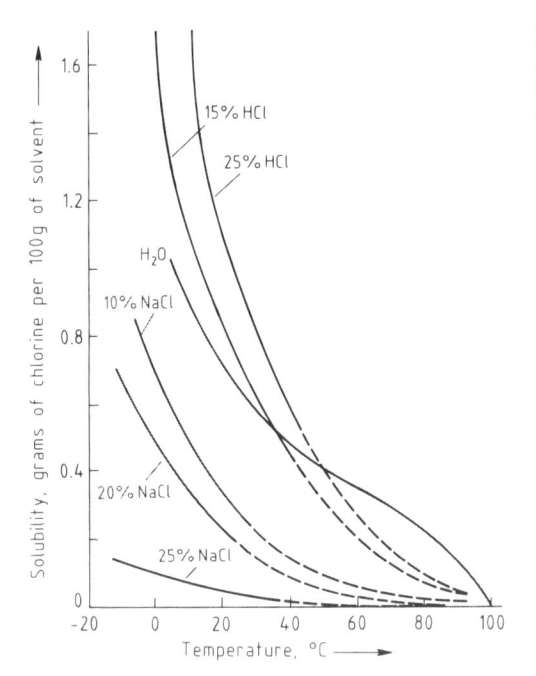

Figure 5. Solubility of chlorine in water, hydrochloric acid (two concentrations), and sodium chloride solutions (three concentrations) All percentages are weight percents.

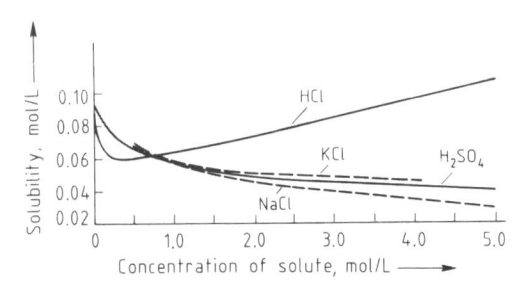

Figure 6. Solubility of chlorine in solutions of KCl, NaCl, H_2SO_4, and HCl at 25 °C

Table 2. Solubility of chlorine in various solvents

Solvent	Temperature, °C	Solubility, wt %
Sulfuryl chloride	0	12.0
Disulfur dichloride	0	58.5
Phosphoryl chloride	0	19.0
Silicon tetrachloride	0	15.6
Titanium tetrachloride	0	11.5
Benzene	10	24.7
Chloroform	10	20.0
Dimethylformamide	0	123*
Acetic acid, 99.84 wt%	15	11.6*

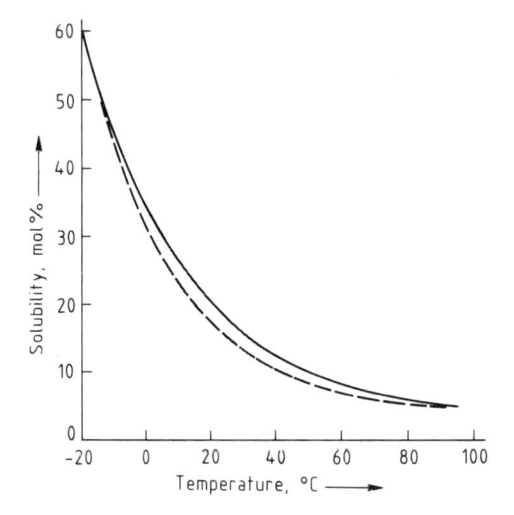

* g/100 cm^3

Figure 7. Solubility of chlorine in hexachlorobutadiene (——) and carbon tetrachloride (– – –) at 101 kPa as a function of temperature

some ionic character to the bond, with chlorine the cation. Chlorine is very reactive, combining directly with most elements but only indirectly with nitrogen, oxygen, and carbon. Excess chlorine in the presence of ammonia salts forms the very explosive nitrogen trichloride, NCl_3. Hypochlorites react with ammonia to produce the chloramines NH_2Cl and $NHCl_2$. Oxygen and chlorine form several chlorine oxides (→ Chlorine Oxides and Chlorine Oxygen Acids).

Chlorine gas does not react with hydrogen gas [1333-74-0] at normal temperatures in the absence of light. In sunlight or artificial light of wavelength ca. 470 nm or at temperatures over 250 °C, the two gases combine explosively to form hydrogen chloride. The explosive limits of mixtures of pure gases lie between ca. 8 vol % H_2 and ca. 14 vol % Cl_2 (the detonation limits). The limits depend on pressure, and the detonation range can be reduced by adding inert gases, such as nitrogen or carbon dioxide (Fig. 8) [9].

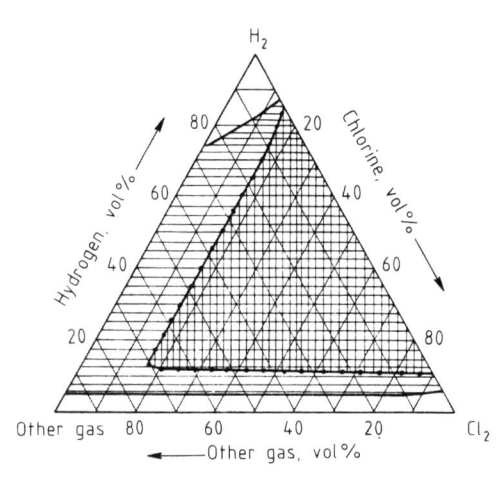

Figure 8. Explosive limits of chlorine–hydrogen–other gas mixture
Horizontally hatched area = explosive region with residue gas from chlorine liquefaction (O_2, N_2, CO_2); checkered area = explosive region with inert gas (N_2, CO_2)

Chlorine reacts vigorously with ammonia

$$3\ Cl_2 + 4\ NH_3 \longrightarrow NCl_3 + 3\ NH_4Cl$$

In the presence of the catalyst bromine, chlorine reacts with nitric oxide to give nitrosyl chloride

$$NO + 0.5\ Cl_2 \longrightarrow NOCl$$

Sulfur dioxide and chlorine in the presence of light or an activated carbon catalyst react to form sulfuryl chloride, SO_2Cl_2. Under these conditions carbon monoxide and chlorine react to produce the colorless, highly toxic carbonyl chloride (phosgene), $COCl_2$.

Chlorine reacts with sodium cyanide and sodium thiocyanate to produce cyanogen chloride and thiocyanogen chloride. The reaction of chlorine with sodium thiosulfate [7772-98-7] (Antichlor) is used to remove free chlorine from solutions.

$$Na_2S_2O_3 + 4\ Cl_2 + 5\ H_2O \longrightarrow 2\ NaHSO_4 + 8\ HCl$$

Chlorine reacts with carbon disulfide to produce carbon tetrachloride and disulfur dichloride.

$$CS_2 + 3\ Cl_2 \longrightarrow CCl_4 + S_2Cl_2$$

The reaction of chlorine with phosphorus produces phosphorus trichloride (PCl_3) and pentachloride (PCl_5). Wet chlorine attacks most metals to form chlorides. Although titanium [7440-32-6] is resistant to wet chlorine, it is rapidly attacked by dry chlorine. Tantalum is resistant to both wet and dry chlorine. Most metals are resistant to dry chlorine below 100 °C, but above a specific temperature for each metal, combustion takes place with a flame. This specific temperature, the ignition temperature, also depends on the particle size of the metal so that the following values are only approximate: iron at 140 °C, nickel at 500 °C, copper at 200 °C, and titanium at 20 °C.

Most metallic chlorides are soluble in water [10, p. 668], notable exceptions being those of silver (AgCl) and mercury (Hg_2Cl_2). Chlorine liberates bromine and iodine from metallic bromides and iodides, but is itself liberated from metallic chlorides by fluorine.

$$0.5\ Cl_2 + KBr \longrightarrow KCl + 0.5\ Br_2$$

Selenium and tellurium react spontaneously with liquid chlorine, whereas sulfur begins to react only at the boiling point. Liquid chlorine reacts vigorously with iodine, red phosphorus, arsenic, antimony, tin, and bismuth. Potassium, sodium, and magnesium are unaffected in liquid chlorine at temperatures below − 80 °C. Aluminum is unattacked until the temperature rises to − 20 °C, when it ignites. Gold is only slowly attacked by liquid chlorine to form the trichloride ($AuCl_3$). Cast iron, wrought iron, carbon steel, phosphor bronze, brass, copper, zinc, and lead are unaffected by dry liquid chlorine, even in the presence of concentrated sulfuric acid.

Organic Compounds. The chlorine – carbon bond is covalent in organic compounds. Chlorine reacts with hydrocarbons either by substitution or by addition. In saturated hydrocarbons, chlorine replaces hydrogen, either completely or partially, to form chlorinated hydrocarbons and hydrogen chloride. Methane can be chlorinated in stages through to carbon tetrachloride.

The reaction of chlorine with unsaturated hydrocarbons destroys the double or triple bond, with decomposition in many cases. In industry the reaction velocity is increased by light (photochlorination), heat (cracker furnace), or catalysts. In aromatic hydrocarbons, both addition and substitution are possible, depending on conditions (light, temperature, pressure, or catalysts). The reaction of chlorine with organic compounds, particularly unsaturated hydrocarbons, can be so violent under some conditions as to lead to complete decomposition and the formation of free carbon.

4. Chlor-Alkali Process

In the chlor-alkali electrolysis process, an aqueous solution of sodium chloride is decomposed electrolytically by direct current, producing chlorine, hydrogen, and sodium hydroxide solution. The overall reaction of the process

$$2\ NaCl + 2\ H_2O \longrightarrow Cl_2 + H_2 + 2\ NaOH$$

takes place in two parts, at the anode and at the cathode. The evolution of chlorine takes place at the anode:

$$2\ Cl^- \longrightarrow 2\ Cl + 2\ e^- \longrightarrow Cl_2 + 2\ e^-$$

There are three basic processes for the electrolytic production of chlorine, the nature of the cathode reaction depending on the specific process. These three processes are (1) the diaphragm cell process (Griesheim cell, 1885), (2) the mercury cell process (Castner–Kellner cell, 1892), and (3) the membrane cell process (1970).

Each process represents a different method of keeping the chlorine produced at the anode separate from the caustic soda and hydrogen produced, directly or indirectly, at the cathode.

These three processes are described in detail in the following three chapters. The basic flow sheets of the three processes are shown in Figure 9. In all three processes, nearly saturated, purified brine is introduced into the electrolysis cell.

The *hydrogen* produced is cooled as it leaves the decomposer or the cathode compartment and is carried through electrically insulated pipework to a vessel fitted with a water seal (Fig. 10). If a hydrogen–air mixture forms because of a shutdown or breakdown, the seal allows the mixture to escape. A demister ensures that the gas is free of spray, whether water or sodium hydroxide solution. The hydrogen is compressed by Roots-type blowers or reciprocating compressors before it passes through coolers on its way to the consuming plants. At no stage is the pressure allowed to fall below ambient pressure.

Electrolytic hydrogen is very pure, > 99.9%; however, unwanted traces of oxygen can be removed by reaction with the hydrogen over a platinum catalyst. The hydrogen is used for organic hydrogenation, catalytic reductions, and ammonia synthesis and to provide hot flames or protective atmospheres in welding technology, metallurgy, or glass manufacture. It is also used in the manufacture of high-purity hydrogen chloride by combustion with chlorine and as a fuel for heating and drying.

In the *mercury cell process*, sodium amalgam is produced at the cathode. The amalgam is reacted with water in a separate reactor, called the decomposer, to produce hydrogen gas and caustic soda solution.

Because the brine is recirculated, solid salt is required for resaturation. The brine, which must be quite pure, first is dechlorinated and then purified by a straightforward precipitation–filtration process.

The products are extremely pure. The chlorine, along with a little oxygen, generally can be used without further purification. The sodium hydroxide solution contains little chloride and leaves the decomposer with a 50 wt % concentration.

Of the three processes, the mercury process uses the most electric energy; however, no steam is required to concentrate the caustic solution. The use of large quantities of mercury demands measures to prevent environmental contamination. In addition, the hydrogen gas and sodium solution must be freed from mercury. Generally, the operation of the cells is not simple.

In the *diaphragm cell process*, the anode area is separated from the cathode area by a permeable asbestos-based diaphragm. The brine is introduced into the anode compartment and flows through the diaphragm into the cathode compartment. Cheaper solution-mined brine can be used; the brine is purified by precipitation–filtration.

A caustic brine leaves the cell, and this brine must be freed from salt in an elaborate evaporative process. Even so, the resultant 50 wt% sodium hydroxide solution contains up to 1 wt% NaCl. The salt separated from the caustic brine can be used to saturate

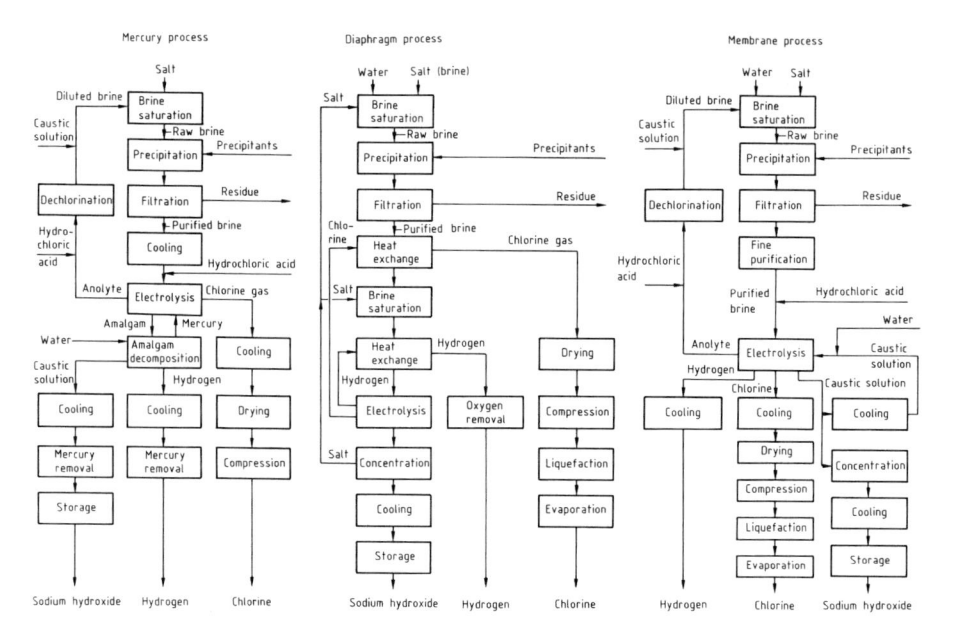

Figure 9. Flow diagram of the three chlor-alkali processes

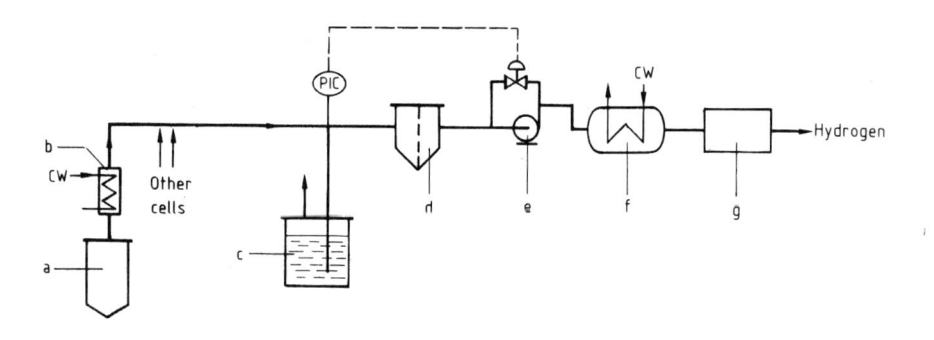

Figure 10. Processing of hydrogen gas from the amalgam decomposer
a) Vertical decomposer; b) Individual cell hydrogen cooler; c) Safety seal; d) Demister; e) Blower; f) Final hydrogen cooler; g) Mercury removal equipment

dilute brine. The chlorine contains oxygen and must be purified by liquefaction and evaporation.

The consumption of electric energy with the diaphragm cell process is ca. 15 % lower than for the mercury process, but the total energy consumption is higher because of the steam required to concentrate the caustic brine (see Fig. 63). Environmental contamination with asbestos must be avoided. Under constant operating conditions, cell operation is relatively simple.

In the *membrane cell process,* the anode and cathode are separated by a cation-permeable ion-exchange membrane. Only sodium ions and a little water pass through the membrane.

As in the mercury process, the brine is dechlorinated and recirculated, which requires solid salt to resaturate the brine. The life of the expensive membrane depends on the purity of the brine. Therefore, after purification by precipitation–filtration, the brine is also purified with an ion exchanger.

The caustic solution leaves the cell with a concentration of 30–35 wt % and must be concentrated. The chloride content of the sodium hydroxide solution is as low as that from the mercury process. The chlorine gas contains some oxygen and must be purified by liquefaction and evaporation.

The consumption of electric energy with the membrane cell process is the lowest of the three processes, ca. 25% less than for the mercury process, and the amount of steam needed for concentration of the caustic is relatively small (see Fig. 63). The energy consumption should be even lower when oxygen-consuming electrodes become common. There are no special environmental problems. The cells are easy to operate and are relatively insensitive to current changes, allowing greater use of the cheaper off-peak-time electric power.

4.1. Brine Supply

The brine used in the mercury cell and membrane cell processes is normally saturated with solid salt although there are some installations that use solution-mined brine on a once-through basis. The brine supply for diaphragm cells is always used on a once-through basis, although the salt recovered from caustic soda evaporators may be recycled into the brine supply.

Salt. The basic raw material for the mercury cell and membrane cell processes is usually solid salt. This may be obtained from three sources: rock salt, solar salt, or vacuum-evaporated salt from purifying and evaporating solution-mined brine.

In the United States and Europe, rock salt is most commonly used. The most important impurities are shown in Table 3. The concentrations of these impurities depend on the method of production and on the different grades: crude rock salt, prepared rock salt, and evaporated salt. Solar salt is used in Japan and many other parts of the world, the most important sources being Pakistan, Egypt, China, Taiwan, India, and Mexico. The salt produced by solar evaporation is usually much less pure than rock salt. In a few cases the salt may be obtained from other processes, such as caustic soda evaporation in the diaphragm process.

Brine Resaturation. In older plants, the open vessels or pits used for storing the salt are also used as resaturators. The depleted brine from the cells is sprayed onto the salt

Table 3. Impurities in rock salt and sea salt, wt %

	Rock salt	Sea salt
Insolubles	≤2	0.1–0.3
Water	≤3	2.0–6.0
Calcium	0.2–0.3	0.1–0.3
Magnesium	0.03–0.1	0.08–0.3
Sulfate	≤0.8	0.3–1.2
Potassium	≤0.04	0.02–0.12

and is saturated, the NaCl concentration reaching 310–315 g/L. Modern resaturators are closed vessels, to reduce environmental pollution [11], which could otherwise occur by the emission of a salt spray or mist. The weak brine is fed in at the base of the resaturator, and the saturated brine is drawn off at the top. If the flow rates of the brine and the continously added salt are chosen carefully, the differing dissolution rates of NaCl and $CaSO_4$ result in little calcium sulfate dissolving within the saturator [12]. Organic additives also reduce the dissolution rate of calcium sulfate [13]. The solubility (g per 100 g of H_2O) of NaCl in water does not increase much with temperature (t, °C), whereas the solubility of KCl does:

t	0	20	40	60	80	100
c_{NaCl}	35.6	35.8	36.4	37.0	38.5	39.2
c_{KCl}	28.2	34.4	40.3	45.6	51.0	56.2

Brine Purification. In mercury cells, traces of heavy metals in the brine give rise to dangerous operating conditions (see p. 1144), as does the presence of magnesium and to a lesser extent calcium [14]. In membrane cells, divalent ions such as Ca^{2+} or Mg^{2+} are harmful to the membrane. The circulating brine must be rigorously purified to avoid any buildup of these substances to undesirable levels [15]. Calcium is usually precipitated as the carbonate with sodium carbonate; magnesium and iron, as hydroxides with sodium hydroxide; and sulfate, as barium sulfate.

The reagents are usually mixed with weak brine and added to the brine stream at a controlled rate. If the magnesium content is high, the Ciba lime–soda process [16] can be used. If solar salt is used, treatment costs may be reduced by prewashing the salt [17]. Brine saturated with calcium sulfate can be used if the brine is acidified to pH 1.5–3 [18]. In order to precipitate calcium at low pH, sodium bicarbonate [19], phosphoric acid [20], or oxalic acid [21] can be added.

The sulfate content can be reduced without the use of expensive barium salts by discharging a part of (purging) the brine [22], by crystallization of $Na_2SO_4 \cdot 10\ H_2O$ on cooling the brine stream [23], or by precipitation of the double salt $Na_2SO_4 \cdot CaSO_4$ [24]. Hoechst [25] has a process for recovering barium sulfate of pure pigment quality by precipitation under acid conditions. Chlorate buildup can be avoided by addition of sodium metabisulfite, $Na_2S_2O_5$ [26], or oxalic acid [27].

After stirring for 1–2 h, the precipitated impurities are removed by filtration alone or by sedimentation followed by filtration. Sedimentation is carried out in large circular settling tanks, from which the slurry is removed by mechanical raking equipment, e.g., Clariflocculator, Cyclator, or Dorr thickener. Filtration is carried out with a sand filter or pressure-leaf filter with filter cloths of chlorine-resistant fabrics. The filter is cleaned by water jets, vibrating, or shaking. The separated filter cake is concentrated to 60–80% solids content in rotary drum vacuum filters or centrifuges before disposal. Any soluble material present may be removed from the sludge by washing with water. Barium salts may be recovered by treating with sodium carbonate under pressure [28]. The purified brine should contain ideally $c_{Ca} < 2$ mg/L, $c_{Mg} < 1$ mg/L, and $c_{SO_4} < 5$ g/L.

In the diaphragm process, the removal of sulfate is not always necessary because SO_4^{2-} can be removed from the cell liquor as pure Na_2SO_4 during the concentration process. In the membrane process, the brine must be purified to a much higher degree to avoid the deterioration of the membrane. The Ca^{2+} and Mg^{2+} concentration must be < 0.05 ppm (50 ppb), so a second, fine purification step is required (see Section 7.4.1).

Before the brine enters the electrolysis cells, it should be acidified with hydrochloric acid to pH < 6, which increases the life of the titanium anode coating, gives a purer chlorine product with higher yield, and reduces the formation of hypochlorite and chlorate in the brine.

Brine Dechlorination. In the mercury and membrane processes, the depleted brine leaving the cells must be dechlorinated before resaturation. Further acidification with hydrochloric acid to pH 2–2.5 reduces the solubility of chlorine by shifting the equilibrium point of hydrolysis and inhibits the formation of hypochlorite and chlorate. Chlorine discharged in the anolyte tank prior to dechlorination may be fed into the chlorine system. The dissolved chlorine of the brine then is still 400–1000 mg/L, depending on pH and temperature. The brine is passed down a packed column or sprayed into a vacuum of 50–60 kPa, which reduces the chlorine concentration in the brine to 10–30 mg/L. The vacuum is produced by steam jet or liquid-ring vacuum pump. The pure chlorine gas obtained is fed into the chlorine stream.

The water that evaporates from the dechlorinated brine is condensed in a cooler. The condensate, which may be chemically dechlorinated, is returned to the brine circulation system if necessary to maintain the volume of the brine circuit. If necessary, the remaining chlorine content can be further reduced by blowing with compressed air, by a second vacuum treatment, by treatment with activated carbon [29], or by chemical treatment with hydrogen sulfite, thiosulfate, sulfur dioxide, ethylene [30], or sodium hydrogen sulfide.

Brine Monitoring. The sodium chloride concentration in the brine is determined by density measured by equipment involving radioactive isotopes, vibration techniques, hydrometry, or weighing.

The pH following alkali or acid additions is determined with glass electrodes, and the redox potential following chlorine removal is determined with metal electrodes.

Excess OH^- and CO_3^{2-} ions ensure adequate precipitation of dissolved calcium, iron, and magnesium. After filtration, a test sample of 100 mL should require 4–6 mL of 0.1 N acid to reach the phenolphthalein end point and a further 0.5–1.5 mL to reach the methyl orange end point. Inadequate filtration is detected by turbidimetry in transmitted light or by the Tyndall effect. Calcium and magnesium are determined hourly, and chlorate and sulfate about once per day, all by titration.

4.2. Electricity Supply

Since 1960 the direct current for electrolysis has been provided exclusively by silicon rectifiers. A set of rectifiers can supply up to 450 000 A. Voltages up to 4.0 kV per diode are feasible, but usually for safety, a peak a.c. voltage of 1500 V, corresponding to a d.c. output of 1200 V, is not exceeded. Liquid cooling of the diodes permits a compact design, and self-contained equipment reduces leakage losses. Modern membrane cell plants also use continuously variable thyristor converters in place of silicon diodes [31].

Rectification equipment is required to provide steady direct current at a voltage determined by the cell room. The current must remain steady even though the voltage is varied both by the operating condition of the cells and by the number of cells operating. The rectifier equipment usually consists of

- transformer capable of variable output voltage with adequate compensation for changing input voltage
- silicon rectifiers
- constant-current control gear
- transducers for metering and control
- control panels
- isolators
- cooling equipment
- ancillary safety and monitoring equipment

Each set of rectifiers is connected through high-voltage switchgear to the three-phase supply [32]. Smaller units use a 10–30 kV supply, but large units can be connected into the high-voltage power system (> 100 kV) [33].

The unit cost of the d.c. supply decreases with increasing voltage and current. A plant is therefore most economical when as many high-current cells as possible are connected in series [34]. Total currents of 450 000 A are achieved. The switches for short-circuiting the cells are designed for 10 000–30 000 A and are operated by compressed air, hydraulically, or by spring action. Erosion of the main contacts is dealt with by using replaceable pre-contacts [35]. The contacts are protected from corrosion by installation in vacuum housings.

The current in the bus bars or in anode rods can be measured by means of iron-free transportable equipment with an accuracy of ca. 1% [36].

5. Mercury Cell Process

The clean separation of chlorine from the cathode products is possible because of the high overvoltage of hydrogen at the mercury electrode. Hydrogen and sodium hydroxide are not produced at the cathode; instead, sodium is produced and dissolves in the mercury as an *amalgam*. The liquid amalgam is removed from the electrolytic cell to a separate reactor, called the decomposer or denuder, where it reacts with water in the presence of a catalyst to form the sodium hydroxide and hydrogen gas. The process may also be used to produce potassium hydroxide by feeding the cell with potassium chloride solution, although this is much less common. The sodium hydroxide is produced from the denuder at a concentration of ca. 50 wt %; the maximum value is 73 wt %. The hydroxide solution is very pure and almost free from chloride contamination.

The process was developed in 1892 almost simultaneously by H. Y. CASTNER and C. KELLNER and used on an industrial scale, although the amount of chlorine produced remained relatively small until 1930, when the rapid growth of the rayon (artificial silk) industry, especially in Germany, increased the demand for pure chloride-free sodium hydroxide solution. At this time, the horizontal high-current cell was developed and output increased rapidly. The development work in Germany was described in the FIAT final reports, published after World War II, and this led to widespread use of the process in Europe and Japan [37]. In the United States, the mercury cell process became more widespread, increasing its share of chlorine production from 3–4% in 1945 to 20% in 1960, reaching a maximum of 27% in 1970.

The development of the mercury cell can be followed in the technical data: the cell current increased from 3.4 kA in 1895 to ca. 30 kA in 1945, 200 kA in 1960, and 450 kA in 1970. The current density rose from 2 kA/m^2 in 1950 to the current maximum of 15 kA/m^2. The cell area increased over the same period from ca. 7 m^2 to 37.5 m^2, while the k-factor (specific voltage coefficient, see p. 1144) was reduced by 50%.

Since 1972 the importance of the mercury cell has decreased. Increasing concern about the effect of mercury on the environment has led to a considerable increase in the number and variation of statutory regulations that affect the mercury cell process.

In particular, widespread concern about cases of mercury poisoning in Japan, which were not related to the mercury cell process [38], caused the process to be legally banned since 1972. However, conversion to the alternative processes was delayed because of demand for low-chloride sodium hydroxide and because of the anticipated advantages of the rapidly developing membrane cell process. The last remaining mercury cell installations were closed in 1986.

In the United States, existing mercury cell plants are still in operation, but official regulations and uncertainty about possible further legal restrictions have hindered expansion.

In Europe, the United States, and Japan, great efforts are being made to develop methods of protecting the environment from mercury (see Section 5.3.5). These measures have greatly reduced emissions of mercury into the atmosphere and into wastewater to the extent that the present levels of emitted mercury are negligible in comparison to those arising from natural sources, such as volcanic action, geological erosion, or other nonnatural sources such as fuel combustion or metallurgical processes. The mercury emissions from modern mercury cell plants are being reduced and are today ecologically justifiable.

At present new plants are being built in the Middle East, India, South America, and the Comecon countries. However, throughout the world the proportion of chlorine produced by mercury cells is decreasing as more plants using the diaphragm or membrane process are installed. In 1984, the mercury cell process accounted for 45% of the world chlorine production [39].

The future of the mercury cell process will depend on energy prices, industrial progress in the developing countries, and the politics and legislation of individual countries.

5.1. Principles

The cathode reaction

$$Na^+ + e^- + Hg_x \longrightarrow NaHg_x$$

forming sodium amalgam, is followed by the decomposition reaction in a separate reactor

$$2\,NaHg_x + 2\,H_2O \longrightarrow 2\,NaOH + H_2(g) + 2\,Hg_x$$

Process Description (Fig. 11). Mercury flows down the inclined base of the electrolytic cell (A). The base of the cell is electrically connected to the negative pole of the d.c. supply. On top of the mercury and flowing cocurrently with it is a concentrated brine with a sodium chloride content of ca. 310 g/L. Anodes are placed in the brine so that there is a small gap between the anode and the mercury cathode. The concentration of the amalgam is maintained at 0.2–0.4 wt % Na, so that the amalgam flows freely (Fig. 12). The chlorine gas and depleted brine (270 g/L) flow out of the cell, either separately or as a two-phase mixture separated later in the process. The amalgam flows out of the cell through a weir and into the decomposer. The amalgam may be passed through a water wash between the cell and the decomposer to remove traces of sodium chloride. The amalgam flows through the decomposer countercurrent to a flow of softened or demineralized water in the presence of a catalyst to produce sodium hydroxide solution and hydrogen. Stripped of its sodium, the mercury flows out of the lower end of the decomposer and is recirculated through a pump back into the cell.

Anode Reactions. The oxidation of chloride ions to chlorine gas has a standard potential of 1.358 V. In a 300 g/L sodium chloride solution at 70 °C, the reversible reaction potential is reduced to 1.248 V [40, p. 339]. Some side reactions occur, such

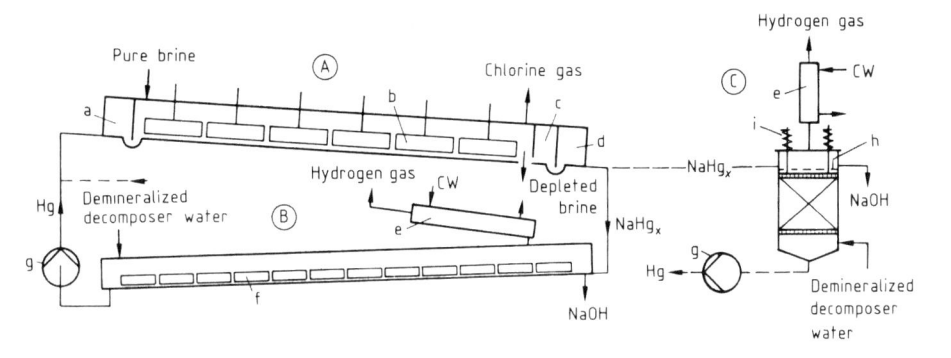

Figure 11. Schematic view of a mercury cell with decomposers
A) Mercury cell: a) Mercury inlet box; b) Anodes; c) End box; d) Wash box
B) Horizontal decomposer: e) Hydrogen gas cooler; f) Graphite blades; g) Mercury pump
C) Vertical decomposer: e) Hydrogen gas cooler; g) Mercury pump; h) Mercury distributor; i) Packing pressing springs

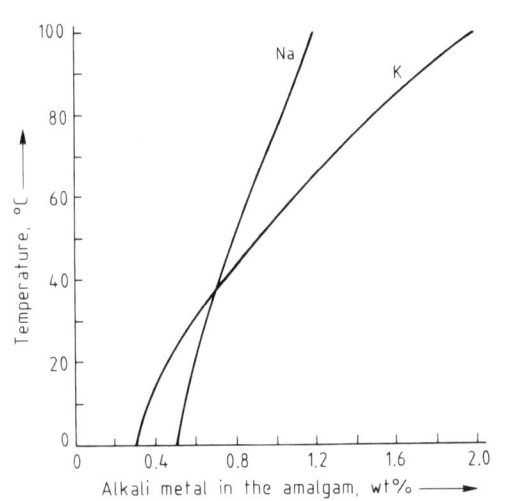

Figure 12. Freezing point curves of sodium amalgam and potassium amalgam

as the oxidation of OH^- and SO_4^{2-} ions and the electrochemical formation of chlorate ions. Nonelectrochemical reactions also take place in the region of the anode, such as hypochlorite formation (because of hydrolysis of chlorine) and chlorate formation. All of these side reactions represent a loss of efficiency.

Cathode Reactions. The standard potential of the hydrogen-liberating reaction is 0 V, which is considerably higher than the potential for the formation of 0.2 wt % sodium amalgam, -1.868 V. However, hydrogen is not liberated at the mercury surface because the reaction is kinetically inhibited. Mainly sodium ions are discharged. At the sodium chloride concentrations used, the reversible potential is reduced by ca. 0.2 V. (Exact values of the discharge potential are given as a function of the sodium concentration in the amalgam, the sodium chloride concentration in the brine, and the temperature [41].) Electrochemical side reactions occur: the reduction of chlorine molecules or hypochlorous acid and the liberation of hydrogen gas. In addition, sodium in the amalgam can react directly with free chlorine, or chlorite and chlorate ions can be reduced to chloride by the action of nascent hydrogen at the cathode. All of these side reactions represent a loss of efficiency, normally ca. 2–4% under good operating conditions.

Contamination of the system by heavy metals can lead to a reduction of the hydrogen discharge potential at the mercury cathode, thus increasing hydrogen liberation, and reducing amalgam formation [42]. The hydrogen concentration in the chlorine can increase to the point at which the cell and downstream chlorine handling equipment contains explosive mixtures. The probability of such problems is estimated by a hazard analysis of an existing plant [43]–[45].

The cell system is sensitive to trace quantities of catalysts in the brine, for example, vanadium, molybdenum, and chromium at the 0.01–0.1 ppm level or iron, cobalt, nickel, and tungsten at the parts per million level. Magnesium, calcium, aluminum, and barium are also active at the parts per million level.

In addition, relatively high concentrations of sodium in the amalgam (> 0.5 wt%) can cause increased hydrogen evolution in the cells. Potassium chloride electrolysis is considerably more sensitive to both catalysts and high concentration in the amalgam than the sodium chloride process.

Current Efficiency. The theoretical electrochemical equivalents representing the materials produced or consumed in the electrolysis of sodium chloride or potassium chloride brines are given in Table 4. In practice, the yield is ca. 95–97% of the theoretical value, owing to side reactions at the electrodes and in the electrolyte. With activated titanium anodes, the yield is largely independent of the distance between the electrodes.

The decrease in salt concentration Δc is determined by the current I, the brine flow rate M, and the electrochemical equivalent f.

$$\Delta c = f I / M$$

The usual units are c in g/L, f in kg kA^{-1} h^{-1}, I in kA, and M in m^3/h.

Cell Voltage. The d.c. voltage across the cell circuit is determined by five factors:

1) The reversible decomposition voltage of the salt
2) The overpotentials of the chlorine and alkali metal at the electrodes
3) The voltage drop in the electrolyte
4) Voltage losses in the bus bars, switches, electrical conductors, anode materials, and cathode
5) The operating current density of the cells

Factor 1. The reversible decomposition potential of NaCl under standard conditions is $E° = 3.226$ V (KCl $E° = 3.234$ V). Under the operating conditions $c_{NaCl} = 290$ g/L, $c_{amalgam} = 0.15\%$, and 70 °C, the reversible decomposition voltage is $E = 3.095$ V [46].

Table 4. Electrochemical equivalents f, kg kA^{-1} h^{-1}

Element	Element produced	Salt required	Alkali produced
Na	0.8580	2.1810 (NaCl)	1.4923 (NaOH)
K	1.4586	2.7816 (KCl)	2.0931 (KOH)
Cl$_2$	1.3228 (0.4115 m^3 STP)		
H$_2$	0.0376 (0.4185 m^3 STP)		

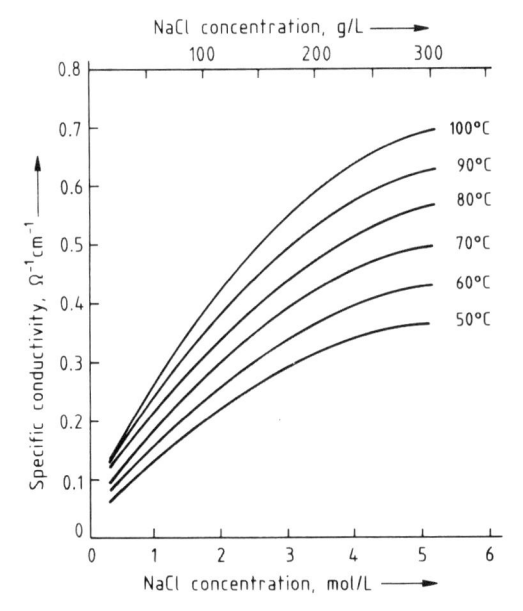

Figure 13. Specific conductivity of sodium chloride solutions

Factor 2. The overpotential of chlorine depends on the material and shape of the anodes. At the high current densities (10 kA/m^2) present in modern cell rooms, the overpotential can reach several hundred millivolts, outweighing the effect of concentration changes in the electrolyte (concentration polarization) [47] and the retarding effect that formation of molecular chlorine has on the process of ion discharging [48]. Chlorine gas bubbles cover part of the anode surface, thereby increasing the current density at the free surface. The anode is designed so that the gas bubbles are liberated as quickly as possible. The rapid removal of these gas bubbles from the reaction zone is one of the advantages of titanium anodes over graphite anodes [49].

The overpotential of sodium on the amalgam cathode is caused by the limited diffusion rate of the liberated sodium atoms into the amalgam, but it is small compared to the chlorine overvoltage.

Factor 3. The specific conductivity of sodium chloride solutions increases with concentration and temperature (Fig. 13), but is independent of pH over the range 2–11. The brine normally enters the cells at 60–70 °C and leaves the cells at 75–85 °C. The conductivity of potassium chloride solutions at 70 °C is 30 % greater than that of sodium chloride solutions. Chlorine gas bubbles in the electrolyte increase the resistance between anodes and cathode. Better circulation of the electrolyte in the gap between electrodes allows more rapid removal of gas bubbles, thus reducing the voltage.

Factor 4. The voltage losses in the cell room are minimized by compactly arranging the cells, which shortens the current path. The relatively low conductivity of steel cell bases can be improved by copper or aluminum fittings. These measures also reduce problems caused by magnetic fields, which occur in wide cells [50].

Factor 5. In practice, cell current density and cell voltage have a linear relationship. The slope of the line is termed the specific voltage coefficient or *k*-factor, a useful measure of the specific energy requirement of cells produced by different manufacturers.

The cell voltage is given by $U_{cell} = 3.15 + kJ$, J = current density, kA/m^2, k = specific voltage coefficient, V kA^{-1} m^{-2}.

Computer-controlled cells with activated titanium anodes are run with *k*-factors from 0.085 to 0.11. The corresponding cell voltages at 10 kA/m^2 are 4.00 – 4.25 V (Fig. 14).

In addition to the d.c. voltages considered above, there are energy losses across the transformer and rectification equipment. All cell installations use a.c. power, which is rectified by silicon diodes in which the energy losses are minimized by operating at greater than 100 V. This voltage is achieved by operating at least 25 cells in series.

Energy Consumption. To operate a cell installation economically, the consumption of d.c. electrical energy per unit mass of product must be minimized. The specific energy consumption w is given by

$$w = 1000\, U_{cell}/af$$

where w = kW h/t, a = yield factor or current efficiency, f = electrochemical equivalent, kg kA^{-1} h^{-1}

For example, if the cell voltage U_{cell} is 4.20 V and the current efficiency is 0.970, then ca. 3275 kW h is required to produce 1 t of chlorine. Since af is almost a constant, the specific energy consumption per tonne of chlorine w is effectively proportional to the cell voltage. In that case, w also depends on the cell current density (see Fig. 14). In the example, w = 3275 kW h corresponds to 10 kA/m^2.

The total energy requirement per tonne of Cl$_2$ must also include the transformer and rectifier losses (30 – 40 kW h/t) and the energy requirements of all of the ancillary equipment (120 – 160 kW h/t).

A mathematical model of the cell has been described [51].

Decomposition of the Amalgam. The amalgam is decomposed in horizontal decomposers, alongside or beneath the cell, or more often since ca. 1960 in vertical decomposers or denuders. The energy stored in the amalgam has an emf of ca. 0.8 V. The hydrogen overpotential at the amalgam prevents spontaneous decomposition in contact with water, and a catalyst (depolarizer) must be used. The overall decomposition reaction is

$$2\,NaHg_x + 2\,H_2O \longrightarrow 2\,NaOH + H_2 + 2\,Hg_x$$

and takes place in two stages, first as an anode reaction at the surface of the amalgam

$$2\,Na \longrightarrow 2\,Na^+ + 2\,e^-$$

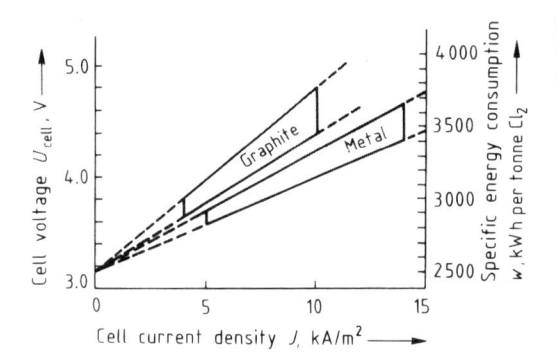

Figure 14. Cell voltage and specific energy consumption per tonne of Cl_2 versus cell current density

and then as a cathode reaction on the catalyst surface, where the water is decomposed

$$2\,H_2O + 2\,e^- \longrightarrow 2\,OH^- + H_2$$

Industrial decomposers are essentially short-circuited electrochemical primary cells (Fig. 15). The most common *catalyst* is graphite [7782-42-5], usually activated by oxides of iron, nickel, or cobalt or by carbides of molybdenum or tungsten. The hydrogen overpotential on graphite (0.5 – 0.6 V at 2 kA/m² and 80 °C) increases with current density and decreases with temperature; therefore, the decomposer should be operated at as high a temperature as possible [52]. Good catalyst material must meet many requirements: resistance to alkali solutions, hydrogen, and mercury; low hydrogen overpotential; good electrical conductivity; long-lasting activity; wettability by amalgam; and incapability of amalgamation.

Attempts to recover some of the energy stored in the amalgam by creating an electrical circuit by using the catalyst as the anode separated from the amalgam or by using the amalgam electrode with an oxygen gas diffusion electrode have so far had no practical outcome [53].

Horizontal decomposers are ducts with a rectangular cross section, which are installed with a 1 – 2.5 % slope near to or underneath the cells. The amalgam flows in a stream ca. 10 mm in depth, and the sodium content is thereby reduced to < 0.02 wt %. The catalyst consists of graphite blades 4 – 6 mm thick, which are immersed in the amalgam in a lengthwise direction (Fig. 16, also see Fig. 11 B). The water for the reaction, which is softened or demineralized by ion exchange, flows in the direction opposite the amalgam and is removed as 50 % caustic alkali solution. The hydrogen gas is cooled as it leaves the decomposer so that any condensed water and mercury run back into the decomposer. Advantages of the horizontal decomposer are serviceability, simple construction, and a pure product that is low in mercury. However, horizontal decomposers require a greater mercury inventory than vertical decomposers.

Vertical decomposers are designed as towers [54] containing packings of activated graphite spheres or other shapes 8 – 20 mm in diameter. The towers are packed 0.6 – 0.8 m high. The cross section of the tower is 0.35 m² per 100 kA of cell current.

Figure 15. Principle of amalgam decomposition

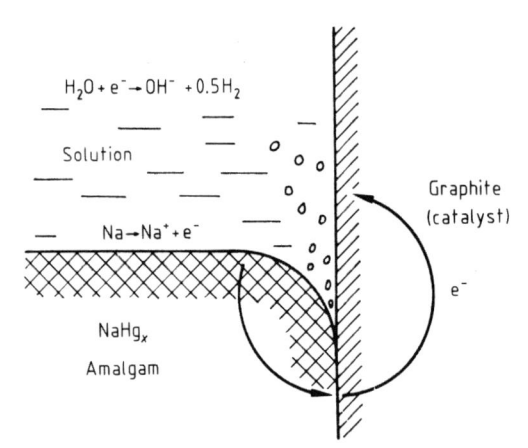

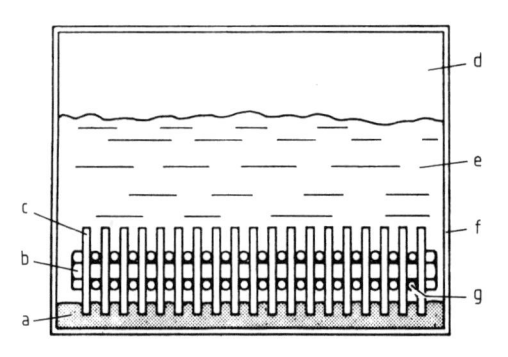

Figure 16. Cross section through a horizontal decomposer
a) Amalgam; b) Bolt; c) Graphite blades; d) Hydrogen; e) Sodium hydroxide solution; f) Decomposer casing; g) Spacers

The amalgam is fed in via an overhead distributor, and the mercury is pumped from the base of the tower back to the cell by a closed centrifugal pump (see Fig. 11C and Fig. 19). The water for the reaction is fed into the base of the tower and flows upward counter to the amalgam. The 50% caustic alkali solution flows out at the top. The smaller volume of the vertical decomposer leads to higher product temperature because of the greater energy intensity of the system. Cooling the hydrogen is essential. Compared with the horizontal decomposer, the amount of space required is small, and the mercury inventory is small, but the caustic alkali contains more mercury.

In alternative decomposition reactions, other products may be obtained from the amalgam in place of sodium or potassium hydroxide solutions [55, p. 518], [56]: sodium sulfide from sodium polysulfide solution, alcoholates from alcohols, sodium dithionite from sodium hydrogen sulfite, hydrazobenzene or aniline from nitrobenzene, adiponitrile from acrylonitrile, and alkali metals by distillation.

5.2. Mercury Cells

During the first decades after the rocking cells of CASTNER and KELLNER were first commissioned, considerable efforts were made to develop suitable materials for the cells and the anodes. A large number of cell configurations were tested, resulting in the development of the continuous cell. Since 1950, the cell areas and the specific load were increased considerably.

In 1972, the changeover from graphite to metallic anodes began, with a parallel development of computer monitoring and control, leading to improved short-circuit protection and a reduction of the specific energy consumption by computer-controlled anode adjustment, of great significance in view of the drastic increase in electricity costs in the late 1970s. In the years following 1972, producers operating the electrolysis plants also concentrated on the development and installation of devices to reduce mercury emissions.

The cells currently available possess a number of common features. The mercury flows over a steel base that has a slope of 1.0 – 2.5 %. The flanged side walls are lined with rubber. The cell covers are mostly steel, lined with rubber on the underside, but they may also be rubberized fabric. The anodes, today almost always of activated titanium, hang in groups from carrying devices that can be varied in height manually, hydraulically, or by motor-operated lifting devices. Each cell can be short-circuited externally by a switch. The cell bus bars are usually copper. The anodes are protected from internal short circuits by means of electronic monitoring systems. The size of the cells can be varied within a broad range to give the desired chlorine production rate. Computer programs optimize the cell size, number of cells, and optimum current density as a function of the electricity cost [57] and capital cost.

For comparison, a list of cells manufactured by leading engineering firms and cell characteristics is given in Table 5. Cathode surface areas are ca. 17 – 30 m^2, and nominal currents are ca. 170 – 300 kA [58, p. 204].

5.2.1. Uhde Cell

The Uhde cells (Fig. 17, also see Fig. 21) are available with a cathode surface area 4 – 30 m^2 for chlorine production rates 10 – 1000 t/d for the complete cell installation. The brine flows in via an inlet box fitted with two pipes for the removal of chlorine. The weak brine is removed at the end of the cell. The solid cover is fixed to the side walls by clamps. The anodes are suspended in groups in carrying frames supported near the cells on transverse girders with lifting gear. The anode rods are raised and lowered within a bellows seal. Short copper bus bars between the cells also serve for shunt measurement of the anode currents. The electric current is brought in above the cell covers via flexible copper straps that run immediately above the anode rods and are bolted to them. The compressed-air switches are situated under the cells. The cell bottom is usually a current conductor when cells are short-circuited, but in wide,

Table 5. Characteristics of modern mercury cells

Characteristic	Manufacturer					
	Uhde	De Nora	Krebs-kosmo	Olin – Mathiesen	Solvay	Krebs Paris
Cell type	300–100	24M2	232–70	E 812	MAT 17	15 KFM
Cathode area, m^2	30.74	26.4	23.2	28.8	17	15.4
Cathode dimensions, $l \times b$, m^2	14.6 × 2.1	12.6 × 2.1	14.4 × 1.61	14.8 × 1.94	12.6 × 1.8	9.6 × 1.6
Slope of cell base, %	1.5	2.0	1.8	1.5	1.7	
Rated current, kA	350	270	300	288	170	160
Max. current density, kA/m^2	12.5	13	13	10	10	10.4
Cell voltage at 10 kA/m^2, V	4.25	3.95	4.25	4.24	4.10	4.30
Number of anodes	54	48	36	96	96	24
Stems per anode	4	4	4	2	1	4
Number of intercell bus bars	36	32	18	24	24	12
Quantity of mercury per cell, kg	5000	4550	2750	3800		1650
Energy requirement per tonne of Cl_2, kW h d.c.	3300	3080	3300	3300	3200	3400

heavily loaded cells the cathode current is carried by copper bus bars to prevent the occurrence of strong magnetic fields, which could interfere with the amalgam flow. The automatic equipment for protection and adjustment of the anodes depends on the shunt measurement of the currents and is controlled by a central computer. In this way, an optimum *k*-factor is selected for each cell. The vertical decomposers are provided with hydrogen coolers and are situated at the end of each cell. The amalgam flows into the decomposer under the force of gravity [59].

5.2.2. De Nora Cell

The size of the De Nora cell (Fig. 18, also see Fig. 22) varies from 4.5 to 36 m^2, corresponding to electric currents from 45 to 400 kA. The cover is a flexible multilayer sheet of elastomer spread over the cell trough. This cover is supported by the anode rods and seals them. The DSA anodes (see Section 8.1) are held rigidly in strong carrying frames, which are automatically adjusted by electric motors. Individual anode adjustment is not provided. The anode rods are individually connected by flexible copper straps to the anode bus bars. The cathode current is carried by copper bus bars. Devices for the improvement of brine circulation and gas removal within the cells reduce specific energy consumption. Consequently, the reduction in brine concentration can be increased from the usual 35–40 g/L to 60–70 g/L, and the brine circulating rate can be reduced by ca. 40%. Separate outlets are present at the inlet box for the normal chlorine gas production and the weak chlorine gas produced during start-up. The graphite catalyst in the vertical decomposer is activated with molybdenum.

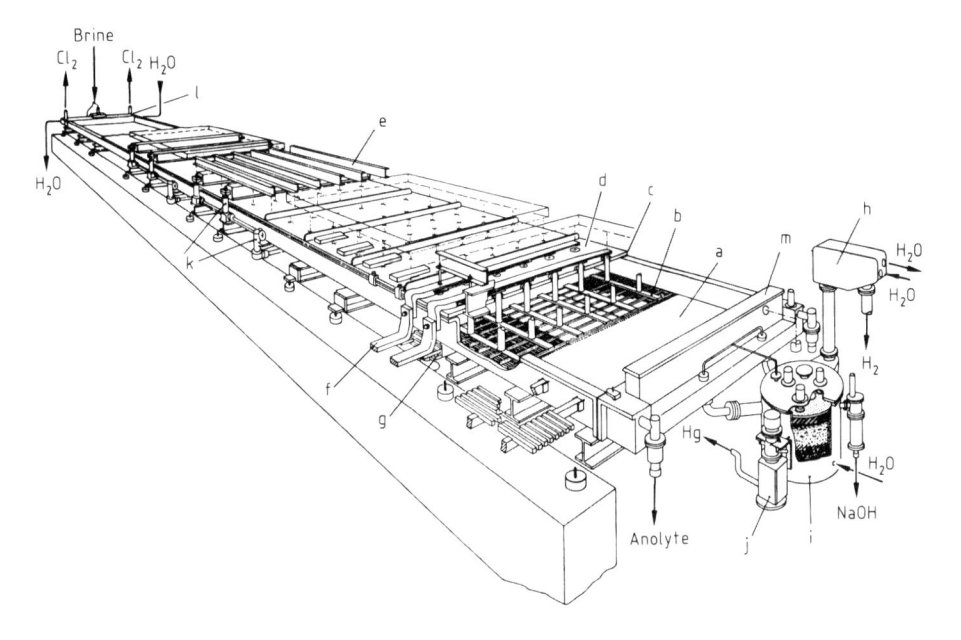

Figure 17. Uhde mercury cell
a) Cell base; b) Anode; c) Cover seal; d) Cell cover; e) Group adjusting gear; f) Intercell bus bar; g) Short-circuit switch; h) Hydrogen cooler; i) Vertical decomposer; j) Mercury pump; k) Anode adjusting gear; l) Inlet box; m) End box

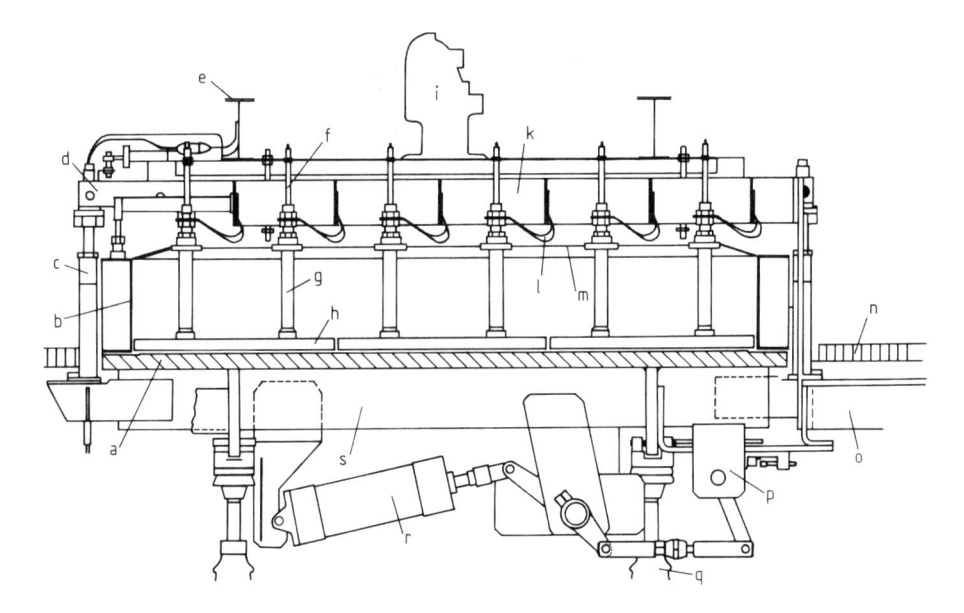

Figure 18. Cross section through the De Nora mercury cell
a) Cell base (steel); b) Side wall (rubber-lined steel); c) Lifting gear; d) Transverse support; e) Lengthwise support; f) Anode carrier; g) Anode rod; h) Anode surface; i) Adjusting motor; k) Bus bar; l) Flexible anode current strap; m) Multilayer cell cover; n) Service walkway; o) Intercell bus bar; p) Switch; q) Insulator; r) Switch drive; s) Support

5.2.3. Krebskosmo Cell

In the Krebskosmo cell (Fig. 19) the brine flows over a flow meter in the inlet box. A device distributes the brine evenly over the width of the cell. The depleted brine is drawn off at the end box. The chlorine gas may be routed either into the main chlorine stream or into a separate absorption system for weak chlorine. The anodes are fixed by copper anode rods to the anode carrier. They are sealed to the rubber-lined steel cell cover by polytetrafluoroethylene (PTFE) bellows. The anodes are adjustable individually or as a group, the latter either by electric motors or by hand. The lifting gear is supported by the cell cover. Pressure sensors in the mercury circulation system shut down the cell immediately in case the mercury pump fails. The electric current is carried to the cell covers by copper or aluminum bus bars and then by flexible copper straps to the anode rods. The short-circuit switches are situated under the cells. In a short-circuited cell, the cell bottom also acts as a conductor [60]. The rectangular vertical decomposer is situated at the end of each cell.

5.2.4. Olin–Mathieson Cell

The special feature of the Olin–Mathieson cell lies in the system of mounting and adjusting the anodes. Above each row of anode rods, a U-shaped copper or aluminum bus bar also serves to support the anode lifting gear. The anode rods are bolted to the U-shaped bus bar. The anodes are adjusted as a group, either manually or by a remote computer with the remote computerized anode adjuster (RCAA) system. The currents are measured independently of the cell potentials by means of reed contacts [61].

5.2.5. Solvay Cell

The bus bars in the Solvay cells are made primarily of aluminum. Above the cells is a cover that also serves as a convenient walkway, giving access to the anode rods. The titanium anodes are specially coated and are automatically adjusted by computer. The tall vertical decomposers are located under the cells.

5.3. Operation

The aspects of the operation of mercury cells that differ from those typical for chloralkali processes enough to justify separate discussion include the brine circulation system, the cell room, treatment of the products, measurement and control, and reduction of mercury emissions. The last is only of importance for the mercury process, since mercury is not used in the diaphragm and membrane processes. A modern mercury cell plant is not a hazard to the environment.

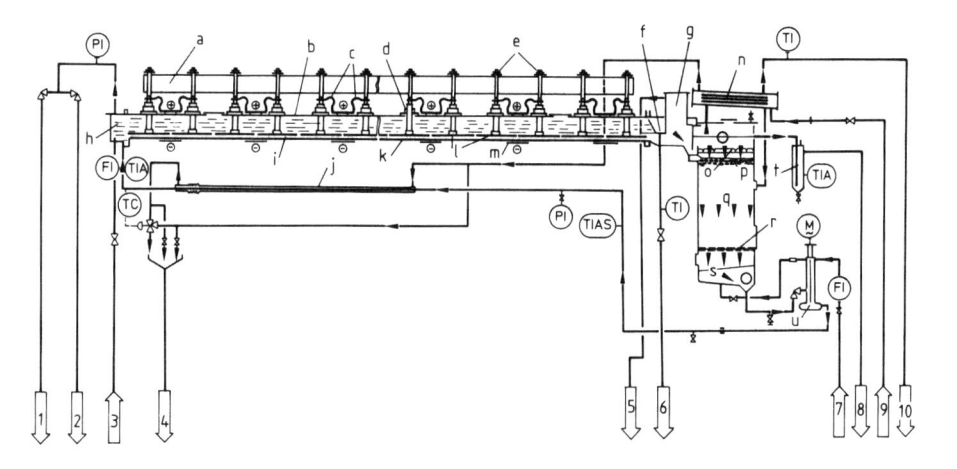

Figure 19. Krebskosmo mercury cell: longitudinal section through the cell and vertical decomposer with connecting pipes

a) Anode group adjustment; b) Cell cover; c) Flexible anode connections; d) Holes in cell cover for anodes; e) Individual anode adjustment; f) Brine outlet; g) Cell end box (wash box); h) Strong brine inlet distributor; i) Flowing mercury cathode; j) Mercury cooler; k) Cell base; l) Titanium anodes; m) Current connections to cell base; n) Hydrogen cooler; o) Amalgam distributor; p) Steel spheres loaded onto graphite packing; q) Graphite packing; r) Packing support grid; s) Wash section; t) NaOH liquid seal; u) Mercury pump

Supply pipes: 1) Concentrated chlorine gas; 2) Dilute chlorine gas; 3) Strong brine; 4) Cooling water exit; 5) End box degassing system; 6) Depleted brine; 7) Demineralized water; 8) Sodium hydroxide solution; 9) Cooling water inlet; 10) Hydrogen gas

5.3.1. Brine System

A typical brine circulation system for the mercury cell process is shown in Figure 20. In the cells the sodium chloride concentration of the brine is reduced by 35 – 60 g/L to 260 – 280 g/L at 70 – 85 °C. To avoid mercury emissions into the air, the resaturators are generally closed vessels. The mercury cathode is very sensitive to poisoning by heavy metals; therefore, a test [62] has been developed that allows rapid determination of the suitability of any particular salt or brine.

5.3.2. Cell Room

The cells are usually situated in a building (Fig. 21), although sometimes they are erected in open air (Fig. 22). Figure 23 shows a bird's eye view, and Figure 24 shows a cross section of cell room. The cells are arranged parallel to each other so that bus bars and supply lines are kept short. The cells stand on supporting structures and are insulated to prevent shorting to the earth. The transformer and rectifiers are situated at one end of the room, and the cell service and repair area is at the opposite end. Ancillary equipment is installed near the cell room in a spillage containment area.

Cell floors, gangways, and spillage containment areas are constructed with smooth, sloping floors so that any mercury can be easily recovered or wash water can be conveniently collected for treatment. The supply pipes run under the cells and are connected to them by flexible, insulating connections. The heat given off by the cells and the decomposers is removed by a ventilation system.

The plant is operated with continuous 24 h/d supervision and control. An additional day-shift team carries out anode changes, repairs, and cleaning.

Occupational Health. Anyone working in the cell area must undergo regular health checks. The Western European chlorine manufacturers in the Bureau International Technique Chlore (BITC) have prepared a *Mercury Code of Practice,* in which rules are recommended for dealing with mercury, including protective measures and medical tests [63]. The U.S. Environmental Protection Agency has established 18 rules relating to cleanliness of the cell room [64]. Adherence to these rules eliminates any danger to the health of personnel caused by mercury. The maximum allowable concentration or threshold limit value (TLV) of mercury in the atmosphere in Western Europe and in the United States is between 0.025 and 0.100 mg/m^3.

5.3.3. Treatment of the Products

Chlorine. See Chapter 11.

Hydrogen. The treatment of the hydrogen gas leaving the demister is described in Chapter 4. It must pass special equipment for the removal of the traces of mercury before it is used (see p. 1159).

Sodium Hydroxide Solution. The great advantage of the mercury cell process is that very pure sodium hydroxide solution is produced (see Table 16) at a suitable concentration. The chloride content is only 5–50 mg/kg.

Sodium hydroxide from the decomposer usually has a concentration of 50% and a temperature of 80–120 °C. It passes through rubber-lined steel pipe work to nickel or Incoloy coolers, where it is cooled to 40–60 °C. Any particles of graphite from the decomposer or traces of mercury are effectively removed by centrifuges, candle filters, or precoated leaf filters (Fig. 25).

The freezing-point and boiling-point curves of sodium hydroxide solutions are shown in Figure 26. The phases separating from the solution, i.e., ice, hydrates, and NaOH, are indicated along the freezing-point curve. Sodium hydroxide is supplied to consumers as aqueous solution, solid block, flakes, prills, or powder. For the processes involved and for uses → Sodium Hydroxide.

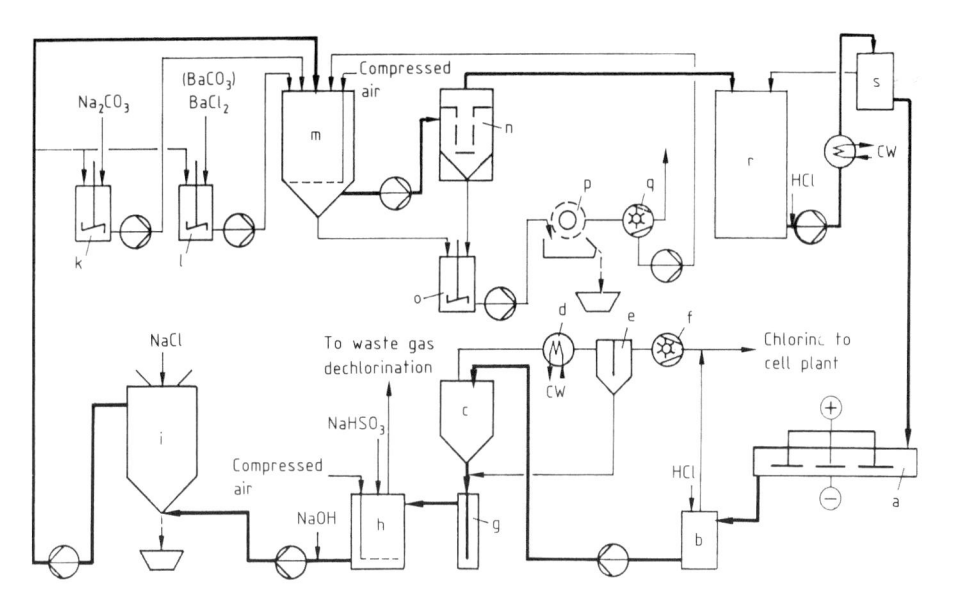

Figure 20. Schematic diagram of a brine circulation system in the mercury cell process
a) Electrolysis cell; b) Anolyte tank; c) Vacuum column dechlorinator; d) Cooler; e) Demister; f) Vacuum pump; g) Seal tank; h) Final dechlorination; i) Saturator; k) Sodium carbonate tank; l) Barium chloride tank; m) Brine reactor; n) Brine filter; o) Slurry agitation tank; p) Rotary vacuum filter; q) Vacuum pump; r) Brine storage tank; s) Brine supply tank

Figure 21. Cell room: Uhde mercury cells

Figure 22. Open-air cell room: De Nora cells

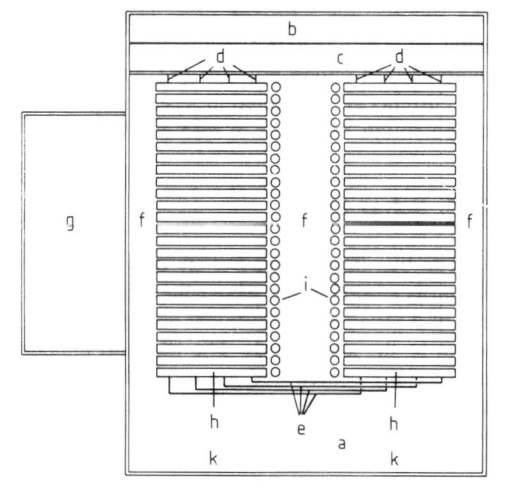

Figure 23. Mercury cell room (bird's-eye view, schematic)
a) Cell room; b) Transformer room; c) Rectifier room; d) Bus bars; e) Turnaround bus bars; f) Service walkways; g) Ancillary equipment; h) Electrolysis cells; i) Vertical decomposers; k) Cell assembly and maintenance area

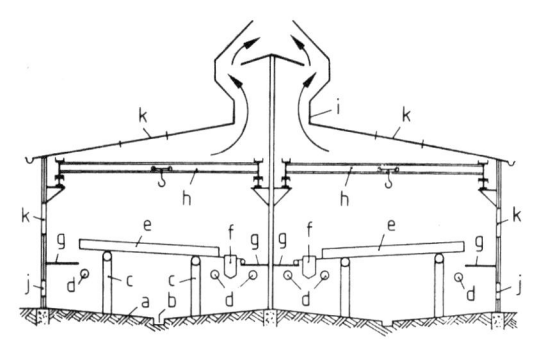

Figure 24. Mercury cell room (cross section, schematic)
a) Basement floor; b) Floor drains; c) Cell supports with insulators; d) Supply pipes; e) Cells; f) Decomposers; g) Service walkways; h) Crane; i) Ridge ventilator; j) Ventilation air supply; k) Windows/lighting

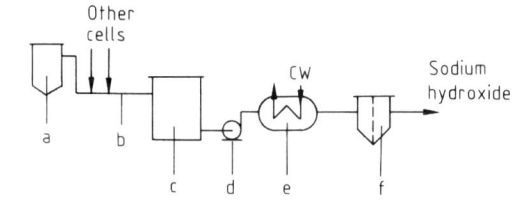

Figure 25. Processing of sodium hydroxide solution from the amalgam decomposer
a) Vertical decomposer; b) Collection main; c) Collecting tank; d) Pump; e) Cooler; f) Mercury removal filter

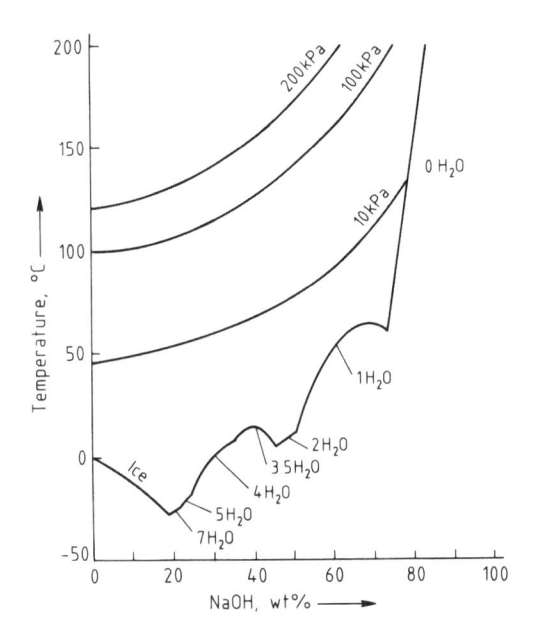

Figure 26. Freezing and boiling point curves of sodium hydroxide solutions

5.3.4. Measurement

The condition of the brine, the cells, and the products must be continuously and carefully monitored, since even small deviations from the correct conditions can increase the hydrogen concentration in the chlorine. The measuring operations are mostly automatic: critical limits are chosen and if these limits are exceeded, alarms are set off.

Cell Operating Conditions. The sodium concentration in the amalgam is determined at the cell inlet (max. 0.05 wt%) and outlet (max. 0.45 wt%).

A 20-g sample of the amalgam is reacted with 30 wt% aqueous sulfuric acid in an absorption pipette. The evolution of 1 cm^3 of gas corresponds to 0.01 wt% Na.

A portable analytical and recording apparatus is available that works electrochemically [65].

If the mercury pump stops, the steel cathode base of the cell is exposed to electrolyte, and hydrogen evolves to form an explosive mixture with the chlorine in the cell.

Failure of the mercury pump or mercury flow automatically short-circuits the cell. Mercury flow failure is detected by monitoring the mercury level at the lowest part of the mercury circulation system or in the inlet box, by direct flow measurement [66], or by loss of pressure at the pump delivery.

The motors for the mercury pumps, the chlorine absorption plant, and the most vital control equipment are all provided with an emergency power supply, ensuring safe shutdown of the plant if a power failure occurs. The installation is protected by a complex system of interlocks so that failure of important equipment, such as the chlorine compressor, shuts down the rectifiers.

A large number of systems have been developed for the protection of cells from short circuits. Titanium anodes are destroyed by short circuits and must be raised before any contact with the amalgam takes place.

The operation of the monitoring system depends on magnetic-field current measurement for individual anodes [61], [67], [68] or on shunt measurement of the supply bus bars [69]. Monitoring is achieved by comparison of the anode–cathode voltage of different cell sections [70] or by following the conductivity of the brine in the electrode gap [71]. The signals from the instruments are fed into central computers or local microprocessors at each cell [72] that control the anode lifting gear. The mercury inventory in each cell may be measured by a radioactive tracer technique once a year without affecting cell operation [73].

Products. The concentration of the sodium hydroxide solution is determined from its density, and the purity its checked by titration to determine hydroxide, carbonate, and chloride contents. The purity of the water for the decomposer is determined from its conductivity.

The oxygen content of the hydrogen gas is determined from the magnetic susceptibility, oxygen being paramagnetic.

5.3.5. Mercury Emissions [38]

Any chlor-alkali plant up to modern technical standards is not a hazard to the environment. The residual emissions are ecologically acceptable according to current knowledge [74]. The mercury in the electrolytic cells circulates in a closed system. All materials that come into contact with the mercury — equipment, products, auxiliary chemicals, wash water, waste gases, other waste materials — are slightly contaminated with mercury, which must be removed. For the exact measurement of these trace amounts and for control of the effectiveness of the measures to reduce the emissions, analytical methods have been developed with sensitivities in the microgram region [75].

Many countries have set legal limits on emissions in waste air and water. The limits on the products of a chlor-alkali plant may depend on their end use, e.g., drinking water treatment or food processing. The sources of contamination are listed, and means of reducing them are described [40], [76], [77]: Mercury cells are sealed vessels, and the products are conveyed in closed pipes. The cell rooms must have smooth joint-free floors with easily cleaned drainage surfaces and irrigated collection gutters (see Fig. 24). Spilled mercury is immediately washed away with water into collecting tanks or sucked up with a vacuum system.

Control of mercury loss is only possible if the mercury content of all the cells is known exactly. The gravimetric and volumetric methods formerly used were cumbersome and led to additional mercury emissions, disadvantages that are avoided by a radioactive tracer method.

Mercury in Products. Hot, moist *chlorine* leaving the cell contains small amounts of mercuric chloride. This is almost completely washed out in the subsequent cooling process and may be fed back into the brine with the condensate. In the cooled and dried chlorine gas, there are only minute traces of mercury: 0.001 – 0.1 mg/kg.

The equilibrium mercury concentration in *hydrogen gas* is a function of temperature and pressure. The mercury concentration (mg of Hg per m^3 of H$_2$ at 101.325 MPa) increases rapidly with temperature:

t, °C	0	20	40	60	80	100
c, mg/m^3	2.36	14.1	66.1	255	836	2404

Subjecting the hydrogen gas to pressure lowers the mercury content of the resulting product gas at atmospheric pressure. For example, at 5 °C

$$c_{Hg}\, p = 0.37\, \mathrm{Pa\, kg\, m^{-3}}$$

c_{Hg}, concentration of mercury in hydrogen gas at atmospheric pressure, mg of Hg per m^3 of H$_2$
p, pressure to which the hydrogen gas is subjected, MPa

When the mixture is cooled to 2–3 °C, the mercury concentration is reduced to ca. 3 mg/m^3 at standard pressure. This mercury content can be reduced by (1) compressing and further cooling, (2) adding chlorine to form mercurous chloride (calomel), which is collected on rock salt or similar material in a packed column, or (3) washing with solutions containing active chlorine. However, it is usually removed by adsorption on activated carbon impregnated with sulfur or sulfuric acid, leaving a mercury concentration in hydrogen of 0.002–0.015 mg/m^3. The highest purity can be achieved by adsorption on copper/aluminum oxide or silver/zinc oxide, < 0.001 mg/m^3.

Centrifugation or filtration in candle filters or in disk filters precoated with charcoal gives *sodium hydroxide solutions* containing mercury concentrations of < 0.1 ppm (mg/kg).

The circulating brine contains mercury concentrations of 2–20 mg/L. Mercury emissions from the brine system can occur through losses of brine into the wastewater, by brine vaporization in the resaturators, or by disposal of the residues from the brine purification filter. These emissions are minimal at a chlorine concentration < 30 mg/L, giving a redox potential > 500 mV vs. NHE. Under these conditions mercury remains dissolved in the brine even if the brine is alkaline.

Mercury in Wastewater. Mercury-containing wastewater has several sources [78]:

1) The process, e.g., condensate and wash liquor from treatment of chlorine, hydrogen, and brine; stuffing-box rinse water from pumps and blowers; brine leakages; ion-exchange eluate from process-water treatment
2) Cell cleaning operations
3) Cleaning of floors, tanks, pipes, and dismantled apparatus

The amount of wastewater can be reduced by separately disposing of the cooling water and process water and by feeding the condensate back into the brine, provided the water balance allows this. A wastewater rate of 0.3–1.0 m^3 per tonne of chlorine is achievable. In older installations, the wastewaters from various parts of the factory must be collected for treatment, and this can cause considerable expense.

There are various methods of making wastewater suitable for discharge:

1) Chemical removal of mercury by reducing any compounds to the metal with hydrazine or sodium borohydride or by precipitating mercuric sulfide with thiourea or sodium sulfide. The mercury metal or sulfide is then filtered off.
2) Oxidation of the mercury by chlorine and adsorption on an ion-exchange medium. Elutriation is done with hydrochloric acid, which is then used to acidify the brine [79].
3) Liquid–liquid extraction [80]. The residual mercury content can be reduced to < 0.01 mg/L by liquid–liquid extraction or a combination of methods.

The Clean Water Act of 1972 (United States) demands the use of the "best available technology economically achievable." Since 1982 each plant has been limited to a maximum of 0.1 g of Hg per tonne of chlorine averaged over 30 d measured at the outlet of the wastewater treatment plant.

In Western Europe, an EC directive has been issued on the subject of the mercury content of wastewater from chlor-alkali plants, following various earlier agreements such as the Rhine protection agreement, the EC guidelines concerning the protection of natural waters, and the Paris Convention [81]. This directive requires plants with circulating brine systems to have a limit of 1.0 g of Hg per tonne of chlorine produced.

Mercury in Process Air. Air from the process, for example, the cell end box ventilation system or the vacuum cleaning systems, can be treated to remove mercury by the methods used for hydrogen.

Ventilation of the Cell Room. The heat produced during electrolysis requires that the air must be changed 10–25 times per hour, depending on the type of building. Mercury spillage can occur during essential operations involving cells or decomposers, for example, opening the cells for anode changing or cleaning, assembling or dismantling equipment, or replacing defective pipes. Spillage leads to small losses in the exhaust air owing to the vapor pressure of mercury. In addition, products that contain mercury, such as the sodium hydroxide solution, hydrogen, or process waste air, can escape via faulty seals in pipes and equipment, leading to emissions. Closed cell construction and special care in handling mercury, i.e., good housekeeping, by adhering to the EPA rules or the *Mercury Code of Practice*, keep the mercury concentrations and, hence, emissions below the allowable work place concentrations (MAK and TLV).

Purification of large volumes of waste air is not feasible. In the United States, the upper limit for the emission of mercury in process waste air and hydrogen is 1 kg per day per facility, and for ventilation air it is 1.3 kg per day per facility [82, p. 372]. In the Federal Republic of Germany the new *TA-Luft* (1985) requires a limit of 1.5 g per tonne of chlorine capacity for new plants and 2 g/t for existing plants [83].

Mercury in Residues. Mercury-containing residues include brine filter slurry, spent decomposer catalyst, discarded cell components, residues from the purification of products, waste material from rinsing media, adsorption materials, ion-exchange media, etc. Mercury can be recovered from these materials by distillation in closed retorts. The residues after distillation must be disposed at special sites.

Summary. In old installations, the procedures described must be adjusted to suit plant conditions. Good operation can restrict total mercury emissions to < 10 g/t. In new plants, total measured emissions (normally less than actual emissions) < 3 g/t are to be expected.

6. Diaphragm Process

The commercial production of chlorine by electrolytic processes began in Europe and the United States in the 1890s. Early cells of the bell-jar type had no diaphragm and relied on the flow of anolyte toward the cathode to prevent the hydroxide ion from back-migrating toward the anode. This method had limited capacity because gas evolution caused mixing and loss of efficiency. The Griesheim cell, another early design, used porous cement as the diaphragm.

E. A. LE SUEUR is credited with the design of a cell incorporating a percolating asbestos diaphragm, which is the basis for all diaphragm chlor-alkali cells currently in use. When brine is caused to flow into the anolyte and subsequently through the diaphragm into the catholyte, continuous operation with much improved efficiency is obtained. This Le Sueur cell, and the similar Billiter cell developed in Germany, incorporated a horizontal asbestos sheet as the diaphragm. During the 1920s, the Billiter cell became the most widely used cell in the world; a few are still in operation today.

Following the invention of synthetic graphite, numerous cells were developed. These fall into three basic types:

1) Rectangular vertical electrode cells
2) Cylindrical vertical electrode cells
3) The vertical electrode bipolar filter press cell developed by Dow Chemical

In 1913, C. W. MARSH developed a cell with finger cathodes and side-entering anodes and cathodes, which greatly increase the electrode area per unit of floor space. About 1928, KENNETH STEWART of Hooker Chemical (now Occidental Chemical Co.) developed a method of depositing asbestos fibers onto the cathode by immersing the cathode in a slurry of asbestos fibers and applying a vacuum. All significant installations of diaphragm cells currently in operation are derived from that development [84].

All diaphragm cells produce cell liquor that contains ca. 11 wt% caustic soda and 18 wt% sodium chloride. To market the caustic soda, its concentration must be increased to 50%. During the evaporation and cooling processes, the salt becomes less soluble in the stronger caustic, and at 50% NaOH the NaCl concentration is ca. 1%.

6.1. Principles

The principles needed to understand the efficient operation of the diaphragm process involve the current efficiency, cell voltage, power consumption, and optimization of the operating conditions [85], [86].

The reaction at the positively charged anode is the same for all three chlor-alkali processes

$$2\,Cl^- \longrightarrow Cl_2 + 2\,e^-$$

The reaction at the negatively charged cathode of the diaphragm cell is

$$2 H_2O + 2 e^- \longrightarrow H_2 + 2 OH^-$$

Figure 27 is a cutaway view of a diaphragm cell that shows the orientation of the various parts of the cell and the various reactions that take place. Figure 27 also shows the location of the diaphragm, which is deposited on the outside of the cathode screen and which separates the cell into two compartments, one containing the anodes and one containing the cathode. The sodium chloride solution (brine) enters the anode compartment and completely covers the anodes and the cathode tubes or fingers. The chlorine leaves the cell through an outlet in the cell head. The anolyte flows through the diaphragm into the cathode compartment because of the difference in liquid level between the two compartments. The catholyte is a solution of sodium chloride and sodium hydroxide because a portion of the water is converted to hydroxide at the cathode. The hydrogen produced at the same time leaves the cell through an outlet on the cathode. The solution of sodium chloride and sodium hydroxide overflows the cell through a level control pipe on the cathode, and is then commonly called cell liquor.

Current Efficiency. Current efficiency is defined as the amount of product actually produced divided by the amount of product that theoretically should have been produced on the basis of the amount of direct-current electrical energy input. The current efficiency is never 100% because of side reactions. The efficiency of a diaphragm cell is usually based on the chlorine production.

The side reactions that lower the efficiency are a result of chlorine that enters the catholyte compartment or of hydroxide ions that enter the anolyte compartment. The amount of chlorine that enters the catholyte compartment is small. The majority of the efficiency losses in a diaphragm cell are due to migration of hydroxide ions from the catholyte through the diaphragm and into the anolyte. This back migration takes place because the negatively charged hydroxide ions are attracted to the positively charged anodes and because of the hydroxide ion concentration gradient across the diaphragm. This migration of hydroxide ions through the diaphragm is in equilibrium with the opposing flow of brine through the diaphragm.

Three factors control the migration of the hydroxide ions into the anolyte:

1) Concentrations of hydroxide and chloride ions at the cathode side of the diaphragm
2) Flow rate of brine through the diaphragm
3) Condition of the diaphragm

These three factors in dynamic equilibrium determine the efficiency of the cell.

Factor 1. The higher the concentration of hydroxide ions in the catholyte, the larger the concentration gradient across the diaphragm, and the higher the probability of hydroxide ions crossing through the diaphragm. As a result, changing cell liquor strength strongly affects cell efficiency. The concentration of chloride ions in the catholyte also affects cell efficiency because some of the chloride ions migrate in place of hydroxide ions.

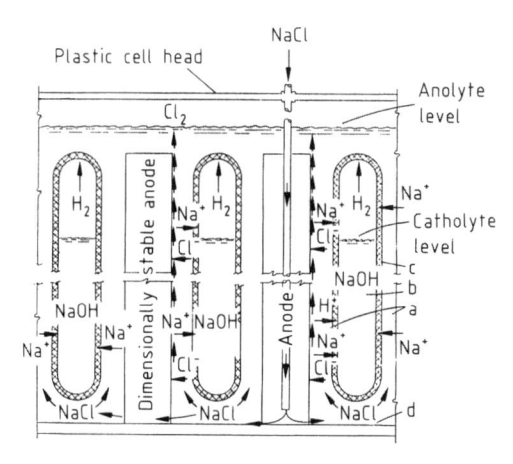

Figure 27. Basic chemical reactions within the cell
a) Anode compartment; b) Cathode compartment; c) Deposited diaphragm on cathode tubes, rims, and end screens

Factor 2. Decreasing brine flow rate to a cell increases the conversion of sodium chloride to sodium hydroxide and raises the hydroxide concentration in the catholyte because of reduced overflow from the cell. The decreased flow rate of brine through the diaphragm allows increased migration of hydroxide ions into the anolyte. These factors decrease cell efficiency.

Factor 3. The condition of the diaphragm is extremely important. Nonuniformity in the diaphragm results in high flow rates of brine through thin or loosely compacted areas and low flow rates through thick or compacted areas. In the areas where there is a low brine flow rate, back migration of hydroxide is increased.

The degree of inefficiency in a cell is indicated by the two products of the side reactions, oxygen in the chlorine and sodium chlorate in the cell liquor. Oxygen in the chlorine gas is the result of hydroxide ions that migrate through the diaphragm into the anolyte, where they are oxidized:

$$2\,OH^- \longrightarrow 1/2\,O_2 + H_2O + 2\,e^-$$

Sodium chlorate in the cell liquor is a result of hydroxide ions that migrate through the diaphragm into the anolyte and react with chlorine before reaching the anode:

$$3\,Cl_2 + 6\,NaOH \longrightarrow NaClO_3 + 5\,NaCl + 3\,H_2O$$

Equations. The simplest equations for calculating cell efficiency are based on the masses of products produced per unit of electrical input. Theoretically, 1.492 kg of sodium hydroxide and 1.323 kg of chlorine are produced per kiloampere-hour. It then follows that

Cathode efficiency, % = (kg of NaOH × 100)/(Q × 1.492 × the number of cells)
Anode efficiency, % = (kg of Cl_2 × 100)/(Q × 1.324 × the number of cells)

where Q is the quantity of electricity in kA h

Unfortunately, the production of a single cell cannot be measured with sufficient accuracy to give meaningful results. To get around this problem, the chlorine industry uses an equation based on the analysis of the chlorine gas, the cell liquor, and the anolyte:

$$\% \text{CE} = [\% \text{Cl}_2 \times 100]/[\% \text{Cl}_2 + 2(\% \text{O}_2) + (\% \text{Cl}_2 \times anox \times F)/c_{\text{NaOH}}]$$

where

% CE = anode current efficiency, %
% Cl_2 = percent chlorine in cell gas (air free)
% O_2 = percent oxygen in cell gas (air free)
$anox$ = oxidizing power of anolyte expressed as grams of NaClO_3 per liter
c_{NaOH} = NaOH concentration in the cell liquor, g/L
F = conversion factor

The denominator is the amount of chlorine produced plus the amount of chlorine consumed in the side reactions. This is equivalent to the amount of chlorine that could have been produced theoretically from the input of current. The conversion factor F is the product of a volume factor, an electric field factor, and a stoichiometric factor. In practice, it is a function of cell liquor strength.

The *SIX equation* is a practical alternative to the previous equation and is often used with computers linked to a gas chromatograph and an automated cell liquor analyzer. The SIX equation is

$$\% \text{CE} = [\% \text{Cl}_2 \times 100]/[\% \text{Cl}_2 + 2(\% \text{O}_2) + (\% \text{Cl}_2 \times 6 \times c_{\text{NaClO}_3})/c_{\text{NaOH}}]$$

This equation also accounts for chlorine lost to the anolyte. However, it approximates the oxidizing potential of the anolyte with the concentration of chlorate in the cell liquor and assumes a fixed conversion factor from anolyte concentration to catholyte, namely *SIX*. The SIX equation approximates the standard equation within 0.5%.

Cell Voltage. The voltage of a cell is the sum of five component voltages: anode potential, cathode potential, cell structure voltage drop, diaphragm voltage drop, and anolyte–catholyte voltage drop. The anode and cathode potentials are sums of the reversible voltages, which are the thermodynamic minimum amounts of work to cause the reactions to take place, and the overpotentials, which are the additional voltages required for nonreversible kinetics. The cell structure voltage drop includes the voltage losses in the cathode, anodes, intercell bus, and all other connectors in the cell. The sum of the diaphragm voltage drop and the anolyte–catholyte (brine) voltage drop is the potential between the electrodes. All of these voltages are functions of current density. Table 6 shows how cell voltage is strongly affected by cell current. Cell temperature, feed brine NaCl concentration, and the cell liquor NaOH concentration also affect cell voltage (Table 7), because they affect the conductivity of the solutions between the electrodes. Table 7 clearly shows that current density is the most important factor.

Table 6. Typical voltage distribution [a]

Component voltages, V	Current density, kA/m^2				
	1.24	1.55	1.86	2.17	2.48
Anode potential [b]	1.30	1.30	1.30	1.30	1.31
Cathode potential [b]	1.12	1.13	1.15	1.16	1.17
Structure loss [c]	0.11	0.14	0.17	0.20	0.22
Brine loss	0.11	0.15	0.19	0.23	0.27
Diaphragm loss	0.24	0.31	0.36	0.41	0.47
Intercell bus	0.02	0.02	0.03	0.03	0.03
Total	2.90	3.05	3.20	3.33	3.47

[a] OxyTech MDC-55 cell with Modified Diaphragm and expandable anodes. Conditions: anolyte temperature 93 °C, anolyte NaCl concentration 250 g/L, catholyte NaOH concentration 130 g/L.
[b] Potential vs. NHE.
[c] Includes anode base, anodes, cathode, cathode screens, copper end connectors, and copper side plates.

Table 7. Factors affecting cell voltage: the change in cell voltage ΔU_{cell} divided by the change in four important cell factors

Factor	Modified cell *	Standard cell *
ΔU_{cell}/change in current density J, mV m^2 kA^{-1}	450	450
ΔU_{cell}/change in cell temperature t_{cell}, mV/°C	−7.7	−10.1
ΔU_{cell}/change in brine concentration c_{NaCl}, mV L g^{-1}	−0.7	−1.8
ΔU_{cell}/change in cell liquor concentration c_{NaOH}, mV L g^{-1}	0.26	0.6

* OxyTech MDC-55 cell. The modified cell is outfitted with the Modified Diaphragm and expandable anodes, whereas the standard cell is outfitted with the standard asbestos diaphragm and the standard DSA anode. Conditions: anolyte temperature of 93 °C, anolyte NaCl concentration of 250 g/L, and catholyte NaOH concentration of 130 g/L.

Excessive brine impurities or other severe operating problems can adversely affect the voltage of the cell.

Power Consumption. The power consumption of a cell, kW h per tonne of Cl_2, may be calculated from the cell voltage by the following equation:

$$\text{Power consumption} = U_{cell} \times 756/\varepsilon$$

where

U_{cell} = cell voltage, V
ε = cell efficiency expressed as a decimal

Optimization. The relationships described in the preceding paragraphs can be used to determine the optimum economical cell operating conditions. The optimizations that must be considered are the following:

1) Higher cell liquor caustic strength and lower steam usage in caustic evaporation versus lower cell efficiency and higher power consumption
2) Lower current density, lower voltage, and lower power consumption versus additional cells and higher capital costs
3) Lower feed brine temperature, thus decreased steam usage for brine heating, versus higher cell voltage, lower efficiency, and higher power consumption
4) High brine pH and reduced acidification costs versus lower chlorine efficiency, higher power consumption, and lower product purity

Each diaphragm cell chlorine plant must determine its own optimum conditions for the most economical operations.

6.2. Diaphragm Cells

Electrolyzers for the production of chlorine and sodium hydroxide, including both diaphragm and membrane cells, are classified as either monopolar or bipolar. The designation does not refer to the electrochemical reactions that take place, which of course require two poles or electrodes for all cells, but to the electrolyzer construction or assembly. There are many more chlor-alkali production facilities with monopolar cells than with bipolar cells.

Bipolar electrolyzers have unit assemblies of the anode of one cell unit directly connected to the cathode of the next cell unit, thus minimizing intercell voltage loss. These units are assembled in series like a filter press, and therefore, the voltage of an electrolyzer is the sum of the individual cell voltages created by the anode of one unit, a diaphragm, and the cathode of the next unit. Bipolar electrolyzers have high voltages and relatively low amperage; therefore, the cost of electrical rectification is higher per unit of production capacity. Bipolar electrolyzers either must be installed in a large number of electrical circuits or be designed with very large individual cell components. Developers have chosen the option for large components.

Dow Chemical was the only early developer to have chlorine production needs large enough to consider the bipolar option [84]. Later, following the development of the DSA anode, PPG Industries and Oronzio De Nora Impianti Elettrochimici designed, and PPG Industries installed Glanor bipolar electrolyzers in a large complex at Lake Charles, Louisiana [87].

The *monopolar* electrolyzer is assembled so that the anodes and cathodes are in parallel. Therefore, the potential difference of all cells in the electrolyzer is the same, and the amperage at any particular current density only depends on the electrode surface area. A monopolar electrolyzer has low voltage and high amperage. The highest amperage rating of the most common modern monopolar cells is ca. 150 kA. Because a monopolar electrolyzer has a voltage of only 3–4 V, circuits of up to 200 electrolyzers have been constructed, producing 900 t of chlorine per day.

Diaphragms. The earliest asbestos diaphragms were made of sheets of asbestos paper. Asbestos was chosen because of its good chemical stability and its ion-exchange properties. Asbestos has been relatively inexpensive, since it is a relatively abundant natural material that was already being mined and processed for other industrial purposes, such as insulation.

The deposited asbestos diaphragm developed by Hooker Chemical in 1928 was the most common diaphragm until 1971, when what is now OxyTech Systems developed the Modified Diaphragm. The Modified Diaphragm is a mixture of asbestos and a fibrous fluorocarbon polymer [88]. The polymer stabilizes the asbestos, which in itself lowers cell voltage and also allows for the use of the expandable DSA anode [89]. In its various formulations, the Modified Diaphragm is the most common diaphragm in use today. The Modified Diaphragm still contains a minimum of 75% asbestos.

6.2.1. Dow Cell [84], [90], [91]

The Dow Chemical Company is the largest chlor-alkali producer, accounting for one-third of the U.S. production and one-fifth of the world capacity. Because Dow's production capacity is large and concentrated in a few sites, Dow's cell development followed a different path than other chlor-alkali technology developers. Dow uses its own cell design of the filter press *bipolar* type. Dow has operated filter press cells for over 80 years. Dow cell development occurred in several stages, characterized by simple rugged construction and relatively inexpensive materials.

The current cell employs vertical anodes of graphite or DSA coated titanium anodes, vertical cathodes of woven wire mesh bolted to a perforated steel backplate, and a vacuum-deposited asbestos diaphragm. A single bipolar element may have 100 m^2 of both anode and cathode active area. The anode of one element is connected to the cathode of the next by copper spring clips. This connection is immersed in the cell liquor during operation. Figures 28 B and 29 show these internal cell parts.

Dow operates at lower current densities than others in the chlor-alkali industry. The electrolyzers are normally operated with 50 or more cells in one unit or series. One electrical circuit may consist of only two of these electrolyzers. Figure 28 A shows a view of six electrolytic cells.

Treated saturated brine is fed to the anolyte compartment, where it percolates through the diaphragm into the catholyte chamber. The percolation rate is controlled by maintaining a level of anolyte to establish a positive, adjustable hydrostatic head. The optimum rate of brine flow usually results in the decomposition of ca. 50% of the incoming NaCl, so that the cell liquor is a solution containing 8–12 wt% NaOH and 12–18 wt% NaCl.

The Dow diaphragm cell, optimized for low current density, consumes less electrical energy per unit of production than the rest of the industry. The cell voltage at these low current densities is only 300–400 mV above the decomposition potential of the cell. However, Dow has a larger investment in the electrolyzers, especially anodes.

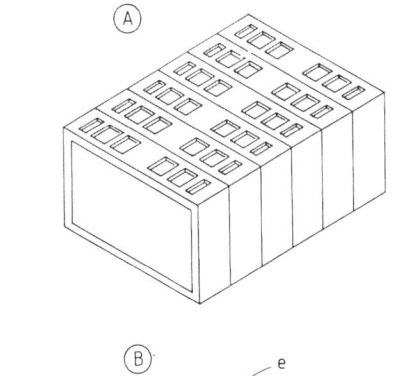

Figure 28. Dow diaphragm cell
A) Six-cell series
B) Internal cell parts: a) Cathode element; b) Cathode pocket elements; c) Copper spring clips; d) Perforated steel backplate; e) Brine inlet; f) Chlorine outlets; g) Copper backplate; h) Titanium backplate; i) Anode element

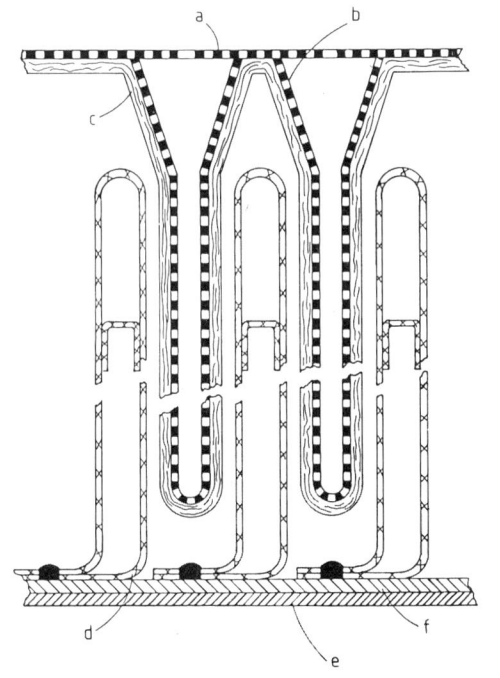

Figure 29. Dow diaphragm cell, section view
a) Perforated steel backplate; b) Cathode pocket; c) Asbestos diaphragm; d) DSA anode; e) Copper backplate; f) Titanium backplate

The electrolyzers are operated at ca. 80 °C, lower than the 95 °C typical of other types of cells. This lower operating temperature allows cell construction with less expensive materials, such as vinyl ester resins and other plastics. Dow uses a mixture of chrysotile asbestos and crocidolite asbestos, whereas the rest of the industry uses chrysotile asbestos only [90].

Operating data have not been published.

6.2.2. Glanor Electrolyzer [87], [92]–[94]

Glanor *bipolar* electrolyzers are a joint development of PPG Industries and Oronzio De Nora Impianti Elettrochimici S.p.A. The Glanor electrolyzer consists of several bipolar cells clamped between two end electrode assemblies by means of tie rods, thereby forming a filter press type electrolyzer (Fig. 30). The electrolyzer is equipped with DSA titanium anodes. Each electrolyzer normally consists of 11 cells. A lower number of cells can, however, be assembled in one electrolyzer. The Glanor electrolyzer was especially designed for large chlor-alkali plants.

The current is fed into the electrolyzer by means of anodic and cathodic end elements. The anodic compartment of each cell is connected to an independent brine feed tank by means of flanged connections.

Chlorine gas leaves each cell from the top through the brine feed tank and then passes to the cell room collection system. Hydrogen gas leaves from the top of the cathodic compartment of each cell, while the catholyte liquor leaves from the bottom through an adjustable level connection.

The V-1144 electrolyzer (Fig. 31) was the first commercial unit, and eight plants utilize this model. The second generation is the V-1161 electrolyzer, which employs Modified Diaphragms, narrower electrode gaps, lower current density, and DSA anodes to achieve lower power consumption than the V-1144 electrolyzer.

The operating characteristics of the Glanor electrolyzers are shown in Table 8.

6.2.3. OxyTech Hooker Cells [84], [87], [95], [96]

The first commercialized deposited asbestos diaphragm cell was the Hooker type S-1 monopolar cell, introduced in 1929. The basic design featured vertical graphite anode plates connected to a copper bus bar and a cathode with woven steel wire cloth or perforated steel fingers between the anodes. The cathode held vacuum-deposited asbestos fiber diaphragms that separated the anode and cathode compartments. The cathode fingers did not extend completely across the cell, but left a central circulation space. In the following 40 years, a family of S series cells with similar characteristics evolved, with over 12 000 having been installed in licensed plants.

In 1973, a new H series of *monopolar* cells was introduced. They incorporated the use of DSA anodes, which had been developed and commercialized in the late 1960s. These cells have significant voltage savings over the S series, thus allowing increases in cell

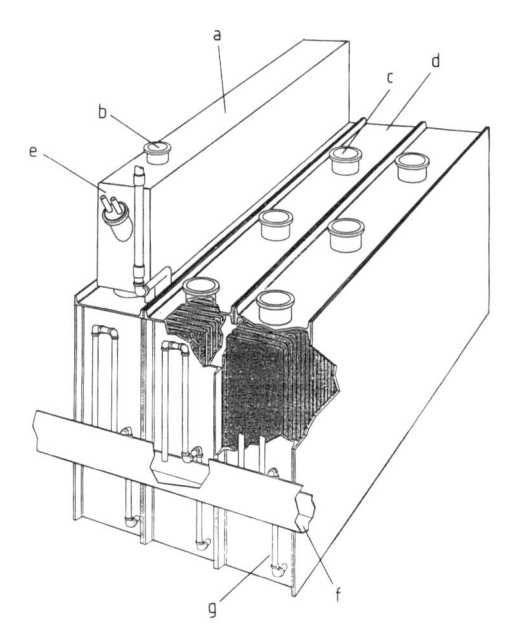

Figure 30. Glanor bipolar electrolyzer
a) Disengaging tank; b) Chlorine outlet; c) Hydrogen outlet; d) Bipolar element; e) Brine inlet; f) Cell liquor trough; g) Cell liquor outlet

Figure 31. Glanor bipolar electrolyzer type V-1144

capacity without corresponding increases in rectification capacity. The H series also incorporate cathode tubes with both ends open, extending across the cell, as the circulation space requirement was satisfied by the change from solid graphite anodes to the open DSA anodes (Fig. 32).

Table 9 is a summary of operating characteristics and current densities of the H-series cells currently available for license.

Table 8. Glanor bipolar diaphragm electrolyzers: design and operating characteristics

Item	Model V-1144	Model V-1161
Cells per electrolyzer	11	11
Active anode area per cell, m^2	35	49
Electrode gap, mm	11	6
Current load, kA	72	72
Current density ϱ at 72 kA, kA/m^2	2.05	1.47
Cell voltage, V	3.50	3.08
Current efficiency, %	95–96	95–96
Power consumption (d.c.), kW h/t *	2500	2200
Anode gas composition (alkaline brine)		
Cl_2, %	97.3–98.0	97.0–98.0
O_2, %	1.5–2.2	1.5–2.2
H_2, %	<0.1	<0.1
CO_2, %	0.4	0.4
Cell liquor		
NaOH, g/L	135–145	135–145
$NaClO_3$, %	0.03–0.15	0.03–0.15
Production per electrolyzer		
Chlorine, t/d **	26.7	26.7
NaOH, t/d	29.8	29.8

* Per short ton of chlorine.
** Short tons.

Table 9. OxyTech Systems Hooker H-series diaphragm cells: design and operating characteristics

	H-2A		H-4	
Operating current, A	80 000		150 000	
Anode area, m^2	36.16		64.52	
in.^{2}	56 050		100 000	
Current density, A/m^2	2212		2325	
A/in.^{2}	1.43		1.50	
Cell voltage, V	3.44		3.44	
Approximate cell dimensions, m	1.87 × 2.66		2.58 × 3.11	
Diaphragm life, days	300–500		300–500	
Anode life, years	5–7		5–7	
Operating NaOH concentration, g/L	140	160	140	160
%	11.35	12.89	11.33	12.87
Current Efficiency, %	96.4	94.6	96.6	94.9
Chlorine output,				
metric ton/day	2.45	2.41	4.60	4.52
short ton/day	2.70	2.65	5.07	4.98
Caustic soda output,				
metric ton/day	2.76	2.71	5.19	5.10
short ton/day	3.05	2.99	5.72	5.62

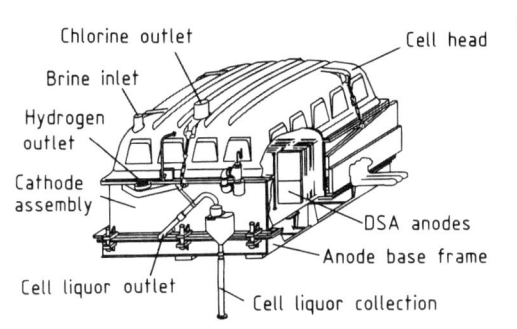

Figure 32. Hooker type H-4 cell

6.2.4. HU Monopolar Cells [96]

The HU type cells were a joint development of Hooker (now OxyTech Systems) and Uhde. The HU-type electrolyzer (Fig. 33) is rectangular, not cubic, and is narrow in the direction of current flow, since anodes are arranged in a single row. The cathode is long and narrow; consequently, the current density is lower through the cathode shell. The long, narrow cathode fabrication lends itself to closer anode – cathode tolerances and spacing. Copper on and around the cathode shell has been eliminated. Another advantage of the long, narrow design is shorter electrolysis current paths through the cell room, resulting in savings in piping and other materials. The HU-type cell incorporates a Modified Diaphragm.

A further novelty of the HU cell system is the design and arrangement of the bypass switch. The HU switch is installed underneath, not next to, the circuit of cells. This is accomplished by raising the cells from the floor, similar to mercury cells, creating a second operating floor. The interconnecting bus bars are flexible and are distributed over the entire length of the cell. The HU cell design incorporates a bus bar for each individual anode. This, as well as the elevation of the cell from the floor below, which allows access, enables connection of facilities for monitoring the current flowing through each anode. During operation of the bypass switch, connection is made for each individual anode, and no additional contact bus bars are required.

The HU-type cells are offered to cover 30 – 150 kA. All of the different cell types are equipped with cathodes and anodes of identical height and width. The only basic difference between the various cell models is the number of elements and consequently the length of the cell (Table 10). Cell voltage and power consumption per tonne of chlorine, identical for all cell types, are shown in Table 11 for the specific current loads of 1.5 and 2.3 kA/m^2.

6.2.5. OxyTech MDC Cells [87], [97]

OxyTech Systems manufactures and licenses the MDC series of *monopolar* diaphragm cells (Fig. 34). The MDC cells feature woven steel wire cathode screen

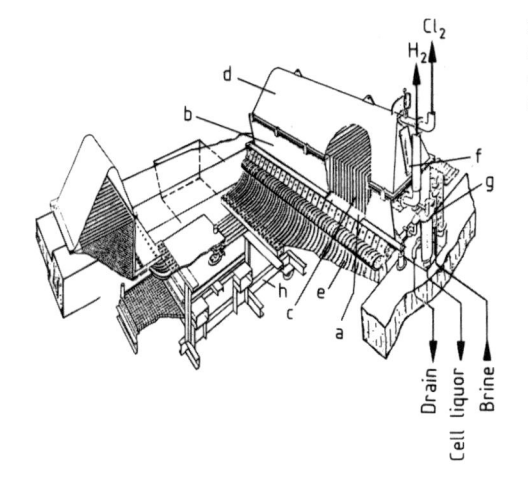

Figure 33. OxyTech/Uhde HU-type cells
a) Cell bottom; b) Cathode; c) Anode; d) Cell cover; e) Bus bars; f) Brine level gauge; g) Brine flow meter; h) Bypass switch

Table 10. HU series diaphragm cells: design and operating characteristics

Item	Cell type						
	HU 24	HU 30	HU 36	HU 42	HU 48	HU 54	HU 60
Number of anodes	24	30	36	42	48	54	60
Anode surface area, m^2	20.6	25.8	31.0	36.1	41.3	46.4	51.6
Load, kA	30–45	40–60	50–70	55–85	60–95	70–105	80–120
Cl$_2$ production, t/d	0.90–1.36	1.19–1.82	1.49–2.12	1.64–2.58	1.79–2.88	2.09–3.18	2.39–3.64
NaOH (100%) production, t/d	1.01–1.54	1.35–2.05	1.68–2.39	1.85–2.91	2.02–3.25	2.36–3.59	2.69–4.10
H$_2$ production, kg/d	25–39	34–52	42–60	47–73	51–82	59–91	68–103
Cell length, m	2.1	2.6	3.0	3.5	3.9	4.4	4.8
Distance, cell-to-cell, m	1.5	1.5	1.5	1.5	1.5	1.5	1.5

Table 11. HU series diaphragm cells: specific load, cell voltage, and power consumption

	Specific load, kA/m^2	
	1.5	2.3
Cell voltage, V	3.12	3.41
Power consumption (d.c., average), kW h/t*	2500	2700

* Per tonne of chlorine.

tubes open at both ends, which are welded into thick steel tube sheets at each end. The tubes, tube sheets, and the outer steel cathode shell form the catholyte chamber of the cell (Fig. 35). Copper is bonded, rather than welded, to the rectangular cathode shell on the two long sides prallel to the tube sheets. Copper connectors attached at the ends of the bonded copper side plates complete the encompassing of the cathode with copper. Anodes are connected to a copper patented cell base, which is protected from the anolyte by a rubber cover or a titanium base cover (TIBAC) [98]. Orientation of the cathode tubes is parallel to the cell circuit, the opposite of a Hooker-type cell. This arrangement accommodates thermal expansion of the cell and circuit without changing the anode-to-cathode alignment.

The combination of the Modified Diaphragm and expandable DSA anodes reduces power consumption 10–15% from that of regular asbestos diaphragms and standard, fixed DSA anodes [99]. Table 12 presents performance data for the two most common MDC cell sizes [97]. The OxyTech MDC-29 is shown in Fig. 36. The licensed chlorine capacity of OxyTech cells now exceeds 20 000 t/d.

6.3. Operation

The process description in this section is intended to provide an overview of typical diaphragm cell process areas. A general block diagram for a diaphragm cell facility is shown in Figure 9. Included on the drawing are many process areas that may be optional, depending on the design of the plant and its end products. The operation of a cell room may be broken down into six areas: the two incoming systems, brine and electrical; the cells; and the three outgoing systems, chlorine, hydrogen, and cell liquor. Some of these are essentially the same for all three chlor-alkali processes and are described in Chapter 4— the brine system (general), the electrical system, and the hydrogen system. The treatment of the chlorine is the subject of Chapter 11. Only aspects that are reasonably specific to the diaphragm cell process are described in this section.

6.3.1. Brine System

Most commonly, diaphragm cells are supplied with well brine on a once-through basis. The treated well brine flows to the treated brine storage tanks, which usually have 12-h capacity. From there the brine is fed to the cell room. The flow to each individual electrolyzer is controlled by a rotameter. If the flow of brine to the cells is suddenly disrupted by failure of the brine feed pump, the rectifiers automatically shut down since an inadequate supply of brine to the cells is potentially dangerous. The specifications for brine for diaphragm cells are given in Table 13.

A brine recovery lagoon is usually available to handle any major upsets in the brine system. Brine sludges or out-of-spec brine can be sent to the lagoon. Supernatant clear brine can be recovered from the lagoon.

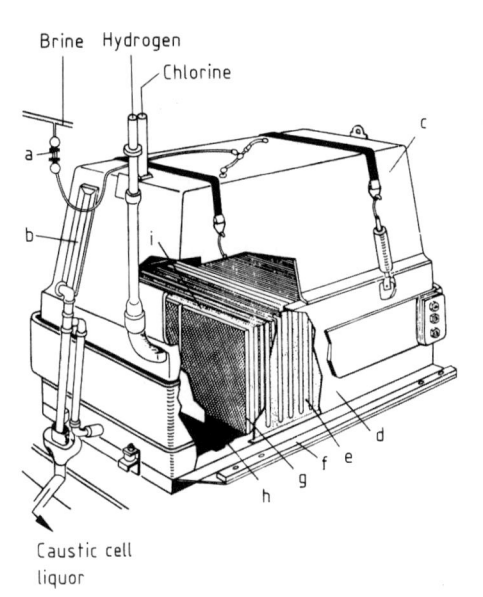

Figure 34. OxyTech Systems MDC cells
a) Brine feed rotometer; b) Head sight glass; c) Cell head; d) Cathode assembly; e) Tube sheet; f) Grid plate; g) Cathode tube; h) Grid protector; i) DSA expandable anode

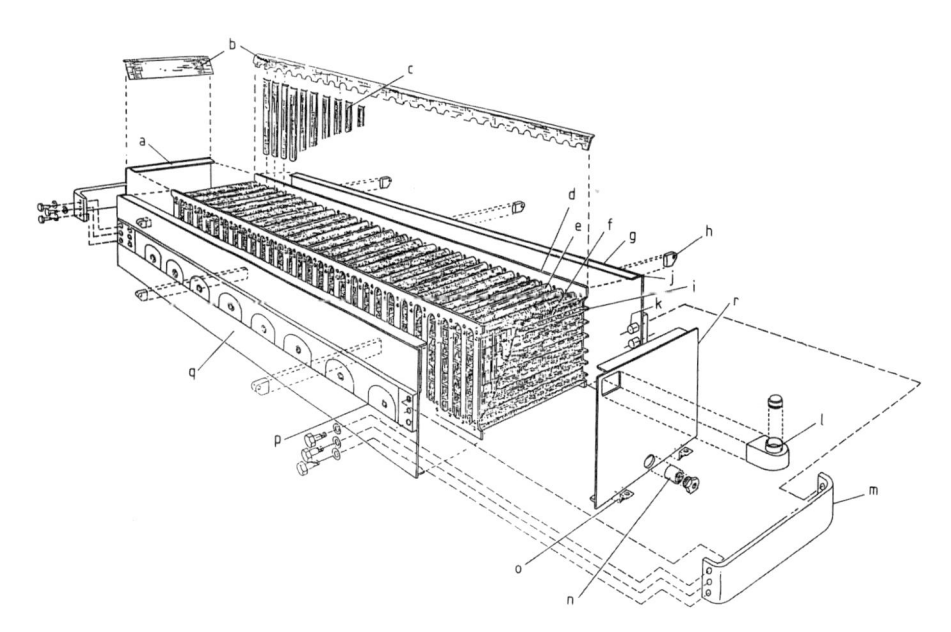

Figure 35. Exploded view of an OxyTech MDC-55 cathode
a) End plate; b) Rim screen; c) Side screens; d) Tube sheet; e) Full cathode tube; f) Half-cathode or end tube; g) Side plate; h) Lifting lug; i) Punched and coined stiffener strap; j) Bosses; k) End plate, operating aisle end; l) Hydrogen outlet; m) Connector bar; n) Caustic outlet; o) Clip angles; p) Grid bar, connector side; q) Side plate

Figure 36. OxyTech MDC-29 with the author (1971)

In most cases, operation with acidic brine is preferred because of the reduced amount of side-reaction products in the chlorine and the cell liquor.

6.3.2. Cell Room

Typical cell rooms are shown in Figures 37 (bipolar cells) and 38 (monopolar cells).

A cell in normal operation requires little attention. The critical requirement is that the *brine flow rate* is sufficient to maintain an anolyte level above the cathode.

Under no circumstances should a cell be operated with an inadequate or excessive anolyte level. Operation with the anolyte level not visible in a sight glass is unsafe. At least one operator should be in the cell room at all times. The cell room operator should inspect the anolyte level and brine flow to each cell at least once per hour. Any change in the anolyte level or brine flow rate should be investigated.

As the cell ages, the diaphragm will undergo changes in porosity because of the following:

1) Electrolysis effect
2) Brine impurities
3) Upsets in operation
4) Gradual wear of the diaphragm

A change in porosity may necessitate a change in brine flow rate. If the increase in porosity is severe, the cell may be replaced or doped with an asbestos slurry or inorganic salts.

Impurities in the brine often lead to decreased porosity. Decreased porosity can be offset to some extent by increasing the anolyte level and, if necessary, by lowering the catholyte level.

A cell operated with the anolyte level at the maximum value and the lowest catholyte level is called a sleeper. To gain additional diaphragm life after a cell has entered the sleeper position, the brine flow

Table 12. OxyTech Systems MDC cells: operating capacities and characteristics

Item	Model number and operating range, kA			
	MDC-29		MDC-55	
	35 to	80	75 to	150
Chlorine capacity,				
metric ton/day	1.05	2.41	2.33	4.53
short ton/day	1.16	2.66	2.48	5.00
Caustic capacity,				
metric ton/day	1.21	2.76	2.59	5.18
short ton/day	1.33	3.04	2.85	5.70
Hydrogen capacity,				
m^3/day	335	765	720	1435
cubic feet/day	11 830	27 010	25 420	50 670
Current density,				
kA/m^2	1.21	2.76	1.37	2.74
A/$in.^2$	0.78	1.78	0.88	1.76
Cell voltage, V[a]				
steel cathode	2.90	3.62	3.00	3.62
activated cathode	2.80	3.51	2.90	3.51
Power consumed (d.c., steel cathode)[b],				
kW h/t	2310	2876	2390	2870
kW h/short ton	2100	2610	2175	2610
Power consumed (d.c., activated cathode)[b],				
kW h/t	2230	2786	2310	2780
kW h/short ton	2025	2530	2100	2530
Diaphragm life, years	1–2	0.5–1.0	1–2	0.5–1.0
Anode life, years	8–10	5–8	8–10	5–8
Cathode life, years	10–15	10–15	10–15	10–15
Distance between cells[c],				
m		1.60		2.13
inches		63		84

[a] Cell voltage includes loss in intercell bus.
[b] Power consumed per ton (metric or short) of chlorine produced.
[c] Distance centerline-to-centerline and side-by-side with bus connecting.

Table 13. Typical brine feed specifications for diaphragm cells

Parameter	Specification
NaCl	≥ 320 g/L
pH	2.5–3.5
Hardness ($Ca^{2+} + Mg^{2+}$)	<5 ppm
Magnesium	<0.4 ppm
Sodium sulfate (Na_2SO_4)	<5 g/L
Organics	<1 ppm
Manganese	<0.01 ppm
Barium	<0.01 ppm
Nickel	<0.1 ppm
Iron	<0.5 ppm
Silicon	<15 ppm
Cobalt	<0.02 ppm
Mercury	<1 ppm
Phosphate	<1 ppm

rate must be decreased below normal. This is not normally a recommended practice because current efficiencies of these cells are usually low.

For safe operation of diaphragm cells, the *header pressures* must be maintained at the proper values.

The chlorine header should be maintained at positive pressure to permit detection and correction of any chlorine piping leaks. The hydrogen header is also maintained at a positive pressure to avoid pulling air into the hydrogen, creating a potentially explosive mixture. The brine header pressure should be maintained to give the desired caustic concentration in the cell liquor. Normal practice is to adjust individual brine feed valves so that each cell receives the correct brine flow rate.

Load changes must be smooth to avoid fluctuations in the header pressures and detrimental effects on the diaphragms.

The brine feed rate to each cell should be increased to the new rate before circuit amperage is increased. The brine feed rate to each cell should be decreased immediately after amperage is decreased. During any period of operation when brine flow rates are being changed, extra attention should be given to the anolyte levels of the cells. Adjustment of the brine feed temperature may also be necessary when a load change occurs.

6.3.3. Diaphragm Aging

Of all the cell components, the diaphragm usually has the shortest life. The ability of a diaphragm to resist the back migration of hydroxide slowly becomes impaired with service life. The performance of the diaphragm deteriorates for the following reasons:

1) Chemical attack
2) Brine impurities
3) Unsteady operating conditions

Figure 37. Cell room: bipolar PPG Industries/De Nora Glanor cells

Figure 38. Cell room: monopolar OxyTech H-4 cells

The major reason for the deterioration is *chemical attack* on the asbestos by the alkaline catholyte and acidic anolyte. The rate of chemical attack can be minimized and diaphragm life maximized by careful operation of the cell. The most important situations to avoid are high concentrations of brine impurities and unsteady operating conditions. High *brine impurities* cause plugging of the diaphragm with insoluble hydroxides, which reduce the diaphragm's separation ability. The most common harmful impurities are calcium, magnesium, iron, nickel, silicates, aluminum, manganese, and barium.

Unsteady operation, such as electrical load changes, cell liquor strength changes, changes in brine concentration or pH, gas-pressure fluctuations, and shutdowns, change the pH of the various regions of the diaphragm, thus accelerating chemical attack on the asbestos. Diaphragm cell plant operators should strive to minimize these changes.

The real importance of the equations in Section 6.1 is as an aid in deciding when the diaphragms should be replaced.

6.3.4. Treatment of the Products

Chlorine. See Chapter 11.

Hydrogen. See Chapter 4.

Sodium Hydroxide Solution. The hydroxide produced at the cathode is associated with sodium ions and water to form a 10–12 wt % sodium hydroxide solution leaving the electrolytic cell. This cell liquor also contains 18 wt % unreacted sodium chloride.

Most large modern chlor-alkali plants have or will soon have an associated cogeneration power plant. In these facilities, the caustic evaporators are an important use for the byproduct steam.

Modern diaphragm cell plants use triple-effect evaporators and, in many cases, quadruple-effect evaporators.

Caustic Soda Evaporation. A flow diagram for a typical triple-effect caustic soda evaporator is shown in Figure 39. The evaporator is of the backward-feed design and concentrates 10–11.3 wt % NaOH cell liquor to 50 wt % NaOH. Liquor flows from the third to the second to the first effect and from the first effect to the liquor flash tank. A cyclone is used for each effect to utilize the pressure drop across the circulating pump to clarify the transfer liquor.

The salt precipitated in the liquor flash tank is isolated from the rest of the salt precipitated in the evaporator and used as seed crystals in the cooling system to help diminish coil scaling and supersaturation of the product liquor with sodium chloride. The sodium chloride and triple salt ($NaOH-NaCl-Na_2SO_4$) precipitated in the liquor flash tank and cooling system is removed from the cooled product liquor with centrifuges. The salt precipitated in the three effects flows countercurrently to the liquor flow so that all of the salt is discharged from the last effect, the effect that has the coldest liquor and the lowest caustic soda concentration.

Two-stage steam-jet air ejectors with a common intercondenser are used to maintain vacuum in the evaporator. In the caustic cooling system, agitated tanks are used to cool the slurry discharged from the liquor flash tank. The slurry flows from the cooling system to the centrifuge feed tank from where it is pumped into centrifuges. Salt discharged from the centrifuges drops into the evaporator feed tank, where it is dissolved in cell liquor. The 50 wt % NaOH concentrate liquor, which flows by gravity to the pressure filter feed tank, contains ca. 1.0–1.5 wt % dissolved NaCl and ca. 0.1 wt % crystalline NaCl.

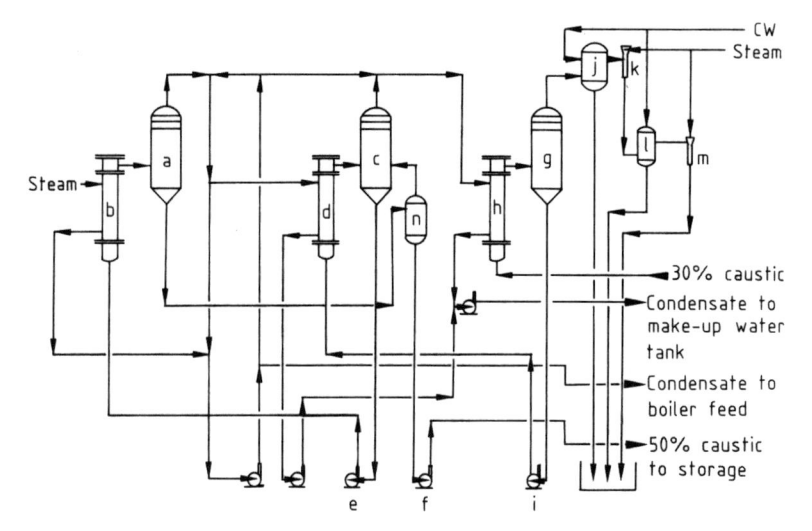

Figure 39. Process flow diagram: triple-effect caustic evaporator
a) First-effect vapor body; b) First-effect heat exchanger; c) Second-effect vapor body; d) Second-effect heat exchanger; e) Second-effect forwarding pumps; f) 50% caustic transfer pumps; g) Third-effect vapor body; h) Third-effect heat exchanger; i) Third-effect forwarding pump; j) Barometric condenser; k) First-stage ejectors; l) Intercondenser; m) Second-stage ejector; n) Liquor flash tank

The liquor is pumped from the pressure filter feed tank into the pressure leaf filters, where the remaining traces of salt crystals are removed. The product caustic flows by gravity to the filtered product tank and then is pumped to storage. Salt removed in the pressure filters is reslurried with cell liquor and pumped to the evaporator feed tank via the filter backwash pump.

The salt discharged from the centrifuges drops into the leaching tank, where it is reslurried with condensate and recycled brine from the Glauber's salt ($Na_2SO_4 \cdot 10 H_2O$) crystallizer. Concentrate from the pusher centrifuges flows by gravity into the evaporator feed tank.

The product salt is discharged from a cyclone into the salt reslurry tank. The overflow from the cyclone is returned to the leaching tank. The product salt is diluted with brine and pumped to the resaturator tank.

Brine containing the dissolved sodium sulfate is separated from the salt crystals in a cyclone. The underflow returns to the leaching tank. The overflow is collected in the feed tank for the Glauber's salt crystallizer. Sodium sulfate is crystallized from the liquor in a continuous vacuum cooled crystallizer.

Mother liquor removed from the crystallizer is pumped under level control to the brine tank. Slurry discharged from the crystallizer is thickened to ca. 50 wt%. The liquor from the thickener is collected in the brine tank and pumped back to the leaching tank. The thickened slurry is redissolved in the Glauber's salt dissolving tank and pumped to a waste treatment system.

Caustic Purification [100]. Diaphragm-cell chlor-alkali producers requiring higher purity caustic than that produced by the diaphragm process can use caustic purification or DH process (Fig. 40). *Salt removal* in the purification unit is effected by contacting the 50 wt% caustic with anhydrous liquid ammonia under pressure sufficiently high to maintain all materials in the liquid state.

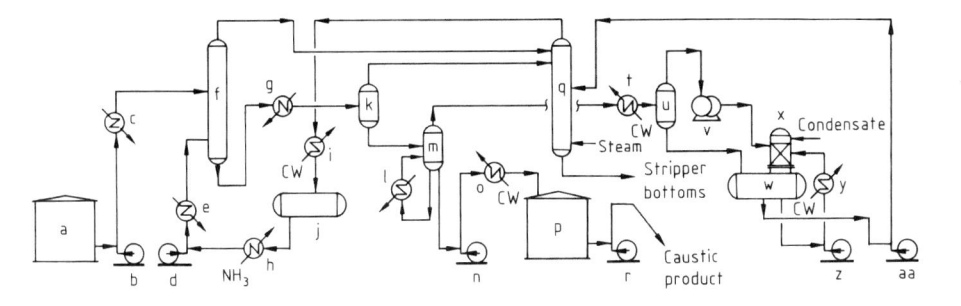

Figure 40. Caustic purification system
a) 50% caustic feed tank; b) 50% caustic feed pumps; c) Caustic feed preheater; d) Ammonia feed pumps; e) Ammonia feed preheater; f) Extractor; g) Trim heater; h) Ammonia subcooler; i) Stripper condenser; j) Anhydrous ammonia storage tank; k) Primary flash tank; l) Evaporator reboiler; m) Evaporator; n) Caustic product transfer pumps; o) Purified caustic product cooler; p) Purified caustic storage tank; q) Ammonia stripper; r) Purified caustic transfer pumps; t) Overheads condenser; u) Evaporator; v) Evaporator vacuum pump; w) Aqueous storage ammonia tank; x) Ammonia scrubber; y) Scrubber condenser; z) Ammonia recirculating pump; aa) Ammonia recycle pump

The liquid ammonia absorbs salt, chlorate, carbonate, water, and some caustic. It is then stripped, concentrated, and returned to the extraction process. The concentrated caustic leaving the extractor is stripped free of ammonia, which is recovered, concentrated, and recirculated. Typical purities before and after caustic purification are shown in Table 16. This process is offered for license by PPG Industries and OxyTech Systems.

In addition, producers and users of diaphragm-cell caustic may wish to *reduce metal impurities* by utilizing the porous cathode cell process (PPG Industries) [101]. The process consists of an electrolysis cell with porous nonmetallic cathodes. The caustic soda (50 wt%) is freed from iron, nickel, lead, and copper, which are deposited on the cathode. The cell must be regenerated periodically with water and hydrochloric acid. Typical feed and product analyses based on anhydrous NaOH are

Metal	Content, ppm	
	Feed	Product
Iron	10.0	2.0
Nickel	3.0	0.2
Lead	4.0	0.4
Copper	0.2	0.1

6.3.5. Measurement

Recorded data is an important tool for determining the operating condition of the plant and diagnosing problems.

The following should be recorded continuously or hourly:

ampere load on each circuit
voltage for each circuit

chlorine header pressure
hydrogen header pressure
brine header pressure or flow rate
brine temperature
brine pH
cell liquor temperature

Samples of brine should be taken every 4 h and combined into a daily composite. In addition, samples of cell liquor should be taken from each cell string, the sodium hydroxide content analyzed, and the temperature taken every 4 h. A daily composite should be made and samples should be analzyed by the laboratory for the following:

NaOH content
NaCl content
salt:caustic ratio
$NaClO_3$ content
NaOCl content
Fe content
average temperature
specific gravity at 25 °C

Chlorine gas from each cell circuit should be analyzed for chlorine and hydrogen content at least twice each 8-h shift. Each day a complete analysis of the chlorine header gas should be made. Additions or extensions to list may be dictated by plant operation.

Each plant must develop a procedure for taking *individual cell data* so that individual cells may be scheduled for renewal. The following is a minimum schedule:

weekly	voltage and cell liquor composition (NaOH, NaCl, $NaClO_3$)
monthly	chlorine composition (Cl_2, O_2, H_2, CO_2, N_2)

6.3.6. Replacement of Asbestos

The hazardous nature of asbestos fibers has been learned in recent years. Modern diaphragm cell chlor-alkali plants have not encountered difficulty in meeting all required exposure limits for asbestos fibers. However, concern for the future supply of asbestos and increased costs for disposal of used asbestos diaphragms have led the industry to search for alternatives to the use of asbestos.

Nonasbestos diaphragms, such as the Hooker/OxyTech Systems Microporous Diaphragm, are based on the use of fluorocarbon polymer sheets to replace the asbestos. OxyTech Systems has developed, but not yet commercialized, a synthetic diaphragm of inorganic ceramic-like fibers bonded as in the Modified Diaphragm with a fluorocarbon polymer. Other firms, such as Dow Chemical, have also reported in the patent literature on developments in this field.

The current results of all of the above work has not yet demonstrated an economical alternative to asbestos. However, new regulations or improved performance of these substitute diaphragms could change the situation.

7. Membrane Process

In the membrane process, the anolyte and catholyte are separated by a cation-exchange membrane that selectively transmits sodium ions but suppresses the migration of hydroxyl ions from the catholyte into the anolyte. A strong caustic soda solution with a very low sodium chloride content can be obtained as the catholyte effluent. The process was a theoretical possibility long before the advent of a suitable membrane. The process was started in the early 1970s with the development of a perfluorosulfonate membrane, Nafion, by Du Pont [102].

Especially in Japan, where the abolition of the mercury process has been promoted by the government, the development of both the membrane and the cell progressed rapidly. In 1975, a perfluorocarboxylate membrane capable of 35 wt% caustic production became available, initially from Asahi Glass in Japan [103].

In addition to the development of a high-performance membrane, the availability of an efficient metal anode was indispensable for the development of the membrane cell.

The cells currently in use are usually of the filter-press type, electrodes connected either monopolarly (parallel) or bipolarly (in series). Based on various designs, membrane sizes range from 0.2 to 5 m^2. The production capacity of one electrolyzer varies from 26 to 1900 t of Cl_2 per month (30–2200 t of NaOH per month).

In one method, used especially in Japan, diaphragm cells are retrofitted as membrane cells by installing ion-exchange membranes between anodes and cathodes in place of the asbestos diaphragm. The developments in the membrane process that have taken place over the past 5 years have been quite significant.

The advantages of the membrane process are its energy efficiency and its ability to produce, with no effect on the environment, a strong caustic of high quality. Chlor-alkali producers worldwide tend to adopt the membrane process in extension or conversion of production facilities.

The production capacity of the chlor-alkali plants using the membrane process is now ca. 10% of the world total production capacity. The development of membranes and cells continues, along with the evolution of new technologies, such as an oxygen depolarized cathode. The membrane process is expected to be dominant in the chlor-alkali industry by the end of the 20th century [104].

7.1. Principles

In a membrane cell a cation-exchange membrane separates the anolyte and catholyte, as shown in Figure 41. Saturated brine is fed into the anode compartment, where chlorine gas is evolved at the anode and sodium ions migrate into the catholyte through the membrane. Depleted brine is discharged from the cell. In the cathode compartment, hydrogen is evolved at the cathode, leaving hydroxyl ions, which

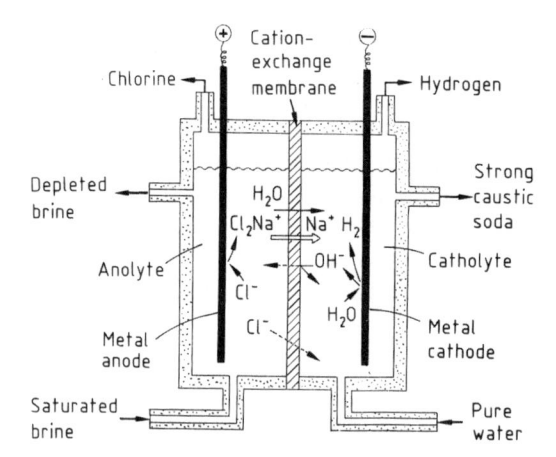

Figure 41. Membrane cell: transfer and transformation of materials

together with the permeating sodium ions constitute the caustic soda. Pure water is added to the catholyte to control the concentration of the caustic soda.

Chloride ions in the anolyte are excluded by the cation-exchange membrane so that the rate of diffusion of chloride from the anolyte to the catholyte is extremely low. As a result, a strong caustic (32–35 wt%) with low salt content (< 100 ppm) can be obtained as the catholyte effluent.

Since the catholyte is a strong caustic, there is some back migration of hydroxyl ions from the catholyte into the anolyte, even with the use of a high-performance cation-exchange membrane. This causes a loss of current efficiency of 3–5% in caustic production.

Anode Reaction. The reactions described in Chapter 4 occur on the anode in a membrane cell. A complicating factor in the membrane process is the back migration of hydroxyl ions into the anolyte. This increases the pH of the anolyte, leading to an increase of oxygen in the evolved gas and of hypochlorites and chlorates in the anolyte. The evolution of oxygen gas can be depressed by selecting an anode coating with suitable characteristics (see Section 8.2).

Because anodes are placed close to or in contact with the membranes, the anode coating should be durable at the high pH of the brine in the vicinity of the membrane. If necessary, hydrochloric acid can be added to the feed brine to neutralize hydroxyl ions in the anolyte.

Cathode Reaction. On the cathode in the membrane cell, hydrogen gas and hydroxyl ions are generated in the caustic solution:

$$2\,H_2O + 2\,e^- \longrightarrow 2\,H + 2\,OH^- \longrightarrow H_2 + 2\,OH^-$$

The required properties for the cathodes are durability in the presence of strong caustic and low hydrogen overpotential. Activated cathodes have been developed for use in membrane cells.

Membrane. Membranes are made of perfluoro polymers containing ion-exchange groups. The carboxylate type is indispensable for obtaining a high current efficiency. The performance of a membrane depends on the operating condition of the cell:

1) Concentration of the anolyte and the catholyte
2) Current density
3) Temperature

Current Efficiency. The current efficiency in caustic production depends on the strength of the caustic solution, as shown in Figure 42. The optimum caustic strength depends on the composition of the membrane polymer (e.g., type A or type B). To achieve stable operation with a high current efficiency, wide fluctuations in operating conditions or upsets must be avoided. Fluctuation in the caustic strength beyond the optimum range decreases current efficiency, as seen in Figure 42 [105], and after a great deviation, some time may be required for complete recovery. Dilution of the anolyte caused by a disorder in the brine feed also decreases the current efficiency. In extreme cases, perfect recovery cannot be attained afterward. The sensitivity of membrane performance to operating conditions is attributed to changes in the water content of the membrane.

In the absence of fluctuations in operation, upsets, or contaminants in the brine feed, stable current efficiency can be maintained at a high level for more than two years.

Equations. The current efficiency for caustic soda can be obtained either by directly measuring the quantity of caustic soda produced or by calculating from the compositions of the anode gas and the anolyte with the following equation:

$$\text{CE}(\%, \text{NaOH}) = 100 - 100\,(\eta O_2 + \eta ClO + \eta ClO_3 - \eta NaOH - \eta NaHCO_3)$$

where ηO_2, ηClO, and ηClO_3 represent the loss of current efficiency due to the generation of oxygen, hypochlorite, and chlorate, respectively, and $\eta NaOH$ and $\eta NaHCO_3$, the equivalent current efficiency of the NaOH and $NaHCO_3$ in the feed brine.

Cell Voltage. The cell voltage of a membrane cell is composed of the following:

1) Theoretical decomposition voltage
2) Electrode overpotentials for chlorine and hydrogen
3) Membrane potential between the anolyte and the catholyte
4) Ohmic drop in the membrane
5) Ohmic drop in the electrolytes
6) Ohmic drop in electrodes and conductors

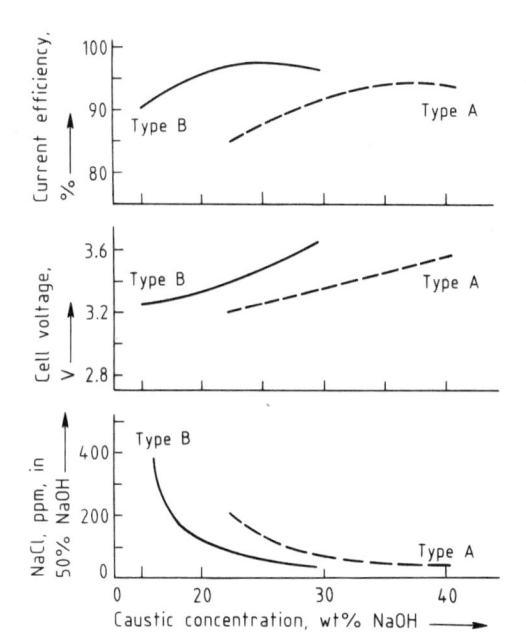

Figure 42. Performance of perfluorocarboxylate membranes

Type A has high ion-exchange capacity, whereas type B has low ion-exchange capacity.
Anolyte concentration 3.5 mol/L
Current density 2 kA/m² Temperature 90 °C

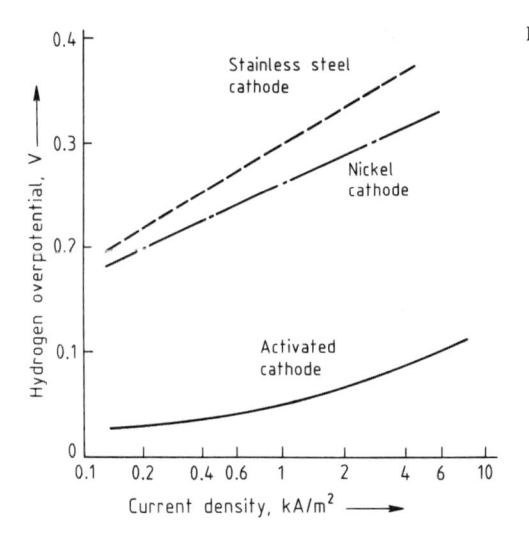

Figure 43. Hydrogen overpotential

Factor 1. Under normal electrolyte conditions, the decomposition voltage in the membrane cell is ca. 2.15 V.

Factor 2. Metal anodes with a chlorine overpotential of ca. 0.1 V are generally used in membrane cells (see Chapter 8.1.1.1, Overpotential and Current-Voltage Relations). The technology of activated cathodes has marked great progress in the last several years. Steel or nickel substrates are coated by electroplating or sputtering with catalytic powdered materials [106], [107]. Coating materials include nickel, cobalt, molybdenum, etc. Hydrogen overpotentials of ca. 0.1 V are attained with these modern activated cathodes (Fig. 43).

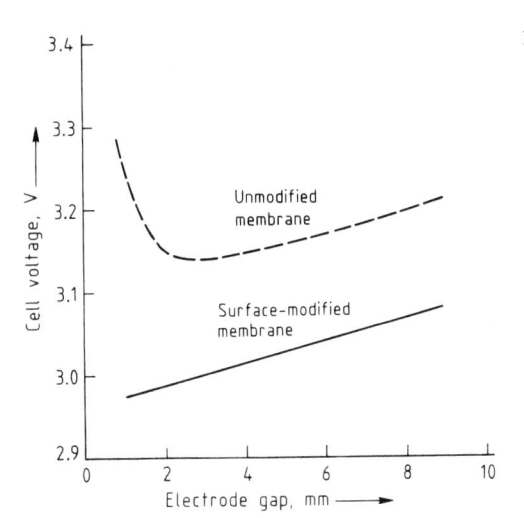

Figure 44. Effect of electrode gap on cell voltage

Factor 3. Under normal conditions, i.e., producing 32–35 wt % caustic solution, the membrane potential is ca. 80 mV.

Factor 4. The ohmic drop of advanced commercial membranes under normal operating conditions is 0.3–0.4 V at 3 kA/m^2. The lower limit of the ohmic drop is estimated to be > 0.2 V. Present membrane technology may well have approached the limit.

Factor 5. To diminish the ohmic drop of an electrolyte, the gaps between membrane and electrodes are minimized in the design of a membrane cell.

However, when the gap is very small, a rise of voltage is observed owing to the entrapment of gas bubbles between the electrodes and the hydrophobic fluoropolymer membrane. However, the development of a membrane with increased hydrophilicity has solved this problem. The surfaces of the membrane are covered with thin layers of a porous inorganic material. This material is an oxide, hydroxide, or carbide of metals belonging to groups 4, 5, 6, and 14 or the iron triad (Fe, Co, Ni) [105], [108]. Figure 44 illustrates the effect of this surface modification. The surface-modified membrane shows a lower cell voltage than the original membrane, and the voltage decreases linearly with the gap size, reaching the minimum value at zero gap.

Other methods of surface modification, such as mechanical roughening, have also been proposed [109]. With advanced membranes, so-called zero-gap cells have been made possible, and the ohmic loss in electrolytes has been nearly eliminated.

Factor 6. The ohmic drop of the conductor is diminished by a bipolar design. In a monopolar cell design, the decrease of the ohmic drop within electrodes and conductors is a major topic.

Energy consumption. During the years since the first commercialized membrane process, a great reduction in power consumption has been achieved. In 1976, d.c. power consumption was ca. 3600 kW h per tonne of chlorine (ca. 3200 kW h per tonne of caustic soda). In 1984, it was less than 2500 kW h per tonne (2200 kW h) at 3 kA/m^2.

7.2. Membranes

The membrane is the key component of the membrane cell. The energy requirement and the quality of solution produced in a membrane cell depend on membrane performance. Requirements for the membrane are as follows:

1) Durability under the conditions of chlor-alkali electrolysis
2) High selectivity for sodium ion transport
3) Low electrical resistance
4) Sufficient mechanical strength for practical use

Materials. The membrane is exposed to chlorine and strong caustic solution at high temperature. Only ion-exchange membranes made of perfluoropolymer can withstand such severe conditions. The first membranes that showed significant potential for use in the chlor-alkali process were made of a perfluorosulfonate polymer. These proved to be durable in chlor-alkali cells, but they were relatively inefficient.

$$-(CF_2CF_2)_{\overline{x}}(CF_2CF)_{\overline{y}}-$$
$$|$$
$$(OCF_2CF)_{\overline{m}}O-CF_2CF_2-SO_2F$$
$$|$$
$$CF_3$$

$$-(CF_2CF_2)_{\overline{x}}(CF_2CF)_{\overline{y}}-$$
$$|$$
$$(OCF_2CF)_{\overline{m}}O-(CF_2)_{\overline{n}}COR$$
$$|\qquad\qquad\qquad\|$$
$$CF_3\qquad\qquad\qquad O$$

$m = 0$ or 1, $n = 1-5$, $R =$ alkyl

In pursuit of greater efficiency, a perfluorocarboxylate polymer membrane was developed. Because of its low water content, the perfluorocarboxylate membrane effectively inhibits the migration of hydroxyl ions from the catholyte, making it possible to obtain a high current efficiency, with a 32–35 wt% caustic soda catholyte. However, the voltage drop for such a highly efficient membrane is relatively large. Further development resulted in a two-layered membrane, composed of a thin, high-efficiency carboxylate membrane on the cathode side and a supporting high-conductivity membrane, either carboxylate or sulfonate, on the anode side [110]–[112].

The ion-exchange groups of the original polymers are in the fluorosulfonate form, $-SO_2F$, or the ester form, $-CO_2R$. These polymers are processable by melt and are formed into membranes by extrusion, heat-pressing, or lamination. In most cases, membranes are embedded with some reinforcement, such as woven fabric made of polytetrafluoroethylene (PTFE), so that even with a thickness of ca. 250 μm, they exhibit sufficient mechanical strength for practical handling. Prior to use, the ion-exchange groups are converted to $-SO_3Na$ or $-CO_2Na$ by caustic solution.

Because the performance of the membrane is the most important element in the efficiency of a membrane cell, many refinements of the technology have been made in membrane manufacturing. The improvement of hydrophilicity by covering the surface with a porous, nonconductive inorganic material (see *Factor 5* in Chap. 7.1, Cell Voltage,) brought about a great reduction in the cell voltage. To reduce the amount of current screening caused by fabrics, membrane reinforcement with dispersed microfibriles or a interwoven fabric of electrolyte-soluble fibers and PTFE have been contrived [113], [114]. The membranes are supplied commercially from Du Pont (Nafion) and Asahi Glass (Flemion).

Performance. The performance of membranes depends on the operating conditions, especially on the caustic strength of the solution (see Fig. 42). Commercially available membranes are designated for use in a specific caustic strength. The membrane must be suitable for the desired caustic strength, and the electrolyzer should be operated under the optimum conditions. For the production of caustic potash, the selection of the appropriate membrane is also essential.

If properly used without mechanical damage or excess impurities in the brine, membranes last for more than two years without chemical degradation. A slow relaxation of the polymer takes place over a long period, but this results in a loss of current efficiency of only 1–2% during the two years. The active life of a membrane should be determined by the economic balance between membrane cost and energy cost in use [115].

7.3. Membrane Cells

The development of membrane cells started in the 1970s, when perfluoropolymer membranes became commercially available.

To obtain the best performance from the membrane, it is essential to design a cell with a homogeneous distribution of electrolyte flow and electric current and to operate the cell under the optimum conditions for the type of membrane. Many chlor-alkali producers and engineering companies in the world designed their own cells, and competition accelerated the technical advance. Early electrolyzer designs were simple structures with gaps between the electrodes. With the introduction of the zero-gap concept, manufacturing tolerances became critical. The emphasis in design was placed on achieving the required degree of flatness over the total electrode area and a reduced cost of construction.

Generally, membranes are clamped vertically between the meshlike metal anodes and cathodes. The cells are filled with electrolytes, and gas-separating means are provided outside the cell. Many cell units can be piled or stacked like a filter press to constitute one electrolyzer with ample production capacity.

Depending on the mode of electrical connection between the cell units, electrolyzers may be monopolar or bipolar (also see Section 6.2). In the monopolar type, all the anodes, and all the cathodes, are connected in parallel, forming an electrolyzer with large current and low voltage. In the bipolar type, the cathode of a cell is connected to the anode of the adjacent cell and so on, that is, in series, resulting in an electrolyzer with small current and high voltage. The *bipolar type* is advantageous for attaining the minimum voltage drop between the cells. However, as the influx and efflux of electrolytes for the cells with different electric potential are gathered in common manifolds, problems of current leakage may arise. However, such problems are prevented by proper design of the bipolar cell.

A *monopolar cell* may have a simple structure. The voltage loss through conductors between the electrolyzers is an inevitable drawback, which must be minimized without extravagant use of conductor materials.

The construction material of the cell must be selected to withstand the corrosive electrolytes, for corrosion gives rise to impurities. Impurities in the anolyte originating from the frame or gasket may be detrimental to the membrane, and any impurities in the catholyte can degrade the caustic.

Various sizes of membranes (Table 14) are used, depending on the cell design. The effective membrane area of a cell unit ranges from 0.2 to 5 m^2. Current density is 2–5 kA/m^2. Cell voltage and current efficiency depend on the performance of the membrane. However, in selecting a membrane cell for use in a plant, it is not sufficient to just compare the cell voltages and current efficiencies of the particular membranes because the performance of a cell depends both on the membrane and on cell as a whole. Several kinds of membranes are now in use in commercial cells. Examples of cells developed by leading companies are described below.

Because of economic or environmental factors, operating mercury cells or diaphragm cells can be replaced by membrane cells. The existing facilities, such as rectifiers, equipment for brine purification, and equipment for product treatment, are utilized as much as possible. The reconstructions are termed diaphragm conversion (diaphragm→membrane) and mercury conversion (mercury→membrane). In these conversions, monopolar electrolyzers are adopted because the capacity of existing rectifiers and bus work can be matched simply by adjusting the number of cell units in each electrolyzer.

As a temporary method of *diaphragm conversion,* ion-exchange membranes replace asbestos diaphragms in existing cells. This type of conversion is termed *diaphragm retrofit.* When a diaphragm cell is retrofitted, the membranes are bent and heat-sealed to enclose the electrodes. Accordingly, membranes with sufficient toughness are a necessity, even at the expense of additional ohmic drop. The retrofit has been promoted especially in Japan. As an example, Chlorine Engineers retrofit of the DS cells is described and illustrated in Section 7.3.10. Kanegafuchi Chemical has retrofitted the OxyTech H Series cells with membranes [123].

Table 14. Characteristics of membrane cells [116]–[122]

	Bipolar type						Monopolar type			
	Asahi Chemical Industry		PPG BIZEC	Tokuyama Soda TSE-270	Hoechst – Uhde BM	Krebskosmo MZB	Asahi Glass AZEC	De Nora K-40	OxyTech MGC	ICI FM-21
	Standard	Super								
Effective membrane area, m^2	2.7	5.08	3.83	2.7	1.2–3	2.5	0.2	0.64	1.5	0.21
Cells per electrolyzer	80–110	80–110	20–50	30–120	up to 100	4 × 18	30–540	20–60	2–30	1–120
Current load, kA	10.8	20.3	15.3	8.1–10.8	3–15		18–340	40–150	6–225	1–100
Current density, kA/m^2	4	4	4	3–4	2–5		3–4	3–4	2–5	1.5–4.1

7.3.1. ACI (Asahi Chemical Industry) Electrolyzer

The ACI electrolyzer (Fig. 45) is of the *bipolar* type, and the standard size of each frame is 1.2 m high and 2.4 m long. The anode and cathode compartments are separated by a partition made of a titanium–steel composite plate. The side facing the anode compartment is covered with titanium, and the cathode side is made of steel. The anode and cathode are welded onto the ribs of their compartments. Each compartment has an inlet nozzle for the electrolyte at the bottom and an outlet nozzle for the electrolyte and gas at the top. Supporting arms are attached to both sides of the cell frame to hang the frame on the side bars of the press. The ion-exchange membrane is installed between the anode and cathode compartments. The standard production capacity of one electrolyzer is 8800 t of chlorine (10 000 t of caustic soda) per year. Electrolysis is conducted under pressure to decrease the size of the gas bubbles in the electrolytes [124].

7.3.2. PPG BIZEC

BIZEC, a *bipolar* zero-gap cell, is a large cell with 1.2 m × 3.7 m membranes (Fig. 46). Asahi Glass cooperated in its development. Titanium is used for the anode compartment and nickel is used for the cathode compartment to ensure corrosion resistance and the quality of caustic soda. Cathodes of coated nickel with a low hydrogen overpotential are employed. Zero-gap design provides minimal voltage drop between electrodes. An electrolyzer contains up to 50 series-connected cells in a filter-press arrangement operating at a current density of 4 kA/m^2.

7.3.3. Tokuyama Soda TSE-270

Bipolar cells of three different sizes, with a unit active area of 2.7, 1.25, or 0.85 m^2, have been developed by Tokuyama Soda (Fig. 47). The TSE-270 is the 2.70 m^2 size, and 30–120 cell units are incorporated in one electrolyzer. A self-developed low-hydrogen overpotential cathode (trade name LHOC) is employed. Titanium is used for the anode compartment, and carbon steel is used for the cathode compartment. With special gaskets and a specially designed filter-press mechanism, a tight seal for gas and liquid is achieved by oil under pressure.

7.3.4. Hoechst–Uhde Bipolar Cell

A *bipolar* cell (Fig. 48) was jointly developed by Hoechst and Uhde. An anode and a cathode, composing a bipolar unit, are assembled into the stack separately and pressed

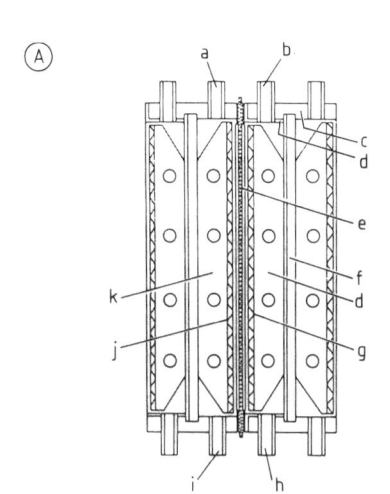

Figure 45. ACI bipolar electrolyzer
A) Structure of cell unit: a) Catholyte outlet; b) Anolyte outlet; c) Steel frame; d) Titanium; e) Membrane; f) Partition wall; g) Anode; h) Anolyte inlet; i) Catholyte inlet; j) Cathode; k) Steel
B) ACI electrolyzer: a) Hydraulic press; b) Header for anolyte discharge; c) Power supply; d) Ion-exchange membrane; e) Cell; f) Header for catholyte discharge; g) Header for anolyte feed; h) Side bar

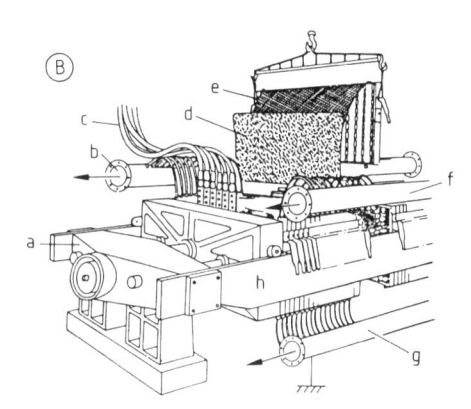

onto each other with a number of pressure bolts at one end of the stack. A gasket is inserted between the membrane and the cathode frame, but the anode frame contacts the membrane directly. Inlet and outlet nozzles to and from each cell are connected to the external pipings with flanges. The effective area of a unit cell is $1.2-3$ m^2. Up to 100 unit cells can be assembled into one electrolyzer [125].

7.3.5. Krebskosmo MZB

The Krebskosmo electrolyzer MZB is of the bipolar type and has internal channels for the distribution of the electrolyte and the collection of the gases. Two special features are

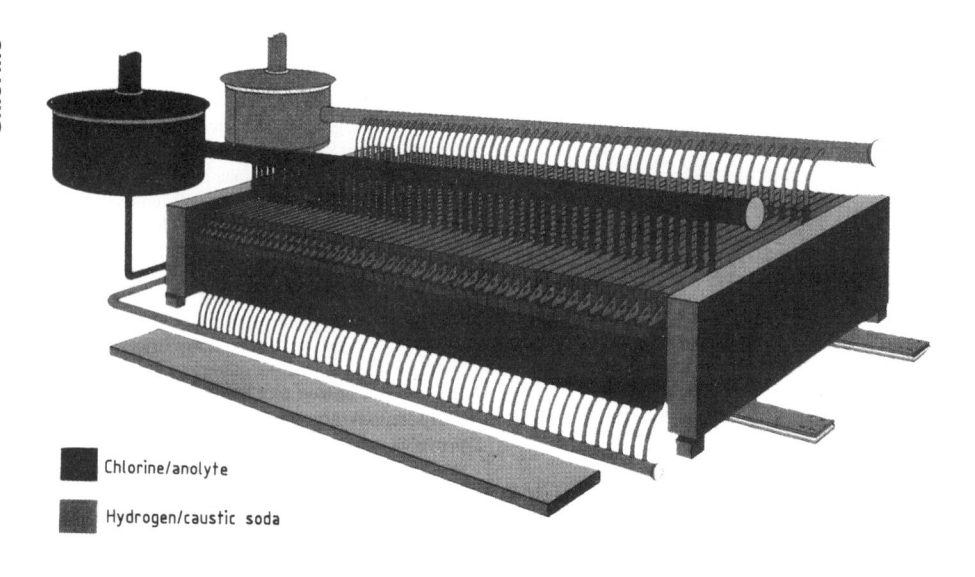

Figure 46. PPG Industries BIZEC

Figure 47. Tokuyama Soda TSE-270

1) the annular shaped passages between electrolysis compartments and distribution/collection channels that allow a strong electrolyte flow with minimum leakage of electrical current
2) the partition wall between anode and cathode compartment made of a polytetrafluoroethylene (PTFE) foil supported by an expanded metal sheet, which reduces the cell weight and cost below those of a cell containing a rigid multilayer partition wall

The cell elements are built up to a size of a 2.5 m^2 membrane. There are 72 cell elements, arranged in four stacks of 18 elements each and connected in parallel. One such cell block corresponds to a unit capacity of ca. 20 t of NaOH per day.

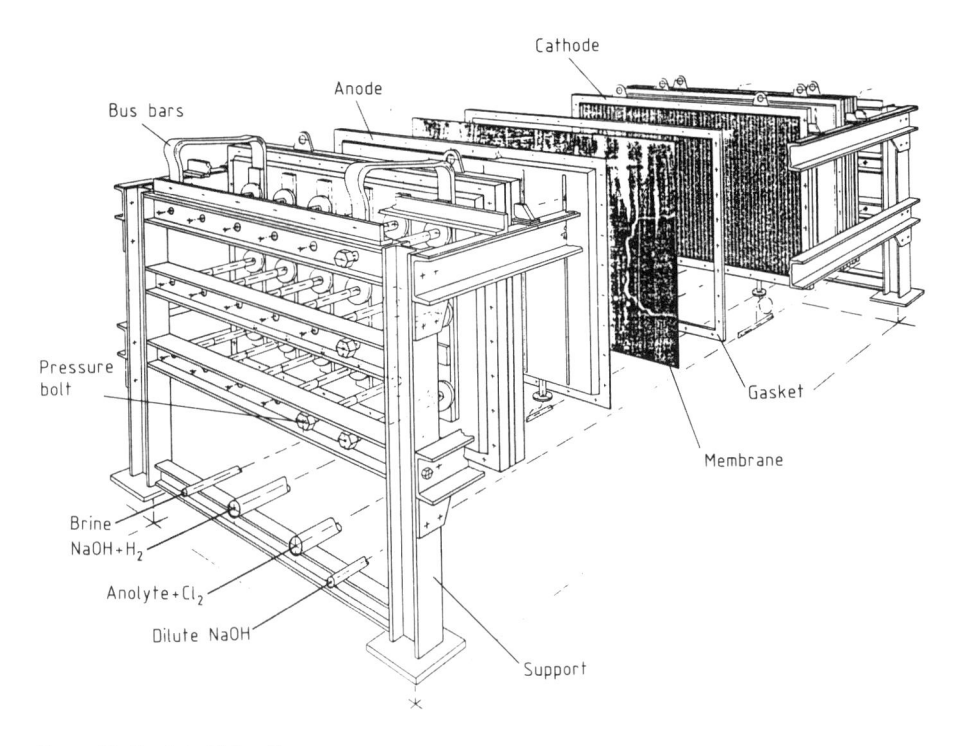

Figure 48. Hoechst – Uhde cell

Electrolyzers of this design have been in operation producing NaOH or KOH since 1980 [126], [127].

7.3.6. Asahi Glass AZEC

AZEC (Fig. 49) is a zero-gap system featuring

1) A Flemion-DX membrane with increased hydrophilicity
2) A proprietary, activated cathode with a low hydrogen overpotential and stability against reverse current
3) Simple, reliable zero-gap structure

The electrolyzer is of the *monopolar* type and is composed of cell units 0.26 m wide and 1.45 m high. Electrolyzers from 30 kA up to 340 kA are operated in commercial plants.

A simple, dependable zero-gap electrode structure, producing a minimal ohmic drop, has been devised, as shown in Figure 49. Manifolds for the influx and efflux of electrolytes are formed in the gaskets. Electrodes and membranes are placed between the gaskets. As the electrolyzer is quite high, gases released from the electrodes induce sufficient natural circulation of the electrolytes to bring about homogeneous distribution. Electrolytes and gases pass through the inner ducts of the electrolyzer

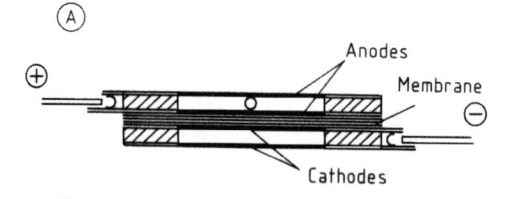

Figure 49. Asahi Glass AZEC monopolar electrolyzer
A) AZEC cell; B) AZEC electrolyzer

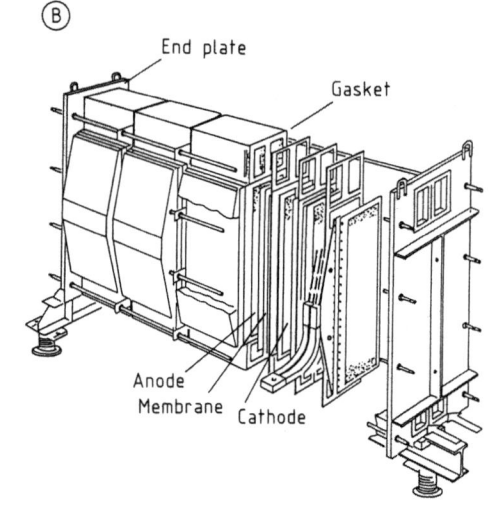

to gas–liquid separators. The electrolytes, after being separated from the gases, are recirculated through the electrolyzers. The electrical connections between adjacent electrolyzers are short.

7.3.7. De Nora K40

The K40 is a *monopolar* electrolyzer (Fig. 50) with an effective area of 0.64 m². A feature of the cell structure is the employment of a resilient, compressible element between the membrane and the cathode assembly. The element is coextensive with the electrode surface. It is compressible against the membrane, while exerting an elastic reaction force on the electrode in contact with the membrane at a number of evenly distributed contact points. It is capable of transferring excess pressure on individual contact points to adjacent points, thus distributing the pressure over the entire electrode surface. The open structure of this element permits gas and electrolyte to flow. Intercell manifolds are composed of pipe elements attached to the electrode assemblies: 20–60 cell units are pressed into one electrolyzer.

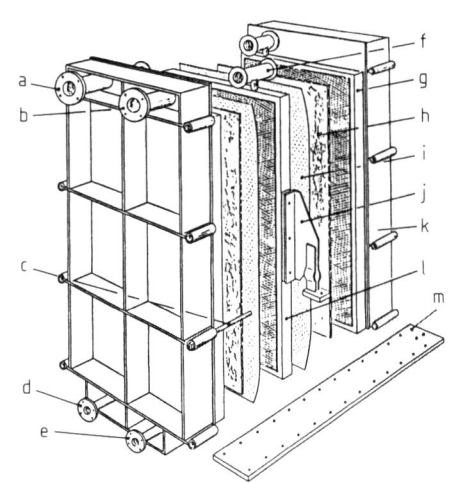

Figure 50. De Nora K40
a) Anolyte outlet; b) Catholyte outlet; c) Tie-rod;
d) Recirculated catholyte and water inlet;
e) Recirculated anolyte and fresh brine inlet; f) Intercell manifold; g) Cathode assembly; h) Resilient compressible element; i) Membrane; j) Individual bus; k) End head; l) Anode assembly; m) Copper bus bar

7.3.8. OxyTech MGC Electrolyzer

The *monopolar* MGC (membrane gap cell) electrolyzer features a design in which the anode and cathode elements are separated only by the thickness of the membrane (Figs. 51 and 52). High-surface-area DSA anodes and activated nickel cathodes are utilized.

The anode and cathode sections are alternated in the electrolyzer sandwich, with the ion-exchange membrane placed between each monopolar section. The anode section is composed of a titanium anode frame with DSA anode, a gasket, interface material, copper current distribution plate, interface material, and a second anode assembly. The nickel cathode section, with a nickel cathode, is similarly composed.

Connecting spool pieces and gaskets are inserted to connect the anode and cathode manifold sections and form a manifold passage attached to the frames. The frames are sealed with O-rings in a staggered gasket design. The cathode O-ring is located closer to the liquid than the anode O-ring. This design protects the anode O-ring from corrosion and provides a double seal, which improves reliability. Exclusive of buses, the size is 1.14 m × 1.96 m. The practical limit is considered to be 30 cells per electrolyzer.

7.3.9. ICI FM-21

The FM-21 (Fig. 53) is a *monopolar* electrolyzer incorporating a simple, pressed electrode structure of relatively small size. The anode assembly is composed of a 1-mm-thick titanium panel between compression-molded joints of ethyle-ne – propylene – diene monomer (EPDM) copolymer, a synthetic cross-linked elastomer. The cathode

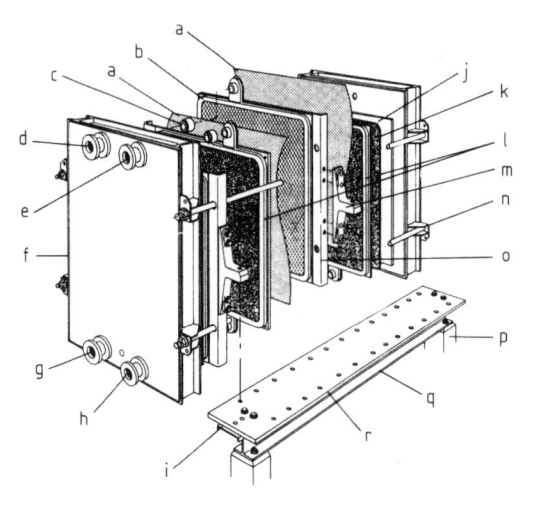

Figure 51. OxyTech MGC
a) Membrane; b) Anode assembly; c) Manifold spacer; d) Anolyte outlet; e) Catholyte outlet; f) Bulkhead; g) Brine inlet; h) NaOH inlet; i) Insulating channel; j) Bulkhead insulator; k) Interface material; l) Cathode assembly; m) Intercell bus; n) Tie-rod; o) Current distributor; p) Electrolyzer support; q) Support beam; r) Connecting bus bar

Figure 52. OxyTech MGC

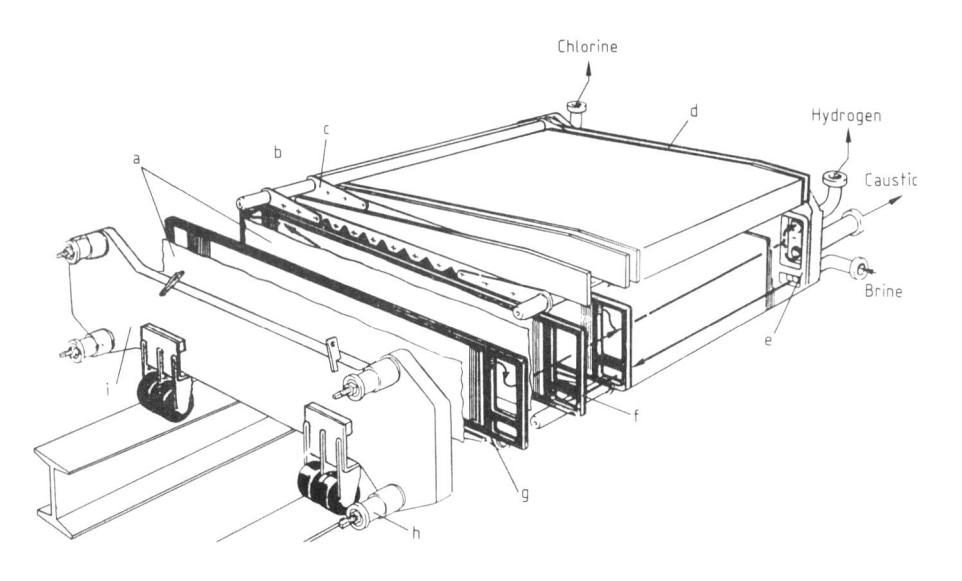

Figure 53. ICI FM-21
a) Membrane; b) Support rail and tie-rod assembly; c) Panel support; d) Copper connections; e) Fixed end-plate; f) Cathode assembly; g) Anode assembly; h) Disk spring assembly; i) Floating end-plate

assembly is composed of a 1-mm-thick nickel panel between compression joints of molded EPDM.

The series is repeated until the number of electrodes required for the desired cell capacity is assembled. The rear end plate is then attached, sandwiching the assembly into a complete cell. External piping to individual cell compartments is eliminated by using internal porting.

The cell features coated titanium anodes. Cathodes are pure nickel, also available with a coating that reduces the hydrogen overpotential. Because the electrodes are pressed from integral sheets of pure metal, their preparation for recoating is simple, and rapid turnaround can be achieved.

7.3.10. Chlorine Engineers MBC

In retrofitting diaphragm cells to make membrane cells, the ion-exchange membrane is installed in the form of a bag that encloses the anodes (membrane bag cell). In the MBC-29, shown in Figure 54, one bag encloses two anodes. In the MBC-55, one bag encloses three anodes. The power-supply rod for each anode passes through a hole in the bottom of the membrane bag for connection to the base plate, as in a diaphragm cell. The open end of the bag, facing upward, is fixed to the partition plate by a sealing plug.

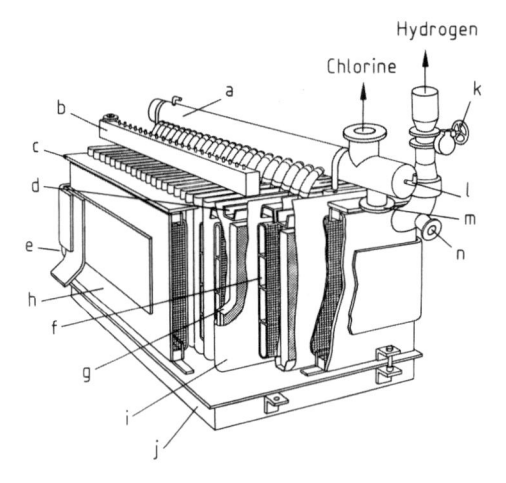

Figure 54. Chlorine Engineers MBC
a) Manifold; b) Frame; c) Partition plate; d) Sealing plug; e) Recirculated NaOH inlet; f) Cathode; g) Anode; h) Cathode can; i) Membrane bag; j) Base; k) Butterfly valve l) Feed brine; m) Depleted brine; n) Caustic outlet

The use of a baglike ion-exchange membrane on the anode side, rather than on the cathode side, simplifies the setting of the membrane and decreases the electrolyte IR drop, resulting in easy assembly and a low operating cost.

Type A utilizes an existing cell head, while type B is equipped with a manifold rather than a cell head. The two types are similar in basic construction and provide the same performance. Type A is lower in cost, since existing cell heads and chlorine gas branch pipes can be reused. However, if existing cell heads are near the end of their service life, conversion to type B with its manifold is preferable [128].

7.4. Operation

The process comprises brine purification, electrolysis, and product handling, as shown in Figure 9. The feed brine is carefully purified to ensure good performance of the membrane. The brine depletion rate is ca. 50%. The caustic soda solution from the electrolyzer can be used as is, or a 50% caustic solution can be produced by a double- or triple-effect evaporator.

7.4.1. Brine Purification

The saturated brine is treated by two-stage purification. The primary purification involves conventional precipitation and sedimentation, while secondary purification is accomplished by the absorption of impurities such as calcium and magnesium ions in a fixed bed of chelating resins.

The membrane comes into contact with a relatively highly concentrated NaOH solution, and Ca^{2+} and Mg^{2+} can react with OH^- within the membrane to form insoluble hydroxides, which then precipitate inside the membrane. This causes serious problems: an increase in cell voltage and a reduction in current efficiency. To maintain the high performance of the ion-exchange membrane for a long period of time, the feed brine must be purified to a greater degree than in the conventional mercury or diaphragm processes [129]. The Ca^{2+} concentration c_{Ca} (mg/L) in feed brine and the Ca^{2+} accumulation rate in the ion-exchange membrane R (µg/hcm^2), while producing 35 wt % NaOH are linearly related:

$$R = a c_{Ca} + b$$
$$a = 3 \text{ cm/h}, \quad b = -0.08 \text{ µg h}^{-1} \text{ cm}^{-2}$$

The equation indicates that if the feed brine is purified to a Ca^{2+} concentration of 0.02 mg/L or less, Ca^{2+} precipitation in the membrane is reduced to zero.

Primary brine purification can reduce the Ca^{2+} in brine to 1–5 mg/L. This brine can then be further purified in the secondary brine purification system. A highly purified feed brine, with a Ca^{2+} concentration < 0.02 mg/L, can be obtained with a *chelating resin* in the secondary brine purification system. The purification system consists of two chelating resin towers: the type and quantity of chelating resin packed in each tower must be adequate to carry out ion-exchange treatment in one of the towers while the other is regenerated. The important properties of chelating resin for brine purification are

large ion-exchange capacity
low pressure drop
resistance to crushing during repeated use
resistance to high temperature

7.4.2. Cell Room

A typical bipolar membrane cell room is shown in Figure 55, and a typical monopolar membrane cell room is shown in Figure 56.

Purified brine is fed to the anode compartment, and water is injected into the catholyte stream to maintain the concentration of catholyte. The catholyte effluent is supplied to the evaporator to create a more concentrated product.

The catholyte and anolyte streams are effectively isolated by the membrane, and operation is simple. There is a little daily maintenance because exclusively corrosion-proof material is used in the cell. Startup or shutdown of the entire plant is quickly accomplished, requiring no special skill or attention. Operation is flexible and easily accommodates a daily current shift, allowing the use of inexpensive off-peak power.

Figure 55. Bipolar cell room

Figure 56. Monopolar cell room

7.4.3. Treatment of Products

Chlorine. See Chapter 11.

Hydrogen. See Chapter 4.

Sodium Hydroxide Solution. If 50 wt% NaOH, rather than the 32–35 wt% membrane cell caustic, is required, the catholyte effluent must be concentrated by evaporation. However, the amount of water to be evaporated is small, and no crystallizer is needed for NaCl. As shown in Figure 57, a 50 wt% NaOH can be obtained by evaporating the catholyte effluent in a double-effect evaporator.

Caustic soda solution obtained by the membrane process contains few impurities, owing to the barrier effect of the membrane. An example of the composition of caustic solution is shown in Table 16.

8. Electrodes

8.1. Anodes

From the beginning of electrolytic chlorine production to about 1913, platinum and magnetite anodes were used. However, platinum was too expensive, and magnetite only allowed an anodic current density of ca. 400 A/m^2. From 1913 until 1970 — nearly 60 years — graphite was used almost exclusively.

Since 1957 attempts have been made to replace graphite by activated titanium because of the excellent corrosion resistance of this metal in chlorinated brine. The activation tested was usually Pt, in a few attempts Pt/Ir, but most of these anodes suffered from either short life or high cost.

In 1965 and 1967, H. B. BEER applied for two patents [130], that changed the situation completely. The first patent described a coated titanium electrode in which the operative material of the coating was one or more platinum group metal oxides, possibly with the addition of some nonprecious metal oxides. The second patent described coatings that were mixed crystals of valve metal oxides and platinum group metal oxides, with usually more than 50 mol% of the valve metal oxide. (The valve metals include titanium, tantalum, and zirconium.)

This type of coating and the titanium structure of the anode were further improved by V. DE NORA [131], who developed titanium anode production on an industrial scale and commercialized the anodes with the trade name Dimensionally Stable Anodes, or DSA.

Figure 57. Two-effect caustic evaporator

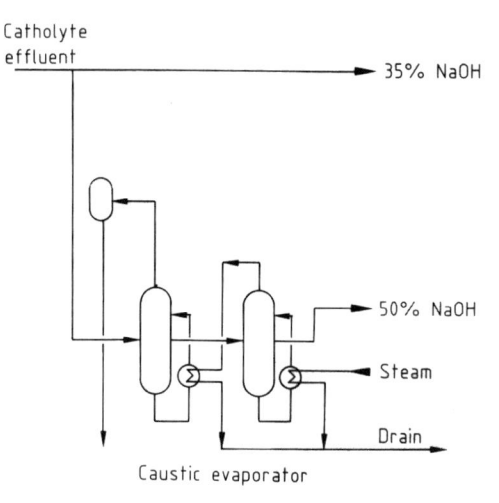

At about the same time, other anode producers developed activated titanium anodes. As a result, more than a thousand patent applications were submitted for titanium anodes.

Most plants have been converted to coated titanium anodes because they allow high current density at a low voltage and have long life. By 1986 only a few plants were still operating with graphite anodes.

8.1.1. General Properties of the Anodes

8.1.1.1. Coating Properties and Preparation

A critical compilation of the literature is given by TRASATTI [132], [133].

Chemical Composition. All coatings that are used on an industrial scale are composed of an oxide of a platinum group metal (usually Ru, in some cases, two or three metals) and an oxide of a nonplatinum metal group (usually Ti, Sn, or Zr). The preferred ratio of platinum group metal oxide to nonplatinum group metal oxide varies from 20:80 to 45:55 by weight. Some of these coatings may contain glassy fibers [134], and some may contain crystals like $Li_{0.5}Pt_3O_4$ [135] or rhodium-containing solids.

Preparation. All of these coatings are prepared from a "paint" that contains the desired metals as salts or organic compounds in an aqueous or organic or mixed solvent. This paint is then applied to the titanium by spraying, brushing, dipping, or any other technique. The solvent is evaporated, and after this drying the paint is converted into the oxidic coating by heating to 350–600 °C, depending on the type of coating. This procedure is repeated several times until the final coating thickness is achieved.

Great care must be taken to ensure the correct pretreatment of the substrate surface, the baking temperature, the amount of metal per single coat, and the final coating thickness. Exact details of all of these aspects are proprietary to the anode producers and, thus, are not available in the literature. The operating parameters of an anode can be affected more by the preparation than by the coating composition.

Crystallographic Composition, Morphology, True Surface Area. The active phase of the coating is of the rutile type and is thermodynamically unstable — even though it may last for many years of operation. The stable phase (anatase TiO_2 in the case of RuO_2/TiO_2) is electrochemically inactive [136]. The final coating looks like cracked mud. The BET surface or the electrochemically determined area is ca. 400-fold the geometrical area, and even 1000-fold surface areas may be prepared [137].

Overpotentials and Current–Voltage Relations. From the data available [138]–[140], the chlorine overpotentials are calculated to be 90–120 mV between 2 and 10 kA/m^2 anodic current density. About 95% of this overpotential is due to diffusion [141], not to electrocatalytic effects.

Side Reactions. During operation of metal anodes for chlorine production, only O_2 and ClO_3^- are generated as byproducts.

Coating Wear and Coating Lifetime. The actual coating lifetime is mainly determined by operating parameters and the particular type of cell — membrane, diaphragm, or mercury cell.

There is, however, an upper limit under ideal conditions, NaCl content > 200 g/L, no contact with amalgam, no operation in strongly alkaline medium, and temperature of ca. 80 °C. This upper limit has been determined by measuring coating thickness in mercury cells after 2 and 3 years of operation and in diaphragm cells after 4 and 5 years of operation. This upper limit is at least 2–3 times greater than that of normal industrial experience of ca. 500 t/m^2 for the same type of electrodes [142].

Radiotracer measurements [140], [143] have shown that initial wear rates (a few hours to a few days) are 1–2 orders of magnitude higher than final wear rates (some months) and that after each shutdown of a cell, the initial wear is almost attained again. The reported initial rates, (ca. 45 mg of Ru per t of Cl_2), are far higher than the ones for upper-limit lifetime.

The wear mechanism has been discussed elsewhere [132], [133].

Other Coatings. Two recent reviews [144] cover the many attempts to find chlorine anode coatings other than ruthenium-based ones. Of these, $Li_{0.5}Pt_3O_4$ [135], Pt/Ir [145], cobalt spinel [146], and PdO-based anodes [147] have been used on a small industrial scale.

Pt/Ir is too expensive for diaphragm or membrane operation and suffers from poor short-circuit resistance in mercury cells, even though it proved to be an excellent chlorate anode and a good anode for sea water electrolysis. Cobalt spinels are cheap, but allow only low current density. Coatings based on PdO suffer from poor resistance to current reversal and high palladium wear, but they have been used in membrane operation to a certain extent [148].

8.1.1.2. Titanium Substrate

The substrate of the anode is usually titanium of ASTM Grade 1 or 2 because the surface of other grades cannot easily be prepared for coating and the anodes cannot easily be flattened. Sintered titanium has also been used for anode production [149]. In this case, the titanium substrate is covered with a layer of Ti_2O, and the porous structure is sealed to prevent the paint from being sucked into the pores.

One of the most important advantages of titanium anodes is that the costly titanium structure can be *recoated* and, therefore, be used probably more than 20 years.

8.1.1.3. Anode and Coating Poisons

Because organic acids (formic acid, oxalic acid, etc.) and fluoride attack titanium, they must be avoided. Organics such as hydraulic oil must be avoided because they cover the anode with a blanket-like inactive coating.

Barium and strontium compounds can precipitate on the anode surface, but if they do, they usually cover the back, rather than the front, of mercury cell anodes. Manganese must be avoided because of MnO_2 formation, which increases anode potential and oxygen level. Even though iron is precipitated as an oxide on the front of diaphragm anodes, it does not affect the voltage or lifetime of the anodes much. Operation in strongly alkaline brine (pH > 11) may destroy the coating rapidly. Operation with NaCl concentration so low that oxygen is evolved with the chlorine must also be avoided.

Current reversal, which occurs during shutdown of a cell, especially an amalgam cell, seems to affect coating lifetime less than believed earlier.

8.1.2. Anodes for Mercury Cells

Structure. Because the typical current density for mercury cells is ca. 10 kA/m^2, heavy structures must be built to minimize ohmic losses. A mercury anode consists of an active surface, one or more current distributing devices welded to the substrate, and one, two, four, or six titanium threads welded to the current distributors (Fig. 58). The greatest part of the electrical current should not pass through the threads; instead, electrical contact is made by a copper shaft, within the threads, that is pressed against a nonthreaded boss on the electrode itself, e.g., at the contacting flat end surfaces or conical sides.

One-stem anodes typically have a size of 0.1 m^2; four- and six-stem anodes may have a size of more than 1 m^2.

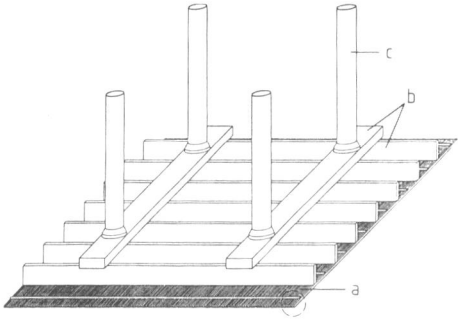

Figure 58. Four-stem anode for amalgam cells
a) Active surface; b) Current distributor; c) Riser tube to protect the copper bar inside

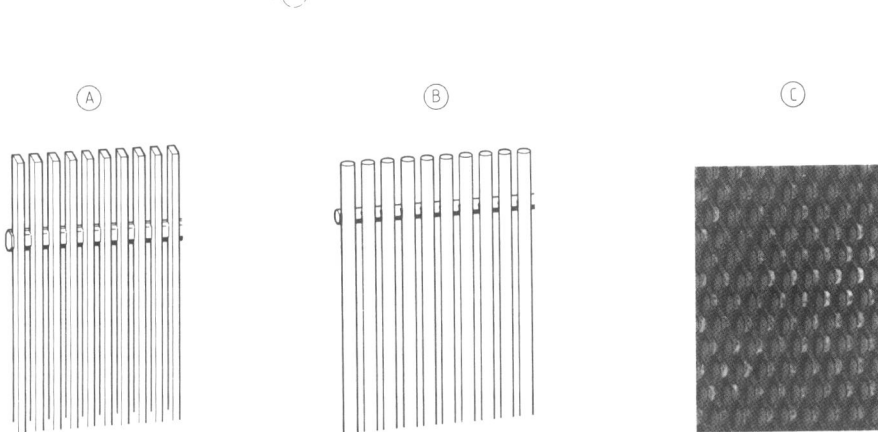

Figure 59. Anode designs for quick gas release
A) Flat profile (channel blades); B) Rod type (3-, 4-, 5-mm diameter); C) Expanded metal

Quick gas release is one of the major requirements for a mercury anode, and many different types of surface structures have been built. The three most common are shown in Figure 59. Rods with oval or inversed delta cross sections and perforated, sintered 1-cm-thick plates have also been built.

At low current density, all of the structures seem to behave similarly, but at high current density, certain surfaces may be better. Enhanced brine circulation and, thus, quicker gas release are claimed if there are baffles on the back of the active surface. These baffles operate like a gas lift to force the brine to flow into the gaps on the surface structure [150].

The surface of the anode must be as flat as possible to be parallel to the cathode. A typical value is ± 0.5 mm; a tolerance of ± 0.25 mm from an ideal plane can be achieved for small anodes.

Coating Life, Short Circuits, Oxygen Content of Cell Gas. If 10% of the anodes in a cell have been damaged or deactivated, the cell can no longer be operated at the desired k-factor (see p. 1146) of $0.08-0.1$ mΩ m^2, and the anodes should be removed to be recoated. Therefore, the practical coating life is not determined by the upper-limit life (see p. 1207), but rather by operating the anodes in alkaline medium (light butter), by massive short circuits [151], and by semi-short-circuits (tickle shorts) [152].

Therefore, 18–30 months (about 180 t of Cl_2 per m^2) is a typical lifetime. The anodes on the outlet side of the cell usually show much higher wear than those on the inlet side. To extend the lifetime, anode coatings have been developed that have an intermediate layer of plasma-sprayed conductive TiO_{2-x} between active coating and titanium substrate (Fig. 60) [139].

The intermediate layer prevents formation of an electrical arc, which spreads out and destroys the anode as a consequence of a small short circuit. The active coating of these anodes is stabilized for operation in low-concentration brine and in alkaline light butter by a second platinum group metal. These Long Life Anodes have been tested on a large scale against standard anodes: repair costs were reduced by more than 50% and their lifetime was twice that of standard coated anodes.

Another way to prevent the formation of an electrical arc following a short circuit is to add glassy fibers to the coating. This prevents a direct contact between the amalgam and the coating due to the high surface tension of the amalgam [134]. Contact with the amalgam may also be minimized with an anode protection device [153].

8.1.3. Anodes for Diaphragm Cells

Structure. The typical current density for a diaphragm anode is ca. 2 kA/m^2. Therefore, light structures may be used since only a low voltage drop occurs inside the anode structure. All diaphragm anodes have flattened, expanded metal, usually 1.5 mm thick. Most of the anodes for monopolar cells (see Section 6.2) are similar to the one shown in Figure 61 [154].

During assembly of the cell, the two expanded metal sides are fixed close together. After the cathode has been positioned, the clamps (not shown) are removed, and the anode is pressed against two plastic rods of 3 mm diameter for separation from the diaphragm. An older type of anode, the box type, which had no springs, has been completely replaced by expandable anodes.

Another type of diaphragm anode is used in the bipolar Glanor cells [155].

Coating Life and Mechanism of Deactivation. The coating lifetime for DSA coatings exceeds 10 years, and production of chlorine may exceed 200 t/m^2. These figures are lower limits — up to now no standard anode has failed. Estimates based on coating wear rate suggest an expected lifetime of more than 12 years, and even higher values are probable. The wear is caused by the relatively high oxygen level in a diaphragm cell [156].

The amount of oxygen in the cell gas depends mostly on operating parameters; an effect of the coating composition can hardly be detected. However, coatings with compositions very different from the ones used industrially do influence oxygen content [156]. In spite of the spacers, the diaphragm

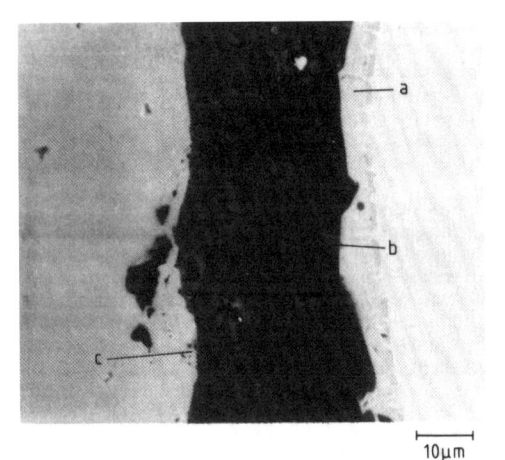

Figure 60. Coating with intermediate layer
a) Active coating; b) Intermediate layer; c) Titanium substrate

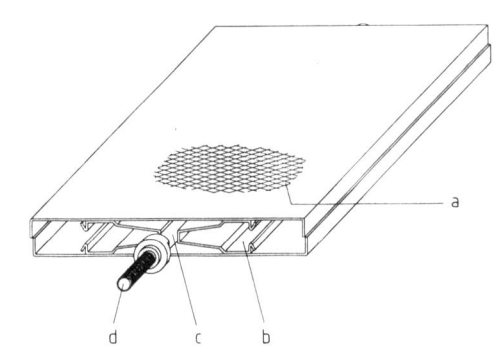

Figure 61. Anode for monopolar diaphragm cells
a) Activated (coated) expanded metal; b) Expanding spring; c) Titanium-clad copper bar; d) Copper thread to fix the anode to the cell base

usually makes contact with the anode. However, if the diaphragm swells and grows into the anode, the coating can be worn off rapidly.

8.1.4. Anodes for Membrane Cells

Structure. Only the structure of the active part of the anode is described here. Because the membrane is usually in contact with the anode, gas is released mainly on the back of the anode. Therefore, thin, flattened, expanded or punched metal is used as the substrate. Louver-like structures are also in use [157]

Punched sheet material is claimed to be superior to flattened, expanded metal [147]. Attempts to find the optimum geometry are discussed in the literature [158]. Sharp peaks or edges that may puncture the membranes must be avoided.

Coating Life Time. Up to now no recoating has been necessary. X-ray determinations of coating thickness show that coating wear is greater in membrane cells than in diaphragm cells, but the reasons for this are unclear. The expected coating lifetime exceeds 7 years. Lifetimes of more than 10 years are possible, depending on coating thickness. Strong caustic flows through holes in the membrane cause rapid coating loss.

Oxygen Content of Anolyte Gas. According to recent measurements, oxygen content of the cell gas varies linearly with active chlorine in the anolyte (Cl_2 + OCl^- + $HOCl$), but is nearly independent of pH of the anolyte (1.7 < pH < 5) provided the concentration of active chlorine is kept constant [147], [156]. In an anolyte with no OH^- back migration through the membrane (acidic catholyte), the oxygen content is ca. 0.2 – 0.3 vol % [156].

If 2 – 4 g/L of active chlorine is present, the oxygen content may climb to 3 – 4 vol %. The chlorate level also increases, but far less so than according to the Foerster mechanism [159] of coproduction of chlorate and oxygen on the anode. About 10 – 20 % of the oxygen comes from water oxidation, 10 – 20 % comes from chlorate formation, and the rest comes from oxidation of active chlorine to oxygen and chloride. Coating composition influences oxygen content [158], but operating parameters — mainly HCl addition — may override this influence for commercially available coatings.

8.1.5. Graphite Anodes

Graphite electrodes were used almost exclusively for 60 years, up to about 1970.

The anode material, which was low in ash and vanadium, was composed of various types of particulate coke mixed with a pitch binder. It was formed by a hydraulic extrusion press, baked at temperatures up to 1000 °C and finally graphitized at 2600 – 2800 °C. The final shape was produced by machining. The horizontally suspended anodes usually had an area of 0.1 – 0.2 m^2 and an initial thickness of 7 – 12 cm.

The chlorine gas bubbles, which were produced on the under side of the anode, were drawn off via a system of vertical slits and from these through holes.

Some oxygen was liberated at the anodes with the chlorine, and this oxygen attacked the graphite forming carbon monoxide and carbon dioxide. This electrode wear was the cause of a graphite consumption of 1.8 – 2.0 kg per tonne of chlorine produced from sodium chloride. The consumption was 3 – 4 kg per tonne of Cl_2 from potassium chloride.

The graphite anodes had to be lowered from time to time, usually once per day. The anodes were replaced when they reached a thickness of 2 – 3 cm. When the anodes were well constructed and frequently adjusted, an average voltage drop coefficient of 0.12 – 0.14 Vm^2/kA was achieved. The shortcomings of graphite anodes, particularly their high energy consumption, led to a change over to titanium, except in a few special cases, such as developing countries with limited foreign exchange.

8.2. Activated Cathode Coatings

Diaphragm and membrane brine electrolyzers use steel or nickel cathodes. Depending on current density and measuring technique, they have a hydrogen overpotential of ca. 300 mV. Activated cathode coatings can lower that by 200–250 mV, thus providing significant energy savings (see Fig. 43). Active coatings have been described often in the literature and used in water electrolysis for over 40 years. They are only now coming into regular use in chlor-alkali production as cell technology is becoming more sophisticated.

The patent literature covers many different types of coatings, and new ones are being published regularly. There are two basic approaches to activation: high-surface-area coatings and catalytic coatings. Both bare nickel and iron show lower overpotential once in operation, and their surfaces roughen. Simply adding high-surface-area iron in the brine can show some savings [160]. More common are porous nickel-type coatings that offer high surface area and good chemical resistance. These coatings consist of two or more components. At least one is usually leached out in caustic to leave the porous, high-surface-area nickel [161]. These coatings are typically nickel–zinc admixtures [162], nickel–aluminum Raney nickel [163], or nickel–aluminum admixtures [164], or nickel–sulfur alloys [165]. A variety of additives are recommended for strength, life, and resistance to poisoning by impurities. Rough coatings of nickel–nickel oxide mixtures [166] and nickel with embedded activating elements such as ruthenium [167] are also used. Sintered nickel coatings are described in patents [168] as well being available from Huntington Alloys. Recently, nickel coatings containing platinum group metals have been sold by Dow [169], Johnson Matthey, and ICI [170].

Cathode coatings can be applied in various ways, the most common of which is electroplating. The largest commercial application facility is OxyTech Systems electroplating line in Texas. Electroless (autocatalytic) plating is used, but is generally quite expensive. Thermal spray procedures work well and can be economical for many types of cathode structures. Baked catalytic coatings use intermediate temperatures, while sintered coatings use even higher temperatures, which can limit the type of cathode structure to be coated.

The coatings used for diaphragm and membrane cells usually differ because of the different operating conditions. The weak 11 % caustic in diaphragm cell liquor is less corrosive than the strong 35 % caustic of a membrane cell. Therefore, the less expensive and more fragile coatings, like nickel–zinc, can be used in diaphragm cells. Membrane cell suppliers tend to favor the plasma spray, sintered, and platinum group metal coatings.

There are other factors affecting the choice of coatings besides the caustic strength. The shape or structure of the cell is a major concern. The complex cathodes of diaphragm cells lend themselves to liquid systems that can coat the entire structure [171]. Membrane cells have flat cathodes, which are easy to access for spray techniques. The lower operating current density of diaphragm cells means more cathodic surface area per unit of production; thus, less expensive coatings are preferred. Most diaphragm cells employ heat-cured polymer–asbestos separators that are vacuum deposited after the cathode coating is applied. This curing operation can destroy the activity of certain coatings.

As experience is gained with activated cathode coatings, a growing concern is damage during operation. Some coatings, especially the platinum group metal coatings, are subject to damage from reverse currents during cell outages; thus, precautions are needed to protect the coatings with reducing agents [172] or by cathodic protection [173]. In membrane cell systems, even where ultrapure brine is used, trace impurities such as iron coming from the caustic recirculation system can deposit on the cathode, thus blinding the coating. Some coatings are claimed to be more resistant to this effect than others because of their microstructure and the presence of additives. Porous nickel coatings in diaphragm cells are less susceptible to iron blinding because the spalling of the brittle coating makes them self-cleaning.

Activated cathode coatings are becoming the standard throughout the chlor-alkali industry. Exceptions are plants having low power costs, usually hydroelectric, and those too remote from a centralized coating facility. Even though the basic technology has been in existence for many years, commercial use has been common only since the late 1970s. However, many aspects of the operation of activated cathode coatings are only now being discovered. Many of these discoveries remain proprietary information of the cell and coating suppliers.

9. Comparison of the Processes

The advantages and disadvantages of the three chlor-alkali processes are summarized in Table 15. The three chlor-alkali processes can be compared in respect to the quality of the chlorine and caustic produced, the capital investments, and the operating costs.

Today the membrane process is clearly the best choice for a new plant.

9.1. Product Qualities

Table 16 shows typical composition values for the chlorine and caustic produced by the diaphragm, mercury, and membrane processes. Chlorine produced by the mercury process can be used directly for most uses. Chlorine produced by the diaphragm or membrane process contains up to 2% O_2, depending on the pH of the anolyte. This oxygen is removed by condensation and evaporation of the chlorine.

The sodium hydroxide solution from the mercury process is the purest of the three; the amounts of NaCl and $NaClO_3$ are especially low. However, the quality of caustic

Table 15. Advantages and disadvantages of the three chlor-alkali processes

Process	Advantages	Disadvantages
Diaphragm process	use of well brine, low electrical energy consumption	use of asbestos, high steam consumption for caustic concentration in expensive multistage evaporators, low purity caustic, low chlorine quality, cell sensitivity to pressure variations
Mercury process	50% caustic direct from cell, high purity chlorine and hydrogen, simple brine purification	use of mercury, use of solid salt, expensive cell operation, costly environmental protection, large floor space
Membrane process	low total energy consumption, low capital investment, inexpensive cell operation, high-purity caustic, insensitivity to cell load variations and shutdowns, further improvements expected	use of solid salt, high purity brine, high oxygen content in chlorine, cost of membranes

Table 16. Product qualities: typical compositions of chlorine, caustic, and hydrogen

Product and contents	Process			
	Diaphragm		Mercury	Membrane
	Unpurified	Purified		
Chlorine gas (from cells), vol %				
Cl_2	96.5–98		98–99	97–99.5
O_2	0.5–2.0		0.1–0.3	0.5–2.0
CO_2	0.1–0.3		0.2–0.5	
H_2	0.1–0.5		0.1–0.5	0.03–0.3
N_2	1.0–3.0		0.2–0.5	
NaOH solution, wt %				
NaOH	50.0	50.0	50.0	50.0
NaCl	1.0	0.025	0.005	0.005
Na_2CO_3	0.1	0.1	0.05	0.04
Na_2SO_4	0.01	0.01	0.0005	0.0001
$NaClO_3$	0.1	0.001	0.0005	0.001
SiO_2	0.02	0.02	<0.001	0.002
CaO	0.001	0.001	0.001	0.0001
MgO	0.0015	0.0015	0.0002	0.0001
Al_2O_3	0.0005	0.0005	0.0005	0.0001
Fe	0.0007	0.0007	0.0005	0.0004
Ni				
Cu	0.0002	0.0002	0.00001	0.0001
Mn				
Hg	none*	none*	0.00003	none*
NH_3		0.001		
Hydrogen gas, vol %				
H_2	>99.9		>99.9**	>99.9

* $< 10^{-6}$ %.
** Hydrogen gas from the mercury process contains mercury: 1 µg/m – 10 mg/m, depending on the purification process. The hydrogen gas from the other two processes is free of mercury.

from the membrane process is almost as good. A main drawback of the diaphragm process is the high concentration of NaCl and $NaClO_3$ in the caustic solution. This sodium hydroxide solution cannot be used for some processes. A chloride-free grade, commonly referred to as rayon-grade caustic, is required for 20–30% of the demand in industrialized countries. Even the use of purification processes (see p. 1181) does not reduce the NaCl content below 0.03 wt%. In addition to the NaCl and $NaClO_3$, the levels of Si, Ca, Mg, and sulfate impurities are higher than for the mercury and membrane processes.

9.2. Economics

The wide variation in the principal cost, that for electrical energy, which varies from region to region by a factor of 3, makes a direct comparison of production costs problematic. Further, the cost of electrical energy is increasing in different regions at drastically different rates, depending on the basic source of energy. For example, in the Federal Republic of Germany, the percentage of the total cost arising from the energy consumed rose from ca. 30% in 1979 to 60% in 1985. The rapidly changing foreign exchange rates also make international comparisons difficult.

Therefore, instead of the absolute costs, the relative capital investment costs for the three processes are discussed and compared. Then the operating costs are compared, and the energy consumptions per tonne of chlorine produced, the most important costs, are given and compared.

9.2.1. Capital Investment

The following estimates are based on a medium-sized plant, one producing 100 000 t of chlorine and 110 000 t of NaOH (as 50% caustic) annually. Electrolysis is in principle a two-dimensional process, for which the scaling-up factor is close to unity (between 0.8 and 0.9). Therefore, the estimates are also reasonably accurate for either smaller or larger capacities.

The expenses for the rectifier, chlorine and hydrogen systems, HCl system, caustic storage, utilities, and engineering and construction overhead are approximately the same for the three processes: together they make up 75% of the capital investment, the largest expense being the 20–25% for utilities. The capital investment is different for the following equipment:

1) *Cells.* The complex mercury cells are considerably more expensive than the simpler diaphragm and membrane cells. However, this disadvantage is largely offset by the higher current density, 8–10 kA/m^2 versus 3–4 kA/m^2 for the membrane process or 2–2.5 kA/m^2 for the diaphragm process, i.e., the cell surface of the mercury cell

need be only 1/3 or 1/4 that of the other processes for the same production capacity.

2) *Brine system.* The brine system for the diaphragm process is the simplest of the three — there is neither sulfate precipitation nor dechlorination — and makes up only 3–4% of the capital investment. The brine system is the most complex for the membrane process, for fine purification by ion exchange is necessary. However, the two- or three-fold greater depletion of the brine in the membrane process allows the brine system to be smaller than that for the mercury process. Therefore, the cost of the brine system for either process is approximately the same, 4–7% of the total.

3) *Caustic concentration.* The elaborate multistage evaporators required for the concentration of the diaphragm-cell caustic and the separation of NaCl and Na_2SO_4 must be nickel plated because of the corrosiveness of the cell liquor containing NaCl and $NaClO_3$. These evaporators cost 20–35% of the total. The evaporators for the membrane process may be constructed of stainless steel and are much smaller because the essentially salt-free cell liquor is more concentrated, costing 3–4% of the total. The mercury process produces 50% caustic directly, evaporation is not required.

4) *Facilities for handling salt.* The mercury and membrane plants require storage and handling facilities for solid salt. If a diaphragm plant uses well brine, such facilities are not needed.

5) *Mercury.* In addition to the capital cost of mercury itself, there is the expense of the equipment to prevent emission of mercury into the environment and to remove mercury from the products (see p. 1159). This equipment costs 10–15% of the total capital investment.

If the capital investment of a medium-sized diaphragm plant is taken as 100%, then the capital investment for a mercury plant is 90–95%, and that of a membrane plant is ca. 80%. This is illustrated in Figure 62.

In 1983 the total capital costs for a 600-t/d (ca. 200 000 t of chlorine annually) plant on the Gulf Coast of the United States was estimated to be $ 118 100 000 for a diaphragm plant but only $ 96 400 000 for a membrane plant [174], corresponding to $ 547 and $ 446 per tonne of chlorine per year.

9.2.2. Operating Costs

The fixed costs for operators and other personnel, taxes, insurance, repairs, and maintenance are about the same for all three processes. The 20% lower depreciation of the membrane process is offset by the additional expense for purchase and replacement of the membranes and for the more elaborate brine purification.

Of the variable costs, the expense for salt, precipitants, and anode reactivation are roughly the same. The difference among the three processes shows up in the consumption of energy, as electricity and steam. If 1 t of steam is taken to be

Figure 62. Comparison of construction costs

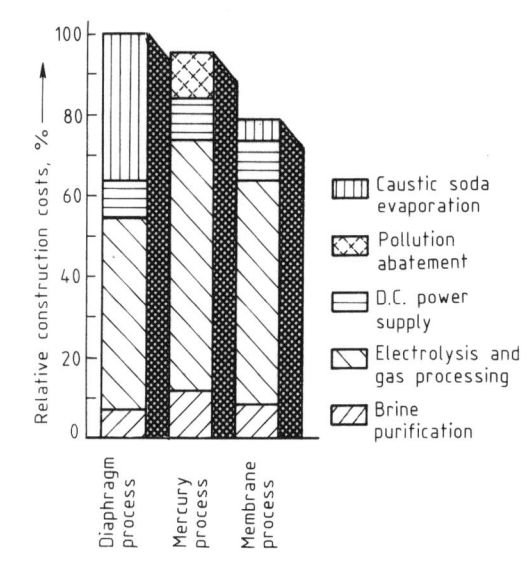

Figure 63. Relative consumption of energy (electricity and steam) in the three chlor-alkali processes in producing 50 wt% NaOH

equivalent to 400 kW h of electrical energy, then the comparison in Table 17 can be made. The differing total energy consumptions are illustrated in Figure 63.

The price of electrical energy varies widely from region to region. The relatively broad range of possible current densities combined with the steep increase in the cell voltage with current density for the diaphragm and membrane cells allows optimization of the current density with respect to the local energy price. That is, if electrical energy is relatively expensive, a greater number of cells, and thus a greater capital investment, can be tolerated to reduce the specific energy consumption and thus minimize total unit production cost [175].

Table 17. Energy consumed to produce 1 t of chlorine in the three chlor-alkali processes

Energy	Process		
	Diaphragm	Mercury	Membrane
Electricity for electrolysis, kW h	2800–3000	3200–3600	2600–2800
Steam equivalent, kW h	800–1000	0	100–200
Total, kW h	3600–4000	3200–3600	2700–3000
Relative energy costs	100%	92%	75%

9.2.3. Summary

For a new plant, the membrane process is the first choice, both the capital investment and the operating cost being the least. The advantage increases as energy prices increase. In special cases a combination diaphragm–membrane plant is advantageous: the salt recovered from the diaphragm cell liquor is used to make the brine for the membrane plant. For large plants, 500 t of chlorine per day, a combination of a diaphragm and mercury plant can be advantageous for the same reason.

There are a large number of mercury and diaphragm plants operating around the world. Continued production from these plants is economic under special circumstances. For the mercury process this is the case where electrical energy is cheap or where the plant is old and completely depreciated. Diaphragm plants are still economic where inexpensive brine is available, where only salt containing much sulfate is available, or where cogeneration of power and steam at the plant provides inexpensive steam.

However, the special circumstances making a mercury or diaphragm plant economic can change, or the price of electrical energy can increase sharply. The plant can perhaps then be kept economic by conversion to the membrane process. Governmental regulation can also effect a change: in Japan, the mercury process is not to be used after 1986, the membrane processes now providing about two-thirds of the total production capacity.

Figure 64 shows the relative capital investment for the three types of conversion (see Section 7.3) for a medium-sized plant, one producing 250 t of caustic soda per day (ca. 100 000 t of chlorine per year) from solid salt. In the case of conversion from the mercury process, the cost for the cells and the membranes accounts for more than 60% of the total investment. In conversion from the diaphragm process, the cost for the cells and the membranes is almost equal to that in mercury conversion, but the other costs for modification are less. In the diaphragm retrofit, the cost for the membranes amounts to 25% of total investment.

On conversion from the mercury process, power consumption is reduced to that of the membrane process. The diaphragm retrofit markedly reduces the cost of steam,

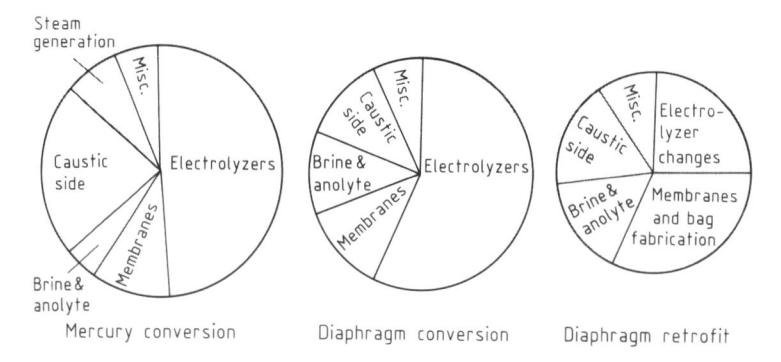

Figure 64. Capital investment for conversion of existing plants to the membrane process, the size of the circles showing the relative total costs.
Solid salt, 4 kA/m² membrane electrolyzers, 2 kA/m² diaphragm cells

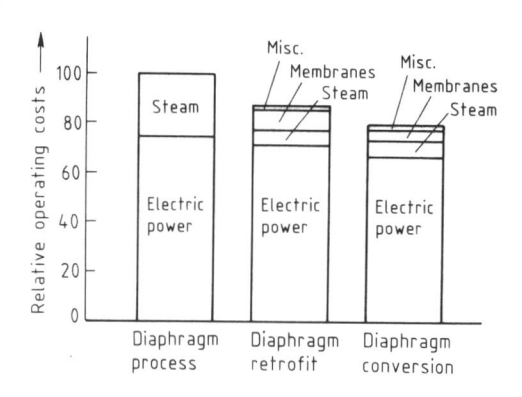

Figure 65. Relative selected operating cost for diaphragm process, diaphragm retrofit, and diaphragm conversion

somewhat reducing the electric power consumption, and the diaphragm conversion reduces power consumption further, to that of the membrane process, as is shown in Figure 65 [175], [176].

10. Other Production Processes

10.1. Electrolysis of Hydrochloric Acid

Electrolytic decomposition of aqueous hydrochloric acid is used to produce chlorine and hydrogen. The first pilot plant was set up by G. MESSNER in 1942 in Bitterfeld, Germany, and since 1964 eight full-scale plants have been commissioned in Europe and the United States, a total capacity of 510 000 t/a [177]. Hydrogen chloride is a byproduct of many organic industrial processes. Electrolysis of hydrochloric acid competes with chemical processes in which either hydrogen chloride is used to produce chlorinated hydrocarbons directly, e.g., by oxychlorination, or where chlorine is produced by chemical reaction, e.g., in the KEL chlorine process (see p. 1223). The advantages of the electrolytic process are very pure products without further treatment, reliability (simple design), ease of operation, flexibility (5:1 turndown ratio), and low energy consumption even with small installations.

Principles. Hydrochloric acid (22 wt% HCl) is fed into the cells in two separate circuits, a catholyte circuit and an anolyte circuit. During electrolysis the concentration is reduced to ca. 17%, and the temperature increases from 65 to 80 °C. A part of the depleted acid is separated from the catholyte stream, concentrated in the absorption plant to ca. 30%, and fed back into the main stream. The electrolyzer is bipolar, with pairs of electrodes arranged like the leaves of a filter press. A diaphragm separates the anode space from the cathode space to prevent mixing of the gaseous products.

The reversible standard decomposition potential of hydrochloric acid is 1.358 V, made up of the anode potential, the discharge of chloride ions with formation of chlorine, and the cathode potential, the discharge of hydroxonium (H_3O^+) ions with formation of hydrogen. In practice (> 15% HCl, 70 °C), the decomposition potential is $\leq$ 1.16 V.

The graphite electrode plates are not attacked by 22% hydrochloric acid. A poly(vinyl chloride) (PVC) fabric constitutes the diaphragm. Chlorine dissolved in the anolyte diffuses through the diaphragm and is reduced at the cathode, causing a loss of 2–2.5% of the theoretical current yield. The increase of cell voltage when current flows is mainly because of the hydrogen overpotential at the graphite cathode and the resistance of the electrolyte. Depolarizing agents (polyvalent metallic ions) in the catholyte reduce the overpotential by $\leq$ 300 mV at 4 kA/m^2 [178].

The conductivity of hydrochloric acid is maximized at a concentration of 18.5 wt%. High temperatures improve the conductivity, but to avoid increased vapor pressure of HCl and material problems, the temperature is kept below 85 °C. Modern cells have a voltage of ca. 1.90 V at 4 kA/m^2, corresponding to an energy consumption of 1400–1500 kW h per tonne of chlorine.

Cells. Hydrochloric acid electrolysis cells are manufactured by Hoechst–Uhde [179]. Each Hoechst–Uhde electrolyzer consists of 30–36 individual cells that are formed from vertical graphite plates connected in series, between which there are diaphragms. To improve gas release, vertical slits are milled in the graphite plates, which are cemented in frames made of HCl-resistant plastics. At the bottom of the frames, channels feed in the electrolyte. The gases rise up the plates and pass through ducts into collection channels in the upper part of the cell. Chlorine leaves the cell with the anolyte, and hydrogen leaves with the catholyte. The end plates of the electrolyzer are made of steel lined internally with rubber and are held together by spring-loaded tension rods. The electric current is supplied via graphite terminals. The unit rests on insulated steel frames. The effective surface of the electrodes is 2.5 m^2, and the current loading can be up to 12 kA.

DeNora and General Electric are developing an electrolyzer with a solid polymer electrolyte (SPE) based on Nafion [180]. In addition to a voltage savings of 20%, it is hoped that completely chloride-free hydrogen gas can be produced.

Operation. A simplified flow diagram of the process as operated by Bayer–Hoechst–Uhde is shown in Figure 66.

In the absorption column, the hydrogen chloride gas is absorbed adiabatically by depleted hydrochloric acid from the catholyte. In the upper section of the column, an absorber removes the remaining hydrogen chloride and the water vapor by absorption in a water stream, which makes up the water balance of the process. The 30 wt % acid that is produced is then cooled, purified if necessary by activated carbon, and supplied to the anolyte and catholyte circulation systems.

The electrolyte is pumped through a filter and heat exchanger to a gravity feed tank for the electrolyzer unit. The gases produced are freed from the electrolytes in separators, and the electrolytes flow back into their respective collecting tanks to be resaturated. The working life of the PVC diaphragms, 1–2 years, depends on the impurities in the acid. The concentrated acid is, therefore, purified carefully [181].

The product gases are saturated with water vapor and hydrogen chloride at the partial vapor pressures of 20% hydrochloric acid. Both product streams are cooled. Sodium hydroxide solution is used to wash the hydrogen, removing chlorine and hydrogen chloride and producing a 99.9% product. The chlorine, which is dried by sulfuric acid, contains ca. 0.5% hydrogen and ca. 0.05% carbon dioxide. The hydrogen overpotential can be reduced by activation of the cathodes.

10.2. Chemical Processes

The chlor-alkali process produces chlorine and sodium hydroxide solution in fixed stoichiometric proportions. Experience has shown that there tends to be a surplus of either chlorine or sodium hydroxide. Chlorine may, however, be produced competitively without the byproduct sodium hydroxide by nonelectrolytic methods. The starting material is usually hydrogen chloride, which is catalytically oxidized to chlorine by oxygen, air, nitric acid, sulfur trioxide, or hydrogen peroxide. Other processes start from ammonium chloride or metal chlorides. Only the KEL chlorine process has been brought to full-scale operation.

10.2.1. Catalytic Oxidation of Hydrogen Chloride by Oxygen

A catalyst is essential for the economic oxidation of hydrogen chloride to chlorine by air or oxygen (Deacon Process), and the catalyst must be active at low temperature and have adequate life. There are many patents claiming improved catalysts and equipment. Most of the catalysts are oxides and/or chlorides of metals on various substrates. Only two processes have been commercialized.

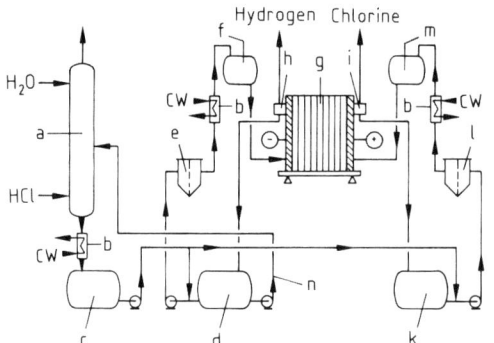

Figure 66. Simplified flow diagram of a hydrochloric acid electrolysis
a) Absorption column; b) Heat exchanger; c) Strong acid tank; d) Catholyte collecting tank; e) Catholyte filter; f) Catholyte supply tank; g) Electrolyzer; h) Hydrogen–catholyte separator; i) Chlorine–anolyte separator; k) Anolyte collecting tank; l) Anolyte filter; m) Anolyte supply tank; n) Weak acid line to absorber

The KEL Chlorine Process. The process developed by KELLOG [182] uses concentrated sulfuric acid (ca. 80%) with ca. 1% nitrosylsulfuric acid as the catalyst. A full-scale plant for recovering up to 600 t of chlorine per day has been operated since 1975 by Du Pont in Corpus Christi, Texas. The raw material, from a fluorinated hydrocarbon plant, consists of waste gases that contain hydrogen chloride [183]. Figure 67 shows a simplified flow diagram.

Sulfuric acid catalyst is fed into the top of the stripper column. The hydrogen chloride gas reacts with the catalyst to form nitrosyl chloride:

$$HCl + NOHSO_4 \longrightarrow NOCl + H_2SO_4$$

The oxygen, the ultimate oxidizing agent, blows the remaining hydrogen chloride out of the sulfuric acid, which becomes more concentrated and also is cooled in a flash vaporizer. This acid is then fed back into the process. Nitrosyl chloride, hydrogen chloride, oxygen, and water vapor flow as a gaseous stream into the oxidizer and react there, increasing the temperature:

$$2\ NOCl + O_2 \longrightarrow 2\ NO_2 + Cl_2$$
$$NO_2 + 2\ HCl \longrightarrow NO + Cl_2 + H_2O$$

In the absorber–oxidizer, the rest of the hydrogen chloride is oxidized. Concentrated sulfuric acid is fed in at the top, reacts with the oxides of nitrogen to form nitrosylsulfuric acid, absorbs the water that has formed, and is conducted back into the stripper:

$$NO + NO_2 + 2\ H_2SO_4 \longrightarrow 2\ NOHSO_4 + H_2O$$
$$NOCl + H_2SO_4 \longrightarrow NOHSO_4 + HCl$$

The cooled, dried chlorine gas still contains ca. 2% hydrogen chloride and up to 10% oxygen. Both are removed by liquefaction.

The net reaction is

$$4\ HCl + O_2 \longrightarrow 2\ Cl_2 + 2\ H_2O$$

The installation at Corpus Christi operates at 1.4 MPa and 120–180 °C. On account of the aggressive nature of the chemicals, expensive materials, such as tantalum-plated equipment and pipes, must be used. For outputs of 250–300 t of chlorine per day, this process can be more economical than the electrolysis of hydrochloric acid, depending on local conditions.

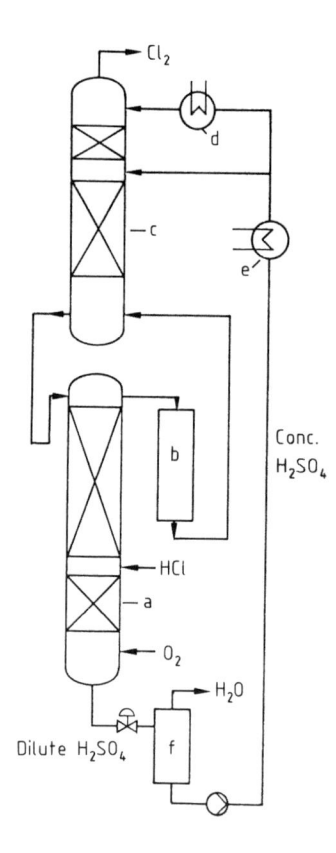

Figure 67. Flow diagram of the KEL chlorine process (simplified)
a) Stripper; b) Oxidizer; c) Absorber – oxidizer; d) Acid chiller; e) Acid cooler; f) Vacuum flash evaporator

The Shell Chlorine Process. The catalyst developed by Shell consists of a mixture of copper(II) chloride and other metallic chlorides on a silicate carrier [184]. The reaction of the stoichiometric mixture of hydrogen chloride and air takes place in a fluidized-bed reactor at ca. 365 °C and 0.1 – 0.2 MPa. The yield is 75 %. The water condenses out from the gas stream, and the hydrogen chloride is removed by washing with dilute hydrochloric acid. After the residual gas has been dried with concentrated sulfuric acid, the chlorine is selectively absorbed, e.g., by disulfur dichloride. After desorption and liquefaction, the chlorine has a purity > 99.95 %.

A manufacturing unit was built by Shell in the Netherlands, 41 000 t/a, and another in India, 27 000 t/a, but both have been closed down owing to the prolonged surplus of chlorine on the market.

10.2.2. Oxidation of Hydrogen Chloride by Nitric Acid

The nitrosyl chloride route to chlorine is based on the strongly oxidizing properties of nitric acid [185]:

$6 \text{ HCl} + 2 \text{ HNO}_3 \longrightarrow 2 \text{ Cl}_2 + 2 \text{ NOCl} + 4 \text{ H}_2\text{O}$
$2 \text{ NOCl} + 2 \text{ H}_2\text{O} + \text{O}_2 \longrightarrow 2 \text{ HCl} + 2 \text{ HNO}_3$

The practical problems lie in the separation of the chlorine from the hydrogen chloride and nitrous gases. The dilute nitric acid must be reconcentrated. Corrosion problems are severe. Suggested improvements include (1) oxidation of concentrated solutions of chlorides, e.g., LiCl, by nitrates followed by separation of chlorine from nitrosyl chloride by distillation at 135 °C or (2) oxidation by a mixture of nitric and sulfuric acids with separation of the product chlorine and nitrogen dioxide by liquefaction and fractional distillation [186].

10.2.3. Production of Chlorine from Chlorides

Alkali-metal chlorides, ammonium chloride, and other metallic chlorides are reacted, usually with nitric acid, to produce nitrate fertilizers [187]. Chlorine is not produced directly, but it can be obtained from the intermediate products nitrosyl chloride or hydrochloric acid.

11. Chlorine Purification and Liquefaction

Chlorine produced by the various processes, especially by electrolysis, is saturated with water vapor at high temperature and also contains brine mist and traces of chlorinated hydrocarbons, and is normally at atmospheric pressure. Before the chlorine can be used, it must be cooled, dried, purified, compressed, and where necessary, liquefied. A simplified flow sheet is shown in Figure 68.

11.1. Cooling

Table 18 shows the volume, water content, and heat content of 1 kg of chlorine gas at 101.3 kPa as a function of temperature. To avoid solid chlorine hydrate formation, the gas is not cooled below 10 °C [188]. Cooling is accomplished in either one stage with chilled water or in two stages with chilled water only in the second stage.

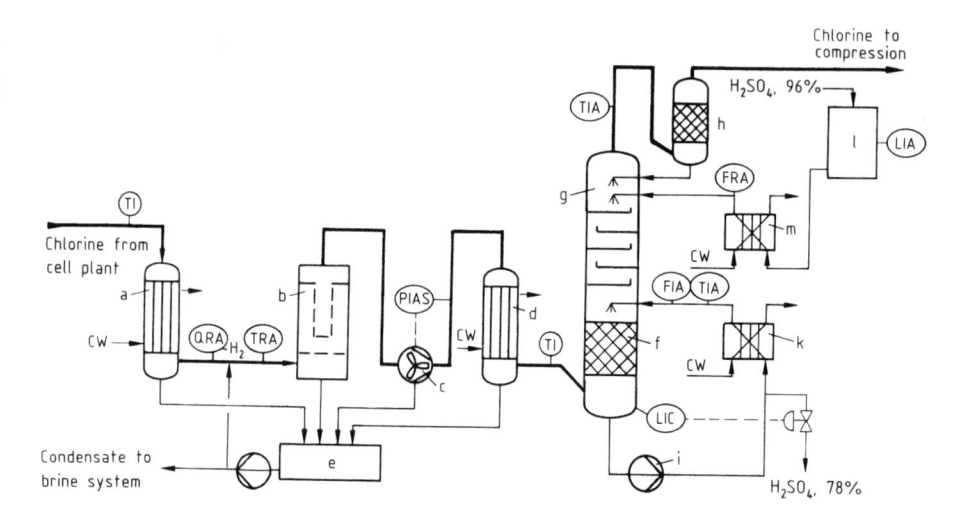

Figure 68. Simplified flow diagram of a chlorine processing plant
a) Chlorine gas cooler (primary); b) Chlorine demister; c) Blower or fan; d) Chlorine gas cooler/chiller (secondary); e) Condensate collection tank; f) Drier, first stage; g) Drier, second stage; h) Sulfuric acid mist separator; i) Sulfuric acid circulation pump; k) Cooler for circulating sulfuric acid; l) Sulfuric acid feed tank; m) Cooler for sulfuric acid feed

Table 18. Volume, moisture content, and enthalpy of 1 kg of chlorine gas at 101.3 kPa as a function of temperature t

t, °C	Volume, m^3		Water content, g/kg**	Heat content, kJ/kg*
	Dry	Saturated *		
0	0.312	0.314	1.54	3.81
20	0.335	0.342	5.95	24.45
40	0.357	0.385	19.7	69.50
60	0.380	0.473	61.5	188.41
70	0.392	0.565	112.0	325.73
80	0.404	0.756	222.0	623.83

* Chlorine gas saturated with water vapor at temperature t.
** Grams of H_2O per kg of Cl_2.

The chlorine gas can be *cooled indirectly* in a tubular titanium heat exchanger so that the cooling water is not contaminated and the pressure drop is small. The resultant condensate is either fed back into the brine system of the mercury process or dechlorinated by evaporation in the case of the diaphragm process.

The chlorine gas can be *cooled directly* by passing it into the bottom of a tower in which the packing is divided into two sections, for two-stage cooling. Water is sprayed into the top and flows countercurrent to the chlorine. This treatment thoroughly washes the chlorine; however, dechlorination of the wastewater consumes a large amount of energy. The cooling water should be free of traces of ammonium salts to avoid the formation of nitrogen trichloride [189]. The cooled chlorine is passed through demisters before it is dried.

Closed-circuit direct cooling of chlorine combines the advantages of the two methods. The chlorine-laden water from the cooling tower is cooled in titanium plate coolers and recycled. The surplus condensate is treated exactly like the condensate from indirect cooling. Spray towers, as well as packed towers, are used. Water carry-over is removed by demisters, which reduce the amount of sulfuric acid used for drying.

11.2. Chlorine Purification

Water droplets and impurities such as brine mist are mechanically removed by special filter elements with glass wool fillings. The efficiency varies with the gas throughput. A commonly used device is the Brink *demister* [190]. Instead of glass wool, porous quartz granules can be used.

In *electrostatic purification*, the wet chlorine gas is passed between wire electrodes in vertical tubes. The electrodes are maintained at a d.c. potential of 50 kV with a current density of 0.2 mA/m^2. The particles and droplets in the chlorine become charged and collect on the tube walls. The resultant liquid is fed back into the brine system or chemically treated before disposal.

Activated carbon filters can adsorb organic impurities and may be regenerated by heating to 200 °C.

Gaseous impurities can be removed by *absorption* of the chlorine in a suitable solvent, such as carbon tetrachloride, water, or disulfur dichloride, followed by *desorption*. This can be coupled with further processes, such as the recovery of chlorine from the waste gas remaining after liquefaction [40, pp. 418–422].

A wash with concentrated hydrochloric acid removes the dangerously explosive nitrogen trichloride [191]. *Scrubbing with liquid chlorine* (see Fig. 70) mainly reduces the content of organic impurities and carbon dioxide, but it can also lower the bromine content. When the chlorine is cooled down to near its dew point, liquid chlorine scrubbing is often combined with compression by turbo or reciprocating compressors.

11.3. Drying

Drying of chlorine is carried out almost exclusively with concentrated sulfuric acid (96–98 wt%). Depending on the desired final concentration of the waste acid, drying can be a two-, three-, or four-stage process. The acid and chlorine flow countercurrently. The final moisture content depends on the concentration and temperature of the acid in the final stage (Fig. 69). An upper limit is 50 ppm H$_2$O.

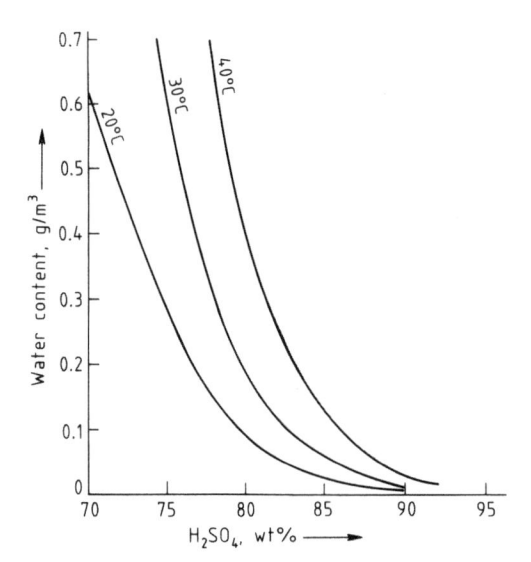

Figure 69. Drying chlorine with sulfuric acid
Attainable moisture content as a function of concentration and temperature of the acid

Low-temperature liquefaction (– 70 °C) demands lower moisture content, which can be achieved with molecular sieves, whereby 2.5 ppm is possible [192].

The *packed towers* usual in the first stages are constructed of rubber-lined steel or glass-fiber-reinforced poly(vinyl chloride). The heat liberated on dilution of the circulating acid is removed by titanium heat exchangers, and the weak acid is dechlorinated chemically or by blowing air. Generally, columns with *bubble cap plates* or *sieve trays* are used at the final stage. The drying is effective, but the pressure drop is great. Occasionally, spray towers are used to dry chlorine.

After drying, the chlorine gas is passed through a demister or a packed bed to remove sulfuric acid mist.

11.4. Transfer and Compression

In all operations involving compression, care must be exercised to prevent the heat of compression from increasing the temperature enough to ignite material in contact with the chlorine.

Wet chlorine gas can be compressed 20–50 kPa by a single-stage *blower* or *fan* with a rubber-lined steel casing and titanium impeller. It can also be compressed in *liquid-ring compressors,* so that further treatment of the chlorine can be accomplished in smaller equipment [193]. Sulfuric acid ring compressors are used for throughputs of 150 t of dry chlorine gas per day per compressor and for pressures of 0.4 MPa or, in

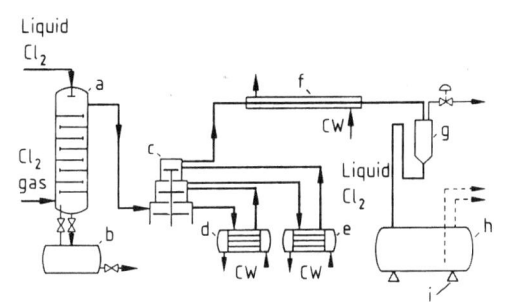

Figure 70. Multistage reciprocating compressor for chlorine liquefaction at 1 MPa with cooling water at 15 °C with liquid chlorine scrubbing
a) Low-temperature cooling and scrubbing column; b) Collection tank for impurities; c) Three-stage compressor; d) Intermediate cooler, stage 1; e) Intermediate cooler, stage 2; f) Liquefier; g) Chlorine collection vessel; h) Chlorine storage tank; i) Chlorine storage tank on load cells

two-stage compressors, 1.2 MPa. The heat of compression is removed by cooling the circulating liquid; cooling of the gas is not necessary. Advantages are simplicity of construction, strength, and reliability, but efficiency is low [194].

Reciprocating compressors were formerly lubricated with sulfuric acid, but are now available as *dry-ring compressors* (no lubrication). They can compress up to 200 t per day. Multistage compressors produce pressures up to 1.6 MPa. The heat of compression of each stage must be removed by heat exchangers or by injection of liquid chlorine (see Fig. 70). Well-purified chlorine gas is essential for trouble-free operation [195].

Turbo compressors are most economical when they operate with large amounts of chlorine. Each unit compresses up to 1800 t/d. In multiple-stage operation, pressures up to 1.6 MPa are reached. Labyrinth seals are used on the high-speed shafts. Requirements for cooling and gas purity are like those of reciprocating compressors. *Screw compressors* handle low rates of chlorine and give pressures up to 0.6 MPa. *Sundyne blowers* are one-stage high-speed centrifugal compressors handling 80–250 t per day and giving pressures up to 0.3 MPa. Liquid chlorine injection is used for cooling [196].

11.5. Liquefaction

The most suitable liquefaction conditions can be selected within wide limits. Important factors are the composition of the chlorine gas, the desired purity of the liquid chlorine, and the desired yield. There are nomograms that give the relationship between the chlorine concentrations of the incoming and residual gases, liquefaction yields, pressures, and temperatures [197]. Increasing the liquefaction pressure increases the energy cost of chlorine compression, although the necessary amount of cooling decreases, resulting in an overall reduction in energy requirement (Table 19) [198].

Any hydrogen is concentrated in the residual gas. To keep the hydrogen concentration below the 6% explosive limit, conversion of gas to liquid should be

Table 19. Electrical energy requirement for compression and liquefaction of 1 t of chlorine gas

	Liquefaction pressure, MPa			
	0.1	0.3	0.8	1.6
Energy for compression, kW h/t	5	23	42	57
Energy for cooling, kW h/t	87	68	27	3
Combined energy, kW h/t	*92*	*91*	*69*	*60*
Starting temperature, °C	−36	−8	25	53
Final temperature, °C	−42	−17	14	40

limited to 90–95 % in a single-stage installation. Higher yields may be obtained by condensing the chlorine from the residual gas in a second stage, which is constructed to reduce the risk from explosion [199]. This is achieved by the use of sufficiently strong equipment to withstand explosions or by the addition of enough inert gas to keep the mixture below the explosive limit. Multistage installations can liquefy over 99.8 % of the chlorine gas.

High-pressure (0.7–1.6 MPa) liquefaction with *water cooling* (Fig. 70) does not require a cooling plant. Therefore, it has the lowest energy cost of all methods; however, the high construction cost must be set against this.

Medium-pressure (0.2–0.6 MPa) liquefaction with *cooling* (−10 to −20 °C) is especially useful when only a part of the chlorine is to be liquefied and the remaining gas is to be reacted at the liquefaction pressure, e.g., with ethylene to form ethylene dichloride. The residual gas can be fed into the compressor suction systems, provided that the increased inert gas content does not interfere with the subsequent process. Otherwise, the residual gas must be scrubbed free of chlorine or liquefied in a second stage.

Figure 71 shows a two-stage liquefaction by the Uhde system, which operates at 0.3–0.4 MPa and −20 °C in the first stage and −60 °C in the second stage, with a yield of 99 % [200]. The refrigerant is difluoromonochloromethane. The gaseous refrigerant is compressed, liquefied by water cooling, and collected in a container. The liquid refrigerant is sprayed into the shell of the chlorine liquefier, where it evaporates, absorbing heat and cooling the chlorine, which flows from the liquefiers at −15 °C (first liquefier) or −55 °C (second liquefier) [201]. The residual gas from the first horizontal liquefier contains < 5 % hydrogen. It is fed into the second liquefier, which is at an angle of 60° and has a strong, low-volume construction. There the gas mixture passes through the explosive concentration limits. In case of an explosion, there is a comprehensive control system to ensure safety:

The explosion pressure is vented by means of a bursting disk to a residual gas absorber. Simultaneously the residual gas from the first stage is passed directly into this absorber. The chlorine gas to the second stage is shut off, and an inert gas purge is introduced. Finally, the liquid chlorine exit valve is closed to prevent back flow of the liquid chlorine into the second liquefier and from there into the absorber.

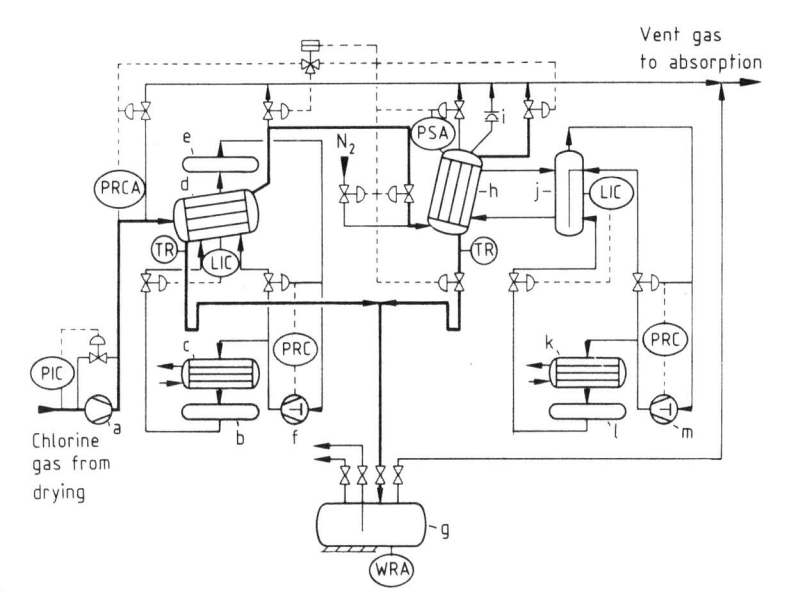

Figure 71. Flow diagram of a two-stage chlorine liquefaction plant at intermediate pressure — Uhde system
a) Chlorine gas compressor; b) Refrigerant collector, stage 1; c) Refrigerant condenser, stage 1; d) Chlorine liquefier, stage 1; e) Refrigerant separator, stage 1; f) Refrigerant compressor, stage 1; g) Liquid chlorine storage tank; h) Chlorine liquefier, stage 2; i) Bursting disk; j) Refrigerant separator, stage 2; k) Refrigerant condenser, stage 2; l) Refrigerant collector, stage 2; m) Refrigerant compressor, stage 2

With *normal-pressure* (ca. 0.1 MPa) liquefaction and *low temperature* ($< -40\,°C$), cryogenic storage of the liquid chlorine is possible. This process is advantageous when large quantities of chlorine must be liquefied as completely as possible. Attention must be paid to the increased solubility of other gases at low temperatures, especially carbon dioxide [188]. This carbon dioxide can be removed from the liquid chlorine by passage of hot chlorine gas [199].

An *absorption–desorption process* by Akzo is based on carbon tetrachloride [202]. It requires little energy and yields over 99.8 % of a pure liquid chlorine that is almost free of carbon dioxide. A similar process by Diamond Shamrock has been described [203].

11.6. Chlorine Recovery

Chlorine can be recovered from the tail gas from liquefaction with a chlorine recovery system.

Tail gas from liquefaction and chlorine from the plant evacuation system together with the snift compressor and stripper recycle streams are supplied to a snift compressor suction knock-out drum. The gas is compressed by the snift gas compressor to 7.0 kg/cm^2 with a discharge temperature of 85 °C.

The snift gas is then cooled by cooling water to 45 °C and then further cooled to −12.2 °C by Freon. Gas is sent to the absorber, whereas liquid is either returned to chlorine storage or is used for reflux at the stripper.

The off-gas enters the bottom of the chlorine absorber and passes upward through the two packed sections of the tower while cold carbon tetrachloride flows downward. All of the chlorine is absorbed in the carbon tetrachloride.

The chlorine-rich carbon tetrachloride leaves the bottom of the chlorine absorber at ca. 10 °C and is forced by pressure difference to the chlorine stripper. Chlorine stripper feed enters the middle of the column and flows downward through two packed sections, releasing chlorine as it is heated. A thermosiphon reboiler is provided at the base of the stripper.

Chlorine boiled off in the stripper passes upward through a packed top section of the column where it is scrubbed and purified by liquid chlorine from the discharge knock-out drum. The stripper overhead stream, a mixture of chlorine and a small amount of inerts, is sent to the chlorine liquefaction system or recycled to the suction knock-out drum to maintain the stripper reflux [202].

12. Chlorine Handling

Both the chlorine industry and governmental organizations are well aware of the risks of chlorine. In the United States and Canada, The Chlorine Institute [204] has established standards and recommendations for safe transport and handling of chlorine since 1924. In Europe, the Bureau International Technique du Chlor (BITC) [205], an association of major Western European chlorine manufacturers, publishes recommendations, codes, and memorandums for chlorine handling and transport concerning European conditions and regulations. Both organizations distribute manuals and pamphlets worldwide. Surveys of existing national and international regulations for the handling and transport of hazardous chemicals are available [84], [206], [207]–[210].

12.1. Storage Systems

Chlorine is liquified and stored at ambient or low temperature [208], [211], [212]. In both cases the pressure in the storage system corresponds to the vapor pressure of liquefied chlorine at the temperature in the stock tank. *Pressure storage* is recommended by the BITC [205] for all usual customers. The BITC recommends a maximum capacity of 300–400 t for individual tanks. For the large storage capacities required by producers, usually a *low-pressure storage* system, operating at a liquid chlorine temperature of ca. −34 °C, is chosen. A low-pressure system needs a cooling or recompression system, and, for this reason, it is basically unsuitable for small chlorine consumers.

A few major design aspects must be mentioned. Any risk of fire or explosion must be eliminated. All tanks having an external connection below the liquid level should be placed in a liquor-tight embankment (bund). In the event of leakage the liquid should be collected in a small area to reduce the rate of vaporization. The outer shell around a double-enveloped low-pressure storage tank can provide such a facility. To vent chlorine, there must be an absorption or liquefaction system. In the course of all operations, the design pressure should not be exceeded. The dimensions of branches and the amount of pipe work should be minimized. Bottom connections from storage tanks are not recommended for small chlorine users. Large branches should always be located in the gas space of a vessel. The pipe work system should be provided with remotely operable valves to permit isolation in case of emergency.

Before being put into service, the whole storage system must be degreased, cleaned, and dried to achieve a dew point of −40 °C in the purge gas at the outlet of the system. No substance that could react with the chlorine can be allowed to enter the storage system. The filling ratio in the tank should never exceed 95 % of the total volume of the vessel; for pressure storage tanks, this corresponds to 1.25 kg of liquid chlorine per liter of vessel capacity at 50 °C (Fig. 72).

Typical measuring and control equipment of a pressure storage tank is shown in Figure 73. The ISO codes for process measurement control functions and instrumentation are explained in Table 20 [213]. The measuring equipment of a low-pressure storage system needs supplementary devices, for example, a temperature indicator with an alarm and, in the case of a double-shell vessel, a device to determine the quality of the purging gas inside the double shell. The vessel and an external envelope should be protected against overpressure or underpressure. In low-pressure systems, the chlorine is removed by vertical submerged pumps, canned pumps below the vessel, or ejector pumps operating with a flow of liquid chlorine produced by external pumps.

Periodic inspection and retesting of the whole system, including a visual examination, a thickness test of the wall of the vessel and pipes, and an examination of the welds and the surfaces under any thermal insulation, is recommended. Hydraulic retesting is accompanied by risk of corrosion and is, therefore, not favored.

12.2. Transport

Within a chemical plant and over distances of several kilometers, chlorine can be transported by pipelines, either as gas or liquid [204], [205], [214]. Every precaution should be taken to avoid any vaporization of chlorine in a liquid-phase system or any condensation in a gas-phase system. Wherever liquid chlorine could be trapped between two closed valves or wherever the system could be overpressurized by thermal expansion, an expansion chamber, a relief valve, or a rupture disk should be provided.

Commercial chlorine is transported as a liquid, either in small containers (cylinders and drums) or in bulk (road and rail tankers, barges, and ISO containers). The design,

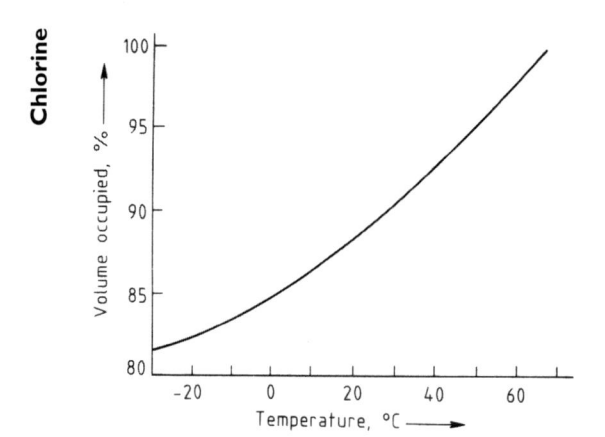

Figure 72. Proportion (%) of a one-liter vessel occupied by 1.25 kg of liquid chlorine as a function of temperature

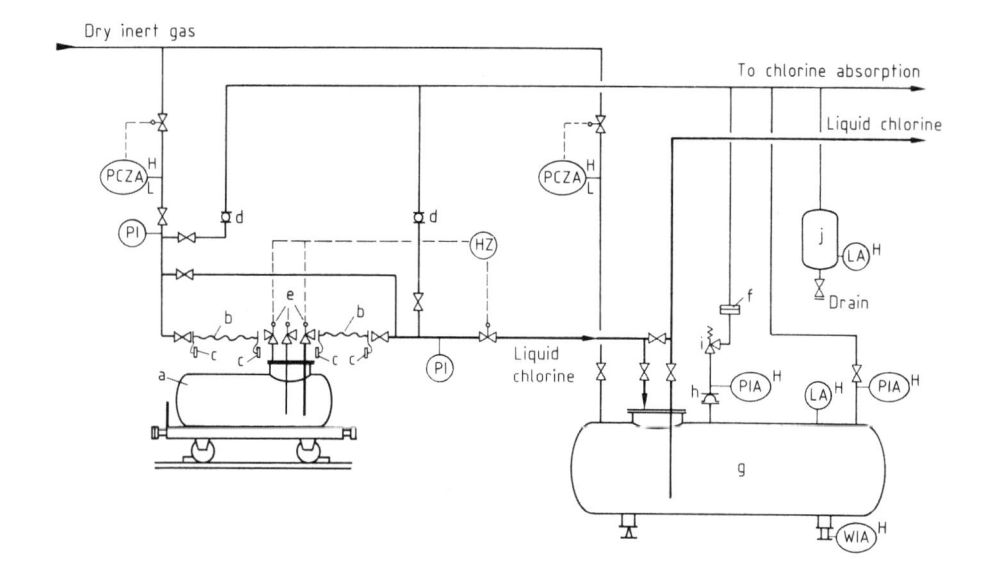

Figure 73. Discharge of liquid chlorine by padding to pressure storage
a) Liquid-chlorine rail tanker; b) Flexible connection; c) Plug; d) Viewing glass; e) Remote-control tank valves;
f) Protective membrane; g) Storage vessel; h) Rupture disk; i) Relief safety valve; j) Buffer vessel for liquid chlorine

construction, system of labeling, inspection, and commissioning are covered by national and international regulations [204], [205], [207]. *Cylinders* have a chlorine content up to 70 kg. A protective hood is provided to cover the valve during transport. The ton containers (drums) have a capacity of 500 – 3000 kg of chlorine. Drums are equipped with two valves near the center of one end and connected with internal eductor pipes.

Table 20. ISO codes and miscellaneous symbols for process measurement control functions and instrumentation

Codes	Function or Instrumentation
AA	analysis alarm
CW	cooling water
dPI	difference pressure indicating
FA	flowrate alarm
FI	flowrate indicating
FIA	flowrate indicating alarm
FICA	flowrate indicating controlling alarm
FRA	flowrate recording alarm
HZ	hand operated emergency acting
H	high
L	low
LA	level alarm
LIA	level indicating alarm
LIC	level indicating controlling
M	moisture analysis
PA	pressure alarm
PCZA	pressure controlling emergency acting alarm
PI	pressure indicating
PIA	pressure indicating alarm
PIAS	pressure indicating alarm switching
PIC	pressure indicating controlling
PRC	pressure recording controlling
PSA	pressure switching alarm
QRA	quality recording alarm
TA	temperature alarm
TC	temperature controlling
TI	temperature indicating
TIA	temperature indicating alarm
TIC	temperature indicating controlling
TRA	temperature recording alarm
WI	weight indicating
WIA	weight indicating alarm
◯	measuring device
⊗	remote control value
- - -	control line

The capacity of *tank cars* (rail tankers) ranges from 15 to 90 t. Special angle valves are mounted on the manhole cover on top of the vessel. In Europe, pneumatic valves are normally used [205], [211]. During loading and unloading, these valves can be closed rapidly and remotely in case of an accident. They have an internal safety plug, providing a tight seal against the passage of gas or liquid chlorine in the event of failure of the body of the valve.

In North America, the eductor pipe inside the vessel has an excess-flow valve at the top, immediately below the manhole cover. This valve closes the eductor pipe when the rate of liquid flow exceeds a set rate [84], [204]. North American tank cars have a spring-loaded safety relief valve, which protects the vessel against overpressure in case of external heat. The tanks have thermal insulation. In Europe thermal insulation and safety relief valves are not used or recommended [205].

Road tankers and ISO containers have a chlorine capacity of 15 – 20 t. The design of and the equipment on chlorine pressure road tankers is similar to these of rail tankers. In North America, large amounts of chlorine are transported by *tank barges* [204]. These barges usually are of the open-hopper type with several cylindrical uninsulated pressure vessels. The total capacity of barges ranges from 600 to 1200 t. In Europe, special ships with a capacity of up to 1000 t are in service for maritime transport of chlorine. The chlorine is transported at low temperature.

12.3. Chlorine Discharge Systems

All containers should be discharged in the same order as received. They must be placed where any external corrosion, risk of fire, explosion, or damage is avoided [204], [205]. At normal room temperature, the discharge rate of chlorine gas from a single 70-kg cylinder is ca. 5 kg/h and the rate of a drum is ca. 50 kg/h. The flow of chlorine gas can be increased by a higher ambient temperature or by connecting two or more containers. A system of two or more containers must be carefully operated and controlled to avoid overfilling by transfer of chlorine from warm to cool containers. Direct heating of containers is not recommended [204], [205]. The best way to determine the flow rate and container content is to observe the weight of the container [208]. A flexible tube is used to connect a mobile container with the fixed piping system. Any reverse suction from the consuming plant must be prevented by a barometric leg or other adequate precaution if the chlorination process runs at atmospheric pressure. Pressurized processes need a pressure controlling system with automatic isolation valves.

Uninsulated tanks have a maximum gaseous discharge rate of ca. 2 t/h. The chlorine gas can be used only for low-pressure chlorination processes and at low rates. This method increases the risk of concentrating nitrogen trichloride and other nonvolatile residues in the liquid phase within the tank [205]. In all other circumstances, the liquid chlorine should be transferred into a fixed storage vessel and then vaporized in a special installation.

Liquid chlorine is discharged by putting the tank under pressure with dry inert gas or dry chlorine gas [208]. The inert transfer gas must have a dew point below – 40 °C at atmospheric pressure and must be clean and free of impurities such as dust or oil. Before closing the valves, the tanks must be vented to avoid the risk of high pressure in the container on account of the additional partial pressure of the inert gas. The use of an inert gas requires the availability of a chlorine absorption or neutralization system. Discharge with pressurized chlorine gas requires a chlorine vaporizer or a special chlorine compressor. Articulated arms, flexible hoses, and steel coils are used for the flexible connections. Remote-control valves installed close to the ends of the flexible connections limit leakage in the event of a failure.

12.4. Chlorine Vaporization

When large amounts of chlorine gas are required or when the chlorination process needs pressurized gas, liquid chlorine must be vaporized and superheated to avoid liquefaction. It is advisable to operate the vaporizer at a sufficiently high temperature to accelerate the decomposition of nitrogen trichloride. As a source of heat, steam with a maximum allowed temperature of 120 °C is used when the vaporizing system is constructed of mild steel. Water above 60 °C is also suitable, as shown in Figure 74. Direct electrical heating is not appropriate because there is always a risk of overheating the steel.

Coil-in-bath vaporizers use a coiled tube or a spiral located in a vessel of hot water (Fig. 74). Generally, they are used for small throughputs; they are simple in design and construction. Double-envelope vaporizers have compact construction and are easy to operate and to maintain. Vertical tube vaporizers have a large surface area and allow a high flow rate. Kettle vaporizers are also constructed for large unit capacities. Every effort must be taken to avoid the reverse suction of water or organic materials into the vaporizer. The recommended water and nitrogen trichloride content [205] of introduced liquid chlorine must not be exceeded. Vaporizers operating at low temperature or with a constant liquid level need to be purged to avoid dangerous concentration of nitrogen trichloride.

12.5. Treatment of Gaseous Effluents

Gaseous effluents containing chlorine arise from various sources and must be treated in such a way to obtain a tolerable concentration of chlorine when they are released into the air. The vent gas may contain other substances, such as hydrogen, organic compounds, CO_2, etc., which must be considered in design and operation of an effluent treatment installation.

Operation of the collection system below atmospheric pressure facilitates the purging of chlorine vessels, pipes, etc. The risk of corrosion in dry chlorine installations by moisture from the treatment system must be excluded. The most commonly used and recommended reagent is caustic soda. The effluents are treated in an absorption system, such as packed absorption towers, venturi scrubbers, etc. An example of a flow sheet for a large plant is shown in Figure 75.

To avoid any formation of solid salts, the recommended concentration of caustic soda is < 22 wt %. The operating temperature should not exceed 55 °C; under normal conditions a temperature of ca. 45 °C is usual [205]. A cooling system may be necessary. In large chlorine absorption units, the sodium hypochlorite solution that is produced can be used in other processes. Where this is not possible, several methods can be used to decompose the hypochlorite: controlled thermal decomposition, catalytic decomposition, acidification, for example, with sulfuric acid

$$NaCl + NaOCl + H_2SO_4 \longrightarrow Na_2SO_4 + Cl_2 + H_2O$$

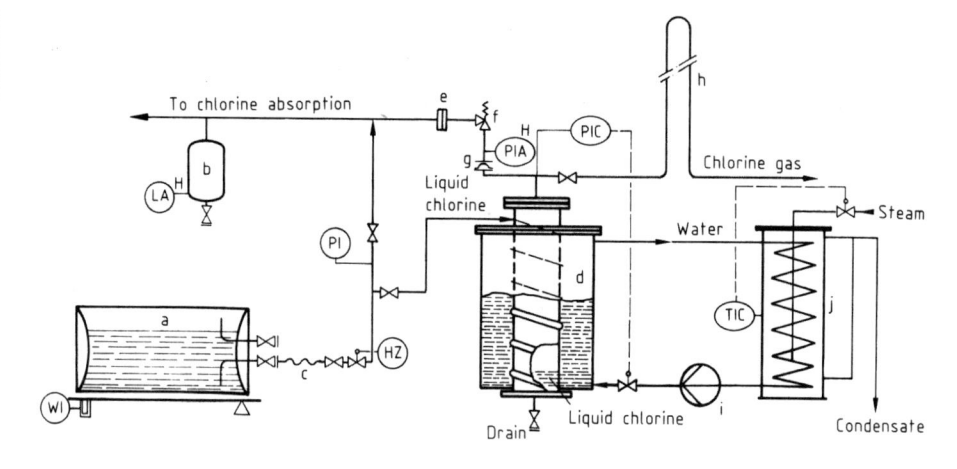

Figure 74. Liquid chlorine vaporizer
a) Liquid-chlorine drum; b) Buffer vessel; c) Flexible coil; d) Chlorine vaporizer; e) Protective membrane; f) Relief safety valve; g) Rupture disk; h) Barometric leg; i) Water pump; j) Water heater

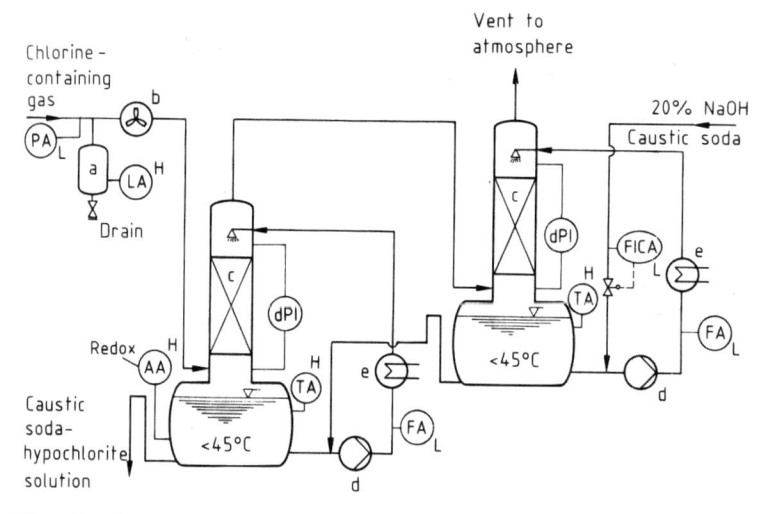

Figure 75. Absorption equipment for the treatment of gases containing chlorine
a) Buffer vessel; b) Vent fan; c) Packed tower; d) Circulating pump; e) Heat exchanger (cooler)

12.6. Materials

The choice of material [204], [205] depends on the design and operating conditions and must take into account all circumstances. A chlorine manufacturer should be consulted to confirm the suitability of a material. Any use of silicone materials in chlorine equipment should be avoided [205].

Dry Chlorine Gas (water, < 40 ppm by weight). Carbon steel is the material most used for dry chlorine gas. It is protected by a thin layer of ferric chloride. For practical purposes the recommended temperature of these materials is $\leq$ 120 °C. High-surface areas, such as steel wool, or the presence of rust and organic substances increase the risk of ignition of steel.

The resistance of stainless steels to chlorine at high temperature increases with the content of nickel. For stainless steels containing less than 10 wt% nickel, the upper temperature limit is 150 °C. High-nickel alloys, such as Monel, Inconel, or Hasteloy C, are suitable up to 350–500 °C. Poor mechanical strength limits the use of nickel. Copper is used for flexible connections and coils, but it becomes brittle when stressed frequently.

Because titanium ignites spontaneously in dry chlorine, it must be avoided. Graphite, glass, and glazed porcelain are used where there is a risk of moisture in the dry chlorine gas, and poly(vinyl chloride) (PVC) or chlorinated PVC and polyester resins are suitable if the temperature limits of these materials are regarded.

Liquid Chlorine. Nonalloyed carbon steel and cast steel are used with liquid chlorine. Low-temperature chlorine systems apply fine-grain steels with a limited tensile strength to guarantee good conditions for welding. To avoid erosion of the protective layer, practice is to limit the velocity of the liquid to less than 2 m/s. Organic materials — rubber lining, ebonites, polyethylene, polypropylene, PVC, chlorinated PVC, polyester resins, and silicone — are dangerous [215]. Zinc, tin, aluminum, and titanium are not acceptable. For certain equipment, copper, silver, lead, and tantalum are appropriate.

Wet Chlorine Gas. Wet chlorine gas rapidly attacks most common metallic materials with the exception of tantalum and titanium. To assure a protective oxide layer on the surface of the titanium, sufficient water must be present in the chlorine gas. If the system does not remain sufficiently wet, titanium ignites spontaneously [205].

Most organic materials are slowly attacked by wet chlorine gas. Rubber-lined iron is successfully used up to 100 °C. At low pressure and temperature the use of plastic materials like PVC, chlorinated PVC, and reinforced polyester resins is advantageous. Polytetrafluoroethylene (PTFE), poly(vinylidene fluoride) (PVDF), and fluorinated copolymers like tetrafluoroethylene–hexafluoropropylene (FEP) are resistant even at higher temperature. Ceramics have been progressively replaced by plastics. Impregnated graphite is suitable up to 80 °C; the impregnation should be resistant to wet chlorine.

Materials for Special Parts. Rubberized, compressed asbestos is used for gaskets in dry gas and liquid chlorine systems. In wet chlorine gas, rubber or synthetic elastomers are acceptable. Even at temperatures up to 200 °C, PTFE is resistant against wet and dry chlorine gas and liquid chlorine.

Materials resistant because of protection by a chloride surface layer are not recommended for protective membranes, rupture disks, and bellows. Suitable materials are tantalum, Hasteloy C, PTFE, PVDF, Monel, and nickel.

12.7. Safety

In hazard and risk assessment studies, the design of chlorine installations and equipment and the operating and maintenance concepts are examined in detail to minimize risks [216]. However, there remains a certain risk, and all efforts must be taken to protect people and the environment in the case of a chlorine emergency. The penetrating odor and the yellow-green color of a cloud indicate chlorine in the air. If around-the-clock surveillance by operators is not possible, automatic leak detectors are available. Safety in handling chlorine depends largely on the education and training of employees. An emergency plan should be brought to the attention of the personnel involved. Computer-assisted systems can be used in certain circumstances [217]. Periodic exercises and safety drills should be carried out.

All people on a chlorine plant are advised to carry escape-type respirators. The use of filter masks is prohibited where there is a risk of a high concentration of chlorine. Anyone who enters an area with high chlorine concentrations should be equipped with self-contained breathing apparatus and full protective clothing suitable for dealing with liquid chlorine. Protective equipment, safety showers, eye-wash facilities, and emergency kits [204], [205] must be quickly accessible.

A means of indicating the actual wind direction should be located near the chlorine installation.

Fixed or mobile water curtains can be used to divert the dispersion of a chlorine gas cloud [218]. However, the direct discharge of water into liquid chlorine and on the area of a chlorine leak must be avoided.

In most countries, chlorine manufacturers have organized groups of experts who are well versed and drilled in handling chlorine and can be called at any time in case of chlorine emergency [204], [205].

13. Quality Specifications and Analytical Methods

13.1. Quality Specifications

Liquid chlorine of commercial quality must have a purity of at least 99.5 wt % [219]. The water content is < 0.005 wt %, and solid residues are < 0.02 wt %. The impurities are mainly CO_2 ($\leq$ 0.5 wt %), N_2, and O_2 (each 0.1–0.2 wt %). There are traces of chlorinated hydrocarbons (originating from rubberized or plastic piping) and inorganic salts such as ferric chloride. The chlorine may also contain small amounts of bromine or iodine, depending on the purity of the salt used in the electrolytic process.

13.2. Analytical Methods

Industrial liquid chlorine is mainly analyzed by the methods in ISO regulations. The liquid chlorine is evaporated at 20 °C, and this gas is then analyzed.

Sampling	ISO 1552 [220]
Moisture	ISO 2121 [221]
	ISO 2202 [222]
Chlorine content	ISO 2120 [223]
Gaseous components	DIN 38 408, part 4 [224]
NCl_3	[225]
Mercury	[226]

The residue is weighed, and the organic constituents are taken up in acetone, hexane, or diethyl ether and determined by gas chromatography. The inorganic residue is analyzed. For quick analysis, liquid chlorine can be introduced directly onto a silica gel column of a gas chromatograph.

Chlorine Gas. The chlorine gas can be analyzed for chlorine content, gaseous impurities, hydrogen, organics, and moisture:

1) Chlorine content. One method for process monitoring and control of chlorine concentration is measurement of thermal conductivity.
2) Gaseous impurities. A known amount of chlorine gas is passed through a solution of potassium iodide or phenol to absorb the chlorine. The residual gases (O_2, N_2, H_2, CO, CO_2) are collected in a gas burette, measured, and analyzed by gas chromatography or with an Orsat apparatus.
3) Hydrogen. A known amount of air is added to the residual gas after removal of the chlorine to ensure excess oxygen, and the volume reduction is measured after the hydrogen is consumed on a heated platinum coil. The hydrogen content can be continuously monitored by thermal conductivity measurement.
4) Organics. Organic components can be determined most conveniently by gas chromatography.

5) Moisture. A known amount of chlorine gas is passed through a drying tube filled with a weighed amount of phosphorus pentoxide. The moisture content is determined from the weight gain of the drying tube (ISO 2121). Continuous determination can be carried out, e.g., by absorption with phosphorus pentoxide and measuring the current required to electrolyze the absorbed water or by the electrical conductivity after absorption in sulfuric acid.

Detection of Chlorine. Chlorine can be recognized by smell or color. Small amounts can be detected by the blue coloration of starch–iodide paper, although other oxidizing agents can produce the same effect. Another method for chlorine detection depends on its ability to combine with mercury. If the unknown gas mixture is shaken with water and mercury, all the chlorine disappears and the remaining water has a neutral reaction. However, if the chlorine contains some hydrogen chloride, the water becomes acidic and reacts with silver nitrate solution to give a white precipitate (AgCl) that is soluble in aqueous ammonia. Leaks in pipes or equipment are detected by testing with the vapor from aqueous ammonia: a thick white cloud of chloride forms.

Quantitative Determination of Free Chlorine. The gas mixture can be shaken with a *potassium iodide solution,* and the liberated iodine can be then determined by titration. Chlorine in alkaline solution can be reduced to chloride by potassium or sodium *arsenite,* and the arsenite can be then oxidized to arsenate. The end point is detected by spot tests with starch–iodide paper. Excess arsenite is back-titrated with acidified potassium bromate solution. Small amounts of chlorine, e.g. in drinking water, can be determined by *photometric measurement* of the yellow color produced by the reaction with *o*-tolidine in hydrochloric acid solution [227].

To determine both *chlorine and carbon dioxide,* the chlorine is absorbed by a solution that contains known amounts of acid and potassium arsenite, and the chlorine is determined by back-titration of the arsenite. The carbon dioxide, which is not absorbed by this solution, is then absorbed by potassium hydroxide solution.

Detection Tubes. Commercial detection tubes (Drägerwerk, Lübeck; Auergesellschaft, Berlin) are available for measuring chlorine in air. They have various ranges: 0.2–3, 2–30, and 50–500 ppm. The Chlorometer (Zeiss-Ikon, Berlin) can determine the free chlorine content of water in a few minutes.

For the protection of the environment and control of working conditions, traces of chlorine as small as 0.01–10 ppm must be determined. There are many types of apparatus on the market for measuring workplace concentrations or emissions. They depend on physicochemical methods, such as conductometry, galvanometry, potentiometry, colorimetry, and UV spectroscopy [228].

14. Uses

The first industrial use of chlorine was to produce bleaching agents for textiles and paper and for cleaning and disinfecting. These were liquid bleaches (solutions of sodium, potassium, or calcium hypochlorite) or bleaching powder (chlorinated lime). Chlorine was then regarded merely as a useful chemical agent.

Since 1950, chlorine has achieved constantly increasing importance as a raw material for synthetic organic chemistry. Chlorine is an essential component of a multitude of end products, which are used as materials of construction, solvents, insecticides, etc. In addition, it is contained in intermediates that are used to make chlorine-free end products. It is these areas of use that allow chlorine production to increase (Table 21). The direct use of elemental chlorine, e.g., for sterilizing water, has declined in some areas, but not in others. For example, in the Federal Republic of Germany, it is < 0.1%, but in the United States it is 6% (1982).

The number of possible reactions of chlorine and, therefore, the number of intermediates and end products, are remarkably large. Some important reactions are shown in Figure 76 along with the areas of application of the end products [229].

Inorganic Products. The use of chlorine to produce pure hydrochloric acid, bleaching agents, and other inorganic products is increasing, although slowly enough that this forms a steadily decreasing proportion of the total consumption. The production of some metallic chlorides is increasing, e.g., titanium tetrachloride, aluminum chloride, and silicon tetrachloride.

Bleaching. Countries with large paper and cellulose industries have a large chlorine consumption for bleaching purposes. In the United States, 13–16% of total production was used for this purpose in 1982/1983; in Canada and Scandinavia, this usage was 60–70%. However, in Western Europe and Japan, the comparable figures are 3–4% and 9%, respectively.

Organic Products. Organic intermediates and end products accounted for the following percentages of total chlorine consumption in 1982/1983:

United States	55–65% [230]
Japan	50–60% [231]
Western Europe	>80% [232]

The single product with the largest chlorine requirement is *vinyl chloride,* accounting for ca. 20% of world chlorine consumption.

Other *chlorinated aliphatic hydrocarbons* are important. Chlorine reacts with methane to give the solvents methyl chloride, methylene chloride, chloroform, and carbon tetrachloride. Butane and pentane are the starting point for the transformer oils hexachlorobutadiene and hexachlorocyclopentadiene, respectively. The latter is also the basis of many pesticides. Chlorinated higher paraffins are the starting materials for detergent raw materials, softening agents, and lubrication additives.

The chlorination of ethylen first gives 1,2-dichloroethane, which is an important starting material for an extraordinary variety of further products, including vinyl chloride, ethylene oxide, ethylene glycol, ethylenediamine, tetraethyl lead, and chloral. Other products of the chlorination of ethylene are trichloroethylene, perchloroethylene, etc.

Table 21. Chlorine consumption, %, of various product groups in 1982 for the United States, Japan, and Federal Republic of Germany

Product group	United States	Japan	Federal Republic of Germany
Vinyl chloride	22	28	23
Solvents	17	16	19
Misc. organic products	27	39	47
Water treatment	6		0
Inorganic products	12	9	8

The following products are obtained from propylene: propylene oxide, propylene glycol, carbon tetrachloride (via propylene dichloride), glycerine (via allyl chloride), and epoxy resins and synthetic rubber (via epichlorhydrin).

Aromatic derivatives are also important. The reaction of chlorine with benzene produces the following compounds, among others: monochlorobenzene, dichlorobenzene, and hexachlorobenzene, which are used as solvents or pesticides. A more important use of monochlorobenzene, however, is as an intermediate in the manufacture of phenol, the insecticide DDT, aniline, and dyes. A further important synthesis route is to start from toluene to produce toluenediamine and to convert this to toluene diisocyanate, which is used in plastics.

15. Economic Aspects

Figure 77 shows how chlorine production has developed worldwide from 1966 to 1983 in the United States, Western Europe, and Japan [233]. The world production figures are only estimates, as many important industrial nations do not publish this information. If the world outputs are taken to be ca. 30×10^6 t of chlorine and 33.9×10^6 t of sodium hydroxide per year and the market prices are taken to be the 1984 U.S. figures [234], the total annual market value is ca. $\$ 15 \times 10^9$. Some of the larger national chlorine production figures are given in Table 22.

Since the beginning of this century, apart from the war years, chlorine production has increased more quickly than that of the other inorganic bulk chemicals. In 1975, as a consequence of the oil crisis, the first slump in output occurred, from which the industry only emerged in 1976. Production has stagnated since 1980 in most industrialized countries. The total annual world output is ca. 75% of world capacity (ca. 40 Mt). Figures are available [235] for world capacities in 1983, categorized according to countries and known producers [236]. The capacities of the various economic regions are shown in Figure 78.

An analysis of consumption in the three countries with the largest production shows that poly(vinyl chloride) (PVC) consumes the largest single share and that organic

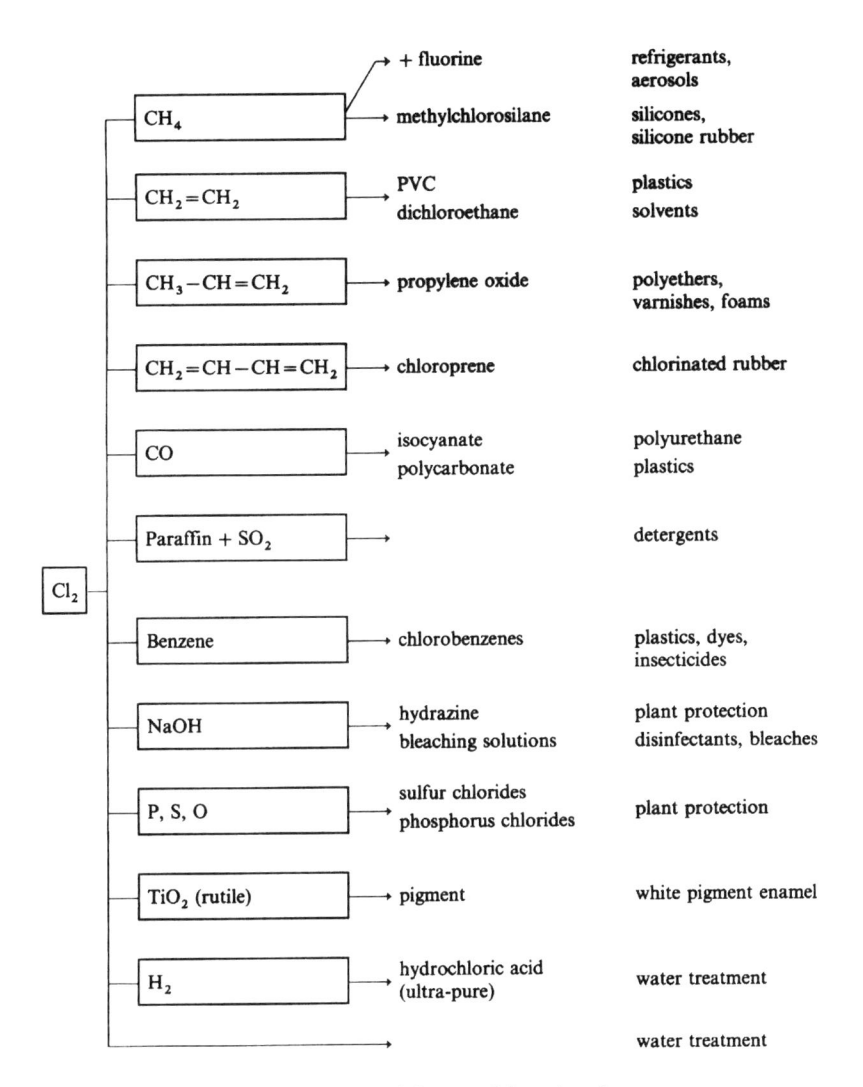

```
                    ┌─→ + fluorine              refrigerants, aerosols
        ┌─ CH₄ ─────┤
        │           └─→ methylchlorosilane      silicones, silicone rubber
        │
        │              PVC                      plastics
        ├─ CH₂=CH₂ ──→ dichloroethane           solvents
        │
        ├─ CH₃–CH=CH₂ ──→ propylene oxide       polyethers, varnishes, foams
        │
        ├─ CH₂=CH–CH=CH₂ ──→ chloroprene        chlorinated rubber
        │
        │              isocyanate               polyurethane
        ├─ CO ───────→ polycarbonate            plastics
        │
        ├─ Paraffin + SO₂ ──→                   detergents
 Cl₂ ───┤
        ├─ Benzene ──→ chlorobenzenes           plastics, dyes, insecticides
        │
        │              hydrazine                plant protection
        ├─ NaOH ─────→ bleaching solutions      disinfectants, bleaches
        │
        │              sulfur chlorides
        ├─ P, S, O ──→ phosphorus chlorides     plant protection
        │
        ├─ TiO₂ (rutile) ──→ pigment            white pigment enamel
        │
        │              hydrochloric acid
        ├─ H₂ ───────→ (ultra-pure)             water treatment
        │
        └─────────────→                         water treatment
```

Figure 76. Important reactions of chlorine and the uses of the end products

products generally are quite important (Table 21). For example, these products, including PVC, account for almost 90 % of the chlorine production of the Federal Republic of Germany.

The Future. The cost of electricity accounts for 40 – 60 % of the total production cost in the electrolysis of sodium chloride. Large regional differences in energy costs account for the considerable regional variations in the location of chlorine producing units. The demand for the coproduct sodium hydroxide affects the amount and regional distribution of chlorine production [237].

Table 22. Chlorine production, 1000 t

Country	Year		
	1978	1980	1982
United States	10 026	10 160	8 272
Federal Republic of Germany	3 011	2 997	2 842
Japan	2 498	2 808	2 471
USSR	2 300	2 663	2 463
Peoples Republic of China*	1 600*	1 700	1 834
France	1 263	1 257	1 254
Canada	941	1 303	1 233
United Kingdom	960	909	878
Italy	883	826	722
World	*28 000*	*31 300*	*32 000*

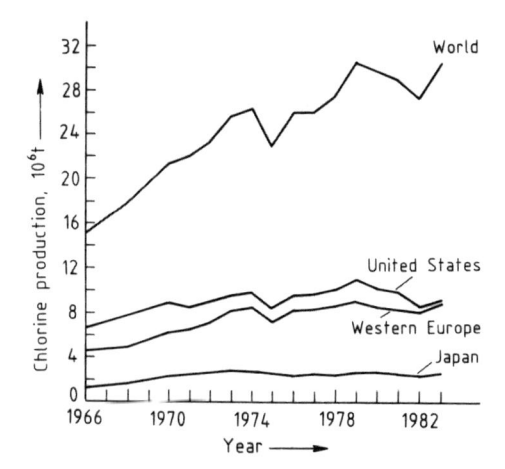

* Estimate.

Figure 77. Chlorine production: Japan, Western Europe, United States, and the World (est.)

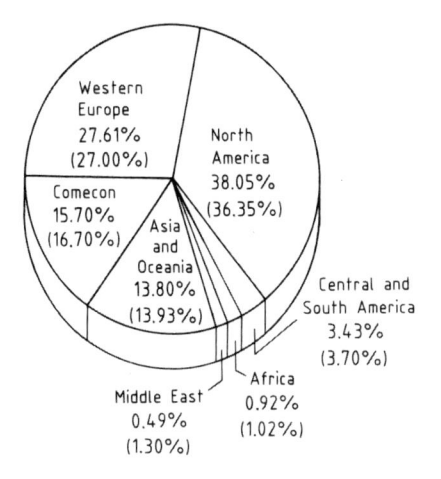

Figure 78. Chlorine production capacity by economic regions in 1983
In parentheses: forecast distribution after completion of known projects.

Legal decisions, e.g., the ban of an insecticide such as DDT or restrictions in the use of such products as fluorinated hydrocarbons, can limit chlorine consumption. The sharp increases in the costs of petroleum derivatives until 1985 caused by the oil crises have meant that such construction materials as PVC have become candidates for replacement by other materials.

Environmental considerations have led to further reductions in chlorine use, e.g., the partial replacement of chlorine by hydrogen peroxide for bleaching.

In the United States, chlorine production is expected to remain fairly static until the 1990s, and the utilization of capacity is expected to fluctuate near 80% [82, pp. 317–365]. In Japan, production is likewise not increasing: it is more economical to buy from time to time intermediates such as dichloroethane on the world market [231]. The quick changeover to the membrane process may bring back an increasingly competitive situation, despite high energy costs. In Europe, a slight increase in chlorine production, up to 5%, is anticipated.

Known projects for new plants should produce 5–20% growth in countries where industry is on the increase, e.g., Brazil, Mexico, India, Taiwan, and Egypt, and a 6–10% increase is expected for the Comecon countries. The greatest increase in chlorine capacity is expected in the oil-producing countries: Saudi Arabia, Iraq, and Libya. They are forging ahead with the production of increasing quantities of bulk chemicals such as dichloroethylene and PVC. The amounts produced considerably exceed their own requirements; thus, these products are predominantly for export.

16. Toxicology

Chlorine gas is dangerous to health because it is a powerful oxidant. In the physiological pH range, it is converted to hypochlorous acid, a cytotoxic substance. The extent to which the cells are damaged depends on the gas concentration, the exposure time, the water content of the tissue, and the health of the person exposed to the gas. Besides getting into the eyes, larynx, and trachea, chlorine also reaches the bronchi and the bronchioles. Because of its moderate solubility in water, chlorine affects the alveoli only at high concentrations. In this respect, chlorine differs from other gases with low water solubility and high lipid solubility, such as phosgene, nitrogen monoxide, and nitrogen dioxide. Initial moderate bronchial irritation is followed by the development of a toxic pulmonary edema because of increased alveolar injury.

The olfactory threshold of chlorine gas is $0.2-3.5$ mL/m^3. Prolonged exposure seems to raise the olfactory threshold. Concentrations of $3-5$ mL/m^3 are tolerated for up to 30 min without any subjective feeling of malaise. At concentrations between 5 and 8 mL/m^3, mild irritation of the upper respiratory tract and the conjunctiva is observed. In addition, running of the eyes and coughing are observed at concentrations of 15 mL/m^3 and higher. Above 30 mL/m^3, the following symptoms are observed: nausea, vomiting, oppressive feeling, shortness of breath, and fits of coughing, sometimes leading to bronchial spasms. Exposure to 40–60 mL/m^3 leads to the development of toxic tracheobronchitis. After a latent period of several hours with fewer symptoms, pulmonary edema may occur because of alveolar membrane destruction. This is indicated by increased shortness of breath, restlessness, and cyanosis. Subsequently, a further complication can occur after several days in the form of pneumonia caused by superinfection of the injured pulmonary tissue.

A clear dose–effect relationship of chlorine gas at different concentrations in humans has not yet been published. On the basis of results obtained from animal experiments, the LC$_{50}$ for healthy humans is assumed to be 300–400 mL/m^3 at a 30-min exposure [238]. No deaths have occurred in animal experiments at 30-min exposures for concentrations below 50 mL/m^3. Death following acute intoxication is caused by a fulminant pulmonary edema.

Many investigations deal with the toxicity of *low chlorine concentrations*. Recent investigations indicate the possibility of reversible damage to the lung function parameters at concentrations starting at 0.5 mL/m^3 [239], [240]. Long-term investigations of workers exposed to chlorine, e.g., in chlor-alkali electrolysis plants or in pulp manufacture, however, do not indicate increased rates of mortality or morbidity caused by pulmonary diseases [241]–[244].

No indications of *carcinogenicity* or *mutagenicity* of chlorine have been detected in animal experiments or encountered in industrial medicine. The MAK value is 0.5 mL/m^3 (1.5 mg/m^3); the TLV is 1.0 mL/m^3 (3.0 mg/m^3) with an STEL of 3.0 mL/m^3 (9 mg/m^3).

17. References

[1] C. W. Scheele, *Sv. Akad. Handl.* **35** (1774) 88.
[2] J. Eidem, L. Lunevall in [245]
[3] R. E. Means, T. R. Beck, *Chem. Eng. (N.Y.)* **91** (1984) no. 22, 46–51.
[4] The Chlorine Institute: *Properties of Chlorine in SI Units*, New York 1981.
[5] J. J. Martin, D. M. Longpre, *J. Chem. Eng. Data* **29** (1984) 466–473.
[6] L. Ziegler, *Chem. Ing. Techn.* **22** (1950) 229.
[7] E. M. Aleta, P. V. Roberts, *J. Chem. Eng. Data* **31** (1986) 51–54.
[8] I. A. A. Ketalaar, *Electrochem. Technol.* **5** (1967) 143.
[9] R. F. Schwab, N. H. Doyle, *Electrochem. Technol.* **5** (1967) 228. Y. E. Frolov, A. S. Maltsera, A. N. Baratov, A. I. Roslovskii, *Khim. Promst (Moscow)* **7** (1977) 530 –532.
[10] *Kirk-Othmer*, vol. **1**, 1978, pp. 799–883.
[11] Hoechst, EP 35 695-B1, 1983.
[12] Solvay & Cie., EP 30 756-A1, 1981.
[13] A. I. Postoronko et al., SU 537 027-A, 1976. Jamestown Chem.: *Calcium Sulphate Inhibitor* .
[14] S. L. Fuks, T. M. Orchinnikova, *Zh. Prikl. Khim (Leningrad)* **56** (1983) no. 8, 1757–1761.
[15] *Winnacker–Küchler*, 3rd ed., vol. **1**, pp. 228–239; *Winnacker–Küchler*, 4th ed., vol. **2**, pp. 379–480.
[16] Ciba, DE 1 209 562-B, 1966 (A. Georg).

[17] K. Harada, Cleveland Meeting: *The Electrochem. Soc.*, May 1966.
[18] Olin Mathieson, US 2 787 591-A, 1957 (W. C. Gardiner, J. L. Wood).
[19] Olin, US 4 277 447-A, 1981 (R. F. Chambers, N. Rachima).
[20] Toa Gosei Chem. Ind., JP 57-106 520-A2, 1982 (K. Suzuki, R. Nagai, H. Goto).
[21] Olin, US 4 303 624-A, 1981 (R. J. Dotson, W. Lynch).
[22] Hoechst, DE 3 037 818-A1, 1982 (S. Benninger).
[23] Bayer, DE 3 216 418-A1, 1983 (R. Schäfer).
[24] Bayer, DE 2 709 728-C2, 1981 (R. Schäfer).
[25] Hoechst, DE 2 905 125-A1, 1980 (M. Schott, J. Russow).
[26] Z. I. Lifatova et al., SU 865 800-A1, 1981.
[27] Olin, US 4 405 465-A, 1983 (S. H. Moore, R. L. Dotson).
[28] Bayer, DE 1 177 122-B, 1964 (E. Zirngiebl).
[29] Ebara Infilco, JP 54-81 174-A2, 1979 (M. Kanekawa, K. Akagi).
[30] Dynamit Nobel, DE 1 467 219-C3, 1973 (K. Hass, R. Cordes).
[31] *Processing*, June 1978.
[32] W. Glas, *Chem. Ing. Tech.* **43** (1973) 832.
[33] F. Lüns, *BBC-Nachrichten*, 1/2 (1974) 36–42.
[34] H. A. Horst, *Chem. Ing. Tech.* **43** (1971) 164.
[35] Siemens, DE 3 123 877-C1, 1983 (G. Achilles). Hundt & Weber Schaltgeräte, EP 11 820-A1, 1980 (E. Stratmann).
[36] F. Lappe, *Chem. Ing. Tech.* **42** (1970) 1228.
[37] Fiat Final Report 732, 797, 816, 834.
[38] *Ullmann*, 4th ed., **19**, 665.
[39] W. Kramer in [246]
[40] J. J. McKetta, W. A. Cunningham: *Encyclopedia of Chemical Processing and Design*, vol. **7**, Marcel Dekker, New York-Basel 1977.
[41] T. Sugino, K. Aoki, *J. Electrochem. Soc. Jpn.* **27** (1959) no. 1–3, E 17.
[42] G. Hauck, W. Dürr, *Chem. Ing. Tech.* **39** (1967) 720. H. J. Antweiler, J. P. Schäfer, *Chem. Ing. Tech.* **43** (1971) 180.
[43] D. S. Nielson, O. Platz, *Reliabil. Eng.* **4** (1983) 1–18.
[44] PPG Ind., US 4 155 819-A, 1979 (W. W. Carlin).
[45] Diamond Shamrock, US 4 073 706-A, 1978 (Z. Nagy).
[46] H. Tsukuda et al., *J. Electrochem. Soc. Jpn.* **32** (1964) 1.
[47] D. W. Wabner, P. Schmittinger, F. Hindelang, R. Huß, *Chem. Ing. Tech.* **49** (1977) 351.
[48] L. J. J. Janssen in [82] p. 271.
[49] Society of Chemical Industry, Electrochemical Technology Group: *Intern. Meeting on Electrolytic Bubbles*, 13–14 Sept. 1984, London.
[50] BASF, US 3 502 561, 1970 (W. Rasche, E. Wygasch, G. Csizi).
[51] A. Zimmer, L. Franke, *Chem. Tech. (Leipzig)* **36** (1984) no. 2, 64–69.
[52] H. Hund, *Chem. Ing. Tech.* **39** (1967) 702.
[53] W. Vielstich, *Chem. Ing. Tech.* **34** (1962) 346. Bayer, DE 1 792 588-C3, 1977 (E. Zirngiebl). Murgatroyd's Salt & Chem., DE 1 915 765, 1969 (M. J. Lockett).
[54] M. M. Jaksic, I. M. Csonka, *Electrochem. Technol.* **4** (1966) 49.
[55] A. Kuhn: *Industrial Electrochemical Processes*, Elsevier, Amsterdam 1971.
[56] G. Barthel, *Chem. Anlagen + Verfahren* **8** (1973) 59–60.
[57] P. Theissing et al., *Chem. Ing. Tech.* **51** (1979) 237. F. Hine: *Topics in Pure and Applied Electrochemistry*, SAEST, Karaikudi, India.

[58] AIChE: *Sympos. Ser. 77, no. 204,* 1981.
[59] Uhde: *Alkalichloridelektrolyse nach dem Quecksilberverfahren.*
[60] Krebskosmo: *Chloralkali-Anlage.*
[61] Olin, US 4 004 989-A, 1977 (R. W. Ralston).
[62] G. Hauck, *Chem. Ing. Tech.* **34** (1962) 369.
[63] Bureau International Technique du Chlore: *The Mercury Code of Practice,* Bruxelles 1984.
[64] Environmental Protection Agency: *Control Techniques for Mercury Emissions from Extraction and Chlor-Alkali Plants,* Publ. No. AP-118, Washington, DC.
[65] A. Gellera, L. Cavalli, G. Nucci, *Analyst (London)* **109** (1984) 1537–1539.
[66] H. Klotz, *Chem. Ing. Tech.* **52** (1980) 444.
[67] Akzo: *Anode Current Control by Computer ACCS-System,* 1972.
[68] A. I. Vasilev et al., SU 1 024 528-A1, 1983.
[69] Bayer, DE 1 767 840-B2, 1975 (R. Schlee, W. Büsing).Hoechst, DE 3 244 033-A1, 1984 (K. Lehmann, H. Valentin, K. Gnann).
[70] Montedison, EP 93 452-A2, 1983 (F. Lo Vullo, E. Malvezzi, P. Balboni).
[71] Solvay & Cie., EP 85 999-A1, 1983 (J. P. Detournay, J. Defourny).
[72] Heraeus Elektroden, EP 68 076-A2, 1983 (P. Fabian).
[73] H. Becker, *Dechema Monographie,* vol. **61,** Verlag Chemie, Weinheim 1968, p. 59.
[74] Umweltbundesamt of the Federal Republic of Germany: *Möglichkeiten zur Verminderung von Quecksilberemissionen bei Alkalichloridelektrolysen,* Abschlußbericht 1980.
[75] BITC Working Group Mercury Analysis, *Anal. Chim. Acta* **72** (1974) 37–48; **84** (1976) 231–257; **87** (1976) 273–281.
[76] A. S. Kulyasova, E. L. Babayan, *Niitekhim,* Moscow 1979.
[77] *Ullmann,* 4th ed., **6,** 174–177.
[78] Federal Republic of Germany: " 42. Verwaltungsvorschrift über Mindestanforderungen an das Einleiten von Abwasser in Gewässer-Alkalichloridelektrolysen nach dem Amalgamverfahren," 1984.
[79] AKZO: *Das IMAC-TMR-Verfahren.*
[80] *Eur. Chem. News,* 18 Feb. 1980.
[81] EEC Directive 82/176, 22. 3. 1982.
[82] C. Jackson: *Modern Chlor Alkali Technology,* vol. **2,** Ellis Horwood, Chichester, Great Britain, 1983.
[83] *TA Luft: Technische Anleitung Zur Reinhaltung der Luft,* 2nd ed., Heider Texte, Bergisch Gladbach 26. 2. 1986.
[84] J. S. Sconce: *Chlorine, Its Manufacture, Properties and Uses,* Reinhold Publ. Co., New York 1962.
[85] ELTECH Systems Corporation Technical Service Department: *Diaphragm Electrolytic Chlorine Cell, Technical Data Manual,* ELTECH Systems Corporation, Chardon, Ohio 1982.
[86] ELTECH Systems Corporation Technical Service Department: *Diaphragm Chlorine Cell, Operating Manual,* ELTECH Systems Corporation, Chardon, Ohio 1983.
[87] D. W. F. Hardie, W. W. Smith: *Electrolytic Manufacture of Chemicals from Salt,* The Chlorine Institute, New York 1975.
[88] ELTECH Systems Corporation, US 4 410 411, 1983 (R. W. Fenn III, E. J. Pless, R. L. Harris, K. J.O'Leary).
[89] ELTECH Systems Corporation, US 3 674 676, 1972 (Edward I. Fogelman).
[90] R. N. Beaver: *The Dow Diaphragm Cell,* Dow Chemical, Freeport, Texas 1985.
[91] Dow Chemical Company, US 4 497 112.

[92] Oronzio De Nora Impianti Elettrochimici S.p.A.: *Design Features of Glanor Diaphragm Electrolyzer*, Oronzio DeNora Impianti Elettrochimici S.p.A., Milano.
[93] PPG Industries: *PPG'sBipolar Diaphragm Electrolyzers*, PPG Industries, Inc., Pittsburgh, P.
[94] PPG Industries: *Licensing From PPG Chemicals*, PPG Industries, Inc., Pittsburgh, Pa.
[95] Occidental Chemical Corporation: *Hooker Chlor-Alkali Systems, A Record of Achievement in Diaphragm Cell Technology*, Occidental Chemical Corporation, Niagara Falls, N.Y.
[96] Uhde GmbH: *Alkaline Chloride Elektrolysis by the Diaphragm Process; System Hooker*, Uhde GmbH, Dortmund.
[97] ELTECH Systems Corporation: *ELTECH Modified Diaphragm Cells*, ELTECH Systems Corporation, Chardon, Ohio 1985.
[98] ELTECH Systems Corporation, US 3 591 483, US 3 707 454, 1971 (Richard E. Loftfield, Henry W. Laub).
[99] ELTECH Systems Corporation, US 3 928 166, 1975 (Kevin J. O'Leary, Charles P. Tomba, Robert W. Fenn III).
[100] ELTECH Systems Corporation: *Caustic Purification System 1985*.
[101] G. A. Carlson, E. E. Estep, "Porous Cathode Cell for Metals Removal from Aqueous Solutions," paper presented at the Electrochemical Society Meeting, May, 1972.
[102] W. G. Grot, *Chem. Ing. Tech.* **44** (1972) 167; Diamond Shamrock, US 4 025 405, 1977 (R. L. Dotson, K. J. O'Leary).
[103] Asahi Glass, US 4 065 366, 1977 (Y. Oda, M. Suhara, E. Endoh).
[104] C. Jackson, S. F. Kelham, *Chem. Ind. (London)* 1984, 397.
[105] M. Nagamura, H. Ukihashi, O. Shiragami, *AIChE Winter Meeting*, Orlando, Florida, March 1–2, 1982.
[106] Tokuyama Soda, US 4 190 514, 1980 (S. Matsuura, T. Oku, M. Kuramatani, N. Kramoto, Y. Ozaki).
[107] PPG, US 4 323 595, 1982 (C. N. Welth, J. O. Snodgrass).
[108] Asahi Glass, JP 59/40 231, 1984 (Y. Oda, T. Morimoto, K. Suzuki).
[109] Asahi Chemical Industry, US 4 323 434, 1982 (M. Yoshida, Y. Masuda, A. Kashiwada).
[110] H. Ukihashi, M. Yamabe, *ACS Polymer Division Workshop on Perfluorinated Ionomer Membranes*, Lake Buena Vista, Florida, Feb. 23–26, 1982.
[111] M. Seko in [110]
[112] Asahi Chemical Industry, US 4 178 218, 1979 (M. Seko).
[113] Asahi Glass, US 4 255 523, 1981 (H. Ukihashi, T. Asawa, T. Gunjima).
[114] Du Pont, US 4 021 327, 1977 (W. G. Grot).
[115] D. Bergner, *Chem.-Ztg.* **107** (1983) 281.
[116] Tokuyama Soda: *Membrane Caustic-chlorine process*.
[117] Oronzio De Nora, US 4 340 452, US 4 343 690, 1982.
[118] PPG, US 4 402 809, 1983 (C. R. Dilmore, C. W. Raetzsch). PPG: *PPG's New BIZEC Electrolyzer Technology*.
[119] Asahi Chemical Industry: *Asahi Chemical Membrane Chlor-alkali Process*.
[120] Asahi Glass: *AZEC Electrolyzer*.
[121] ICI: *FM21 Membrane Cell*.
[122] ELTECH Systems Corp.: *Development of the ELTECH MGC Membrane Gap Electrolyzer*.
[123] Kanegafuchi Chemical Industry, US 4 268 365, 1981 (T. Iijima, T. Yamamoto, K. Kishimoto, T. Komabashiri, T. Kano).
[124] Asahi Chemical Industry, US 4 105 515, 1978 (S. Ogawa, M. Yoshida).
[125] Uhde, DE-OS 3 130 742, 1983; EP 71 740.

[126] Krebskosmo, DE-OS 2 940 120.
[127] Krebskosmo, DE-OS 2 940 121.
[128] Chlorine Engineers Corp: *Membrane Bag Cell.*
[129] Asahi Glass, US 4 202 743, 1980 (Y. Oda, M. Suhara, S. Goto, T. Hukushima, K. Miura, T. Hamano).
[130] H. B. Beer, GB 1 147 442, 1965. H. B. Beer, GB 1 195 871, 1967.
[131] V. de Nora, J.-W. Kühn von Burgsdorff, *Chem. Ing. Tech.* **47** (1975) 125–128. Electronor Corporation, US 3 616 445, 1967 (G. Bianchi, V. de Nora, P. Gallone, A. Nidola).
[132] S. Trasatti, G. Lodi in S. Trasatti (ed.): *Electrodes of Conductive Metallic Oxides*, part A, Elsevier Scientific Publishing Company, Amsterdam-Oxford-New York 1980, pp. 301–358.
[133] S. Trasatti, W. E. O'Grady in H. Gerischer, C. W. Tobias (eds.): *Advances in Electrochemistry and Electrochemical Eng.*, vol. **12**, J. Wiley & Sons, New York-Chichester 1981, pp. 177–261.
[134] GB 1 402 414, 1971 (N. W. J. Pumphrey, B. Hesketh). GB 1 484 015, 1973 (B. Hesketh, C. Pownall, N. W. J. Pumphrey).
[135] C. Conradty, DE-OS 1 813 944, 1968 (G. Thiele, D. Zöllner, K. Koziol); C. Conradty, DE 2 255 690, 1972 (K. Koziol, K.-H. Sieberer, H.-C. Rathjen).
[136] C. Modes, W. C. Heraeus: abstract and poster during the *13th Int. Congress and General Assembly*, Int. Union of Crystallography, Hamburg, August 1984.
[137] D. V. Kokoulina, T. V. Ivanova, Y. I. Krasavitskaya, Z. I. Kudryavtseva et al., *Elektrokhimiya* **13** (1977) no. 10, 1511–1515.
[138] J. E. Currey in *Encycl. of Chem. Proc. and Design*, vol. **7**, Marcel Dekker, New York 1978, pp. 305–450.
[139] D. Bergner, S. Kotowski, *J. Appl. Electrochem.* **13** (1983) 341–350.
[140] Y. M. Kolotyrkin, *Denki Kagaku* **47** (1979) no. 7, 390–400.
[141] D. W. Wabner, P. Schmittinger, F. Hindelang, R. Huß, *Chem. Ing. Tech.* **49** (1977) 351.
[142] S. Kotowski, B. Busse in [247], p. 310.
[143] V. V. Gorodetskii, M. M. Pecherskii, V. B. Yanke, D. M. Shub et al., *Sov. Electrochem. (Engl. Transl.)* **15** (1979) 471.
[144] J. P. Randin in J. O'M. Bockris, B. E. Conway, E. Yeager, R. E. White (eds.): *Comprehensive Treatise of Electrochemistry*, vol. **4**, Plenum Press, New York 1981, pp. 473–537. D. M. Novak, B. V. Tilak, B. E. Conway in J. O'M. Bockris, B. E. Conway, R. E. White (eds.): *Modern Aspects of Electrochemistry*, no. 14, Plenum Press, New York 1982, pp. 195–318.
[145] P. C. S. Hayfield, W. R. Jacob in [248], pp. 103–120.
[146] D. L. Caldwell, M. J. Hazelrigg in [248] pp. 121–135.
[147] S. Saito in [248], pp. 137–144.
[148] Asahi Glass Comp, EP 0 032 819, 1981 (K. Saito, H. Shibata).
[149] F. Brandmeir, H. Böder, G. Bewer, H. Herbst, *Chem. Ing. Tech.* **52** (1980) 443.
[150] Oronzio De Nora Impianti Elettrochimici S.p.A., GB 2 051 131, 1980 (A. Pellegri).
[151] A. Nidola in S. Trasatti (ed.): *Electrode of Conductive Metallic Oxides*, part B, Amsterdam-Oxford-New York 1981, pp. 627–659.
[152] C. Traini, *Proc. of O. De Nora Symposium*, 15–18 May 1979.
[153] W. Hartmann, W. Hofmann, D. Bergner, *Chem. Ing. Tech.* **52** (1980) no. 5, 433–435.
[154] Diamond Shamrock Corp., US 3 674 676, 1970 (E. I. Fogelman).
[155] V. de Nora, *Chem. Ing. Tech.* **47** (1975) no. 4, 141.
[156] S. Kotowski, B. Busse, paper presented at the Annual DECHEMA Meeting 1984, *Anodenbeschichtungen in Membranzellen zur Chlor- und Laugeproduktion.*
[157] Uhde GmbH, DE 3 219 704, 1982 (H. Schmitt, H. Schurig, W. Strewe).

[158]   G. Kreysa, H.-J. Külps, *J. Electrochem. Soc.* **128** (1981) no. 5, 979–984.
[159]   D. Bergner, *Chem. Ztg.* **104** (1980) 215.
[160]   Olin Corp., US 4 160 704, 1979 (H. C. Kuo, B. K. Ahn, R. L. Dotson, K. E. Woodard Jr.).
[161]   N. P. Fedot'ev, N. V. Berezina, E. G. Kruglova, *Zh. Prikl. Khim. (Leningrad)* **21** (1948) 317–328.
[162]   Toa Gosei Chemical Industry Co. Ltd., JP 31-6611, 1956 (K. Sasaki).
[163]   E. I. DuPont de Nemours & Co., US 4 116 804, 1978 (C. R. S. Needes).
[164]   Diamond Shamrock Corp., US 4 024 044, 1977 (J. R. Brannan, I. Malkin).
[165]   F. Hine, M. Yasuda, M. Watanabe: "Studies of the Nickel–Sulfur Electrodeposited Cathode," *Denki Kagaku* **47** (1979) no. 7, 401–408.
[166]   Asahi Chemical, EP-A 0 031 948, 1981 (M. Yoshida, H. Shiroki).
[167]   Chlorine Engineers Corp. Ltd., US 4 465 580, 1984 (K. Kasuya).
[168]   Tokuyama Soda, US 4 190 516, 1980 (Y. Kajimaya, T. Kojima, Y. Murakami, S. Matsura).
[169]   Dow Chemical, EP-A 0 129 734, 1985 (N. R. Beaver, L. E. Alexander, C. E. Byrd).
[170]   ICI, EP-A 0 129 374, 1984 (J. F. Cairns, D. A. Denton, P. A. Izard).
[171]   Diamond Shamrock Corp., US 4 104 133, 1978 (J. R. Brannan, I. Malkin, C. M. Brown).
[172]   Kanegafuchi Kagaku Kogyo Kabushiki Kaisha, EP-A 0 132 816, 1985 (Y. Samejima, M. Shiga, T. Kano, T. Kishi).
[173]   Olin Corp., US 4 169 775, 1979 (H. C. Kuo).
[174]   E. Means, T. R. Beck, *Chem. Eng. (N.Y.)* **91** (1984) no. 21, 46–51.
[175]   M. Esayian, J. H. Austin: "Membrane Technology for Existing Chloralkali Plants," E. I. du Pont de Nemours Co., Wilmington, Del., 1984.
[176]   S. Higuchi, *Soda to Enso* **35** (1984) 383.
[177]   H. Isfort, W. Stockmans in [246]
[178]   K. Kerger, *Chem. Ing. Tech.* **43** (1971) 167.
[179]   Uhde: *Chlor und Wasserstoff aus Salzsäure durch Elektrolyse.*
[180]   General Electric, DE 2 856 882-A1, 1979 (R. M. Dempsey, A. B. Conti).
[181]   H. Klotz, P. Orth in [246]
[182]   The M. W. Kellogg Co.: *Hydrogen Chloride to Chlorine, The Kel-Chlor Process,* presentation at the 20th Chlorine Plant Managers Seminar, New Orleans, Feb. 9, 1977.
[183]   The M. W. Kellogg Co., *Hydrocarbon Process.* 60 (1981) no. 11, 143.
[184]   Shell, BE 599 241-A, 1961 (W. S. Engel, F. Wattimena).
[185]   Dynamit Nobel, DE 1 245 922-B, 1967 (W. Schmidt, K. Hass, H. Epler).
[186]   Institut Français du Pétrole, FR 1 294 706-A, 1962 (P. Bedagne et al.).
[187]   C. P. van Dijk, *Chem. Econ. Eng. Rev.* **4** (1972) no. 12, 42.
[188]   H. Schmidt, F. Holzinger, *Chem. Ing. Tech.* **35** (1963) 37.
[189]   A. A. Krashennikova, A. A. Furmann, G. S. Ulyankina, *J. Appl. Chem. USSR (Engl. Transl.)* **44** (1971) 2232.
[190]   Monsanto Enviro-Chem. Systems: *Fiber Bed Mist Eliminator,* 1979.
[191]   Dow Chemical, US 3 568 409-A, 1971 (B. Mac Ferguson, J. F. Gilbert, D. N. Glew).
[192]   Koninkl. Nederlandsche Zoutindustrie, US 3 534 562-A, 1970 (D. Meyer, T. Thijssen).
[193]   ICI, DE 1 946 096-C3, 1973 (K. J. Howliston).
[194]   G. Alberti, *DECHEMA Monogr.* **66** (1971) 1193–1211. R. Gabbioneta: *Garo* . Nash International: *HC-7 Chlorine Compressor.*
[195]   Sulzer Burckhardt, Prospectus No. 1393 a.
[196]   Sundstrand: *Sundyne Zentrifugalverdichter,* 1980.
[197]   H. Abel, F. Özvegyi, *Kälte-Klima-Rundsch.* **7** (1969) 100.

[198] *Inf. Chim.* **98** (1971) 55.

[199] H. Hagemann, *Chem. Ing. Tech.* **39** (1967) 744.

[200] R. F. Schwab, W. H. Doyle, *Electrochem. Technol.* **5** (1967) 228.

[201] B. Liggenstorfer: "Kälteanlage zum Verflüssigen von Chlor," *Tech. Rundsch. Sulzer* **63** (1981) no. 2, 53–56.

[202] AKZO: *Chlorine Liquefaction by Absorption and Distillation*, 1974.

[203] T. A. Liederbach, *Chem. Eng. Prog.* **70** (1974) 64.

[204] The Chlorine Institute: *Chlorine Manual*, New York 1969.

[205] GEST Récommendations, Bureau International Technique Chlore BITC, Bruxelles.

[206] Joint Chlorine Institute: *BITC Meeting, London, June 1982*, The Chlorine Institute, New York.

[207] R. Dandres: *Le Chlore*, 3rd ed., L'Institut National de Recherche et de Sécurite pour la Prevention des Accidents du Travail et de Maladies Professionelles, Paris 1978. H. Dorias: *Gefährliche Güter*, Springer Verlag, Berlin-Heidelberg-New York-Tokyo 1984.

[208] G. Payne: "Safe Handling and Storage of Liquefied Gases, in Particular Chlorine and Chlorofluorohydrocarbon Refrigerants," *Trans. Inst. Chem. Eng.* **42** (1964) no. 4, 92–99.

[209] M. Reuter: *Stand und Entwicklungstendenzen der deutschen und westeuropäischen Druckbehälterregeln und Prüfpraxis*, Technische Überwachung 19 (1978) no. 6, 181–187.

[210] D. P. Meinhardt, "Chlorine," *Chem. Eng.* (1981) Oct. 5, 125–133.

[211] N. C. Harris, J. P. Shaw, *European Chlorine Storage Practice*, paper presented at the Chlorine Institute's 23rd Plant Manager Seminar, 1980. R. Papp, *Chlorine '74*, Chemistry and Industry, 15 March (1975) 243–246.

[212] *Winnacker-Küchler:* 4th ed., pp. 428–429.

[213] ISO 3511/1.

[214] A. F. Timofeer et al.: "Steps to Prevent Failures of Chlorine Containers," *Khim. Promst* (Moscow) 10 (1978) no. 3, 202–205; Sov. Chem. Ind. (Engl. Transl.) 10 (1978) no. 3, 222–225. M. I. Bereshovski et al.: "Transportation of Liquid Chlorine," *Sov. Chem. Ind.* **2** (1970) no. 10, 776–780.

[215] W. A. Statesir: "Explosive Reactivity of Organics and Chlorine," *Chem. Eng. Prog.* **69** (1973) no. 4, 52–54.

[216] T. B. Meslin: "Assessment and Management of Risk in the Transport of Dangerous Materials: The Case of Chlorine Transport in France," *Risk Anal.* **1** (1981) no. 2, 137–142. G. B. Frame: "Determination of Risk from a Chlorine Spill or Major Release," *Chem. Can.* **32** (1980) no. 8, 27–29.

[217] C. L. Melancon, *Dow Chemicals Emergency Response System*, paper presented at The Chlorine Institute's 22nd Plant Manager's Seminar 1979. R. Sklarew et al., *Emergency System for Toxic Chemical Releases*, Pollution Engineering, July 1982.

[218] J. M. Buchlin, *Aerodynamic Behaviour of Liquid Spray-Design Method of Water Spray Curtain*, von Karman Institute, Rhode Saint Genese, Belgium 1980, Test Report No. 171.

[219] British Standard Institution: *Specification for Liquid Chlorine*, BS 3947, 1965. DIN 19 607: Chlorgas zur Wasseraufbereitung, 1985. E. J. Laubusch, *Jour. AWWA* **51** (1959) 742–748.

[220] ISO 1552: Liquid Chlorine for Industrial Use — Method of Sampling.

[221] ISO 2121: Liquid Chlorine for Industrial Use — Determination of the Water Content — Gravimetric Method.

[222] ISO 2202: Liquid Chlorine for Industrial Use — Determination of the Water Content — using an Electrolytic Analyzer.

[223] ISO 2120: Liquid Chlorine for Industrial Use — Determination of the Content of Chlorine by Volume in the Vaporized Product.

[224] DIN 38 408 Teil 4: Gasförmige Bestandteile (Gruppe G) Bestimmung von freiem und Gesamtchlor (G 4).
[225] *Anal. Chim. Acta,* **156** (1984) 221–233.
[226] *Anal. Chim. Acta,* **87** (1976) 273–281.
[227] *Deutsche Einheitsverfahren zur Wasseruntersuchung,* Verlag Chemie, Weinheim 1971.
[228] M. L. Langhorst, *Am. Ind. Hyg. Assoc. J.* **43** (1982) 347–360. The Chlorine Institute: *Atmospheric Monitoring Equipment for Chlorine,* 2nd ed., March 1982, CI Member Information Report 138.
[229] W. König, K. Krinke, H. Kron et al., *Chem. Ind.* **36** (Feb. 1984) 69.
[230] Oil Paint Drugs Chemical Marketing Reporter, OPD 11 Apr. 83.
[231] K. Niki, *Jpn. Chem. Annu.* 1982/83, 56.
[232] *Chem. Rundsch.* **37** (1984) no. 24, 3.
[233] E. Weise: *Recent Developments in the Field of Alkali Chloride Electrolysis,* Lecture on the 20th IUPAC Congress 1983, Cologne, FRG.
[234] *Chem. Eng. News,* Apr. 2 (1984).
[235] *Inf. Chimie* **238** (Spécial Juin 1983) 177–194.
[236] The Chlorine Institute: Pamphlet no. 16, Jan. 1984.
[237] *Chem. Mark. Rep.* (1981) no. Aug. 24, 27–35.
[238] P. Davies, I. Hymes: "Chlorine toxicity criteria for hazard assessment," *Chem. Eng.,* June 1985, 30–33.
[239] H. H. Rotman, M. J. Fliegelman, T. Moore, R. G. Smith, D. M. Anglen, C. J. Kowalski, J. G. Weg: "Effects of low concentrations of chlorine on pulmonary function in humans," *J. Appl. Physiol.: Respir. Environ. Excercise Physiol.* **54** (1983) no. 4, 1120–1124.
[240] "Chronic Inhalation Toxicity Study on Chlorine in Non-Human Primates," *CITT Activities Chemical Insitute of Toxicology* **4** (1984) no. 8, 1–3.
[241] B. Grenquist-Norden: "Respiratory effects of industrial chlorine and chlorine dioxide exposure," Institute of Occupational Health, University of Helsinki, Finland, 1983.
[242] F. Schuckmann (Occupational Health Department of Hoechst AG): "Pulmonary Function Study on Workers with Long-Term Exposure to Chlorine," in: *Medichem proceedings, XI. International Congress, Calgary, Alberta, Canada,* 26.–29. 9. 1983, pp. 475–483.
[243] B. G. Ferris, Jr., S. Puleo, H. Y. Chen: "Mortality and morbidity in a pulp and a paper mill in the United States: ten-year follow-up," *Br.J. Ind. Med.,* **36** (1979) no. 2, 127–134.
[244] L. R. S. Patil, R. G. Smith, A. J. Vorwald, T. F. Mooney: "The health of diaphragm cell workers exposed to chlorine," *Am. Ind. Hyg. Assoc. J.* **31** (1970) 678–686.
[245] The Electrochemical Society: *Extended Abstracts of Industrial Electrolytic Division,* The Chlorine Institute, Spring Meeting, San Francisco, May 12nd–17th, 1974.
[246] Technische Elektrolysen, *Dechema Monographie* vol. **98,** Verlag Chemie, Weinheim 1985.
[247] K. Wall, *Modern Chlor-Alkali Technology,* vol. **3,** Ellis Horwood, Chichester, Great Britain, 1986.
[248] M. O. Coulter: *Modern Chlor-Alkali Technology,* Ellis Horwood, London 1980.

Chlorine Oxides and Chlorine Oxygen Acids

HELMUT VOGT, Technische Fachhochschule Berlin, Berlin, Federal Republic of Germany (Chaps. 1, 4.2–4.6, 7, and 9)

JAN BALEJ, Ingenieurbüro für chemische Technik, Jülich, Federal Republic of Germany (Chap. 1)

JOHN E. BENNETT, Eltech Systems Corp., Fairport Harbor, Ohio 44077, United States (Chaps. 4.2–4.6)

PETER WINTZER, Cellulosefabrik Attisholz AG, Luterbach, Switzerland (Chaps. 7 and 9)

SAEED AKBAR SHEIKH, Davy McKee AG, Frankfurt, Federal Republic of Germany (Chaps. 2, 3, 4.1, 5, 6, and 9)

PATRIZIO GALLONE, Politecnico di Milano, Milano, Italy (Chaps. 8 and 9)

1.	Introduction	1258	5.2.5. Other Processes	1283
2.	Hypochlorous Acid	1262	5.3. Economic Aspects	1284
3.	Solid Hypochlorites	1263	5.4. Uses	1284
3.1.	Properties	1263	6. Sodium Chlorite	1285
3.2.	Production	1263	6.1. Properties	1285
3.3.	Quality Specifications	1265	6.2. Production	1286
3.4.	Uses	1265	6.3. Uses	1287
4.	Hypochlorite Solutions	1265	7. Chloric Acid and Chlorates	1287
4.1.	Chemical Production	1266	7.1. Properties	1287
4.1.1.	From Chlorine	1267	7.2. Production Fundamentals	1291
4.1.2.	From Bleaching Powder	1268	7.2.1. Chlorate-Generating Reactions	1291
4.2.	Electrosynthesis	1269	7.2.2. Loss Reactions	1293
4.2.1.	Reaction Fundamentals	1269	7.3. Industrial Electrosynthesis Systems	1295
4.2.2.	Industrial Cells	1271	7.3.1. Electrolysis Cell Types	1297
4.3.	Storage	1276	7.3.2. Electrodes	1301
4.4.	Uses	1276	7.3.3. Operational Parameters	1302
4.5.	Economic Aspects	1277	7.3.4. Brine Purification	1304
4.6.	Plant Safety	1278	7.4. Crystallization	1304
5.	Chlorine Dioxide	1278	7.5. Construction Materials	1305
5.1.	Properties	1278	7.6. Environmental Protection	1306
5.2.	Production	1279	7.7. Quality Specifications	1307
5.2.1.	Day–Kesting Process	1280	7.8. Storage, Transportation, and Safety	1307
5.2.2.	R2 Process	1281		
5.2.3.	Mathieson Process	1282		
5.2.4.	Solvay Process	1283	7.9. Uses	1308

7.10.	Economic Aspects 1308	8.3.	Environmental Protection ... 1315
8.	Perchloric Acid and Perchlorates 1308	8.4.	Chemical Analysis 1315
8.1.	Physical and Chemical Properties 1309	8.5.	Storage, Transportation, and Safety 1316
8.1.1.	Perchloric Acid 1309	8.6.	Uses 1317
8.1.2.	Perchlorates 1311	8.7.	Economic Aspects 1318
8.2.	Production 1311	9.	Toxicology and Occupational Health 1318
8.2.1.	Perchloric Acid 1311
8.2.2.	Perchlorates 1312	10.	References 1319

1. Introduction

Numerous chlorine oxides are known. However, only two anhydrides of chlorine oxygen acids, *dichlorine oxide*, Cl_2O, and *chlorine dioxide*, ClO_2, and two mixed anhydrides, *dichlorine hexoxide*, Cl_2O_6, and *dichlorine heptoxide*, Cl_2O_7, are fairly stable under certain conditions. Table 1 shows some important properties of these chlorine oxides. Other chlorine oxides, such as *dichlorine dioxide* [12292-23-8], Cl_2O_2, *dichlorine trioxide* [17496-59-2], Cl_2O_3, or *dichlorine tetroxide* [27218-16-2], Cl_2O_4, are unstable.

The chlorine oxygen acids are formed by reaction of the corresponding chlorine oxides with water.

Hypochlorous acid [7790-92-3], HClO:

$Cl_2O + H_2O \longrightarrow 2\,HClO$

Chlorous acid [13898-47-0], $HClO_2$:

$2\,ClO_2 + H_2O \longrightarrow HClO_2 + HClO_3$

Chloric acid [7790-93-4], $HClO_3$:

$Cl_2O_6 + H_2O \longrightarrow HClO_3 + HClO_4$

Perchloric acid [7601-90-3], $HClO_4$:

$Cl_2O_7 + H_2O \longrightarrow 2\,HClO_4$

Table 2 lists the thermodynamic properties of chlorine oxides, chlorine oxygen acids, and their sodium salts [1].

The *oxidation power* of individual chlorine oxygen compounds is characterized by the changes in the standard enthalpy $\Delta H°$ and Gibbs free energy $\Delta G°$ of the decomposition

Table 1. Properties of chlorine oxides

	Dichlorine oxide [7791-21-1], Cl_2O	Chlorine dioxide [10049-04-4], ClO_2	Dichlorine hexoxide [12442-63-6], Cl_2O_6	Dichlorine heptoxide [12015-53-1], Cl_2O_7
Oxidation state	+ 1	+ 4	+ 6	+ 7
M_r	86.91	67.45	166.91	182.90
Melting point, °C	− 116	− 59	+ 3.5	− 90
Boiling point, °C	+ 2	+ 11	−	82
Appearance	yellow-brown gas, red-brown liquid	orange-yellow gas, red liquid	red liquid	oily, colorless liquid
Stability	decomposes at 100 °C, explodes on heating or shock	unstable at ambient temperature	decomposes at melting point	decomposes slowly at ambient temperature

Table 2. Thermodynamic properties of chlorine oxides, chlorine oxygen acids, and their salts at 25°C

Compound	Formula	CAS registry number	State*	$\Delta H_f°$, kJ/mol	$\Delta G_f°$, kJ/mol
Dichlorine oxide	Cl_2O	[7791-21-1]	g	80.3	97.9
Chlorine dioxide	ClO_2	[10049-04-4]	g	102.5	120.5
Chlorine trioxide	ClO_3	[13932-10-0]	g	155	−
Dichlorine heptoxide	Cl_2O_7	[12015-53-1]	l	238.1	−
			g	272.0	−
Hydrochloric acid	HCl	[7647-01-0]	ao	− 167.2	− 131.2
Hypochlorous acid	HClO	[7790-92-3]	ao	− 120.9	− 79.9
Chlorous acid	$HClO_2$	[13898-47-0]	ao	− 51.9	+ 5.9
Chloric acid	$HClO_3$	[7790-93-4]	ai	− 104.0	− 7.95
Perchloric acid	$HClO_4$	[7601-90-3]	ai	− 129.3	− 8.52
Sodium chloride	NaCl	[7647-14-5]	ai	− 407.3	− 393.1
Sodium hypochlorite	NaClO	[7681-52-9]	ai	− 347.3	− 298.7
Sodium chlorite	$NaClO_2$	[7758-19-2]	ai	− 306.7	− 244.7
Sodium chlorate	$NaClO_3$	[7775-09-9]	ai	− 344.1	− 269.8
Sodium perchlorate	$NaClO_4$	[7601-89-0]	ai	− 369.5	− 270.4

* ao: undissociated solute in aqueous ideal solution at unit molality; ai: electrolyte in the hypothetical ideal solution at unit activity, dissociated into ions.

Table 3. Thermodynamic data of oxygen-forming decomposition reactions of chlorine oxygen compounds at 25°C

Reaction	$\Delta H°$, kJ/mol	$\Delta G°$, kJ/mol
Cl_2O (g) + H_2O $\longrightarrow$ 2 HCl (aq) + O_2 (g)	− 128.8	− 123.2
4/5 ClO_2 (g) + 2/5 H_2O $\longrightarrow$ 4/5 HCl (aq) + O_2 (g)	− 101.4	− 106.5
4/7 ClO_3 (g) + 2/7 H_2O $\longrightarrow$ 4/7 HCl (aq) + O_2 (g)	− 102.4	
1/4 Cl_2O_7 (l) + 1/4 H_2O $\longrightarrow$ 1/4 HCl (aq) + O_2 (g)	− 71.6	
1/4 Cl_2O_7 (g) + 1/4 H_2O $\longrightarrow$ 1/4 HCl (aq) + O_2 (g)	− 80.1	
2 HClO (aq) $\longrightarrow$ 2 HCl (aq) + O_2 (g)	− 92.5	− 102.7
$HClO_2$ (aq) $\longrightarrow$ HCl (aq) + O_2 (g)	− 115.3	− 137.1
2/3 $HClO_3$ (aq) $\longrightarrow$ 2/3 HCl (aq) + O_2 (g)	− 42.1	− 82.2
1/2 $HClO_4$ (aq) $\longrightarrow$ 1/2 HCl (aq) + O_2 (g)	− 18.9	− 61.4

reactions forming molecular oxygen (Table 3). As shown in Table 3, the thermodynamic stability of chlorine oxygen acids and chlorine oxides increases with increasing oxidation state of the chlorine atom. Therefore, concentrated perchloric acid can be isolated, whereas all other oxygen acids are stable only in diluted form. The instability of the chlorine oxides and their acids determines their industrial significance. All chlorine oxygen compounds are strong oxidants; the strongest are those with the lowest oxidation state of the chlorine atom.

Dichlorine oxide, Cl_2O, chlorine dioxide, ClO_2, all oxygen acids, and their salts, particularly those of sodium and potassium, are used industrially.

History. Soon after the discovery of chlorine in 1774, scientific and commercial interest was directed to the chlorine oxides, the chlorine oxygen acids, and their salts.

Hypochlorite was first prepared in 1787 by C. L. BERTHOLLET by feeding chlorine into potash lye. This bleach liquor (eau de Javel) was soon applied in bleaching textiles and in papermaking. LABARRAQUE replaced potash lye by the cheaper soda lye (eau de Labarraque).

Hypochlorite was prepared by electrolysis of sodium chloride as early as 1801. However, commercial electrochemical production did not start for a long time.

At the beginning of the 20th century, the traditional routes to form bleach liquors [2] fell out of use by the rapidly expanding chlorine – caustic industry that made large quantities of cheap waste chlorine available. Within the last 20 years, however, electrosynthesis has found a widespread revival and is today an alternative to chemical hypochlorite production wherever safety risks of small plants are decisive.

In 1799 C. TENNANT and C. McINTOSH developed a process for the production of *bleaching powder* by absorbing chlorine onto dry calcium hydroxide. This bleaching powder was much more stable than previously obtained bleaching products. In 1906, G. PISTOR succeeded in producing highly concentrated bleaching powder with more than 70% available chlorine.

R. CHEVENIX first made *chlorine dioxide* in 1802 by the reaction of concentrated sulfuric acid and potassium chlorate, but the product was first identified by H. DAVY and F. VON STADION in 1815/1816. CALVERT and DAVIES used oxalic acid instead of sulfuric acid in 1859 and obtained chlorine dioxide together with carbon dioxide, thus eliminating the explosion risk. This and the investigations of E. SCHMIDT in 1921 – 1923 [3] laid the ground to today's extensive use of chlorine dioxide as a bleaching agent in the textile, pulp, and paper industry.

J. R. GLAUBER probably prepared *chlorate* for the first time, but C. L. BERTHOLLET first made chlorate in 1787 by the reaction of chlorine with potassium hydroxide and identified it as the salt of chloric acid. The first electrochemical chlorate preparation was performed by W. VON HISINGER and J. J. BERZELIUS in 1802. In 1851 a cell patent was granted to C. WATT that had the essential features of later cells, but industrial production did not start before 1886. Originally, all cells were operated with an alkaline electrolyte until J. LANDIN added chromic acid to the electrolyte and solved the problem of cathodic hypochlorite reduction [4], [5]. Within the few years before and after 1900, scientific research focused on the mechanism of chlorate formation [6]–[12]; this was supplemented by numerous investigations within the last 30 years. However, the detailed reaction mechanism is still a subject of scientific debate.

The most important technological improvements were (1) the separation of the electrochemical reactor from the chemical reactor in 1933 (A. SCHUMANN-LECLERCQ) [13], (2) the utilization of the hydrogen formed at the cathode for "stirring" the electrolyte [14], and (3) the introduction of dimensionally stable titanium anodes 20 years ago [15].

F. VON STADION first made *perchlorates* by oxidation of chlorate at platinum anodes in 1816; he also obtained perchloric acid by the reaction of concentrated sulfuric acid with potassium perchlorate and by anodic oxidation of hydrochloric acid. Preparation of perchloric acid by electrolysis of dilute chloric acid was first carried out by J. J. BERZELIUS in 1835. Like all other electrochemical production methods, electrosynthesis of perchlorate and perchloric acid did not gain industrial significance before the end of the 19th century. In 1890, O. CARLSON was granted a patent for electrochemical perchlorate production, and in 1895 he operated the first commercial plant in Sweden. The first methodic studies of the reaction fundamentals were initiated by German researchers in 1898 [16]–[18].

General Significance. The importance of chlorine oxygen compounds is based predominantly on their oxidizing power. For 200 years, after the lawn bleaching of textiles became obsolete, the textile and paper industry has invariably been a main consumer of chlorine dioxide and hypochlorite. The importance of these products has steadily grown in accordance with the expansion of the pulp and paper industry. Traditional uses of chlorate as a herbicide or explosive became less important. On the other hand, the increasing demand for disinfection of process and drinking water and for sanitation in general has strongly favored the use of chlorine oxygen compounds. The importance of chlorinated lime as a disinfectant has declined. However, other chlorine oxygen compounds, above all hypochlorite, have grown. This trend was particularly favored by the danger of handling chlorine. Moreover, some of the chlorine oxygen compounds have found numerous novel uses that are reported in detail in the following chapters.

Today, chlorate production is one of the most important inorganic electrosyntheses. In addition, perchlorates are preferably made by electrochemical processes. On the basis of a deeper understanding of the chemical and electrochemical fundamentals and by introducing dimensionally stable anodes, design and operation of electrochemical plants have been revolutionized to such an extent that a modern electrochemical reactor has very little resemblance to a cell 20 years ago.

2. Hypochlorous Acid

Hypochlorous acid [7790-92-3], HOCl, M_r 52.5, is only moderately stable in aqueous solution. It is colorless when dilute and yellowish at higher concentrations. Hypochlorous acid is one of the most powerful oxidizing agents known.

Hypochlorous acid solution decomposes exothermically. The main decomposition products are hydrochloric acid and oxygen:

$$2\ HOCl \longrightarrow 2\ HCl + O_2$$

Minor amounts of chlorine and chloric acid are also formed.

Production. Hypochlorous acid is produced by the reversible reaction of chlorine and water:

$$Cl_2 + H_2O \rightleftharpoons HOCl + HCl$$

For efficient conversion, hydrochloric acid must be removed from the equilibrium mixture. This is achieved by limestone, $CaCO_3$, soda ash, Na_2CO_3, or calcium hypochlorite, $Ca(OCl)_2$.

One of the common ways to produce hypochlorous acid is to pass chlorinated water through towers packed with powdered limestone [19]. The overall reaction is as follows:

$$2\ Cl_2 + 2\ H_2O + 2\ CaCO_3 \longrightarrow 2\ HOCl + CaCl_2 + Ca(HCO_3)_2$$

Calcium hydrogen carbonate present in the resulting solution provides the buffer needed to stabilize the product. Any hydrochloric acid still produced is removed by the reaction

$$Ca(HCO_3)_2 + 2\ HCl \longrightarrow CaCl_2 + 2\ H_2O + 2\ CO_2$$

Hypochlorous acid can also be prepared by passing a mixture of dry chlorine and air through a column packed with yellow mercuric oxide, HgO. The resulting dichlorine oxide is dissolved in water to produce hypochlorous acid [20], [21].

Storage. Because of its limited stability, hypochlorous acid is best used soon after production. Its long-distance transport or intermediate storage should be avoided.

A solution containing less than 1% hypochlorous acid can be stored in the dark over fairly long periods, provided that such metals as copper, nickel, or cobalt are absent. Solutions containing 30% hypochlorous acid can also be stored, but at temperatures of −20 °C or lower.

Uses. Because of its instability, hypochlorous acid is not used extensively as an oxidizing or bleaching agent. It was used mainly in the water treatment industry for slime control, for treatment of drinking water, and for sterilization of swimming pools. These applications have now been taken over by the more stable hypochlorites.

3. Solid Hypochlorites

3.1. Properties

All solid hypochlorites are soft, white, dry powders. Some are almost odorless; others smell more or less strongly of chlorine or hydrochloric acid because of decomposition during storage (Eqs. 1–3). The stability of hypochlorites depends primarily on their water content, which is usually less than 1%; tropical bleach contains even less than 0.3%. They are stable up to 80 °C, tropical bleach even up to 100 °C. When heated to 180 °C, they decompose into chloride and oxygen. Such metals as iron, nickel, or cobalt decrease the stability of hypochlorites. Therefore, the raw materials used for the production of hypochlorites must be free of such metals. *Tropical bleach* (< 0.3% water) free of heavy metals has a shelf life of more than 2 years, if properly stored.

If not properly stored in air-tight containers, hypochlorites suffer loss of available chlorine because of reaction with water:

$$Ca(OCl)_2 + CaCl_2 + 2\,H_2O \longrightarrow 2\,Ca(OH)_2 + 2\,Cl_2 \qquad (1)$$

or reaction with carbon dioxide:

$$Ca(OCl)_2 + CaCl_2 + 2\,CO_2 \longrightarrow 2\,CaCO_3 + 2\,Cl_2 \qquad (2)$$

or reaction with both:

$$Ca(OCl)_2 + CO_2 + H_2O \longrightarrow CaCO_3 + 2\,HOCl \qquad (3)$$

3.2. Production

Bleaching Powder. Standard bleaching powder is a mixture of calcium hypochlorite [7778-54-3], $Ca(OCl)_2$, calcium chloride, and calcium hydroxide containing varying amounts of water. It is made by passing chlorine over hydrated lime. The Rheinfelden bleaching powder process of Dynamit Nobel is a batch operation [22]. Dry, powdered lime hydrate is chlorinated at 45 °C and low pressure (5.3 kPa) in a horizontal reaction drum. Chlorine is injected as a liquid. The reaction mass is permanently mixed by a slowly rotating rake. The reaction of solid lime hydrate and chlorine leads to the formation of a mixture of dibasic calcium hypochlorite [12394-14-8], $Ca(OCl)_2 \cdot 2\,Ca(OH)_2$, and basic calcium chloride, corresponding to 40% conversion of the available calcium hydroxide. The following equation characterizes the reaction [23], [24]:

$$5\ Ca(OH)_2 + 2\ Cl_2 \longrightarrow Ca(OCl)_2 \cdot 2\ Ca(OH)_2 + CaCl_2 \cdot Ca(OH)_2 \cdot H_2O + H_2O$$

On further chlorination, hemibasic calcium hypochlorite [62974-42-9], $Ca(OCl)_2 \cdot 1/2\ Ca(OH)_2$, and neutral calcium chloride hydrate are formed. After ca. 60% of the available calcium hydroxide has been converted, the bleaching powder reaction stops; this can be represented by the following equation:

$$10\ Ca(OH)_2 + 6\ Cl_2 \longrightarrow Ca(OCl)_2 \cdot 2\ Ca(OH)_2 + 2\ Ca(OCl)_2 \cdot 1/2\ Ca(OH)_2$$
$$+ CaCl_2 \cdot Ca(OH)_2 \cdot H_2O + 2\ CaCl_2 \cdot H_2O + 3\ H_2O$$

The reaction is strongly exothermic, generating 1100 kJ of heat per kg of chlorine converted. This heat and the low pressure cause the water formed during the reaction and the liquid chlorine to evaporate. Consequently, the reaction mass is dried completely under vacuum at a maximum temperature of 85 °C. The product is standard bleaching powder of 35–37% available chlorine content (for definition, see Section 3.3).

Gaseous chlorine can also be used for this reaction, but then the reaction takes 2–3 times longer, chlorine losses are higher, and the available chlorine content of the product is smaller.

Tropical Bleach. To reduce the water content further, finest ground quicklime, CaO, is added to the bleaching powder. It absorbs any water still present and is converted into calcium hydroxide. Although this operation decreases the available chlorine content by 1–2%, the extra drying makes the resulting bleaching powder, known as tropical bleach, stable up to temperatures of 100 °C.

ICI has developed a continuous process for the production of bleaching powder, in which countercurrents of calcium hydroxide and chlorine react in a rotating drum [25]. The heat of reaction is removed by spraying the drum externally with water and by diluting the chlorine with cooled air; this gas stream also removes the water formed during the chemical reaction.

High-Percentage Hypochlorite. Solid hypochlorites with 70% and higher available chlorine contents can be prepared by chlorinating slurries of such calcium compounds as calcium hydroxide, or bleaching powder. Initially, hemibasic calcium hypochlorite, $Ca(OCl)_2 \cdot 1/2\ Ca(OH)_2$, is formed. When further chlorinated, this gives neutral calcium hypochlorite dihydrate, $Ca(OCl)_2 \cdot 2\ H_2O$, which is then dried to the desired high-percentage hypochlorite. In all of these reactions, calcium chloride is formed as a byproduct [26]–[28]. Some processes recover the calcium values by adding sodium hypochlorite to the slurries:

$$2\ NaOCl + CaCl_2 \longrightarrow Ca(OCl)_2 + 2\ NaCl$$

In such cases, the product primarily consists of calcium hypochlorite, sodium chloride, and water, which is then removed [29].

Barium and Magnesium Hypochlorite. Barium hypochlorite [*13477-10-6*], Ba(OCl)$_2$, can be produced in large crystals with a maximum available chlorine content of 59%. It is more stable than calcium hypochlorite but also more expensive because of the high cost of raw materials.

Magnesium hypochlorite [*10233-03-1*], Mg(OCl)$_2$, is extremely unstable and decomposes when dried.

3.3. Quality Specifications

The term *available chlorine content*, also called *active chlorine*, represents the mass fraction of liberated chlorine in bleaching powder when bleaching powder reacts with hydrochloric acid.

The following qualities of solid hypochlorites are available on the market today (content of available chlorine, wt %, in parentheses): tropical bleach (34 – 35), bleaching powder (35 – 37), and high-percentage hypochlorite (70).

3.4. Uses

In the paper industry, calcium hypochlorite is used in single-stage bleaching. The more expensive sodium hypochlorite is used in the multistage process, which involves chlorination, caustic extraction, and hypochlorite oxidation. Kraft pulp is processed to higher brightness and greater strength when sodium hypochlorite is used instead of calcium hypochlorite.

For bleaching in laundry operations, bleaching powder is first suspended in water and then decanted. Only the solution is used because the insolubles could damage the fibers.

4. Hypochlorite Solutions

The reaction of gaseous chlorine with a slight excess of alkali produces highly concentrated hypochlorite solutions. Available chlorine concentrations of 170 – 220 g/L can be obtained, or even more if the residual chloride concentration is lowered extremely [30]. Hypochlorite solutions are relatively safe and are often chosen instead of chlorine for bleaching, disinfection, biofouling control, and odor control. In recent years, the concern regarding the safety hazards associated with liquid chlorine has grown. Several major cities now restrict transportation of chlorine within their boundaries, and a great deal of attention has focused on accidents caused by the handling of liquid chlorine by unskilled labor [31]. This has increased the popularity of hypochlorite solutions in spite of their relatively high cost.

An attractive alternative to the chemical production of hypochlorite solutions described above is the on-site electrolysis of brine or seawater. Such processes are described in Section 4.2.

Stability. Hypochlorite solutions are more stable than solutions of hypochlorous acid, but they are active enough to be used as disinfectants or bleaching agents. The factors that affect the stability of the parent acid also affect the stability of hypochlorite solutions: concentration, presence of such metals as copper, nickel, or cobalt, pH, temperature, and exposure to light.

Available Chlorine. When chlorine reacts with caustic soda, half of the chlorine is lost because inert sodium chloride is formed:

$$Cl_2 + 2\ NaOH \longrightarrow NaOCl + NaCl + H_2O$$

However, as an oxidant, sodium hypochlorite decomposes to sodium chloride and oxygen:

$$NaOCl \longrightarrow NaCl + [O]$$

The oxidizing power of one oxygen atom is equivalent to that of two chlorine atoms. Therefore, the complete oxidizing power of the original chlorine is available in the hypochlorite solution; it is expressed in grams of available chlorine per liter of finished solution. Thus, the "available chlorine" of hypochlorite solutions compares the oxidizing power of the agent to that of the equivalent amount of elemental chlorine used to make the solution.

4.1. Chemical Production

A hypochlorite unit is attached to each chlor-alkali plant to render harmless the dilute chlorine that cannot be recovered economically. These units make most of the industrially produced hypochlorite solutions [32]–[34].

Beyond that, hypochlorite solutions with available chlorine contents higher than 5 g/L are made by passing chlorine gas through dilute solutions of sodium hydroxide or potassium hy-droxide. Calcium hydroxide suspensions may also be used, from which, after filtration or decantation, clear calcium hypochlorite solutions are obtained [35].

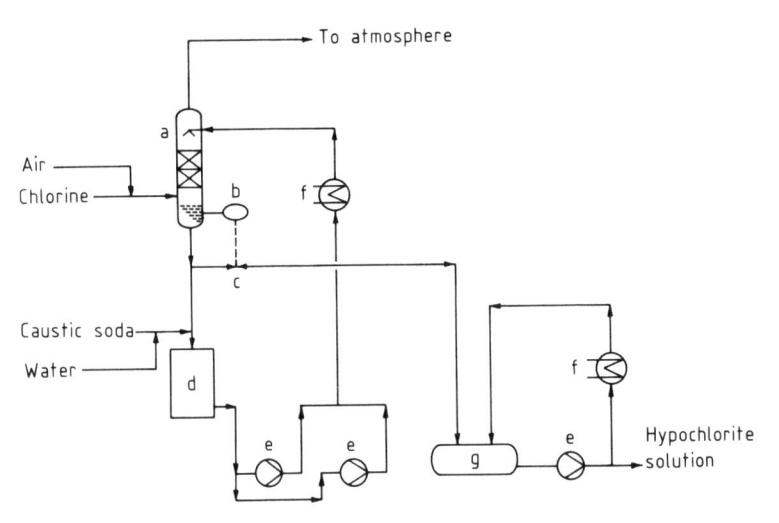

Figure 1. Production of hypochlorite solution from chlorine and caustic soda
a) Chlorination column; b) Analyzer; c) Control valve; d) Buffer tank; e) Pumps; f) Heat exchangers; g) Storage tank

4.1.1. From Chlorine

In the commercial production of hypochlorite solutions from chlorine gas and alkali solutions of various concentrations, the following conditions must be maintained.

1) The temperature must be controlled at 30–35 °C.
2) The solution must be alkaline at any stage.
3) The equipment must provide for thorough mixing and escape of inert gas.
4) Such heavy metals as manganese, iron, cobalt, nickel, or copper must be avoided in the system.
5) Available chlorine contents of more than 150 g/L should be avoided. The high decomposition rate of such concentrated solutions more than offsets any savings in transportation [36].

Sodium or Potassium Hypochlorites. Figure 1 shows a schematic diagram of a typical process for the commercial production of hypochlorite solutions. The continuous process can produce hypochlorite solutions of any available chlorine content between 0 and 150 g/L and any amount between zero and the designated capacity. The only parameters to be adjusted are the redox potential of the analyzer (b) and the desired dilution of the caustic soda solution. For a given amount of chlorine coming in, the system automatically adjusts the caustic soda, process water, and hypochlorite solution flows into and out of the system.

Chlorine gas, diluted with air, is introduced into the chlorination column (a), packed with Raschig rings. Caustic soda is diluted to the desired concentration with water. Tank (d) provides for buffer capacity and homogenization. The pumps (e) circulate the mixture of caustic soda and sodium hypochlorite through the titanium heat exchanger (f), the chlorination column (a), and the tank (d).

Any amount of chlorine between zero and the designated capacity is absorbed in the circulating caustic, producing sodium hypochlorite. When the available chlorine concentration reaches the desired value, the analyzer (b) signals to open control valve (c) for the withdrawal of sodium hypochlorite to the storage tank (g). In this case, the liquid in tank (d) must be replenished by fresh caustic.

The storage tank (g) is equipped with a circulation pump and a heat exchanger to keep the temperature of the stored hypochlorite solution below 35 °C. If necessary, the inert gas leaving column (a) can be scrubbed before entering the atmosphere.

Materials of Construction. All pumps have rubber-lined steel casings and titanium propellers. The heat exchangers are plated with titanium. The tanks (d) and (g) are made from fiberglass-reinforced plastic [37]. The same material is used for the chlorination column and the pipes; chlorinated poly(vinyl chloride) is an alternative material for the pipes.

Calcium Hypochlorite. If a solution of calcium hypochlorite is desired, milk of lime — a suspension of calcium hydroxide in water — is chlorinated. (The solubility of calcium hydroxide in water is 1.3 g/L at 20 °C.) In that case, the plant described in Figure 1 must be modified to handle the solid phase.

In addition to design changes in the equipment, no packing is needed in the chlorination column and the piping must be designed to ensure that pockets are avoided where solid particles can settle and block the pipes; a settler or filter is needed to remove all insoluble or suspended particles before a clear solution of calcium hypochlorite flows to the storage tank.

4.1.2. From Bleaching Powder

To avoid long-term storage of hypochlorite solutions, laundries prefer to store solid bleaching powder and then prepare the sodium hypochlorite solutions by using sodium carbonate, sodium sulfate, or caustic soda:

$CaCl(OCl) + Na_2CO_3 \longrightarrow NaOCl + NaCl + CaCO_3$
$CaCl(OCl) + Na_2SO_4 \longrightarrow NaOCl + NaCl + CaSO_4$
$CaCl(OCl) + 2\,NaOH \longrightarrow NaOCl + NaCl + Ca(OH)_2$

A disadvantage of this method is the precipitation of $CaCO_3$, $CaSO_4$, or $Ca(OH)_2$, which requires filtering or settling before the hypochlorite solution can be used.

4.2. Electrosynthesis

On-site, electrochemical production of dilute hypochlorite solution has long been recognized as an option wherever long-term storage of hypochlorite is unnecessary; it is now rapidly gaining popularity [38]. Currently, hypochlorite solutions with an available chlorine content of up to 10 g/L are commonly produced on the site in electrochemical cells by using either prepared brine or natural seawater as feed.

4.2.1. Reaction Fundamentals

Electrolysis of sodium chloride yields chlorine at the anode:

$$2\,Cl^- \longrightarrow Cl_2 + 2\,e^-$$

The final product depends on the operational conditions of the cell. In the production of chlorine gas, special care is taken to prevent mixing of anode and cathode products ($\rightarrow$ Chlorine). Chlorine hydrolyzes and hypochlorous acid dissociates, forming hypochlorite and chloride in solution [39], [40]:

$$Cl_2 + H_2O \rightleftharpoons HClO + Cl^- + H^+$$
$$HClO \rightleftharpoons ClO^- + H^+$$

The formation of hypochlorous acid and hypochlorite ceases when the electrolyte is saturated with chlorine and chlorine gas evolves at pH 2–3.

However, in the electrosynthesis of hypochlorite, as well as of chlorate, anolyte and catholyte are vigorously mixed. The hydroxyl ions formed at the cathode

$$2\,H_2O + 2\,e \longrightarrow 2\,OH^- + H_2$$

maintain the electrolyte near neutrality (pH 7–9). Under this condition, the concentration of dissolved chlorine near the anode surface remains too low to permit evolution of gaseous chlorine, and hypochlorite is the main product.

There are four main loss reactions that complicate industrial operation; they have been studied recently to optimize the operating conditions [41], [42] and are described in the following paragraphs in more detail.

Cathodic Reduction. Hypochlorite is reduced at the cathode to form chloride:

$$ClO^- + H_2O + 2\,e^- \longrightarrow Cl^- + 2\,OH^-$$

The rate of this reaction is controlled by mass transfer and is, thus, linearly proportional to the overall hypochlorite concentration; it increases with flow rate and temperature [42]–[45].

Loss by cathodic reduction also occurs in chlorate production; in that case it is minimized by adding dichromate to the electrolyte (see Section 7.2.2). This remedy is

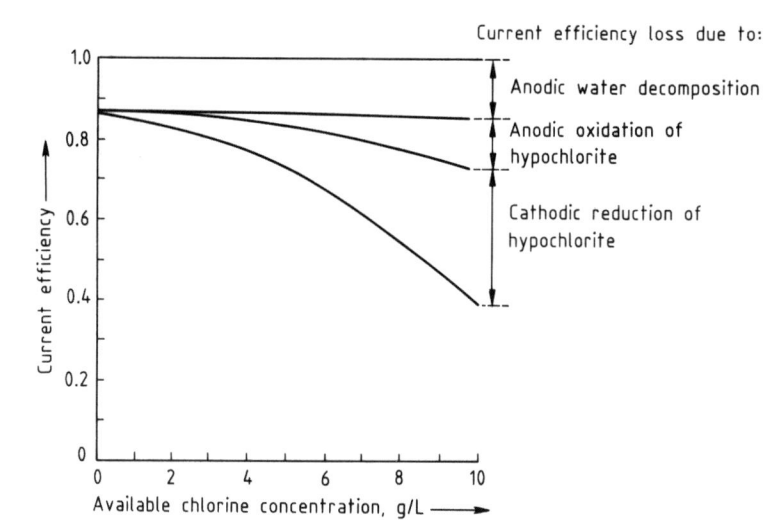

Figure 2. Current efficiency loss vs. available chlorine concentration (NaCl: 28 g/L, 25 °C, 1550 A/m²) [46]

not possible in the simple flow-through hypochlorite cells, and loss in commercial cells is considerable, as Figure 2 shows.

Several means are used to lower reduction loss. Smooth cathode surfaces are superior to rough surfaces [45]. Loss is also lowered by decreasing the active area of the cathodes, thereby increasing the cathode current density. Cathodic reduction was greatly suppressed by decreasing the active area to ca. 1% with a synthetic resin coating [42]; however, the attendant voltage increase makes this approach industrially unattractive. Bubbles always cover a portion of the cathode surface and contribute to lower cathodic reduction. Hydrogen evolution increases cathodic reduction because its stirring action enhances the mass transfer coefficient. Efforts to lower the mass transfer coefficient by covering the cathode with a porous plastic have also been successful at suppressing cathodic reduction [47]. Similarly, the inhibitory action of calcium chloride was related to the formation of insoluble compounds on the cathode surface [48].

Anodic Oxidation. Anodic oxidation of hypochlorite to chlorate is used in industry and is described in detail in Section 7.3:

$$3\ ClO^- + 1.5\ H_2O \longrightarrow ClO_3^- + 3\ H^+ + 2\ Cl^- + 0.75\ O_2 + 3\ e^-$$

This reaction is also controlled by mass transfer, and its rate increases with the hypochlorite concentration. Because hypochlorite is decomposed at the anode and cathode, commercial on-site production is restricted to a maximum concentration of available chlorine of ca. 10 g/L; this is usually sufficient for disinfecting and deodorizing water. The contribution of anodic oxidation to the overall loss can further be lowered by increasing the anodic current density; however, the anode lifetime may then decrease. The loss reaction may also depend on the anode material: dimensionally stable anodes based on RuO_2 were found to be more efficient than platinized titanium or graphite [42].

Anodic Water Electrolysis. This loss reaction competes with chlorine discharge, and its rate depends on the chloride concentration, various mass transfer coefficients, and the nature of the anode material.

$$H_2O \longrightarrow 1/2\, O_2 + 2\, H^+ + 2\, e^-$$

When NaCl concentrations are greater than 100 g/L, the loss is small. However, a typical NaCl concentration in a modern brine cell is ca. 30 g/L, at which the rate of water decomposition is significant, as shown in Figure 2.

When salt must be purchased, the conversion of chloride into hypochlorite must be maximized. A typical on-site electrosynthesis from prepared brine consumes 3–5 kg of salt per kg of available chlorine produced. The question of chloride consumption is meaningless where natural seawater is used as a feed. However, in cases of low salinity or very cold seawater, oxygen evolution caused by anodic water electrolysis may decrease current efficiency by as much as 40% [49].

As in the case of chlorate synthesis, the nature of the anode material strongly influences the amount of water decomposition. Platinized titanium and dimensionally stable anodes based on RuO_2 are more selective for chlorine evolution, whereas PbO_2 and graphite anodes have a greater tendency to evolve oxygen.

Chemical Chlorate Formation. Autoxidation of hypochlorite to chlorate is the preferred route of commercial chlorate production (Section 7.3). The rate of this reaction depends on pH and temperature; in industrial hypochlorite production, it is minimized by keeping the temperature below 40 °C and the pH above 7 [41].

4.2.2. Industrial Cells

The number of companies offering hypochlorite cells has grown quickly in the past few years. Currently, over 20 suppliers offer a wide variety of cell designs. They can all be classified into three basic types: (1) tube cells, (2) parallel-plate cells, and (3) rotating or mechanical scraper-type cells. Of the modern hypochlorite cells, the tube cells were developed first. They generally consist of two concentric pipes, one being the anode and the other the cathode, with the annular space serving as the electrode gap. This type of cell may be operated under pressure and is well-suited for small applications. Parallel-plate cells achieve a much better packing of electrode area, and most large industrial installations use cells of this type. Only two of the manufacturers offer rotating or mechanical scraper-type cells. In theory, such cells should be able to operate indefinitely without deposit buildup, and would therefore result in low

Table 4. Typical operational data of electrolytic cells for hypochlorite generation

Parameter	Brine feed	Seawater feed
Current density, A/m^2	1500	1500
Current efficiency, %	65	90
Temperature, °C	25	5–25
Concentration		
NaCl (cell entrance), g/L	30	15–30
Available chlorine (cell exit), g/L	8–10	1.0–3.0
Energy consumption, kW/h/kg of available chlorine	4.5–5.0	3.3–4.1
Sodium chloride consumption, kg of NaCl per kg of available chlorine	3–3.5	

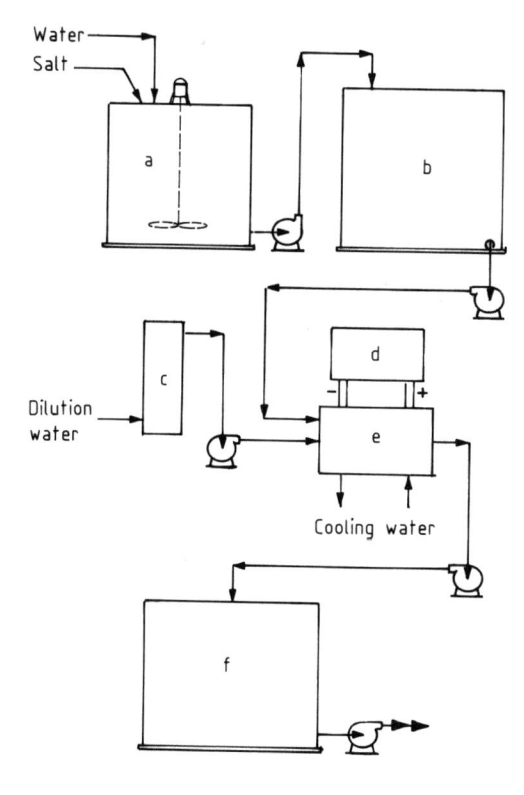

Figure 3. Typical layout for a brine hypochlorite cell system
a) Automatic brine makeup; b) Brine storage tank; c) Water softener; d) Rectifier; e) Hypochlorite cell; f) Hypochlorite storage tank

maintenance cost. In practice, however, their extra mechanical action is difficult to maintain, and such cells have not yet captured a significant market share.

Seawater Cells. Cells designed to operate by using natural seawater are different from those using prepared brince. Cells using seawater tend to operate at a higher electrolyte flow rate and a wider electrode gap than brine cells to minimize problems arising from cathode deposits. Seawater cells also produce a lower product concentration, typically solutions with an available chlorine content of 0.5–4.0 g/L. Hypochlorite concentration is kept low to maximize current efficiency whenever high salt usage is not a concern (see also Table 4).

Brine Cells. Brine cells usually produce solutions with an available chlorine content of 7–10 g/L to keep salt cost low. Most manufacturers offer brine cells of the same design as their seawater cells, but electrolyte flow rate is lowered to maximize current efficiency at the higher concentration of available chlorine. Figure 3 shows a flow diagram for a typical system designed for electrolysis using prepared brine. Although brine cells were developed first, they now have only a small share of the total on-site hypochlorite cell market.

Table 4 shows typical operational data of seawater and brine feed cells.

Sanilec System. Eltech System Corp. first offered the Sanilec system for seawater electrolysis in 1973. The cells are of a parallel-plate design (Fig. 4) and feature once-through operation without recycle [31]. Cells producing 30, 70, 140, or 155 kg/d can be banked in series to produce hypochlorite solutions of available chlorine concentrations up to 3.0 g/L. Full-load a.c. power consumption is as low as 4.1 kW h per kg of chlorine, partly because dimensionally stable anodes are used. These cells use cathodes made of nickel alloys and remove hydrogen periodically to lower power consumption. Typical operational data are shown in Table 4.

Chloropac System. Englehard Minerals & Chemical Corp. produces tube cells with concentric titanium pipes as anode and cathode under the trade name of Chloropac System [50]. These cells are designed for an operating pressure of 1000 kPa. They utilize a high seawater flow rate to minimize deposit formation.

As shown in Figure 5, the electrodes are assembled in a bipolar arrangement, which is particularly suited for smaller capacities. The platinized titanium anodes are coated with 5 µm of platinum and have good current efficiency for chlorine evolution. The power consumption is claimed to be 3.5–5.0 kW h per kg of chlorine.

M. G. P. S. System. Mitsubishi Heavy Industries offers monopolar plate-type cells arranged in series under the name of the M. G. P. S. system. The cell body is made from mild steel lined with rubber, and the system is well-suited for large industrial applications. Cathodes are made from titanium and anodes from platinized titanium; precious metal oxide anodes are also used. The available chlorine concentration ranges from 0.2 to 1.0 g/L, and the power consumption is reported to be 5.8 kW h per kg of chlorine.

Seaclor System. The Oronzio de Nora Seaclor system features a bipolar parallel plate-type seawater cell [51] shown in Figure 6. Cathodes are made from titanium and anodes are based on RuO_2. Hypochlorite solutions with an available chlorine content of up to 2.5 g/L are produced; power consumption is cited to be 3.4–4.5 kW h per kg of chlorine. The most notable feature of the Seaclor system is the size of the individual cells, which may be large enough to produce nearly 1000 kg/d.

Figure 4. Sanilec seawater electrolysis cell
a) Molded polypropylene cell body; b) O-ring seal; c) Dimensionally stable anodes; d) Seawater inlet; e) Clear acrylic cover (not shown)

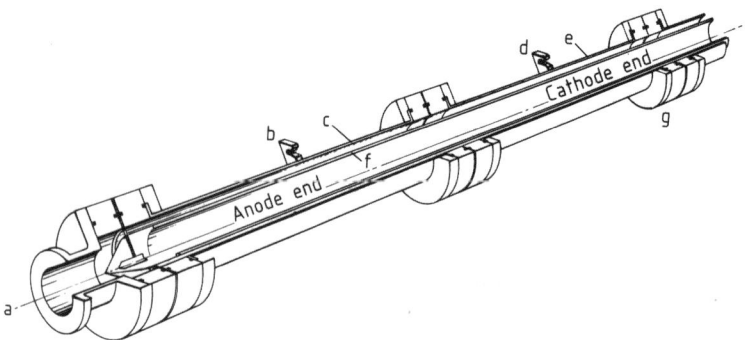

Figure 5. Chloropac seawater cell
a) Inlet; b) Cathode connector; c) Cathode; d) Anode connector; e) Anode; f) Bipolar electrode; g) Insulating flange

Pepcon System. The Pepcon system made by Pacific Engineering & Production Co. of Nevada is distinctly different from other commercial cells because it employs less expensive PbO_2 anodes. Pacific Engineering uses a tube-type design, with a steel or titanium pipe forming the outside of the cell and a PbO_2-plated graphite rod at the center. Steel has a small overpotential for hydrogen evolution, but it requires careful protection to prevent corrosion during shutdown.

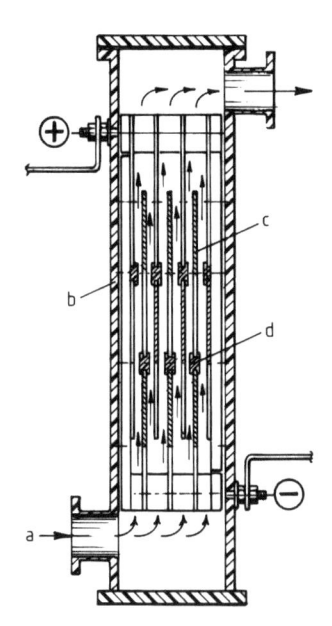

Figure 6. Seaclor hypochlorite cell
a) Inlet; b) PVC cell body with overlay; c) Bipolar electrode; d) Insulator

Maintenance. The major problem of on-site hypochlorite cells, especially those using natural seawater as feed, is that of deposit formation. Because the electrolyte adjacent to the cathode is strongly alkaline, magnesium hydroxide and calcium hydroxide deposit at the cathode surface. If allowed to build up, these deposits may bridge the electrode gap, reducing cell efficiency and ultimately causing anode failure [52].

Most manufacturers of industrial hypochlorite cells will tolerate deposit formation, but simultaneously attempt to minimize the problem by controlling the current density, turbulence, and cathode surface. In the case of seawater the deposit — chiefly $Mg(OH)_2$ — is soft and can generally be scoured from the cathode surface and flushed from the cell by using a high flow rate. Although at least two suppliers suggest that acid cleaning of deposits is never necessary, it seems likely that occasional removal of deposits is required on a cycle that varies from a few days to a few months. Usually, the deposit is removed by flushing the cell with hydrochloric acid. This dissolves the deposit quickly and consumes very little acid. Large industrial installations usually provide for convenient acid washing.

Brine cells may also form deposits as a result of hardness ions introduced from either feed water or from impurities in the salt. In this case, the deposit will usually consist of a hard calcium carbonate, which is also easily removed with acid. Unlike in seawater cells, however, it is often practical to soften the cell feed; brine cells can then be designed to operate with very little deposit formation.

A second major maintenance item is occasional replacement of anodes. The lifetime of both precious metal oxide and platinized titanium anodes is adversely affected by high current density, low salinity, low electrolyte temperature, and severe deposit formation. Manufacturers should be consulted on the expected anode lifetime and cost of replacement. Typically, anodes must be replaced or recoated after 2–5 years from startup.

4.3. Storage

Hypochlorite solutions slowly decompose if catalytic amounts of cobalt, nickel, or copper are present; iron and magnesium do not act catalytically [53]. Therefore, long-term storage and transport over long distances must be avoided. Hypochlorite solutions prepared from seawater are especially unstable and should be consumed directly after on-site preparation.

Hypochlorite solutions for household use have an available chlorine concentration of ca. 40 g/L. Special attention must be paid to minimize oxygen evolution. Improper storage of hypochlorite bottles may cause stoppers to be blown out and bottles to explode. Vented stoppers are used to avoid pressure buildup in the bottles.

4.4. Uses

Concentrated hypochlorite solutions are used primarily in the paper and textile industries for bleaching. Since chlorine dioxide (Chap. 5) produces a brighter product and is less harmful to fibers, the use of a hypochlorite solution as bleaching agent in these industries has declined.

On the other hand, the use of electrochemically produced dilute hypochlorite solutions has increased largely over the past several years. Electrolytic generators can be used whenever hypochlorite is needed for disinfection or for bleaching; their advantages are economy, safety, and convenience.

Currently, the broad area of biofouling control accounts for over half of the market for such equipment, especially at locations remote from chlor-alkali plants. The largest seawater electrolysis plants for the production of hypochlorite in the world, a 60 000-kg/d plant in Kuwait and a 48 000-kg/d plant in Saudi Arabia, were commissioned in 1980 as parts of large desalination projects. The hypochlorite is used to control slime and algae in piping and tubes and to eliminate odor in the desalted water.

Another important and quickly growing use for hypochlorite cells is disinfection of seawater for secondary oil recovery; otherwise, slime growth would clog the oil-bearing strata. The largest plant for this purpose is a 5400-kg/d plant at Qurayyah, Saudi Arabia, where disinfected seawater is filtered, deaerated, and pumped 100 km inland for injection.

Shipboard applications account for many seawater electrolysis installations worldwide. In this case, smaller units are used to inhibit marine growth in seawater systems for sanitary services and for distilled water treatment. Fishing fleets also use hypochlorite as a disinfectant for storage.

Coastal utilities and industrial plants use hypochlorite from seawater electrolysis to control mollusks, algae, and slime, which may block seawater intakes, clog piping, and reduce heat transfer efficiency. Inland utility and industrial plants use brine hypochlorite cells when handling and safety are of primary concern. Nuclear power plants are particularly sensitive to the hazards of liquid chlorine.

Both coastal and inland wastewater treatment plants are major users of chlorine for disinfection before discharge. This was first started on the island of Guernsey in 1966 and was operated for 6 years [54].

On-site hypochlorite production from brine is also used for drinking water treatment, but it is only economical at remote locations or where safety is a major concern.

A number of companies have introduced on-site generators to supply chlorine demand for swimming pools. Many of these have been small chlor-alkali cells with separated anolyte and catholyte; but recently unseparated hypochlorite generators have gained acceptance and now represent an important portion of that market.

On-site hypochlorite generators are sometimes selected for other less common applications, such as textile manufacturing, industrial and laundry bleaching, cyanide destruction, odor control, ocean aquariums, and food processing [41], [55].

4.5. Economic Aspects

The Middle East now represents about half of the total market for electrolytic hypochlorite generation. The rest of the market is roughly balanced between East Asia, South America, and the United States. The total installed capacity has grown from ca. 200 t/d in 1979 to estimated 700 t/d in 1984. This probably represents total system sales approaching U.S. $ 200×10^6 through 1984.

On-site hypochlorite generation is difficult to compare with purchased chlorine or hypochlorite because chemical costs vary widely from nation to nation. Bulk liquid chlorine prices vary from ca. $ 150/t in the United States to well over $ 1000/t in remote locations, and hypochlorite prices vary from $1500/t of available chlorine content in the United States to over $ 2500/t where transportation cost is high. Therefore, a direct economic comparison with purchased chemicals can only be made for each location individually. There are many cases in which the selection of on-site generation is motivated primarily by economics, rather than by safety or convenience.

4.6. Plant Safety

Dilute sodium hypochlorite solution is much safer than liquid or gaseous chlorine. Sodium hypochlorite solution, as produced by on-site electrolytic generators (with an available chlorine content of 0.5 – 10.0 g/L), is regarded as corrosive and as an irritant when ingested or inhaled. It is also a mild skin irritant, and prolonged exposure may result in a burn or rash [56].

The primary safety concern associated with the electrolysis equipment is the explosion and fire hazard from byproduct hydrogen. This hazard is increased if oxygen is present in the cell gas. Oxygen content ranges from 3% to 9% in efficiently operating cells, but it varies widely in practice. Cells with anodes having poor selectivity for chlorine evolution or with poor design for operation at low temperature and low salinity may result in an oxygen concentration of 40% or more in the cell gas.

The explosion limit for hydrogen–oxygen mixtures is 6.0% O_2. Explosive impact will be soft at this point, but will increase quickly in severity as oxygen content increases. Two measures are taken in commercial cells against this: (1) dilute the byproduct hydrogen with sufficient air to less than 4.0%, the explosion limit for hydrogen in air, or (2) allow the cell gas to vent with proper precaution, e.g., the installation of rupture disks.

A special hazard arises when cell deposits are washed with dilute hydrochloric acid. In addition to the hazards of handling hydrochloric acid, chlorine gas evolves when HCl is accidentally mixed with stored hypochlorite solution, or when the cells are started without first flushing out the acid.

5. Chlorine Dioxide

5.1. Properties

Chlorine dioxide [*10049-04-4*], ClO_2, M_r 67.45, is a yellowish-green to orange gas. It can be condensed to a reddish-brown liquid at 11 °C and solidified to orange red crystals at − 59 °C. The density of liquid ClO_2 is as follows [57]:

t, °C	− 33	− 21	− 17	5
ϱ, g/cm³	1.907	1.788	1.735	1.635

Chlorine dioxide has an irritant, pungent odor that resembles that of a mixture of chlorine and ozone. Chlorine dioxide is an extremely unstable gas, readily decomposing into chlorine and oxygen even on mild heating. It is explosive as a gas or liquid at high concentration [58], [59]. However, it can be handled easily when it is diluted with air to less than 15 vol%.

Table 5. Solubility of chlorine dioxide in water [60]

25 °C		40 °C		60 °C	
p, mbar	c, g/L	p, mbar	c, g/L	p, mbar	c, g/L
46	3.01	74.9	2.63	141.2	2.65
29.5	1.82	45.7	1.6	71.6	1.18
17.9	1.13	25.2	0.83	28.4	0.58
11.2	0.69	13.2	0.47	16.0	0.26

Chlorine dioxide is easily soluble in water (heat of solution: − 26.8 kJ/mol); at 10 °C its solubility in water is 5 times that of chlorine (Table 5).

Chlorine dioxide can be easily driven out of aqueous solutions with a strong stream of air. When an aqueous solution is cooled, $ClO_2 \cdot 8\,H_2O$ crystals precipitate. Chlorine dioxide is also soluble in carbon tetrachloride, sulfuric acid, or acetic acid [61].

Other important properties of chlorine dioxide are its photochemical [62] and thermal [63] decomposition. Solutions of chlorine dioxide are relatively stable in the dark, but they decompose into chlorine and oxygen when exposed to light. Therefore, these solutions are very strong oxidants and extremely reactive and corrosive. They attack all metals except platinum, tantalum, and titanium.

In dilute aqueous solutions, chlorine dioxide oxidizes hydrocarbons to ketones and alcohols [64]. In the absence of water, chlorine dioxide loses its bleaching power.

5.2. Production

Because of the explosion risk, chlorine dioxide is manufactured on site. Its industrial production is based on the reduction of chlorate [65]–[67]. Undesirable byproducts are chlorine and chloride. For analysis of chlorine dioxide, see [68]–[71].

Reaction Mechanism [72], [73]. The basic reaction mechanism of chlorine dioxide formation is the same for all known processes. All processes use chlorate as the raw material; in all processes chlorine dioxide formation takes place in strong acidic solutions; and byproduct chloride is found in all generator solutions.

The presence of chloride ions is essential for the formation of chlorine dioxide [74], [75]; this is evident from the following facts:

1) No significant amount of chlorine dioxide is formed in acidified chlorate solutions with various reducing agents when chloride has previously been removed from the reaction system by adding silver sulfate.
2) If chlorine dioxide generators are allowed to stand overnight, minor ClO_2 formation consumes all of the chloride present in the reaction mass. When the supply of fresh raw materials is started the next morning, a specific chloride level must build up before chlorine dioxide can be produced at the desired rate. This chloride buildup

phase can be eliminated by deliberately adding fresh chloride to the generator in the beginning.

Independent on the choice of reducing agent, the primary reaction for chlorine dioxide production is the reaction between chloric acid and hydrochloric acid to form chlorine dioxide and chlorine. Traces of Mn^{2+} and Ag^+ ions catalyze the reaction [76].

$$HClO_3 + HCl \longrightarrow HClO_2 + HClO$$
$$HClO_3 + HClO_2 \longrightarrow 2\,ClO_2 + H_2O$$
$$HClO + HCl \longrightarrow Cl_2 + H_2O$$
$$\overline{2\,HClO_3 + 2\,HCl \longrightarrow 2\,ClO_2 + Cl_2 + 2\,H_2O}$$

In industrial chlorine dioxide production, sulfur dioxide, hydrochloric acid, or methanol are used as reducing agents. Other reducing agents are not economical.

5.2.1. Day–Kesting Process [74], [77], [78]

This process combines the production of chlorine dioxide from sodium chlorate, with hydrochloric acid as reducing agent, and electrochemical sodium chlorate production from sodium chloride:

$$2\,NaCl + 6\,H_2O \xrightarrow{Electrolysis} 2\,NaClO_3 + 6\,H_2$$
$$2\,NaClO_3 + 4\,HCl \longrightarrow 2\,NaCl + 2\,ClO_2 + Cl_2 + 2\,H_2O$$

The efficiency of chlorine dioxide production depends on how far the rate of the competing chlorine production can be decreased:

$$NaClO_3 + 6\,HCl \longrightarrow 3\,Cl_2 + NaCl + 3\,H_2O$$

This can be achieved by maintaining a high chlorate concentration and a low hydrochloric acid concentration in the reaction system.

The following description incorporates recent major improvements :

Chlorate solution is rapidly circulated from a large storage tank through the electrolytic cells, which oxidize chloride to chlorate. From the same tank the chlorate solution is also circulated slowly through the chlorine dioxide generator. The reaction takes place in a heated multicompartment column. The reactant solutions, chlorate and hydrochloric acid, are added from the top and air is introduced from the bottom of the column [79]. Chlorate is reduced with hydrochloric acid to chlorine dioxide and chlorine. Both gases are stripped from the system with air; at the same time, chlorine dioxide is diluted below the explosion limit [80]. The depleted chlorate solution is then returned to the storage tank and recycled to the chlorate electrolysis. The mixture of chlorine dioxide and chlorine is stripped with water to give a solution rich in chlorine dioxide [80], [81] and a gas phase rich in chlorine. Chlorine can then be recycled to the process as hydrochloric acid by reduction with hydrogen produced in the electrolytic cells or neutralized with NaOH solution to produce hypochlorite.

Although it has always been cheaper to produce chlorine dioxide by this process, it never gained momentum because of the high initial capital expenditure involved.

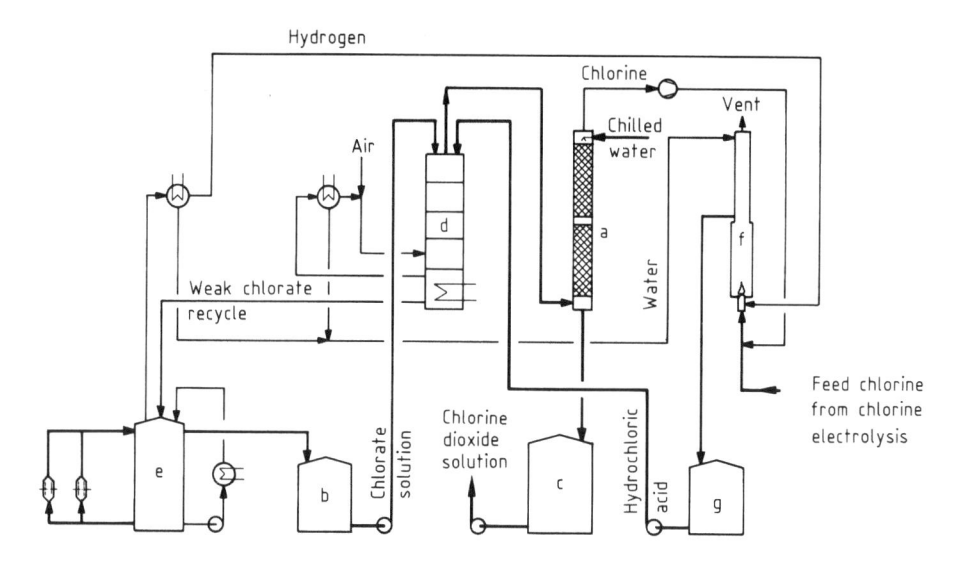

Figure 7. Schematic diagram of the integrated Day–Kesting process [82]
a) Chlorine stripper; b) Sodium chlorate solution; c) Chlorine dioxide solution storage; d) Chlorine dioxide generator; e) Chlorate electrolysis; f) Hydrochloric acid furnace; g) Hydrochloric acid storage

Lurgi [79], [82]–[86], Chemetics [87], [88], and others [89]–[91] have developed integrated systems that combine this process with the production of chlorine, caustic soda, and hydrochloric acid. Such combined systems use sodium chloride and electric power as the raw materials and produce chlorine dioxide solution and caustic soda (Fig. 7).

5.2.2. R2 Process [92]–[100]

Hooker Chemical Corporation, together with Electric Reduction Company, developed the R2 process, starting from the single-vessel process [101]–[103], and Electric Reduction Company further improved the details. Chlorine dioxide is produced according to the following overall reaction:

$$2\ NaClO_3 + 2\ NaCl + 2\ H_2SO_4 \longrightarrow 2\ ClO_2 + Cl_2 + 2\ Na_2SO_4 + 2\ H_2O$$

An undesirable side reaction produces additional chlorine:

$$NaClO_3 + 5\ NaCl + 3\ H_2SO_4 \longrightarrow 3\ Cl_2 + 3\ Na_2SO_4 + 3\ H_2O$$

In continuous industrial processes, the molar ratio of chlorine dioxide to chlorine is nearly 2:1, indicating that the ClO_2 production efficiency is almost 100% (based on chlorate).

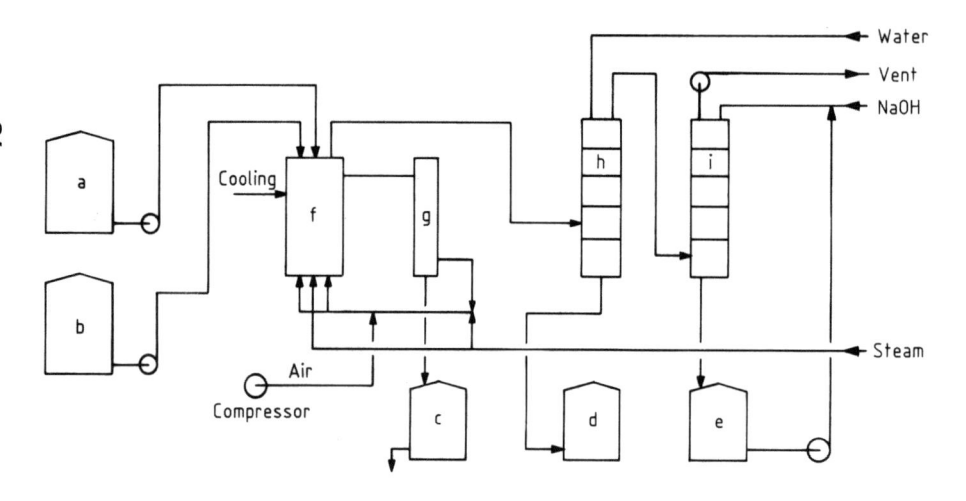

Figure 8. Schematic diagram of the R2 process [106]
a) NaCl + NaClO$_3$ storage; b) Acid storage; c) Spent acid; d) ClO$_2$ solution storage; e) NaOCl solution storage; f) ClO$_2$ generator; g) Stripper; h) ClO$_2$ absorption tower; i) Cl$_2$ absorption tower

Figure 8 shows a schematic diagram of the R2 process. Concentrated solutions of sodium chlorate, sodium chloride, and sulfuric acid (equimolar ratio) are metered to a vigorously agitated reaction vessel. Air is blown into the reactor through porous plates. Under optimum conditions, the reaction mass contains 0.1–0.2 mol/L of sodium chlorate, 4.5–5 mol/L of sulfuric adic and 0.02–0.08 mol/L of sodium chloride. Chlorine dioxide is absorbed from the gas phase in packed towers in cold water, and chlorine leaves the system as byproduct. The liquid effluent from the reactor is a mixture of sodium sulfate and sulfuric acid. The process can also be operated in such a way that sodium hydrogen sulfate crystallizes from the effluent solution [104], [105]. It provides sulfate for kraft pulping; the sulfuric acid is recovered for reuse.

The R2 process has been modified [105] and optimized [103], [107], [108] to suit the requirements of the industry. Currently, it is probably the most extensively used process for the production of chlorine dioxide.

5.2.3. Mathieson Process [67], [74], [96], [98], [106]

Sulfur dioxide is the reducing agent in the Mathieson process. The main overall reaction is as follows:

$$2\ NaClO_3 + H_2SO_4 + SO_2 \longrightarrow 2\ ClO_2 + 2\ NaHSO_4$$

Solutions of sodium chlorate and sulfuric acid are added continuously to a relatively large, cylindrical, lead-lined tank (primary reactor) from the top. Sulfur dioxide, diluted with air, is introduced through gas diffusion plates at four points in the bottom of the tank. The reaction mass overflows to a smaller secondary reactor of similar construction. The generated chlorine dioxide still contains some unreacted sulfur dioxide; it is stripped from the reaction mass by air and then washed in a scrubber packed with Raschig rings. The scrubber is installed on top of the primary generator, and the fresh sodium chlorate solution serves as a washing liquid on its way down to the primary generator. The mixture of chlorine dioxide, air, and chlorine coming out of the scrubber goes to the absorption tower.

5.2.4. Solvay Process [67], [74], [96], [98], [109]

The Solvay process uses methanol as reducing agent. The main overall reaction is as follows:

$$2\,NaClO_3 + CH_3OH + H_2SO_4 \longrightarrow 2\,ClO_2 + HCHO + Na_2SO_4 + 2\,H_2O$$

The reaction between sodium chlorate, methanol, and sulfuric acid takes place in two jacketed, lead-lined steel reactors. The chemicals are added to the bottom of the first vessel and flow by gravity from one reactor to the other. Each vessel has its own supply of methanol and air. The two reactors are operated at different temperatures and chlorate concentrations. Up to 70% of the reaction is completed in the first reactor. Additional methanol and up to 10% of the total acid requirement is fed to the second reactor; the conversion of chlorate is then more than 95%.

5.2.5. Other Processes

The *Holst process* [110] – [112] is a batch version of the Mathieson process and has not been exploited because of its low chlorine dioxide yield. When solid sodium chlorate is successively added, the unreacted sulfuric acid from the previous operating period can be utilized and the conversion efficiency of chlorate increases to 83%.

The *Persson process* [74], [113] has only historical value. The main raw materials were sodium chlorate and sulfur dioxide, but the latter was used to reduce chromic acid to chromic sulfate, and this in turn reduced chlorate to chlorine dioxide.

The *CIP process* [114], [115] was in operation long before it was published, maintaining secrecy for several years. Concentrated sodium chlorate solution is carefully metered and distributed to a packed reaction tower. Sulfur dioxide gas, diluted with air or nitrogen, enters from the bottom, reacts with the chlorate, and is oxidized to sulfuric acid. Chlorate is reduced to chlorine dioxide. The additional inert gas dilutes the ClO_2 formed to a safe concentration, and the mixture is led to the absorption tower.

Small-Scale Production from Sodium Chlorite [116]–[119]. Small-scale consumers produce chlorine dioxide by passing chlorine gas through a sodium chlorite solution:

$$2\ NaClO_2 + Cl_2 \longrightarrow 2\ NaCl + 2\ ClO_2$$

The reaction is almost quantitative.

The alternative acidification of sodium chlorite solution with hydrochloric acid is also commonly used:

$$5\ NaClO_2 + 4\ HCl \longrightarrow 4\ ClO_2 + 2\ H_2O + 5\ NaCl$$

5.3. Economic Aspects

The economy of chlorine dioxide production depends on the efficient utilization of the byproducts of the process [97], [104]. When the chlorine dioxide production is combined with chlor-alkali membrane cells and hydrochloric acid synthesis (Lurgi concept), chlorate cells work most efficiently [82]–[90]. Byproducts from one unit can then be used as raw materials or make-up chemicals for the others. In this way, high initial capital cost is paid back in the long run by a far lower operating cost. The R2 process offers similar possibilities if it is combined with a paper mill [95], [107], [120].

5.4. Uses

Chlorine dioxide is the most widely used bleaching agent today, in particular for high-quality cellulose. It destroys lignin without attacking cellulose, yielding a characteristically white cellulose. The general trend is to eliminate chlorine and hypochlorite as bleaching agents altogether [121], [122] and replace them with chlorine dioxide [81], [123], [124].

Chlorine dioxide is used in the pulp and paper, textile, and food industries. In the pulp and paper industry, a unique whiteness can be achieved in kraft pulp, sulfite pulp, and soda pulp [96]. In the textile industry, chlorine dioxide produces high-quality fibers with additional special advantages. Shrinkproof wool owes its quality to the reaction of chlorine dioxide with the cross-linking sulfur atoms of the wool.

Chlorine dioxide is also used in sanitization, e.g., of industrial and municipal waters, sewage, algae, or decomposed vegetables. Waterworks use chlorine dioxide to handle taste and odor problems of household water [68], [118], [119], [125].

6. Sodium Chlorite

6.1. Properties

Sodium chlorite [7758-19-2], $NaClO_2$, M_r 90.45, is the sodium salt of the unstable chlorous acid; it exists as an anhydrous and a trihydrated form (transition point 38 °C). Very pure $NaClO_2$ crystals are white, but they usually have a greenish tint because traces of chlorine dioxide are present.

The stability of sodium chlorite lies between that of hypochlorite and chlorate. For further information, see [126]. Sodium chlorite is not sensitive to impact if organic matter is excluded. It can be struck with a clean metal surface without detonation. However, in the presence of organic matter — the film usually occurring on a hammer suffices — the impact may result in a spontaneous puffing. As a strong oxidizing agent, solid sodium chlorite forms explosive mixtures with such oxidizable materials as sulfur, powdered coal, metal powders, or organic compounds. Sodium chlorite solutions should never be allowed to dry on fabrics because this would result in a flammable combination.

Sodium chlorite is soluble in water: [126]

t, °C	5	17	20	30	40	45	50	60
Solubility, wt %	34	39	40.5	46	50.7	53	53.7	55

More important is its solubility in the presence of caustic soda (Table 6), sodium chlorate [127], sodium chloride [128], [129], and sodium carbonate [130]. For the system $NaClO_2 - NaCl - NaClO_3 - H_2O$, see [131], [132].

Aqueous solutions of sodium chlorite must be protected from light. At low pH (approximately 2), chlorite solutions contain chlorous acid that decomposes to form chlorine dioxide and chlorate:

$$4\ HClO_2 \longrightarrow 2\ ClO_2 + HClO_3 + HCl + H_2O$$

At pH 3–4, decomposition slows down. Alkaline solutions are stable, and dilute solutions can even be boiled without decomposition. Concentrated alkaline solutions of sodium chlorite slowly decompose when heated [133]–[135]:

$$3\ NaClO_2 \longrightarrow 2\ NaClO_3 + NaCl$$

The reaction of sodium chlorite with hypochlorite depends on pH [135]. At low pH, the reaction produces chlorine dioxide, whereas at high pH, chlorate is formed. The reaction with chlorine produces chlorine dioxide and sodium chloride:

$$2\ NaClO_2 + Cl_2 \longrightarrow 2\ ClO_2 + 2\ NaCl$$

Crystalline sodium chlorite is slightly hygroscopic without caking; it is stabilized with alkali for long-term storage. When heated to 180–200 °C, it decomposes partially

Table 6. Solubility in the $NaOH-NaClO_2-H_2O$ system at 30°C

d_4^{30}	c_{NaOH}		c_{NaClO_2}		Solid phase
	wt%	g/L	wt%	g/L	
–	55.3	–	0	–	
–	51.1	–	1.5	–	$NaOH-H_2O$
1.568	51.2	802.8	1.7	26.7	
1.575	50.6	797.0	2.45	38.6	$NaOH + NaClO_2$
1.546	48.9	756.0	2.5	38.7	
1.519	42.3	627.3	5.1	77.5	
1.496	35.9	537.1	7.65	114.4	
1.474	31.5	464.3	10.5	154.8	
1.453	26.8	389.4	14.9	216.5	$NaClO_2$
1.457	23.6	343.9	18.7	272.5	
1.447	20.3	293.7	22.4	322.7	
1.451	16.0	232.2	28.2	409.7	
1.457	15.4	224.4	29.2	425.4	
1.469	10.3	151.3	36.8	540.6	
1.469	9.5	139.6	38.0	558.2	
1.471	9.0	132.4	38.8	570.7	
1.473	7.86	115.7	40.5	596.6	
1.473	6.2	91.3	42.9	631.9	unstable
1.468	5.8	85.1	43.1	632.7	unstable
1.476	6.1	90.0	42.6	628.8	
1.441*	3.55	51.2	43.5	626.8	$NaClO_2 \cdot 3 H_2O$
1.410**	0.0	–	45.7	644.7	

* The solid phase was analyzed.
** The solutions contain traces of NaCl (0.06 wt%).

to sodium chlorate and sodium chloride or completely to sodium chloride and oxygen [126]. For the analysis of sodium chlorite, see [69], [70], [136]–[138].

6.2. Production

Sodium chlorite is produced by treating chlorine dioxide with caustic soda [139]–[141]:

$$2\ ClO_2 + 2\ NaOH \longrightarrow NaClO_2 + NaClO_3 + H_2O$$

The reaction products, sodium chlorite and sodium chlorate, have nearly the same solubility in water and are difficult to separate. Industrial sodium chlorite production uses the following procedure [142]. The absorption of chlorine dioxide in caustic soda solution and the simultaneous reduction of chlorate to chlorite are achieved by a suitable reducing agent, such as hydrogen peroxide [143]:

$$2\ ClO_2 + 2\ NaOH + H_2O_2 \longrightarrow 2\ NaClO_2 + 2\ H_2O + O_2$$

The product is a 33 wt% solution of sodium chlorite, which is then converted to a dry solid containing ca. 80 wt% of sodium chlorite, the rest being stabilizers.

Numerous complex chlorites have been prepared [144], but only sodium chlorite has proven to be of any commercial value.

Efforts to produce sodium chlorite by electrolysis, similar to hypochlorite or chlorate, have not yet been successful. Its synthesis from chlorine dioxide and sodium amalgam [145] could not be realized because redox potential [146] and pH [147] were difficult to control. Other processes for the direct reduction of chlorine dioxide were not efficient [148], [149].

6.3. Uses

Sodium chlorite is a very efficient bleaching agent. Its oxidation potential allows a controlled bleaching that is not attainable with other bleaching agents. Therefore, it is widely used as a bleaching agent in the textile industry [133]–[135], [150]–[153].

Another important use of sodium chlorite is the small-scale production of chlorine dioxide; see Section 5.2.5 [116]–[119].

7. Chloric Acid and Chlorates

Chloric acid is not being produced on an industrial scale. However, sodium chlorate and potassium chlorate have outstanding industrial significance. Sodium chlorate is produced on a very large scale by one of the most important inorganic electrosyntheses. All other chlorates are produced in much smaller amounts for special purposes, usually from sodium chlorate.

7.1. Properties

Physical Properties. Anhydrous *chloric acid* [7790-93-4], $HClO_3$, M_r 84.46, is unstable and explosive. Dilute aqueous solutions are colorless and odorless; they are stable at low temperature if catalytically active contaminants are excluded. In the presence of such catalysts, the solutions may decompose vehemently, particularly at elevated temperature. Addition of polyphosphates or hydrogen peroxide lowers the decomposition tendency. Above 95 °C, pure chloric acid solutions decompose to form chlorine dioxide, chlorine, oxygen, and perchloric acid, but in the presence of hydrochloric acid, the products are chlorine and chlorine dioxide.

Concentrated chloric acid is a strong oxidant. In addition to the noble metals, only Hastelloy C exhibits satisfactory resistance against corrosion. At low temperature, dilute chloric acid may be kept in containers made of poly(vinyl chloride). At 18 °C, the

Table 7. Physical properties of sodium chlorate and potassium chlorate

	NaClO$_3$ [7775-09-9]	KClO$_3$ [3811-04-9]
M_r	106.44	122.55
Crystal system	cubic	monoclinic
mp, °C	260	356
Enthalpy of fusion, kJ/mol	21.3	
Density, g/cm^3	2.487 (25 °C)	2.338 (20 °C)
	2.385 (252 °C)	
Molar heat capacity, J mol^{-1} K^{-1}	54.7 + 0.155 T*	99.8 (20 °C)
	(298–533 K)	
Standard enthalpy of formation, kJ/mol	− 365.8 (cryst.)	− 391 (cryst.)
	− 344.1 (ai)**	
Standard entropy, J mol^{-1} K^{-1}	123.4 (cryst.)	143 (cryst.)
	221.3 (ai)**	
Enthalpy of dissolution (200 mol of H$_2$O/mol of chlorate, 25 °C), kJ/mol	+ 21.6	+ 40.9

* T = temperature, K.
** ai = ideal solution of unit activity.

density of an aqueous 13 % HClO$_3$ solution is 1.080 g/cm^3 while that of a 25 % solution is 1.166 g/cm^3.

Physical properties of *sodium chlorate* and *potassium chlorate* are shown in Table 7. Figure 9 shows the densities of aqueous chloride–chlorate solutions. Figure 10 shows the relative vapor pressure depression of aqueous solutions of chloride and chlorate. The electric conductivity of pure chlorate solutions is given in Table 8; for further data, see [154]. Solubility data of the aqueous chloride–chlorate system are shown in Figures 11 and 12 and in Table 9. The mass fraction of saturated sodium chlorate solutions in the range from 0 to 100 °C can be calculated from

$$w = 0.445 + 0.00226\, t$$

where

t = temperature, °C
w = mass fraction, kg of NaClO$_3$/kg of solution

The freezing point depression ΔT (in K) of aqueous solutions of sodium chlorate is given by

$$\Delta T = 33.64\, \zeta - 115.1\, \zeta^2$$

where ζ = mass ratio, kg of NaClO$_3$/kg of H$_2$O

Table 8. Electric conductivity of pure sodium chlorate solutions [155]

Concentration, g/L	Conductivity, Ω^{-1} m^{-1}		
	20 °C	40 °C	60 °C
100	6.2	8.9	11.8
200	10.4	14.9	19.7
300	13.4	18.9	25.0
400	15.0	21.5	28.5
500	15.7	22.7	30.3
600	15.5	23.1	30.8
750		21.7	29.7

Table 9. Mass ratio (kg of salt/kg of H_2O) of the solution in equilibrium with crystalline chloride and chlorate [154]

t, °C	NaCl–NaClO$_3$ solution		KCl–KClO$_3$ solution	
	NaCl	NaClO$_3$	KCl	KClO$_3$
− 9.8	0.270	0.360	0.2466	0.0056
+ 10	0.249	0.499	0.3123	0.0144
30	0.2125	0.706	0.3703	0.0321
50	0.1785	0.958	0.4226	0.0635
70	0.1495	1.238	0.4651	0.1162
100	0.1245	1.85	0.518	0.2588

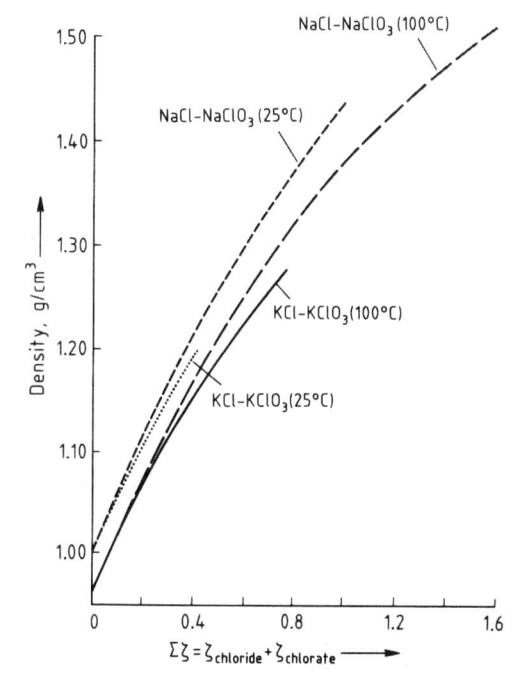

Figure 9. Densities of aqueous solutions of NaCl–NaClO$_3$ and KCl–KClO$_3$ [156]
$\Sigma \zeta$ is the sum of mass ratios ζ_{chloride} (kg of chloride/kg of H_2O) and ζ_{chlorate} (kg of chlorate/kg of H_2O)

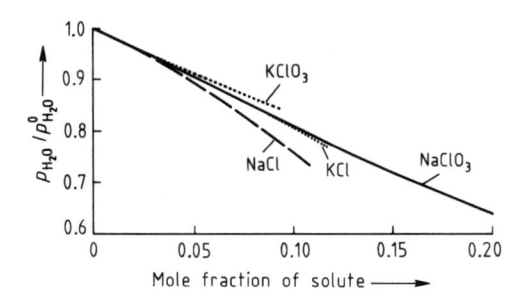

Figure 10. Vapor pressure depression of aqueous chloride and chlorate solutions

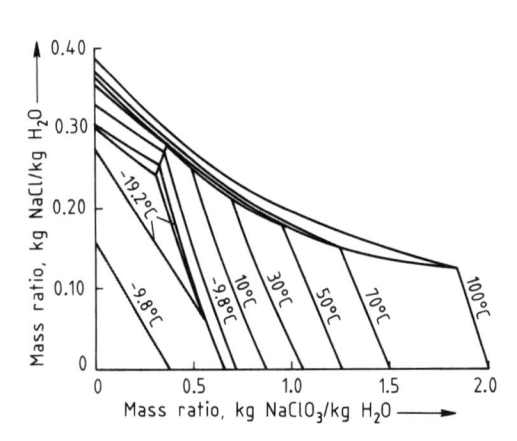

Figure 11. Solubility of aqueous solutions of NaCl and NaClO₃ [157]

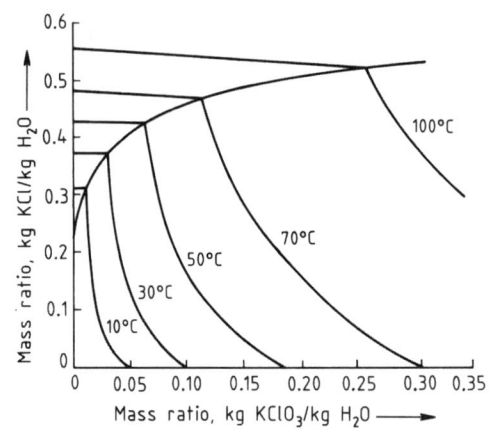

Figure 12. Solubility of aqueous solutions of KCl and KClO₃ [157]

Chemical Properties. Chlorates decompose to yield oxygen. They form flammable and explosive mixtures with organic substances, phosphorus, ammonium compounds, some sulfur compounds, and some metal salts, oxidizable solvents, or other oxidizable substances. Potassium chlorate decomposes below the melting temperature. Alkaline chlorate solutions do not exhibit strong oxidizing properties. With decreasing pH, however, the oxidizing activity of chlorate solutions increases. Concentrated acidic solutions are vigorous oxidants as a result of chloric acid formation.

Solutions containing more than 30% $HClO_3$ decompose spontaneously. In the presence of organic matter or reducing agents the reaction may be violent, especially at elevated temperature.

7.2. Production Fundamentals

Sodium chlorate (and to a minor extent potassium chlorate) is produced by electrolysis of an aqueous sodium chloride (potassium chloride) solution. Hypochlorite forms as an intermediate that is further oxidized to chlorate along two competing reaction paths. The concentrated chlorate solution is either submitted to crystallization (see Section 7.4) or it is used directly, particularly in the production of chlorine dioxide.

An aqueous solution containing 450–550 g/L of sodium chlorate and 90–100 g/L of sodium chloride is generated directly by electrosynthesis; it can be used as feed in the Kesting process (Section 5.2.1) [158]. The spent solution from that process, containing 140 g/L of sodium chlorate and 220 g/L of sodium chloride, is then fed back to the electrolysis system.

The other, less common chlorates are chemically formed by conversion of sodium chlorate with the corresponding chloride. Other production methods, such as the chemical formation of chlorates by introducing gaseous chlorine into a warm hydroxide solution, are now obsolete.

7.2.1. Chlorate-Generating Reactions

An aqueous solution of sodium chloride is electrolyzed, usually in cells without a diaphragm. Hydrogen and sodium hydroxide are formed at the cathode, while chloride is discharged at the anode. Chlorine does not evolve as a gas, but undergoes hydrolysis (Fig. 13):

$$Cl_2 + H_2O \rightleftharpoons HClO + H^+ + Cl^- \tag{4}$$

Chlorate then forms simultaneously by two competing reactions: (1) predominantly by autoxidation of hypochlorite in the bulk electrolyte, and (2) to a small extent (ca. 20%) by anodic chlorate formation. The detailed mechanism of the reaction was

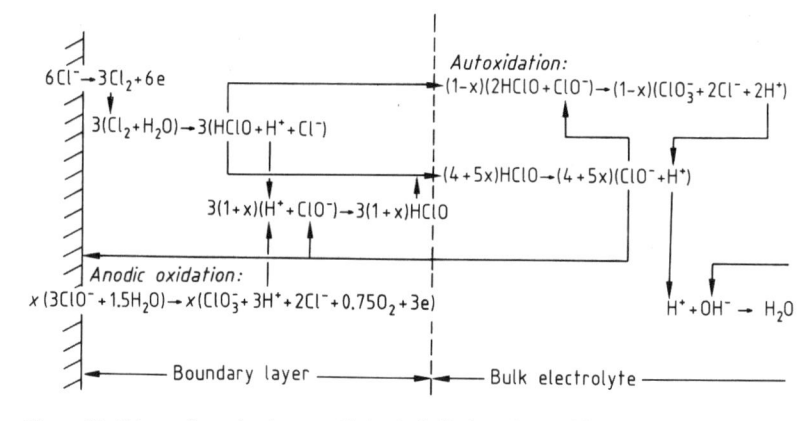

Figure 13. Chlorate formation by autoxidation in bulk electrolyte and by anodic oxidation

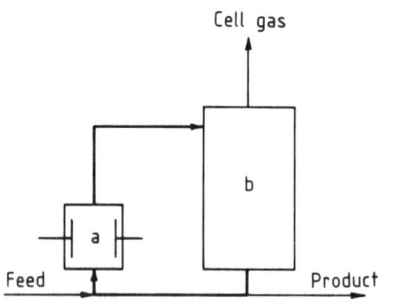

Figure 14. Concept of separate electrochemical reactor (a) and chemical reactor (b) for chlorate electrosynthesis

essentially clarified at the turn of the century, but continues to be the subject of intense studies [39] b.

Autoxidation. Autoxidation of hypochlorous acid, also called chemical chlorate formation, is a homogeneous reaction that proceeds according to the following overall equation, which gives no information on the individual steps:

$$3\,HClO \longrightarrow ClO_3^- + 2\,Cl^- + 3\,H^+ \tag{5}$$

Autoxidation is preceded by dissociation of a part of the total hypochlorous acid involved in autoxidation:

$$HClO \longrightarrow ClO^- + H^+ \tag{6}$$

This dissociation can occur to a significant degree only at some distance from the anode, where the electrolyte is sufficiently buffered by hydroxyl formed by the cathode reaction. The hypochlorite formed then reacts with the complementary amount of hypochlorous acid:

$$2\,HClO + ClO^- \longrightarrow ClO_3^- + 2\,Cl^- + 2\,H^+ \tag{7}$$

Equation (7) shows the classical form of FOERSTER [159], [160], which was attacked [161] but has recently been confirmed [162], [163].

Anodic Chlorate Formation. The mechanism of anodic chlorate formation was established by LANDOLDT and IBL [164]. Hydrolysis of chlorine (Eq. 4) is considered to be fast. Because of the formation of H^+ ions, one might expect the electrolyte in the anodic boundary layer to be strongly acidic. However, this is the case only at low chloride concentration. Large chloride concentrations, as they occur in industrial processes, shift the hydrolysis equilibrium (Eq. 4) to the left. In the electrolyte layer adjacent to the anode, the H^+ concentration is too small to permit noticeable diffusion of H^+ into the inner electrolyte. Therefore, hydrogen is transported away from the anode as hypochlorous acid rather than H^+. In the bulk electrolyte where the pH is high, hypochlorous acid is largely dissociated. The hypochlorite ion diffuses back to the anode and more than two-thirds of it is consumed by buffering before reaching the anode. However, less than one-third of that hypochlorite is discharged at the anode to form chlorate and oxygen [164]–[166]:

$$3\,ClO^- + 1.5\,H_2O \longrightarrow ClO_3^- + 3\,H^+ + 2\,Cl^- + 0.75\,O_2 + 3\,e^- \quad (8)$$

The stoichiometry of the anodic chlorate formation (Eq. 8) was recently reviewed [39 b].

As shown in Figure 13 the discharge of 6 mol of chloride yields 1 mol of chlorate; this is independent of the reaction route, either autoxidation or anodic discharge of hypochlorite. However, the anodic oxidation (Eq. 8) requires 50% additional electric energy. Therefore, industrial processes endeavour to suppress the anodic oxidation in favor of the autoxidation. The effective means is a short residence time of the electrolyte solution inside the electrochemical reactor by applying large flow rates using rather short electrode lengths. Thereby the average bulk concentration of hypochlorite (ClO^- + $HClO$) is lowered and the undesired anodic chlorate formation controlled by mass transfer is minimized. However, increasing flow rates increase the mass transfer coefficient, thus counteracting the beneficial effect of short residence time. The optimum is at elevated values of the flow rate; this method to suppress anodic oxidation is used in all modern industrial chlorate systems. Under these conditions, chlorate is formed predominantly outside the interelectrode space, e.g., in a separate chemical reactor where the route is necessarily restricted to autoxidation (Fig. 14).

7.2.2. Loss Reactions

In addition to anodic chlorate formation, further loss reactions decrease current efficiency and must be suppressed in industrial systems.

Cathodic Reduction of Hypochlorite. The main loss occurs by cathodic reduction of hypochlorite, as described in Section 4.2.1:

$$ClO^- + H_2O + 2\,e^- \longrightarrow Cl^- + 2\,OH^-$$

The reaction competes with hydrogen generation. Without special precautions, its rate is controlled by mass transfer and, therefore, linearly proportional to the hypochlorite concentration in the bulk electrolyte. The reaction can widely be suppressed by adding a small amount of dichromate (1–5 g/L) to the electrolyte solution. By cathodic deposition a porous surface film of chromium hydroxide forms [167]. Inside the pores of this film, the effective current density increases, and the resulting potential gradient across the film impedes the diffusion of any kind of anions to the cathode, whereas the access of cations to the cathode and their reduction are facilitated. The protective film ceases to grow after a certain thickness is reached; the increase in cell voltage caused by the ohmic resistance of the film is more than compensated by the gain in current efficiency [163], [168]. Increasing the electrolyte temperature favors mass transfer of hypochlorite to the cathode and, therefore, requires more dichromate. When the current is interrupted, the surface layer dissolves. It takes some time to reestablish a new film after the current is switched on, but cathodic reduction immediately takes place. Therefore, cells should be operated continuously, and a minimum cathodic current density is recommended to prevent dissolution of the film [169]; from industrial experience, a current density of 20 A/m^2 is considered satisfactory.

Cathodic Reduction of Chlorate. Chlorate is also reduced at the cathode, but this reaction is less significant than the reduction of hypochlorite described in the preceding paragraph:

$$ClO_3^- + 3\,H_2O + 6\,e^- \longrightarrow Cl^- + 6\,OH^-$$

The rate of reduction is strongly affected by the cathode material; reduction at iron is much faster than at nickel and platinum. Chlorate reduction is also restrained by the chromium hydroxide layer on the cathode [16], [169]–[172]. In closed-loop electrolyte systems (e.g., Fig. 14), the presence of residual dichromate in the solution has a beneficial effect because its additional buffering capacity stabilizes the various reaction equilibria [163], [173], [174].

Catalytic Decomposition of Hypochlorite. Hypochlorite solutions decompo e in the presence of catalytically active contaminants:

$$2\,ClO^- \longrightarrow 2\,Cl^- + O_2$$

Oxides of nickel, cobalt, and copper are very effective; their catalytic activity decreases in that order. The catalytic action of manganese, iron, lead, and tin has been demonstrated, but is at least one order of magnitude smaller [53], [175]. Autocatalytic decomposition is also possible, but its extent is negligible at ambient temperature [175], [176].

Chlorine Desorption. Hydrogen and oxygen gas bubbles evolved at the cathode and anode, respectively, tend to desorb some chlorine from the solution. This amount of

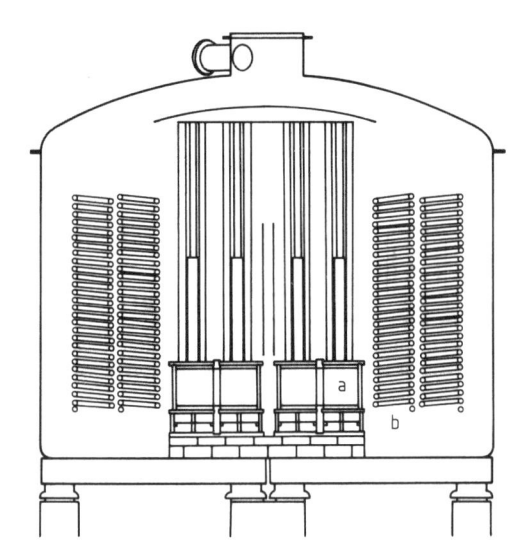

Figure 15. Huron electrosynthesis system with integrated electrolysis cells
a) Chemical reactor; b) Cooling coils

chlorine is lost for chlorate production unless it is recovered from the cell gas by absorption outside the cell. Water vapor contained in the gas bubbles increases their volume and, thus, the desorption loss. Therefore, the operation temperature is limited to a maximum of 80 – 90 °C. Chlorine loss results not only in a corresponding loss in current efficiency, but also in a steady increase in pH. Therefore, chlorine gas or, more usually, hydrochloric acid must be permanently added to maintain the pH.

Further loss reactions are of minor importance: *perchlorate* may be formed by anodic oxidation of chlorate, particularly at low chloride concentration, provided the oxygen overvoltage is large enough (see Section 8.2.2). With the industrial anodes used in chlorate electrosynthesis, perchlorate formation is negligible. The *anodic decomposition of water*, which plays a major role in hypochlorite production, is negligible as long as the chloride concentration is larger than 100 g/L.

7.3. Industrial Electrosynthesis Systems

Chlorate cell design has changed much within the last 25 years by the introduction of coated titanium anodes [177] and by the systematic application of the results of chlorate formation theory. Hypochlorite autoxidation occurs at high temperature outside the interelectrode gap or even outside the electrochemical reactor to decrease the hypochlorite concentration before the electrolyte reenters the cell. The first industrial plant of this type was started in 1969 in Finland.

The requirement for large electrolyte flow rate resulting from the considerations of Section 7.2 is taken into account in all modern cells. Hydrogen evolved at the cathode provides for a natural flow of the electrolyte and is used for its recirculation through

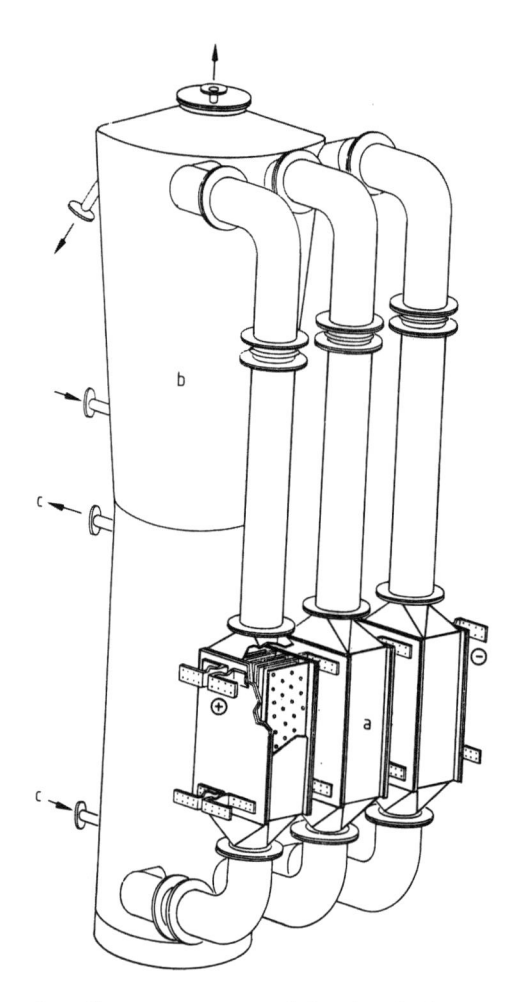

Figure 16. Electrosynthesis system with natural convection of electrolyte (Krebs, Paris)
a) Electrolysis cell; b) Chemical reactor; c) Cooler

the cell compartment and the chemical reactor without requiring additional pumping. Other systems use a mechanical pump to provide the required circulation rate. Hydrogen gas is released at the top of the unit. For safety reasons, the gas volume in the equipment must be minimized. To stabilize the pH value, the solution is continuously acidified (cf. Section 7.2.2).

Some electrosynthesis systems combine the cell and the chemical reactor in a single unit (Huron, Fig. 15, and Atochem). Other industrial systems are composed of an electrochemical reactor and a separate chemical reactor [178] interconnected by electrolyte circulation pipes with integrated heat exchangers for the removal of excess heat (Krebs, Fig. 16, and Pennwalt, Fig. 17) or separate heat exchangers (Krebskosmo, Chemetics, and Fröhler–Lurgi–Uhde). None of these electrosynthesis systems is clearly superior from an engineering point of view.

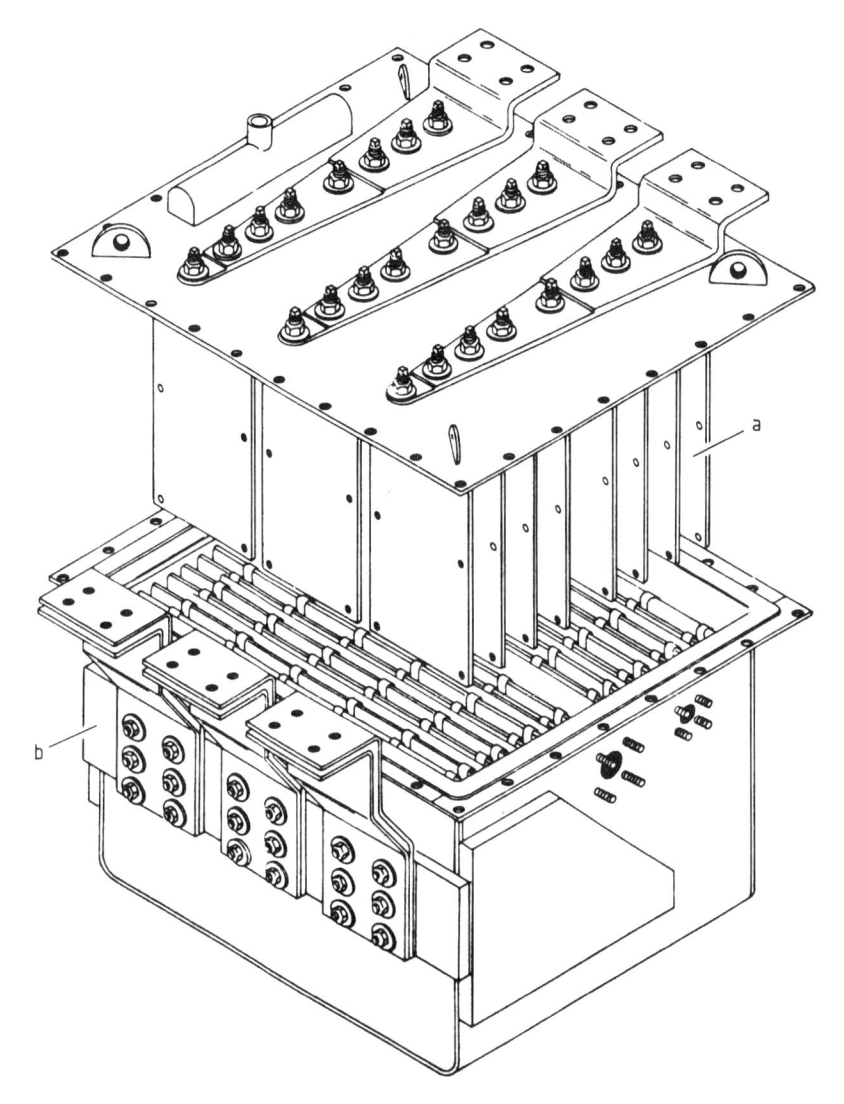

Figure 17. Pennwalt electrochemical reactor
a) Anodes; b) Cathodes

7.3.1. Electrolysis Cell Types

Currently, most cells are equipped with coated titanium anodes [179], and only some 5–10% of all chlorate cells still use graphite anodes. Older cells have also been operated with anodes of magnetite (Fe_3O_4) [180], [181], platinum [182], graphite [183]–[186], or lead dioxide [187], [188]; reviews on these older anodes can be found in [39] b and [189]–[191]. The following survey is restricted to modern cells with coated titanium anodes. Typical operational data are shown in Table 10.

Chlorine Oxides and Chlorine Oxygen Acids

Table 10. Typical operational data of sodium chlorate cells

Parameters	Krebs, Paris	Chemetics, Vancouver	Fröhler Lurgi–Uhde	Pennwalt, Philadelphia	Atochem, Paris	Huron Chemicals, Kingston	Krebs, Zürich	Krebskosmo, Berlin
Current per cell, kA	75	15–85	6–50	30	30–75	480	30	17.5
Cell type	NC	DC	TDK	P8-3	TA2	cell-in-tank	ZMA	CZB 50–150
Cell connection	unipolar	multipolar	unipolar	unipolar	unipolar	multipolar	unipolar	bipolar
Anode surface, m^2	30	7–25	7.1–17.8	8.92	20/30	proprietary	10	30.6/91.8
Current density, kA/m^2	2.5	1.8–3.5	2.8	3.36	1.5–2.5	1.5–4	3.0	3.0
Operating voltage,								
min., V	2.85	2.70	2.9	3.4	2.92	3.1 at 3 kA/m^2	3.1	3.1
max., V	3.00	3.00	3.0	3.7	3.10	3.4 at 3 kA/m^2	3.3	3.4
Energy consumption (d.c.), kW h/t of NaClO$_3$	4500–4750	4250–4750	4610–4770	5650	4600–4900	5000–5500	5100	5000–5500
Current efficiency, %	95.5	96	95	96	96	> 93.5	95	94–96
Current concentration, A/L	30	25–30	22	20	13–22	6–20	20	25
Operating temperature, °C	80	80	75	90	70–80	65	80	80
pH of cell liquor	6.2	6–7	6.4	5.5–6.5	6.2–6.4	6.3–6.9	6.5	6.5
Interpolar distance, mm	4	–	3.5	4	6	proprietary	5	4
Kind of coating	Pt/Ir/RuO$_2$	IMI	platinate	Pt/Ir	RuO$_2$	Pt/Ir or RuO$_2$	Pt/Ir/RuO$_2$	noble metal oxide
Anode coating life, years	6 8–10	> 10	8–10	7–10	7–10	dep. on coating	> 6	4
Cathode material	spec. carbon steel	steel	mild steel	carbon steel	spec. steel	titanium	steel	steel
Cell liquor composition:								
Min. NaCl concentration, g/L	100	100	90	70–110	100	50	100	80
Max. NaClO$_3$ concentration, g/L	600	640	550	500–650	625	700	600	650
NaOCl concentration, g/L	1.3–1.8	–	3	1–2	1	–	2	2–3
Na$_2$Cr$_2$O$_7$ concentration, g/L	3–4	3–4	3	3–5	5	1.1	5	3
Cell gas composition:								
H$_2$, vol%	97.7–98.2	97.5	97.5	97.5	> 98	97–99	97.5	97.5
O$_2$, vol%	1.5–2	2	2	1.5	1–2	1.0–2.5	2	2
Cl$_2$, vol%	0.3	0.5	0.5	1.0	0.1–0.2	0.1–0.2	0.5	0.3

Table 10. (continued)

Parameters	Krebs, Paris	Chemetics, Vancouver	Fröhler Lurgi–Uhde	Pennwalt, Philadelphia	Atochem, Paris	Huron Chemicals, Kingston	Krebs, Zürich	Krebskosmo, Berlin
Requirements per t of $NaClO_3$ (100%):								
NaCl, kg	560	560	–	560	550	–	550	570
HCl (33%), kg	35	–	–	Cl_2	30	–	33	50
NaOH (30%), kg	24	–	–	–	10	–	–	60
$Na_2Cr_2O_7 \cdot 2\,H_2O$, kg	0.05–0.1	–	–	5	0.3	–	neglig.	0.3
Production:								
$NaClO_3$ (100%), kg/d	1137	–	–	446	–	7628	453	1370/4110
H_2, m³ (STP)/d	753	–	–	291	–	4760	300	910/2730

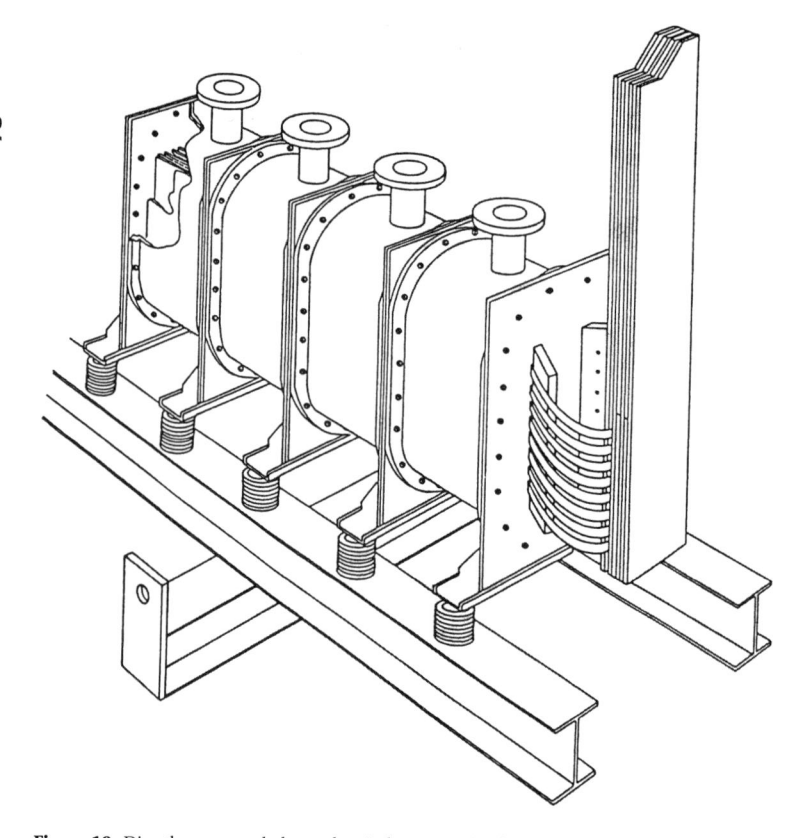

Figure 18. Directly connected electrochemical reactors (multipolar arrangement, Chemetics)

Modern electrochemical reactors [192]–[194] cover the complete spectrum of unipolar and bipolar electrodes, including direct coupling of cells with unipolar electrodes, an arrangement sometimes called multipolar (Fig. 18).

Cells with *unipolar* electrodes (Krebs, Pennwalt, Fröhler–Lurgi–Uhde, and Atochem) are suitable for large currents up to 100 kA. Repair of one cell does not necessitate shutdown of the entire electrolyzer. Parasitic currents [195], corrosion, and malfunction of a cell unit are easy to locate. Pennwalt and Atochem use two cathodes per anode. This design allows circulation within the cell [192], [193]. Cells with *bipolar* electrodes (Krebskosmo) seldom exceed 30 kA (in charge of the voltage), thus lowering the rectifier cost. *Multipolar cells* (Chemetics and Huron) combine some of the advantages of cells with unipolar and bipolar electrodes and require a minimum of bus bars, resulting not only in savings in investment cost, but also in cell voltage. However, the entire electrolyzer system must be shut down for repair of a single unit.

7.3.2. Electrodes

Anodes. N. B. BEER made a decisive breakthrough in anode technology in the mid-1960s by introducing anode coatings based on ruthenium dioxide, RuO_2 [15], [196]–[199]. In fact, the anode surface is composed of a mixture of RuO_2 with other metal oxides, primarily titanium dioxide. These anodes maintain a low overvoltage for chlorine formation at high current densities for long periods without being consumed and changing their mechanical dimensions. Independent of the detailed nature of the material, they are called *dimensionally stable anodes* (DSA) (→ Chlorine) [200]–[202]. In the chlorate field, dimensionally stable anodes have been selected for 80% of the total capacity installed since 1978.

Sintered titanium anodes (STA) have been developed by Sigri. In these electrodes, an intermediate layer of sintered titanium suboxide is applied between the titanium substrate and the coating, which gives improved mechanical stability, excellent adhesion to sintered titanium structures, and increased active surface area [203], [204]. Imperial Chemical Industries (ICI) has used these anodes in large-scale production since 1971. Other novel anodes have been developed by Imperial Metal Industries (IMI) and Marston Excelsior (Marex Electrode).

Titanium anodes normally consist of 1.5–4-mm sheets or of expanded mesh, activated by platinum group metal oxides. In troublefree operation, the durability of the coating based on ruthenium dioxide [205] is between 6 and 10 years. The literature on coated titanium anodes for chlorate electrosynthesis is reviewed in [191].

Cathodes. Mild steel has been the industrial standard for many years. It has a low hydrogen overvoltage and is cheap and stable under the operating conditions. However, steel corrodes when the cells are shut down, and it absorbs hydrogen, which tends to make the metal brittle and to form fissures in the welding areas and blisters [206]. The cathodes are usually 3 – 8 mm thick. The sheets are often perforated with holes of 3 – 6 mm diameter or they are slotted, which facilitates the release of hydrogen bubbles from the interelectrode gap. Mild steel cathodes tend to collect carbonate deposits, which increase the potential of the cathodes. This requires periodic acid washing unless the process water is deionized to contain less than 0.05 mg/kg of impurities.

The application of steel cathodes in bipolar cells is complicated because dissimilar metals must be joined. However, no connection problems arise by explosion bonding of titanium and steel, and recent developments look promising [207].

Attempts to coat the steel cathodes or to replace them by other metals have not brought satisfactory results. Titanium has not been generally accepted because of its tendency to form titanium hydride, TiH_2, which gradually makes the cathode break off and limits the lifetime to 2 years. Attempts to lower the overvoltage by coating the cathodes with nickel, molybdenum, or their oxides [208] have not yet been successful because these materials catalyze the decomposition of hypochlorite during shutdown. Cathodes made from titanium alloys, e.g., Ti – 0.2% Pd, remain stable for 2 years at a temperature of 95 °C, whereas the durability is much longer at 75 °C.

7.3.3. Operational Parameters

Temperature. The operation of electrochemical reactors benefits from temperature increase in two ways: (1) the cell potential decreases by lowering the resistivity of the electrolyte, and (2) autoxidation of hypochlorite to chlorate is more strongly favored than anodic oxidation. Consequently, the current efficiency increases with temperature. On the other hand, the temperature should not be near the boiling temperature of the electrolyte because then the rate of chlorine and water desorption increases. The introduction of dimensionally stable metal anodes has permitted raising the operating temperature from 40–45 °C (previously with graphite anodes) to 80–90 °C.

Pressure. Increasing the electrolyte pressure lowers not only the rate of chlorine desorption, but also the volume of the gas bubbles released from the electrolyte; thus, it acts favorably on the interelectrode resistivity. However, serious engineering problems arise when the pressure is increased. Industrial electrochemical reactors are usually operated at a gage pressure of less than 0.1 MPa (1 bar).

Interelectrode Distance. The electric interelectrode resistance strongly depends on the quantity of gas bubbles dispersed in the electrolyte [209]. When the electrolyte flow rate is increased and the electrode distance decreased, the resistance decreases. However, a very small electrode distance leads to a large frictional pressure drop in the reactor. Therefore, the minimum spacing of industrial reactors is ca. 3 mm, which roughly agrees with the results of optimizations [210]; distances of 3–5 mm are generally used between metal electrodes.

pH Value. According to Equation (7) (see p. 1292), autoxidation of hypochlorite requires the simultaneous presence of hypochlorous acid and hypochlorite. The concentration of both species is very dependent on pH. A carefully controlled pH value, therefore, is a prerequisite of an industrially acceptable current efficiency. The optimum pH depends on the temperature and is maintained in the range of 6.1–6.4 in modern industrial electrochemical reactors operated at 80–90 °C.

Chloride Concentration. The brine feed to the cells is always close to being saturated to obtain a large final chlorate concentration of the product solution. The chloride concentration should not fall below 80–100 g/L because water decomposition is favored at low chloride concentration [211]. If necessary, the brine is resaturated with salt, but solubility limits must be observed (Figs. 11 and 12).

Current Density. High current density increases the acidity near the anode, thus depressing anodic hypochlorite oxidation. In addition, cathodic hypochlorite reduction is hampered. However, current density of industrial cells is limited by the concurrent increase in cell potential. Current densities of industrial cells are optimized on the basis of these considerations and today permit operation at higher current efficiencies than 20 years ago.

Current Efficiency. If x is the fraction of the hypochlorous acid (or chlorine) involved in the anodic oxidation and $(1-x)$ is the fraction that produces chlorate along the autoxidation reaction, the current efficiency ε is (cf. Fig. 13)

$$\varepsilon = \frac{1}{1 + 0.5\,x} \tag{9}$$

Exclusive anodic chlorate formation would result in a current efficiency of $\varepsilon = 66.7\,\%$, provided the contribution of loss reactions is negligible.

The short survey of the operational parameters given in the previous sections highlights the complex system that governs current efficiency in chlorate electrosynthesis. The problem has been dealt with in detail in [39]b and is the object of methodical investigations [212], [213]. Mathematical models have been established to provide means to theoretically predict current efficiency. A model assuming ideal stirred-tank-reactor conditions for both the electrochemical and the chemical reactor [214] was compared with industrial results [178]. A recently developed model combines a plug flow electrolyzer with either a stirred tank or a plug flow chemical reactor [215]. The results show satisfactory agreement with data from two industrial systems [216].

The current efficiency of operating industrial cells with external recirculation of the electrolyte (Fig. 14) can be assessed by measuring the difference Δc of total hypochlorite concentration (mol/L) at the cell outlet and inlet. Because the difference of total hypochlorite concentration represents the amount of hypochlorite converted to chlorate by autoxidation outside the cell, the current efficiency is calculated to be

$$\varepsilon = \frac{2}{3}\left[1 + \frac{\dot{V}_L \cdot \Delta c \cdot F}{I}\right] \tag{10}$$

where

F = Faraday constant (96 487 A s/mol)
$\dot{V}_L$ = volume flow rate, m³/s
I = total current, A

Equation (10) does not take account of the autoxidation in the interelectrode gap, and it neglects all loss reactions; both factors tend to balance.

In all electrosynthesis systems, including those without external electrolyte recirculation, the current efficiency may be estimated by analyzing the gas mixture leaving the electrochemical reactor. The amount of anodic chlorate formation (together with the catalytic decomposition) can be assessed from the volume fraction of oxygen (φ_{O_2}). The loss caused by chlorine desorption is represented by the volume fraction of chlorine (φ_{Cl_2}). Moreover, the ratio of the true hydrogen flow rate N_{H_2} to the flow rate N'_{H_2} as calculated from Faraday's law indicates the loss through cathodic reduction of hypochlorite and chlorate [217]:

$$\varepsilon = \frac{1 - 3\varphi_{O_2} - 2\varphi_{Cl_2}}{1 - \varphi_{O_2} - \varphi_{Cl_2}} \cdot \frac{N_{H_2}}{N'_{H_2}} \qquad (11)$$

In industrial operation, the current efficiency is predominantly lowered by anodic chlorate formation; when neglecting any other loss reaction, one obtains a very simple approximation equation:

$$\varepsilon \approx 1 - 2\varphi_{O_2} \qquad (12)$$

The error of Equation (12) as compared to Equation (11) is within the accuracy of gas analysis in the current efficiency range of modern cells. Automatic measurement of the oxygen concentration in the cell gas, therefore, serves as a simple and reliable means to control cell operation.

7.3.4. Brine Purification

Solar salt from seawater, rock salt, or very pure "vacuum salt" are used as raw material. The brine must not contain large amounts of magnesium and calcium, which could form deposits on the cathodes. Sodium carbonate and sodium hydroxide are added to the brine to raise its pH to at least 10. The sulfate concentration must be lower than 10–20 g/L. The brine must be free of heavy metals that favor the decomposition of hypochlorite (see Section 7.2.2). Maximum impurity concentrations are listed below:

SO_4^{2-}	10 g/L
SiO_2	10 mg/kg
Al, $P_2O_7^{4-}$, Ca	5 mg/kg
Fe, Mn, Sn	1–5 mg/kg
Cu, Mg, Pb, Ti	1 mg/kg
Cr, Mo	0.5 mg/kg
Ir, Co	0.2 mg/kg
Ni	0.05 mg/kg

7.4. Crystallization

For the crystallization of chlorate, a solution of 560–630 g/L of sodium chlorate and 90–120 g/L of sodium chloride is produced at 70–90 °C. The temperature at the cell outlets is first raised to 85–95 °C to convert the remaining 1.5 g/L of hypochlorite into chlorate. Urea, ammonia, hydrogen peroxide, or sodium formate is added to complete the hypochlorite decomposition. The solution is then rendered slightly alkaline with NaOH; this reduces corrosion in the crystallizer. The NaClO$_3$ solution is subsequently fed to a vacuum crystallizer, based on flash cooling with subsequent crystallization.

Crystalline chlorate is separated from sodium chloride solution by cooling or evaporation [218]. Crystallization conditions depend on the concentration of NaCl and NaClO$_3$, on solution temperature, and on mixing intensity. The cooling rate has little influence on the crystal size; ca. 90% of the crystals have 120–260 µm diameter. The solution temperature, initially raised to 90–95 °C, drops to 48 °C in the separator and to 35 °C in the vacuum crystallizer at 2.9 kPa. The concentrated slurry contains 15–20 wt% of crystals; it is passed through a hydrocyclone and then a pusher centrifuge. Energy consumption, efficiency and economic optimization of the overall process were studied [219]; an energy flow diagram is given in [220].

Potassium chlorate is generally made from potassium chloride and sodium chlorate:

$$NaClO_3 + KCl \longrightarrow KClO_3 + NaCl$$

Solid KCl is added to the sodium chlorate cell liquor in stoichiometric amounts. The mixture is then transferred to a crystallizer and the potassium chlorate slurry is removed as described in the preceding paragraph. The mother liquor is recycled to the cells, where the salt is converted to chlorate and the process is repeated. Quality requirements for solid KCl are high, because no purification is possible after it is added to the solution.

7.5. Construction Materials

Chlorate solutions are kept slightly alkaline except during electrolysis, where a specific acidity must be maintained.

Mild steel and cast iron are suitable for equipment if the liquor is alkaline and free of active chlorine. In acidic media, poly(vinyl chloride) (PVC) tubes and tanks made from fiber-reinforced PVC are used up to 50 °C. Poly(vinyl chloride) stabilized with Ba or Cd (PVC-C) is much better suited, but the consumption rate is nevertheless 1 mm per year at 60 °C [221].

Titanium, polytetrafluoroethylene (PTFE), glass, and poly(vinylidene fluoride) (PVDF) have excellent resistance at all temperatures and under all conditions. Evaporators are made from yellow brass, rubber-lined steel, monel, stainless steel (AISI: 316; DIN: 1.4401) with cathodic protection, or titanium. Stainless steel should be used for the dryer because it is imperative to avoid Fe$_2$O$_3$ contamination of the chlorate, which would act as an explosion catalyst. Pump seals must be of the noncombustible type. Titanium is increasingly used in all equipment as the material for anodes, heat exchangers, reaction vessels, pumps, and tubes.

Table 11. Quality specifications for chlorates of sodium and potassium

	Sodium chlorate			
	"White crystals", wt %	Crystalline, wt %	Powder, wt %	Solution at 20 °C, g/L
NaClO$_3$	96.5–97.5	99.75–99.95	99.4–99.75	500
NaCl, max.	0.05	0.03	0.06	0.16
Na$_2$SO$_4$, max.	0.04	traces	0.04	0.11
CaO, max.	nil	nil	nil	0.11
NaBrO$_3$, max.	0.02	traces	0.02	0.004
Fe, max.	traces	nil	traces	traces
Na$_2$Cr$_2$O$_7$, max.	nil	nil	nil	0.004
Insolubles, max.	0.02	0.02	0.02	0.02
Humidity, max.	2.00	0.02	0.02	–
Bulk density, g/cm^3				
not vibrated	1.15	1.31	1.49	1.317
vibrated	1.55	1.58	1.65	

	Potassium chlorate	
	Typical specification, wt %	Aragonesas S.A., for matches, wt %
KClO$_3$, min.	99.7	99.8
KCl, max.	0.032	0.05
NaCl, max.	0.2	
KClO$_4$, max.	0.1	
K$_2$CrO$_4$	nil	
KBrO$_3$, max.	0.07	0.04
Fe, max.	0.01	nil
Insolubles, max.	0.01	0.02
Humidity, max.	0.05	0.02

7.6. Environmental Protection

The gas released from the electrochemical reactor always contains some chlorine in addition to hydrogen and oxygen. The mixture passes a cooler, where the major amount of chlorine condenses. Residual amounts of chlorine (0.3%) are lowered to $(1-3) \times 10^{-4}$ % by absorption in alkaline solution (15% NaOH) and to $(1-10) 10^{-7}$ % by subsequent adsorption on activated carbon. The resulting gas mixture of hydrogen and some oxygen can either be vented to the atmosphere or used as a fuel. The oxygen concentration (1.5–2.5%) depends on the reaction conditions of the cell (see Section 7.2). In trouble-free operation, environmental pollution problems do not arise. Pure hydrogen can be produced by passing the residual gas mixture over a noble metal catalyzing combustion and subsequently purified by selective permeation in palladium at elevated temperature (350 °C).

Wastewater from cleaning tanks and building floors requires appropriate treatment to avoid damage of the water flora by chlorates acting as herbicides. A process to remove almost 100% of dichromate from chlorate solutions by use of a fixed bed of standard ion-exchange resin has been operated for several years in a pilot plant and is about to be commercially exploited [222].

7.7. Quality Specifications

The purity requirements of chlorate depend on the intended purpose and are usually negotiated between producer and user. Various products differ considerably in quality. Standard values for white and yellowish sodium chlorate and potassium chlorate are shown in Table 11. The given potassium chlorate quality is satifactory for many purposes, but may be recrystallized for special requirements (Table 11).

7.8. Storage, Transportation, and Safety

Chlorates should be stored in a cool, dry, fireproof building. Preferably, a separate storage building should be provided with cement floors and metal catwalks.

Wooden construction must be avoided because of the combustibility of chlorate-impregnated wood. Crystalline chlorate may be stored in concrete, lined steel bins, or glazed tile silos. Dried air should be supplied to prevent caking from atmospheric moisture. Drums and packages containing chlorates should not be stored where such noncompatible chemicals as acids, solvents, oils, organic substances, sulfur, or powdered metals could be spilled [223]. Ventilation must be provided for operations where fine chlorate dust arises. The ventilation system should discharge to a water scrubber and be designed for easy cleaning. Pumps handling chlorate solutions should be of the "packless" type. Electrical supply and distribution points are to be inspected periodically for dust. Motors and switches must not be housed in the plant or store and must conform to the National Electrical Code (United States). Metal or plastic pallets are recommended for handling containers.

Safety showers and water supply should be available to workers. The containers (barrels) must be kept closed and not be stored on top of each other. Chlorates are handled in polyethylene bags, in metal drums, or metal-lined fiberboard or plywood drums (25–170 kg). Bulk chlorates can be delivered by tank trucks or railroad tank cars either in the dry or wet state. They are usually unloaded as a slurry by recirculation of hot water (60 °C) between the car and the dissolver tank. The recommended material for storage tank construction is stainless steel (AISI 316). Tanks lined with rubber or plastics are not recommended because of possible fire or explosion hazards. For more details, see references [224]–[235].

7.9. Uses

Sodium Chlorate. The rapid growth of sodium chlorate production is mainly due to the widespread use of the chlorate-derived chlorine dioxide bleach by the pulp and paper industry. The once predominant use of sodium chlorate as a nonselective herbicide has strongly declined. Second in importance is its use as an intermediate in the production of other chlorates, mainly potassium chlorate, and of sodium perchlorate for conversion to ammonium perchlorate, which is used as an oxidizer in solid propellants. Sodium chlorate is further used as oxidizing agent in uranium refining and other metallurgical operations, as an additive to agricultural products and dyes, in textile and fur dyeing, metal etching, and in chemical laboratories and throughout the chemical industry as an oxidizing agent.

Potassium Chlorate. Potassium chlorate is used mainly in the manufacture of matches and in the pyrotechnics, explosives, cosmetics, and pharmaceutical industry; it is superior to sodium chlorate because of its smaller hygroscopicity.

Barium Chlorate. Barium chlorate is prepared by reaction of barium chloride with sodium chlorate solution. It is precipitated by cooling, purified by recrystallization, and used in pyrotechnics.

Calcium Chlorate. Solutions of calcium chlorate are used as herbicide.

7.10. Economic Aspects

Because of the steadily increasing demand for chlorine dioxide in the pulp industry, production of sodium chlorate has doubled in the past 10 years. Figure 19 shows annual production data. Canada is the world's leading producer with 0.5 Mt $NaClO_3$ in 1984, closely followed by the United States with 0.41 Mt $NaClO_3$. The United States productivity has grown by the factor of 25 within the past 40 years [236].

Since electric power cost amounts to ca. one-third of the total production cost of sodium chlorate, cheap energy is the key economic factor.

8. Perchloric Acid and Perchlorates

Perchloric acid [7601-90-3] and its salts, particularly ammonium perchlorate [7790-98-9], NH_4ClO_4, sodium perchlorate [7601-89-0], $NaClO_4$, and potassium perchlorate [7778-74-7], $KClO_4$, find many applications because of their strong oxidizing power; their chemical stability is sufficient to permit high-energy oxidation under controlled

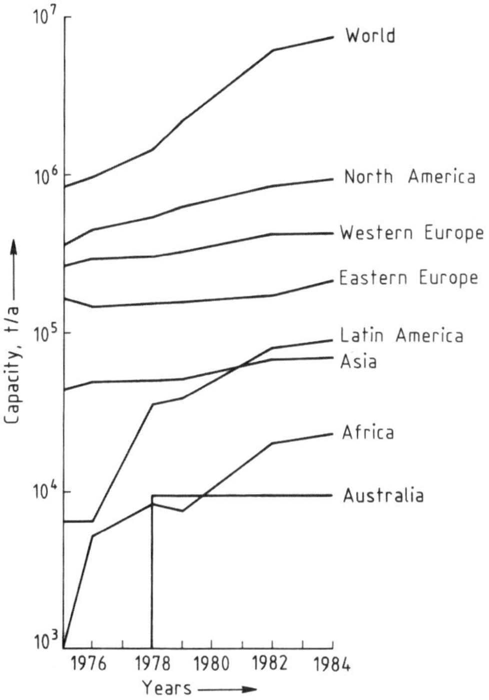

Figure 19. Chlorate production capacities

conditions. Perchloric acid is used on a limited scale mainly as a reagent for analytical purposes; its production has recently increased because of its use as a starting material for pure ammonium perchlorate, a basic ingredient of explosives and solid propellants for rockets and missiles. Minor amounts of potassium perchlorate occur in Chile, in natural deposits of sodium nitrate.

8.1. Physical and Chemical Properties

8.1.1. Perchloric Acid

Anhydrous perchloric acid, M_r 100.5, can be obtained by distilling at reduced pressure a mixture containing 1 part of 20% $HClO_4$ and 4 parts of 20% oleum. It is a colorless, strongly hygroscopic liquid, melting at -102 °C. At atmospheric pressure, it decomposes at 75 °C. Its density and vapor pressure vs. temperature are given in Table 12.

Perchloric acid is miscible with water in all ratios and forms a series of hydrates as shown in Table 13. Boiling points and densities vs. concentration are given in Table 14. For other physical properties, see [237], [238].

Table 12. Density, ϱ, and vapor pressure, p, of anhydrous perchloric acid

	t, °C			
	0	10	20	25
ϱ, g/cm^3	1.808	1.789	1.770	1.761
p, kPa	1.546	2.506	3.913	4.826

Table 13. Perchloric acid and its hydrates, HClO$_4 \cdot n$ H$_2$O

	n					
	0	1	2	2.5	3	3.5
CAS registry no.	[7601-90-3]	[60477-26-1]	[13445-00-6]	[34099-94-0]	[35468-32-7]	[41371-23-1]
M_r	100.46	118.47	136.49	145.5	154.51	163.5
c_{HClO_4}, wt%	100	84.8	73.6	69.1	65.0	61.5
mp, °C	− 112	+ 50	− 17.5	− 29.8	− 37 (α) − 43.2 (β)	− 41.4

Table 14. Boiling points and densities of aqueous perchloric acid solutions

	c_{HClO_4}, wt%							
	24.34	38.9	50.67	56.65	61.2	65.20	70.06	72.5
bp, °C	105.8	114.8	132.4	148.0	162.3	189.2	198.7	203
ϱ^{25}, g/cm^3	1.154	1.280	1.4058	1.4799	1.5413	1.5993	1.6748	1.7150

Table 15. Solubility of some perchlorates in water at 25°C (grams in 100 g of water)

NH$_4$ClO$_4$	24.922	Sr(ClO$_4$)$_2$	309.67
LiClO$_4$	59.71	Ba(ClO$_4$)$_2$	198.33
NaClO$_4$	209.6	Cu(ClO$_4$)$_2 \cdot$ 2 H$_2$O	259
KClO$_4$	2.062	AgClO$_4$	540
RbClO$_4$	1.338	Cd(ClO$_4$)$_2 \cdot$ 6 H$_2$O	478
CsClO$_4$	2.000	Mn(ClO$_4$)$_2 \cdot$ 6 H$_2$O	268
Mg(ClO$_4$)$_2$	99.601	Co(ClO$_4$)$_2 \cdot$ 6 H$_2$O	292
Ca(ClO$_4$)$_2$	188.60	Ni(ClO$_4$)$_2 \cdot$ 6 H$_2$O	267

Anhydrous perchloric acid is a very strong oxidizing agent. Some metals like nickel, copper, silver, and gold are only slightly oxidized at ambient temperature. Platinum is not attacked, but it decomposes the acid by catalytic action. In contact with combustible materials, the perchloric acid reacts violently, forming explosive mixtures with paper, charcoal, ethanol, acetic anhydride, and gelatin.

Perchloric acid is more stable in aqueous solution than in anhydrous form. The azeotropic mixture (72.5 wt% HClO$_4$) decomposes in the absence of oxidizable matter only above the boiling point. Concentrated perchloric acid is a strong oxidizing agent, especially at high temperature; it forms explosive mixtures with organic compounds, which can detonate on heating, percussion, or exposure to sparks or a flame.

Perchloric acid is a strong acid and reacts in aqueous solution with metals, metal oxides, and hydroxides, as well as with salts of volatile acids, forming the corresponding perchlorates. Some oxides, e.g., CuO, catalyze the decomposition of perchloric acid. This occurs through a chain of reactions that give chlorine, oxygen, and water as ultimate products.

8.1.2. Perchlorates

At least one element in each group of the periodic table, except the noble gases, forms perchlorates; this includes not only metals, but also nonmetallic elements, such as nitrogen in hydrazine perchlorate [13762-80-6] or fluorine in fluorine perchlorate [37366-48-6]. Organic ammonium, diazonium, and sulfonium perchlorates form another large class. Perchlorates are generally colorless and well (or fairly) soluble in water and such organic solvents as alcohols, ketones, or esters. Table 15 lists the solubilities of some metal perchlorates. Other physical properties can be found in [225], [239], and [240].

When heated, alkali metal and alkaline-earth metal perchlorates decompose before reaching the melting point, with the exception of $LiClO_4$ (*mp* 247 °C). The hydrates of the perchlorates generally have definite melting points, but they decompose on further heating, after liberation of water.

The outstanding chemical property of the perchlorates is their strong oxidizing power. This is exploited either in mixtures of perchlorates with combustible materials or by using organic perchlorates as explosives or propellants. On the other hand, all perchlorates, also in the pure state, are liable to undergo explosive decomposition if they are heated above a critical temperature, with liberation of chlorides and oxygen; ammonium perchlorate has been investigated extensively in this respect in view of its importance for jet propulsion [241]–[246].

8.2. Production

8.2.1. Perchloric Acid

Perchloric acid is made from sodium perchlorate and hydrogen chloride [237], [247]. A saturated $NaClO_4$ solution reacts with an excess of HCl, and NaCl precipitates. The dilute per-chloric acid produced after filtration (32% $HClO_4$) is concentrated in three stages up to 70–71% $HClO_4$; the last stage is vacuum distillation.

A continuous, electrochemical process for the production of perchloric acid has been developed in the Federal Republic of Germany. It involves the anodic oxidation of gaseous chlorine dissolved in chilled 40% $HClO_4$ [248], [249]:

$$Cl_2 + 8\,H_2O - 14\,e^- \longrightarrow 2\,HClO_4 + 14\,H^+$$

The electrolyzer is of the filter press type; it is composed of poly(vinyl chloride) (PVC) frames and the electrodes are separated by a diaphragm of PVC fabric. The anodes are made from platinum foil, spot-welded to tantalum rods. The cathodes are horizontally slitted silver plates. A 40% $HClO_4$ solution is circulated from the electrolyzer to an external cooler, which keeps the temperature within -5 to $+3$ °C. The operating conditions are as follows: current 5000 A; current density $2.5-5$ kA/m^2; voltage 4.4 V; current efficiency 0.60; and chlorine concentration at inlet 3 g/L. The outflowing solution is distilled to remove residual chlorine and byproduct hydrogen chloride and to obtain high-purity perchloric acid. Platinum consumption is 0.025 g/t of 70% $HClO_4$; it dissolves at the anode, but partly redeposits at the cathode and can thus be recovered. Per ton of 70% $HClO_4$, 9600 kWh of electricity (d.c.) is used. The high purity of the product, directly obtained by a continuous process not requiring separation stages, is an advantageous tradeoff versus the relatively low current efficiency. The process further allows unusual perchlorates to be prepared by direct conversion with perchloric acid, thus avoiding the route via sodium chlorate.

8.2.2. Perchlorates

Various methods for the production of perchlorates are described in [237]. In practice, commercial production is based on the following steps:

1) Electrochemical production of sodium chlorate from sodium chloride (Chap. 7)
2) Electrochemical oxidation of sodium chlorate in aqueous solution to sodium perchlorate
3) Conversion of sodium perchlorate into another perchlorate, e.g.:

$NaClO_4 + KCl \longrightarrow KClO_4 + NaCl$
$NaClO_4 + NH_4Cl \longrightarrow NH_4ClO_4 + NaCl$

In the last step, the corresponding sulfate, which causes fewer corrosion problems, can be used instead of the chloride.

This indirect method for the production of perchlorates other than $NaClO_4$ is advantageous over their direct electrosynthesis from the corresponding chlorates, because of their relatively small solubility in water (Table 15). Attempts have been made to make sodium perchlorate directly from sodium chloride, by a single electrolysis [250]–[252]. The advantage, however, seems debatable, because the resulting current efficiency is only slightly better than 50% under optimum conditions.

Anodic Oxidation. The anodic oxidation of chlorate occurs according to the following overall reaction:

$$ClO_3^- + H_2O - 2\,e^- \longrightarrow ClO_4^- + 2\,H^+$$

The standard potential of the anodic process (1.19 V) is very close to that of water oxidation (1.228 V), which gives rise to oxygen evolution. Both reactions compete strongly, independent of the acidity of the solution. To obtain a satisfactory current efficiency, the anode potential should be as large as possible, because a high polarization enhances the chlorate oxidation rate more than the oxygen evolution [253]. This is achieved by selecting a suitable anode material and a high current density. Several reaction mechanisms have been proposed for chlorate oxidation to perchlorate [39] c.

Cell design and operating conditions must be selected to optimize electricity consumption, which depends on cell voltage and current efficiency. Both are affected by a number of variables in conflicting ways. In particular, the current efficiency is improved by increasing the current density, but this also increases the cell voltage. A temperature rise operates in reverse in that not only the voltage, but also the current efficiency is diminished. A high sodium chlorate concentration in the effluent liquor improves current efficiency as well as voltage, but requires more expensive procedures and higher chlorate consumption to obtain a sufficiently pure perchlorate solution.

Industrial Cells. Contrary to electrochemical chlorate production, the residence time of the electrolyte in the perchlorate cell is not a critical parameter. The cell is undivided and hydrogen gas evolution provides sufficient agitation to decrease the sodium chlorate concentration gradient between the cell inlet and outlet. Because of the relatively high heat dissipation, the cell is normally provided with internal cooling.

Smooth platinum anodes yield the highest oxygen overpotential and, hence, the highest current efficiency (95 – 97 %). Platinum is either used in foils or cladded onto rods of some other metal, such as tantalum or copper. Lead dioxide anodes are also used [254]; they consist of a lead dioxide deposit plated on some conductive substrate, such as graphite rods or plates [255] – [258]. They are less expensive than platinum, which more than offsets their lower current efficiency (85 %). When a small amount of sodium fluoride (2 g/L) is added to the electrolyte, the current efficiency increases [255]. More recently, platinate-coated titanium has been claimed to substitute profitably for massive smooth platinum in perchlorate manufacture [259].

Cathodes are made from carbon steel, chromium – nickel steel, nickel, or bronze. As in chlorate electrosynthesis, sodium dichromate is added to the solution to minimize the current efficiency loss caused by cathodic reduction of perchlorate. At the same time, dichromate inhibits corrosion of the steel parts that are directly exposed to the electrolyte, in addition to their cathodic protection. However, dichromate cannot be used with lead dioxide anodes, because it catalyzes oxygen evolution [253]. In that case, the most suitable cathode material is nickel or chromium – nickel steel.

Figure 20 shows some typical cell models [260]. In the *Bitterfeld* cell, each anode, made from platinum foil, works between a pair of cathodes made from perforated steel plates. The cell is kept at 35 °C by cooling water running through a steel pipe bundle.

In the *Cardox* cell, the anodes consist of platinum cladded onto copper rods 1.3 cm in diameter. Each anode is surrounded by a cathodic tube 7.6 cm in diameter, which is

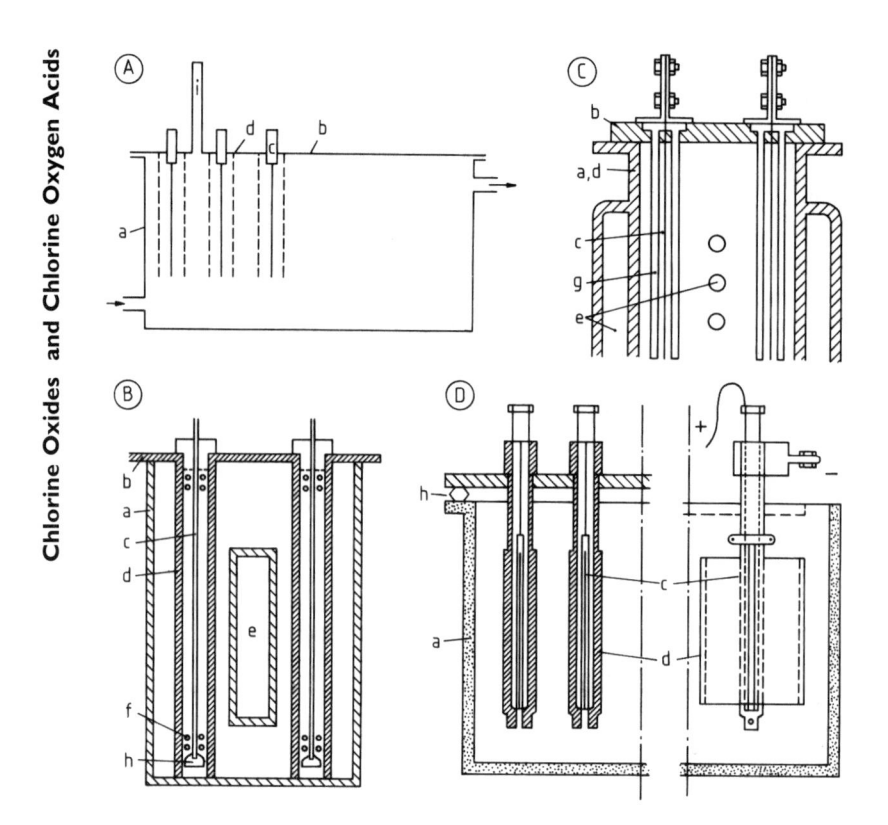

Figure 20. Some typical models of perchlorate cells
A) I. G. Farbenindustrie, Bitterfeld; B) Cardox Corp.; C) American Potash and Chemical Corp.; D) Pechiney, Chedde
a) Cell tank; b) Cover; c) Anode; d) Cathode; e) Cooling system; f) Circulation holes; g) Glass rods; h) Porcelain insulators; i) Gas outlet

made from steel and perforated at both ends. This allows hydrogen gas to escape into the interelectrode space and ensures convective circulation.

In the model of *American Potash and Chemicals* (formerly Western Electrochemical Co.), a set of anodic platinum foils is arranged around the inner wall of the cylindrical cell body performing as the cathode; the interelectrode gap is stabilized by a number of glass rod spacers. The temperature is kept at 40–45 °C by internal cooling and by a water jacket surrounding the cylindrical cell tank.

In the *Pechiney* model, the anodes are platinum foils and the cathodes are plates made from bronze.

Operation. Table 16 shows typical operational data [39 c], [237], [260]. Sodium perchlorate can be separated from the cell effluent either as hydrate or as anhydrous salt. Depending on concentration, separation is carried out by cooling alone or by evaporation, followed by cooling. The monohydrate precipitates from the solution below ca. 52 °C; above this temperature, the perchlorate crystallizes as the anhydrous

Table 16. Typical operational data for perchlorate cells

Current	500–12 000 A
Current density	1500–5000 A/m^2
Cell potential	5–6.5 V
Current efficiency	
Platinum anodes	90–97%
Lead dioxide anodes	85%
Electric energy (d.c.) per ton of NaClO$_4$	2500–3000 kWh
Electrolyte temperature	35–50 °C
pH	6–10
Concentrations, g/L	
Na$_2$Cr$_2$O$_7$	0–5
Cell inlet	
NaClO$_3$	400–700
NaClO$_4$	0–100
Cell outlet	
NaClO$_3$	3–50
NaClO$_4$	800–1000
Platinum consumption per ton of NaClO$_4$	2–7 g

salt. In either case, the mother liquor still contains a large amount of perchlorate. After enrichment with sodium chlorate, this solution is recycled to the electrolysis process.

When the sodium perchlorate is used for conversion to other perchlorates, it need not be separated from the cell effluent. Sodium chlorate is destroyed by chemical treatment and dichromate precipitated as insoluble chromic hydroxide. The purified and concentrated solution is then ready for conversion to other salts [261].

8.3. Environmental Protection

On account of the limited volume produced and their slight toxicity, perchlorates do not constitute an environmental pollution problem. Aquatic life (fish, leeches, and tadpoles) survives indefinitely even when the perchlorate concentration in water exceeds 500 mg/L [262]. However, because of the antithyroid effect of perchlorate, some chronic symptoms may appear at low levels.

Some bacteria, such as *Vibrio dechloraticans,* can metabolize perchlorates; this is exploited for perchlorate destruction in the sanitation of contaminated sewage waters.

8.4. Chemical Analysis

Perchlorate is analyzed by its decomposition to chloride. This occurs by melting the perchlorate salt in a platinum dish with sodium carbonate, which accelerates the thermal decomposition. Sodium carbonate can be replaced with ammonium chloride if alkali-metal perchlorates are analyzed that decompose rapidly at 450–550 °C. The cooled melt is then dissolved in water or dilute nitric acid, and the total chloride is determined by normal volumetric or gravimetric methods. The chlorate or chloride content initially present in the sample must be subtracted to obtain the amount of perchlorate in the sample.

Quantitative precipitation of perchlorate from an aqueous solution by one of the following reagents is also possible: methylene blue [263]; potassium, rubidium, or cesium salts in cold ethanol–water solution [264]; or tetraphenylphosphonium chloride [265], [266]. The most sensitive quantitative determination of perchlorate (10^{-6}–10^{-7}) is possible by means of ion-specific electrodes [267], [268].

8.5. Storage, Transportation, and Safety

The same storage and shipping regulations that apply to chlorates (Section 7.8) are also applicable to perchlorates.

According to the U.S. Department of Transportation (DOT) regulations for hazardous materials, perchloric acid and perchlorates are shipped in glass containers or metal drums as specified in [269]. They are classified as oxidizing substances and require oxidizer shipping labels. Transportation is forbidden on passenger-carrying aircraft or railcars. Shipment of perchloric acid in concentrations greater than 72 % is forbidden. Ammonium perchlorate may be shipped in steel drums with plastic liners; large quantities are transported in portable aluminum containers holding up to 2.27 t. Lower-side or hopper-type product discharge openings are not permitted.

Sodium or magnesium perchlorate may be shipped wet in tank cars with a minimum of 10 % water, which must be equally distributed. All perchlorates not specifically covered by DOT regulations should carry special precautionary labels indicating the specific fire or explosion hazards expected from the individual perchlorate.

Despite their limited toxicity, perchloric acid and perchlorates must be considered, like chlorates, as hazardous chemicals during fabrication, transportation, and handling. Because of its tendency toward spontaneous explosion, anhydrous perchloric acid should be prepared only in very small batches and in the absolute absence of impurities. It can be stored for a limited time at low temperature and must be protected from any kind of contamination. Distillation must be carried out under vacuum (2.4 kPa at 16 °C) and with protective shielding. One must never try to obtain the acid in anhydrous form by treatment with a drying agent. It should be stored in glass containers with glass stoppers, possibly embedded in kieselguhr or glass wool for protection.

Ammonium perchlorate and alkali-metal and alkaline-earth metal perchlorates are relatively insensitive to rubbing, heating, and shock. They also require special precautions in manufacturing and further handling [223], [224]. Clean work clothing must be worn each day and laundered afterward; it must not be taken home. Clothing wet with perchlorate solution should be changed before drying. No smoking can be permitted in perchlorate working areas or while wearing work clothes. Deluge type safety showers or jump tanks should be provided. Only rubber shoes and rubber or rubberized gloves are permitted.

Perchlorates should not be stored close to flammable materials, reducing agents, or other hazardous substances. Buildings in working areas should be fireproof. Dust control or dust prevention in perchlorate solutions is particularly important. Since mixtures of such solutions with oil or grease are violently explosive, motor and pump bearings must be provided with special lubricating and washing devices [223]. Fires must be extinguished with water, but carbon dioxide may provide sufficient cooling to extinguish small fires. Dry powder is ineffective because it cannot smother a self-sustaining fire. For burns, cold water treatment should be used as quickly as possible. For other first aid measures, consult [224].

A second class of more hazardous compounds is formed by inorganic perchlorates containing nitrogen or heavy metals, organic perchlorates, or mixtures of inorganic perchlorates with organic substances, finely divided metals, or sulfur. They are all very sensitive to rubbing, shock, percussion, sparks, and heating and must, therefore, be handled with the same precautions as high explosives.

8.6. Uses

Perchloric Acid. The commercial product is the azeotropic, aqueous solution of 72.5% $HClO_4$ (*bp* 203 °C). In analytical chemistry, it serves to determine the metallic elements present in oxidizable substances, such as organic compounds [270]–[275]. Perchloric acid is an acetylation catalyst for cellulose and glucose and is used in the preparation of cellulose fibers.

Perchlorates. The most outstanding property of ammonium perchlorate is its high oxygen concentration (54.5% O_2) and the fact that it decomposes without leaving a solid residue. Therefore, it is used as an oxidizing component in solid rocket propellants.

Lithium perchlorate is used in lithium–nickel sulfide dry batteries [276], and sodium perchlorate in electrochemical machining [277]; potassium perchlorate is a component of pyrotechnics and an ignition ingredient in flash bulbs [278]. Magnesium perchlorate is known as a very effective drying agent.

Ammonium perchlorate mixed with an epoxy resin forms a temporary adhesive between two metallic surfaces, such as two steel plates. These can be separated whenever desired by heating at ca. 300 °C because of the self-sustained combustion of the adhesive layer [279]. Ammonium perchlorate has been tried as a feed supplement to increase the weight of livestock, with an optimum dose of 2–5 mg/kg [280]–[284]. Perchlorates have also proven to be helpful in oxygen-regenerating systems to be used in enclosed environments, such as submarines and spacecraft, and in breathing equipment [285]. Potassium values from enriched bittern, obtained from the Dead Sea or the Great Salt Lake, can be recovered by precipitating potassium perchlorate with sodium perchlorate [286], [287].

8.7. Economic Aspects

Statistical data about the production and use of perchlorates are not easily accessible because of their strategic importance. The United States' production in 1974 was reported to be ca. 25 000 t, or ca. 12% of the chlorate production [288]. On account of the increasing developments of artificial satellites and space shuttle programs, perchlorate production will increase, due to their major use in solid propellants.

9. Toxicology and Occupational Health

Hypochlorous acid and hypochlorites are toxic because they liberate chlorine on contact with acid. Sodium hypochlorite solution produced by on-site electrolyzers (available chlorine concentration 0.5–10 g/L) is regarded as corrosive and as an irritant when ingested or inhaled. It is also a mild skin irritant, and prolonged exposure may result in a burn or rash [56]. However, concentrated hypochlorous acid burns human skin in seconds.

Chlorine Dioxide. Chlorine dioxide gas is the most toxic and hazardous of all chlorine oxides. Even when small amounts of chlorine dioxide are inhaled, the respiratory system is severely damaged [289]. The symptoms of chlorine dioxide intoxication depend on its concentration and on the exposure time; they include lacrimation, headache, vomiting, severe cough, asthmatic bronchitis, dyspnea, and even death. Such defects as dyspnea or asthmatic bronchitis only heal slowly after the exposure to chlorine dioxide has ceased.

Exposure to 5 ppm of chlorine dioxide in air during several hours severely irritates the mucous membranes [290]; short-term exposure to 20 ppm may cause death. According to animal experiments and observations of employees in chlorine dioxide production plants, the human and animal organisms are extremely sensitive to chlorine dioxide, in particular when the concentrations are higher than 1 ppm [289]–[292].

Chlorine dioxide levels below 0.1 ppm are relatively harmless [293], [294]. Exposure of rats to 0.1 ppm of chlorine dioxide (5 h/d; 10 weeks) showed no toxic effects whereas repeated inhalation of 10 ppm of chlorine dioxide caused death.

The odor threshold of chlorine dioxide is less than 0.1 ppm [295], [296]. The MAK and the TLV of chlorine dioxide have been established at 0.1 ppm (0.3 mg/m^3).

Chlorine dioxide production plants must be operated at slightly subatmospheric pressure. The plant design must also account for the high explosiveness of chlorine dioxide; therefore, chlorine dioxide is diluted with air to below 15% before it leaves the generator. No attempt should ever be made to transport chlorine dioxide over long distances in any form.

Chlorates. The major health hazard of the chlorates arises from their extreme danger of flammability on contact with oxidizable substances. Irritation of the skin, mucous membranes, and eyes may occur from prolonged exposure to dusty atmospheres. Contrary to chlorides, the chlorates are strong blood toxins. Chlorates are readily absorbed by the mucous membranes; doses of a few grams of chlorate are lethal for humans. Abdominal pain, nausea and vomiting, diarrhea, pallor, blueness, shortness of breath, and unconsciousness are the immediate symptoms when toxic amounts of chlorates are ingested. In workplaces with dust formation, dust masks must be worn. Cases of chronic toxicity have not been reported. The instructions for industrial safety must carefully be observed [224], [225], [227]. For detailed information on precautionary and first aid procedures, see [226], [231]–[235].

Perchlorates. Sodium perchlorate appears in the urine of humans within 10 min after ingestion and is largely eliminated within 5 h. During many years of large-scale production of alkali-metal perchlorates, no case of perchlorate intoxication has been reported [237]. Industrial experience also indicates that these salts are not particularly irritating to the skin or mucous membranes, although inhalation should be avoided. A sodium perchlorate dose of 2–4 g/kg is lethal to rabbits.

10. References

[1] D. D. Wagman et al., *J. Phys. Chem. Ref. Data* **11** (1982) suppl. no. 2.
[2] V. Engelhard: *Hypochlorite und elektrische Bleiche. Technisch-konstruktiver Teil,* Knapp, Halle (Saale) 1903.
[3] E. Schmidt, E. Graumann, *Ber. Dtsch. Chem. Ges.* **54** (1921) 1861–1873.
[4] H. Vogt, *J. Electrochem. Soc.* **128** (1981) 29C–32C.
[5] W. Wallace, *J. Electrochem. Soc.* **99** (1952) 309C–310C.
[6] F. Oettel, *Z. Elektrotech. Elektrochem.* **1** (1894) 354–361; **1** (1895) 474–480.
[7] H. Wohlwill, *Z. Elektrochem. Angew. Phys. Chem.* **5** (1898) 52–76.
[8] F. Foerster, E. Müller, *Z. Elektrochem. Angew. Phys. Chem.* **8** (1902) 515–540.
[9] F. Foerster, E. Müller, *Z. Elektrochem. Angew. Phys. Chem.* **8** (1902) no. 633–665, 665–672.
[10] F. Foerster, E. Müller, F. Jorre, *Z. Elektrochem. Angew. Phys. Chem.* **6** (1899) 11–23.
[11] F. Foerster, *Trans. Electrochem. Soc.* **46** (1924) 23–50.
[12] J. B. C. Kershaw: *Die elektrolytische Chloratindustrie,* Knapp, Halle (Saale) 1905.

[13] A. Schumann-Leclercq, FR 772 326, 1933.
[14] F. Oettel, *Z. Elektrochem. Angew. Phys. Chem.* **7** (1901) 315–320.
[15] H. B. Beer, *J. Electrochem. Soc.* **127** (1980) 303C–307C.
[16] F. Foerster, *Z. Elektrochem. Angew. Phys. Chem.* **4** (1898) 386–388.
[17] F. Haber, S. Grinberg, *Z. Anorg. Chem.* **16** (1898) no. 198–228, 329–361.
[18] F. Winteler, *Chem.-Ztg.* **22** (1898) 89–90.
[19] Shell-Development Co., US 2 347 151, 1944.
[20] D. S. Davis, *Ind. Eng. Chem.* **34** (1942) 624.
[21] *Gmelin,* System no. 6, Chlor, p. 228.
[22] I. G. Rheinfelden, DE 656 413, 1931.
[23] J. Ourisson, *Atti Congr. Int. Chim.* **4** (1938) 40–51.
[24] C. W. Bunn, L. M. Clark, J. L. Clifford, *Proc. R. Soc., London, Ser. A* **151** (1935) 141–167.
[25] ICI, DE 541 821, 1930.
[26] Olin Mathieson, US 1 787 048, 1930.
[27] Olin Mathieson, US 1 713 650, 1929.
[28] Olin Mathieson, US 2 195 755, 1940; US 2 195 757, 1940.
[29] Olin Corp., EP 0 086 914 A 1, 1982.
[30] W. H. Sheltmire in J. S. Sconce (ed.): *Chlorine, Its Manufacture, Properties and Uses,* Reinhold, New York 1962.
[31] G. L. Culp, R. L. Culp: *New Concepts in Water Purification,* Van Nostrand Reinhold, New York 1974, pp. 188–196.
[32] Murgatroyd's Salt and Chemical Co., GB 984 378, 1962.
[33] R. E. Stanton, US 3 222 269, 1962.
[34] R. E. Stanton, *Manuf. Chem.* 1962, no. 10, 400–404.
[35] B. M. Milwidsky, *Manuf. Chem.* 1962, no. 10, 400–404.
[36] H. B. H. Cooper, *Chem. Process (Chicago)* **33** (1969) 28.
[37] B. Alt, *Chem. Tech. (Heidelberg)* **4** (1975) no. 4.
[38] M. I. Ismail, T. Z. Fahidy, H. Vogt, H. Wendt, *Fortschr. Verfahrenstech.,* vol. **19**, VDI-Verlag, Düsseldorf 1981, pp. 381–414 (Engl.).
[39] N. Ibl, H. Vogt in J. O'M. Bockris, B. E. Conway, E. Yeager, R. E. White (eds.): *Comprehensive Treatise of Electrochemistry,* vol. **2**, Plenum Publishing, New York 1981, a) pp. 201–208; b) pp. 169–201; c) p. 210.
[40] L.R. Czarnetzki, L.J.J. Janssen, *J. Appl. Electrochem.* **19** (1989) 630–636.
[41] G. R. Heal, A. T. Kuhn, R. B. Lartey, *J. Electrochem. Soc.* **124** (1977) 1690–1697.
[42] P. M. Robertson, W. Gnehm, L. Ponto, *J. Appl. Electrochem.* **13** (1983) 307–315.
[43] L. Hammar, G. Wranglén, *Electrochim. Acta* **9** (1964) 1–16.
[44] F. Hine, M. Yasuda, *J. Electrochem. Soc.* **118** (1971) 182–183.
[45] A. T. Kuhn, H. B. H. Hamzah, *Chem. Ing. Tech.* **52** (1980) 762–763.
[46] J. E. Bennett, *Chem. Eng. Prog.* **70** (1974) no. 12, 60–63.
[47] P. M. Robertson, R. Oberlin, N. Ibl, *Electrochim. Acta* **26** (1981) 941–949.
[48] V. I. Skripchenko, E. P. Drozdetskaya, K. G. Il'in, *Zh. Prikl. Khim. (Leningrad)* **49** (1976) 887–889.
[49] J. E. Bennett, *Int. J. Hydrogen Energy* **5** (1980) 383–394.
[50] Engelhard Minerals & Chemicals Corp., US 3 873 438, 1975 (E. P. Anderson, T. L. Lamb).
[51] Oronzio de Nora Impianti Elettrochimici S.p.A., US 4 032 426, 1977 (O. de Nora, V. de Nora, P. M. Spaziante).
[52] C.-Y. Chan, K. H. Khoo, T. K. Lim, *Surf. Technol.* **15** (1982) 383–394.

[53] M. W. Lister, *Can. J. Chem.* **34** (1956) 479–488.
[54] *Chem. Eng. (N.Y.)* **76** (1966), 9th May, 98.
[55] A. T. Kuhn, R. B. Lartey, *Chem. Ing. Tech.* **47** (1975) 129–135.
[56] G. D. Muir: *Hazards in the Chemical Laboratory*, The Chemical Society, London 1977, p. 388.
[57] D. S. Davis, *Ind. Eng. Chem.* **34** (1942) 624.
[58] F. Göbel, *Dissertation* TH Dresden (1961).
[59] R. A. Crawford, B. I. Dewitt, *Tappi* **50** (1967) no. 7, 138.
[60] T. F. Haller, W. W. Northgraves, *Tappi* **38** (1955) no. 4, 199.
[61] J. Kepinski, J. Trzeszczynski, *Chemii*, Ann. Soc. Chim. Pol. 38 (1964) 201.
[62] W. Finkelnburg, H. J. Schumacher, *Z. Phys. Chem. Bodenstein-Festband* (1931) 704.
[63] H. J. Schumacher, G. Steiger, *Z. Phys. Chem. Abt. B* **7** (1930) 363.
[64] A. S. C. Chen, R. A. Larson, V. L. Snoeyink, *Environ. Sci. Technol.* **16** (1982) no. 5, 268–273.
[65] P. L. Gilmon, *Tappi* **51** (1968) no. 4, 62A–66A.
[66] R. Swindells, G. Cowley, R. A. F. Latham, *Pulp Pap. Can.* **81** (1980) no. 4, 49–52, 54, 56, 58.
[67] W. H. Rapson, *Tappi* **37** (1954) no. 4, 129–137.
[68] H. Karge, *Z. Anal. Chem.* **200** (1964) 57.
[69] T. Ozawa, T. Kwan, *Chem. Pharm. Bull.* **31** (1983) no. 8, 2864–2867.
[70] E. T. Eriksen, J. Lind, G. Merenyi, *J. Chem. Soc. Faraday Trans. 1* **77** (1981) no. 9, 2115–2123.
[71] E. Asmus, H. Garschagen, *Z. Anal. Chem.* **138** (1953) 404.
[72] W. H. Rapson, *Tappi* **39** (1956) no. 8, 554–556.
[73] C. C. Hong, F. Lenzi, W. H. Rapson, *Can. J. Chem. Eng.* **45** (Dec. 1967) 349–355.
[74] W. H. Rapson, *Can. J. Chem. Eng.* **36** (1958) 262.
[75] Hooker Chem. Corp., US 2 936 219, 1961.
[76] R. Hirschberg, *Papier (Darmstadt)* **17** (1963) 25.
[77] E. Kesting, *Tappi* **36** (1953) no. 4, 166.
[78] E. Kesting, DE 831 542, 1948.
[79] Metallgesellschaft AG, EP 0 094 718 A1, 1983.
[80] R. S. Saltzmann, *Adv. Instrum.* **30** (1975) Pt. 2, 609.
[81] C. A. Lindholm, *Pap. Puu* **60** (1978) no. 5, 359–362, 365–366, 369–371.
[82] "Production of Bleaching Chemicals – Chlorine Dioxide, Chlorine, Caustic Soda," *Chem. Econ. Eng. Rev.* **15** (1983) no. 10 (no. 172).
[83] Lurgi, Express Information, Bleaching Chemicals, 1983.
[84] Metallgesellschaft AG, EP 0 011 326 B1, 1979.
[85] K. Lohrberg, G. Klamp, *Pulp Pap.* **56** (1982) no. 11, 114–116.
[86] K. Lohrberg, *Pulp Pap.* **54** (1980) no. 10, 185–187.
[87] T. D. Hughes, *Pulp Pap. Can.* **81** (1980) no. 4, 30–32, 34.
[88] T. D. Hughes, *Chemetics, Integrated Chlorine Dioxide System*, International Pulp Bleaching Conference, June 13, 1979, Toronto, Canada.
[89] P. Wintzer, *Chem.-Ing.-Techn.* **52** (1980) no. 5, 392–398.
[90] "Clean Process for Making Chlorine Dioxide," *Environ. Sci. Technol.* **13** (1979) no. 7, 788–790.
[91] Uhde: *Bleaching Chemicals, Chlorine Dioxide, Chlorine, Sodium Hypochlorite, Sodium Hydroxide*, Dortmund 1984.
[92] W. H. Rapson, *Tappi* **41** (1958) no. 4, 181–185.
[93] J. W. Robinson, *Tappi* **46** (1963) no. 2, 120–123.
[94] P. Orlich, R. Simonette, *Pulp Pap.* **40** (1966) no. 7.
[95] Hooker Chem. Corp., 1184-D.

[96] F. M. Ernest, *Pap. Trade J.* **143** (1959) no. 34, 46–50.
[97] E. S. Atkinson, R. Simonette, *Pulp Pap.* **42** (1968) no. 4.
[98] E. S. Atkinson, R. Simonette, *Pulp Pap.* **42** (1968) no. 8.
[99] H. D. Partridge, W. H. Rapson, *Tappi* **44** (1961) no. 10, 698.
[100] ERCO, Chlorine Dioxide Processes, R2, R3, R3H, R5, R6, R7, R8, 1983.
[101] E. S. Atkinson, *Chem. Eng. (N.Y.)* **81** (1974) no. 3, 36–37.
[102] H. de V. Partridge, E. S. Atkinson, *Pulp. Pap. Mag. Can.* **73** (1972) no. 8, 203–206.
[103] H. D. Partridge, E. S. Atkinson, A. C. Schulz, *Tappi* **54** (1971) no. 9, 1484–1487.
[104] D. W. Reeve, W. H. Rapson, *Pulp Pap. Mag. Can.* **74** (1973) no. 1, 19–27.
[105] R. K. Bradley, G. N. Werezak, *Tappi* **57** (1974) no. 4, 98–101.
[106] J. F. Serafin, H. C. Scribner, *Pulp Pap. Mag. Can.* **62** (1961) 473–476.
[107] E. S. Atkinson, *Tappi* **61** (1978) no. 8, 61–64.
[108] D. D. Koch, L. L. Edwards, *Tappi* **55** (1972) no. 5, 752–756.
[109] J. Schuber, W. A. Kraske, *Chem. Eng. N.Y.* **60** (1953) no. 9, 205.
[110] G. L. Cunningham, US 2 089 913, 1937.
[111] T. G. Holst, *Sven. Papperstidn* **47** (1944) 537–546.
[112] T. G. Holst, US 2 373 830, 1945.
[113] S. H. Persson, US 2 376 935, 1945.
[114] US 2 481 241, 1949, (W. H. Rapson, M. Waymann).
[115] US 2 598 087, 1952, (M. Waymann, W. H. Rapson).
[116] E. R. Woodward, G. P. Vincent, *Trans. Am. Inst. Chem. Eng.* **40** (1944) 271.
[117] W. J. Masschelein, *J. Am. Water Works Assoc.* **76** (1984) no. 1, 70–76.
[118] J. Valenta, W. Gahler, O. Mohr, *BBR, Brunnenbau, Bau Wasserwerken, Rohrleitungsbau* **32** (1981) no. 3, 106–109.
[119] R. W. Jordan, J. A. Kosinski, R. J. Baker, *Water Sewage Works* **127** (1980) no. 10, 44–45, 48.
[120] M. C. Fredette, *Pulp Pap. Can.* **85** (1984) no. 1, 52–56.
[121] W. H. Rapson, K. A. Hakim, *Pulp Pap. Mag. Can.* **151** (July 1957) 151.
[122] W. H. Rapson, C. B. Anderson, G. F. King, *Tappi* **41** (1958) 442.
[123] M. Gummerus, C. A. Lindholm, N. E. Virkola, *Pap. Puu* **59** (1977) no. 5, 335–338, 341–344, 347–348, 353–355.
[124] F. Kocevar, H. Talovic, *Melliand Textilber. Int.* **54** (1973) no. 10, 1064–1066.
[125] C. Rav-Acha, R. Blits, E. Choshen, A. Serri, B. Limoni, *J. Environ. Sci. Health, Part A* **A18** (1983) no. 5, 651–671.
[126] M. C. Taylor, J. E. White, G. P. Vincent, G. L. Cunningham, *Ind. Eng. Chem.* **32** (1940) 890.
[127] G. Cunningham, T. S. Oey, *J. Amer. Chem. Soc.* **77** (1955) 4498.
[128] *Ullmann*, 3rd ed., **5,** 521.
[129] G. Cunningham, T. S. Oey, *J. Amer. Chem. Soc.* **77** (1955) no. 3, 799.
[130] V. Popjankow, N. Kolarow, *Z. Anorg. Allg. Chem.* **346** (1966) 322.
[131] I. Nakamori, Y. Nagino, K. Hideshima, T. Hirai, *Kogyo Kagaku Zasshi* **61** (1958) 147–149; *Chem. Abstr.* **53**, 21322 a.
[132] T. S. Oey, G. Cunningham, J. Koopman, *Chem. Eng. Data* **5** (1960) no. 3, 248.
[133] H. Korte, W. Kaufmann, *Melliand Textilber.* **23** (1942) 234–239.
[134] M. C. Tayler et al., *Ind. Eng. Chem.* **32** (1940) 899–903.
[135] J. F. White, M. C. Tayler, G. P. Vincent, *Ind. Eng. Chem.* **34** (1942) 782–792.
[136] L. F. Ynterna, T. Fleming, *Ind. Eng. Chem. Anal. Ed.* **11** (1939) 375–377.
[137] I. F. Haller, S. Listek, *Analytic.* **20** (1948) 639–642.
[138] I. F. White, *Am. Dyest. Rep.* **31** (1942) 484–487.

[139] W. C. Gardiner, E. H. Karr, *Office Tech. Serv. Rept.*, **PB-33** (1946) 218.
[140] I. G. Farbenind., DE 744 369, 1944 (F. Erbe).
[141] Mathieson Alkali Works, US 2 092 944, Sept. 1937.
[142] Potasse et Produits Chimiques, FR 1 019 027, 1953.
[143] A. R. Reychler, *Bull. Soc. Chim. Fr.* **25** (1901) 663.
[144] G. R. Levi, *Atti V Congr. Nazl. Chim. Pura Applicata, Rome 1935*, 1936, part I, 382–386; *Chem. Abstr.* **31**, 5290^5.
[145] Solvay + Cie., CH 258 842, 1946.
[146] Nederld. Zoutindustrie, DE 963 505, 1954.
[147] Farbwerke Hoechst, DE 1 059 891, 1958.
[148] Mathieson Alkali Works, DE 668 140, 1935.
[149] Farbw. Hoechst, DE 906 447, 1942.
[150] E. Elöd, A. Klein, *Melliand Textilber.* **27** (1947) no. 191, 263.
[151] H. Baier, *Melliand Textilber.* **32** (1951) no. 2, 141–146.
[152] W. Hundt, K. Vieweg, *Text.-Prax.* **6** (1951) 439.
[153] S. K. Kapoor, Y. V. Sood, T. K. Roy, A. K. Kohli, R. Pant, *IPPTA*, **20** (1983) no. 1, 27–32.
[154] R. A. Crawford, W. B. Darlington, L. B. Kliever, *J. Electrochem. Soc.* **117** (1970) 279–282.
[155] N. V. S. Knibbs, H. Palfreeman, *Trans. Faraday Soc.* **16** (1920) 402.
[156] H. Vogt, *Chem. Age India* **26** (1975) 540–544.
[157] A. Nallet, R. A. Paris, *Bull. Soc. Chim. Fr.* 1956, 488–494.
[158] P. Wintzer, *Chem.-Ing.-Tech.* **52** (1980) 392–398.
[159] F. Foerster, *J. Prakt. Chem.* **63** (1901) 141–166.
[160] F. Foerster, E. Müller, *Z. Elektrochem.* **9** (1903) 171–185.
[161] I. E. Flis, M. K. Bynyaeva, *Zh. Prikl. Khim. (Leningrad)* **30** (1957) 339–345; *J. Appl. Chem. USSR (Engl. Transl.)* **30** (1957) 359–365.
[162] I. Taniguchi, T. Sekine, *Denki Kagaku* **43** (1975) 715–720.
[163] B. V. Tilak, K. Viswanathan, C. G. Rader, *J. Electrochem. Soc.* **128** (1981) 1228–1232.
[164] D. Landolt, N. Ibl, *Electrochim. Acta* **15** (1970) 1165–1183.
[165] N. V. S. Knibbs, *Trans. Faraday Soc.* **16** (1920) 424–433.
[166] V. A. Shlyapnikov, *Zh. Prikl. Khim. (Leningrad)* **42** (1969) no. 2182–2188, *J. Appl. Chem. USSR (Engl. Transl.)* **42** (1969) 2051–2056.
[167] I. Taniguchi, T. Sekine, *Denki Kagaku* **43** (1975) 632–637.
[168] C. Wagner, *J. Electrochem. Soc.* **101** (1954) 181–185.
[169] V. A. Shlyapnikov, T. S. Filippov, *Elektrokhim.* **5** (1969) 866–868; *Sov. Electrochem. (Engl. Transl.)* **5** (1969) 806–807.
[170] F. Hine, M. Yasuda, *J. Electrochem. Soc.* **118** (1971) 182–183.
[171] J. S. Booth, H. Hamzah, A. T. Kuhn, *Electrochim. Acta* **25** (1980) 1347–1350.
[172] I. E. Veselovskaya, E. M. Kuchinski, L. V. Morochko, *Zh. Prikl. Khim.* **37** (1964) 76–83; *J. Appl. Chem. USSR (Engl. Transl.)* **37** (1964) 85–91.
[173] I. Taniguchi, T. Sekine, *Denki Kagaku* **43** (1975) 532–534.
[174] V. de Valera, *Trans. Faraday Soc.* **49** (1953) 1338–1351.
[175] M. W. Lister, R. C. Petterson, *Can. J. Chem.* **40** (1962) 729–733.
[176] J. M. González Barredo, *An. Fis Quim.* **37** (1941) 123–157.
[177] J. R. Newberry, W. C. Gardiner, A. J. Holmes, R. F. Fogle, *J. Electrochem. Soc.* **116** (1969) 114–118.
[178] T. R. Beck, R. Brännland, *J. Electrochem. Soc.* **119** (1972) 320–325.
[179] J. Horacek, S. Puschaver, *Chem. Eng. Prog.* **67** (1971) no. 3, 71–74.

[180] T. Matsumura, R. Itai, M. Shibuya, G. Ishi, *Electrochem. Technol.* **6** (1968) 402–404.
[181] T. Nagai, T. Takei, *J. Electrochem. Soc. Japan* **25** (1957) no. 2, 11.
[182] J. Billiter: *Die technische Elektrolyse der Nichtmetalle*, Springer-Verlag, Wien 1954.
[183] M. M. Jakić, *J. Appl. Electrochem.* **3** (1973) 219–225.
[184] M. M. Jakić, A. R. Despić, B. Z. Nikolić, *Elektrokhim.* **8** (1972) 1573–1584 (Russ.); *Sov. Electrochem. (Engl. Transl.)* **8** (1972) 1533–1542.
[185] D. G. Elliott, *Tappi* **51** (1968) no. 4, 88A–90A.
[186] M. M. Jakić, B. Z. Nikolić, D. M. Karanović, C. R. Milovanović, *J. Electrochem. Soc.* **116** (1969) 394–398.
[187] *Chem. Eng. (N.Y.)* **72** (19 July 1965) 82–83.
[188] K. C. Narasimham, H. V. K. Udupa, *J. Electrochem. Soc.* **123** (1976) 1294–1298.
[189] *Ullmann*, 3rd ed., **5**, 539.
[190] M. Y. Fioshin, *Uspekhi v oblasti elektrosinteza neorganiceskikh soedinenii. (Progress of electrosyntheses of inorganic compounds)* Isdatelstwo "Khimia," Moscow 1974.
[191] W. Schreiter, H. Wendt, H. Vogt, *Fortschr. Verfahrenstechnik,* **vol. 17,** VDI-Verlag, Düsseldorf 1979, pp. 379–412.
[192] J. E. Colman, *AIChE Symp. Ser.* **77** (1981) no. 204, 244–263.
[193] K. Viswanathan, B. V. Tilak, *J. Electrochem. Soc.* **131** (1984) 1551–1559.
[194] C. Oloman: *Electrochemical Processing for the Pulp and Paper Industry,* The Electrochemical Consultancy, Romsey, Hants, 1996.
[195] I. Rouar, V. Cezner, *J. Electrochem. Soc.* **121** (1974) 648–651.
[196] H. B. Beer, DE 1 571 721, 1966.
[197] H. B. Beer, *J. Electrochem. Soc.* **127** (1980) 303C–305C.
[198] H. B. Beer, DE 1 671 422, 1968.
[199] H. B. Beer, CH 492 480, 1966.
[200] S. Trasatti, W. E. O'Grady, *Adv. Electrochem. Electrochem. Eng.* **12** (1981) 177–261.
[201] H. B. Beer, *Chem. Ind. (London)* 1978, 491–496.
[202] O. De Nora, *Chem.-Ing.-Tech.* **42** (1970) 222–226.
[203] G. Bewer, H. Debrodt, H. Herbst, *J. Met.* **34** (1982) 37–41.
[204] F. Brandmeir, H. Böder, G. Bewer, H. Herbst, *Chem.-Ing.-Tech.* **52** (1980) 443–444.
[205] L. M. Elina, V. M. Gitneva, V. I. Brystow, N. M. Shmygul', *Elektrokhim.* **10** (1974) 68–70 (Russ.); *Sov. Electrochem. (Engl. Transl.)* **10** (1974) 59–61.
[206] *Ullmann*, 4th ed., **22,** 41, 69.
[207] Heraeus, DE 3 239 535, 1982.
[208] N. V. Krstajić, M. D. Spasojević, R. T. Atanasoski, *J. Appl. Electrochem.* **14** (1984) 131–134.
[209] H. Vogt, in *Fortschritte der Verfahrenstechnik,* vol. **20,** VDI-Verlag, Düsseldorf 1982, pp. 369–404.
[210] W. Thiele, M. Schleiff, *Chem. Tech. (Leipzig)* **31** (1979) 623–624.
[211] F. Hine, M. Yasuda, T. Noda, T. Yoshida, J. Okura, *J. Electrochem. Soc.* **126** (1979) 1439–1445.
[212] M. M. Jakić, B. Z. Nikolić, M. D. Spasojević, *Chem. Tech. (Leipzig)* **27** (1975) no. 158–162, 534–538.
[213] M. M. Jakić, *Electrochim. Acta* **21** (1976) 1127–1136.
[214] T. R. Beck, *J. Electrochem. Soc.* **116** (1969) 1038–1041.
[215] H. Vogt, *J. Appl. Electrochem.* **22** (1994) 1185–1191.
[216] H. Vogt, *Electrochim. Acta* **39** (1994) 2173–2179.

[217] M. M. Jakić, A. R. Despić, I. M. Csonka, B. Z. Nikolić, *J. Electrochem. Soc.* **116** (1969) 1316–1322.
[218] Pennwalt Corp., US 3 883 406, 1973.
[219] H.-P. Neumann, J. Kardos, *Chem. Tech.* **39** (1987) 205–207.
[220] T. R. Beck, R. T. Ruggeri, *Adv. Electrochem. Electrochem. Eng.* **12** (1981) 263–354.
[221] J. Fleck, *Chem.-Ing.-Tech.* **43** (1971) 173–177.
[222] J. R. Hodges, *J. Electrochem. Soc.* **131** (1984) 96C.
[223] C. M. Olson, *J. Electrochem. Soc.* **116** (1969) 33C–35C.
[224] *Safety Data Sheet SD-42*, Manufacturing Chemists Ass., Washington D.C., 1952.
[225] *Data Sheet 371, Chlorates*, National Safety Council, Chicago, Ill., 1977.
[226] *Merkblatt für Arbeiten mit Kalium-Natriumchlorat*, Verlag Chemie, Weinheim 1965.
[227] I.C.C. Regulations for Transportation of Explosives and other Dangerous Articles by Rail Freight, Rail Express, Rail Baggage and Motor Carrier.
[228] RID/ADR International Rules for the Transportation of Dangerous Goods by Rail and Road, Office Central des Transports par chemins de fer, Bern 1985.
[229] *Accident Prevention Manual for Industrial Operations*, National Safety Council, 7th ed., Chicago, Ill., 1978.
[230] *Unfallverhütungsvorschriften der Berufsgenossenschaft der chemischen Industrie*, Carl Heymanns Verlag, Köln, published annually.
[231] *Merkblätter Gefährliche Arbeitsstoffe* (R. Kühn, K. Birett eds.), 4th ed., Verlag Moderne Industrie, München 1979, p. N04.
[232] *Hygienic Guide Series, Sodium Chlorate*, American Industrial Hygiene Association, Detroit, Mich., United States.
[233] L. Roth, M. Daunderer: *Toxikologische Enzyklopädie*, l. 1–3, Verlag Moderne Industrie, München 1976.
[234] G. Hommel: *Handbuch der gefährlichen Güter*, vol. **1**, no. 142, Springer-Verlag, Berlin 1980.
[235] L. Roth, U. Weller: *Gefährliche chemische Reaktionen*, ecomed, Landshut 1982, pp. III-N, 1–4.
[236] W. C. Gardiner, *J. Electrochem. Soc.* **125** (1978) 22C–29C.
[237] J. C. Schumacher (ed.): *Perchlorates – their Properties, Manufacture and Uses*, Am. Chem. Soc. Mon. no. 146, Reinhold Publ. Corp., New York 1960.
[238] *Gmelin*, System no. 6, Chlor, Verlag Chemie, Berlin 1927, pp. 362–391.
[239] H. V. Casson, G. J. Crane, G. E. Styan, *Pulp Pap. Mag. Can.* **69** (1968) 39.
[240] G. K. Loseva, *Tr. Novocherkassk, Politekh. Inst.* **266** (1972) 78–81.
[241] F. Shadman-Yazdi, E. E. Petersen, US Nat. Tech. Inform. Serv., AD Rep. 1972, no. 746 728; *Gov. Rep. Announc.* **72** (1972) 172.
[242] A. V. Nikolaev, G. G. Tsurinov, L. A. Khripin, *Izv. Sib. Otd. Akad. Nauk SSSR, Ser. Khim. Nauk.* 1972, 98–99.
[243] P. Jacobs, W. McCarthy, H. M. Whitehead, *Chem. Rev.* **69** (1969) 551.
[244] P. Barret, *Cah. Therm.* **4** (1924) 13.
[245] C. N. R. Rao, B. Prakash, *Natl. Stand. Ref. Data Ser. (U.S. Natl. Bur. Stand.)* **56** (1975) 28.
[246] F. Solymosi, *Acta Phys. Chim.* **19** (1973)∥67.
[247] Oldbury Electrochem. Co., US 2 392 861, 1945 (J. C. Pernert).
[248] E. Merck AG, DE 1 031 288, 1956.
[249] W. Müller, P. Jönck, *Chem. Ing. Tech.* **35** (1963) 78–80.
[250] Y. Kato, K. Sugino, K. Koizumi, S. Kitahara, *Electrotech. J.* **5** (1941) 45–48.
[251] M. Nagalingam, R. P. Govinda, R. C. J. Narasimham, S. Sampath, H. V. K. Udupa, *Chem. Ing. Tech.* **41** (1969) 1301–1308.

[252] H. V. K. Udupa, K. C. Narasimham, M. Nagalingam, N. Thiagarajan et al., *J. Appl. Electrochem.* **1** (1971) 207–212.

[253] O. de Nora, P. Gallone, C. Traini, G. Meneghini, *J. Electrochem. Soc.* **116** (1969) 146–151.

[254] J. C. Grigger in C. A. Hampel (ed.): *The Encyclopedia of Electrochemistry*, Reinhold Pub. Corp., New York 1964, pp. 762–764.

[255] T. Osuga, S. Fujii, K. Sugino, T. Sekine, *J. Electrochem. Soc.* **116** (1969) 203–207.

[256] Pacific Eng. and Prod. Co., US 2 945 791, 1960.

[257] J. C. Schumacher, D. R. Stern, P. R. Graham, *J. Electrochem. Soc.* **105** (1958) 151–155.

[258] J. C. Schumacher, *J. Electrochem. Soc.* **129** (1982) 397C–403C.

[259] K. R. Koziol, K. H. Sieberer, H.-C. Rathjen, J. B. Zenk, E. F. Wenk, *Chem. Ing. Tech.* **49** (1977) 288–293.

[260] A. Legendre, *Chem. Ing. Tech.* **34** (1962) 379–386.

[261] T. W. Clapper in C. A. Hampel (ed.), ref. [252], pp. 886–890.

[262] D. Burrows, J. C. Dacre, U.S. NTIS, AD-A, Rep. no. 010 660, Army Med. Bioeng. Res. Dev. Lab. Fort Detrick, Md., 1975.

[263] F. D. Snell, C. T. Snell: *Colorimetric Methods of Analysis*, vol. **2**, Van Nostrand Co., Princeton, N.J., 1957, pp. 718–719.

[264] H. H. Willard, H. Diehl: *Advanced Quantitative Analysis*, Van Nostrand Co., Princeton, N.J., 1944, pp. 254–257.

[265] G. M. Smith, *Ind. Eng. Chem. Anal. Ed.* **11** (1939) 186.

[266] G. M. Smith, *Ind. Eng. Chem. Anal. Ed.* **11** (1939) 269.

[267] R. J. Bacquk, R. J. Dubois, *Anal. Chem.* **40** (1968) 685.

[268] M. J. Smith, S. E. Manahan, *Anal. Chim. Acta* **48** (1969) 315.

[269] R. M. Graziano: *Hazardous Materials Regulations*, Tarif No. 31, Dept. of Transportation, Washington DC, 1977.

[270] G. F. Smith: *Mixed Perchloric, Sulfuric and Phosphoric Acids and their Applications in Analysis*, G. F. Smith Chem. Co., Columbus, Ohio, 1942.

[271] G. F. Smith, *Anal. Chim. Acta* **8** (1953) 397.

[272] G. F. Smith, *Anal. Chim. Acta* **17** (1957) 175.

[273] E. Kahane: *L'Action de l'Acide Perchlorique sur les Matières Organiques*, Herman et Cie, Paris 1934.

[274] H. Diehl, G. F. Smith, *Talanta* **2** (1959) 209.

[275] T. T. Gorsuch: *The Destruction of Organic Matter*, Pergamon Press, Oxford 1970.

[276] L. Gaines, R. Jasinski, US Natl. Tech. Inform. Serv., AD Rep. no. 749 861 (1972).

[277] J. P. Hoare, K. W. Mao, A. J. Wallace, *Corrosion (Houston)* **27** (1971) 211–215.

[278] General Electric Company, US 3 724 991, 1971.

[279] Nissan Motor Co., US 3 993 524, 1976 (Y. Okada, S. Kensho).

[280] P. N. Razumovskii, *Khim Sel'sk Khoz.* 1976, no. 14, 71.

[281] P. N. Razumovskii, G. S. Semanin, G. I. Balk, *Kompleksn. Ispol'z Biol. Akt. Veshchestv. Korml. S-kh Zhivotn. Mater. Vses. Soveshch.* **1** (1973) 370.

[282] A. S. Solan, *Zhivotnovodstvo* 1974, no. 11, 63.

[283] V. I. Mikhailov, B. R. Gotsulenko, V. P. Kardivari, *Zhivotnovodstvo* 1976, no. 5, 83.

[284] H. G. Pena, *Poult. Sci.* 1976, 188.

[285] Thiokol Corp., US 3 993 514 (E. J. Pacanovsky, E. A. Martino).

[286] J. A. Epstein, *Hydrometallurgy* **1** (1975) 39.

[287] D. R. George, J. M. Riley, J. R. Ross, paper presented at 62nd Nat. Meet. Inst. Chem. Eng., Salt Lake City, May 21–24, 1967.

[288] W. C. Gardiner, *J. Electrochem. Soc.* **125** (1978) 22C–29C.
[289] H. Petry, *Arch. Gewerbepath. Gewerbehyg.* **13** (1964) 363.
[290] H. B. Elkins: *The Chemistry of Industrial Toxicology,* 2nd ed., J. Wiley &Sons, New York 1958.
[291] J. Gloemme et al., *AMA Arch. Ind. Health* **16** (1957) 169.
[292] A. Graefe, *Dtsch. Med. Wochenschr.* **8** (1902) 191.
[293] L. T. Fairhall: *Industrial Toxicology.* Williams & Wilkins Co., Baltimore 1957.
[294] T. Dalhamm, *AMA Arch. Ind. Health* **15** (1957) 101.
[295] "Hygienic Guide, Chlorine Dioxide," *Ind. Hyg. Ass. J.* **21** (1958) 381.
[296] M. N. Gleason, R. E. Gosselin, H. C. Hodge: *Clinical Toxicology of Commercial Products,* Williams & Wilkins Co., Baltimore 1963.

Chlorosulfuric Acid

JOACHIM MAAS, Bayer AG, Leverkusen, Federal Republic of Germany
FRITZ BAUNACK, Wuppertal, Federal Republic of Germany

1.	Introduction 1329	5.	Safety Precautions, Transportation 1333
2.	Physical Properties 1329	6.	Uses and Economic Aspects . . 1334
3.	Chemical Properties 1331	7.	Toxicology and Occupational Hygiene 1335
4.	Production 1332	8.	References 1335

1. Introduction

Chlorosulfuric acid [7790-94-5], M_r 116.53, is also referred to in the literature as chlorosulfonic acid, sulfuryl hydroxychloride, sulfuric chlorohydrin, sulfonic acid monochloride, and chlorohydrated sulfuric acid.

$$HSO_3Cl \quad \text{or} \quad H-O-\overset{\overset{O}{\|}}{\underset{\underset{O}{\|}}{S}}-Cl$$

Chlorosulfuric acid is a colorless, mobile, extremely reactive liquid. It fumes very strongly in air and, consequently, has been used as a military smoke-generating agent, being mixed with oleum for this purpose (ca. 50–65 wt% HSO_3Cl, 50–35 wt% SO_3). A very thick, opaque mist is produced when 2 L of chlorosulfuric acid is mixed with 1000 m^3 of air.

Chlorosulfuric acid is highly corrosive and hygroscopic. It decomposes explosively in the presence of water, forming sulfuric acid and hydrogen chloride. A. WILLIAMSON first prepared chlorosulfuric acid in 1854 [1].

2. Physical Properties

The values of physical constants quoted in the literature should not be accepted uncritically. Chlorosulfuric acid is difficult to purify and has frequently been used in an impure state for experimental determinations. Table 1 shows the best values available. Very pure chlorosulfuric acid can be obtained by fractional crystallization [2].

Table 1. Physical properties of chlorosulfuric acid

Property	Value
Freezing point fp	-80 to -81 °C
Boiling point bp(decomp.)	$151-152$ °C
Vapor pressure	
(p in mbar at T in K)	$\log p = 9.495 - 2752/T$
(p in Pa at T in K)	$\log p = 11.495 - 2752/T$
Density*	$d_4^{20} = 1.753$
	$d_4^{-70} = 1.90$
	$d_4^{-10} = 1.80$
	$d_4^{\,t}$ (0–100 °C) $= 1.7847 - 1.616\,t \times 10^{-3}$
	$+ 1.21\,t^2 \times 10^{-6} - 4.1\,t^3 \times 10^{-9}$
Specific heat capacity (15–80 °C)	1.18 kJ kg^{-1} K^{-1}
Heat of vaporization (at boiling point)	53.6 kJ/mol (460 J/g)
Heat of solution in water (at 18 °C)	168.7 kJ/mol
Heat of formation from elements (at 25 °C)	-597.5 kJ/mol
Viscosity at -31.6 °C	10 mPa s
at -17.8 °C	6.4 mPa s
at 15.6 °C	3.0 mPa s
at 49 °C	1.7 mPa s
Refractive index n_D^{14}	1.437
Dielectric constant (at 15 °C)	60 ± 10
Electrical conductivity (at 25 °C)	$0.2-0.3 \times 10^{-3}\,\Omega^{-1}$ cm^{-1}
Specific magnetic susceptibility	-0.402

* Because chlorosulfuric acid decomposes at the boiling point, its vapor density cannot be determined precisely.

The large heat of vaporization is explained by the strong association of chlorosulfuric acid molecules, which results from hydrogen bonding [3]. Chlorosulfuric acid is soluble in methylene chloride, chloroform, 1,1,2,2-tetrachloroethane, 1,2-dichloroethane, acetic acid, acetic anhydride, trifluoroacetic acid, trifluoroacetic anhydride, sulfuryl chloride, liquid sulfur dioxide, and sparingly soluble in carbon tetrachloride and carbon disulfide. It reacts with alcohols, ketones, ethers, and dimethyl sulfoxide.

Whereas sulfur trioxide is miscible in all proportions with chlorosulfuric acid, the limit of solubility of free hydrogen chloride is 0.5 wt% at 20 °C and atmospheric pressure [4].

Warning: Care must be taken when working with chlorosulfuric acid in the presence of solvents; an increase in temperature can lead to violent reactions, particularly when catalysts are also present.

3. Chemical Properties

Chlorosulfuric acid is a strong acid, stable under normal conditions. The tetrahedral molecule [5] contains a relatively weak S–Cl bond [6]. Its force constant is only about one-third that of the single bond in S–O(H) [7]. Depending on the co-reactant, the solvent (if any), the temperature, and the amount of excess HSO_3Cl present, the first step of the reaction will be the formation of free chlorine atoms [8], chlorine ions, or protons [6], [9]. Prolonged heating or distillation, especially in vacuo, leads to partial decomposition to hydrogen chloride, chlorine, sulfur dioxide, sulfuryl chloride, pyrosulfuryl chloride, and sulfuric acid [2].

If chlorosulfuric acid is stored in the presence of an excess of sulfur trioxide, the following reaction occurs [2]:

$$2\ HSO_3Cl + SO_3 \longrightarrow H_2SO_4 + S_2O_5Cl_2$$

In freshly prepared and cooled solutions chloropyrosulfuric acids are formed by the following equilibrium reaction:

$$H(SO_3)_nCl + SO_3 \rightleftharpoons H(SO_3)_{n+1}Cl$$

where n = 1, 2, or 3 [10].

Chloropyrosulfates are obtained from the reaction of alkali metal chlorides with sulfur trioxide in liquid sulfur dioxide at −12 °C [11].

Many reactions with chlorosulfuric acid can be carried out at low temperatures, which is facilitated by its low melting point. Sulfonic acids are produced with equimolar quantities of aromatic hydrocarbons under mild conditions; the reaction usually does not affect other substituents present in the molecule. However, in some cases, amino groups must be protected by acylation from reacting with chlorosulfuric acid. Side reactions lead to the formation of small quantities of sulfonyl chlorides, which usually react further with aromatic hydrocarbons to form sulfones. The sulfonyl chlorides are used as intermediates in the manufacture of pharmaceuticals or dyes.

Depending on the reaction conditions, sulfonyl chlorides, sulfones, or chlorinated aromatic hydrocarbons can be obtained by using an excess of chlorosulfuric acid [12], [13]. Sulfonyl chloride formation takes place in two steps: formation of a sulfonic acid followed by an equilibrium reaction [6], [14], [15]:

$$ArH + ClSO_3H \rightleftharpoons ArSO_3H + HCl$$
$$ArSO_3H + ClSO_3H \rightleftharpoons ArSO_2Cl + H_2SO_4$$

This equilibrium can be shifted to the right by reacting the sulfuric acid formed with pyrosulfuryl chloride or with thionyl chloride. The reaction with pyrosulfuryl chloride leads to the formation of chlorosulfuric acid and sulfur trioxide; thionyl chloride reacts with the nascent sulfuric acid to form chlorosulfuric acid, hydrogen chloride, and sulfur dioxide [16].

The reaction of chlorosulfuric acid with aliphatic hydrocarbons is slow, provided there are no double bonds or other reactive groups present. Therefore, long-chain saturated fatty alcohols react with chlorosulfuric acid to form the corresponding sulfates without chain cleavage or discoloration of the product. This type of reaction is used on a large scale in the manufacture of detergents:

ROH + HSO$_3$Cl $\longrightarrow$ HCl + ROSO$_3$H

Both chlorosulfuric acid and fluorosulfuric acid have been proposed as polymerization catalysts [17]. Many of the reactions of chlorosulfuric acid have been reviewed by JACKSON [18]. Special information on this subject can be found in refs. [19]–[27].

4. Production

All the industrial routes to chlorosulfuric acid are based on the reaction of hydrogen chloride with sulfur trioxide:

HCl + SO$_3$ $\longrightarrow$ HSO$_3$Cl

Earlier processes made use of contact process gas which contains 6–7% SO$_3$ [28], but today it is more usual to use pure sulfur trioxide. Waste-gas problems are thereby dealt with more easily, and a more compact plant design is possible. Production processes vary both in the manner of bringing the two raw materials into contact and in the methods of removing the heat of reaction.

The reactor can be a packed column with a chlorosulfuric acid spray at the top and hydrogen chloride and sulfur trioxide entering at the bottom [29]. In this case, the reaction and the heat removal both take place in one piece of equipment. Alternatively, it is possible to separate these steps by first mixing the components intensively in a separate mixer such as a mixing nozzle [30]; the hot reaction product is then quickly cooled with cold chlorosulfuric acid in a packed column or other suitable unit. Cooling has also been suggested by means of a water-cooled condenser [31]. In one modification of the process, the first product is chlorosulfuric acid which contains a small excess of sulfur trioxide; this is then cooled and saturated with hydrogen chloride in a bubble column [32]. The off-gas is normally scrubbed, first with 98% sulfuric acid and then with water.

Several patents describe the recycling of recovered hydrogen chloride and sulfur trioxide [33]. The heat of formation of chlorosulfuric acid can be used to evaporate sulfur trioxide from low-percentage oleum. The sulfur trioxide then reacts with hydrogen chloride at sub-atmospheric pressure [34].

Construction Materials. Because commercial chlorosulfuric acid should contain very little or no iron, suitable construction materials for reactors, containers, etc., are enamel, glass, aluminum, or steel lined with polytetrafluoroethylene (PTFE) [35].

Analysis and Quality Requirements. The usual method for analyzing chlorosulfuric acid involves hydrolysis of the acid to form SO_3 and HCl, followed by determination of the total acid-ity and chloride content. When the content of HSO_3Cl, H_2SO_4, and free SO_3 or HCl is calculated, the amount of water that is absorbed by SO_3 to form H_2SO_4 must be taken into account. Variations of this method are described in ref. [27]. The presence of sulfuryl and pyrosulfuryl chloride (SO_2Cl_2 and $S_2O_5Cl_2$) must be determined by special methods [28]. Specific color reactions in combination with too low a melting point identify chlorosulfuric acid producing a cherry red color with tellurium powder or a moss green color with selenium [2].

A product low in iron is normally required, with the following analysis:

HSO_3Cl	98–99.5%
H_2SO_4	0.2–2%
Free SO_3	0–2%
Free HCl	0–0.5%
Fe	5–30 ppm

5. Safety Precautions, Transportation

Because chlorosulfuric acid is very aggressive, the regulations dealing with hazardous materials apply [36]. Chlorosulfuric acid reacts violently with water and, with moist air, decomposes to form hydrogen chloride and sulfuric acid mist. All vessels, pipes, metallic hoses, etc., should be dry. Even small leakages must be prevented. Containers should be handled with care: they must never be dropped because a slight internal pressure may build up. The stopper should be opened only with a wrench and not with hammer and chisel. Containers must be emptied by siphoning and not by the use of compressed air. The access of moist air can be avoided by tight connection to a gas bypass. After each removal of chlorosulfuric acid the container must be tightly closed.

Although chlorosulfuric acid itself is not combustible, it can induce ignition of combustible materials. Reaction with moist metals can produce hydrogen. Naked flames must therefore be avoided in the neighborhood of containers or pipes. If welding must take place, the safety precautions pertaining to hydrogen should be observed. Enclosed spaces in which chlorosulfuric acid is being handled must always be well ventilated. Acid-proof clothing, goggles, and gloves must be worn. A full face gas mask with a filter element for absorbing acid fumes, and safety boots must be at hand. Polychloroprene coated with a copolymer of vinylidene fluoride and hexafluoropropylene would be a suitable material for gloves, protective clothing, and full-face gas masks. Boots could be made of poly(vinyl chloride). Contaminated clothing must be changed without delay.

Small spillages of chlorosulfuric acid must be washed away with water, and all necessary safety measures should be observed; the spilled material should be downwind and the area cordoned off. A high-velocity jet of water should not be used; the spillage should rather be worked from the outside toward the middle, preferably by means of a hose with a spray nozzle attachment. The area can then be neutralized with sodium carbonate, sodium hydrogen carbonate, or lime.

Chlorosulfuric acid must not be allowed to leak into the sewerage system or the soil. Spilled acid may be blanketed with paraffin oil or FEP film (perfluorinated ethylene–propylene copolymer). The contaminated area will reveal itself by fog formation; and this can be prevented from spreading by shielding with water. Personnel involved in such cleaning operations must wear full protective clothing and use compressed air breathing apparatus.

Transportation. Differing regulations apply to transportation by road, rail, or sea [37]. Road transportation may utilize barrels made from aluminum or iron.

Chlorosulfuric acid should, however, be carried by rail whenever possible. The construction material for tanks of rail or road trucks is usually stainless steel, but enameled steel may also be used.

6. Uses and Economic Aspects [38]

The breakdown of chlorosulfuric acid production according to end use is as follows:

Detergents	40%
Pharmaceuticals	20%
Dyes	15%
Crop protection	10%
Ion-exchange resins [39], plasticizers, and others	15%

The following companies are published as manufacturers of chlorosulfuric acid:

Europe [38]:	BASF, Ludwigshafen	D
	Bayer, Leverkusen	D
	UCB, Ostende	B
	EniChem, Pieve-Vergonte	I
	Säurefabrik, Schweizerhalle	CH
	ICI, Runcorn	GB
	Tiszamenti, Szolnok	H
USA [39]:	DuPont, Wurtland, Kent.	US
	Gabrieil Chem., Geismar, Lous.	US
Asia [40]:	Mitsubishi, Kitakyushu	J
	Nippon Soda, Toaoka	J
	Nissan Chem., Neigun	J
	Takuyama, Tokuyama	J
	Choheung, Seoul	KOR
	Chen Yeh, Taipei	Taiwan
	Chung Hwa, Yingko	Taiwan
	Kuo Kai, Taipei	Taiwan

With regard to the annual production of chlorosulfuric acid in recent years, no reliable figures are available in the relevant literature.

7. Toxicology and Occupational Hygiene [41]

Chlorosulfuric acid is a strong acid and dehydrating agent that produces severe burn of the skin. The acid fog formed in moist air strongly irritates the respiratory tract and eyes. Contact with the skin or inhalation of the vapor must be avoided under any circumstances.

If there is accidental contact with the skin or eyes, the affected part must be washed immediately with large amounts of water; medical help should then be sought as soon as possible.

No MAK or TLV has been established for HSO_3Cl but the values for its principal decomposition products are:

H_2SO_4: MAK and TLV 1 mg/m^3 (1995)
HCl: MAK and TLV 7 mg/m^3 (1995)

8. References

[1] A. Williamson, *Proc. Royal Soc. VII, London* **1854/55**, 11–15.
[2] C. R. Sanger, E. R. Riegel, *Z. Anorg. Chem.* **76** (1912) 79–128.
[3] A. Simon, H. Kriegsmann, *Z. Phys. Chem.* **204** (1955) 369–392.
[4] E. Korinth, *Angew. Chem.* **72** (1960) 108–109.
[5] R. Vogel-Högler, *Acta Physic. Austriaca* **1** (1948) 323–338.
[6] R. J. Gillespie, E. A. Robinson, *Can. J. Chem.* **40** (1962) 644–657.
[7] Unpublished results, R. Seelemann, Bayer AG, Leverkusen.
[8] G. E. Chivers, R. J. W. Cremlyn, T. N. Cronje, R. A. Martin, *Aust. J. Chem.* **29** (1976) 1573–1582.
[9] M. Goehring, in *Scientia Chimica*, vol. **9**, Akademie-Verlag, Berlin 1957, pp. 110–113.
[10] H. Gerding, *J. Chim. Phys. Phys. Biol.* **46** (1949) 118–119. R. J. Gillespie, E. A. Robinson, *Can. J. Chem.* **40** (1962) 675–685.
[11] G. H. Weinreich, *Bull. Soc. Chim. Fr.* 1960, 791. W. Traube, *Ber. Dtsch. Chem. Ges.* **46** (1913) 2513–2524.
[12] R. J. W. Cremlyn, T. N. Cronje, *Phosphorus Sulfur* **6** (1979) 459–504.
[13] E. Gebauer-Fuelnegg, H. Figdor, *Monatsh. Chem.* **48** (1927) 627–638.
[14] *Ullmanns Encyklopädie*, 4th ed., vol. **8**, p. 414, Verlag Chemie, Weinheim 1975. J. C. D. Brand, W. C. Horning, *J. Chem. Soc. (London)* 1952, 3922.
[15] G. Schroeter, DE 634 687, 1933.
[16] Chemische Fabrik von Heyden, DE 752 572, 1941 (E. Haack, R. Jacob).

[17] H. Staudinger, H. A. Bruson, *Liebigs Ann. Chem.* **447** (1926) 110–122.
[18] K. E. Jackson, *Chem. Rev.* **25** (1939) 67–119.
[19] O. Ruff, *Ber. Dtsch. Chem. Ges.* **34** (1901) 3509–3515.
[20] D. J. Salley, *J. Amer. Chem. Soc.* **61** (1939) 834–838.
[21] P. Claesson, *J. Prakt. Chem. 2,* **19** (1879) 231–265.
[22] R. Levaillant, L.-J. Simon, *C. R. Acad. Sci.* **169** (1919) 140–143.
[23] P. Baumgarten, *Ber. Dtsch. Chem. Ges.* **64** (1931) 1505.
[24] M. Müller, *Ber. Dtsch. Chem. Ges.* **6** (1873) 227–231.
[25] K. Fuchs, E. Katscher, *Ber. Dtsch. Chem. Ges.* **60** (1927) 2288–2296; DE 505 687, 1928.
[26] H. Beckurts, R. Otto, *Ber. Dtsch. Chem. Ges.* **11** (1878) 2058–2066.
[27] I. G. Farbenind. AG, DE 582 265, 1930 (E. Konrad, H. Kleiner).
[28] Ullmanns Encyklopädie, 4th ed., vol. **9,** pp. 583–585, Verlag Chemie, Weinheim 1975.
[29] BASF, DE 1 226 991, 1964 (H. Wolf, W. Klingler).
[30] BASF, DE 2 059 293, 1970 (H. Wolf, W. Gösele, S. Schreiner).
[31] The Sulfur Institute, Washington, D.C., DE 1 767 169, 1968 (M. Schmidt).
[32] Hoechst, DE 2 730 011, 1977 (R. Borger, E. Malow, A. Renken, G. Riess).
[33] Ciba-Geigy, CH 597 094, 1978 (M. Buerli, A. Henz).
[34] Bayer, DE 3 129 976, 1983 (K. H. Schultz, F. Baunack).
[35] *DECHEMA-Werkstofftabelle,* DWT 475–476.
[36] Kühn, Birret (eds.): Merkblätter Gefährliche Arbeitsstoffe; Registerband, p. 56; Bd. 2, p. 72 (GefStoffV); Bd. 5, VI–4, Wassergefährdende Stoffe, p. 36; Bd. 7, C24; Ecomed Verlag, München–Landsberg 6/1994. Merkblätter der Berufsgenossenschaft der Chemischen Industrie, M004 (1987), M051 (1985), M053 (1989), Jedermann Verlag, Heidelberg. StörfallV (Bundesgesetzblatt Teil 1, Nr. 54 of 28.09.1991, p. 1891) and 1st, 2nd and 3rd VwV zur StörfallV. *European Community:* Guideline of 14. 7. 1976 (76/907 EEC), EG-Nr. 016-017-001.
United States: Manufacturing Chemists' Assoc., Inc., Washington, D.C., 1949: Chemical Safety Data Sheet SD-33.
Japan: Chem. Sub. Control Law, Existing Chem. Sub. 1–327.
[37] IMDG Code; class 8, UN-No. 1754; RID/ADR/ADNR: class 8, Ziffer 12 A; GGVSee/GGVE/GGVS: Class 8, Ziffer 12 A, Warntafel: Gefahr-Nr. 088; ICAO/IATA-DGR: not allowed; Post and Express transportation in Germany not allowed; Kühn, Birret (eds.): Merkblätter Gefährliche Arbeitsstoffe; Bd. 5, Merkblatt 175400 (6/1994); Merkblatt T015 (1990) der Berufsgenossenschaft der chemischen Industrie.
[38] *Directory of Chemical Producers – Europe – Vol. 2.*
[39] *Chemical Economics Handbook – 10/94.*
[40] *Directory of Chemical Producers – East Asia*
[41] L. Arzt, *Dermatol. Wochenschr.* **97** (1933) 995–997. F. Roulet, O. Straub, *Arch. Gewerbepathol. Gewerbehyg.* **10** (1941) 451–459. G. R. Cameron, *J. Pathol. Bacteriol. (Edinburgh)* **68** (1954) 197–204. Chlorschwefelsäure, Nr. 248, Toxikologische Bewertung, Ausgabe 12/1995, ISSN 0927–4248, Berufsgenossenschaft der chemischen Industrie, Jedermann Verlag, Heidelberg.

Chromium Compounds

GERD ANGER, Leverkusen, Federal Republic of Germany (Chap. 2)

JOST HALSTENBERG, Bayer AG, Leverkusen, Federal Republic of Germany (Chaps. 3–4.1, 4.3–8, and 14)

KLAUS HOCHGESCHWENDER, Bayer AG, Leverkusen, Federal Republic of Germany (Chaps. 9–13)

CHRISTOPH SCHERHAG, Bayer AG, Uerdingen, Federal Republic of Germany (Chaps. 9–13)

ULRICH KORALLUS, Bayer AG, Leverkusen, Federal Republic of Germany (Chap. 13)

HERBERT KNOPF, Bayer AG, Leverkusen, Federal Republic of Germany (Chaps. 3–4.1, 4.3–7, and 14)

PETER SCHMIDT, Bayer AG, Leverkusen, Federal Republic of Germany (Chaps. 3–4.1, 4.3–7, and 14)

MANFRED OHLINGER, BASF Aktiengesellschaft, Ludwigshafen, Federal Republic of Germany (Chaps. 4.2 and 13)

1. Introduction 1338
2. Chromium Ores 1339
2.1. Ore Deposits 1340
2.2. Ore Beneficiation 1344
3. Production of Sodium Dichromate 1345
3.1. Alkaline Roasting 1345
3.2. Leaching of the Roast 1348
3.3. Acidification 1349
3.4. Crystallization 1350
4. Chromium Oxides 1350
4.1. Chromium(III) Oxide and Chromium Hydroxide 1350
4.2. Chromium(IV) Oxide (Chromium Dioxide) 1353
4.3. Chromium(VI) Oxide 1355
5. Chromium(III) Salts 1357
5.1. General Properties 1357
5.2. Chromium(III) Sulfates and Chrome Tanning Agents 1359
5.3. Other Chromium(III) Salts ... 1362
6. Chromic Acids and Chromates(VI) 1364
6.1. Chromic Acids 1364
6.2. Alkali Chromates and Dichromates 1366
6.3. Other Chromates 1369
7. Other Chromium Compounds . 1371
8. Analysis 1372
9. Transportation, Storage, and Handling 1373
10. Environmental Protection ... 1373
11. Ecotoxicology 1374
12. Nutrition 1376
13. Toxicology and Occupational Health 1376
14. Economic Aspects 1380
15. References 1383

Table 1. Uses of chromium compounds

Branch of industry	Product	Use
Building industry	chromium(III) oxide	pigment for coloring building materials
Chemical industry	dichromates, chromium(VI) oxide	oxidation of organic compounds, bleaching of montan waxes, manufacture of chromium complex dyes
	chromium(III) oxide	catalysts
Printing industry	dichromates	photomechanical reproduction processes
	chromium(VI) oxide	chromium plating of printing cylinders
Petroleum industry	chromates(VI)	corrosion protection
Paints and lacquers	chromates, chromium(III) oxide	pigments
Refractory industry	chromium(III) oxide	additive for increasing slag resistance
Electroplating	chromium(VI) oxide	bright and hard chromium plating
Wood industry	chromates, chromium(VI) oxide	in mixtures of salts for protecting wood against fungi and insects
Leather industry	basic chromium(III) sulfates	tanning of smoothed skins
Metal industry	chromium boride, chromium carbide	flame sprays
	chromium(III) oxide	polishing agents
Metallurgy	chromium(III) oxide	aluminothermic extraction of pure chromium metal
Textile industry	dichromates	dyeing with chrome dyes
	basic chromium(III) acetates and chromium(III) fluorides	mordanting of textiles
Recording industry	chromium(VI) oxide	magnetic information storage
Pyrotechnics industry	dichromates	additive to igniting mixtures

1. Introduction

Historical. Chrome iron ore (chromite) was discovered in 1798. A few decades later this ore was being subjected to oxidative roasting in the presence of soda and lime in manually operated furnaces to produce water-soluble sodium dichromate. This was processed further to yield yellow, red, and green chromium pigments which were used, among other things, for dyeing wallpaper; they replaced the toxic arsenic dyes that had been used until then. Chromium salts soon found their way into the textile industry as mordants for the dyeing of wool.

The importance of dichromates increased considerably in the period following 1870 when the rising coal tar dye industry needed large quantities for the oxidation of chemical intermediates. With the advent of the 20th century, chrome tanning was introduced in leather factories and in many areas replaced vegetable tanning.

The manufacture of chromium compounds received a further boost after 1930, when metallic chromium was successfully precipitated from chromic acid solutions by special additives. Since then this possibility has been used extensively in electroplating for bright and hard chromium plating.

Chromium compounds are used in numerous fields. In addition to the applications mentioned, chromates have long been used in printing as an aid in photomechanical reproduction. For some time, chromium dioxide has been a component of magnetic tapes for information storage. Table 1 lists important applications of chromium chemicals.

Table 2. Estimated reserves of chromium ore [14]

	Reserves, 10^6t			
	Total	Metallurgical[a], > 45% Cr_2O_3	Chemical[a], > 40% Cr_2O_3	Refractory[a], > 20% Al_2O_3
Republic of South Africa	2000[b]	100 (5%)	1900 (95%)	–
Zimbabwe	600	300 (50%)	300 (50%)	–
Turkey	10	9 (90%)	–	1 (10%)
Philippines	7.5	1.5 (20%)	–	6 (80%)
United States	8	0.4 (5%)	7.4 (92.5%)	0.2 (2.5%)
Canada	5	–	0.5 (100%)	–
Finland	7.5	–	7.5 (100%)	–
Others	11.35	8.175 (72%)	0.2 (2%)	2.975 (26%)
Total	2649.35	419.075 (16%)	2220.1 (84%)	10.175 (0.4%)
USSR and other Eastern bloc countries	51.5	26.5 (51%)	15 (29%)	10 (20%)
Total worldwide (rounded off)	2701	446 (17%)	2235 (83%)	20 (1%)

[a] Graded according to Cr_2O_3 or Al_2O_3 contents.
[b] Ores containing 30–50% Cr_2O_3.

2. Chromium Ores [1]–[13]

The distribution of chromium in terrestrial rocks is closely linked to magmatic intrusions and their crystallization. The average content in the ten-mile crust of the earth is 100 ppm of chromium [7]. Table 2 contains a worldwide estimate of chromium ore resources.

The most important applications of chromium ores are in the manufacture of stainless steel, grey cast iron, iron-free high-temperature alloys, and chromium plating for surface protection. In the nonmetallic mineral industry, chromite is processed in conjunction with magnesite (sintered magnesia, calcined magnesia) and binders (clay, lime, gypsum, bauxite, corundum). The products are intended to have good resistance to pressure, fire, and temperature change, as well as good insulating properties between basic and acidic masonry. The chemical industry uses chromium ores in the production of chromium compounds (Chaps. 3–7). Table 3 shows quality requirements of chromium ores for different areas of application.

Minerals. Of the many minerals that contain chromium only the chromium spinels are of economic importance. The formula for the series of isomorphous mixtures of chromium spinels that form geological deposits is

$(Fe^{II}, Mg)O \cdot (Cr, Al, Fe^{III})_2O_3$

The proportion of Cr_2O_3 in the chromium spinels varies widely, causing the Cr:Fe ratio (also known as the Cr–Fe factor) to vary as well; this can have a profound effect

Table 3. Quality requirements (mass fractions in %) for chromium ores (according to U.S. Bureau of Mines)

	Metallurgical[a] (high-chromium chromite)	Refractory[b] (high-aluminum chromite)	Chemical[c] (high-iron chromite)
Cr:Fe ratio	3:1 or higher	–	–
Cr_2O_3	>48	>31	>44
$Cr_2O_3 + Al_2O_3$	–	>58	–
Fe	–	<12	–
SiO_2	<8	<6	<5
S	<0.08	–	–
P	<0.04	–	–
CaO	–	<1	–

[a] Zimbabwe, South Africa, Turkey, USSR.
[b] Turkey, Philippines.
[c] South Africa, USSR.

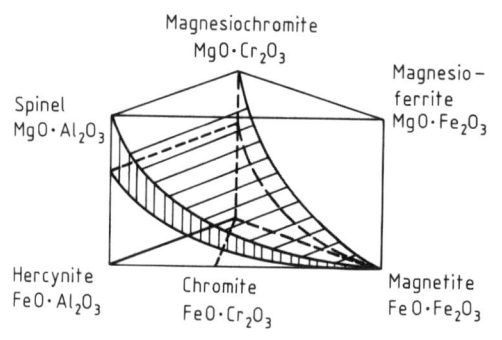

Figure 1. Ternary spinel system showing main isomorphous region

on the evaluation of a deposit. In an ideal chromium spinel ($FeO \cdot Cr_2O_3$; 67.8% Cr_2O_3, 32.2% FeO) the chromium:iron ratio is 2. As a result of the isomorphous inclusion of MgO, the Cr:Fe ratio may rise to between 2.5 and 5. Figure 1 shows the region of isomorphism with varying composition of the spinels. Natural chromium spinels usually contain 33–55% Cr_2O_3, 0–30% Fe_2O_3, 0–30% Al_2O_3, 6–18% FeO, and 10–32% MgO. Table 4 lists some physical properties of chromium spinels.

Chromium also occurs in all groups of silicates where chromium replaces Al^{3+}, Fe^{3+}, and Mg^{2+}. Sulfidic chromium ores do not occur on earth. Chromates and chromium iodates are described, which originate from the weathering zone of sulfidic lead deposits (e.g., crocoite, $PbCrO_4$).

2.1. Ore Deposits

Chromium ore deposits can be divided into two genetically different types:

1) Seam-like deposits, also called stratiform or anorogenic deposits. Main representatives are Bushveld, Great Dyke, and Stillwater.

Table 4. Physical properties of chromium spinels

Properties	Notes
Specific density: 3.8–4.8	increases as Fe and Cr contents increase
Hardness (Mohs): 4.5–8	increases with increase in the ferrochromite component, very high for Al spinels
Melting point: 1545–1730 °C	inclusion of Mg raises melting point, inclusion of Fe^{2+} reduces it
Color: dark brown to black	reddish with high Cr_2O_3 content
Streak on porcelain plate; hammer striking mark: brown	important feature for differentiating from serpentine

2) Deposits which are shaped like sacks or tubes; they are called podiform or orogenic deposits. Main representatives are Selukwe, Guleman, and Tiébaghi.

Various intermediate types such as adjacent "seam" pockets, striated chromite slabs, mottled ores, and vein-like deposits also occur. "Placer" deposits, i.e., enrichment due to chromite lumps and grains on or near primary deposits, are now achieving economic importance.

The seam-like deposits reveal layers or strata of chromite enrichment, with thicknesses ranging from centimeters to decimeters; the layers are regularly interlaminated with banded series of olivine-rich or pyroxene-rich rocks. The Main Seam of the Western Bushveld is, for example, 1.10–1.30 m thick and can be traced for over 65 km without any significant change in the mineral composition or thickness.

The demarcation between the chromite enrichment and the underlying bed is usually razor-sharp; in the direction of the overlying layer, disintegration into layers or mottled ores as a result of increased silicate content is observed.

The chromite bodies that are sack-like to tube-like in appearance are usually aligned with the direction of the magmatic stratification, i.e., the lowest sections are massive chromite ores; in the direction of the overlying layer, these merge into striated slabs or mottled ores.

The internal texture of the chromite ore bodies varies widely. The closest chromite crystal packing results in the formation of massive ores containing 75–85 vol% of chromite. Sphere or leopard ores, which consist of round chromite crystal aggregates 0.5–2 cm in diameter in a silicate matrix (olivine, pyroxene, serpentine), are also characteristic. Banded ores are closely related to the massive ores, but they are frequently richer in silicate and then form a link with the mottled ores (chromite single crystals in silicate matrix).

During transformation (serpentinization), the silicate content within the chromite ore bodies has resulted in the formation of friable and pulverizable masses (friable ore) which are encountered not only near the surface but also at depths of several hundred meters below the present-day land surface.

Chromite transformation in the course of more recent tectonic superficial modification under pneumatolytic or hydrothermal conditions has resulted in the

Table 5. Chemical analyses (mass fractions in %) of some chromium ores (crude ores, concentrates)

Country	Cr_2O_3	FeO	SiO_2	MgO	Al_2O_3	CaO	V_2O_5	Cr:Fe ratio
South Africa								
Rustenburg (c)[a]	44.5	26.4	3.5	10.6	14.4	n.d.[b]		1.7:1
Lydenburg (c)	44.3	24.6	2.3	11.2	16.1	0.4		1.8:1
Zimbabwe								
Great Dyke (m)[a]	48.5	18.3	5.6	13.4	11.5	0.8		2.6:1
Great Dyke (r)[a]	50.7	16.4	4.3	13.2	13.0	0.8		3.1:1
Selukwe (m)	47.0	12.0	5.7	15.5	12.6	1.8		3.9:1
Selukwe (r)	42.0	15.7	8.6	15.8	13.8	0.3		2.7:1
Turkey								
(m)	48.3	14.1	5.1	16.8	13.0	0.9		3.4:1
(r)	37.0	15.2	4.3	17.7	24.3	0.2		2.4:1
Philippines								
(Masinloc) (r)	33.3	13.2	4.6	19.6	28.2	0.4		2.5:1
Finland (Kemi)								
Crude ore	26.5	15.0	18.5	19.5	9.5	–	0.04–0.1	1.8:1
Concentrate	45.7	33.8	0.4	2.9	13.6	n.d.	0.1	1.4:1
Albania								
(m, r)	43.0	16.2	9.8	22.2	7.9	0.1		2.6:1
USSR								
(m)	53.9	12.6	5.8	13.3	9.6	1.1		4.3:1
(r)	39.1	14.0	9.4	16.1	17.4	0.7		2.8:1

[a] (m) = metallurgical, (r) = refractory, (c) = chemical.
[b] n.d. = not determined.

striking colors of recent uvarovite, smaragdite, and kammererite formations which act as pathfinders in prospecting and exploring for chromite deposits. Table 5 shows some analyses of selected chromium ores.

Soviet Union. The Soviet Union is one of the most important producers of chromium ore in the world. All the deposits are distributed in ultrabasite massifs in the Central and Southern Urals. The deposits that are most important at present were found in the late 1930s in the Akhtiubinsk region (North Kazakhstan). The Donskoye deposit, which is associated with the mining settlement of Khrom Tau, contains high-grade chromium ores for ferrochromium production and low-grade ores for chemical purposes. Mining is carried on in numerous open-pit mines, which implies that the ore bodies are small. A new open-cast mine was put into production near Donskoye, as is a processing plant with a throughput of 10^6 t/a. Strong prospection effort for new occurences is being made in the Northern Urals, but because of the rough climate no mine has been opened up to 1986.

Bushveld. In the Bushveld (Republic of South Africa) mining began in the 1920s in two districts: the Lydenburg district (Eastern Bushveld) and the Rustenburg district (Western Bushveld).

From a geological and petrological point of view this is a large intrusion of 500×250 km with a thickness of over 5 km. The chromite "seams" are located in the pyroxenite–norite zone of the basal section of the intrusion, always below the platinum-bearing Merensky Reef. In the case of Rustenburg there are up to 25 chromite seams on top of each other. The thickness of the individual seams varies from a few centimeters to 1.80 m. The seams are workable from 0.35 m upward, especially if they can be combined into mining units (Cr_2O_3 content in the crude ore 30–40%; Cr:Fe ratio = 1.6–2.3). In the Lydenburg district, which is genetically very similar to the Rustenburg district, only two seams are being mined; the Cr_2O_3 content is 44% and the chromium:iron ratio 1.6–1.7. The iron content is frequently high, which may cause difficulties in the case of metallurgical ores; however, these ores are highly valued as chemical ores.

Great Dyke. The Great Dyke (Zimbabwe) is an intrusion which is 610 km long and 6–9 km thick — a remarkable length:thickness ratio which is unique in the world. The internal structure is similar to that of the Bushveld. From north to south, the individual complexes are Musengezi, Hartley, Selukwe, and Wedza. Selukwe consists of sack-like deposits containing 48% Cr_2O_3 and even more, with a chromium:iron ratio greater than 2.8 (a highly valued metallurgical ore). In the Hartley region, on the other hand, numerous bands and seams 2–75 cm thick are being mined; these are separated by serpentinized peridotite layers, some of which are very thick and make mining very difficult. However, the Cr_2O_3 content varies between 48 and 57%, and the chromium:iron ratio is over 2.8 (Table 6).

According to conservative estimates, 1 km^2 of the Great Dyke contains around 1.4×10^6 t of crude ore, which corresponds to assured reserves of 600×10^6 t (geologically 4.6×10^9 t are possible).

Other Ore Deposits Madagascar. On the island of Madagascar, chromium ores are being mined since 1967 with an annual production of around 60 000 t of metallurgical grade ore. Total output is calculated to be almost 2×10^6 t since the beginning of the operation (50–52% Cr_2O_3). The reserves are said to be around 5.5×10^6 t.

Turkey. Turkey still is the traditional country for chromite deposits of metallurgical quality, but because of falling prices on the world market and exhaustion of reserves, many mines have been forced to close. The most important regions belong essentially to the alpidic era, e.g., Bursa, Mugla district, and Elazig, including the Guleman chrome ore field. Open-pit mining, and in some places underground mining at shallow depths, are employed.

Iran. Chromite deposits are described in two regions of Iran: northwest of Sabzawar near Mashad and 200 km northeast of the Gulf port of Bandar Abbas. These deposits are pocket-like, sometimes

Table 6. Analysis of chromium ores from Zimbabwe

	Cr_2O_3 content, wt%	Cr:Fe ratio	Proportion, wt%
Metallurgical	over 48	over 2.8	80
Chemical	45–48	2.2–2.5	17
Refractory	42–46	1.8–2.0	3

containing only 500 t of ore. Extraction is by open-pit mining and by primitive underground mining. Only hard lump ore for metallurgical applications is sometimes exported.

Philippines. On the Philippine island of Luzon the most important chromite deposits are to be found in the Coto region (near Masinloc, province of Zambales). These are chromite seams and pockets within layered dunites and harzburgites. They are classical metallurgical and refractory ores. More recently a new type of chrome ore has been put into production: chromite from lateritic soils. The concentrates are suited for the chemical industry.

Finland. In 1959 a fairly large deposit of chromite was discovered near Kemi, which has been developed into a productive mine. Chromium ore occurs in a serpentinite–anorthosite massif 12 km long and 1–2 km wide; the ore zone, however, is only 15–100 m thick and dips at an angle of 60° toward the north. The Cr_2O_3 content of the crude ore of the various ore bodies varies between 17.5 and 21.9% (locally even up to 30.5% Cr_2O_3); the chromium:iron ratio is low (0.81–1.87).

Yugoslavia. The deposits of chromite in Yugoslavia are restricted to the Radua massif near Skopje, but the mining of metallurgical ores there has fallen considerably. Native ores are being processed in a new plant, whereas ores imported from Albania are being processed in the old one. The chromium ores are always associated with serpentinized ultrabasites. The striated slab type predominates, but massive chrome ores are encountered in some places.

Albania. Since 1960, Albania has become the third largest producer in the world. All actual data are based onestimates because the Albanian government withholds production and export figures. Albanian deposits belong to the podiform type and normal grades are reported to be 43% Cr_2O_3 with a Cr:Fe ratio of 3:1. The largest chrome ore mines are Bulquize and Matanesh with concentration plants of 300 000 t/a each.

2.2. Ore Beneficiation

The simplest method of concentrating chromite is by hand picking; this is still employed today at many pits, including those in Turkey, Brazil, Iran, and the Philippines. Because the mining of richer ores continues to decline, concentrating procedures, chiefly using the gravity method, have been developed to separate the serpentine from the chromite. For example, in South Africa or Brazil the chromite ores are enriched by crushing, milling, screening, and sophisticated gravity procedures. In South Africa, spirals and diamond pans are standard equipment. A combination of

Reichert cones and Reichert spirals has also been employed. Although the costs are higher, the use of hydrocyclones for separating the fine chromite grains from waste is of great importance in the recycling of tailing dumps. Some chromium ores contain magnetite which can be removed by means of magnetic separation. However, if the magnetite is present as an individual phase within the chromite grains or as a fringe around the grains, this method is only suitable if Cr_2O_3 is further enriched. Flotation and electrostatic processes have so far enjoyed little success in the concentration of chromium ores. If the Cr_2O_3 content or the chromium:iron ratio is sufficient, fine-grained chromite concentrates can be briquetted or pelleted with the aid of binders.

The yield (65–85% of the chromite actually contained in the crude ore) depends on many factors including the nature of the chromite–serpentine intergrowth, grain size, and Cr_2O_3 content of the ore or individual grain.

At the chromite concentration plant at Kemi in Finland a fraction of the crude ore (70%) is crushed to below 10 mm in a primary crusher plant at the open-pit mine. After further grinding (rod and ball mill) and removal of sludge, the intermediate product is dried in a rotary kiln. The magnetic separation (a combination of weak and strong fields) produces two concentrates: concentrate 1 containing 45.9% Cr_2O_3, which is sold or used as molding sand, and concentrate 2 containing 42.0% Cr_2O_3 for the production of ferrochromium.

3. Production of Sodium Dichromate

Directly or via several intermediate stages, sodium dichromate [7789-12-0], $Na_2Cr_2O_7 \cdot 2 H_2O$, is the starting material for the production of all chromium compounds and pure chromium metal.

Sodium dichromate is made in a three-step process: (1) alkaline roast of chromite under oxidizing conditions (Eq. 1), (2) leaching, and (3) conversion of sodium monochromate to sodium dichromate by means of an acid (Eq. 2).

$$4 \, FeO \cdot Cr_2O_3 + 8 \, Na_2CO_3 + 7 \, O_2 \xrightarrow{ca.\ 1000\,°C} 8 \, Na_2CrO_4 + 2 \, Fe_2O_3 + 8 \, CO_2 \quad (1)$$

$$2 \, Na_2CrO_4 + 2 \, H^+ \longrightarrow Na_2Cr_2O_7 + H_2O + 2 \, Na^+ \quad (2)$$

3.1. Alkaline Roasting

Soda ash (sodium carbonate) is generally used as the alkali component but sodium hydroxide may also be employed [15]–[17]. The degree of solubilization of chromites by the roasting process depends on their composition. For optimum results, the process is controlled by adding so-called carrier materials. These ensure sufficient porosity of the material so that oxygen can diffuse into the roast. Porosity is maintained by means of such materials as iron oxide, bauxite, or dried leach residue; CO_2-emitting additives include lime and/or dolomite. The inert additives dilute the sodium carbonate and sodium chromate, which both melt at the reaction temperature. In the *low-lime process* the carbonates evolve CO_2, decrease the reaction temperature to below 1000 °C, and raise the melting point of the reaction products; the amount of lime added must be controlled so that the compound $5\,Na_2CrO_4 \cdot CaCrO_4$ [18] is produced in the roast. Temperatures above 1150 °C must be avoided because they result in the subsidiary components of the ore being attacked. At still higher temperatures the degree of conversion is markedly decreased. The optimum temperature range is very narrow and depends strongly on the type of ore used and the composition of the mixture.

Process Description. A typical roast mixture contains 100 parts of ore, 60–75 parts of sodium carbonate, 0–100 parts of lime or dolomite, and 50–200 parts of inert materials. The components are first finely ground, then mixed, and fed into the furnace. Annular hearth furnaces or rotary kilns are commonly used in large plants today.

The *annular hearth furnace* (Fig. 2) is made from steel with a refractory lining (inner diameter ca. 20 m; outer diameter ca. 30 m); it is driven by a gear wheel underneath and has rails running on rollers. The rotating hearth is sealed from the stationary parts of the furnace by sheets of metal dipping into annular water troughs. The furnace is heated by several burners from the side or from the top with gas, coal dust, or oil. The exhaust gas can be utilized to preheat the burner air or to generate steam. The mixture is fed to the outer edge of the annular hearth by a feed screw. A water-cooled ribbon screw transports the mixture inward, each time the annular hearth revolves, and finally removes it in the middle.

In the annular hearth furnace, the mixture is uniformly heated to the reaction temperature and made to travel toward the center of the hearth with a well-defined layer thickness. The furnace process is fairly independent of the sintering of the roast; it allows the production of melts that contain 40 wt % of water-soluble sodium chromate. The yield is 80–95%, based on the chromite feed. The roast takes 2–6 h, depending on the composition of the mix.

Most of the kiln tube of the *rotary kiln* (Fig. 3) between the feed point and the reaction zone is used to heat the mixture. Shortly before the mix reaches the actual reaction zone, the soda melts and calcines. At this point the mixture bakes, and pellets or wreath-shaped cakes may be formed. If the furnace is operated inexpertly (temperature too high) or the composition of the mixture is wrong (too little carrier material) the kiln tube may get substantially clogged. In such cases, the constriction can be cleared by using an industrial gun.

The roast from the rotary kiln contains up to 30 wt% of water-soluble sodium chromate. The yield is 75–90%, based on the chromite feed. The roast takes 3–8 h, depending on the composition of the mixture. The hot exhaust gas from the rotary kiln can be used to preheat the burner air.

Gas Purification. Exhaust gas purification systems able to achieve a high degree of separation are required for dust collection. They essentially consist of two components: (1) an exhaust gas cooling system, with optional energy recovery, for example, steam generation; and (2) the exhaust gas purification system, usually an electrostatic

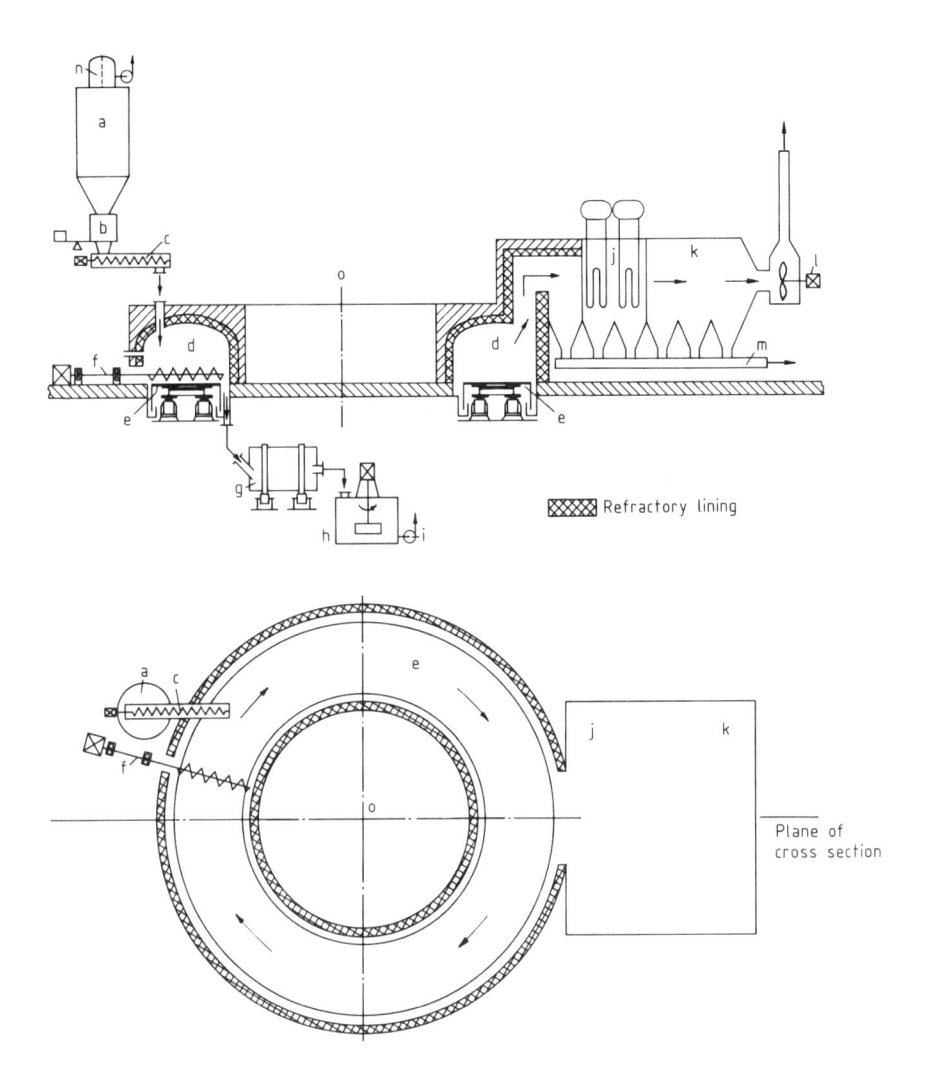

Figure 2. Annular hearth furnace
a) Mixture silo; b) Scales; c) Feed screw; d) Furnace; e) Annular hearth; f) Water-cooled ribbon screw; g) Wet tube mill; h) Stirred vessel; i) Pump for filtering system; j) Waste heat boiler; k) Electrostatic gas purification; l) Exhaust gas fan; m) Dust drag chain; n) Bin filter; o) Rotation axis

separator. The size and the design of the purification system depend on the type of furnace. In the rotary kiln, ca. 10% of the feed mixture is carried off by the exhaust gas, whereas in the annular hearth furnace, less than 1% is carried off. However, operation of the annular hearth furnace necessitates a considerable expenditure on gas cooling.

Other Processes. In the literature, other processes are proposed but so far these have not achieved any industrial significance. Thus, attempts have been made to roast

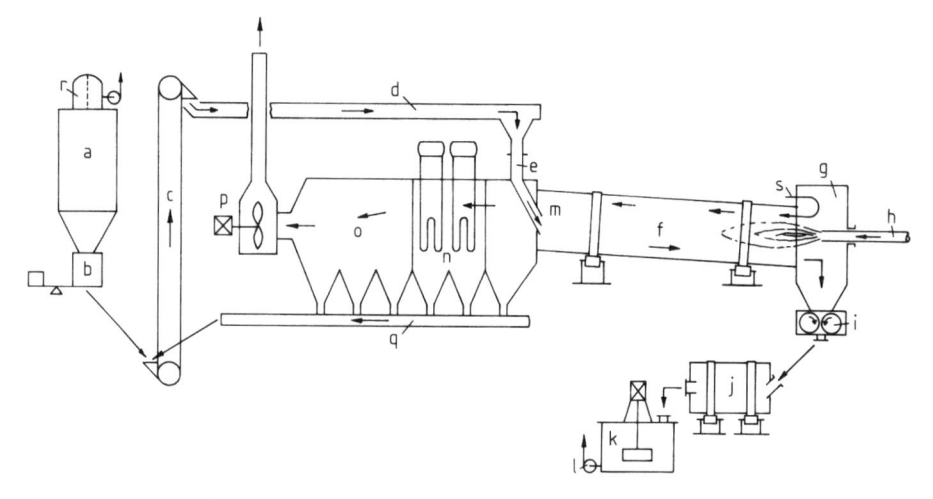

Figure 3. Rotary kiln for roasting chromium ores
a) Mixture silo; b) Scales; c) Elevator; d) Drag chain; e) Inlet tube; f) Rotary kiln; g) Combustion chamber; h) Burner; i) Crusher; j) Wet tube mill; k) Stirred vessel; l) Pump for filtering system; m) Kiln inlet; n) Waste heat boiler; o) Electrostatic gas purification; p) Exhaust gas fan; q) Dust drag chain; r) Bin filter; s) Air inlet

chromium ore in a shaft furnace [19] or in a fluidized-bed reactor [20]. A fundamentally different process involves the reaction of the chromium ore and soda in a molten salt mixture with oxygen-containing gases being injected [21].

3.2. Leaching of the Roast

After the oxidative process, the roast is a mixture of soluble salts and insoluble components. It contains sodium chromate, sodium aluminate, magnesium oxide, sodium vanadate(V), iron(III) oxide, unused alkali, unchanged chromite, and small amounts of sodium chloride originating from the soda.

When the roast is extraced with hot water, a pH of 10.5–11.2 results. The pH is controlled by adding acids or carbonates so that all chromate dissolves, whereas the alkali-soluble impurities hydrolyze and form a readily filterable precipitate along with the iron hydroxide and the unchanged ore components.

The roast is first cooled on a Fuller grate or in a cooling drum. Then it is either ground in a wet tube mill after addition of water or wash solution (see below) with carbonates or acids added, or it is dissolved in a stirred vessel. The insoluble residue is separated from the sodium chromate solution and thoroughly washed with a countercurrent of water. Nowadays continuous multistage Dorr plants or rotary filters are used; after separation, the insoluble residue is extracted two to three more times in counterflow. Dorr plants only exhibit satisfactory separation of residue and solution if the sodium chromate concentration is not too high. Rotary filter plants can be employed without difficulty for nearly saturated hot sodium chromate solutions; such filters are frequently preferred because the higher water consumption of the Dorr plant results in unnecessary steam costs in the subsequent evaporation process.

After removal of residual aluminum hydroxide and other undissolved components in a final purification process (e.g., thickener), the concentrated sodium chromate solution is acidified (Section 3.3).

Some of the filter residue is dried and added to the roasting mix (Section 3.1). The remainder is subjected to reducing treatment to convert the residual chromate content into an ecologically harmless form. To do this, the residue is suspended in water and treated with sulfuric acid and sodium hydrogen sulfite or iron(II) sulfate; chromate residues are converted into chromium(III) compounds in this way. Subsequent addition of alkali precipitates trivalent chromium (and iron(III), if present) as hydroxide. The suspension is then filtered and the cake (optionally after further removal of water) is dumped.

3.3. Acidification

The sodium chromate solution is converted into sodium dichromate solution by acidification with sulfuric acid (Eq. 3) or carbon dioxide (Eq. 4). The sequence of individual steps depends on the acid.

$$2\,Na_2CrO_4 + H_2SO_4 \longrightarrow Na_2Cr_2O_7 + Na_2SO_4 + H_2O \tag{3}$$
$$2\,Na_2CrO_4 + 2\,CO_2 + H_2O \rightleftharpoons Na_2Cr_2O_7 + 2\,NaHCO_3 \tag{4}$$

Sulfuric Acid Acidification. Sulfuric acid is added to the concentrated sodium monochromate solution in an agitated vessel until the pH is about 4. The sodium dichromate solution is then concentrated in a continuous evaporation plant. Each liter of sodium dichromate solution yields 400–500 g of anhydrous crystalline sodium sulfate. The sulfate is removed by centrifugation. The clear, dark-red sodium dichromate solution contains 900–1200 g of $Na_2Cr_2O_7 \cdot 2\,H_2O$ per liter and additional small amounts of sodium sulfate; it is dispatched in tanks, e.g., of steel. The solution is either used directly as an oxidizing agent or processed to yield dichromate crystals.

Carbon Dioxide Acidification. After filtration the sodium monochromate solution from the first filtration stage is concentrated to ca. 850 g of Na_2CrO_4 per liter. The saturated sodium chromate solution is then acidified with a countercurrent of carbon dioxide at 0.5–1.5 MPa (5–10 bar) to yield sodium dichromate and sodium hydrogen carbonate. A series of stirred autoclaves is preferred for this reaction. They must be cooled to remove the heat of neutralization; the slurry leaves the last reactor at room temperature. The degree of conversion is about 80–90%.

The sodium hydrogen carbonate is removed by centrifugation or filtration, preferably under pressure to prevent reaction (4) from being reversed. After it has been washed, water can be removed in a pusher centrifuge. Still moist, the sodium hydrogen carbonate is then transferred to a calcining furnace. The carbon dioxide produced in the furnace may be fed back to the acidification autoclaves [22]; the sodium carbonate obtained is recycled for alkaline ore roasting [23].

To obtain commercial sodium dichromate solution, further evaporation to a concentration of 1000 g of sodium dichromate per liter is required. This is then followed by a second acidification with carbon dioxide or sulfuric acid.

Production of soda is the main advantage of carbon dioxide acidification. However, this is offset by a number of difficulties, particularly the formation of deposits on the evaporator during the concentration of the sodium monochromate solution. Other problems include mastering the pressure technology and cooling, separation of the sodium hydrogen carbonate, yield loss due to reverse reaction, and clogging of the calcination furnace.

3.4. Crystallization

For the purpose of crystallization the sodium dichromate solution (950 – 1200 g/L) is further concentrated and may, if necessary, be filtered while hot to remove additional sodium sulfate or sodium chromate. It is then slowly cooled to 30 – 35 °C with constant stirring to obtain orange-red crystals of $Na_2Cr_2O_7 \cdot 2\,H_2O$.

Today continuous vacuum crystallization is carried out to an increasing extent. The initial difficulties of this process, particularly in obtaining coarse, dust-free crystals, have largely been overcome. The crystalline slurry is continuously separated from the mother liquor and dried. Precise control of the drying temperature is important because hydrated sodium dichromate is converted into anhydrous sodium dichromate above 84.6 °C and, therefore, cakes if overheated.

For reasons of occupational health, care should be taken to ensure that workplaces and production plants are dust free when sodium dichromate, especially the dried product, is being handled. In such locations, extensive ventilation and dust removal systems (wet scrubbers, electrostatic separators) are necessary.

4. Chromium Oxides

4.1. Chromium(III) Oxide and Chromium Hydroxide

Chromium(III) oxide [1308-38-9], Cr_2O_3, M_r 151.99, ϱ 5.2 g/cm^3, is green in finely dispersed form, whereas fairly large crystals have a blackish green hue and a metallic luster. The crystals have a hexagonal rhombohedral structure of the corundum type. The compound melts at 2435 °C, but begins to evaporate at 2000 °C to form clouds of green smoke; the boiling point is estimated to be 3000–4000 °C. The enthalpy of formation is −1141 kJ/mol. Macrocrystalline chromium(III) oxide has a hardness of 9 on the Mohs scale. An amorphous form of the oxide is also known; this crystallizes on heating.

Chromium(III) oxide does not dissolve in water, acid, alkali, or alcohols. It is converted by a molten bath of sodium peroxide into soluble sodium monochromate(VI). Chromium(III) oxide and chromates(III) are used in organic chemistry as catalysts, e.g., in the hydrogenation of esters or aldehydes to form alcohols and in the cyclization of hydrocarbons. They also catalyze the formation of ammonia from hydrogen and nitrogen.

Production. The industrial production of chromium(III) oxide involves the reduction of solid sodium dichromate, generally with sulfur. The finely divided components are thoroughly mixed, fed into a brick-lined furnace, and brought to dark-red heat. The reaction proceeds exothermically. After the reaction mass has cooled, it is broken up and the sodium sulfate produced is leached out with water. The remaining solid is separated, rinsed, dried, and ground. To obtain 100 kg of chromium(III) oxide, 200 kg of sodium dichromate must react with at least 22 kg of sulfur; usually an excess of sulfur is used. Additives such as ammonium chloride or starch in the crude mixture affect the pigment properties. When sodium dichromate is replaced by the corresponding potassium salt, the hue of the pigment becomes more bluish.

The compound can also be prepared by a wet route involving reduction of sodium chromate by sulfur [24], with sodium thiosulfate being produced as a coproduct. The hydrate initially obtained is washed by decanting, filtered, and calcined to form the oxide.

Chromium(III) oxide destined for aluminothermic production of pure chromium metal must be heated additionally at 1000 °C to increase its grain size. If products particularly low in sulfur are to be produced for this purpose, charcoal can be used for the reduction instead of sulfur. High-purity oxides can also be obtained by thermal decomposition of chromium(VI) oxide or ammonium dichromate(VI), the latter yielding a material of very low density.

Chromium(III) oxide pigments contain 99.1–99.5% Cr_2O_3. The aluminum oxide and silicon dioxide impurities each amount to ca. 0.1%; the annealing loss at 1000 °C is about 0.3%. The individual particles are spherical, with a diameter of 0.3 µm predominating. Chromium(III) oxide finds widespread application as a green pigment resistant to atmospheric conditions and heat. In addition, it is used as a colorant in glass products and printing inks, as a vitrifiable pigment in the ceramics industry, and as a polishing agent because of its considerable hardness.

Chromium(III) Aquoxides [12292-46-5], [12182-82-0]. Pure chromium(III) hydroxide [1308-14-1], $Cr(OH)_3$, M_r 103.02, can only be prepared with difficulty because the hydrates initially obtained by precipitation are subject to aging.

After drying in air, specimens prepared by precipitation with alkali in the cold from violet chromium(III) salt solutions have a composition corresponding to $Cr_2O_3 \cdot 9\,H_2O$

[*41993-26-4*], and are usually formulated as $Cr(H_2O)_3(OH)_3$ [*41993-26-4*]. They are bright bluish green powders with limited life. All three hydroxyl groups react immediately with acids. Upon careful heating, dehydration occurs in steps and compounds containing 8, 5, 3, and 1 mol of water are formed. The density increases as the water content falls. Above 50 °C, conversion to a gelatinous green aging product occurs, and the solubility and chemical reactivity decrease; oxygen bridges are formed through the elimination of water, and polynuclear complexes are produced. The composition approaches that of chromium(III) oxide hydroxide, $CrO(OH)$ [*20770-05-2*]. Aging is accelerated by the presence of hydroxide ions.

In freshly precipitated hydroxides a crystalline phase isomorphic with bayerite [$Al(OH)_3$] is observed (→ Aluminum Oxide), whereas aged compounds are X-ray amorphous. The chromium(III) hydroxide hydrates are amphoteric compounds. With acids they form Cr^{3+} salts, whereas they dissolve in strong hydroxide solution to form chromates(III), e.g., the deep green sodium chromate(III), $Na_2Cr_2O_4$ (previously known as sodium chromite). When ammonium hydroxide is added, red solutions are formed. Oxidizing agents in the presence of alkali produce chromates(VI). With halogens this takes place immediately on gentle heating, but with oxygen several hours are required at a pressure of 4 MPa (40 bar) at 175 °C.

Chromium(III) hydroxide forms a stable colloid solution, whose isoelectric point is at pH 5. At higher pH the sol becomes negatively charged and adsorbs cations, whereas below pH 5 the charge is positive and anions are adsorbed. The adsorption capacity of chromium(III) hydroxide sols is higher than that of aluminum or iron(III) hydroxide sols. Sols containing 127 g of Cr_2O_3 per liter have been obtained from concentrated chromium(III) chloride solution by addition of ammonium carbonate and dialysis while the solution is hot [25].

Production. In industry chromium(III) hydroxide hydrates are usually produced from solutions of chromium(III) sulfate or chromium alum by precipitation with soda, sodium hydroxide solution, or ammonium hydroxide. Production by reduction of sodium chromate with sodium sulfide [26] has also been proposed.

For production from potassium chromium alum 54 kg of soda is dissolved in 300 L of water and a solution of 180 kg of alum in 900 L of water is added slowly. After the evolution of carbon dioxide has subsided, about 220 kg of moist chromium hydroxide containing 12% Cr_2O_3 is obtained by filtration. Chromium(III) hydroxide hydrates are used for the production of chromium(III) salts by reaction with the corresponding acids.

Hydrated chromium(III) oxide, $Cr_2O_3 \cdot x\,H_2O$, is a brilliant emerald-green pigment known as Guignet's green [*12001-99-9*] that consists of very finely divided chromium(III) oxide to which water is bonded by adsorption. It is produced by heating a ground mixture of one part by weight of potassium dichromate and three parts by weight of boric acid in a muffle furnace to a faint red heat, which results in the formation of chromium(III) and potassium tetraborates. The molten mass still contains 6–7% water and, after cooling, already has a deep green color. When this mass is boiled with water, it decomposes into chromium(III) oxide hydrate and boric acid. The product is coarse-grained and difficult to grind. Use of sodium dichromate as raw material results in a more yellowish color, whereas addition of thiourea or polysulfide to the reaction mixture produces a pigment with a bluish hue. The composition of the commercial products varies; typical values are: Cr_2O_3 79.3–82.5%, H_2O 16.0–18.0%, B_2O_3 1.5–2.7%.

Hydrated chromium oxide green is a pigment with properties similar to those of Guignet's green but with a somewhat less intensive coloration. This pigment is prepared by reducing sodium chromate or sodium dichromate in aqueous solution with sulfur or sodium formate in a stirred autoclave or pressure tube [27]. The temperature required is 250–270 °C. The solid is separated by filtration, washed, dried, and ground. The finished pigment consists of fine needles, with a particle size of 0.02 × 0.1 μm predominating. The product contains 79–80% Cr_2O_3, the annealing loss is about 19%, and the density 3.7 g/cm³. The coloration changes at elevated temperature. Because of its high reflecting power at infrared wavelengths the product had at times been of special importance in camouflage paints.

4.2. Chromium(IV) Oxide (Chromium Dioxide)

WÖHLER discovered ferromagnetic chromium dioxide in 1859 when he decomposed chromyl chloride [28]. About 100 years later, Du Pont produced it in pure form by decomposition of chromic acid under hydrothermal conditions [29], [30]. Industrial exploitation began after the morphological and magnetic properties had been modified by doping chromium dioxide with heavy metals [31], [32] in order to meet the requirements for a magnetic pigment [33]. The marketing of chromium dioxide at the beginning of the seventies initiated the development of cobalt-doped iron oxides which are nowadays used as an alternative to chromium dioxide for information storage.

Physical and Chemical Properties [34]. Chromium dioxide [*12018-01-8*], CrO_2, M_r 84.00, crystallizes in black tetragonal needles. The lattice is of the rutile type and belongs to the space group 4/mmm. The dimensions of the unit cell are $a = b = 44.21$ nm and $c = 29.16$ nm. The X-ray density is 4.89 g/cm³, and the phase width is between $CrO_{1.89}$ and $CrO_{2.02}$ [35]. The enthalpy of formation is -590 kJ/mol [36]. The temperature coefficient for the c-axis is negative [37]. At 100 °C, agglomerated blocks have a linear coefficient of expansion of -6×10^{-6} K^{-1} [38].

At room temperature chromium dioxide is ferromagnetic, the magnetic moment being 2 Bohr magnetons. The Curie temperature is 120 °C and increases to 155 °C as a result of doping with iron [39]. Finely crystalline needle-shaped chromium dioxide has a specific magnetic saturation M_s/ϱ of 77–92 A m²/kg, whereas in single crystals M_s/ϱ rises to 100 A m²/kg [31]. The magnetocrystalline anisotropy constant is 22×10^3 J/m³

[40]. The coercivity H_c depends on the size of the crystals and on their shape and magnetocrystalline anisotropies. The coercivity is affected to differing extents by various heavy metals [41]. Iron, antimony, and tellurium increase H_c from 35 kA/m to 60 kA/m, whereas iridium increases it to 220 kA/m [42].

Chromium dioxide behaves as a metallic conductor [34], [43], with a specific electrical resistivity between 2.5×10^{-4} and 4×10^{-2} Ω cm [37], [44].

At room temperature and normal pressure chromium dioxide is metastable; when heated to temperatures above 350 °C, it decomposes into chromium(III) oxide and oxygen. Chromium dioxide has an oxidizing action on reactive organic compounds [45]; the reactivity is considerably decreased by enveloping it with iron(III) and chromium(III) oxides [46]–[48]. Chromium dioxide is insoluble in water. Reaction with water occurs at the crystal surface, with disproportionation to chromate and Cr^{3+} ions. The aqueous suspension has a pH of 3. However, chromium dioxide is soluble in concentrated sulfuric acid or concentrated alkali solution.

Production [49]. Chromium dioxide is made by decomposition of chromyl chloride, chromic acid anhydride [49], and chromium(III) chromate [50], or by oxidation of chromium(III) compounds with oxygen, hydrogen peroxide, chromic acid anhydride [51], or ammonium perchlorate [42].

Industrial production employs a process originally carried out under licence from Du Pont [41], which involves hydrothermal oxidation of chromium(III) oxide with excess chromic acid:

$$Cr_2O_3 + 3\,CrO_3 \longrightarrow 5\,CrO_2 + O_2$$

Iron(III) oxide and antimony(III) oxide are used as a dopant. Finely divided chromium(III) oxide is obtained either by thermal decomposition of ammonium dichromate or by dehydration of chromium(III) hydroxide [51].

A highly viscous paste (50–100 Pa · s) is produced by intensively homogenizing the starting materials. This paste is heated at 300 °C and 35 MPa (350 bar) to form a hard agglomerate of fine chromium dioxide needles which must be drilled out of the reactor trays, broken, and carefully ground. If the residual moisture exceeds 5%, the product is reheated in a rotary kiln. The chromium dioxide is deagglomerated in an aqueous sodium sulfite suspension, and the crystal surface is simultaneously reduced; this forms a chromium(III) oxide hydroxide layer about 1 nm thick. To do this, the suspension is circulated through a mill which generates intense shear fields and the fine component is removed by using a hydrocyclone. After filtration and washing, drying is carried out in a spray tower. The chromium dioxide obtained has a bulk density of 0.8 g/cm³.

Production is carried out to a large extent in closed equipment. Less than 2 mg of dust is emitted per 1 m³ of exhaust air (STP). The chromium-containing wastewater from the production is worked up by reduction.

Use and Economic Importance. Chromium dioxide is used as a magnetic pigment. So far no other uses have achieved any significance [32].

Table 7. Powder data for typical chromium dioxide pigments

Application	Particle geometry				Magnetic data [d]		
	SSA [a], m^2/g	$\bar{l}$ [b], µm	$\bar{l}/d$ [c]	V [c], 10^{-4} µm	H_c, kA/m	M_r/ϱ [e], A m^2/kg	M_s/ϱ [f], A m^2/kg
Audio	28	0.29	9	2.5	41	35	77
Video	35	0.29	11	1.5	49	34	74
Data storage	24	0.32	8	3.5	39	35.5	79

[a] Specific surface area (SSA) determined by N_2 adsorption using the BET method (1-point measurement).
[b] Mean length $\bar{l}$ determined by electron microscope photography with a magnification of 20 000 times.
[c] Diameter d and volume V calculated from SSA and $\bar{l}$.
[d] Measured with a vibration magnetometer, H_m = 800 kA/m.
[e] M_r/ϱ = specific remanent magnetization.
[f] M_s/ϱ = specific saturation magnetization.

In the audio field, chromium dioxide is used in mono- and multilayer tapes, the layers in the latter containing chromium dioxide of different coercivity. In video tapes, it is employed either on its own or mixed with cobalt-doped iron oxides. Because of its low Curie point, chromium dioxide allows high-speed thermomagnetic copying of prerecorded audio and video tapes [52]. Chromium dioxide has been used in digital data storage since 1985. Since the magnetostriction of chromium dioxide is low, repeated playing results in virtually no level losses [53]. Table 7 contains data on chromium dioxide powder intended for various applications.

In 1984 the demand for magnetic pigments was about 50 000 t, about 10% of this being chromium dioxide. The most important manufacturers are Du Pont (United States) and BASF (Federal Republic of Germany). In addition, there is a chromium dioxide plant in the Soviet Union.

4.3. Chromium(VI) Oxide

Chromium trioxide [1333-82-0], chromic acid anhydride, chromic acid, CrO_3, M_r 99.99, ϱ 2.7 g/cm^3, forms dark red crystals which deliquesce in air. The enthalpy of formation is – 594.5 kJ/mol. The oxide melts at 198 °C and starts to decompose, giving off oxygen and brownish red vapors with a pungent smell. The rate of decomposition reaches a maximum at 290 °C, chromium(III) oxide, Cr_2O_3, being produced as the final product via various intermediate stages. Chromium(VI) oxide dissolves in water to form chromic acids (see 6.1); the solubility depends only slightly on temperature. A saturated solution contains 166 g of CrO_3 at 20 °C and 199 g of CrO_3 at 90 °C per 100 mL of water. The compound also dissolves in sulfuric acid and nitric acid. Chromium(VI) oxide is a powerful oxidizing agent, particularly in the presence of acids. Reactions with alkali metals and numerous organic compounds, e.g., low-boiling hydrocarbons, acetone, or benzene and its derivatives, proceed explosively with considerable heat being produced. Esters of chromic acid are also known, e.g., with such cyclic tertiary alcohols as methylfenchol and methylborneol.

Chromium(VI) oxide is made by the reaction of sodium dichromate with sulfuric acid. The reaction can be carried out with solid sodium dichromate or with solutions or suspensions. Both methods are in use industrially. The reaction proceeds rapidly and completely after the components have been mixed, with heat being evolved. Isolation of chromium(VI) oxide from the reaction mixture or purification of the crude product obtained from the aqueous solution is difficult. Quantitative separation of sodium hydrogen sulfate is possible only if the chromium(VI) oxide is melted, but at this temperature the product begins to decompose. The melting process must therefore be controlled very precisely.

Dry Process. Even today the old *discontinuous* process is in use to some extent. The reaction vessels are made from carbon steel or stainless steel. The conical containers are equipped with stirrer, exhaust facilities, and external heating. Sulfuric acid and sodium dichromate are added simultaneously with stirring. The paste heats up to 80 °C during mixing and is heated further to evaporate water. At 170 °C sodium hydrogen sulfate melts, followed by chromium(VI) oxide at 198 °C. As soon as the reaction products are liquid, the heating and the stirrer are turned off. After a few minutes the heavier chromium(VI) oxide (ϱ 2.2 g/cm^3) settles at the bottom, and is covered by a layer of the lighter sodium hydrogen sulfate (ϱ 2.0 g/cm^3). Liquid chromium(VI) oxide is drawn off the bottom and conveyed to a cooling drum where it solidifies to form scales. Sodium hydrogen sulfate is subsequently drained. The yield of chromium(VI) oxide is about 85 %; 175 kg of sodium dichromate and 140 kg of 96 % sulfuric acid are required to obtain 100 kg of chromium(VI) oxide. About 150 kg of sodium hydrogen sulfate is obtained as byproduct.

Figure 4 shows a *continuous* dry process [54]. The raw materials are fed to a mixing screw for intimate mixing. A viscous paste of chromium(VI) oxide, sodium hydrogen sulfate, and water forms which is fed into a heated rotary kiln of stainless steel where it is melted. The heating must be controlled very carefully. The melt flows into a separator, where the heavier chromium(VI) oxide collects at the bottom of the trough, is removed by means of a rising pipe, and is converted into scales on cooling drums. The upper sodium hydrogen sulfate layer leaves the separating cell via an overflow and is also cooled on drums. Exhaust air from the various pieces of equipment is purified in a wash tower. The yield of this process is over 90 %.

If a highly concentrated solution is used instead of dichromate crystals, the kiln can be heated directly [55].

Wet Process. A hot saturated solution of sodium dichromate, which may still contain dichromate crystals, reacts with sulfuric acid [56]. In the course of 30 – 60 min the chromium(VI) oxide precipitates from the hot solution. On filtration a crude product is obtained, with a yield of about 80 %. Sodium hydrogen sulfate must be removed from the crude product by fusion. The filtrate can be recycled for converting sodium monochromate into dichromate; it can also be used in a fresh reaction mixture [57] if sodium hydrogen sulfate is first crystallized and removed at 20 – 25 °C. The crude chromium(VI) oxide is purified by continuous successive fusion and decanting [58].

As an alternative to production by reaction with sulfuric acid, Diamond – Shamrock developed an electrochemical process [59] – [61] in which chromic acid is produced from sodium dichromate in a two- or three-compartment cell.

Chromium trioxide is usually sold in the form of flakes, but the coarsely or finely ground product is also marketed. Good commercial products contain 99.5 – 99.7 % CrO$_3$ and a

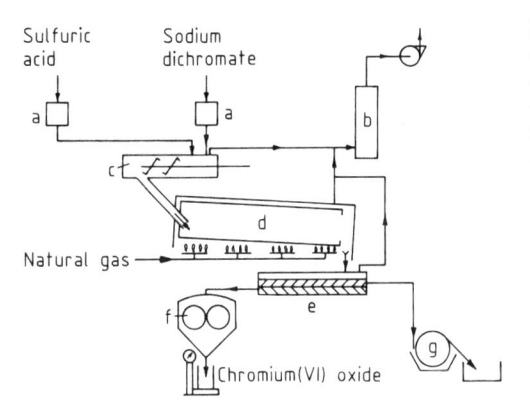

Figure 4. Continuous production of chromium(VI) oxide
a) Metering device; b) Wash tower; c) Mixer; d) Rotary kiln; e) Separation cell; f) Cooling drum for chromium(VI) oxide; g) Cooling drum for sodium hydrogen sulfate

maximum of 0.1% of sulfate. In the form of flakes, the product has a bulk density of 1.1 kg/L whereas that of the ground product is 1.4 kg/L. Steel drums must be used as containers and they must be tightly sealed because the product absorbs moisture from the air.

Chromium trioxide is classified as a dangerous substance in the EEC list and must be marked as fire-promoting and corrosive. In the IMDG code chromic acid has been put in class 5.1., UN No. 1463. The MAK of CrO_3 is 0.1 mg/m^3.

Electroplating is the most important field of application of chromium(VI) oxide. Numerous mixtures containing chromium trioxide are on the market; these "compounds" often contain hexafluorosilicates which improve the properties of the chromium coatings. Chromic acid solutions are also used for passivating zinc, aluminum, cadmium, and brass. Proprietary mixtures predominantly contain additions of fluoride, nitrate, and phosphate ions. Other uses for chromic acid are in the production of chromium dioxide and in wood preservation.

5. Chromium(III) Salts

5.1. General Properties

Water Content. In contrast to many other inorganic salts, chromium(III) salts occur in a variety of forms that depend on water content and on the particular conditions under which they are formed. Anhydrous compounds do not dissolve in pure water.

However, some of them, e.g., chromium(III) chloride or chromium(III) sulfate, dissolve in the presence of chromium(II) ions. In this process, one dissolved divalent ion transfers an electron via an anion bridge to a trivalent chromium ion in the solid crystal. Having become divalent, this ion detaches itself, acts in a similar manner on another chromium(III) ion in the crystal array, and reverts again to the trivalent state.

Complex Formation. Dissolved chromium-(III) ions are always coordinated by various ligands. In the simplest case of the hexaaquochromium(III) ion, $[Cr(H_2O)_6]^{3+}$, six water molecules surround the central chromium ion in an octahedral arrangement as ligands. In addition to the aquo complexes, numerous coordination compounds with other molecules are known, and research on them, particularly studies of the ammine complexes (NH_3 as ligand), has played an important part in the development of coordination compound chemistry [62].

When negatively charged ligands enter the chromium complex, the charge is decreased appropriately. If the sum of the negative charges is four or more, the complex becomes anionic, an example of this being the diaquodisulfatochromium(III) ion $[Cr(SO_4)_2(H_2O)_2]^-$. In this case, each sulfate radical with a double negative charge occupies the position of two ligands.

The tendency of negatively charged ligands to form complex compounds with chromium increases in the following order:

$$Cl^- < SO_4^{2-} < CH_3CO_2^- < HCO_2^- < OH^-$$

Nitrato complexes are unknown.

Hydrate Isomerism. Chromium(III) complexes exhibit hydrate isomerism due to the positioning of anions and water molecules.

Thus chromium(III) chloride hexahydrate [*10060-12-5*] is known in three different forms:

1) Hexaaquochromium(III) chloride, $[Cr(H_2O)_6]Cl_3$, bluish grey
2) Pentaaquochlorochromium(III) chloride hydrate $[Cr(H_2O)_5Cl]Cl_2 \cdot H_2O$, bright green
3) Tetraaquodichlorochromium(III) chloride dihydrate $[Cr(H_2O)_4Cl_2]Cl \cdot 2\,H_2O$, dark green

The anions that are directly bound to the central atom do not dissociate in water and consequently do not react with the common precipitating agents; therefore, in the two green chlorides, only one-half and one-third of the chloride ions, respectively, are precipitated by silver nitrate.

Basic Salts. Hydroxide ions form coordinate bonds, with central ion of the hexaaquochromium complex being hydrolyzed. In this process, the pentaaquohydroxochromium(III) ion with a double positive charge is first formed:

$$[Cr(H_2O)_6]^{3+} \xrightarrow{OH^-} [Cr(H_2O)_5OH]^{2+} + H^+$$

When more alkali is added, chromium hydroxides are precipitated immediately. Finally, with a strong hydroxide solution, a soluble deep green hydroxo salt is produced:

$$Cr(OH)_3 + 3\,NaOH \longrightarrow Na_3[Cr(OH)_6]$$

Pentaaquohydroxochromium(III) complexes are very weak bases. Their salts hydrolyze and the pH of aqueous solutions usually is 2. The basicity of these salts is defined as the ratio of hydroxyl groups (in percent) bound to chromium to the number of hydroxyl groups in chromium(III) hydroxide, that could theoretically be bound to chromium. Pentaaquohydroxochromium(III) complexes therefore have a basicity of 33%. When a second hydroxyl group enters the complex the basicity increases to 67%. However, from a basicity of 60% onward, chromium(III) hydroxide precipitates and these compounds are not used in practice.

5.2. Chromium(III) Sulfates and Chrome Tanning Agents

Chromium(III) Sulfate. *Anhydrous chromium sulfate* [10101-53-8], $Cr_2(SO_4)_3$, M_r 392.18, ϱ 3.0 g/cm^3, is a violet powder which is insoluble in water but dissolves to form complexes when reducing agents are added. For its preparation, chromium metal or chromite is heated over 250 °C with sulfuric acid.

The *octadecahydrate* [13520-66-6], $[Cr(H_2O)_6]_2(SO_4)_3 \cdot 6\,H_2O$, M_r 716.45, ϱ 1.86 g/cm^3, forms cubic crystals. The violet compound gives off water on heating and above 70 °C it is converted, with further loss of water, into a dark-green crystalline *pentadecahydrate* [10031-37-5]. As the water content diminishes, the solubility decreases.

Solutions of chromium(III) sulfates can be made by treating chromite with sulfuric acid in the presence of chromium(VI) compounds [63]. Since other components of the ore are solubilized at the same time, the solutions are strongly contaminated; separation of magnesium, aluminum, and iron presents such great difficulties that this process has not yet gone beyond the experimental scale. Chromium(III) sulfate solutions are also obtained by dissolving ferrochromium in sulfuric acid, a process in which iron(II) sulfate is obtained as a coproduct. So far, the economical preparation of a pure product has been only partially successful.

Large quantities of chromium(III) sulfate solution are produced in the oxidation of organic substances with chromic acid or sodium dichromate in sulfuric acid solution. Examples of this are the preparation of anthraquinone from anthracene, the preparation of benzoquinone from aniline, or the bleaching of montan wax. These solutions are used to produce other chromium products; an electrolytic regeneration to dichromate is also possible.

Tanning Agents. *Basic chromium(III) sulfates* are used on a large scale as tanning agents for leather. Industrially, two processes are available for the reduction of sodium dichromate: (1) reaction with organic compounds (molasses, sugar) in the presence of sulfuric acid and (2) reduction with sulfur dioxide.

Reduction with molasses is carried out in aqueous solution; about 30 kg of molasses or 15 kg of cane sugar is required for 100 kg of sodium dichromate dihydrate. The amount of sulfuric acid required depends on the desired basicity. To adjust the basicity to 33%, about 103 kg of 96% acid is needed. The reaction is strongly exothermic; water evaporates in abundance and must be continuously replenished. Lead-lined vats have proved successful as reaction vessels. The exhaust gases have an unpleasant smell, but this can be eliminated by scrubbing with water in a wash tower or by heating them after condensing the water vapor.

The properties of the final product are, to a certain extent, dependent on how the reaction is performed. If sulfuric acid is added first to the dichromate solution and the reducing agent is then added slowly, relatively few organic acids are produced as result of side reactions. The proportion of these acids becomes considerably larger if the dichromate solution is mixed first with the reducing agent and the sulfuric acid is added last. The organic acids form chromium complexes and mask the tanning agent. This masking delays the tanning process.

In the *reduction with sulfur dioxide* sulfuric acid is generated in such proportions that the tanning agent produced has a basicity of 33%. For the reduction of 100 kg of sodium dichromate dihydrate, 65 kg of SO_2 is theoretically required.

The reaction is carried out in lead-lined or brick-lined absorption towers containing ceramic packing material. Sulfur dioxide is produced by combustion of liquid sulfur which yields a gas containing 8–18% SO_2; SO_2-containing gases from other manufacturing processes are also suitable for the reaction.

For the production of chrome tanning agents, chromium(III) sulfate solutions which are generated in the manufacture of organic intermediates may also be used. Impurities must be removed from such solutions, and the solutions are then converted to the correct basicity by acidification or neutralization. They must also be concentrated by evaporation.

To manufacture *solid tanning agents*, the concentrated solutions are dried in spray driers made of stainless steel. With a basicity of 33%, the amorphous green powder obtained generally contains 24–26% Cr_2O_3, 25–27% SO_3, 22–25% Na_2SO_4, and 22–25% water. In air, the powder absorbs moisture and the particles stick together. Under the microscope, the individual particles, which are often hollow spheres or fragments of such spheres, have a glassy appearance.

The product obtained by spray drying is sold under numerous trade names, e.g., Chrometan (British Chrome and Chemicals, UK); Chromitan (BASF, Federal Republic of Germany); Chromosal (Bayer, Federal Republic of Germany); Salcromo (Stoppani, Italy); Tanolin (Hamblett & Hayes Co., Mass., USA).

Paper or jute sacks with watertight polyethylene liners or wrappings, or plastic sacks are used for packing.

Solutions of basic chromium(III) sulfate containing 12–18% Cr_2O_3 are also available commercially. Solutions of higher concentration must be kept warm because sodium sulfate precipitates at room temperature. The solutions are transported in rubber-lined rail or road tankers; lead-lined tankers are also suitable.

Besides the standard 33% basic type, a large number of products of higher basicity are available. To improve their stability toward alkali, these contain various quantities of organic acids. The market importance of chrome tanning agents containing 30% chromium oxide and having a basicity of 50% has increased.

In addition to the standard products, mixtures of chrome tanning agents have established themselves on the market. These contain basifying agents which react slowly and eliminate the need for the tedious basifying process. Mixed products containing special organic masking agents and having a high total basicity have recently been developed. These are used in combination with conventional chrome tanning agents and afford a high degree of chromium exhaustion in the liquors.

Potassium chromium(III) sulfate [10141-00-1], [10279-63-7], [7788-99-0], potassium chrome alum, $KCr(SO_4)_2 \cdot 12\,H_2O$, M_r 499.11, ϱ 1.813 g/cm^3, crystallizes in the cubic system forming violet regular octahedra which decay in air and melt at 89 °C, the color changing to green; the enthalpy of formation is -5788 kJ/mol. The solubility in water at 25 °C is 11.1 wt%. The solution is violet when cold, but becomes green above 50 °C, this change being accompanied by a decrease in the molar conductance. For 0.125 M solution at 50 °C, the molar conductance is 221 Ω^{-1} cm^2 mol^{-1} for the violet form and 202 Ω^{-1} cm^2 mol^{-1} for the green form. This change is reversible and its rate is increased by acids. The green form always occurs as an amorphous solid and crystals are unknown.

In addition to the alum containing 12 molecules of water of crystallization, potassium chromium(III) sulfates containing one [35177-45-8], two [35177-44-7], and six [35177-43-6] molecules of water are known.

For the preparation of potassium chromium alum, a saturated potassium dichromate solution is reduced with sulfur dioxide in the presence of sulfuric acid.

During the reaction, the temperature must be kept below 40 °C by cooling to prevent the green modification being produced. Apart from sulfur dioxide, such organic compounds as formaldehyde, methanol, or starch are suitable as reducing agents. Crystallization starts after sulfuric acid has been added and the temperature is kept further below 40 °C. If the solution is allowed to settle in vats, large crystals are produced, whereas fine ones result if the solution is stirred. The industrial product contains 15% Cr$_2$O$_3$ and 0.01–0.03% Fe. The mother liquor may be reused in the production of the alum solution, but after several cycles it is so enriched in magnesium and sodium sulfate that crystallization of the alum is retarded. The trivalent chromium is then precipitated from the solution with alkali and may be recycled. Lead-lined equipment is used for the production.

The alum can also be prepared from ferrochromium. The reaction with sulfuric acid first results in a solution of chromium(III) and iron(II) sulfates. The majority of the iron(II) sulfate can be removed by crystallization. With potassium sulfate added, the filtrate yields the alum which is contaminated with 0.1–0.2% Fe.

Potassium chrome alum was formerly used on a large scale as a tanning agent in the leather industry, but its importance in this field has receded with the introduction of the basic chromium sulfates. Other fields of application are the textile industry and the film and photographic industry.

Ammonium chromium(III) sulfate [*10141-00-1*], ammonium chrome alum, $NH_4Cr(SO_4)_2 \cdot 12\,H_2O$, M_r 478.47, ϱ 1.72 g/cm^3, crystallizes in the cubic system forming bluish violet octahedra which appear ruby red when held against the light and slowly decay in air. When heated to 70 °C they turn green, and at 94 °C the compound melts in its water of crystallization. Ammonium chrome alum is obtained from chromium(III) sulfate solutions by addition of a stoichiometric amount of ammonium sulfate. Ammonium chrome alum crystallizes much more easily than potassium chrome alum. The preparation of ammonium chrome alum from carbon-rich ferrochromium has assumed relatively great importance in the electrochemical production of pure chromium metal [64].

5.3. Other Chromium(III) Salts

Chromium(III) fluoride [*7788-97-8*] CrF_3, M_r 108.99, ϱ 3.8 g/cm^3. The anhydrous compound forms highly refractive rhombohedral crystals. It melts above 1000 °C and is distinctly volatile between 1100 and 1200 °C. Chromium(III) fluoride is insoluble in water if no divalent chromium is present. Double compounds are formed with other metal fluorides, e.g., green $CrF_3 \cdot 2\,KF \cdot H_2O$.

Hydrates are known which contain three [*16671-27-5*] to nine [*68374-27-6*] molecules of water. The violet hexaaquochromium(III) fluoride, $[Cr(H_2O)_6]F_3$, and its trihydrate, $[Cr(H_2O)_6]F_3 \cdot 3\,H_2O$, can be obtained from hexaaquochromium(III) salt solutions and alkali fluorides. Products containing less water are green. The composition of the industrial product corresponds approximately to $CrF_3 \cdot 3.5\,H_2O$ and the product contains about 30% chromium.

For the production of chromium(III) fluoride hydrate, chromium(III) oxide hydrate is dissolved in hot aqueous hydrofluoric acid and the green salt crystallizes. Chromium(III) fluoride is used in the textile industry for mordanting wool, for chromating dyestuffs, and in vigoureux printing. Recently, chromium(III) fluorides have also found application in rust-prevention paints as corrosion inhibitors.

Chromium(III) Chloride. Anhydrous chromium(III) chloride [*10025-73-7*], $CrCl_3$, M_r 158.36, forms hexagonal reddish violet flakes which sublime at 950 °C yielding a vapor that dissociates above 1300 °C. The enthalpy of formation is −554.8 kJ/mol. Chromium(III) chloride is insoluble in water if no reducing agent is present. On roasting in air, chromium(III) oxide is produced.

Anhydrous chromium(III) chloride is obtained along with iron(II) chloride by chlorinating roasting of chromite in the presence of carbon at 900–1050 °C [65]. Oxygen is added to the chlorine to prevent nonvolatile residues, in particular minor constituents of the ore, from sintering together. Fractionating condensation between 400 and 640 °C has been suggested for separating the chloride vapors [66]. The compound can also be obtained by chlorinating chromium(III) oxide in the presence of reducing agents or by treating ferrochromium with chlorine [67].

Chromium(III) chloride can be prepared readily from chromyl chloride by reaction with carbon monoxide and chlorine. The reaction proceeds rapidly in the gas phase at 750–850 °C [68]. Since chromium(III) chloride evaporates only at a higher temperature, a considerable portion of the product, which varies as a function of the partial pressure, is produced in the form of fine crystals. As the smoke cools down, these act as crystallization nuclei for any gaseous chromium(III) chloride still present. This procedure prevents the deposition of solid on the cooling surfaces.

Anhydrous chromium(III) chloride has been suggested for chromizing steel parts by surface diffusion; it can be used for the production of high-purity ductile chromium metal by reduction with magnesium and for the synthesis of organic chromium compounds.

Chromium(III) chloride hexahydrate [10060-12-5] is obtained in pure form by introducing hydrogen chloride and methanol into an aqueous solution of chromic acid. The reaction proceeds exothermally, and adequate cooling must be provided. Of the three isomeric hydrates (see 5.1) the dihydrate of the tetraaquodichlorochromium(III) chloride crystallizes in the cold. This is used as an intermediate in the production of chromium complex dyes and other chromium salts, e.g., chromium stearates, which are of interest as impregnating agents for textiles or paper. The solution is also used as a mordant in the textile industry.

Chromium(III) Acetate [1066-30-4]. The bluish violet hexaaquo salt [66851-10-3], $[Cr(H_2O)_6](CH_3COO)_3$, M_r 337.22, forms needle-shaped crystals. Basic chromium(III) acetates are green. For their preparation chromium(III) hydroxide hydrate is dissolved in dilute acetic acid, and the solid is obtained by drying on drums or in a spray drier. Basic chromium acetates are used as mordants in calico printing and worsted top printing, and also for fixing vigoureux dyes. Combinations of basic chromium(III) acetate and basic chromium(III) formate also find application in the textile industry as mordants. In addition, chromium(III) acetate is used as a starting compound in the production of organic chromium dyes.

Chromium(III) Nitrate [13548-38-4]. Normally, chromium(III) nitrate crystallizes with nine molecules of water, $[Cr(H_2O)_6](NO_3)_3 \cdot 3 H_2O$ [7789-02-8], M_r 400.15, ϱ 1.8 g/cm^3. The dark violet rhombic prisms become green above 36 °C and melt at 66 °C. The nitrate group is not bound to the trivalent chromium in a coordination compound. The salt is readily soluble in water, acid, alkali, and alcohol. To prepare chromium(III) nitrate, chromium(III) oxide hydrate is dissolved in nitric acid and the nitrate is allowed to crystallize. The compound is also produced by reduction of chromic acid with methanol in the presence of nitric acid. During the exothermal reaction, the temperature is kept between 55 and 65 °C by cooling. Chromium(III) nitrate is used to a limited extent as a mordant in cotton printing, usually together with basic chromium acetates. In addition, the salt is suitable for producing alkali-free catalysts.

Chromium(III) Phosphate. *Anhydrous chromium(III) phosphate* [7789-04-0], $CrPO_4$, M_r 146.97, ϱ 2.99 g/cm³, is a black powder which belongs to the orthorhombic crystal system. It is insoluble in water, hydrochloric acid, and aqua regia but is attacked by boiling sulfuric acid. It is obtained by calcining its hydrates. The violet *hexahydrate* [13475-98-4], $[Cr(H_2O)_6]PO_4$, M_r 255.06, ϱ 2.12 g/cm³, precipitates from chromium(III) salt solutions upon addition of phosphoric acid and disodium hydrogen phosphate. The crystals are triclinic and only sparingly soluble in water. Green chromium(III) phosphates precipitated hot contain two to three molecules of water after drying. A salt containing four molecules of water is also known. The green products are virtually insoluble in water. They are used to a small extent as pigments and have corrosion-inhibiting properties.

Chromium lignosulfonates [9066-50-6] are prepared by reaction of sulfite waste liquor from the pulp industry with sodium dichromate solution, the hexavalent chromium being reduced by the organic material to trivalent chromium. After filtration, the pH is adjusted by means of alkali and the product is dried in spray driers. It is a free-flowing, water-soluble powder and normally contains cations in the following quantities: ammonium 1.5–2.5%, chromium 2.5–4.2%, iron 0–5%, and sodium 1.5–2.5%. These saltlike compounds are used to a large extent in the petroleum industry as additives for lowering the viscosity of drilling muds and decreasing liquid loss [14].

6. Chromic Acids and Chromates(VI)

6.1. Chromic Acids

Chromic acids are not known in the free state. Depending on the method of preparation, mono-, di-, tri-, or tetrachromic acids are formed in aqueous solution. In alkaline or dilute solution, formation of the yellow monochromate ion is favored, but in acid solution or at high concentrations, the orange-red dichromate ion is formed preferentially. Aqueous solutions of chromic acids are, therefore, yellow or red depending on their concentrations. Figure 5 shows the density as a function of concentration.

Dissociation constants (at 25 °C) are as follows:

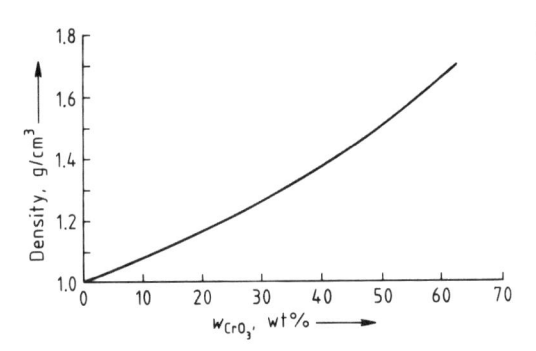

Figure 5. Density of aqueous chromic acid solutions at 15.6 °C (60 °F)

$$H_2CrO_4 \rightleftharpoons H^+ + HCrO_4^- \quad K_1 = 1.21$$
$$HCrO_4^- \rightleftharpoons H^+ + CrO_4^{2-} \quad K_2 = 3.7 \times 10^{-7}$$
$$H_2Cr_2O_7 \rightleftharpoons H^+ + HCr_2O_7^- \quad K_3 = 1$$
$$HCr_2O_7^- \rightleftharpoons H^+ + Cr_2O_7^{2-} \quad K_4 = 0.85$$
$$Cr_2O_7^{2-} + H_2O \rightleftharpoons 2H^+ + 2CrO_4^{2-} \quad K_5 = 3 \times 10^{-15}$$
$$Cr_2O_7^{2-} + H_2O \rightleftharpoons 2HCrO_4^- \quad K_6 = 0.023$$

The standard redox potential for the reaction

$$Cr_2O_7^{2-} + 14H^+ + 6e \rightleftharpoons 2Cr^{3+} + 7H_2O$$

is 1.36 V.

Chromic acid solutions are strong oxidizing agents with a strongly acidic character; they form salts with metals and bases. The monochromates(VI), M_2CrO_4, which are derived from chromic acid, hydrolyze in aqueous solution:

$$CrO_4^{2-} + H_2O \rightleftharpoons HCrO_4^- + OH^-$$

The easiest method of preparing chromic acid solutions is to dissolve chromium(VI) oxide in water. In industry, chromic acid is often produced from sodium dichromate(VI) and sulfuric acid.

Chromic acid solutions can also be prepared by anodic oxidation of chromium(III) sulfate solutions [69]; lead-lined cells with a diaphragm and lead electrodes are employed. To keep the concentration of sulfuric acid constant, the chromium(III) sulfate solution is introduced first into the cathode space, where it becomes depleted of sulfuric acid, and then into the anode space. Here, oxidation to chromic acid takes place and the concentration of sulfuric acid is restored to its original value.

In practice, several electrolytic cells are combined to form a unit. At a current density of 3 A/dm² the voltage is 3.5 V. Current efficiency is 80%. Lost chromium is periodically replenished by adding chromium(III) oxide. The electrolytic preparation of chromic acid can also start from chromium hydroxide hydrate with chromic acid as electrolyte [70].

Aqueous chromic acid solutions are used as pickling and chromium-plating baths in the metal processing and plastics processing industries.

Table 8. Solubility (in wt%) of various chromates in water

	Temperature, °C								
	0	20	25	40	50	60	75	80	100
Sodium chromate		44.3		48.8		53.5		55.8	56.1
Sodium dichromate dihydrate	70.6	73.18		77.09		82.04		88.39	91.43
Potassium chromate		39.96							45.0
Potassium dichromate	4.3	11.7		20.9		31.3		42.0	50.2
Ammonium chromate	19.78		27.02		34.4		41.2		
Ammonium dichromate	15.16	26.67		36.99		46.14		54.10	60.89
Silver chromate		0.0025			0.0053				0.0041

6.2. Alkali Chromates and Dichromates

Sodium chromate [7775-11-3] Na_2CrO_4, M_r 161.97, mp 792 °C, ϱ 2.723 g/cm^3, ΔH^0_{298} −1329 kJ/mol, crystallizes in the orthorhombic system in small yellow needles or columns; transformation to the hexagonal form takes place at 413 °C. The bulk density of the powder is 0.7 g/cm^3; that of the crystals is 1.67 g/cm^3. For solubility, see Table 8. The compound is hygroscopic and forms several hydrates: below 19.5 °C the decahydrate [13517-17-4]; between 19.5 and 25.9 °C, the hexahydrate; and between 25.9 and 62.8 °C, the tetrahydrate [10034-82-9] which undergoes transformation into anhydrous sodium chromate above 62.8 °C.

To prepare the salt, sodium dichromate solution is usually mixed with a stoichiometric amount of sodium hydroxide, and the salt solution is then crystallized or spray dried. The 96.5–98.5% product (0.4% NaCl, 2% Na_2SO_4) is stored and dispatched in watertight steel drums.

Sodium chromate is used as a corrosion inhibitor in the petroleum industry and as a dyeing auxiliary in the textile industry.

Sodium Dichromate. The *dihydrate* [7789-12-0], $Na_2Cr_2O_7 \cdot 2\,H_2O$, M_r 298.0, ϱ 2.348 g/cm^3, $\Delta H_{298}°$ − 2194 kJ/mol, forms orange-red, monoclinic, translucent needles which are converted into the anhydrous salt above 84.6 °C. The bulk density is 1.2 g/cm^3. For solubility, see Table 8. The heat of solution is − 118 kJ/kg. The compound is very hygroscopic and deliquesces in air; in acid solution, it is a strong oxidizing agent.

The preparation of sodium dichromate is described in Chapter 3.

Sodium dichromate is the most important of the industrial chromium chemicals and is used as the starting compound for almost all chromium compounds. Large quantities are used in numerous industrial fields. In the textile industry (wool, cotton, silk, and synthetics), sodium dichromate is used in mordanting and in aftertreatment baths. The leather industry virtually no longer uses sodium dichromate. Sodium dichromate has a variety of uses in the surface treatment of metals, e.g., in the pickling of steel, aluminum, magnesium, and other metals and their alloys. The ability of the chromates to convert gelatin or protein into an insoluble form on exposure to light is exploited on a large scale in printing technology (lithography). A further field of application for the dichromates is in corrosion protection; they are added to crude oil in pipelines and to water in closed cooling systems as direct corrosion inhibitors. Sodium dichromate is also used for the manufacture of wood preservatives.

In the chemical industry, sodium dichromate is used as a strong oxidizing agent in numerous cases; the most important include oxidation of anthracene to anthraquinone (dyes), of aniline to quinone (hydroquinone for the photographic industry), of camphene to camphor, and of contaminants in oils, fats, tallow, and waxes (soap industry, wax bleaching).

To these classical applications, the wide field of catalysts and catalyst carriers containing chromium(III) oxide or chromates has been added over the past thirty years. These are important in a variety of oxidation and carbonizing processes.

Anhydrous sodium dichromate [10588-01-9], $Na_2Cr_2O_7$, M_r 261.96, mp 356.7 °C, ϱ 2.52 g/cm^3, bulk density 1 g/cm^3, heat of solution ca. − 33.5 kJ/kg, forms light-brown to orange-red plates which are strongly hygroscopic. They decompose above 400 °C with the formation of sodium monochromate(VI), chromium(III) oxide, and oxygen.

Anhydrous sodium dichromate can be prepared by melting down sodium dichromate dihydrate, by crystallizing aqueous dichromate solutions above 86 °C, or by drying sodium dichromate solutions in spray driers.

Anhydrous sodium dichromate is required for cases in which the water content of the dihydrate has an interfering action. Thus, for example, the energy liberated in the oxidation with anhydrous sodium dichromate is greater than that liberated in the case of sodium dichromate dihydrate. Anhydrous sodium dichromate is, therefore, used in the preparation of chromium(III) oxide by the dry process (4.1), in pyrotechnics, and in anhydrous oxidation processes where it replaces the more expensive potassium dichromate. Compared with sodium dichromate dihydrate, anhydrous sodium dichromate has the advantage that it can first absorb two molecules of water (13 wt %) instead of deliquescing immediately when moisture is admitted.

Potassium chromate [7789-00-6], K_2CrO_4, M_r 194.2, mp 968.3 °C, ϱ 2.73 g/cm^3, $\Delta H_{298}°$ − 1383 kJ/mol, occurs as the stable β-modification. The lemon-yellow, nonhygroscopic prisms are isostructural with K_2SO_4. At 666 °C they are converted into hexagonal α-potassium chromate. For solubility, see Table 8. The heat of solution is −71.3 kJ/kg. The salt crystallizes from aqueous solution in anhydrous form and is thermally stable.

Potassium chromate is obtained by reacting potash with potassium dichromate. The potassium salt has been supplanted nearly completely by the cheaper sodium chromate and is used only for very specific purposes such as in the photographic industry.

Potassium dichromate [7778-50-9], $K_2Cr_2O_7$, M_r 294.19, mp 397.5 °C, occurs in two modifications. α-$K_2Cr_2O_7$, tabular or prismatic, bright orange-red triclinic crystals, ϱ 2.676 g/cm^3, has a bulk density of about 1.3–1.6 g/cm^3; at 241.6 °C α-$K_2Cr_2O_7$ transforms to β-$K_2Cr_2O_7$. For solubility, see Table 8. The heat of solution is −258.3 kJ/kg. The thermodynamic data are as follows: c_p 219.7 J mol^{-1} K^{-1}, $\Delta H_{298}°$ −2033 kJ/mol, $S°$ 291.2 J mol^{-1} K^{-1}, heat of fusion 36.7 kJ/mol. The substance is not hygroscopic and, above the melting point, decomposes into potassium chromate, chromium oxides, and oxygen.

Today potassium dichromate is obtained primarily by conversion of sodium dichromate with potassium chloride. Potassium dichromate has largely been supplanted by the cheaper sodium dichromate but is still used whenever its advantage of being nonhygroscopic is important, for example, in the match, firework, film, and photographic industries. Potassium dichromate is of interest in the preparation of yellow and green zinc pigments.

Ammonium chromate [7788-98-9], $(NH_4)_2CrO_4$, M_r 152.07, smells of ammonia and forms golden yellow needles, ϱ 1.886 g/cm^3, $\Delta H_{298}°$ −1152 kJ/mol, $S_{298}°$ 656 J mol^{-1} K^{-1}. For solubility, see Table 8. In air, it decomposes into ammonia, water, and ammonium dichromate. On heating it ignites and decomposes into chromium(III) oxide, ammonia, and nitrogen. Ammonium chromate is prepared from ammonium dichromate with the addition of ammonia.

Ammonium dichromate [7789-09-5], $(NH_4)_2Cr_2O_7$, M_r 252.06, forms large, bright orange-red crystals, ϱ 2.155 g/cm^3, bulk density ca. 1.0–1.3 g/cm^3. For solubility, see Table 8. The heat of solution is −230.9 kJ/mol. Ammonium dichromate crystallizes in anhydrous form from aqueous solution and is not hygroscopic. Decomposition, which is not preceded by melting, sets in on heating to 180 °C; this becomes self-maintaining at 225 °C and above. Decomposition proceeds with displays of fire and heat, and large amounts of gas are developed. The products of decomposition are chromium(III) oxide, nitrogen, and water vapor. Ammonium dichromate reacts very violently with organic solvents.

Ammonium dichromate is prepared by reaction of sodium dichromate with ammonium chloride or, less frequently, ammonium sulfate. Ammonium dichromate is used as the starting material for preparing very finely divided chromium(III) oxide and, in addition, finds application primarily in pyrotechnics, wood preservation, and photography (lithography) for the preparation of light-sensitive solutions of gelatin or proteins. Ammonium dichromate is also used to prepare catalysts for organic syntheses. A further field of application is the production of magnetic chromium(IV) oxide. Because of its self-ignition properties and explosiveness, ammonium dichromate is subject to the German Explosives Law and the IMDG code, class 5.1, UN No. 1439. It is also distributed moist.

6.3. Other Chromates

Barium chromate [10294-40-3], BaCrO$_4$, M_r 253.33, mp 1400 °C (decomp.), ϱ 4.498 g/cm^3, $\Delta H_{298}°$ −1156 kJ/mol, crystallizes as light-yellow transparent rhombic crystals which are isomorphous with barium sulfate. Barium chromate is only sparingly soluble in water but dissolves readily in acids.

In the presence of excess alkali chromate or dichromate, barium chromate has a tendency to form double salts, among which special mention may be made of potassium barium chromate [13819-19-7], K$_2$CrO$_4$ · BaCrO$_4$, and ammonium barium chromate [13819-20-0], (NH$_4$)$_2$CrO$_4$ · BaCrO$_4$, both of which are light yellow.

These barium chromate double salts can be prepared by reaction of soluble barium salts or barium hydroxide with alkali chromate(VI) or dichromate(VI). In weakly acid solutions, as is the case, for example, if dichromates(VI) are used, the precipitation is incomplete. Quantitative precipitation is achieved by adding sodium acetate.

Yellow barium chromate and its double salts can be used for the production of chrome pigments for paints; the double salts, in particular, are excellent corrosion protection paints for all metals. They form sparingly soluble metal chromates, which prevent attack by moisture (condensation, seawater) even more readily than zinc chromate.

Calcium chromate [13765-19-0], CaCrO$_4$, M_r 156.07, mp 1020 °C (decomp.), ϱ 3.12 g/cm^3, is a yellow powder which is sparingly soluble in water (4.3 wt% at 0 °C, 0.42 wt% at 100 °C). The compound has acquired no industrial importance, but its preparation directly from chromium ores is described in several patents [71], [72].

Calcium dichromate [14307-33-6], CaCr$_2$O$_7$, M_r 256.06, $\Delta H_{298}°$ −1821 kJ/mol, is thought not to constitute a uniform crystalline phase but to be a mixture of phases consisting of calcium monochromate and chromium(VI) oxide. The compound forms a series of readily soluble hydrates. Thus, below 10 °C, the hexahydrate CaCr$_2$O$_7$ · 6 H$_2$O exists; between 20 and 40 °C, the pentahydrate [61204-19-1]; between 50 and 60 °C, the tetrahydrate; and above 70 °C, the monohydrate [85752-77-8]. A red deliquescent calcium dichromate trihydrate has also been prepared from the tetrahydrate.

Calcium dichromate can be made industrially by oxidative roasting of chromium-containing ores with calcium carbonate or calcium oxide and subsequent leaching of the cake with an acid, e.g., chromic or sulfuric acid [72].

Copper Chromates. Neutral copper chromate [13548-42-0], CuCrO$_4$, M_r 179.53, is produced as a yellowish brown, water-containing compound by precipitation from a copper sulfate solution with sodium or potassium dichromate. The compound is used as the starting material for "chrome black" which is prepared by calcining the neutral salt under oxidizing conditions and subsequently leaching with hydrochloric acid.

The double salt *copper ammonium chromate* (cupric ammonium chromate) is required in dyeworks along with logwood and fustic extracts to obtain olive green wool or cotton dyes. Copper chromate is used as such or in reduced form as a catalyst in a number of petrochemical reactions.

Iron Chromates. No anhydrous iron(II) chromate is known. On the other hand, the watersoluble double salts, $KFe^{III}(CrO_4)_2 \cdot 2\,H_2O$ [20161-12-0] and $NH_4Fe^{III}(CrO_4)_2 \cdot 2\,H_2O$ [20161-14-2] do exist, and potassium iron(III) chromate is a corrosion inhibitor. A similar corrosion-inhibiting compound is also thought to be formed on steel surfaces that are treated with chromate solutions. Industrially, the compound is produced by the reaction of iron(III) chloride solution with potassium dichromate in an autoclave at 130–160 °C. The precipitate is filtered and dried.

Lead chromate [7758-97-6], $PbCrO_4$, M_r 323.18, mp 844 °C (with evolution of oxygen), ϱ 6.123 g/cm^3, $\Delta H_{298}°$ −910 kJ/mol, is a yellowish orange powder which occurs in three modifications:

$$\text{monoclinic} \xrightarrow{707\,°C} \text{orthorhombic} \xrightarrow{783\,°C} \text{tetragonal}$$
$$\text{yellowish orange} \qquad\qquad \text{yellow} \qquad\qquad \text{red}$$

The solubility product in water is 1.5×10^{-14} at 18 °C. Lead chromate forms mixed crystals with lead(II) oxide, lead sulfate, and lead molybdate. All these salts are virtually insoluble in water; the molybdenum-containing compound is particularly well-known as molybdenum red. Lead chromate is prepared in a manner similar to that used for barium chromate. The composition, color, and quality of lead chromates depend on the conditions of precipitation; they are used widely as yellow to red pigments in the lacquer and paint industry.

Silver Chromate and Silver Dichromate. Ag_2CrO_4 [7784-01-2], M_r 331.73, ϱ 5.625 g/cm^3, $\Delta H_{298}°$ −711.7 kJ/mol, $S_{298}°$ 216 J mol^{-1} K^{-1}, c_p 142.3 J mol^{-1} K^{-1} (for solubility, see Table 8), and $Ag_2Cr_2O_7$ [7784-02-3], M_r 431.73, ϱ 4.770 g/cm^3, $\Delta H_{298}°$ −1218 kJ/mol, form dark red crystals and are soluble in acids, ammonia, and potassium cyanide solutions but not in water. The salts can be precipitated from a silver salt solution with chromate or dichromate solution and are used in the photographic industry.

Zinc chromate [13530-65-9], $ZnCrO_4$, M_r 181.36, is sparingly soluble in water but dissolves readily in acids. A series of zinc chromate hydrates exists having the composition $n\,ZnO \cdot m\,CrO_3 \cdot x\,H_2O$. In industry zinc chromate is known as zinc yellow. Its color may be controlled by the mode of preparation. It is made by the reaction of either a suspension of finely ground zinc white (ZnO) in concentrated sulfuric acid or water-soluble zinc salts ($ZnCl_2$, $ZnSO_4$) with potassium or ammonium dichromate. The zinc chromates prepared in this manner always incorporate potassium or ammonium ions into their lattice. Zinc chromate is used in lacquer primers as a corrosion inhibitor instead of minium (red lead) because of its ability to form insoluble iron(III) chromates with iron or to passivate metal surfaces by oxidation.

7. Other Chromium Compounds

Chromyl chloride [14977-61-8], CrO_2Cl_2, M_r 154.90, mp −96.5 °C, bp 116.7 °C, ϱ 1.912 g/cm^3, is a blood red oily liquid with a pungent smell, c_p 545 J kg^{-1} K^{-1}, $\Delta H_{298}°$ −567.8 kJ/mol, $S_{298}°$ 510 J kg^{-1} K^{-1}, heat of fusion 268 kJ/kg. In the temperature range 178–390 K the vapor pressure obeys the equation

$$\log p = -3340\, T^{-1} - 9.08 \log T + 35.06$$

where p is in hPa (mbar) and T in K. The liquid is electrically conducting.

Chromyl chloride is easily hydrolyzed; it is an extremely powerful oxidizing and chlorinating agent and, therefore, reacts with organic solvents, often violently. It decomposes in daylight within a week via CrO_2 and Cl_2 to form a series of chromium oxides and chlorides with a low degree of oxidation.

Industrially chromyl chloride is produced from chromium(VI) oxide and hydrogen chloride gas, with concentrated sulfuric acid (> 68%) being used primarily to bind the water of reaction. The higher density chromyl chloride is drained, distilled, and collected in cooled receptacles.

Chromyl chloride can also be prepared by using mixtures or molten baths of sodium chloride and alkali chromates or dichromates with fuming sulfuric acid. An elegant route involves the reaction of chromium(VI) oxide with liquid thionyl chloride; these reactants are converted quantitatively into chromyl chloride by elimination of SO_2.

Hexacarbonylchromium [13007-92-6], chromium hexacarbonyl, $Cr(CO)_6$, M_r 220.06, ϱ 1.77 g/cm^3, forms colorless, highly refractive crystals which belong to the orthorhombic system. The compound sublimes slowly even at room temperature; when heated in a sealed tube it melts at 149–150 °C. The boiling point has been calculated to be 147 °C. At 210 °C, explosive decomposition occurs. The enthalpy of formation is −1077 kJ/mol. Hexacarbonylchromium is somewhat soluble in chloroform and carbon tetrachloride but insoluble in benzene, ether, alcohol, and acetic acid. It is resistant to water and dilute acids at room temperature; there is no sign of attack even by concentrated hydrochloric acid or sulfuric acid in the cold, but concentrated nitric acid causes decomposition. No reaction with alkali occurs, but derivatives are produced with ammonia, pyridine, cyclopentadiene, and other organic ligands.

Attempts to synthesize hexacarbonylchromium directly from chromium and carbon monoxide have been unsuccessful. The action of carbon monoxide and a solution of phenylmagnesium bromide in ether on anhydrous chromium(III) chloride suspended in a mixture of benzene and ether produces intermediates that lead to hexacarbonylchromium upon addition of acid and distilling the ether in vacuo. Finely divided sodium can also be used for reduction instead of phenylmagnesium bromide. Yields of up to 80% are reported when very finely divided sodium reacts with anhydrous chromium(III) chloride below 0 °C with carbon monoxide at a pressure of about 6 MPa (60 bar); the reaction proceeds in the presence of diethylene glycol dimethyl ether; the mixture is subsequently hydrolyzed while the carbon monoxide pressure is maintained [73].

Applications quoted for hexacarbonylchromium include the tempering and hardening of metal surfaces by chromizing, use as a fuel additive, an intermediate in the preparation of organic chromium compounds, and a catalyst for oxo syntheses.

Chromium(II) Compounds. In general, compounds of divalent chromium are extremely unstable in air and, therefore, are of only minor importance in industry. Gaseous chromium(II) halides (for preparation, see [74]) are used to produce chromium diffusion layers on iron and nickel parts. Air stable sodium fluorochromate(II) is said to be suitable for the precipitation of metals in electroplating operations and for corrosion-protection coatings [75].

8. Analysis

Compounds that are insoluble in water and acids, for example, chromium ore, are solubilized by roasting with sodium peroxide [76] in the presence of soda. Chrome-tanned leathers are incinerated before being treated with sodium peroxide. Under these conditions, chromium is completely converted into water-soluble chromate(VI). The roasted mass is leached with hot water and after cooling sulfuric acid is added. Compounds of lower oxidation state that are soluble in water or acids are oxidized with ammonium persulfate to chromate(VI) in dilute boiling sulfuric acid in the presence of silver ions as catalysts [77]. Chloride ions must be removed beforehand. Traces of chloride ions are precipitated by adding a few drops of silver nitrate solution. The chromate(VI) solutions obtained in this way or solutions of chromium (VI) compounds that are soluble in water or sulfuric acid, are then titrated with iron(II) sulfate solution after phosphoric acid is added [78]. The end point of this titration can be determined potentiometrically or by means of a redox indicator, e.g., sodium diphenylamine-4-sulfonate [*6152-67-6*], [78]. Iodometric titration of chromate(VI) is also possible [79]. For very low concentrations, e.g., in the analysis of water, atomic absorption spectrometry (AAS) and, in recent years, inductively coupled plasma atomic emission spectrometry (ICP-AES) have proved successful [80]. Trace amounts of chromium are also detected photometrically at 540 nm after oxidation to chromate(VI) and addition of diphenylcarbazide [*140-22-7*] (formation of a reddish violet complex) [81].

9. Transportation, Storage, and Handling

Both hexavalent and trivalent chromium compounds are of commercial significance. Whereas the trivalent compounds and chromium(IV) oxide are not subject to regulations governing transport, the hexavalent compounds are classified internationally as dangerous goods. According to the IMDG code [82], the following classifications are applicable for hexavalent chromium compounds:

Compound	Hazard class	UN No.
$Na_2Cr_2O_7 \cdot 2\,H_2O$, crystals	6.1	3288
$Na_2Cr_2O_7$ solution	6.1	3287
$Na_2Cr_2O_7$ anhydr., crystals	5.1	3087
$K_2Cr_2O_7 \cdot 2\,H_2O$, crystals	6.1	3288
$(NH_4)_2Cr_2O_7$	5.1	1439
CrO_3, solid chromic acid	5.1	1463
Chromic acid solution	8	1755

During handling and storage, chromates, dichromates, and chromic acid must not be brought into contact with readily oxidizable substances.

A limit of 50 µg/L of chromium recommended by the WHO has been adopted in drinking water regulations. This figure was estimated by the U.S. Public Health Services from toxicity data. The value considers the No Observable Adverse Effect Level (NOAEL) from animal tests and the calculated Average Daily Intake (ADI) of drinking water consumers [83].

10. Environmental Protection

Wastewater. In Germany limits for the chromium content of wastewaters have been set up. The permitted concentrations depend on the type of industry. For the leather tanning industry, for instance, the concentration of Cr(IV) must not exceed 0.5 mg/L, and that of total chromium, 1 mg/L [84].

To avoid the pollution of water supply facilities, the risk of a release to surface and ground water has to be minimized during transportation, storage, and handling. In Germany, soluble hexavalent chromium compounds are classified in the highest water pollution class [85].

Exhaust Gas. In Germany, the emission of such carcinogenic chromates as calcium chromate, strontium chromate, chromium(III) chromate, and zinc chromate is limited to 1 mg/m^3 max. (as Cr) [86]; the emission of the other chromium compounds is limited to 5 mg/m^3 (as Cr).

Waste. Within the European Community, waste containing Cr(VI) compounds is considered hazardous [87]. It can only be deposited in dumps with particular safety measures and drainage-water treatment.

In Germany, sewage sludge containing not more than 900 mg per kilogram of chromium may be used as fertilizers in soils for agricultural purposes provided the chromium content of the soil does not exceed 100 mg chromium per kilogram soil before the application [88]. According to the U.S. Environmental Protection Agency, the chromium content of the soil should not be regarded as a limiting factor for the application of sewage sludge [89]. Studies carried out in Germany on the use of sewage sludge containing a few percent of chromium(III) hydroxide as fertilizer did not reveal any adverse effects on soil and plants [90].

11. Ecotoxicology

Insoluble inert chromium(III) oxide, Cr_2O_3, is the stable mineral end product into which chromium compounds are converted in the environment as a result of natural processes.

Inland Waters. The natural concentration of dissolved trivalent chromium in surface waters is $< 1-10$ µg/L [91]–[93]. Hexavalent chromium does not occur in natural fresh waters. If dissolved trivalent chromium from anthropogenic sources enters an inland body of water, it precipitates under neutral conditions as chromium hydroxide. This ages and becomes increasingly insoluble, and only a small proportion remains in solution. Chromium(VI) compounds entering inland waters are reduced to chromium(III) compounds by the natural content of organic substances in water and sediments [94], [95].

Table 9 summarizes toxicity data for chromium(III) and chromium(VI) compounds in relation to fish, bacteria, algae, daphnia, and plants.

Limits for Chromium in Fresh Water. Various countries have set guidelines for the content of chromium in fresh water which can be considered tolerable [110]–[112]. Whereas the U.S. recommendations distinguish between trivalent and hexavalent chromium, both the Canadian and English guidelines consider total chromium.

Soil. The chromium content of the earth's crust varies from a few mg/kg to ca. 50 % in chromite deposits. The average chromium content of rock is 83 mg/kg. The chromium content of soil varies very widely, depending on the geological conditions. The range is from 5 to 1500 mg/kg, and the average concentration about 50 mg/kg [113]. The highest chromium contents are found in soils over basaltic rock. Chromium in soils is linked to the relicts of primary minerals such as chromite or olivine or bound by adsorption to clay minerals. Chromium is found everywhere in the

Table 9. Ecotoxicology of chromium compounds

Species or medium	Chromium(III) compounds	Chromium(VI) compounds
Freshwater fish	$CrCl_3$: LC_{50} (48 h) minnows, static, 400 mg/L LC_{50} (48 h) trout, static, >1000 mg/L [98] LC_{50} (48 h) ides, static, 300 mg/L $Cr_2(OH)_2(SO_4) \cdot Na_2SO_4$ LC_0 (30 days) Brachydanio rerio, static >1000 mg/L [99], [100]	$K_2Cr_2O_7$: 14 day no observable effect level (NOEL) for zebra fish (Brachydanio rerio) at 50 mg/L [96] or 80 mg/L [97]
Bacteria	$KCr(SO_4)_2 \cdot 12 H_2O$: 100 mg/L proved not to be poisonous for bacteria of the genus Escherichia [99], [100]	$Na_2Cr_2O_7 \cdot 2 H_2O$: toxic limiting concentration for Pseudomonas putida 0.78 mg/L [101]
Algae	$KCr(SO_4)_2 \cdot 12 H_2O$: incipient injurious effect at 4–6 mg/L (Scenedesmus) [99]	$K_2Cr_2O_7$ (Scenedesmus subspicatus): EC_{10} (96 h) 0.5 mg/L [96]; 0.3–1.3 mg/L [102]; 1.8 mg/L [97] EC_{50} (96 h) 1.4 mg/L [96]; 1.6–4.7 mg/L [102]
Daphnia	$KCr(SO_4)_2 \cdot 12 H_2O$ marked injurious effect at 42 mg/L [99]	$K_2Cr_2O_7$ (21-day test): concentrations of > 0.1 mg/L markedly b decreased the production of offspring [97]; swimming ability of Daphnia magna: $EC_0 = 0.3$ mg/L; $EC_{50} = 0.9$ mg/L; $EC_{100} = 2.2$ mg/L [103]
Mammals (oral, rat)	$Cr(NO_3)_3$: LD_{50} 3250 mg/kg [104]	$Na_2Cr_2O_7 \cdot 2 H_2O$: LD_{50} 160 mg/kg [105]
Soil mobility	$Cr(NO_3)_3$: low, only certain complex compounds are biologically available [106]	$K_2Cr_2O_7$: low since chromate is strongly adsorbed on the podsol type of soil and is also reduced by organic soil constituents [97]
Bioaccumulation	$Cr(NO)_3$: plants, particularly the parts above ground, do not absorb much chromium; accumulation of Cr is prevented at the very beginning of the food chain [107], [108]	$K_2Cr_2O_7$: carp (Cyprinus carpio): no bioaccumulation detected [97]
Higher plants EC_{50} * **(14 d)**	$CrCl_3 \cdot 6 H_2O$: Avena sativa (oat) 560 mg of Cr/kg of soil [109] Brassica rapa (turnip) 230 mg of Cr/kg of soil [109]	$K_2Cr_2O_7$: oat: 27 mg/kg [96]; 32 mg/kg [97]; 96 mg/kg [102] turnip: 23 mg/kg [96]; 22 mg/kg [97]; 24 mg/kg [102]
Earthworm LC_{50} **(28 d)**	$CrCl_3 \cdot 6 H_2O$: not known	$K_2Cr_2O_7$: > 2000 mg/kg of soil [96] no lethal effects at 1000 mg/kg [102]

* EC = effective concentration; at EC_{50} growth is retarded by 50% compared with control.

soil of Germany [114], and levels of over 1100 mg/kg may be reached. Such levels are caused solely by geological conditions. All natural chromium is trivalent. Minerals of hexavalent chromium (eg. Crokoite) are extremely rare and can be found only in very few places worldwide.

Plants. As chromium occurs in soil only in the form of sparingly soluble chromium(III) compounds, which are available to plants only to a small extent, chromium is not enriched in the food chain [107], [108], [115]–[117].

Animals. Grazing animals are not subjected unduly to chromium since only small amounts enter the parts of plants above the ground [108]. Less than 1% of the chromium contained in plants is available to animal and human organisms. To cover essential needs, the chromium must be in a special, biologically suitable form [118].

12. Nutrition

Chromium is an essential trace element both for the animal and human organism. A number of physiological functions in animal and humans can no longer proceed normally in a state of chromium deficiency [119], [120]. Examples include the role of chromium in glucose-tolerance factor, a cofactor for the action of insulin [121], in cytochrome C reductase [122], and in other metabolic processes [123].

The chromium content in food is in the range of 0.01–10 mg Cr/kg dry substance. Consumers in the western world ingest an average of 60 µg Cr per day in their food. The U.S. Food and Nutrition Board (1980) recommends [117] as the safe, but also adequate amount (average dietary intake): 50 to 200 µg per day.

The amount of chromium necessary to humans is not always guaranteed in the dietary intake and even with additional uptake from the environment. Therefore, nutrition science literature contains more references to deficiencies in chromium supply than any toxic effects due to chronic excess supply of chromium [124], [125].

For chromium the gap between essential and toxic concentrations is particularly large. Mammals can tolerate 100 to 200 times the normal chromium content of their bodies without adverse effects [126].

13. Toxicology and Occupational Health

The properties of chromium and its compounds which determine the toxicity and the impact on the environment differ greatly depending on the valency state. Only hexavalent chromium compounds are biologically active. Metallic chromium and the trivalent compounds, including those in chromium ores, are neither irritating, mutagenic, nor carcinogenic [126].

Reviews of the toxicity of chromium and its compounds are given in [127]–[129].

Chromium Dioxide. In a single administration, chromium dioxide is resorptively nontoxic; the lethal dose is greater than 17000 mg/kg (rats, oral). In rabbits, chromium dioxide causes slight primary irritation of skin and mucous membranes.

Since any chromium dioxide in the normal atmosphere may contain traces of hexavalent chromium, care is to be taken during production and further processing to ensure that the concentration of chromium(VI) compounds at the workplace does not exceed the TLV for chromium(VI).

Employees of chromium dioxide plants are examined annually for chromium exposure. So far no indications have been found of any disease caused by chromium in chromium dioxide production plants.

Hexavalent Chromium Compounds. Particularly chromic acid and the alkali chromates corrode and irritate the skin and mucous membranes.

Acute Effects. Sodium monochromate and dichromate, potassium dichromate, and ammonium dichromate are soluble chromium(VI) compounds. These substances are toxic when swallowed, very toxic when inhaled, and dangerous to health when they come into contact with the skin. In humans, inhalation of these compounds, even at low doses, leads to coughing, chest pain, difficult breathing and fever.

Sodium monochromate and dichromate, potassium dichromate, and ammonium dichromate irritate the skin, eyes and respiratory tract. The risk of serious eye damage cannot be ruled out. In humans, repeated contact of hexavalent chromium compounds with the skin may lead to so-called chromium ulcers, but only on previously damaged parts of the skin.

Potassium dichromate and sodium monochromate show a skin-sensitizing effect in animal studies. There is evidence that soluble chromium(VI) compounds have a skin-sensitizing effect in humans. Skin allergy in humans is generally tested with potassium dichromate.

Hexavalent chromium compounds are capable of sensitizing the skin strongly, which may lead to chronic eczemas, particularly when the source is cement dust containing chromium(VI). In contrast, no sensitization of the respiratory tract occurs.

Uptake by the digestive system (mostly suicidal cases) causes serious intestinal inflammation, sometimes with loss of blood. Damage to the renal tubules occurs mainly after dermal absorption. This can lead to kidney failure if the spontaneous reduction capacity of plasma of about 2 ppm (20 min) is not sufficient to reduce chromium(VI) to nontoxic chromium(III). The administration of high doses of ascorbic acid facilitates this reduction, and chromium(III) is then excreted in the urine without causing kidney damage [130].

Chronic Effects. Repeated inhalation of ammonium dichromate and sodium dichromate primarily affects the respiratory tract, while no corresponding studies are available with sodium monochromate and potassium dichromate. There are no significant data on repeated oral intake of ammonium dichromate or sodium dichromate, while liver and kidney damage have been described for sodium monochromate and potassium dichromate. In humans, prolonged inhalation of the dust damages the nasal mucosa or even the nasal septum (septal perforation) [131].

Chronic irritation of the nasal mucous membrane can lead to its atrophy. Chronic bronchitis has been reported following long-term exposure to hexavalent chromium compounds, but this does not occur normally under today's manufacturing conditions [126]. Impairment of breathing has not been observed.

Mutagenicity. Sodium monochromate and dichromate and potassium dichromate show mutagenic activity in tests with bacteria, microorganisms, cell cultures, and in animal studies. Ammonium dichromate shows mutagenic activity in tests with bacteria.

All hexavalent chromium compounds are mutagenic, but trivalent chromium compounds do not have any mutagenic potential. In some cases, the hexavalent compounds must first be solubilized to show a mutagenic effect. This effect is counteracted by the reduction of chromium(VI) to chromium(III), for example, by adding body fluids or organ homogenates [132], [133].

Carcinogenicity. Many animal experiments have been carried out to test the carcinogenicity of chromium compounds. These include subcutaneous injection, inhalation, intratracheal instillation, and introduction of pellets into the bronchial tree [134]–[138].

Depending on solubility, chromium(VI) compounds show carcinogenic activity after long-term inhalation. For this reason, the MAK Commission classed them under Group III A2 (substances with carcinogenic potential in animal studies).

Chromium(VI) compounds have been classed by IARC under Group I (carcinogenic in humans) based on early epidemiological studies, although clear classification of the findings to the category of carcinogenic substances was not possible.

The EU has classified chromic acid and zinc chromates as category 1 carcinogens (human carcinogens), other chromates/dichromates ($SrCrO_4$, $CaCrO_4$, $Na_2Cr_2O_7$, $K_2Cr_2O_7$, K_2CrO_4) as category 2 carcinogens (animal carcinogens), and $PbCrO_4$ as category 3 carcinogen (suspected carcinogens) [139].

In the United States [140], water-soluble Cr(VI) compounds and insoluble Cr(VI) compound—both "not otherwise classified"—are considered A1, i.e., "confirmed humans carcinogens"

Reproductive Toxicity. In animal studies, sodium dichromate and potassium dichromate do not induce deformities at nonmaternotoxic doses. An embryotoxic effect has been described for potassium dichromate. There are no data on an embryotoxic effect of ammonium dichromate or sodium monochromate.

In animal studies, inhalation of sodium dichromate does not impair fertility. Animal studies with potassium dichromate and sodium monochromate at doses that are nontoxic to the animals yield no evidence of a fertility-impairing effect. There are no data on a fertility-impairing effect of ammonium dichromate.

Epidemiology. Epidemiological studies among workers in chrome-tanning plants and chrome(III) pigment manufacturing plants have proved trivalent chromium compounds to be noncarcinogenic [141].

Earliest indications of increased incidences of lung cancer among workers in the chromate manufacturing industry were observed in the 1930s. More evidence in support of this observation came from a large number of epidemiological studies carried out after 1948 [142], [143]. Studies carried out between 1948 and 1956 showed a 25- to 29-fold increase in the incidence of lung cancer. In contrast, investigations performed since 1979 reflect the effects of improved hygienic working conditions and production processes, for example, by avoiding the use of lime in the oxidative roasting of chromium ores (low-lime process). These measures have led to a convergence between the incidences of lung cancer observed in the workers and that expected in unexposed persons [131], [144] – [147].

An increased incidence of lung cancer was observed among persons employed in the chromate pigment industry only after exposure to zinc chromate. Lead chromate showed no carcinogenic effect [148] – [150]. Results of epidemiological studies carried out during the handling of chromic acid, particularly during chromium plating, are blurred by confounding factors; a statistically significant increase in the rate of lung cancer was not observed [134], [151]. An increased risk of lung cancer has also not been clearly established in the manufacture of ferrochromium [152], [153]. Occupational health care has been provided in the chromate-producing industry for several decades.

Classification. According to the EU List of Dangerous Substances [139] the following labelling has to be applied for

$K_2Cr_2O_7$, $Na_2Cr_2O_7 \cdot 2 H_2O$

T+	very toxic
N	dangerous for the environment
R49	may cause cancer by inhalation
R46	may cause heritable genetic damage
R21	harmful in contact with skin
R25	toxic if swallowed
R26	very toxic by inhalation
R37/38	irritating to respiratory sytem and skin
R41	risk of serious damage to eyes
R43	may cause sensitization by skin contact
R50/53	very toxic to aquatic organisms; may cause long-term adverse effects in the aquatic environment
S53	avoid exposure; obtain special instructions before use
S45	in case of accident or if you feel unwell, seek medical advice immediately (show label where possible)
S60	this material and its container must be disposed of as hazardous waste
S61	avoid release to the environment; refer to special instructions / safety data sheet

$(NH_4)_2Cr_2O_7$

E	explosive
T	very toxic

N	dangerous for the environment
R49–46	(see above)
R1	explosive when dry
R8	contact with combustible material may cause fire
R21–25–26–37/38–41–43–50/53	(see above)
S53–45–60–61	(see above)

Chromic acid (anhydride), CrO_3

O	oxidizing
T	toxic
C	corrosive
N	dangerous for the environment
R49	may cause cancer by inhalation
R8	contact with combustible material may cause fire
R25	toxic if swallowed
R35	causes severe burns
R43	may cause sensitization by skin contact
R50/53	very toxic to aquatic organisms; may cause long-term adverse effects in the aquatic environment
S53–45–60–61	(see above)

Occupational Exposure Limits. *Germany* [154]. The TRK values for Cr(IV) compounds, except those which are practically insoluble in water, are (as CrO_3 in total dust) for

Arc welding	0.1 mg/m^3 (corresponding to 0.05 mg/m^3 as Cr)
Manufacturing of soluble Cr(IV) compounds	0.1 mg/m^3 (0.05 as Cr)
Others	0.05 mg/m^3 (0.025 as Cr)

United States [140]. The TLV values (as Cr) are as follows;

Cr metal and Cr(III) compounds	0.5 mg/m^3
Water-soluble Cr(IV) compounds	0.05 mg/m^3
Insoluble Cr(IV) compounds	0.01 mg/m^3

Biological Monitoring. In addition to the classical assay for chromium in blood or urine, the degree of previous exposure to chromium can be estimated by determining the extent of bound chromium in erythrocytes [155].

14. Economic Aspects

The output of chromium ore is subject to considerable variations. Rich ores which are obtained easily by handpicking have declined. This shrinkage in output has to be counterbalanced by exploiting low-grade ores and upgrading them to saleable concentrates. The statistical documents are incomplete and contradictory. Table 10 shows the output of chromium ores according to various sources of information.

Table 10. Output of chromium ore and concentrates (in 1000 t) [156]–[159]

	1901–1982	1982	1983	1984
USSR*	55 000	3 400	3 400	3 400
South Africa	50 000	2 200	2 250	2 450
Zimbabwe	22 000	430	450	450
Turkey	21 000	370	360	400
Philippines	18 000	350	360	400
Albania*	15 000	1 200	1 350	1 350
India	7 900	340	360	410
Brazil	4 100	400	370	400
Finland	3 700	400	360	400
Cuba*	3 700	30	40	30
Iran	3 400	40	40	30
Greece	2 000	40	50	50
Madagascar	2 000	60	40	40
Remainder	800	40	70	90
Total worldwide	208 600	9 300	9 500	9 900

* Estimated.

Table 11. Estimated consumption of chromium ore in terms of products in 1984 [157], [160]

Product	World	United States
Ferrochromium, FeSi chromium, chromium metal, and foundry sands	72%	60%
Refractory bricks	12%	20%
Chemicals	17%	20%

Even the most pessimistic estimates of the worldwide reserves and the extraction possibilities in individual countries do not predict a chromium shortage before the end of this century. In addition, the huge lean-ore deposits in South Africa, Southeast Asia, and the United States, which are still completely untouched, afford a replacement for the rich ores being mined at present.

In Table 11 the consumption of chromium ore is broken down according to products. The refractory brick share is declining because Siemens–Martin steel is being increasingly supplanted by oxygen-blown steel produced in converters with a basic lining.

Table 12 shows chromium ore imports into the Federal Republic of Germany and Table 13 the share of individual applications in the consumption of chromium chemicals.

The starting material for virtually all chromium chemicals is sodium dichromate prepared from chromium ore. Table 14 summarizes the production capacities for $Na_2Cr_2O_7 \cdot 2 H_2O$.

In Table 15 the chromic acid production capacities are broken down according to countries.

Table 12. Imports of chromium ore* into the Federal Republic of Germany [161]

	1981	1982	1983
1000 t	268	244	247
Millions of DM	48.8	43.8	42.3
Price, DM/t	181.90	179.10	171.30

* Includes all types of chromium ore.

Table 13. Estimated share (in %) of individual applications in the consumption of chromium chemicals, 1985*

	United States	Worldwide
Pigments	22	19
Metal processing	15	24
Tanning	12	32
Wood preservation	26	11
Corrosion protection	5	3
Petroleum industry	5	2
Textile dyes	2	1
Catalysts	2	1
Video tapes	2	2
Remainder	9	5

* Excluding countries with state trading organizations.

Table 14. Estimated 1985 sodium dichromate capacities in 1000 t of $Na_2Cr_2O_7 \cdot 2 H_2O$ per year

Western Europe (EEC)	260
State trading countries	250
United States	220
Asia, Africa	110
Worldwide	840

Table 15. Estimated 1985 chromic acid capacities in 1000 t/a

Western Europe (EEC)	42
United States	40
State trading countries	32
Japan	15
Others	8
Worldwide	137

15. References

[1] Minerals Yearbook US Bureau of Mines, Washington, D.C., 1983.
[2] *Chromium minerals. World Survey of Production and Consumption.* ROSKILL Information Services Ltd., London 1972.
[3] *Mining J. (London),* Mining annual review 1985.
[4] Mineral Facts and Problems, US Bureau of Mines Bull. 650, 1970.
[5] H. Borchert: "Principles of the genesis and enrichment of chromite ore deposits," in *Methods of prospection for chromite,* OECD, Paris 1964, pp. 175–202.
[6] S. Janković: *Wirtschaftsgeologie der Erze,* Springer, Wien 1967.
[7] H. Kern: "Zur Geochemie und Lagerstättenkunde des Chroms und zur Mikroskopie und Genese der Chromerze," *Clausthaler Hefte zur Lagerstättenkunde und Geochemie der Mineralischen Rohstoffe,* no. 6, Borntraeger, Berlin 1968.
[8] T. Lukkarinen, L. Heikkilä: "Beneficiation of chromite ore, Kemi, Finland,". *10th International Processing Congress 1973,* Inst. Min. Met., London 1973.
[9] H. Schneiderhöhn: *Die Erzlagerstätten der Erde,* vol. **I,** Gustav Fischer, Stuttgart 1958.
[10] T. Shabad: *Basic industrial resources of USSR,* Columbia University Press, New York 1969.
[11] H. Strunz: *Mineralogische Tabellen,* Akadem. Verlagsges., Leipzig 1966.
[12] A. Sutulov: *Mineral resources and the economy of the USSR,* McGraw-Hill, New York 1973.
[13] H. D. B. Wilson (ed.): *Magmatic ore deposits, a symposium,* monograph 4, Economic Geology Publ. Co., Lancaster (Pa., USA) 1969.
[14] W. F. Rogers: *Composition and Properties of Oil Well Drilling Fluids,* Gulf Publ. Co., Houston, Texas, 1963, pp. 459–464.
[15] Bayer AG, DE-AS 1 533 076, 1966.
[16] Bayer AG, FR 1 531 069, 1967.
[17] Bayer AG, DE-OS 1 926 660, 1969.
[18] S. K. Cirkov, *Zh. Prikl. Khim. (Leningrad)* **13** (1940) 521–527.
[19] Associated Chemical Comp., DE-OS 1 467 298, 1963.
[20] G. N. Bogachov et al., *Khim. Promst. (Moscow)* 1961, 655.
[21] Produits Chimiques Ugine Kuhlmann, DE-OS 2 329 925, 1972.
[22] M. J. Udy: *Chromium,* vol. **1,** Reinhold, New York 1956.
[23] Bayer AG, DE-AS 1 533 077, 1966.
[24] C. K. Williams, US 2 695 215, 1950.
[25] R. N. Mittra, N. R. Dhar, *J. Indian Chem. Soc.* **9** (1932) 315–327.
[26] Pacific Bridge Co., US 2 431 075, 1945/47.
[27] I. G. Farben, DE 492 684, 1927.
[28] F. Wöhler, *Justus Liebigs Ann. Chem.* **111** (1859) 117–121.
[29] DuPont, US 2 885 365, 1959 (A. L. Oppegard).
[30] DuPont, US 2 956 955, 1960 (P. Arthur).
[31] T. J. Swoboda, P. Arthur, N. L. Cox, J. N. Ingraham, A. L. Oppegard et al., *J. Appl. Phys.* **32** (1961) 374–375.
[32] F. Hund, *Farbe + Lack* **78** (1972) 11–16.
[33] E. Köster, "Neue Werkstoffe für magn. Speicherschichten," in H. H. Mende (ed.): *Neuere magn. Werkstoffe und Anwendungen magn. Methoden,* Verlag Stahleisen, Düsseldorf 1983, pp. 149–170.
[34] B. L. Chamberland, *CRC Crit. Rev. Solid State Mater. Sci.* **7** (1977) 1–31.

[35] D. S. Chapin, J. A. Kafalas, J. M. Honig, *J. Phys. Chem.* **69** (1965) 1402–1409.
[36] L. Brewer, *Chem. Rev.* **52** (1953) 9.
[37] F. J. Darnell, W. H. Cloud, *Bull. Soc. Chim. Fr.* 1965, 1164–1166.
[38] BASF, unpublished, 1983.
[39] DuPont, US 3 034 988, 1962 (J. N. Ingraham, T. J. Swoboda).
[40] F. J. Darnell, *J. Appl. Phys.* **32** (1961) 1269.
[41] H. Y. Chen, D. M. Hiller, J. E. Hudson, C. J. A. Westenbroek, *IEEE Trans. Magn.* **MAG-20** (1984) 24–26.
[42] CNRS, DE-OS 3 209 739, 1982 (G. Demazeau, P. Maestro, M. Pouchard, P. Hagenmuller et al.).
[43] J. B. Goodenough, *Bull. Soc. Chim. Fr.* 1965, 1200–1207.
[44] D. S. Rodbell, J. M. Lommel, R. C. DeVries, *J. Phys. Soc. Jpn.* **21** (1966) 2430.
[45] J. Boháček, *J. Signalaufzeichnungsmater.* **8** (1980) 55–60.
[46] DuPont, US 3 512 930, 1970 (W. G. Bottjer, H. G. Ingersoll).
[47] BASF, DE-OS 2 749 757, 1977 (G. Väth, M. Ohlinger, H. J. Hartmann, M. Velic et al.).
[48] Montedison, EP 29 687, 1980 (G. Basile, G. Boero, E. Mello Ceresa, F. Montino).
[49] W. Ostertag, W. Stumpfi, R. Falk, M. Ohlinger, *Elektronik-Anzeiger* **11** (1972) 225–227.
[50] Montedison, DE 2 648 305, 1976 (G. Basile, A. Mazza, M. Spinetta).
[51] DuPont, US 3 278 263, 1966 (N. L. Cox).
[52] G. R. Cole, L. C. Bancroft, M. P. Chouinard, J. W. McCloud, *IEEE Trans. Magn.* **MAG-20** (1984) 19–23.
[53] P. J. Flanders, *IEEE Trans. Magn.* **MAG-12** (1976) 348–355.
[54] Bayer AG, DE 1 203 748, 1961.
[55] Aktjubinskij Sawod Chromowych Isdelij, DE-OS 2 119 450, 1971.
[56] Bayer AG, DE 1 065 394, 1958.
[57] Diamond Alkali Co., US 2 993 756, 1959.
[58] Diamond Alkali Co., DE-OS 2 018 602, 1970.
[59] Diamond-Shamrock Corp., DE-OS 3 020 260, 1980.
[60] Diamond-Shamrock Corp., DE-OS 3 020 261, 1980.
[61] Diamond-Shamrock Corp., DE-OS 3 020 280, 1980.
[62] A. Werner, P. Pfeiffer: *Neue Anschauungen auf dem Gebiet der anorganischen Chemie*, 5th ed., Vieweg, Braunschweig 1923.
[63] Griesheim Elektron, DE 369 816, 1920.
[64] R. B. Norden, *Chem. Eng.* **63** (1956) 308–311.
[65] Degussa, FR 817 502, 1936.
[66] Pittsburgh Plate Glass Co., US 2 185 218, 1938.
[67] I. G. Farben, DE 514 571, 1925.
[68] Bayer AG, DE 1 467 327, 1963.
[69] Farbenfabriken Meister Lucius Brüning, DE 103 860, 1898.
[70] M. J. Udy, US 1 739 107, 1925.
[71] M. J. Udy, GB 546 681, 1940.
[72] Ass. Chem. Comp., DE-AS 1 467 298, 1964.
[73] Ethyl Corp., DE-AS 1 159 913, 1959.
[74] G. Brauer: *Handbuch der präparativen Anorganischen Chemie*, 2nd ed., vol. **2**, Enke Verlag, Stuttgart 1962.
[75] DuPont, DE-AS 1 272 909, 1960.

[76] I. M. Kolthoff, P. J. Elving (eds.): *Treatise on Analytical Chemistry*, Part II, vol. **8**, Interscience, New York-London 1963, pp. 301 ff.

[77] I. M. Kolthoff, P. J. Elving (eds.): *Treatise on Analytical Chemistry*, Part II, vol. **8**, Interscience, New York-London 1963, p. 326.

[78] I. M. Kolthoff, P. J. Elving (eds.): *Treatise on Analytical Chemistry*, Part II, vol. **8**, Interscience, New York-London 1963, pp. 327–328.

[79] I. M. Kolthoff, P. J. Elving (eds.): *Treatise on Analytical Chemistry*, Part II, vol. **8**, Interscience, New York-London 1963, pp. 330

[80] C. Veillon, K. Y. Patterson, N. A. Bryden, *Analytica Chim. Acta* **164** (1984) 67–76. A. G. Cox, G. Cook, C. W. McLeod, *The Analyst* **110** (1985) 331–333.

[81] I. M. Kolthoff, P. J. Elving (eds.): *Treatise on Analytical Chemistry*, Part II, vol. **8**, Interscience, New York-London 1963, pp. 338–339.

[82] *International Maritime Dangerous Goods Code (IMDG Code)*.

[83] EPA: *Ambient Water Quality Criteria for Chromium*, Washington, D.C., NTIS No. PB 81-117647, 1980.

[84] Anhang 25 der Abwasserverwaltungsvorschrift nach § 7A des Wasserhaushaltsgesetzes, veröffentlicht als Beilage 164a des Bundesanzeiger vom 31.7. 1996, p. 30.

[85] Verwaltungsvorschrift wassergefährdende Stoffe (VwVwS), Verwaltungsvorschrift vom 18. April 1996 zum Wasserhaushaltgesetz über die Einstufung wassergefährdender Stoffe in Wassergefährdungsklassen.

[86] *Technische Anleitung zur Reinhaltung der Luft (TA Luft)*, Feb. 28, 1986.

[87] *EEC Directive on poisonous and hazardous wastes*, 1991 (91/689/EEC).

[88] Klärschlammverordnung, April 15, 1992, *BGBl* 1992, p. 912–934.

[89] EPA, Municipal Construction Div.: "Application of Sewage Sludge to Cropland. Appraisal of Potential Hazards of the Heavy Metals to Plants and Animals," NTIS PB-264015, EPA-430/9-76-013, Nov. 15, 1976.

[90] G. Schmid, W. Pauckner, *Leder* **35** (1984) 165–171.

[91] C. N. Durfor, E. Becker, *U.S. Geol. Survey Water Supply, Paper No.1812*, 1964.

[92] NAS-U.S. *Geochemistry and the Environment*, Washington, D.C., U.S. Government Printing Office, 1974.

[93] RIWA annual report 1983, Part A: *Der Rhein*, Aug. 1984, p. 60/61.

[94] Industrial Health Foundation (IHF): "Kinetic Reactions of the Fate of Chromium in the Environment," IHF, Pittsburgh 1988.

[95] C. A. Harzdorf, *Int. J. Environ. Analyt. Chem.* **29** (1987) 249. C. A. Harzdorf: *Spurenanalytik des Chroms*, Thieme Verlag, Stuttgart 1990.

[96] P. Friesel et al. (Bundesgesundheitsamt): *Überprüfung der Durchführbarkeit und der Aussagekraft der Stufe 1 und 2 des Chemikaliengesetzes*, Umweltbundesamt (UBA) Forschungsbericht 1984.

[97] H. W. Marquart et al. (Bayer AG), same publication as [96].

[98] B. Hamburger, H. Häberling, H. R. Hirtz, *Arch. Fischereiwiss.* **28** (1977) 45–55.

[99] G. Bringmann, R. Kühn, *Gesund. Ing.* **80** (1959) no. 4, 115–120.

[100] D. M. M. Adema, A. de Ruiter: "The influence of "Basisches Chromsulfat" on the early life stages of *Brachydanio rerio* (Els-test)," Netherlands Organization for Applied Scientific Research, Report R 89/399, Delft 1990.

[101] G. Bringmann, R. Kühn, *Haustech. Bauphys. Umwelttech. Gesund. Ing.* **100** (1979) 250.

[102] W. Kördel et al. (Fraunhofer-Institut für Toxikologie und Aerosolforschung), same publication as [96].

[103] G. Bringmann, R. Kühn, *Z. Wasser Abwasser Forsch.* **15** (1982) 1–6.

[104] NIOSH, Registry of Toxic Effects of Chemical Substances 1985.
[105] Unpublished results, Bayer AG, 1984.
[106] R. A. Griffin et al., *J. Environ. Sci. Health Part A* **A 12** (1977) 431–449.
[107] H. Rasp, *Leder* **32** (1981) 188–203.
[108] D. Sauerbeck, *Zeitschrift des Verbandes Deutscher Landwirtschaftlicher Untersuchungs- und Forschungsanstalten (VDLUFA)*, Landwirtschaftliche Forschung, special issue no. 39, Kongreßband 1982, pp. 108–129.
[109] Unpublished results, Bayer AG, May 1985.
[110] U.S. Water quality criteria documents, U.S. Federal Register 45, No. 231, Nov. 28 1980, p. 79331.
[111] A. V. Pawliz, R. A. Kent, U. A. Schneider, C. Jefferson, "Canadian Water Quality Guidelines for Chromium," *Environmental Toxicology and Water Quality* **12** (1997) 123–183.
[112] S. M. Hunt, S. Hedgecott: Revised Environmental Quality Standards for Chromium in Water, WRc Medmenham, Report DoE 2858(P), 1994.
[113] H. J. M. Bowens: *Environmental Chemistry of the Elements*, Academic Press, London 1979.
[114] *Chrom, Bachsedimente (creek sediments)*, Bundesanstalt für Geowissenschaften und Rohstoffe, Hannover 1984.
[115] Umweltbundesamt, Jahresbericht 1982, Berlin 1983, pp. 33, 35.
[116] D. J. Swaine, R. L. Mitchell, *J. Soil Sci.* **11** (1960) no. 2, 347–368.
[117] S. Langard: *Biological and Environmental Aspects of Chromium*, Elsevier, Amsterdam 1982, p. 58.
[118] S. Langard, see [117], p. 60, p. 62.
[119] R. A. Anderson, *Sci. Total Environ.* **17** (1981) 13–28.
[120] W. Mertz, *Proc. Nutr. Soc.* **33** (1974) 307–313.
[121] W. Merz, K. Schwarz, *Am. J. Physiol.* **196** (1959) 614–618.
[122] B. L. Horecker, E. Stotz, T. R. Hogness, *J. Biol. Chem.* **128** (1939) 251.
[123] W. Mertz, *Physiol. Rev.* **49** (1969) 163.
[124] F. J. Diehl in *Zeitschrift des Verbandes Deutscher Landwirtschaftlicher Untersuchungs- und Forschungsanstalten (VDLUFA), Landwirtschaftliche Forschung* (1982) special issue no. 39, 58.
[125] Deutsche Forschungsgemeinschaft: *Forschungsbericht Schadstoffe im Wasser*, Deutsche Forschungsgemeinschaft, Bonn 1982, p. 411.
[126] D. Henschler: *Gesundheitsschädl. Arbeitsstoffe*, Verlag Chemie, Weinheim 1973.
[127] S. Fairhurst, C. A. Minty: "Toxicity of chromium and Inorganic Chromium Compounds," *Toxicity Review*, vol. **21**, HSE, London 1989.
[128] Environmental Health Criteria 61, Chromium, WHO, Geneva 1988.
[129] Toxicological Profile for Chromium, TP-92/08, U.S. Department of Health and Human Services, Public Health Service, Agency for Toxic Substances and Disease Registry, Atlanta, 1993.
[130] U. Korallus, C. Harzdorf, J. Lewalter, *Int. Arch. Occup. Environ. Health* **53** (1984) 247–250.
[131] U. Korallus, H. Lange, A. Neiss, E. Wüstefeld, T. Zwingers, *Arb. Med., Soz. Med., Praev. Med.* **17** (1982) 159–167.
[132] F. L. Petrilli, S. de Flora, *Appl. Environ. Microbiol.* **33** (1977) 805–809.
[133] F. L. Petrilli, S. de Flora in: *Mutagens in our Environment*, A. R. Liss, New York 1982, pp. 453–464.
[134] EPA Health Assessment Document for Chromium, August 1984 (EPA-600/8-83-D14F).
[135] W. C. Hueper, W. W. Payne, *Arch. Environ. Health* **5** (1962) 445–462.
[136] L. S. Levy, Fac. of Sc., University of London, 1975.

[137] NIOSH (1975) Criteria for a recommended standard. Occ. exposure to chromium(VI), US DHEW, Washington, D.C.
[138] D. Steinhoff, C. Gad, G. T. Hatfield, U. Mohr, *Exp. Pathol.* **30** (1986) no. 3, 129–141.
[139] EC Commission Directive 96/64/EC of July 30, 1996 on the classification, packing and labelling of dangerous substances; 22nd adaptation of EEC Directive 67/548.
[140] American Conference of Governmental Industrial Hygienists (ACGIH): Threshold Limit Values, Biological Exposure Indices, ACGIH, Cincinatti, OH 1996.
[141] U. Korallus, H. Ehrlicher, E. Wüstefeld, *Arb. Med., Soz. Med., Praev. Med.* **3** (1974) 51–54;76–79; 248–252.
[142] W. Machle, F. Gregorius, *U.S. Publ. Hlth. Rep.* **63** (1948) no.no. 35, 1114–1127.
[143] T. Mancuso, W. C. Hueper, *Ind. Med. Surg.* **20** (1951) 358–363.
[144] M. R. Alderson, N. S. Rattan, L. Bidstrup, *Br. J. Ind. Med.* **38** (1981) 117–124.
[145] R. B. Hayes, A. M. Lilienfeld, L. M. Snell, *Int. J. Epidemiol.* **8** (1979) 365–374.
[146] W. J. Hill, W. S. Ferguson, *JOM J. Occup. Med.* **21** (1979) 103–106.
[147] H. Satoh, Y. Fukuda, K. Teric, N. Chike Katsumo, *JOM J. Occup. Med.* **23** (1981) 835–838.
[148] J. M. Davies, *Br. J. Ind. Med.* **41** (1984) 148–169.
[149] R. Frentzel-Beyme, *J. Cancer Res. Clin. Oncol.* **105** (1983) 183–188.
[150] S. Langard, T. Norseth, *Br. J. Ind. Med.* **32** (1975) 62–65.
[151] H. Royle, *Environ. Res.* **10** (1975) 141–163.
[152] G. Axelsson, R. Rylander, A. Schmidt, *Br. J. Ind. Med.* **37** (1980) 121–127.
[153] S. Langard, A. Andersen, B. Gylseth, *Br. J. Ind. Med.* **37** (1980) 114–180.
[154] Deutsche Forschungsgemeinschaft: *MAK- und BAT-Werte-Liste 1996*, Mitteilung 32, VCH Verlagsgesellschaft, Weinheim 1996.
[155] J. Lewalter, U. Korallus, C. Harzdorf, H. Weidemann, *Internat. Arch. Occ. Environ. Health* **55** (1985) 305–318.
[156] U.S. Geological Survey Circular 930-B, 1984.
[157] U.S. Bureau of Mines: Mineral Commodity Summaries 1984.
[158] U.S. Bureau of Mines: Mineral Commodity Summaries 1985.
[159] T. Power: "Chromite – the non metallurgical markets," *Industrial Minerals* 1985, April, 17–51.
[160] Minerals Bureau, International Report No. 86, Braamfontein.
[161] Statistisches Bundesamt Wiesbaden, Fachserie: Außenhandel Reihe 2, Außenhandel nach Waren und Ländern (Spezialhandel) 1983.

Clays

HAYDN H. MURRAY, Indiana University, Department of Geology, Bloomington, Indiana 47405, United States

1.	Introduction 1389	4.3.	Hormite 1415
2.	Structure and Composition of Clay Minerals 1390	5.	Properties and Uses 1417
		5.1.	Kaolin 1417
3.	Geology and Occurrence of Major Clay Deposits 1394	5.1.1.	Physical Properties 1417
		5.1.2.	Paper 1419
3.1.	Kaolin 1396	5.1.3.	Paint 1423
3.2.	Smectite 1400	5.1.4.	Ceramics 1424
		5.1.5.	Rubber 1424
3.3.	Hormite 1404	5.1.6.	Plastics 1425
3.4.	Miscellaneous Clays 1406	5.1.7.	Ink 1425
4.	Mining and Processing 1407	5.1.8.	Catalysts 1425
4.1.	Kaolin 1407	5.1.9.	Fiberglass 1425
4.1.1.	Dry Process 1408	5.2.	Smectite 1426
4.1.2.	Wet Process 1408		
4.1.3.	Combined Dry and Wet Processing 1412	5.3.	Hormite 1428
		6.	Environmental Aspects 1429
4.1.4.	Special Processes 1412	7.	Production and Consumption . 1430
4.2.	Smectite 1414	8.	References 1431

1. Introduction

The term *clay* has a double meaning and, therefore, should be defined when it is used. Clay is used both as a rock term and as a particle size term. As a rock term, clay is used for a natural, earthy, fine-grained material composed largely of a limited group of crystalline minerals known as the clay minerals. As a particle size term, clay is used for the category that includes the smallest particles. Soil investigators and mineralogists generally use 2 µm as the maximum size, although the Wentworth scale [1] defines clay as material finer than 4 µm. GRIM [2] uses the term clay material for any fine-grained, natural, earthy, argillaceous material; in this way, the term can include clays, shales, or argillites, and some soils if they are argillaceous.

Clay is an abundant raw material and has an amazing variety of uses and properties that depend on the clay mineral composition and other factors enumerated by GRIM [2]. These factors are clay mineral composition, nonclay mineral composition, presence of organic material, the type and amount of exchangeable ions and soluble salts, and texture.

Clays are comprised of certain groups of hydrous aluminum, magnesium, and iron silicates that may contain sodium, calcium, potassium, and other ions. These silicates are called the clay minerals, and the major clay mineral groups are kaolins, smectites, illites, chlorites, and hormites. The specific clay minerals are identified by several techniques, including X-ray diffraction [3], differential thermal analysis [4], electron microscopy [5], and infrared spectrometry [6]. Identification and quantification of the clay minerals and nonclay minerals present in a clay material is important because the uses and engineering properties are controlled largely by these two factors.

Clays and clay minerals are important geologically, industrially, and agriculturally; they are important to the engineer in building foundations, tunnels, road cuts and fills, and dams. There are several scientific groups whose work is largely devoted to clays, including AIPEA (Association Internationale pour l'Étude des Argiles), Clay Minerals Group of the Mineralogical Society of Great Britain, CMS (Clay Minerals Society, United States), European Clay Minerals Society, Clay Minerals Society of Japan, and the Australian Clay Minerals Society. Major publications include the *Proceedings of International Clay Conferences*, published by AIPEA; *Clays and Clay Minerals*, published by CMS; *Clay Minerals*, published by the Mineralogical Society of Great Britain; and *Clay Mineral Science*, published by the Japanese clay group.

2. Structure and Composition of Clay Minerals

GRIM [2] proposed a tentative classification based on the structural attributes of the various clay minerals. This classification is relatively simple:

I) Amorphous
  Allophane group [*12172-71-3*]

II) Crystalline

  A) Two-layer types (sheet structures composed of units of one layer of silica tetrahedrons and one layer of alumina octahedrons)
   1) Equidimensional
     Kaolin group [*1332-58-7*]
       kaolinite [*1318-74-7*], dickite [*1318-45-2*], nacrite [*12279-65-1*]
   2) Elongate
     halloysite [*12244-16-5*]

  B) Three-layer types (sheet structures composed of two layers of silica tetrahedrons and one central dioctahedral or trioctahedral layer)

1) Expanding lattice
   Smectite group [*12199-37-0*]
   a) Equidimensional
      montmorillonite [*1318-93-0*], sauconite [*12424-32-7*], vermiculite [*1318-00-9*]
   b) Elongate
      nontronite [*12174-06-0*], saponite [*1319-41-1*], hectorite [*12173-47-6*]
2) Nonexpanding lattice
   Illite group [*12173-60-3*]

C) Regular mixed-layer types (ordered stacking of alternate layers of different types)
   Chlorite group

D) Chain-structure types (hornblende-like chains of silica tetrahedrons linked together by octahedral groups of oxygens and hydroxyls containing aluminum and magnesium ions)
   Hormite group
      palygorskite [*12174-11-7*] (attapulgite [*1337-76-4*]), sepolite [*15501-74-3*]

The materials described as *allophane* are amorphous [7] and contain variable proportions of silica, alumina, and water. Allophane is commonly associated with halloysite and generally has a glassy appearance.

The kaolin minerals [8] are hydrous aluminum silicates and have the approximate composition $2\,H_2O \cdot Al_2O_3 \cdot 2\,SiO_2$. *Kaolinite* is the most common of the kaolin minerals. The structure of kaolinite consists of a single silica tetrahedral sheet and a single alumina octahedral sheet combined to form the kaolin unit layer (Fig. 1). These unit layers are stacked on top of each other. Variations in orientation of the unit layers in stacking cause differences in the kaolin mineral itself and lead to the differentiation of *nacrite* and *dickite* in the kaolin group [2]. Platy and vermicular forms of kaolinite are shown in the electron micrograph in Figure 2.

Halloysite is an elongate mineral that occurs in a hydrated and dehydrated form [7]. BATES et al. [9] proposed that the hydrated form of halloysite consists of curved sheets of kaolin unit layers (Fig. 3), which is consistent with electron micrographs (Fig. 4).

Smectite is composed of units consisting of two silica tetrahedral sheets with a central alumina octahedral sheet. The lattice has an unbalanced charge because of substitution of alumina for silica in the tetrahedral sheet and iron and magnesium for alumina in the octahedral sheet. Because of this and because oxygen layers are adjacent when these unit layers are stacked, the attraction holding these layers together is weak, and cations and polar molecules can enter between the layers and cause expansion (Fig. 5). Sodium *montmorillonite* is a smectite in which sodium and water molecules are the interlayer material, and calcium montmorillonite is a smectite in which calcium and water molecules are the interlayer material. Figure 6 shows an electron micrograph of a sodium montmorillonite.

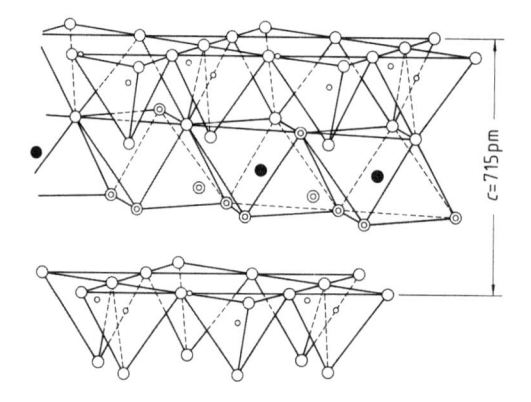

Figure 1. Structure of kaolinite
○ Oxygen; Hydroxyl; ○ Silicon tetrahedrally coordinated; ● Aluminum octahedrally coordinated

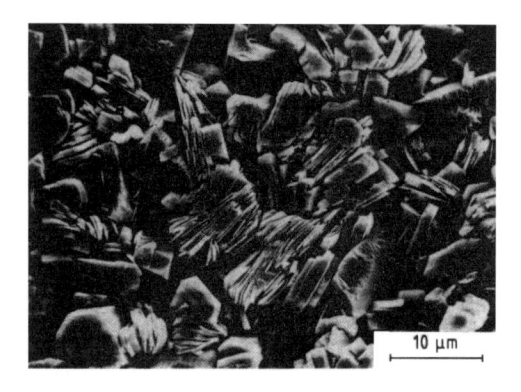

Figure 2. Scanning electron micrograph of kaolinite stacks and plates

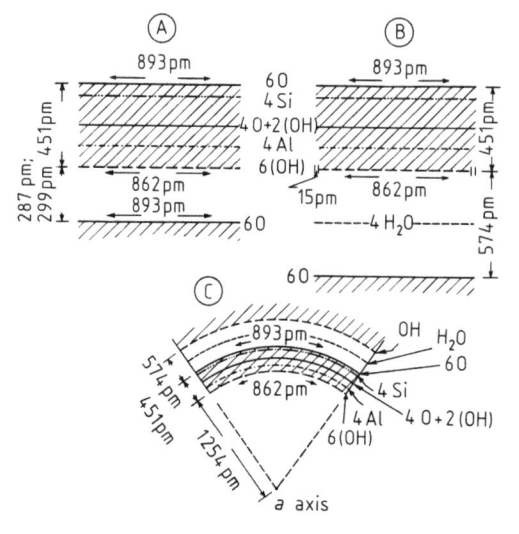

Figure 3. Structure of halloysite
A) Dehydrated halloysite; B) Hydrated halloysite; C) Proposed curved structure of halloysite

Figure 4. Scanning electron micrograph of elongate halloysite growing on feldspar

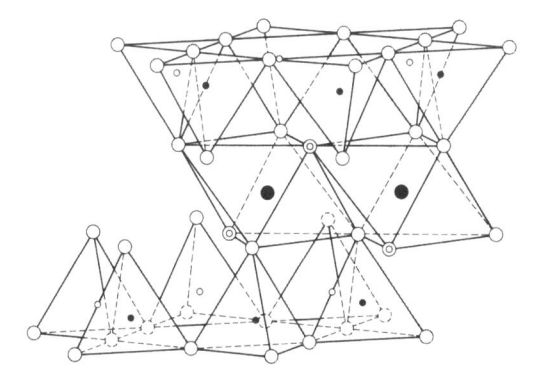

Figure 5. Structure of smectite
○ Oxygen; Hydroxyl; ● Aluminum, iron, magnesium; ○● Silicon, occasionally aluminum

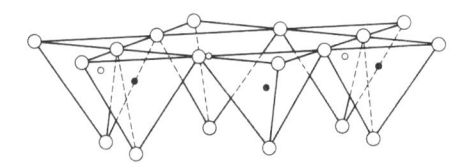

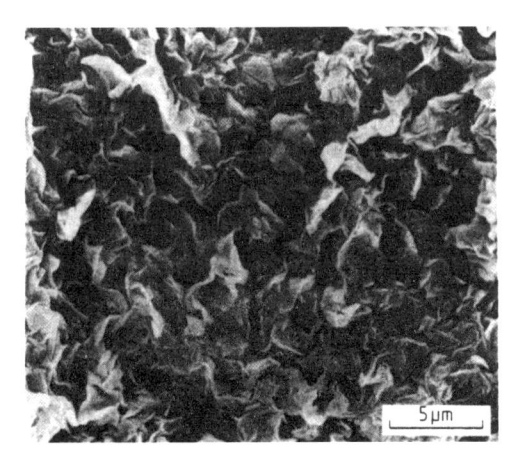

Exchangeable cations, nH_2O

Figure 6. Scanning electron micrograph of a sodium montmorillonite (Wyoming)

Sauconite is a smectite in which zinc has replaced some of the aluminum. *Nontronite* results when iron replaces the aluminum, and *saponite* results when magnesium replaces aluminum. *Hectorite* results when magnesium replaces aluminum and lithium is the ion between the layers.

Vermiculite is an equidimensional expandable clay mineral. The structure [10] consists of a layer made up of two silica tetrahedral sheets with a central magnesium–iron octahedral sheet separated by two layers of water molecules and adsorbed magnesium and calcium ions. Vermiculite differs from montmorillonite in that the layers do not expand as much and show less random stacking [2].

Illite is a general term for the micalike clay mineral [11]. The basic structural unit is a layer composed of two silica tetrahedral sheets and a central octahedral sheet. The unit is similar to that of montmorillonite except that more aluminum ions replace silicon in the tetrahedral sheet, which causes a charge deficiency. This charge deficiency is balanced by potassium ions that act as a bridge between the unit layers and bind them together so that illites are nonexpandable (Fig. 7).

Another common group of clay minerals is the *chlorite* minerals. The structure consists of alternate mica layers and brucite layers (Fig. 8). The brucite-like layer can be a magnesium, aluminum, or iron hydroxide, or a combination of all three. The various members of the chlorite group are differentiated on the kind and amount of substitution in the brucite and the mica layers [12].

The hormite clay minerals *palygorskite* (attapulgite) and *sepiolite* have a chainlike structure (Fig. 9). Palygorskite and attapulgite [13] are synonymous, palygorskite being the preferred name. The differentiation between sepiolite and palygorskite is the substitution of iron for magnesium, which causes some minor structural differences. These minerals consist of double silica tetrahedral chains linked together by octahedral oxygen and hydroxyl groups containing aluminum and magnesium ions. An electron micrograph of palygorskite is shown in Figure 10.

Mixed-layer clays are mixtures of the unit layers on a layer-by-layer basis and are relatively common in nature. Illite–smectite, illite–chlorite, smectite–chlorite, vermiculite–illite, and even smectite–kaolinite have been described. Mixed-layer clays with both random and regular layering were treated by REYNOLDS [14].

3. Geology and Occurrence of Major Clay Deposits

Certain clays, such as kaolins, smectites, and hormites, can be found in relatively pure occurrences or can be beneficiated to relatively high purity. These particular clays have special industrial applications. Illites and chlorites are usually in shales or clays that are used for structural clay ceramics; they are not utilized where the application depends on the particular physical or chemical properties of the individual clay mineral.

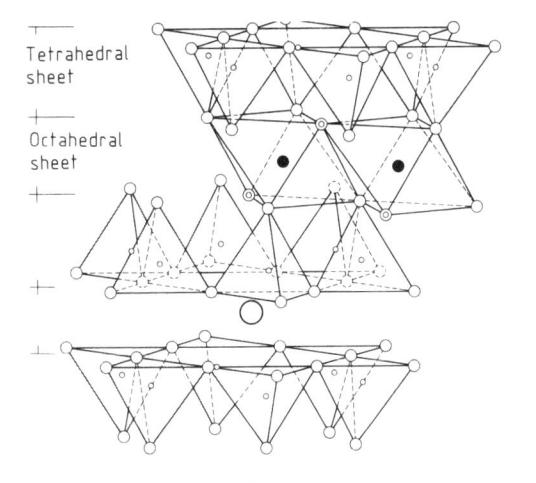

Figure 7. Structure of illite
◯ Potassium; ◯ Oxygen; ● Aluminum (only 2/3 of available positions filled); Hydroxyl; ◯ Silicon (1/4 replaced by aluminum)

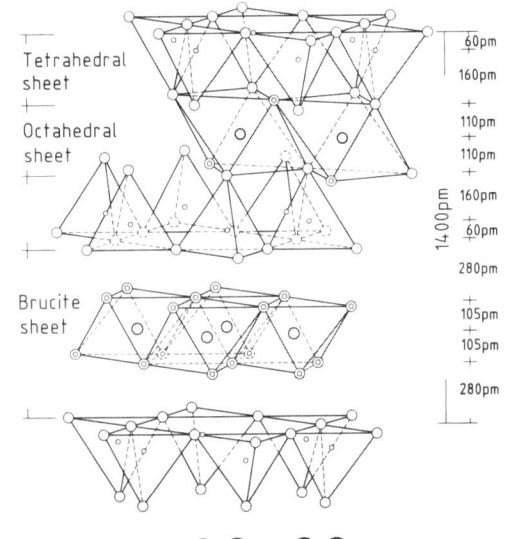

Figure 8. Structure of chlorite
◯ Magnesium, iron (all available positions filled); Hydroxyl; ◯ Oxygen; ° Silicon (1/4 replaced by aluminum)

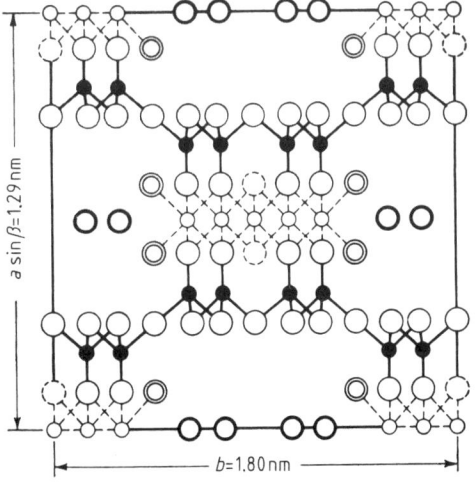

Figure 9. Structure of palygorskite (attapulgite), $(OH_2)_4 (OH)_2 Mg_5 Si_8 O_{20} \cdot 4 H_2O$
◯ H_2O; Hydroxyl; ◯ Magnesium or aluminum; OH_2; ◯ Oxygen; ● Silicon

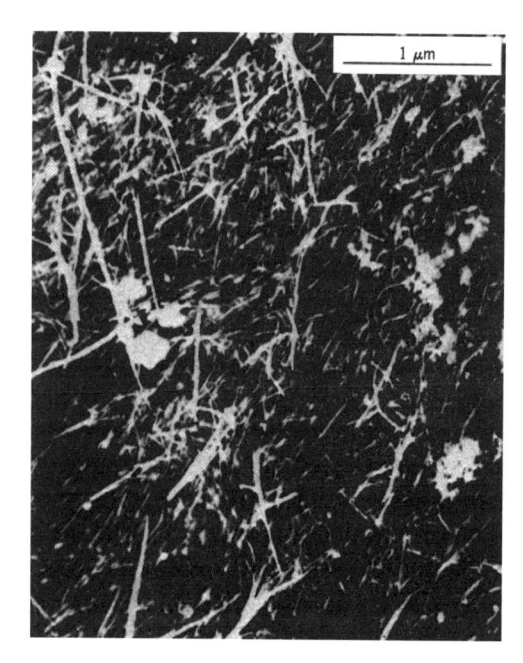

Figure 10. Scanning electron micrograph of palygorskite (attapulgite)

3.1. Kaolin

Kaolin is a rock term, a clay mineral group, and an industrial mineral commodity. It is used interchangeably with the term china clay. The term kaolin is derived from the Chinese word *kauling*, meaning high ridge, the name for a hill near Jauchau Fu in China where a kaolin clay was mined for many centuries. Kaolin is a clay that

1) consists of substantially pure kaolinite or related clay minerals;
2) is naturally or can be beneficiated to be white or nearly white;
3) fires white or nearly white; and
4) is amenable to beneficiation by known methods to make it suitable for use in whiteware, paper, rubber, paint, and similar uses.

The term is applied without direct relation to the purity of deposits. Many large kaolin deposits are essentially pure and require little concentration during preparation for market. Most, however, are slightly off-color and require beneficiation to improve the brightness and whiteness and to control the particle size. Other kaolins contain as little as 10% kaolinite; these must be washed and concentrated to recover a marketable product.

Kaolin deposits are classed as primary or secondary. Primary or residual kaolins are those that have formed by the alteration of crystalline rocks such as granite and remain in the place where they formed. Secondary deposits of kaolin are sedimentary and have been transported from their place of origin and deposited in beds or lenses associated

Table 1. Geologic eras, periods, and epochs

Eras	Periods	Epochs	Years ago
Cenozoic	Quaternary	Holocene	
		Pleistocene	
	Tertiary	*Pliocene*	
		Miocene	
		Oligocene	
		Eocene	
		Paleocene	70×10^6
Mesozoic	*Cretaceous*	*Upper*	
		Lower	
	Jurassic		160×10^6
	Triassic		
Paleozoic	Permian		230×10^6
	Pennsylvanian		
	Mississippian		
	Devonian		
	Silurian		
	Ordovician		
	Cambrian		

The terms in italics are mentioned in the text.

with other sedimentary rocks such as sands. Kaolins occur on every continent, but only a few (Georgia, England, Brazil, Czechoslovakia, East Germany, Spain, the Soviet Union, and Australia) can be mined and beneficiated to meet the rigid industrial specifications required for use as a filler, extender pigment, or ceramic raw material. Only those in Georgia, England, and Brazil can meet specifications for coating clays for paper. Over half of the world's kaolin is produced in the United States and England.

The largest kaolin deposits in the world are the sedimentary deposits that occur in Georgia and South Carolina in the United States. These deposits are late Cretaceous and early Tertiary [15] (Table 1) and occur in lenses and beds in relatively coarse sands (Fig. 11). The kaolin was derived from granitic rocks on the Piedmont plateau that were altered by weathering over a long period of geologic time. During the Cretaceous period, the Piedmont was uplifted, and the weathered residue consisting of kaolinite, quartz, feldspar, and mica was transported by streams onto a deltaic platform where it was deposited as sinuous lenticular deposits in lakes, oxbows, and swamps along the Cretaceous shoreline. The environment in which the deposition took place was similar to the one now existing along the coast of Georgia and South Carolina, with deltas, offshore bars, swampy, muddy lagoons, and winding estuaries. The deposits are concentrated in a belt ca. 30 km wide extending from central Georgia northeastward into South Carolina south of the fall line (Fig. 12).

These minable sedimentary deposits range in thickness from 3 to 15 m and extend laterally from a few hundred meters to kilometers in length and width. There are hundreds of millions of tons of reserves in this area. In 1984 ca. 7 000 000 t of kaolin were produced from this area. The Georgia kaolins generally contain 85–95% kaolinite. The other minerals present are quartz, muscovite, biotite, smectite, ilmenite,

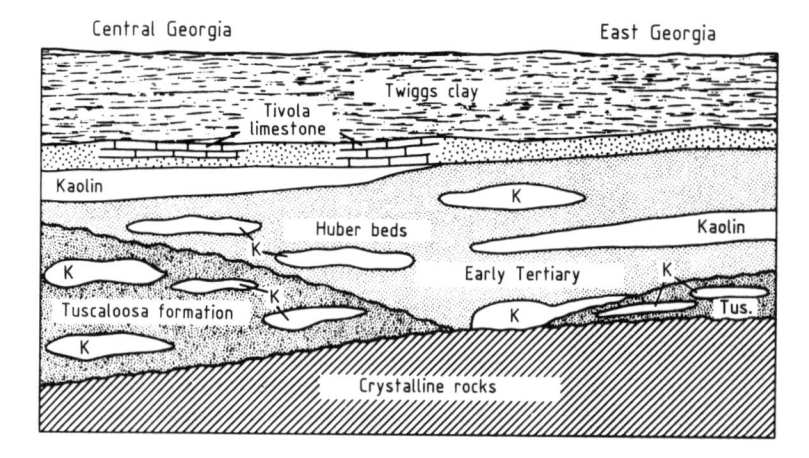

Figure 11. Diagrammatic sketch showing kaolin beds (K) in sands of Cretaceous(Tuscaloosa formation(Tus.)) and Tertiary periods

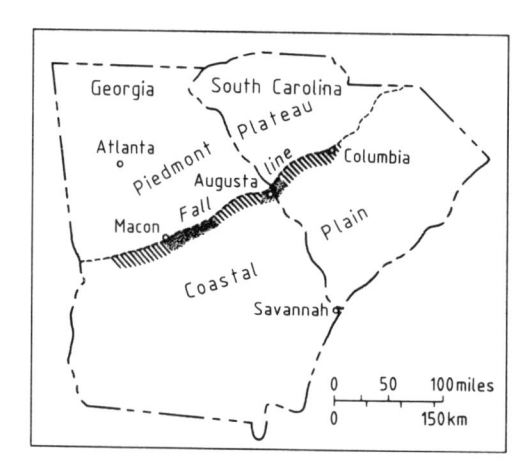

Figure 12. Location of kaolin deposits in Georgia and South Carolina

anatase, rutile, leucoxene, goethite, and traces of zircon, tourmaline, kyanite, and graphite.

The English kaolins that occur in southwest England in Cornwall (Fig. 13) are primary. These kaolins are the result of alteration of large granite bodies by warm acidic hydrothermal solutions migrating upward along structural channels such as faults, fractures, and joints [16]. These warm solutions attacked the granite, altering most of the feldspar and leaving quartz, mica, and tourmaline relatively unaltered. Superimposed on the hydrothermally altered granites is additional alteration by surface weathering. Over 3 000 000 t of kaolin is produced annually from the Cornwall area. The mines are located in the area around St. Austell. Drilling has shown that kaolinization exists at depths over 800 feet (> 250 m) [17].

Figure 13. Location of kaolin deposits in southwestern England
° Working china clay pit

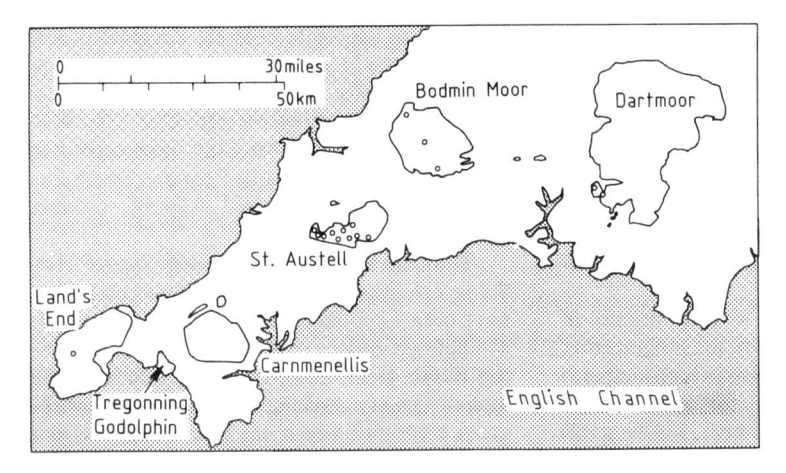

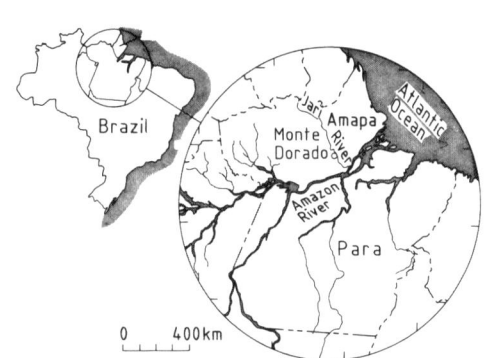

Figure 14. Location of Jari kaolin deposit at Monte Dorado

A large sedimentary deposit of kaolin of Pliocene age [18] is mined along the Jari River, a tributary of the Amazon, on the border of the states of Para and Amapa (Fig. 14). This deposit is 40 m thick and extends over an area of 60 km^2. The deposit is thought to be lacustrine in origin, and the kaolin was derived from weathering of the feldspathic granites on the Guyana shield. Kaolinite is the major mineral present, comprising 80–98% of the deposit. The major nonclay mineral present is quartz, along with small amounts of gibbsite, zircon, rutile, anatase, ilmenite, tourmaline, and goethite.

Czechoslovakia is the principal producer of kaolin in eastern Europe. The major kaolin deposits are located in the vicinity of Karlovy-Vary, Pilzen, and Podborany. The kaolin is primary in origin, and the deposits are extensive, occurring as a weathering crust on crystalline rocks of the Bohemian Massif [19]. The parent rocks are granites and gneisses. The feldspar in these rocks is almost completely kaolinized. The major nonclay minerals are quartz and mica.

In East Germany, the kaolin deposits are primary and are part of the same weathering crust on crystalline rocks as in Czechoslovakia [20]. The largest deposits, near Caminau and Kemmlitz, are the result of kaolinization of granodiorites and porphyrites. The major nonclay minerals present are quartz and muscovite.

The major kaolin deposits in the western part of Germany are located in Bavaria near the village of Hirschau. This kaolin is sedimentary and is an altered arkose of the Triassic period [21]. The deposit extends ca. 15 km and is 3–8 m thick. The major minerals present are quartz and feldspar. Kaolinite constitutes only 5–15% of the total.

Kaolin is mined in Spain in two locations. A primary kaolin is mined in the northwestern corner of Spain near Galicia. This deposit is a residual deposit resulting from the weathering of a large granitic body [22]. A secondary or sedimentary kaolin of Cretaceous age is mined in the state of Guadalajara, east of Madrid. This kaolin is a minor constituent (10–20%) in a friable white sand. There are extensive deposits of these kaolinitic sands extending from Guadalajara to Valencia.

In the Soviet Union, the largest kaolin deposit is located in the Ukraine and is primary [23]. Reserves are very large. The kaolin is washed to remove quartz and mica, which are the major nonclay minerals present.

In Australia, both primary and secondary kaolin deposits are mined [24]. The largest deposit being mined is north of Melbourne near Pittong and is primary. The kaolin is a residual weathering product of altered granite. The major nonkaolin minerals are quartz and muscovite. A large sedimentary kaolin deposit of late Cretaceous or early Tertiary occurs below bauxite on the Cape York Peninsula near Wiepa. This deposit is being developed and is to be in production in 1986. The major nonclay mineral is quartz (40–50%), which is relatively coarse in size (> 44 µm).

Other kaolin deposits that are smaller and have low production are listed in the literature [25].

Ball clays and refractory clays are special types of kaolinitic clays that are used primarily for ceramics. Ball clays are secondary clays characterized by the presence of organic matter, high plasticity, high dry strength, long vitrification ranges, and light color when fired. Kaolinite is the principal mineral constituent of ball clay and typically makes up more than 70% of this type of clay. Ball clays are found in the United States [26], England [17], the Federal Republic of Germany [21], and the German Democratic Republic [20].

Refractory clays include several varieties of kaolinitic clays that are used in the manufacture of products requiring resistance to high temperatures [26]. Fire clay and refractory clay are used interchangeably by many authors, but NORTON [27] excludes white burning clays in his definition of fire clay. Refractory clays are common and are found on every continent except Antarctica, where they are probably present but have not yet been discovered.

3.2. Smectite

Smectite is the name for a group of sodium, calcium, magnesium, iron, lithium aluminum silicates, which include the individual minerals sodium montmorillonite, calcium montmorillonite, nontronite, saponite, and hectorite. The rock in which these smectite minerals are usually dominant is *bentonite*. The name bentonite was first suggested in 1898 by KNIGHT [28] and is the term commonly used to describe the industrial mineral. The term bentonite was defined by ROSS and SHANNON [29], who restricted it to a clay material altered from a glassy igneous material, usually a tuff or volcanic ash. WRIGHT [30] suggested that bentonite was any clay composed dominantly of a smectite clay mineral and whose physical properties are dictated by this clay mineral. GRIM and GUVEN [31] used WRIGHT's definition in their book on bentonites because there are many clays designated as bentonite that did not originate by the alteration of volcanic ash or tuff. Therefore, the term bentonite usually does not include the mode of origin.

Bentonites in which sodium montmorillonites are the major mineral constituent commonly have a high swelling capacity [32]. The largest and highest quality sodium bentonite deposits in the world are located in South Dakota, Wyoming, and Montana [33]. These clays are commonly called Western or Wyoming bentonites. The bentonite beds are Cretaceous and are part of the Mowry formation [33]. There are several bentonite beds in the Mowry, but the thickest and most extensive is the Clay Spur bentonite bed (Fig. 15). This Clay Spur bentonite bed extends west from Belle Fourche, South Dakota, and then north into Alberta, Canada (Fig. 16).

The Clay Spur bentonite usually ranges from 0.5 to 2 m thick, although thicknesses up to 4 m have been reported [34]. The bentonite is generally light yellowish green on the outcrop, becoming bluish green away from the outcrop. The highest quality swelling clay is the yellowishgreen bentonite, which has been weathered and oxidized. This Western bentonite was formed by alteration of volcanic ash. The major nonclay minerals present in the Clay Spur bentonite bed are quartz, cristobalite, feldspar, mica, and some zeolite. Locally calcite and gypsum may also be present.

Bentonites in which calcium montmorillonites are the major mineral constituent commonly have a low swelling capacity [32]. These clays are generally referred to as Southern or subbentonites. Large deposits of these calcium bentonites occur in Texas and Mississippi [31]. In Mississippi, the calcium bentonite occurs in the Eutaw formation of the Upper Cretaceous [35]. It is commercially produced in Itawamba and Monroe Counties, where some bentonites reach a thickness of 4 m. The bentonite is waxy and varies from blue when fresh to yellow when weathered. Quartz, feldspar, and micas are the major nonclay minerals present. Another calcium bentonite has been produced from the Tertiary Vicksburg formation in Wayne and Smith Counties of Mississippi [36]. Several beds have been identified and are generally 0.5–1 m thick. This Tertiary bentonite ranges from white to gray to yellow. Quartz and micas are the major nonclay minerals. This central Mississippi bentonite is often referred to as the Polkville, Mississippi, bentonite.

The Texas calcium bentonites are found in a belt paralleling the present gulf coast in Cretaceous and Tertiary sediments. They are best developed in the Tertiary Jackson and Gueydan formations in Gonzales and Fayette Counties [37]. The mined bentonites range from 0.5 to 3 m thick. They vary from white to yellow to green to dark brown. Some of the white bentonites in Gonzales County contain kaolinite [38]. The major nonclay minerals are quartz, cristobalite, feldspar, and mica.

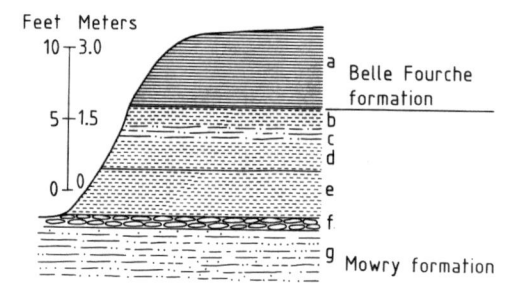

Figure 15. Diagrammatic sketch of the Clay Spur bentonite bed in the Mowry formation a) Dark shale, nonsiliceous; b) Dark bentonite; c) Interlaminated siliceous shale; d) Dark bentonite; e) Lightcolored bentonite; f) Chertlike floor; g) Siliceous shale

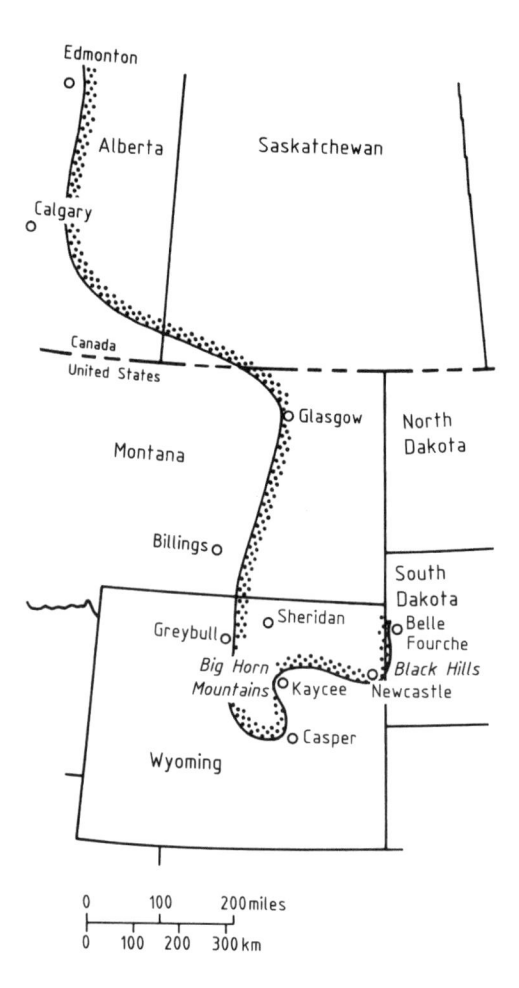

Figure 16. Outcrop area of the Mowry bentonites

Other calcium bentonites are produced in the United States in Arizona, Alabama, and Nevada. The best known, most extensively mined deposit is at Cheto, Arizona, where a thick (5 m) ash bed in the Pliocene Bidahochi formation has altered to calcium montmorillonite [39]. The bentonite is white to light gray. This clay also contains a small amount of kaolinite. Quartz, micas, and feldspars are the common nonclay impurities. A small amount of calcium bentonite is produced from the Amargosa Valley in Nevada [40]. This bentonite is Pliocene in age and is white to light gray. The major nonclay minerals include quartz, mica, calcite, and dolomite. In Alabama, a calcium bentonite is produced at Sandy Ridge from the Ripley formation [41].

Some special bentonites, including hectorite and saponite, are produced in the United States. Hectorite, a lithium-bearing smectite, is mined at Hector, California. The hectorite was formed where siliceous lithium–fluorine solutions reacted with carbonates [42]. The hectorite deposits are Tertiary. Saponite is mined in the Ash Meadows area of the Amargosa Valley in Nevada. It is called Pliocene [40], although DENNY and DREWES [43] mapped these rocks as Pleistocene.

Bentonites are produced in many other areas of the world and are commonly of the calcium montmorillonite variety. In England, there are two well-known occurrences of smectites, which are of the calcium montmorillonite variety. In the English literature, they are referred to as *fuller's earth*. The term fuller's earth refers to any natural material that has the capacity to decolorize oil to an extent that is of commercial value (also see 3.3) [31]. At Redhill in Surrey, near London, is a calcium bentonite in the Upper Cretaceous [44]. The clay is waxy and blue. Another calcium bentonite, of the Jurassic period, is found in Somerset [45].

In the Federal Republic of Germany, there are extensive deposits of calcium bentonite in the vicinity of Moosberg and Landshut in Bavaria [46]. These Bavarian bentonites occur in a marine section of marls and tuffaceous sands in the Upper Miocene molasse. Some beds attain a thickness of 3 m. The bentonite is light yellow to gray and has a waxy appearance. These deposits are an important source of commercial bentonite in European and world markets.

In Greece, several of the volcanic islands contain bentonite. The most extensively developed deposit, a Pliocene calcium bentonite, is on the island of Milos [31].

In Hungary, bentonite has been produced for about 50 years and is now an important product, especially recently. The most important deposits are located at Istenmezo, Nagyteteny, Band, Komloska, Mad, and Vegardo [47]. The last three are the result of hydrothermal alteration of rhyolites and pyroxene andesites associated with Tertiary volcanism; the first three are the result of alteration of tuff that fell into the sea and altered under marine conditions. These altered tuff deposits are Surmatian (Tertiary).

In Italy, important calcium bentonite deposits are located on the islands of Sardinia and Ponza. The Ponza bentonite is ivory to blue white and is the result of hydrothermal alteration of rhyolitic glassy tuff [48]. The Sardinian bentonites are of two types, hydrothermally altered trachytic tuffs and sedimentary bentonite of the Miocene, which is a marine-altered ash [49], [50].

In Yugoslavia, there are several calcium bentonites produced commercially. These are located near Pula, Kutina, Laktasi, Kriva Palanka, and Petrovac. The Petrovac deposit is Triassic, and the others are Middle to Upper Tertiary [31]. These bentonites are altered ashes and tuffs [51].

In India, bentonite is produced in the Barmer district of Rajasthan near Akli and Hathi-ki-Dhani [52]. These are quite pure calcium montmorillonites up to 3 m thick. They are Lower Tertiary and are associated with calcareous sands and conglomerates.

In Japan, commercial bentonites are located in the Yamagata, Gumma Niigata, and Nagano Prefectures of northern Honshu and on Hokkaido. The deposits are Miocene and Pliocene, formed by alteration of ash, pumice, and tuff [53]. Some of the bentonites in the Gumma Prefecture are up to 5 m thick.

In the Soviet Union, bentonites are found in many areas. The most important deposits are in Georgia, Azerbaijan, Armenia, the Ukraine, and Uzbekistan. The best known deposits are the Askana and Gumbri deposits in Georgia. The Askana deposits are found in a series of Eocene tuffs and breccias ranging from trachyte to basalt. Both sodium and calcium montmorillonites are produced from the Askana deposit. RATEEV [54] believes the Askana bentonites are hydrothermal. The Gumbri bentonite occurs in beds up to 8 m thick of Cretaceous Cenomanian and Turonian [55]. It was formed by the alteration of volcanic glass in a submarine environment.

Other bentonite deposits along with their geologic age and occurrence are described in [31].

3.3. Hormite

Hormite is a group name for clay minerals with a chain structure. These include palygorskite and sepiolite. Attapulgite is the same as palygorskite, but the more recent name attapulgite has been discredited as a mineral name. However, it is so prevalent in the literature that the term still persists.

These hormite clays are sometimes referred to as sorptive clays because they have a large surface area and can absorb and adsorb many materials. The term *fuller's earth* is more or less a catchall term for clays or other fine-grained earthy materials suitable for bleaching and sorbent uses. It has no compositional or mineralogical meaning. The term was first applied to earthy material used in cleansing and fulling wool, thereby removing the lanolin and dirt; thus, it acquired the name fuller's earth. Industrially the term fuller's earth is used quite often for the hormite minerals and other sorptive clays.

Palygorskite (attapulgite) is produced in the southeastern United States near the Georgia–Florida border in the vicinity of Quincy, Florida, and Attapulgus, Georgia, in Senegal near Dakar, and in the Urals in the Soviet Union. Sepiolite is mined commercially in Spain, which has the largest deposits, in Turkey, and in Nevada in the United States. These hormite clays are lathlike or elongate in shape (see Fig. 10). Sepiolite is structurally similar to palygorskite except the latter has a slightly larger unit cell and the individual laths are ca. one-third wider [56]. This may be due to the presence of more iron substituting for magnesium in the lattice, because generally palygorskite has a higher iron content.

The palygorskite deposits in southeastern United States are of the Miocene age and were deposited in marine lagoons [57]. The clay ranges from 0.5 to 3 m thick and is gray to tan or brown. These deposits are mixtures of palygorskite and smectite with the palygorskite content higher in the southern part of the district (Fig. 17).

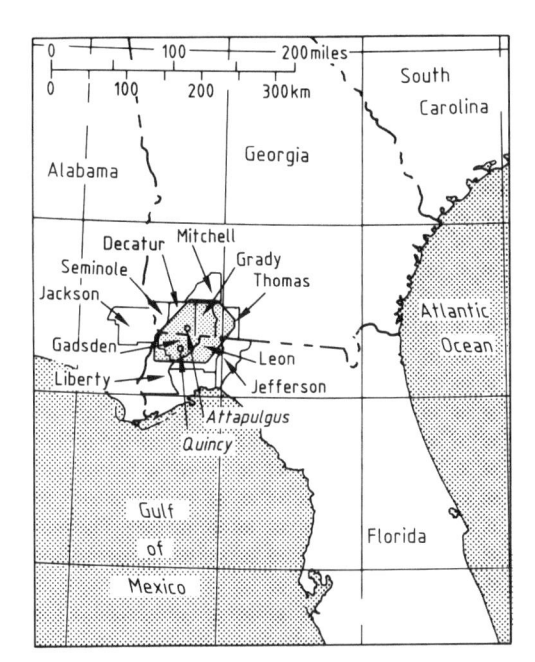

Figure 17. Fuller's earth deposits in Florida and Georgia

The palygorskite in Senegal [58] is a large deposit that extends from Theis, where it immediately overlies a commercial aluminum phosphate deposit, south-southwest about 100 km to the Senegal–Barrie border. This deposit has only been in production for ca. 10 years. The age of the deposit is early Eocene, and the thickness ranges from 1.5 to 6 m. The overburden thickness is low, usually less than 2 m. The color of the palygorskite is light tan to gray green. There are calcite, dolomite, and chert impurities in some areas. Smectite is also a common constituent, but in some areas it is barely detectable on X-ray diffraction patterns.

In the Soviet Union, palygorskite was first described in the Urals in the Palygorsk Range [59]. Palygorskite is mined in the central part of the Ukrainian crystalline massif along the borders of the Cherkassay and Kiev regions. The palygorskite occurs in the middle section of a productive bentonite and is Lower Miocene [59]. The thickness of the palygorskite layer is ca. 2 m. Smectite, hydromica, and quartz are the other major constituents.

Sepiolite is mined in Spain near Toledo from a large Tertiary lacustrine deposit [60]. It is light tan and ranges from 1 to 5 m thick. Palygorskite is mined at Trujillo in the province of Caceres in western central Spain.

Sepiolite is mined in small quantities in the Ash Meadows area in the Amargosa Valley in Nevada [61]. The sepiolite occurs as 0.3–1 m stringers in saponite clays of lacustrine origin. The sepiolite is lightweight, very porous, and brown to brownish gray. The age of these sepiolite beds is Pliocene or Pleistocene [40], [43].

Sepiolite is produced in Turkey in the vicinity of Eskisehir in Anatolia [56]. The sepiolite is white and is interbedded with palygorskite shales. The sepiolite contains up to 3 wt% organic carbon in the form of humic acid, which enhances the dispersability of the clay.

Palygorskite is produced in the Mudh District in India [62]. The deposits are Tertiary and occur between fossiliferous limestones, thus apparently marine in origin. The thickness averages ca. 1.5 m.

Palygorskite is produced in small quantities near Xuyi (near Nanjing) in China. The palygorskite occurs beneath a smectite-rich layer 1–2 m thick. The palygorskite is light gray and averages ca. 6 m in thickness. There are lenses and beds of a light tan dolomitic palygorskite interspersed in the palygorskite. The bed contains up to 80% palygorskite and ca. 20% quartz. The dolomitic palygorskite contains ca. 50% dolomite and 50% palygorskite. These beds are Tertiary. Several million tons are present in two known deposits. The author visited these deposits in 1985.

3.4. Miscellaneous Clays

Illite or chlorite deposits are not mined because they are composed of the mineral illite or chlorite. Many clays and shales contain these clay minerals and are used for structural clay products, lightweight aggregate, and cement. Generally, clays and shales that contain illite and chlorite in major proportions have properties that are important in the manufacture of structural clay products. Structural clay products are a broad category that includes brick, drain tile, sewer piper, conduit tile, floor tile, wall tile, terra cotta, and other items. The properties that are important in the manufacture of structural clay products are plasticity, green strength, dry strength, drying and firing shrinkage, vitrification range, and fired color. These physical properties depend on the constituents of the clay and their grain size [63].

Illitic and chloritic shales and clays are not restricted geologically and can be found in many areas of the world. In the United States, shales and clays with high illite and chlorite content are found in all geologic periods with a preponderance in Pennsylvanian, Cretaceous, and Tertiary formations. Usually these clays and shales are relatively thick with low overburden.

Lightweight aggregate is a material produced from clays and shales. The material is fired rapidly in rotary kilns or sintering machines up to the temperature between incipient and complete fusion. At this temperature, substances that release gas cause bloating and vesiculation after fusion has developed a molten jacket around the particle, which prevents the escape of the gas. The mineral content is important, i.e., the clay minerals and the minerals such as calcite or dolomite that produce a gas on firing [64]. Clays and shales used for lightweight aggregate production are found in many areas of the United States and again are not restricted geologically or geographically.

Large quantities of clays and shale are used in the manufacture of cement. These clays and shales provide silica and alumina for the cement charge. Practically all clay or shale with an average alumina and silica content can be used in the cement industry.

Usually these clays and shales can be found in the area in which the cement is being manufactured (→ Cement and Concrete).

4. Mining and Processing

Almost all clay in the world is mined by open pit. There are a few underground mines; however, in most instances the cost of production and poor roof conditions prohibit underground mining of clay deposits unless the clay has exceptional properties.

An understanding of the geology and the origin of a particular type of clay plays an important part in clay exploration. After a clay deposit is discovered, it must be evaluated to determine the quantity and quality of the clay. In most instances, this is done by core drilling and testing the recovered cores. The spacing of the drill holes depends on the geologic and surface conditions associated with the deposit. For example, the drilling pattern used to evaluate a sedimentary kaolin deposit is much different than that used to evaluate a residual or hydrothermal kaolin deposit.

A small section that is contaminated with a deleterious material can affect the results of the total length of the core if it is composited. For this reason, the core is normally tested in increments of a meter or less. If the core is 5 m long, the core is composited for testing in five increments. In this way, a section that does not meet a particular specification can be isolated. Further drilling can corroborate the thickness and extent of the material that is out of specification, and a mining plan can probably be designed to handle the particular problem. Drilling is normally done on a pattern grid so that maps indicating thickness of overburden and thickness and quality of clay can be prepared.

4.1. Kaolin

All kaolins that are used as paper coating clays and most that are used as filler clays are processed because practically no deposits are naturally pure. Usually beneficiation is accomplished by one of two basic processes, a dry process or a wet process [65]. The higher quality grades of kaolin, except for some kaolins with special ceramic properties, are prepared by wet processing because the product is more uniform, relatively free from impurities, and of better color.

Major kaolin processing developments and their significance to the growth of the kaolin industry have been summarized [66]. There has been a steady growth of the kaolin industry since 1930, and the only physical property that has not been significantly improved by process developments is viscosity.

4.1.1. Dry Process

The dry process [65] is relatively simple and yields a lower cost, lower quality product than the wet process. In the dry process, the properties of the finished kaolin reflect the quality found in the crude kaolin. Therefore, deposits containing the desired qualities of brightness, low grit content, and satisfactory particle-size distribution must be located by drilling and testing.

The overburden on the kaolin is stripped by motorized scrapers and bulldozers or draglines. Once the overburden is removed, the kaolin is mined by using draglines, shovels, or front-end loaders. The kaolin is transported to the processing plant in trucks. The kaolin is usually shredded or crushed and stored in large storage sheds. Usually the kaolin is segregated into two or three bins according to specific properties, such as brightness and grit percentage. (Grit is material, usually quartz, that is coarser than 44 µm, i.e., the material that is retained on a 325-mesh screen.) The general flow sheet utilized in dry processing is shown in Figure 18.

The material, as it enters the storage shed or enters the process, is shredded or crushed to particles equal to or less than the size of an egg. Normally, the crude kaolin contains ca. 20–25% moisture and, therefore, must be dried. This is commonly done in a rotary dryer. After drying, the kaolin is disintegrated in a roller mill, hammer mill, disk grinder, or some other disintegrating device. Commonly, heat is applied to the kaolin during the disintegration to further reduce the moisture content. The disintegrated kaolin may then be air-classified to separate the fine particles from the coarse. The top size can be controlled to a degree so that various grades, based upon particle size, can be produced. The material can be shipped in bulk or in bags by truck or railcar.

4.1.2. Wet Process

The general flow sheet for wet processing Georgia kaolin [67] to produce both filler and coating grades is shown in Figure 19. The kaolin is mined with shovels, draglines, motorized scrapers, or front-end loaders and can be either transported to the processing plant or fed into a stationary or mobile blunger (Fig. 20). The blunger separates the kaolin into small particles, which are mixed with water and a dispersing chemical to form a clay–water slip or slurry. The dispersing chemical can be sodium polyphosphate, sodium silicate, or sodium polyacrylate, each of which can be blended with soda ash to reduce cost.

The percent solids of the slurry is normally between 30 and 40%, but in some special circumstances it can be in excess of 60% and as high as 70% solids (see Section 4.1.3). This clay–water slurry is pumped from the blunger to rake classifiers or hydrocyclones and screens to remove the grit. The grit is removed from the kaolin slurry and discarded into waste ponds or into mined-out areas. The degritted slurry is collected

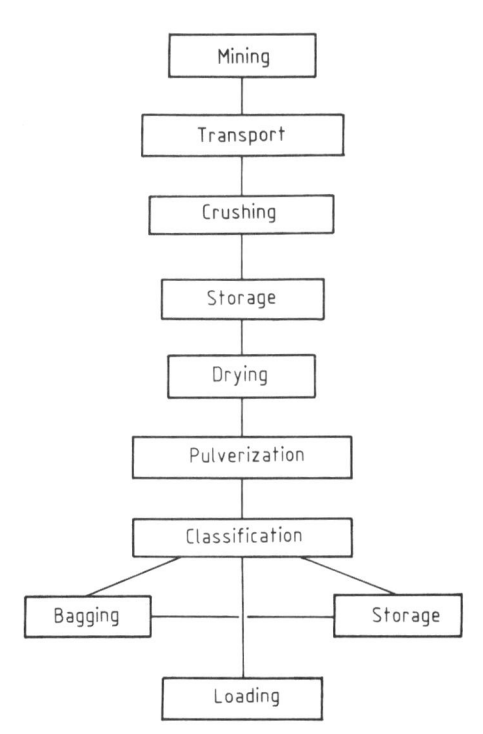

Figure 18. Dry-process flow sheet for kaolin production

```
Mining → Transport → Crushing → Storage → Drying → Pulverization → Classification → Bagging / Storage → Loading
```

into large storage tanks with agitators and is then pumped to the processing plant, which may be several miles away.

The kaolin slurry is collected in large storage tanks at the plant before it is processed. Although each kaolin producer may process the kaolin slightly differently, the steps in the processing are generally as shown in Figure 19. The first step is to separate the kaolin particles into a coarse and fine fraction through continuous centrifuges. Some operators pass the degritted slurry through a high-gradient magnetic separator prior to centrifugation to upgrade the crude clay by removing iron and titanium minerals (see Section 4.1.4).

The coarse fraction can be used as a filler in paper, plastics, paint, and adhesives, as a casting clay for ceramics, or as a feed to delaminators, which can be a subsequent processing step to produce a special paper coating grade (see 4.1.4). The fine particle size fraction is used in paper coatings, high-gloss paints, inks, special ceramics, and rubber. As shown in Figure 19, there are many different steps that can be taken after the clay is fractionated. Normally, the coarse kaolin takes one of two routes:

1) It can go directly to the leaching department, where it can be chemically treated to solubilize some of the iron if the brightness needs to be upgraded, or it can be flocculated so that the dewatering step is facilitated.
2) It can go through magnetic separation and into the delaminators or directly to the delaminators and then to the magnetic separator or to the delaminators without any magnetic separation step.

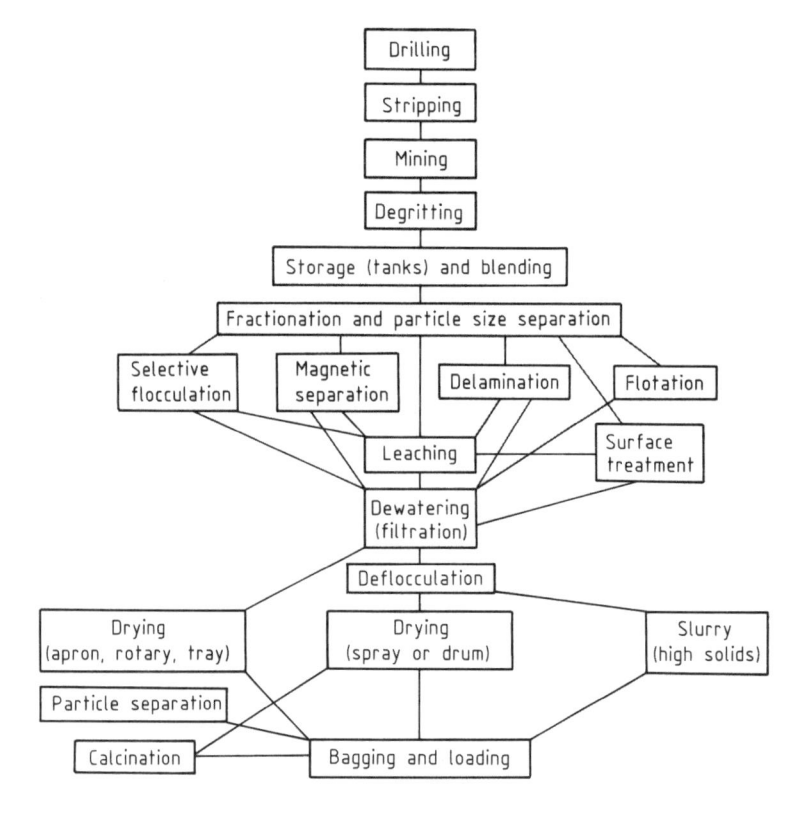

Figure 19. Wet-process flow sheet for kaolin coating and filler production

Figure 20. Dragline feeding kaolin into a mobile blunger

In the leaching operation, the kaolin slurry is acidified with sulfuric acid to a pH of ca. 3.0 to solubilize some of the colloidal iron. Zinc or sodium dithionite, a strong reducing agent, is then added to reduce the iron to the iron(II) state, which is more soluble and forms a clear iron sulfate that is removed with the filtrate in the dewatering step. In some cases, the iron in the kaolin is in a reduced state rather than the normal oxidized state. Usually such kaolins are slightly gray. These gray kaolins are treated with a strong oxidizing agent, such as ozone or sodium dithionate, prior to the acidification and normal reducing leach procedure. To facilitate flocculation prior to dewatering, in many cases alum is added along with the sulfuric acid.

At this point in the process, the kaolin slurry is between 20 and 30% solids and is flocced. This flocced slip is then dewatered, which raises the solids to $\geq 60\%$. The dewatering is accomplished by large rotary vacuum filters or plate and frame filter presses. The dewatered filter cake is then prepared for drying by using rotary, apron, drum, or spray dryers. Most of the clay that is shipped in dry bulk form is spray dried in a dispersed state. The usual process is to take the filter cake, which is flocced, and disperse it by adding a long-chain sodium polyphosphate or sodium polyacrylate. This dispersed high-solids slurry ($\geq 60\%$) is fed into the spray dryer, where it is sprayed or atomized and dried to meet the final moisture specifications, which normally range from less than 1% to as high as 6%. This dispersed clay slurry can also be dried by rotary drum dryers for some special uses that require the flake form.

Slurry clay, which is normally shipped in tank cars or trucks at 70% solids, is produced by mixing the dispersed clay from the filters with enough spray-dried clay to bring the final solids to 70%. About 70% of all shipments of coating and fine filler clays to the paper industry is now being shipped in slurry form.

If the filter cake is to be dried with rotary or apron dryers, then the filter cake is extruded into noodles ca. 3/8 in. (1 cm) in diameter and fed into the rotary dryer or onto an apron dryer. The final moisture from these dryers is generally ca. 6%. This type of clay can be shipped in lump form, or if necessary, the clay can be further dried and disintegrated, using hammer mills swept with hot air. The final moisture of this disintegrated kaolin can be $\leq 0.5\%$. It is commonly referred to in the kaolin industry as acid clay because the pH is ca. 4.5.

The wet processes are used on the sedimentary kaolins in Georgia and Brazil and are to be used on the Australian kaolin at Weipa when it comes on stream in 1986. Primary kaolins such as the English china clays produced in the Cornwall district of southwest England are wet processed in much the same manner as the Georgia kaolins except for the mining methods. The kaolin content of the weathered granites ranges from 10 to 20%. Mining uses high-pressure hydraulic monitors that play a stream of water on the mine face, washing out the fine particle kaolin and leaving the coarse quartz and mica residue. The fine kaolin is suspended and is transported in small rivulets to a collecting basin in the bottom of the open pit. This slurry is pumped to large thickeners, where the clay slurry is concentrated, and then processed generally following Figure 19.

4.1.3. Combined Dry and Wet Processing

A sort of hybrid development, a blend of dry processing and wet processing, is the whole clay slurry produced from fine particle size East Georgia Tertiary crude kaolins. Two processes are used to produce this high-solids whole clay slurry. One is to dry process the kaolin and then slurry the product at 70% solids and degrit the clay slurry with a scalping cut on a centrifuge. A second method is to blunge the crude clay at high solids (65–68%), degrit with a centrifuge, and add dry clay to bring the solids to 70%.

4.1.4. Special Processes

Several special processes are used to produce unique and special quality grades of kaolin. One of these special processes is *delamination*, which is a process that takes a large coarse kaolin stack and separates it into several thin, large-diameter plates (Fig. 21). These thin, large-diameter plates have excellent covering power on rough base sheets of paper and are also used to produce high-quality lightweight coatings.

The process of delamination involves attrition mills into which fine media, such as glass beads or nylon pellets, are placed along with the coarse kaolin stacks and intensely agitated. The fine media impact the kaolin stacks, separating them into thin plates [68]. The brightness and whiteness of the delaminated kaolin is very good: the clean newly separated basal plane surfaces are white because they have been protected from groundwater and iron staining. The delamination process has enabled the kaolin industry to convert low-priced coarse clay into a higher priced special coating and filler clay.

Another special process is *ultraflotation*, which is a special flotation process to remove iron and titanium contaminants to make a 90% brightness kaolin product [69]. *Selective flocculation* [70]–[72] is another special process that produces a 90% brightness kaolin product by selectively removing iron and titanium impurities.

High-gradient magnetic separation (HGMS) is another special process, although it has become a standard processing technique in the kaolin industry. The high-gradient magnetic separation process [73] uses a canister filled with a fine stainless steel wool that removes iron, titanium, and some mica minerals from the kaolin slurry as it passes through the canister (Fig. 22). The process can be used to upgrade marginal low-brightness crude kaolins and to produce high-brightness coating clays. It also reduces the amount of chemicals that are needed to leach out the iron in a subsequent processing step. The development of HGMS has dramatically increased the usable reserves of kaolin in Georgia. A new superconducting magnet is being developed that will reduce power costs and improve the removal of fine paramagnetic mineral particles [74].

Calcining is another special treatment that is used to produce special grades. One grade is thermally heated to a temperature just above the point where the structural hydroxyl groups are driven out as water vapor, which is between 650 and 700 °C. This

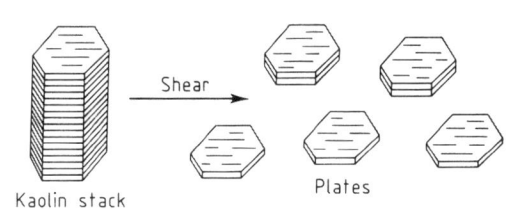

Figure 21. Diagrammatic representation of delamination of a kaolin stack into large thin kaolin plates

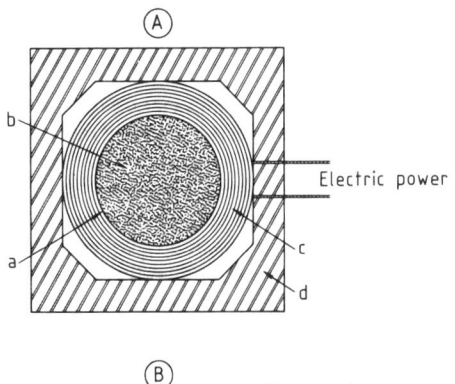

Figure 22. Diagrammatic sketch of a high-gradient magnetic separator
A) Top view: a) Canister wall; b) Steel-wool matrix; c) Magnet coil; d) Magnet steel
B) Side view section through magnet:
a) Canister; b) Magnet coil; c) Matrix; d) Magnet steel

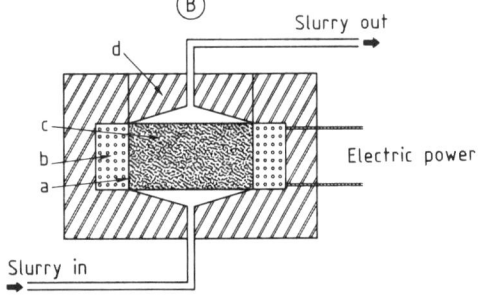

produces a bulky product that is used as a paper coating additive to enhance resiliency and opacity in low basis weight sheets. A second grade is thermally heated to 1000–1050 °C. By proper selection of the feed kaolin and careful control of the calcination and the final processing [75], the abrasiveness of the calcined product can be reduced to acceptable levels. The brightness of this fully calcined, fine-particle kaolin is 92–95%, depending on the feed material. It is used as an extender for titanium dioxide in paper coating and filling and in paint and plastic formulations.

Another special process is *surface treatment*. Kaolin is hydrophilic and can be easily dispersed in water. Because of the nature of the kaolinite surface, it can be chemically modified to produce hydrophobic and organophilic characteristics. Generally, an ionic or a polar nonionic surfactant is used as a surface treating agent. These surface-modified kaolins are used in paper [76], paint [77], plastics [78], and rubber [79].

4.2. Smectite

Virtually all bentonite, which is comprised mainly of a smectite mineral, is strip-mined. Bulldozers and motorized scrapers are most commonly used for removing overburden. In a typical open pit, the overburden is removed in panels. The exposed bentonite is then mined by loading the material into trucks with draglines, shovels, or front-end loaders. Once the bentonite is removed, the overburden from the next panel is shifted to the mined-out panel.

Thicknesses of removed overburdens vary considerably. Most Wyoming-type bentonite has 10 m or less overburden. The thickness of overburden removed from some Mississippi bentonite deposits is as much as 30 m [80], about the same as that on some of the Arizona bentonites near Chambers. In Bavaria, some overburden on the bentonites is up to 50 m thick.

Bentonite beneficiation and processing involves relatively simple milling techniques that involve crushing or shredding, drying, and grinding and screening to suitable sizes (Fig. 23). The high-swelling Wyoming or sodium bentonites contain ca. 30% moisture when delivered to the plant. The Southern or calcium bentonites have ca. 25% moisture. The processed bentonite generally contains only 7–8% moisture, although because bentonite is hygroscopic, it may contain considerably more moisture when used.

The raw bentonite is passed through some sort of crushing or shredding device to break up the large chunks before drying. Drying in most plants is accomplished by gas- or oil-fired rotary dryers, but in one bentonite plant in Wyoming, a fluidized-bed dryer is used. The bentonite properties can be seriously affected by overdrying, so the drying temperature must be carefully controlled.

The dried bentonite is ground and sized in several ways. Granular bentonite is cracked by using roll crushers and screened to select the proper size range of granules. Most powdered bentonite is ground with roll or hammer mills to ca. 90% finer than 200 mesh. The bentonite is shipped mainly in bags, but some is shipped in bulk.

Special processing is used in some plants to produce special products. Extruders can be used to pug the clay and mix it with additives that may improve the viscosity and dispersion properties. Some hectorite and Southern bentonites are beneficiated wet. In this process, the bentonite is dispersed in water and degritted by centrifuges, hydrocyclones, settling devices, or screens. The slurry is either filtered and then flash-dried or sent directly to rotary drum or spray dryers.

Organic-clad bentonite is bentonite that is processed to replace the inorganic exchangeable ions with alkylamine cations, which produces a hydrophobic clay [81]. These organic-clad bentonites are used in paints, greases, oil-base drilling muds, and to gel organic liquids.

Acid-activated bentonites are special sorptive clays used for bleaching edible oils and decolorizing special lubricating oils. The process flow sheet for acid activation is shown in Figure 24. The process involves slurrying the bentonite in warm water, removing the grit by hydrocyclones, reacting the clay with either sulfuric or hydrochloric acid at

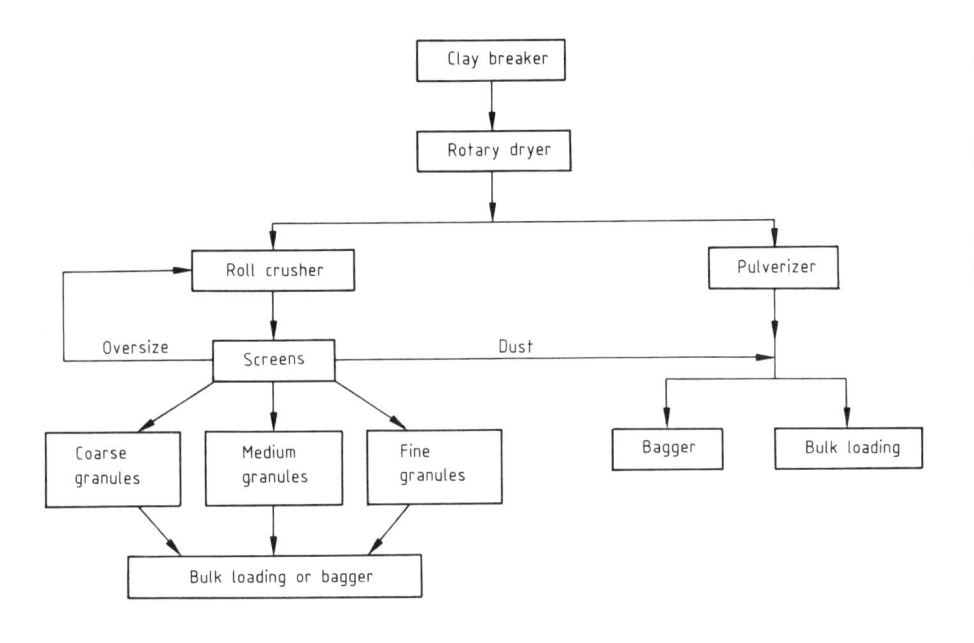

Figure 23. Flow sheet for bentonite production

elevated temperature, dewatering in plate and frame filter presses, and flash drying. Acid treatment removes some of the aluminum, magnesium, and iron octahedral ions, resulting in a highly charged particle saturated with hydrogen ions, which makes for a good bleaching and decolorizing clay.

4.3. Hormite

The mining of hormite clays is no different than mining bentonites. The processing is also similar (Fig. 25). The hormite clay is shredded or crushed, dried, screened, air-classified, and packaged. Some of this hormite clay is thermally treated to harden the particles to prevent disintegration of the granules during bagging, transport, and handling prior to arrival at the customer. The temperature of this thermal treatment is less than 400 °C so that no structural modification results, but all of the adsorbed water is driven from the surface and interior of the particles.

When palygorskite (attapulgite) is to be used for salt-water drilling mud, the clay is extruded at high pressure through small orifices. This shears apart the bundles of palygorskite into individual elongate particles, which produces a higher viscosity drilling fluid.

Figure 24. Flow sheet for producing acid-activated bentonite

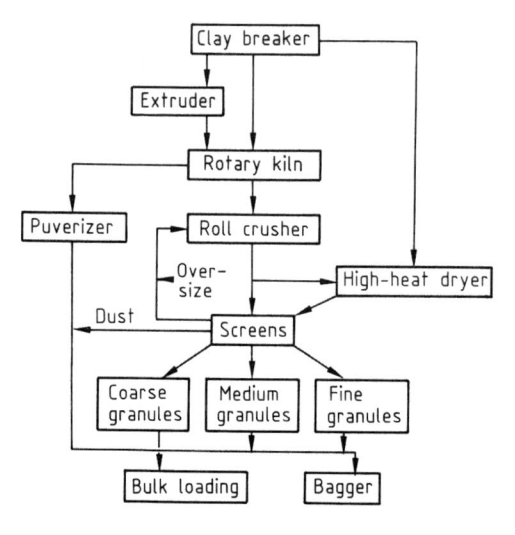

Figure 25. Flow sheet for producing hormite clays

5. Properties and Uses

Kaolins, smectites, hormites, and miscellaneous clays form an important industrial minerals group. Because of their fine particle size, particle shape, and surface chemistry, they have unique colloidal and physical properties.

5.1. Kaolin

Kaolin is one of the most versatile of the industrial minerals [82] and is used extensively for many applications [83]. It is a unique industrial mineral because it

1) is chemically inert over a relatively wide pH range (except for catalytic activity in some organic systems);
2) is white or near white;
3) has good covering or hiding power when used as a pigment or extender in coating and filling;
4) is soft and nonabrasive;
5) has low conductivity of both heat and electricity; and
6) is lower in cost than most materials with which it competes.

Some uses of kaolin require rigid specifications, including particle size, brightness, color, and viscosity. On the other hand, some uses have no critical specifications, e.g., cement, where the only concern is light color and chemical composition. Ceramic specifications are variable in that individual users may have different requirements as to strength, plasticity, fired color, shrinkage, and pyrometric cone equivalent (PCE). The major process industries in which kaolin finds a substantial use are paper, paint, ceramics, rubber, plastics, ink, catalyst, and fiberglass, as well as some special uses.

5.1.1. Physical Properties

Diagnostic tests are used to evaluate kaolin physical properties [84]. Some properties that can be used as screening tests for the many potential uses for kaolin are as follows:

mineralogy
screen residue (grit)
particle-size distribution
brightness
viscosity
pH

Mineralogy. The mineral content of a kaolin is important in assessing the uses, the results of many tests determining the physical properties, and the beneficiation processes that may be needed to produce a saleable product. X-ray diffraction of the whole sample gives a quick assessment of the gross mineralogy, but in many instances minor quantities of illite and smectite are not detected. Separation of the 2-μm and 0.5-μm fractions and preparation of oriented slides for X-ray enhances the basal reflections of the clay minerals and enables the detection of very small quantities of smectite and illite.

The presence of quartz, cristobalite, alunite, smectite, illite, muscovite, biotite, chlorite, gibbsite, feldspar, anatase, pyrite, iron oxides and hydroxides, and halloysite affects the beneficiation processes and the possible uses of the kaolin. In many instances, the presence of halloysite in a kaolin cannot be detected by X-ray diffraction; therefore, electron micrographs and differential thermal analysis must be used.

Screen Residue. The grit test gives the percentage of particles that are retained on a 325-mesh screen (44-μm openings). Quartz and mica, along with hard agglomerates of kaolin, are the most common minerals retained on the screen. The amount of screen residue is important in estimating the recovery of the minus 325 mesh material since this is the fraction that is the usable portion of the kaolin. Generally, if the kaolin contains more than 5–8% grit, it cannot be dry processed.

Particle-Size Distribution. Particle-size distribution (Fig. 26) is one of the more important properties of a kaolin because it affects viscosity, brightness, opacity, gloss, ceramic strength and shrinkage, and many other properties. Sedimentation methods based on Stoke's law of settling are used for measuring particle size. The 2-μm content of a kaolin is particularly important because it is this fraction that constitutes the major portion of paper coating clays and high-glossing paint clays. The particle-size distribution also determines the beneficiation procedures that are used in processing a kaolin for industrial uses.

Brightness. The brightness of a kaolin is an important property because it determines the potential uses and prices. In general, the higher the brightness of the kaolin the more value the kaolin has. Standard brightness values are measured at a wavelength of 457 nm, and a common standard against which the samples are compared is smoked magnesium oxide. As a general rule, filler clays have a brightness of 80–85%, and paper coating clays and high-glossing paint clays have a brightness of 85–90%.

Viscosity. The flow properties of kaolin used for paper coating are very important. Both low-shear and high-shear viscosity measurements are made at 70–71% solids. Low-viscosity kaolins are required for paper coating because the kaolin coating color must flow easily as it is applied to the paper surface, not leaving streaks or blotches, which can be caused by high-viscosity coating colors.

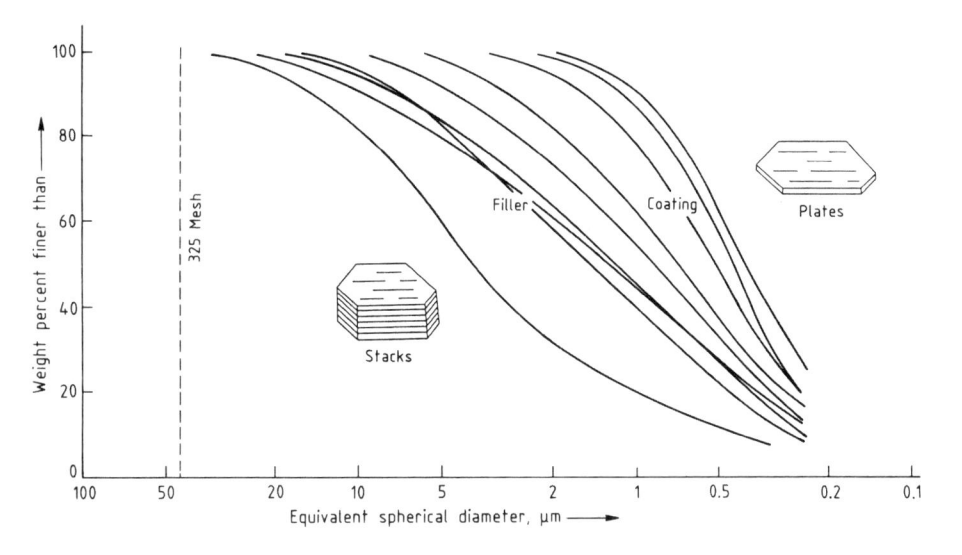

Figure 26. Particle-size distribution of some typical kaolin coating and filler grades

pH. The pH of untreated kaolin slurries normally ranges from 4.5 to 6.5. A high pH generally indicates the presence of soluble salts that can cause problems in many applications.

5.1.2. Paper

The mineral kaolinite has many physical properties that are advantageous to the paper industry. It is white, fine, soft and nonabrasive, and relatively inert. In the United States, ca. 70% of all of the wet processed kaolin is used by the paper industry as a coating clay and as a filler. Some physical constants that are representative of kaolins from Georgia are the following:

Specific gravity d_4^{20}	2.62
Index of refraction	1.57
Hardness (Mohs' scale)	1.5–2
Fusion temperature, °C	1850
Valley abrasion number	4–10
Dry brightness (at 457 nm), %	75–91
Crystal system	triclinic

The Georgia kaolins and English kaolins are the most widely used kaolins in the paper industry. Table 2 gives typical chemical analyses of two Georgia kaolins and an English kaolin.

Table 2. Typical chemical analyses, wt%

Component	Cretaceous middle Georgia kaolin	Tertiary east Georgia kaolin	English kaolin	Theoretical kaolin
SiO_2	45.30	44.00	46.77	46.3
Al_2O_3	38.38	39.5	37.79	39.8
Fe_2O_3	0.30	1.13	0.56	
TiO_2	1.44	2.43	0.02	
MgO	0.25	0.03	0.24	
CaO	0.05	0.03	0.13	
Na_2O	0.27	0.08	0.05	
K_2O	0.04	0.06	1.49	
Ignition loss	13.97	13.9	12.79	13.9

Important Properties. *Particle-size* and *shape distribution* are probably the most important factors in controlling the many physical properties of kaolin important to the papermaker. Figure 26 shows the range of particle-size distributions available commercially. The 2-µm size is used as the commercial control point. The coarse particle size kaolins are normally used as filler clays, and the fine particle size kaolins are used as coating clays.

Some of the properties of importance to the papermaker are dispersion, rheology, brightness and whiteness, opacity, gloss and smoothness, orientation, adhesive demand, film strength, and ink receptivity.

Complete *dispersion* of the kaolin particles is necessary to obtain maximum efficiency in a coating or filler clay. Kaolin is easily dispersed because of its hydrophilic surface. In the natural state, kaolin is acid and is flocculated. This flocced state is the result of the presence of both positive and negative charges on the particles, which attract each other and form aggregates. Therefore, dispersing agents must be used to deflocculate the kaolin by making the particles all the same charge order, thus obtaining maximum fluidity. The most efficient dispersing agents for kaolin in aqueous systems are alkali-metal polyphosphates, polyacrylates, and silicates [85].

The *rheology* or flow properties of coating clays and coating colors are complex and exert a controlling influence on the coating operation. Such factors as coat weight, smoothness, and freedom from pattern on the coated sheet are related to viscosity. Low-viscosity kaolins can be dispersed (made down) readily at high solids concentration (71%). In addition, these low-viscosity suspensions can be pumped and screened with relative ease. The coating process (i.e., trailing blade, roll coater, air knife, etc.), the coating color formula, the type and amount of adhesive, and the percent solids of the coating color determine the high- and low-shear viscosity values that can be tolerated for the coating clay.

*Brightness and whiteness*and of kaolin coatings are controlled primarily by the presence of iron and titanium minerals in the kaolin. However, the coating structure is also important. Significant differences in refractive indices between coating components are necessary to obtain maximum light scattering and, thus, optimum brightness and whiteness. The coating structure must have a maximum of uniformly distributed air-filled pores. Differences in brightness of one percentage point of the dry kaolin powder may result in a difference of less than 0.2% brightness of the coated sheet. Coating structure and weight, particle-size distribution, particle shape, and brightness and whiteness of the dry kaolin powder all affect the coated sheet brightness and whiteness.

Opacity of kaolin coatings is strongly influenced by particle packing, which depends largely on particle size and shape and particle-size distribution. Particle size and particle-size distribution must be optimal to develop efficient scattering of light and consequent opacity. The presence of ultrafine particles of ca. 0.1 µm seriously lowers the opacity [86]. The opacity of the coating layer depends on the difference in the refractive indices and the relative proportions of the various components, including air, pigment, and adhesive.

Gloss and smoothness are related to particle size and shape. Superior gloss is attained with kaolins of fine particle size. Smoothness is developed with fine, thin, small-diameter particles. The object of coating paper is to smooth over the fine structure of the base paper to make high-quality ink transfer possible.

Parallel *orientation* of the kaolin particles results in conformation to substrate irregularities, thus reflecting imperfections of the substrate surface. The method of application of the coating affects the orientation and, hence, the smoothness. For optimum effects on irregular substrates, the particle orientation should be random adjacent to the substrate and parallel at the top of the coating [87].

Coating clays are characterized by a low *adhesive demand*, ranging between 10 and 18%. Finer particle size clays require slightly higher amounts of adhesive. The introduction of adhesive can disrupt orientation and reduce smoothness, gloss, and other optical qualities. The selective drainage of adhesive into the substrate or the concentration at the coating surface strongly affects the coating properties [87]. The adhesive requirement is directly related to the surface area of the clay. The amount of adhesive in the coating affects the opacity, brightness, color, and smoothness, as well as film strength. If the amount of adhesive is excessive, the printability is poor.

Film strength is related to preferential adhesive migration into the substrate and to the preferred orientation of the kaolin particles [88]. Maximum film strength is attained with well-oriented, dense films, which imply a highly bonded surface area. If pick occurs within the coating layer, it is generally caused by insufficient adhesive, excessive adhesive migration into the base sheet, or poorly dispersed pigment.

Kaolins are considered to have low ink absorption even though they readily accept printing inks. Substantial differences exist between the *ink receptivity* of coatings made from different grades of kaolin. This is related to film permeability, which is influenced by void volume. Thus, small-diameter particles that are randomly oriented give excellent ink receptivity. The interrelationship of the kaolin and adhesive is a major control over ink receptivity and holdout.

There are many grades of coating and filler clays available to the paper industry. The properties that define these coating and filler clays are based on brightness, particle size and particle-size distribution, and particle shape. Table 3 shows the major coating grades and their particle size and brightness.

Coating Clays. The no. 3 and no. 2 grades are generally used in high-solids formulations for publication-grade paper and medium-finish lower cost enamels. Slick sheet magazines with large circulation are generally coated with one of these two regular grades often mixed with the regular delaminated grade. The no. 1 grade is used to coat high-quality boxboard and enamel papers. This grade gives excellent gloss and brightness and relatively uniform ink receptivity and good smoothness. The fine no. 1 grade is used where maximum gloss and very good ink holdout is desired.

The regular delaminated grade is used on lightweight paper and paper that has a rough surface where good printability must be maintained. This clay gives coatings that are exceptionally smooth with excellent ink and varnish holdout. The fine delaminated grade is used to achieve maximum gloss, good runnability, and high opacity.

The high-brightness grades, having a blue-white tint, are very bright. The coatings that use these grades are very smooth and give excellent gloss ink holdout and good printability. These high-brightness grades are used to coat premium-grade paper.

The special coating grades were developed to meet specific coating needs. The low-glossing coating clays are used when a dull or matte finish is required. The calcined kaolin extender is used to replace titanium dioxide in the coating to reduce the cost of the coating formulation and still maintain the desired optical properties.

Filler Clays. Fillers are highly desirable in printing papers because they increase the opacity, raise the brightness, increase the smoothness or finish, and improve the ink receptivity and printability. Filler clays also improve the appearance and absorbency of paper, as well as increase the density. A perfect filler, if available, would have the following characteristics [89]:

1) Reflectance of 100% at all wavelengths of light
2) High index of refraction
3) Grit-free and a particle size close to 0.3 µm, approximately half the wavelength of light
4) Low specific gravity, soft, and nonabrasive
5) Ability to impart to paper a surface capable of taking any finish, from the lowest matte to the highest gloss
6) Complete retention in the paper web
7) Completely inert and insoluble
8) Reasonable in price

Obviously, such a perfect filler is not available, and filling paper is a matter of compromise. Kaolin is the usual filler in white or essentially white papers such as newsprint, printing grades, uncoated book, to develop certain desirable properties, and in paper may be the only filler used. In some grades, kaolin may be partially replaced by other mineral fillers to develop certain desirable properties, or may be used alone without clay.

Table 3. Particle size and brightness of Georgia coating kaolins

Grades	Mass fraction with particle size < 2 µm, %	Brightness, %
Regular coating clays		
No. 3	73	85.0–86.5
No. 2	80–82	85.5–87.0
No. 1	90–92	87.0–88.0
Fine no. 1	95	86.0–87.5
Delaminated coating clays		
Regular	≈ 70–80	88.0–90.0
Fine	≈ 90–95	87.0–88.0
High-brightness coating clays		
No. 2	80	89.0–91.0
No. 1	92	89.0–91.0
Fine no. 1	95	89.0–91.0
Special coating clays		
Low-glossing clays	≈ 50	87.0–88.0
Calcined extender	85	92.5–94.0

In general, the lighter the weight of the paper, the greater the need for a filler to improve its opacity and brightness. Cost reduction is an important criterion for using fillers. In most cases, the filler is much less expensive than the pulp it replaces. Table 4 gives the filler grades of kaolin that are available.

In general, filler grades are produced to meet brightness and grit specifications. The lowest price filler grade is air-floated, whereas the highest price filler grade is the calcined kaolin extender. A good review of the properties of paper affected by fillers is presented by CASEY [90].

5.1.3. Paint

Kaolins are used extensively in water-based paints. Kaolin is a functional extender pigment that has good covering or hiding power, imparts desirable flow properties to the paint, and is inexpensive. *Brightness* and *particle size* are important physical properties, as is a *resistivity specification*. A resistivity test gives an index of the amount of residual soluble salts that remain in the clay. Higher resistivity values reflect a low soluble salt content, which is important because high soluble salt content adversely affects the dispersion of the paint ingredients and alters paint properties.

Another important physical property is *oil absorption*, which gives an indication of surface area and which is related to a parameter called vehicle demand. A test commonly used by the paint manufacturer to measure dispersion is a fineness of grind measurement using a Hegman gage. Actual paint formulations are made, and many performance tests are run, including gloss, color, smoothness, flow characteristics, dispersion, stability, weathering characteristics, aging, washability, and hiding power.

Table 4. Filler grades of kaolin

Type	Brightness, %	Screen residue, %
Air-floated filler	80–81	0.30
Whole clay filler	81–83	0.10
Whole clay filler	83–85	0.10
Water-washed filler	81–83	0.09
Water-washed filler	83–84	0.09
Water-washed filler	84–86	0.05
Delaminated filler	86–87	0.005
High-brightness delaminated filler	87–89	0.005
Calcined kaolin extender	91–94	0.01

The kaolins used in paint vary from very coarse, which produces a matte finish, to very fine, which gives a high-gloss finish. Calcined kaolins are used in paint as a titanium dioxide replacement and extender. Delaminated clays are used to give a high-gloss finish with good covering power, good color, and excellent washability.

5.1.4. Ceramics

The ceramic industry is a large user of kaolin clays in whiteware, sanitary ware, insulators, and refractories. Several tests are used to evaluate kaolins for use in ceramics, including plasticity, shrinkage, modulus of rupture, absorption, fired color, casting rate, PCE (see 5.1), and chemical analysis.

Coarse-particle-size clays are used for casting clays in sanitary ware, and fine-particle kaolins are used when high strength is important, such as in whiteware and as a bond in certain refractory applications. Ball clays, which are a fine-particlesize, very plastic kaolinitic clay, are used in whiteware, stoneware, enamels, etc. Calcined kaolin called grog is marketed to the refractory industry for use in high-alumina refractories. Both dry-process and wet-process kaolins are used by the ceramic industry (→ Ceramics, General Survey).

5.1.5. Rubber

Kaolin is used by rubber manufacturers because of its reinforcing and stiffening properties and because of its low cost. Fine-particle-size kaolins give good resistance to abrasion and are used extensively in nonblack rubber goods. The kaolins that are used in rubber contain a maximum of 1% free moisture. Both dry-process and wet-process kaolins are used, but by far the largest tonnage are dry-process grades. Brightness, particle size, and grit percentage are important physical properties. Other tests that are important in evaluating kaolin for use in rubber are water settling characteristics and oil absorption. The water settling test gives a good idea of the reinforcing characteristics of the clay, and oil absorption also correlates with reinforcement. Tests on the clay-filled rubber itself are stress–strain, tear resistance, abrasion resistance, heat generation, energy rebound, extrusion and plasticity, hardness, aging characteristics, and water absorption [91].

5.1.6. Plastics

Kaolin is used in plastics because it aids in producing a smooth surface finish, reduces cracking and shrinkage during curing, obscures the fiber pattern when fiberglass is used as reinforcement, contributes to a high dielectric strength, improves resistance to chemical action and weathering, and helps control the flow properties. Kaolin is a functional filler in that it contributes some beneficial effects. Loadings in various plastic compositions vary from ca. 15% to as high as 60%.

5.1.7. Ink

Kaolin clays are used extensively in printing inks. In black inks the pigment is normally carbon black, but in white and other colored inks kaolin is a common extender pigment. The kaolins used in ink must be fine particle size and relatively white. A major disadvantage to the use of kaolin in ink is its hydrophilic nature. However, coating the kaolin particles with hydrophobic material increases its usefulness in inks. Surface-modified kaolins have found a significant market in printing inks.

5.1.8. Catalysts

Kaolin clays are used extensively in the production of cracking catalysts and molecular sieves. The kaolin clay should have relatively low iron and sulfate content. Although the surface of the kaolin does contribute some catalytic activity, the kaolin serves largely as a diluent for the silica–alumina gels [92]. Partially calcined kaolins, which are heated to a temperature just above the dehydroxylation temperature of ca. 600 °C, are used in the manufacture of molecular sieves [93].

5.1.9. Fiberglass

Kaolin clays with a low iron content are used as a source of alumina and silica in the manufacture of fiberglass. A good grade of air-float kaolin is generally used.

5.2. Smectite

Smectite is a group name, and many minerals comprise the smectite group, including sodium montmorillonite, calcium montmorillonite, saponite, nontronite, hectorite, and sauconite. Each of these minerals has unique properties, and therefore, the individual smectite minerals and the other minerals that are present in the bentonite must be determined. This is done by X-ray diffraction, electron microscopy, differential thermal analysis, infrared spectrometry, and selective chemical analysis. Once the smectite mineral has been identified, then some of the properties can be established, for example, viscosity, thixotropy, plasticity, shrinkage, bonding strength, shear strength, and water impedance. These properties are not the same for all smectites and, in fact, are quite different.

The exchangeable cations present between the silica and alumina sheets (see Fig. 5) have a strong influence on the use and properties of the smectite. The major uses of smectites (bentonites) are in drilling muds, foundry sand bond, and iron-ore pelletizing. In addition to these three major uses, bentonite is used for many miscellaneous products, including filtering agents, water impedance, cosmetics, animal feed, pharmaceuticals, paint, ceramics, slurry trenching, catalysts, and decolorizing.

Drilling Fluids. Sodium montmorillonite, which is the major constituent of the Wyoming or Western bentonites, is a high-swelling clay with some unique properties. It is used throughout the world as an ingredient in drilling fluids because commonly this Western bentonite yields 100 barrels per ton (20 m^3/t) [31]. Only 5% of this bentonite is necessary to produce the high viscosity, thixotropy, and low filter-cake permeability required in the drilling fluid. Hectorite, the lithium smectite, also gives high yields and has excellent properties as an ingredient in drilling fluids. Some calcium montmorillonites can be treated with a sodium salt, usually soda ash, to produce a bentonite suitable for use in drilling fluids, but in general such treatment does not substantially improve the water loss property of the bentonite.

Foundry Molding Sands. The molding sands used in foundries are composed of sand and clay. The clay provides bonding strength and plasticity. A small amount of tempering water is added to the mixture to make it plastic. It can then be molded around a pattern and be cohesive enough to maintain the shape after the pattern is removed and while molten metal is poured into it. The important foundry properties are green compression strength, dry compression strength, hot strength, flowability, permeability, and durability. These properties vary greatly with the amount of tempering water [31]. Both sodium and calcium bentonites (montmorillonites) are used as bonding clays. Calcium bentonite has a higher green strength, lower dry strength, lower hot strength, and better flowability than sodium bentonite. Final evaluations of the bonding clay can only be made after the bentonite is actually used in foundry practice so that flowability, durability, ease of shakeout of the casting from the mold, and cleanness of the surface of the cast metal can be determined. Commonly, blends of sodium and calcium bentonites are used to gain the optimum properties from each.

Iron-Ore Pelletizing. Bentonites are used extensively in pelletizing iron ores [94]. Finely pulverized ore concentrate must be pelletized into units of ca. 1 in. (2.5 cm) or more in diameter before it is used as blast furnace feed. Pelletized ore is a superior blast furnace feed. Bentonite is ca. 0.5 wt% of the ore. Because of its superior dry strength sodium bentonite is the preferred bentonite for pelletizing ores.

Other Uses. Acid-activated bentonites are widely used to *decolorize* mineral, vegetable, and animal oils. The clay also serves to deodorize, dehydrate, and neutralize the oils. In addition, the clay must have low oil retention and good filtration characteristics. It must not change the character of the oil nor give an objectionable taste to the oil. Some bentonitic clays naturally possess adequate decolorizing ability. Such materials are called fuller's earth. The bentonites that work best as both natural and acid-activation decolorizers are the calcium variety.

Some clay *catalysts* for cracking petroleum are produced from bentonites with very low iron content. The bentonite is reacted with sulfuric or hydrochloric acid at elevated temperatures followed by washing, drying, and calcining at ca. 500 °C. This process removes the adsorbed alkali metals and alkaline-earth metals along with the removal of iron and partial removal of magnesium and aluminum from the lattice. This increases the surface area and alters the pore-size distribution of the product. Calcium bentonite is the preferred material for making a smectite clay catalyst.

Sodium bentonites are used in *water impedance*. The clay is used to impede the movement of water through earthen structures such as dams, to stop seepage of water from ponds and irrigation ditches, and to stop water from entering basements of homes. One use of high-swelling sodium bentonite is in the slurry-trench or diaphragm-wall method of excavation in construction in areas of unconsolidated rock or soil [95]. In this method, the trench being excavated is filled with bentonite slurry, and the earth being excavated is removed through it. A thin filter cake on the walls of the excavation prevents loss of fluid, and the hydrostatic head of the slurry prevents caving and running of loose soil, which makes costly shoring unnecessary. Small amounts of sodium bentonites are added to cement to improve workability, lessen aggregate segregation, and enhance impermeability.

Sodium bentonites are widely used as *emulsifying* and *suspending agents* in many oil and water systems and in medicinal, pharmaceutical, and cosmetic formulations [96]. Sodium bentonites are used in both oil- and water-based paints. In water-based paints, the bentonite acts as a suspending and thickening agent. In oil-based paints, it is used as an emulsifying agent.

In *paper*, sodium bentonite is sometimes used in the deinking process to recover the cellulose fibers. These deinking processes generally involve heating the paper to be deinked in a caustic soda solution to break down the ink, thus freeing the ink pigment.

Next a detergent is added to force the pigment from the paper fiber. Smectite is then added to disperse the pigment particles and adsorb them. Washing removes the smectite, which carries the ink pigment with it. Sodium bentonite is also used in papermaking to prevent the agglomeration of pitch, tar, wax, and resinous material. The addition of 0.5% smectite, based on the dry weight of the paper stock, prevents agglomerates so that the globules will not stick to screens, machine wires, and press rolls and thus cause defects in the paper.

Sodium montmorillonite is the base for the organoclays described earlier. The organoclad clays are used in lubricating greases, inks, paints, and oil-based drilling fluids. These clays improve pigment suspension, viscosity, and thixotropy control. They also improve brushability and spraying characteristics of paints.

Calcium montmorillonite is used in adhesives; as granular pet litter; to clarify wines, cider, and beer; as floor absorbents; carriers and diluents in pesticide preparations; and in water clarification. White calcium smectites are used in detergents, paints, enamel suspensions, pharmaceuticals, cosmetics, food, pelletized animal feed, and in no-carbon-required copy paper [25].

Both calcium and sodium smectites are used as additives in ceramics, particularly because of their plasticity and relatively high green and dry strengths. For example, many brick clays are silty materials. The addition of a small amount of smectite improves the plasticity and green and dry strengths. In addition, both types are used as a liner for toxic and low-level radioactive waste disposal sites and as additives to latex for thickening and stabilizing.

5.3. Hormite

The use of the hormite clays is almost as varied as that of the kaolins and smectites. Because of the particle shape of the hormite clays, i.e., palygorskite (attapulgite) and sepiolite are elongate (see Fig. 10), they have some special applications. The specific type of hormite mineral can be identified with X-ray diffraction and is generally characterized by using scanning electron microscopy to determine the crystal size and length. In addition to the crystal shape, another set of important properties is the absorption and adsorption characteristics.

One of the more important, larger uses is *drilling fluids*. Considerable quantities of palygorskite and sepiolite are used in drilling fluids because the viscosity and gel strength of the mud are not affected by variations in electrolytic content as are those of muds comprised of bentonite. Thus, palygorskite and sepiolite drilling muds can be used with salt water or when formation brines become a serious problem. In recent years, the offshore drilling industry has used salt-water-based muds, so that the quantities of palygorskite and sepiolite used have increased dramatically.

Because palygorskite and sepiolite are not easily flocculated on account of their particle shape, these clay minerals are used as *suspending agents* in paint, medicines, pharmaceuticals, and cosmetics. In paint, palygorskite is used because it improves the thickening and thixotropic properties. Another major use is as *floor sweep compounds*, for absorbing oil and grease spills on factory floors, in service stations, and other areas where oil and grease spills are a problem. Granular particles of palygorskite are the most effective floor absorbent.

Palygorskite and sepiolite are also used extensively in *agriculture* as absorbents and adsorbents for chemicals and pesticides. The chemical is mixed onto the granular hormite clay particle, and the treated particle is placed in the ground with the seed particle. The pesticide or fertilizer chemical is slowly released to provide the necessary fertilizer or protection for the growing plant. Finely pulverized hormite clays are also used to adsorb chemicals, which can then be dusted or sprayed on the plant or on the surface of the ground.

Another large use for hormite clays is *pet litter*. Granular particles of hormite are an effective litter for absorbing animal waste, particularly for domestic cats. The clay effectively absorbs the waste and odors.

HADEN [97] described the properties and uses of palygorskite and pointed out both the colloidal and noncolloidal characteristics that make this mineral important. Other uses of hormites include foundry sand binder, polishes, wax emulsions, metal drawing lubricants, laundry washing powders, bonding agents for granulation of powders, decolorizing oils, anti-caking agents, flattening agent in paints, catalyst in no-carbonrequired paper, and catalyst carrier.

6. Environmental Aspects

Almost all clays are surface mined, so the industry is required to reclaim the disturbed land in most countries. Common practice is to open a cut and then spoil the overburden from the following panels or cuts into the mined out areas. The land is leveled or sloped to meet the governmental requirements and then planted with grasses or trees.

In the processing plants the waste materials, which include chemicals and separated clay particles or other minerals, are collected in impounds. The clay and other particles are flocculated with alum or other chemical flocculants, and the clear water is released into streams after adjusting the pH to 6–8.

Air quality is maintained in the processing plants by using dust collectors on the dryers and by enclosing transfer points where dry clay may cause dust. In areas such as the bagging departments, the workers may be required to wear dust masks. None of the clays, kaolins, smectites, or hormites are health hazards. Any dust, if inhaled in large quantities, can be a problem, but the industries use dust collectors and require workers to wear masks. Therefore, no serious lung ailments result. The clays themselves contain no deleterious trace elements or chemicals that are hazardous to human health.

7. Production and Consumption

Clays are important industrial minerals in most parts of the world. As indicated in Chapter 5 they are utilized in many process industries and are necessary ingredients in a great many products: kaolins are used in paper, ceramics, paint, etc.; bentonites, in drilling fluids and foundries; and hormites, in special drilling fluids and as suspending agents in paints and pharmaceuticals. The total economic value for this group in 1984 from U.S. Bureau of Mines statistics was 40×10^6 t (44 million short tons) at a value of $ 1.1×10^9.

The total kaolin production in the world is ca. 25×10^6 t/a. The three largest producing countries are the United States with ca. 8×10^6 t, the United Kingdom with ca. 4.5×10^6 t, and the USSR with ca. 3.5×10^6 t. The paper and the ceramic industries are the largest consumers. In the United States the average price per short ton (0.9 t) for processed kaolin is ca. $ 80, with a range in price from $ 34 for some filler clays to $ 150 for high-quality coating clay. A fine-particle-size, high-brightness calcined kaolin sells for ca. $ 350 per short ton. There is a large export market for kaolins, especially coating grade clays produced in the United States and England. The largest markets for these clays are in Western Europe and Japan.

Bentonite production is ca. 6×10^6 t/a at the present time. In 1981 the production was ca. 8×10^6 t, primarily because of the extensive drilling programs for oil and gas around the world. With the present decline in drilling for new oil and gas reserves, the usage of bentonite has severely decreased. The United States is by far the largest producing country at 3.5×10^6 t, with the USSR next with ca. 750 000 t, and Greece third with about the same production. The average price per short ton ranges from ca. $ 30 for the low-swelling varieties to $ 45 for the high-swelling quality. High-swelling drilling mud quality is exported from the United States to most countries in the world where deep oil well drilling is taking place.

Hormite clay production in the world is ca. 3×10^6 t/a. Again, the United States is the largest producing country with ca. 2×10^6 t shipped per year. Spain is the second largest producer with 350 000 t/a, and third is Senegal with ca. 150 000 t/a. The average price per short ton for hormite clay is ca. $ 60. The largest consumption of hormite is for pet litter with the second being oil and grease absorbents.

No world figures are available for the production of common clay and shale, but the United States produces ca. 16×10^6 t annually at an average value of $ 3.50 per short ton. The figure for the world is probably many times that because common clays and shales are used in greater quantities for brick and structural-clay products in many parts of the world than in the United States.

The production and consumption of clays in the world are large tonnage figures with fairly high values. Clays are important in the economy of many countries, and their use enhances our standard of life. The export markets for kaolin, swelling bentonite, palygorskite, and sepiolite are important and continue to grow because clays are about the least expensive ingredient that goes into many products — but the clays perform necessary functions.

8. References

[1] C. K. Wentworth: "A Scale of Grade and Class Terms for Clastic Sediments," *J. Geol.* **30** (1922) 377–392.
[2] R. E. Grim: *Clay Mineralogy*, 1st ed., McGraw-Hill, New York 1953, p. 384.
[3] G. W. Brindley, G. Brown (eds.): *Crystal Structures of Clay Minerals and Their X-ray Identification*, Monograph 5, Mineralogical Soc., London 1980, p. 495.
[4] W. J. Smothers, Y. Chiang: *Handbook of Differential Thermal Analysis*, Chem. Pub. Co., New York 1966, p. 633.
[5] H. Beutelspacher, H. W. Van Der Marel: *Atlas of Electron Microscopy of Clay Minerals and Their Admixtures*, Elsevier, Amsterdam 1968, p. 333.
[6] H. W. Van Der Marel, H. Beutelspacher: *Atlas of Infrared Spectroscopy of Clay Minerals and Their Admixtures*, Elsevier, Amsterdam 1976, p. 396.
[7] C. S. Ross, P. F. Kerr: "Halloysite and Allophane," *Geol. Surv. Prof. Paper (U.S.)* no. 185-G (1934) 135–158.
[8] C. S. Ross, P. F. Kerr: "The Kaolin Minerals," *Geol. Surv. Prof. Paper (U.S.)* no. 165-E (1931) 151–180.
[9] T. F. Bates et al.: "Morphology and Structure of Endellite and Halloysite," *Am. Mineral.* **35** (1950) 463–484.
[10] J. W. Gruner: "Vermiculite and Hydrobiotite Structures," *Am. Mineral.* **19** (1934) 557–575.
[11] R. E. Grim et al.: "The Mica in Argillaceous Sediments," *Am. Mineral.* **22** (1937) 813–829.
[12] S. W. Bailey: "Chlorites" in J. E. Gieseking (ed.): *Soil Components*, vol. **II.**, Springer-Verlag, New York 1975, pp. 191–263.
[13] W. F. Bradley: "The Structural Scheme of Attapulgite," *Am. Mineral.* **25** (1940) 405–410.
[14] R. C. Reynolds in [3], pp. 249–303.
[15] H. H. Murray in S. Shimod (ed.): *7th Int. Kaolin Sym. Proc.*, Univ. of Tokyo 1976, pp. 114–125.
[16] E. M. Durrance et al.: "Hydrothermal Circulation and Post-Magnetic Changes in Granites of Southwest England," *Proc. Ann. Conf., Usher Soc.*, 1982, pp. 304–320.
[17] C. M. Bristow: "Kaolin Deposits of the United Kingdom of Great Britain and Northern Ireland," *Int. Geol. Congr. Rep. Sess.* 23rd (1969) 275–288.
[18] H. H. Murray, P. Partridge: "Genesis of Rio Jari Kaolin," *Devel. Sedimentol.* **35** (1982) 279–291.
[19] M. Kuzvart: "Kaolin Deposits of Czechoslovakia," *Int. Geol. Congr. Rep. Sess.* 23rd (1969) 25–32.
[20] M. Storr: *Kaolin Deposits of The GDR in The Northern Region of the Bohemian Massif, Guidebook*, 5th Int. Kaolin Sym., Ernst-Moritz-Arndt Univ., Greifswald 1975, p. 243.
[21] H. M. Koster in S. Shimoda (ed.): *7th Int. Kaolin Sym. Proc.*, Univ. of Tokyo, 1976, p. 83.
[22] L. Martin-Vivaldi: "Kaolin Deposits of Spain," *Int. Geol. Congr. Rep. Sess.* 23rd (1969) 225–261.
[23] V. P. Petrov: "Kaolin Deposits of the USSR," *Int. Geol. Congr. Rep. Sess.* 23rd (1969) 289–319.
[24] H. H. Murray, C. C. Harvey: *Australian and New Zealand Clays*, Preprint 82–83, 1st Int. SME-AIME Fall Mtg., Honolulu, Hawaii, 1982, p. 5.
[25] H. H. Murray in R. W. Hagemeyer (ed.): *Paper Coating Pigments*, Tappi Press, Atlanta, Ga., 1984, pp. 95–141.
[26] S. H. Patterson, H. H. Murray: "Kaolin, Refractory Clay, Ball Clay, and Halloysite in N. America, Hawaii, and the Caribbean Region," *Geol. Surv. Prof. Pap. U.S.* no. 1306 (1984) 56.
[27] F. H. Norton: *Refractories*, 4th ed., McGraw-Hill, New York 1968, p. 450.

[28] W. C. Knight: "Bentonite," *Eng. Min. J.* **66** (1898) 491.
[29] C. S. Ross, E. V. Shannon: "Minerals of Bentonite and Related Clays and Their Physical Properties," *J. Am. Ceram. Soc.* **9** (1926) 77–96.
[30] P. C. Wright: "Meandu Creek Bentonite — A Reply," *J. Geol. Soc. Aust.* **15** (1968) 347–350.
[31] R. E. Grim, N. Guven: "Bentonites, Geology, Mineralogy, Properties and Uses," *Devel. Sedimentol.* **24** (1978) 256.
[32] S. H. Patterson, H. H. Murray, in S. J. LeFond (ed.): *Ind. Minerals and Rocks,* 5th ed., AIME 1983, pp. 585–651.
[33] M. M. Knechtel, S. H. Patterson: "Bentonite Deposits of the Northern Black Hills District, Wyoming, Montana, and South Dakota," *U.S. Geol. Surv. Bull.* **1082** (1962) 893–1029.
[34] M. Slaughter, J. M. Early: "Mineralogy and Geological Significance of the Mowry Bentonites," *Spec. Pap. Geol. Soc. Am.* **83** (1965) 116.
[35] R. E. Grim: "Preliminary Report on Bentonite in Mississippi," *Miss. State Geol. Surv. Bull.* **22** (1928) 14.
[36] R. E. Grim, G. Kulbicki: "Montmorillonite: High Temperature Reaction and Classification," *Am. Mineral.* **46** (1961) 1329–1369.
[37] A. F. Hagner: "Adsorptive Clays of the Texas Gulf Coast," *Am. Mineral.* **24** (1939) 67–108.
[38] P. Y. Chen: *Geology and Mineralogy of the White Bentonite Beds of Gonzales County, Texas,* Ph.D. Thesis, University of Texas 1968.
[39] R. L. Sloan, J. M. Gilbert: "Electron-Optical Study of Alterations in the Cheto Clay Deposit," *Clays Clay Miner.* **15** (1966) 35–44.
[40] K. G. Papke: "Montmorillonite Deposits in Nevada," *Clays Clay Miner.* **17** (1969) 211–222.
[41] W. H. Monroe: "Notes on Deposits of Selma-Ripley Age in Alabama," *Ala. Geol. Surv. Bull.* **48** (1941) 150.
[42] L. L. Ames, Jr., et al.: "A Contribution on the Hector, California, Bentonite Deposit," *Econ. Geol.* **53** (1958) 22–37.
[43] C. S. Denny, H. Drewes: "Geology of the Ash Meadows Quadrangle, Nevada–California", *U.S. Geol. Surv. Bull.* **1181-L** (1965) 56.
[44] I. A. Cowperthwaite et al.: "Sedimentation, Petrogenesis, and Radioisotopic Age of the Cretaceous Fuller's Earth of Southern England," *Clay Miner. Bull.* **9** (1972) 309–327.
[45] A. Hallam, B. W. Selwood: "Origin of the Fuller's Earth in the Mesozoic of Southern England," *Nature (London)* **220** (1968) 1193–1195.
[46] R. Fahn: "The Mining and Preparation of Bentonite," *InterCeram* **12** (1965) 119–122.
[47] V. Szeky-Fux: "The Hydrothermal Genesis of Bentonites on the Basis of Studies in Komloska," *Acta Geol. Acad. Sci. Hung.* **4** (1957) 361–382.
[48] G. C. DeAngeles, G. Novelli: "Bentonite from Ponza," *Geotectonia* **4** (1957) 1–7.
[49] V. Annedda: "Deposits of Bentonite on Sadali and Vallanova Tulo Territories, Sardinia", *Resoconti Assoc. Min. Sarda* **60** (1956) 5–9.
[50] T. E. Wayland: "Geologic Occurrence and Evaluation of Bentonite Deposits," *Trans. AIME* **250** (1971) 120–132.
[51] E. Lukas: "State of Exploration of Zaluska Gorica Bentonites in Comparison with other Yugoslavian Bentonites," *Rud. Metal. Zb.* 1968, no. 3, 277–286.
[52] N. N. Siddiguie, D. P. Bahl: "Geology of the Bentonite Deposits of the Barmer District, Rajasthan," *Mem. Geol. Surv. India* **96** (1965) 36.
[53] S. Iwao: "The Clays of Japan," *Int. Clay Conf. Proc. 1969,* 209 pp.
[54] M. A. Rateev: "Sequence of Hydrothermal Alteration of Volcanic Rocks into Bentonite Clays of the Askana Deposit in the Georgian S.S.R.," *Dokl. Akad. Nauk SSSR* **175** (1967) 675–678.

[55] G. S. Dsotsenidze, E. A. Matchabeley: "The Genesis of Bentonites of Georgian S.S.R.," *Int. Clay Conf. Proc. 1963*, pp. 203–210.
[56] "Bentonite, Sepiolite, Attapulgite," *Industrial Minerals Magazine* 1978, no. 126, 50.
[57] S. H. Patterson: "Fuller's Earth and Other Industrial Mineral Resources of the Meigs-Attapulgus-Quincy District, Georgia and Florida," *Geol. Surv. Prof. Pap. (U.S.)* no. 828 (1974) 45.
[58] L. Wirth: *Attapulgites du Sénégal Occidental*, Rapport No. 26, Laboratoire de Géologie, Faculté des Sciences, Université de Dakar 1968, p. 55.
[59] F. D. Ovcharenko et al.: *The Colloid Chemistry of Palygorskite*, Israel Program for Scientific Translations, Jerusalem 1964, p. 101.
[60] E. Galan, J. J. Brell, A. La Iglesia, R. H. S. Robertson: "The Caceres Palygorskite Deposit," *Proc. Int. Clay Conf. 1975* (1976) 81–94.
[61] H. N. Khoury, D. D. Eberl, B. F. Jones: "Origin of Magnesium Clays from the Amargosa Desert, Nevada," *Clays Clay Miner.* **30** (1982) 327–336.
[62] M. K. H. Siddiqui: *Bleaching Earths*, Pergamon Press, New York 1968, p. 86.
[63] H. H. Murray: "Clay," in J. L. Gilson (ed.): *Industrial Minerals and Rocks*, AIME, New York 1960, pp. 259–284.
[64] H. H. Murray: "Mining and Processing Industrial Kaolins," *Ga. Miner. Newsl.* **16** (1963) 3–11.
[65] H. H. Murray: "Dry Processing of Clay and Kaolin," *InterCeram.* **31** (1982) no. 2, 108–110.
[66] H. H. Murray: "Major Kaolin Processing Developments," *Int. J. Miner. Process.* **7** (1980) 263–274.
[67] H. H. Murray: "Equipment for Wet Processing of Clay Material," *InterCeram* **31** (1982) no. 3, 196–197.
[68] F. A. Gunn, H. H. Morris, US 3 171 718, 1965.
[69] E. W. Greene, J. B. Duke: "Selective Froth Flotation of Ultrafine Minerals or Slimes," *Trans. Soc. Min. Eng. AIME* **223** (1962) 389–395.
[70] W. M. Bundy, J. P. Berberich, US 3 477 809, 1969.
[71] R. N. Maynard, B. R. Skipper, N. Millman, US 3 371 988, 1968.
[72] V. V. Mercade, US 3 701 417, 1972.
[73] J. Iannicelli: "New Developments in Magnetic Separation," *IEEE Trans. Magn., MAG-12* (1976) 436–443.
[74] J. A. Wernham, T. H. Falconer: *The Eriez Commercial Superconducting High Gradient Separators (Abstract)*, Conf. on Superconducting Magnetic Separation, Imperial College, London 1985, p. 2.
[75] H. R. Franselow, D. A. Jacobs, US 3 586 523, 1971.
[76] W. M. Bundy, J. L. Harrison, J. G. Ishley: "Chemically Induced Kaolin Floc Structures for Improved Paper Coating," *Proc. Tech. Assoc. Pulp Pap. Ind.* 1983, 175–187.
[77] R. F. Grim: *Applied Clay Mineralogy*, McGraw-Hill, New York 1962, p. 375.
[78] P. G. Nahin, US 3 248 314, 1966.
[79] P. W. Libby, J. Iannicelli, C. R. McGill: "Elastomer Reinforcement with Amino Silane Grafted Kaolin," *Abstr. Pap. Am. Chem. Soc. 154th 1967* (Div. of Rubber Chem.).
[80] K. H. Teague, "Southern Bentonite," SME Preprint No. 72 H 328 (1972), 7 pp.
[81] J. W. Jordan, US 2 531 440, 1950.
[82] H. H. Murray: "Kaolin — A Versatile Pigment, Extender, and Filler," *Prepr. Pap. Annu. Meet. Tech. Sect. C.P.P.A. 63rd*, 1977, A29–34.
[83] H. H. Murray: *Industrial Applications of Kaolin, Clays and Clay Minerals*, vol. **10,** Pergamon Press, London 1963, pp. 291–298.

[84] H. H. Murray: "Diagnostic Tests for Evaluation of Kaolin Physical Properties," *Acta Mineral.-Petrogr.*, **23** (1980) 67–76.

[85] A. S. Michaels: "Deflocculation of Kaolinite by Alkali Polyphosphates," *Ind. Eng. Chem.* **59** (1958) 951–958.

[86] W. M. Bundy, W. D. Johns, H. H. Murray: "Physico-Chemical Properties of Kaolinite and Relationship to Paper Coating Quality," *Tappi* **48** (1965) 688–696.

[87] G. A. Hemstock, J. W. Swanson: "The IGT Printability Tester as an Instrument for Measuring the Adhesive Strength of Pigment-Coated Paper," *Tappi* **40** (1957) 794–801.

[88] A. C. Eames: "The Transverse Tensile Strength of Clay-Starch Coatings," *Tappi* **43** (1960) 2–10.

[89] W. R. Willets: *Paper Loading Materials*, Introduction-Tappi Monograph 19, New York 1958, p. 5.

[90] J. P. Casey: *Pulp and Paper. Papermaking* (vol. 2), Interscience, New York 1960, pp. 985–1020.

[91] *Kaolin Clays and Their Industrial Uses*, J. M. Huber Corp., New York 1955, p. 214.

[92] J. S. Magee, J. J. Blazek in "Preparation and Performance of Zeolite Cracking Catalysts", *ACS Monogr.* **171** (1976) 615–679.

[93] D. W. Breck: *Zeolite Molecular Sieves*, J. Wiley & Sons, New York 1974, pp. 314–320.

[94] F. D. DeVaney, US 2 743 172, 1956.

[95] W. J. Lang: *Bentonite: The Demand and Markets for the Future*, SME Preprint No. 71 H 29, AIME 1971, p. 9.

[96] G. Novelli, "Properties and Possible Use of Montmorillonite in Pharmaceutical and Cosmetic Preparation," *Riv. Ital. Essenze, Profumi, Piante Off., Olii Veg., Saponi* **54** (1972) 267–270.

[97] W. L. Haden, Jr.: "Attapulgite: Properties and Uses," *Clays Clay Miner.* **10** (1963) 284–290.

Coal

JOHN C. CRELLING, Southern Illinois University at Carbondale, Carbondale, Illinois 62901, United States (Chaps. 1–4 and 10–12)

DIETER H. SAUTER, Lurgi GmbH, Frankfurt, Federal Republic of Germany (Chaps. 5 and 9)

RAJA V. RAMANI, The Pennsylvania State University, University Park, Pennsylvania 16802, United States (Chap. 6)

DIETER LEININGER, Bergbauforschung GmbH, Essen, Federal Republic of Germany (Chap. 7)

BERNHARD BONN, Bergbauforschung GmbH, Essen, Federal Republic of Germany (Chap. 8)

UDO BERTMANN, Bergbauforschung GmbH, Essen, Federal Republic of Germany (Chap. 8)

RAINER REIMERT, Lurgi GmbH, Frankfurt, Federal Republic of Germany (Chap. 9)

WOLFGANG GATZKA, Ruhrkohle AG, Essen, Federal Republic of Germany (Chap. 13)

1.	Coal Petrology	1436	6.2.	Surface or Underground Mining ... 1456
1.1.	Coal Characterization	1436	6.3.	Surface Coal Mining Methods ... 1458
1.2.	The Maceral Concept	1437	6.4.	Underground Coal Mining Methods ... 1462
1.3.	Maceral Types and Properties	1437	7.	Hard Coal Preparation ... 1471
2.	Coalification	1441	7.1.	Preliminary Treatment and Classification of Raw Coal ... 1471
3.	Occurrence	1443	7.2.	Wet Treatment ... 1473
3.1.	Coal Seam Occurrence	1443	7.2.1.	Cleaning in Jigs ... 1473
3.2.	Coal Seam Structures	1443	7.2.2.	Cleaning in Dense-Media Separators ... 1474
3.3.	Coal Seam Distribution	1445	7.2.3.	Cleaning by Flotation ... 1474
4.	Classification	1445	7.2.4.	Other Cleaning Methods ... 1475
5.	Chemical Structure of Coal	1449	7.3.	Dewatering ... 1475
5.1.	Characterization of Coals	1451	7.4.	Decantation and Thickening of the Process Water ... 1476
5.1.1.	Standard Analytical Methods	1451	7.5.	Dosing and Blending ... 1477
5.1.2.	Laboratory and Bench-Scale Simulation Tests	1453	7.6.	Removal of Pyritic Sulfur ... 1477
5.1.3.	Petrographic Studies	1453	7.7.	Thermal Drying ... 1478
5.2.	Structural Deductions from Analytical and Bench-Scale Data	1454	8.	Coal Conversion (Uses) ... 1478
5.3.	Bonding of Elements in Coal	1454	8.1.	Preparation ... 1478
5.4.	Structural Evidence of Coals	1455	8.2.	Briquetting ... 1480
6.	Coal Mining	1456	8.3.	Carbonization and Coking ... 1480
6.1.	General Considerations	1456		

8.4.	Pyrolysis	1481
8.5.	Coal Liquefaction	1482
8.5.1.	Direct Liquefaction	1482
8.5.2.	Indirect Liquefaction	1483
8.6.	Coal Gasification	1483
8.7.	Coal Combustion	1486
8.8.	Conversion of Coal for Purposes Other Than the Generation of Energy	1486
8.8.1.	Coal Tar Chemical Industry	1486
8.8.2.	Production and Use of Activated Carbon	1486
9.	Agglomeration	1487
10.	Transportation	1488
11.	Coal Storage	1489
12.	Quality and Quality Testing	1490
12.1.	Chemical Analyses	1490
12.2.	Mineral Matter	1492
12.3.	Thermal Properties	1493
13.	Economic Aspects	1493
13.1.	World Outlook	1493
13.2.	Some Major Coal Producing Countries	1494
13.2.1.	United States of America	1494
13.2.2.	Australia	1497
13.2.3.	Federal Republic of Germany	1498
13.2.4.	Poland	1498
13.2.5.	Republic of South Africa	1499
14.	References	1500

1. Coal Petrology

1.1. Coal Characterization

Coal is an extremely heterogeneous, complex material that is difficult to characterize. Coal is a rock formed by geological processes and is composed of a number of distinct organic entities (macerals) and lesser amounts of inorganic substances (minerals). Each of the coal macerals and minerals has a unique set of physical and chemical properties; these in turn control the overall behavior of coal. Although much is known about the properties of minerals in coal, for example, the crystal chemistry, crystallography, and magnetic and electrical properties, surprisingly little is known about the properties of individual coal macerals. Even though coal is composed of macerals and minerals, it is not a uniform mixture of these substances. The macerals and minerals occur in distinct associations called lithotypes, and each lithotype has a set of physical and chemical properties that also affect coal behavior.

Coal seams, the basic units in which coal occurs, are in turn composed of layers of coal lithotypes, and individual coal seams may also have their own sets of physical and chemical properties. For example, even if two coal seams have the same maceral and mineral composition, the seams may have significantly different properties if the maceral associations in lithotypes in the two seams are different. The enclosing rocks immediately adjacent to a coal seam can also affect the properties of the coal. This aspect is particularly important in mine design, production, and strata control. The compositional characterization of a coal seam must cover the nature of the macerals, the lithotypes, the entire seam, and the association of the seam with the neighboring strata.

In addition to compositional factors, coal properties also change with the rank or the degree of coalification of a given sample of coal. Coal is part of a metamorphic series ranging from peat, through lignite and subbituminous and bituminous coal, to anthracite. Temperature, pressure, and time alter the original precursors of coal through this metamorphic series. As the rank of the coal changes, the properties of the coal macerals change progressively and, therefore, so also do the properties of lithotypes and the entire seam.

Because of these factors, coal characterization requires a detailed knowledge of both the maceral composition and the rank of the coal. All coal properties are ultimately a function of these two factors.

1.2. The Maceral Concept

The term maceral was introduced by STOPES to distinguish the organic components of coal (macerals) from the inorganic components (minerals) [1]. The term is now interpreted in two conflicting ways. The interpretation of the International Committee for Coal Petrology (ICCP) is that a maceral is the smallest microscopically recognizable component of coal [2]. This conception, generally held by most European coal petrographers, is based on morphology and other criteria such as size, shape, botanical affinity, and occurrence. The ICCP concept implies that the properties of the macerals change with rank. Thus, the same maceral, vitrinite, can exist in a bituminous coal as well as in an anthracite.

In contrast, the maceral concept of SPACKMAN (commonly held in North America) is based on the idea that macerals are substances with distinctive sets of properties [3]. Thus, vitrinite in a bituminous coal is viewed as a different maceral than vitrinite in an anthracite because the two materials have different properties.

The basis of the petrographic study of coal composition is the idea that coal is composed of a number of distinct macerals. The entire body of coal petrographic literature supports this idea and is in direct contrast to the earlier chemical concept of coal being composed of a single unique molecular substance. Although the concept and the term "coal molecule" may once have been useful, modern chemical studies of coal recognize coal's heterogeneous maceral composition.

1.3. Maceral Types and Properties

A large number of different macerals have been named and classified in various systems. The ICCP system is given in part in Table 1. Although this system is useful for some purposes, it is impractical for routine maceral analysis because of the large number of terms. For such routine analysis, classification systems with a limited number of terms are needed. In the standard method for maceral analysis D 2799 of

Table 1. Survey of the macerals of hard coal, according to the ICCP system

Maceral group	Maceral	Maceral type*	Maceral variety*	Kryptomaceral*
Vitrinite	telinite		cordaitotelinite fungotelinite xylotelinite	
	collinite	telocollinite gelocollinite desmocollinite corpocollinite		kryptotelinite kryptocorpocollinite
	vitrodetrinite			
Exinite	sporinite		tenuisporinite crassisporinite microsporinite macrosporinite	kryptoexosporinite kryptointosporinite
	cutinite			
	resinite			
	alginite			
	liptodetrinite			
Inertinite	micrinite			
	macrinite			
	semifusinite			
	fusinite	pyrofusinite degradofusinite		
	sclerotinite		plectenchyminite corposclerotinite pseudocorposclerotinite	
	inertodetrinite			

* Incomplete, can be arbitrarily extended.

the American Society for Testing and Materials (ASTM), only six terms are required, although some additional terms are defined in ASTM Standard D 2796 (see Table 2). Although there is no standard method for the analysis of fluorescent macerals, some additional terms, listed in Table 2, are used for this type of analysis.

As shown in Table 2, coal macerals fall into three distinct groups: vitrinite, liptinite, and inertinite. *Vitrinite macerals* are generally the most abundant, commonly making up 50–90% of North American coals. This group is not as abundant in coals that originate in the southern hemisphere.

Most vitrinite macerals are derived from the cell wall material (woody tissue) of plants. Although the details of the vitrinization process are not well understood, it is generally believed that during the coalification process the plant cell wall material is chemically altered and broken down into colloidal particles that are later deposited and desiccated. This process commonly homogenizes the components so that the resulting macerals are structureless. The variation in vitrinite macerals is usually thought to be due to differences in the original plant material or to different conditions of alteration at the peat stage or during later coalification.

Table 2. Classification of macerals, according to the ASTM system, for routine analysis

Maceral group	Maceral ASTM D 2799	Additional ASTM D 2796	Terms used in fluorescence analyses
Vitrinite	vitrinite	pseudo-vitrinite	fluorescing vitrinite
Liptinite	exinite resinite	sporinite cutinite alginite	fluorinite bituminite exudatinite
Inertinite	micrinite semifusinite fusinite	macrinite sclerotinite	

Under the microscope, vitrinite macerals have a reflectance (brightness) between that of the liptinite and inertinite macerals. Because the reflectance of the vitrinite macerals shows a more or less uniform increase with coal rank, reflectance measurements for the determination of rank are always taken exclusively on vitrinite macerals. The reflectance of vitrinite macerals is also anisotropic, so that in most orientations a particle of vitrinite will display two maxima and two minima with complete rotation. Two types of vitrinite are usually distinguished in North American coals. Normal vitrinite is almost always the most abundant maceral present and makes up the groundmass in which the various liptinite and inertinite macerals are dispersed. It has a uniform gray color and is always anisotropic. With UV excitation, some normal vitrinites will fluoresce. Pseudovitrinitealways has a slightly higher reflectance than normal vitrinite in the same coal. It also tends to occur in large particles that are usually free of other macerals and pyrite. Pseudovitrinite particles commonly show brecciated corners, serrated edges, wedge-shaped fractures, and slitted structures. Pseudovitrinite does not fluoresce under UV light.

In international practice, the terms desmocollinite, heterocollinite, and vitrinite B are used to describe normal vitrinite, and telocollinite, homocollinite, and vitrinite A are used for pseudovitrinite.

*Liptinite macerals*are derived from the waxy and resinous parts of plants, i.e., the spores, cuticles, and resins. This group generally makes up 5–15% of most North American coals, although it dominates some unusual types of coal such as cannel and boghead. In any given coal, liptinite macerals have the lowest reflectance. Liptinite macerals are the most resistant to alteration or metamorphism in the early stages of coalification; thus, the reflectance changes are slight up to the rank of medium-volatile coal. In this range the reflectance of liptinite macerals increases rapidly until it matches or exceeds the reflectance of vitrinite macerals in the same coal and, thus, essentially disappears. Sporiniteis the most common of the liptinite macerals and is derived from the waxy coating of fossil spores and pollen. It generally has the form of a flattened spheroid with upper and lower hemispheres compressed until they fuse. The outer surfaces of the sporinite macerals often show various kinds of ornamentation. In Paleozoic coals, two sizes of spores are common. The smaller ones, usually < 100 μm in diameter, are called microspores, and the larger ones, ranging up to several millimeters in diameter, are called megaspores. Cutinite, found as a minor component in most coals, is derived from the waxy outer coating of leaves, roots, and stems. It occurs as long stringers, which often have one fairly flat surface and another that is crenulated. Cutinite usually has a reflectance equal to that of sporinite. Occasionally the stringers of cutinite are distorted. Resiniteis also common in most coals and usually occurs as ovoid bodies with a reflectance slightly greater than that of sporinite and cutinite but still less than that of vitrinite. Some of the larger pieces of resinite may appear to be translucent with an orange color. In some coals, particularly in those from the western United States, a number of different forms of resinite may be distinguished by using fluorescent microscopy. Alginiteis derived from fossil algae colonies. It is rare in most coals and is often difficult to distinguish from mineral matter. However, in UV light it fluoresces with a brilliant yellow color and can display a distinctive flower-like appearance.

Newly Defined Fluorescent Macerals. With the use of fluorescence microscopy, TEICHMULLERdefined three new macerals: fluorinite, bituminite, and exudatinite [4]. Although these macerals have some characteristic features in normal white-light microscopy, they can be properly identified only by their fluorescence properties.

Fluoriniteusually occurs as very dark lenses that may show internal reflections. Fluorinite is also commonly associated with cutinite. Fluorinite fluoresces with a very intense yellow color.

Bituminiteis difficult to detect in white light and is often mistaken for mineral matter. It is common in vitrinite-poor detrital coals. It occurs as stringers and shreads and fluoresces weakly with an orange to brown color. This material is similar to what organic petrologists call amorphous organic matter (AOM).

Exudatiniteis a secondary maceral which appears as an oil-like void filling. It has no shape of its own, and can usually only be detected by its weak orange to brown fluorescence in UV light.

The *inertinite* maceral groups are derived from plant material, usually woody tissue, that has been strongly altered by charring in a forest fire or by biochemical processes such as composting either before or shortly after deposition. These macerals can make up 5–40% of most North American coals, with the higher amounts occurring in Appalachian coals. In southern hemisphere coals, this group commonly is more abundant than vitrinite. The inertinite maceralshave the highest magnitude and greatest range of reflectance of all the macerals. They are distinguished by their relative reflectances and by the presence of cell texture.

Fusiniteis seen in most coals and has a charcoal-like structure. It is always the highest reflecting maceral present and is distinguished by a cell texture that is commonly broken into small shards and fragments.

Semifusinitehas the cell texture and general features of fusinite except that it is of lower reflectance. In fact, semifusinite has the largest range of reflectance of any of the various coal macerals. Semifusinite is also the most abundant of the inertinite macerals in most coals.

Macriniteis a very minor component of most coals and usually occurs as structureless ovoid bodies with the same reflectance as fusinite.

Micriniteoccurs as very fine granular particles of high reflectance. It is commonly associated with the liptinite macerals and sometimes gives the appearance of actually replacing the liptinite.

2. Coalification

A unique and often troublesome feature of coal that distinguishes it from other fuels and bulk commodities is its property of rank. All coal starts out as peat, which is then changed into progressively higher ranks of coal. This transformation is generally divided into two phases. The first, which occurs in the peat stage, is called diagenesis or biochemical coalification. In this phase most of the plant material making up the peat is biochemically broken down. Specifically, most of the cellulose in the plant material is digested away by bacteria, and the lignin in the plant material is transformed into humic acids and humic compounds, humins. Some plant material is also thermally altered by partial combustion or biochemical charring. Still other plant material, such as spores and pollen, survive the diagenesis stage without much change. After diagenesis is complete and the altered peat is buried, geological forces begin to act in the geological or metamorphic phase of coalification.

Because of the common occurrence of high-rank coals in geologically deformed areas, for example, anthracite in the eastern Appalachian Mountains and low-volatile bituminous coal in the folded and faulted Canadian Rocky Mountains, it was assumed that tectonic pressure was responsible for the high rank of most coals. However, this position has now lost most of its support because recent studies in the folded coal seams of the Ruhr Basin show that isorank lines (isovols) follow the folds in the seams [5]. This shows that the coal achieved its rank before the structural deformation and was not altered during or after such deformation. Similar studies of stratigraphic relationships in the Canadian Rocky Mountains showed that the higher rank coals correlated with increasing depth of burial and not tectonism [6]. Laboratory experiments using increased confining pressure have not led to increased coalification [7].

The majority of the geological evidence suggests that temperature is the major factor in coalification and that the temperature range in which most coalification takes place is 50–150 °C. The depth at which these temperatures occur is a function of the natural geothermal gradient (dT/dZ), which ranges from 0.8 °C/100 m to 4 °C/100 m. Therefore, at these two geothermal gradient extremes, the depth at which 150 °C would be reached, if a surface temperature of 20 °C is assumed, is 16.25 km (53 300 feet, 10.1 miles) for the lowest gradient and 3.25 km (10 700 feet, 2.02 miles) for the highest. The major lines of evidence for the importance of temperature in coalification are as follows:

1) An increase in coal rank and temperature with depth is confirmed by thousands of bore holes.
2) Lines of equal rank contour igneous intrusions and show an increase in rank in the direction of the intrusion. In some cases, natural coke is found at the contact of a coal seam with an intrusion.
3) Laboratory experiments show that wood can be coalified by heating in an inert atmosphere. Increasing the pressure alone has no effect on the coalification of wood.
4) When thermodynamic factors are considered, it is clear that temperature should have more of an effect than pressure. While an increase in temperature will encourage most coalification reactions by increasing the available energy, an increase in pressure may often inhibit such reactions by raising the energy requirements for such reactions to take place.

It was long thought that time was not a factor in coalification, and the example of the Russian paper coals of lignite rank and Carboniferous age was often cited as evidence. However, it is now believed that if the peat has been buried deeply enough to effect coalification, then time does have a "soaking effect." For example, in Venezuela there is a subbituminous coal of Eocene age that has been buried at a depth of 1120–1220 m at 125 °C for $(10-20) \times 10^6$ years. In Germany there is a coal that has been at the same depth and temperature for 300×10^6 years and is of anthracite rank [5]. Numerous other examples have been found and a number of correlation charts relating depths of burial, temperature, rank, and time are based on this type of data.

Changes in the coal rank correspond to changes in most of the properties of the coal. For example, as rank increases, moisture, volatile matter, and ultimate oxygen and hydrogen decrease, whereas fixed and ultimate carbon, calorific value, and reflectance increase. All of these measurements and even some combinations of these have been used as measures of coal rank. However, with the exception of reflectance, they all suffer from two major drawbacks. First, none actually change uniformly across the rank range of coal. Second, they are all bulk properties of coal and, thus, can be significantly affected by changes in maceral composition having nothing to do with rank. For example, a coal with a higher-than-normal content of liptinite macerals can have a higher-than-normal hydrogen content and, therefore, appear to have a lower rank than it actually does on the basis of a rank parameter independent of composition such as reflectance. The reflectance parameter of coal is based on the amount of light reflected from the vitrinite macerals in a coal compared to a glass standard of known refractive index and reflectance. The vitrinite reflectance of coal changes uniformly across most of the coal rank range. However, it is not very sensitive in the lowest rank range (lignite to lower subbituminous).

3. Occurrence

3.1. Coal Seam Occurrence

Although coal particles are scattered throughout many rock units, most coal occurs in seams, which can range in thickness from a few millimeters to over 30 m. In North America and, indeed, most of the world, anthracite and bituminous coals occur in thinner seams of 1–2 m while the lower rank coals, lignite and subbituminous, commonly occur in thicker seams of up to 20–30 m. While many seams, even minable seams, are of a limited size area, some seams continuously underlie large areas. The Pittsburgh coal seam extends over 77 700 km^2 in Pennsylvania, Ohio, West Virginia, and Maryland, and it is of minable thickness for over 15 540 km^2.

Although there are hundreds of named coal seams in the United States, relatively few are of the quality, thickness, and size of area to be extensively exploited commercially. In the Appalachian anthracite region, the Mammoth seam is the major source of production. In the northern Appalachian field, the major minable seams are the Pittsburgh, Lower Kittanning (No. 5 Block), Upper Freeport, Campbell Creek (No. 2 Gas), Upper Elkhorn (No. 3), Fireclay, Pocahontas, and Sewell seams. All of these seams are suitable for use in making coke. In the southern Appalachians, the Sewanee, Mary Lee, and Pratt seams are the most important, and the latter two are used extensively as coking coals. In the Illinois Basin, the Harrisburg–Springfield No. 5 and the Herrin No. 6 are of a minable thickness of over 38 800–51 800 km^2. In the Oklahoma, Kansas, Missouri, and Arkansas region, the Weir–Pittsburg and Lower Hartshorne seams are the most important. In the Powder River Basin, the Anderson–D–Wyodak seam is continuously exposed for 193 km and is estimated to contain at least 91 Gt of coal. In Utah the Lower Sunnyside and Hiawatha seams have produced the most coal. The former is of coking quality, and the latter is characterized by a high resinite content, which itself has been exploited. In Colorado the Wadge, Raton–Walsen, and Wheeler A, B, C, and D are the most important coal seams and some production from the Raton –Walsen and Wheeler seams has been used for coking. Finally, in Washington state the Roslyn (No. 5) is the most mined coal seam. Production from these seams accounts for ca. 75–80% of the cumulative past United States coal production [8].

3.2. Coal Seam Structures

All coal seams have a number of structural features, including partings, splits, rolls, cutouts, cleats, faults, folds, and alterations caused by igneous intrusions. These features strongly affect the mining and economical recovery of the coal. *Partings* are layers of rock, usually shale or sandstone, that occur within a coal seam and were caused by an influx of sediment into the original coal swamp. In commercial seams, the partings must, of necessity, be few and thin (a few centimeters). However, even in such seams the partings can be very extensive. For example, near the base of the Herrin No. 6 seam, there is a parting known as the blue band that is found throughout the entire Illinois Basin. When the thickness of a layer of rock within a seam increases to the point where it is no longer practical to mine the parting with the coal, the seam can be considered to have *split*. Although some seams like the Hiawatha in Utah split into three or more minable seams, the splitting of one seam into multiple seams can present serious problems in mining, including correlation (i.e., tracing a given seam across a zone, e.g., a valley, where it is missing) and loss of minable thickness.

In addition to partings and splits within the coal seam, the upper and lower surfaces of a seam often pinch and swell to change the coal seam thickness. Such features have been called *pinches, rolls, horsebacks,* and *swells,* and their occurrence at the base of a seam is usually attributed to differential compaction. Roof rolls are common and the protrusion of rock into the coal causes some serious problems with mine roof stability. *Cutouts* or stream washouts are the extreme cases where the protrusion actually eliminates the coal seam. These features are clearly the result of nondeposition or erosion by ancient streams associated with the coal swamps.

Another important feature of coal seams, especially in bituminous coals, is the presence of closely spaced fractures within the coal. These fractures, called *cleats*, are usually perpendicular to the bedding plane of the coal and commonly occur in two sets perpendicular to each other. This gives the coal a tendency to break into blocks. The cleat controls the ease with which the coal breaks up, and it has long been used in coal mining. The most prominent cleat is called the face cleat because the working face of a mine is often parallel to it. The other cleat is called the butt cleat. Cleats give coal high permeability to gas and groundwater and also act as sites of mineral deposition. Calcite and pyrite are the most common cleat-filling minerals, although other minerals including gypsum can occur.

Faulting and folding of coal seams by geological forces can also be important features. Seams that are faulted or folded usually cannot be mined by surface methods and are more expensive to mine than flat seams. When faulting is extensive, even thick seams of high quality may be too discontinuous to mine. On the other hand, folding can double the minable thickness of some seams, thereby increasing their value, as with the anthracite seams in eastern Pennsylvania.

The *alteration* of coal seams by igneous intrusions is widespread and is a serious problem in some coal fields, such as those in the Rocky Mountains of the United States. The alteration is thermal and causes an increase in carbon content and a decrease in hydrogen, moisture, and volatile matter. In the contact zone, the coal may be transformed into natural coke. Although such contact zones are usually small, extensive amounts of natural coke are known and have been commercially exploited in some areas such as the Raton Mesa of Colorado.

3.3. Coal Seam Distribution

Although coal seams are found in rocks of all geologic ages since the Devonian, the age distribution is not even. Major coal deposits of Carboniferous age occur in eastern and central North America, in the British Isles, and on the European continent. Major deposits of Permian age occur in South Africa, India, South America, and Antarctica. In Jurassic times the major coal accumulation was in Australia, New Zealand, and parts of Russia and China. The last great period of coal deposition was at the end of the Cretaceous period and the beginning of the Tertiary period. Coals originating at this time are found in the Rocky Mountains of North America, in Japan, Australia, New Zealand, and in parts of Europe and Africa. Because they are younger, these coals tend to be of lower rank, usually subbituminous, than the Carboniferous coals. Since the Cretaceous some coal has been deposited in scattered locations more or less continuously and tends to be lignite or brown coal.

The distribution of coal seams throughout the world is also not uniform. As shown in Figure 1, most of the world's coal is located in only three countries, the United States, the Soviet Union, and China. Although the figures vary from source to source, each of these countries has about 25% of the total coal resources, while the rest of the world shares the remaining 25%. In the United States, bituminous coal seams are concentrated in the Appalachian and Illinois Basins (Fig. 2). Most of the subbituminous coal occurs in the various smaller basins in the Rocky Mountain region, and the lignite seams are concentrated in the northern Great Plains and the Gulf Coast area.

4. Classification

Coal is combustible and should be composed of more than 50 wt% carbonaceous material [9]. Commercially, coal is classified in a number of ways on the basis of (1) the original plant or maceral composition, sometimes called coal type, (2) the degree of maturity or metamorphism, called coal rank, (3) the amount of impurities such as ash or sulfur, called coal grade, and (4) the industrial properties such as coking or agglomeration.

One of the main classifications by composition used by the United States Bureau of Mines is based on the relative amounts of petrographic entities detected in thin-section analysis, including anthraxylon (translucent material roughly equivalent to vitrinite), translucent attritus (roughly equivalent to liptinite), and opaque attritus and fusain (roughly equivalent to inertinite) [10], [11]. Under this system, coals are divided into two groups: **banded coals,** with > 5% anthraxylon, and **nonbanded coals,** with < 5% anthraxylon. The banded coals are subdivided into three types: *bright coal,* consisting mainly of anthraxylon and translucent attritus with < 20% opaque matter; *semisplint coal,* consisting mainly of translucent and opaque attritus with 20–30% opaque matter; and *splint coal,* consisting mainly of opaque attritus with > 30% opaque

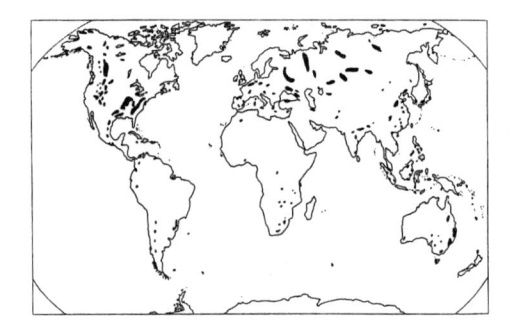

Figure 1. Geographic location of the world's coal

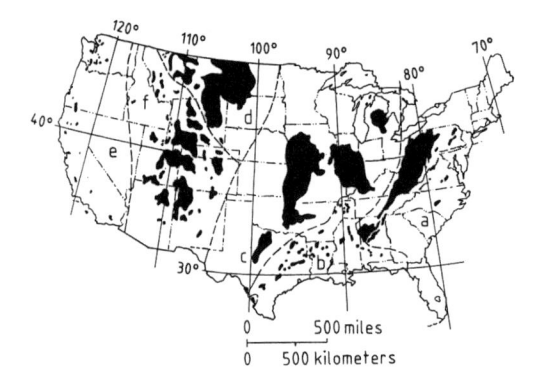

Figure 2. Coal provinces of the conterminous United States
a) Eastern; b) Gulf; c) Interior; d) Northern Great Plains; e) Pacific Coast; f) Rocky Mountains

matter. The nonbanded coals are divided into *cannel coal*, consisting of attritus with spores, and *boghead coal*, consisting of attritus with algae.

The various bands or layers in coal evident to the unaided eye have also been classified into four types [12]. *Vitrain* layers appear bright and vitreous; *clarain* appears as relatively less bright, striated layers; *durain* is dull and featureless; *fusain* layers are dull gray and like charcoal. Although these terms (all ending in ain) are megascopic terms meant to be applied to handspecimen samples, they do have some compositional implications at the microscopic level. For example, vitrain layers contain mainly vitrinite macerals, fusain layers contain mainly inertinite macerals, and clarain and durain are mixtures of all three maceral types.

The most important classification for commercial purposes in the United States is the ASTM classification by rank. It is the basis on which most of the coal in the United States is bought and sold. This classification, ASTM Standard D 388 shown in Table 3, divides coals into 4 classes, anthracite, bituminous, subbituminous, and lignitic, which are further subdivided into 13 groups on the basis of fixed carbon and volatile matter content, calorific value, and agglomerating character. The fixed carbon and volatile matter values are on a dry, mineral-matter-free basis and the calorific values are on a moist, mineral-matter-free basis. In this system, coals with $\geq 69\%$ fixed carbon are

Table 3. Classification of coal by rank [a]

Class	Group	Fixed carbon limits, % (dry, mineral-matter-free basis)		Volatile matter limits, % (dry, mineral-matter-free basis)		Calorific value limits, Btu/lb (moist, mineral-matter-free basis) [b]		Agglomerating character
Anthracite	metaanthracite	≥ 98	—	—	≤ 2	—	—	nonagglomerating
	anthracite	≥ 92	<98	>2	≤ 8	—	—	nonagglomerating
	semianthracite [c]	≥ 86	<92	>8	≤ 14	—	—	nonagglomerating
Bituminous	low-volatile bituminous coal	≥ 78	<86	>14	≤ 22	—	—	commonly agglomerating [e]
	medium-volatile bituminous coal	≥ 69	<78	>22	≤ 31	—	—	commonly agglomerating [e]
	high-volatile A bituminous coal	—	<69	>31	—	$\geq 14\,000$ [d]	—	commonly agglomerating [e]
	high-volatile B bituminous coal	—	—	—	—	$\geq 13\,000$ [d]	<14 000 [d]	commonly agglomerating [e]
	high-volatile C bituminous coal	—	—	—	—	$\geq 11\,500$	<13 000 [d]	
						$\geq 10\,500$	<11 500	agglomerating
Subbituminous	subbituminous A coal	—	—	—	—	$\geq 10\,500$	<11 500	nonagglomerating
	subbituminous B coal	—	—	—	—	$\geq 9\,500$	<10 500	nonagglomerating
	subbituminous C coal	—	—	—	—	$\geq 8\,300$	< 9 500	nonagglomerating
Lignitic	lignite A	—	—	—	—	$\geq 6\,300$	< 8 300	nonagglomerating
	lignite B	—	—	—	—	—	< 6 300	nonagglomerating

[a] This classification does not include a few coals, principally nonbanded varieties, which have unusual physical and chemical properties and which come within the limits of fixed carbon or caloric value of the high-volatile bituminous and subbituminous ranks. All of these coals either contain < 48% dry, mineral-matter-free fixed carbon or have > 15 500 Btu/lb moist, mineral-matter-free.
[b] Moist refers to coal containing its natural inherent moisture but not including visible water on the surface of the coal.
[c] If agglomerating, classify in low-volatile group of the bituminous class.
[d] Coals having 69% or more fixed carbon on the dry, mineral-matter-free basis shall be classified according to fixed carbon, regardless of caloric value.
[e] It is recognized that there may be nonagglomerating varieties in these groups of the bituminous class, and there are notable exceptions in high-volatile C bituminous group.

classified by fixed carbon content and those with <69% fixed carbon content are classified by calorific value. Thus, all lignitic and subbituminous coals and the lower rank bituminous coals are classified by their calorific value. It is also important to note that not all coals can be fitted into this system. This is especially true of coals with a high liptinitic maceral content, such as cannel and boghead types.

The other important classification system is the international system of the ISO. In this system, coals are divided into two types: hard coalswith greater than 23.86 MJ/kg (10 260 Btu/lb) and brown coals and lignites with calorific values less than that amount. In the hard coal classification shown in Table 4, the coals are divided into classes, groups, and subgroups. The classes are similar to ASTM groups and based on dry, ash-free volatile matter and moist, ash-free calorific value. The classes are

Table 4. International classification of hard coal by type[a]

Groups (caking properties)			Code numbers[d]										Subgroups (coking properties)		
Group number	Free-swell-ing index	Roga index											Subgroup number	Dilatation number	Gray-King
3	>4	>45				435	535	635					5	>140	>G_8
						434	534	634					4	50–140	G_5–G_8
					334	433	533	633	733				3	0–50	G_1–G_4
					333	432	532	632	732	832			2	≦0	E–G
					332[b]										
2	2.5–4	20–45				423	523	623	723	823			3	0–50	G_1–G_4
					323	422	522	622	722	822			2	≦0	E–G
					322	421	521	621	721	821			1	contraction	B–D
					321										
1	1–2	5–20		212	312	412	512	612	712	812			2	≦0	E–G
				211	311	411	511	611	711	811			1	contraction	B–D
0	0–0.5	0–5	100	200	300	400	500	600	700	800	900		0	nonsoftening	A

Class number[e]		0	1A	1B	2	3	4	5	6	7	8	9
Class parameters	Volatile matter, % (dry, ash-free)	0–3	3–6.5	6.5–10	10–14	14–20	20–28	28–33	>33	>33	>33	>33
	Calorific parameter[c]	—	—	—	—	—	—	—	>13950	12960–13950	10980–12960	10260–10980

[a] Where the ash content of coal is too high to allow classification according to the present system, it must be reduced by laboratory float-and-sink method (or any other appropriate means). The specific gravity selected for flotation should allow a maximum yield of coal with 5–10% of ash.
[b] 332a: volatile matter 14–16% and 332b: volatile matter 16–20%.
[c] Gross calorific value on moist ash-free basis (30°C, 96% relative humidity) in Btu/lb.
[d] First digit class, second digit group, third digit subgroup.
[e] Approximate volatile matter content 33–41% (class 6), 33–44% (class 7), 35–40% (class 8), and 42–50% (class 9).

Table 5. Classification of brown coal according to ISO 2950-1974-(E)

Group parameter: yield of tar on a dry, ash-free basis, %	Group number	Code number					
> 25	4	14	24	34	44	54	64
20 – 25	3	13	23	33	43	53	63
15 – 20	2	12	22	32	42	52	62
10 – 15	1	11	21	31	41	51	61
≤ 10	0	10	20	30	40	50	60
Class number		1	2	3	4	5	6
Class parameter: total moisture content of mined coal on an ash-free basis, %		≤ 20	20 – 30	30 – 40	40 – 50	50 – 60	60 – 70

numbered as 0 to 9. The classes are divided into four groups, numbered 0 to 3 on the basis of the swelling properties. These groups are further broken down into six subgroups numbered 0 – 5 on the basis of their Audibert – Arnu dilatation number and Gray – King coke type. The system is set up in such a way that all coals are classified with a three-digit number, in which the first digit is the class, the second digit is the group, and the third digit is the subgroup.

The lignites and brown coals are only divided into classes and groups. The classes, numbered from 1 to 6, are based on ash-free moisture; the groups, based on dry, ash-free tar yield, are numbered from 0 to 4. This classification is shown in Table 5.

Although the ASTM and International Systems are different, there is a reasonable correspondence between the ASTM group names and the International System class numbers. This is shown in Table 6.

5. Chemical Structure of Coal

The *organic components* of coal consist of a complex mixture of macromolecular carbon compounds of varied chemical constitution. The heterogeneity of coal is well-known from optical studies of thin coal sections. The microscopically distinct and distinguishable maceral groups of vitrinite, liptinite, and inertinite, their proportions, and the association of macerals of several groups with each other are essentially responsible for the physical, chemical, and technological properties of a particular coal. Preformed macerals derived from transformation of the original lipidic, woody, and waxy plant matter under varying oxidizing and reducing conditions and processes are present in *peat* and *soft lignites*. Temperature-induced chemical changes of these macerals and increased pressures influencing the physical structure of the material undergoing coalification have led to the formation of *lignitic, subbituminous, bituminous,* and *anthracitic* coals, forming the basis of coal classification by *rank* [13], [14].

Table 6. Comparison of class numbers and boundary lines of the International System with group names and boundary lines of the ASTM system

International classification, class number	0	1	2	3	4	5	6	7	8	9
		5	10	15	20	25	30			
				Volatile-matter parameter[a]				14 000	13 000	12 000	11 000	10 000
									Calorific-value parameter[a]		
ASTM classification, group name	Meta-anthracite	anthracite	semianthracite	low-volatile bituminous coal	medium-volatile bituminous coal		high-volatile A bituminous coal [b]	high-volatile B bituminous coal	high-volatile C bituminous coal and subbituminous A coal	subbituminous B coal

[a] Parameters in the International System are on ash-free basis; in the ASTM system, they are on mineral-matter-free basis.
[b] No upper limit of calorific value for class 6 and high-volatile A bituminous coals.

The *minerals* of coal play a considerable part in the industrial use of coals of all ranks. The mineral content, its distribution within the coal, and its composition may affect carbonization, gasification, combustion, and liquefaction processes by modifying the process of coal depolymerization and influencing hydrogenation and the thermal behavior of the resulting ash under both oxidizing and reducing conditions.

Besides microscopically unidentifiable plant ash, coals contain varying amounts of *syngenetic* and *epigenetic* minerals, which entered the organic coal substance during the first or second phase of coalification, respectively. They have either grown together with the organic substance or are physically separable. Generally, medium to high ash coals contain epigenetic minerals. Structure and hardness bear considerable influence on the distribution in the various fractions of the crushed coal from coal preparation plants. The mineral structure of coal ash is extremely difficult to determine. Low-temperature combustion in oxygen plasma produces an ash that can be analyzed by X-ray diffraction and IR absorption spectrometry. Deductions must also be made from the chemical analysis of the oxides present in the ash. The major constituents of coal ash as determined by this chemical analysis are SiO_2, Al_2O_3, Fe_2O_3, and CaO, in addition to MgO, Na_2O, K_2O, BaO, TiO_2, P_2O_5, and SO_3.

These oxides may be derived from minerals of the following types originally present in the coal: kaolinite, illite, and bentonite as major clay minerals; calcite or siderite as major carbonates; pyrite or marcasite as the sulfides; and many other minor minerals such as apatite, hematite, rutile, quartz, etc.

5.1. Characterization of Coals

5.1.1. Standard Analytical Methods

Coals are characterized by various chemical and physical methods that have been established as national and international standards. These include proximate analysis, carbonization assay, ultimate analysis, heating value (calorific value) determination, sulfur distribution, caking and coking behavior, ash composition, and fusibility.

The above tests are carried out on representative prepared coals, using either as-received (by the laboratory) or predried samples. The results are commonly expressed on an as-received, dry (d), dry ash-free (daf), or dry mineral-matter-free (dmmf) basis.

Figure 3 shows the comparison and relationship between the proximate and ultimate analyses as well as the carbonization assay.

In Figure 4 the sulfur forms occurring in coals are illustrated. Figure 5 shows selected typical distributions of mineral components.

The coalification series, expressed by plotting atomic ratios of H/C vs. O/C, is graphically represented in Figure 6. As one moves from right to left, i.e., from peat to bituminous coals, the diagram shows only a slight shift to somewhat lower H/C ratios while the O/C ratios steadily decrease. At an O/C level of ca. 0.05, a steep decline in the

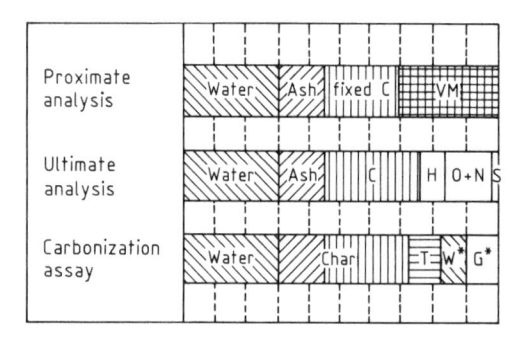

Figure 3. Comparison of proximate and ultimate analysis with carbonization assay
C, H, O, N, S = elements; T = tar; W* = decomposition water; G* = carbonization gas; VM = volatile matter

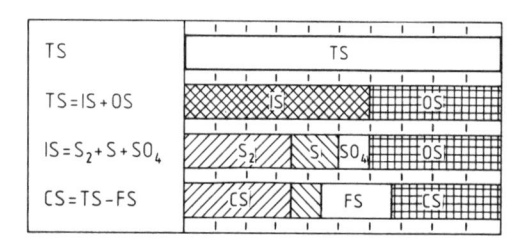

Figure 4. Scheme of sulfur distribution in coal
TS = total sulfur; IS = inorganic sulfur; OS = organic sulfur; S_2 = pyrite sulfur; S = sulfide sulfur; SO_4 = sulfate sulfur; CS = combustible sulfur; FS = fixed sulfur

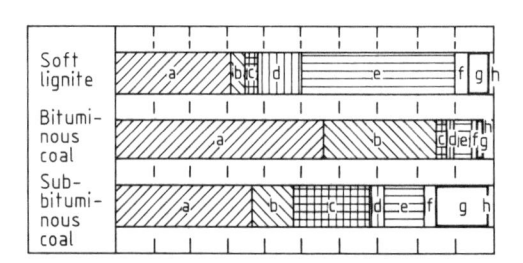

Figure 5. Distribution of mineral components of ash
a) SiO_2; b) Al_2O_3; c) Fe_2O_3; d) MgO; e) CaO; f) $Na_2O + K_2O$; g) SO_3; h) Balance

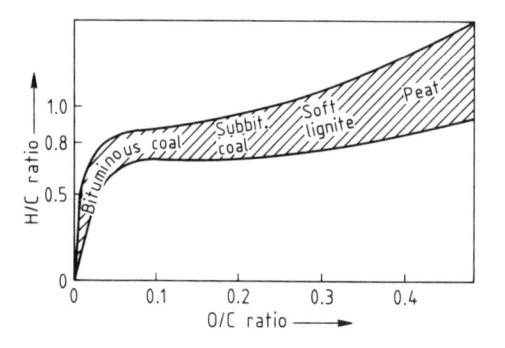

Figure 6. Coalification series: atomic ratios H/C vs. O/C

Table 7. Analytical data for coals of different degree of coalification

Analytical parameter	Peat	Soft lignite	Lignite	Subbituminous coal	Bituminous coal	Anthracitic coal
Moisture (as received), wt%	>75	56.7	38.7	31.2	3.7	1.0
Ultimate analysis (daf)						
Carbon, wt%	58.20	70.30	71.40	73.40	82.60	92.20
Hydrogen, wt%	5.63	4.85	4.79	4.86	4.97	3.30
Nitrogen, wt%	1.94	0.74	1.34	1.16	1.55	0.15
Sulfur, wt%	0.21	0.27	0.60	0.31	1.50	0.98
Oxygen (difference), wt%	34.02	23.84	21.87	20.27	9.38	3.37
Elemental ratio						
H/C	1.15	0.82	0.80	0.79	0.72	0.43
O/C	0.44	0.25	0.23	0.21	0.09	0.03
Heating value						
(dry, ash-free), kJ/kg	23 500	27 500	28 500	29 400	30 600	35 700

H/C ratio occurs when the anthracitic coals are reached. Examples of actual compositions of various coals of different degrees of coalification are given in Table 7.

5.1.2. Laboratory and Bench-Scale Simulation Tests

Becauseof the complex structure of coal and its vastly varying composition, its depolymerization, liquefaction, and combustion characteristics in thermal and extractive applications are somewhat unpredictable. In commercially operated processes, they depend very much on the prevailing conditions, such as heating and flow rates, pressure, and gas or solvent composition.

Therefore, it is often necessary to evaluate coal properties further by subjecting samples to thermal, extractive, and mechanical tests, assessing on a small scale the expected behavior and at the same time providing for study of the types and compositions of specific reaction products.

Such experiments that yield fragmental decomposition compounds can also be employed for the structural identification of coals.

The determinations include, e.g., pyrolytic or progressive devolatilization tests, extraction or solubility studies, performance of selected coking tests, reactivity measurements on specially prepared cokes, grain stability evaluation, and ash slagging behavior.

5.1.3. Petrographic Studies

The determination of the petrographic composition of coals by maceral group analysis is a valuable tool for assessing structural aspects and predicting coal behavior.

Usually it also incorporates the determination of the degree of coalification by measuring the reflectance of the vitrinite as the most important and abundant group of macerals. This method allows a ranking or classification of coals by their degree of coalification and, hence, identifies the type of coal. It also reveals specific phenomena that may have induced structural coal changes within the geochemical stage of coal formation, such as aging or contact coalification due to inflowing magma.

5.2. Structural Deductions from Analytical and Bench-Scale Data

The resultsof coal analyses predict thermal decomposition products (proximate analysis, carbonization assay, caking and coking tests) and the element composition (ultimate analysis). Furthermore, the moisture level of run-of-mine coals is indicative of the capillary structure, allowing a first tentative classification of a particular coal. The lack of moisture holding capacity of highly coalified coals is proof of an increasingly densified structure. The element ratio, particularly that of H/C and its change to lower values as coalification progresses, points toward the aromatization of structural groups. Likewise, the dramatic decrease in oxygen within the series peat to anthracite strongly suggests the loss of functional end groups.

The softening properties of bituminous coals (caking and coking tests) reveal the mobility of fragmented decomposition products, observed particularly in coals with high hydrogen and, hence, high vitrinite and liptinite contents. On the other hand, oxygen-rich coals, i.e., low-rank coals with an associated appreciable inertinite content or highly aromatic coals with a high degree of cross-linkage, yield recondensed solid depolymerization products. The tar and gas composition (carbonization assay, devolatilization, caking, and coking tests) shows that hydrogen preferentially enters liquid and gaseous components due to rearrangement [15].

Stable fragments, obtained from specific pyrolysis or hydropyrolysis studies, indicate the major types and the volatility of aromatic molecules present in a particular coal. Liquid-phase treatment produces larger unit fragments, which can be determined analytically. A petrographic survey permits selective composition analyses and structural identification of individual major maceral groups such as vitrinite.

5.3. Bonding of Elements in Coal

Thermaldepolymerization by controlled slow or flash pyrolysis of coals, oxidation, hydrogenation, liquefaction of coals, supercritical and selective solvent extraction, as well as the application of modern physical techniques such as X-ray diffraction, IR spectroscopy, nuclear magnetic resonance, or pyrolysis mass spectrometry have furnished extensive data on the bonding of elements in coals. Major findings from these investigations are [16]:

Carbon: The aromaticity increases from 40–50% C in subbituminous coals to 70–80% C in bituminous coals and is over 90% C in anthracites.

Hydrogen: Aromatic bonding to carbon and as aliphatic hydrogen in, e.g., methylene ($-CH_2-$) and methyl ($-CH_3$) groups occurs.

Oxygen: Major functional groups are hydroxyl ($-OH$), carboxyl ($-COOH$), carbonyl ($>C=O$), etheric ($-O-$) and heterocyclic oxygen.

Sulfur: Bonding is predominantly as thiophenes, also in the thiolic form ($R-SH$), decreasing with higher rank coals; up to ca. 25% S is present as aliphatic sulfides ($R-S-R$). Heterocyclic sulfur compounds are also known to exist.

Nitrogen: Very little information on organic nitrogen bonding is available. It appears to occur in heterocycles.

5.4. Structural Evidence of Coals

Muchdata on the structural aspects of coal has been accumulated and published by numerous authors, with the aid of the chemical and physical techniques available to modern science [17], [18].

The overall picture, however, is still very incomplete. The difficulty is that coal is a highly heterogeneous mixture and, as such, is not a clearly defined macromolecular aromatic compound of uniform molecular mass and structure. Thus, evaluated structural parameters refer to isolated fragments, components (vitrinite), or constituents. According to the current level of understanding, however, coal may be described structurally as a three-dimensional skeleton of generally four to five highly stable condensed aromatic and hydroaromatic units with cross-linkage by weaker short-chain aliphatic groups. Reactive functional end groups and aliphatic structures are attached to the aromatic skeleton. The average molecular mass distribution is ca. 500–800 for low-rank coals, increasing to ca. 3000–6000 for bituminous coals, and possibly values in excess of 100 000 for anthracites.

Various overall models for the complex organic structure of coals, combining available experimental data on structural fragment analysis, have been tentatively proposed. However, none as yet fully accounts for all the phenomenological characteristics [16].

6. Coal Mining

6.1. General Considerations

A coal deposit is the final result of the cumulative effects of decaying vegetation, deposition of sediments, movements of the earth crust, and forces of erosion. The transformation of the vegetation in swamps into peat, lignite, anthracite, and bituminous deposits is associated with high pressure or high temperature. Movements of the earth crust (folding and faulting), igneous intrusions, and wind and water erosion have played important roles in determining the inclination, continuity, and quality of the seams (see Section 3.2). These have also resulted in (1) changing the position of the coal seams relative to the surface of the earth, (2) altering the condition of the roof and floor of the coal seams, (3) eroding away of entire seams or portions of seams, (4) outcropping of the surface of otherwise concealed seams, and (5) determining the economic and technical feasibility of mining these coal seams [19], [20].

The various methods of mining a coal seam can be classified as surface or underground, depending on whether the coal seam is exposed and recovered from operations on the surface of the earth or from operations in underground openings, approached from suitably located surface openings. The results of an analysis of the world coal production by mining method and rank of coal are summarized in Table 8. Bituminous and lignite seams account for over 65 and 25% of the world production, respectively. Ten countries account for nearly 75% of the total production. Contributions from surface and underground mining to worldwide production are approximately equal; however, by country the production distributions are quite variable. In general, most anthracite production is from underground mines, and almost all lignite production is from surface mines.

Decisions on whether a coal seam will be mined and the mining method to be practiced on a given coal seam depend on many technological, economic, and social factors. Very often, the technological factors, the most important of which are associated with geology, constrain the mining engineer to a single mining method. However, at any given time, economic and social factors dictate whether a coal seam can be mined at all.

6.2. Surface or Underground Mining

The key to determining whether to mine a coal seam by a surface or underground method is the *stripping ratio.* Distinction is often made between several stripping ratio calculations. The first is the calculation of the stripping ratio of the areas over the coal seam. It is generally defined as the number of cubic meters of the strata overlying the

Table 8. Salient statistics on leading coal-producing countries and world coal production by method of mining and coal rank[a] (1983 production; all figures are in 10^6 short tons)

Country	Bituminous			Anthracite[e]	Lignite[e]	Total	Proportion of total, %
	Underground[b]	Surface[b]	Total				
Australia	54.2[c]	88.5[c]	133.4		38.6	172.0	3.9
China	555.5	23.2	578.7	209.4		788.1	17.8
West Germany (FRG)	98.8		98.8	6.6	136.9	242.3	5.5
India	75.1	75.1	150.2		8.1	158.3	3.6
Poland	180.7	29.9	210.6		46.8	257.5	5.8
South Africa	100.1	42.9	143.0	2.4		145.4	3.3
United Kingdom	112.3	16.2	127.1	2.9		130.1	3.0
United States	294.2	420.0	719.6	3.8	58.3	781.0	17.7
Soviet Union	380.9	142.7	523.6	87.1	178.5	789.4	17.8
Others			204.0	77.9	669.7	951.7[d]	21.6
Total			2889.0	390.2	1136.9	4416.1	100
Proportion of the total coal production, %			65.4	8.8	25.8	100	

[a] Compiled from *Statistics in International Coal*, 1985 Edition, National Coal Association, Washington, D.C., USA. Numbers may not add up due to independent rounding off.
[b] Estimated, except for the United States.
[c] Fiscal 1983–1984 data.
[d] Estimated on the basis of world production.
[e] Anthracite production is mostly from underground mining and lignite production is mostly from surface mining.

coal seam for every ton of coal in the seam. The basic inputs are the thicknesses of the coal seam and the strata overlying the coal seam.

$$\text{stripping ratio} = \frac{\text{cubic meters of strata material over a specified area of coal seam}}{\text{tons of coal in the same specified area of coal seam}} \quad (1)$$

The second is the calculation of a stripping ratio at which recovery by surface mining is equivalent in cost to recovery by the underground method. The basic inputs are the costs of mining that reflect the mining technology to be used.

$$\text{break-even stripping ratio} = \frac{\left\{\begin{array}{c}\text{underground}\\ \text{mining cost}\\ \text{per ton of coal}\end{array}\right\} - \left\{\begin{array}{c}\text{surface mining and}\\ \text{reclamation cost}\\ \text{per ton of coal}\end{array}\right\}}{\text{surface mining cost per cubic meters of strata above the coal seam}} \quad (2)$$

Only those parts of the coal seam for which the stripping ratio calculated by Equation (1) does not exceed the stripping ratio shown in Equation (2) should be mined by surface mining methods.

A third stripping ratio, also called break-even stripping ratio, incorporates the technology of mining and the market value of coal produced. It is calculated as follows:

$$\text{break-even stripping ratio} = \frac{\left\{\begin{array}{l}\text{sales realiza-}\\\text{tion per ton}\\\text{of coal}\end{array}\right\} - \left\{\begin{array}{l}\text{total cost of produc-}\\\text{ing and marketing}\\\text{per ton of coal}\end{array}\right\}}{\text{surface mining cost per cubic meters of strata above the coal seam}} \qquad (3)$$

6.3. Surface Coal Mining Methods

Surfacecoal mining is a very broad term and refers to the removal of the soil and strata over a coal seam and the removal of the seam itself. It is conducted in a relatively simple sequence of operations, which includes the unit operations of (1) scalping the surface, (2) drilling, (3) blasting, (4) stripping or overburden removal, including the hauling of waste, (5) coal loading, (6) coal hauling, and (7) reclaiming. Figure 7shows a typical sequence of operations in an area surface mine. Other operations include building drainage systems, handling waste, and storing top soil. Associated with the mining operations are plants for such operations as machinery maintenance, coal cleaning, and water treatment.

The mining techniques for a particular region are dictated by the geological and topographical conditions, i.e., the thickness and inclination of the coal seam, the thickness of the overlying strata, and the topographic relief. Figure 8shows schematically how geological factors influence the design of a surface mine. The chosen method may not include all of the unit operations previously mentioned. Even where the techniques are generally comparable, equipment choices and operational procedures may differ. If the surface topography is steep, then at least initially the overburden material must be transported to an outside dump.

A variety of equipment is used in a surface mining operation. Bulldozers and scrapers find extensive applications in scalping the surface, removing the topsoil, and preparing the mining area for subsequent unit operations. Some strata can be prepared for mining and loading through the use of ripper bulldozers. Where the strata is hard, the unit operations of drilling and blasting are necessary. Drilling for fragmentation is commonly done with rotary-type units capable of hole diameters of 14–40 cm, with vertical drilling being more common. However, in the United States horizontal sidewall drilling has been introduced in several coal stripping operations.

Overburden removal is the most important unit operation in the system. The equipment used varies from fleets of mobile units such as front-end loaders and trucks to giant draglines, shovels, and bucket-wheel excavators. There are two types of shovels currently used as mining tools: the stripping shovel and the quarry-mine (or loading) shovel. The common method of shovel application in stripping overburden for coal overcasting is shown in Figure 9. The shovel usually sits in the pit, digs the bank, and

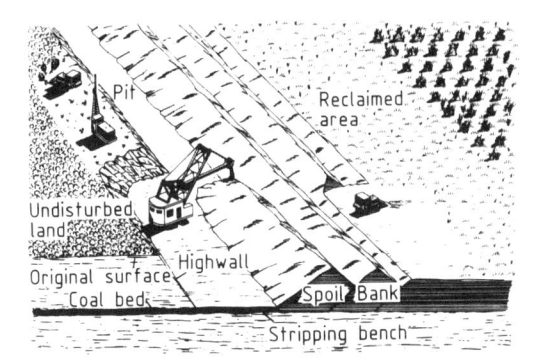

Figure 7. Typical sequence of operations in an area surface mine

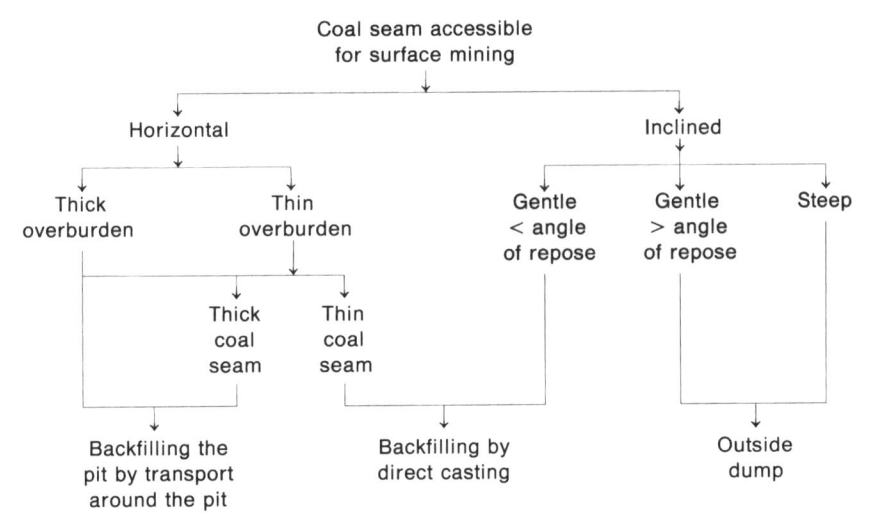

Figure 8. Influence of geological factors in surface mine design

deposits the overburden in the adjacent mined-out area. The quarry-mine shovel is carefully sized according to the capacity of the haul units into which it will be dumping its load, as shown in Figure 10. The haul units usually are large off-highway trucks.

Draglines are the most common stripping equipment used in the surface mining of coal. As opposed to the shovel, they operate on top of the overburden and can disperse the dug material over greater distances than a shovel, as is shown in Figure 11. Draglines provide greater flexibility, work on higher bench heights, move more material per hour, and are more popular than shovels. The largest shovel and dragline in operation today have bucket capacities of 140 and 170 m^3, respectively.

While shovels and draglines are cyclic equipment, bucket-wheel excavators are continuous extraction machines capable of removing up to 12 000 m^3/h. Wheel excavators hold considerable promise where conditions are favorable. Ideally, this machine has the capability for continuous overburden removal and for selective

Figure 9. Shovel pit
Note also the vertical coal drill and the horizontal overburden drill.

Figure 10. Quarry mine shovel loading the blasted overburden onto a rear-dump off-highway truck

Figure 11. Dragline pit
Note the vertical overburden drill on the surface; also note the coal loading shovel and a coal hauling truck in the pit.

placement of the soil and strata. A bucket-wheel excavator is shown in Figure 12. In the United States, wheel excavators usually operate in tandem with another major stripping machine, as shown in Figure 13. In applications around the world, wheel excavators are widely used in lignite mining.

Coal loading is usually done by loading shovels, end loaders, hydraulic excavators, and small-wheel excavators. Coal transport is typically accomplished by bottom-dump trucks. Reclamation operations, which include backfilling the last cut after coal

Figure 12. Bucket-wheel excavator

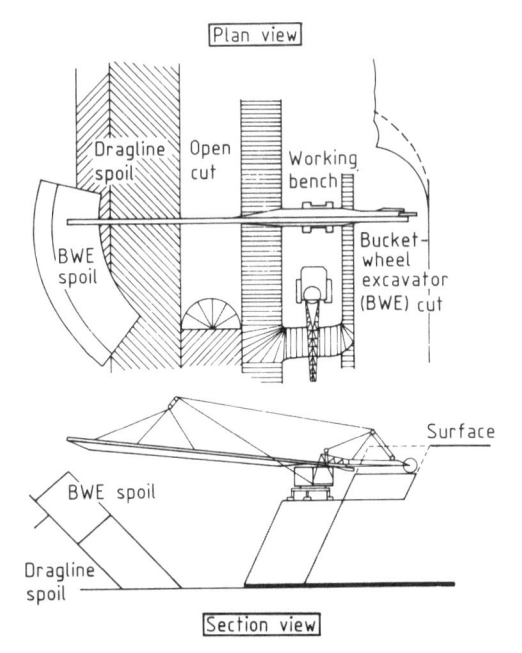

Figure 13. Bucket-wheel excavator – dragline tandem operation in an Illinois surface coal mine
Note that both machines are sitting on the working bench. The dragline digs the hard strata below it, whereas the wheel excavates the soft material above the working bench. This kind of operation has been very beneficial for reclamation and environmental conservation.

removal, regrading the final surface, and revegetating and restoring the land for future use, are integrated with the mining operations. Use of bulldozers, scrapers, graders, and agricultural equipment is common.

Surface mining systems can be broadly classified as continuous or noncontinuous mining systems. Continuous mining systems are applicable where the overlying strata requires little or no drilling and blasting before removal by overburden removal equipment. Usually the bucket-wheel excavator is the principal stripping equipment.

There are several methods of transporting the overburden material from the pit. With continuous mining systems, it is common to use belts, which discharge to railroad cars or dumping areas. In noncontinuous systems, off-highway trucks are most commonly used. These vehicles range in capacity from 35 to 300 t and have bottom-dump, rear-dump, or side-dump capability. Whereas rear-dump trucks are commonly employed for overburden handling, bottom-dump trucks are more common for coal transport out of the pit. The various possible combinationsof the different equipment available for the extraction of a coal seam are shown in Figure 14.

Surface mining techniques are used in open-cast mining, open-pit mining, strip mining, contour mining, pit mining, quarrying, and dredging. The most commonly applied methods in coal are open-pit mining with shovels and trucks or bucket-wheel excavators and belts and area strip mining with shovels or draglines.

The controversy with regard to the environmental damages associated with surface mining of coal has resulted in an increased appreciation of the problems and potential for reclaiming surface mined lands. Modern surface coal mining methods are engineered to reflect this concern and to reduce, if not to eliminate, the environmental damages.

6.4. Underground Coal Mining Methods

Underground coalmining is a complex operation where the working environment of the miners and equipment is completely enclosed by the geological medium overlying and underlying the coal seam. Therefore, the correct assessment of the geological factors is more critical for an underground mine than for a surface mine. The depth to the coal seam, its thickness and quality, the conditions of roof and floor strata, and the problems associated with methane and water inflows must be analyzed in detail to select the appropriate applicable methods and equipment.

An underground mine can be classified as a shaft mine, a drift mine, or a slope mine, depending on the type of access from the surface to the coal seam, as shown in Figure 15. Where applicable, drift is the obvious choice for access to a coal seam. In other cases, the choice between slope and shaft for deep mines can be difficult. A mine with large daily production and a long life would benefit from a slope opening for hauling coal.

Two distinct activities in underground coal mining are the unit operations in the production and hauling of coal and the auxiliary operations that are essential to maintaining a safe productive operation. Specifically, the unit operations include cutting, drilling, blasting, loading, and hauling. Auxiliary operations include ventilation, drainage, power, communications, and lighting. Roof support operations in the face area (active face where coal is being mined) can be considered as a unit operation. Support of other areas in the mine can be included in the auxiliary operations. The roof support used in most coal mines is a 1.5–2 m long, 15–25 mm

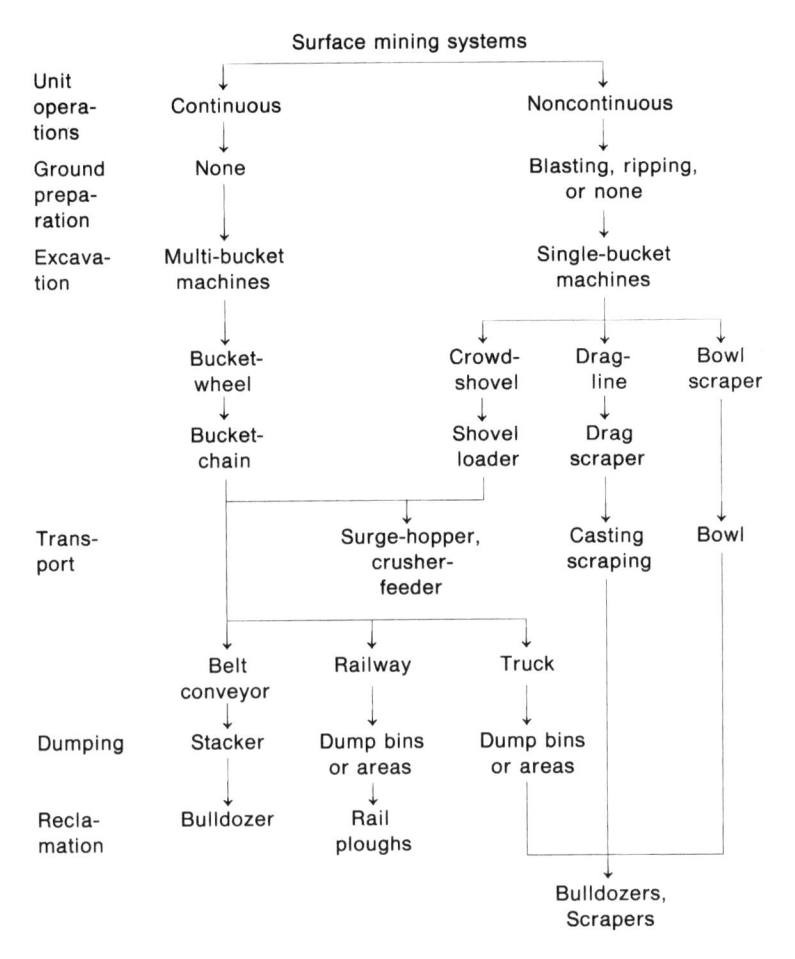

```
                        Surface mining systems
                        ↓                    ↓
Unit          Continuous              Noncontinuous
opera-            ↓                        ↓
tions
Ground          None              Blasting, ripping,
prepa-                                 or none
ration            ↓                        ↓
Excava-      Multi-bucket           Single-bucket
tion          machines                machines
                  ↓              ↓         ↓         ↓
               Bucket-        Crowd-    Drag-      Bowl
               wheel          shovel    line     scraper
                  ↓              ↓         ↓
               Bucket-        Shovel     Drag
               chain          loader   scraper
                  ↓              ↓         ↓
Trans-                      Surge-hopper, Casting    Bowl
port                         crusher-    scraping
                             feeder
                  ↓              ↓         ↓
                Belt         Railway    Truck
              conveyor
                  ↓              ↓         ↓
Dumping        Stacker      Dump bins  Dump bins
                            or areas   or areas
                  ↓              ↓
Recla-        Bulldozer        Rail
mation                        ploughs
                                           ↓
                                      Bulldozers,
                                       Scrapers
```

Figure 14. Equipment applications for surface mining systems [21]

diameter steel bolt, usually set in rows such that the bolts are 1.2–1.8 m apart in a row, with the rows 1.2–1.8 m apart. Additional supports may be used as required.

There are several theories which explain how rock bolts hold the roof. These include the beam theory (roof bolts tie together several weak roof strata into one strong beam), the suspension theory (the weak members are suspended from a strong anchor horizon), and the keying effect theory (the roof bolts act much like the keystone in an arch).

There are two major systems of underground coal mining: the room and pillar and the long-wall. There are many variations in each of the systems. Several types of equipment — cutting machines, drills, loaders, roof bolters, shuttle cars, belts, etc. — are used in the mining process, depending on the method chosen to extract a coal seam; the choice is determined by local geological conditions and operating practices.

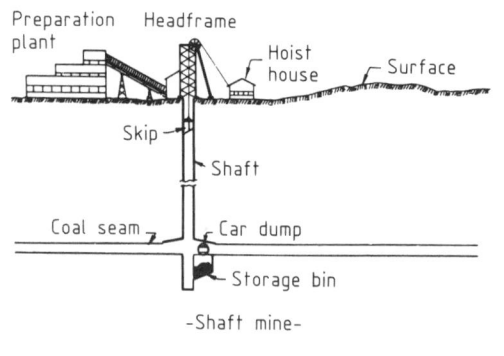

Figure 15. Types of access to a coal seam Deep coal seams can be accessed only by sinking vertical shafts.

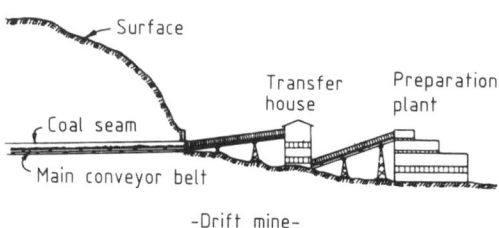

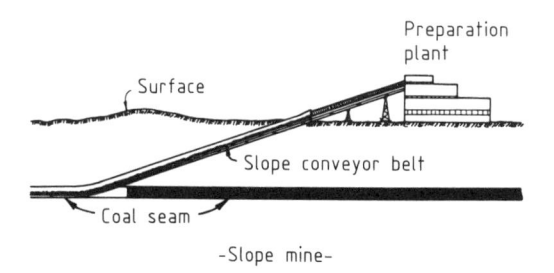

Two main types of room and pillar systems are the conventional and continuous mining systems, shown in Figure 16. In the former, the operations of cutting, drilling, blasting, and loading are performed by separate machine units. In the latter, a continuous miner rips the coal from a coal seam and loads it directly into a hauling unit, all in one step. In either system, the objective is to create blocks of coal, usually rectangular or square, during the first stage of the mine, called "development." These blocks of coal may be left in place. However, they may also be recovered during the second stage of mining, usually called "pillaring." Where the surface features (such as houses, roads, streams, etc.) have to be protected, pillars underneath these features are not recovered. In the United States, the average recovery from a coal seam, when room and pillar methods are used, is ca. 50%. The sequence of operations for the various machines in a conventional mining development plan is shown in Figure 17. After finishing its operation in a working area (or face), a machine moves to the next higher numbered working area. Because of the large number of machines, there is a need to have a large number of working faces to avoid excessive delays.

Continuous mining incorporates the unit operations of cutting, drilling, blasting, and loading into one machine: the continuous miner. There is no need for a large number of working faces. While there are many types of continuous miners, the milling-head miner, shown in Figure 18, is by far the most popular machine in operation. The sequence of operations for the continuous miner in a five-entry development is shown in Figure 19. The only other machine unit that cycles with the miner is the roof bolter. The ventilation plan has been laid out such that the miner and the roof bolter operations can have their own split of ventilating air current. To a large extent, continuous mining has replaced conventional mining. However, in thick and hard coal seams, conventional mining has great potential for increased production.

The methods that are practiced to extract the coal left in the pillars depend greatly on the nature of roof and floor and the manner in which the increased pressure generated due to pillaring is transferred to other pillars and the gob. The necessary evaluations are specific to each site and must ensure that sudden bumps and coal eruptions do not result. In addition, the unrecovered pillars should not be crushed to a point where loss of life, equipment, or coal is possible. Two methods are shown in Figures 20 and 21.

In longwall miningsystems, a block of coal 200–300 m wide and 2000–3000 m long is extracted by a machine that runs up and down the width of the face and takes a 0.75–1 m slice of the block as shown in Figure 22. A modern mechanized longwall system is a true continuous coal production system. It offers great productivity and safety. The system is most common in Europe; however, it is gaining popularity in the United States and elsewhere. There are essentially four machines: the cutting machine (usually a shearer or a plough), the armored face conveyor, the line of roof supports, and the stage loader. The cutting machine rides on the conveyor, cutting and transferring coal to it (Figs. 23 and 24). Both of these machines and all of the men in the face are under a protective canopy of steel, formed by a continuous line of roof supports. In modern longwall systems, shields are most common for supporting the roof over the immediate face area.

There are three main types of longwall system layout. The retreating longwall, shown in Figure 25, is most common in the United States. The advancing longwall, shown in Figure 26, is most common in Europe. The third method, alternate retreat and advance, is not in common practice. A method combining the longwall layout with continuous miners is known as the shortwall method and is shown in Figure 27. It is also not in common practice, but where conditions are favorable, this method may have great potential.

Mention has already been made about thick and inclined coal seams. Thick coal seams are a special gift of nature. However, thick and steeply pitching seams present challenges to the ingenu-ity of miners to devise safe and efficient methods for their extraction. A generalized classification of thick-seam mining methods is shown in Table 9.

The cutting action of a water jet has a long history (at least 40 years) of successful application in the Soviet Union and other European countries for the extraction of coal.

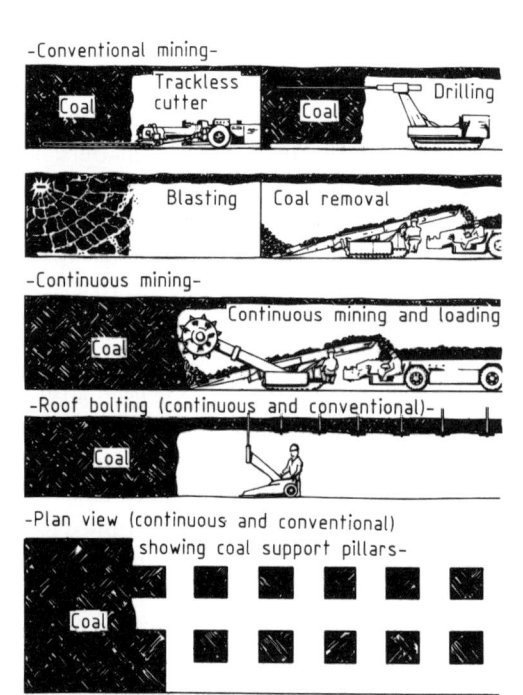

Figure 16. Conventional and continuous room and pillar underground mining systems

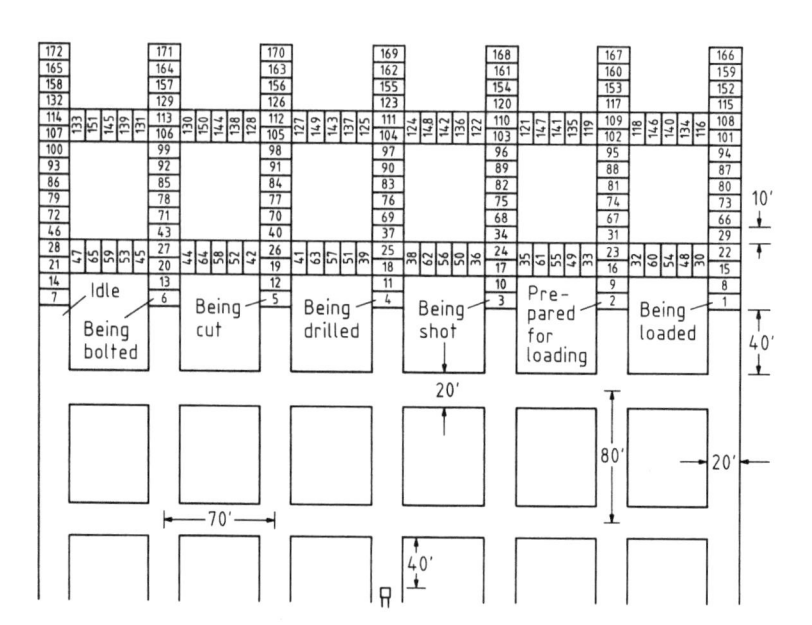

Figure 17. Development plan for a conventional room and pillar mining section (lengths in feet; 1' = 0.305 m)

Figure 18. The milling-head continuous miner
The coal is cut by the picks mounted on a number of oscillating and rotating disks. The cut coal is loaded onto a chain conveyor in the machine by a gathering arm in the loading head.

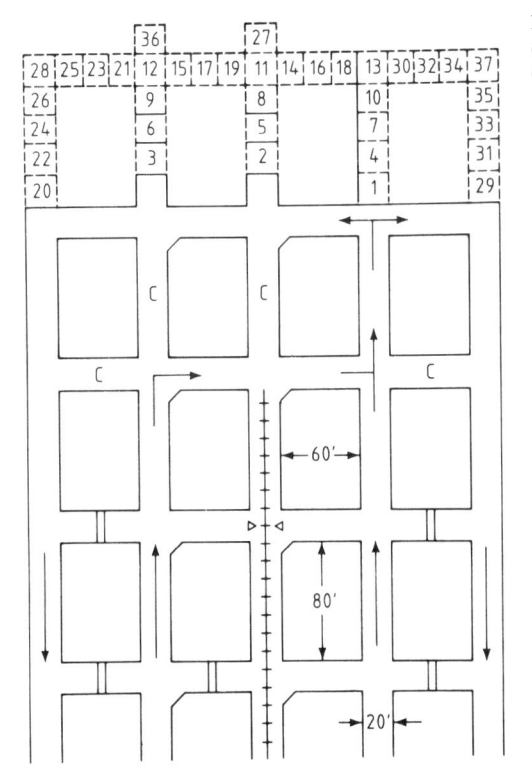

Figure 19. Development plan for a five-entry continuous room and pillar mining section (lengths in feet; $1' = 0.305$ m)

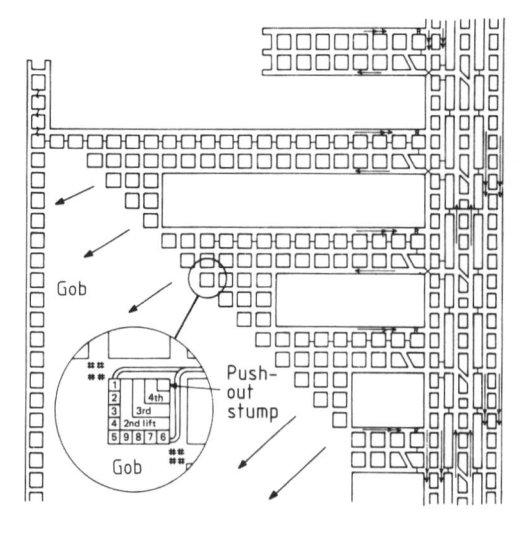

Figure 20. Open-end method of pillar extraction [22]
The pillar extraction must be sequenced to maintain the 45° line.

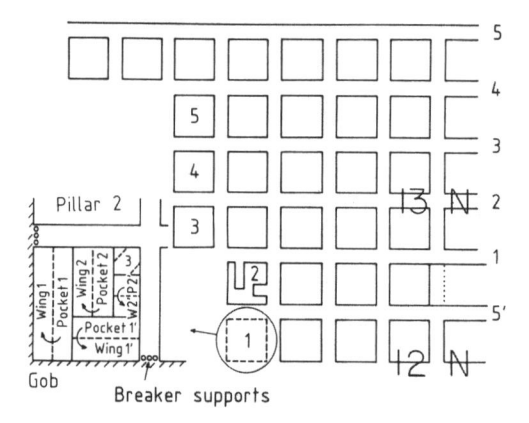

Figure 21. Pocket and wing method of Pittsburgh block room and pillar slicing system [22]

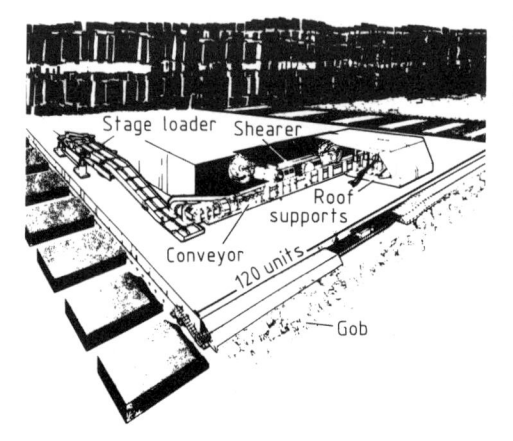

Figure 22. Perspective and close-up view of longwall system showing the equipment and its arrangement
Note that this is a retreating longwall system.

Figure 23. Plough longwall face in a German coal mine

Figure 24. Shearer longwall face in an American coal mine

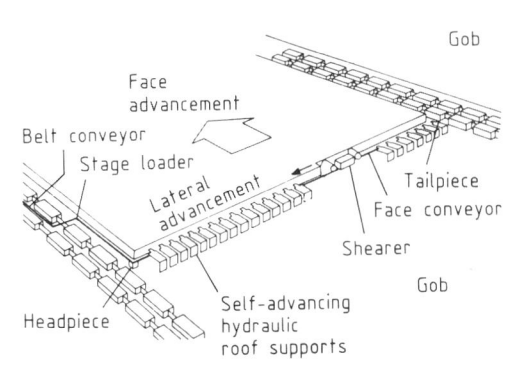

Figure 25. Equipment arrangement in a retreating longwall face [22]

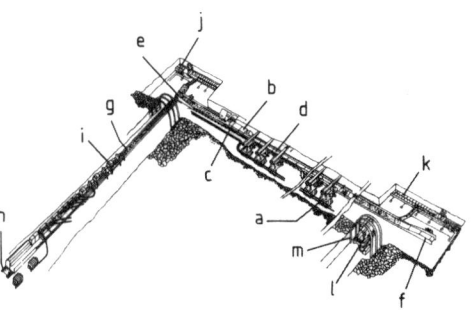

Figure 26. Equipment arrangement in an advancing long-wall face [22].
a) Powered roof supports; b) Shearer mining machine; c) Cable-handling equipment; d) Face chain conveyor; e) Chain conveyor head drive; f) Face chain conveyor tail drive; g) Stage loader; h) Belt conveyor; i) Hydraulic power pack for face supports; j) Head gate stable driving equipment; k) Tail gate stable driving equipment; l) Ripping machine; m) Yieldable arches

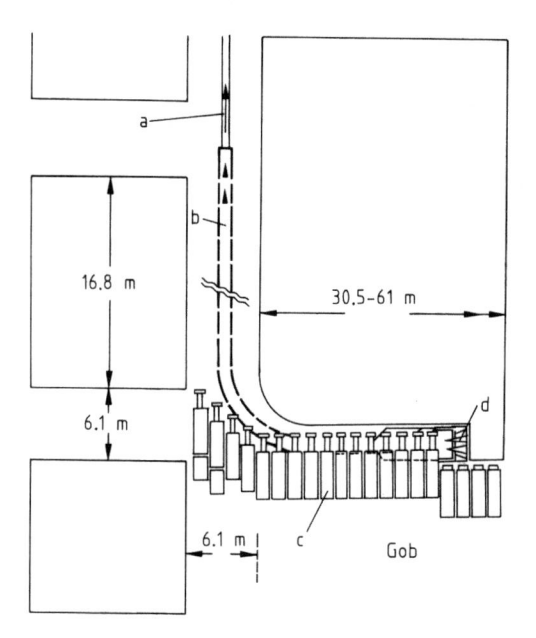

Figure 27. Shortwall mining method
Shortwall is generally 40–60 m wide. The cut width is usually 3 m. The roof support equipment (chocks) must be specially designed for the longer cantilever support and is more costly. a) Panel belt conveyor; b) Flexible conveyor train; c) Chock; d) Continuous miner

Table 9. A generalized classification of thick-seam mining methods

Method	Variations	Operational dip constraint	Thickness constraint	Ground control	Haulage constraint	Practicing countries
Full face	longwall	0–15°	5 m	caving stowing	conveyors	Germany (FRG) Poland
	conventional	depends on haulage	7 m	caving stowing	shuttle cars < 12° conveyors < 15°	United States India
Slicing	longwall	0–15°	3 m/pass (max. 14 m)	caving stowing	chain conveyors	France Poland
	continuous	depends on haulage	3 m/pass (max. 7 m)	caving	shuttle cars < 12° conveyors < 15°	United States
	conventional	depends on haulage	3 m/pass (max. 7 m)	caving stowing	chain conveyors	India
Caving	longwall	0–15°	< 14 m	caving	chain conveyors	France
	conventional	dip is used	< 7 m	caving	none	Soviet Union
	hydraulic	> 2.5°	penetration up to 20 m	caving	fluming > 2.5°	Canada Soviet Union

Some important requirements for the application of such hydraulic mining underground are that the coal should be soft or have friable bands to aid in the breaking and that the roof should be competent and part easily from the coal. This allows the maximum extraction of coal before the roof collapses and also minimizes dilution. The floor must resist degradation by water. Where the coal is hard, blasting can aid subsequent water jetting. Under proper conditions, hydraulic mining is very efficient, safe, and productive.

7. Hard Coal Preparation

The ROM(run of mine) coal brought to the surface contains various types of accompanying minerals and interstratifications, which must be removed to obtain a coal that complies with market demands for the different types and grades. Grain size, ash content, and moisture content can be controlled within a narrow range by applying mechanical and physical cleaning methods. The ROM coal is subjected almost exclusively to mechanical wet treatments that are more difficult and expensive if the feed is intimately intergrown.

Figure 28shows that the characteristics of raw coals have increasingly deteriorated in recent years due to the mechanization of coal retrieval and to the measures adopted to suppress dust. Large mining complexes concentrating on few face areas produce coal with characteristics that fluctuate within a wide range. A raw coal homogenization step can minimize investment and operational costs for new or extended washery buildings needed to handle such diverse coals. It can also improve the precision of separation and, thus, the yield of saleable product and provide a more consistent product.

The sequence of hard coal cleaning procedures is depicted in the flow sheet of a modern washery shown in Figure 29.

7.1. Preliminary Treatment and Classification of Raw Coal

The uppergrain size of raw coal is defined by the dimensions of the coal cleaning equipment. Before being fed to that equipment, the coal is subjected to preliminary screening and crushing to a diameter of 120–150 mm; foreign matter is simultaneously removed. Coarse dirt is normally reduced in particle size along with the coal, and only in exceptional cases is it removed by preliminary treatment. Such preliminary removal of coarse dirt is done exclusively in drum or inclined separators. Dual eccentric and resonance screens are used for lump coal; each of these machines has a throughput of up to 2 kt/h.

After preliminary screening and removal of foreign matter, the coal in the 0–150 (or 0–120) mm diameter range is defined as raw feed coal. To an increasing extent, this coal is fed to a storage and homogenization plant without further treatment. Homogenization then produces a consistent grain size and uniform contents of moisture, dirt, and volatile materials. The uniform quality of the coal feed makes possible a uniform product and optimal utilization of the washery capacity [23].

Homogenizationof the raw feed coal can be carried out in blending silos or on blending yards. Figure 30shows the homogenization results of a silo. A critical factor in homogenization is the size of the storage location for containing one day's output.

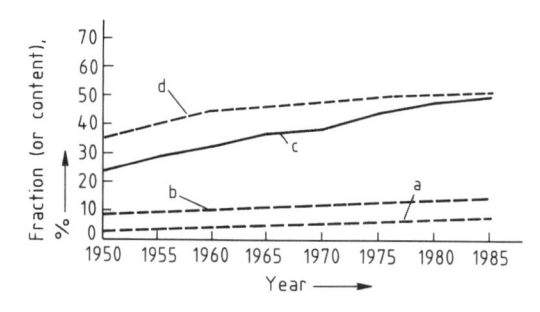

Figure 28. Change in the characteristics of ROM coal
a) Moisture content in particles < 6 mm;
b) Fraction of particles < 0.5 mm; c) Refuse content of the ROM coal; d) Fraction of particles < 10 mm

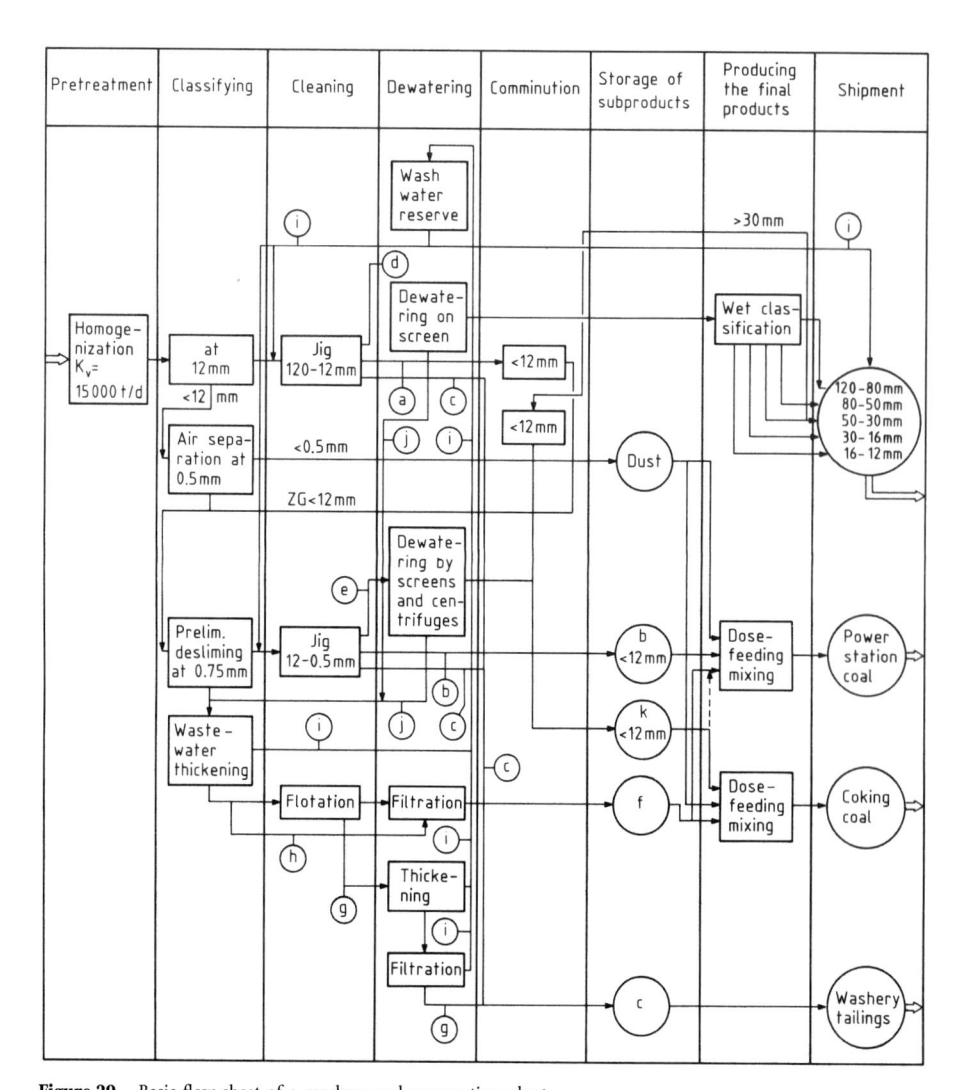

Figure 29. Basic flow sheet of a modern coal preparation plant
K_v Capacity of homogenization; a) Intermediate products; b) Middlings; c) Refuse; d) Cleaned coarse; e) Cleaned smalls; f) Froth fines; g) Tailings; h) Filtrate; i) Clarified water; j) Dirty water; k) Cleaned coal

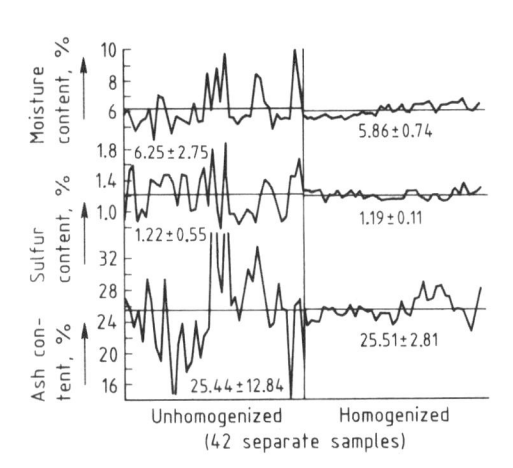

Figure 30. Homogenization of raw coal in a blending silo

In the washery the coal is first classified to ensure optimal utilization of the subsequent cleaning equipment for coarse coal and smalls. Normally the raw feed coal is screened off between 10 and 12 mm and subdivided into coarse raw coal and raw smalls. Such classification is carried out either on vibrating screens or on tensioned screens equipped with varying types of bottoms of perforated plate or tissue. The throughput for a unit may be as high as 1 kt/h [24]. The raw coarse fraction is then subjected to wet mechanical cleaning, which completely removes adhering smalls and ultrafines by spraying.

Prior to cleaning of the raw smalls 0–12 (or 0–10) mm, the ultrafines are removed by a complex method. Normally an air separation method, e.g., dry classifying, or wet desliming on screens is used. Among the air separators, cyclone separators with mechanical dust discharge have proven to be most successful. These are designed for throughputs of up to 700 t/h. Vibrating separators are used in some exceptional cases where dust must be removed from all of the raw feed 0–150 (or 0–120) mm [25]. With some rare exceptions, the separator dust is added unscreened to the saleable products; in the future, however, an increasing proportion of the dust will need to be cleaned.

7.2. Wet Treatment

Cleaning of the raw feed is done by wet treatment in jigs and dense-medium separators, which separate according to differences in density or flotation (surface) properties.

7.2.1. Cleaning in Jigs

Almost all of the jigs used in the Federal Republic of Germany are of the air-pulse type or, in exceptional cases, of the wash-box type [26], [27]. They usually yield three products. Separation is accomplished by density, under the action of a pulsating flow of water. Pulsation may be applied either by a simple stroke or a superimposed double stroke.

During recent years, the Batac jig with compressed air chamber underneath the bed proved to be the most successful machine. With such a system, all of the machine width, including the chamber width, is available for jigging. An electronic valve control or a mechanical rotational valve is used to adjust the compressed-air load. Air pulsation coupled to the hydraulic inductive discharge controls enables the operator to modify the cut point by simply actuating a selector. This allows the cleaning of two different coal types successively in the same jig. The jigs usually accept coarse coal 10–150 (or 10–120) mm in diameter and smalls < 10 mm.

7.2.2. Cleaning in Dense-Media Separators

The coarse fraction is also cleaned by dense media, particularly in star-wheel extractors or inclined separators [28].

Whereas previously baryte, clay, ultrafine dirt, loess, and pyrite were used as dense media, the more recent equipment runs almost exclusively on magnetite and ilmenite. Selection of the dense medium depends on the method of regeneration (magnetic separation or gravity). The regeneration method also depends on the grain size of the dense medium.

A jig can be controlled by using either floaters or a layer of a given weight. It can also be fully automatic, controlled by gamma-ray counters. Dense-media equipment is controlled by maintaining the densities constant via automatic addition of the dense liquid. This adjustment can be set so precisely that it approximates separation by a true solution. While the cut point with dense-media cleaning is, in general, more precise than that with jigs, the latest achievements in jig cleaning have resulted in the widely expanded use of jigs within the Federal Republic of Germany because the coal normally is highly amenable to jig cleaning and because machine expenditure and costs are lower.

7.2.3. Cleaning by Flotation

Currently, some 14% of the raw coal feed in the Federal Republic of Germany is subjected to flotation. The diverging degrees of wettability of different mineral components allow cleaning of very fine particles by means of air bubbles. The bubbles are created by dispersion of the air feed in the dense medium; they remove the coal particles that have been rendered hydrophobic by flotation agents [29]. Flotation is applied to slurries if they contain an excessive percentage of mineral components so that their inclusion in the high-grade product is economically impossible.

Flotation plants consist of five to seven separate cells or flotation troughs with the corresponding number of stirrers. Each cell is between 5 and 14 m^3 and has a throughput up to 100 t/h particulate matter corresponding to 1000 m^3/h of pulp. Flotation is done either in one or two steps. The current flotation agent is of the collecting and foaming type. A ton of particulate matter requires 0.3–0.8 kg of flotation agent for the process.

7.2.4. Other Cleaning Methods

Cleaning on tables and in dense-media cyclones has been used in certain cases to reduce the sulfur content of the product and to obtain a coal extremely low in ash.

7.3. Dewatering

Moisture must be removed after wet cleaning [29]–[32]. There are static and dynamic methods. Among the dynamic methods are screen dewatering of nut-size coal and preliminary dewatering of small products on stationary screening tables or in refuse elevators. The static methods are being replaced by dynamic ones (centrifuge, vacuum, and pressurized dewatering) because of the demand for low final moisture content. The concurrent higher ultrafine and moisture contents in the raw coal have necessitated additional process steps.

Smalls are normally dewatered on vibrating screen centrifuges. Several improvements in their design—particularly the conversion from vertical to horizontal arrangements—have allowed high throughputs and easy repair and maintenance. Centrifuges with drum diameters of 1.3 m and rated capacities of 250 t/h (water-free) are the best. The smalls are dewatered down to 5 or 7% at centrifugal forces of 80–100 g, depending on grain size distribution.

Coarse slurries are dewatered in pusher screens where the centrifuge forces (400–500 g) are higher than in vibrating screen centrifuges and where residences are considerably longer. Figure 31 shows a pusher screen. A first (comparatively short) step provides preliminary dewatering and an even distribution of particles on the drum periphery. This is followed by the secondary dewatering step. Recent pusher screen designs have drum diameters up to 1.2 m with a rated throughput of coarse slurrying of more than 40 t/h. The coarse slurry is dewatered to 9–15%, depending on grain size distribution.

Drum filters with surface areas up to 120 m^2 and disk filters of up to 500 m^2 are available for ultrafine dewatering. With this equipment, specific feed rates of 100–800 kg m^{-2} h^{-1} (wf) and moisture contents of 18–28% are attainable.

The flotation concentrate is also dewatered by means of a solid-bowl screen centrifuge where a solid-bowl section is followed by a cylindrical screen section, which

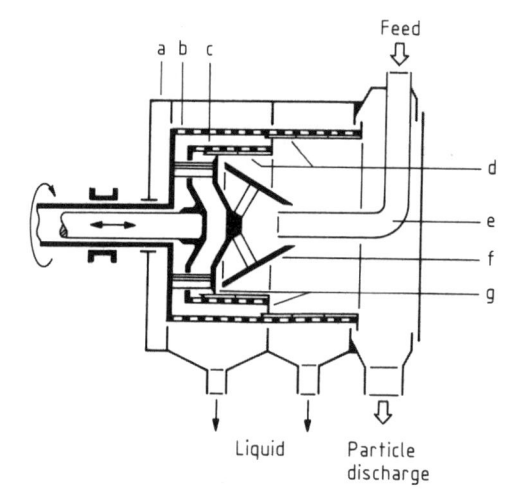

Figure 31. Two-step pusher screen
a) Enclosure; b) External screen drum; c) Internal screen drum; d) Slot screens; f) Vortex; g) Pusher rings

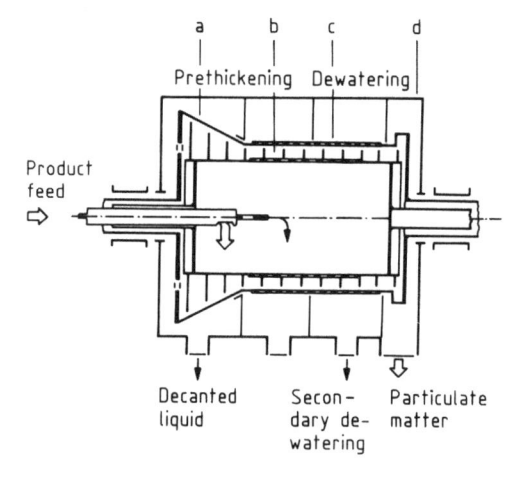

Figure 32. Solid-bowl centrifuge
a) Solid bowl; b) Conveyor screw; c) Screen bowl; d) Enclosure

dewaters the prethickened feed under favorable conditions independent of the particle concentration in the feed medium. Such a solid-bowl centrifuge is depicted in Figure 32. Flotation tailings are dewatered either by chamber filter presses or by solid-bowl centrifuges. More recently, screen belt presses have also been used. The selection of these alternatives depends on the desired moisture content of the waste.

7.4. Decantation and Thickening of the Process Water

Wet-type coal preparation plants involve a number of particle–water circuits where up to 6000 m³/h medium are circulated. If the water is to be reused for further cleaning steps, the particulate matter must be removed from it as far as possible. Decantation units, which usually consist of circular thickeners with diameters up to 40 m, are used. Rectangular thickeners are used less often. Water circulation varies, depending on the mechanical equipment, and can be as high as 6 m^3 per ton of raw feed coal. Water consumption oscillates near 0.2 m^3/t feed for a closed circuit. It can rise to as high as 1 m^3/t feed if, e.g., salt content is excessive or if the flotation waste must be discharged to an external disposal site.

The decantation units are meant to bring about simultaneously a desirable degree of thickening for subsequent dewatering or cleaning procedures. Thickener outputs fluctuate between final concentrations of 120 g/L for flotation feed and 650 g/L for pusher screen dewatering. About 0.5 g of organic sedimentation accelerators are added per cubic meter to promote sedimentation for wash-water decantation.

7.5. Dosing and Blending

The products from coal preparation (nuts, dewatered smalls, air separation dust, flotation concentrate, and middlings, i.e., product that has a high content of intergrown material) are stored separately in bunkers. The demand for nut-size particles has been declining, and they are reduced to < 10 mm diameter by impact or hammer mills. They then serve as a constituent of high-grade coals. The cleaned products are withdrawn from the bunkers and blended according to a set program. During this operation, the ash contents can be monitored or adjusted by automatic rapid-measuring instruments. The addition of those components that are highest in ash (air separation dust) is adjusted, depending on the measurements, so that the final product is brought to the desired ash content. Instruments for rapid and continuous measurement of moisture content will soon be introduced. Other mechanical equipment used for dose-feeding and blending includes filling-level indicators, dose-feeding devices, and belt weighers.

7.6. Removal of Pyritic Sulfur

Sulfur may be present in hard coal as elementary sulfur, sulfate, organic sulfur, or sulfide (pyrite).

Pyritic sulfur is removed by modern cleaning procedures [33], [34]. The raw coal is screened off at ca. 40 mm. The dirt of the > 40 mm product is removed in a jig. The jig floatings are reduced to < 40 mm, fed along with the < 40 mm product from sizing to a blending yard, and homogenized. The product is then cleaned in another jig; the coal and intergrowth of this step are reduced to a diameter of < 10 mm and fed to dense-media cyclones to yield, at low separation density, a raw coal low in sulfur. The intergrowth from the cyclone is reduced to a diameter < 3 mm and then classified in cyclones at 0.063 mm and subjected to secondary cleaning on tables. The < 0.063 mm product is then subjected to flotation. Dewatering of the intermediate products is done in vibrating screen centrifuges, pusher screens, and solid-bowl screen centrifuges.

7.7. Thermal Drying

Thermal drying is restricted to a few special cases in the treatment of coking and power station coals because the procedure involves high costs. The continuing increase of the ultrafine proportion, which is difficult to dewater, and the increasingly stringent quality demands necessitate more advanced dissociation and cleaning of ultrafines, as well as novel techniques of coal extraction, haulage, and treatment. These factors have led to increasing reliance on thermal drying in the aforementioned sectors [34].

Drum-type dryers are usually superior to the gas suspension dryers in terms of energy balance. The current quality requirements for coals for coking and power stations require that only part of the coal volumes be dried to stay within the moisture tolerances required for the final smalls product. Less drying also reduces problems of dust.

8. Coal Conversion (Uses)

Coal conversion processes may be divided into three groups. *Mechanical conversion processes* include coal preparation and briquetting. *Processes for transforming coal into secondary fuels* include coking, gasification, liquefaction, and combustion. *Processes for the conversion of coal for purposes other than the generation of energy* include the recovery of byproducts during coking, the production of active carbon, and the preparation of coal-based materials.

In Figure 33 the individual conversion processes are shown by ellipses, and the intermediates and end products are depicted by rectangles. Because there are so many variations of the processes, only the most important routes are depicted. The dark lines show processes and common applications. The thinner lines represent less common routes, and the broken lines show those that are still under development or have not yet found commercial application.

8.1. Preparation

Coal preparation is part of the colliery operation and is intended to turn the raw coal into more saleable products with defined characteristics. The choice of preparation

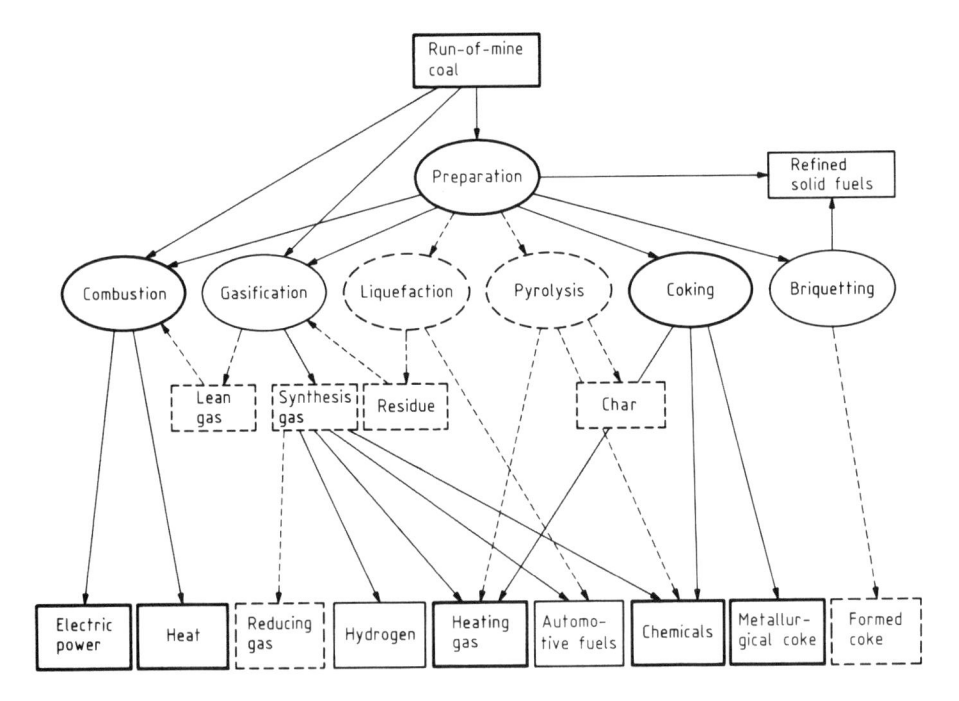

Figure 33. Coal conversion processes and end uses

process depends on the *raw material* and its technological properties, the *market requirements* for quantity, type, and quality, and *economic and environmental considerations*.

Coal preparation is used to differing extents in different countries. In the Federal Republic of Germany, all of the coal obtained by underground mining is subjected to preparation, whereas ca. 85–90 % of the lignite, which is obtained exclusively by opencast mining, is used for power generation without preparation. The lignites and subbituminous coals in the United States and Australia are also used without preparation. These two countries only submit coal to full preparation if it is to be used for coking. Probably about half of the coal used throughout the world undergoes preparation.

The water, ash, and sulfur contents, the size, and the volatile constituents are all important properties of coal that can be affected by preparation. Reduction of the sulfur content is especially significant because there is a worldwide concern about pollution.

Comminution, screening, grading (the true focal point of the preparation process), dewatering, settling, thickening, storage, proportioning, and mixing to blend the finished products are all important preparation processes.

The production and stabilization of suspensions of coal in water and coal in oil, which can be used to replace fuel oil in power stations and industrial boilers, are not considered to lie within the normal scope of preparation. The application of suspensions of coal in water to the transport of coal in pipelines is being examined as an alternative to other types of transport, see Chap. 10, [35]–[37].

8.2. Briquetting

Lignites or low-volatile coals of a particle diameter < 6 mm that are not suitable for coking and are not used in power stations can be burned in grate stokers only if they are converted to lump form. This can be carried out commercially by briquetting. Binding agents are used for low-volatile coals, but these are not required for lignites. Less common methods of compacting include hot briquetting, in which a coking coal is used as the binding agent, and pelletizing. With some exceptions, such as in the German Democratic Republic, briquetting is of only minor significance in the sales of coal to private and industrial users. Since the mid-1950s, oil and natural gas have effectively replaced coal as a means of heating in other countries because of their cost and convenience.

8.3. Carbonizationand Coking

Low-temperature carbonization and coking involve heating of coal with the exclusion of air. This process removes the condensable hydrocarbons (pitch, tar, and oil), gas, and gas liquor, leaving a solid residue of coke. Low-temperature carbonization (up to 800 °C) and coking (> 900 °C) are differentiated by the final temperature. The two processes also differ sharply in the rate of heating of the coal and in the residence time in the reactor. These parameters have a direct effect on the product yields. Low-temperature carbonization produces fine coke and fairly large quantities of liquid and gaseous products, whereas high-temperature coking is used primarily for the production of a high-temperature lump coke. After the oil crisis of 1973, attempts were made to submit the coal used in power stations to preliminary low-temperature carbonization. The coal byproducts were then to be refined while the coke was used for firing. However, the process could not be made profitable.

Currently, high-temperature coking of coal is carried out entirely in batch-operated coke ovens, of which the majority are of the horizontal chamber type. The feedstock is a coking coal of given size composition. The coking properties depend chiefly on softening and resolidification temperatures and on swelling behavior. Coking takes place at a temperature of 1000–1400 °C. The coking time of 15–30 h depends on the operating conditions and type of oven. The main product is metallurgical coke required for the production of pig iron. It is characterized by its suitable size and high resistance to abrasion even under blast furnace conditions. Coke oven gas and liquid byproducts are also produced. In Western Europe, these have considerable influence on the economy of coking and, therefore, are reprocessed. However, in many coke oven plants in the United States, the byproducts are burned.

Considerable technical improvements in coke production have occurred in recent years. These have led to greater cost effectiveness. They include mechanization and automation of oven operation, reduction of coking time and increase of specific throughput by the use of thinner bricks of higher thermal conductivity, and increased oven sizes.

Coking in horizontal chamber ovens is becoming increasingly difficult for the long term due to the deteriorating quality of coal feedstocks. Therefore, coking technology must be assisted with additives and auxiliary technology. Potential additives include petroleum coke, bitumen, and oil. Among the auxiliary technologies to be considered are the ramming operation, such as is used in the Federal Republic of Germany and some East European countries, and the various modifications of coal preheating, which are used in Western Europe, the United States, and Japan.

Dry cooling of coke has recently found increased use in Western Europe to recover stored heat. For climatic reasons this technique has been used for a considerable time in the Soviet Union. Dry cooling of coke leads to considerable savings in energy and pollution, especially when used in combination with preheating.

At the start of the 1960s, many countries considered developing a continuous production process for formed coke in expectation of a further worldwide increase in demand for coke for steel production. Numerous process developments, such as briquette coking or hot briquetting of a mixture of char and caking coal, have been demonstrated on an industrial scale. Even so, these processes have not been able to gain acceptance. This is partly because unexpected improvements in chamber coking have been realized and partly because the steel industry has been having difficulty for more than a decade, and thus, the anticipated shortage of coking coal has not yet occurred.

8.4. Pyrolysis

Pyrolysis includes carbonization and coking. It is also the starting reaction in gasification, combustion, and direct liquefaction processes. As the modern coal conversion processes have come to involve higher pressures, e.g., to increase the reaction rates and reduce the vessel dimensions, pyrolysis under pressure and in different gas atmospheres has been systematically studied.

Laboratory-scale trials show that under high hydrogen pressure, coal can be converted to gaseous and liquid products, especially BTX aromatics (benzene, toluene, and xylene), with good yield. In addition, the process results in a residue char that must be either burned or gasified. The technology is called *hydropyrolysis*. It is classified as gasification or liquefaction, according to temperature and pressure (800 °C, 100 MPa). An extended research and development program on hydropyrolysis is funded by the International Energy Agency. After successful laboratory trials, a process development unit has begun operation.

8.5. Coal Liquefaction

Coal liquefactioncan be accomplished in two ways. Treating coal suspended in suitable oils with hydrogen in the presence of a catalyst or with hydrogenating solvents yields oil products and some unreactive residue. This technology is called *direct liquefaction* or *coal hydrogenation*. In addition, coal can be gasified with steam and oxygen to yield a mixture of hydrogen and carbon monoxide (synthesis gas) from which liquid products can be synthesized. This technology is usually called *indirect liquefaction*. Both routes were developed into industrial-scale processes during the 1930s. However, (indirect) coal liquefaction is currently employed on an industrial scale only in South Africa (Sasol plants I, II, and III). Further developments took place in 1975 – 1985 mainly in the Federal Republic of Germany, the United States, and Japan.

8.5.1. Direct Liquefaction

Coal hydrogenation is a hydrogenating digestion of the coal molecule. Hydrogenation using a solvent brings about depolymerization at ca. 15 MPa and 430 – 460 °C to yield an asphalt-like product. This product can have the ash removed, e.g., by hot filtration, and it can serve as a boiler fuel or as a feedstock, e.g., for the production of high-quality carbon products. However, many development schemes went on to further hydrogenate the extract in a separate stage and thereby produce a synthetic oil that can be distilled and refined into marketable products [38], [39]. New examples of this technology are the American EDS (Exxon donor solvent) [40] and SRC (solvent refined coal) [41] processes and several Japanese developments on a pilot-plant scale [42]. In the new American development of ITSL (integrated two stage liquefaction), extraction and further hydrogenation are combined in one process [43].

Catalytic hydrogenation of coal usually requires a pressure of tens of megapascals and a temperature of 450 – 480 °C. These strong reaction conditions yield a light oil fraction that can be further refined. This technology is used, e.g., in the Kohleöl process (Federal Republic of Germany), in the H-Coal process (United States), and in Japanese developments. For the Kohleöl process a demonstration plant of 200 t/d coal throughput has been in operation since 1981. The Japanese are just completing a demonstration plant in Australia. These new processes surpass the prewar Bergius – Pier process by increased selectivity and yield of useful liquid products, by a considerable reduction in operating pressure to 20 – 30 MPa from ca. 70 MPa, by increased volumetric reactor throughput, by improved availability of the plant, and last but not least, by having the unreacted residue conditioned in such a way as to make it accessible to steam gasification for hydrogen production. The very recent German development of integrated raffination in which part of the downstream processing of the Kohleöl has been incorporated into the hydrogenation process has led to further substantial improvements and may even constitute an entirely new liquefaction process.

Direct liquefaction also reduces the sulfur and nitrogen contents of the liquid products compared to those of the reactant feed coal. Thus, extractive hydrogenation is also considered as a means of obtaining a clean boiler fuel from coal. However, the economic prospects that complete hydrogenation will produce a synthetic crude oil are more promising. Further refinement into light fuel oil, automotive fuels, and chemical feedstocks essentially follows petrochemical technology although coal oils require an adaptation of the refining catalysts.

8.5.2. Indirect Liquefaction

The synthesis of hydrocarbons from synthesis gas is called indirect liquefaction. This technology is also named the Fischer–Tropsch synthesisafter its discoverers. The process was developed in prewar Germany and was used extensively during World War II. After the war, the process could not compete with cheap oil from the Middle East. New capacity was erected only in South Africa, where it was set up downstream of Lurgi fixed-bed coal gasifiers. In the Sasol I plant, two different Fischer–Tropsch process technologies are employed, the fixed-bed Arge process (Lurgi–Ruhrchemie) and the entrained-flow Synthol process (Kellogg–Sasol) (→ Carbon Monoxide). The Arge process produces mainly gasoline, fuel oil, and waxes, whereas the Synthol process yields a larger fraction of gasoline and low molecular mass products (methane and propylene). The Sasol II and III plants, built in the late 1970s, use only the Synthol process. On an energy basis the total liquid product yield obtained by indirect liquefaction of coal in the Sasol II plant is ca. 32%.

8.6. Coal Gasification

Systematicdevelopment of coal gasification began in the first half of the 19th century. A mixture of carbon monoxide and hydrogen was produced. This was generally used for chemical purposes, which remained the main application of coal gasification for nearly 100 years. Many communities relied on coke oven gas to supply town gas for illumination, cooking, and heating into the second half of this century. However, coal or coke gasification was also employed mainly for peak shaving in large gas schemes and also for the supply of gas to industrial furnaces. These applications provided strong development incentives. Natural gas and mineral oil in large amounts and at low cost replaced gas manufactured from coal not only in the heating market, but also in the chemical industry in most countries. The oil crisis of 1973–1974, with its sudden steep increase in prices of mineral oil and natural gas and the subsequent supply shortages, triggered new interest in coal gasification.

Gasification can be accomplished by reacting coal with oxidizing or reducing agents. Using oxygen (or air) and steam as gasifying agents yields water gas.

$$C + H_2O \xrightarrow{O_2} CO + H_2$$

The composition chosen for the blast depends on individual process conditions and governs the carbon monoxide:hydrogen ratio of the product. Varying the steam:oxygen ratio in the blast is a very sensitive and convenient means of controlling the reaction temperature.

Gasification with hydrogen (hydrogasification) leads primarily to methane.

$$C + 2 H_2 \longrightarrow CH_4$$

However, hydrogasification is usually incomplete and produces a char residue, which cannot economically be converted with hydrogen. Thus, reductive gasification is considered as only a preliminary step in the complete gasification of coal. However, it is a very favorable step if a gas with a high heating value, e.g., substitute natural gas (SNG), is to be produced. Hydrogasification is also involved in most fixed-bed countercurrent-flow gasification processes, contributing to the methane content of the product. Several other reactions take place simultaneously, including the Boudouard reaction, complete combustion of carbon, and methane formation from carbon monoxide and hydrogen.

In coal gasification processes, it is necessary to optimize heat transfer, mass transfer, and chemical reaction conditions between huge flows of solid and gaseous agents and to achieve a high throughput and optimal energy economy at the same time. Many varieties of coal gasification processes have been developed, reflecting compromises of the many constraints and conditions.

First consideration must be given to the intended use of the product gas. Current interest focuses on the preparation of the following:

1) a gas with a high heating value, e.g., SNG
2) a fuel gas, e.g., a clean boiler or turbine fuel
3) synthesis gas, for the production of commodity chemicals
4) hydrogen, for ammonia production or for energy purposes
5) reducing gas, for direct reduction of iron ore

A second, equally important, consideration applies to the available coal. Of particular importance are coal rank, grain size, caking properties, ash content, and ash melting behavior. Third, environmental regulations, site specific infrastructure, etc. have to be taken into account. Considering all these factors, a decision in favor of fixed-bed, fluidized-bed, or entrained-flow gasification can be made.

In a *fixed-bed* gasifier, coal and blast are usually contacted in countercurrent flow. This leads to the splitting of the overall gasification reaction into several zones with favorable conditions for the various reaction steps involved. Fixed-bed gasification usually results in excellent carbon utilization and high efficiency. It is applicable to coals of all ranks. However, it requires lump, noncaking or weakly caking coal with a high ash softening temperature. The product contains tars and other liquid byproducts that complicate the gas treatment. Atmospheric fixed-bed gasifiers of various design are still occasionally found in small-scale industrial use. On a large scale, some Lurgi fixed-bed pressurized gasification plants are operating commercially, e.g., in the Republic of South Africa (Sasol I, II, and III).

Fluidized-bed gasification, invented in 1922 by WINKLER at BASF, has the advantage of fairly simple reactor design and very high reactor throughputs even at atmospheric pressure. However, reaction temperature is limited by the need to avoid ash agglomeration so that fluidized gasification is restricted to very reactive fuels such as subbituminous coals or lignite. Winkler gasifiers operating at atmospheric pressure produce synthesis gas in the German Democratic Republic and in India. In the Federal Republic of Germany, further development to Winkler gasification at high pressure has taken place, and a first demonstration unit has been completed [44], [45].

Entrained-flow gasification takes place in a flamelike reaction zone, usually at a very high temperature to produce a liquid slag. For economical operation, a high-standard heat recovery system is mandatory, but the product gas is free of methane, tars, etc., thereby considerably simplifying gas and water treatment.

Entrained-flow gasifiers of the Koppers –Totzek design operated at atmospheric pressure are used industrially in many countries to produce hydrogen or synthesis gas. The Texaco partial-oxidation process recently has been adapted to coal gasification in a pressurized entrained-flow reactor. Industrial-scale plants operate or are under construction in the United States, the Federal Republic of Germany, and Japan.

Apart from these well-established or highly developed processes, further intense development is under way on existing processes, as well as on new concepts. During the last decade most of the existing processes have been modernized or modified by incorporating new process engineering, materials, and control technology. Such new concepts as pressurized gasification in various modes, hydrogasification, multistage processes, slag bath and molten iron processes, and allothermal processes, e.g., using process heat from gas-cooled nuclear reactors, have been proposed and tested [46]–[51].

Some of the new development projects have subsequently been terminated or shelved because of the present availability of mineral oil and natural gas. However, the long-term incentive of securing and diversifying the world energy supply is still valid. Environmental considerations may favor converting coal into a clean fuel gas that can be used either conventionally or in highly efficient technologies like fuel cells or combined cycle power plants. Furthermore, low-quality coals, which are located in remote places and are therefore not marketable as such, can be gasified and converted on site into transportable products such as commodity chemicals or hydrocarbons and thereby contribute to worldwide supplies.

8.7. Coal Combustion

In combustion the energy content of the fuel is completely released and can be recovered as sensible heat. The combustion products are usually considered to be useless waste products. Therefore, coal combustion is not classically considered to be a coal conversion or refining process, but rather an energy conversion process. However, modern combustion systems, e.g., two-stage combustion or low-nitrogen oxide burners, involve clearly distinguishable reaction steps such as pyrolysis and partial oxidation (gasification), which are classical conversion reactions. In the fluidized-bed combustion of coal, sulfur dioxide is removed from the flue gas during the combustion process by adding limestone sorbent to the feed coal. This enables low-grade coal to be burned cleanly, eliminating the need for preliminary upgrading steps.

8.8. Conversion of Coal for Purposes Other Than the Generation of Energy

8.8.1. Coal Tar Chemical Industry

Considerable quantities of aromatic chemical feedstocks are still prepared by reprocessing the coal tar that is produced during coking at a worldwide rate of ca. 16 Mt/a (tar and pitch). The tar components are reprocessed into dyes, pesticides, varnishes, vitamins, and textile auxiliaries. Tar oils are used for the production of carbon black, e.g., for tire manufacture, and for the production of impregnating oils. Special pitches recovered from coal tar are used for the production of special coke, e.g., that used for electrodes.

Until the early 1950s, industrial production of aliphatic hydrocarbons was based on acetylene produced from calcium carbide. This coal-based chemical feedstock was subsequently replaced by the more economical ethylene produced from crude oil. Research into the production of acetylene is again being carried out worldwide, and new methods using carbide and plasma arcs are being investigated. Special attention is given to reducing the high energy requirement.

8.8.2. Production and Use of Activated Carbon

Precise thermal and chemical treatment of lignite or coal can produce activated carbon with a specific pore system. Activated carbons made from coal are particularly noted for their high ignition temperatures and good resistance to abrasion. Therefore, they are suitable for use in industrial processes for removing sulfur and nitrogen compounds from the flue gases emitted by coal-fired power stations, for wastewater purification, for separation of various mixtures of gases, and for solvent recovery, etc. (→ Carbon).

9. Agglomeration

Numerous industrial processes require coal in its coarse state for handling and transportation and as a feed for such thermal applications as combustion, gasification, iron ore reduction, and carbonization. However, highly mechanized mining operations, coal preparation, and beneficiation yield a high proportion of fines. Upgrading of such coal fines by agglomeration is frequently applied and considered [52]–[57].

The choice of agglomeration process and of a possible binder depends very much on the degree of coalification of the coal and the intended application.

The agglomeration ability of a coal is mainly influenced by its moisture-holding capacity, ash content and composition, hydrophobicity, size distribution, and plastic or elastic behavior.

Fundamental coal agglomeration methods are grouped into

1) briquetting and extrusion [53], [57]–[61];
2) balling and tumbling [53], [55], [56], [62]–[65].

In the processes of the first group, a compressive force is applied to a prepared coal or a coal and binder mixture.

The double-roll hydraulic press and extrusion presses of various designs are the most common mechanical devices used. Binderless briquetting or extrusion requires that a high pressure be exerted on the coal. Therefore, its applicability is generally limited to predried peat or soft lignites sized to particle diameters < 1 or 0.5 mm that exhibit defined plastic properties. Ash contents should be low to prevent excessive erosion [61].

Binder briquetting or extrusion exercised at low to medium pressures is applied to coal with insufficient plastic properties, but essentially elastic properties, i.e., subbituminous, bituminous, and anthracitic coals usually of particle sizes < 3 mm. Organic binders such as pitch, bitumen, starch, molasses, lignosulfonates, or inorganic additives, e.g., clay minerals, are used in quantities of 2–8 wt %. Briquettes produced with water-soluble binders require a subsequent drying step to attain final strength [57], [66], [67].

Hot briquetting techniques utilizing the softening of bituminous coals at a temperature of 350–450 °C are employed for the production of hot briquettes from mixtures of coal and coke.

Balling of finely ground prepared coal generally smaller than 0.315 mm in particle diameter with 2–6 wt % of organic or inorganic binders has only recently evolved as a commercially viable coal agglomeration technique by using pelletizing disks designed for light bulk materials [55], [65], [68]. Both moist and dry pellets may be used industrially. Although the disk produces pellets of almost uniform size, tumbling in, e.g., rotary drums or mixers yields pellets of a wide size distribution. This mode of agglomeration is preferentially applied to subbituminous, bituminous, and anthracitic coals of both low and high ash contents, but predried lignites can also be pelletized [53], [62], [63]. Drying and thermal treatment of these pellets, however, cause severe cracking and significant losses in strength and stability.

Coal agglomerates must withstand mechanical and thermal treatment when used in industry. The type and severity of stress depend on the particular use and mode of treatment which mainly include transportation, drying, thermal shock, pyrolysis, gasification, and combustion [57], [60], [62]. Suitable procedures for the agglomeration step may be developed and selected from bench-scale simulation studies. Subsequent handling and thermal experiments with these agglomerates lead to a final, optimized product, which may be tested on a larger scale in pilot or commercial plants.

10. Transportation

The large amount of coal that is mined every year is a major commodity for the transportation industry and one that requires a significant capital investment. In the United States, ca. 50% of the 700–800 Mt of coal produced per year is shipped by rail; 18% is shipped by river barges; and the rest is handled by trucks, conveyor belts, and pipelines. The actual cost of long-distance shipping is usually quite modest, but the loading and unloading of the coal adds greatly to the total cost of shipping. In a typical surface mining operation, for example, coal is moved from the mine by truck or conveyor belt to a nearby preparation plant, where it is loaded onto a unit train for rail shipment to a utility or port facility. At the port it may then be loaded onto river barges or ocean-going bulk carriers.

Although some of the coal transported by *rail* is shipped at single car or bulk rates, most is carried at lower rates in unit trains of up to 100 cars that travel continuously to a single destination and return. The unit train carries only coal and each car can carry up to 100 t. The cars are of either the bottom-dumping or roll-over type, and each car can be unloaded in 1.5–5 min in modern processing facilities, which have a capacity of 2000–6000 t/h.

The least expensive means of transporting coal is by *river barges*, whereby the cost is as low as $ 0.005 per mile ($ 0.003 per km). The barges have open tops and commonly hold up to 1000–1500 t each. They are moved in groups or tows of 20–30 barges powered by a single towboat. Although there can be some problems with delivery schedules due to congestion, the ease and economy of river transport are major factors in the location of many coal-burning utilities along inland waterways in the United States and in Europe.

In the United States, ca. 12% of coal production is carried by *trucks*. Much of this is done at surface mines, where the coal is hauled from the working face to preparation plants. This is usually done with dedicated high-capacity vehicles; however, significant amounts of coal are hauled by trucks on public highways. *Conveyor belts* are also used for transporting coal from mines to preparation plants, as well as for short cross-country hauls of over 10 miles (16 km) to loading points. Because they are dependable and economical, conveyor belts are also widely used for in-mine coal transport.

Although > 80% of the coal mined in the United States is used domesticly as steam coal, some of the metallurgical coal production is shipped overseas. This coal is shipped in dry bulk carriers that can be loaded and unloaded in modern port facilities at rates of up to 100 000 t/h. These *ocean-going ships* are usually in the 60 000 t range, although some are larger. In addition, some smaller vessels in the 30 000 t range that are self-unloading are also in service. These ships have the advantage of being able to use a much greater number of ports.

The transportation of coal by *slurry pipeline* has been demonstrated to be dependable and economical, even though there are only a few such pipelines in use around the world. In this technique the coal is reduced to a diameter of < 1 mm, treated with chemicals to prevent corrosion and improve flow characteristics, and mixed with water. It is then pumped as a slurry at a velocity of 1.5 – 2 m/s. The fact that pipelines can be built above ground or buried eliminates some environmental problems, but their high water requirements can be a serious drawback. Sustained delivery rates of > 600 t/h have been demonstrated.

11. Coal Storage

The use of any bulk commodity requires the storage of sufficient material to ensure efficient operations. In the case of coal, this requirement varies from a few days supply at the mine to a few months supply at a power plant. Although some covered storage in bins or silos is practical at mines and small operations, most of the larger volumes of coal are stored in open piles, where it is exposed to the air and precipitation. When coal is exposed to air, it loses moisture and begins to oxidize. This oxidation can degrade coal quality, and more important, it can cause spontaneous combustion that destroys the coal. The main coal properties that influence oxidation are the particle size, pyrite content, and rank. Coal oxidation leading to spontaneous combustion is enhanced as the particle size decreases and as the pyrite content increases. Low-rank coals such as lignite and subbituminous coal are very difficult to store because of their strong tendency toward spontaneous combustion, and care must be taken even with the lower rank bituminous coals. Repeated wetting and drying also exacerbate oxidation. The continued oxidation generates more heat than can be dissipated, and this leads to hot spots, which eventually ignite.

In a stockpile even mild oxidation can lead to a degradation of coal quality. For steam coals there can be some loss in calorific value, but in coking coals there can be a total loss of quality. The fluid properties can be destroyed, and the heat transfer properties and coke production yields can also be reduced. The quality of coke made from oxidized coal is also seriously altered. The prime coke property, coke strength or stability, is reduced, and this can result in the loss of iron production in the blast furnace. Oxidized coal also increases the reactivity of coke and leads to higher coke consumption in the blast furnace. Because all of the effects of oxidation are undesirable, every effort should be made to prevent it.

The main steps to prevent oxidation involve limiting the access of air and protection from wind and precipitation. Stacking the coal to eliminate size segregation and compaction of the pile with earth-moving equipment to reduce pore space have proven helpful. Coating the pile with a sealant also helps protect it from air and moisture, although this may be expensive. Another costly but useful step is to provide the storage piles with some kind of wind protection. The wind not only enhances oxidation, but it also removes the fine particles from the coal pile and can thereby create environmental problems.

Another aspect of coal storage is the need to mix and blend the stockpile coal to homogenize the product coal reclaimed from it, so that a consistent, uniform product can be delivered to the plant. This can be a serious concern in cases where coal is coming into an operation from a number of sources. The usual way to solve this problem is to develop a bedding and blending system. A typical system consists of a plan to build a stockpile with long, thin layers of the various incoming coals and to reclaim the coal from the pile with vertical cuts across these layers. The key design factors are the variability of a target blending parameter such as ash or sulfur content and the desired degree of homogeneity. These are evaluated to determine the necessary pile parameters such as individual layer thickness.

12. Quality and Quality Testing

Because coal is so variable in its maceral and mineral composition, its suitability for a given use must be determined by a variety of tests. Although some tests such as chemical analysis are quite general in nature, others such as ash fusion are specific for particular uses.

12.1. Chemical Analyses

Although coal is composed of a large number of organic chemical components, no true standard organic chemical test exists for coal. The two major kinds of chemical analyses used are the proximate analysis and ultimate analysis, which are standard tests defined by the American Society for Testing and Materials. The *proximate analysis* (ASTM D 3172) consists of a determination of the moisture, ash, and volatile matter, and a calculation of the fixed carbon value. The *moisture* is determined by heating the sample at 104–110 °C to a constant weight. The percent weight loss is reported as the moisture value. The *ash* in this analysis is the incombustible residue after the coal is burned to a constant weight. It should be noted that the ash value is not a measure of the kinds or relative amounts of the minerals in the coal. The *volatile matter* is a measure of the amount of gas and tar in a coal sample. It is reported as the weight loss minus the moisture after the coal is heated in the absence of air at 950 °C for 7 min. The *fixed carbon* is not a distinct chemical entity. It is reported as the difference between 100% and the sum of the moisture, ash, and volatile matter values. While the proximate analysis as described above is a rather simple assay of the chemical nature of coal, its value as a quality parameter is well established and is widely used in commerce.

The *ultimate analysis* (ASTM D 3176) consists of direct determinations of ash, carbon, hydrogen, nitrogen, sulfur, and an indirect determination of the oxygen. The ash is determined as in the proximate analysis, and because all of the values are reported on a moisture-free basis, moisture must also be determined.

Both proximate and ultimate analyses are reported in a number of different ways, and care must be taken to be certain of the method of reporting. On the "as-received" basis, the results are based on the moisture state of the coal sample as it was received for testing. With the "dry" basis, the results are calculated back to a condition of no moisture, and with the "dry, ash-free" basis, the results are calculated to a condition of no moisture and no ash. These calculations are done so that different coals can be compared on their inherent organic nature. The reporting basis can cause significant changes in the values reported, and it is essential that a target value and the value of a sample in question be on the same basis. For example, for a given coal with a moisture of 10%, an ash value of 15%, a volatile matter of 30%, and a fixed carbon content of 45%, the volatile matter content would be 33.3% on a dry basis and 40% on a dry, ash-free basis. The corresponding fixed carbon values are 50% and 60%, respectively.

Because some components of the minerals such as water from the clays and carbon dioxide from calcite are lost in the high-temperature ashing process, the ash value determined is less than the actual mineral matter in the raw coal. A number of corrections for this loss are in use, but the one most used in the United States is the Parr formula, where the corrected mineral matter is equal to 1.08 times the ash percentage plus 0.55 times the sulfur percentage. Results reported with this correction are considered to be on a dry, mineral-matter-free basis.

12.2. Mineral Matter

The mineral matter content of a coal is the actual weight percent of the minerals present. It is the best measure of the inorganic content of a coal, but it is difficult to determine. Although there is no standard test for mineral matter, low-temperature ashing ($< 150\ °C$) with an oxygen plasma device is widely used for this purpose. The X-ray diffraction analysis of the low-temperature ash is used to identify the actual minerals in the ash. However, with present techniques it is impossible to accurately determine the amounts of the various minerals present in a given coal.

Although a large number of different minerals have been identified in coal, the four most common are clays, pyrite, calcite, and quartz. The minerals get into the coal in a variety of ways. Some can be part of the original plant material itself. Silica (quartz) is present in some of the saw grasses of modern swamps, and similar types of plants are known from ancient swamps. Certainly much of the mineral matter is transported into the coal-forming swamp from the environment. Much of the clays and quartz minerals were brought into the coal swamp as clastic material in streams or in airborne dust. Most of the common minerals could be chemically precipitated from solution under conditions that can exist in swamps. Some minerals such as calcite and pyrite can also be precipitated into cleats and fractures in coal after it has formed and coalified beyond the peat stage.

The presence of sulfur in coal is of great interest because of the problems it causes in utilization, especially air pollution. There are three commonly recognized forms of sulfur in coal. Pyrite sulfur is the sulfur tied up in the mineral pyrite, FeS_2; it is determined by leaching with nitric acid. Sulfate sulfur is of minor importance and thought to be formed by the weathering of pyrite; it is determined by leaching with hydrochloric acid. Organic sulfur is that portion of the total sulfur that is organically bound with the various coal macerals; it is determined indirectly by difference. There also may be some minor elemental sulfur present in coal. The forms of sulfur are determined in accordance with the ASTM standard D 2492.

The occurrence of pyrite in coal is of special interest because it is the only form of sulfur that can usually be removed by mechanical cleaning methods and because it is a major source of air and water pollution. Although some pyrite is formed by the chemical combination of the sulfur and iron that occur naturally in peat, most of the pyrite found in coal is thought to form from Fe(III) ions absorbed on clay minerals that are transported into coal swamps, and from sulfate ions that are introduced into the swamp in seawater. Marine water usually has about two orders of magnitude more sulfate ions than fresh water. In the swamp the ferric Fe(III) is reduced to Fe(II), which combines with the sulfur in the sulfate ions to form pyrite. While this process is clearly not the only way that pyrite can form in coal seams, it is a good model for many United States coal seams, especially those in the Illinois Basin. In these seams it is reported [69]–[71] that both the total sulfur and pyritic sulfur contents are controlled by the nature of overlying rocks; they are both high under marine rocks and low under nonmarine rocks.

Pyrite can occur in coal in a number of forms such as single crystals, void fillings, irregular and dendritic masses, and framboids (raspberry-like clusters). The most significant forms are the more massive ones, which are easier to remove from the coal by washing, and the framboidal forms, which are more chemically active because of their high surface area. When pyrite is exposed to air and water, it oxidizes in a series of chemical reactions that generate sulfuric acid and cause acid mine drainage. In fact, the reaction of 1 mol of pyrite with air and water results in the generation of 2 mol of sulfuric acid.

12.3. Thermal Properties

The most widely used thermal property is the calorific value, which is a measure of the heat produced by combustion of a unit quantity of coal under given conditions (ASTM D 3286). It is usually reported on a moist, mineral-matter-free basis and is used this way in the ASTM classification of coals by rank. The major use of the calorific value is in the evaluation of coals for use in steam generation. In many utility company contracts, coal is bought on the basis of total calorific value per unit mass.

13. Economic Aspects

13.1. World Outlook

Reserves. The world's reserves of coal are an estimated 6.9 Tt of hard coal and 6.5 Tt of brown coal, as shown in Table 10. Coal deposits are found distributed over the whole world, and there are not yet any internationally standardized criteria for a clear, quantitative classification of coal deposits. The assignment of terms such as "technically recoverable" or "economically recoverable" depends largely on prevailing technological and economic conditions. With current technology only ca. 7–8% of the total reserves are economically recoverable.

Production. Totalproduction of hard coal worldwide has been increasing since 1970. However, the individual areas of production have developed to different extents. For instance, production in the EEC has been reduced, but at the same time, the coal output of countries such as the People's Republic of China, South Africa, Australia, Canada, and Poland has increased, in some cases considerably. In 1983, production of hard coal worldwide reached 2.9 Gt, as seen in Table 11. World production of brown coal has been increasing continuously since 1966. In 1983, the brown coal output amounted to ca. 1.1 Gt, as Table 12shows. This amount yields the same heat as 0.75 Gt of hard coal.

Table 10. World reserves of coal as of 1984 [72]

Region	Quantity, Mt	
	Hard coal	Brown coal
Western Europe	430 365	85 314
Canada and United States	1 385 335	2 688 734
Latin America	20 085	35 607
Africa	216 057	2 426
Eastern Europe*	2 767 388	3 410 593
Asia**	1 567 990	86 035
Australia	548 800	231 100
Total	6 936 020	6 539 809
Economically recoverable total	549 234	431 252

* The Soviet Union is included with Eastern Europe.
** The Soviet Union is excluded from the estimate for Asia.

Consumption. Approximately 20% of the 2.9 Gt of hard coal used per year in the world is allocated to coking coal and 80% to steam coal. Virtually all of the brown coal produced in the world, ca. 1.1 Gt/a, is used in power plants.

Trade. Trade in hard coal has assumed worldwide proportions, as seen in Table 13. This applies, to the same extent, to both coking coal and steam coal. International trade in coal coke is predominantly carried out within the EEC, between COMECON countries, and between Canada and the United States. The same applies to trade in brown coal.

13.2. Some Major Coal Producing Countries

Coal reserves, coal production, and coal consumption in some of the world's most important coal-producing nations can be described in detail.

13.2.1. United States of America

Reserves. The United States possesses reserves of hard coal amounting to 1286.4 Gt and reserves of brown coal to the extent of 2313.3 Gt, but only 9.7% and 5.7%, respectively, of these reserves are economically recoverable. The United States contains 23% of the economically recoverable hard coal reserves of the world and 31% of its brown coal reserves.

Table 11. World hard coal production in 1983 [73]

Region or country	Quantity, kt	Proportion, %
Europe		
EEC		
Belgium	6 097	0.21
Federal Republic of Germany	89 620	3.10
France	17 022	0.59
United Kingdom	116 448	4.02
Ireland	62	0.00
EEC total	229 249	7.92
Other Western European countries	16 264	0.56
Yugoslavia	380	0.01
COMECON		
Bulgaria	243	0.01
Poland	191 090	6.60
Romania	7 250	0.25
Czechoslovakia	26 915	0.93
Soviet Union *	553 700	19.11
Hungary	2 827	0.10
COMECON total	782 025	27.00
Europe total	1 027 918	35.49
Asia		
People's Republic of China	675 130	23.31
India	132 800	4.59
Japan	16 880	0.58
Turkey**	7 000	0.24
Other Asian countries	69 380	2.39
Asia total	901 190	31.11
Americas		
United States	664 500	22.94
Canada	22 563	0.78
Latin America	23 780	0.82
The Americas total	710 843	24.54
Africa		
South Africa	145 838	5.03
Africa total	151 325	5.22
Australia and New Zealand	105 537	3.64
World	2 896 813	100.00

* The USSR total includes the part in Asia.
** The Turkey total includes the part in Europe.

Table 12. World brown coal production in 1983 [73]

Region or country	Quantity, kt	Proportion, %
Europe		
EEC		
Federal Republic of Germany	124 335	11.38
France	2 612	0.24
Greece	30 026	2.75
Italy	1 738	0.16
EEC total	158 711	14.53
Other Western European countries	26 974	2.47
Albania	1 650	0.15
Yugoslavia	57 700	5.28
COMECON		
Bulgaria	32 120	2.94
German Democratic Republic	277 968	25.45
Poland	42 532	3.89
Romania	32 800	3.00
Czechoslovakia	102 416	9.38
Soviet Union *	162 200	14.85
Hungary	22 386	2.05
COMECON total	672 422	61.56
Europe total	917 457	83.99
Asia		
People's Republic of China	25 200	2.31
India	7 000	0.64
Japan	23	0.00
Turkey**	21 500	1.97
Other Asian countries	16 850	1.54
Asia total	70 573	6.46
Americas		
United States	47 500	4.35
Canada	21 645	1.98
The Americas total	69 145	6.33
Africa		
South Africa		
Africa total		
Australia and New Zealand	35 224	3.22
World total	1 092 399	100.00

* The USSR total includes the part in Asia
** The Turkey total includes the part in Europe.

Production. Although coal production had been increasing (542 Mt in 1973; 760 Mt in 1982), production in 1983 dropped to the 1979 level (712 Mt). The production of coke has steadily declined (85 Mt in 1973, 34 Mt in 1983).

Consumption. Domestic consumption reached a new peak of 668 Mt in 1983. In fact, consumption had steadily increased until 1981 (665 Mt) before it dropped to 641 Mt in 1982. As much as 84% of the hard coal production is used in power plants and heating plants. The second largest consumer is industry, accounting for 10%. The

Table 13. World trade in hard coal in 1983 [74]

Country	Quantity, kt	
	Exports	Imports
United States	70 135	1 083
Canada	16 794	13 823
Federal Republic of Germany	10 990	9 613
France	556	18 420
Belgium and Luxembourg	606	7 678
The Netherlands	587	7 631
United Kingdom	6 332	4 277
Italy		17 678
Denmark		8 601
Spain		5 908
Finland		4 390
German Democratic Republic		4 198
Poland	35 148	738
Czechoslovakia	2 849	4 929
Bulgaria		7 087
Romania		4 087
Yugoslavia		4 546
Soviet Union	21 545	11 564
Japan		74 315
Australia	61 304	
South Africa	27 500	
Others	4 449	48 229
Total	258 795	258 795

remaining 6% is absorbed by coking plants and gasworks, by home owners, small consumers, and others. The United States imported 1.1 Mt of coal in 1983, which corresponded approximately to the amounts imported in 1981 and 1982.

Exports, on the other hand, fell drastically from a peak of 101.7 Mt in 1981 to 96.2 Mt in 1982 and to 70.1 Mt in 1983. In 1984 exports grew slightly (75 Mt), but the National Coal Association anticipates further declines because of aggressive competition. The main importers of American coal are the EEC, Japan, and Canada, which account for 75% of total United States exports.

13.2.2. Australia

Reserves. Australia possesses geological deposits of 548.8 Gt of hard coal and 231.1 Gt of brown coal. Of these deposits, 27.4 Gt of hard coal and 38.3 Gt of brown coal are economically recoverable. These respresent 5.0% and 8.9%, respectively, of the world's economically recoverable reserves.

Production. The production of coal has steadily risen since 1973 (79 Mt in 1973, 140.8 Mt in 1983), and will further increase by more than 70% in the course of the next 10 years. The planned outputs of hard coal and brown coal for the year 1987 are 116.4 Mt and 39.5 Mt, respectively. This means that the hard coal production will have increased by 10.3% from 1983 to 1987 and the brown coal output by 24.1% in the same period of time.

Consumption. In 1983, domestic consumption accounted for 41.4 Mt of hard coal. Of this amount, 64% went to power and heating plants, 12.8% to coking plants and gasworks, and only 8.2% was used by industry. The remaining 15% was used by home owners, small consumers, and others. More than 90% of the brown coal output was used for the generation of electricity. A full 5% was used in the preparation of briquettes, and the remainder was consumed by industry. In 1973–1974, coal accounted for 81.3% of total production of electricity (hard coal 48.6%, brown coal 32.7%). An increase is planned for the period 1991/1992.

Australia's exports rose by 15 Mt in 1984, reaching an all-time peak of 76 Mt. As a result, Australia's share of international trade increased from 22% to 24%. In 1984, Australia became the world's foremost hard coal exporter. Future exports of coal are expected to show a steady increase. Japan is the main buyer of Australian coal, consuming more than 60% of its exports.

13.2.3. Federal Republic of Germany

Reserves. The Federal Republic of Germany possesses geological deposits to the extent of 230 Gt of hard coal and 55 Gt of brown coal. Of these deposits, 24 Gt of hard coal and 35 Gt of brown coal are economically recoverable.

Production. Output of hard coal has fallen from a peak of 150 Mt in 1957 to 90 Mt in 1983. Brown coal production dropped from a record 135 Mt in 1976 to 124 Mt in 1983.

Consumption. The largest portion of the hard coal, 51%, used in the Federal Republic of Germany is channeled into the production of electricity. The second largest consumer (of coke) is the steel industry (20%). Even the exported coal goes primarily to the steel industry. Hard coal imports have not changed much in recent years, and the country remains a net exporter of hard coal. Domestic power plants consume ca. 90% of the brown coal output.

13.2.4. Poland

Reserves. Polish deposits of coal amount to 144 Gt of hard coal and 40 Gt of brown coal. Of these reserves, 19% and 30%, respectively, are economically recoverable. Many of the seams that have not yet been mined lie either too deep underground (more than

1 km) or are overlain with strata that make them geologically almost inaccessible. This applies, for instance, to seams found near Lublin, which contain an estimated 40 Gt of coal.

Production. Hard coal production reached a low of 163 Mt in 1981, but Poland boosted its production to 191 Mt in 1983. Brown coal production rose to a record 42.5 Mt in 1983. Poland is expected to increase its hard coal production to 205 Mt and its brown coal output to 100 Mt by the end of this century.

Consumption. The total consumption of hard coal was 152.1 Mt in 1983, a drop by 4.5 Mt from the previous year. In 1983, the power industry, the largest consumer in Poland, absorbed 56.5% of the total domestic consumption and 20.4% went to home owners and small consumers. Coking plants and gasworks used a further 14.9% and industry consumed no more than 3.8% of the total. The remaining 4.4% was absorbed by briquette factories and others. Brown coal consumption increased by 19% over 2 years, from 35.6 Mt in 1981 to 42.5 Mt in 1983. It is used almost exclusively in power plants.

Poland imported 1 Mt of hard coal in 1983 from other socialist countries and exported 35.1 Mt. In 1984, it was able to increase its exports to 45 Mt, thereby attaining a 14% share of the total volume of international trade. Poland has set its intermediate export goal at 42 Mt of hard coal per year. Of this amount, 60% will be sold to the West for hard currency.

13.2.5. Republic of South Africa

Reserves. South Africa possesses hard coal reserves to the extent of 92.5 Gt, 56% of which is economically recoverable.

Production. Hard coal production has been increasing (65.0 Mt in 1974, 145.8 Mt in 1983), and South Africa plans to continuously expand production.

Consumption. Domestic consumption amounted to 113 Mt in 1983. Power plants and heating plants consumed 53% of the hard coal output, and 37.2% was absorbed by industry (including coal liquefaction). Other consumers, including coking plants, gasworks, home owners, and small consumers, had a 9.8% share of the total consumption.

In South Africa, coal is the main raw material used in the production of electrical energy (95%). In addition, it is becoming increasingly important as a basis for gaseous and liquid fuels (substituted for mineral oil), for chemical products, and for the manufacture of steel. The government expects the demand for coal to grow, on an average, by 5.8% per year. South Africa exported 29 Mt of coal in 1983 and 38 Mt in 1984. It expects to continue to increase exports in the future.

14. References

[1] M. C. Stopes: "On the petrology of banded bituminous coals," *Fuel* **14** (1935) 4–13.

[2] International Committee for Coal Petrology: *Handbook of coal petrography*, 2nd ed., Centre National de la Recherche Scientifique, Paris 1963.

[3] W. Spackman: "The maceral concept and the study of modern environments as a means of understanding the Nature of Coal," *Trans. N.Y. Acad. Sci.* **20** (1958) no. 5, 411–423.

[4] M. Teichmuller: "Über neue Macerale der Liptinit-Gruppe und die Entstehung von Micrinit," *Fortschr. Geol. Rheinld. Westfalen* **24** (1974) 37–64.

[5] M. Teichmuller, R. Teichmuller: "Geological cause of coalification," in R. F. Gould (ed.): "Coal Science," *Adv. Chem. Ser.* **55** (1966) 133–163.

[6] P. A. Hacquebard, J. D. Donaldson: "Rank studies of coal in the Rocky Mountains and Inner Foothills Belt Canada," in R. R. Dutcher, P. A. Hacquebard, J. M. Schopf, J. A. Simon (eds.): "Carbonaceous materials as indicators of metamorphism," *Spec. Pap. Geol. Soc. Am.* **153** (1974) 75–93.

[7] E. Hryckowian, R. R. Dutcher, F. Dachille: "Experimental studies of anthracite coals at high pressures and temperatures," *Econ. Geol.* **62** (1967) no. 4, 517–539.

[8] P. Averitt: "Coal resources of the United States," *U.S. Geol. Surv. Bull.* **1412** (1975).

[9] J. M. Schopf: "A definition of coal," *Econ. Geol.* **51** (1956) 521–527.

[10] R. Thiessen: "What is coal?" *Int. Circ. U.S. Bur. Mines* **7397** (1947) 48 p.

[11] B. C. Parks, H. J. O'Donnell: "Petrography of American coals," *Bull. U.S. Bur. Mines* **550** (1956) 193 p.

[12] M. C. Stopes: "On the four visible ingredients in banded bituminous coals," *Proc. R. Soc. London Ser. B* **90** (1919) 470.

[13] M. L. Gorbaty, K. Ouchi (eds.): "Coal Structure," *Adv. Chem. Ser.* **192**, American Chemical Society, Washington, D.C., 1981.

[14] R. C. Neavel: "Origin, Petrography and Classification of Coal," in M. A. Elliot (ed.): *Chemistry of Coal Utilization*, 2nd suppl. vol., Chap. 3, J. Wiley & Sons, New York, Chichester, Brisbane, Toronto 1981, pp. 91–158.

[15] D. W. van Krevelen: *Coal and Its Properties Related to Conversion*, Int. Conference on Coal Conversion, Pretoria, RSA, 16–20 August 1982.

[16] R. M. Davidson: *Molecular Structure of Coal*, Rep. No. ICTIS/TR 08, Jan. 1980, IEA Coal Research, London.

[17] G. R. Gavalas, P. H.-K. Cheong, R. Jain, *Ind. Eng. Chem. Fundam.* **20** (1981) 113–122.

[18] G. J. Pitt, G. R. Millward: *Coal and Modern Coal Processing, An Introduction*, Academic Press, London, New York, San Francisco 1979.

[19] R. V. Ramani et al.: "User's Manual for Premining Planning of Eastern Surface Coal Mining, vol. 2: Surface Mine Engineering," *National Technical Information Service Report No. PB81-109-415*, U.S. Dept. of Commerce, Springfield, VA, 1980.

[20] L. J. Thomas: *An Introduction to Mining*, Hicks, Smith & Sons, Sydney 1973.

[21] T. Atkinson: "Selection of Open-Pit Excavating and Loading Equipment," *Trans. Inst. Min. Metall. Sect. A* **80** (1971) 101–129.

[22] R. Stefanko: *Coal Mining Technology: Theory and Practice*, Society of Mining Engineers of AIME, New York 1983.

[23] R. von der Gathen, K.-H. Kubitza, B. Bogenschneider: "Die Auswirkungen der Vergleichmäßigung von Rohförderkohle auf Kosten und Produkte der Aufbereitung," *Glückauf* **119** (1983) 22–27.

[24] M. Hampel: "Großsiebmaschinen für die Vorklassierung in Steinkohlenaufbereitungsanlagen," *Glückauf* **113** (1977) 80–85.

[25] K.-H. Kubitza, P. Wilczynski: "Trockene Feinstkornabscheidung in Sichtern: neuere Entwicklungen bei der Ruhrkohle AG," *Glückauf* **110** (1974) 480–484.

[26] E. Fellensiek: "Das Setzverhalten unterbettgepulster Feinkorn-Großsetzmaschinen," *GlückaufForschungsh.* **39** (1978) 207–212.

[27] E. Fellensiek: "Feinkornsortierung auf einer neuartigen, in Doppelfrequenz gepulsten Durchsetzmaschine," *Glückauf-Forschungsh.* **42** (1981) 130–136.

[28] S. Heintges: "Die Entwicklung der Schwertrübesortierung von Steinkohlen in der Bundesrepublik Deutschland," *Glückauf* **109** (1973) 955–960.

[29] M. Becker: "Der Weg zu den Großraumflotationsanlagen bei der Ruhrkohle AG," *Glückauf* **113** (1977) 952–955.

[30] W. Blankmeister, B. Bogenschneider, K.-H. Kubitza, D. Leininger, L. Angerstein, R. Köhling: "Optimierung der Fein- und Feinstkohlenentwässerung im Bereich unter 10 mm," *Glückauf* **112** (1976) 758–762.

[31] W. Erdmann: "Neuere Entwicklungen bei der Entwässerung von fein- und feinstkörnigen Steinkohlenerzeugnissen," *Aufbereit. Tech.* **26** (1985) 249–258.

[32] D. Leininger, P. Wilczynski, R. Köhling, W. Erdmann, T. Schieder: "Behandlung und Verwertung von Flotationsbergen in der Bundesrepublik Deutschland," *Glückauf* **115** (1979) 467–472.

[33] R. von der Gathen: "Möglichkeiten und Grenzen der Entschwefelung von Steinkohle," *Glückauf* **115** (1979) 112–118.

[34] W. P. Bethe, G. Koch: "Neue Bauformen für Aufbereitungsanlagen," *Glückauf* **119** (1983) 368–373.

[35] D. Rebb: "Latest Design in Coal Pipelining," *Can. Min. J.* **104** (1983) no. 3, 20–22.

[36] T. Wheeler: "Hydraulic Transport of Coal: Slurry Pipelines," *Mine Quarry,* **14** (1985) 35–38.

[37] J. R. Siemon: "Economic Potential of Coal-Water Mixtures," *IEA Coal Res.*, London, Sept. 1985.

[38] E. Ahland, F. Friedrich, I. Romey, B. Strobel, H. Weber, *Erdöl Erdgas* **102** (1986) no. 3, 148–154.

[39] A. G. Comalli, J. B. MacArthur, H. H. Stotler: "H-Coal Process Demonstrations, Development and Research Activities," *Prepr. Pap. Am. Chem. Soc. Div. Fuel Chem.* **27** (1982) no. 3/4, 104–113.

[40] D. T. Wade, L. L. Ansell, W. R. Epperly: "Coal liquefaction," *CHEMTECH* 1982, no. 4, 242–249.

[41] W. G. Schützendübel: "SRC II," *Energie* **32** (1980) no. 6/7, 254–259.

[42] K. Uesugi: "The Status of Coal Liquefaction Technology Development in Japan," *Int. Working Forum of Coal Liquefaction,* Atlanta, Ga., Apr. 4–9, 1986.

[43] J. M. Lee, R. V. Nalitham, C. W. Lamb: "Recent developments in Two-Stage Coal Liquefaction at Wilsonville," *Prepr. Pap. Am. Chem. Soc. Div. Fuel Chem.* **31** (1986) no. 2, 316–324.

[44] G. Franken, W. Adlhoch, W. Koch, *Chem. Ing. Tech.* **52** (1980) 324–327.

[45] K. A. Theis, E. Nitschke: "Make Syngas from Lignite," *Hydrocarbon Process.* **66** (1982) no. 9, 233–237.

[46] H.-D. Schilling, B. Bonn, U. Krauß: *Kohlenvergasung*, 3rd ed., vol. **22**: "Bergbau, Rohstoffe, Energie," Verlag Glückauf, Essen 1982.
[47] H.-D. Schilling, B. Bonn, U. Krauß: "Coal Gasification," Graham & Trotmann, London 1981.
[48] R. Specks, *Glückauf* **119** (1983) no. 23, 1147–1159.
[49] P. Nowacki (ed.): "Coal Gasification Processes," *Energy Technology Review No. 70*, Noyes Data Corp., Park Ridge, N.J., 1981.
[50] H. Teggers, H. Jüntgen, *Erdöl Kohle Erdgas Petrochem.* **37** (1984) no. 4, 163–173.
[51] J. M. Caffin: "Industrial Coal Gasification; Technology, Applications and Economics," *Energy Prog.* **4** (1984) no. 3, 131–137.
[52] M. Teper, D. F. Hemming, W. C. Ulrich, EAS-Report E2/80, London, January 1983.
[53] S. A. Elmquist et al., DOE/FE/05147-1488, May 1983.
[54] M. K. Schad, C. F. Hafke, CEP May 1983, 45–51.
[55] K. V. S. Sastry, D. W. Fuerstenau, EPRI CS-2198, Project 1030-1, January 1982.
[56] F. P. Calhoun, *Min. Congr. J.* **49** (1962) 38–39.
[57] N. Galbenis, O. Abel, J. Lehmann, W. Peters, *Erdöl Kohle Erdgas Petrochem.* **34** (1981) 59–65.
[58] F. H. Beckmann, *Stahl Eisen* **100** (1980) 803–813.
[59] E. Dunger, P. Dittmann, H. Reißmann, *Neue Bergbautech.* **10** (1980) 427–430.
[60] H. Krug, W. Naundorf, *Energietechnik* **31** (1981) no. 62–68, 228–232.
[61] R. Kurtz, *Braunkohle (Düsseldorf)* **32** (1980) no. 368–372, 443–448.
[62] A. F. Baker, R. E. McKeever, A. W. Deurbrouck, DOE, RI-PMTC-12 (82), March 1982.
[63] M. A. Colaluca, TENRAC/EDF-043, July 1981.
[64] H. P. Hudson, J. E. Landon, J. H. Walsh: *The Canadian Mining and Metallurgical Bull. for January 1964*, Montreal, pp. 52–58.
[65] K. V. S. Sastry, V. P. Mehrotra, 3rd Int. Symp. Agglomeration, Nürnberg, May 1981, H 36–38.
[66] E. Ahland, J. Lehmann, *Erdöl Kohle Erdgas Petrochem.* **34** (1981) 402–407.
[67] H.-G. Schäfer, 3rd Int. Symp. Agglomeration, Nürnberg, May 1981, S 53–55.
[68] M. Schad, D. Sauter et al., BMFT-FB-T-85, Oct. 1985, pp. 108–109.
[69] G. H. Cady, *Illinois Geological Survey Report of Investigations*, no. 35, 1935, pp. 25–39.
[70] H. J. Gluskoter, J. A. Simon, *Illinois Geological Survey Circular*, no. 432, 1968, p. 28.
[71] C. N. Kravits, J. C. Crelling, *Int. J. Coal Geol.* **1** (1981) 195–212.
[72] World Energy Conferences 1980 (München) and 1983 (New Delhi): *Oil Gas J.*, 26. Dec. 1983.
[73] Statistical Office of the European Communities; United Nations, *Monthly Bulletin of Statistics*; ECE, Statistical Indicators of short term economic changes in ECE-countries.
[74] Statistisches Bundesamt, Wiesbaden, *Außenhandelsstatistik 1983*; Eurostat, *Monatsberichte Kohle*; Eurostat, *Nimexe 1983 Import/Export*; United Nations, *Annual Bulletin of coal statistics*; ECE, *Statistical indicators in ECE-countries*; OECD, Paris, *OECD-Energiebericht 1983*; Verein deutscher Kohlenimporteure, *Jahresbericht 1983*; Department of Energy, Washington, *Coal Statistics International*; Ministère de l'Energie, Bruxelles, *Statistiques*.

Cobalt Compounds

JOHN DALLAS DONALDSON, The City University, London, United Kingdom

1.	**Occurrence** 1503	2.3.5.	Electronics and Solid-State Devices 1522
2.	**Individual Compounds** 1506	2.3.6.	Agriculture, Nutrition, and Medicine 1522
2.1.	**Chemistry** 1506	3.	**Cobalt Powders** 1523
2.2.	**Commercially Important Cobalt Compounds** 1511	3.1.	**Production** 1523
2.3.	**Industrial Applications** 1515	3.2.	**Uses** 1523
2.3.1.	Glasses, Ceramics, and Refractories 1515	3.2.1.	Cemented Carbides 1523
2.3.2.	Driers, Paints, Varnishes, and Dressings 1517	3.2.2.	Cobalt Powders in Powder Metallurgy and Cobalt-containing Metal 1524
2.3.3.	Catalysts 1518	4.	**Physiology and Toxicology** . . . 1524
2.3.4.	Electroplating 1520	5.	**References** 1525

1. Occurrence [1], [2], [3]

Cobalt occurs in nature in a widespread but dispersed form in trace quantities in many rocks, soils, and plants. It is also found in sea water and in manganese-rich marine nodules. The cobalt content of the earth's crust is about 20 mg/kg, whereas its concentration in sea water has been reported as being 0.1 – 1 part in 10^9. The concentration of the element in marine nodules is usually 0.1 – 1 %. The largest concentrations of cobalt are found in mafic and ultramafic igneous rocks; the concentration of the element and the nickel – cobalt ratio decreases from ultramafic to acidic rocks, as shown in Table 1. The nickel – cobalt ratio changes because cobalt enters the lattice of early crystallizing magnesium silicates less readily than nickel. Sedimentary rocks contain varying amounts of cobalt, with average values of 4, 6, and 40 mg/kg being reported for sandstone, carbonate rocks, and clays or shales, respectively. During the formation of metamorphic rocks, very little movement or concentration of cobalt took place; thus, the levels of cobalt found in metamorphic rocks depend essentially on the amount of the element in the original igneous or sedimentary source. Those formed from ultramafic or mafic sources contain an average cobalt content of ca. 100 mg/kg, whereas gneissic granites and metasedimentary rocks contain an average of 16 and 8 mg/kg, respectively.

Under oxidizing conditions, cobalt shows a strong tendency to concentrate with manganese oxides. In the weathering process of mafic and ultramafic rocks to form

Table 1. Average cobalt content of igneous rocks

Rock type	Cobalt content, mg/kg	Ni–Co ratio
Ultramafic	270	7
Gabbro	51	2.6
Basalt	41	2.5
Diabase	31	2.3
Intermediate igneous	14	1.9
Felsic	5	1.1

laterites, nickel tended to be leached downward with magnesia and silica, whereas cobalt with manganese oxides was residually enriched near the surface of the deposit. The concentration of cobalt and other metals in marine nodules has been attributed to the strong ion-exchange properties of the submicroscopic particles of colloidal manganese dioxide from which the nodules were formed.

Cobalt is a major constituent of about 70 minerals [4] and is a minor or trace constituent of several hundred more, particularly those containing nickel, iron, and manganese. The minerals that have been mined or concentrated for their cobalt content and those that are relatively high in cobalt are listed in Table 2. The sulfide minerals include the copper-containing carrollite, which is one of the main sources of cobalt in Zaire, linnaeite found in Zaire, Zambia, and the United States, and cattierite, which is also present in ores from Zaire. The arsenide ores include smaltite, which is found in the silver–copper ores from Cobalt, Ontario and in Morocco, and skutterudite, the main cobalt mineral in Canadian and Moroccan deposits. The sulfoarsenide, cobaltite, is found in ore bodies in Zaire, Canada, and the United States. The oxide mineral heterogenite is a hydrated metal oxide containing varying amounts of cobalt and copper and is one of the main cobalt-bearing components of the Zaire deposits. The hydrated manganese–cobalt mineral, asbolite, is the source of most of the cobalt in ores from New Caledonia.

Ore Deposits. Cobalt is produced mainly as a byproduct of the mining and processing of the ores of other metals, particularly those of copper, nickel, and silver, but also those of gold, lead, and zinc.

The deposits of cobalt can be classified under the following headings:

1) *Hypogene deposits* associated with mafic intrusive igneous rocks. The massive and disseminated iron–nickel–copper sulfides containing cobalt are important examples of this type of deposit. They include the ore from the Sudbury district of Ontario, which has an average cobalt content of 0.07%. The suite of ore minerals consists of pyrrhotite, pentlandite, pyrite, marcasite, cobaltite, and gersdorffite in veins stringers and disseminated grains within an igneous host rock of Precambrian age.
2) *Contact metamorphic deposits* associated with mafic rocks. Deposits of magnetite, chalcopyrite, and cobalt-containing pyrite, formed by contact metamorphism of

Table 2. Cobalt minerals

Mineral	CAS reg. no.	Cobalt content, %
Cattierite, CoS_2 (pure)	[12017-06-0]	47.8
Linnaeite, Co_3S_4 (pure)	[1308-08-3]	58.0
Siegenite, $(Co,Ni)_3S_4$	[12174-56-0]	20.4–26.0
Carrollite, $(Co_2Cu)S_4$	[12285-42-6]	35.2–36.0
Cobaltite, (Co,Fe)AsS	[1303-15-7]	26.0–32.4
Safflorite, $(Co,Fe)As_2$	[12044-43-8]	13.0–18.6
Smaltite, $(Ca, Ni)As_2$	[12006-41-6]	ca. 21
Glaucodot, (Co,Fe)AsS	[12198-14-0]	12.0–31.6
Skutterudite, $(Co,Fe)As_3$	[12196-91-7]	10.9–20.9
Heterogenite, CoO(OH) (pure)	[12323-83-0]	64.1
Asbolite	[12413-71-7]	0.5–5.0
Erythrite, $(CoNi)_3(AsO_4)_2 \cdot 8\ H_2O$	[149-32-6]	18.7–26.3
Gersdorffite, (Ni,Co)AsS	[12255-11-7]	(low)
Pyrrhotite, $(Fe,Ni,Co)_{x-1}S_x$	[1310-50-5]	up to 1.00
Pentlandite, $(Fe,Ni,Co)_9S_8$	[53809-86-2]	up to 1.50
Pyrite, $(Fe,Ni,Co)S_2$	[1309-36-0]	up to 13.00
Sphalerite, Zn(Co)S	[12169-28-7]	up to 0.30
Arsenopyrite, Fe(Co)AsS	[1303-18-0]	up to 0.38
Manganese oxide minerals		0.10–1.0

carbonate rock by sills and dikes of diabase, gave the ore deposits at Cornwall and Morgantown in the United States.

3) *Lateritic Deposits.* The weathering of peridotite and serpentine generally gives laterite that is rich in iron, nickel, cobalt, and chromium. Commercially valuable deposits contain 40–50% iron, 1–2% nickel, and 0.01–0.1% cobalt. Major lateritic deposits occur in Cuba, New Caledonia, Australia, the United States, and Russia.

4) *Massive sulfide deposits* in metamorphic rocks, largely of volcanic sedimentary origin. These deposits consist mainly of pyrite and pyrrhotite and are mined in the United States.

5) *Hydrothermal deposits* are subdivided into two classes: vein deposits and replacement deposits. Some of these deposits are the only ones that have been mined specifically as sources of cobalt. In Canada, veins that contain as much as 10% cobalt occur in the Cobalt-Gowganda region of Ontario, whereas veins in the Bou-Azzer area of Morocco contain an average of 1.2% cobalt. Hydrothermal replacement deposits containing 0.5% cobalt as gersdorffite are found in Burma and at Outokumpu in Finland; a copper-rich sulfide deposit contains 0.2% cobalt mainly in linnaeite.

6) *Strata-bound deposits:* the copper–cobalt deposits of Zaire and Zambia are of this type. They occur in folded shale and dolomite and contain a number of minerals, including chalcopyrite, bornite, chalcosite, linnaeite, and carroleite. The ore bed is 6–24 m thick and currently provides the world's major source of cobalt.

Table 3. Cobalt reserves

Country	Reserves, t	Country	Reserves, t
Australia	295 000	New Caledonia	385 000
Brazil	>9 000	New Guinea	18 000
Burma	16 000	Philippines	159 000
Canada	250 000	Puerto Rico	68 000
Columbia	22 500	Solomon Islands	22 500
Cuba	1 048 500	Soviet Union	181 500
Dominican Republic	89 000	Uganda	8 000
		United States	>764 000
Finland	22 500	Venezuela	60 000
Guatemala	45 500	Zaire	1 920 000
Japan	2 500	Zambia	370 000

7) *Deposits formed as chemical precipitates* usually contain chemically precipitated cobalt in associated marine manganese nodules.

Reserves. As generally accepted, there are $3-5 \times 10^6$ t of minable cobalt reserves and a further $4-5 \times 10^6$ t of potential reserves. Table 3 lists the reported reserves in a number of countries. However, the minable reserves do not have the same degree of profitability. Cuba has minable resources of cobalt, but provides only 5% of the world production because its ore bodies are laterites with only small quantities of nickel and cobalt that are costly to refine. Zaire produces more than half of the world's cobalt because its ores can be treated more profitably.

In addition to the land-based reserves listed in Table 3, almost 6×10^9 t of copper is available in marine nodules if it can be recovered and worked economically.

2. Individual Compounds [5]

2.1. Chemistry

Cobalt is stable to atmospheric oxygen unless heated. When heated, it is first oxidized to Co_3O_4 and then, above 900 °C, to CoO, which is also the product of the reaction between the red-hot metal and steam. The activation energy of the oxidation above 900 °C has been calculated as 155–160 kJ/mol. The metal does not combine directly with hydrogen or nitrogen, but it combines with carbon, phosphorus, and sulfur on heating. The reaction with sulfur is influenced by the formation of a low melting eutectic (877 °C) between the metal and the Co_4S_3 phase; the reaction between cobalt and sulfur is rapid above this temperature. Below 877 °C, a protective layer of sulfide scale is formed. In an atmosphere of hydrogen sulfide, cobalt also forms a scale of sulfide, but in air containing sulfur dioxide, a mixed oxide–sulfide scale is formed.

Table 4. Corrosion of cobalt in aqueous media at 25°C

Corrosive medium	Rate of cobalt corrosion, mg dm^{-2} d^{-1}
Distilled water	1.1
5 vol% Ammonia	5.3
10 vol% Sodium hydroxide	5.6
Conc. phosphoric acid	7.4
50% Phosphoric acid (aqueous)	65.1
5 vol% Acetic acid	12.5
5 vol% Sulfuric acid	56.8
50% Hydrofluoric acid (aqueous)	178.6

In bulk form, cobalt is resistant to many mild corrosive agents, but it is more readily attacked when it is finely divided. Table 4 gives the corrosion rates of cobalt in some aqueous media. Cobalt is strongly attacked by concentrated nitric acid at ambient temperature. The metal dissolves slowly in dilute mineral acids, the Co^{2+}/Co potential being -0.277 V.

The solubility of oxygen in cobalt is 0.006, 0.013, 0.125, and 0.4 wt % (at 600, 1200, 1510, and 1700 °C, respectively). Nitrogen is only slightly soluble (40 mg/kg) in cobalt at its melting point; the solubility rises to 60–70 mg/kg at 1750 °C. The solubility of hydrogen in cobalt increases with temperature from 1 mg/kg at 700 °C to ca. 8 mg/kg at the melting point. The solubility of hydrogen in the liquid metal increases to ca. 20 mg/kg at 1500 °C and to ca. 27 mg/kg at 1750 °C.

The main oxidation states of cobalt are Co^{2+} [22541-53-3] and Co^{3+} [22541-63-5]. In acid solution and in the absence of complexing agents, Co^{2+} is the stable oxidation state, with oxidation to Co^{3+} being difficult.

$$[Co(H_2O)_6]^{3+} + e^- \longrightarrow [Co(H_2O)_6]^{2+} \quad \Delta E° = 1.84 \text{ V}$$

The oxidation can be achieved by electrolysis or ozone, but the Co^{3+} is very unstable and rapidly reduces to Co^{2+}, even at room temperature, with evolution of oxygen from the water. The solution chemistry of cobalt in acid solution in the absence of complexing agents is dominated by Co^{2+}. The most common species present is $[Co(H_2O)_6]^{2+}$, although the ions $[CoX_3]^-$, $[CoX_4]^{2-}$, and $[CoX_6]^{4-}$ (X = halide) are also found in solutions of hydrogen halides. Cobalt is often removed from solution as its sulfide, and it is interesting to note that although the precipitates of cobalt sulfide obtained by using H_2S are not readily soluble in dilute acids, those obtained by using Na_2S or $(NH_4)_2S$ are soluble. All sulfide precipitates become less soluble as they age in the atmosphere to form Co(OH)S. The cobalt dihalides (except fluoride) are also readily soluble in some organic solvents such as alcohol, acetone, and methyl acetate.

In alkali, Co^{2+} is more readily oxidized to Co^{3+}

$$CoO(OH)\ (s) + H_2O + e^- \longrightarrow Co(OH)_2\ (s) + OH^- \quad \Delta E° = 0.17 \text{ V}$$

Table 5. $\Delta E°$ for some Co^{3+}/Co^{2+} couples in acid solution

Redox reaction	$\Delta E°$
$[Co(C_2O_4)_3]^{3-} + e^- \rightleftharpoons [Co(C_2O_4)_3]^{4-}$	0.57
$[Co(EDTA)]^- + e^- \rightleftharpoons [Co(EDTA)]^{2-}$	0.37
$[Co(bipy)_3]^{3+} + e^- \rightleftharpoons [Co(bipy)_3]^{2+}$	0.31
$[Co(en)_3]^{3+} + e^- \rightleftharpoons [Co(en)_3]^{2+}$	0.18
$[Co(CN)_6]^{3-} + H_2O \rightleftharpoons [Co(CN)_5(H_2O)]^{3-} + CN^- + e^-$	−0.8

EDTA = ethylene diaminetetraacetic acid; bipy = bipyridyl; en = ethylene diamine.

In the presence of complexing agents, oxidation is very easy in any solution because Co^{3+} has a particularly high affinity for complex formation. With N-donor ligands the redox reaction in the presence of ammonia is

$$[Co(NH_3)_6]^{3+} + e^- \longrightarrow [Co(NH_3)_6]^{2+} \quad \Delta E° = 0.1 \text{ V}$$

Oxidation of Co^{2+} solutions containing complexing agents can be achieved with air or hydrogen peroxide. Thus, the solution chemistry of cobalt in the presence of complexing agents is dominated by the complex chemistry of Co^{3+}. The sensitivity of the reduction potential of the Co^{3+}/Co^{2+} couple to different ligands whose presence renders Co^{2+} unstable to air oxidation is shown by the data in Table 5.

Cobalt(II) Compounds. Cobalt combines with oxygen to form cobalt(II) oxide [1307-96-6], CoO, which is stable above 900 °C. This oxide has the sodium chloride structure and is antiferromagnetic at ordinary temperature. When cobalt(II) oxide is heated at 400–500 °C in an atmosphere of oxygen, the cobalt(II,III) oxide, Co_2O_3 [1308-06-1], is formed. This mixed-valence oxide has the spinel structure with cobalt(II) in tetrahedral sites and cobalt(III) in octahedral sites. Cobalt hydroxide [1307-86-4], $Co(OH)_2$, is a product of the hydrolysis of solutions containing Co^{2+} ions. The hydroxide is amphoteric, dissolving both in alkali, to give blue solutions containing the $[Co(OH)_4]^{2-}$ ion, and in acids. Cobalt(II) forms an extensive range of simple and hydrated salts with all of the common anions including acetate, bromide, carbonate, chloride, fluoride, nitrate, perchlorate, and sulfate. Many of the hydrated salts and their solutions contain the pink octahedral $[Co(H_2O_6)]^{2+}$ ion. Although complexes of cobalt(II) are generally unstable to oxidation, a number of octahedral species are formed, including (1) a series of $[Co(N-N)_3]^{2+}$ complexes with neutral bidentate donor ligands, such as ethylenediamine and bipyridyl, and (2) the acetylacetonates $[Co(acac)_2 \cdot 2 H_2O]$ and $Co(acac)_2$. Tetrahedral complexes $[CoX_4]^{2-}$ are usually formed with monodentate anionic ligands, such as chloride, bromide, iodide, thiocyanate, azide, and hydroxyl. Tetrahedral $[CoX_2L_2]$ complexes are formed by a combination of two such ligands with two neutral ligands (L). Addition of cyanide to a solution of Co^{2+} produces a dark green color attributed to the $[Co(CN)_5]^{3-}$ ion.

The colors produced by cobalt(II) complexes in aqueous media have been used to distinguish between octahedral and tetrahedral coordination in the complexes. In general, octahedral species are pink to violet, whereas tetrahedral species are blue. This is not an infallible distinction, but it does provide a useful guide that can be improved by analyzing the electronic spectra of the complexes.

Cobalt(III) Compounds. In addition to the mixed-valence oxide Co_2O_3, impure forms of the unstable cobalt(III) oxide, Co_2O_3, have been prepared. Very few simple cobalt(III) salts are known. The blue sulfate, $Co_2(SO_4)_3 \cdot 18\ H_2O$ [13478-09-6], which contains the $[Co(H_2O)_6]^{3+}$ species, can be obtained by the electrolytic oxidation of cobalt(II) in 4M sulfuric acid solution. It is stable when dry, but decomposes in the presence of moisture. The alums $MCo(III)(SO_4)_2 \cdot 2\ H_2O$ (M = K, Rb, Cs, NH_4) are also known, and a hydrated fluoride $2\ CoF_3 \cdot 7\ H_2O$ has been reported.

The Cobaltchemistry of cobalt(III) is dominated by complex formation. Cobalt(III) complexes are kinetically inert, and for this reason, indirect methods of syntheses are used to obtain them. Usually the ligand is added to a solution of Co^{2+}, which is then oxidized with some convenient oxidant, often in the presence of a catalyst such as activated charcoal. A wide range of cobaltammines have been prepared and studied; the species identified in solution or as solid derivatives include $[Co(NH_3)_6]^{3+}$, $[Co(NH_3)_5H_2O]^{3+}$, $[Co(NH_3)_5X]^{2+}$ (X = Cl, Br, NO_2, NO_3), and cis- and trans-$[Co(NH_3)_4X_2]^+$ (X = Cl, NO_2).

In addition to these simple cobaltammines, a number of polynuclear species containing bridging groups such as NH_2^-, NH^{2-}, NO_2, OH^-, and O_2 have been prepared. Polynuclear cobaltammines that have been identified include:

$$\left[(NH_3)_3Co\underset{(O_2)}{\overset{NH_2}{<>}}Co(NH_3)\right]^{3+} \quad \left[(NH_3)_4Co\underset{OH}{\overset{OH}{<>}}Co(NH_3)_4\right]^{4+}$$

$$\left[(NH_3)_5Co-NH_2-Co(NH_3)_5\right]^{5+} \quad \left[(NH_3)_4Co\underset{OH}{\overset{NH_2}{<>}}Co(NH_3)\right]^{4+}$$

$$\left[(NH_3)Co\underset{OH}{\overset{NH_2}{\underset{OH}{<>}}}Co(NH_3)_3\right]^{3+} \quad \left[(NH_3)_5Co-(O_2)-Co(NH_3)_5\right]^{3+}$$

Complexes of cobalt(III) with O-donor ligands are generally less stable than those with N-donor species, although $[Co(acac)_3]$ [14024-48-7] and $[Co(C_2O_4)_3]^{3-}$ are known. Apart from the O_2-bridged compounds mentioned above, the octahedral fluoro complexes $[CoF_6]^{3-}$ and $[CoF_3(H_2O)_3]$ are the only known high-spin cobalt(III) complexes, the paramagnetic moment of $[CoF_6]^{3-}$ at room temperature being about 5.8 BM. All other cobalt(III) complexes are low-spin and diamagnetic, including the hexacyanocobaltates (III), $[Co(CN)_6]^{3-}$, and the hexanitritocobaltates(III), $[Co(NO_2)_6]^{3-}$.

In addition to the two most stable oxidation states, 2+ and 3+, cobalt also forms compounds in the 1−, 0, 1+, and 4+ oxidation states. There are only a few reported examples of Co(IV) [20499-79-0] compounds, mainly fluoro complexes, CoF_6^{2-}, and mixed-metal oxides (the purity of compounds in this oxidation state is questionable).

The formation of cobalt compounds in oxidation states lower than 2+ requires the presence of stabilizing π-acceptor ligands. The 1+ state [16610-75-6] is not common for cobalt, and most of the known examples are pentacoordinated complexes of the type $[Co(NCR)_5]^+$ (NCR = organic nitrile). Cobalt forms a wide range of complexes in which its formal oxidation state is 0 or 1 − [16727-18-7]. Many of these contain ligands such as CO, CN^-, NO^+, and RNC, but other ligands such as tertiary phosphanes also stabilize lower oxidation states to give compounds like $[Co(P(CH_3)_3)_4]^+$, which is prepared by reducing an ethereal solution of cobalt(II) chloride with sodium or magnesium amalgam in the presence of trimethylphosphane.

Cobalt Carbonyls [5], [6]. The carbonyl complexes of cobalt are important because of their uses as hydroformylation catalysts (see Section 2.3.3). Because cobalt has an odd number of valence electrons, it can satisfy only the eighteen-electron rule in its carbonyls if Co–Co bonds are formed. For this reason, the principal binary carbonyls of the element are octacarbonyldicobalt [15226-74-1], $[Co_2(CO)_8]$, dodecacarbonyl-tetracobalt [19212-11-4], $[Co_4(CO)_{12}]$, and hexadecacarbonylhexa- cobalt [12182-17-1], $[Co_6(CO)_{16}]$.

Octacarbonyldicobalt is prepared by heating the metal to 250–300 °C at 200–300 bar of carbon monoxide or by heating cobalt carbonate under similar conditions in the presence of hydrogen. It is an air-sensitive orange-red solid with a melting point of 51 °C. The compound can be reduced with sodium amalgam in benzene to give the tetrahedral monomeric ion $[Co(CO)_4]^-$, acidification of which leads to tetracarbonylhydridocobalt [16842-03-8], $[HCo(CO)_4]$. This hydride is a yellow liquid that forms a colorless vapor. It melts at −26 °C and decomposes above this temperature to H_2 and $[Co_2(CO)_8]$. It is partly soluble in water to give a strong acid solution containing $[Co(CO)_4]^-$ ions. The hydrogen atom in $[HCo(CO)_4]$ is bound directly to the cobalt, giving a Co–H infrared stretching frequency of ca. 1934 cm^{-1}. The $[Co(CO)_4]^-$ ion is reoxidized to $[Co_2(CO)_8]$ by carbon tetrachloride, and further reaction with this reagent leads to a triply bridged chloromethynyl derivative, from which a carbidocarbonyl compound, disodium carbidopentadecacarbonylhexa- cobaltate, $Na_2[Co_6C(CO)_{15}]$, can be obtained:

$$10 [Co_2(CO)_8] + 6 CCl_4 \xrightarrow[90\% \text{ yield}]{40 °C} 4 [Co_3(CO)_9(CCl)] + 8 CoCl_2 + 44 CO + C_2Cl_4$$

$$[Co_3(CO)_9(CCl)] + 3 Na[Co(CO)_4] \xrightarrow[\text{Propylene oxide}]{25 °C} Na_2[Co_6C(CO)_{15}] + NaCl + 6 Co$$

Other carbidocarbonyls are obtained by similar routes, including $[Co_6C(CO)_4]^-$, $[Co_8C(CO)_{18}]^{2-}$, and $[Co_{13}(C)_2(CO)_{24}H]^{4-}$. Dodecacarbonyltetracobalt is obtained as a green black solid by heating $[Co_2(CO)_8]$ in an inert atmosphere at 50 °C.

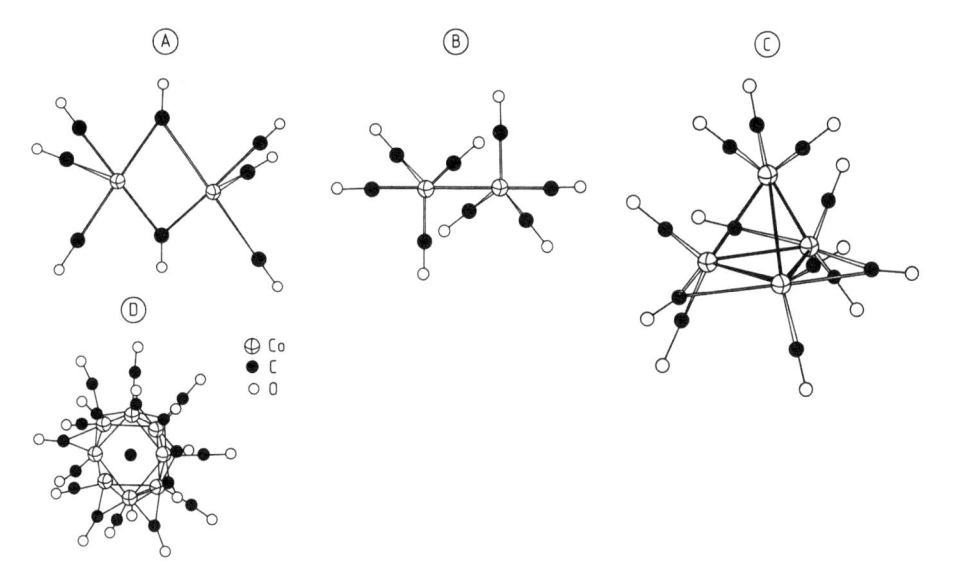

Figure 1. Structures of carbonyl complexes of cobalt
A) [Co$_2$(CO)$_8$] (solid state); B) [Co$_2$(CO)$_8$] (solution); C) [Co$_4$(CO)$_{12}$]; D) [Co$_8$C(CO)$_{18}$]

The structure of [Co$_2$(CO)$_8$] in the solid state is different from that in solution. The solid-state structure shown in Figure 1 A involves two bridging carbonyl groups and can be best rationalized in terms of the formation of a bent Co–Co bond. In solution, however, this structure is in equilibrium with the form shown in Figure 1 B, in which the dimer is held together by a Co–Co bond. The structure of [Co$_4$(CO)$_{12}$] is shown in Figure 1 C. The structure of the carbidocarbonyl (Fig. 1 D) shows the presence of a carbon atom at the center of a distorted square antiprismatic cobalt cluster.

In addition to the carbonyls, cobalt also forms complexes with N$_2$ and NO$^+$, which are isoelectronic with CO. Examples of this type of complex are [CoH(PR$_3$)$_3$(N$_2$)] and [Co(CO)$_3$NO], where R is an alkyl group.

2.2. Commercially Important Cobalt Compounds [7]

Table 6 lists the known applications of cobalt compounds. The most important commercially available compounds are the oxides, as well as hydroxide, chloride, sulfate, nitrate, phosphate, carbonate, acetate, oxalate, and other carboxylic acid derivatives.

Table 6. Industrial uses of cobalt compounds

Compound		Formula	Uses
Acetate(III)	[917-69-1]	$Co(C_2H_2O_2)_3$	catalyst
Acetate(II)	[71-48-7]	$Co(C_2H_2O_2)_2 \cdot 4\,H_2O$	driers for lacquers and varnishes, sympathetic inks, catalysts, pigment for oil cloth, mineral supplement, anodizing, stabilizer for malt beverages
Acetylacetonate	[14024-48-7]	$Co(C_5H_7O_2)_3$	vapor plating of cobalt
Aluminate	[13820-62-7]	$CoAl_2O_4$	pigment, catalysts, grain refining
Ammonium sulfate	[13586-38-4]	$CoSO_4(NH_4)_2SO_4 \cdot 6\,H_2O$	catalysts, plating solutions
Arsenate	[24719-19-5]	$Co_3(AsO_4)_2 \cdot 8\,H_2O$	pigment for paint, glass, and porcelain
Bromide (II)	[7789-43-7]	$CoBr_2$	catalyst, hydrometers
Carbonate	[513-79-1]	$CoCO_3$	pigment, ceramics, feed supplements, catalyst
Carbonate (basic)	[7542-09-8]	$2\,CoCO_3 \cdot Co(OH)_2 \cdot H_2O$	chemicals
Carbonyl	[15226-74-1]	$Co_2(CO)_8$	catalyst
Chloride	[7791-13-1]	$CoCl_2 \cdot 6\,H_2O$	chemicals, sympathetic inks, hydrometers, plating baths, metal refining, pigment, catalyst
Chromate	[24613-38-5]	$CoCrO_4$	pigment
Citrate	[18727-04-3]	$Co_3(C_6H_5O_7)_2 \cdot 2\,H_2O$	therapeutic agents, vitamin preparations
Ferrate	[12052-28-7]	$CoFe_2O_4$	catalyst, pigment
Fluoride	[10026-17-2]	CoF_2	fluorinating agents
	[10026-18-3]	CoF_3	
Fluoride	[13817-37-3]	$CoF_2 \cdot 4\,H_2O$	catalyst
Fluorosilicate	[15415-49-3]	$CoSiF_6 \cdot 6\,H_2O$	ceramics
Formate	[6424-20-0]	$Co(CHO_2)_2 \cdot 2\,H_2O$	catalyst
Hydroxide	[1307-86-4]	$Co(OH)_2$	paints, chemicals, catalysts, printing inks
Iodide	[15238-00-3]	CoI_2	moisture indicator
Linoleate	[14666-96-7]	$Co(C_{18}H_{31}O_2)_2$	paint and varnish drier
Manganate	[12139-69-4]	$CoMn_2O_4$	catalyst, electrocatalyst
Naphthenate		$Co(C_{11}H_{10}O_2)_2$	catalyst, paint and varnish drier
Nitrate	[10026-22-9]	$Co(NO_3)_2 \cdot 6\,H_2O$	pigments, chemicals, ceramics, feed supplements, catalyst
2-Ethylhexanoate	[136-52-7]	$Co(C_8H_{15}O_2)_2$	paint and varnish drier
Oleate	[14666-94-5]	$Co(C_{18}H_{33}O_2)_2$	paint and varnish drier
Oxalate	[814-89-1]	CoC_2O_4	catalysts, cobalt powders
Oxide	[1307-96-6]	CoO	chemicals, catalysts, pigments
Oxide	[1308-06-1]	Co_3O_4	enamels, semiconductors
Oxides		(mixed metal)	pigments
Dilanthanum tetroxide	[39449-41-7]	La_2CoO_4	catalyst, anode
Tricobalt tetralanthanum decaoxide	[60241-06-7]	$La_4Co_3O_{10}$	catalyst
Lithium oxide	[12190-79-3]	$LiCoO_2$	battery electrode
Sodium oxide	[37216-69-6]	$NaCoO_2$	battery electrode
Dicobalt manganese tetroxide	[12139-92-3]	$MnCo_2O_4$	catalyst
Dicobalt nickel tetroxide	[12017-35-5]	$NiCo_2O_4$	catalyst, anode
Lanthanum trioxide	[12016-86-3]	$LaCoO_3$	oxygen, electrode
Phosphate	[10294-50-5]	$Co_2(PO_4)_2 \cdot 8\,H_2O$	glazes, enamels, pigments, steel pretreatment

Table 6. (continued)

Compound		Formula	Uses
Potassium nitrite	[17120-39-7]	$K_3Co(NO_2)_6 \cdot 1.5\ H_2O$	pigment
Resinate		$Co(C_{44}H_{62}O_4)_2$	paint and varnish drier, catalyst
Succinate	[3267-76-3]	$Co(C_4H_4O_4) \cdot 4\ H_2O$	therapeutic agents, vitamin preparations
Sulfamate		$Co(NH_2SO_3) \cdot 3\ H_2O$	plating baths
Sulfate	[10026-24-1]	$CoSO_4 \cdot x\ H_2O$	chemicals, ceramics, pigments
Sulfide	[1317-42-6]	CoS	catalysts
Tungstate	[12640-47-0]	$CoWO_4$	drier for paints and varnishes

Cobalt Oxides. Two main types of cobalt oxide distinguishable by their colors are available: *grey* cobalt(II) oxide, containing 75–78% cobalt, and *black* cobalt(II) dicobalt(III) oxide, containing 70–74% cobalt.

Cobalt(II) oxide, CoO (78.66% Co), is usually prepared by the controlled oxidation of the metal above 900 °C followed by cooling in a protective atmosphere to prevent partial oxidation to Co_3O_4. Cobalt(II) oxide has a cubic unit cell with $a = 0.425$ nm. It is insoluble in water, ammonium hydroxide, and alcohol, but dissolves in strong acids in the cold and in weak acids on heating. Cobalt oxide (CoO) absorbs a large amount of oxygen at room temperature. Cobalt(II) dicobalt(III) tetroxide, Co_3O_4 (73.44% Co), can be prepared by the controlled oxidation of cobalt metal or CoO or by the thermal decomposition of cobalt(II) salts at temperatures below 900 °C. The cubic spinel lattice has $a = 0.807$ nm; the solid material readily absorbs oxygen at room temperature, but never transforms into cobalt(III) trioxide, Co_2O_3. The mixed-valence oxide is insoluble in water and only slightly soluble in acids.

Cobalt(II) Hydroxide. $Co(OH)_2$ (63.43% Co) is prepared commercially as a pink solid by precipitation from a cobalt(II) salt solution with sodium hydroxide. It has a hexagonal crystal structure with $a = 0.317$ nm and $c/a = 1.46$. It is insoluble in water and alkaline solutions, but dissolves readily in most inorganic and organic acids; for this reason, it is commonly used as a starting material in the synthesis of cobalt chemicals. $Co(OH)_2$ decomposes thermally by loss of water, starting at 150 °C, to give the anhydrous oxide at 300 °C. Care must be taken to store the hydroxide in the absence of air because slow air oxidation leads to a product that is poorly soluble in weak acids.

Cobalt(II) Chloride. Cobalt(II) chloride hexahydrate, $CoCl_2 \cdot 6\ H_2O$ (24.79% Co), is a dark red deliquescent crystalline compound with a monoclinic unit cell ($a = 0.886$ nm, $b = 0.707$ nm, $c = 1.312$ nm, $\beta = 97°17'$). It is prepared by concentrating a hydrochloric acid solution of cobalt oxide or carbonate. The chloride is very soluble in water, alcohols, and a number of other organic solvents. The solubility of the chloride in aqueous media does, however, decrease with increasing hydrochloric acid content. The hexahydrate dehydrates thermally in three stages, giving the dihydrate at 50 °C, the monohydrate at 90 °C, and the anhydrous chloride at 130–140 °C.

The anhydrous chloride and the lower hydrates are very hygroscopic and transform to the hexahydrate in a moist atmosphere. The fact that the anhydrous chloride is blue and the hexahydrate red is used as a humidity indicator in silica gel desiccants. The chloride is also used in the electroplating, ceramics, glass, chemical, agricultural, and pharmaceutical industries.

Cobalt(II) Sulfate. Cobalt(II) sulfate hexahydrate, $CoSO_4 \cdot 6\,H_2O$ (20.98 % Co), is a brownish-red crystalline compound with a monoclinic unit cell ($a = 1.545$ nm, $b = 1.308$ nm, $c = 2.004$ nm, $\beta = 104°40'$). It is prepared by concentrating a sulfuric acid solution of cobalt oxide or carbonate. It is an efflorescent substance and loses one molecule of water when exposed to dry air or when heated gently. The hexahydrate obtained loses water in two stages to give the monohydrate at 100 °C and the anhydrous sulfate above 250 °C. The monohydrate, $CoSO_4 \cdot H_2O$ (34.08 % Co), has been manufactured and sold commercially. Cobalt sulfates are very soluble in water and are generally more stable than cobalt(II) chlorides or nitrates; for this reason they have been widely used as sources of cobalt(II) in solution for the manufacture of chemicals and for the electroplating industries. The sulfates are also used in the ceramics, linoleum, and agricultural industries.

Cobalt(II) Nitrate. Cobalt(II) nitrate hexahydrate, $Co(NO_3)_2 \cdot 6\,H_2O$ (20.26 % Co), is a red-brown crystalline compound with a monoclinic unit cell ($a = 1.509$ nm, $b = 0.612$ nm, $c = 1.269$ nm, $\beta = 119°$). It is prepared by concentrating a nitric acid solution of cobalt oxide or carbonate. The hexahydrate loses water rapidly at 55 °C to give the trihydrate; the monohydrate can also be prepared. The nitrates are very soluble in water, alcohols, and acetone. Cobalt nitrate is an important source of high-purity cobalt for use in the electronics and related industries, and the compound has also found uses in the chemical and ceramics industries.

Cobalt(II) Phosphate. Cobalt(II) phosphate octahydrate [10294-50-5], $Co_3(PO_4)_2 \cdot 8\,H_2O$ (34.63 % Co), is obtained as a purple flocculent precipitate when an alkali-metal phosphate is added to the solution of a cobalt(II) salt. The phosphate is soluble in inorganic acids and particularly in phosphoric acid, but is insoluble in water or alkaline solutions. It is used in the paints and ceramics industries and is a component of some steel phosphating formulations.

Cobalt(II) Carbonate. Although pure $CoCO_3$ has a cobalt content of 49.57 % and a pseudohexagonal unit cell, the material available commercially is a mauve basic carbonate, $[CoCO_3]_x\,[Co(OH)_2]_y \cdot z\,H_2O$, of indeterminate composition and a cobalt content of 45–47 %. The basic carbonate is insoluble in water, but dissolves easily in most inorganic and organic acids. For this reason it is often used as a starting material for the manufacture of other chemicals. The basic carbonate is also used in the ceramics and agricultural industries.

Cobalt(II) Acetate. Cobalt(II) acetate tetrahydrate, $Co(CH_3CO_2)_2 \cdot 4 H_2O$ (23.68% Co), is a deliquescent mauve-pink crystalline compound with a monoclinic unit cell ($a = 0.847$ nm, $b = 1.190$ nm, $c = 0.482$ nm, $\beta = 94°18'$). It is prepared by concentrating solutions of cobalt hydroxide or carbonate in acetic acid. The tetrahydrate is soluble in water, alcohols, inorganic, and organic acids including acetic acid. The compound loses its water of crystallization at ca. 140 °C. It is used in the manufacture of drying agents for inks and varnishes, fabrics adressings, catalysts, and pigments, as well as in the anodizing and agricultural industries.

Cobalt(II) Oxalate. Cobalt(II) oxalate dihydrate, $Co(C_2O_4) \cdot 2 H_2O$ (32.23% Co), is obtained as a pink precipitate when oxalic acid or an alkali-metal oxalate is added to the solution of a cobalt(II) salt. The dihydrate loses water when it is heated gently in air, and at 200 °C it decomposes to cobalt(II) oxide. The oxalate is insoluble in water, slightly soluble in acids, but soluble in solutions containing ammonia or an ammonium salt. The main use of cobalt oxalate is as a starting material for the preparation of cobalt metal powders.

Cobalt(II) Carboxylates. The cobalt salts of carboxylic acids can be made by the direct reaction of cobalt powder, oxide, or hydroxide with the organic acid or by precipitation reactions involving the addition of the sodium salt of the acid to an aqueous solution of a cobalt salt, such as the sulfate. Cobalt(II) resinate, oleate, linoleate, soyate, naphthenate, ethylhexanoate, formate, acetylacetonate, and citrate have been produced for use in a variety of applications including catalysts, drying agents, metal film production, and medical uses.

2.3. Industrial Applications [7]

The more important applications of cobalt chemicals can be considered under the following headings: glasses, ceramics, and refractories; driers, paints, varnishes, and dressings; catalysts; electroplating; electronics and solid-state devices; and agricultural, nutritional, and medical uses.

2.3.1. Glasses, Ceramics, and Refractories

The addition of cobalt oxides to provide a blue pigment for glass, ceramics, and enamels has been used for many centuries. The level of cobalt in the final product depends on the color intensity required, but is typically 0.4 – 0.5% for colored ceramic

Table 7. Colors of cobalt-containing pigments

Metals in mixed-metal oxide			Color
Co	Al		blue
Co	P		violet
Co	Zn		green
Co	Sn	Si	light blue
Co	Cr	Al	turquoise
Co	Mg		pink
Co	Fe		brown

Table 8. Compositions of cobalt-containing pigments

Color	Content of Co_3O_4, %	Content of other components (% in parentheses)
Mazarine blue	68.0	SiO_2(12), cornish stone (16), $CaCO_3$(4)
Willow blue	33.3	$CaCO_3$(50), SiO_2(16.7)
Dark blue	44.6	Al_2O_3(55.4)
Matt blue	20.0	Al_2O_3(60), ZnO (20)
Blue-green	41.8	Al_2O_3(39), Cr_2O_3(19.2)
Black	20.6	Fe_2O_3(41.1), Cr_2O_3(32.4), MnO_2(5.9)

bodies, 0.5 % for the blue glass used in welder's goggles, and a few mg/kg in camera lens glass. A range of cobalt-containing mixed-metal oxides and cobalt tripotassium hexanitrite (a yellow pigment) is also available for use as pigments. An indication of the colors obtained from mixed-metal oxide pigments is given in Table 7. The shades of the colors also depend on the exact composition of the oxide and on the method of preparing the pigment. The compositions of some ceramic pigments are given in Table 8.

The pigments are normally prepared by mixing the ingredients as oxides or as readily decomposable salts and then calcining the mixture at 1100–1300 °C prior to milling the product to obtain the pigment as a fine powder. In ceramic applications, the pigments can be added to the ceramic base materials to give a body color or, after being mixed with suitable fluxes, applied as an underglaze (as on Delft china), or overglaze decoration. The final shade after the ceramic has been fired may be modified by reaction between the pigment and the clay base.

Cobalt pigments are used for decolorizing both glass and pottery articles that contain iron oxide, which would give a yellow coloration to the products in the absence of a decolorizer. The yellow color is masked by the complementary blue color of the cobalt pigment added. The levels of cobalt required for the compensation of iron oxide colorations in typical pottery bodies, enamels, and glasses are 0.003–0.02 %, 0.002–0.01 %, and 1–2 mg/kg, respectively.

The color of the pigments and of the compounds used for decolorizing arise from the optical absorption spectra of cobalt atoms modified by lattice effects in the pigments and in the final products.

In the vitreous enameling of steel sheet, a small amount of cobalt oxide (0.15 – 1.0 %) is included in the ground coat mixture of feldspar, sand, borax, and soda ash to improve the adhesion of the enamel to the steel. As generally accepted, the cobalt does not contribute directly to the adhesion process, but it is involved in creating suitable conditions for the development of good adhesion.

Ceramic coatings for low-carbon and low-alloy steels for high-temperature use can consist of a mixture of aluminum oxide and two cobalt coats, one of which is high-firing and the other low-firing. The purpose of the latter is to seal off oxygen at low temperatures to protect the base metal from oxidation from the start of the firing. These coatings are claimed to withstand combustion products and corrosive vapors up to 750 °C and have been used to protect flue gas and gas turbine exhaust pipes.

The metal – glass coatings used to protect plain carbon steels typically contain 80 % powdered chromium and 20 % of a borosilicate glass containing 5 % cobalt oxide. Superalloys have also been coated with a ceramic consisting of high-firing cobalt ground coats with high aluminum oxide content or with boron nitride containing a lithium cobalt oxide binding agent.

2.3.2. Driers, Paints, Varnishes, and Dressings

A number of cobalt pigments have been used in oil paintings, including Cobalt blue smalt [*1345-16-0*] (a silicate), Thenard's blue [*13820-62-7*] (an aluminate), Cerulean blue [*68187-05-3*] (a stannate), Cobalt violet [*82196-89-2*] (a phosphate), and Aureolin yellow [*13782-01-9*] (potassium cobalt(III) nitrite). Some of the pigments have also been used for inks, for printing on fabrics and paper, and for coloring plastics.

Cobalt acetate is used for coloring the oxide layers in anodized aluminum to give shades varying from bronze to black. The anodized metal is immersed in a 20 % solution of cobalt acetate and then either in a solution of ammonium sulfide to produce cobalt sulfide (black) or in a solution of potassium permanganate (bronze). The different depths of bronze shades are achieved by different numbers of alternate immersions in the cobalt acetate and permanganate solutions. In addition to its use as a coloring agent, cobalt acetate is also used to improve the lightfastness of anodized aluminum dyed with organic pigments.

The cobalt(II) salts of a number of organic acids have been used as drying agents for nonsaturated oils in paints, varnishes, and inks. The salts used for this purpose include the oleate, ethylhexanoate, naphthenate, soyate, linoleate, resinate, and tallate. These cobalt salts are either soluble in the nonsaturated oils or they react with them to form soluble compounds that then act both as oxidation accelerators of the oil and as polymerization catalysts, forming a film with increased stability, resistance, and flexibility. The drying action is due entirely to the redox behavior of the cobalt species present, the carboxylate moieties simply act as oleophilic groups to solubilize the cobalt compounds in the oils. The amounts of cobalt added to the oils as drying agents are in the range 0.01–0.6%; the salts can be added as solutions in organic solvents or as dispersions of ultrafine powders. Cobalt-containing drying agents are also used in low-temperature curing processes for silicone resins. The cobalt species promote drying at the surface of the film in contact with atmospheric oxygen. Its action is so fast that a hard polymeric film can be formed that prevents deep drying; for this reason cobalt is often used in conjunction with other metals such as zinc and calcium that do not have a drying action but slow down and control the effects of cobalt.

2.3.3. Catalysts [3]

Cobalt compounds are versatile catalysts. Cobalt-containing materials have been used to catalyze the following types of reactions: hydrogenation; dehydrogenation; hydrogenolysis including hydrodenitrification and hydrodesulfurization; hydrotreatment of petroleum products; selective oxidation; ammonoxidation; oxidation; hydroformylation; polymerization; selective decomposition; and ammonia synthesis. In addition to these more general reactions, there are a number of reactions for which cobalt chemicals are among the best catalysts. These include their uses as driers, described in Section 2.3.2., their use in the oxidation of ammonia to nitric acid, and the syntheses of fluorocarbons.

The effectiveness of cobalt as a catalyst is related to the ease with which the element forms complexes and particularly to the large variety of ligands in these complexes.

The main industrial processes using cobalt catalysts are the removal of sulfur from various petroleum-based feedstocks (hydrodesulfurization), selective liquid-phase oxidation, and hydroformylation. The catalysts used in the hydropurification reactions including *hydrodesulfurization* consist of cobalt and molybdenum oxides supported on an inert material such as alumina and contain 2.5–3.7% cobalt. The homogeneous catalysts used in the *liquid-phase oxidation* processes are soluble cobalt(II) carboxylates such as acetate, naphthenoate, and oleate. Relatively large concentrations of cobalt (0.1–1%) are used as cobalt salts in the catalysis of *hydroformylation* reactions in which $HCo(CO)_4$ is the active material. The amounts of cobalt catalysts used in hydrodesulfurization, oxidation, and hydroformylation have been estimated at 800–1800, 200, and 800 t/a, respectivly.

In *hydrotreating*, the cobalt–molybdenum catalysts function in the sulfide form and catalyze two reactions, hydrogenation and hydrogenolysis of carbon–heteroatom bonds. Two theories have been used to explain the enhanced catalytic activity obtained by having both cobalt and molybdenum oxides in the catalysts: the *pseudointercalation model* and the *remote control model*. The first model assumes an association between molybdenum sulfide (MoS_2) and cobalt to produce active sites for the catalysis. The remote control model, on the other hand, assumes that active centers are created in the MoS_2 lattice by a mobile species of hydrogen produced at a cobalt sulfide component.

The cobalt in the catalyst thus controls the nature and the number of catalytically active sites formed.

The types of *oxidations* catalyzed by cobalt salt solutions include *p*-xylene to terephthalic acid, cyclohexane to adipic acid, and hydrocarbons or acetaldehyde to acetic acid. All of these liquid phase oxidations follow classical radical chain mechanisms. The following reaction mechanism for cyclohexane oxidation shows the role of cobalt in decomposing the hydroperoxides formed in the reaction and thus avoiding unwanted side reactions:

Initiation:

$$C_6H_{12} + R\cdot \longrightarrow C_6H_{11}\cdot + RH$$

Chain reaction:

$$C_6H_{11}\cdot + O_2 \longrightarrow C_6H_{11}OO\cdot$$

$$C_6H_{11}OO\cdot + C_6H_{12} \longrightarrow C_6H_{11}OOH + C_6H_{11}\cdot$$

Hydroperoxide decomposition:

a) $C_6H_{11}OOH \longrightarrow$ cyclohexanone + H_2O

b) $Co^{2+} + C_6H_{11}OOH \longrightarrow Co^{3+} + C_6H_{11}O\cdot + OH^-$
   $Co^{3+} + C_6H_{11}OOH \longrightarrow Co^{2+} + C_6H_{11}OO\cdot + H^+$
   $\longrightarrow$ chain

c) $C_6H_{11}OOH$ + all radicals $\longrightarrow$ degradation products, cyclohexanol, and cyclohexanone

In *hydroformylation* and *hydroesterification*, oxo reactions such as those listed below occur:

$$CH_3OH + CO \longrightarrow CH_3COOH$$

$$RCH=CH_2 + CO + H_2 \longrightarrow RCH_2CH_2CHO$$

$$RCH=CH_2 + CO + 2H_2 \longrightarrow RCH_2CH_2CH_2OH$$

$$RCH=CH_2 + CO + R'OH \longrightarrow RCH_2CH_2COOR'$$

The cobalt salts added to the reaction mixture are converted into $HCo(CO)_4$, which catalyzes the oxo reaction by the following types of mechanism:

$$HCo(CO)_4 + RCH=CH_2 \longrightarrow RCH_2CH_2Co(CO)_4$$

$$RCH_2CH_2Co(CO)_4 + CO \longrightarrow RCH_2CH_2CO-Co(CO)_4$$

$$RCH_2CH_2CO-Co(CO)_4 + H_2 \longrightarrow RCH_2CH_2COH + HCo(CO)_4$$

In addition to the main types of reactions catalyzed by cobalt species, cobalt compounds have also been used as catalysts for automobile exhaust gas purification, for the manufacture of nitric acid, for heterogeneous oxidation reactions, and for the oxidation of toluene to benzoic acid. Cobalt(II) dicobalt(III) tetroxide is one of the most effective catalysts for the oxidation of carbon monoxide and also has a high activity in oxidizing nitrogen monoxide; these properties have led to the use of Co_3O_4 and the $LaCoO_3$ (perovskite phase) as components of exhaust gas purification catalysts. The oxide Co_3O_4 is also a highly selective catalyst in the oxidation of ammonia to nitric acid, although it loses activity due to sintering after repeated use and due to poisoning by sulfur dioxide.

Cobalt is often cited in the patent literature as an additive in nearly all heterogeneous oxidation catalyses, including the oxidations of propene, butane, butenes, and methanol. New systems containing cobalt can also be used to catalyze the conversion of thiols to disulfides in gasoline sweetening.

Cobalt compounds have considerable potential in *homogeneous catalytic systems* and are used with copper in the oxidation of toluene to benzoic acid; good selectivities for the conversion of methanol to ethanol have been achieved by using cobalt cluster catalysts.

Cobalt fluoride can be used as a catalyst in the *fluorination* of hydrocarbons. These reactions utilize the cobalt fluoride as a fluorine carrier and take place in two steps – oxidation of CoF_2 to CoF_3 at 150–200 °C followed by fluorination of the hydrocarbon at 250–350 °C. Both stages of the reaction are exothermic, and careful control of the reaction is needed to obtain optimum yields.

Cobalt has been a component of the catalysts used in the *Fischer–Tropsch synthesis* of hydrocarbons from synthesis gas. The mixed cobalt–thorium oxide–magnesium oxide–kieselguhr catalyst (100:5:8:200), for example, is generally prepared by boiling solutions of sodium carbonate containing cobalt and thorium nitrates to precipitate the metals as carbonates prior to adding magnesium oxide and kieselguhr. The mixture of metal carbonates and support materials is then filtered off, washed, dried, milled, and reduced in a hydrogen atmosphere to produce the Fischer–Tropsch catalyst.

Raney cobalt can be prepared by leaching a finely powdered aluminum–cobalt alloy containing 40–50% cobalt with sodium hydroxide at 15–20 °C, which dissolves the aluminum to leave a very active porous cobalt residue. The properties of Raney cobalt are similar to those of Raney nickel, and the materials can be used to catalyze the same types of reaction.

2.3.4. Electroplating [8]

Cobalt is readily electrodeposited from a number of electrolyte solutions. Pure plated layers of cobalt are of little commercial value, but cobalt is an important component of a number of alloy electrodeposits. The electrochemical equivalent of cobalt is 1.099 g A^{-1} h^{-1} and, with cathode efficiencies of 90–100% for most electrolytes, the deposition rates are ca. 1 g A^{-1} h^{-1}. Solutions of cobalt sulfate,

Table 9. Compositions of cobalt-containing plating baths

Electrolyte composition			Content, g/L	Metal deposited
Metal salts	Content, g/L	Additives		
$CoSO_4 \cdot 7 H_2O$	332	H_3BO_3	30	Co
$CoCl_2 \cdot 6 H_2O$	300	H_3BO_3	30	Co
$CoSO_4(NH_4)_2SO_4 \cdot 6 H_2O$	200	H_3BO_3	25	Co
$Co(SO_3NH)_2 \cdot 4 H_2O$	450	$HCoNH_2$	30*	Co
$Co(BF_4)_2$	116–154	H_3BO_3	15	Co
$NiSO_4 \cdot 7 H_2O$	240	H_3BO_3	30	Co–Ni
$NiCl_2 \cdot 6 H_2O$	45	$NaHCO_2$	35	Co–Ni
$CoSO_4 \cdot 7 H_2O$	3–15			
$CoSO_4 \cdot 7 H_2O$	150	NaCl	28	Co–W
$NaWO_3$	15–50	H_3BO_3	40	Co–W
		Na heptanoate	100	Co–W

* mL/L.

chloride, sulfamate, or fluoroborate have been used as electrolytes. The compositions of typical plating baths are listed in Table 9. To obtain a smooth metal deposit at a cathode efficiency of 90–100% and a current density of 5 A/dm^2, the bath should have a high cobalt ion concentration (65–100 g/L) and should be operated at a temperature of 25–50 °C and a pH of 4–5.

Additions of cobalt sulfate to nickel plating baths have been used to produce bright nickel plates consisting of nickel–cobalt alloys. The proportion of cobalt in the electrolyte controls the composition of the alloy deposited. Since the deposition potential of cobalt is less than that of nickel, the cobalt–nickel ratio is higher in the deposit than in the electrolyte. A typical bath composition is given in Table 9. If this bath is operated at 55–60 °C, pH 4–4.3, and with a current density of 2.5–4 A/dm^2, an alloy deposit containing 18% cobalt is produced. The cobalt content of the bath must be maintained by the addition of the cobalt salt when nickel anodes are used. Cobalt sulfamate is used as an alternative to cobalt sulfate in some chloride –sulfate nickel plating baths.

Cobalt–tungsten alloys with tungsten contents of up to 10% can be deposited from baths containing cobalt sulfate and sodium tungstate (typical electrolyte composition given in Table 9). A temperature of 70–90 °C and a pH of 7–9 are used to obtain smooth crack-free cobalt–tungsten deposits. Cobalt–molybdenum alloys can be obtained under similar conditions when sodium molybdate replaces the sodium tungstate in the electrolyte. A number of other cobalt-containing alloys can be obtained as electrodeposits, including cobalt–iron, cobalt–platinum, cobalt–gold, and cobalt–phosphorus.

Electrodeposited cobalt and cobalt alloys have been used as matrices for composite wear-resistant coatings in systems that codeposit suspended particles of materials such as alumina and silicon carbide with the plating metal.

Table 10. Properties of cobalt oxide coated solar panels

Coating method	Substrate	Absorbance, %	Emittance, %
Plated	Cu, Ni	93	28
Plated	Ni-plated steel	90	7
Plated/thermal decomposition	Ag-coated steel	> 90	20
Plated	Ni	95	–
Plated	Cu-coated steel	96	13
Oxidized paint	Al	91	30
Paint	Ni alloy	85	10
Spray pyrolysis	Al, galvanized Fe	91, 92	13, 12
Spray pyrolysis	Stainless steel	93	14

2.3.5. Electronics and Solid-State Devices

There has been recent interest in black cobalt oxide, Co_3O_4, as a selective coating material for high-temperature solar collectors [9]. The cobalt oxide coatings are superior to the black chrome coatings often used. The function of cobalt oxide as a coating intended to operate up to 1000 °C is to concentrate the solar radiation on the collector surface by a factor of up to 2000 times. For this purpose, a high solar absorbance with low infrared emittance is required. Cobalt oxide coatings can be prepared by plating, pyrolysis of cobalt salts on heated substrates, vacuum deposition, and by use of Co_3O_4-based paints. Table 10 gives absorbance and emittance data for cobalt oxide coated solar panels prepared in a variety of ways on different substrates.

Cobalt(II) oxide has been used in thermistors to improve both the resistivity and the temperature coefficient of the resistivity of the device. A number of mixed-metal cobalt oxides have been used in devices, including $NiCoO_4$ and La_2CoO_4 as anodic materials in water electrolysis, $LiCoO_2$ as an electrode in lithium batteries, $NaCoO_2$ as cathodes in sodium batteries, and $CoMoO_4$ as a fuel cell electrode. Iron oxides doped with cobalt are used extensively in thin-film coatings for magnetic recording tapes.

2.3.6. Agriculture, Nutrition, and Medicine

Cobalt chemicals are used to correct cobalt deficiencies in soils and in animals. Soil treatments usually involve top dressings containing cobalt sulfates, whereas treatment of ruminant animals involves the use of either salt licks containing ca. 0.1% of cobalt as sulfate, or concentrated feeds, or pellets of cobalt oxide bound in an inert material such as china clay.

The medicinal uses of cobalt are dominated by the use of vitamin B_{12}. The vitamin is obtained from the mother liquors of the microbial formation of antibiotics such as streptomycin, aureomycin, and terramycin after the removal of the antibiotic. Vitamin B_{12} and cobalt treatments are used as remedies in certain types of anemia. Cobalt salts, administered intravenously, have been used as an antidote in cyanide poisoning.

3. Cobalt Powders [10]–[12]

3.1. Production

Cobalt powder can be produced by a number of methods, but those of industrial importance involve the reduction of oxides, the pyrolysis of carboxylates, and the reduction of cobalt ions in aqueous solution with hydrogen under pressure. Very pure cobalt powder is prepared by the decomposition of cobalt carbonyls.

Reduction of Oxides. Grey cobalt(II) oxide (CoO) or black cobalt(II) cobalt(III) oxide (Co_3O_4) is reduced to metal powder with carbon monoxide or hydrogen. The reactions occur under conditions well below the melting point of the oxides or the metal. The purity of the powder obtained is 99.5% with a particle size of ca. 4 µm, although the density and the particle size of the final product depend on the reduction conditions and on the particle size of the parent oxide. The finely powdered metal is stored at very low temperature.

Pyrolysis of Carboxylates. The thermal decomposition of such cobalt carboxylates as formate or oxalate in a controlled reducing or neutral atmosphere produces a high-purity (about 99.9%), light, malleable cobalt powder with a particle size of ca. 1 µm that is particularly suitable for the manufacture of cemented carbides. The particle size, form, and porosity of the powder grains can be changed by altering the pyrolysis conditions.

Reduction of Cobalt Ions in Solution. Purified leach solutions containing cobalt pentammine complex ions can be treated in autoclaves with hydrogen under pressure and at high temperature to give an irregular chainlike form of the powder that is suitable for consolidation by direct strip rolling.

3.2. Uses

3.2.1. Cemented Carbides [13], [14]

One of the most important uses of metallic cobalt is as a bonding agent in cemented carbides that are used extensively as cutting tools for metals, rocks, and other high-strength materials. No suitable substitute has been found for cobalt as a cementing agent for carbides. The most commonly used cemented carbide, tungsten carbide containing 2–30% cobalt, is manufactured as follows: pure cobalt powder is added to tungsten carbide (particle size ca. 1 µm), and the mixture is ball-milled for a long period by using hard metal balls. The mixed powders are then consolidated by cold pressing, by hot pressing at ca. 1300–1400 °C, by cold extension, or by slip casting to produce small artifacts, large artifacts, constant-section rods, and complex-shaped articles, respectively. The consolidated parts are sintered in a reducing atmosphere at 1300–1600 °C. The sintering process involves the formation of a liquid phase; the cobalt liquefies at ca. 1320 °C and dissolves tungsten and carbon from the carbide. When the mixture is cooled, most of the tungsten and carbon reprecipitates, but the binder phase is much stronger than pure cobalt because enough tungsten and carbon are present to stabilize the metal in its cubic form. During the sintering process, the consolidated powders shrink by 20 %. The binding mechanism depends on (1) the liquid metal phase to wet the carbide particles so that surface tension forces help densify the sintered phases and (2) on the ability of the liquid to dissolve and retain traces of carbide impurities.

The properties of cemented carbides are controlled by the amount of cobalt present and the particle size of the carbide used. The hardness of the cemented carbides increases with decreasing particle size and with increasing cobalt content. The impact strength of cemented carbides is directly proportional to the cobalt content; it rises from less than 1 kg m for a sample with 6 % cobalt to ca. 2.5 kg m for a sample with 25 % cobalt in tungsten carbide with a particle size of 1.4–3.1 µm. The properties of cemented tungsten carbides can be improved by the addition of other carbides such as those of niobium, tantalum, or titanium.

3.2.2. Cobalt Powders in Powder Metallurgy and Cobalt-containing Metal [14]

Cobalt powders have been used in the formation of a number of alloy phases including maraging steel by hot extension of prealloyed powders, cobalt-based superalloys by consolidation of powders involving liquid phase sintering, fine-particle magnetic alloys, and bearing materials impregnated with low-friction substances such as graphite, lead, nylon, and molybdenum disulfide.

4. Physiology and Toxicology [15]–[19]

The wasting disease in sheep and cattle known as pine in Britain, bush sickness in New Zealand, coast disease in Australia, and salt sick in the United States is treated with vitamin B_{12}. This vitamin is a coenzyme in a number of biochemical processes, the most important of which is the formation of red blood cells.

The acute rat toxicity LD_{50} for cobalt powder is 1500 mg/kg and the TLV is 0.1 mg/m^3. Prolonged exposure to the powder may produce allergic sensitization and chronic bronchitis. In 1977 NIOSH recommended that when the cobalt content of cemented carbide dusts exceeds 2 %, the dust level should not exceed 0.1 mg/m^3 in air.

Occupational cobalt poisoning is caused primarily by inhalation of dust containing cobalt particles and by skin contact with cobalt salts. Very few people have been affected by cobalt dust, but those liable to be affected must be protected adequately. Skin irritation and diseases caused by cobalt are extremely rare, but it is possible to distinguish two types: one appears as erythema, which is normally found on the hands shortly after contact with cobalt especially during warm weather; the other appears as eczema. However, this form of allergy does not appear until after many years of contact with cobalt compounds.

Toxic effects have been observed from cobalt therapy for certain types of anemia. A single oral dose of up to 500 mg of cobalt chloride has resulted in severe toxic effects, but milder symptoms have been caused by daily intravenous injection of 5–10 mg. In some patients, cobalt has been shown to have a toxic effect on thyroid function.

Over 100 cases of serious heart ailments appeared between 1966 and 1969 in Canada, the United States, and Belgium. They were finally traced to cobalt sulfate that several brewers added to beer (0.075–1 mg/kg) to stabilize the froth. Apparently, chronic alcoholism combined with a lack of nourishment sensitized the hearts of those affected. The acute oral toxicities (LD_{50}, rats) for cobalt oxides, cobalt carbonate, cobalt chloride hexahydrate, cobalt nitrate hexahydrate, and cobalt acetate tetrahydrate are 1750, 630, 766, 691, and 821 mg/kg, respectively [19].

5. References

[1] *Cobalt Monograph*, Centre d'Information du Cobalt, Brussels 1960.
[2] R. S. Young: "Cobalt," *ACS Monogr. Ser.* **149,** New York 1960.
[3] *Proceedings International Conference on Cobalt: Metallurgy and Uses,*Brussels Nov. 10–13, 1981, vol. I and II, ATB Metall. (1981).
[4] R. W. Andrews: *Cobalt: Overseas Geol. Surveys*, Mineral. Resources Div, H. M. Stationary Office, London 1962.
[5] *Gmelin*, vol. 58.
[6] B. F. G. Johnson, R. E. Benfield in G. L. Geoffroy (ed.): *Topics in Inorganic Stereochemistry*, Wiley, New York 1981, pp. 253–335.
[7] E. DeBie, P. Doyen, *Cobalt* **1962**, no. 15, 3–13; no. 16, 3–15.
[8] F. R. Morral, W. H. Safrawek in F. A. Lowenheim (ed.): *Modern Electroplating*, 3rd ed., Wiley, New York 1974, pp. 152–164.
[9] S. J. Clark, J. D. Donaldson, S. M. Grimes, *Cobalt News* 1983, no. 4, 6.
[10] H. Hagon, *Metall (Berlin)* **29** (1975) no. 11, 1157–1158.
[11] H. Hagon, *Met. Powder Rep.* **31** (1976) no. 1.
[12] P. Doyen, H. Hagon, Rev. Univ. Mines **115** (1972) 205.
[13] H. E. Exner, *Int. Met. Rev.* **24** (1979) no. 4, 137–176.
[14] J. D. Donaldson, S. J. Clark: *"Cobalt in Superalloys,"* The Monograph Series, Cobalt Development Institute, London 1985.
[15] R. S. Young: *Cobalt in Biology and Biochemistry*, Academic Press, London 1979.
[16] E. Browning: *Toxicity of Industrial Metals*, Butterworths, London 1969.

[17] F. Caudrolle, D. Meynier, *Rev. Pathol. Gen. Physiol. Clin.* **55** (1959) 245.
[18] H. E. Harding, *J. Ind. Med.* **7** (1950) 76.
[19] J. D. Donaldson, S. J. Clark, S. M. Grimes: *"Cobalt in Medicine, Agriculture and the Environment,"* The Monograph Series, Cobalt Development Institute, London 1986.

Composite Materials

BERNHARD ILSCHNER, Ecole Polytechnique Fédération de Lausanne, Lausanne, Switzerland (Chaps. 1–3)

JOSEPH K. LEES, E.I. Du Pont de Nemours & Co., Wilmington, DE 19898, United States (Chaps. 4–7)

ASHOK K. DHINGRA, E.I. Du Pont de Nemours & Co., Wilmington, DE 19898, United States (Chaps. 4–7)

R. L. MCCULLOUGH, University of Delaware, Newark, DE 19716, United States (Chaps. 4–7)

1.	Definition	1527
2.	Classification	1530
2.1.	Geometric Classification	1530
2.2.	Material Classification	1532
3.	Principles of Fiber Reinforcement	1533
3.1.	High-Strength Fiber Composites	1533
3.2.	Behavior in Compliant Matrices	1534
3.3.	Behavior in Brittle Matrices	1537
4.	Materials Used in Advanced Composites	1538
4.1.	Reinforcing Fibers	1539
4.1.1.	Carbon Fibers	1539
4.1.2.	Organic Fibers	1543
4.1.3.	Inorganic Fibers	1546
4.1.4.	Reinforcement Geometry	1551
4.2.	Matrix Materials	1552
5.	Processing and Fabrication Technology	1559
5.1.	Thermoset Resin Matrix Composites	1560
5.1.1.	Classic Low-Volume Production Methods	1560
5.1.2.	Automation of Traditional Production Methods	1564
5.1.3.	Alternate Production Methods	1565
5.2.	Thermoplastic Resin Matrix Composites	1566
5.3.	Design Integration	1567
6.	Properties of Advanced Composite Materials	1568
6.1.	Properties of Composite Structures	1569
6.2.	Laminate Analysis	1571
6.3.	Micromechanical Models	1572
7.	Applications	1580
8.	References	1588

1. Definition

Structure as a Basis for Definition. The word *composite* is used in the technical sense to describe a product that arises from the incorporation of some basic structural material into a second substance, the *matrix*. That which is incorporated can be in the form of particles, whiskers, fibers, lamellae, or a mesh. Its task is to impart its own advantageous mechanical characteristics to the matrix material. Composites can be produced either by thermomechanical mixing of individual components or by means of chemical reactions that lead to phase separations within homogeneous starting mixtures (cf. next paragraph).

To obtain a composite with desirable mechanical properties, it is necessary that the separation between the incorporated elements be in the range of 0.01 – 1 mm and that the incorporation show directional characteristics. The extent of mutual adhesion and interpenetration of the components must also be favorable.

The key characteristics sought in an incorporated substance, which are in turn conferred on the composite, are above all elastic rigidity, tensile and fatigue strength, and hardness, as well as appropriate electrical and magnetic properties. Glass fiber reinforced resins are typical examples of composites, and these are probably the best known.

Composite materials and glass fiber reinforced plastics, respectively, are known in German as "Verbundwerkstoffe" and "glasfaserverstärkte Kunststoffe," and in French as "matériaux composites" and "matière plastique armée aux fibres de verre."

For further information in the form of introductory textbooks, monographs, and review articles, see [1]–[6].

Modes of Preparation as a Basis for Definition. The mostcommon way to produce composites entails thermomechanical introduction of one or more additive components R_1, R_2, R_3,..., into a preexisting matrix. Particularly suitable techniques include winding and infiltration (see the literature sources cited above). Extrusion [7] and thermal injection [8] methods have also been described. The variety of possible forms that composite materials can take and the number of possible component combinations are large. Finishing of the materials presents no major problems; the appropriate technology is both proven and cost effective. Rather, it is the preparation of composites in suitable form that tends to be costly.

An interesting alternative to thermomechanical mixingis the possibility of using a phase-separationprocess within a homogeneous starting material to create the desired heterogeneity. The following processes lend themselves to this approach: eutectic crystallization from a melt, eutectic decomposition of solid solutions, or phase separation within a supersaturated solid solution [9], [10]. Composites that form to some extent in this way within partially prefabricated construction elements are known as *in situ composites*.

The method is a particularly elegant one, and it affords an opportunity to produce exceedingly fine geometric distribution of the phase components. Moreover, constituents that are difficult to prepare or handle can sometimes be introduced by this method. Nonetheless, the following are disadvantages:

1) The number of multicomponent systems suited to the approach and at the same time capable of yielding desirable properties is extremely limited.
2) The resulting ratio of matrix : included phase is often a fixed property of the system.
3) Reactions of this kind proceed under diffusion control and with a finite number of transport paths to the growing front. Moreover, both of these parameters change

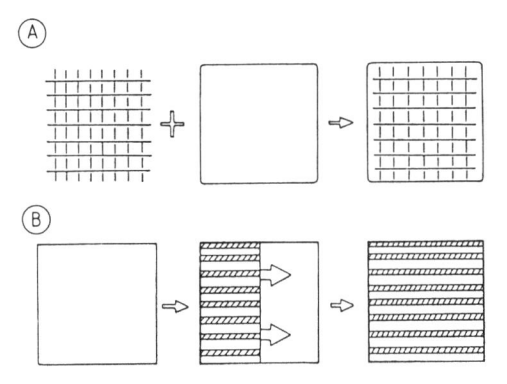

Figure 1. Symbolic comparison of the two principal approaches to the manufacture of composite materials
A) Thermomechanical mixing; B) In situ preparation by phase separation

during the course of the reaction. To introduce a measure of selectivity on the directional characteristics of the composite, the reaction must be correspondingly controlled, for example, by provision of a steep temperature gradient in the direction of desired growth. Thus, production is slow, highly subject to interference, and limited to geometrically simple forms.

A material prepared in the latter way through controlled solidification of a eutectic mixture is referred to as a *directionally solidified eutectic*, commonly abbreviated DSE.

Current state of the art permits only the use of the thermomechanical method for the manufacture at modest cost of large or mass-produced objects (e.g., in the automotive, sporting goods, and boat building industries, and for large containers, etc.). The method of phase separation is best suited to the manufacture of small items that are subject to rigid demands, where cost plays a minor role (e.g., for turbine blades or microcomponents) [11].

Both approaches are represented schematically in Figure 1.

Distinction from Other Classes of Materials. The definition provided above is occasionally criticized as being too narrow, where the objections arise from the point of view of the material sciences. Thus, it is argued that it is the nature of the interaction between two or more finely divided materials that determines measurable macroscopic properties of a composite, not the way the mixture is prepared or the purpose it is intended to serve. The very fact that oriented materials resulting from eutectic crystallization are called in situ composites demonstrates that the eutectic mixture itself determines the properties of the composite. This line of reasoning suggests that all eutectic mixtures (not only those that are oriented) should be regarded as composites — including cast iron, Al – 12 % Si alloys, and soft solder. It is further asserted that no fundamental difference exists between an inclusion comprised of Ti/W carbides in a Co matrix and the product resulting from distribution of Cr and Nb carbides in a ferrite matrix, whereas only the former system (cemented carbides) is normally referred to as a composite. Similar argumentation can lead to the conclusion that paper and even the natural product wood are composite materials.

The term "composite" is not normally used in as broad a sense as this interpretation would suggest. Therefore, it is preferable to employ a more general term such as "multiphase" material (or solid) when one is discussing the general relationship between microstructure and macroscopic properties, and to reserve the word "composite" for cases in which some specific characteristic of an incorporated material is intentionally being exploited. The latter criterion does not apply to most eutectic or precipitation-hardened alloys. It is true that the hardness of a carbide or intermetallic compound that is introduced into a metal matrix does contribute strengthening characteristics, but this is not a property that is exploited as such. In glass fiber reinforced resins, on the other hand, the elastic rigidity of the glass fibers is the desired property. Similarly, the hardness of carbides is the reason these are added in the preparation of hardened alloys, and asbestos fibers are what supply the tensile strength of asbestos cement.

Cast iron represents an interesting borderline case. Gray cast iron with eutectically oriented graphite lamellae is often employed in applications in which the damping properties of the graphite lamellae are advantageous. White cast iron, on the other hand, is preferred for purposes that require hardness, where eutectic carbide is performing the same function as in hardened alloys. Nevertheless, the historical development of both forms of cast iron was unrelated to such considerations, so that it is reasonable to conform to tradition and refrain from designating cast iron as a "composite."

It is also inappropriate to describe as a composite a macroscopic material into which whole structural elements (wires, meshes), themselves homogeneous, have been introduced for the purpose of improving mechanical or other properties (e.g., steel-reinforced concrete and steel-belted tires). The same is true even if the union of materials is conducted in a layered fashion (e.g., plywood) or by coating. Similarly, sheet steel or pipe stock that has been electrolytically nickel plated, galvanized, plated in a rolling mill, or plastic coated is not a composite material.

2. Classification

2.1. Geometric Classification

A reinforcing agent is introduced into a given matrix in the form of particles. These are understood to exist as dispersed entities with planar (usually crystalline) or rounded boundary surfaces. For a particular particle shape, the relationship of particle radius r to the distance between centers of adjacent particles d determines the volume fraction f of the composite material, relative to the volume V:

$$f = V_R/(V_M + V_R) \tag{1}$$

Throughout this section the subscript R is used to refer to the incorporated reinforcing material, wheras the index M signifies the matrix. The maximum volume fraction that can be obtained by closest spherical packing of spherical particles of uniform radius is $f_{max} = 0.680$. Higher values of f can be attained by using particles of varying diameter if these are arranged (e.g., by applying vibration) so that smaller spheres occupy the spaces between larger ones. According to [12], space filling that approaches 95%, for example, results from the layering of spherical particles of radius ratio of $1:7:40:320$ in a volume relationship of $6:10:25:59$. Even higher degrees of admission are possible if the spherical shape is abandoned, as shown by the polycrystalline structure of metals and ceramics. Thus, cubes and hexagonal dodecahedra would, in theory, lead to values of 100%. Other particle shapes (ellipsoids) reduce the potential packing density, but at the same time they open the way to establishing preferential orientation.

Whiskers are monocrystalline fibers that are largely free of defects and are thus subject to elastic deformation even under extreme tension. Such whiskers can be obtained by the chemical vapor deposition (CVD) process, which entails reactive deposition from the gas phase. Whisker diameter ranges from 0.1 to 1 µm, and is thus an order of magnitude below that of ordinary fibers. Their length reaches 0.1–1 mm. Whiskers exhibit very high mechanical strength as a result of their lack of defects. Tensile strengths ca. 10% as large as the elastic modulus have been measured, a value 5 times that associated with a high-quality polycrystalline fiber and 100 times as large as that for massive polycrystalline material. Unfortunately, the preparation and handling of whiskers is both difficult and costly. Despite 20 years of developmental effort, it has still not proven possible to mass produce them, although well over 100 substances have been prepared as whiskers on a laboratory scale. Expectations were high — probably unrealistically so — in the 1960s for future applications of whiskers, but these have been largely unfulfilled [13]. In principle, whiskers should be ideal components for use in composites, since this application is one that would take full advantage of their remarkable strength while at the same time making available in the form of the matrix a suitable container and protective coating.

Fibers are incorporated into composites in both short and continuous forms, although even the so-called short fibers rarely have length: radius ratios under 20. In most cases the fiber's high rigidity or tensile strength is to be conferred upon the composite material. Thus, a key role is played by the transfer of forces out of the matrix and, therefore, by the microgeometry and wettability of the boundary surfaces. For this reason, the fibrous materials are often either provided with a coating or in some other fashion "activated."

Most fiber-reinforced composites are distinguished by the rigid orientation of their fibers ("unidirectional composites," UDC). This leads to a distinctive directionality with respect to macroscopic (e.g., mechanical) properties of the materials as well — they show high strength in the direction of the fibers, but show poor characteristics in the transverse direction. Weakness of this kind intolerable is in many applications (e.g., in automotive components). One way to compensate for this weakness is through introducing fibers in the form of a mesh (laminates) rather than unidirectional single fibers. Resulting improvements in strength transverse to the principal axis of stress are obtained at a price, however: the effective volume fraction in the direction of primary stress is substantially reduced, as is the corresponding carrying capacity. The same holds true for irregular "fleecy" inclusions.

2.2. Material Classification

Since the volume fraction of fibers in a composite rarely exceeds 0.5, the general appearance of the material is conferred by the matrix. This fact is reflected in the terminology that is employed (e.g., fiber-reinforced resin). Almost all of the known structural material types can play the part of the *matrix*.

High polymers are used as building materials [14], [15]. Most often encountered are unsaturated polyester (UP) and epoxy (EP) resins. Other examples include polyacrylate dental materials containing silicate particles and polypropylene to which glass fibers have been added for use in shatterproof safety glass.

Metals are less commonly used for ordinary temperature composite materials because of their undesirable weight. Exceptions include aluminum and other light-metal alloys strengthened by the incorporation of boron, SiC, or carbon fibers [16], [17]. Al_2O_3 whiskers are, of course, also utilized, although the economics involved are far from ideal. Strengthening through the use of steel threads has also been studied [18].

One important domain of fiber-reinforced metal composites is the area of high-temperature applications. The use of organic matrices is here excluded, although ceramic reinforcing materials represent an alternative. A special case of growing importance is the category of in situ composites (oriented eutectic mixtures, already mentioned above). Examples of the latter include Ni–Co-based materials containing fibrous TaC, or ones based on Fe and reinforced with $(Cr,Mn)_7C_3$.

Cobalt-based alloys containing particulate WC/TiC (so-called hardened alloys) are also employed on a large scale for cutting tools.

A third domain of metal-based composites includes materials for electrical engineering applications, such as contacts [19], [20].

Ceramics containing fibrous materials (Al_2O_3, SiC, and B_4C) have recently come into use as lightweight bulletproof shields (e.g., helmets, vests, and helicopter seats), though civilian applications exist as well.

Cement containing asbestos, glass, plastic, or steel threads has numerous uses in the construction industry, especially for poured concrete structures. Also noteworthy is the high impact resistance of concrete reinforced with polypropylene fibers [21]–[23].

Rubber can also serve as the basis for composites, as in preparations containing magnetic particles ("magnetic rubber").

Materials with proven value as *fibrous reinforcing agents* for composites include the following:

Glass fibers, especially those derived from E-type glass (aluminum–boron–silicate glass) and, for better elastic properties, S-type glass ($SiO_2 - Al_2O_3 - MgO$) — these materials are also introduced in fabric or fleece form.

Carbon fibers, prepared by the pyrolysis of cellulose or polyacrylonitrile fibers at high temperature and under inert atmosphere — if the final stage of the thermal treatment is carried out at ca. 2800 °C, the material is converted largely into a graphite possessing exceptional tensile strength [24], [25].

Tungsten-core fibers, including boron, B_4C, and SiC — these are prepared by thermal decomposition of halides on a tungsten core.

Metal fibers can be prepared from the extremely light but toxic beryllium, as well as from steel, niobium, molybdenum, and tungsten. Their advantage is their plastic ductility (albeit restricted within specific limits), but they have the disadvantage of great weight.

Ceramic fibers — these include important natural fibers (asbestos, especially in its chrysotile modification), but synthetic ceramic fibers are also known (though largely on the laboratory scale) [26].

Hybrid composite materials containing two or more additives (e.g., carbon and glass fibers) are increasingly important because they represent a means of creating a balance between the advantages and the disadvantages shown by individual components.

3. Principles of Fiber Reinforcement [1]–[6]

3.1. High-Strength Fiber Composites

Despite the complications attending their preparation, composite materials represent an interesting alternative to metals whenever there is a demand for great strength with minimal mass. To characterize this property, it is common to employ either the term *specific strength* $R_m/(\varrho \cdot G)$ (ϱ = density in g/m³, $G = 9.80665$ m/s² = gravitational acceleration) or else the *specific modulus of elasticity* (Young's modulus) $E/(\varrho \cdot G)$. Since $\varrho \cdot G$ has the unit N/m³, whereas R_m and E are in N/m², the resulting specific quantity has the unit m. In practice, data are normally reported in km; e.g. $E/(\varrho \cdot G)$ for graphitized carbon fibers is 2000 km. This measure corresponds to the breaking length data commonly reported for textile fibers.

High values of this parameter are only attainable with materials having high tensile strength and low density. These criteria apply mainly to the fibrous component, since the strength of the composite is determined almost exclusively by the fibers. The matrix serves to protect fiber surfaces from damage, distribute applied forces, and compensate for microscopic fiber fracturing. Thus, it is more important that the matrix material have a low density than that it be particularly strong.

Strong binding and high rigidity are achieved primarily by maximizing the number of covalent bonds per unit cross-section. This high concentration of bonds ("valences") can only be realized without undue increase in fiber weight if the constituent atoms are of low atomic mass.

For these reasons only a few elements can be considered candidates for the construction of high-strength fibers: those with low atomic number and covalent bonding character [3]. Beryllium can be largely excluded on the basis of its toxicity. Thus, there remain (apart from Al_2O_3) only boron, carbon, and silicon, as well as a few of their compounds with oxygen and nitrogen. Hydrocarbon and silane fibers generally possess low Emoduli; their constituent hydrogen atoms occupy much space, but contribute nothing to material strength. The best results within this category of substances have been achieved with aramide fibers, which combine benzene rings and – CONH – groups in rather strict alignment.

When one considers applications entailing the strengthening of metals, especially for use as high-temperature materials, the thermal stability of the fibrous component becomes another criterion, which limits the choice further. Currently, the best solution in a qualitative sense is graphitized carbon fibers [24], although B_4C and BN fibers of very high quality and with melting points above 2500 °C can also be prepared (by high-temperature boration of carbon fibers and high-temperature nitration of B_2O_3, respectively).

The choice is further influenced by price and production costs, factors that are highly susceptible to change in response to technological developments. Materials exist whose costs are more favorable than those of the components described above, but their properties are also less ideal. An example is mullite fiber ($Al_2O_3 - SiO_2 - B_2O_3$) produced by the centrifugal melt spinning process. Glass fibers fall at the low end of both the price scale and the thermal stability spectrum. It should be noted again that optimization of the balance of technical and economic factors is often possible with hybrid composites, such as those containing combinations of carbon and glass fibers.

Strength factors as they are actually observed, particularly in the case of graphite fibers, come very close to the theoretical values predicted from atomic bond strength data. Thus, it is unrealistic to anticipate that further significant increases will be obtained.

3.2. Behavior in Compliant Matrices
[27] – [30]

Themost easily described situation consists of a fiber composite material containing infinitely long fibers oriented in a single direction. If this material contains n_F fibers of cross-section a_F within an overall cross-sectional unit area, then the corresponding fiber volume fraction can be calculated as $f = n_F a_F$. When such a composite is placed under elastic stress, the extent of elongation is identical for both fiber and matrix (apart from small regions near the ends of the sample). Application of Hooke's law under these considerations illustrates the role of the modulus of elasticity E_F of the fibers. Thus, the average tension that the composite is capable of sustaining for a given elongation is

$$\sigma_c = f\sigma_F + (1-f)\sigma_M = [fE_F + (1-f)E_M]\varepsilon \qquad (2)$$

This can be rewritten as

$$\sigma_c = E_c \cdot \varepsilon$$

or, if the effective modulus of elasticity for the composite is expressed as E_c, it can be written as

$$E_c = fE_F + (1-f)E_M \qquad (3)$$

Equation (3) contains the so-called "rule of mixtures," which provides a first approximation to the behavior of a composite material. It is known, however, that this approximation can, for a number of reasons, lead to errors of $\pm 30\%$ or more in individual cases. Notice that if one assumes a fiber volume fraction of ca. 50%, then $E_c \approx 1/2 E_F$, provided that (as is usually the case) $E_F \gg E_M$. Thus, the elastic properties of the matrix are irrelevant to the overall behavior of the composite. The fibers alone determine the strength of the whole.

Nevertheless, glass fibers and most of the other fibers that lend themselves to practical application are brittle; that is to say, they break rather than undergoing plastic deformation upon reaching some maximum load σ_F^*. This behavior is of considerable importance in the development of a composite material, because the total elongation ε (see above) at maximum load (the breaking point) remains very low due to the high E modulus of the fibers:

$$\varepsilon^* = \sigma_F^*/E_F \qquad (4)$$

At this elongation the amount of stress assumed by the matrix, determined by its elastic resilience, is correspondingly small, namely $\sigma_M = E_M \varepsilon^*$. The load supported by the matrix at the breaking point is thus

$$\sigma_c^* = f\sigma_F +^* (1-f)\tilde{\sigma}_M \qquad (5)$$

This composite strength increases in a linear manner from $\tilde{\sigma}_M$ as the degree of admission increases. Nonetheless, it is also necessary to take into account the fact that a nonreinforced matrix with the same cross-section $(1-f)$ has, in the region $f = 0$, a greater strength than the composite (if it is assumed that the breakage of a high proportion of the fibers also leads to destruction of the matrix). Thus:

$$\sigma_{MN}^* = (1-f)E_M \varepsilon_M^* \qquad (6)$$

where ε_M^* is the breaking elongation of a sample of matrix from which a certain amount of material has been removed to create the channels that would otherwise be occupied by fibers. Figure 2 is a graph of Equations (5) and (6), clearly showing that at low volume fractions there is a decrease in overall tensile strength caused by reduction in the matrix cross-section.

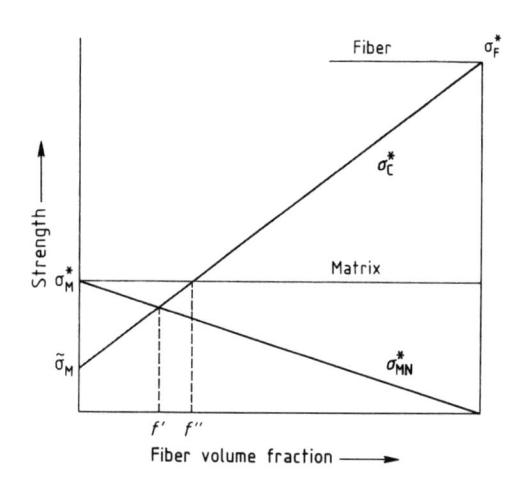

Figure 2. Influence of the volume fraction on strength σ_c^* for a composite whose fibers are oriented in a single direction as compared to the strength σ_{MN}^* of a nonreinforced matrix of the same cross-section. Only if $f > f''$ is there a true strengthening of the composite relative to matrix material.

Only after a *critical volume fraction* f' has been reached does the strength begin to rise. True strengthening by fibers can only be said to occur after a volume fraction f'' has been exceeded, where f'' is defined such that above this value, the strength of the composite is greater than that of a pure matrix of comparable cross-section. Thus:

$$f'' = (\sigma_M^* - \tilde{\sigma}_M)/(\sigma_F^* - \tilde{\sigma}_M) \qquad (7)$$

In the event that the *E*modulus of the fibers is very large, $\tilde{\sigma}_M \ll \sigma_F^*$ and, in fact, $\ll \sigma_{MN}^*$. Equation (7) can then be approximated by:

$$f'' \approx \sigma_M^*/\sigma_F^* \qquad (7a)$$

Thus, if the fibers have a tensile strength 10 times that of the matrix, the degree of admission must be at least 10% for a strengthening effect to be achieved. This would appear readily attainable, but certain limitations must also be recognized. For example, the tensile strength of a composite is reduced when, in the interest of improving transverse strength, uniaxially incorporated fibers are replaced by a fiber network. Moreover, defects are always present in the fibers of a real material. It is true that premature breakage of individual fibers will be somewhat confined as a result of localized crack branching and plastic microdeformation of the matrix. At the same time, however, the risk of crack spreading is increased if breakage of one fiber causes transmission of critical loads through the matrix to adjacent fibers. This risk makes it mandatory that an ample volume of matrix material be present even at high fiber volume fractions in order that it may serve to maintain the integrity of the composite and prevent the spreading of breaks from one fiber to another. Furthermore, the Weibull statistical probability of fiber breakage and the rate of subcritical expansion of cracks within the fibers must be known if one is to make a realistic estimate of the value of σ_F^*. In other words, account must be taken of the distribution of microdefects both in the fibers and at their surfaces.

All of these aspects of the subject have been thoroughly discussed in the recent literature [27]–[30]. The same is true for metal fibers, which undergo plastic deformation when subjected to stress above the flow limit, consequently become stiffened, and then break only when subjected to relatively high degrees of elongation. The matrix, too, is considerably stretched in the process, so that its fracture properties must also be taken into account. It is fundamentally true that the inherent strength of the matrix is here better utilized. The transmission of forces through a matrix by short fibers is thoroughly discussed in the literature [31], [32]. In this case, transverse stress along the fiber encasement surfaces plays a decisive role.

3.3. Behavior in Brittle Matrices

A brittlematrix represents a different situation, one in which the task of fibers is not to increase elastic strength or to raise the tensile strength relative to the matrix material. Instead, fibers are expected to counter what is the principal disadvantage of a brittle matrix: the tendency for cracks to be rapidly propagated and — unlike metals — for breakage to occur due to the absence of damage-limiting plastic deformation in the vicinity of the crack front.

The "ductilization" that results from incorporating fibers into ceramic materials is a consequence of the fact that any cracks that form in the matrix transverse to the direction of stress fail to penetrate into the fibers: the lower Emodulus (greater ductility) of the fibers allows them initially to undergo elongation (Fig. 3). In other words, those fibers that happen to lie in the direction of stress effectively "bridge" the crack and exert a force counterbalancing the opposing stress. Elastic elongation does lead automatically to cross-sectional contraction of the fibers, however, just as with the plastic stretching of metal fibers. Consequently, fiber separation from the matrix can occur in strain-free regions at the edge of a crack. This results in additional fiber length becoming available to accommodate further energy-consuming length changes, thus permitting the crack to open wider and increasing the potential for its further growth.

Whenever short fibers are utilized, the probability (which can be easily quantified) exists that ends of fibers will be present in the vicinity of the walls of a crack. Thus, there arises the possibility that fibers will be pulled out of the matrix as the crack grows and cross-sectional contraction occurs. This process, known as "pullout," also consumes energy in the form of friction with the walls. The extent of the extra work expended in broadening of main cracks is clearly demonstrable with or without pullout. It appears as an increase in the effort required to break a sample (equal to the integral under the corresponding stress/elongation curve), and leads to an increase in ductility.

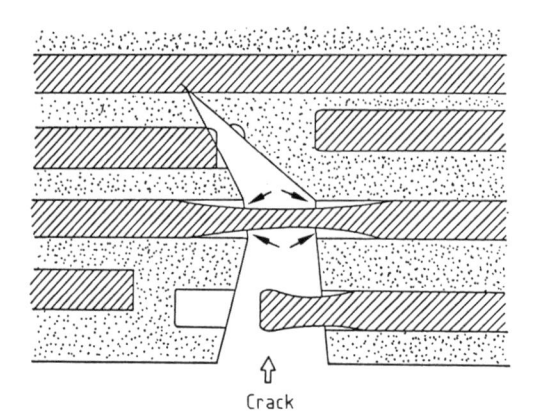

Figure 3. Elastic–plastic bridging of a crack and "pull-out" of fibers within a matrix at the onset of crack formation

4. Materials Used in Advanced Composites

Approximatelytwo decades ago, to meet the requirements of advanced military systems, a number of low-density fibers were produced that exhibited both high specific stiffness and high specific strength. In the ensuing years, the category of high-performance fibers has grown to include such diverse materials as carbon (graphite), p-aramid, boron, silicon carbide, alumina, silica, and glass. Incorporating one or more classes of these high-performance filaments in a suitable matrix to achieve properties that approach those of steel or aluminum produces a specialized class of composite materials, herein called advanced composite materials, that exhibit physical and structural properties not attainable with conventional engineering materials. The matrix or bonding agent can be a thermosetting resin, a thermoplastic resin, a metal, or a ceramic material.

On a functional basis, advanced composite materials may be described as a specialized class of composite materials that contain a high volume concentration of continuous high-performance fibers and, therefore, exhibit very high specific strength (strength/weight) and very high specific modulus (stiffness/weight).

The choice of production methods and most of the environmental properties of advanced composite materials are a function of the matrix material rather than the high-performance fiber. In this respect, resin matrix composites have the general characteristics of reinforced plastics, metal matrix composites behave like metals, and so forth. While there is much technical interest in the development of metal, ceramic, and carbon matrix advanced composite materials, resin matrix materials dominate the technology and commercial use of advanced composite materials. The predominance of resin matrix systems is due to a number of reasons, which include

1) the relatively low temperatures involved in the fabrication of resin matrix materials as compared to carbon, ceramic, or metal matrix materials, thus reducing the problems of internal stresses induced by differences in the coefficients of thermal expansion of the components and/or reaction products between components
2) the direct applicability or adaptability of fabrication technology developed for glass fiber reinforced plastics to the fabrication of advanced composite materials systems.

Carbon, ceramic, or metal matrix composites are being developed principally for use in high-temperature environments hostile to resin matrix materials, or in applications that require characteristics not attainable with organic resins, such as electrical conductivity or toughness of the level attainable with metal matrix materials.

4.1. Reinforcing Fibers

In general, reinforcing fibers used in advanced composite materials can be characterized as being light, stiff, and strong. These properties are usually exhibited by materials on the basis of the elements Be, Mg, B, Al, C, Si, Ti. Most of these relatively light elements form thermally stable covalent bonds, which are inherently stiffer and stronger than compounds formed through metallic or ionic bonds. However, many of these materials are quite brittle in bulk form because of the presence of internal or surface flaws. The probability of internal flaws, which detract from the potential strength of these materials, is minimized when the material is used in a fibrous form.

Of the elements listed above, carbon and silicon are the most important in terms of the composition of commercially available high-performance fibers. These fibers can be considered to belong to one of the three following broad families of materials: carbon, organic resins, and inorganic compounds. Within each family, several classes of high-performance fiber materials have been developed that satisfy the basic criteria of low density, high strength, and high stiffness to varying degrees. Each of these classes of fiber materials is unique in that it offers a different combination of properties and characteristics. The various fiber classes exhibit different combinations of mechanical properties, physiochemical properties, electromagnetic properties, and economic characteristics (cost and availability).

4.1.1. Carbon Fibers

Carbon fibers are currently the predominant high-strength, high-modulus reinforcing fiber used in the manufacture of advanced composite materials. Carbon fibers are made by the pyrolytic degradation of a fibrous organic precursor. In the process, an organic polymeric fiber is heated under tension and in an inert atmosphere to a very high temperature to drive off the volatile constituents. The residual carbon atoms tend to orient themselves along the fiber axis into graphitic crystallites to form a high-strength, high-modulus fiber. The composition of the precursor and the time–temperature history of the fiber establish the properties of the carbon fiber product. In general, as the processing temperature to which the fiber is exposed increases, the extent of crystallite orientation parallel to the fiber axis increases, and thus the fiber modulus increases. Because of greater sensitivity to flaws, this increase in modulus is usually accompanied by a decrease in strength. Because of the high degree of internal structure orientation, graphite fibers are strongly anisotropic. Their transverse tensile and shear moduli are usually an order of magnitude lower than the axial modulus.

A good carbon fiber precursor has the following characteristics:

1) Its chemical structure favors the formation of an aligned graphitic carbon structure during pyrolysis
2) A high carbon fiber yield is obtained
3) Fiber integrity is maintained during the pyrolysis process
4) The fiber is strong enough to be handled during all phases of the pyrolysis process
5) It should be inexpensive and readily available from various commercial sources.

The first characteristic is essential to obtain the required mechanical properties. The others, particularly achievement of a high carbon yield, are important for reasons of practicality and cost. Obtaining a high carbon yield is necessary not only to minimize raw material consumption, but also to reduce the volume of off-gas that is generated and must be treated.

The organic materials that have best satisfied these criteria to date are rayon (cellulose acetate), polyacrilonitrile (PAN), and petroleum pitch.

Rayon-Based Carbon Fibers. Carbonfibers based on rayon are now mainly of historical interest. The commercial use of carbon filaments produced by the pyrolysis of cellulose dates back to 1880 when THOMAS EDISON patented the incandescent electric lamp [33]. The development of methods of producing rayon-based carbon fibers that warranted their use as reinforcements in structural composites is a much more recent event, dating back to the mid-1950s and early 1960s. Key discoveries included recognition of the need to apply stress to the fiber prior to and during carbonization, which resulted in the initial development of high-modulus carbon fibers. These efforts led to the commercial production of Thornel 25 by Union Carbide. This material exhibited an average modulus of 25×10^6 psi (170 GPa) and an average tensile strength of 184×10^3 psi (1.27 GPa). As the process became more refined with time, Union Carbide was able to introduce rayon-based carbon fibers with improved performance characteristics, such as Thornel 50, and Thornel 75, which respectively exhibited an average modulus of 57×10^6 psi (380 GPa) and 75×10^6 psi (520 GPa) and average tensile strengths of 315×10^3 psi (2.17 GPa) and 385×10^3 psi (2.65 GPa)

[34]. High-modulus rayon-based carbon fibers are, however, expensive to produce because of the low carbon yields (10–30%) and because of the extremely high temperatures (> 2800 °C) required for the "stretch graphitization" operation needed to attain moduli in excess of 100 GPa. Because of their high price (> $ 600/kg in the early 1970s), they have not been able to compete with lower cost carbon fibers that are based on other precursor materials.

PAN-Based Carbon Fibers. Most of the carbon fibers that are currently commercially available are obtained by the pyrolysis of polyacrylonitrile(PAN), an atactic, linear polymer that contains nitrile pendant groups attached to a

$$\begin{array}{c}H_2C\diagdownH_2C\diagdownH_2C\diagdownH_2C\diagdown\\ CHCHCHCH\\ ||||\\ C\!\!\equiv\!\!NC\!\!\equiv\!\!NC\!\!\equiv\!\!NC\!\!\equiv\!\!N\end{array}$$

carbon–hydrogen backbone. This technology is principally based on the work of SHINDO et al. in Japan [35] and of WATT and JOHNSON of the Royal Aircraft Establishment in Great Britain [36]. The essential elements of the production of PAN-based carbon fibers are outlined in Fig. 4. In the process, bundles (tows) that may contain from ca. 1000 to 320 000 individual fibers, each ca. 10 µm in diameter, are passed, while under tension, through a series of ovens where the temperature and atmosphere are closely controlled. The time–temperature history of the fiber establishes the characteristics of the carbon fiber produced: the higher the temperature, the higher the degree of graphitization and the higher the modulus of the resulting carbon fiber.

Worldwide, there are currently over a dozen producers of PAN-based carbon fibers, each of whom offers a number of specific products. Most manufacturers offer a high-strength product (type I carbon fiber) and a high-modulus product (type II carbon fiber), with some manufacturers also offering an ultrahigh-modulus product (type III carbon fiber). Representative properties for these materials are given in Table 1. As can be noted from this table, there is a close correlation between carbon content and physical properties. Strain to failure is low for carbon and all types of carbon fibers, but especially for the higher modulus products. This makes the handling of carbon fibers difficult and is one of the major limitations of the products listed in Table 1. Carbon fibers differ from other high-perfomance reinforcing fibers because they are good conductors of electricity. They also exhibit unusual thermal expansion characteristics, since the coefficient of thermal expansion is negative along the fiber axis while being positive in the transverse directions.

The price of commercial PAN-based carbon fibers ranges from ca. $ 40/kg to $ 1000/kg (1986). These prices are mainly a function of fiber modulus and the number of filaments in a yarn bundle (end count). Type I (high-strength) fibers are less expensive than type II (high-modulus) fibers, in part because processing costs are lower and in part because they are manufactured in higher volumes. Type III (ultrahigh-modulus) fibers are expensive speciality materials. For any fiber type, price is also a

Figure 4. Graphite fiber process outline

```
Precursor
preparation
   │
   ▼
Stabilization
(preoxidation)
200-300°C
air
2-4 hours
   │
   ▼
Carbonization
1200-1500°C        ──► High strength product
nitrogen
45-60 seconds
   │
   ▼
Graphitization         High modulus and
2000-3000°C        ──► ultrahigh-modulus
nitrogen/argon         products
15-20 seconds
   │
   ▼
Surface treatment
various media
and conditions
   │
   │                   Resin impregnation and
   └───────────────►   intermediate product
                       forms
```

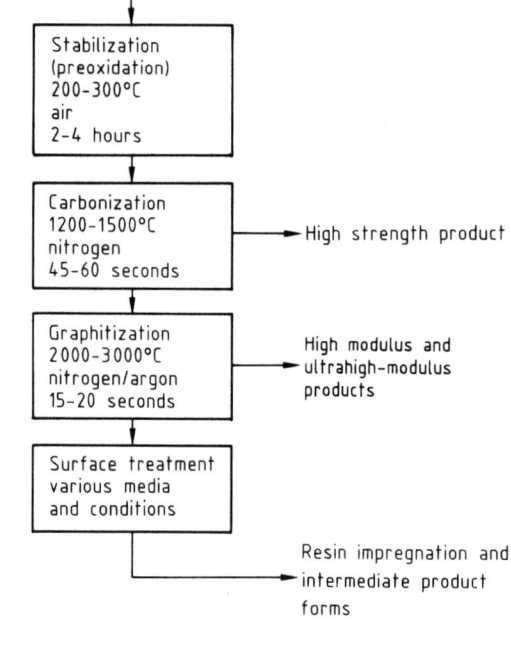

Table 1. Representative properties of standard PAN-based carbon fibers

Fiber type Nominal designation	Type I High strength	Type II High modulus	Type III Ultrahigh modulus
Carbon content, wt%	92–94	>99	>99.9
Specific gravity	1.7–1.8	1.8–1.9	1.9–2.1
Filament diameter, µm	7–8	7–8	8–9
Tensile modulus, GPa	220–250	340–380	520–550
Tensile strength, GPa	2.5–3.5	2.2–2.4	1.8–1.9
Tensile elongation, %	1.2–1.4	0.6–0.7	0.3–0.4
Toughness, MPa	20	7.5	3
Electrical resistivity, µΩ/m	15–18	9–10	6–7
Longitudinal CTE*, $10\ mm^{-1}K^{-1}$	–0.5	–0.7	–0.9 (est.)
Range of available tow counts, 1000 fibers/bundle	1–320	1–12	<1
Price range, 1985 $/kg	40–300	90–1000	>300

* CTE = coefficient of thermal expansion.

strong function of end count. High end count material (40 K or more) is significantly less expensive than low end count material (12 K or less). This price differential is a reflection of the fact that plant production capacity is directly proportional to end count. Thus, production costs increase as the end count decreases. However, pyrolyzing low end count bundles results in a more uniform product. In most critical applications, high end count material is rarely used.

Table 2. Representative properties of high-strain type I PAN-based carbon fibers

Carbon content, wt%	92–94
Specific gravity	1.7–1.8
Filament diameter, μm	5–8
Tensile modulus, GPa	240–270
Tensile strength, GPa	4.0–4.7
Tensile elongation, %	1.7–1.8
Toughness, MPa	34–42
Longitudinal CTE, $10\ m\ m^{-1}K^{-1}$	–0.5
Range of available tow counts 1000 fibers/bundle	6–12
Price range, 1985 $/kg	90–100

The low strain to failure characteristics have been the major limitation of the standard PAN carbon fibers listed in Table 1. High-strain carbon fibers are now available that exhibit tensile elongations of ca. 1.7–1.8%, as shown in Table 2. These fibers are not only unusually strong, but also much easier to handle, which facilitates the fabrication of composite parts.

Pitch-Based Carbon Fibers. Petroleumpitch is that fraction of the petroleum barrel that is uneconomical to further process into light hydrocarbons because of its tendency to carbonize and interfere with processing operations such as catalytic cracking. Because petroleum pitch is a high carbon content, readily available, and low cost material, interest has been growing in its use as a precursor material in an attempt to reduce the manufacturing cost of carbon fibers.

Petroleum pitch must be purified and preconditioned to produce high-performance carbon fibers. An important step is the development of a liquid crystalline phase (mesophase) prior to spinning to form fibers with a high degree of preferred orientation. These preliminary operations add substantially to the fabrication costs, and the current prices of pitch-based carbon fibers are not significantly lower than those of PAN-based carbon fibers of comparable modulus. Pitch-based carbon fibers that are currently available exhibit a much lower strain to failure at equal modulus.

4.1.2. Organic Fibers

The measured tensile strength and modulus of most natural and synthetic organic fibers are typically less than 1 and 10 GPa, respectively. These values are significantly lower than the theoretical values of the maximum tensile strength (> 20 GPa) and modulus (> 100 GPa) calculated on the basis of force constants of bonds in the polymer chain and crystal extensions observed by X-rays on stressed fibers [37]. A number of polymeric fibers are now commercially available, whose strength and modulus more closely approach these theoretical values and which are high enough for these materials to be considered as reinforcements for advanced composites. As compared to other textile fibers, these high-strength, high-modulus materials all exhibit a high degree of molecular orientation and chain alignment. High-performance organic fibers of this class that have been commercialized to date include p-aramids, ultrahigh molecular mass polyolefins, and certain classes of polyesters.

***p*-Aramid Fibers.** Aromatic polyamidesor aramids were introduced initially in the early 1960s to meet the need for fibers with improved heat and flammability resistance. The first example of this class of compounds was poly(*m*-phenyleneisophthalamide), which is sold commercially by Du Pont under the trademark Nomex. In the early 1970s, Du Pont introduced aramid fibers under the trademark Kevlar, based on para-substituted aromatic polyamides, most notably poly(*p*-phenyleneterephthalamide) (PPD-T), which is a condensation product of *p*-phenylenediamine and terephthalic acid.

$$H_2N-\text{C}_6H_4-NH_2 + Cl-\overset{O}{\underset{\|}{C}}-\text{C}_6H_4-\overset{O}{\underset{\|}{C}}-Cl \longrightarrow$$

p-Phenylene- Terephthaloyl
diamine chloride

$$\text{--}(HN-\text{C}_6H_4-\underset{H}{N}-\overset{O}{\underset{\|}{C}}-\text{C}_6H_4-CO\text{--})_n$$

Poly(*p*-phenyleneterephthalamide)

Other aramid fibers have been more recently introduced to the market by Enka (Netherlands) and Teijin (Japan).

Because of its rigid chain structure, PPD-T can exhibit liquid crystalline behavior in solution. The rodlike molecules of these materials can aggregate in nematic, ordered domains. When liquid crystalline solutions of these materials are subjected to shear, these ordered domains tend to be oriented in the direction of flow. When passed through a spinneret, these liquid crystalline solutions retain the high degree of orientation imparted by the spinning operation, leading to as-spun fibers with extraordinary levels of crystallinity and orientation.

As-spun fibers obtained by the spinning of a 20% solution of PPD-T in 100% sulfuric acid exhibit a crystalline orientation angle of ca. 12° (as determined from wide angle X-ray diffraction) and a modulus of ca. 72 GPa. Heat treatment increases the degree of crystalline alignment. Heat-treated fibers have an orientation angle of ca. 9°, and a tensile modulus of ca. 120 GPa.

Du Pont markets several types of fibers under the Kevlar aramid trademark that offer different combinations of properties to satisfy the requirements of a variety of end use requirements. The major products are Kevlar, Kevlar 29, and Kevlar 49 aramid fibers. The major end uses of these materials are outlined in Table 3. Selected physical properties of Kevlar 49 and of Kevlar 29, the two products of interest in terms of plastics reinforcements, are presented in Table 4. The stress–strain graph for Kevlar 29 is nonlinear. At failure, its tensile modulus is about 20% higher than the initial value reported in the table.

Table 3. Commercial applications of aramid fibers

Material	Kevlar	Kevlar 29	Kevlar 49
Tires	×		
Mechanical rubber goods	×		
Ropes and cables		×	(×)
Ballistics protection		×	(×)
Coated fabrics		×	
Tapes and webbing		×	
Friction products and gaskets		×	
Reinforced plastics			×

Supplier: E. I. du Pont de Nemours & Co., Inc.

Table 4. Representative properties of Kevlar *p*-aramid fibers

Fiber type	Kevlar 29	Kevlar 49	Kevlar 49
Nominal designation	Standard	Standard	Improved
Specific gravity	1.44	1.44	1.44
Filament diameter,			
denier	1.5	1.5–2.25	2.25
μm	13	13–16	16
Tensile modulus, GPa	62*	131	124
Tensile strength, GPa			
Unimpregnated yarn**	2.8	2.7	
Impregnated strand***	3.8	3.8	4.1
Tensile elongation, %	4.0	2.8	3.3
Toughness, MPa	87	55	68
Longitudinal CTE, 10 m m^{-1} K^{-1}		−2	
Range of available tow counts			
1000 fibers/strand	0.13–10	0.13–2	2

* Initial modulus,
** As per ASTM test D885,
*** As per ASTM test D2343.

Kevlar 49, because of its higher modulus, is the member of the Kevlar aramid product family normally used in advanced composites. Although Kevlar 49 behaves elastically in tension, it exhibits nonlinear characteristics under compressive stress. The application of compressive stress results in formation of structural defects called kink bands and eventual ductile failure [38]. The onset of this behavior occurs at compressive yield strains of ca. 0.3–0.5%. This phenomenon has been attributed to a molecular rotation of the amide carbon–nitrogen bond from the normal extended trans configuration to a kinked cis configuration. This results in about a 45° bend in the chain, which propagates to form a kink band across the fiber.

As a result of this unusual behavior under compression, the use of *p*-aramid fibers in applications subject to high-strain compressive or flexural loads is limited. At the same time, however, because of their ductile response under compression, *p*-aramid fibers are inherently tough materials that exhibit good damage tolerance and energy absorption characteristics.

Polyolefin Fibers. The high degree of molecular orientation that results in a well-defined crystalline structure, and the corresponding high-modulus, high-strength properties, can also be obtained by the imposition of a very high degree of plastic extensional deformation on a polymer in the solid state. When this method is applied to ultrahigh molecular mass polyethylene, which exhibits a very high degree of crystallinity, very strong and stiff fibers can be prepared [39]. Allied has developed a high-strength, high-modulus polyethylene fiber, trademarked as Spectra 900, for tensile applications such as ropes and cordage. It may also have potential use in polymer matrix composites. Room-temperature properties of this fiber are presented in Table 5. While this fiber exhibits superior mechanical properties at ambient temperature, its utility decreases rapidly with increasing temperature above 50 °C due to the relatively low melting point of the polymer. A noticeable decrease in tensile properties occurs above 50 °C. At 120 °C, the tensile strength is approximately half the value at ambient temperature, with elongation increasing proportionally.

4.1.3. Inorganic Fibers

Even though the densities of inorganic fibers are significantly higher than those of carbon and polymeric fibers, these materials are of interest in the preparation of advanced composite materials because of their stability at high temperature and their compatibility with metal and ceramic matrices, which are also inorganic materials. In addition, any listing of reinforcing fibers for advanced composite applications must include glass fibers because of their low cost and economic importance as the primary fibrous reinforcing material currently used by the plastics industry.

Glass Fibers. Glass fibers, also called fiberglass, have become a major product since their commercial introduction by Owens-Corning Fiberglass Corp. in 1939. Worldwide, the reinforced plastics and composites industry is currently estimated to use over 500 000 metric tons annually of glass fibers.

Glass fibers are prepared by melting a suitable mixture of raw materials (silica sand, limestone, boric acid, alumina, etc.), extruding the molten glass through a platinum spinneret and cooling the streams to form a multitude of individual glass fibers, which are then lubricated and mechanically gathered into strands. Glass fibers produced in this manner are amorphous and isotropic.

The principal grades of glass fibers used in composite applications are E-glass (E: electrical) and S-glass (S: strength). A third type, C-glass (C: corrosion), is used where resistance to acid corrosion is particularly important. The nominal compositions of E-glass and S-glass are presented in Table 6. Representative properties of these grade of glass fibers are presented in Table 7. Glass fibers generally have a high strength : weight ratio, but their moduli, which are in the range of aluminum alloys, are significantly lower than those of most of the other high-performance fibers considered in this chapter. Glass fibers are extremely sensitive to abrasion damage. They are coated with

Table 5. Representative properties of Spectra 900 polyolefin fibers

Specific gravity	0.97
Filament diameter, µm	38
Tensile modulus, GPa	117
Tensile strength, GPa	2.6
Tensile elongation, %	2.2
Toughness, MPa	29
Range of available tow counts, 1000 filaments/strand	0.06–0.12%

Supplier: Allied Corp.

Table 6. Chemical composition (in wt%) of inorganic oxide fibers

Material	Trade-name	Suppliers	Components						
			SiO_2	Al_2O_3	FeO_3	CaO	MgO	$Na_2O + K_2O$	B_2O_3
E-Glass		various	53.2	14.8		21.1	0.3	1.3	9.1
S-Glass		Owens-Corning	64.3	24.8	0.2	<0.01	10.3	0,27	<0.01
Quartz			>99.95						
Ceramic	Nextel 312	3M	24	62					14
Ceramic	Nextel 440	3M	28	70					2
Alumina		Sumitomo/Avco	15	85					
Alumina	Saffil	ICI	3	97					
Alumina	Fiber FP	Du Pont		>99					

protective sizings to minimize this problem. Sizing materials also lubricate the fiber and promote its resin adhesion.

E-glass is the most common grade of fibrous glass used by the plastics industry. While originally introduced for electrical applications because of its dielectric and loss tangent properties, E-glass is used in many nonelectrical applications because of its combination of mechanical properties and low cost. E-glass is the only reinforcing fibrous material available at a price of $ 1/kg.

S-glass is a high-strength glass initially developed for military applications. Its modulus is about 20% greater than that of E-glass, and it is about one-third stronger. The creep rupture resistance of this material is significantly better than that of E-glass, and it is about one-third stronger. It is significantly less susceptible to acid corrosion than E-glass. The failure energy of S-glass is high, making it a very tough fiber.

S-2 glass is a fiber of the same chemical composition as S-glass that was developed for commercial applications. The primary difference in the two products is the sizing on the fibers. The cost of S-2 glass fibers is, however, significantly lower than that of S-glass fibers because they are not subject to the stringent and expensive quality control procedures imposed by military procurement requirements.

High-Silica and High-Quartz Fibers. A number of high-silica (> 95% silica) and high-quartz (> 99.95% silica) fibers, which melt at a higher temperature than glass fibers, have been developed for extreme-temperature textile applications. At lower

Table 7. Representative properties of inorganic oxide fibers

Material designation Trade name Supplier	Unit	E-Glass Various	S-Glass S2-Glass Owens-Corning	Quartz Various	Ceramic Nextel 312 3M	Ceramic Nextel 440 3M	Alumina Sumitomo/Avco	Alumina Saffil ICI	Alumina Fiber FP Du Pont
Specific gravity		2.60	2.49	2.2	2.7	3.1	3.25	3.3	3.95
Filament diameter	μm	9	9	10	8–12	8–12	17	3	20
Range of available tow counts	1000 filaments/strand	0.2–2	0.2	0.3	0.39	0.39	1		0.21
Price range	1985 $/kg	1	6–30	100–400	180	440	760		440
Tensile modulus	GPa	72	87	69	152	220	200	300	377
Tensile strength	GPa	3.45	4.6	0.9	1.5	1.7	1.5	2	1.4
Tensile elongation	%	4.8	5.4	1.3	1	0.8	0.8	0.67	0.36
Toughness	MPa	83	124	5.9	7.7	6.6	5.6	13.3	2.5
Dielectric constant at 10^{10} Hz		6.1	5.2	4.5	4.7				9.5
Longitudinal CTE	10^{-6} m m^{-1} K^{-1}	5	5.6						6.8
Melt temperature	°C	1260		1650	1800			>2000	2045
Max. use temperature	°C	600	760	900	1200			1600	
90% Tensile strength retention temperature	°C	220	275		1100	1400	1250		1100

temperatures, the mechanical properties of this family of fibers are not substantially different from those of glass fibers. They exhibit superior electric properties, in terms of exhibiting an even lower dielectric constant and a lower power dissipation factor than E-glass. The high-temperature properties of silica fibers are of little value in resin matrix composites, where temperature limitations are imposed by the thermal resistance of the matrix material. Because they are significantly more expensive than glass fibers, high-silica and high-quartz fibers are rarely used as composite reinforcing fibers, except in specialized electronic applications, where an extremely low dissipation factor is required.

Alumina-Based Fibers. A numberof fibers with a high alumina content (> 60 wt% alumina) have been developed both for high-temperature textile applications, where they compete with high-silica and high-quartz fibers, and as reinforcements for metal matrix and ceramic matrix composites. The nominal composition and representative properties of five such materials that are currently commercially available are listed in Tables 6and 7. These fibers are generally prepared by extruding an aqueous or organic precursor gel through spinnerets, drying the resulting fibers to remove most of the liquid phase, and then subjecting the resulting filaments to high-temperature (> 1200 °C) heat treatment to form a continuous refractory yarn. ICI's Saffil fibershave a δ-alumina crystal phase, whereas Du Pont's FP fibers are polycrystalline α-alumina. Saffil alumina is not currently available as continuous filaments, but only as fibers 1–5 cm long (in mat form) or less (in bulk and milled form).

The Sumitomo and 3M Nextel fibers are alumina-rich mixtures of alumina and silica and exhibit properties intermediate to those of pure silica and pure alumina fibers.

The high-alumina fibers all exhibit superior high-temperature properties, as exemplified by the fact that for all these fibers, increasing temperature from ambient to 1100 °C results in a reduction in tensile strength of less than 10%. These fibers are claimed to be able to sustain exposure temperatures that range from 1200 °C to 1600 °C. At 1100 °C, these fibers are wetted by and stable in molten light metal alloys, such as those of aluminum and magnesium. Thus, these fibers are of interest as reinforcing materials for metal matrix composites.

Silicon Carbide Fibers. Sincethe 1960s, there has been significant interest in development of continuous carbide, nitride, and boride fibers and filaments as reinforcing fibers for hightemperature structural materials. Of these materials, silicon carbide fibers offer the most promise and are currently the most fully developed. Silicon carbide filaments are of interest because of their thermal stability, oxidation resistance, superior mechanical properties at elevated temperatures, and compatibility with most molten metals, as well as with many ceramic materials. Silicon carbide is currently available in two different types of continous filaments, as well as in whisker form. These fibers, as well as isotropic SiC powder, are all being actively examined as reinforcements for a variety of metal matrix and ceramic matrix applications.

Because of the refractory nature of silicon carbide, silicon carbide fibers cannot be produced by consolidation from the melt or by sintering. The silicon carbide fibers that

Table 8. Representative properties of SiC fibers

Manufacturing process		CVD	Pyrolysis
Trade name			Fiber FP
Supplier		Avco	Nippon Carbon Co./Dow-Corning
Specific gravity		3	2.55
Filament diameter	μm	140	10–20
Range of available tow counts		1000 fibers/bundle	0.5
Price range	1985 $/kg	not applicable	500
Tensile modulus	GPa	1100	190
Tensile strength	GPa	430	2
Tensile elongation	%	2.4	1.1
Toughness	MPa	0.56	11
Electrical resistivity	μΩ/m	13	0.1
Longitudinal CTE	10^{-6} m m^{-1} K^{-1}		3.1
Melt temperature	°C	2700 *	2700 *
Max. use temperature	°C	1150	
90% Tensile strength retention temperature	°C	800	300

* Sublimes

are currently available are produced by very different process routes. Avco Corp. produces large-diameter silicon carbide filaments by a vapor chemical deposition process, Nippon Carbon Corp. produces finer diameter continuous filaments by the pyrolysis of polycarbosilane polymers, and Arco Chemical Co. produces silicon carbide whiskers by the pyrolysis of rice hulls.

Avco's CVD process is very similar to the one it developed for the production of boron filaments. Boron filaments, which were major reinforcing fibers about a decade ago, have been displaced by carbon and aramid fibers in resin matrix composites, principally because these fibers are expensive ($ 700/kg) and difficult to handle. In the Avco process, silicon carbide filaments are produced by reducing a blend of methyltrichlorosilane and methylhydrogendichlorosilane with hydrogen in the presence of an electrically heated, continuously moving carbon substrate filament. Representative properties of these filaments are given in Table 8.

The original impetus for developing SiC filaments was to obtain a fiber better suited than boron for incorporation in metal matrix composites. Boron rapidly loses its strength at temperatures above 600 °C and is unsuitable for high-temperature applications. Furthermore boron reacts with molten metals, such as aluminum, which makes the fabrication of boron-reinforced metal matrix composites by liquid infiltration or standard casting techniques not feasible. Silicon fibers made by this process are potentially less expensive than boron fibers because the rate of deposition is faster and lower cost raw materials are required. The price of $ 1100/kg (1986) reflects the developmental status of these fibers. Avco projects that in production quantities, the price of these fibers would be less than $ 300/kg.

Nippon Carbon Co.'s silicon carbide filaments are produced by the pyrolysis of a carbosilane polymer precursor, a process originally developed by YAJIMA of Tohoku University [40]. YAJIMA discovered that polydimethylsilane polymers, obtained by condensing dimethyldichlorosilane in an autoclave at ca. 450 °C, could be melt spun into fibers, which when subsequently pyrolyzed at ca. 1300 °C in an inert atmosphere were transformed into high-strength, high-modulus fibers that were predominantly β-silicon carbide (> 80 wt%) and carbon. In spite of their carbon content, these fibers were observed to be oxidation resistant and wetted by molten metals. It was subsequently observed that the physical properties of the silicon carbide fibers were improved by slight modifications in the composition of the polycarbosilane precursor. In particular, the addition of diphenyldichlorosilane and of boron-modified silanes to the condensing reaction mixture was observed to have a beneficial effect on fiber properties. The properties of these silicon carbide fibers, which are marketed under the trademark Nicalon, are also given in Table 8. According to the Dow Corning Co., the United States distributor of these fibers the price of these fibers in production quantities could drop from the present level of $ 500/kg to ca. $ 220/kg.

4.1.4. Reinforcement Geometry

The fiberdiameter of the various fibers discussed in the previous sections of this chapter is in the 5 – 20 µm range. Exceptions to this are the silicon carbide and boron filaments produced by chemical vapor phase deposition, which have a diameter of ca. 140 µm (Avco). As a result, these filaments are neither combined in strands nor treatable by standard textile fiber processes. Fibers of small diameter range are flexible enough to be amenable to processing by many standard textile operations in spite of their high modulus. At the same time, fibers of this diameter range are too large to be aspirated into the human respiratory tract, and thus, do not present the environmental health hazards associated with fibrous materials, such as asbestos. Most of these fibers also have a circular cross-section. Notable exceptions are rayon-based carbon fibers, which are crenulated, and certain PAN-based ultrahighmodulus carbon fibers (e.g., Celion 70), which are bilobar.

Fibers in the above diameter range are too fragile to be handled individually and are only available as multifilament strands. The reinforcing materials used in most advanced composite applications typically have a tow countof 200 – 12 000 filaments per strand. It should be noted that carbon fibers are also available in much higher tow counts, as noted in Table 1. Carbon fiber tows with a high tow count are not commonly used in the manufacture of advanced composite parts, even though they are less expensive than low end count carbon tows. In general, low end count strands are preferred as reinforcements for critical applications of advanced composite materials. The higher the tow count, the greater the variation in filament properties, and the more difficult it becomes to assure uniform contact between the matrix and the individual filaments.

Strands of reinforcing fibers can also be woven into flat fabrics, or three-dimensional woven or knitted fabrics, or even braided shapes, by slight modifications of standard textile machines. So-called structural fabrics, which contain one or more reinforcing fibers, are available in a wide variety of weaves and fabric weights.

4.2. Matrix Materials

The matrix of an advanced composite is the continuous phase that binds the fibers together so that they can act in concert. The major characteristics required of a matrix material are that it be

1) chemically compatible with the reinforcing fibers
2) compatible with the manufacturing methods used to fabricate the desired advanced composite component
3) environmentally stable under conditions of use of the advanced composite component.

The matrix material is the primary influence on mechanical properties of interlaminar shear strength and compression and flexural strength, especially at high temperature. The matrix also dictates the processibility of the composite, and its environmental resistance. Since the processibility and the environmental properties of advanced composite components depend to a significant extent on the characteristics of the matrix material, no single ideal matrix material exists that satisfies the multitude of requirements imposed by different applications. Thus, advanced composites have been prepared with a diversity of matrix materials, which include organic polymers, carbon, metals, and ceramics.

Polymeric materials historically have been the predominant matrix materials used in advanced composites. Currently, high-performance fiber-reinforced epoxy resins are nearly synonymous with the term "advanced composite materials." This situation is changing rapidly, and many high-performance fibers are now being combined with many other thermosetting and thermoplastic resins.

Thermosetting Resins. Thermosets are those resins which in the presence of a catalyst, heat, radiation, and/or pressure undergo an irreversible chemical reaction, or cure. Prior to cure, thermosets may be liquid or made to flow under pressure to any desired form. Once cured, they cannot be returned to the uncured state, and can no longer flow. This cure reaction can be either an addition reaction, such as the free radical polymerization of unsaturated groups, or a condensation reaction, which occurs as a result of the stepwise combination of different functional groups. If the volatile byproducts can be removed, superior properties can be attained with condensation polymers. This generally requires long, high-temperature cure cycles.

Epoxy Resins Epoxies are, by far, the matrix materials most commonly used in advanced composite materials. The main reason is the good balance of properties obtained with relative ease of handling and processing.

Epoxies are generally obtained by reacting epichlorohydrins (ECH) with molecules containing active hydrogen, such as phenols, amines or acids, followed by dehydrochlorination:

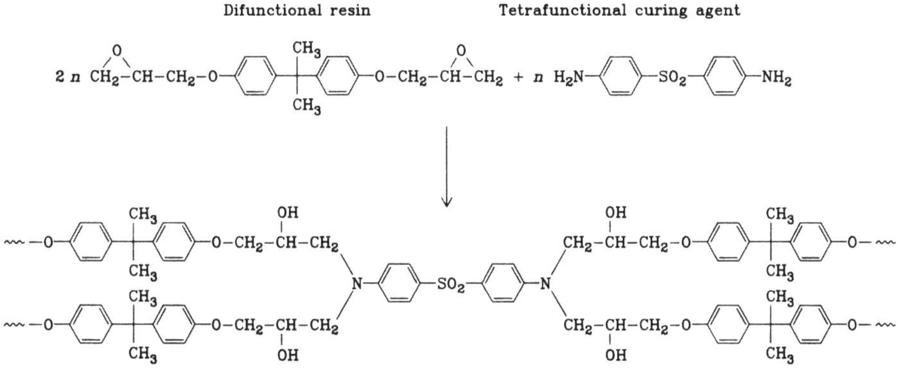

Difunctional resin   Tetrafunctional curing agent

$$2n\ CH_2\text{-}CH\text{-}CH_2\text{-}O\text{-}\langle\text{Ph}\rangle\text{-}C(CH_3)_2\text{-}\langle\text{Ph}\rangle\text{-}O\text{-}CH_2\text{-}CH\text{-}CH_2 + n\ H_2N\text{-}\langle\text{Ph}\rangle\text{-}SO_2\text{-}\langle\text{Ph}\rangle\text{-}NH_2$$

↓

(crosslinked network structure with bisphenol-A and diaminodiphenylsulfone units, bearing OH groups on the CH-carbons)

Most of the commercial epoxies are based on bisphenol A; higher functionality resins for higher temperature are based on aminophenols, diaminodiphenyl methane, triphenol methane, or novolaks.

Epoxies can be cured either by ionic homopolymerization of the epoxy groups or by reaction with suitable polyfunctional curing agents, such as primary or secondary amines, anhydrides, phenols, acids, or dicyandiamide. Some typical curing agents are:

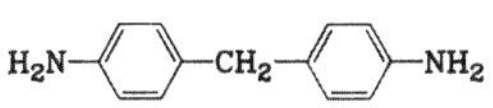

$H_2N\text{-}\langle\text{Ph}\rangle\text{-}CH_2\text{-}\langle\text{Ph}\rangle\text{-}NH_2$
4,4'-Diaminodiphenylmethane

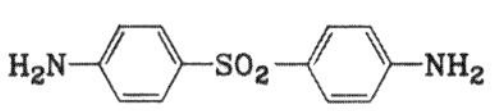

$H_2N\text{-}\langle\text{Ph}\rangle\text{-}SO_2\text{-}\langle\text{Ph}\rangle\text{-}NH_2$
4,4'-Diaminodiphenylsulfone

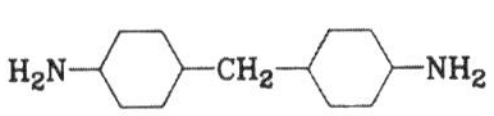

$H_2N\text{-}\langle\text{Cy}\rangle\text{-}CH_2\text{-}\langle\text{Cy}\rangle\text{-}NH_2$
4,4'-Diaminodicyclohexylmethane

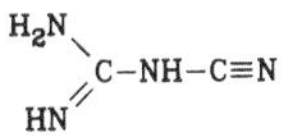

$H_2N\text{-}C(=NH)\text{-}NH\text{-}C\equiv N$
Dicyandiamide

Imidazole, R = -H, -CH$_3$, -C$_2$H$_5$, -C$_6$H$_5$

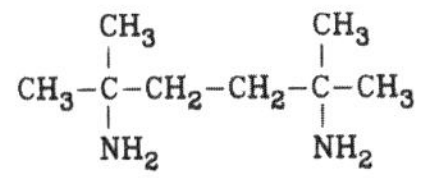

$CH_3\text{-}C(CH_3)(NH_2)\text{-}CH_2\text{-}CH_2\text{-}C(CH_3)(NH_2)\text{-}CH_3$
2,5-Dimethyl hexan-2,5-diamine

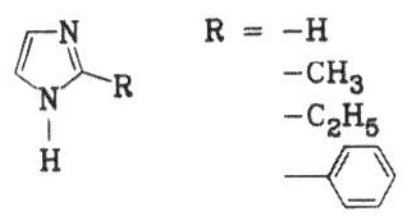

Cure and processing behavior are governed by structure and choice of components. Aliphatic amines allow ambient temperature curing, whereas slow-reacting, aromatic amines, such as diaminodiphenyl-sulfone (DDS), require a high temperature to cure.

The range of properties attainable for neat (unfilled), cured epoxy resins of different structure are summarized in Table 9. In general, increasing the concentration of aromatic rings and cross-links tends to improve resin properties such as thermal stability, chemical resistance, and rigidity, but lowers the strain to failure (increased brittleness).

Most of the epoxy resins used by the aerospace industry in advanced composite structures, which must withstand exposure to high temperature (350 °F = 177 °C), are highly cross-linked aromatic resins. The properties of a commercial epoxy resin, which has been accepted for high-temperature composite applications, are given in Table 10. The major performance limitations of materials of this class are their low toughness and the significant loss in mechanical properties at high temperatures, particularly in the presence of moisture, as shown in Figure 5.

Bismaleimides. Bismaleimides are synthesized by the reaction of maleic anhydride with primary aromatic diamines and subsequent cyclodehydration of the intermediate thus formed. The double bonds can be reacted further by copolymerization with suitable comonomers or by a Michael-type addition of amines, for example, [46].

Bismaleimides offer high glass transition temperatures and good aging stability and flammability characteristics. These resins are still fairly brittle and exhibit limited high-temperature (>350 °C) stability. Significant interest in these resins exists because they may be modified or formulated to achieve better environmental resistance (e.g., better properties under hot/wet conditions) than accepted epoxy resins while retaining most of the processibility characteristics of epoxies. Representative properties of a commercial bismaleimide resin are given in Table 10.

Polyimides. Polyimides are a class of resins that can withstand continuous exposure in air at temperatures above 300 °C. These materials have become of practical interest because they are more processable than most other resin systems that are stable at high

Table 9. Property values of unfilled cast epoxy resin systems [41]

Property	Typical value at 23 °C (73 °F)
Specific gravity	1.2 – 1.3
Rockwell hardness	M100 – M110
Impact strength (notched bar, Izod test) J/m of notch	0.1 – 1.0
Coefficient of thermal conductivity, $Wm^{-1} K^{-1}$	0.17 – 0.21
Coefficient of thermal linear expansion, $m\ m^{-1} K^{-1}$	$(5-8) \times 10^{-5}$
Specific heat, $Jkg^{-1} K^{-1}$	$(1.25-1.80) \times 10^3$
Volume resistivity, $\Omega \cdot m$	$(1-50) \times 10^{17}$
Dielectric constant (at 60 Hz)	2.5 – 4.5
Tensile strength, MPa	55 – 130
Tensile modulus of elasticity, MPa	2800 – 4200
Poisson's ratio	0.20 – 0.33
Flexural strength, MPa	125

Table 10. Representative properties of selected thermosetting resin matrix materials [42] – [45]

Polymer	Unit	Epoxy Narmco 5208	BMI * Ciba-Geigy XU-922	Polyimid Nat. Starch thermid 600	Rigid P'ester generic
Specific gravity		1.265	1.35	1.37	1.10 – 1.46
Glass transition temperature	°C	238	310	350	70 – 140
Processing temperature	°C	177	250	370	20 – 150
Mechanical properties at ambiant					
Tensile strength	MPa	50	94	83	40 – 90
Tensile modulus	GPa	39	39	39	20 – 44
Tensile elongation	%	1.4	3.0	2.0	< 2
Fracture toughness	kJ/m²	0.08	0.21		
Solvent resistance		+	+	+	+
Price range	$/lb	10			1

temperature. Polyimide resins have evolved from early amic-acid type condensation products in high-boiling solvents (such as N-methylpyrrolidone). These were difficult to process and tended to form porous laminates because of the subsequent condensation of volatile reaction byproducts which could not be totally removed from the resin. Subsequent developments have led to resins in which final cure is obtained by addition-type reactions, which do not emit volatile byproducts [47]. Examples include PMR-type resins, initially developed by NASA Lewis Research Center. These entail the in situ condensation of an imidized prepolymer on the reinforcing fibers at a temperature of 150 – 220 °C. The composite is then cured by cross-linking via a retro-Diels – Alder reaction at 300 – 350 °C. More recently, a family of condensation polyimides with a flexible diamine has been developed by Du Pont. These materials have the potential to be prepolymerized and handled like thermoplastics; thus, they should provide a good processing balance [48].

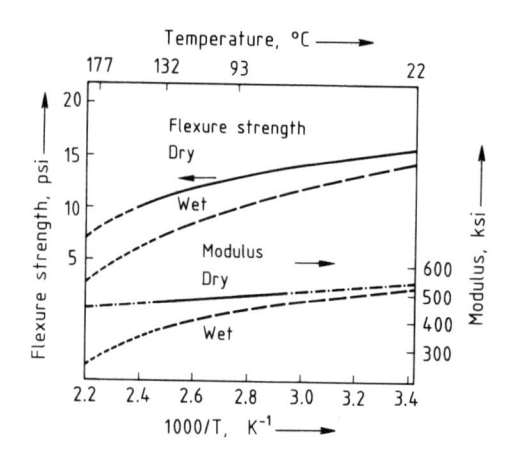

Figure 5. Environmental effects on mechanical properties of cured high-temperature epoxy resin (NEAT) [42]
System = Narmco 5208, wet: after 50 h water boil (3.1% moisture pickup)
Note: 1 psi = 6.895 × 10³ Pa; 1 ksi = 1000 psi

Ethynyl-Terminated Resins. Ethynyl-terminated resins are another promising class of high-temperature thermosetting resins. A very stable thermosetting resin is obtained by the high-temperature polymerization and cyclotrimerization of terminal ethynyl groups on resin oligomers. This intramolecular cyclization (IMC) [44]results in the formation of benzene rings as the cross-link sites. Ethynyl-terminated resins have been prepared with a wide range of backbones, including polyimides and epoxies. These materials produce thermosets with high toughness and thermal and oxidative stability, low water absorption, and versatile handling and processing characteristics. Representative properties of a commercial resin of this type (National Starch and Chemical Corps., Thermid 600) are given in Table 10.

Other Thermosets. Some of the oldest and most widely used thermosets are generally not considered as suitable matrix materials for advanced composite materials. Unsaturated polyesters are obtained by condensation of unsaturated dicarboxylic acids with polyethylene or polypropylene diols. These resins are then cross-linked by radical copolymerization of the double bonds along the chain with suitable comonomers such as styrene, divinyl benzene, etc. Because of their low cost, unsaturated polyesters are extensively used in the manufacture of glass fiber reinforced plastic parts. However, this class of polymers finds little application in advanced composites because of its relatively poor thermal resistance (i.e., low glass transition temperature) and low mechanical properties, as indicated by the data presented in Table 10. Vinyl ester resins are obtained by reacting epoxies with methacrylic acid to form a bismethacrylate. Although the double bonds of this oligomer can be cured without comonomers, it is generally diluted with comonomers similar to those used in conjunction with unsaturated polyester resins. Properties and the cost of the vinyl esters generally lie between those of unsaturated polyesters and epoxy resins, and can be varied to meet specific requirements in this range. Phenolic resins are made by condensation of phenols with aldehydes, such as formaldehyde, to form mixtures of hydroxymethylphenols. Resol or novolak type resins are obtained, depending on the stoichiometric ratio of monomers. These resins can be cross-linked by heat, by acidic catalysts, or by the addition of cross-linking agents, such as hexamethylenetetramine. During curing, the material condenses further to form a methylene-bridged three-dimensional network. Water, which is given off as a byproduct, tends to form voids, which result in relatively poor mechanical properties. However, phenolic resins are important in the manufacture of carbon fiber reinforced carbon matrix composites, which are obtained by the pyrolysis of carbon fiber reinforced phenolic resins.

Thermoplastic Resins. Thermoplastics are materials which are normally solids at room temperature, which melt or soften when heated to a sufficiently high temperature, and which become solid again when cooled. This cycle can be repeated. As a result, it is usually easier and less expensive to fabricate a complex part out of a thermoplastic resin than with a thermosetting resin. Thermoplastic resins lend themselves to rapid processing and to the formation of complex parts by processes such as injection molding, extrusion, and thermoforming. The mechanical properties at ambient temperature of many commercial thermoplastics are comparable to those of thermosets. Thermoplastics are generally more ductile and tougher than thermosets. Indeed, for many common applications, thermoplastics are the material of choice, as evidenced by the high-volume use of polyolefins, vinyls, polyamides, polyacrylics, polyesters, etc.

The ability of thermoplastics to be deformed by heat also has deleterious effects. The creep resistance of most thermoplastics at an elevated temperature is lower than that of thermosets. This has been a serious impediment to their wider use in structural applications. Furthermore, as a class, thermoplastics also tend to be more susceptible than thermosets to solvent attack. Representative properties of selected standard thermoplastics are presented in Table 11.

Short-fiber-reinforced thermoplastics have been in use for a number of years, but there have been relatively few applications of continuous fiber reinforcement. As a consequence, there has been relatively little use made of thermoplastics in the formulation of advanced composites. This situation is now changing for two reasons. First, thermoplastic materials are now available which exhibit good high-temperature properties and are reasonably resistant to solvent attack. Second, there has been a mutual adaptation of these materials and advanced composite fabrication methods. This combination has resulted in advanced composite products whose performance and cost are competitive with those made with thermoset matrices.

High-temperature thermoplastics that are being actively considered for use in advanced composites include aromatic polyesters, poly-(phenylene ether), polyphenylene sulfide, polysulfone, aromatic polyetherether ketone (PEEK), and polyimides, such as ether–, ester–, or amide–imides. These materials all contain an aromatic backbone, but differ in the way the aryl groups are linked, as indicated in Table 12. Representative properties of these materials are given in Table 11. These materials can be either amorphous or crystalline. In general, crystalline polymers are more resistant to creep and solvents than the amorphous polymers, but are more difficult to process because they are more subject to shrinkage and warpage.

Table 11. Representative properties of selected thermoplastic resin matrix material [49] – [52]

Polymer Supplier Designation	Unit	Polyimide Du Pont K-I	Polyimide Du Pont K-II	Polyamide Du Pont	Polyamide- imide Amoco Torlon C	PPS * Phillips Ryton	PEEK ** ICI AFC-2	Nylon 66 Du Pont	PET *** Du Pont
Specific gravity		1.37	1.31	1.14	1.38	1.36	1.30	1.14	1.38
Glass transition temperature	°C	195	255	145	135	88	145	50	75
Melting temperature	°C	n.a	n.a		n.a	290	334	260	255
Processing temperature	°C	371	371	315	350	325	390	290	290
Mechanical properties at ambiant									
Tensile strength	MPa	103	110	103	138	76	83	83	69
Tensile modulus	GPa	2.38	2.86	3.17	3.30	3.31	3.65	3.10	2.83
Tensile elongation	%	7	11	27	25	2–20	46	60	50
Fracture toughness	kJ/m²	6.3	14		3.3				
Solvent resistance		+	+	–	+	+	+	+	–
Price range	$/lb	20	20	2	20	3	28	2	2

* PPS: Polyphenylene sulfide;
** PEEK: Polyetheretker ketone;
*** PET: Polyethelene terephthalate

Table 12. High-temperature thermoplastics [46]

Polymer	Structure	Glass temp., °C	Melting temp., °C
Poly(phenylene ether)		210	
Polyarylate		192	
Polyphenylene sulfide		88	290
Polysulfone		187	
Polyether sulfone		225	
Polyetherimide		200	
Polyetherketone		145	334

5. Processing and Fabrication Technology

The manufacture of advanced composite components entails the following basic operations:

1) Surrounding the fibers with the matrix material while the latter is in a nonrigid state
2) Arranging the reinforcing fibers in selected patterns and orientations
3) Solidifying the matrix material under constrained conditions that prevent fibers from moving during the solidification of the matrix
4) Integrating the advanced composite material into the finished component or system.

Most advanced composite components are made by laying up precoated fibers. However, the order of the first three steps can be changed, depending on the type of resin system and process technology selected.

Advanced composite components can be manufactured by a variety of fabrication processes. The particular process choice made by a manufacturer depends on both the composition and the structure of the composite materials used to fabricate the part, as well as on the complexity of the design and part. Each step in the forming process is affected by a combination of fiber characteristics, the resin type, and the position in the fabrication process. The orientation process (step 2), if carried out with bare fiber, is dominated by the fiber handling properties such as yarn quality and toughness. If this step is carried out with matrix resin impregnated yarn, resin viscosity can significantly alter the process rate. In the choice of consolidation technology (step 3), resin characteristics such as cure chemistry, resin viscosity, heat transfer, and reaction kinetics will control the process. In the final step of joining the composite section to other components (step 4), the process, while affected by resin type, is primarily application specific.

It is self-evident that different processes are required to fabricate a polymeric matrix composite part and one made with a metallic matrix. In fact, even small differences in the properties of the matrix material significantly influence the choice of processing methods used to fabricate composite components. For this reason, the rest of this chapter is organized by the type of matrix material.

Processing technology choices are also influenced by non-material-product characteristics, such as part geometry, the number of parts required and their rate of production, reliability requirements, the usual economic considerations of the relative costs of capital, labor, and raw materials and the desire to minimize the total costs of producing a part to meet a given set of specifications.

5.1. Thermoset Resin Matrix Composites

5.1.1. Classic Low-Volume Production Methods

Traditionalmethods of fabrication of advanced composite structures in the aerospace industry have entailed cutting and manual lay-up of preimpregnated tapes of collimated fibers in a prearranged design, which are then cured by the application of heat and pressure to form the desired composite structure.

Combining the Matrix and the Reinforcement. Thermoset "prepreg" (preimpregnated)tape is the basic starting material now used in fabricating most advanced composite structures. These are most commonly unidirectional tapes that contain parallel strands of fibers that are bonded to each other by a thin film of partially cured resin. As indicated in Figure 6nonwoven unidirectional prepreg tape is made by wetting parallel strands of filaments in a resin tank, passing the impregnated

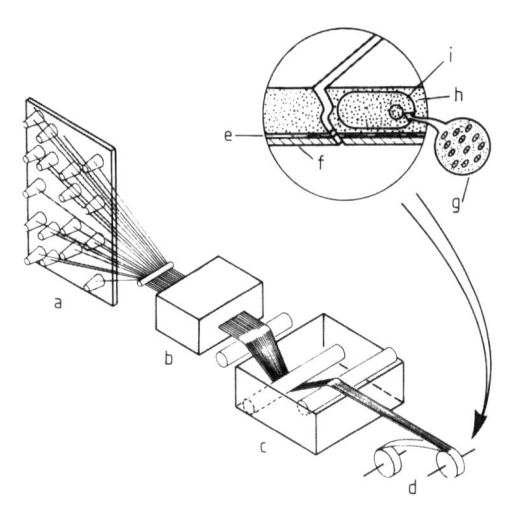

Figure 6. Typical steps in the fabrication of composite tapes
a) Creel; b) Surface treatment; c) Resin; d) Backing paper; e) Carrier cloth; f) Removable backing; g) Glass or carbon filaments; h) Resin; i) Glass or carbon yarns

filaments through a set of doctor blades to control the resin content, collimating the filaments on a removable paper substrate, and then driving off the volatiles that are present by passing the tape through controlled-temperature ovens to attain a fiber concentration of 65 vol%. The resulting B stage product is a dry, relatively tack-free material that is relatively easy to handle. There are two categories of prepreg tapes: (1) continuous tapes, generally less than 3 in. (7.5 cm) in width, and (2) "broad goods," which are sheets whose dimensions are measured in meters.

If the fiber can be woven, an alternate form of prepreg is obtained by drawing a woven fabric through the impregnating resin solution. The fiber content of woven prepregs is somewhat less than that of unidirectional tape, from ca. 55 vol% to 65 vol%, depending on the fabric weave.

The various reinforcing fibers and thermoset resins discussed above have been combined by a number of suppliers to provide composite molders with a wide range of commercially available prepreg materials. While the exact compositions of prepreg materials are proprietary, prepreg suppliers certify their physical and processing properties, and their compliance to military and industry specifications and quality control standards. Specifications include the following:

1) Composition: fiber type and matrix type
2) Content: volatiles content and volume fraction of resin matrix or reinforcement in cured composite
3) Processing characteristics: shelf life as a function of storage temperature, tack (a measure of adhesion of the prepreg), flow (resin flow during curing), and gel time as a function of temperature.

Prepregs are attractive to molders of advanced composite components because they are relatively easy to handle and because their use generally results in cured laminate

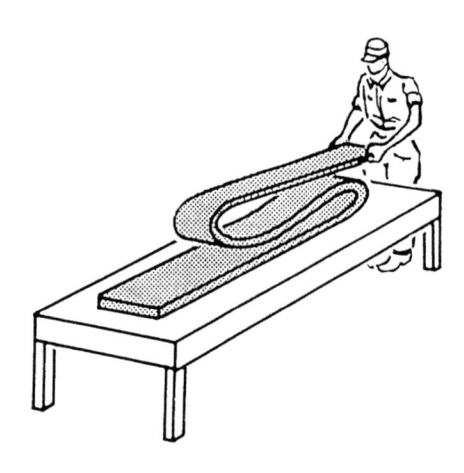

Figure 7. Elements of lay-up forming

structures that exhibit reasonably reproducible properties. While these are significant advantages, there are also certain inherent disadvantages associated with thermoset prepregs. These include a very short shelf life (less than 24 h) at ambient conditions, necessitating refrigerated shipment and storage. A second limitation is that each prepreg material on the market is unique and generally not compatible with similar prepregs (i.e., prepregs that result in a composite with similar properties) offered by other suppliers. In general, the unit price of a prepreg is at least equal to the unit price of the contained fiber.

Fiber Orientation. The advantage of using prepreg tapes is the ability to cut them into specific patterns and then arrange the pieces individually to define the internal structure and contours of the composite part. This lay-up method, which is illustrated in Figure 7, is best suited to flat or slightly curved surfaces. More convoluted parts can be layed up by stacking woven prepregs over a suitably shaped preform. This forming method offers the advantages of versatility in selecting fiber orientation and in allowing the incorporation of different types of fiber (e.g., carbon and aramid, or aramid and glass) to achieve an optimized structure. The disadvantages of manual lay-up as a fabrication method are that it is slow, extremely labor intensive, subject to human error, and consequently an expensive operation.

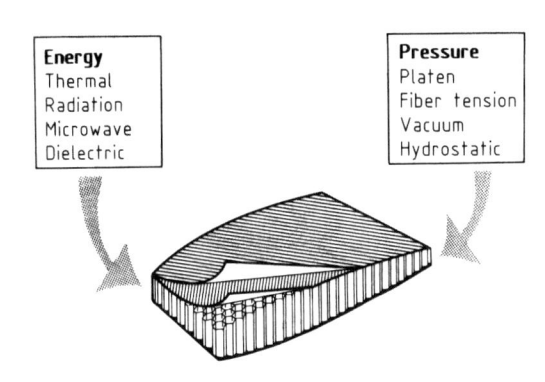

Figure 8. General methods of curing

Energy
Thermal
Radiation
Microwave
Dielectric

Pressure
Platen
Fiber tension
Vacuum
Hydrostatic

Consolidation of the Composite. Consolidation, or curing, of the layed-up prepreg structure is most commonly achieved by the application of heat and pressure, as outlined in Figure 8. As previously noted, during the cure of thermosetting polymers, the molecules react with each other to form a more rigid network of cross-linked molecules. Epoxy matrix composites are cured at temperatures of 250 °F (121 °C) to 350 °F (177 °C), depending on their formulation. High-temperature resins, such as polyimides, require cure temperatures in excess of 500 °F (260 °C). Alternately, curing can be achieved by the application of other energy sources, such as X-ray, microwave, or dielectric radiation.

Pressure is also applied to ensure compaction of the fibers and resin and to establish the dimensions of the part. This is most easily achieved by bag molding. In this method, the prepreg lay-up is encased in a flexible airtight sheet or blanket and a vacuum is applied to the material in the bag so that the atmosphere exerts pressure. The pressure differential may be increased by placing the bag in an autoclave. Evacuation is a useful method for reducing or eliminating volatiles prior to or during cure, thereby reducing the formation of voids in the composite structure. Bag molding is used for curing large and complex parts. However, it is a relatively slow process, which may entail the application of pressure to the part for a period of 2–4 h, followed by up to a day in a postcure oven.

Matched die compression molding and elastomeric tooling are also commonly used to cure prepreg lay-ups. In compression molding, prepreg is layed up in the cavity of a two-part metal mold. The mold assembly is then placed between the platens of a hydraulic press, and pressure and heat are applied until the resin cures. Elastomeric tooling is a variation of matched metal tooling, which does not require the use of a hydraulic press. In this instance, the prepreg layup is placed in a metal mold with a silicone rubber mandrel. The mold is then bolted shut and placed in an oven. Thermal expansion of the rubber provides the pressure necessary to consolidate the composite. The advantages of these methods over bag molding techniques are that they provide a rigid, predefined envelop for enclosing the composite part and that they result in shorter curing times because of higher heat transfer rates.

System Integration — Joining Methods. Not all functional characteristics required of an advanced composite material can be attained in a molded part. In spite of the fact that it is possible to mold advanced composites into very complex geometrical shapes, it is sometimes necessary to machine a cured composite material or to attach it to another material. Because of the directional characteristics of the mechanical properties of advanced composites and the brittle nature of thermoset resins, proper joint design is critical to prevent stress concentration around joints and premature failure of the assembly. In general, factors of safety used in the design of attachments of advanced composite parts are usually more conservative than those used in metallic designs.

Since thermoset resins cannot be joined by the equivalent of welding or brazing, adhesive bonding is the most commonly used method of forming permanent joints. This method is of limited value if coefficients of thermal expansion of the materials to be joined differ greatly or are dissimilar to that of the adhesive. Advanced composites can be fastened mechanically by drilling holes and joining with rivets, bolts, or pins. The fiber discontinuities caused by machining can be eliminated by molding in suitable inserts.

5.1.2. Automation of Traditional Production Methods

Themajor disadvantages of the traditional approach to the production of advanced composite structures described above have been its low productivity and high production costs. To overcome these economic disadvantages, while maintaining versatility in selecting fiber orientations and flexibility in terms of being able to produce a variety of composite structures in a given manufacturing facility, the aerospace in-dustry has invested heavily in advanced manufacturing methods, such as robotics and other computer-aided manufacturing tools. Machine tool builders and material handling suppliers have developed a diversity of process automation modules to support the fabrication of composite structures, from the removal of the prepreg from refrigerated storage to the placement of the composite part on an airframe. These modules include automated, guided vehicles to transfer material from one production site to another, integrated composite cutting centers, automated tape-laying machines, "intelligent" autoclaves which incorporate computer control of the major process parameters, automated on-line inspection methods based on ultrasonic scanning, and more recently on CAT (computer-aided tomography) scanning, and automation of secondary processes such as trimming, drilling, and fastening [53]–[63].

Currently, specific modules are in operation at different manufacturing facilities. The next advance in composite manufacturing technology will be to integrate all of these modules into a fully automated manufacturing process to form what is currently being called the factory of the future.

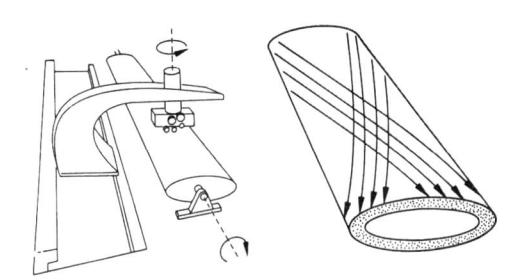

Figure 9. Elements of winding methods

5.1.3. Alternate Production Methods

Advanced composite structures can also be produced directly from roving and resin. Filament winding and pultrusion are the most important alternate processes. These are continuous or nearly continuous automated processes that are better suited to high-volume production than the manufacturing methods based on hand lay-up of prepregs.

Filament winding is a method of fabricating hollow composite structures that have one or more axes of rotation from fiber roving or tape. As outlined in Figure 9, continuous fibers or rovings are fed from a creel, drawn through a resin bath, and wound under tension over a rotating mandrel in the shape of the finished part. By controlling the rate of rotation of the mandrel and the position of the carriage that deposits the filaments onto the mandrel, layers of reinforcing fibers can be placed in a variety of predetermined patterns over the length of the mandrel. Originally, filament winders were mechanical machines in which the relative motion between the carriage and the mandrel was determined by gearing. With these machines, wind patterns were those that had a constant wind angle and an adjustable dwell (no carriage motion while the mandrel turns). Other limitations of the mechanical machines were the difficult gearing calculations and the often lengthy gearing changes needed to change patterns. A major technological advance was the application of dedicated computer control to filament winding. Motion of the carriage relative to the rotating mandrel is controlled by software, which allows wind patterns to be modified without changing gears. There are many types and forms of filament winders available, including horizontal winders, vertical winders, two-axis winders, four-axis winders, and advanced machines with seven or more axes of controlled direction. Filament winders can be used to wind very large parts. For example, the mandrel of a large filament winding machine may be 4 m (160 in.) in diameter and 11.5 m (460 in.) long, and weigh in excess of 45 000 kg (100 000 lbs) [64].

Because of the need to deposit the filaments under tension and because of the rapid motion of the carriage, the filaments being wound into a composite part need to be somewhat ductile. The process has been used successfully with more ductile fibers such as aramid, high-strength carbon, and glass, which exhibit a strain to failure of 1% or

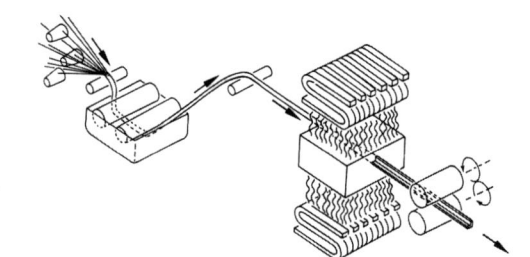

Figure 10. Pultrusion — an illustration of continuous forming and curing

more. Very brittle high-performance fibers, such as silicon carbide and very high modulus carbon, are not compatible with filament winding.

After the filament is deposited on the mandrel, the resin is then cured either at room temperature or in an oven, depending on the size of the part and the characteristics of the resin matrix. The composite part is then stripped from the mandrel.

Pultrusionis a method of producing continuous reinforced composites of constant cross-section, as outlined in Figure 10. In pultrusion, continuous filaments are drawn through a resin bath for impregnation and then pulled through a hardened steel die. The die orients the reinforcement, sets the final shape of the laminate, and controls its resin content. Cure may be completed within the die or may require additional heating after the formed composite leaves the die. Many structural shapes such as I-beams, channels, solid rods and bars, hollow rectangular beams, and flat plates can be pultruded. Some manufacturers have used postforming of partially cured pultruded stock to produce pultrusions of variable cross-section.

5.2. Thermoplastic Resin Matrix Composites

Thevarious production methods used to manufacture thermoset resin matrix composites are also applicable, with certain modifications, to the production of thermoplastic resin matrix composites. The key difference in the processing of these two classes of materials is in the applied time–temperature profile. Although long periods of time at elevated temperatures are required to cure properly thermosetting resin matrix composites, consolidation of thermopastic resin composites is achieved by heating to the fusion temperature of the resin and then cooling rapidly. The use of thermoplastic resin matrices eliminates the cure time delays that can stretch into days for large thermoset parts. Because of the resultant short cycle time, mass production applications of advanced composites are being enabled by the use of thermoplastic matrix materials. Other inherent manufacturing advantages of thermoplastic resin composites are that the raw materials have an inherently long storage life at room temperature, that molded parts can be postformed, and that scrap parts can be reclaimed.

Tape lay-up, filament winding and pultrusion are all adaptable to the production of thermoplastic resin matrix composites. A variety of uniform thermoplastic resin–fiber prepregs is available from commercial sources. Developmental products include tapes of glass, aramid, carbon, or other reinforcing fibers, which are combined with engineering thermoplastics such as polyetheretherketones, polyphenylene sulfide, polyamides, or thermoplastic polyester, as well as lower cost thermoplastic materials such as polypropylene, polyethylene, and ABS. These prepregs are made by a variety of processes, which include pulling the reinforcing fiber through a molten resin bath and a forming die, interweaving the reinforcing fiber with a thermoplastic monofilament of equal diameter, impregnating the reinforcement with minute particles of resin in a fluidized bed, or coextruding resin-sheathed fibers. It is also claimed that thermoplastic resin prepregs are better adapted to automatic tape laying systems than thermoset prepregs because of their lack of tack at ambient. Once the tapes are layed up, as with thermoset resin matrix composites, a thermoplastic resin matrix composite part can then be consolidated, if needed, by bag molding or compression molding, with the appropriate modifications of the time–temperature profile as mentioned above.

Thermoplastic filament winding systems are not radically different from thermosetting resin filament winding systems. The major difference is in providing means of preheating resin-coated rovings to the resin melt temperature prior to their being deposited on the mandrel. By heat loss to the surroundings the resin is resolidified and the composite structure is consolidated once the filaments are laid down. Heat sources that can be used range from ultrasonics and infrared for low-temperature materials to focused plasma guns and laser beams for high-melting engineering thermoplastic matrices.

The major difference between pultrusion systems for thermoplastic resin matrix composite materials and thermoset matrix composites is in the temperature of the resin impregnation bath and pultrusion die. Thermoplastic pultrusions are made by drawing filaments through a hot bath of molten resin and then through a chilled die, which shapes and cools the material to its final form.

5.3. Design Integration [65], [66]

Thecomplexities of these materials force a different approach to design. The traditional design approach to any complex system was to treat it as an assemblage of relatively isolated systems to minimize interdependencies. The limitations of the "open loop" process have been recognized, and a significant change is taking place. The emphasis is shifting to "configuration integration," by which technological advances, new products, or system requirements are permitted to impact the total configuration. Capabilities and advances in diverse disciplines can be coupled to achieve the valuable benefits of a multidisciplinary interaction and the full potential of these anisotropic materials. Thus, groups of independent specialists are being replaced by multidisciplinary teams.

The part performance depends not only on the geometric envelope, but also on the fiber orientation. In this way the material properties can be treated as a design parameter that can be varied to best meet the overall system requirements. The reduction to practice is best realized through the development of computer-aided design (CAD) systems that allow these very complex calculations to be made. This can then be coupled with a computer-aided manufacturing program that allows a process such as filament winding and tape lay down to be directly controlled by the design. The result is both an improved efficiency in the use of the materials and an efficient composite manufacturing process limited only by the limits of the available hardware.

6. Properties of Advanced Composite Materials

Thedesign of load-bearing structures requires knowledge of a large number of material properties. Resistance to deformation is characterized by the stiffness of the material as described by Young's moduli, shear moduli, and Poisson's ratio. The ability to support loads is described by the tensile or compressive strength. Coefficients of thermal expansion characterize changes in dimensions under thermal loads.

The properties of the polymeric components of composite materials are dependent on the duration, rate, and frequency of applied loads or deformations. This viscoelastic behavior is translated into the response characteristics of the composite and is manifested as creep, stress relaxation, and energy dissipation, which may result in internal heating. Metals and ceramics may also exhibit such behavior near the melting point; however, at normal operating temperature, these effects are negligible and are not usually included in traditional design considerations. If the viscoelastic behavior of polymer composites is ignored, the consequences may be disastrous. Consequently, mechanical properties should be characterized as functions of time, frequency, and temperature. The dependence of polymer composites on time, frequency, and rate of loading is reflected in quantities such as fatigue life, fracture toughness, and impact strength.

In addition to mechanical requirements, composite materials may be employed in applications involving the shielding of electric or magnetic fields, electrical and heat conduction, and permeation by fluids or gases. These characteristics are described by the appropriate transport properties: dielectric constant, magnetic permeability, thermal conductivity, electrical resistivity, and diffusion coefficient.

All of the above properties may be modified by environmental effects such as infrared and ultraviolet radiation, moisture, etc.

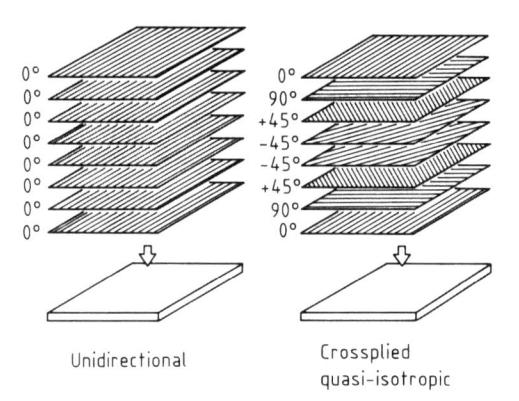

Figure 11. Representative composite lay-up configurations

6.1. Properties of Composite Structures

A unique feature of composite materials is the directional dependence (or anisotropy) of the various properties. The feature is intuitively clear for a unidirectional lamina; i.e., properties along the direction of the fiber reflect the higher performance characteristics of the fiber while properties perpendicular to the fiber tend toward the behavior of the matrix material. This directional dependence of the properties introduces the opportunity to further tailor the material to specific applications. The alignment of stiff and strong fibers along anticipated load paths can yield highly efficient structures.

The most common method of controlling fiber directions is shown in Figure 11. In this manual lay-up procedure, sheets of continuous fiber prepregs can be stacked in a variety of prescribed orientations. The effects of fiber orientation, developed through different stacking sequences, are illustrated in Table 13. Properties of aluminum and cold rolled steel are included for comparison.

To compensate for the low transverse properties of unidirectional material (0°), laminates of unidirectional fiber may be cross-plied at right angles to each other (0, 90°). The resulting structure has improved transverse properties compared to the unidirectional structure, but poorer longitudinal properties. Furthermore, the in-plane shear strength is not significantly improved over that of the unidirectional structure. Shear strength can be improved, but at the expense of longitudinal and transverse properties, by a stacking sequence of $\pm 45°$. A stacking sequence of 0°, 45°, and 90° plies (also illustrated in Fig. 11) yields a quasi-isotropic structure in which the in-plane properties are independent of direction.

As shown in Table 13, woven fabric structures are similar to crossplied laminate structures; however, they are usually less stiff and strong due to the various weaving patterns. Fabric geometries offer the advantage of drapability and handling characteristics that facilitate the fabrication of geometrically complex shapes.

Composite Materials

Table 13. Effect of fiber orientation on mechanical properties of various fiber reinforced composites [67]

Material	Fiber lay-up geometry	Density g/cm³	Elastic moduli, GPa			Ultimate strength, MPa			Yield strength as % of ultimate strength	Fatigue strength as % of ultimate strength
			Longi-tudinal	Trans-verse	Shear	Longi-tudinal	Trans-verse	Shear		
A type graphite/epoxy	unidirectional (0°)	1.57	138	6.9	4.5	1517	41	97		70
	crossply (0°, 90°)	1.57	74	74	4.5	838	838	97		59
	crossply (±45°)	1.57	17.2	17.2	31.0	138	138	345		
	isotropic (0°, 90°, ±45°)	1.57	48	48	17.8	604	604	221		78
	harness satin weave cloth (warp)	1.57	62	62		462	476			
IM type graphite/epoxy	unidirectional (0°)	1.60	221	6.9	4.8	1206	34	69		
	crossply (0°, 90°)	1.60			4.8			69		
	crossply (±45°)	1.60	17.2	17.2	44.8	124	124	290		
	isotropic (0°, 90°, ±45°)	1.60	73	73	24.8	345	345	179		
UHM type graphite/epoxy	unidirectional (0°)	1.68	303	6.9	6.6	758	28	48		
	crossply (0°, 90°)	1.68	159	159	6.6	402	402	48		
	crossply (±45°)	1.68	20.7	20.7	79.3	96.5	96.5	207		
	isotropic (0°, 90°, ±45°)	1.68	103	103	42.8	242	242	128		
Kevlar 49/epoxy	unidirectional (0°)	1.38	86 (tens) 41 (comp)	5.5	2.1	1517 (tens) 276 (comp)	28	41		70
	crossply (±90°)	1.38								
	crossply (±45°)		7.6	7.6	20.7	207	207	221		
	isotropic (0°, 90°, ±45°)									
	181 fabric (warp)(50v/e)	1.33	31	31	2.0	517 (tens) 172 (comp)	517 (tens) 172 (comp)	110		
S-Glass/epoxy	unidirectional (0°)	1.88	48	6.9	3.4	1730	40	10		
	crossply (0°, 90°)	1.88	31	31		980	900			
	crossply (±45°)	1.88	16	16		170	170			
	isotropic (0°, 90°, ±45°)	1.88	25	25		730	730			
E-Glass/epoxy	unidirectional (0°)	1.80	39	9.6	2.1	1104	20		23	
	crossply (0°, 90°)	1.80	25	25		518	518		32	22
	crossply (±45°)	1.80	11	11		152	152			
	isotropic (0°, 90°, ±45°)	1.80	18	18		330	330		42	25
Aluminum 6061-T-6		2.70	72	72	26.2	310	310	207	88	31
Steel, cold rolled (0.25 carbon)		7.85	207	207	83	552	552	414	75 (tension) 60 (shear)	40–45

The structures described in Table 13 consist of a single-fiber component. However, the material properties may be further altered through hybrid composites consisting of two (or more) types of fibers in a common matrix. By judicious positioning of different fibers, it is possible to greatly expand the range of properties that can be achieved with composite materials systems [68].

The benefits of tailoring materials through control of composition and reinforcing geometry greatly complicate the direct characterization of the properties of composite structures. For homogeneous materials, such as aluminum and steel, it is feasible to develop testing programs to generate design data concerning stiffness, strength, fracture toughness, etc. In contrast, the properties of a composite structure are strongly dependent on the composition and the reinforcing geometries associated with the various methods of fabrication. The labor and expense required to acquire direct characterization data for the directionally dependent properties for each variation in composition and geometry prohibit the use of this traditional approach. Consequently, analytical techniques that predict behavior in terms of composition and reinforcing geometry have emerged as an important component of the technology of composite materials. These analyses proceed at two levels, laminate analysis and micromechanics, as further discussed below.

6.2. Laminate Analysis

In this treatment, as in manufacturing, the unidirectional lamina is taken as the basic structural element. The microstructure of the individual plies is ignored and each ply is treated as a homogeneous material with different properties along and perpendicular to the fiber direction. Building on techniques inherited from plywood, tires, and cloth reinforced plastics, analysts have constructed sophisticated techniques for defining the optimum orientation of plies for specific applications.

Laminate Theory. The starting point for laminate analysis is a knowledge of the properties of a unidirectional ply. The ply is treated as a homogeneous orthotropic material with different properties along and perpendicular to the fiber direction. The direction-dependent properties are formulated as matrix arrays according to the methods of anisotropic elasticity theory (an introduction to anisotropic elasticity is given in [69]). The structural member is considered to consist of variously oriented orthotropic plies, the boundary conditions on the structural member are formulated, and laminated plate (or laminated shell) theory is used to determine stresses on individual plies. Relative orientations of the plies, which best match the properties of the plies to the loading conditions imposed on the structural member, are determined.

The execution of laminate analysis requires mathematical treatments beyond the scope of this review. An introduction to laminate analysis is given in [70]. An important result of laminate analysis shows that a structure made from a sufficient number of plies stacked in the sequence illustrated in Figure 11 approaches the behavior of a quasi-isotropic material. For this structure, the in-plane properties are independent of direction. For example, the in-plane Young's modulus (E_0) is uniform and is related to the longitudinal modulus (E_L) and transverse modulus (E_r) of the ply through the following relationship: $E_0 = (3/8) E_L + (5/8) E_r$. TSAI[71] recommends that comparisons between different compositions of composite materials should be made in terms of these properties rather than the common practice of comparing longitudinal properties and ignoring the possibility that the transverse properties might not be acceptable.

Typical Properties of Unidirectional Laminae. The application of laminate analysis requires a complete set of properties for the individual lamina. Some properties of representative, unidirectional materials are presented in Tables 14–19. The influence of various reinforcing fibers on the room-temperature properties of otherwise comparable epoxy matrix composites is illustrated in Table 14. Conversely, the effects of various matrix materials for comparable groups of carbon fiber and p-aramid fiber reinforced composites are shown in Tables 15 and 17, respectively, as well as in Figures 12 and 13. The impact of various matrix materials on selected properties of carbon fiber reinforced composites is illustrated in Table 18. Tables 16 and 19 provide data on the effects of temperature and humidity on p-aramid and glass composites in epoxy matrix resins.

Although the utilization of laminate analyses reduces the characterization effort by focusing attention on the characterization of unidirectional specimens, the wide range of possible compositions and effects illustrated in Tables 14, 15, 17, 18 tends to preclude a total reliance on direct laboratory characterizations.

6.3. Micromechanical Models

Starting ca. 1960, emphasis was given to predicting the properties of the individual unidirectional plies in terms of the properties and concentration of the fiber and matrix components. These models became a necessary supplement to laminate analysis as the choices of components became sufficiently numerous to make a strictly empirical approach economically impractical. In this approach, the inhomogeneous nature of the ply at a microscopic level is recognized. Various models have been developed by introducing approximations concerning the packing geometry and/or the response fields within the ply. By these models the mathematical analyses can be performed to relate the properties and concentrations of the fibers and matrix to the effective properties of the ply.

A detailed treatment of the various micromechanical models is beyond the scope and intent of this review. Introductory treatments are given in [77]–[79]; more advanced treatments are given in [80] and [72].

Table 14. Properties of unidirectional epoxy matrix composites with varying reinforcing fibers (normalized to 60 vol% fiber) [72]

Property	Unit	Standard carbon	1.5% Strain carbon	1.8% Strain carbon	Int mod carbon	HI mod carbon	UL HI mod carbon	Pitch 100 carbon	Kevlar 49 aramid	EGlass	S-2 Glass	Nicalon SiC
Specific gravity		1.58	1.60	1.61	1.60	1.80	1.83	1.83	1.45	1.90	2.02	1.90
Ultimate tensile strength	MPa	1516	1895	2584	2756	779	758	1034	1364	1034	1688	1378
Tensile modulus	GPa	131	134	138	165	239	314	420	65	41	52	110
Tensile strain	%	1.16	1.42	1.88	1.67	0.33	0.24	0.25	2.11	2.50	3.22	1.25
Ultimate compressive strength	MPa	1309	1585	1585	1378	345	338	255	207	827	827	1654
Compressive modulus	GPa	131	131	134	145	227	316	310	41	41	60	–
Transverse shear strength	MPa	65	65	69	65	34	37	34	59	–	–	–
Interlaminar shear	MPa	110	110	110	110	35	65	31	52	76	76	96

Table 15. Mechanical properties of 60 vol% Kevlar unidirectional p-aramid-reinforced composites of varying matrix composition [73]

Property	Unit	Thermoset resin matrix		Thermoplastic resin matrix					
		Polymer K-I	Polymer K-II	Polymer J-1	Polymer J-2	PPS	Nylon 6-6	PET
Tensile strength at 25 °C	MPa	1380	1070	945	1380	1366	1056		
at 121 °C	MPa	1172							
Compressive strength, dry									
at 25 °C	MPa	276	642	607	676	710	566	642	634
at 121 °C	MPa	221			277				
Flexure strength, dry	MPa	621	62	69	55	69	35	58	47
Short beam shear, dry	MPa	53							
Fracture toughness	kJ/m²				1.40		0.9		

Properties of Advanced Composite Materials

Table 16. Hot, humid aging of p-aramid fabric epoxy composites*

Property	Conditioning	Temperature, °C	p-Aramid resin		E-Glass resin	
			Narmco 5208	Hexcel F-161	Narmco 5208	Hexcel F-161
Flexural strength MPa	dry	23	434	426	500	639
		100	391	405	485	563
		160	343	345	465	530
Flexural strength MPa	wet**	23	374	341	470	466
		100	344	234	420	396
		160	174	136	315	275
Short beam shear, dry MPa	dry	23	32	37	56	53
		100	29	32	54	45
		160	25	28	46	38
Short beam shear MPa	wet**	23	31	36	54	51
		100	27	23	40	36
		160	19	13	28	26

* S-281 Fabric – Kevlar 49;
** 21-day exposure at 82.2 °C and 95% relative humidity

Table 17. Mechanical properties of unidirectional graphite-reinforced composites of varying matrix composition [73]–[82]

Property	Unit	Thermosetting resins				Thermoplastic resins						
		H.T. Epoxy	BMI	Thermid 600	Polymer K-I	Polymer K-II	Polymer J-1*	Polymer J-2**	PEEK (APC-2)	PPS	Nylon 6-6	PET
		AS 4	AS 4	HTS	AS 4	AS 4	AS 4	AS 4	AS 4	AS 4	AS 4	AS 4
Fiber, vol%		60	65	60	60	60	60	60	60	60	60	60
Tensile strength	MPa	1860					2432		2439	2274	2322	2150
Compressive strength, dry	MPa	1378	1860	1344	1034	1530	1068	1040	1040			
Flexure strength, dry	MPa	1654	123	83	1364	96	1240	1447	1495	1226	1350	1220
Short beam shear, dry	MPa	110			110		83	103	117	83	83	83
Fracture toughness	kJ/m²	0.26			1.05		2.10		1.75	1.40		

Table 18. Selected high-temperature mechanical properties of unidirectional graphite-reinforced composites of varying matrix composition [73]–[76]

Property	Temperature °C	Unit	Thermosetting resins			Thermoplastic resins							
			H.T. Epoxy AS 4 60	BMI AS 4 65	Thermid 600 HTS 60	Polymer K-I AS 4 60	Polymer K-II AS 4 60	Polymer J-1 AS 4 60	Polymer J-2 AS 4 60	PEEK (APC-2) AS 4 60	PPS AS 4 60	Nylon 6-6 AS 4 60	PET AS 4 60
Flexure strength, dry	23	MPa	1654	1860		1364	1530	1240	1447	1495	1226	1350	1220
	93	MPa	1557		1344	1261	1412	971	1116		827	923	951
	117	MPa	1192	1509		1123	1096						
	232	MPa					909						
	315	MPa			1020								
Flexure strength, wet *	23	MPa	1764			1426	1412			1323			
	93	MPa	1392		641	1151			717				
	177	MPa	648	1123		806	923						
	232	MPa					847						
	315	MPa			350								
Short beam strength, dry	23	MPa	110	123	83	110	96	83	103	117	69	83	83
	93	MPa	90	82		90	83	62	83		42	69	48
	177	MPa	65	79	53	54	63						
	232	MPa					56						
	315	MPa			55								
Short beam strength, wet *	23	MPa	103			95	86						
	93	MPa	75			73							
	177	MPa	40			52	65						
	232	MPa					47						

* 2-Week exposure at 71 °C and 95 % relative humidity

Table 19. Air aging of *p*-aramid unidirectional epoxy composites* at 150°C

Hours exposure at 150 °C	Tensile strength (in MPa)** at composite thickness		
	14 mils	42 mils	73 mils
0	1234	1207	1289
1000	1165	1165	1254
2000	1110	1138	1241
8760 (1 year)			1082
19600	676	883	

* SP-306 Epoxy, 50 vol% fiber;
** Measured at room temperature after exposure

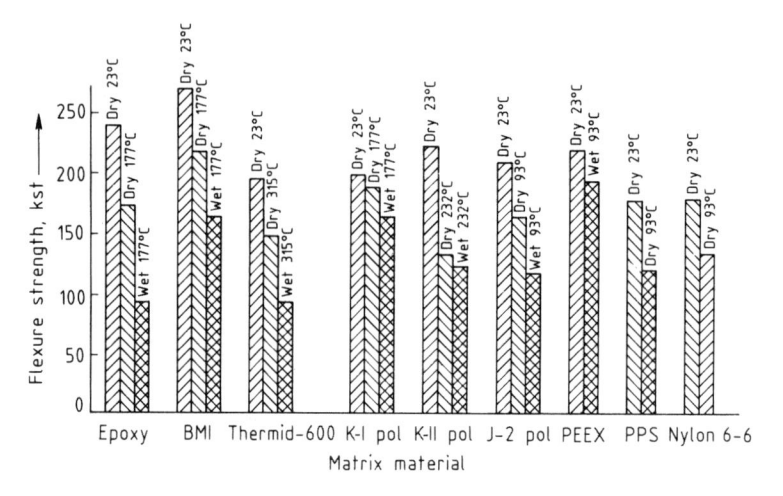

Figure 12. Effect of environment on flexure strength of unidirectional graphite reinforced composites

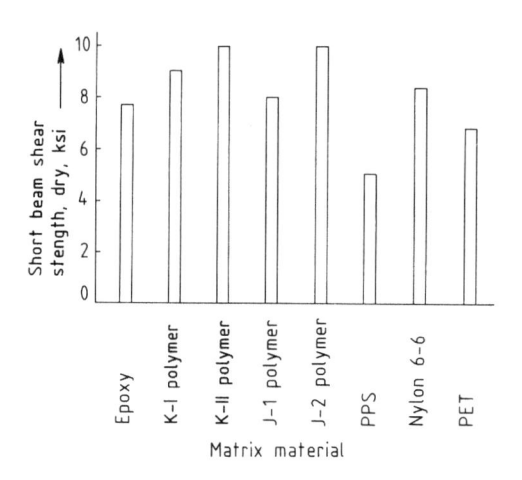

Figure 13. Effect of matrix material on interlaminar shear strength of *p*-aramid reinforced composites
1 ksi = 6.895 × 10^6 Pa

For the purposes of model development properties may be grouped under the classifications of averaged or localized properties. The class of properties within the first group can be treated in terms of average properties and average response characteristics. Thermoelastic properties, certain viscoelastic properties, and transport properties fall within this group. The class of the properties belonging to the localized group is strongly dependent on the statistical features of the material properties as well as fluctuations in microstructure; thus, properties belonging to this group are much more sensitive to the nature of the local environment. Properties such as tensile strength, fatigue life, and fracture toughness fall in this group. For such properties, the identification of modes of local failure (i.e., the weak links) is a prerequisite to the application of model relationships.

Averaged (Effective) Properties. Numerousimportant properties of advanced composites can be predicted by micromechanical models. They include thermoelastic properties such as Young's modulus, Poisson's ratio, shear modulus, and the thermal coefficient of expansion; transport properties such as magnetic permeability, electrical conductivity, thermal conductivity, and diffusion coefficients; as well as expansional strains induced by environmental factors such as moisture adsorption.

All of the above properties share the common feature that they can be treated in terms of average properties and the overall (average) response for the composite system. In principle, the average thermoelastic and transport properties of a ply can be obtained by specifying (1) the details of the geometry (fiber shape, packing geometry, and spacing), (2) the distribution of the thermal, mechanical, or electrical loads on the surfaces of the ply, and (3) the connectivity between the fiber and the matrix phase. The average bulk response can be determined by taking the appropriate volume averages, thereby relating the volume fraction of the components and their properties to the average properties of the ply.

The simplest models for predicting thermoelastic and transport properties may be developed intuitively by treating response to mechanical, thermal, and electrical loads by analogy to springs (for mechanical loads and thermal expansion) or resistors (for thermal or electrical transport), which can be arranged either in parallel or in series. The analogies are illustrated in Figure 14.

The longitudinal responsecharacteristics are treated as if the components of the ply react to the loads by parallel response. For mechanical loads, the central feature of this analogy is the assumption that each component is subject to the same strain. This analogy yields the following simple linear relationship:

$$P = \sum_{i=1}^{n} v_i P_i \qquad (8)$$

where P_i is any one of the pertinent longitudinal mechanical, thermal, or transport properties of the *i*th component. The concentration terms, v_i, are the volume fractions of the *i*th component given by

$$v_i = V_i/V \qquad (9)$$

Figure 14. Simple composite models

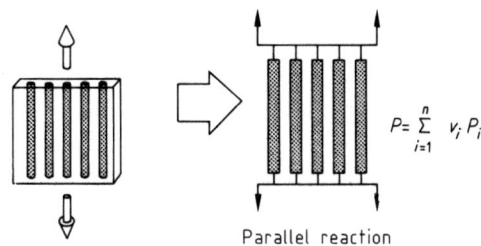

Longitudinal response — Parallel reaction

$$P = \sum_{i=1}^{n} v_i P_i$$

Transverse response — Series reaction

$$\frac{1}{P} = \sum_{i=1}^{n} \frac{v_i}{P_i}$$

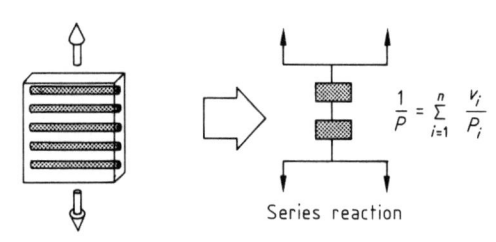

where V_i is the overall volume of the ith component and V is the total volume of the ply. The volume fractions must satisfy the following relationship:

$$\sum_{i=1}^{n} v_i = 1 \qquad (10)$$

The transverse response of the ply to mechanical and other loads is modeled by considering that the components react to these loads as though they were connected in series. The central feature of this analogy (for mechanical loads) is the assumption that each component is subject to the same stress. This analogy yields the following simple reciprocal relationship:

$$\frac{1}{P} = \sum_{i=1}^{n} \frac{v_i}{P_i} \qquad (11)$$

For a two-component ($n = 2$) fiber–resin composite, Eq. (10) becomes

$$v_f + v_r = 1$$

or

$$v_r = 1 - v_f$$

and Eqs. (8) and (11) can be expressed as follows:

$$P = v_f P_f + v_r P_r$$

and

$$1/P = v_f/P_f + v_r/P_r$$

If the property of the fiber (P_f) is considerably greater than the corresponding property of the matrix (P_r), these equations are reduced to the following:

$$P \approx v_f P_f$$

for longitudinal properties and

$$P \approx P_r / v_r$$

for transverse properties.

These simple relationships, which have become popularly known as the "Rule of Mixtures," indicate that for advanced composite materials in which fiber properties are considerably greater than matrix properties, the longitudinal properties of a ply are dominated by the fiber contributions and the transverse properties are dominated by the contributions of the matrix.

While it has been demonstrated that the Rule of Mixtures relationship for longitudinal properties gives excellent agreement between the predicted and observed properties of continuous fiber composites, prediction of transverse properties is usually underestimated. Considerable effort has been devoted to developing more accurate and comprehensive models and methods of analysis. Improved models for predicting shear moduli and transverse properties are discussed in [70]–[80].

Strength of Composites(Localized Properties). In contrast to the thermoelastic properties, an understanding of the average response fields is insufficient for a description of strength. The strength of a system is dependent on the behavior at local levels, with the weaker sections of the material exerting more influence than the strong sections. Because of this weighting of the weak elements, the strength of the assembly is less than the average strength of the individual components. The appropriate weighting of the weak elements is dependent on the mechanism or mode of failure.

The weak elements or flaws can be due to characteristics either of the fibers or of the matrix. Even in the case of longitudinal tensile strength, while the direct contribution of the strength of the matrix phase to the tensile strength of the composite is negligible, an improper choice for the matrix phase could seriously detract from the realization of the full strength potential of the reinforcing fibers.

The tensile, compressive, and shear strengths of a composite are determined by a complex interplay between several factors:

1) The mechanical properties of the fiber system, taking into account the statistical distribution in the tensile and/or compressive strengths of individual fibers
2) Statistical variations in the fiber–resin geometry in a ply
3) Coupling between the fiber and the resin phase, taking into account not only the bond strength at the fiber/resin interface, but also stresses induced by the difference in the responses of the fiber and resin to environmental changes; for example, a mismatch in the coefficients of thermal expansion of the fiber and resin phases can cause strains to develop near the fiber–resin interface when the composite is

subjected to significant changes in temperature, as often occurs in the composite manufacturing process.

4) The viscoelastic properties of the resin phase, taking into account possible spatial variations in these properties due to presence of the fibers; one of the unique features of composite materials is the extremely large surface area over which interfacial contact is made. Because of the large surface : volume ratio and the heterogeneous nature of the matrix resins currently used, particularly the multicomponent thermosets and semicrystalline thermoplastics, the probability is high that certain components of the resin phase are preferentially adsorbed on the fiber surface, thus resulting in spatial variations in both the composition and the properties of the matrix phase.

5) Flaws and discontinuities in the matrix phase.

In summary, the strength of a composite is determined by a complex interplay between all of the above factors; the achievement of the full capabilities potentially available from high-performance reinforcing fibers requires careful control over the extent to which each of the above factors exerts its influence. As reviewed and discussed in more detail in [68]–[80], current theoretical models do not adequately relate all the various factors and, therefore, can offer only qualitative guidelines as to their relative significance.

7. Applications

Historically, advanced composite materials have been used for reasons of performance in applications where their value in use is the driving economic consideration. The major impetus has been their potential for reducing the mass of a structure due to their high specific stiffness and specific strength. More recently, however, the focus has shifted to the need for cost reduction through parts consolidation and total part count reduction.

Advanced composite materials have found greatest use in applications where mass savings result in significantly improved performance, thereby allowing the product to be value priced. This criterion applies to high-performance military aircraft, where lower mass translates into higher speed, greater acceleration, and/or greater payload, which are the essential elements of survival, much less victory in combat. It also applies to wide classes of equipment that use human muscle power as the motive force. Better control and ease of handling, and thus improved performance, have been the driving factors behind the use of advanced composites in sports equipment or medical prostheses.

Advanced composite materials are also being used in applications where mass reduction results in significantly lower life cycle costs. This criterion applies broadly to all forms of transportation equipment, where reduced system mass results in directly increased payload or reduced fuel consumption.

Missiles and rockets are extreme examples of transportation vehicles in which any small reduction in mass is extremely valuable. For example, the Space Shuttle System weighed 1.9×10^6 kg at liftoff. This was many times larger than the weight of the Orbiter, which was 68 000 kg, or of its low earth orbit altitude payload of ca. 30 000 kg [81]. On the basis of the most recent cost of a Space Shuttle launch of ca. $ 80×10^6, it cost ca. $ 2700/kg to send an object into low earth orbit. Sending an object into geosynchronous orbit is even more constrained in terms of maximum payload, ca. 2000 kg, and costs, which range from ca. $ 20 000 to $ 60 000/kg.

The effects of mass reduction on the fuel consumption of various common modes of transportation are summarized in Table 20. The data presented in this table are based on published fuel consumption and vehicle utilization data. An economic value of mass savings for vehicles used to serve these various modes of transportation is obtained by assigning a net present value to the fuel that otherwise would have been consumed, currently ca. $ 0.25/L ($ 1/gal). Not surprisingly, mass reduction has greater value in aircraft, particularly commercial transport aircraft, than it does in ground transportation equipment.

The value of providing tailored anisotropic material properties for reasons other than mass savings is now also being recognized and is being applied in developmental aircraft, as discussed further below. Other general technical characteristics of advanced composite materials that have led to their use include corrosion resistance and their relatively high transparency to electromagnetic radiation.

In some instances, use of an advanced composite has been based on the unique characteristics of a specific fiber–matrix combination in addition to the general characteristics described above. In some applications, advanced composites reinforced with carbon fibers are used specifically because of (1) the high electrical and thermal conductivity of these fibers or (2) their unique thermal expansion characteristics, which makes these composites the materials of choice for constructing low-mass structures that do not distort under thermal gradients, such as telescope reflectors and antennae. Conversely, glass or p-aramid fiber reinforced composites are used in applications, such as radomes, that must be transparent to electromagnetic radiation and, thus, require a material with a low dielectric constant.

Advanced composite materials are now also used for reasons of cost as well as performance, as in the case of helicopter fuselages. In these applications, complex structures, normally made from a large number of metal parts, are now being fabricated from advanced composite materials as a single, integrated component. Cost savings derive from the smaller number of parts required and the lower assembly costs.

In the concluding sections of this chapter, a select number of representative applications of various composite materials are presented. These examples are chosen to illustrate various materials, manufacturing processes, or reasons for using advanced composites. They are not meant to be representative of all possible applications of advanced composite materials.

Table 20. Reduction in lifetime fuel consumption per pound of mass savings for various modes of transportation

Mode of transport	Nominal gross mass, kg	Nominal life of equipment, km of use	Liters per km and kg of mass	Lifetime fuel reduction per kg of mass saved, L
Aircraft				
Light pleasure craft	1 400	3×10^5	1.3×10^{-4}	39
Business jet	5 500	3×10^6	1.4×10^{-4}	420
Commercial transport	95 000	3×10^7	8×10^{-5}	2 400
Turbine helicopter	1 400	1×10^6	3.3×10^{-4}	330
Automotive vehicles				
Passenger auto, gasoline	1 400	2×10^5	7.0×10^{-5}	14
Passenger auto, diesel	1 400	2×10^5	5.0×10^{-5}	10
Intercity bus	16 000	2×10^6	1.0×10^{-5}	20
Commercial truck	27 000	2×10^6	7.0×10^{-6}	14

Satellite Systems. Satellite systems are now commonly utilized to perform a wide variety of military and civilian purposes. Commercial satellites are used for telecommunications and TV broadcasting. Government satellites are used to gather data for weather forecasting and geophysical studies, to conduct a variety of scientific experiments, and to perform a variety of military missions.

The generic technology issues and specific materials and structures criteria that address these issues are as follows [82]:

> Good dimension stability by low thermal expansion, high thermal conductivity, and no outgassing
> High structural rigidity (specific stiffness) by high modulus and low density
> Noncontaminating by no outgassing
> Rapid track and point capability by good vibrational damping characteristics
> Deployable/erectable in space by low mass (shuttle transportable) and joining and fastening capabilities
> Survivable in laser/nuclear natural environments by high strength, no induced outgassing, high-temperature integrity, controllable reflectivity, and good temperature dissipation control
> Minimal space charging by good electrical conductivity.

The specific criteria include minimum mass, low thermal distortions, well-separated natural frequencies, high damping, environmental stability, as well as good electrical conductivity and a low thermal coefficient of expansion. Carbon fiber reinforced composites with their high specific strength and specific stiffness, their conductivity, and their controlled coefficient of thermal expansion that approaches zero (or even a negative value, depending on their structure), have become the materials of choice for satellite structures.

Military Aircraft. TheUnited States Department of Defense has supported the development and use of advanced composite materials since the early 1960s. Use of advanced composite materials has increased with each new type of aircraft. Use of advanced composite materials has progressed from replacement of metal in a few selected secondary structures, whose failure in flight would not endanger the aircraft, to current designs in which advanced composites are used in primary structures critical to airworthiness and safety, and account for over 20% of the mass of the airframe.

The first production part made of an advanced composite to reach flight status, in 1970, was a boron–epoxy horizontal stabilizer on the Grumman/U.S. Navy F-14. This 85-kg part, which is still in service, represents only 0.8% of the structural mass of this airplane. In contrast, the McDonnell-Douglas/U.S. Navy AV-8B Advanced Harrier VSTOL aircraft contains over 450 kg of advanced composites, which is ca. 20% of its structural mass. Major components made from carbon–epoxy composites include the forward fuselage, as well as wing substructures, such as spars, ribs, and skins, which account for over 70% of the wing mass.

By the mid-1990s, it is anticipated that advanced performance military aircraft must be capable of maintaining supersonic flight for long missions without refueling and at the same time demonstrate superior agility, VSTOL capability, and superior firepower (e.g., high payload). To meet these requirements, aircraft size and mass will be severely constrained, and advanced composites will be extensively used. For example, not only will proposed United State Air Force Advanced Tactical Fighter (ATF) have a structure made largely (40–60%) of advanced composites and, thus, weigh 25% less than a comparable one made of aluminum or titanium, but also the total configuration of the aircraft will be influenced by advanced composite materials.

The ATF could have a forward-swept wing, which would be aerodynamically more efficient than the traditional rear-swept wing. Advantages of forward-swept wings are reduced drag, improved control, and greater resistance to spin leading to smaller engines and lighter, less costly aircraft. Despite its advantages, the forward-swept wing has been sparingly employed because the phenomenon of divergence makes it a structurally unstable design, as discussed in more detail in [83]. Conventional metallic structures of sufficient strength and stiffness to counter these divergence effects at desirable sweep angles are also very heavy, and result in unacceptable mass penalties.

The problems of divergence can be overcome by using advanced composite materials without incurring a major structural mass penalty. Since laminates can be fabricated with fiber directions selected to produce a wide variety of twisting characteristics, wing skins can be tailored to twist the leading edge down while bending, so that the wing is then structurally stable.

Radomes. Glass–epoxy and, more recently, *p*-aramid–epoxy composites are extensively used in radomes for both military and civilian aircraft. The combination of low mass, damage tolerance, and low dielectric constant (needed for signal transparency) uniquely qualifies these composites as the materials of choice.

Helicopters. Since the introduction of the first production helicopters, damage caused by fatigue of critical flight structures has had an adverse effect on all aspects of rotorcraft operations. Thus, helicopter designers have been interested in using advanced composite materials since their commercialization, and the current designs of all major manufacturers of rotorcraft make extensive use of advanced composites for reasons of cost as well as performance.

The CH 53-E Blackhawk developed by Sikorsky for the U.S. Army was the first military helicopter to make extensive use of advanced composites. This rotorcraft contains ca. 350 kg of advanced composites, mainly p-aramid–epoxy composite. p-Aramid composites are also extensively used on the Boeing CH-47 Chinook helicopter and on the Hughes AAH-64A close ground support helicopter. The high impact resistance and toughness of p-aramid fibers led to these composites being preferred over carbon fiber reinforced composites, as well as over aluminum, for military helicopter structures. These properties result in a rotorcraft that is much less sensitive to damage from exploding shells and shrapnel.

Helicopter manufacturers are using the technology developed for military programs on the latest generation of commercial helicopters. For example, both the Sikorsky S-76 and the Boeing Commercial Chinook make extensive use of p-aramid–epoxy composites. By using p-aramid–epoxy composites, the frame of the S-76 weighs 30% less than that of a structurally equivalent aluminum frame, resulting in a 20% improvement in efficiency.

The U.S. Army has further pursued the advancement of using composites by supporting the development of the all-composite helicopter (ACAP). The Sikorsky ACAP candidate, designated the S-75, has met or exceeded all the objective and goals that the Army had set for the ACAP program. These included a mass reduction of 22%, a cost savings of 17%, and improved crashworthiness, survivability, reliability, and maintainability. The primary materials of construction on the S-75 are carbon and aramid reinforced epoxy.

In addition to a significant mass reduction, one of the major benefits of using advanced composites instead of metals was a sizable reduction in manufacturing costs. Previously complex metal assemblies can now be fabricated as a single, integrated composite part.

Commercial Transport Aircraft. In the past decade, the average price of aviation fuel has increased by an order of magnitude, from 10 /gal to > $ 1/gal. As a result of this escalation, fuel costs have become a dominant cost factor and can now account for 60% of the direct costs of operating a commercial transport jet. Since a marginal kilogram represents an additional consumption of 600 gal (2400 L) of fuel over the life of the aircraft (see Table 20), a strong economic incentive exists for developing more fuel-efficient transport aircraft. Technical advances that have led to more fuel-efficient transport aircraft have included improved propulsion systems, aerodynamically more efficient designs, and improved structural materials, particularly advanced composites.

Lockheed set the pace in the early 1970s by adapting 2500 lb (1140 kg) of p-aramid–epoxy composites on the secondary exterior structures of their L-1011-500 transport. This resulted in a structural mass reduction of 805 lb (366 kg), which contributed to the extended range of this model.

The current generation of commercial transport aircraft, such as the new Boeing 757 and 767 and the Airbus A-310, long-range, wide-bodied aircraft, makes somewhat more extensive use of advanced composite materials. In these aircraft, advanced composites are used in a variety of interior and secondary exterior structures and account for < 5% of the structural mass. Typical applications include fairing, flaps, and parts of the empennage structure.

It is anticipated that the next generation of commercial transport aircraft will make more extensive use of advanced composites, and that these applications will include important primary structures, such as the horizontal tail plane and elevator, the fin box, and the flaps, in addition to the components already in use. Such increased use is predicated on statisfactory performance of the secondary components that are currently flying and on the development of lower cost, more damage-tolerant composite structures than the epoxy matrix materials now in use.

General Aviation. Some of the most innovative concepts for the integrated use of advanced composite materials have been developed by manufacturers of fixed-wing aircraft for general aviation. Their incentive is to build a relatively inexpensive (ca. $ 1×10^6), fuel-efficient (> 10 mpg), 6–10-passenger aircraft capable of transcontinental flight.

The Lear Fan 2100 is the first fixed-wing aircraft to make intensive use of advanced composites. This airplane has an all-composite primary structure that makes extensive use of graphite–epoxy composite materials. This aircraft is currently undergoing testing and evaluation by the FAA and is likely to be the first all-composite airplane to be certified.

The Avtec 400 represents a more radical design approach that is enabled by the use of advanced composites. This aircraft, which is currently undergoing development, uses a canard concept to improve lift and reduce drag. The skins are made of aramid–epoxy for damage tolerance, while carbon–epoxy is used in the spars and other parts that require maximum stiffness and compressive strength.

Another business aircraft of unusual design that makes extensive use of carbon–epoxy composites is the Beech Starship. This airplane has a forward wing that adjusts forward for takeoff and landing and back for flight. Beech is considering a filament wound structure.

Finally, a very special part of general aviation, human-powered flight, owes its realization to advanced composite materials. Without the availability of advanced composite materials, an aircraft light enough to be effectively powered by a human being simply could not have been built.

Ballistic Protection. *p*-Aramid-reinforced composites are being increasingly used in military helmets, for fragment protection on ships, and for military shelters. They are replacing fiberglass and metals in mass-critical applications [84]. The ability of *p*-aramid fabrics to stop high-speed objects is controlled by their dynamic energy-absorbing properties, primarily high tensile strength and specific modulus, complemented by good thermal resistance and high fracture toughness. Special woven fabrics need to be designed to balance positive and negative aspects of fabric cross-overs. Cross-overs help spread the impact by involving more yarns, but in excess, can reduce fabric deflection causing fiber shearing and loss in performance.

Automotive Transportation. Advanced composite materials have been accepted for primary body structures and chassis parts in speciality and racing vehicles, in which light mass and performance take precedence over cost. Their use in production automobiles, however, has been extremely limited for reasons of both high materials costs and lack of a suitable technology for mass production of composite parts and components. The most notable automotive application of advanced composites to date has been the continuous glass–epoxy composite transverse rear leaf spring on the Chevrolet Corvette. This single leaf spring, which weighs only 8 lb (3.6 kg), has a fatigue life that is an order of magnitude greater than that of the much heavier (45 lb, 22 kg) multileaf steel spring that it replaced [85].

This situation may change radically within the next decade because of recent developments in process and materials technology. Computer-controlled filament winding of thermoplastic matrix composites has generated major interest in the automobile industry as a means of producing large intricate parts, such as the frame, the suspension system, and the steering column. In particular, it has been reported that filament wound structural frames for cars with plastic bodies are currently under development by General Motors [86].

There is also active interest in the application of advanced composite materials to automotive engine parts. Prime candidates are those highly stressed parts that oscillate or rotate. High-speed reciprocating parts such as wrist pins, connecting rods, and valve lifters made from advanced composite materials would weigh significantly less than the metallic components now used, and their use would significantly improve engine performance, in terms of vibration, noise, and fuel efficiency.

General Industries/Electronics. Whilegeneral use of advanced composite materials in the industrial sector on the basis of their mechanical properties has been limited to selected high-speed rotating or reciprocating parts, such as textile machinery components as picking sticks, use of advanced composite materials is increasing in applications that require tailored electromagnetic properties as well as superior mechanical properties. For example, X-ray tables are currently made of advanced composites because of the combination of stiffness and X-ray transparency. As the electronic industry moves to direct surface mounting of chip carriers, the ability to match the coefficients of expansion of the board and of the chip carrier, coupled with the desired electrical properties, make p-aramid reinforced composites attractive materials of construction for circuit boards.

Sporting Goods. The sporting goods and recreational equipment markets were the first significant commercial uses of advanced composite materials. Advanced composites have been successfully used in applications such as golf club shafts, fishing rods, tennis (and other) racquets, archery equipment, hand gliders, windsurfer boards and masts, etc. The successful penetration of advanced composites in these applications is based on psychological as well as technical reasons. While use of advanced composite materials does result in lighter mass and more responsive equipment that can marginally improve performance during play, the ego gratification that comes from using the most expensive, as well as the most technologically innovative, equipment available cannot be discounted in these markets.

Prospective Developments. In less than 30 years, advanced composite materials have evolved from their experimental beginnings to become well-established speciality products used in a wide variety of high-performance applications, where mass savings has a substantial value in use. Since a higher level of complexity is inherently needed to understand and properly use advanced composites than is needed with traditional materials, the successful development of this technology was greatly enabled by changes in the design environment brought about by the dramatic increases in computing power and computerized methods of analysis, design, and control that also occurred during this period.

The value of using advanced composite materials for reasons other than mass savings are now being recognized. These reasons include

1) The ability to provide tailored anisotropic properties, which allows otherwise unfeasible structural design concepts to be realized by providing an additional degree of freedom in material properties
2) The ability to manufacture complex structures previously manufactured as assemblies of numerous metal parts as single, integrated components of advanced composite materials, which may result in a simpler and less expensive overall manufacturing process

Areas of future advanced composites research include the development of materials that can result in higher performance composites, as well as those that will allow advanced composites to be used more economically.

In terms of improved performance, the emphasis will be on developing composites that exhibit

1) superior mechanical properties at elevated temperatures
2) improved three-dimensional properties
3) improved toughness and ductility.

Penetration of advanced composite materials into mass production applications, in which their value in use would be modest, will depend on the development of suitable manufacturing processes, as well as lower cost materials. The most significant development in this regard may well be the filament winding of composites of high-strain fiber in a thermoplastic matrix.

8. References

[1] D. Hull: *An Introduction to Composite Materials*, Cambridge University Press, Cambridge 1981.
[2] S. W. Tsai, H. T. Hahn: *Introduction to Composite Materials*, Cambridge University Press, Cambridge 1981.
[3] A. Kelly: *Strong Solids*, Clarendon Press, Oxford 1973.
[4] A. Kelly, G. J. Davies: "The Principles of Fibre Reinforcement of Metals," *Met. Rev.* **38** (1965) 1–77.
[5] T. W. Chou, A. Kelly: "Fibre Composites," *Mater. Sci. Eng.* **25** (1976) 35–40.
[6] Deutsche Gesellschaft für Metallkunde (eds.): *Verbundwerkstoffe (Symposium 1980)*, DGM Informationsges. mbH, Oberursel 1980.
[7] D. Ruppin, K. Müller, *Z. Werkstofftech.* **12** (1981) 263–271.
[8] H. D. Steffens, H. Kayser, K. N. Müller, *Z. Werkstofftech.* **2** (1971) 44.
[9] W. Kurz, D. Sahm: *Gerichtet erstarrte eutektische Werkstoffe*, Springer Verlag, Berlin 1975.
[10] H. Mangers, E. Blank, *Z. Werkstofftech.* **11** (1980) 367–373.
[11] P. R. Sahm, U. W. Hildebrandt, *Z. Werkstofftech.* **10** (1979) 257–262.
[12] A. Kelly, unpublished presentation.
[13] C. A. Calohr, A. Moore: "No Hope for Ceramic Whiskers or Fibres as Reinforcement of Metal Matrices at High Temperatures," *J. Mater. Sci.* **7** (1972) 543.
[14] P. H. Seldan: *Glasfaserverstärkte Kunststoffe*, Springer Verlag, Berlin-Heidelberg-New York 1967.
[15] G. Mengers, H. Brintrup, R. Jonas, *Z. Werkstofftech.* **9** (1978) 164–171.
[16] R. T. Pepper, G. W. Upp, R. S. Rossi, E. G. Kendall: "The Tensile Properties of a Graphite-Fiber Reinforced Al-Si-Alloy," *Metall. Trans.* **2** (1971) 117.
[17] J. P. Giltrow, J. K. Lancaster: "Friction and Wear Properties of Carbon-Fiber Reinforced Metals," *Wear* **12** (1968) 91–105.
[18] K. E. Saeger, *Aluminium (Düsseldorf)* **46** (1970) 681–686; see also W. Dawihl, W. Eicke, *Z. Werkstofftech.* **9** (1978) 27–30.
[19] G. Rau: *Metallische Verbundwerkstoffe*, Werkstofftechnische Verlagsgesellschaft, Karlsruhe 1977.
[20] G. Rosenkranz, *Z. Werkstofftech.* **12** (1981) 289–296.
[21] A. Neville (ed.): *Fibre Reinforced Cement and Concrete*, Construction Press, Lancaster, United Kingdom, 1975.
[22] G. Batson: "Steel Fibre Reinforced Concrete," *Mater. Sci. Eng.* **25** (1976) 53–58.
[23] J. A. Manson: "Modifications of Concretes with Polymers," *Mater. Sci. Eng.* **25** (1976) 41–52.
[24] H. Böder, D. Gölden, P. Rose, H. Würmseher, *Z. Werkstofftech.* **11** (1980) 275–281.
[25] R. M. Gill: *Carbon in Composite Materials*, Iliffe Books, London 1972.
[26] E. Fitzer, J. Schlichting, *Z. Werkstofftech.* **11** (1980) 330–341.
[27] R. M. Chaddell: *Deformation and Fracture in Solids*, Chap. 9: "Composites," Prentice-Hall, Englewood Cliffs 1980, pp. 252–270.
[28] B. W. Rosen in ASM (eds.): Fibre Composite Materials, Metals Park, Ohio, USA, 1965, p. 37.
[29] C. Sweben, B. W. Rosen, *J. Mech. Phys. Solids* **18** (1970) 189–196.
[30] G. Meder, *Z. Werkstofftech.* **12** (1981) 366–374.
[31] C. C. Chamis: "Mechanics of Load Transfer in the Interface," in E. P. Plueddemann (ed.): *Interface in Composite Materials*, Academic Press, New York-London 1974, pp. 32–78.

[32]   M. J. Folkes: *Short Fibre Reinforced Thermoplastics,* J. Wiley/Research Studies Press, London 1982.
[33]   T. Edison, US 223, 898, 1880.
[34]   W. E. Chambers, "Low Cost High Performance Fibers," *Mech. Eng.* December (1975).
[35]   A. Shindo, *J. Ceram. Assoc. Japan,* **69 C** (1961)   195.
[36]   W. Watt, W. Johnson, *Nature (London)* **220** (1960)   835.
[37]   E. E. Magat, *Philos. Trans. R. Soc. London A* **294** (1980)   463–472.
[38]   D. Tanner, *Future of Aramid Fiber Composites as a General Engineering Material* E. I. Du Pont de Nemours & Co., Textile Fibers Department, Wilmington, Del 19898, 1985.
[39]   I. M. Ward, *Philos. Trans. R. Soc. London, A* **294** (1980) 473–481.
[40]   S. Yajima, *Philos. Trans. R. Soc. London A* **294** (1980)   419–426.
[41]   L. S. Penn, T. T. Chiao in G. Lubin (ed.): "Epoxy Resins," *Handbook of Composites,* Chap. 5, Van Nostrand Reinhold Company, New York, 1982, p. 73.
[42]   "Rigidite 5254C, *Carbon Fiber Prepreg System,*" Narmco Materials Inc., Anaheim, Calif., April 1984.
[43]   M. Chaudhari et al., Nat. SAMPE Symp. Exhib. Proc. **30** March (1985) 735–746.
[44]   *Technical Bulletins,* National Starch and Chemical Corp., 1985.
[45]   *Mod. Plast. Encycl.* **62** (1985) no. 10a, 449–480.
[46]   D. Nissen, H. Stutz, *Advanced Composite Matrices* CCM Annual Workshop, Center for Composite Materials, University of Delaware, Newark, Del. 19716, May 1985.
[47]   J. A. Harvey, *Proc. ICCM V,* 1985, pp. 1623–1632.
[48]   H. H. Gibbs, *Proc. ICCM V,* 1985, pp. 971–993.
[49]   I. Y. Chang, *Composites Science & Technology,* **24,** 198561–79 .
[50]   H. H. Gibbs, *Proc. ICCM V,* 1985, pp. 971–993.
[51]   J. E. O'Connor, et al, *Proc. ICCM V,* 1985, pp. 963–969.
[52]   B. Cole, Proc. 30th National SAMPE Symposium, 799–808, 1985.
[53]   G. W. Ewald, "Two Stage Tape Placement System," *Proc. Nat. SAMPE Symp.* 30 March 1985, 565–578.
[54]   E. L. Bonahan Jr., "F/A 18 Composite Wing Automated Drilling System," *Proc. Nat. SAMPE Symp.* 30 March 1985, 579–585.
[55]   R. A. Postier, "Factory Automation for Composite Structures Manufacturing," *Proc. Nat. SAMPE Symp.* 30 March 1985, 586–594.
[56]   O. Weingart, "Flexible Assembly Subsystems," *Proc. Nat. SAMPE Symp.* 30 March 1985, 820–831.
[57]   R. H. Sievers Jr., "Robotic Opportunities in Advanced Material Processing," *Proc. Nat. SAMPE Symp.* 30 March 1985, 832–842.
[58]   J. F. Wagner, "Intelligent Methodology for Automation and Robotics," *Proc. Nat. SAMPE Symp.* 30 March 1985, 843–854.
[59]   J. F. Kober, "Automated Fiber Placement – Process Creativity," *Proc. Nat. SAMPE Symp.* 30 March 1985, 1238–1245.
[60]   T. G. Gutowski, et al., "Advanced Composites Automation – Process Modelling," *Proc. Nat. SAMPE Symp.* 30 March 1985, 1275–1283.
[61]   D. J. McNally, J. R. Mitchell: "Inspection of Composite Rocket Motor Case Using AE," *Soc. Plast. Ind. Proc. Reinforced Plastics/Composites Institute* **40** Paper 5A (1985).
[62]   J. W. Saveriano: "Automated Contour Tape Laying of Composite Materials," *Nat. SAMPE Tech. Conf.* 16 October 1984, 176–182.

[63] G. Gugliotta: "Computer Aided Laser Gauging and Inspection," *Nat. SAMPE Tech. Conf.* 16 October 1984, 183–190.

[64] T. M. Harper, J. S. Roberts: "Advanced Filament Winding Machines for Large Structures," *Society of the Plastics Industry, Proc. Reinforced Plastics/Composites Institute* **39** Paper 9E (1984).

[65] R. H. Tolson, J. Sobieszczanski-Sobieski: "Multi-disciplinary Analysis and Synthesis: Needs and Opportunities," *AIAA Pap.* **85-0584** (1985) Apr. 15, 1–12 (Proc. 26th Structures, Structural Dynamics and Materials Conference).

[66] R. R. Emery, F. W. Crossman, *Proc.* 29th Nat. SAMPE Symp. April 1985, pp. 1465–1476.

[67] R. Kaiser: *Automative Applications of Composite Materials*, Prepared for U.S. Dept. of Transportation, NHTSA, Report No. DOT HS-804745, July 1978.

[68] H. S. Kliger: *Design of Cost Effective Hybrid Composites for Automotive Structures*, Paper 18-D, Proc. 33rd Annual Technical Conference, Reinforced Plastics/Composites Institute, The Society of the Plastics Industry, Washington, D.C., January 1978.

[69] R. F. S. Hearmon: *An Introduction to Applied Anisotropic Elasticity*, Oxford University Press, London 1961.

[70] J. E. Ashton, J. C. Halpin, P. W. Petit: *Primer on Composite Materials: Analysis*, Technomic Publ., 1969.

[71] S. W. Tsai, N. J. Pagano: *Invariant Properties of Composite Materials*, Composite Materials Workshop, Technomic Publ., 1968, p. 233.

[72] *Composite Materials Brochure*, Fiberite Corporation, Winona, Minn, 1985.

[73] I. Y. Chang, *Compos. Sci. Technol.* **24** 1985, 61–79.

[74] M. Chaudhari et al., Proc. 30th National SAMPE Symposium, March 1985, 735–746.

[75] *Technical Bulletins*, National Starch and Chemical Corp., 1985.

[76] H. H. Gibbs, *Proc. ICCM V*, 1985, pp. 971–993.

[77] R. L. McCullough: *Concepts of Fiber-Resin Composites*, Marcel Dekker, 1971.

[78] R. M. Jones: *Mechanics of Composites Materials*, Scripta Book Company, 1975.

[79] R. M. Christensen: *Mechanics of Composite Materials*, J. Wiley and Sons, New York 1979.

[80] C. T. D. Wu, R. L. McCullough in G. S. Holister, (ed.): *Constitutive Relationships for Heterogeneous Materials*, Chap. 7 in "Developments in Composite Materials-1," Applied Science Publishers, 1977.

[81] C. Teixeira: *Space Shuttle/High Energy Upper Stage Capabilities for the 1990's*, Proc. 1982 National Telesystems Conference, IEEE Cat No 82CH1824-2, November 1982.

[82] Raj N. Gounder: *Structures and Materials Technologies for Spacecraft Systems – An Overview*, Proc.29th National SAMPE Symposium, April 1984, pp. 1–7.

[83] S. Dastin, H. L. Eidinoff, H. Armen: *Some Engineering Aspects of the X-29 Airplane*, Proc. 29th National SAMPE Symposium, April 1984, pp. 1438–1449.

[84] D. Tanner et al., *Future of Aramid Fiber Composites as a General Engineering Material*, Du Pont, Wilmington, Del., 1985.

[85] *A Leaf with Lots of Spring*, Exhibited Products Guide, 34th Annual Conference, Reinforced Plastics/Composites Institute, Society of the Plastics Industry, February 1979, p. 8.

[86] Joseph A. Sneller, "Thermoplastics Open the Way to Mass-Produced RP Composites," *Mod. Plast.* **62** (1985), no. 2, 44–47.

Construction Ceramics

DIETER HAUCK, Institut für Ziegelforschung Essen e. V., Essen, Federal Republic of Germany (Chap. 1)

ERNST HILKER, Institut für Ziegelforschung Essen e. V., Essen, Federal Republic of Germany (Chap. 1)

ENNO HESSE, Ibbenbueren, Federal Republic of Germany (Chap. 1)

JOCHEN HARTMANN, Cremer + Breuer Keramische Betriebe GmbH, Frechen, Federal Republic of Germany (Chap. 2)

P. SCHUSTER, Cremer + Breuer Keramische Betriebe GmbH, Frechen, Federal Republic of Germany (Chap. 2)

HANSHEINZ VOGEL, Friedrichsfeld GmbH, Steinzeug- und Kunststoffwerke, Mannheim, Federal Republic of Germany (Chap. 2)

1.	Bricks and Structural Tiles	1591
1.1.	Classification	1592
1.2.	History	1594
1.3.	Raw Materials	1594
1.4.	Preparation	1598
1.5.	Molding	1600
1.5.1.	Processing	1600
1.5.2.	Machinery	1601
1.6.	Drying	1605
1.7.	Firing	1607
1.8.	Properties of the Fired Body	1609
1.9.	Coloration and Surface Effects	1610
1.10.	Quality Control and Standardization	1611
1.11.	Economic Aspects	1612
2.	Stoneware	1613
2.1.	History	1613
2.2.	Raw Materials	1615
2.3.	Composition	1616
2.4.	Preparation	1618
2.5.	Molding	1621
2.6.	Drying	1622
2.7.	Firing	1624
2.8.	Glazing	1626
2.9.	Stoneware Products	1626
2.9.1.	Clinker Ware	1626
2.9.2.	Sewer Pipe	1627
2.9.3.	Stable Ware	1637
2.9.4.	Products for the Chemical Industry	1637
2.9.5.	White Chemical Stoneware	1642
2.9.6.	Tiles and Slabs	1643
3.	References	1646

1. Bricks and Structural Tiles

Bricks and structural tiles are mass-produced ceramic products requiring large amounts of raw materials for manufacture. Brick and tile making is mostly a low-value industry, but is nevertheless often problematic because of the dependence on suitable clay close to the manufacturing facility. These raw material problems, the increasingly higher quality standards, and competition from other building materials necessitate costly but economical manufacturing techniques.

1.1. Classification

In order to give an idea of the essential features of the various products, some porosity and strength values specified in the German DIN standards are given in the following.

Bricks. Ordinary bricks constitute the largest fraction of total brick production. They have a porous body. If there are no special requirements in regard to appearance or resistance to weathering, they are called *inner-wall bricks*, or *backing bricks*, and can be solid or hollow. For hollow bricks there are requirements for strength (4–28 MPa) and a brick bulk density of 1.0–2.2 kg/dm^3. (The brick bulk density is calculated by dividing the mass of the dried brick by its volume *including* all channels, finger holes, and mortar pockets, or frogs.)

One special variety of brick is the *insulating brick* used to reduce heat flow [1]. For these the strength specification is 2–28 MPa. The brick density must be only 0.6–1.0 kg/dm^3. The porosity is increased by adding materials that burn during firing. Insulating brick can be as large as 49 × 30 × 23.8 cm.

Progress is being made in producing building components with height equal to the height of the room. These are known as *brick planks* [2] and are produced as monolithic pieces with an extrusion press.

If the brick has surface and color suitable for facing and the brick is able to withstand frost, the brick is classified as *face brick*, or *facing brick*. If the average compression strength is ≥ 45 MPa, the brick is ranked as *extra strong*.

Clinker Bricks. To achieve clinker classification, the bricks must have a specified degree of sintering (water absorption below ca. 7 wt%), average compression strength > 35 MPa, and a *body density* > 1.9 kg/dm^3. (The body density is calculated by dividing the mass of the dried brick by its volume *excluding* all channels, finger holes, and mortar pockets.) Extra strong clinker bricks must have an average compression strength > 45 MPa.

Ceramic clinker bricks, which are manufactured from high-quality nonporous sintering clays, are face bricks having a water absorption capacity ≤ 6%. The net density must average ≤ 2.0 kg/dm^3, and the average compression strength must be > 75 MPa. Furthermore, the surface must have a specified abrasive hardness and colorfastness in light. The clinker materials also include bricks for sewer construction, paving stones, and sintered flagstones, for which the water absorption must be below 3–6%, depending on the intended use.

Roofing Tiles. Roofing tiles are manufactured to produce their particular shape, sometimes with their natural color and sometimes with a slip coating or a glazed finish (see Section 1.9). They are manufactured by extrusion or stamping. They usually have a porous, unsintered structure, and to be weatherproof they must be accurately shaped, water impermeable, and able to bear sufficient load.

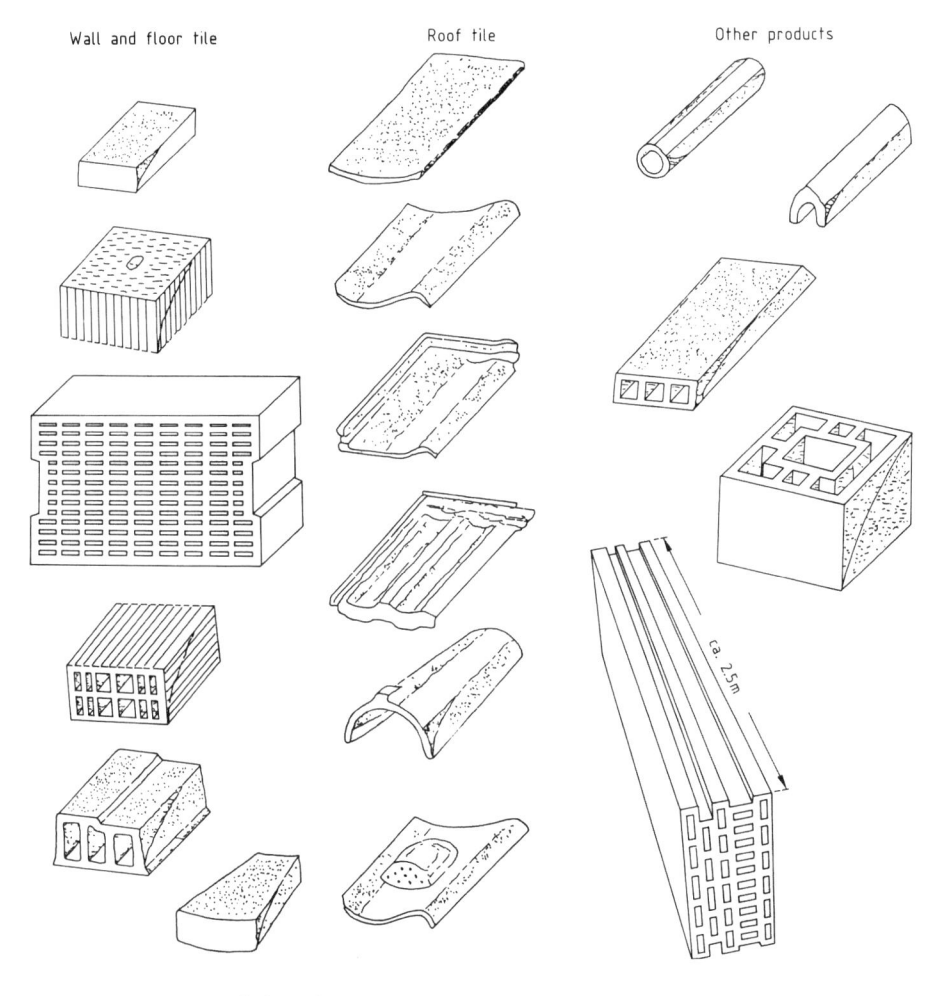

Figure 1. Typical brick and tile products

Special Products. There are a number of porous products other than roofing tiles: wall tiles, floor tiles, drain pipe, and cable conduits.

Expanded clay is a specialproduct [3], [4], [5], usually consisting of spherical particles with a diameter of 1–16 mm, having a densely sintered surface and an expanded finely porous structure. The quality of this product depends critically on the packing density of the particles, the density of the outer skin and the uniformity of the pore structure. The bulk density varies from 0.3 to 0.8 kg/dm^3, permitting use as an aggregate in light concentrate or as an insulator for walls and floors.

Typical brick and tile products are shown in Figure 1.

1.2. History

Brick manufacture is one of the oldest mass-production trades. Excavations along the Nile have revealed the earliest bricks, estimated to be 15 000 years old, which were made by hand from Nile mud, sand, and powdered straw. The dimensions were approximately the same as those of the standard bricks used in Germany. They were air dried and used to construct small dwellings. The oldest description of brickmaking (ca. 1500 B.C.) was also discovered in Egypt, all stages of the process being described, from clay extraction to air drying of the product [3].

Bricks and brick fragments 6000 years old have been excavated in Mesopotamia. In Uruk (Sumeria) both air-dried and fired bricks have been found, as well as colored mosaic pieces of fired clay, these dating from 3100 to 2900 B.C. The magnitude of the industry may be appreciated from the 9.5-km brick city wall with 900 towers.

In the Punjab region of India there are remains of brick towers 8 m in height and dating from 2500 B.C. Other well-known examples are the glazed and embossed brick walls of Babylonia and Persia, and the 5000-year-old houses of Habuba Kabira, on the upper Euphrates, constructed of air-dried, die-formed bricks and provided with a sewer system of fired clay (Fig. 2), including sockets, pipes, and U-sections [6].

The 800-year-old pueblos of the American Southwest are recent by comparison. They were built four or five stories high from dried adobe bricks. A complex of buildings could contain up to 800 rooms [7].

As reported by Pindar (5th century B.C.), roofing tiles were invented in Corinth: later, these were adopted and used over the entire western European part of the Roman Empire. The words tile (English), tuile (French), Ziegel (German), and tegel (Dutch) are all derived from the Latin word for roofing tile, *tegula*.

Eaves trough tiles, antefix tiles, and ridge tiles up to 50 cm × 100 cm were produced. The tile shapes and methods of attaching them have changed little since Roman times [3].

1.3. Raw Materials

A great variety of clay sediments are used, ranging from loose to compact in consistency. The principal raw materials are loams, clays, marls, mudstone, shale, and slate together with smaller quantities of boulder clay, loess, and sand [8]. They can contain all types of suitable clay minerals (layered silicates → Clays), such as illite, sericite, kaolinite, chlorite, and montmorillonite. Quartz is an important component and is always present, and there can also be considerable amounts of calcite, dolomite, and iron compounds. The composition and particle size are the two principal factors that determine the nature of the products.

Some constituents are harmful, mainly coarse pyrites and limestone inclusions, which can cause spalling; sulfur compounds, which can cause efflorescence [9]; and large carbonaceous particles and plant remains.

Whereas common bricks may be manufactured from fairly low-quality brick clays, large hollow bricks require adequate plasticity of the clay body. For hollow bricks, clays

Figure 2. Ceramic sewer pipe and stone-covered sewers in a 5000-year-old house in Habuba Kabira (Courtesy of the Deutsche Orient-Gesellschaft)

and marls are used primarily, sometimes mixed with loam and shale. Other products with a porous structure are produced from these raw materials, e.g., face bricks, paving bricks, roofing tiles, drainage pipes, and cable conduits. Clinker, with its impervious or near impervious structure is produced mainly from shale and stoneware clays. Shale clay tends to be used for the manufacture of clinker face bricks, high-compressive-strength solid or hollow bricks, paving bricks, and sewer channel bricks. Stoneware clays are the first choice for high-quality face bricks and for split clinker flags and flooring tiles (see Section 2.7).

Additives. Additives have increased in importance in recent years, improving both the manufacturing process and the products. A variety of combustible organic materials are used, such as foamed polystyrene, sawdust, powdered straw, powdered brown coal, and materials with organic content including oily filtration aids (fuller's earths), oily metallic hydroxide slimes and coal washery refuse. These additives may increase the porosity and thus the insulating quality of the brick. Furthermore, the amount of fuel required for firing may be reduced [10]–[12].

Powdered limestone or chalk may improve firing properties, increase porosity, and lighten the color [13], [14]. Powdered slags from blast furnaces or power stations are opening agents that add strength, while making the plastic mass leaner; glass powder is also a suitable additive [11]. Iron oxide powder and various sludges provide color [15].

As a consequence of ever more stringent quality requirements, some high-quality clay is commonly used for brick manufacture.

Raw-Materials Testing. Owing to the wide variety of raw materials, effective tests for suitability and effective methods for optimizing the proportions are needed.

Raw materials are characterized by their geological origin and condition as well as by mineralogical analysis, particle size analysis, and to a limited extent chemical analysis. Harmful constituents, e.g., coarse inclusions that could cause defects, salts that could cause efflorescence, and compounds that could produce noxious substances on firing, must be revealed by testing [9], [16], [17].

To assess the workability of the unfired mass, the water requirement, the relationship between water content and stiffness (Pfefferkorn method), the tensile and shear strength (Linseis method), the shrinkage on drying, and the transverse strength after drying are needed. Also needed is knowledge about how sensitive the material is to the drying process (Bigot curves showing shrinkage versus water loss, see Sections 1.6 and 2.6) and the water permeability.

The expected firing characteristics and properties of the fired body can be obtained by dilatometry, differential thermal analysis (DTA), thermogravimetric analysis (TGA), and test firings. Firing at various peak temperatures allows determination of the tendency to form reduction cores, the ability to retain shape (extent of softening at high temperature), the shrinkage on firing, water absorbancy (vapor and liquid) of the fired body, density, and color [3], [18].

Examples are given in Figure 3 of dilatometer curves for three different raw materials, showing the expansion and contraction during firing.

These testing methods are especially useful for working out optimum proportions of various clays and additives. In addition, there are some tests used for production control that are designed individually for particular manufacturing processes [19].

Clay Winning. The raw materials for brick manufacture are nearly always obtained by opencut mining. To reduce costs, all the various strata are usually mined, and these strata are mixed to the extent possible during the extraction process. The higher the quality of the products, the more worthwhile it is to separate the valuable clay strata from the lower value inclusions and strata.

For the extraction of soft or crumbly materials, especially when the strata are horizontal, bucket excavators are suitable because each bucket scrapes away thin layers of material from top to bottom of the strata, thoroughly mixing them in the process (Fig. 4). These excavators are, however, not especially mobile and are unsuitable for selective extraction of materials. For mainly vertical strata, bulldozers or scrapers achieve the best mixing effect (Fig. 5). These can also be used to transport the material over short distances.

Dragline excavators mix somewhat less effectively, but can be used where the geological formations are complex. A dredging shovel suspended by a cable from a

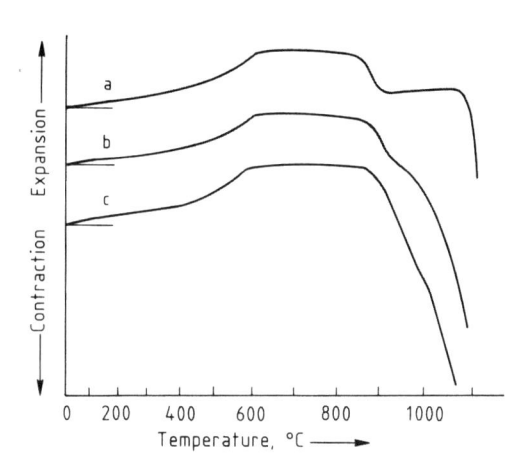

Figure 3. Dilatometer curves
a) Marl clay (calcareous); b) Clay (noncalcareous); c) Slate clay

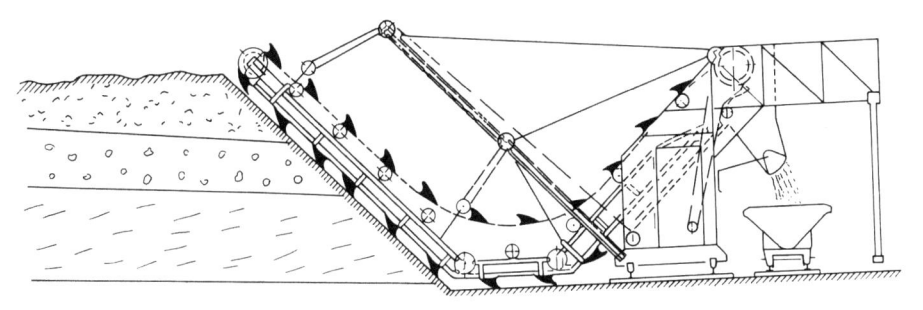

Figure 4. Bucket excavator

Figure 5. Scraper

boom is thrown outwards and then dragged back upwards through the material being extracted.

Power shovel excavators (Fig. 6) are widely used. These were formerly cable operated, but are now usually hydraulic and normally fitted with a grab. An advantage of shovel machines is their great mobility. Unwanted strata (gravel, chalk, etc.) are removed relatively easily, and lenticular deposits of clay can be extracted. Heavy power shovels have been used to extract diagenetically solidified raw materials such as shale. Hard clay stone must first be broken by explosives. Transportation of the materials within the mining area and to the processing area is by rail, belt conveying, or truck, depending on the distance [3].

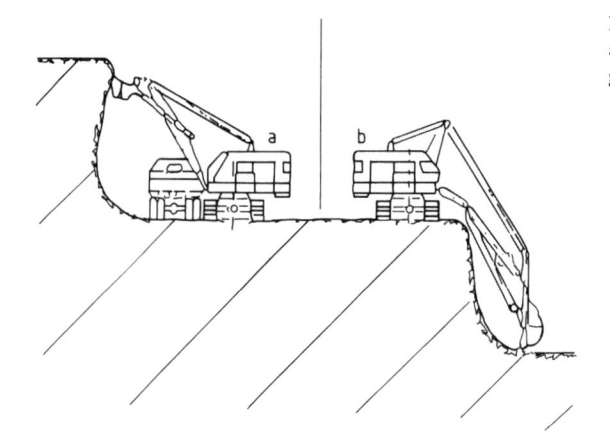

Figure 6. Power shovels
a) Excavating a face; b) Digging below ground level

1.4. Preparation

Raw materials and mixtures that are variable in consistency and composition are converted into homogeneous, easily moldable masses of uniform quality [20]. If the available equipment is unable to achieve adequate mixing, and therefore the quality required, a premix should be carried out with a suitable extraction process in the mine or with a large-scale mixer, e.g., one for mixing stocks of materials. Coarse, hard materials, such as limestone, pyrites, and quartz, should be removed as completely as possible; otherwise, the cost of producing a fine product is too high. Diagenetically hardened clays, such as shale, can be stored in the open air; the weathering that takes place favors deflocculation of the clay.

The choice of the preparation process is determined by the type and condition of the raw material and the end product. There are three processes: wet, semidry, and dry. The wet process is used most often.

Wet Processing. Most raw materials for brick manufacture have a high natural moisture content as mined. This moisture may have to be carefully supplemented during the preparation process with gauging water if the product is intended for wet molding. The constituents are precisely measured out with charging boxes (Fig. 7 A and 7 B) and wet-milled. Edge runner mills are still commonly used for this purpose, their heavy rollers (Fig. 7 C, b) effectively grinding quite hard clay stone. Only the hardest raw materials need an initial crushing operation with a roll crushing mill. The edge runner mill also acts as a mixer to some extent, incorporating most of the water needed to give good molding properties. The intensity of the size reduction and mixing processes is determined by the number and size of the slits in the sieve plate (c).

If necessary, the product may be further treated with coarse and fine crushing rollers, which reduce the particle size to < 1 mm (Fig. 7 D). A few installations

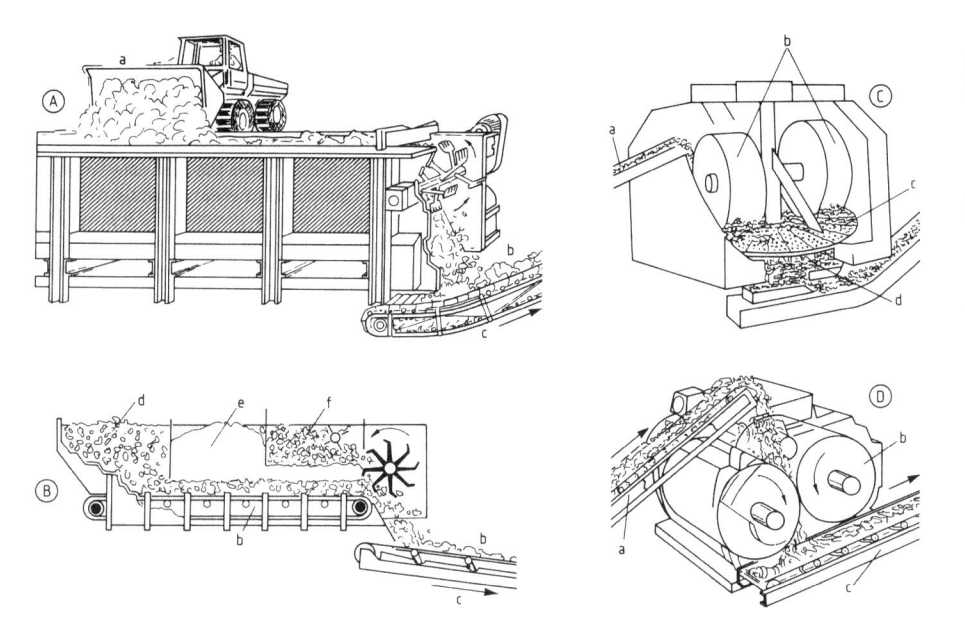

Figure 7. Mixing and grinding machines
A) Box feeder: a) Feed; b) Conveyer belt; c) Direction of the next processing machine
B) Box feeder (cutaway side view): b) Conveyer belt; c) Direction of the next processing machine; d) Clay; e) Sand; f) Sawdust
C) Kollergang (edge runner): a) Feed; b) Rollers (e.g., 1.8 m in diameter); c) Perforated plate; d) Collecting plate
D) Rolling mill: a) Conveyer belt from the kollergang; b) Rollers; c) Conveyer belt to the next processing step

include a series of about three roll crushers, which progressively reduce the particle size of the raw material mixture. Another technique is to treat coarse material, while water is being added simultaneously with cone-shaped rim gears, which produces a fine, soft plastic mass. Lean mixtures, which break down more easily, may usually be molded directly after mechanical crushing. Very plastic materials are better processed intensively with mixing and homogenizing machines to break up the compressed clay mass into small pieces.

The material is often put into temporary storage in large containers, bunkers, or aging towers in a moist condition. This procedure encourages deflocculation and homogenization of the clay, and tends to prevent operating problems. It also ensures a stable supply and good cross mixing of material. However, prolonged storage of moist material is not advisable if pyrite is present, because this slowly oxidizes to sulfate, which produces severe efflorescence on drying.

There are mixing and homogenizing machines specially designed for the molding process, e.g., circular feeders and single and double trough mixers. They facilitate the addition of measured amounts of powders (e.g., barium carbonate), materials to promote porosity (e.g., polystyrene), and water or steam to heat up the mass (hot preparation). These machines, like the aging tower, fulfill two separate functions [3], [1].

Semidry and Dry Processing. These procedures are mainly used in the production of clinker flagstones, more rarely for face bricks, pipes, and clinker paving slabs. The best raw material for these products is shale, which has a low natural moisture content, < 10 wt %, and therefore does not need to be dried. If shale has a high quartz content it is very abrasive and perhaps should be wet ground in a heavy edge runner mill. Dry processing is used from time to time in the manufacture of high-quality face bricks and floor tiles when highly plastic, moist clays, e.g., stoneware clays, are used. However, a large proportion of the water must be removed before processing.

Depending on the type and hardness of the raw material, various types of size reduction machinery are used: jaw crushers, kollergangs, impact mills, hammer mills, pendulum roller mills, and centrifugal grinders.

Moist raw materials can be dried during milling with the waste hot air from driers or kilns. When harmful inclusions interfere with the manufacture of high-quality products, these combined grinding/drying machines can also pay for themselves by finely milling these particles in the presence of high clay moisture contents [21].

An advantage of dry processing is that the material being milled can be mixed with the precise proportions of powdered additives required, e.g., powdered clay, brick dust, sand, metallic oxide pigments, etc. Then the product can be transported satisfactorily by screw conveyor and is easy to remove from the storage bunkers in a uniform manner.

Dry processing is especially suitable when dry pressing is to be used. If necessary, the material is carefully moistened and granulated on trays. In order to obtain the correct granule size and to cause densification, pan granulators may also be used.

1.5. Molding

1.5.1. Processing

The molding process most commonly used for brick is *wet pressing*. Other processes are stiff mud pressing for low water additions and the semidry and dry pressing processes. The choice of pressing process is determined by the raw materials and products [3], [22], [23]. The most suitable moisture content depends mainly on the type of raw material to be treated.

Wet Pressing. Materials with sufficient plasticity may be molded by wet pressing using extrusion presses or stamping presses to form almost all brick products, including wall slabs and flagstones. The correct water content (tempering water) is between 20 and 30 wt % and depends on the ratio of clay to nonclay material. Whereas ordinary bricks tended to be soft pressed, recently there has been a tendency toward stiff pressing. The reasons are improved green strength and therefore dimensional stability, and a saving in energy for drying.

Stiff Pressing. Good plastic masses and fairly soft shale are sometimes molded by stiff pressing. The gauging water is then ca. 15 wt%. Mixtures that contain lean and silty minerals and low-quality clays are unsuitable because their molding properties are extremely sensitive to slight variations in moisture content. Stiff pressing, particularly for hollow bricks, requires high pressures and powerful presses. The main advantage is that the molded pieces are stacked on the kiln car immediately after forming and then can be passed through the dryer and kiln without repeated handling. This method of molding is not so suitable for making products with many holes, e.g., with thin webs or thin walls, nor can it be considered for pressing insulating bricks whose porous materials would be destroyed by high pressure.

Soft Mud Brick Making. By means of strickling machines, lean or silty clay mixtures that are unsuitable for the extrusion process may be molded to give face bricks with a very rough (rustic) surface to fulfill special architectural requirements or for restoration work (sometimes even hand molding is employed). These processes require soft plastic masses with gauging water contents of 25 – 35 wt%.

Semidry Pressing. Semidry pressing, which is, for example, commonly used in Great Britain to make bricks [3], requires a gauging water content of 10 – 15 wt%. As with stiff pressing, the molded pieces may be handled without difficulty and stacked directly onto the kiln cars without predrying. Compared to other molding processes, preparation of raw materials is less costly, normally only involving particle-size reduction and incorporation of the required tempering water.

Dry Pressing. Materials for dry pressing, which consist of shale, require as a rule 5 –10 wt% gauging water. Stoneware clays and other rich clay materials, or those with complicated shapes, can have a water content exceeding 10 wt%. Diagenetically solidified clay stones, e.g., shale, are especially suitable for dry pressing with low moisture content because they can be fairly easily milled to the correct particle size. If plastic clays are used, they must, after drying, be brought to the appropriate grain size by granulation. Molding is then carried out by various stamping presses.

1.5.2. Machinery

Extrusion Molding. The most popular machine for extrusion molding is the screw or helical extruder(Fig. 8).

The prepared mass is usually mixed first in a double trough mixer (b) with a sieve or annular discharge into the screw feed of the extruder (d). The molding head (e) forms an integral part of the

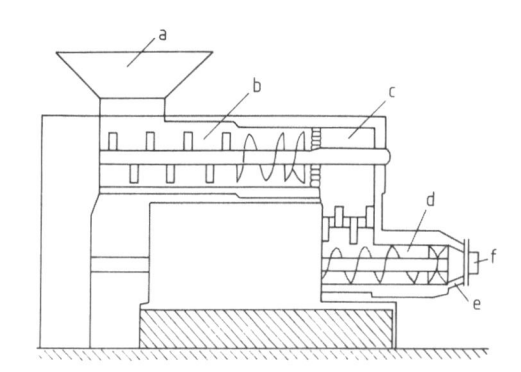

Figure 8. Screw extruder
a) Feed; b) Double shaft mixer; c) Vacuum chamber; d) Extrusion tube with screw; e) Die

die (f). The final pressed shape is formed in the die. Between the double trough mixer and the screw there is usually a vacuum chamber (c) in which the air is extracted from the shredded mass. This improves the binding and flow properties, and allows raw material of naturally poor molding properties to be molded. In addition, the formation of the extrudate by the die is improved by water or steam treatment.

Although the extrusion press has the advantages of continuous operation, high throughput, and varied forms, it has the disadvantage of texture formation, which can be aggravated by drying and firing. Lean mixtures also have a tendency to surface tearing and formation of "dragon's teeth" on the edges. The ring-shaped and S-shaped textures [24]caused by the action of the screw feeder cannot be wholly avoided. However, altering the granule characteristics (e.g., increasing as much as possible the proportion of easily fractured coarse granules) and modifying the screw, the die, and the molding head reduces the number of these textures so that the final product quality is not excessively affected.

Apart from the screw-fed extruder, the disk press (Europress, Fig. 9) is also used. In this press the screw is replaced by a roller feeder (c) resembling a V-belt pulley.

The extrusion method makes possible the production of a wide range of surfaces on the molded shapes by the use of devices such as shaped rollers and brushes located near the die outlet. The surface of the extrudate may be coated with an aqueous sealant to avoid dry efflorescence.

Forming the extrudate with a suitably designed press and die (double die) allows the use of lower quality raw material for the molded shape with a surface layer of high-quality material a few millimeters thick.

After leaving the die, the extrudate is carried by rollers or band conveyors to the cutting machinery, where it is cut into separate pieces. Each type of product requires its own *cutting system*,the most common being stretched steel wires. For roofing tiles, split clinker flags, and large blocks formed from soft material, knives are also used. With a ribbon harp (i.e., an array of parallel ribbon cutters) a long length of extrudate can be cut into a number of pieces in one cutting action. Even though this is quite slow, a high output may be obtained. For extrudates that are difficult to cut, e.g., stiff plastic masses

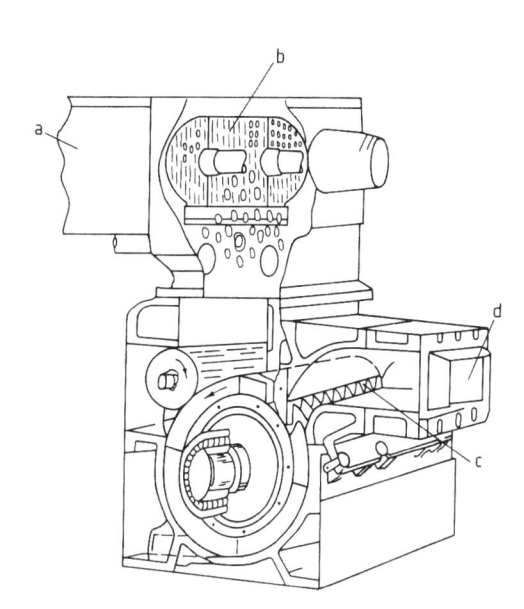

Figure 9. The Europress
a) Feed; b) Vacuum chamber; c) Roller; d) Die

or molded pieces with low green strength, a vibrating or saw action cutter is an improvement [10].

Soft Mud Molding. Before the brick industry became mechanized, the usual process was hand molding (Fig. 10), and this is still used to a small extent. A lump of soft plastic clay was beaten into a sanded mold, the excess material was removed by strickling, and the mold was inverted to release the product. These techniques are usually carried out with mass-production strickling machines.

Press Molding. Products with complex shapes cannot be produced by extrusion; complex shapes are produced by stamping.

Articles such as special types of roofing tiles, wall and floor tiles with embossed surfaces, and face bricks with a "handmade" surface are made on a machine known as a *revolver press*(Fig. 11). The clots of clay are produced on a screw-fed extrusion press, either with or without a deairing stage. All stages of the process are automatic: placing the clots in the pressure mold, the various steps of the compression process, the trimming of the shaped pieces, their removal from the mold, and placement on a carrying frame or band conveyor.

For molding ridge tiles, ventilation bricks, and other special shapes, semi and fully automatic turntable presses, as well as high lift revolver presses, are used.

The molds for stamping presses are still mainly made from high-strength plaster, sometimes bonded with plastics. Other mold materials used are lubricated cast iron, steel, molds with rubber sheeting, and metal molds with vulcanized rubber coating [3].

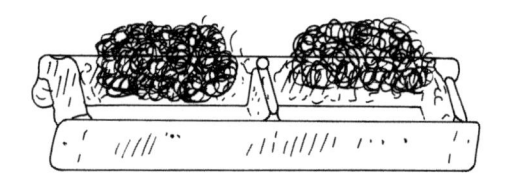

Figure 10. Soft mud molding by hand

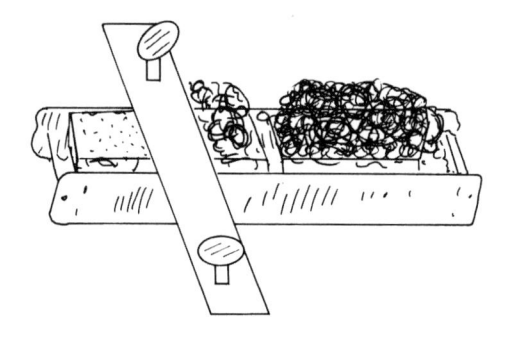

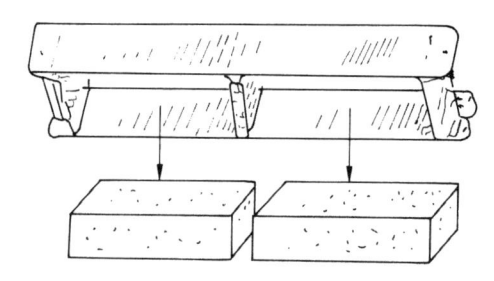

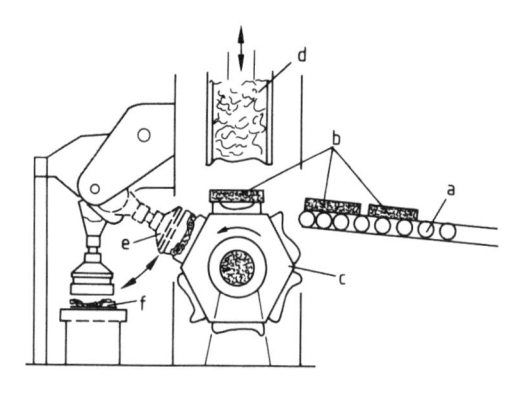

Figure 11. Revolver press (schematic)
a) Conveyer belt from the clot press; b) Clots; c) Drum with lower molds; d) Plunger with the upper mold; e) Green body removal; f) Conveyer belt to the dryer

Dry press material is usually molded with *rotating table presses*, which are driven mechanically, hydraulically, or by a combination of the two. Mechanical or hydraulic toggle lever presses or friction presses are also used. These last two give the highest compression force.

With dry presses, the prepared, loose feed is loaded into the mold, where it occupies about twice the volume of the final molded piece. Exactly reproducible amounts of material must be charged so that the pieces are of constant size, and they must be compressed to the same extent, a task that can be difficult.

While the production of thin wall and floor tiles is relatively easy, thick pieces and pieces containing holes can present more problems. There are large frictional forces between the material and the die walls, and between the grains of the material itself, and these effects reduce the densification produced by the applied pressure. Steps can be taken to reduce these density differences, e.g., lowering the viscosity of the added water with flow-promoting agents such as lubricating gums (sulfite liquor, etc.). Another approach is to even out the differences by applying pressure from both sides.

An important method of preventing delamination is to deair the mixture. It can be improved by stepwise compaction of the grains by vibration processes and also by carrying out the compression in several stages [23].

The machines for the semidry pressing are fairly simple and require little maintenance. The pieces are preformed in molding boxes between two mechanically operated rams and then compacted again in a second shallower molding box (double pressing process).

1.6. Drying

The gauging water, which was needed to make a plastic mass, is removed during drying, the process that immediately follows forming. This drying is associated with marked shrinkage, which varies considerably in extent, generally ranging between 10 and 25% by volume. When the green bricks are formed under conditions of flow, as in an extrusion press, the amount of shrinkage in the direction of the flow can be strikingly different from that in the perpendicular direction as a result of the orientation of the clay particles.

Drying occurs in two stages. In the first stage, during which ca. 1/3–2/3 of the water evaporates, depending on the material, nearly all the shrinkage takes place. A large proportion of the pore water and the water film around each solid particle evaporates, and the particles move closer together. In the second stage, the remaining water evaporates without appreciable shrinkage. The shrinkage can be measured as a function of the water removed with a Barelettograph, which requires only a small sample, and can be displayed as Bigot curves.

Incorrect drying of green bricks can produce distortion and cracking. The drying conditions are determined by both the type of brick and the material, which can vary with respect to mineral composition and grain-size distribution. Fine-grained, unmixed clay minerals, especially montmorillonite, can easily cause defects on drying, even in the case of small amounts. Products made from mudstone or shale, however, give few problems.

The most important factor affecting the drying properties of ceramic masses is the moisture diffusion through the mass. Reducing the sensitivity to drying, and hence drying times, therefore depends largely on improving moisture-diffusion properties. Suitable additives are nonclay inorganic materials, such as sand, slag, and brick dust, or organic fibrous materials, such as paper pulp or sawdust. In some cases, flocculating agents, such as calcium hydroxide, may be used, provided that they do not cause excessive deterioration in the molding properties [25]. The effects of nonclay minerals, and especially their particle size, on drying have been fully discussed [26].

On an industrial scale, brick products are currently nearly always dried with an air current, usually preheated. In times past bricks were almost exclusively dried in the open air or in large rooms above an oven. This is still true in developing countries, where land and labor are less scarce than capital. Modern drying equipment is almost always capable of year-round operation, and the air temperature, moisture content, and throughput are controlled to give optimum product quality [27]. Open-air drying is no longer common.

The systems most frequently used are chamber dryers and tunnel dryers. The *chamber dryers*, which have single or double chambers, are periodically charged and discharged by means of fork lift trucks, with the green bricks stacked in several layers on grids or frames. The whole of the chamber contents are in contact with the drying air.

In the *tunnel dryer*, the feed material, which is carried by frame or pallet cars, is transported through a tunnel so that the drying air contacts the green bricks, either countercurrently or cocurrently. The principal advantage of a tunnel dryer is that it is a continuous process, thus providing a continuous feed to the next stage.

In both types of dryer, the air can be circulated horizontally and vertically, improving the drying effect [28]. Uniformity of drying, and therefore reduction of the time required, can be achieved with batch driers by judicious direction of the air stream [29]. One special characteristic is the rhythmic drying of the Rotomixair system, a system provided with wide currents of air or with longitudinal slit nozzles in the walls. The warm air flows intermittently for brief periods at a high velocity onto the material [3].

The evenness of drying of a charge or a chamber is affected considerably by the arrangement of the green bricks; this determines the permeability of the stack. As a general rule, drying time exceeds 24 h, because even green bricks with many holes are mainly dried by the stream of air flowing around the outside of the bricks [26], [30], [31].

So-called *rapid dryers* achieve crack-free drying of bricks containing holes in less than 6 h. A large number of air jets are used, whereby every single brick is subjected to an air stream over and through it. Whether or not a rapid dryer can be used in an individual case depends both on the shape and on the raw materials of the brick. Raw materials that are particularly sensitive to drying conditions are unsuitable for rapid dryers. The type and significance of the drying stress when air jets are used has been investigated [32].

The tunnel cars that contain the dried bricks are usually kept in heat-retaining tunnels before going into the kilns. Even if the raw materials contain a high proportion of swelling clay, this avoids most reabsorption of moisture from the atmosphere, and the consequent reduction in quality.

1.7. Firing

Brick products are fired predominantly in tunnel kilns because these require fewer personnel, are easily automated, and may be operated under uniform conditions. The stacking of the green brick on the cars is often done by stacking machines, not by hand. The earlier ring kilns (Fig. 12), zigzag kilns, and chamber-ring kilns are now used only rarely. For firing special products in small quantities and for influencing the firing colors of roofing tiles and clinker bricks by means of a reducing atmosphere, batch chamber kilns are occasionally used because they permit individual firing regimes.

As a rule, bricks are direct-fired at peak temperatures of 900–1200 °C. The fuel can be gas, oil, or coal. Clinker bricks and slabs that should not come into contact with combustion products (which might affect the fired colors) are heated indirectly in tunnel kilns. High-quality wall and floor tiles are occasionally fired in electrically heated kilns.

Bricks are usually heated by overhead firing, but sometimes also from the sides. The temperature distribution can be improved by high-velocity jet burners [3].

The energy requirement varies with the type of raw material, firing temperature, and kiln operating conditions. Ordinary bricks fired in a tunnel kiln require ca. 1 MJ/kg of fired material [31]. For the firing of insulating bricks, a considerable proportion of the heat energy is supplied by the combustible substances, e.g., sawdust, that are added to increase the porosity of the material [11], [12].

In order to fire at higher temperatures without causing distortion, e.g., to increase frost resistance or density of the fired body, roofing and wall tiles are fired inside boxes (saggars) to avoid deformation.

Firing Conditions. Because of the variety of raw material mixtures, as well as the sizes and shapes of the products, there are no universal firing conditions: each product needs its own firing regime [33].

The firing rate allowed at each stage of firing depends on the physical and chemical reactions of the raw materials: swelling, shrinkage, evaporation, combustion and decomposition processes, solid-state reactions, and formation of liquid phases, all depending on the thickness, size, and shape of the molded piece [34].

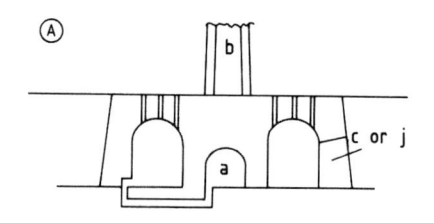

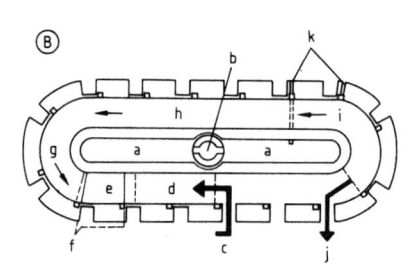

Figure 12. Ring kiln
a) Flue-gas collector; b) Chimney; c) Entrance; d) Stacks of the green bricks; e) Prefiring zone; f) Damper; g) Heating-up zone; h) Burning zone; i) Cooling zone; j) Exit; k) Flue-gas skimmers
The small arrows show the direction of motion

In the prefiring zone, with temperatures up to 300 °C, the rapid evaporation of the water can give rise to significant shrinkage, and consequent high distortion and crack formation, particularly with highly plastic masses containing large proportions of swelling clays.

Another critical temperature range lies between 450 and 650 °C. If the material contains quartz, this quartz can undergo large volume changes producing considerable internal stresses and therefore cracking.

For many materials, the rate at which the temperature may be increased above 600 °C is limited owing to a tendency to give *black cores*, the mottling of the outer surfaces or intumescence. The basic cause is the presence in the raw material of bituminous substances, plant remains, wood, coal, or combustible materials that have been added to increase porosity of the fired body or to economize on expensive fuels. The oxidation of these substances must be complete before the diffusion of gases is obstructed by vitrification of the material. This may be achieved by correct dwell times at set temperatures, which must be determined for each case [35].

Materials rich in limestone, such as marl clays, are useful for the manufacture of light-weight insulating bricks. However, significant shrinkage, causing cracks, can take place in the range of 800–900 °C because of silicate formation, especially in the case of large-sized bricks.

Cooling of brick products through 573 °C is a problem because at this temperature any free quartz changes from the high-temperature (β) to the low-temperature (α) form, with an accompanying reduction in volume (quartz transition). In products that require a high firing temperature, such as clinker bricks and floor tiles, some quartz can change to cristobalite, and this too gives rise to a contraction on cooling, in this case at 230 °C [36].

Figure 13 shows the typical firing curves for four different raw materials or product types.

Possibilities of controlling the emission of harmful substances, e.g., fluorine, during firing has been described [17], [37].

Expanded clay products occupy a special position. The wet-formed granules, which consist of pellets from a pan granulator, or fragments produced from shale by size reduction [38], are usually (ca. 90%) fired in rotary kilns [3]. The feed material is heated to ca. 1150 °C in 30–45 min, with the firing controlled so that the outer skin is densely sintered before the gas begins to form in the interior, which causes the clay spheres to swell [4], [39].

1.8. Properties of the Fired Body

The fired body acquires the properties that render it suitable for its particular use mainly in the finishing burn, i.e., the time that the piece spends near the peak temperature. At the same time the density and strength of the body increase. In general, a longer residence time around the peak temperature improves product quality.

In the course of this sintering process, new mineral phases, mainly silicates, are formed as well as the glassy phases that bind the whole body together. The final properties that are desired are produced not only by the nature of the raw materials, the pretreatment, and composition but also by the firing process [40]. Porosity data, such as water absorption and strength, are used to assess the degree of sintering. The designation *sintered* denotes in practice that a definite, prescribed degree of densification and hardening has taken place.

Most brick products, such as inner-wall bricks, face bricks, roofing tiles, floor tiles, and drainage pipes, can be and even should be porous. For these products then, the required fired body strength, weather resistance, etc., can be attained before the sintering process is so advanced that softening and deformation occur as the result of the excessive formation of liquid phase.

Lightweight insulating bricks are made especially porous (body density of ca. 1.2 kg/dm^3) by incorporating substances that burn out during firing. Convenient raw materials are those that on firing produce a porous structure without additives, either because of their naturally high organic content (e.g., the overburden clays in the brown coal regions) or because of a high content of finely divided chalk. The relatively high porosity of fired products from chalk-containing raw materials (marl clays) is produced not only by the formation of CO_2 from calcium carbonate but also by the formation of calcium silicates, which prevent shrinkage at 900–1000 °C, sometimes even leading to expansion. The fired strengths of lightweight bricks made from clays rich in chalk are comparatively high; therefore, calcium carbonate as limestone or

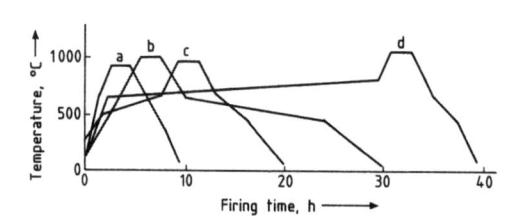

Figure 13. Typical firing curves

powdered chalk, is sometimes added to clays poor in calcium carbonate [13]. A further feature of chalk-containing bodies is that they must not be heated above 1060 °C, or they soften when calcium iron silicates suddenly melt.

Clinker products are manufactured from materials low in chalk. These materials, on account of the minerals and potential fluxes present, can withstand higher temperatures, reducing atmospheres, and prolonged dwell times at the peak firing temperature, becoming dense without deformation.

A special firing method, known as the Hydrite process, was developed in the German Democratic Republic. This process involves firing the green bricks in indirectly heated kilns with maximum temperatures of ca. 780 °C. The dehydratation of the clay minerals produces a moist atmosphere, which irreversibly strengthens the fired brick, and, it is reported, giving higher compression strengths than by normal firing methods [41].

1.9. Coloration and Surface Effects

The color of fired brick products is determined mainly by the ratio of iron (III) oxide to aluminum oxide if the lime content is low. If the lime content is high, the color is determined mainly by the iron oxide/calcium oxide ratio [14]. Other raw material properties and firing methods play a role. Iron oxide usually gives red brick; lime tends to produce whitish-yellow brick. Incorporating a reducing stage in the high temperature and cooling zones produces varied shades, from yellow to green, blue, and gray black, in place of the usual red. This is mainly caused by the reduction of iron (III) to iron (II). If the reducing stage is prolonged and is continued until the product has completely cooled, small particles of carbon deposit to give a silver-gray appearance. This process is mostly used for the production of silver-gray roofing tiles. The use of raw materials containing adequate lime combined with a firing in a reducing atmosphere produces light colors.

Glazing and Engobing. A broaderproduct range may be achieved by the use of a glaze coating on face bricks, wall and floor tiles, and similar items. These glazes are fused onto the already fired clay body; they consist of a mixture of silica sand, clay, alkali-metal and alkaline-earth metal oxides, lead and boron oxides, and colored metallic oxides, such as CoO, Cr_2O_3, MnO_2, and Fe_2O_3.

Sometimes roofing tiles are also given a glaze. Usually, however, they are engobed with a porous layer, i.e., they are coated with a fine clay that gives a color on firing [42]. A shiny surface may be obtained if these coatings also contain lead silicates, alkali-metal compounds, etc., which melt at low temperatures (sintered engobing). Because glazing and sintered engobing both have the effect of densifying the surface it may be necessary to allow for the loss of moisture at the back of the brick to avoid frost damage [43], [44].

Solid Coloring. Occasionally, face bricks, tiles, and roofing tiles are colored throughout. Some examples of substances used are as follows:

iron (III) oxide	red
manganese clays or manganese/iron ores	brown
mixtures of manganese/iron ores with chromium oxide	black

To reduce the cost, the oxides used are ores or byproducts from other branches of industry, if possible [15]. The use of chromium oxide to produce a green color is economical only for high-value floor tiles.

Other methods of modifying the surface appearance of face bricks are treatment with rollers or brushes, removal of the surface skin, application of sand, surface treatment with combustible materials, and rock facing, i.e., mechanical removal of the outer surface of the fired brick.

1.10. Quality Control and Standardization

Continuous quality control of construction ceramics by recognized institutes is gaining in importance because over the years the quality requirements have increased considerably. In addition, most producers have their own quality control. The recent trend is to control the quality in-house and to rely on external testing only to monitor in-house quality control [5], [44].

In the *Federal Republic of Germany*, for example, the product requirements vary significantly with the end use but are standardized:

DIN 105		Mauerziegel (bricks and tiles)
	Part 1	Vollziegel, Hochlochziegel (facing and inner-wall brick)
	Part 2	Leichthochlochziegel (insulating brick)
	Part 3	Hochfeste Ziegel und Klinker (high-strength bricks and clinker)
	Part 4	Keramikklinker (ceramic clinker)
	Part 5	Leichtlanglochziegel und Leichtlangloch-platten (brick planks)
DIN 456		Dachziegel (roof tiles)
DIN 4 159		Ziegel für Decken- und Wandtafeln (wall and floor tiles)
DIN 4 160		Ziegel für Decken- und Wandtafeln (wall and floor tiles)
DIN 278		Tonhohlplatten und Hohlziegel (hollowbricks and tiles)

DIN 4 051   Kanalklinker (bricks for sewer construction)
DIN 1 057   Mauersteine für freistehende Schornsteine (bricks for freestanding stacks)
DIN 18 503   Pflasterklinker (paving clinker)
DIN 18 158   Bodenklinkerplatten (flooring clinker)
DIN 1 180   Dränrohre aus Ton (ceramic drain pipes)

European Standards. For several years now European standards (EN) have tended to replace the various national standards. For example, the quality control for split flags has been established as an European standard.

International Standards. The International Organization for Standardization (ISO) consists of representatives of national standardization organizations. Some non-European countries, e.g., Australia, Brazil, Canada, China, India, the Soviet Union, and the United States, are also represented. The technical commission ISO/TC 179 has published "Materials Testing Methods and Requirements," which is important for ceramic products. In addition, the design and construction of masonry structures, both reinforced and non-reinforced, has been standardized. For developing countries, the *simple rules* have been made available. A technical commission standard on "Construction and Design" is currently being formulated.

Table 1 shows the standard lengths of bricks in several countries. The testing of ceramic raw materials and products is treated in more detail under → Ceramics, General Survey.

1.11. Economic Aspects

In 1984, brick production in the Federal Republic of Germany was ca. 5×10^9 units (one unit equaling the standard format $240 \times 115 \times 71$ mm). This production is equivalent to ca. 82 000 homes. In 1981, the fraction of brick wall construction in homes in the Federal Republic of Germany was 43%. Sand-lime blocks and expanded-concrete blocks accounted for 34% and 9%, respectively, of wall construction. Ceramic tile is used for 22% of the nonflat roof surface in the Federal Republic of Germany.

The number of brick producers in the Federal Republic of Germany decreased from ca. 1000 to ca. 320 over the 20-year period 1965–1985, the number of employees decreasing from 58 000 to 12 500. In the same period, the total production of bricks decreased by 22%; that of roof tiles, by 32%. In countries such as Belgium, France, Finland, Great Britain, Holland, and Sweden, the number of producers has decreased 75–80% [5].

In 1982, a modern brick plant in the Federal Republic of Germany could produce ca. 120 000 units (ca. two homes) per day, i.e., 44×10^6 units per year at full capacity. The capital cost of such a plant was ca. 25×10^6 DM in 1982. The sales were ca. 18×10^6 DM, wages accounting for ca. 20% of the sales.

Table 1. Dimensions of standard bricks [1]

Country	Nomenclature	Dimensions, mm
France	most usual size	220 × 110 × 60
Germany, Federal Republic of	NF	240 × 115 × 71
United Kingdom	BS	210 × 102.5 × 65
United States	standard	203 × 95 × 57

Table 2 compares brick and roof tile production in several European countries [45].

Brick production is reported by number of bricks or by cubic meters of brick. Tile production is reported by number of tile or square meters of tile.

2. Stoneware

Stoneware is a traditional, clay-based ceramic with a densely sintered body, which is not translucent and has a conchoidal, stonelike fracture. Its external characteristics distinguish stoneware from other ceramic products: The densely sintered body is not absorbent and does not adhere to the moist tongue, which clearly differentiates stoneware from brick, grog (chamotte), and earthenware. Stoneware differs from porcelain mainly in the color of the body and often also in its coarser structure. Both unglazed stoneware and unglazed porcelain are impermeable to liquids. Light-colored stoneware is a type of whiteware.

The highly plastic nature of the clay mass for stoneware, the many possible methods of forming the shape, and the useful physical properties of the sintered body have led to variegated stoneware products during the course of the historical development. These include tableware, art objects, construction ceramics, and industrial ceramics. The division into coarse and fine ceramic products is based on the visible macrostructure of the body. Among the coarse ceramics are clinker, sewer pipes, and stable equipment; fine stoneware includes all densely fired ceramic products that are not porcelain.

2.1. History

The earliest stoneware is Chinese and dates back to the 1st and 3rd centuries A. D. All earlier ceramic products were porous, underfired pottery. The Chinese were the first to make a completely nonporous, brown stoneware. The Koreans and Japanese followed. In the 9th century A. D. the Chinese succeeded in producing white stoneware, representing the transition to porcelain. The famous China red glazes were the model for many European developments.

Table 2. Production of European countries in 1984

Country	Plants	Employees	Bricks, 10^6	Roof tiles, 10^6
Austria	62*	2 150**	969	1.7
Belgium	61**	2 275**	1.4 m^3	
Denmark	47	1 170**	409	20.6
Finland	29**	930**	136	
France	242	8 400	4.6 m^3	42.2 m^2
Germany, Federal Republic of	325	12 500	11.7 m^3	23.0 m^2
Great Britain	200	14 500**	5.2 m^3	2.0 m^2
Italy	513	27 000**	11.6 m^3	33.0 m^2
Netherlands	86	2 150	1 600	41.5
Switzerland	31	2 255	667	131.8

* 1982.
** 1983.

In Europe, where only porous earthenware had been produced in ancient times, there was a course of development similar to that in the Far East. Roman water pipes show that the ancient Romans were capable of manufacturing a fairly nonporous fired body. These high-quality pipes can be regarded as the first forerunners of industrial stoneware. With the fall of the Roman Empire, however, the art of stoneware manufacture was lost.

Stoneware reappeared in Middle Europe in the 11th century, favored by the availability of suitable clay, mainly in the Rhineland in the Eifel region (northwest of the Moselle) and the Westerwald (between the Lahn and Sieg rivers). Higher firing temperatures than was usual produced densely sintered bodies of great hardness and strength. At first these ceramics were unglazed. Stoneware manufacture was established in many other localities during the Middle Ages, e.g., Thuringia, Saxony, Bavaria, Silesia, and the Brandenburg March. The salt glazing process was discovered in the Rhineland in the 11th century.

In the main centers of stoneware manufacture in the 14th–16th centuries, especially in the Raeren/Aachen region, the product was brown. Colored oxides, such as cobalt blue enamels and manganese dioxide allowed brown to violet colors to be obtained. The earliest manufacture of stoneware with a whitish to light-yellow body was in Siegburg. Typical Siegburg ware of ca. 1400 was decorated with detailed raised reliefs, which required a very fine, plastic clay body. Westerwald ware of the 16th century, especially from the Höhr-Grenzhausen region, may be recognized by its bluish-green body, incised patterns, and blue, violet, and brown decorations. Salt-glazed ware with red-brown colors was also common by this time. Very dark brown types of stoneware made their appearance in the early 17th century in the Creussen region (near Bayreuth). These were the precursors of Böttger ware (1708–1709), which led to the development of porcelain in Germany.

The Bunzlauer brown tableware of Silesia was not salt glazed, but had a clay glaze that was made from a mixture of several locally available earths. The establishment of technical colleges specializing in ceramics in Höhr-Grenzhausen (1879) and Bunzlau (1897) gave new impetus to the pottery industry and was a turning point in the industrial development of these regions.

Dense stoneware was first manufactured in England in the 17th century. In 1688, the Ehler brothers from Nuremberg founded a large pottery in Staffordshire; they also founded the stoneware industry in Lambeth, where their company became known as Doulton. From the 18th century onwards, English stoneware attained a worldwide reputation through the Wedgwood Company.

In the 19th century, industrial products became important. The manufacture of sewage pipes led the way. In Germany, materials were developed that were watertight, acid proof, abrasion resistant, and mechanically strong, and these replaced the porous pipes of English origin. In England, the chemical industry had attained importance during the first half of the 19th century, and thus it was here that Wedgwood and Doulton produced the first vessels, apparatus, and cooling coils from stoneware for chemical use.

In 1904 several German stoneware factories amalgamated, forming the Deutsche Ton- und Steinzeug-Werke AG in Charlottenburg. In 1922 they combined with the company Deutsche Steinzeugwarenfabrik für Kanalisation und chemische Industrie in Friedrichsfeld (Baden). The two companies were world leaders in the field of stoneware for industrial chemical application.

Friedrichsfeld became famous for its Hoffmann stoneware body, a gas tight material that was relatively insensitive to temperature changes, and for another product, *Korundsteinzeug*(corundum stoneware), which had extremely high mechanical strength. The Deutsche Ton- und Steinzeugwerke AG provided a completely densified stoneware, *DTS-Sillimanite*,which was suitable for high-voltage insulators.

2.2. Raw Materials

The most important raw materials for all types of stoneware are the stoneware clays that contain kaolinite and fireclay (25–40% total), together with other main components, illite (25–75%), and sericite (→ Clays). Their relatively high alkali-metal content gives them good sintering properties. Montmorillonite can be present, although only in small amounts. The feldspar content is also small. The quartz content is 20–30%, and the quartz is usually very finely divided. Quartz is beneficial in the formation of a good salt glaze. Stoneware clays in general have a very small particle size, and this is closely related to their most important properties. Between 60 and 95% of the particles are less than 2 µm in size; consequently, stoneware clays are very plastic.

Typical stoneware clays with no added flux must be completely sintered when fired at Seger cones 4a–10 (1160–1300 °C). At the same time they must have as wide a sinter interval as possible: there must be at least five Seger cones between sintering point and melting point. Good stoneware clays sinter between 1200 and 1300 °C (Seger cones 6a–10) and have a cone "squat" temperature of 1580–1750 °C (Seger cones 26–34). Therefore, there is excellent rigidity on firing up to the sintering point, in contrast to porcelain, and thus large articles can be manufactured. The sinter interval can be influenced by the type and amount of clay minerals or fluxes present in the clay. Illite contains alkali metals, and consequently promotes the widening of the sinter interval, but lime, magnesia, and iron oxide, especially iron oxide in a reducing atmosphere, act as fluxes and can significantly reduce it.

Depending on the iron oxide and titanium oxide content, and also on whether the kiln atmosphere is oxidizing or reducing, stoneware color is gray white, stone gray, blue gray, yellowish, or red to brown.

Typical stoneware clays can also be supplemented by plastic clays with a high quartz content, such as kaolin clays and kaolins, depending on the type of product and desired properties. However, there is no need to make use of kaolins for the purpose of achieving fine particle size with consequent high plasticity, these properties being more easily attained through the stoneware clays. Stoneware clays of various qualities (Table 3) are extracted in Germany, Czechoslovakia (Wildstein), and England (Devon).

The amount of emphasis to be placed on clay purity depends on the type of ware. In any case, the clays must be free from coarse particles of sand and quartz, which could have a harmful effect on molding and drying. More important is the effect of these particles in producing stresses and structural damage caused by volume changes associated with quartz transitions during firing and cooling.

Good quality stoneware clays can also sometimes contain pyrites or other iron compounds such as marcasite, hematite, and siderite. These can give rise to melting defects at high firing temperatures unless they are finely ground and thoroughly mixed into the clay body. Calcium carbonate must not be present in stoneware clays because the very small particles of quicklime produced during firing slake subsequently and give rise to spalling.

Nonplastic, nonclay minerals include sand and grog, and some substances that also have a fluxing action, e.g., feldspar, feldspar sands, porphyry, and basalt. Fluxes differ from grog in that they must be finely divided or finely ground. Too much coarse material can cause porosity. The flux content must not be too high, or it can lead to distortion and formation of melting defects, blistering, and pitting.

2.3. Composition

Today a single stoneware clay is generally not used to make pottery, the exceptions being art and decorative tableware. Usually the desired clay properties are obtained by mixing several clays, sometimes with nonplastic fluxes. The proportions of the various clays and nonclay additions are chosen so that the plasticity and sintering properties of the clay body are suitable for the molding and firing process.

Clay bodies, normally quite plastic, are rendered less so by the addition of fired bodies ground to a definite particle size between 0.2 and 1.8 mm. This addition reduces drying and firing shrinkage and, therefore, sensitivity to the drying process. It also improves the rigidity of the mass during firing. Suitable additives are fired stoneware, fired porcelain, or grog produced by firing stoneware clay in a shaft furnace [50]. The amount, particle size, and particle shape of the added grog affect the fired density and strength of the body of the ware (Fig. 14).

The chemical composition of the clay body (→ Ceramics, General Survey, for the Seger convention) lies within the limits given by

$RO \cdot 0.33 - 7.0\ Al_2O_3 \cdot 4.0 - 44\ SiO_2$

with RO varying between 0.7 (CaO + MgO + FeO) + 0.3 K_2O/Na_2O and 0.3 (CaO + MgO + FeO) + 0.7 K_2O/Na_2O

Table 3. Chemical composition of stoneware clays, weight percent

Clay	Region	SiO_2	Al_2O_3	Fe_2O_3	TiO_2	CaO	MgO	Alkalis		Ignition loss	Reference
								K_2O	Na_2O		
Ball clay	England	59.80	26.40	1.00	1.40	0.20	0.50	2.40	0.40	7.90	[46]
Duinger stoneware clay	Lippe*	56.96	26.58	2.57	0.91	0.91	1.38	1.65		9.06	[47]
Lämmersbach clay 101/W	Westerwald*	62.95	27.95	0.55		trace		0.48	0.67	7.42	[48]
Klardorf bonding clay	Oberpfalz*	50.50	32.80	1.67	1.10	0.14	0.40	1.52	0.08	11.30	[48]
Niesky stoneware clay	Oberlausitz	51.80	32.20	1.60	1.70	0.05	trace	1.22		11.80	[47]
Pfalz glass pot clay	Rheinpfalz*	56.39	37.18	1.54	0.96	0.28	0.77	1.46	1.42	11.52	[48]
Pfalz yellow clay	Rheinpfalz*	60.31	31.28	3.76	1.00	0.11	0.13	1.58	1.83	8.18	[48]
Westerwald standard clay FT-W	Westerwald*	64.60	23.00	1.00	1.30	0.20	0.50	2.70	0.20	6.50	[49]
Westerwald stoneware clay 204	Westerwald*	53.60	30.00	2.50	1.60	0.40	0.50	2.10	0.10	9.20	[49]
Wildstein clay AGB	Czechoslovakia	53.30	29.90	3.90	1.20	1.40	0.20	3.20		6.90	[48]

* Federal Republic of Germany.
** German Democratic Republic.

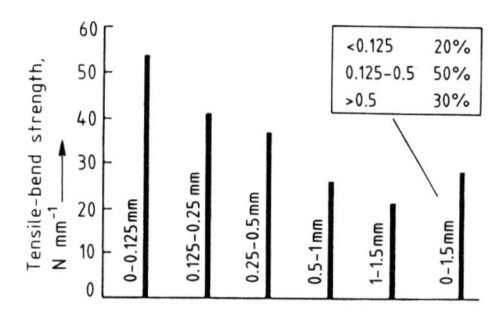

Figure 14. Effect of particle size of added grog (chamotte, chamotte:clay = 30:70) on strength of stoneware [51]

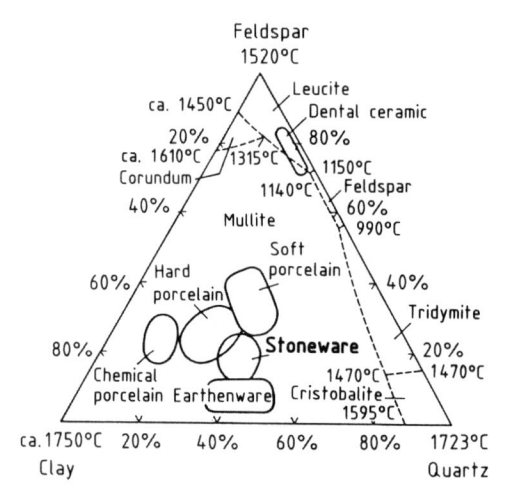

Figure 15. Some ceramic products in the three-phase system clay–feldspar–quartz (% = wt%) [53]

Table 4 shows the chemical compositions of stoneware clay bodies. The stoneware bodies lie, along with porcelain bodies, in the mullite precipitation zone (Fig. 15).

2.4. Preparation

The preparation process brings the raw materials into a workable form, giving a clay body mixed to the correct consistency for molding [54]. Sometimes these processes begin in the clay pit, where harmful impurities, such as pyrite or limestone, are separated as far as possible. The actual preparation process comprises the following steps:

1) Breakdown of the plastic raw material
2) Breakdown, milling, and screening of the nonclay minerals
3) Proportioning
4) Mixing of the constituents

Table 4. Chemical composition of important stoneware bodies, weight percent [52]

Source	SiO_2	Al_2O_3	Fe_2O_3	CaO	MgO	Alkalis K_2O	Na_2O
Ancient Roman	65.62	27.94	1.60	1.25	1.33	0.39	1.42
Vauxhall	74.00	22.04	2.00	0.60	0.17	1.06	
Helsingborg	74.60	19.00	4.25	0.62	trace	1.30	
Voisonlieu	74.30	19.50	3.90	0.50	0.80	0.50	
Baltimore	67.40	29.00	2.00	0.60	trace	0.60	
Wedgwood	66.49	26.00	6.12	1.04	0.15	0.20	
China	62.00	22.00	14.00	0.50	trace	1.00	
China	62.04	20.30	15.58	1.08	trace	trace	
Japan, gray	71.29	21.07	1.25	2.82	1.98	1.03	0.44
Japan, brown	73.68	19.20	4.37	0.70	0.32	1.41	0.32
Bitterfeld	71.24	25.25	2.11	0.11	0.21	0.64	
Krauschwitz	53.77	41.34	3.34	0.03	0.01	1.40	0.10
Moskow	68.05	29.22	1.31	0.13	0.08	0.91	0.24
Rhineland	62.60	34.20	1.70	0.30	0.10	0.90	0.40

The breakdown of the clay must disperse the clay into its finest particles to produce good plastic properties.

The type and quality of the products determine the choice of the preparation process. As the structure of the fired body becomes finer and the specifications more demanding, the preparation process generally becomes more costly. Basically, there are four preparation processes, dry, semidry, hot, and wet, with the last the one used most often.

Dry Preparation [55]. All raw materials are broken down or milled in a dry state to the required fineness. The clay and nonclay materials may be milled separately or together. The clay minerals must sometimes be predried to a moisture content of 5% to facilitate the size reduction process. The nonclay minerals are ground to a predetermined particle size before mixing.

The coarse size reduction of the clay (down to 10–50 mm diameter) is done by gyratory crushers; medium size reduction is by edge runner mills, roll crushers, or cutters; and for the finest product, double hammer mills, pin disk mills, ball mills, centrifugal roll mills, single roller ring mills, or clay disintegration mills are used. A particle size of 0.5 mm can be obtained. If still greater fineness is required, wet processing is necessary.

The grog (chamotte) often needed for stoneware manufacture must first be broken down roughly with jaw crushers prior to medium and fine grinding. Cross-beater mills and pin mills reduce the particle size to 0.5–3 mm. Swing hammer mills, centrifugal roll mills, single roller ring mills and Maxecon ring roll mills reduce the particle size below 1 mm. The sieved milled material can be used directly. The milled material may be classified so that the predetermined particle-size distribution gives the greatest packed density. Classification of the grog is carried out with vibration, resonance, and sonic sieves or with Mogensen sizers. The sieving equipment is generally fitted with magnetic separators (drum magnets), which remove the iron particles produced by abrasion in the milling equipment.

The various plastic and nonplastic constituents are measured by weight or volume [56]. Mixing is the final stage of the preparation process. The breaking down of the clay mix is further improved, and at the same time the finely powdered nonclay components are mixed in homogeneously. Continuous mixers may be used, e.g., single and double trough mixers or sieving mixers. The countercurrent intensive mixer, sometimes with a high-speed agitator, is a batch mixer.

The pure mixing stages can be followed by a souring process, in which the plasticity of the moist clay body is improved by prolonged storage. The clay–water reaction is promoted by the action of microorganisms such as algae and bacteria. This process can be considerably accelerated by raising the temperature, e.g., by steam heating. Improved plasticity can be achieved in a shorter time with souring towers.

Semidry and Moist Preparation. Another approach involves size reduction in which the clays are treated in the dry state or moist as mined. By later addition of the finely ground nonclay minerals and the water, a plastic mass is obtained with a water content of 15–25%.

Hot Preparation. A temperature increase is produced by direct condensation of steam on the clay-body mixture, completely breaking down the clay and making molding easier to carry out. Drying time is also reduced. However, this method of preparation is used less frequently than the others.

Wet Preparation. The usual process for fine ceramics, wet preparation gives the most complete clay breakdown and finest milling, thus producing especially homogeneous mixtures. Milling and mixing take place in the presence of water. Nonclay minerals, such as feldspar and silica sand together with any hard or impure clays, are milled in rotating cylindrical mills with flint pebbles or balls made of porcelain, steatite, or aluminum oxide. Clays and kaolins sufficiently pure and finely divided not to require milling are slurried with water in blungers and then mixed with the wet-milled nonplastic materials. Both raw material streams are treated before mixing with magnets and vibration sieves to remove iron or remaining coarse impurities such as wood, coal, or coarse sand [57]. A filter press may be used to convert the slurry to a plastic mass with a water content of 20–25%. Afterwards, grog or other material may be incorporated in a mixer. Alternatively, the filter cake may be dried, milled, and used in a dry press process or converted into a slip. The slurry can also be converted to granules in a spray dryer or simply to a plastic mass [58].

Table 5 sets out the preparation and molding processes used in the manufacture of various products [59].

Table 5. Preparation processes for various products [59]

	Fine stoneware → ← Coarse stoneware					
	Floor tiles	Kiln linings	Split flags	Stoneware pipes	Clinker	Roofing tiles and bricks
Consistency and granule size	plastic 0–2 mm	plastic	plastic	plastic	plastic 0–3 mm	plastic
% H$_2$O	4–7	18–25 or 28–40	17–22 or 28–40	16–20	3–10 or 16–22	16–24
Molding process	dry pressing	plastic pressing, modeling, slip casting	extrusion, slip casting	pipe pressing (vertical or horizontal)	extrusion, dry pressing	revolver press, extrusion
Preparation *						
Dry	2	2	2	1	1	1
Semidry	0	2	2	2	2	2
Wet	2	1	0	0	0	0

* 2 = commonly used; 1 = restricted use; 0 = not used.

2.5. Molding

Various molding methods are used to produce the shapes from prepared body [60], [61]. The usual processes are plastic molding, dry molding, semidry molding, and slip casting. The most important factor affecting the choice of process is the shape of the molded piece:

Small hollow bodies with uniform wall thickness by slip casting in a mold; larger ones by jollying

Bodies of any complicated shape with variable wall thickness by slip casting with a core; larger ones by molding

Small, symmetrical, thin-walled bodies by turning in plaster molds; larger ones by molding or extruding

Highly plastic, typical stoneware clay bodies are well suited for throwing, jollying, jiggering, or extruding, but unsuitable for slip casting, for which a leaner mass must be prepared. The method is often the most economical shaping method.

The most important forming process for continuous production of stoneware articles is *extrusion pressing* with a continuously operating vacuum screw press. The shaping process in the press, and especially the extrusion of the mass through the die, give rise to flow and the particles align, causing layering and "texturing" [62]. This effect can be reduced by adding grog and thus increasing particle size or by using clays with as large a proportion as possible of particles between 11 and 38 µm [63]. In the intermittent *piston press* the texturing is avoided, but the problem can occur at the die exit.

Moist or wet pressing of plastic masses is possible by using stamping presses, friction screw presses, and hydraulic presses. Heavy metal molds or porous plaster molds are filled with excess plastic mass. Surplus material becomes flowable during the pressing operation and comes away.

An important shaping method is *trimming*, in which preformed, extruded or slip-cast pieces are affixed to molded shapes or complicated apparatus.

Dry or semidry molding requires dry or nearly dry, crumbly or flowable granules with a moisture content of 5 – 8%. Molding is done under high pressure by partly or wholly automatic hydraulic presses, e.g., friction screw presses, and turntable or toggle presses. Large blocks can be formed by stamping a crumbly mass containing ca. 6% moisture through use of compressed-air stamping machines.

A slip with the lowest attainable water content is required for the slip-casting process. For this reason, 0.05 – 0.2% electrolyte, based on dry clay [64], is added to produce a good, flowable slip. Sodium carbonate and/or silicate can be used. The viscosity can be adjusted to suit the wall thickness and corresponding standing time for a casting operation in a mold or around a core. The water content should, if possible, be no more then 3 – 6% greater than that of the corresponding plastic mass, so that the drying process is not too prolonged. Stoneware clays are generally difficult to make into a liquid slip. They are slip cast in porous plaster molds with a water content of 20 – 26%.

2.6. Drying

The water added to produce a moldable mass must be removed by drying before firing. The drying process must not distort the shape of the molded clay or crack it, which demands great care because the highly plastic stoneware clays require much water to make them moldable. The amount of shrinkage on drying is larger for the more plastic, finely divided clays containing larger amounts of water than for the less plastic clays.

In Bourry's diagram (Fig. 16), the volume change of the mass is shown during the course of drying. Although the rate of water loss remains fairly constant during the whole drying period, the volume reduction takes place mainly in the early stages, i.e., while the water film surrounding the particles is being lost. Pores are formed when the particles touch each other and cannot move any closer. Further water loss then merely increases the pore volume.

The correlation between shrinkage on drying and water loss is shown in the Bigot curve, which has a characteristic form for each material (Fig. 17). The clay body, which is initially moldable, is brittle when the drying is complete. The drying operation must be carried out so that the changes in the state of the body take place evenly, and cause neither distortion nor cracking.

Figure 16. Bourry drying diagram for clay bodies [53]

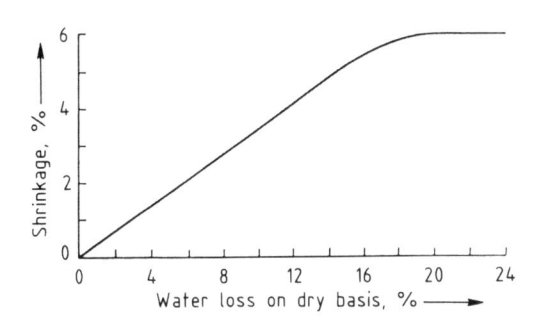

Figure 17. Bigot curve for drying a stoneware clay body

Inhomogeneities or texturing in the molded piece, sometimes produced by extrusion, causes drying stresses and raises the possibility of cracking.

The moist air drying process (→ Ceramics, General Survey) can be used for drying stoneware safely and economically [65].

After shrinkage is complete, usually when the residual water content has reached 6–8%, drying can be safely speeded up by further increasing the temperature and reducing the relative humidity (Fig. 18). Total drying time and the course of each individual drying period must be established for each clay body and each type of product. The value of the lowest permissible residual moisture content should be set. The thicker the walls of the pieces and the finer the clay body, the lower this value must be, e.g., for stoneware pipes it must be < 1%.

Depending on the type of product, batch dryers (floor dryers, chamber dryers, channel dryers) or continuous dryers (tunnel dryers) may be used [66].

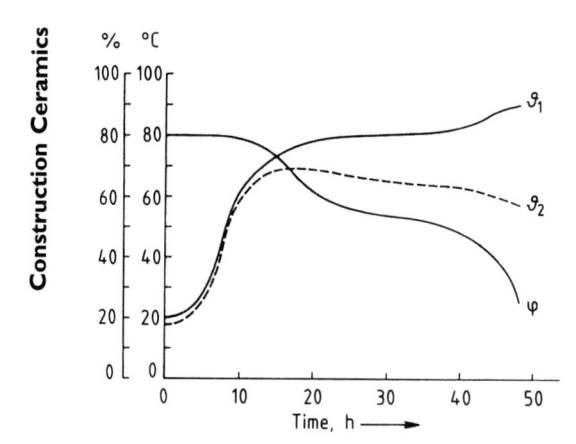

Figure 18. Moist air drying process for stoneware
φ = relative humidity, %; ϑ_1 = dry-bulb chamber temperature, °C; ϑ_2 = wet-bulb chamber temperature, °C

2.7. Firing

The clay body or molded pieces after drying possess sufficient strength to undergo glazing, transportation, or setting in the kiln. During firing, processes take place within the body that are known collectively as sintering. Stoneware undergoes liquid sintering: a glassy phase is formed from the components of the body, which dissolves and binds together the other nonmelting components. The firing temperature must be chosen so that it is high enough to cause strengthening. However, it must not be so high that an excessive deformation occurs in addition to the normal uniform size reduction. The firing temperature of stoneware is 1100–1300 °C. The distinctive feature of the sintering of stoneware is that strengthening takes place while shape is retained.

The sintering properties of the stoneware body are dependent on a large number of factors: the chemical and mineralogical composition of the body; the condition of the raw materials, e.g., particle size, standard of preparation, and packed density; firing temperature; time; and atmosphere. The porosity of the fired body is increased by loss of free water at the start of firing, combustion of organic components of the raw materials, subsequent splitting off of bonded water, breakdown of kaolinite, and decomposition of carbonates. It reaches its maximum at 900 °C. An oxidizing kiln atmosphere in this temperature region is advantageous because the remaining combustible components within the body are burned away. Above 900 °C the firing shrinkage begins, and porosity decreases. Increased flux content gives more of the glassy phase and thus a denser body (Fig. 19). This process takes place more readily when the clay body is finely milled, especially the fluxing component. Higher temperatures reduce the viscosity of the molten phase and cause the quartz to begin to dissolve, more readily if it is finely divided. Another factor promoting the sintering process is the packed density of the raw material. Every addition of grog reduces or delays sintering. All these relationships require that for each stoneware product the composition and fineness of the materials must be specified to suit the product.

Figure 19. Drying and firing shrinkage

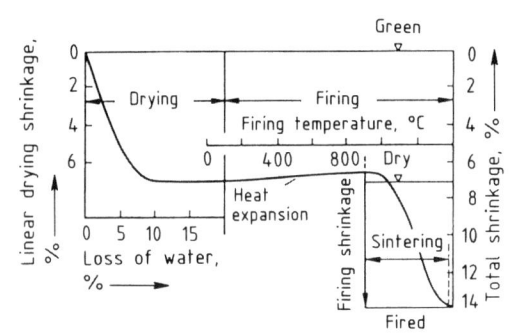

The firing time is also important. There are two possible ways to achieve densely fired stoneware: either by terminating the firing process after reaching a high temperature, or by holding at a low temperature for a longer time, until the optimum degree of sintering is reached. The second method produces less distortion and eliminates blistering caused by overfiring.

Rigidity and strength development during firing are partially a consequence of reactions between the decomposition product of kaolinite, metakaolinite, and molten feldspar to produce mullite crystals, which strengthen the structure.

After the first (oxidative) firing stage, which produces the greatest porosity of the body, the final sintering can be carried out in either an oxidizing or a reducing atmosphere. The choice depends on the desired product. In a reducing atmosphere, i.e., firing with an oxygen deficiency, the reduction must take place between 900 and 1000 °C, so that it can still be effective before the pores in the interior of the body become closed by sintering. If the reduction starts at a lower temperature, below 800–850 °C, the Boudouard reaction can take place:

$$2\,CO \rightarrow CO_2 + C$$

and this can give carbonaceous deposits and blistering at higher temperatures. Reduction causes the conversion of Fe_2O_3 in the raw material to FeO, which reduces the viscosity of the molten phase and promotes sintering by acting as a flux.

The depth of the color of the stoneware body depends on the iron content. Oxidatively fired bodies are light yellow to brown; reductively fired bodies are light to dark gray or blue gray.

Firing is followed by cooling. The rate of cooling is determined by the type and size of products. The stoneware body, because of its composition after firing, still contains free quartz, and must be cooled slowly in the regions of the quartz transition (575 °C) and cristobalite change (200–240 °C).

The furnaces may be intermittent, e.g., chamber kilns, bogie hearth kilns, and top hat kilns [67], or continuous, e.g., tunnel and fast firing kilns [68]. The fuel may be coal, heavy oil, light oil, or gas.

Setting the ware in the kiln is an important operation. Shrinkage must be trouble-free, with no quality problems resulting from deformation. If necessary, the ware may be placed on some supporting material that shrinks at the same or a similar rate to the ware. Because the setting arrangement is often open, i.e. without saggars, fired refractory materials are often used, such as grog, cordierite, or silicon carbide, so that the ware may be equally distributed over the whole of the firing space.

To prevent ware softened through sintering from sticking to the supporting surface, the surface is covered with a suitable engobing, or it may be covered with sand, quartz, or alumina-fiber paper.

2.8. Glazing

The traditional stoneware glaze is *salt glazing*, which is formed at the end of the firing. When the ware is already partly sintered, common salt is thrown onto the burning bed of coal and blown into the kiln by a blast of air, or sprayed into the kiln as an aqueous solution. The process is repeated several times during the firing. The salt is decomposed by the steam in the kiln atmosphere:

$2 NaCl + H_2O \rightarrow Na_2O + 2 HCl$

The Na_2O reacts with the SiO_2 and Al_2O_3 on the surface of the body and forms a fused coating, which on cooling solidifies to a glass.

This glaze formation requires temperatures $\geq 1100\ °C$ and a well-sintered body to avoid penetration of the glaze into a porous surface. It also requires that the body have a sufficiently high silica content. The final color is determined by the Fe_2O_3 content of the body and by the kiln atmosphere during firing and cooling. It can be brown to red-brown, light brown, or gray. Salt glazing is rarely used today because of changed kiln technology and the need to reduce environmental pollution.

When stoneware is to be glazed today, it is coated with the glaze before firing by spraying, dipping, pouring, or brushing. The glaze may simply be a clay, or a feldspar, which may be colored with metallic oxides or colored bodies. The color range is limited owing to the high firing temperature, but an attractive range of effects is obtainable with the help of glazing technology: running glazes, craquelure glazes, matt glazes, ash glazes, crystal glazes, and shrinkage glazes.

2.9. Stoneware Products

2.9.1. Clinker Ware

The term clinker includes relatively thick slabs for floors and façades, wall and street surfacing, and clinker bricks, all having a sintered body and made from usually strongly colored clays (brown to reddish-black when fired). They are extremely hard, strong, and abrasion resistant. The method of manufacture is similar to that for bricks. For shaping, the extrusion press is usual for bricks, and the dry press is usual for slabs. Firing temperatures are 1150–1250 °C (see 446).

Standards for unglazed floor slabs are DIN 105 and Preliminary Standard DIN 18 158 (Dec. 1978).

2.9.2. Sewer Pipe

Stoneware is the classical material for pipes carrying domestic and industrial sewage. In the Federal Republic of Germany between 1980 and 1985, the consumption of stoneware pipes was ca. 300 000–350 000 t/a. This extensive use of stoneware is a tribute to outstanding properties: impermeability, strength, and corrosion and abrasion resistance.

The first sewer systems, which were based on clay ceramics, were constructed by 6000 B.C. in Turkey, although, of course, they were not completely impermeable (cf. Fig. 2). The first sewer system in Frankfurt am Main was built in 1200 A.D.; the first in England, in 1840. Hamburg's sewer system, begun in 1842, first used clinker pipes, then in 1875 stoneware, and in 1900 some concrete. The first factory for stoneware pipes was founded in Germany in 1852.

Preparation. The manufacture of stoneware pipe requires a plastic clay body made up of as many as five stoneware clays, so that any negative effects caused by variability of the properties of a single clay are averaged out. Clays low in chalk may be added, and other additives include basalt, sand, porphyry, and feldspar. In addition, 20–35% powdered grog of grain size 0–2 mm may be added to reduce shrinkage, thus improving dimensional accuracy, reducing sensitivity to drying, and increasing rigidity of the product during firing [69]. The grog can be made from waste kiln furniture (support rings) of the same material, pottery fragments, and specially fired clay grog, e.g., bought-in material made in a shaft furnace. Fine-grained porphyry (0.8 mm) may also be used, either alone or with other additives.

A stoneware grog for sewer pipes has approximately the following size distribution:

<0.063 mm	3%	0.2–0.5 mm	30%
0.063–0.1 mm	5%	0.5–1.0 mm	35%
0.1–0.2 mm	12%	>1.0 mm	15%

Coarse grog should first be broken down in a jaw or hammer crusher to a size of ca. 150 mm. Final size reduction to ca. 10 mm is carried out in a Symons cone crusher or an impact and gyratory crusher. For uniform fine grinding to ≤ 3.5 mm, the Hazemag impact mill may be used or, more often, the Maxecon ring roll mill. The Mogensen vibratory sieve, which cleans itself by electrostatic repulsion, is the most suitable for separation into definite size ranges. With this equipment the mesh size of the top sieve should correspond to the size of the coarsest feed material.

The ingredients for the clay body could be 30% extremely plastic stoneware clay, 22% plastic stoneware clay, 10% lean clay, 8% loam, and 30% grog.

The *clays*for stoneware pipes are not always usable in the raw state, but must undergo preparation. The quality requirements for stoneware, particularly in regard to strength, require the semidry or moist processes. The dry preparation process is hardly used.

In the *semidry*process [70], the various clays first arrive at a circular feeder, which measures out the individual clays in definite proportions. The moist clay mixture produced is then squeezed through the small sieve holes of a purifier, which removes gross contaminants such as coal, stones, and iron minerals. After passing through a roll crusher, the clay mixture is carried by band conveyor to a double trough mixer, where the grog is added. Finally it goes to a sieve mixer.

Molding. Stoneware pipes with sockets are formed by vacuum extrusion mainly by vertical screw presses. The water content of the stiff plastic mass is $< 15-17\%$. Although stiff clay bodies have a tendency to texturing, they may be used for manufacturing straight pipes up to 2 m in length and placed immediately on palettes or trucks where they keep their shape unsupported. They also dry more quickly. Of course, such pipes must be formed with heavy, powerful presses, e.g., 125–150 kW with a compression cylinder diameter of 0.45 m.

Both pipe and socket are formed by the extrusion press. In order to make the socket, a former, fixed to a vertically movable table, is brought up to the die exit before the start of the extrusion process and locked in position, where it exerts the necessary back-pressure against the thrust created by the screw press(Fig. 20). When the socket has been formed, the table is lowered to correspond to the rate of extrusion of the body of the pipe. Grooves are made in the plain end of the pipe, the socket end having been grooved during its formation. The pipe is then cut off and picked up by a gripping or suction device and placed on a palette or drying-chamber bogie. The density of the extruded article is influenced greatly by the form of the screw, i.e., the helix angle and the shape of the part nearest the mouth of the extruder.

The entire operation of the extrusion press, including groove making and cutting off and removing the finished pipe, is usually fully automatic [71]. So-called *top clay* pipes(Cremer + Breuer, Frechen) constitute an interesting development in the automation of stoneware pipe manufacture. These pipes receive an inner glaze coating during the extrusion process and are immediately placed vertically on the tunnel kiln car. The setting ring, which is necessary to prevent deformation, is extruded at the same time in the form of an extra length of pipe. Firing in the tunnel kiln is carried out immediately after the drying process, and the usual transfer from drying cars to firing cars is thus omitted. *Top clay* pipes are unglazed on their outer surface.

The German Standard DIN 1230, Part 1, specifies that stoneware pipes be manufactured with a nominal diameter of 0.1–1.0 m and in lengths of 1.0, 1.25, 1.5, and 2.0 m. Larger diameters are available by special order only. Pipes with extra

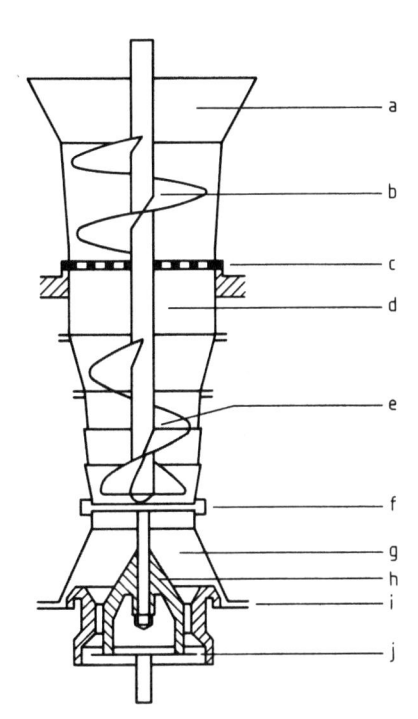

Figure 20. Screw press
a) Feed hopper; b) Prepress; c) Grid; d) Vacuum chamber; e) Press cylinder; f) Holder; g) Bell; h) Former; i) Die; j) Table

carrying capacity are specially strengthened (1.5 times normal wall thickness). The grooves formed in the pipes and sockets enable them to be joined tightly together.

In the United States, England, Belgium, and Holland, smaller diameter pipes are both extruded and dried horizontally. This technique is very popular in the United States, where the shale clays may be extruded in a rigid form and do not deform to oval.

Curved pipes are formed by using horizontal vacuum presses. After the socket has been formed, the emerging pipe is bent around a suitable guide. The process is automatic. The manufacture of branched pipes, at one time a matter of hand joining separate pipes, is now mechanized.

Drying. Artificial drying is carried out today by using modern techniques of measurement and control. Chamber or, more often, tunnel dryers are generally used. The pipes, usually standing vertically, are dried in 24–48 h to < 1% residual moisture by the moist air process. In tunnel dryers with so-called wandering cell drying, the pipes, standing on the grid floors of tunnel cars, pass through a drying tunnel with no blown or circulated air but with finned tubes at floor level carrying hot water. This hot water can be produced in the cooling zone of a tunnel kiln [72]. The heat rises from below through the vertical pipes and the moisture can escape through the roof of the tunnel, the permeability of which increases along the length of the tunnel.

Glazes and Glazing. Although the pipe body after firing is fully sintered and impervious, the surface is usually glazed, at least on the inside, to give a smooth, abrasion-resistant finish. The roughness factor K is thus decisively reduced, values between 0.02 and 0.15 mm being obtained. The earlier process of salt glazing of pipes in tunnel kilns is no longer carried out. Instead, the pipes are glazed by dipping in loam or feldspar glaze immediately after forming or drying [73]. With small- and medium-sized pipes the entire contents of a drying car can be glazed in one operation. The loam glaze consists of iron oxide-containing loam, marl, and flux-rich clay. Loam glazing is done at Seger cone 3a–12 (1140–1350 °C). Because the melting interval is only 3–4 Seger cones, the melting point is reduced by adding other fluxes such as pumice, basalt, calcite, dolomite, wollastonite, zinc oxide, feldspar, or nepheline syenite. The brown color produced by the loam or clay is itself insufficient, and is intensified with iron oxide and/or manganese dioxide.

The composition of a glaze for Seger cone 6a–10 (1200–1300 °C) corresponds to

0.4	CaO
0.25	MgO
0.25	K_2O
0.1	Na_2O
0.35–0.5	Al_2O_3
0.2	Fe_2O_3
3.5–4.5	SiO_2

A glaze composition for Seger cone 7–9 (1230–1280 °C) is 20% glazing loam, 25% Niederahr clay, 30% feldspar K 40 (Mandt), 20% dolomite, and 5% calcite.

The glaze is ground in a well ball mill. This glaze must not have more than 5–10% oversize on a standard 0.063 DIN 4188 sieve.

Firing. Chamber kilns and circular chamber kilns have now been replaced by bogie hearth kilns and tunnel kilns. Because of their greater economy of operation, tunnel kilns have generally taken over. Where chamber kilns are still in operation, they are fired with grid gas or oil. The heat consumption is 8–15 MJ/kg of ware. In contrast, large tunnel kilns specially constructed for stoneware pipe manufacture and having a width of 3–4 m, even 6 m (equal to the car width), require only 3–4 MJ per kilogram of ware. Tunnel kiln cars carrying 0.5 t/m^2 can give a daily output of 100–150 t. The kilns are 100–140 m in length and have a height that will accommodate the pipes in a vertical position. Pipes 1–1.5 m long are fired with additional curved and branched pipes placed on top, but 2-m pipes are fired alone.

To prevent deformation and loss of the circular cross section, the pipes either are fired with setting rings which contract along with the pipes, or are made extra long but scored to enable easy separation after firing.

During firing there must be equalization of temperature over the whole cross section, from top to bottom, to avoid stresses in the pipes, and the same is true during cooling. However, cooling can be rapid — 400 °C/h — until the 800–700 °C region is reached, when cooling must be slowed, becoming as slow and steady as possible in the region between 600 and 500 °C, where the quartz transition takes place.

The cooling system of a reliable tunnel kiln consists of water-cooled tubes in the roof of the tunnel. The heated water can be cooled in a heat exchanger. The cooling of the kiln may be made to follow a predetermined cooling curve by automatic control of dampers under the cold-water tubes. Other cooling systems remove the heat by injecting air into the cooling zones and extracting it at a higher temperature from other locations.

A particularly well-established system for firing stoneware pipes is Cremer's tunnel kiln system, which uses side firing, a hot water cooling system, and continuous operation (Fig. 21).

The tunnel kiln often has a preceding predrying space, where the pipes are heated to ca. 150 °C. During the following heating period, all the remaining water, including absorbed water, is completely driven off. However, the rate of heating must be low enough to avoid spalling. Also, in this zone all organic components should be burned off. The maximum firing rate is between 800 and 950 °C.

The firing zone must densify the body thoroughly. The vitrification, dissolution of quartz, and mullite formation must take place evenly over the entire cross section of the body.

Skillful arrangement of the burners, optimal air flow pattern, correct setting of the pipes on the kiln cars, and a suitable throughput time can ensure an even distribution of temperature across the whole tunnel cross section.

The processes taking place within the cooling zone determine the properties of the vitreous and quartz phases. As long as the vitreous phase still contains viscous components (the case down to ca. 750 °C), rapid cooling may be used, because the resulting stresses are relieved by plastic deformation. Immediately below this temperature, cooling must be very slow, especially in the region of the quartz transition at 573 °C, where a quiescent zone is necessary.

During firing, the reactions shown in Figure 22 take place.

Kaolinite, which has a layer structure, loses adsorbed water and water of crystallization to form metakaolinite, which then is converted to a spinel phase, a type of mullite, and finally, to primary mullite. During these reactions amor-phous silica is liberated in several steps, and largely dissolves into the molten phase, but can also be changed to cristobalite. Illite gives up its adsorbed interlayer and crystallization water, and forms a molten phase at 920 °C and illite–mullite at 1000–1050 °C.

The chemical composition of fired stoneware materials can vary between the following limits:

60–70%	SiO_2	0–1%	CaO
20–30%	Al_2O_3	0–1%	MgO
1–4%	Fe_2O_3	0.1–0.5%	Na_2O
1–2%	TiO_2	1.3–2.5%	K_2O

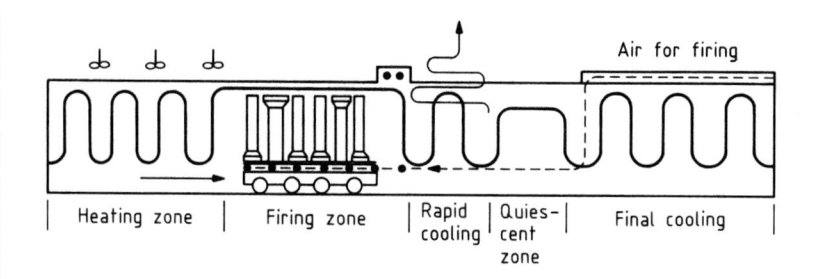

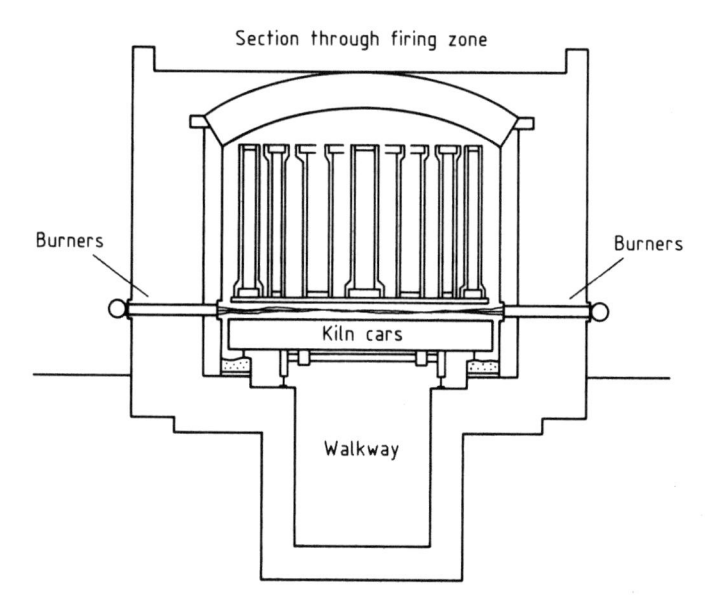

Figure 21. Cremer's tunnel kiln system

However, the properties of the product depend less on the chemical analysis than on the mineralogical composition:

35–50% vitreous phase
10–25% quartz
15–30% mullite
 0–15% cristobalite

If the proportion of vitreous phase falls below 25%, the strength is low, while too much vitreous phase means a brittle product. The mechanical strength is also affected by the mullite, which is formed on firing, especially its particle size and the way in which it is embedded in the vitreous phase. Small crystals of mullite improve the properties of stoneware.

The density of stoneware sewer pipe is ca. 2.5 g/cm^3, the unfired density being 2.1–2.3 g/cm^3 and the total porosity (from texturing) between 2 and 15%.

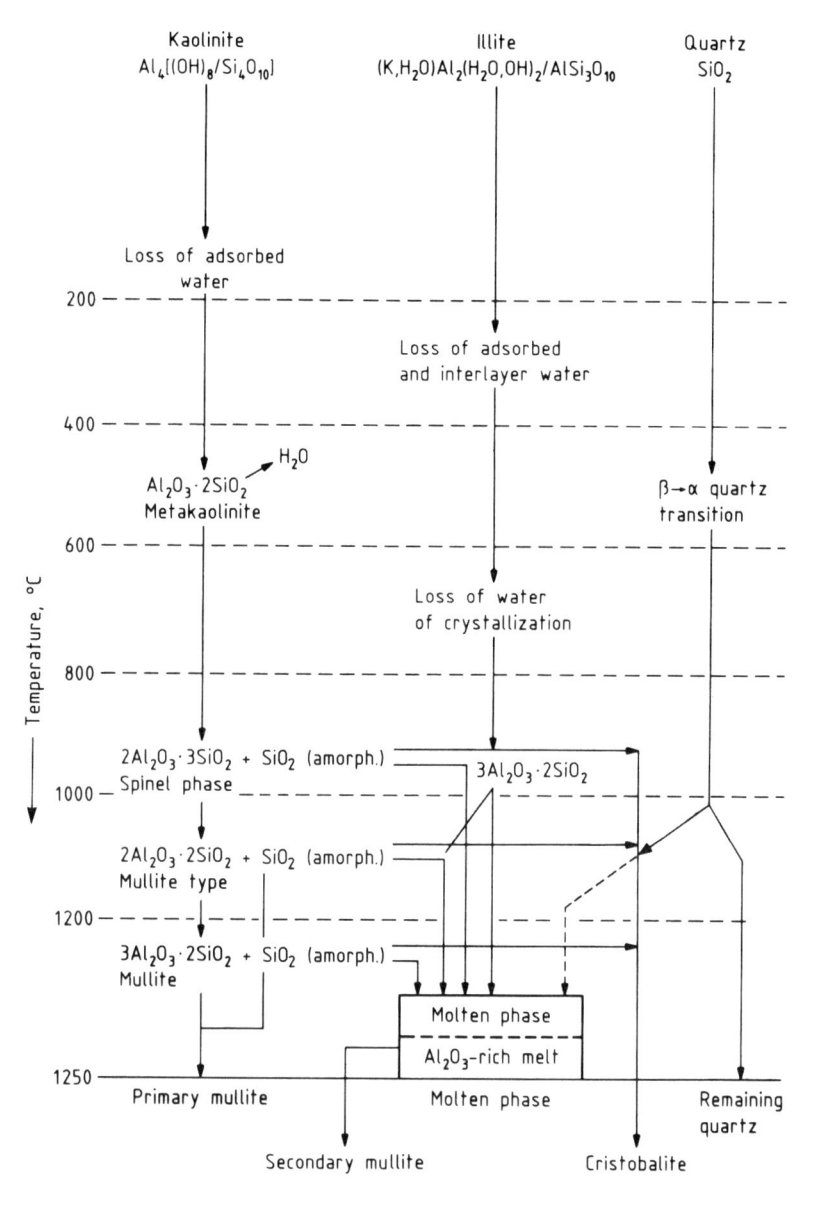

Figure 22. Reactions during firing

Jointing. The componentparts of a stoneware sewer system (pipes, fittings, and special shapes) must be connected together by long-lasting, flexible, corrosion- and temperature-resistant, watertight joints. The seal must also resist tree roots and pressure from inside or outside the piping system. At one time common bituminous materials, which were melted and poured, and tarred rope, both of which were brought to the construction site, were used. This was a time-consuming task and did not always guarantee a leak-proof seal. These have been replaced by elastic seals that are tightly bonded to the pipe. The development of these prefabricated seals (by the Friedrichsfeld GmbH, Mannheim-Friedrichsfeld) was an important contribution to the continued existence of stoneware pipes in their competition with other piping materials. The socket seals were introduced into Germany under the names *Steckmuffe K* and *Steckmuffe L*.

Compression provides the seal. Pipes ≤ 0.2 m in diameter often need to be shortened and, therefore, the seal only comprises a solid ring of synthetic polyester resin cast within the socket (*Steckmuffe L*, Fig. 23). There is an additional seal consisting of a rubber ring.

Pipes over 0.2 m in diameter are sealed with *Steckmuffe K* (Fig. 24), consisting of a silica-filled solid polyester ring inside the socket and an elastic polyurethane ring cast onto the outside of the plain end.

The compression gives increased chemical resistance.

Standard Dimensions. The standard dimensions in Germany are given for pipes with sockets and for seals in DIN 1230, Part 1 (1979) and Part 2 (1979). Testing conditions and methods of control are also given.

Testing. The most important properties to be tested, apart from dimensions, are watertightness and corrosion resistance.

Strength. Tests are carried out on pipe sections 0.3 m in length (method A; Fig. 25) or on complete pipes. In order to allow for the large effect of the socket on the compression strength, the results must be calculated by method A:

$$F_N = \frac{F_B \cdot 1000}{l} \cdot 1.09 \left(1 - \frac{295}{l + 1440}\right)$$

F_N = compression strength, kN/m
F_B = breaking force of pipe section, kN
l = length of pipe section, mm

The minimum values of compression strength that pipes must withstand when tested are given in DIN 1230 (Fig. 26). The ASTM standards are also given in Figure 26. The BSI standards are shown in Figure 27.

Watertightness. To prevent pollution of the groundwater, the pipe seals of a sewer system must be watertight. This requirement is tested with an excess pressure (0.5 bar)

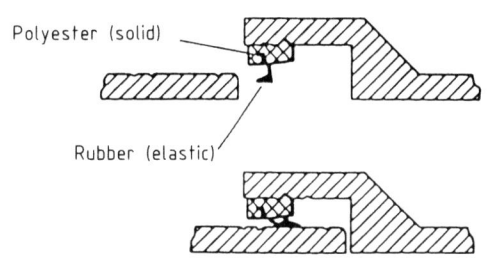

Figure 23. Steckmuffe L

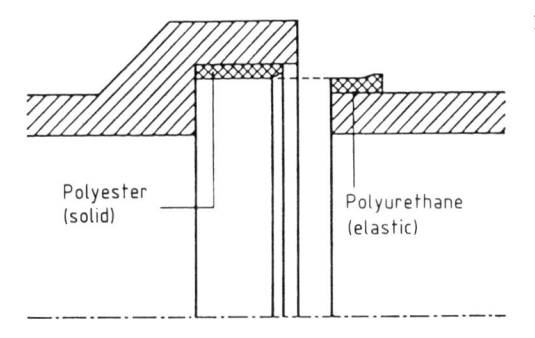

Figure 24. Steckmuffe K

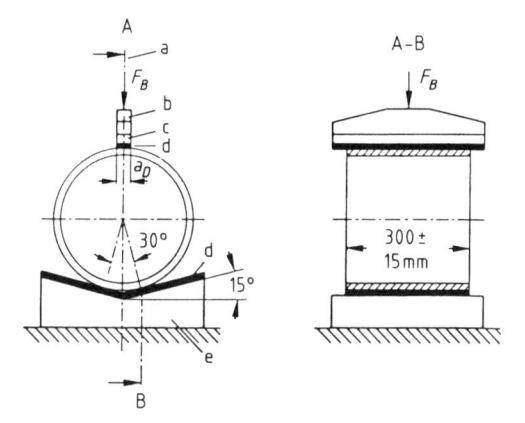

Figure 25. Compression strength testing of pipe sections

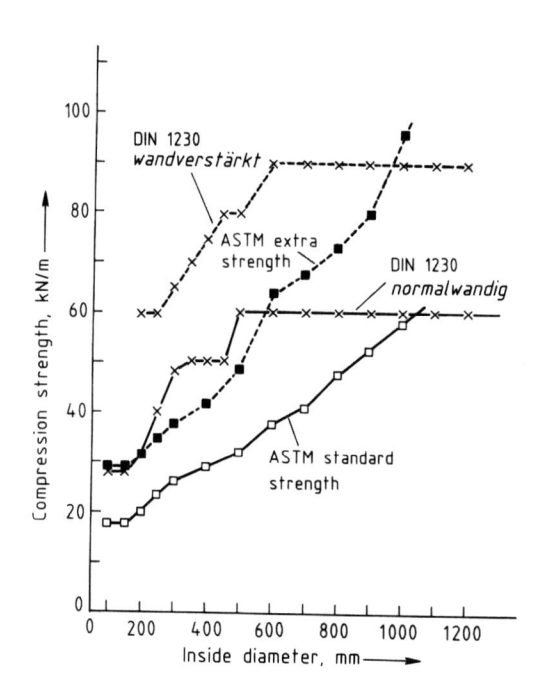

Figure 26. Minimum compression strength — DIN and ASTM

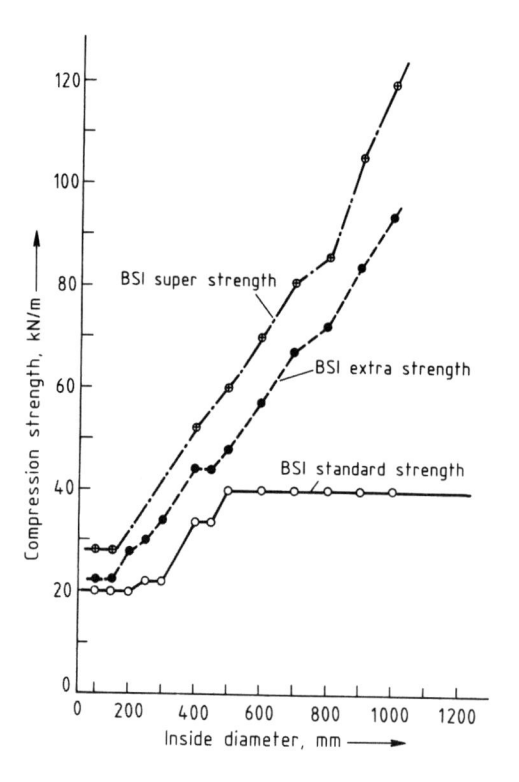

Figure 27. Minimum compression strength — BSI

inside the pipe, corresponding to the maximum pressure of water at 5 m depth. The test takes place over a 15 min period, after 1 h during which the pipe stands full of water at the test pressure. The water loss is measured and the loss factor calculated:

$$W_{15} = \frac{V_{15}}{\pi d_1 l_1}$$

W_{15} = water loss factor, L m^{-2} (15 min)$^{-1}$
V_{15} = water loss, L/(15 min)
d_1 = inside diameter, m
l_1 = length of piping, m

The water-loss factor must not exceed 0.07 L/m² of inner piping surface, and there must be no formation of droplets or wet areas.

Corrosion Resistance. According to DIN 1230, Part 1, stoneware pipes must not be corroded by wastewater, groundwater, or earth material, with the exception of hydrofluoric acid; DIN 51 102, Part 1, describes the determination of corrosion resistance using test pieces of prescribed dimensions that are subjected to the action of 70% sulfuric acid for 6 h in a boiling water bath. The loss must not exceed 0.5 wt%.

Other required properties of pipes relate to roughness and abrasion resistance of inner surfaces. The testing of seals relates to their sealing properties, mechanical properties, chemical resistance, and temperature properties. The observance of all required properties is controlled either internally or by an outside body such as the Güteschutzgemeinschaft Steinzeugindustrie, Köln (Cologne). The internal supervision comprises daily or weekly testing of the quality of the pipes and seals, dimensions, load-bearing capacity, and watertightness. The outside supervision involves carrying out the same tests twice a year without previous notice.

2.9.3. Stable Ware

Stable ware is made from the same material as sewer pipes. The molding processes are extrusion pressing and slip casting in plaster molds. The glazed stoneware articles such as mangers, troughs, drinking basins, and gutters fulfill all hygienic requirements. They may be cleaned either roughly or thoroughly, they are not attacked by acids or fermenting materials produced by decaying fodder, and they do not allow penetration by harmful fungi or yeasts.

2.9.4. Products for the Chemical Industry

Thestoneware in the chemical industry has a variety of uses. Its properties closely resemble those of the material used for sewer pipes, split flags, and floor tiles, and resemble industrial porcelain. Chemical stoneware may have a coarse or a fine structure, although the latter is more common. Products include containers, heating vessels, filters, columns, troughs, basins, working surfaces, pipes, valves, rollers, cyclones, extractor fans, pumps, and bleaching equipment. They must resist corrosion and, very often, mechanical and thermal stresses and abrasion.

Raw Materials. High-quality raw materials such as pure, uniform stoneware clays or standardized clays are offerd by suppliers, as well as kaolins, nonclay minerals, and fluxes. If quartz is necessary, it is added as ground silica sand or quartz powder. Feldspar acts as a flux and also makes the mix leaner. Potassium feldspar and potassium sodium feldspar are used.

Composition. The composition of the clay body for chemical stoneware depends on the molding method and the desired properties. The proportion of clays with more or less plasticity is occasionally supplemented by kaolin, which can improve firing properties but can at the same time lead to greater shrinkage. The addition of fired broken stoneware ground to a particle size of 0.2–1.2 mm, sintered clay grog, or porcelain fragments can have several useful effects, including a leaner clay body that reduces shrinkage on drying and firing, reduced drying sensitivity, and increased rigidity during firing.

Chemical stoneware consists mainly of SiO_2 (40–70%) and Al_2O_3 (25–50%). The 40–60% vitreous phase contains crystalline components such as mullite, quartz, and/or corundum. The approximate makeup of the clay body is

Component	Proportion, %
Stoneware clay, very plastic	20–35
Stoneware clay, plastic	10–15
Stoneware clay, lean	10–15
Kaolin	5–15
Feldspar	10–20
Stoneware grog (0.2–1.2 mm)	20–30

Special Bodies. The use of selected ingredients and especially crystalline materials that may be added or that are formed during firing, such as corundum, indialith (cordierite), and β-eucryptite, which are embedded in the vitreous phase, encourages the development of special properties: mechanical strength, abrasion resistance, thermal conductivity, and resistance to thermal shock. The enrichment or presence of the following oxides tend to produce the physical properties stated:

MgO	Better resistance to alkalis and molten metals; with other suitable additives, formation of cordierite, which has low conductivity and hence good resistance to thermal shock
SrO	Properties intermediate between those of calcium and barium stoneware
BaO	Better resistance to alkalis and molten metals
ZnO	Better resistance to molten metals without affecting resistance to acids
Al_2O_3	Better mechanical strength
Cr_2O_3	Better alkali resistance without loss of acid resistance
ZrO_2	Better acid resistance

Additions of corundum and argillaceous earth increase mechanical strength and hardness. The development of these bodies by the Deutsche Steinzeugwarenfabrik für Kanalisation und Chemische Industrie, Friedrichsfeld (Baden), dates back only to the early 1920s. They are currently manufactured in an improved form as "corundum stoneware" by the Friedrichsfeld GmbH, Steinzeug- und Kunststoffwerke in Mannheim-Friedrichsfeld. They are used for mechanically strong acid-resistant pumps and apparatus.

A low coefficient of expansion is necessary to achieve good thermal shock resistance. This usually incurs penalties such as difficulties in manufacture or in attaining a good standard for other properties. Quartz or vitreous silica may be added as a grog grain owing to its low thermal expansion coefficient (0.55×10^{-6}) [74]; however, below a grain size of 0.4 mm, vitreous silica can be converted on firing to cristobalite, which has poor thermal shock properties. Also, quartz with a grain size over 0.8 mm can lower impermeability. Another problem with quartz additions to the body can be the formation of hairline cracks in the glaze, giving poor densification. By using steatite, talc, or magnesium carbonate in clay bodies and firing over 1150 °C, indialith (cordierite, $2\,MgO \cdot 2\,Al_2O_3 \cdot 5\,SiO_2$), can be formed. Such cordierite bodies have the low thermal expansion coefficient of $(1-2) \times 10^{-6}$ between 20 and 100 °C. Furthermore, as a consequence of the small firing interval, production of densely sintered products with precise dimensions is difficult. Their use is still further restricted in the chemical industry by their poor acid resistance [75].

Products made from bodies containing barium oxide have lower expansion coefficients than mullitic stoneware, although not as low as those of cordierite stoneware. However, the products designed for good alkali resistance are better than mullitic stoneware in this respect [76].

The use of lithium-containing minerals and salts leads to the formation of β-eucryptite crystals, which have a very low and sometimes even negative thermal coefficient of expansion. The body has the extremely low expansion coefficient of $(0.6-2.0) \times 10^{-6}$. The sinter interval is, however, small, and dimensionally accurate products are difficult to make. The thermal shock resistance of these products depends on size, shape, and wall thickness because of the anisotropy of the thermal expansion in different axial directions of the lithium aluminum silicate crystals.

Bodies with increased heat conductivity are obtained by the addition of silicon carbide, ferrosilicon, or silicon.

Preparation. The clay bodies are prepared by the semidry or the wet process (see Chap. 2.4). The latter is preferred if superior mechanical or thermal properties are required or if the surface properties are especially important, e.g., for polished rollers, dense surfaces, or products whose glazed surface has aesthetic qualities.

Molding. Plastic molding and slip casting are the usual methods (see Chap. 2.5). The choice depends on type, shape, size, wall thickness, service conditions, and number of pieces to be produced. Rectangular vessels up to 1000 L in volume are extruded as box sections, and the side pieces are attached with slip. Larger containers are made entirely from vacuum-pressed plates (up to 2 m long) which are also attached with slip. Circular molded pieces up to a diameter of 1 m can likewise be extruded. Larger circular containers and vessels over 1000 L in volume are made by joining pre-extruded pieces in plaster molds, or they can be made on a rotating wheel in a plaster mold. For slip casting with plaster shapes, the clay body is prepared with a water content of 22–26% with the aid of electrolytes such as sodium carbonate, sodium silicate, or humic acid. The method of casting around a core is used most often.

Drying. See Chap. 2.6.

Glazing. See Chap. 2.8. A glazing process is frequently unnecessary, e.g., with polished ware, especially since the acid resistance and liquid impermeability are not produced by the glaze, but by the combination of densely sintered components.

Firing. See Chap. 2.7. The rate of heating is determined by the size and shape of the ware, the firing properties, the weight of the charge, and the kiln and burner construction.

The preconditions for an economical method of firing with rapid kiln turnaround, as well as consistent and improved quality, were provided by new circulatory and jet burners and corresponding kiln construction. The ware is now fired in gas-heated preprogrammed, rapid-firing kilns such as chamber kilns, bogie hearth kilns, and top hat kilns at Seger cone 8–10 (1250–1300 °C). One firing including cooling takes 4–5 days. Tunnel kilns are preferred for large batches of similar articles where the batch weight does not change continuously.

Finishing. Although shrinkage is taken into account at the molding stage, dimensional variations of up to ± 3% are possible. For long production runs, tolerances of ± 0.5% may be achieved. Demands for dimensions accurate within 0.01 mm may be met by machining, and this finishing process can also achieve a polished surface. The grinding medium can be a silicon carbide or often a diamond wheel. Diamond tools are suitable for drilling and cutting.

Properties. Chemical stoneware is resistant to corrosion by reactive media such as acids, solvents, and solutions of salts at all concentrations and temperatures, with the exception of hydrofluoric acid, which will damage ceramic materials even in trace amounts. Resistance to alkalis is strongly dependent on concentration and temperature (Fig. 28). If unglazed or surface ground bodies are to be corrosion resistant, they must be impermeable to liquids. Like all ceramics, chemical stoneware has high compression strength but low tensile strength. It is possible by structural means and by variation of the composition to produce special properties to suit the conditions in which pumps and apparatus operate in the chemical industry.

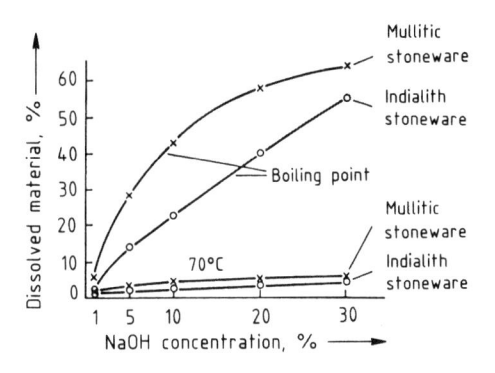

Figure 28. Solubility after 6 h of mullitic and indialith stoneware (1 g, grain size 0.25–0.60 mm) in sodium hydroxide solutions (DIN 51 103) [76]

Table 6. Properties of chemical stoneware types [61]

Property	Symbol	Unit	DKG 150 chemical stoneware	DKG 152 chemical stoneware	DKG 154 corundum stoneware
Al_2O_3 content		wt%	30–35	25–35	40–50
Water absorbency (DIN 51 056, Section 5.2)	W_g	wt%	0–3	0–0.5	0–0.1
Green density	ϱ_R	kg/dm	2.2	2.3	2.5
Tensile strength	σ_{zB}	MPa	10–20	15–30	25–35
Compression strength	σ_{dB}	MPa	100–250	200–300	250–500
Transverse strength	σ_{bB}	MPa	30–40	45–65	50–90
Modulus of elasticity	E	kPa	50	55	60–70
Coefficient of linear expansion	α_t	$10^{-6} K^{-1}$			
20–100 °C			4	4–5	5
20–600 °C			4–5	4–5	5–5.5
Acid resistance (relative weight loss, DIN 51 102, Sheet 2)		wt%	0.5–0.8	0.3–0.6	0.2–0.6

Chemical stoneware is hard wearing, the hardness being 7–8 on the Mohs scale, equal to that of quartz or topaz. Corundum stoneware in which the corundum is very fine (< 50 μm), evenly distributed, and well bonded within the vitreous phase is especially hard wearing. Stoneware products are especially suitable for simultaneously corrosive and abrasive conditions [77]. In the "DKG-Werkstoffkennblätter für technische keramische Werkstoffe" ("German Ceramic Association Information Sheets on Industrial Ceramic Materials") [61] the areas of application, properties, and manufacturers of all industrial ceramic materials are described. Chemical stoneware materials are included under the headings DKG 150, DKG 152 and DKG 154 (see Table 6).

Use. Chemical stoneware can be used for pipes and shutoff devices, ventilation equipment [78], pumps, hydrocyclones, rollers, laboratory equipment (acid resistant basins and large working surfaces), and kitchen ceramics (sink units, working surfaces and cooking appliances).

Standard Specifications. Standard dimensions are given in DIN 7000–7032 for pipes, valves, faucets, vessels etc. and in DIN 12 915 and 12 916 for laboratory basins and large working surfaces in chemical stoneware, this information being also available in Werkstoffblatt (Materials Sheet) 71 of the DECHEMA-Werkstofftabelle (Materials Register). Special standard specifications govern the testing of resistance to acids and alkalis (DIN 51 102, Part 2, and 51 103).

2.9.5. White Chemical Stoneware

The special group of white chemical stoneware arose from a requirement for a material with the same properties as the brown salt-glazed stoneware but with a light-colored porcelain-like body. The *raw materials* are clay bodies of similar composition to the usual stoneware, but with clays as pure as possible, which fire to a light color. To achieve this aim, iron-free raw materials are used and prepared by the wet method. Molding processes must take into account the low plasticity of the clay body. Feldspar glazes opacified with tin or zircon are used for glazing. Flow properties and adherence (for dipping, brushing, or spraying) may be improved with organic thickeners, e.g., carboxymethylcellulose (Tylose from Kalle of Wiesbaden or Relatin from Henkel of Düsseldorf) [79]. The following Seger formula (soft porcelain) is suitable:

0.5	CaO
0.1	MgO
0.15	ZnO
0.25	K_2O
0.3	Al_2O_3
2.8	SiO_2

The ingredients are

44% Norwegian feldspar
6% dolomite
4% zinc oxide

12% calcite
4% kaolin
30% silica flour

White stoneware is fired only in pure, oxidizing atmospheres at Seger cone 8–10 (1250–1300 °C) in chamber or bogie hearth kilns heated with gas or light oil.

It is only a small step from white stoneware to so-called industrial porcelain, which is produced from fine clay bodies without added grog, is fired at Seger cone 8–12, and resembles soft porcelain bodies. It is made into industrial articles such as kettles, columns, pipes, rollers, valves, and faucets whose outstanding property is mechanical strength. The clay bodies may be fired in oxidizing or reducing conditions.

2.9.6. Tiles and Slabs

Ceramic tiles of all kinds will soon be subject to a unified European standard laid down by the Comité Européen de Normalisation (CEN), and stoneware wall and floor coverings occupy an important place in this. According to the proposed CEN definition, tiles are regarded as building materials when they are used to cover floors and walls both indoors and outdoors, irrespective of size or shape. They can be unglazed, glazed, or engobed [80]. In the future these products will be classified according to molding method and water absorption. The following will thus count as stoneware: split clinker flags, extruded slabs, and dry pressed or slip-cast tiles, all with a water absorption of not more than 3%.

Split Clinker Flags. The name comes from the method of manufacture. Two slabs that are weakly bonded together back to back are split apart after firing. The parting line is formed by knives set in the die of an extrusion press. This makes a scratch on the core. The single slabs have grooves with a dovetail shape in the direction of extrusion on the back, and these give a good bond when mortar is applied. Split clinker flags are used as frost-resistant building components for indoor and outdoor wall coverings and for swimming pools. They are manufactured in various sizes, shapes, and colors, both glazed and unglazed.

The body corresponds to typical stoneware compositions with mixtures of stoneware clays of various plasticities:

30–50%	stoneware clays of varying plasticity
10–25%	grog or powdered fired body
10–20%	silica flour
5–20%	feldspar
0–5%	talc

The required grain size of the grog or the powdered fired body depends on the desired surface structure. Fine, smooth surfaces require a grog milled to a particle size < 60 μm. For a rustic effect, which at the present time accounts for 60–65% of production, the grog can be ≤ 1 mm.

For open-air swimming pools a grade made with fine grog is usually used. The raw materials are often prepared in high-output installations by the dry and semidry processes. The vacuum extrusion molding process is automated, including the cutting and setting of drying racks. The slabs are stacked on their longitudinal edges in chamber dryers with good air circulation or dried in continuous dryers with adjustable lateral air injection. Then they are carried on conveyor belts through one or more automatic glazing stages, where they are coated with feldspar glazes by spraying, pouring, or hosing. After this, they are automatically set onto kiln cars and fired at Seger cone 6 a–10 (1200–1300 °C) in an oxidizing atmosphere in a tunnel kiln, where the energy consumption is 2.7–3.0 MJ/kg. The final processes of splitting, sorting, and packing are all automated.

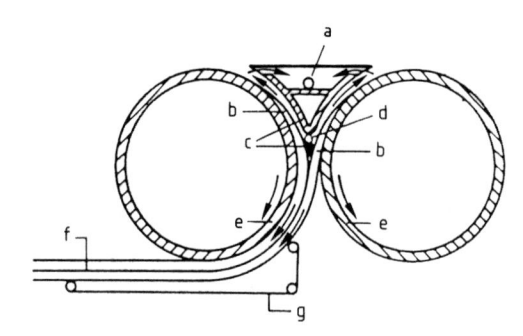

Figure 29. Schematic diagram of electrophoretic formation of clay body sheets
a) Outlet for reusable excess slip; b) Clay body particles deposited by electrophoresis; c) Counterelectrode (cathode); d) Inlet for slip; e) Zinc anode; f) Two-layer sheet of clay body; g) Belt conveyor

Quality requirements and testing procedures are summarized in DIN 18 166. The following important individual test methods are planned: water absorption (DIN 51 056), transverse strength (DIN 51 090), acid and alkali resistance of the glaze (DIN 51 092) and acid and alkali resistance of the unglazed slabs (DIN 51 091).

Single Extruded Slabs. These are also formed by a vacuum extrusion press and may be given their exact rectangular or square shape by a later punch press operation. Sometimes this can also include a dry glazing step to give a rustic effect. Slabs are also produced in a punch press by a process still being developed. Continuous sheets of clay body are produced from a slip by an electrophoretic process using the "Elephant" machine of the firm Karl Händle & Söhne, Mühlacker (Fig. 29). The desired shapes and sizes (from small mosaic tiles up to 60 × 60 cm slabs) can be punched out from this by means of the Lingl punch press [81]. The output of clay body sheet is 60–100 m^2/h, with a thickness of only 4–6 mm.

Large Slabs. Since the early 1970s, large Keraion slabs have been manufactured in stoneware by the company Buchtal, Keramische Betriebe, Schwarzenfeld. The largest size produced is 1.25 × 1.60 m and normal sizes are 0.6 × 0.6 m and 0.3 × 0.6 m. All the slabs are 8 mm thick, weighing only 18 kg/m^2. Again, molding is done with an extrusion press, but the extruded slabs must be passed between rollers until the desired size and thickness are obtained. Keraion slabs may be used for covering walls, floors, or ceilings, and can be used for interiors or façades.

Test methods suitable for Keraion slabs are given in DIN 18 166, and the properties meet or exceed the specifications: compression strength ca. 200 N/mm^2, transverse strength 30–35 N/mm^2 if the glazed surface is under compression and 20–22.5 N/mm^2 if it is under tension, modulus of elasticity 65 000 N/mm^2.

Dry-Pressed Tiles and Slabs. The European tile industry underwent a change of direction in the 1950s after far-reaching developmental work in Italy and Spain. Completely automated factories came into being, with a completely new concept. Mass production of this kind requires that the greatest importance be attached to consistency and standardization of raw materials. These are stored in large silos with sophisticated weighing equipment. The preparation method often depends on the particular types of raw material available. Thus, the wet process with spray drying may be used or the semidry or dry process. The product is a flowable granular clay body with 5–8% moisture, suitable for dry pressing in high-power presses. Grain-size distribution and an even moisture distribution are vital for defect-free molded pieces. The usual automatic presses used today are hydraulic presses, friction screw presses, or combinations of the two systems. They operate at a maximum pressure of 500 t and 15–18 strokes/min, up to four 15×15 cm tiles being produced per stroke. All subsequent operations of removal, transport, drying, glazing, and setting in the kiln are completely automatic. Drying occurs in a rocking dryer.

Glazing is by various means, e.g., squirting, dripping, spraying, sheepskin-covered rollers, brushing equipment, etc. Dry glazes may be applied either evenly or unevenly to produce special effects. Flowable spray-dried glaze can be applied during the pressing operation by feeding it onto the clay body in the mold from a separate push feeder and, after the bottom molding plate has been lowered slightly, pressing the glaze onto the molded piece. Another possibility is to apply the glaze preprepared in the form of sheets or foil.

For firing in tunnel kilns there is automatic equipment to load and unload saggars stacked one on top of another. Flat flame kilns for so-called fast firing are energy saving and highly automated. The throughput times are up to 60 min. The output of a 60-m-long flatflame kiln, for shapes $200 \times 300 \times 10$ mm, with throughput times of 65–100 min and firing temperatures of 1180–1200 °C, is said to be 1200–1500 m^2/d with a heat consumption of 2.5–2.9 MJ/kg. Flat-flame kilns are designed as single and multilayer roller kilns and as bogie kilns. In the latter, the charge is placed on solid supports of aluminum oxide, Pythagoras body, or special steel fixed to the flat upper surface of the kiln car.

The Standard Specification DIN 18 155 (March 1976) now applies to stoneware tiles; in Part 1, the concept and constitution of fine ceramic tiles are defined and their applications are explained. Part 2 gives shapes and measurements; Part 4, quality requirements and testing. Stoneware tiles, whether glazed or unglazed, must have a transverse strength of at least 25 N/mm^2 (for test method see DIN 51 090). Test methods for thermal shock resistance are given in DIN 51 090, and for household chemicals, acids, and alkalis they are given in DIN 51 092.

3. References

[1] C. O. Pels Leusden: *Ziegeleitechnisches Jahrbuch,* Bauverlag, Wiesbaden-Berlin 1977, pp. 277–359.
[2] C. O. Pels Leusden, H. B. Weber, *Ziegelindustrie* 1975, no. 7, 254.
[3] W. Bender, F. Händle (eds.): *Handbuch der Ziegelindustrie, Verfahren und Betriebspraxis in der Grobkeramik,* Bauverlag, Wiesbaden-Berlin 1982.
[4] G. Piltz: *Ziegeleitechnisches Jahrbuch,* Bauverlag, Wiesbaden-Berlin 1972, pp. 218–261.
[5] H. Kromer, W. Potschigmann: "Blähton" in *Handbuch der Keramik,* group II M, Verlag Schmid, Freiburg 1977.
[6] E. Strommenger: *Habuba Kabira – Eine Stadt vor 5000 Jahren,* Verlag Philipp von Zabern, Mainz 1980, pp. 39–46.
[7] K. Benesch: *Auf den Spuren großer Kulturen.* Lexikothek-Verlag, Gütersloh 1979, pp. 176–183.
[8] G. Piltz, *Ziegelindustrie* 1964, no. 13, 493–498.
[9] E. Schmidt: *Ziegeleitechnisches Jahrbuch,* Bauverlag, Wiesbaden-Berlin 1973, pp. 373–411.
[10] W. Köther, E. Hilker, E. Hesse: *Ziegeleitechnisches Jahrbuch,* Bauverlag, Wiesbaden-Berlin 1981, pp. 177–207.G. Piltz„ E. Hilker, *Ziegelindustrie* **1973**, no. 12, 453; **1974**, no. 1, 15; **1974**, no. 2, 60.
[11] D. Hauck, E. Hilker: *Ziegeleitechnisches Jahrbuch,* Bauverlag, Wiesbaden-Berlin 1984, pp. 15–60.
[12] G. Piltz, E. Hilker, *Ziegelindustrie* 1974, no. 9, 374.
[13] W. Köther, E. Hilker, E. Hesse: *Ziegeleitechnisches Jahrbuch,* Bauverlag, Wiesbaden-Berlin 1981, pp. 369–377.
[14] G. Piltz: "Untersuchung der Möglichkeiten der Aufhellung der Brennfarben von Ziegelrohstoffen," Forschungsberichte NRW, no. 1323, Westdeutscher Verlag, Köln-Opladen 1964.
[15] E. Hilker, D. Hauck: *Ziegeleitechnisches Jahrbuch,* Bauverlag, Wiesbaden-Berlin 1986, pp. 17–48.
[16] W. Köther, E. Hilker: *Ziegeleitechnisches Jahrbuch,* Bauverlag, Wiesbaden-Berlin 1982, pp. 198–211. H. Schmidt, H. Scholze, G. Tünker, *Sci. Ceram.* **11** (1981) 333–339. E. Schmidt:
Ziegeleitechnisches Jahrbuch, Bauverlag, Wiesbaden-Berlin 1964, pp. 349–368.
[17] W. Köther, N. Pauls: *Ziegeleitechnisches Jahrbuch,* Bauverlag, Wiesbaden-Berlin 1982, pp. 212–244.
[18] G. Piltz: "Vergleiche in der Grobkeramik angewandter Untersuchungsmethoden," Forschungsberichte NRW, no. 1351, Westdeutscher Verlag, KölnOpladen 1964.
[19] U. Troje, H.-G. Hopp, *ZI Ziegelind. Int.* 1980, no. 5, 288–291.
[20] C. O. Pels Leusden: *Ziegeleitechnisches Jahrbuch,* Bauverlag, Wiesbaden-Berlin 1974, pp. 172–226.
[21] W. Unger, *Keram. Z.* **31** (1979) 26–27.
[22] C. O. Pels Leusden, *Keram. Z.* **29** (1977) 665–668; **30** (1978) 93–97. R. Grätz, *Sprechsaal Keram. Glas Email Silik.* **102** (1969) no. no. 18, 764–787; **102** (1969) no. 22, 990–998.
[23] R. Grätz: *Ziegeleitechnisches Jahrbuch,* Bauverlag, Wiesbaden-Berlin 1972, p. 262–287.
[24] R. Grätz, *Ziegelindustrie* 1969, no. 9/10, 197–203.
[25] E. Hilker, *Ziegelindustrie* 1974, no. 8, 333.

[26] C. O. Pels Leusden, E. Hilker: "Erhöhung der Maßhaltigkeit von keramischen Produkten insbesondere durch Zuschlagstoffe zur Verbesserung der Formgebung und des Trocknungsablaufs," Forschungsberichte NRW, no. 2960, Westdeutscher Verlag, Köln-Opladen 1980.

[27] F. R. Stupperich, *Ziegelindustrie* 1975, no. 11, 400.

[28] K. Junge, *ZI Ziegelind. Int.* **38** (1985) no. 1, 10–22; **38** (1985) no. 4, 227–235.

[29] C. O. Pels Leusden, *ZI Ziegelind. Int.* 1979, no. 7, 384–397.

[30] H.-B. Weber, *Ziegelindustrie* 1973, no. 2, 46–54.

[31] C. O. Pels Leusden: *Ziegeleitechnisches Jahrbuch*, Bauverlag, Wiesbaden-Berlin 1979, pp. 202–352.

[32] W. Köther, *ZI Ziegelind. Int.* 1979, no. 7, 409–416.

[33] C. O. Pels Leusden, *Ber. Dtsch. Keram. Ges.* **46** (1969) 529–533.

[34] E. Hilker: *Ziegeleitechnisches Jahrbuch*, Bauverlag, Wiesbaden-Berlin 1973, pp. 213–253.

[35] D. Hauck, E. Hilker: *Ziegeleitechnisches Jahrbuch*, Bauverlag, Wiesbaden-Berlin 1985, pp. 46–95.

[36] G. Piltz, *Ziegelindustrie* 1965, no. 20, 751–778.

[37] D. Hauck, E. Hilker: *Ziegeleitechnisches Jahrbuch*, Bauverlag, Wiesbaden-Berlin 1986, pp. 58–116.

[38] W. Bender, *Ziegelindustrie* 1968, no. 8, 196–199.

[39] W. Schellmann, H. Fastabend: *Ziegelindustrie* 1963, no. 24, 899–905.

[40] E. Hesse, E. Hilker, *ZI Ziegelind. Int.* 1978, no. 5, 256–265.

[41] H. Hohmann, H.-G. Krüger, J. Geilich, *Silikattechnik* **22** (1971) no. 4, 115–120. H. Hohmann, S. Plüschke, *Baustoffindustrie* **24** (1981) no. 2, 47–50.

[42] E. Hesse, *Ziegelindustrie* 1969, no. no. 11/12, 249–253.

[43] E. Hesse, *Ziegeleitechnisches Jahrbuch*, Bauverlag, Wiesbaden-Berlin 1982, pp. 118–145. G. Piltz, E. Hesse, *Ziegelindustrie* 1972, no. 9, 432–436. G. Schellbach, G. Piltz, E. Hilker: *Ziegeleitechnisches Jahrbuch*, Bauverlag, Wiesbaden–Berlin 1977, pp. 360–426.

[44] G. Schellbach: *Ziegeleitechnisches Jahrbuch*, Bauverlag, Wiesbaden-Berlin 1982, pp. 14–51.

[45] G. Schellbach: *Ziegeleitechnisches Jahrbuch*, Bauverlag, Wiesbaden-Berlin 1983, pp. 101–144.

[46] Firmenschrift Watts, Blake, Bearne & Co. Ltd., Devon, England.

[47] O. Reumann: "Eigenschaften der keramischen Rohstoffe," in: *Handbuch der Keramik*, group 1 A 3, Verlag Schmid, Freiburg 1968.

[48] *Databook 1975*, Sprechsaal-Verlag, Coburg 1975.

[49] Firmenschrift Fuchssche Tongruben KG, Ransbach-Baumbach.

[50] E. Gugel, K. Schröder, E. Frank, *Ber. Dtsch. Keram. Ges.* **49** (1972) 179–184.

[51] P. Fischer, H. A. Müller, *Marsdorfer Tech. Mitt.* **1** (1965) 5–22.

[52] F. Singer, S. Singer: *Industrielle Keramik*, vol. **2,** Springer Verlag, Berlin-Heidelberg-New York 1968, p. 46.

[53] H. Salmang, H. Scholze: *Die Keramik. Physikalische und chemische Grundlagen*, 5th ed., Springer Verlag, Berlin-Heidelberg-New York 1968.

[54] H. B. Ries: "Aufbereitung keramischer Massen," in: *Handbuch der Keramik*, group I C, Verlag Schmid, Freiburg 1968.

[55] H. Zimmermann, *Keram. Z.* **23** (1971) 381–384.

[56] K. Suchowski, *Keram. Z.* **27** (1975) 401–402.

[57] G. Lengersdorf, H. Röhr, *Keram. Z.* **21** (1969) 428–431.

[58] H. B. Ries, *Keram. Z.* **25** (1973) 454–460.

[59] H. B. Ries, *Euro-Ceram.* **13** (1963) 249.

[60] S. Lenk, *Keram. Z.* **25** (1973) 134–136. K. Krahl, C. Richter, H. Hässlich, *Silikattechnik* **29** (1978) 151–153.
[61] DKG-Werkstoffkennblätter für technische keramische Werkstoffe, Dtsch. Keram. Ges. Fachausschußber. no 23, 1978.
[62] G. Teubner, *Sprechsaal* **101** (1968) 752–758.
[63] H. G. F. Winkler, F. Freund, *Ber. Dtsch. Keram. Ges.* **35** (1958) 375.
[64] W. Weiand, *Sprechsaal* **109** (1976) 332–335.
[65] A. Bergholz, K. Herdt, Silikattechnik **19** (1968) 150–154.
[66] F. Rüb, *Keram. Z.* **21** (1969) 98–106.
[67] R. Lenz, *Keram. Z.* **21** (1969) 438–443.
[68] I. Gatzke, *Keram. Z.* **21** (1969) 219–224.
[69] P. Fischer: "Kanalisations-Steinzeug" in: *Handbuch der Keramik*, group II D 1, Verlag Schmid, Freiburg 1972. F. Gorn, *Sprechsaal* **105** (1972) 533–535.
[70] H. B. Ries, *Euro-Ceram.* **10** (1960) no. 2.
[71] W. Richter, *Keram. Z.* **26** (1974) 638–642.
[72] G. Cremer, *Ber. Dtsch. Keram. Ges.* **32** (1955) 365–368. G. Cremer, *Ber. Dtsch. Keram. Ges.* **39** (1962) 175–180.
[73] W. Richter, *Keram. Z.* **28** (1976) 581–583.
[74] R. Masson, *Chimia* **8** (1954) 7.
[75] E. Gugel, H. Vogel, *Ber. Dtsch. Keram. Ges.* **41** (1964) 197–205.
[76] E. Gugel, H. Vogel, O. Osterried, *Ber. Dtsch. Keram. Ges.* **43** (1966) 587–594.
[77] E. Dörre, A. Lipp, D. Rauschert, K.-H. Schüller, H. Vogel, *Ber. Dtsch. Keram. Ges.* **50** (1973) 4.
[78] K. Pfeifer, R. Roth, *VFDB Z.* **16** (1967) no. 2.
[79] W. Weiand, *Keram. Z.* **31** (1979) 148–151.
[80] M. Drews: "Fliesen und Platten" in: *Handbuch der Keramik*, group II H 2, Verlag Schmid, Freiburg 1979.
[81] E. W. Schmid, *ZI Ziegelind. Int.* 1978, 217–220. F. Händle, *Keram. Z.* **32** (1980) 185–188. H. Lingl, *Keram. Z.* **33** (1981) 41–42.

Copper Compounds

H. Wayne Richardson, CP Chemicals Inc., Sumter, South Carolina 29150, United States

1.	Introduction 1649	4.4.1.	Copper(II) Sulfate Pentahydrate . 1666
2.	The Copper Ions 1650	4.4.2.	Anhydrous Copper Sulfate 1671
3.	Basic Copper Compounds 1652	4.4.3.	Copper(II) Sulfate Monohydrate 1672
3.1.	Copper(I) Oxide 1652	4.4.4.	Basic Copper(II) Sulfates 1673
3.2.	Copper(II) Oxide 1654	5.	Compounds and Complexes of Minor Importance 1674
3.3.	Copper(II) Hydroxide 1656	5.1.	Copper Compounds 1674
3.4.	Copper(II) Carbonate Hydroxide 1658	5.2.	Copper Complexes 1680
4.	Salts and Basic Salts 1659	6.	Copper Reclamation 1681
4.1.	Copper(I) Chloride 1659	7.	Copper and the Environment . 1682
4.2.	Copper(II) Chloride 1662	8.	Economic Aspects 1684
4.3.	Copper(II) Oxychloride 1665	9.	Toxicology and Occupational Health 1686
4.4.	Copper(II) Sulfates 1666	10.	References 1687

1. Introduction

Copper compounds, although they represent a small fraction of total copper production, play an important and varied role in industry and agriculture. One of the oldest known fungicides was copper-based and was used extensively in the early part of the century. The last 20 years have brought about a resurgence in the use of copper-based fungicides partly because of a lack of tolerance by fungi to copper and also because of its relatively low toxicity to higher plants and animals (see, however, Chaps. 7 and 9). Although copper is an essential trace element for higher plants and animals, it is acutely toxic in higher doses. Copper compounds are used as nutritional additives in animal feeds and fertilizers, and are found in a variety of dietary supplements. Copper salts are used in the control of algae in lakes and ponds, and the oxides are used in antifouling paints and coatings. Copper acetoarsenite, Paris green, is used as an insecticide, and copper chromium arsenate is an effective alternative to creosote for the preservation of wood.

Copper and its compounds are used catalytically in numerous organic reactions, e.g., polymerization, isomerization, and cracking reactions. They are used in the textile and dye industries in the preparation of rayon and acrylonitrile, as mordants and oxidants in textile dyeing and printing, and in the preparation of azo dyes. Copper compounds are used as pigments in glass, ceramics, porcelains, varnishes, and artificial gems, and in the manufacture of the copper phthalocyanine pigments. Copper salt solutions are used for electroplating, in brazing and burnishing preparations, and as brighteners for aluminum. Solutions of copper(I) complexes are used to selectively absorb carbon monoxide, butadiene, and alkenes from gas streams. In the petroleum industry, copper compounds are used as deodorizing (desulfurization) and purifying agents. Copper(II) carbonate is used in drilling muds to protect against release of poisonous hydrogen sulfide gas, and copper(I) iodide is used in acid muds to bind corrosion inhibitors to the iron drills. More recent applications of copper compounds are in pollution control and solar technology.

The multitude of applications of copper compounds in the biosphere is largely responsible for the extent of academic interest in them. Also, the facile reduction–oxidation of the copper(I)–copper(II) couple, the ease of theoretical treatment of the d^9 copper(II) system, and the varied stereochemistries and magnetic behaviors associated with copper ions enhance their theoretical appeal. Since this treatment of copper compounds is primarily from an industrial perspective, the reader whose interest is academic is referred to the classic references [1]–[4] as well as to more recent materials [5]–[11] that are attune to the subtle chemistries of copper and its compounds. Compounds of primary industrial importance (Chaps. 3 and 4) are distinguished from compounds of minor importance (Chap. 5) in this article.

2. The Copper Ions

Copper, Cu [7440-50-8], M_r 63.546, [Ar]$3d^{10}4s^1$, is a member of the first transition series. It is classified as a transition element in the broader definition because the copper(II) valence state, [Ar]$3d^9$, comprises such a large proportion of the defined chemistry of copper. Copper in its compounds exists primarily in two oxidation states, + 1 and + 2. Although copper(0) and copper(III) compounds have been identified, they are not presently of commercial importance. The stabilities of the various valence states of copper are illustrated by the following standard reduction potentials:

		E_0
$Cu^+ + e^-$	$\longrightarrow Cu^0$	+ 0.52 V
$Cu^{2+} + 2\,e^-$	$\longrightarrow Cu^0$	+ 0.34 V
$Cu^{2+} + e^-$	$\longrightarrow Cu^+$	+ 0.15 V
$Cu^{3+} + e^-$	$\longrightarrow Cu^{2+}$	+ 1.80 V

From the above, it is seen that

$$2\,Cu^+ \longrightarrow Cu^0 + Cu^{2+} \quad \Delta E_0 = -0.37\text{ V}$$

with $pK = -5.95$

In other words, the free copper(I) ion does not exist to any appreciable extent in aqueous solution. In the presence of ligands such as ammonia, chloride, or cyanide, solutions of copper(I) can be prepared that are stable with respect to disproportionation. For example, the colorless solution of tetraamminedicopper(I) sulfate is prepared readily by contact of blue tetraamminecopper(II) sulfate with metallic copper in the absence of air. Upon acidification with sulfuric acid, copper powder and a copper(II) ammonium sulfate solution are produced. The insoluble copper(I) chloride can be produced by sulfuric acid acidification of a copper(I) ammine chloride solution. If the solution is acidified with hydrochloric acid, a solution of $[CuCl_2]^-$, $[CuCl_3]^{2-}$, or $[CuCl_4]^{3-}$ species is produced, depending on chloride concentration. Copper(I) chloride is stable to water because of its insolubility, which is a result of the polymeric structure that arises from the chloride's ability to bridge copper. This contrasts with the sulfate's inability to coordinate or bridge strongly. Consequently, copper(I) sulfate can be produced only in nonaqueous media.

The electronic structure of the copper(I) ion is $[Ar]3d^{10}$. The compounds are diamagnetic and colorless except where charge-transfer bands arise. Copper(I) is isoelectronic with zinc(II) and the preferred stereochemistries are similar. As a result of the filled $3d$ level, no ligand field stabilization occurs and electronic distortions are minimized. The stereochemistry around the copper(I) ion is determined mainly by the size of the anions, as well as by the electrostatic and covalent bonding forces. The preferred stereochemistry is tetrahedral, with linear and trigonal planar compounds also being common.

The majority of copper(II) compounds exhibit square planar or distorted octahedral configurations about the copper ion. The $3d^9$ electronic structure gives rise to the classic example of Jahn–Teller distortion in which the four planar metal–ligand distances are smaller than the two axial distances. Copper(II) ions are also found in distorted tetrahedral and various five-coordinate environments.

The copper(II) ion, $[Ar]3d^9$, is predominantly blue or green, and the unpaired $3d$ electron results in magnetic phenomena. In most copper compounds, the unpaired electrons of the copper ions are sufficiently isolated from each other so that the compounds exhibit paramagnetic behavior. However, there are many polynuclear copper compounds in which the spins are coupled, which lowers the magnetic moment. The coupling may be so weak that it must be observed near the absolute zero of temperature, or it may be strong enough to render the compound diamagnetic at room temperature or above.

3. Basic Copper Compounds

3.1. Copper(I) Oxide

Cu_2O [1317-39-1], M_r 143.09, mp 1235 °C, d_4^{25} 5.8–6.2, decomposes above 1800 °C. It occurs in nature as the red or reddish brown mineral cuprite with a cubic or octahedral crystal morphology. Depending on the method of preparation and particle size, the synthetic material is yellow, orange, red, or purple. The yellow material has erroneously been referred to as copper(I) hydroxide, but X-ray diffraction patterns indicate that there are no differences in the crystal structures of the colored forms. Their thermodynamic data are as follows: c_p (298 K) 429.8 J kg^{-1} K^{-1}, c_p (290–814 K) 519.2 J kg^{-1} K^{-1}, c_p (290–1223 K) 565.2 J kg^{-1} K^{-1}, $\Delta H°$ (25 °C) −166.6 kJ/mol. Copper(I) oxide is stable in dry air but slowly oxidizes to copper(II) oxide [1317-38-0] in moist air. It is practically insoluble in water but dissolves in aqueous ammonia. In excess hydrochloric acid, soluble copper(I) chloride complexes are formed; however, in dilute sulfuric or nitric acids, disproportionation to the soluble copper(II) salts and copper powder results.

Production. Copper(I) oxide is produced easily by a variety of methods; its instability with respect to oxidation requires careful consideration. Copper(I) oxide produced pyrometallurgically is usually coated with isophthalic acid or pine oil to preserve its integrity [12]. Hydrometallurgically produced material can be stabilized by mixing the particle slurry with glue, gelatin, casein, or dextrin before drying [13]–[16].

Pyrometallurgical Processes. Copper(I) oxide is formed when copper powder is heated above 1030 °C in air; to prevent further oxidation, it must be cooled quickly in an inert atmosphere. To allow for lower temperature production of copper(I) oxide, carbon can be blended with copper(II) oxide and heated to 750 °C in an inert atmosphere. The material must be stabilized by coating the formed particles with isophthalic acid or pine oil [12]. A more stable copper(I) oxide results when stoichiometric amounts of copper powder and copper(II) oxide are blended, heated to 800–900 °C in an inert atmosphere, and allowed to cool. The production can be effected at lower temperature if ammonia or certain ammonium salts are added to the blend [17]–[19]. The autoclave oxidation of copper metal at 120 °C and about 0.6 MPa gauge pressure with air in the presence of water and small amounts of sulfuric and hydrochloric acids produces a red, pigment-grade product [20]. By varying the pressure and temperature, considerable differences in particle size, coloring, apparent bulk density, and buoyancy have been found.

Hydrometallurgical Processes. Tetraamminedicopper(I) sulfate, $Cu_2(NH_3)_4SO_4$, prepared by leaching an excess of copper with a solution of ammonia and ammonium sulfate, with air as the oxidant, yields a red copper(I) oxide upon acidification to pH 3–5 [21]. The less corrosive ammonium carbonate leach system in which $Cu_2(NH_3)_4CO_3$ is produced is more common (see p. 1654). Upon vacuum distillation, a very stable red Cu_2O product remains [21]. If sodium hydroxide is added to the leach liquor, a yellow microcrystalline powder is precipitated [22]. When the yellow Cu_2O is heated in an excess of sodium hydroxide, it is converted to an orange material of somewhat larger particle size.

Steam stripping of the copper(I) ammine carbonate solution yields a brown, impure product [23] which can be converted to a red material by washing it in an organic acid, e.g., formic or acetic acid [24]. An impure, brown product can also be converted to a red material by boiling it in 20% sodium hydroxide solution [25].

If a saturated solution of copper(I) ammine carbonate is agitated over copper metal, a layer of red copper(I) oxide is continuously produced which can be broken loose and recovered [26]. When copper salts are leached with chelating agents such as ethylenediaminetetraacetic acid [27] or ammonia [28] under pressure of carbon monoxide or hydrogen, and sodium hydroxide is subsequently added, a relatively stable, yellow copper(I) oxide is obtained; the reaction is catalyzed by an alkali metal iodide [29].

The reduction of a boiling slurry of basic copper(II) sulfate with sulfur dioxide at a pH of about 3 produces a reddish product [30]. Red copper(I) oxide has also been prepared by mixing a slurry of basic copper(II) sulfate with neutral copper(II) sulfate and adding sodium sulfite to a pH of 5.2. The mixture is then acidified to pH 3.5–5 and heated to boiling. The intermediate copper(I) sulfite slurry is decomposed to copper(I) oxide and sulfurous acid. Alkali is subsequently added to maintain a pH of 2.6–2.8 [31].

When a solution of copper(I) chloride and sodium chloride is neutralized with sodium hydroxide and then heated to 138 °C under pressure, a red copper(I) oxide is obtained which has an average particle diameter of about 2.5 µm [32]; an orange product (about 1-µm particles) is prepared by neutralizing the solution to pH 8.5 at 60 °C [33]. Simultaneous mixing of copper(I) chloride solutions with sodium chloride and sodium hydroxide solution in the presence of copper(I) oxide seed crystals at a controlled pH of 10.0, 55 °C, and under nitrogen, gives a reddish purple material (average diameter 48 µm). At pH 7.0, a yellow material is obtained with an average particle size of 0.4 µm [34].

The *electrolytic production* of copper(I) oxide between copper electrodes in brine yields a yellow product at room temperature. At higher temperature, an orange or red material is produced.

Uses. The largest commercial use of copper(I) oxide is in antifouling paints for boat and ship bottoms; it is an effective control for barnacles and algae. The yellow or orange copper(I) oxide is used as a seed and crop fungicide, and the red material is used as a pigment in ceramic glazes and glass. Copper(I) oxide is also used in rectifiers and in brazing. Numerous organic reactions are catalyzed by copper(I) oxide, and it is an effective absorbent for carbon monoxide.

Table 1. Specifications for pigment-grade copper(I) oxide

Assay	Mass fraction, %	
	Navy I [36], [37]	Navy II [36]
Copper(I) oxide	97.0	90.0
Total copper (min.)	86.0	80.0
Reducing power (min.)	97.0	90.0
Nitric acid-insolubles (max.)	0.3	0.3
Chloride (max.)	0.4	0.4
Sulfate (max.)	0.1	0.1
Zinc oxide (max.)	–	10.0
Other metals (max.)	0.5	0.5
Acetone-soluble material (max.)	0.5	0.5

Analysis and Specifications. The ASTM approved analysis and specification for pigment-grade copper(I) oxide [35] and the military specification for the pigment grade [36] are listed in Table 1.

3.2. Copper(II) Oxide

CuO [1317-38-0], M_r 79.54, mp 1330 °C, d_4^{25} 6.48, occurs in nature as the black minerals tenorite (triclinic crystals) and paramelaconite (tetrahedral, cubic crystals). Commercially produced copper(II) oxide is usually black, although a brown product (particle size $< 10^{-6}$ m) can also be produced. Thermodynamic data: c_p (298 K) 531.1 J kg^{-1} K^{-1}, c_p (290–1253 K) 682.4 J kg^{-1} K^{-1}, $\Delta H°$ (25 °C) −155.3 kJ/mol. Copper(II) oxide is stable to air and moisture at room temperature. It is virtually insoluble in water or alcohols. Copper(II) oxide dissolves slowly in ammonia solution but quickly in ammonium carbonate solution; it is dissolved by alkali metal cyanides and by strong acid solutions. Hot formic acid and boiling acetic acid solutions readily dissolve the oxide. Copper(II) oxide is decomposed to copper(I) oxide and oxygen at 1030 °C and atmospheric pressure; the reduction can proceed at lower temperature in a vacuum. Hydrogen and carbon monoxide reduce copper(II) oxide to the metal at 250 °C and to copper(I) oxide at about 150 °C. Ammonia gas reduces copper(II) oxide to copper metal and copper(I) oxide at 425–700 °C [17].

Production. Copper(II) oxide can be prepared *pyrometallurgically* by heating copper metal above 300 °C in air; preferably, 800 °C is employed. Molten copper is oxidized to copper(II) oxide when sprayed into an oxygen-containing gas [37]. Ignition of copper(II) nitrate trihydrate [10031-43-3] at about 100–200 °C produces a black oxide. Basic copper(II) carbonate [12069-69-1], when heated above 250 °C, produces a black oxide if a dense carbonate is employed; a brown material is produced when the light and fluffy carbonate is used. An alkali-free oxide can be prepared by ignition of

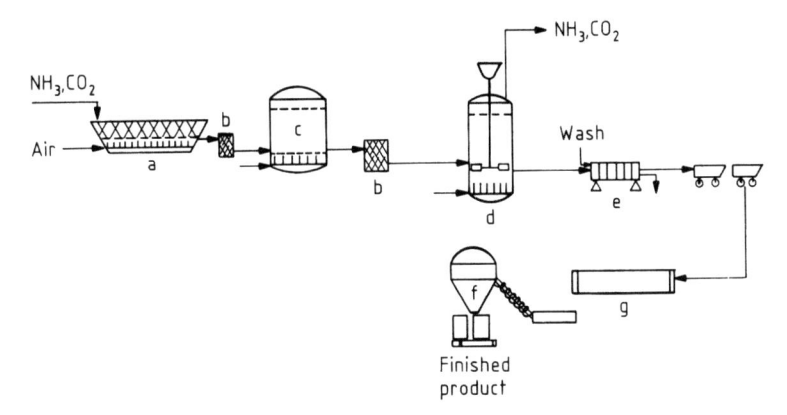

Figure 1. Process flow diagram for production of copper(II) oxide from ammonia–ammonium carbonate leach
a) Leach vat; b) Filter; c) Treatment tank; d) Strip tank; e) Press; f) Bag house; g) Drying kiln

copper(II) carbonate produced from ammonium carbonate and a copper(II) salt solution. Copper(II) hydroxide [20427-59-2], when heated above 100 °C, is converted to the oxide.

Hydrometallurgy is the most common method for the production of copper(II) oxide. A solution of ammonia and ammonium carbonate in the presence of air effectively leaches metallic copper; the process is represented by the following reactions:

$$2\ Cu + 1/2\ O_2 + 2\ NH_3 + (NH_4)_2CO_3 \longrightarrow [Cu_2(NH_3)_4]CO_3 + H_2O$$
$$2\ NH_3 + (NH_4)_2CO_3 + [Cu_2(NH_3)_4]CO_3 + 1/2\ O_2 \longrightarrow 2\ [Cu(NH_3)_4]CO_3 + H_2O$$
$$Cu + [Cu(NH_3)_4]CO_3 \longrightarrow Cu_2(NH_3)_4CO_3$$

The second and third reactions proceed readily; the first is slow. Consequently, in batch operations the leach is usually begun with a small charge of the copper solution, but continuous operations offer significantly improved rates. The leach liquor is then filtered to remove iron impurities and metallic copper, and is subsequently oxidized by air sparging. If necessary, lead and tin are removed by treatment with strontium, barium, or calcium salts [38]–[41]. The solution is filtered again and stripped of ammonia and carbon dioxide by steam injection or pressurized boiling to produce a black copper(II) oxide [38]. The ammonia and carbon dioxide are recycled for further use. The process is illustrated in Figure 1. Alternatively, the leach liquor can be treated with strong alkali to precipitate the intermediate copper(II) hydroxide, and then boiled to remove ammonia, with subsequent decomposition of the hydroxide to the black oxide.

The copper(II) ammine sulfate and chloride systems are treated similarly although they are utilized much less frequently because of their highly corrosive nature. However, the copper ammine chloride system, a byproduct of circuit board etching, is recycled out of economic and environmental necessity to copper(II) oxide [42] (see Chap. 6).

Table 2. Typical analysis of commercial copper(II) oxide

Assay	Mass fraction, %
Copper	78.5
Iron	0.09
Lead	0.08
Water	0.10
Nitric acid-insolubles	0.10
Zinc	0.05
Surface area, m^2/g	10.0

Uses. Copper(II) oxide is used as a precursor to a number of copper(II) salts. One of the largest commercial applications is in the production of compounds for wood preservation. Copper(II) oxide is also used extensively as a feed additive and as a pigment in glass, ceramic, and porcelain enamels [43]. In combination with manganese dioxide, it is used as an oxidative catalyst for exhaust gas [44], in the removal of NO$_x$, CO, and O$_3$ [45], [46], and in the purification of formaldehyde-containing waste gas [47]. Supported on aluminum phosphate, copper(II) oxide is active in reducing tar and polycyclic hydrocarbons in smoke by absorption and catalytic conversion [48]. Copper(II) oxide is used as a catalyst in the preparation of acrylates [49] and in the production of magnetic storage devices [50]. It has limited application in the petroleum industry as a gas sweetener and is used in welding fluxes for bronze.

Analysis and Specifications. Electrolytic deposition from a solution or iodometric analysis with sodium thiosulfate is standard for copper determination [51]. The technical specification is shown in Table 2.

3.3. Copper(II) Hydroxide

Cu(OH)$_2$ [20427-59-2], M_r 97.54, d_4^{25} 3.37, $\Delta H°$ (25 °C) 446.7 kJ/mol, decomposes over 100 °C or over 50 °C in the presence of an excess of alkali.

Copper(II) hydroxide is virtually insoluble in water (0.003 mg/L), and decomposes in hot water to the more stable copper(II) oxide and water:

$$\underset{\text{blue}}{\text{Cu(OH)}_2} \longrightarrow \underset{\text{green to brown}}{\text{CuO} \cdot \text{H}_2\text{O}} \longrightarrow \underset{\text{brown to black}}{\text{CuO}} + \text{H}_2\text{O}$$

Copper(II) hydroxide is readily soluble in mineral acids and ammonia solution. When freshly precipitated, it is soluble in concentrated alkali, with the formation of [Cu(OH)$_3$]$^-$ or [Cu(OH)$_4$]$^{2-}$. Copper(II) hydroxide is inherently unstable but can be kinetically stabilized by a suitable production method.

Production. There are two classes of copper(II) hydroxide. The first is stoichiometrically rather precise, with a copper content as high as 64%; the theoretical copper content of $Cu(OH)_2$ is 65.14%. This class is produced by the ammonia process [52]–[55], which yields a pure product of relatively good stability and large particle size. The best product results from the addition of strong alkali to the soluble copper(II) ammine complex [52], [55]. A relatively large particle-size product, deep blue in color and high in copper content, is precipitated below 35 °C. The resulting material is fairly stable or can be coated with gelatin to enhance its stability [56].

In the copper(II) hydroxide made by the ammonia process, the solubility of the copper(II) ammine complexes provides for crystallite growth. This affords a large particle size, a limited surface area (point of dehydration), and hence a relatively stable product, in contrast with the unstable product (variable assay) that results from the addition of hydroxide solutions to copper(II) salt solutions at 20 °C or above. The reaction with hydroxide is diffusion-controlled, allowing essentially no time for crystallite growth. The product is obtained as a gelatinous, voluminous precipitate with a large surface area, which is quite unstable and difficult to wash free of impurities. If the same reaction, with the same order of addition, is allowed to occur at 0–10 °C, a product of defined particle size and measurable surface area results, with greater stability but low assay.

The second class of copper(II) hydroxide, which represents a "stable" product but has lower assay and greater impurity, is produced from an insoluble precursor such as basic copper(II) carbonate or copper(II) phosphate. The first stable copper(II) hydroxide of this kind was made from copper(II) phosphate with alternate additions of copper(II) sulfate and sodium hydroxide solutions [57]. The process is illustrated by the following series of reactions:

$$3\ CuSO_4 + 2\ Na_3PO_4 \longrightarrow Cu_3(PO_4)_2 + 3\ Na_2SO_4$$
$$Cu_3(PO_4)_2 + 6\ NaOH \longrightarrow 3\ Cu(OH)_2 + 2\ Na_3PO_4$$

The alternate copper(II) sulfate–sodium hydroxide addition is continued through 15 or 20 cycles and yields a stable product with 58–59% copper and 3–5% phosphate. The product has a small particle size and a high surface area, and is used as an agricultural fungicide.

Other stable copper(II) hydroxides of high surface area and fine particle size have been produced more recently [58]–[60]; the processes include the use of a copper(II) oxychloride precursor in the presence of an anionic surfactant [59] or a magnesium sulfate-precipitated precursor [58]. An electrolytically produced material has also been made by using trisodium phosphate as the electrolyte [61].

The classic Bordeaux slurry of copper(II) sulfate and lime in water has been replaced by a powdered stabilized product. This is obtained by mixing copper(II) nitrate solution and lime, adding cellulose pitch liquor (a waste product of the paper industry), and drying to yield a powder which is effective as a fungicide [62]. Sodium carbonate can be used as the precipitating agent instead of lime [63].

Uses. Copper(II) hydroxide is used as an active precursor in the production of copper(II) compounds. Ammonia-processed copper(II) hydroxide is used in the production of copper(II) naphthenate, copper(II) 2-ethylhexanoate, and copper soaps. Ammonia-processed copper(II) hydroxide is also used in the production of rayon (Schweitzer's reagent) and in the stabilization of nylon. Copper(II) hydroxides of the second class are often used as fungicides because of their small particle size. Copper(II) hydroxide is also used as a feed additive, a catalyst in the vulcanization of polysulfide rubber, and an antifouling pigment.

Analysis and Specifications. Copper is determined iodometrically with sodium thiosulfate [51]. The analysis of typical ammonia-processed copper(II) hydroxide (mass fractions in %) is copper 63.0, iron 0.05, zinc 0.05, lead 0.05, naphthenic acid-insoluble material 2.0, sulfate 0.3.

3.4. Copper(II) Carbonate Hydroxide

Copper(II) carbonate hydroxide, also called basic copper(II) carbonate, occurs in nature as the metastable mineral *azurite* [12070-39-2], also called chessylite, a blue, monoclinic crystalline or amorphous powder with a formula approximating $2\,CuCO_3 \cdot Cu(OH)_2$, M_r 344.67, d_4^{24} 3.8, $\Delta H°$ (20 °C) -87.4 kJ/mol, and *malachite* [12069-69-1], green, monoclinic crystals with a formula approximating $CuCO_3 \cdot Cu(OH)_2$, M_r 221.12, d_4^{25} 3.9–4.0, $\Delta H°$ (20 °C) -57.7 kJ/mol. The copper(II) carbonate of commerce, malachite, is also known as Bremen green. Pure copper(II) carbonate, $CuCO_3$, has not been isolated. Copper(II) carbonate is virtually insoluble in water but dissolves readily in aqueous ammonia and alkali metal cyanide solutions. Copper(II) carbonate dissolves quickly in mineral acid solutions and warm acetic acid solution, with the formation of the corresponding copper(II) salt. Malachite is much more stable than copper(II) hydroxide but slowly decomposes to the oxide according to the following reaction:

$$CuCO_3 \cdot Cu(OH)_2 \longrightarrow 2\,CuO + H_2O + CO_2$$

Malachite is rapidly decomposed to the oxide above 200 °C.

Production. Two grades of copper(II) carbonate are available commercially, the light and the dense. The light grade is a fluffy product of high surface area. It is precipitated by adding a copper(II) salt solution, usually copper(II) sulfate solution, to a concentrated solution of sodium carbonate at 45–65 °C. Azurite is formed initially, and complete conversion to malachite usually occurs within two hours. The conversion is accelerated by the addition of malachite nuclei to the reactor.

A dark green, dense product results when a copper(II) salt solution is added to a solution of sodium hydrogen carbonate at 45–65 °C; conversion to malachite requires

Table 3. Typical analysis of commercially available copper(II) carbonate hydroxides

Assay	Mass fraction, %	
	Light	Dense
Copper	55.5	55.0
Sulfate	0.6	0.6
Iron	0.1	0.1
Zinc	0.01	0.02
Lead	0.003	0.005
Hydrochloric acid insoluble material	0.05	0.05
Water	1.0	2.0

about one hour in this case. The density is maximized if the reactor is washed with acid prior to the precipitation to prevent premature nucleation on malachite nuclei. (A less dense product would be produced if malachite nuclei are added to the slurry of azurite.) Solutions of copper(II) salt and sodium carbonate can also be added simultaneously at a pH of 6.5–7.0 and a temperature between 45 and 65 °C; conversion to malachite is usually complete within one hour.

When a solution of copper(II) ammonium carbonate is boiled, ammonia and carbon dioxide are expelled from the solution, and a deep green, dense copper(II) carbonate precipitates [38].

Uses. Copper(II) carbonate is used as a precursor in the production of copper salts and soaps. It is used in animal feeds as a source of copper, in the sweetening of petroleum, and in electroplating for the control of pH. Copper(II) carbonate is used as a hydrogenation catalyst and as an accelerator in polymerization reactions. The light grade is somewhat effective as a fungicide and is used as a seed protectant.

Analysis and Specifications. Copper is analyzed by iodometric titration with sodium thiosulfate solution [51]. Table 3 gives typical analyses of light and dense technical-grade copper(II) carbonates.

4. Salts and Basic Salts

4.1. Copper(I) Chloride

Copper(I) chloride [7758-89-6], CuCl, M_r 99.00, mp 422 °C, bp 1367 °C, d_4^{25} 4.14, $\Delta H°$ (25 °C) −134.6 kJ/mol, occurs in nature as the colorless or gray cubic-crystal nantokite. The commercially available product is white to gray to green and of variable purity. Copper(I) chloride is fairly stable in air or light if the relative humidity is less than about 50%. In the presence of moisture and air, the product is oxidized and

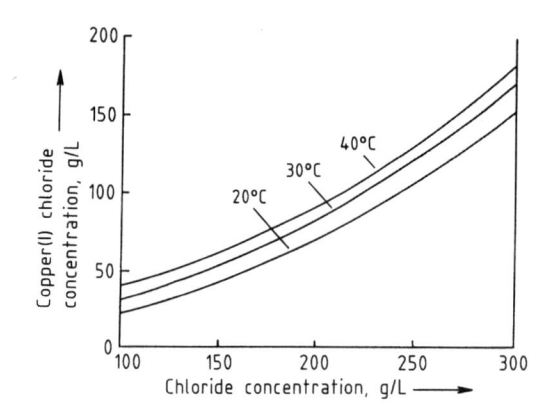

Figure 2. Solubility of copper(I) chloride in excess chloride ion solution at different temperatures

hydrolyzed to a green product that approaches copper(II) oxychloride, $CuCl_2 \cdot 3\,Cu(OH)_2$ [12356-86-4]. In the presence of light and moisture, a brown or blue product is obtained. Copper(I) chloride is slightly soluble to insoluble in water, with values from 0.001 to 0.1 g/L being reported. It is readily hydrolyzed to copper(I) oxide by hot water. Copper(I) chloride is insoluble in dilute sulfuric and nitric acids, ketones, alcohols, and ethers, but it quickly dissolves in hydrochloric acid, alkali halide, or ammonia solutions with the formation of complex compounds that are readily oxidized by air. Copper(I) chloride is soluble in solutions of alkali metal cyanides or thiosulfates and of coordinating amines, pyridines, and nitriles, notably, acetonitrile [64]. The increase in solubility of copper(I) chloride with chloride ion concentration illustrated in Figure 2 [65]. When the chloride concentration is decreased by dilution with water, the pure white copper(I) chloride precipitates.

Production. The direct combination of the elements is the most common method of production. The reaction of copper metal and chlorine is not spontaneous at ambient temperature. Once the metal is heated to red heat in the presence of chlorine, the reaction is self-sustaining and requires external cooling to prevent the metal from melting. A number of similar patents exist [66]–[69] in which copper metal is reacted with chlorine gas to produce a molten copper(I) chloride that is cast and pulverized. The primary difference in conditions is the use of a shaft furnace [66] as opposed to the use of crucibles. The recommended process temperature varies from 450 to 800 °C [66], [67] or 500 to 700 °C [68], [69]. The conditions required for the high-temperature production of a pure copper(I) chloride can be illustrated by the following:

$$Cu + 1/2\,Cl_2 \longrightarrow CuCl \qquad \Delta H = -134.6 \text{ kJ/mol}$$
$$Cu + Cl_2 \longrightarrow CuCl_2 \qquad \Delta H = -247.2 \text{ kJ/mol}$$
$$CuCl + 1/2\,Cl_2 \longrightarrow CuCl_2 \qquad \Delta H = -122.6 \text{ kJ/mol}$$
$$CuCl_2 + Cu \longrightarrow 2\,CuCl \qquad \Delta H = -22 \text{ kJ/mol}$$

Higher temperature and excessive contact with copper metal favor the production of a very pure copper(I) chloride. The lowest possible temperature is obviously 422 °C, the melting point of copper(I) chloride. As the temperature approaches the decomposition temperature for the reaction

$CuCl_2 \longrightarrow CuCl + 1/2\ Cl_2$ (993 °C)

a product of higher purity is obtained. Operationally, a temperature between 750 and 900 °C is ideal, and results in a product of > 98 % purity. The process of Degussa [67] illustrates a commonly used commercial process for the high-temperature production of copper(I) chloride; Figure 3 shows a suitable crucible for the production of technical-grade copper(I) chloride. Once the reaction is initiated, chlorine and copper metal (shot, chopped wire, or briquettes) are added continuously. As the molten product is formed on the surface of the upper metal layer, it flows by gravity down through the porous copper bed to effect further reduction of any copper(II) chloride, and out the exit port onto a rotating table where the product is allowed to cool and solidify. The flakes that form are packaged as is or ground to a powder and packaged. Because of the high temperature during the reaction and the volatility of copper(I) chloride, the exit port must be vented to a caustic scrubber. When the molten product is allowed to fall onto a high-speed, horizontally rotating disk constructed of quartz, graphite, or porcelain, small prills of uniform size are produced [70]. The product is spun out onto a water cooled diaphragm and collected. If the copper is contaminated with oxides, hydrogen chloride gas should be added to the chlorine gas stream to prevent the production of basic copper(II) chlorides that would contaminate the product [66].

If the product is packaged quickly and sealed properly, no extreme precautions are required. Otherwise, the product must be stored under nitrogen to preserve its integrity or coated with mineral oil as a barrier to moisture.

Copper(I) chloride is also produced hydrometallurgically by the reduction of copper(II) in the presence of chloride ions [71]:

$2\ CuCl_2 + Na_2SO_3 + H_2O \longrightarrow 2\ CuCl + Na_2SO_4 + 2\ HCl$

Other reducing agents can be used, such as metallic copper, sulfurous acid, hydroxylamine, hydrazine, or phosphorous acid. The copper(I) chloride solution is produced, for example, by mixing a copper(II) chloride solution with metallic copper in the presence of hydrochloric acid or sodium chloride. The colorless to brown solution is stable only in the absence of air. Continuous preparations of copper(I) chloride solutions have been developed [72], [73]. When they are diluted with water, a white crystalline material precipitates which can be vacuum dried or washed with sulfurous acid, then with alcohol and ether, and carefully dried. Zinc has also been used as a reducing agent in a more recent process [74].

Production of copper(I) chloride by treatment of ores with iron(III) chloride solutions [75], [76] and recovery of the product through chlorination in pit furnaces above 800 °C [77] have also been attempted.

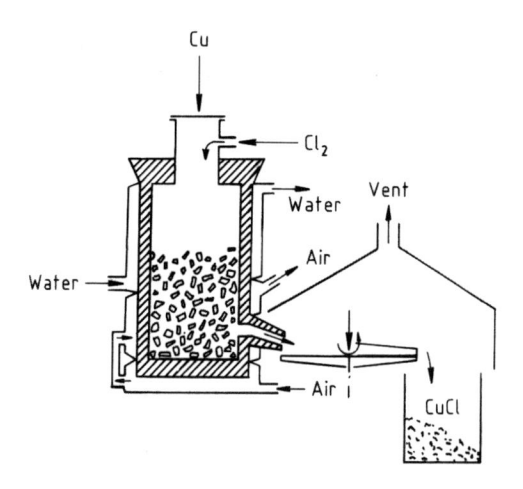

Figure 3. Crucible used in the production of copper(I) chloride

Table 4. Specifications for copper(I) chloride

Assay	Mass fraction, %	
	Technical grade	Reagent grade
Copper (min.)	97.0	90.0
Acid-insolubles (max.)	0.1	0.02
Iron (max.)	0.01	0.005
Sulfate (max.)	0.3	0.10
Arsenic (max.)	–	0.001
Not precipitated by H_2S as sulfate (max.)	–	0.2

Uses. Copper(I) chloride is used as a precursor in the production of copper(II) oxychloride and copper(I) oxide, as well as fine copper powder [78]. The production of silicone polymers, the vulcanization of ethylene–propene rubbers (EPDM) [79], and acrylonitrile production are other applications. Copper(I) chloride is also used in the purification of carbon monoxide gas [80]–[83] and the production of phthalocyanine pigments [84], [85]. More recently, copper(I) chloride has been found to be an effective catalyst in the production of dialkyl carbonates [86]–[88].

Analysis and Specifications. Copper(I) chloride is analyzed according to [51]. Table 4 lists specifications for technical- and reagent-grade material.

4.2. Copper(II) Chloride

Copper(II) chloride [7447-39-4], $CuCl_2$, M_r 134.45, mp (extrapolated) 630 °C, d_4^{25} 3.39, begins to decompose to copper(I) chloride and chlorine at about 300 °C. The often reported melting point of 498 °C is actually a melt of a mixture of copper(I)

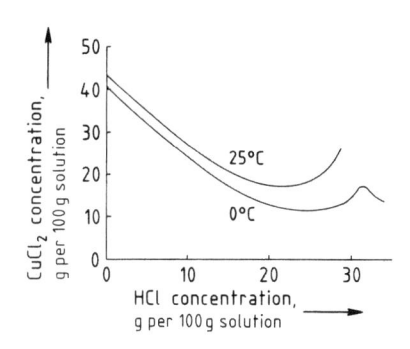

Figure 4. Solubility of copper(II) chloride in hydrochloric acid solutions at different temperatures

chloride and copper(II) chloride. Decomposition to copper(I) chloride and chlorine is complete at 993 °C. The deliquescent monoclinic crystals are yellow to brown when pure; their thermodynamic data are as follows: c_p (298 K) – 579.2 J kg^{-1} K^{-1}, c_p (288–473 K) – 621.7 J kg^{-1} K^{-1}, c_p (288–773 K) – 661.9 J kg^{-1} K^{-1}, $\Delta H°$ (25 °C) – 247.2 kJ/mol. In moist air, the dihydrate is formed. Figure 4 shows the solubility of copper(II) chloride in water and hydrochloric acid at two temperatures [89]. At higher concentrations of hydrogen chloride, [CuCl$_3$]$^-$ and [CuCl$_4$]$^{2-}$ complexes are formed. Copper(II) chloride is easily soluble in methanol and ethanol and moderately soluble in acetone.

The more common commercial form of copper(II) chloride is the dihydrate [10125-13-0], CuCl$_2$ · 2 H$_2$O, M_r 170.45, mp around 100 °C (with decomposition to the anhydrous form). This occurs in nature as blue-green orthorhombic, bipyramidal crystals of eriochalcite, $d\,^{25}_{4}$ 2.51. Its solubility characteristics are proportionally similar to those of the anhydrous form. In moist air the dihydrate deliquesces, and in dry air it effloresces.

Production. Because of the relative stabilities of copper(I) chloride and copper(II) chloride at high temperature, it is improbable that a pure anhydrous copper(II) chloride can be prepared by excessive chlorination of copper in a melt, even though such methods have been reported. The most common method for the production of anhydrous copper(II) chloride is by dehydration of the dihydrate at 120 °C. The product must be packaged in air-tight or desiccated containers.

The dihydrate can be prepared by the reaction of copper(II) oxide, copper(II) carbonate, or copper(II) hydroxide with hydrochloric acid and subsequent crystallization. Commercial production of copper(II) chloride dihydrate uses a tower packed with copper. An aqueous solution is circulated through the tower. Sufficient chlorine is passed into the bottom of the tower to oxidize the copper completely [72], [73]; to prevent hydrolysis [precipitation of copper(II) oxychloride] of concentrated copper(II) chloride solutions, they are kept acidic with hydrochloric acid. The tower can be operated batchwise or continuously; Figure 5 shows the continuous operation. A hot, concentrated liquor is circulated continuously through the tower, and the overflow

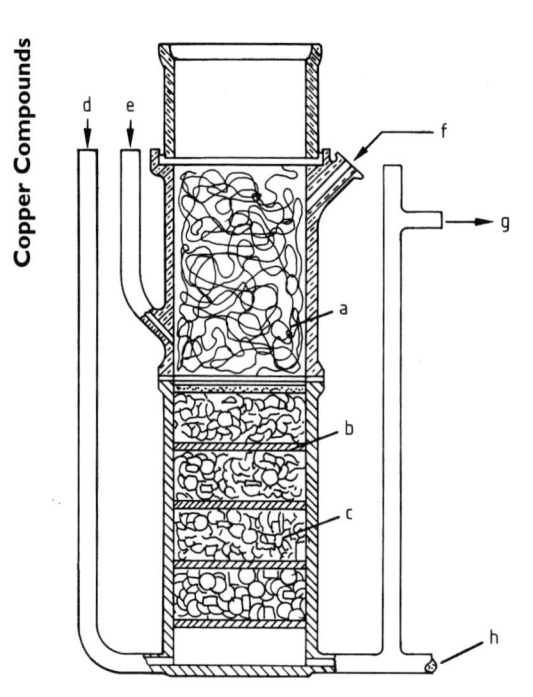

Figure 5. Reactor for the production of copper(II) chloride solutions
a) Copper metal; b) Porous plates; c) Raschig rings; d) Chlorine inlet; e) Steam inlet; f) Solution recycle; g) Solution to crystallizer; h) Drain

from the tower is passed through a crystallizer where the liquor is cooled; the product is then centrifuged, dried, and packaged. The addition of hydrogen chloride is pH controlled; the addition of water is controlled by specific gravity. Copper is added daily or twice daily, as needed.

Uses. Copper(II) chloride dihydrate is used in the preparation of copper(II) oxychloride [90], [91]. It serves as a catalyst in numerous organic chlorination reactions such as the production of vinyl chloride [92] or 1,2-dichloroethane [93]. Copper(II) chloride dihydrate is used in the textile industry as a mordant and in the petroleum industry to sweeten sulfidic crude oil. Copper(II) chloride solutions are used for plating copper on aluminum, and in tinting baths for tin and germanium. Copper(II) chloride dihydrate is used as a pigment in glass and ceramics, as a wood preservative, and in water treatment.

Analysis and Specifications. Copper is analyzed iodometrically with sodium thiosulfate [51]. A typical analysis of technical-grade copper(II) chloride dihydrate (mass fractions in %) is as follows: copper(II) chloride dihydrate 99.0 (min.), iron 0.02, zinc 0.05, sulfate 0.05, water-insoluble material 0.01, material not precipitated by hydrogen sulfide as sulfate 0.15.

4.3. Copper(II) Oxychloride

Copper(II) oxychloride, $Cu_2Cl(OH)_3$ [1332-65-6], M_r 213.56, is usually written as $CuCl_2 \cdot 3\,Cu(OH)_2$. The trade name is copper oxychloride or basic copper chloride; the internationally accepted name (IUPAC) is dicopper chloride trihydroxide. Copper oxychloride is found in nature as the minerals paratacamite, green hexagonal crystals, and atacamite, green rhombic crystals, d_4^{25} 3.72–3.76 [94]. It is virtually insoluble in water, dissolves readily in mineral acids or warm acetic acid, and is soluble in ammonia and alkali-metal cyanide solutions. The green oxychloride is converted into blue copper(II) hydroxide in cold sodium hydroxide solution [59] and into the oxide in hot sodium hydroxide solution. With a lime suspension, copper oxychloride is converted into the blue calcium tetracuproxychloride, calcium tetracopper(II) chloride tetrahydroxide, $CaCl_2 \cdot 4\,Cu(OH)_2$ [95]. It is decomposed to the oxide at 200 °C.

Production. Copper(II) oxychloride is most often prepared commercially by air oxidation of copper(I) chloride solutions [90], [91], [96], [97]. For this purpose, a concentrated sodium chloride solution containing about 50 g/L copper(II) is contacted with copper metal to produce a solution containing about 100 g/L copper(I). The copper(I) chloride–sodium chloride solution, with or without the copper metal, is then heated to 60–90 °C and aerated to effect oxidation:

$$CuCl_2 + Cu \longrightarrow 2\,CuCl$$
$$6\,CuCl + 3\,H_2O + 3/2\,O_2 \longrightarrow CuCl_2 \cdot 3\,Cu(OH)_2 + 2\,CuCl_2$$

The mother liquor is separated from the precipitate and recycled into the process. The particles that are obtained by the above process are generally less than 4 μm in diameter and are suitable as crop fungicides. The particle size can be reduced further by increasing the agitation during oxidation and by utilizing a lower temperature [96]. Also, spray-drying of the product slurry gives a micronized product [98] as a result of deagglomeration of the material.

Copper(II) oxychloride can also be prepared by reaction of a copper(II) chloride solution with sodium hydroxide [94]:

$$4\,CuCl_2 + 6\,NaOH \longrightarrow CuCl_2 \cdot 3\,Cu(OH)_2 + 6\,NaCl$$

or by reaction of a copper(II) chloride solution with freshly precipitated, hydrated copper(II) oxide [99]:

$$CuCl_2 + 2\,NaOH \longrightarrow CuO \cdot H_2O + 2\,NaCl$$
$$CuCl_2 + 3\,CuO \cdot H_2O \longrightarrow CuCl_2 \cdot 3\,Cu(OH)_2$$

Copper oxychloride has also been produced as a byproduct in the electrolytic production of copper(I) chloride [100].

Uses. Copper oxychloride is used primarily as a foliar fungicide [90], [101]–[105]; it is also used as a pigment.

Analysis and Specifications. A typical analysis of technical-grade copper(II) oxychloride (mass fractions in %) is as follows: copper 56.0, chloride 14.0 – 15.0, sulfate 2.0 – 2.5, hydroxide 22.5 – 23.5, water 3.0 – 6.0. Copper is analyzed by iodometric titration with sodium thiosulfate [51].

4.4. Copper(II) Sulfates

4.4.1. Copper(II) Sulfate Pentahydrate

$CuSO_4 \cdot 5 H_2O$ [7758-99-8], M_r 249.61, d_4^{25} 2.285, c_p (273 – 291 K) 1126 J kg^{-1} K^{-1}, $\Delta H°$ (25 °C) – 850.8 kJ/mol, bluestone, blue vitriol, is found in nature as the mineral chalcanthite, blue triclinic crystals that can be ground to a light blue powder. Copper(II) sulfate pentahydrate slowly effloresces in dry air or above 30.6 °C with the formation of the trihydrate, $CuSO_4 \cdot 3 H_2O$. At 88 – 100 °C the trihydrate is produced more quickly. Thermal analysis of the pentahydrate gives the following:

$$CuSO_4 \cdot 5 H_2O \xrightarrow{88\,°C} CuSO_4 \cdot 3 H_2O \xrightarrow{114\,°C} CuSO_4 \cdot H_2O$$
$$\xrightarrow{245\,°C} CuSO_4 \xrightarrow{340\,°C} 3\,Cu(OH)_2 \cdot CuSO_4 \xrightarrow{600-650\,°C} CuO$$

Above about 114 °C, the monohydrate is formed, and between about 245 and 340 °C, the anhydrous product $CuSO_4$, results. Figures 6 A and 6 B show solubility of $CuSO_4$ in water and density of the solution as a function of temperature and sulfuric acid concentration [106], [107]. The pentahydrate can be crystallized from the solution either by addition of sulfuric acid or by evaporation. Although the addition of sulfuric acid appears to be more economical, the concentration of the solution by evaporation is preferred because the crystals obtained from a "neutral" (pH 3.5 – 4.0) medium are less prone to hard cake formation than the acid crystals. As the particle size of the pentahydrate decreases, the tendency toward hard cake formation increases, and the necessity to increase the pH during crystallization is enforced. When a free-flowing, commercial product of fine particle size is required, such alkaline additives as calcium oxide or calcium stearate must be incorporated into the final product to assure flowability. The incorporation of excess acid into the product accelerates the in situ dehydration of the pentahydrate and promotes hard cake formation. Lower temperature and lower humidity slow the caking process.

Copper(II) sulfate pentahydrate is soluble in methanol (15.6 g/100 mL solution) but insoluble in ethanol. It readily forms soluble alkaline complexes at sufficiently high

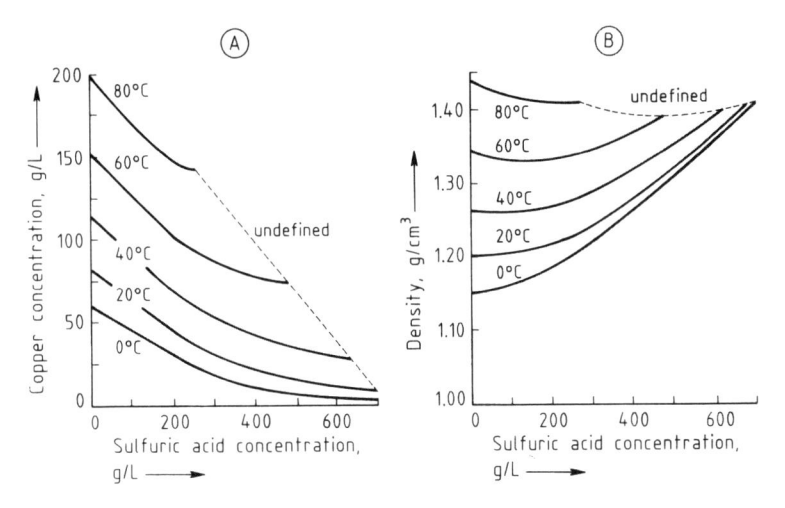

Figure 6. Saturated solutions of copper(II) sulfate
A. Solubility of copper(II) sulfate (as copper) as a function of sulfuric acid concentration and temperature; B. Density of copper(II) sulfate solutions as a function of sulfuric acid concentration and temperature
Reprinted with permission [107]

concentrations of amines or alkali cyanides (see Chap. 5), but basic sulfates are precipitated from solution by ammonia at an intermediate pH (about 4.2–6.8).

Copper(II) sulfate pentahydrate is the most commonly used copper compound because of the economics and the availability of starting materials, the ease of production, and the extent of byproduct utilization (primarily copper electrowinning liquors).

Production. Copper(II) sulfate pentahydrate is prepared most easily by the reaction of a basic copper(II) compound with a sulfuric acid solution (100–200 g/L H_2SO_4), e.g.:

$$CuO + H_2SO_4 \longrightarrow CuSO_4(aq) + H_2O$$

Copper metal, sulfuric acid, and air are the most common starting materials for the production of copper sulfate pentahydrate:

$$Cu + H_2SO_4 + 1/2\,O_2 \xrightarrow{65-100\,°C} CuSO_4(aq) + H_2O$$

Harike Process. The Harike process [108] is the best commercial example of the preceding reaction (Fig. 7). Blister shot copper up to 25 mm in diameter is added to the reaction tower (2.9 m^2 cross–sectional area) to give a bed of copper metal 2.74-m high. Two different reaction conditions are given in Table 5. Condition A allows for the production of concentrated copper(II) sulfate solutions and subsequent crystallization from acid media. Condition B gives a concentrated neutral solution of copper(II) sulfate that can be crystallized or diluted with water and used for direct production of other copper products. The rate of oxygen consumption by the system is directly

Table 5. Production of copper(II) sulfate by the Harike method

	A*	B*
Initial concentration, Cu (g/L)	100	100
H_2SO_4 (g/L)	160	80
Final concentration, Cu (g/L)	160	160
H_2SO_4 (g/L)	80	0
Temperature, °C	85	85
Circulation rate, m³/h	34–45	34–45
Air to tower, m³ h⁻¹ m⁻²	46	46
Entering oxygen, vol%	20.9	20.9
Exiting oxygen, vol%	2.6	7.3
Oxygen consumed, vol%	90	70
Production rate ($CuSO_4 \cdot 5 H_2O$)		
t/m² of tower cross section	4.65	3.6
t/tower	13.2	10.3

* For processes A and B, see text.

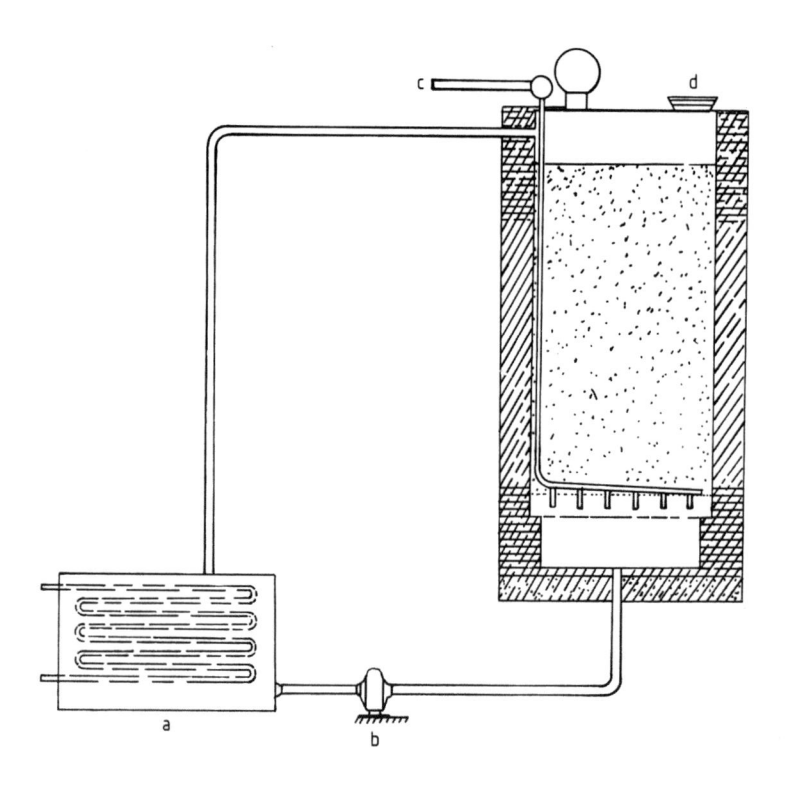

Figure 7. Harike tower process for production of copper(II) sulfate solutions
a) Heat exchanger; b) Circulation pump; c) Air inlet; d) Copper input

proportional to the mass of copper metal dissolved. Air flows of 46 m^3 h^{-1} m^{-2} are used in order to fill with air the voids that are created by the packing of the copper shot. The solution is circulated cocurrently with the air flow to wet the particles continuously and enhance mixing of the solution with the copper metal. If the air flow is decreased and the voids in the copper metal bed are not filled sufficiently, the hot acid solution oxidizes the copper metal in the absence of air with the formation of a copper sulfide film. This film renders the copper metal inert to further oxidation and would lower the efficiency of the tower. On the other hand, an increase in air flow above 46 m^3 h^{-1} m^{-2} decreases the fraction of oxygen consumed and would also result in a lowered efficiency of the tower. If an increased production (greater oxygen consumption) with good tower efficiencies is required, the height of the tower must be increased. This would also be necessary if a copper metal of lower surface area was used such as scrap copper instead of the blister shot copper. The use of cement copper and fine chopped wire creates insufficient voids for adequate air passage; such fine copper must be leached in agitated vessels.

Two-tower Process. Another commercial method for the preparation of copper(II) sulfate pentahydrate utilizes two towers filled with copper shot. One of the towers is filled with a sulfuric acid solution, whereas the other is sparged with air and steam to oxidize the surface of the copper metal and form a layer of copper(II) oxide. The solution is pumped alternately to the other tower to dissolve the layer of oxide while the now drained tower is sparged with air and steam. This process is continued until the desired copper(II) sulfate concentration is reached.

The trickle method is also used commercially: a copper-filled tower is continuously sprayed from the top with leach liquor from an adjacent reservoir. The liquor drains through the tower and is pumped back to the reservoir. Steam and air are sparged continuously into the tower from the bottom. The process is continued until the copper sulfate solution has the desired concentration.

Solvent Extraction. Copper(II) sulfate pentahydrate is produced from alkaline ammoniacal copper(II) solutions by solvent extraction [107], [109]. The alkaline copper(II) solution is first contacted with an organic extractant that is selective to copper. One of the most common extractants can be represented by the following general formula:

$$R-\underset{\underset{R''}{|}}{\overset{\overset{OH}{|}}{C}}-\overset{\overset{NOH}{\|}}{C}-R'$$

where R and R' are preferably unsaturated alkyl groups and R'' is preferably hydrogen [110].

$$[Cu(NH_3)_4]^{2+} \text{ (aq)} + 2 \text{ RH} \longrightarrow CuR_2 + 2 \text{ NH}_3 \text{ (aq)} + 2 \text{ NH}_4^+ \text{ (aq)}$$

The now copper-barren aqueous layer is continuously separated from the copper-containing organic layer which is subsequently contacted with a sulfuric acid solution:

$$CuR_2 + H_2SO_4 \longrightarrow CuSO_4 + 2 \text{ RH}$$

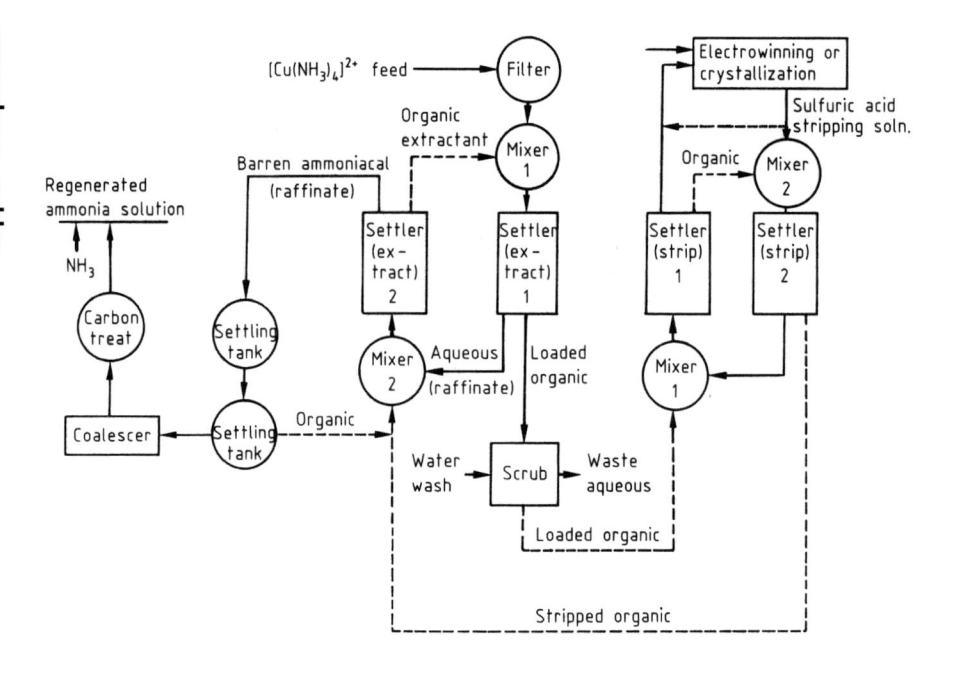

Figure 8. Flow diagram for a solvent extraction plant

The saturated copper(II) sulfate liquor produced is cooled or concentrated by evaporation to crystallize the pentahydrate. The copper-barren ammonia–ammonium salt solution is continuously returned to the copper leach tank. Figure 8 shows a flow diagram for a typical solvent extraction ammonia leach circuit utilizing mixer–settlers. With the exception of small quantities of wash water and loss of ammonia, the plant produces no effluent. Also, this process allows for lower overall energy input than acid dissolution, primarily as a result of the more facile oxidation of copper metal in ammoniacal solutions. Ion-exchange resins are used to recover and concentrate copper from the solution [111], [112]. Elution of the resins with sulfuric acid produces copper(II) sulfate solutions.

Byproduct Recovery. About 20–30% of the total copper(II) sulfate pentahydrate produced comes from the electrowinning industry as a byproduct. When the impurity content of the electrowinning liquors is too high to make cathode copper, they are concentrated and crystallized [113]. The product is suitable for most agricultural applications. Acidic copper(II) sulfate etching wastes are neutralized with a copper sludge, filtered, and concentrated to produce an agricultural-grade pentahydrate [114], [115].

Uses. The major uses of copper(II) sulfate pentahydrate have been grouped as follows [113]:

Agriculture (feed supplement, soil nutrient, fungicide)	41 %
Industrial algicides	27 %
Mining (flotation activator)	10 %
Electroplating	5 %

In many instances, the copper(II) sulfate pentahydrate is used as an intermediate, e.g., to produce active foliar fungicides such as Bordeaux mixture, tribasic copper(II) sulfate, or copper(II) hydroxide. A substantial amount of copper(II) sulfate pentahydrate is used in combination with sodium dichromate and arsenic acid for the preservation of wood. Copper(II) sulfate pentahydrate is an effective and economical algicide for lakes and ponds. In the mining industry, it is used as a flotation activator for lead, zinc, and cobalt ores. Copper(II) sulfate solutions are often used in the electroplating industry. The compound is also used as a mordant in textile dyeing, in the preparation of azo and formazan dyes, as a pigment in paints and varnishes, for preserving hides and tanning leather, in pyrotechnic compositions, and in synthetic fire logs.

Analysis and Specifications. Typical analyses for various commercially available crystal sizes are shown in Table 6. Table 7 lists specifications for an electroless copper(II) sulfate pentahydrate. The analytical methods for copper(II) sulfates are given in [51].

4.4.2. Anhydrous Copper Sulfate

$CuSO_4$ [7758-98-7], M_r 159.61, occurs in nature as the mineral hydrocyanite [14567-54-5]. It is gray to white and has a rhombic crystal morphology. It decomposes to the green basic copper(II) sulfate at 340 °C, and at 600–650 °C it decomposes to copper(II) oxide. Some of the properties of the anhydrous salt are as follows: d_4^{25} 3.6, c_p (273–291 K) 631.8 J kg^{-1} K^{-1}, c_p (273–373 K) 657.3 J kg^{-1} K^{-1}, $\Delta H°$ (25 °C) −771.6 kJ/mol. The compound is soluble in water (Section 4.4.1), somewhat soluble in methanol (1.1 g/100 mL), but insoluble in ethanol. It readily dissolves in aqueous ammonia and excess alkali metal cyanides, with the formation of complexes. The material is hygroscopic, with conversion to the pentahydrate in moist air below 30 °C.

The anhydrous salt is prepared by careful heating of the pentahydrate salt to 250 °C. It can also be prepared by the reaction of hot concentrated sulfuric acid with copper metal, but purification is rather difficult.

The compound has limited commercial use, but it can be used as a desiccant for removing water from organic solvents and is a sensitive indicator of the presence of moisture in such solvents.

Table 6. Typical analysis of copper(II) sulfate pentahydrate

Assay	Mass fraction, %			
	Industrial, granular	Large, medium	Snow, superfine	Powdered
Copper	25.2	25.2	25.2	25.2
Iron	0.008	0.007	0.008	0.17
Silicon dioxide	0.006	–	0.42	–
Water-insolubles	0.008	0.06	0.11	0.11
Calcium oxide	–	–	0.001	0.15

Large crystals:	on 9.5 mm × 19.0 mm through 28.6 mm screen;
Medium crystals:	on 3.2 mm × 12.7 mm through 19.0 mm screen;:
Granular crystals:	on U.S. 20 mesh through U.S. 6 mesh;
Industrial crystals:	on U.S. 14 mesh;
Snow crystals:	through U.S. 18 mesh;
Superfine crystals:	through U.S. 25 mesh;
Powdered:	85% through U.S. 100 mesh

Table 7. Specifications of electroless-grade copper(II) sulfate pentahydrate

Assay	Mass fraction, %
Copper	25.1
Iron	<0.0015
Lead	<0.003
Nickel	<0.001
Manganese	<0.0005
Chromium	<0.0005
Chlorine	nil
Water-insolubles	0.01

4.4.3. Copper(II) Sulfate Monohydrate

$CuSO_4 \cdot H_2O$ [10257-54-2], M_r 175.63, d_4^{20} 3.25, is a whitish powder. The solubility of the monohydrate in water is identical to the pentahydrate on a copper basis. The product is hygroscopic and must be packaged in containers with moisture barriers. It is commercially produced by dehydration of the pentahydrate at 120 –150 °C. A novel preparation is to triturate stoichiometric ratios of copper(II) oxide and sulfuric acid:

$$CuO + H_2SO_4 \longrightarrow CuSO_4 \cdot H_2O$$

The uses of the monohydrate are analogous to those of the pentahydrate. Although there can be slight economic advantages (primarily freight costs) in using the monohydrate, market acceptance has not been great. Presently, less than 5% (copper basis) of copper(II) sulfate is marketed in the monohydrate form.

4.4.4. Basic Copper(II) Sulfates

Four distinct basic copper(II) sulfates can be identified by potentiometric titration of a copper(II) sulfate solution with sodium carbonate or sodium hydroxide solution [115], [116]: $CuSO_4 \cdot 3\,Cu(OH)_2 \cdot H_2O$ (langite), $CuSO_4 \cdot 2\,Cu(OH)_2$ (antlerite), $CuSO_4 \cdot 3\,Cu(OH)_2$ (brochantite), and $CuSO_4 \cdot CuO \cdot 2\,Cu(OH)_2 \cdot x\,H_2O$. Their unique crystal morphologies are confirmed by X-ray diffraction.

The most important commercial basic copper(II) sulfate, commonly referred to as *tribasic copper sulfate* [12068-81-4], is $CuSO_4 \cdot 3\,Cu(OH)_2$, M_r 452.27, mp ca. 380 °C (decomp.), d_4^{25} 3.78, occurs in nature as the green monoclinic mineral brochantite. It is readily soluble in mineral acids, acetic acid solution, and ammonia solution; it is insoluble in water. Above 650 °C it decomposes to copper(II) oxide. If anhydrous copper(II) sulfate is heated cautiously to 650 °C, another naturally occurring mineral is obtained, dolerophane, $CuSO_4 \cdot CuO$, which reacts readily with water at 20 °C to form $CuSO_4 \cdot 3\,Cu(OH)_2 \cdot H_2O$ or at 100 °C to form $CuSO_4 \cdot 2\,Cu(OH)_2$ [117].

Production. Tribasic copper(II) sulfate, the commercially available basic copper(II) sulfate, is most often prepared by the addition of sodium carbonate solution to hot solutions of copper(II) sulfate:

$$4\,CuSO_4 + 3\,Na_2CO_3 + 3\,H_2O \longrightarrow CuSO_4 \cdot 3\,Cu(OH)_2 + 3\,Na_2SO_4 + 3\,CO_2$$

Great variation of particle size can be achieved by control of the precipitation temperature. As the temperature is increased, larger particles with greater bulk density are formed. If a small particle size is desired, lower precipitation temperature should be used. However, as the precipitation temperature is lowered, multiple hydrates and larger amounts of sulfate are incorporated into the product; e.g., around 50 °C, the dried product contains only about 50% copper. A precipitation temperature around 90 °C is required to make a pure tribasic copper(II) sulfate.

The single largest use of the compound is as a crop fungicide, and small particles are preferred because they give a greater degree of coverage. These small particles can be obtained by high-energy attrition of the product or by carefully controlled precipitation conditions.

Another method of making tribasic copper(II) sulfate is the aeration of a suspension of copper(I) oxide in the presence of stoichiometric quantities of sulfuric acid or copper(II) sulfate [118]:

$$2\,Cu_2O + H_2SO_4 + O_2 + 2\,H_2O \longrightarrow CuSO_4 \cdot 3\,Cu(OH)_2$$
$$6\,Cu_2O + 4\,CuSO_4 + 3\,O_2 + 12\,H_2O \longrightarrow 4\,[CuSO_4 \cdot 3\,Cu(OH)_2]$$

A purified, concentrated solution of copper(II) sulfate containing ammonium sulfate can be neutralized to pH 6.0–6.5 with ammonia to give a blue precipitate of approximate stoichiometry $4\,CuSO_4 \cdot 12\,Cu(OH)_2 \cdot 5\,H_2O$. When this is dried, a green tribasic copper(II) sulfate is formed with up to 1 mol of hydration water per mole of tribasic salt [119]; the water content depends on the drying temperature.

A continuous process has been developed in which copper(II) sulfate solutions are neutralized to pH 5.9 at 30 °C by addition of gaseous ammonia to an agitated vessel [120]. A unique, germicidally active, basic copper(II) sulfate whose X-ray diffraction pattern differs from langite and brochantite has been prepared in this way [121].

Reaction of copper(II) sulfate with $Na_3PO_4 \cdot 12\,H_2O$ and sodium hydroxide in aqueous solution gives a fine, blue, basic copper(II) sulfate powder (16% SO_4, 2.6% PO_4, and 54.9% Cu) that is formulated into a germicidal powder.

Tribasic copper(II) sulfate can also be prepared by cautiously heating copper(II) sulfate to 340 °C or by the aeration of a hot copper(II) sulfate solution in contact with copper metal.

Other basic copper(II) sulfates of commercial interest are the classic fungicial mixtures, Bordeaux and Burgundy slurries, in which copper(II) sulfate solution is mixed with lime and soda ash, respectively. They are of variable stoichiometry. More recently, the aqueous Bordeaux suspension has been dried to yield a stable basic copper(II) sulfate powder of variable composition [122].

Uses. As stated earlier, basic copper(II) sulfate is primarily used as a crop fungicide [123], [124]. It has also been utilized as a precursor in the separation of copper from metallic impurities [125].

Analysis and Specifications. The analysis of a typical technical-grade tribasic copper(II) sulfate is (mass fractions in %) as follows: copper 53.5, iron 0.08, sulfate 19.0, carbonate 2.0, water 2.0–5.0. This is analyzed according to [51].

5. Compounds and Complexes of Minor Importance

5.1. Copper Compounds

Copper(I) acetate [598-54-9], $CuCH_3COO$, M_r 122.6, is obtained as white crystals which are stable when dry but decompose on exposure to water. The product is obtained by reducing an ammoniacal solution of copper(II) acetate; on acidification with acetic acid, the crystals precipitate. Ammoniacal solutions of copper(I) acetate are used commercially to absorb olefins.

Copper(II) acetate monohydrate [6046-93-1], neutral verdigris, $Cu(CH_3COO)_2 \cdot H_2O$, M_r 199.65, mp 115 °C, d_4^{25} 1.88, decomposes at 240 °C; it forms dark green, monoclinic crystals; its solubility in water at 25 °C is 6.8 g/100 g solution; the compound is slightly soluble in methanol, diethyl ether, and acetone.

Copper(II) acetate monohydrate is produced by the reaction of copper(II) carbonate or copper(II) hydroxide with a solution of acetic acid or by the reaction of copper(II) oxide with hot dilute acetic acid. Alternatively, the material can be made by refluxing aqueous acetic acid in the presence of copper metal and air. The technical product is generally 99% pure.

Copper(II) acetate monohydrate is used in textile dyeing and as a ceramic pigment. It is used as a fungicide, as a precursor in the production of Paris green [copper(II) acetoarsenite], and as a polymerization catalyst in organic reactions.

Copper(II) acetate, basic [52503-63-6], $Cu(CH_3COO)_2 \cdot CuO \cdot 6\,H_2O$ (variable), M_r 369.26, exists as blue to green salts, depending on the amount of water of hydration. Blue verdigris has the above formula, while green verdigris has fewer molecules of water of hydration. The salts are slightly soluble in water or ethanol, and soluble in dilute mineral acid or aqueous ammonia.

The basic salts are produced by neutralizing copper(II) acetate solutions. They can also be prepared by refluxing acetic acid over copper in the presence of air until the basic salt precipitates.

The basic copper(II) acetates are used as precursors in the manufacture of Paris green [copper(II) acetoarsenite] and as fungicides. They are used as pigments in oil- and water-based paints and in textile dyeing.

Copper(II) acetoarsenite [12002-03-8], (acetato)trimetaarsenitodicopper(II), $Cu(CH_3COO)_2 \cdot 3\,Cu(AsO_2)_2$ (variable), M_r 1013.77, also known as Paris green, is an emerald green, poisonous powder. It is virtually insoluble in water and ethanol but dissolves in dilute mineral acids and aqueous ammonia.

Copper(II) acetoarsenite is primarily made by the reaction of a solution of copper(II) sulfate with arsenic(III) oxide, sodium carbonate, and acetic acid

$$4\,CuSO_4 + 3\,As_2O_3 + 4\,Na_2CO_3 + 2\,CH_3COOH \longrightarrow$$
$$Cu(CH_3COO)_2 \cdot 3\,Cu(AsO_2)_2 + 4\,Na_2SO_4 + H_2O + 4\,CO_2$$

or by the reaction of copper(II) oxide with a hot solution of acetic acid and arsenic(III) oxide. A solution of copper(II) acetate or a suspension of basic copper(II) acetate can also react with arsenic(III) oxide. The product is used as an insecticide, in the preservation of wood, and as an antifouling pigment.

Copper(II) arsenate [7778-41-8], $Cu_3(AsO_4)_2 \cdot 4\,H_2O$ (variable), M_r 540.52, is a blue-green to blue insoluble powder. It is most often prepared by reaction of copper(II) sulfate solutions with arsenic(V) oxide and sodium hydroxide. Copper(II) arsenate is used as an insecticide, fungicide, rodenticide, wood preservative, and antifouling pigment.

Copper(II) arsenite, $CuHAsO_3$ (variable), M_r 187.47, Scheele's green, an insoluble green powder, is prepared by the reaction of a copper(II) sulfate solution with arsenic(III) oxide and sodium hydroxide and is used as pigment and insecticide.

Copper(I) bromide [7787-70-4], CuBr, M_r 143.45, mp 504 °C, bp 1345 °C, d_4^{25} 4.72, forms white cubic crystals that slowly decompose on exposure to light or to moist air; it is soluble in hydrochloric and hydrobromic acids and in aqueous ammonia, but very slightly soluble in water. This compound is prepared pyrometallurgically [see copper(I) chloride, Section 4.1] or by reducing a copper(II) sulfate solution in the presence of sodium bromide, usually with sulfur dioxide or metallic copper. Copper(I) bromide is used as a polymerization catalyst for organic reactions.

Copper(II) bromide [7789-45-9], $CuBr_2$, M_r 223.36, mp 498 °C, d_4^{25} 4.77, crystallizes from warm solution as black, deliquescent, monoclinic crystals. Below 29 °C, the green tetrahydrate, $CuBr_2 \cdot 4\,H_2O$, results. The anhydrous product is very soluble in water (55.7 g/100 g solution) and soluble in ethanol and acetone. It is insoluble in diethyl ether, benzene, or concentrated sulfuric acid. Copper(II) bromide is most conveniently prepared by dissolving copper(II) oxide in hydrobromic acid; it can also be prepared by the direct action of bromine water on metallic copper. Copper(II) bromide is used as a brominating reagent and catalyst in organic synthesis and as an intensifier in photography.

Copper(II) chromate(VI) [13548-42-0], $CuCrO_4$, M_r 179.55, is obtained as reddish brown crystals. Numerous insoluble basic salts can be formed from solutions. Neutral copper(II) chromate is prepared by direct heating of a mixture of copper(II) oxide and chromium(VI) oxide. It decomposes around 400 °C, with the formation of copper(II) chromate(III). Copper(II) chromate(VI) is used in wood preservation and in the weatherproofing of textiles.

Copper(II) chromate(III) [12018-10-9], $CuCr_2O_4$, M_r 231.56, forms black, tetragonal crystals which are insoluble in water. It is prepared by heating neutral copper(II) chromate(VI) to 400 °C. Copper(II) chromate(III) is used as a hydrogenation catalyst.

Copper(I) cyanide [544-92-3], CuCN, M_r 89.56, mp 474 °C (in nitrogen), d_4^{25} 2.92, is usually a white to cream-colored powder, which is practically insoluble in water, cold dilute acids, and ethanol, but dissolves in ammonia and alkali cyanide solutions with the formation of complexes. Copper(I) cyanide is produced by the addition of an alkali cyanide and sodium hydrogensulfite to a solution of copper(II) sulfate:

$$2\,CuSO_4 + 2\,NaCN + NaHSO_3 + H_2O \longrightarrow 2\,CuCN + 3\,NaHSO_4$$

Copper(I) cyanide is very poisonous and must be handled cautiously. It is used in electroplating and in organic reactions as a polymerization catalyst or as a means of introducing the cyanide moiety. Copper(I) cyanide has also been used as an antifouling pigment for marine paints and is an active fungicide and insecticide.

Copper(II) formate [544-19-4], Cu(HCOO)$_2$, M_r 153.58, mp ca. 200 °C (decomp.), d_4^{25} 1.831, is soluble in water and slightly soluble in ethanol. The reaction of copper(II) oxide, carbonate, or hydroxide with formic acid is used to prepare the product. The most common commercial form, a royal blue material, is prepared by crystallization from water at 75–85 °C. A metastable dihydrate is produced by crystallization at 50–60 °C. Copper(II) formate tetrahydrate is prepared by crystallization from cool solutions. Copper(II) formate is used to prevent bacterial and mildew growth in cellulosic materials.

Copper(II) gluconate [527-09-3], Cu(C$_6$H$_{11}$O$_7$)$_2$, M_r 453.85, forms light blue to blue-green crystals from water; it is soluble in water and slightly soluble in ethanol. Copper(II) gluconate is prepared by the reaction of gluconic acid solutions with basic copper(II) carbonates or copper(II) hydroxide. It is used as a dietary supplement for copper deficiency and in mouth deodorants or breath fresheners.

Copper(I) iodide [7681-65-4], CuI, M_r 190.49, mp 605 °C, bp 1290 °C, d_4^{25} 5.62, occurs in nature as the mineral marshite, white to reddish brown cubic crystals. It is virtually insoluble in water but dissolves in ammonia solution, alkali iodide and cyanide solutions, and dilute hydrochloric acid. Copper(I) iodide is manufactured pyrometallurgically by the reaction of hot copper with iodine vapor. Alternatively, copper(II) salt solutions are reacted with alkali iodides to precipitate copper(I) iodide.

Copper(I) iodide is used as a heat and light stabilizer in polymers, photographic emulsions, and light-sensitive papers, and in oil drilling to aid in corrosion inhibition in highly acid environments. It is also used as a feed additive, in cloud seeding, and as a double salt with mercury(II) iodide as a temperature indicator.

Copper(II) nitrate trihydrate [10031-43-3], Cu(NO$_3$)$_2$ · 3 H$_2$O, M_r 241.59, mp 114.5 °C, d_4^{25} 2.32, is a blue, deliquescent salt that crystallizes as rhombic plates. It is very soluble in water (77.4 g/100 g solution) and ethanol. When crystallized from solution below the transition point (26.4 °C), the hexahydrate is produced, Cu(NO$_3$)$_2$ · 6 H$_2$O [13478-38-1], M_r 295.64, mp 26.4 °C (with loss of 3 mol of water of hydration), d_4^{24} 2.07, blue, deliquescent, prismatic crystals. The anhydrous nitrate is not produced by heating the hydrates; instead, decomposition to the basic copper(II) nitrate [12158-75-7], Cu$_2$(NO$_3$)(OH)$_3$, begins around 80 °C. Conversion to copper(II) oxide is complete at 180 °C.

Copper(II) nitrates are obtained by dissolving copper(II) oxide, basic copper(II) carbonate, or copper(II) hydroxide in nitric acid solution. The various hydrate crystals obtained depend on the conditions of crystallization. Basic copper(II) nitrate can be precipitated directly from solution by neutralization. Copper(II) nitrate salts are often produced by dissolving copper metal in nitric acid solution. The reaction is vigorous, and oxides of nitrogen are evolved:

$$Cu + 4\,HNO_3 \longrightarrow Cu(NO_3)_2 + 2\,H_2O + 2\,NO_2$$
$$3\,Cu + 8\,HNO_3 \longrightarrow 3\,Cu(NO_3)_2 + 4\,H_2O + 2\,NO$$

The second reaction is favored by a lower temperature and dilute acid. Anhydrous copper(II) nitrate [3251-23-8] can be produced in ethyl acetate:

$$Cu + N_2O_4 \longrightarrow Cu(NO_3)_2 \cdot N_2O_4 \xrightarrow{90\,°C} Cu(NO_3)_2$$

It is not available commercially.

The hydrates of copper(II) nitrate are used in the textile industry as oxidants and mordants in dyeing. They are used to prepare a copper(II) oxide of high surface area, which is useful as a catalyst in numerous organic reactions. The nitrate salts are used as colorants in ceramics, in the preparation of light-sensitive papers, and in the burnishing of iron, as well as the browning of zinc and brightening of aluminum. They are also used in pyrotechnics and as an oxidative component in solid rocket fuels.

Copper(II) oxalate [814-91-5], CuC_2O_4, M_r 151.56, is a blue-white powder, which is very slightly soluble in water. Copper(II) oxalate is prepared by precipitation from a mixture of a copper(II) salt and sodium oxalate solution. The product is used as a stabilizer for acetylated polyformaldehyde and as a catalyst in organic reactions.

Copper(II) phosphate trihydrate [7798-23-4], $Cu_3(PO_4)_2 \cdot 3\,H_2O$, M_r 434.61, is a light blue powder which is insoluble in cold water, slightly soluble in hot water, and soluble in ammonia solution and mineral acids. This compound is prepared by reaction of copper(II) sulfate solution with soluble alkali phosphates; it precipitates as a voluminous, almost gelatinous, product. Copper(II) phosphate trihydrate is used as a fungicide and corrosion inhibitor.

Copper(II) diphosphate hydrate [10102-90-6], $Cu_2P_2O_7 \cdot x\,H_2O$ (variable), has the following typical analysis: Cu 33–36%, P_2O_7 45–49%. Copper(II) pyrophosphate is a light blue powder which is insoluble in water, but soluble in solutions containing an excess of diphosphate. It is prepared by the precipitation reaction of solutions of copper(II) sulfate and alkali diphosphate. Solutions are used for plating copper on plastics, aluminum, and zinc.

Copper(II) selenide [1317-41-5], CuSe, M_r 142.5, mp not defined, d_4^{25} 6.0, decomposes at dull red heat; it is insoluble in water but dissolves in hydrochloric acid with evolution of H_2Se. Copper(II) selenide is prepared by reducing copper(II) selenite with hydrazine. It is used as a catalyst in the digestion of organic chemicals by the Kjeldahl method.

Copper(II) soaps are water-insoluble copper salts of long-chain fatty acids. They are usually sold as 6 – 10 % copper solutions in a kerosene diluent. The copper(II) soaps are commonly prepared by direct reaction of the fatty acid with copper(II) hydroxide or basic copper(II) carbonate in an organic diluent, or by precipitation from aqueous media when copper(II) sulfate solution is mixed with the sodium salt of the respective fatty acid. The common commercially available soaps, concentrations, and uses follow.

Copper (II) naphthenate [1338-02-9] contains 8% copper and is used as a mildewcide in textiles, woods, and paints, and to prevent barnacle growth on ship bottoms.

Copper (II) oleate (9-octadecenoate)[1120-44-1], M_r 626.43, contains between 6 and 9% copper, and is used as a combustion improver in fuel oils, as an emulsifier and dispersant, and as an antifouling coating for fish nets and lines.

Copper (II) stearate (octadecanoate)[660-60-6], M_r 630.46, contains 10% copper, and is used in antifouling paints and in the preservation of wood and textiles.

Copper(I) sulfide [22205-45-4], Cu_2S, M_r 159.15, mp 1100 °C, d_4^{25} 5.6, occurs in nature as the mineral chalcocite [21112-20-9], blue to gray rhombic crystals, also known as copper glance. It has a lustrous metallic look. Copper(I) sulfide is virtually insoluble in water, but is soluble with decomposition in nitric acid and concentrated sulfuric acid, and soluble in alkali cyanide solution through complex formation. It is decomposed to copper(II) oxide, copper(II) sulfate, and sulfur dioxide by heating in air, and to copper(II) sulfide and copper by heating in the absence of air. Copper(I) sulfide is prepared by heating copper and sulfur in a hydrogen atmosphere or by precipitation with hydrogen sulfide from an ammoniacal copper(II) salt solution. It is used in luminous paints, lubricants, solar cells, thermoelements, and semiconductors.

Copper(II) sulfide [1317-40-4], CuS, M_r 95.60, d_4^{25} 4.6, decomposes at 220 °C and occurs in nature as the mineral covellite [19138-68-2], as blue-black hexagonal or monoclinic crystals. It is virtually insoluble in water, but soluble in alkali cyanides and in ammonia solution with complex ion formation. This compound is decomposed by hot nitric acid. It is stable in dry air, but is slowly oxidized to copper(II) sulfate in moist air. Copper(II) sulfide can be prepared by melting an excess of sulfur with copper(I) sulfide or by precipitation with hydrogen sulfide from a solution of anhydrous copper(II) chloride in anhydrous ethanol. It is used in the dye industry for the preparation of aniline black dyes and as an antifouling pigment.

Copper(II) tetrafluoroborate [38465-60-0], $Cu(BF_4)_2$, M_r 237.15, is prepared commercially by the neutralization of tetrafluoroboric acid with copper(II) hydroxide or basic copper(II) carbonate. It is generally produced as a concentrated solution and is used in the production of printed circuits and in copper plating.

Copper(I) thiocyanate [1111-67-7], CuSCN, M_r 121.62, mp 1084 °C (under nitrogen), d_4^{25} 2.84, is a white to yellow amorphous powder, which is very slightly soluble in water and dilute mineral acids, and soluble in ammonia solution, alkali

Table 8. Dissociation constants of copper(I) and copper(II) complexes (pK = –log K)

Complex ion	Dissociation products		pK
$[CuCl_3]^-$	$CuCl$	+ 2 Cl^-	–2.0
$[CuCl_2]^-$	Cu^+	+ 2 Cl^-	4.7
$[CuBr_3]^-$	$CuBr$	+ Br^-	–3.3 to –2.3
$[CuBr_2]^-$	Cu^+	+ 2 Br^-	5 to 6
$[Cu(CN)_3]^{2-}$	$[Cu(CN)_2]^-$	+ CN^-	11.3
$[Cu(CN)_2]^-$	Cu^+	+ 2 CN^-	16
$[Cu(NH_3)_2]^+$	Cu^+	+ 2 NH_3	10.8
$[Cu(NH_3)]^{2+}$	Cu^{2+}	+ NH_3	4.1
$[Cu(NH_3)_2]^{2+}$	Cu^{2+}	+ 2 NH_3	3.5
$[Cu(NH_3)_3]^{2+}$	Cu^{2+}	+ 3 NH_3	2.9
$[Cu(NH_3)_4]^{2+}$	Cu^{2+}	+ 4 NH_3	2.1
$[Cu(NH_3)_5]^{2+}$	Cu^{2+}	+ 5 NH_3	–0.5
$[CuCl_4]^{2-}$	$[CuCl_3]^-$	+ Cl^-	0.01
$[CuCl_3]^-$	$CuCl_2$	+ Cl^-	0.06
$CuCl_2$	$CuCl^+$	+ Cl^-	0.4
$CuCl^+$	Cu^{2+}	+ Cl^-	1
$[Cu(OH)_4]^{2-}$	$[Cu(OH)_3]^-$	+ OH^-	0.9
$[Cu(OH)_3]^-$	$Cu(OH)_2$	+ OH^-	–5

thiocyanate solutions, and diethyl ether. It is stable to dry air but slowly decomposes in moist air. The material is prepared by reacting a copper(I) chloride solution with an alkali metal thiocyanate at 80–90 °C. The recovered product is dried under nitrogen. Copper(I) thiocyanate is used as an antifouling pigment.

5.2. Copper Complexes

Copper(I) and copper(II) ions form many stable complexes with halides, amines, azo compounds, cyanides, and other complexing media. This, in part, accounts for the fact that more X-ray crystal structures have been determined for copper complexes than for any other first-row transition metal ion. Many of these complexes are of great commercial significance. The dissociation constants for a number of copper complexes are given in Table 8.

Copper Ammine Complexes. Copper(II) salts form complexes of the type $[Cu(NH_3)_n]^{2+}$ where $n = 1-5$. The pentaammine complex is favored only in concentrated ammonia solution (Table 8). The hexaammine complex is formed only in anhydrous ammonia. The tetraammine complex of copper(II) is favored at low ammonia concentrations. For the copper(I) complexes of ammonia, the diammine complex is favored. Tetraammine copper(II) hydroxide, $[Cu(NH_3)_4](OH)_2$, Schweitzer's reagent, is used in the dissolution of cellulose and in the production of rayon. Copper(I) diammine salt solutions are used in the absorption of olefins and carbon monoxide. Many copper circuit boards are etched with ammoniacal ammonium salt solutions; copper ammine solutions, which can be used in the production of copper compounds, are byproducts of this process.

Copper Chloride Complexes. Copper(I) and copper(II) ions form complexes with hydrochloric acid or soluble metal chlorides. Dilute solutions of copper(II) chloride are blue. As the chloride concentration increases, the color of the solution shifts toward green, intensifying as the concentration of the distorted tetrahedral $[CuCl_4]^{2-}$ ion increases. Copper(II) compounds have been isolated that contain either the $[CuCl_4]^{2-}$ anion or the $[CuCl_3]^-$ anion, e.g., $CsCuCl_4$ and $KCuCl_3$, respectively.

The insoluble copper(I) chloride can be solubilized by chloride-containing solutions, with the formation of the $[CuCl_2]^-$ ion. The benzene-insoluble copper(I) chloride dissolves readily in the presence of anhydrous aluminum chloride, with formation of the $Al[CuCl_4]$ complex. In acetonitrile, copper(I) chloride is solubilized with the formation of the isolable $[Cu(CH_3CN)_4]Cl$ complex.

Copper(I) Cyanide Complexes. Alkali cyanides readily dissolve copper(I) cyanide, with the formation of $[Cu(CN)_2]^-$ and $[Cu(CN)_3]^{2-}$ complexes. Solutions of the complexes are of great importance in the electroplating industry because of their versatility. From saturated copper and sodium cyanide baths, crystals of the double salt $Na_2Cu(CN)_3 \cdot 2\,H_2O$ can be obtained readily.

Other Copper Complexes of Industrial Significance. Solvent extraction of copper(II) ions relies on the ability of copper to form complexes and to break those complexes as a function of pH [110], [126], [127]. The most common reagents used commercially are substituted salicylaldoximes, 1,8-hydroxyquinolines, and α-hydroxyoximes.

Copper(II) is solubilized in alkaline media by complex formation with tartrates for use as Fehling's solution, which is utilized in the analysis of reducing sugars. Copper(II) bis(1,8-dihydroxyquinoline) is used as a textile fungicide and as a pigment. Copper phthalocyanines [128], [129] are exceptionally stable blue and green pigments, as are the azo dye complexes of copper.

6. Copper Reclamation

The reclamation or recycling of copper wastes and byproducts is environmentally imperative and, in many cases, economically sound. Better technologies have been developed to produce useful materials from wastes and byproducts, many of which are now considered hazardous wastes. The establishment of the Resource Conservation and Recovery Act (RCRA) by the EPA defined procedures to assure proper disposal or reclamation of hazardous wastes [130]. However, no distinction is made between waste buried in an approved landfill and waste recovered as useful products. Unfortunately, in many cases this encourages the burial of recyclable waste materials.

The largest sources of reclaimable byproduct copper are the electronics and plating industries, as well as spent electrowinning baths [113]. Electrowinning baths are most often crystallized (Section 4.4.1) or precipitated as tribasic copper(II) sulfate for use in agriculture.

In the manufacture of circuit boards, a copper laminate is selectively dissolved (etched). This dissolution process yields the desired circuit board and a byproduct copper solution, the spent etchant. The earliest etching solutions used iron(III) chloride–hydrogen chloride or chromium trioxide–sulfuric acid solutions which were difficult to recycle. Later, the sulfuric acid etchants that used hydrogen peroxide [131], [132] or persulfate [133]–[135] as oxidants were developed. Recycling of the copper is achieved through crystallization, precipitation [135], ion exchange [136], or cementation [137], [138]. Alkaline etchants utilized a mixture of ammonia and ammonium sulfate with persulfate as the oxidant [139] or a mixture of ammonia and ammonium chloride with chlorite as the oxidant [140]. The primary etchants used today are the ammonia–ammonium chloride system with air as the oxidant and the hydrochloric acid system with stabilized hydrogen peroxide as the oxidant [141].

Copper is recovered from the acid chloride etchants (1) by neutralization and precipitation of copper(II) oxychloride or copper(II) oxide [42]; (2) by reduction with sodium hydrogensulfite, for example, to precipitate copper(I) chloride which can then be converted to copper(I) oxide by sodium hydroxide treatment [142]; or (3) by cementation with scrap iron [143]. From the alkaline etchants, copper is recovered (1) by distillation to strip ammonia and precipitate copper(II) oxide [144]; (2) by addition of base to precipitate copper(II) hydroxide, followed by steam stripping of the ammonia [145]; (3) by reduction to copper with formaldehyde [146] or $Na_2S_2O_4$ [147]; or (4) by electrolysis [148].

One of the more elegant methods for the recovery of copper from ammonia–ammonium chloride etchants utilizes solvent extraction [149] (Fig. 8). Not only is the copper recovered, but the etchant is simultaneously regenerated:

$$[Cu(NH_3)_4]Cl_2\,(aq) + 2\,RH \longrightarrow CuR_2 + 2\,NH_3\,(aq) + 2\,NH_4Cl\,(aq)$$

Fresh etchant

$$CuR_2 + H_2SO_4\,(aq) \longrightarrow 2\,RH + CuSO_4\,(aq)$$

Reclaimed copper

where R is an α-hydroxyoxime as shown in 1667.

The copper can be further reclaimed by crystallization or electrowinning.

7. Copper and the Environment

The reclamation of copper byproducts and wastes and the regulation of plant effluents and sanitary waters have resulted in lower indiscriminate release of copper into natural waters [150]. Where large quantities of copper in high concentration exist (see Chap. 6), recovery of the copper can, in many cases, be economical. Regulations have been established to control or limit altogether the introduction of copper into waters [151]–[155]. The Federal Water Pollution Control Act Amendments of 1972 set up a comprehensive program to "restore and maintain the chemical, physical, and biological integrity of the nation's waters" [151].

Limitations of copper in drinking water have also been established: the World Health Organization recommends 0.05–1.5 mg/L [156]. No specific drinking water standards for copper have been established by the Environmental Protection Agency [157].

A variety of technologies presently exist for the removal of copper from wastewaters and drinking water [158], [159]. The most common treatment methods include direct precipitation [151]–[153], [160]–[163] or electrolysis [164], [165].

Although copper is toxic in exceedingly low concentrations to certain lower life forms, notably fungi and algae, it is a necessary constituent of higher plants and animals (see, however, Chap. 9). Copper plays a necessary role as an oxidation catalyst and oxygen carrier, probably second only to iron in importance [166]. Copper aids plants in photosynthesis and other oxidative processes. In higher animals, it is responsible for oxidative processes and is present in many proteins such as phenolase, hemocyanin, galactose oxidase, superoxide dismutase, ceruloplasmin, tyrosinase, monoamine oxidase, and dopamine β-hydroxylase [167], [168].

Copper compounds are regularly applied for their nutrient value to agricultural crops in European countries [169]–[171]. In the United States, more emphasis is now being placed on the value of copper as a nutritional feed additive [169], [172], [173]: it increases the rate of gain and feed efficiencies of the animals. The increased use of copper as a feed additive has caused concern about the environmental impact of high levels of copper in manure [174], [175]. The recently approved maximum total copper content in animal feeds for Western European countries is shown in Table 9 [169].

The largest single application of copper compounds is as an agricultural fungicide (see Chap. 8). Although copper is a necessary constituent of higher plants and animals, it is highly toxic to certain fungi. In human medicine, the importance of copper as a nutrient in proper development and growth is receiving increased study [167], [176]. Also, copper is being used in the production of new copper antibiotics and copper anti-inflammatory drugs [176]. The recommended daily requirement of copper is as follows [176]:

Humans	2.5–5.0 mg
Cattle	50–70 mg
Horses	50–60 mg
Sheep	10–15 mg
Swine	10–20 mg

Table 9. Maximum copper levels in Western European animal feeds

Use	Total copper, mg/kg
Fattening pigs, up to 16 weeks	175
Fattening pigs, 17th week to 6th months	100
Fattening pigs, over 6 months	50
Breeding pigs	50
Calves, fed milk-based products	30
Calves, other feeds	50
Sheep	20
Other animals	50

Table 10. Production and valuation of copper(II) sulfate pentahydrate in the United States

Year	Production		Price of copper(II) sulfate pentahydrate, U.S. $/kg
	Quantity, t	Copper content, t	
1978	31 880	8551	1.04
1979	35 005	9286	1.18
1980	31.010	8445	1.23
1981	35 640	9413	1.14
1982	32 230	8385	1.08

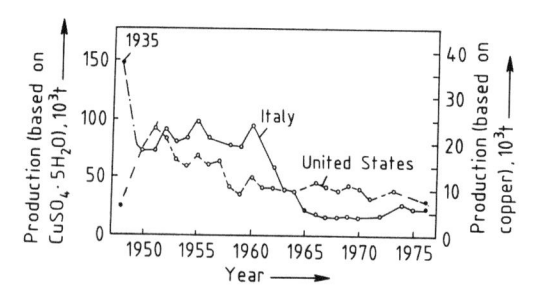

Figure 9. Production of copper(II) sulfate pentahydrate in Italy and the United States (1950–1976)

8. Economic Aspects

The world's production of copper compounds represents less than 2% of the total production of primary metal [65]. In 1974, 7.65×10^6 t of primary copper was produced, with 1.3% (100 000 t) used in the production of copper compounds. Of these 100 000 t, 80% is used to prepare copper(II) sulfate pentahydrate.

Historically, the United States and Italy have led the world in producing copper sulfate pentahydrate (Fig. 9) [65]. The U.S. production and valuation of copper sulfate pentahydrate are given in Table 10 [177].

Table 11. Estimate of current annual use of copper in agriculture [169]

Country	Total, t	Fungicide, %	Animal feed, %	Crop nutrient, %
Europe				
France	5 380	83.6	14.9	0.9
Federal Republic of Germany	2 800	35.7	28.6	35.7
Greece	1 410	97.9	1.4	0.7
Italy	6 220	96.5	3.2	0.3
Portugal	1 120	89.3	8.9	1.8
Spain	2 800	89.3	10.7	0.0
Others	15 563	49.8	34.5	15.8
Asia and Australia				
Australia	1 556	19.3	6.4	74.3
India	3 500	0.0	0.0	100.0
Japan	2 040	98.0	1.2	0.7
Others	1 811	78.7	0.6	20.3
Africa				
Algeria	1 750	100.0	0.0	0.0
Kenya	1 000	100.0	0.0	0.0
Tanzania	2 000	100.0	0.0	0.0
Others	2 635	96.4	0.6	3.0
America				
Brazil	7 160	97.8	1.4	0.8
Mexico	1 650	100.0	0.0	0.0
United States	6 400	46.9	9.4	43.8
Others	3 540	40.7	2.8	56.6
Total	70 305	76.4	12.1	11.4

According to estimates about 70% (70 000 t) of the total world's production of copper compounds is used in agriculture [169]. Table 11 gives a breakdown by region, country, and use. Copper sulfate pentahydrate can be blended directly with feed or fertilizers for animal feeds or crop nutrients, respectively. For application as a foliar fungicide, it must be rendered insoluble to prevent it from being easily washed off the leaf. The first copper fungicide on the market was Bordeaux mixture (1882), a combination of lime and copper sulfate [178]. The inherent phytotoxicity of Bordeaux mixture led to the development of the "fixed coppers" in the 1920s and 1930s. The fixed coppers are grouped into four categories: (1) basic sulfates, (2) basic chlorides, (3) oxides, and (4) miscellaneous — silicates, phosphates, etc. [179], [180]. To enhance their performance, the fixed coppers are usually blended with wetting agents, dispersants, sticking agents, and diluents. Since better coverage and enhanced efficacy are obtained with smaller particles, the basic coppers are usually ground before packaging [181]. Table 12 gives a partial list of available fixed copper fungicides.

Table 12. Commercially available fixed coppers

Manufacturer/representative	Trade name	Active ingredient	Cu content, %
BASF	BASF-Grünkupfer	copper oxychloride	45
	Kauritil	copper oxychloride	47
Bayer	Cupravit-Spezial	copper oxychloride	45
	Cupravit-Forte	copper oxychloride	50
CP Chemicals	Basic Copper Sulfate	tribasic copper sulfate	53
	Champion WP	copper hydroxide	50
Cuproquim	Hydrox	copper hydroxide	50
Hoechst	Vitigran, Conc.	copper oxychloride	45
Kocide/Griffin	Kocide 101	copper hydroxide	50
Merck	Perenox	copper(I) oxide	50
Norddeutsche-Aff.	Cobre Nordox	copper(I) oxide	50
Phelps-Dodge	Tribasic Copper Sulfate	tribasic copper sulfate	53
Sandoz	Cobre Sandoz	copper(I) oxide	50
Simplot	Blue Shield	copper hydroxide	50
Spiess Urania	Funguran	copper oxychloride	45
	Cuprasol	copper oxychloride	50
Tennessee Copper	Tribasic Copper Sulfate	tribasic copper sulfate	53
Wacker	Wacker-Kupferkalk	copper oxychloride	15–18

9. Toxicology and Occupational Health

Copper is an essential trace element in humans. In larger quantities it can be lethal. The salts are usually considered to be more toxic than the metal. The acute oral toxicity in humans, LD_{Lo}, is about 100 mg/kg; however, recovery has occurred after ingestion of up to 600 mg/kg. The symptoms of copper poisoning include nausea, vomiting, gastric disturbances, apathy, anemia, cramps, convulsions, coma, and death.

Many reported cases of illness were once attributed to chronic copper poisoning, but they are now thought to be caused by impurities in the refining of copper, e.g., lead, arsenic, and selenium [182], [183]. The question of chronic poisoning is still open to debate. Prolonged exposure to copper dust can cause skin irritation and discoloration of skin and hair. Whether pathological changes occur is uncertain, although there is evidence of accumulation of copper in the liver. Attempts to induce chronic copper poisoning in animals have been unsuccessful.

The inhalation of dusts of copper compounds irritates the upper respiratory tract [184], [185]. Ulceration and perforation of the nasal septum have occurred. Workers exposed to dusts of copper salts complained of metallic tastes and irritation of the oral and nasal mucosa. Smokers complained of an intense sweet taste during inhalation of the smoke. Long-term exposure to copper-containing dust has resulted in atrophic changes in the mucous membranes. Mild nasal discomfort was noted in workplace concentrations as low as 0.08 mg/m^3. A very small fraction of the workers exhibited allergic skin reactions from exposure to copper-containing dust [186], [187].

Copper compounds embedded in the eye produce a pronounced foreign body reaction with discoloration of the ocular tissue. Conjunctivitis, ulceration, and turbidity have been reported [188], [189].

The current workplace standard for copper dust concentration in air, MAK and TLV, is 1 mg/m^3 [190]–[192]. Copper fume standards are 0.2 mg/m^3 [190] and 0.1 mg/m^3 [191], [192]. A number of reviews and books of a more specific nature report detailed animal and workplace studies [193]–[197].

10. References

[1] *Gmelin*, System No. 60 "Kupfer," Teil A (1955), Teil B (1958–66), Teil C (1978), Teil D (1963).

[2] P. Pascal in *Nouveau Traité de Chimie Minérale*, vol. **3**, Masson et Cie., Paris 1957, pp. 155–421.

[3] J. Mellor: *A Comprehensive Treatise on Inorg. and Theoretical Chemistry*, vol. **3**, Longman's and Green, Co., New York 1923, pp. 49–294.

[4] A. Butts (ed.): *Copper — The Science and Technology of the Metal, its Alloys and Compounds, ACS Monograph 122*, Reinhold Publ. Co., New York 1954.

[5] W. Hatfield, R. Whyman in R. Carlin (ed.): *Transition Metal Chemistry*, vol. **5**, Marcel Dekker, New York 1969, pp. 47–179.

[6] M. Kato, H. Jonassen, J. Fanning, *Chem. Rev.* **64** (1964) 99.

[7] B. Hathaway, D. Billig, *Coord. Chem. Rev.* **5** (1970) 1–43.

[8] H. Sigel, *Angew. Chem. Int. Ed. Engl.* **13** (1974) 394.

[9] R. Doedens, *Prog. Inorg. Chem.* **21** (1976) 209.

[10] M. Bruce, *J. Organomet. Chem.* **44** (1972) 209.

[11] F. Jardine, *Adv. Inorg. Chem. Radiochem.* **17** (1975) 116.

[12] C. K. Williams Co., US 2 554 319, 1953 (J. Ayers).

[13] Rohm and Haas, US 2 184 617, 1939 (L. Hurd).

[14] Rohm and Haas, US 2 273 708, 1942 (L. Hurd).

[15] K. Hauffe, P. Kofstad, *Z. Elektrochem.* **59** (1955) 399.

[16] Merck, US 2 409 413, 1951 (H. Becker).

[17] Calumet and Hecla, US 3 466 143, 1969 (H. Day).

[18] Glidden, US 2 758 014, 1956 (J. Drapeau, P. Johnson).

[19] Glidden, US 2 891 842, 1959 (J. Drapeau, P. Johnson).

[20] Norddeutsche Aff., GB 772 846, 1957; DE 1 020 010, 1957 (E. Klumpp).

[21] S. Mahalla, US 3 492 115, 1970.

[22] Lake Chem. Co., US 2 474 497, 1949 (P. Rowe).

[23] G. E. Co., US 3 186 833, 1965 (R. Cech).

[24] Lake Chem. Co., US 2 474 533, 1949 (L. Klein).

[25] Glidden, US 2 817 579, 1957 (J. Drapeau, P. Johnson).

[26] Tennessee Corp., US 3 457 035, 1969 (J. Barker).

[27] USA, Secretary of the Interior, US 3 716 615, 1973 (D. Bauer, P. Haskett, R. Lindstrom).

[28] USA, Secretary of the Interior, US 3 833 717, 1974 (P. Haskett, D. Bauer, R. Lindstrom, C. Elges).

[29] S. Titova, V. Golodov, D. Sokol'skii, SU 488 788, 1975; *Chem. Abstr.* **84** (1976) 124 057 c.

[30] Mountain Copper, US 2 665 192, 1954 (P. Rowe).
[31] Mountain Copper, US 2 977 195, 1961 (C. Matzinger).
[32] Nippon Chem., JP-Kokai 80/71 629, 1980.
[33] ICI, GB 936 922, 1963 (A. Campbell, A. Taylor).
[34] Nippon Chem., JP-Kokai 78/133 775, 1978; *Chem. Abstr.* **93** (1981) 116 652 p.
[35] ASTM D283-52: Standard Methods of Chemical Analysis of Dry Cuprous Oxide and Copper Pigments, (reapproved 1978) Philadelphia, pp. 89–94.
[36] Military Specification: *Pigment, Cuprous Oxide* MIL-P-15169B (Ships), 1963; Amendment – 1 MIL-P-15169B (Ships), 1969.
[37] Metallgesellschaft, DE 1 007 307, 1957 (J. Dornauf).
[38] W. Kunda, H. Veltman, D. Evans, *Copper Met. Proc. Extr. Met. Div. Symp.* 1970, 27–69.
[39] Fluor Corp., US 2 927 018, 1960 (C. Redemann).
[40] Fluor Corp., US 2 923 618, 1960 (C. Redemann, H. Tschimer).
[41] Sherritt Gordon Mines, US 3 127 264, 1964 (H. Tschimer, L. Williams).
[42] Ruehl, Erich, Chem. Fabrik, DE 3 115 436, 1982 (W. Jagusch, H. Reichelt).
[43] Jap. Bur. Ind. Tech., DE 2 047 372, 1971 (Y. Moriya).
[44] Kachita Co., DE 2 028 791, 1970 (T. Tamura, A. Sakamoto, T. Kato).
[45] TDK Electronics, JP-Kokai 75/131 668, 1975 (S. Adachi, T. Miyakoshi, M. Hattori).
[46] Fordwerke, DE 2 046 180, 1971 (J. Jones, E. Weaver).
[47] Mitsui Toatsu Chem., JP-Kokai 74/72 179, 1974 (T. Kobayagawa, Y. Nakajima).
[48] Oxy. Catalyst, FR 1 402 087, 1965.
[49] Lonza, DE 2 012 250, 1970 (T. Voelker, K. Hering).
[50] Philips Gloelampen Fabrik., DE 2 103 538, 1971 (C. Esveldt, N. Slijkerman).
[51] C. Freedenthal: *Copper Compounds, Encyclopedia of Industrial Chemical Analysis*, J. Wiley & Sons, New York 1970, pp. 651–680.
[52] Cellocilk, Co., US 1 800 828, 1931 (W. Furness).
[53] D. Marsh, B. Marsh, US 2 104 754, 1938.
[54] Mountain Copper, US 2 536 096, 1951 (P. Rowe).
[55] Lake Chem. Co., US 2 525 242, 1950 (P. Rowe).
[56] H. Neville, C. Oswald, *J. Phys. Chem.* **35** (1931) 60–72.
[57] Copper Research, US 24 324, 1957 (W. Furness).
[58] Rohm and Haas, US 3 231 464, 1966 (E. Dettwiler, J. Filliettaz).
[59] Cuproquim S.A., US 4 418 056, 1983 (M. Gonzalez).
[60] Giulini Aldolfomer Ind., BR 83/01 912, 1983 (J. Giulini, A. Meyer).
[61] Kennecott Copper, DE 1 592 441, 1965 (W. Furness); US 3 194 749, 1965.
[62] Hoechst, DE 824 199, 1951 (K. Pfaff, A. Voigt).
[63] Hoechst, DE 824 200, 1951 (K. Pfaff, A. Voigt).
[64] E. Gimsey, D. Muir, A. Parker, *Proc. Australas. Inst. Min. Metall.* **273** (1980) 21–25.
[65] *Ullmann*, 4th ed., **15**, 564.
[66] Degussa, DE 1 000 361, 1955 (F. Bittner).
[67] Degussa, FR 2 009 852, 1969.
[68] Norddeutsche Affinerie, DE 1 813 891, 1958.
[69] Norddeutsche Affinerie, US 3 679 359, 1972 (E. Haberland, W. Perkow).
[70] Mitsubishi Metal Corp., JP-Kokai 80/60 003, 1980.
[71] R. Keller, H. Wycoff, *Inorg. Synth.* **2** (1946) 1–4.
[72] Schering, DE 1 080 088, 1958 (H. Niemann, K. Herrmann).
[73] Harshaw, US 2 367 153, 1945 (C. Swinehart).

[74] Goldschmidt, DE 3 305 545, 1983 (E. Mack, L. Witzke).
[75] Cyprus Metallurg., DE 2 607 299, 1976 (D. Goens, P. Kruesi).
[76] Cyprus Metallurg., US 3 972 711, 1976 (D. Goens, P. Kruesi).
[77] Metallgesellschaft, DE 1 174 996, 1963 (K. Meyer, H. Ransch, H. Pietsch); DE 1 180 946, 1963; DE 1 160 622, 1963.
[78] Duisburger Kupferhütte, US 3 353 950, 1967 (H. Junghauss); DE 1 200 545, 1965.
[79] Nitto Elec. Ind., JP-Kokai 78/65 348, 1978.
[80] Jap. Pure Hydrogen Co., JP-Kokai 70/48 693, 1970 (T. Yamamoto).
[81] Phillips Petro. Co., US 3 658 463, 1972 (W. Billigs).
[82] Tenneco Chem. DE 2 414 800, 1974 (R. Turnbo, D. Keyworth).
[83] H. Allgood, *Fert. Sci. Technol. Ser.* **2** (1974) 289–309.
[84] DuPont, US 3 230 231, 1966 (H. Burtolo, J. Braun, C. Winter).
[85] DuPont, US 4 035 383, 1977 (R. Sweet).
[86] Anic S.p.A., DE 2 743 690, 1978 (U. Romano, R. Tessi, G. Cipriani, L. Micucci); US 4 218 391, 1980.
[87] Ube Industries, JP-Kokai 79/106 429, 1979 (H. Itatani, S. Dano).
[88] Anic S.p.A., DE 3 045 767, 1981 (U. Romano, F. Rivetti, N. Muzio); US 4 318 862, 1983.
[89] as reported in A. Siedell (ed.): *Solubilities of Inorganic and Metal Organic Compounds*, vol. **I**, D. Van Nostrand, New York 1940, p. 478.
[90] Sudhir Chem. Co., US 3 202 478, 1965 (A. Hindle, S. Raval, S. Damani, H. Damani, K. Damani); GB 912 125, 1962.
[91] N. V. Koninklijke Ned. Zoutindustric, FR 957 457, 1947.
[92] Hoechst, DE 1 931 393, 1969 (H. Kreckler, H. Kuckertz).
[93] Distillers Co., BE 616 762, 1962; GB 932 130, 1962.
[94] W. Feitknecht, K. Maget, *Helv. Chim. Acta* **32** (1949) 1639–1653.
[95] Wacker, DE 1 083 794, 1960 (H. Baumgartner).
[96] Bayer, DE 1 159 914, 1963 (E. Podschus).
[97] Instytut Przemyslu Org., PL 55 953, 1968.
[98] Bayer, DE 1 161 871, 1963 (E. Podschus, W. Joseph); US 3 230 035, 1965.
[99] Ciba, Ltd., CH 243 271, 1946.
[100] Sirco, A.G., CH 256 414, 1947.
[101] Fisons Pest Control, US 2 907 691, 1959 (G. Hartley, P. Park).
[102] Fahlberg-List GmbH, DE 1 920 963, 1970 (G. Guenter, F. Schuelde, K. Schoeyen).
[103] Riedel-de-Haen, DE 1 142 724, 1963.
[104] P. Ramos, E. Esteban, V. Callao, *Ars. Pharm.* **12** (1971) 471–477.
[105] W. Feitknecht, K. Maget, *Helv. Chim. Acta* **32** (1949) 1639–1653.
[106] as reported in W. Linke (ed.): *Solubilities of Inorganic and Metal Org. Compounds*, 4th ed., vol. **I**, Amer. Chem. Soc., Washington, D.C. 1958, pp. 965–968.
[107] H. Moyer, *AIME Annual Meeting*, New Orleans, Feb. 18–22, 1979.
[108] Tennessee Copper, US 2 533 245, 1950 (G. Harike).
[109] C. Merigold, D. Agers, J. House, *Int. Solvent Extr. Conf.* 1971.
[110] General Mills, US 3 224 873 (Swanson).
[111] Dow, US 4 031 038, 1977 (R. Grinstead); US 4 098 867, 1978.
[112] S. Tataru, RO 67 946, 1979.
[113] F. Lowenstein, M. Moran in *Faith, Keyes and Clark's Industrial Chemicals*, 4th ed., John Wiley and Sons, New York 1975, pp. 280–283.

[114] Hitachi Chem., JP-Kokai 76/03 396, 1976 (T. Senda, T. Umehara); JP-Kokai 76/03 395, 1976.
[115] L. Markov, K. Balarev, *Izv. Khim.* **15** (1982) 472–481.
[116] H. Weiser, W. Milligan, E. Cook, *J. Am. Chem. Soc.* **64** (1942) 503–508.
[117] A. Binder, *Compt. Rend.* **198** (1934) 653, 2167.
[118] Cities Service, US 3 725 535, 1973 (J. Barker).
[119] Vereinigte Metallwerke, DE 2 701 253, 1977 (J. Bertha).
[120] Compagnie de Saint Gobain, FR 1 302 445, 1962 (M. Provoost).
[121] Takeda Chem. Ind., FR 1 452 608, 1966.
[122] INCRA, US 3 846 545, 1970 (S. Foschi); FR 2 001 428, 1969 (E. Hess, D. Kennedy, C. Whitham).
[123] J. Horsfall, J. Heuberger, *Phytopathology* **32** (1942) 226–232.
[124] J. Horsfall: *Fungicides and their Action*, Chronica Botanica Co., Waltham, Mass. 1945.
[125] Duisburger Kupferhütte, DE 2 602 448, 1977 (W. Roever, K. Lippert).
[126] A. Biswas, W. Davenport: *Extractive Metallurgy of Copper*, Pergamon Press, Oxford–New York 1976.
[127] G. Ritcey, A. Ashbrook: *Solvent Extraction*, Part II, Elsevier, Amsterdam 1979, pp. 197–248.
[128] DuPont, US 3 230 231, 1966 (H. Bartolo, J. Braun, C. Winter).
[129] DuPont, US 4 035 383, 1977 (R. Sweet).
[130] USEPA, *Environmental Protectin Agency*, Code of Federal Regulation 40CFR, parts 240–299, 1983.
[131] Mitsubishi, US 3 597 290, 1971 (A. Naito, Y. Masuda, S. Osawa).
[132] Dart Ind., US 4 158 593, 1979 (J. Allan, P. Readio).
[133] Denka, US 3 770 530, 1973 (O. Fujimoto).
[134] Denka, US 3 936 332, 1976 (A. Matsumoto, K. Itani); US 3 939 089, 1976; Tokai Electro-Chem., US 3 839 534, 1974 (A. Matsumoto, O. Fujimoto).
[135] FMC, GB 873 583, 1959 (P. Margulies, J. Kressbach, H. Wehrfritz); DE 1 160 270, 1959.
[136] K. Jones, R. Pyper, *J. Met.* **31** (1979) no. 4, 19–25.
[137] Micro Copper Corp., US 3 679 399, 1972 (D. Linton, A. Zinkl, J. Ballam).
[138] Showa Electric Wire and Cable, JP-Kokai 75/84 415, 1975 (Y. Kogane).
[139] FMC, US 3 837 945, 1974 (J. Chiang).
[140] FMC, US 3 844 857, 1974 (J. Chiang).
[141] P. Fintschenko, *Met. Finish. J.* **20** (1974) 138–139.
[142] Nippon Chem. Ind., JP-Kokai 83/140 321, 1983.
[143] A. Celi, DE 3 208 609, 1983.
[144] D. Klein, DE 3 204 815, 1983.
[145] J. Leroy, Contribution to International Recycling Congress (IRC), Berlin 1982, 592–595.
[146] Western Electric, US 4 428 773, 1984 (K. Krotz).
[147] Hitachi Cable, JP-Kokai 80/159 895, 1980.
[148] Western Electric, US 3 843 504, 1974 (B. Nayder).
[149] Criterion, US 4 083 758, 1978 (W. Hamby, M. Slade).
[150] F. Harrison, *Gov. Rep. Announce. Index (U.S.)* **84** (1984) 42–139.
[151] USEPA: *Development Document for Effluent Limitations Guidelines and Standards for the Inorganic Chemicals Point Source Category*, Phase II, 440/1-83/007-6 (1983).
[152] USEPA: Nonferrous metals manufacturing point source category; *effluent limitations guidelines, pretreatment standards, and new source performance standards, Fed. Regist.* **49** (1984) 8742–8831.

[153] USEPA: *Clean water;* Inorganic chemicals manufacturing point source category, effluent limitations, guidelines, pretreatment standards and new source performance standards, Fed. Regist. 49 (1984) 33 402 – 33 429.

[154] *Wasserhaushaltsgesetz* – WHG – vom 27. 7. 1957 in der Fassung vom 26. 4. 1976, BGBl. I (1976).

[155] EG – *Gewässerschutzrichtlinie* vom 4. 5. 1976 – Amtsblatt der EB Nr. L 129/23 vom 18. 5. 1976, hier Anhang Stoffliste II u. a. mit Kupfer.

[156] WHO: *International Standards for Drinking Water,* Monograph Series No. 3 (1971) 40.

[157] USEPA: "Interim Primary Drinking Water Standards," *Fed. Regist.* **40** (1975) 11 994.

[158] J. Hallowell, J. Shea, G. Smithson, A. Tripler, B. Gonser, *Gov. Rep. Announce (U.S.)* **74** (1974) 89 – 161.

[159] F. Pierrot, D. Vanel, *Galvano-Organo Trait. Surf.* **52** (1983) 726 – 728.

[160] C. Roy, US 3 816 306, 1974.

[161] Hoellmueller Maschinenbau, DE 2 257 364, 1974 (E. Lendle, G. Wartenberg).

[162] BASF, DE 2 255 402, 1974 (H. Hiller, A. Schuhmacher, H. Schmidt).

[163] M. Manafont, ES 397 046, 1974.

[164] Nippon Electric, JP-Kokai 84/22 694, 1984.

[165] R. Spearot, J. Peck, *Environ. Prog.* **3** (1984) 124 – 128.

[166] D. Kertesz, R. Zito, F. Ghiretti in O. Hayaishi (ed.): *Oxygenases,* Academic Press, London 1962.

[167] Ciba Foundation Symposium: *Biological Roles of Copper,* no. 79, Excerpta Medica and Elsevier-North Holland, Amsterdam 1980, pp. 343.

[168] H. Sigel (ed.): *Metal Ions in Biological Systems,* vol. **12,** Properties of Copper, Marcel Dekker 1981, pp. 384.

[169] INCRA: V. Shorrocks, Planning Study, *Use of Copper as a Micronutrient for Crops,* Part I, New York, July 1984.

[170] W. Ernst, *Schwermetallvegetation der Erde,* G. Fischer Verlag, Stuttgart 1974.

[171] *Micronutrients in Agriculture,* Soil Sci. Soc. of Amer. Inc., Madison, Wisc. 1972.

[172] G. Cromwell, T. Stahly, W. Williams, *Feedstuffs* **53** (1982) 30, 32, 35 – 36.

[173] L. DeGoey, R. Wahlstrom, R. Emerick, *J. Anim. Sci.,* **33** (1971) 52 – 57.

[174] INCRA: *Environmental Impact Analysis Report,* Copper Salts in Animal Feed, Int. Copper Res. Assoc. Inc., New York – London 1974.

[175] USEPA: "Registration on the Level of Copper in Animal Feed," *Fed. Regist.* **38** (1973) 178.

[176] INCRA: *A Critical Review of Copper in Medicine,* Report No. 234, Int. Copper Res. Assoc., Inc., Goteborg – New York 1975.

[177] U.S. Dept. of Interior: *Minerals Yearbook 1982,* vol. **I,** U.S. Govt. Printing Office, Washington, D.C., 1983, p. 209.

[178] G. Johnson, *Agricult. History* **9** (1935) 67 – 79.

[179] D. Frear in *Chemistry of Insecticides, Fungicides, and Herbicides,* 2nd ed., D. Van Nostrand, New York 1948, pp. 211 – 228.

[180] E. R. de Ong in *Chemistry and Uses of Pesticides,* 2nd ed., Reinhold Publ. Co., New York 1956, pp. 31 – 47.

[181] S. Gunther, *Chem. Ing. Tech.* **31** (1959) 731 – 734.

[182] E. Browning: *Toxicity of Industrial Metals,* 2nd ed., Butterworths, London 1969.

[183] R. Fabre, R. Truhaut: *Précis de Toxicologie,* Centre Doc. Univ., Paris 1960.

[184] A. Askergren, M. Mellgren, Scand. J. Work, *Environ. Health* **1** (1975) 45 – 49.

[185] R. Gleason, *Am. Ind. Hyg. Assoc. J.* **29** (1968) 461 – 462.

[186] Y. Kambe, Matsushima, T. Kuroume, S. Suzuki, *Proc. Fourth Int. Congr. Rural Med.* Zürich 1970, 55–57.
[187] E. Saltzer, J. Wilson, *Arch. Dermatol.* **98** (1968) 375–376.
[188] *Documentation of Threshold Limit Values for Substances in Workroom Air*, 3rd ed., Cincinnatti 1974.
[189] W. Grant: *Toxicology of the Eye*, 2nd ed., C. C. Thomas, Springfield, Il. 1974.
[190] NIOSH: *Manual of Industrial Chemicals*, U.S. Dept. of Labor, September 1978.
[191] D. Henschler: *Gesundheitsschädliche Arbeitsstoffe, Toxikologisch-arbeitsmedizinische Begründung von MAK-Werten*, Verlag Chemie, Weinheim 1972/77.
[192] DFG: *Maximale Arbeitsplatzkonzentration*, Verlag H. Boldt, Boppard 1976.
[193] H. E. Stokinger in *Patty*, vol. 2A, pp. 1620–1630 (1981).
[194] S. Cohen, *JOM, J. Occup. Med.* **16** (1974) 621–624.
[195] Int. Labor Office: *Encyclopedia of Occupational Health and Safety*, McGraw-Hill, New York 1971.
[196] N. Sax: *Dangerous Properties of Ind. Materials*, 4th ed., Van Nostrand, New York 1982.
[197] W. Dreichmann, H. Gerarde: *Toxicology of Drugs and Chemicals*, Academic Press, New York 1969.

Cyanates, Inorganic Salts

GERHARD DÜSING, Degussa AG, Frankfurt am Main, Federal Republic of Germany

1. Constitution 1693
2. Properties 1693
3. Production and Uses 1695
4. References 1696

1. Constitution

Cyanates are salts of an acid having the empirical formula HNCO, M_r 43.03, which exists in two tautomeric forms [1], [2]:

H–O–C≡N   ⇌   H–N=C=O
Cyanic acid   Isocyanic acid
[420-05-3]   [75-13-8]

At ambient temperature, isocyanic acid clearly predominates. However, esters of both isomers exist, e.g., ethyl cyanate, C_2H_5OCN, and ethyl isocyanate, C_2H_5NCO. Isocyanic acid can be represented by three mesomeric structures:

$$H-\bar{N}=C=\bar{O} \longleftrightarrow H-\overset{-}{N}-C\overset{+}{=}\bar{O} \longleftrightarrow H-\overset{+}{N}=C-\bar{\bar{O}}^{-}$$

Likewise, the anion of the inorganic salts is a resonance hybrid and may be derived from either tautomeric form of the acid.

$$[\bar{\bar{O}}-C\equiv N]^{-} \longleftrightarrow [O=C=\bar{N}]^{-}$$

Cyanate anion   Isocyanate anion

2. Properties

Only sodium and potassium cyanate are manufactured commercially (for their physical properties, see Table 1).

In aqueous solution, both salts undergo hydrolysis according to Equations (1) and (2) [3]. The rate of hydrolysis depends strongly on temperature, pH, and concentration.

Table 1. Physical properties of sodium and potassium cyanate

	Sodium cyanate	Potassium cyanate
CAS registry number	[917-61-3]	[590-28-3]
Molecular mass	65.01	81.12
Melting point, °C	550	315
Density, g/cm^3	1.893	2.056
Solubility (g/100 g solvent)		
in water	10.68 at 16 °C	75 at 25 °C
in ethanol	0.5 at bp	0.53 at bp
in benzene	0.13 at bp	0.18 at bp
in liquid ammonia	1.72 at −19.8 °C	1.70 at 25 °C
	0.72 at 45 °C	

$$OCN^- + 2 H_2O \longrightarrow CO_3^{2-} + NH_4^+ \tag{1}$$
$$OCN^- + NH_4^+ \longrightarrow CO(NH_2)_2 \tag{2}$$

If the pH of the solution is less than 1.2, hydrolysis follows Equation (1) exclusively and no urea is formed [4]. This reaction, in dilute, strong acid, is sufficiently rapid and complete to be used in the chemical analysis of cyanate. In alkaline solution, cyanates also undergo hydrolysis, but much more slowly. Concentrated hydrochloric acid, however, causes cyanic acid to trimerize, and cyanuric acid is formed.

When heated, the solid, anhydrous cyanates decompose by different routes, depending on the cation and the temperature. The mechanism and course of the thermal decomposition of potassium cyanate are not well established.

$$5\ NaOCN \xrightarrow{500-700\ °C} 3\ NaCN + Na_2CO_3 + N_2 + CO_2$$

$$4\ NaOCN \xrightarrow{>700\ °C} 2\ NaCN + Na_2CO_3 + N_2 + CO$$

$$KOCN \xrightarrow{700-900\ °C} KCN, K_2CO_3, N_2, C, CO, CO_2$$

Many useful reactions of alkali cyanates with organic compounds are known. Urea derivatives, for example, are obtained by reaction with amines or imines.

$$NaOCN \xrightarrow{H^+} HOCN \xrightarrow{+RNH_2} [RNH-\underset{\underset{\text{OH}}{|}}{C}=NH]$$
$$\longrightarrow R-NH-\underset{\underset{\text{O}}{\|}}{C}-NH_2$$

With a suitable choice of amines, eterocyclic compounds can be synthesized, e.g., imidazoles and hydantoins. Both sodium and potassium cyanate may be used in the above reactions. The only difference is in their solubility (which, in water, is much better for KOCN) and cyanate content (which is slightly lower in KOCN than in NaOCN).

Table 2. Specifications of technical-grade cyanates (mass fractions in %)

	Sodiumcyanate "90%"	Sodiumcyanate "98%"	Potassiumcyanate "99%"
MOCN*	89–91	min. 98	min. 98.8
OCN$^-$	min. 57.5	min. 62.3	min. 51.2
M_2CO_3	max. 10	max. 0.9	max. 0.7
MCN	0	max. 0.2	max. 0.05
Moisture (loss at 110 °C)	max. 0.1	max. 0.02	max. 0.02
pH of solution in water (5 wt %)	ca. 10.9	ca. 10.4	ca. 10.1

* M = Na or K.

When alkali cyanates react with alcohols, alkyl carbamates, $H_2N-COOR$, and allophanates, $H_2N-CO-NH-COOR$, are formed. The reaction of alkali cyanates with alkyl halides, sulfates, or phosphates leads to alkyl isocyanates. For example, allyl isocyanate is obtained from the reactions of alkali cyanates with allyl chloride. The product trimerizes readily to form triallyl isocyanurate.

3. Production and Uses

Commercially, alkali cyanates are manufactured by melting urea with sodium or potassium carbonate. Yields of about 98%, based on urea consumption according to the following equation, are obtained [5]–[7]:

$$2\ OC(NH_2)_2 + M_2CO_3 \longrightarrow 2\ MOCN + 2\ NH_3 + CO_2 + H_2O$$
(M=Na or K)

The reaction with potassium carbonate proceeds readily in the molten state at 400 °C. With sodium carbonate, the melting temperature of the mixture is much higher (600 °C), and if the mixture of urea and sodium carbonate is heated rapidly to this temperature, a violent reaction takes place with significant decomposition and loss of urea. Therefore, in practice, the reaction is initiated by prolonged heating of the mixture at 150–180 °C. During this stage, some of the urea polymerizes and some decomposes; the loss resulting from decomposition must be compensated to complete the reaction in the molten phase above 600 °C. Alkali cyanates are obtained with a purity of 98–99% (see Table 2).

Cyanates are used as herbicides, mainly to destroy weeds in lawns and onion crops. Simultaneously, they act as fertilizers because of their high content of available nitrogen. Alone or in combination with cyanides, cyanates are used on a large scale for the "nitrification" and "carbonitrification" of steel [8]–[11]. They are also used as building blocks in the synthesis of pesticides, detergents, and plastic additives.

4. References

[1] J. Goubeau, *Ber. Dtsch. Chem. Ges.* **68** (1935) 912–919.
[2] *Gmelin,* Syst. No. 14 "carbon", part DI, (1971) p. 327.
[3] O. Masson, J. Masson, *Z. Phys. Chem.* **70** (1910) 290.
[4] C. E. Vanderzee, R. A. Myers, *J. Phys. Chem.* **65** (1961) 153.
[5] Degussa, DE 597 956, 1929.
[6] Du Pont, US 1 915 425, 1930.
[7] Sakai Food Stuff Industry Co., DE-AS 1 205 065, Nov. 18, 1965.
[8] Degussa, DE 1 191 655, 1958.
[9] Degussa, DE 1 149 035, 1959.
[10] Degussa, DE 1 237 872, 1961.
[11] Degussa, DE 1 280 018, 1964.

Cyano Compounds, Inorganic

HERBERT KLENK, Degussa, Frankfurt am Main, Federal Republic of Germany
ANDREW GRIFFITHS, Degussa, Frankfurt am Main, Federal Republic of Germany
KLAUS HUTHMACHER, Degussa, Frankfurt am Main, Federal Republic of Germany
HANS ITZEL, Degussa, Frankfurt am Main, Federal Republic of Germany
HELMUT KNORRE, Degussa, Frankfurt am Main, Federal Republic of Germany
CARL VOIGT, Degussa, Frankfurt am Main, Federal Republic of Germany
OTTO WEIBERG, Degussa, Frankfurt am Main, Federal Republic of Germany

1.	Hydrogen Cyanide 1698	2.3.2.	Cyanides of Copper, Zinc, and Cadmium 1722
1.1.	Properties 1698	2.3.3.	Cyanides of Mercury, Lead, Cobalt, and Nickel 1724
1.2.	Production 1701	2.3.4.	Cyanides of Precious Metals . . . 1725
1.2.1.	Andrussow Process 1701	2.4.	Cyanide Analysis 1726
1.2.2.	Methane – Ammonia (BMA) Process 1703	3.	Detoxification of Cyanide-Containing Wastes 1727
1.2.3.	Shawinigan Process 1704	3.1.	Wastewater Treatment 1727
1.3.	Storage and Transportation . . 1705	3.2.	Solid Wastes 1730
1.4.	Economic Aspects and Uses . . 1706	4.	Cyanogen Halides 1730
2.	Metal Cyanides 1706	4.1.	Properties 1730
2.1.	Alkali-Metal Cyanides 1707	4.2.	Production 1732
2.1.1.	Properties 1708	4.3.	Storage and Transportation . . 1733
2.1.2.	Production 1711	4.4.	Uses 1733
2.1.3.	Commercial Forms, Specifications, and Packaging . . . 1712	5.	Cyanogen 1734
2.1.4.	Transportation, Storage, and Handling 1713	5.1.	Properties 1734
		5.2.	Production 1735
2.1.5.	Economic Aspects and Uses 1714	5.3.	Storage, Transportation, and Uses 1736
2.2.	Alkaline-Earth-Metal Cyanides 1715		
2.3.	Heavy-Metal Cyanides 1716	6.	Toxicology and Occupational Health 1736
2.3.1.	Iron Cyanides 1716		
2.3.1.1.	Properties 1716		
2.3.1.2.	Production 1719	7.	References 1739
2.3.1.3.	Commercial Forms, Specifications, and Packaging . . . 1721		

1. Hydrogen Cyanide

Hydrogen cyanide [74-90-8] (hydrocyanic acid, prussic acid, formonitrile), HCN, M_r 27.03, is a colorless liquid with the characteristic odor of bitter almonds. Hydrogen cyanide in aqueous solution was first prepared by SCHEELE in 1782 [1]. The acid occurs naturally in combination with some glucosides, such as amygdalin.

Hydrogen cyanide is currently produced by direct reaction of alkanes and ammonia, or indirectly as a byproduct in the manufacture of 2-propenenitrile [107-13-1] (by the ammoxidation of propene – acrylonitrile, Sohio process. Major end uses include the production of adiponitrile, methyl methacrylate, cyanuric chloride, chelating agents, sodium cyanide, nitrilotriacetic acid, and methionine.

1.1. Properties

Physical properties [2], [3]–[6]:

Melting point	−13.24 °C
Boiling point	25.70 °C
Vapor pressure (0 °C)	35 kPa (0.35 bar)
(20 °C)	83 kPa (0.83 bar)
(50 °C)	250 kPa (2.5 bar)
Enthalpy of formation, ΔH	3910 kJ/kg
Critical temperature	183.5 °C
Critical density	0.20 g/cm^3
Critical pressure	5 MPa (50 bar)
Density (20 °C)	0.687 g/cm^3
Specific heat, liquid (20 °C)	2630 J kg^{-1} K^{-1}
Specific heat, gas (25 °C)	1630 J kg^{-1} K^{-1}
Heat of fusion	310 kJ/kg
Heat of vaporization	935 kJ/kg
Heat of polymerization	1580 kJ/kg
Explosive range in air	5.5 – 46.5 vol%
Flash point	−17.8 °C
Ignition temperature	535 °C
Dynamic viscosity, η (20 °C)	0.192 mPa · s
Surface tension (20 °C)	18.33 mN/m
Dielectric constant (0 °C)	158.1
(20 °C)	114.9
Dissociation constant aqueous solution, pK (20 °C)	9.36

As a result of its high dielectric constant, HCN has acquired some importance in preparative chemistry as a nonaqueous, ionizing solvent [7].

Chemical Properties. Some comprehensive reviews on hydrogen cyanide and cyanogen compounds have been published [2]–[9]. The acid is found only in the nitrile form. Although isomeric isonitrile, HNC, has been detected in interstellar space, all efforts to isolate this compound have failed. As the nitrile of formic acid, HCN undergoes many typical nitrile reactions. For example, hydrogen cyanide can be hydrolyzed to formic acid by aqueous sulfuric acid or hydrogenated to methylamine. Hydrogen cyanide adds on to carbon–carbon double bonds, and it forms cyanohydrins with carbonyl groups of aldehydes or ketones. The most important uses of this type are in the manufacture of acetone cyanohydrin (an intermediate in the production of methyl methacrylate) and in the production of adiponitrile from butadiene and hydrogen cyanide. Another example is the multistep synthesis of amino acids via hydantoins. Hydrogen cyanide can be oxidized by air over silver or gold catalysts at 300–650 °C to yield cyanic acid (HOCN) and cyanogen $(CN)_2$ in an approximate 2:1 ratio.

The reaction of hydrogen cyanide with chlorine gives cyanogen chloride (see Section 4.2). For industrial purposes, the latter compound is usually directly trimerized to cyanuric chloride, the starting material for the chemistry of s-triazines.

In the presence of oxygen or air, hydrogen cyanide burns with a very hot flame. For the reaction

$$2\,HCN + 1.5\,O_2 \longrightarrow N_2 + 2\,CO + H_2O$$

the heat of formation is calculated to be −723.8 kJ/mol and the adiabatic flame temperature is 2780 °C. Pure liquid or gaseous HCN is inert to most metals and alloys such as aluminum, copper, silver, zinc, and brass.

At higher temperatures (> 600 °C), the acid reacts with metals that can form carbides and nitrides (titanium, zirconium, molybdenum, and tungsten). Therefore, containers of unalloyed steel are used for storage and transportation of liquid anhydrous hydrogen cyanide.

Pure liquid hydrogen cyanide has a tendency to polymerize to brown-black, amorphous polymers, commonly called azulmic acid [26746-21-4]. The reaction is accelerated by basic conditions, higher temperature, UV light, and the presence of radicals. Since the decomposition is exothermic, the polymerization reaction is autocatalytic and can proceed with explosive violence to form the HCN dimer, iminoacetonitrile [1726-32-5], and the HCN tetramer, diaminomaleonitrile [1187-42-4], as intermediates [10], [11]. Both compounds are presumed to be important in the evolution of life, and cyano compounds may play a role in prebiotic syntheses [12], [13]. In the liquid phase, hydrogen cyanide is stabilized by the presence of small amounts of acids (0.1 wt% H_3PO_4, 1–5% HCOOH or CH_3COOH) or of 0.2 wt% SO_2 in the gaseous phase.

The HCN–H_2O System. At 25 °C, hydrogen cyanide is miscible with water in all ratios. The solution is a weak acid, with a dissociation constant of the same order of magnitude as amino acids. The relation between total cyanide concentration ($c_{HCN} + c_{CN^-}$) and dissociated cyanide (c_{CN^-}) in a dilute aqueous solution, as a function of pH, is illustrated in Figure 1.

Figure 2 shows the liquid–vapor equilibrium diagram at atmospheric pressure. Because of the high HCN partial vapor pressure it is difficult to separate the acid from

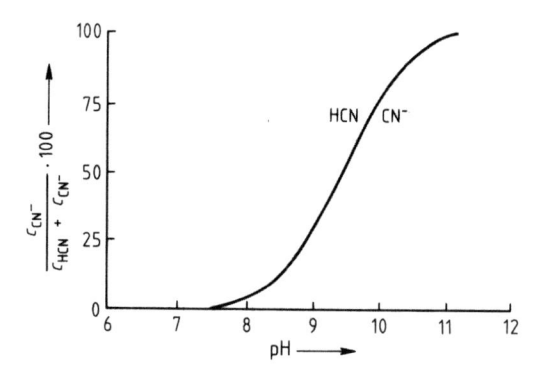

Figure 1. Dissociation of HCN in aqueous solution as a function of pH, based on total cyanide concentration $c_{HCN} + c_{CN^-}$

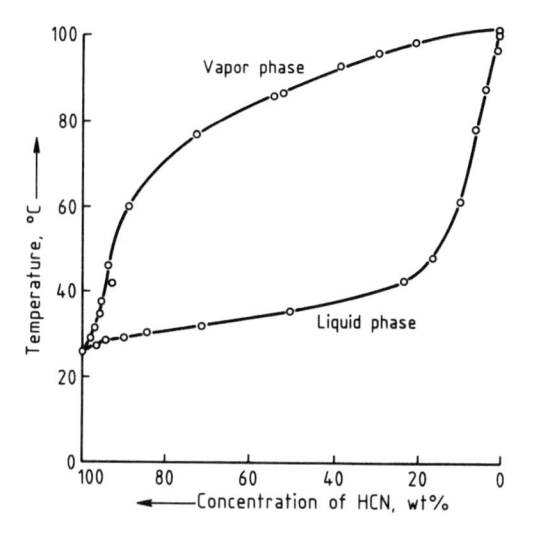

Figure 2. Liquid–vapor diagram of the system H_2O–HCN at atmospheric pressure

a gas mixture by absorption in water, which is an important consideration for industrial practice [14].

The stability of hydrogen cyanide solutions depends on the degree of dilution: at concentrations below 0.1 mol/L HCN, the acid is stable; addition of traces of acid prevents decomposition. Like pure HCN, the exothermic polymerization of an aqueous solution is accelerated by the presence of alkali. In addition to azulmic derivatives, small amounts of amino acids and purine bases are formed. This fact is of some importance to biological chemistry. The C≡N triple bond is hydrolyzed by strong alkali or acid to give formic acid and ammonia. Higher temperatures or hydrothermal conditions favor these reactions.

1.2. Production [2], [15], [16]

Hydrogen cyanide can be produced when sufficient energy is supplied to any system containing the elements hydrogen, nitrogen, and carbon. Generally, only processes starting from hydrocarbons and ammonia are of economic importance today; however, the production of hydrogencyanide from formamide [17] is carried out in one plant at Ludwigshafen in the Federal Republic of Germany. Recent patents indicate interest in alternative raw materials such as methanol, carbon, or carbon monoxide [18] [19]–[21] [22]–[24] [25]–[27]; however, these routes have little chance of competing with present technology. The reaction of hydrocarbons and ammonia

$$C_xH_{2x+2} + x\,NH_3 \longrightarrow x\,HCN + 2 \cdot (2x+1)\,H_2$$

is highly endothermic and needs a continuous heat supply. The means of providing this energy requirement are manifold and characteristic of the different processes [28]. Only three of them are currently used to make hydrogen cyanide: the Andrussow ammoxidation process, which involves the reaction of ammonia, methane, and air over a catalyst gauze; and the two ammonia dehydration routes, the methane–ammonia (BMA) and the Shawinigan processes, developed by Degussa and Gulf Oil Company, respectively. The latter are performed in externally heated ceramic tubes or in an electrically heated fluidized coke bed. Twenty percent of the hydrogen cyanide in the United States and nearly 45% in Western Europe is obtained as a byproduct in the manufacture of acrylonitrile (2-propenenitrile) by the oxidation of propene in the presence of ammonia (Sohio technology).

1.2.1. Andrussow Process

The Andrussow process was developed around 1930 by L. ANDRUSSOW of I.G. Farben [29]–[31] and is currently the most important method for direct synthesis of hydrogen cyanide. The average capacity of single commercial installations is $5-30 \times 10^3$ t/a; large units are operated by Du Pont, Rohm and Haas, and Monsanto in the United States; ICI, Butachemie, and Monsanto in Western Europe; and Kyowa Gas in Japan.

Figure 3 is a flow diagram of the Andrussow process. Natural gas, essentially methane purified from sulfur, is mixed with ammonia which flows from an evaporator. Compressed air is added in a volume ratio, corresponding closely to the theoretical reaction [32]:

$$CH_4 + NH_3 + 1.5\,O_2 \longrightarrow HCN + 3\,H_2O \qquad \Delta H = -474\ \text{kJ/mol}$$

The mixture is passed over a platinum–rhodium or platinum–iridium gauze [33] catalyst; temperature and upper flammable limit should be monitored carefully [34], [35]. The reaction takes place at > 1000 °C, at around atmospheric pressure, and with a gas velocity through the catalyst zone of about 3 m/s. To avoid decomposition of

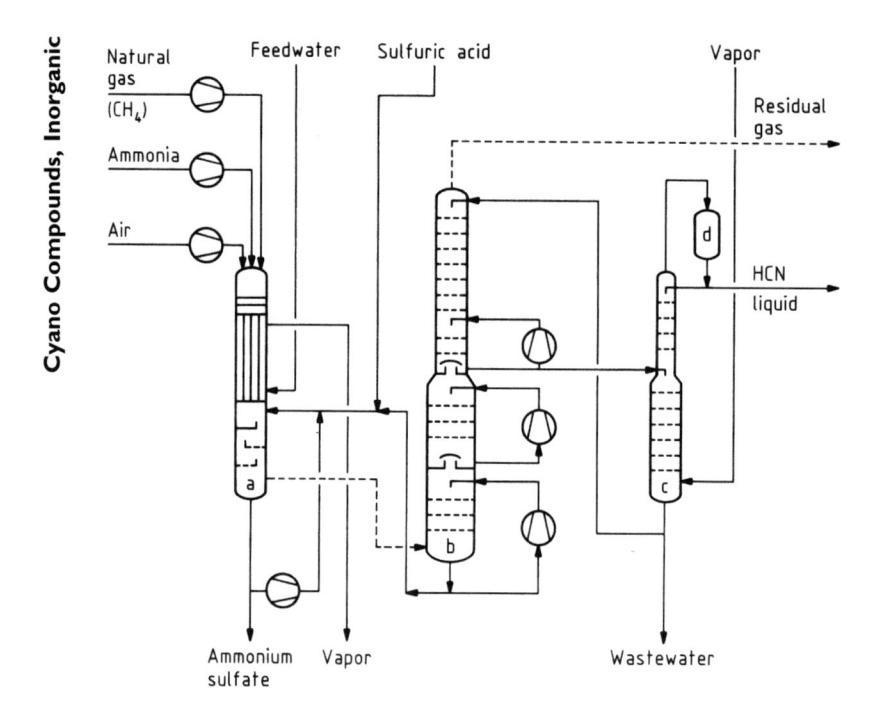

Figure 3. Simplified diagram of the Andrussow process
a) Reactor and ammonia scrubber; b) HCN absorption tower; c) HCN rectifier; d) Condenser

HCN, the effluent gas from the reactor is quickly cooled in a waste-heat boiler, which produces the steam used in the process.

After the waste-heat boiler, the gas is washed with dilute sulfuric acid to remove unreacted ammonia: this is necessary to prevent polymerization of HCN. Because disposal of the resulting ammonium sulfate solution is expensive, other systems have been patented [36]–[38]. Alternatively, the off-gas from the reactor is passed through a monoammonium phosphate solution [39], [40], where the ammonia reacts to form diammonium phosphate. To effect thermal reversal of the phosphate equilibrium, the absorption solution is boiled in a stripper by injection of steam. The ammonia released is condensed and recycled to the reactor, while the regenerated monoammonium phosphate solution is pumped back to the absorber.

After the ammonia scrubber, the gas is passed through a countercurrent column where the hydrogen cyanide is absorbed in cold water and the resultant solution is stabilized by adding acid (ca. 0.1%). The hydrogen cyanide is stripped from the aqueous solution in a rectifier and condensed. The end product is highly pure and has a water content of less than 0.5%. The aqueous absorber solution, containing traces of HCN, is cooled and fed back to the absorption tower. The residual gases, H_2, CO, and N_2, can be used for heating or methanated in a separate unit and recycled as feedstock for HCN manufacture [41].

Table 1. Composition of off-gases and residual gases (volume fractions, %) on the basis of pure methane and ammonia in the BMA and Andrussow processes

Compound	BMA		Andrussow*	
	After reaction	Residual	After reaction	Residual
HCN	22.9	$<10^{-2}$	8.0	$<10^{-2}$
NH_3	2.5	$<10^{-2}$	2.5	>0
H_2	71.8	96.2	22.0	24.6
N_2	1.1	1.5	46.5	51.9
CH_4	1.7	2.3	0.5	0.6
CO			5.0	5.6
H_2O			15.0	16.8
CO_2			0.5	0.6

* Calculated because no consumption figures are available so far.

The advantages of the Andrussow process include (1) long catalyst life, up to 10 000 h; (2) well-tested technology, with a simple and safe reaction system; and (3) high-purity HCN. One disadvantage is the fact that the process is uniquely dependent on pure methane as a raw material as a result of severe carburization of the platinum catalyst; for example, a small percentage of higher hydrocarbon impurities rapidly causes trouble in the catalyst system and reduces conversion rates. Other important poisons for platinum are sulfur and phosphorus compounds. A further problem is the relatively low yield based on carbon (60–70%) and ammonia (70%), as well as the low hydrogen cyanide concentration in the product gas, so that recovery equipment must be built to handle large volumes of gas.

The compositions of reaction and residual gases in the Andrussow and BMA processes are compared in Table 1.

1.2.2. Methane–Ammonia (BMA) Process

The basis of the Degussa BMA process is the formation of hydrogen cyanide in the absence of oxygen [42]–[46]. The reaction

$$CH_4 + NH_3 \longrightarrow HCN + 3 H_2 \qquad \Delta H = +252 \text{ kJ/mol}$$

is endothermic, requires temperatures above 1200 °C, and is performed in externally heated, alumina tube bundles, which are coated with a thin layer of a special platinum catalyst [47], [48]. Several of these bundles are fixed in a reaction furnace unit. A mixture of ammonia and methane (natural or refined gas with a content of 50–100 vol% methane) is passed through the tubes and quickly heated to 1300 °C at normal pressure. To avoid the formation of any disturbing deposits of carbon black, the $NH_3:CH_4$ ratio is kept between 1.01 and 1.08. After leaving the reaction tubes, the product gas is cooled to 300 °C by passage through a water-cooled chamber made of aluminum. A kinetic study has shown that a particular temperature profile is essential for this process [49].

The subsequent reaction steps, ammonia absorption and hydrogen cyanide isolation, are similar to those of the Andrussow process. A distinct advantage is the higher HCN content (Table 1) of the off-gas, so that the number of steps, and the size and cost of recovery equipment, are greatly reduced. The tail gas consists mainly of pure hydrogen. If this is not needed for other syntheses, it can be used as fuel gas for heating the furnace. About 80–85% of the ammonia and 90% of the methane are converted to hydrogen cyanide. The specific energy consumption of ca. 4×10^6 kJ/100 kg HCN reported thus far has been considerably decreased by recent developments [50], [51]. A large part of the heating energy is recovered and used in the air preheater or steam generator.

If the methane supply is limited or costly, the process can be carried out directly with liquefied hydrocarbons or ethanol, or in a three-step reaction starting from methanol [52]–[54]. Both Degussa and Lonza utilize the BMA route to produce hydrogen cyanide.

1.2.3. Shawinigan Process

In the Shawinigan process, hydrocarbon gases are reacted with ammonia in an electrically heated, fluidized bed of coke. The process, sometimes called the Fluohmic process, was developed in 1960 by Shawinigan Chemicals [55]–[58], now a division of Gulf Oil Canada.

In a circular reaction cavity constructed from alumina and silicon carbide, the mixture of ammonia and hydrocarbon (N:C-ratio slightly >1) passes through a fluidized bed of coke, heated by electrodes immersed in the bed. The chemical reaction is similar to the methane–ammonia process, but no catalyst is required and temperatures are kept above 1500 °C. Other carbon compounds, such as naphtha or lighter hydrocarbons, can also be converted. Propane is usually the main feedstock. The reaction can be described as

$$3\,NH_3 + C_3H_8 \longrightarrow 3\,HCN + 7\,H_2 \qquad \Delta H = +634 \text{ kJ/mol}$$

Unreacted feed gas is almost completely decomposed to the elements. This reduces the quantity of ammonia to be removed from the product gas and leads to the formation of coke particles. The control of bed coke size is an important operating parameter.

The reactor effluent contains up to 25 vol% HCN, 72 vol% H_2, 3 vol% N_2, and only 0.005 vol% NH_3. Coke is removed in a water-cooled, cyclone-entrained bed. The gas is further cooled and enters the absorption equipment where HCN is removed. The residual gas, nearly pure hydrogen, can be used for other chemical processes. Some of the hydrogen is recycled to the reaction unit to inhibit the formation of soot.

Coke from the cyclone is screened, and three fractions are separated, stored, and then batched back to the reactor system in the desired proportions to control the particle size distribution. By regulating the rate of coke recycling, the level of the fluidized bed and the reaction temperature can be controlled.

In practice, at least 85% of the ammonia and up to 94% of the hydrocarbon are converted to hydrogen cyanide. Because of the high electric power consumption (6.5 kW h per kilogram of HCN) the Shawinigan process would probably only be attractive where low-cost electricity is available.

This process is employed by South African Explosives Chemical Industries and by EIASA in Spain.

1.3. Storage and Transportation [59], [60]

Handling, storage, and transportation of hydrogen cyanide are determined by its low boiling point, high toxicity, and instability in the presence of moisture, bases, or other impurities. The liquid acid is relatively uncorrosive. Materials compatible with HCN at normal temperatures are stainless steel, Hastelloy, and Monel. To prevent polymerization, stabilizing agents, such as sulfuric acid, phosphoric acid, oxalic or acetic acid, and sulfur dioxide are used. The type and quantity of stabilizer (usually < 0.5%) depend on storage capacity, temperature, and residence time in a container. A combination of H_2SO_4 and SO_2 prevents the decomposition of HCN in the liquid phase as well as in the vapor phase. Larger quantities of hydrogen cyanide are stored at a maximum temperature of 5 °C and must be permanently recirculated.

Additionally, the color of the liquid is monitored and should not exceed APHA 20. To keep the concentration of gas below the danger level, good ventilation of buildings in which HCN is stored and handled is of primary importance. Hydrogen cyanide is usually classified by governmental authorities as an extreme poison, requiring special packaging and transportation regulations. Similar strict procedures exist for solutions with an HCN content of 5% or more. Smaller quantities of the stabilized acid are transported in metal cylinders of < 56 kg nominal water capacity in the United States and < 60 kg in the Federal Republic of Germany. Cylinders cannot be charged with more than 0.55–0.60 kg of liquid HCN per 1-L bottle, and resistance to deformation must be tested up to 10 MPa before first filling. The water content should not exceed 3%, and storage time should be less than one year. According to national regulations, transportation of quantities > 100 kg requires special permission. Procedures covering details of tank car size (up to 50 t), shipping, loading, and handling must be obeyed.

International transportation is regulated by IMDG, code no. 2172, class 2, UN no. 1051. In the United States, hydrogen cyanide is classified as a poison and a flammable gas, and its transportation is governed by the U.S. Department of Transportation (DOT) Safety Act, title CFR 172.101. Transportation in Europe is regulated by RID, ADR, and ADNR: class 6.1, no. 1 a, RN 601, 2601, and 6601 and Blue Book (United Kingdom), Poison IMDG-E 2075.

A filled container must always be tested before transport and must show absolutely no leakage. This can be done, for example, with special indicator paper [61] or with an HCN-sensitive detector system [62].

1.4. Economic Aspects and Uses

In 1984 the production capacity of synthetic hydrogen cyanide was about 600 × 10³ t/a in the United States, 220 × 10³ t/a in Europe, and 25 × 10³ t/a in Japan. Without detailed knowledge of the type and age of catalyst used in various acrylonitrile plants, an estimation of the additional amount of byproduct HCN is difficult. With a 12% HCN formation estimated in the acrylonitrile processes, an additional 120 × 10³ t/a (United States), 160 × 10³ t/a (Europe), and 70 × 10³ t/a (Japan) are available. Currently, between 65 and 75% of the existing production capacity is being utilized. The variety of uses has changed enormously in the past 15 years. Acrylonitrile production starting from acetylene was formerly the major consumer of hydrogen cyanide; in contrast, the Sohio process now in use yields HCN as a byproduct.

Today, the main outlet for HCN is the manufacture of methyl methacrylate. The raw materials — acetone and hydrogen cyanide — are reacted to produce acetone cyanohydrin which is treated with sulfuric acid and methanol to form methyl methacrylate. Another use for HCN is in adiponitrile production, by adding 2 mol HCN to 1 mol butadiene in a two-stage process. Virtually all the adiponitrile is then converted to hexamethylenediamine, a nylon precursor. Hydrogen cyanide reacts with chlorine to form cyanogen chloride (see Section 4.2) which is usually directly trimerized to cyanuric chloride. Herbicides based on cyanuric chloride have been successfully employed in recent years.

D,L-Methionine, one of the largest volume amino acids produced commercially, is manufactured in a rather complicated, multistep synthesis which employs hydrogen cyanide as one of the raw materials. Other uses of HCN are in the production of chelating agents such as ethylenediaminetetraacetic acid (EDTA) or nitrilotriacetic acid (NTA), starting from aldehydes, amines, and hydrogen cyanide. A wide variety of organic intermediates can also be made from HCN and have considerable potential utility. The most important derivatives of hydrogen cyanide in inorganic chemistry are alkali cyanides and cyanide complexes of iron.

2. Metal Cyanides

Metal cyanides can be considered compounds of metals and one or more cyanide ions acting as monodentate ligands which prefer carbon as the donor atom. Simple cyanides are represented by the formula $M(CN)_x$, where M is a metal and x is the number of cyano groups, depending on the valency of M. Depending on the type of metal, simple cyanides dissolve more or less readily in water, forming metal ions and cyanide ions:

$$M(CN)_x \rightleftharpoons M^{x+} + x\,CN^-$$

The solubility is influenced by pH and temperature, especially because hydrogen cyanide is formed by the hydrolysis of cyanide:

$$CN^- + H_2O \rightleftharpoons HCN + OH^-$$

The cyanide complexes can generally be described by the formula $A_y[M(CN)_x]$, where A is an alkali, alkaline-earth, or heavy metal, and y is the number of ions of A present; M is normally a transition metal, and x is the number of CN groups, depending on the valencies of A and M. Most cyanide complexes in which A is an alkali or alkaline-earth metal are highly soluble in water and form alkali or alkaline-earth metal ions and complex transition metal cyanide anions, e.g.,

$$A_y[M(CN)_x] \rightleftharpoons y A^+ + [M(CN)_x]^{y-}$$

The complex cyanide anion may then undergo further dissociation and release cyanide ions:

$$[M(CN)_x]^{y-} \rightleftharpoons z CN^- + [M(CN)_{x-z}]^{(y-z)-}$$

The dissociation of the cyanide complex depends strongly on the type and valency of the complexed metal ion, as well as on the pH and concentration of the solution. Insoluble or slightly soluble cyanide complexes are formed when the alkali or alkaline-earth metal ligand A is replaced by a heavy metal ion. In this case, more complicated cyanide complexes are often formed, which cannot be described by the above-mentioned formula. Furthermore, most heavy metal ions allow the formation of mixed cyanide complexes in which one or more of the cyanide groups can be replaced by other ligands such as halides, pseudohalides, nitrogen oxides, sulfur compounds, or dipolar ligands like water or ammonia [63]. The properties and toxicities of these compounds depend on the structures of the metal cyanides, but discussion of the various correlations is not possible here. In the following sections, only the most important compounds are described.

2.1. Alkali-Metal Cyanides

Since the introduction of the Castner process, sodium cyanide [143-33-9] has been the metal cyanide with the greatest commercial importance. Until about 1900, potassium cyanide [151-50-8] was the more common compound, because it could be produced more easily by melting potassium carbonate with potassium hexacyanoferrate(II), one of the oldest known cyano compounds.

The properties and reactions of both alkali cyanides are very similar and are described together.

Table 2. Physical properties of NaCN [143-33-9] and KCN [151-50-8]

Physical quantity	NaCN	KCN
M_r	49.015	65.119
mp	561.7 °C (98 wt%)	634.5 °C
bp	1500±10 °C	
ϱ, density	1.620 g/cm³ (6 °C, rhombic)	1.553 g/cm³ (20 °C, cubic)
	1.595 g/cm³ (20 °C, cubic)	1.56 g/cm³ (25 °C, cubic)
	1.19 g/cm³ (850 °C, fused)	
c_p, specific heat capacity	1.667 kJ kg⁻¹K⁻¹ (273.1 K)	1.00 kJ kg⁻¹K⁻¹ (25–72 °C)
	31.630 kJ kg⁻¹K⁻¹ (288.5 K)	
	1.402 kJ kg⁻¹K⁻¹ (298.6 K)	
H, enthalpy (25 °C)	−89.9 kJ/mol	−112.63 kJ/mol
ΔH_f, heat of fusion	314 kJ/kg	225 kJ/kg
ΔH_v, heat of vaporization	3185 kJ/kg	
p_v, vapor pressure	0.10 kPa (800 °C)	
	1.65 kPa (1000 °C)	
	11.98 kPa (1200 °C)	
	39.10 kPa (1350 °C)	
Solubility in 100 g of		
Water	see Figure 4	see Figure 4
Ethanol, 100%	1.235 g (25 °C)	0.57 g (19.5 °C)
Ethanol, 95%	2.445 g (25 °C)	
Methanol	6.44 g (15 °C)	4.91 g (19.5 °C)
	4.58 g (25 °C)	
	4.10 g (67.4 °C)	
NH_3, liquid	58 g (−31 °C)	4.55 g (−33.9 °C)

2.1.1. Properties

Sodium and potassium cyanide are colorless, hygroscopic salts with a slight odor of hydrogen cyanide and ammonia in moist air. They are fairly soluble in water (Fig. 4), and the sodium salt forms two hydrates: NaCN · 2 H₂O, at temperatures below 35 °C, and NaCN · 1/2 H₂O at higher temperatures. Further physical properties are listed in Table 2. In the absence of air, carbon dioxide, and moisture, the alkali-metal cyanides are stable, even at fairly high temperature, and can be stored indefinitely [8].

Dry CO₂ does not react with dry alkali-metal cyanides; however, in moist air a slow decomposition takes place, even at normal temperature, releasing HCN:

$$2\ NaCN + CO_2 + H_2O \longrightarrow Na_2CO_3 + 2\ HCN$$

When this occurs, the salt sometimes becomes brownish because of the formation of polymerization products of HCN (azulmic acid). Alkali cyanides are totally decomposed to HCN by the action of strong acids, e.g.,

$$2\ NaCN + H_2SO_4 \longrightarrow 2\ HCN + Na_2SO_4$$

At elevated temperatures, oxygen reacts with alkali-metal cyanides, to form the metal cyanate and carbonate, nitrogen, and carbon dioxide [64]:

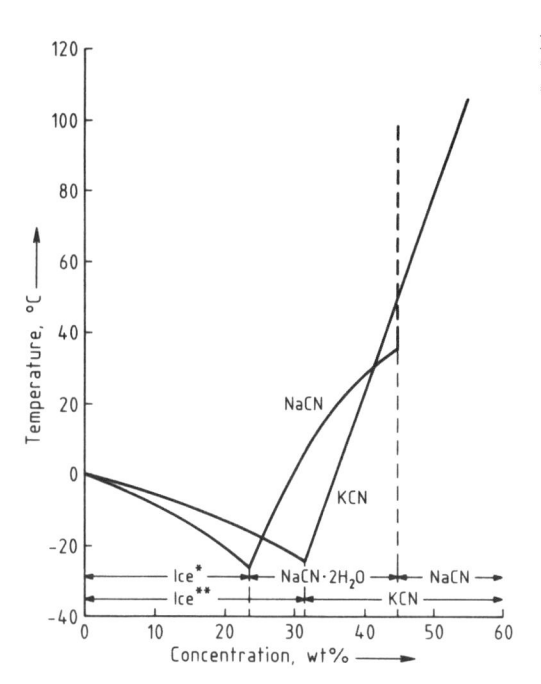

Figure 4. Solubility of NaCN and KCN in water
* Solid phase: ice–NaCN · 2 H$_2$O or ice–NaCN
** Solid phase: ice–KCN

$$2\,NaCN + O_2 \longrightarrow 2\,NaOCN$$
$$2\,NaOCN + 3/2\,O_2 \longrightarrow Na_2CO_3 + N_2 + CO_2$$

The oxidation of alkali-metal cyanides to cyanates by bubbling an air stream through an alkaline melt is used for the bath nitriding of steel [65], [66]. Oxides of lead, tin, copper, nickel, and iron react with alkali-metal cyanides above 560 °C to form the appropriate cyanate and carbonate, nitrogen, carbon dioxide, and the corresponding metal [67], [68]. Very violent oxidation of alkali-metal cyanides can result from the addition of strong oxidants such as nitrate, nitrite, or chlorate to a melt:

$$NO_3^- + CN^- \longrightarrow N_2 + CO_3^{2-}$$
$$5\,NO_2^- + 3\,CN^- \longrightarrow 4\,N_2 + O^{2-} + 3\,CO_3^{2-}$$

These reactions have been utilized for the destruction of waste salts from hardening shops [69], [70].

When alkali-metal cyanides are dissolved in water, a pH-dependent, reversible equilibrium is established between hydrocyanic acid and alkali-metal hydroxide [71], [72]:

$$NaCN + H_2O \rightleftharpoons HCN + NaOH$$

For example, in an aqueous solution of NaCN, at pH 9.4, half of the total cyanide is present as HCN [73] (compare with Fig. 1). As a result of this hydrolysis, solutions of alkali-metal cyanides in water are always strongly alkaline. Furthermore, commercial products always contain small amounts of alkali hydroxide to enhance stability; thus,

Table 3. pH values of aqueous solutions with varying concentrations of NaCN and KCN [56]

	Concentration, mol/L					
	1	10^{-1}	10^{-2}	10^{-3}	10^{-4}	10^{-5}
NaCN	11.64	11.15	10.67	10.15	9.6	8.9
KCN		11.37	10.91	10.35		

the actual pH values of solutions are usually higher than those calculated on the basis of the cyanide concentration and the hydrolysis reaction (Table 3).

Nevertheless, when alkali-metal cyanides are dissolved in tap water, discoloration and precipitation of brownish-black amorphous polymerization products of HCN (azulmic acid) may occur when the water contains too much CO_2 or other acidic components. When solutions of alkali-metal cyanides are stored for a long time or heated, slow hydrolysis of the $C \equiv N$ bond takes place to yield the alkali-metal formate and ammonia [74], [75]:

$$NaCN + 2 H_2O \longrightarrow HCOONa + NH_3$$

The hydrolytic decomposition of concentrated NaCN solutions may be suppressed, even at 60–80 °C, by the addition of 0.5–2% NaOH [76]. However, the hydrolysis is fast and complete when the reaction is carried out at 170 °C under pressure [77], [78]. This process has been investigated for the destruction of cyanide wastes from metal hardening plants [79]. The oxygen in air brings about only partial oxidation of cyanide in aqueous solution at elevated temperature [80]; however, the reaction can be enhanced by the use of catalysts such as activated carbon [81], [82]. A fast conversion of cyanide to cyanate is brought about by ozone at pH 10–12; this is followed by a slower oxidation to nitrogen and carbonate [83], [84].

Cyanide is also oxidized to cyanate by hydrogen peroxide [85] and by peroxomonosulfate, peroxodisulfate [86], and permanganate ions [87]. The anodic oxidation of cyanide leads, via cyanogen and the cyanate, to the carbonate, nitrogen, ammonia, and urea [88]. Halogens react with cyanide in aqueous solution, forming cyanogen halides, XCN (X = Cl, Br, I), that hydrolyze more or less quickly, depending on the pH and temperature of the solution [89] (see Section 4.1).

Reaction with chlorine or hypochlorite above pH 11.5 is the most widely used process for detoxification of solutions and wastewaters containing cyanide at concentrations < 1 g/L.

The reaction of alkali cyanides with sulfur or polysulfides in aqueous solution at elevated temperature is used for the production of alkali thiocyanates [90]. The reaction with thiosulfate to give thiocyanate is used therapeutically in the treatment of cyanide poisoning [91]. Heavy metal ions react in aqueous solution with cyanide to yield insoluble cyanides all of which dissolve in excess cyanide, forming very stable complex salts [92]. These compounds play an important role in metallurgy and electroplating [93]. Some base metals (e.g., zinc or nickel) and, in the presence of oxygen or oxidizing agents, even precious metals (e.g., gold or silver) are dissolved by aqueous solutions of alkali cyanides [94]. This property of alkali cyanides has been utilized for about 100 years in the leaching of ores that contain precious metals [95]:

$$4\ NaCN + 2\ Au + 1/2\ O_2 + H_2O \longrightarrow 2\ Na[Au(CN)_2] + 2\ NaOH$$

Tantalum, titanium, and tungsten are not attacked by alkali cyanide solutions at room temperature.

Alkali cyanides react with carbonyl groups of organic compounds in aqueous solution to form cyanohydrins [96], [97]:

$$\underset{R^2}{\overset{R^1}{>}}C=O + NaCN + H_2O \rightleftharpoons \underset{R^2}{\overset{R^1}{>}}C(OH)CN + NaOH$$

Organic compounds with labile halogen atoms can be converted by reaction with alkali cyanides in aqueous solution to nitriles, which can be further processed to give carboxylic acids, amines, etc.:

$$RCl + NaCN \longrightarrow RCN + NaCl$$

2.1.2. Production

The large-scale production of alkali cyanides began in the second half of the 19th century, when J. S. MACARTHUR and the FORREST Brothers patented their process for the extraction of gold and silver from ores. The manufacturing method was based on work of F. and E. RODGERS [98], who fused potassium or sodium carbonate together with potassium hexacyanoferrate (II), for which animal blood and waste were used as raw materials. A large number of modified processes were patented up to 1900, but the production of cyanides on a major industrial scale finally began with the development of a process based on a patent of H. J. CASTNER, which employed the reaction of metallic sodium with charcoal and ammonia [99]:

$$Na + C + NH_3 \longrightarrow NaCN + 3/2\ H_2$$

The Castner process was originally a twostage process, which was developed technically through work carried out mainly at Degussa [100], [101] and was used there until 1971. Today, sodium cyanide and potassium cyanide are produced exclusively by the so-called neutralization process. This process is based on the stoichiometric reaction of a solution of sodium or potassium hydroxide with liquid or gaseous hydrogen cyanide [102]–[106] and is described in more detail below.

$$MOH + HCN \longrightarrow MCN + H_2O$$

Although the process appears very simple, it involves a number of technical difficulties, which result from the tendency of hydrogen cyanide to polymerize, as well as from the hydrolysis of the alkali cyanides and their reactivity with carbon dioxide and air. High-grade products can be obtained only when the production process is controlled very closely. In principle, all plants employ the following process steps, which differ only in some details.

Gaseous or liquid HCN is first introduced continuously into highly concentrated solutions of NaOH or KOH at elevated temperature. The heat of neutralization is used for the subsequent vaporization of water. The use of special mixing devices and the addition of 0.2–3% excess NaOH prevents polymerization [107]–[109]. Simultaneously or in a second step, water is vaporized under reduced pressure and the alkali cyanide is precipitated below 100 °C. The shorter the residence time of the cyanide in solution, the cleaner the product and the lower the content of hydrolysis products and iron impurities [110], [111].

The cyanide crystals are separated by filtration or centrifugation, and various methods are used to dry the wet salt, which still contains about 12% water. A two-step drying process with hot, CO_2-free air at 200–430 °C is advantageous because the product remains free flowing [112], [113]. Sometimes fluidized-bed dryers are utilized [114]. In any case, the used air has to be washed and the wastewater detoxified to avoid pollution problems. Standard equipment can be employed for further processing of the salt, which still contains up to 1% moisture at this stage [115]. The final commercial forms are powder, granulate, briquettes, or tablets [116], [117].

The production of NaCN from nitrogenous waste from the manufacture of sugar [118] is the precursor of the "neutralization process." This old process no longer has any commercial importance.

2.1.3. Commercial Forms, Specifications, and Packaging

Most alkali cyanide is supplied in the solid form. Common commercial forms for NaCN and KCN are granular products composed of irregularly shaped particles ca. 0.1–5 mm in size, nearly free of dust; pillow-shaped, molded briquettes of ca. 15–40 g; cylindrical tablets ca. 25–40 mm in diameter and weighing 20–40 g.

Briquettes or tablets are preferred as dust-free products. Smaller amounts of NaCN and KCN are used as coarse-grained powders for the preparation of mixtures, e.g., in the electroplating industry, or in organic synthesis.

The specifications according to the German standards [119] for the electroplating industry (DIN 50 971) for NaCN and KCN are shown in Table 4. The quality of commercial products normally exceeds the requirements of DIN 50 971.

Special products containing such other compounds as cyanate, carbonate, or barium salts are sometimes used in the metal surface hardening industry. Preferred application forms are tablets or cast eggs which are often designated by color. The most frequently used type of packaging for solid alkali cyanides is the steel-plate drum containing 50 or 100 kg of cyanide. The lid of the drum is fixed with a locking ring and hermetically sealed. Larger quantities of solid cyanides are supplied in steel-plate or stainless steel "flow bins" or in plywood boxes with polyethylene liners, which contain 1 t or more of

Table 4. Specifications and average analysis of NaCN and KCN (data given in wt% or ppm = mg/kg; M = Na or K)

	NaCN		KCN	
	DIN 50 971	Average analysis	DIN 50 971	Average analysis
MCN, %		> 98.0		98–99
CN^-, %	51.5	> 52.0	39.0	> 39.0
MOH, %		≈ 0.5		≈ 0.1
M_2CO_3, %		≈ 0.5		≈ 1.0
Cl^-, %		≈ 0.02		≈ 0.0015
SO_4^{2-}, %		≈ 0.05		< 0.01
SO_3^{2-}, %		< 0.05		< 0.01
S^{2-}, ppm		< 1		< 1
MOCN, %		≈ 0.1		≈ 0.1
SCN^-, %		< 0.01		< 0.01
HCOOM, %		≈ 0.5		≈ 0.3
H_2O, %		≈ 0.2		≈ 0.2
Sb, ppm	< 5	< 5	< 5	< 5
As, ppm	< 5	< 5	< 5	< 5
Pb, ppm	< 50	< 50	< 50	< 50
Cd, ppm	< 10	< 10	< 10	< 10
Fe, ppm	< 50	< 50	< 50	< 50
Cu, ppm	< 10	< 10	< 10	< 10
Ni, ppm	< 50	< 50	< 50	< 50
Zn, ppm	< 20	< 20	< 20	< 20
Sn, ppm	< 80	< 80	< 30	< 30
Na, %		< 48		< 0.5	< 0.3
Insolubles, %	< 0.005	< 0.001	< 0.005	< 0.001

cyanide. All types of packaging must meet the official requirements for the transportation of toxic substances and must carry labels bearing a skull and crossbones, and advice on safe handling, and first aid procedures.

Sodium and potassium cyanides can be supplied as stable aqueous solutions [76] with about 30 wt% NaCN or 40 wt% KCN in containers with a capacity of at least 20 m³. The concentration of the solution can be adjusted to ensure that even in winter no danger of precipitation exists. A solution is the most economical and technically preferred form when large quantities of cyanide are required and aqueous solutions can be used.

2.1.4. Transportation, Storage, and Handling

For the transportation of cyanides by road, rail, or sea freight, the regulations for dangerous goods according to the following legal agreements must be considered: EEC Guidelines no. 6.7., Transportation Class 6.1, no. 31 a, GGVE, GGVS, RID, ADR; IMCO 6.1; UN no. 1689 for NaCN and no. 1680 for KCN. Cyanides must not be sent by mail. They must never be transported or stored with foods, beverages, smoking materials, or anything used for human or animal consumption, or with acids or acid salts.

Steel drums that contain cyanide must be stored in a dry place and protected against corrosion. Rooms in which cyanides are stored and processed must be well ventilated. A full mask with a filter, type B (identifying color gray, shelf life unused max. five years), must be worn when HCN might be present. Unauthorized persons must never be allowed access to rooms in which cyanides are stored or processed.

Iron, carbon steel, and stainless steel can be used as construction materials for equipment that comes in contact with cyanide solutions up to 100 °C [94]

Concentrated cyanide solutions must be stored in steel or stainless steel containers, which are protected against leakage by a collecting basin constructed of alkali-resistant concrete. The drums must be tightly closed immediately after removal of cyanide and should not be allowed to stand open. Spilled solid cyanide must be swept up immediately with a broom and shovel, and returned to the container. If it has become dirty and unusable, the cyanide must be destroyed or disposed of in accordance with legal regulations. In addition, spilled cyanide solutions and wastewater that contains cyanide must be detoxified (see Section 3.1) before discharge into public sewage systems or open water. Containers, equipment, and any objects or floor areas that have become contaminated with cyanides must be washed thoroughly with water, which must then also be detoxified.

Work with cyanides requires the greatest possible caution and strictest adherence to all safety regulations. When handling cyanide, do not eat, drink, smoke, take snuff, or chew tobacco. Never touch cyanides with unprotected hands. If a danger of cyanide contamination exists, rubber gloves, a rubber apron, and safety goggles or protection for the face and head must be worn. After handling cyanide, wash any contaminated parts of the body thoroughly with large amounts of water. Cyanides must never be brought into contact with acids or acid salts because the highly poisonous, gaseous hydrogen cyanide is generated immediately.

Further safety information is compiled in the Chemical Safety Data Sheets of the Manufacturers Association of the Chemical Industry [120], [121].

2.1.5. Economic Aspects and Uses

Sodium cyanide is cheaper than potassium cyanide and is, therefore, more commonly used. Occasionally, potassium cyanide has some advantages, e.g., in electroplating, because the potassium ion has greater mobility than sodium.

Ore Dressing. The leaching of precious metals such as gold and silver from ores is the main application of NaCN. All efforts to replace cyanide with other complexing agents that are less toxic and less dangerous to the environment, remain largely unsuccessful. Some attempts have been made to use thiourea instead of cyanide, but this product is much more expensive and requires quite different leaching conditions. On the other hand, new methods for the recovery of cyanide [122] and for the detoxification of tailings that contain cyanide have been developed. The use of cyanide as a depressant in the selective flotation of sulfide ores of zinc, lead, copper, and iron [123] is, on the other hand, decreasing as less toxic substitutes become more popular.

Electroplating. Today, just as 100 years ago, large quantities of alkali cyanides, together with the complex cyanides of copper, zinc, and cadmium manufactured from alkali cyanides, are used in several electroplating processes. In particular, the plating of iron, steel, and zinc is carried out by using electrolytes that contain cyanide. Very good corrosion protection resulting from the even deposition of layers, ease of use, and low sensitivity of the bath to impurities are the main advantages of electrolytes containing cyanide. In this application, potassium cyanide is preferred, particularly for the deposition of gold and silver. Generally, electrolytes based on potassium salts have the advantages of better performance and less sensitivity to impurities [124].

When the highest quality is required, cyanide pretreatment and degreasing baths are still in use today, especially since attempts to employ other strong complexing agents have failed because of insurmountable wastewater problems.

Metal-Surface Hardening in Molten Salts. Sodium and potassium cyanides are important because they release nitrogen or carbon (or both) onto the surface of the material being treated at relatively low temperatures. The Tufftride process is employed primarily for the hardening of steel parts used as tools or in engines where they are highly stressed [125]–[128].

Organic Synthesis [96]. Large quantities of sodium cyanide are used to introduce cyanide groups into organic substances. This is the first step in the development of a variety of well-known compounds used in pharmacy, complex chemistry, and polymer chemistry.

Inorganic Chemistry. A very important application of sodium and potassium cyanide is the production of heavy metal cyanides, particularly complex iron cyanides (see Section 2.3).

2.2. Alkaline-Earth-Metal Cyanides

Properties and Uses. The properties of the alkaline-earth-metal cyanides differ from those of the alkali-metal cyanides. They are less stable and decompose at elevated temperatures to release HCN. Even when alkaline-earth-metal cyanides are in contact with moist air or dissolved in water, hydrolysis takes place, e.g.,

$$Ca(CN)_2 + 2\,H_2O \longrightarrow Ca(OH)_2 + 2\,HCN$$

This property is utilized in the application of calcium cyanide as a pesticide [129]. Large quantities of calcium cyanide in the form of "black cyanide" are still used for the

Table 5. Physical properties of the alkaline-earth-metal cyanides

Property	Mg(CN)$_2$ [4100-56-5]	Ca(CN)$_2$ [592-01-8]	Sr(CN)$_2$ [52870-08-3]	Ba(CN)$_2$ [542-62-1]
M_r	76.37	92.12	139.67	189.40
mp, °C		> 350		≈ 600
Decomp., °C	> 300	≈ 640	> 500	> 600
Other crystal forms	Mg(CN)$_2$ · 2 NH$_3$	Ca(CN)$_2$ · 2 NH$_3$	Sr(CN)$_2$ · 4 H$_2$O	Ba(CN)$_2$ · 2 H$_2$O

leaching of precious metals from ores [130]. Apart from this compound, only barium cyanide has any commercial importance as a result of its use in electroplating to separate carbonate from cyanide-containing electrolytes [124]. Some of the physical properties of these compounds are shown in Table 5.

Production. The alkaline-earth-metal cyanides can be obtained by reaction of the corresponding metal hydroxides with hydrogen cyanide [131]. "Black cyanide" (also called "Aerobrand Cyanide") is produced commercially in large quantities by heating crude calcium cyanamide with carbon, in electric furnaces above 1000 °C and in the presence of sodium chloride [132]. The reaction product is a dark-colored mixture of Ca(CN)$_2$, NaCN, CaCl$_2$, NaCl, and excess carbon [133]; the cyanide content corresponds to about 42–44 % NaCN.

2.3. Heavy-Metal Cyanides

The chemistry of heavy metals is characterized by the fact that they usually form very stable, insoluble or slightly soluble simple cyanides, M(CN)$_x$, which are dissolved by excess alkali cyanide to form very stable complex salts [92]:

$$M(CN)_x + y\,CN^- \rightleftharpoons [M(CN)_{x+y}]^{y-}$$

where $x = 1-3$; $y = 1-4$.

When dissolved in water, the complex cyanide anions tend to dissociate, and various dissociation equilibria are set up. The dissociation properties of the most important heavy-metal cyanides are shown in Table 6.

2.3.1. Iron Cyanides

2.3.1.1. Properties

The most important iron cyanides are the hexacyanoferrate(II), ferrocyanide, [Fe(CN)$_6$]$^{4-}$, and the hexacyanoferrate(III), ferricyanide, [Fe(CN)$_6$]$^{3-}$, anion complexes, which have an octahedral configuration. They belong to the most stable inorganic complex compounds and are practically nontoxic because of the strong

Table 6. Stability of heavy-metal cyanide complexes [63], [92]

Dissociation equilibria		$pK_{dissociation}$
$[Pb(CN)_4]^{2-} \rightleftharpoons Pb^{2+} + 4\,CN^-$		10.3
$[Cd(CN)_4]^{2-} \rightleftharpoons [Cd(CN)_3]^- + CN^-$		2.5
$[Cd(CN)_3]^- \rightleftharpoons Cd^{2+} + 3\,CN^-$		14.7
	Σ	17.2
$[Zn(CN)_4]^{2-} \rightleftharpoons [Zn(CN)_3]^- + CN^-$		1
$[Zn(CN)_3]^- \rightleftharpoons Zn^{2+} + 3\,CN^-$		17.9
	Σ	18.9
$[Ag(CN)_2]^- \rightleftharpoons Ag^+ + 2\,CN^-$		20.9
$[Ni(CN)_4]^{2-} \rightleftharpoons Ni^{2+} + 4\,CN^-$	$\approx$	22
$[Cu(CN)_4]^{3-} \rightleftharpoons [Cu(CN)_3]^{2-} + CN^-$		1.5
$[Cu(CN)_3]^{2-} \rightleftharpoons [Cu(CN)_2]^- + CN^-$		5.3
$[Cu(CN)_2]^- \rightleftharpoons Cu^+ + 2\,CN^-$		23.9
	Σ	30.7
$[Fe(CN)_6]^{3-} \rightleftharpoons Fe^{3+} + 6\,CN^-$	$\approx$	36
$[Au(CN)_2]^- \rightleftharpoons Au^+ + 2\,CN^-$	$\approx$	37
$[Hg(CN)_4]^{2-} \rightleftharpoons Hg^{2+} + 4\,CN^-$		40.5
$[Fe(CN)_6]^{4-} \rightleftharpoons Fe^{2+} + 6\,CN^-$	$\approx$	42
$[Co(CN)_6]^{4-} \rightleftharpoons Co^{2+} + 6\,CN^-$	$\approx$	64
$[Pt(CN)_4]^{2-} \rightleftharpoons Pt^{2+} + 4\,CN^-$	$\approx$	40
$[Pd(CN)_4]^{2-} \rightleftharpoons Pd^{2+} + 4\,CN^-$	$\approx$	42

bonding between iron and cyanide (for LD$_{50}$ values, see p. 1737). Almost none of the known cyanide reactions can be observed when these compounds are dissolved in water. Only when irradiated with UV light, does the hexacyanoferrate(II) ion decompose slowly at normal temperature to form cyanide ions [134]:

$$[Fe(CN)_6]^{4-} + H_2O \underset{dark}{\overset{UV}{\rightleftharpoons}} [Fe(CN)_5H_2O]^{3-} + CN^-$$

In the absence of UV light, only very strong oxidants like ozone are able to oxidize the cyanide and destroy the hexacyanoferrate(II) complex slowly at elevated temperature [135]:

$$2\,[Fe(CN)_6]^{4-} + 13\,O_3 + 4\,OH^- + H_2O \longrightarrow 2\,Fe(OH)_3 + 12\,OCN^- + 13\,O_2$$

Permanganate, peroxodisulfate, hypochlorite, chlorine, iodine, cesium(IV), and other oxidants oxidize hexacyanoferrate(II) to hexacyanoferrate(III) at moderate temperatures [136], e.g.,

$$5\,[Fe(CN)_6]^{4-} + MnO_4^- + 8\,H^+ \longrightarrow 5\,[Fe(CN)_6]^{3-} + Mn^{2+} + 4\,H_2O$$

This reaction is used for the determination of hexacyanoferrate(II) in water. Hydrogen peroxide oxidizes hexacyanoferrate(II) to hexacyanoferrate(III) below pH 3 and reduces hexacyanoferrate(III) to hexacyanoferrate(II) above pH 7:

$$2\,[Fe(CN)_6]^{4-} + H_2O_2 + 2\,H^+ \longrightarrow 2\,[Fe(CN)_6]^{3-} + 2\,H_2O$$
$$2\,[Fe(CN)_6]^{3-} + H_2O_2 \longrightarrow 2\,[Fe(CN)_6]^{4-} + 2\,H^+ + O_2$$

When acidified with mineral acids, solutions of hexacyanoferrates decompose slowly at room temperature and more rapidly when heated, releasing HCN and forming precipitates of iron(II) [14460-02-7] [136]:

$$3\,[Fe(CN)_6]^{4-} + 12\,H^+ \longrightarrow 3\,H_4[Fe(CN)_6] \longrightarrow 12\,HCN + Fe_2[Fe(CN)_6]$$

Insoluble or slightly soluble metal hexacyanoferrates(II) are formed by the addition of aqueous solutions to neutral or weakly acidic solutions of metal salts, particularly heavy metal salts [136]:

$$4\,M^+ + [Fe(CN)_6]^{4-} \longrightarrow M_4[Fe(CN)_6]\ \text{(slightly soluble)}$$
$$2\,M^{2+} + [Fe(CN)_6]^{4-} \longrightarrow M_2[Fe(CN)_6]\ \text{(slightly soluble)}$$

These reactions are used to separate heavy metals from solutions for the purpose of refining and for the production of colored pigments, e.g., Prussian blue pigments, which contain the hexacyanoferrate(III) anion:

$$M^+ + Fe^{2+} + [Fe(CN)_6]^{3-} \longrightarrow MFe[Fe(CN)_6]$$

where M = K, Na, NH_4, Cu

Multinuclear complexes are formed by the thermolysis of hexacyanoferrate(II) complexes with complex cations [131]. Many anionic and neutral ligands can be introduced into the hexacyanoferrate anion by substitution of one CN group, to form pentacyano complexes [63], [131].

$$[Fe(CN)_6]^{3-/4-} + X \longrightarrow [Fe(CN)_5 X]^{2-/3-} + CN^-$$

where X = H_2O, NH_3, NO, NO_2, NO_5, CO

$$[Fe(CN)_6]^{3-/4-} + X^- \longrightarrow [Fe(CN)_5 X]^{3-/4-} + CN^-$$

where X = SCN, NCS.

Some of these compounds are used in analytical chemistry. Hexacyanoferrate(II) reacts with a variety of organic amino compounds to form insoluble or slightly soluble salts of general composition BA, B_3A_2, B_2A, B_3A, and B_4A, where B is the cationic nitrogen-compound and A the anionic hexacyanoferrate (II) [63]. These reactions are used in analytical and pharmaceutical chemistry. Hexacyanoferrate(III) is a strong oxidant which can dissolve metals; it is used in photographic processes [137] and etching techniques [138]:

$$[Fe(CN)_6]^{3-} + M \longrightarrow [Fe(CN)_6]^{4-} + M^+$$

The hexacyanoferrates(II) of sodium, potassium, and calcium crystallize with different amounts of water of crystallization, whereas potassium hexacyanoferrate(III) is obtained in anhydrous form. The hexacyanoferrates(II) start to lose water of crystallization above 30 °C and become anhydrous white powders above 80 °C. All hexacyanoferrates decompose above 400 °C to form alkali-metal or calcium cyanide, elemental iron, carbon, and nitrogen, e.g.,

Table 7. Solubilities of hexacyanoferrates in water (wt% of anhydrous salt)

Compound	Temperature, °C		
	20	50	80
$Na_4[Fe(CN)_6]$	16	26	38
$K_4[Fe(CN)_6]$	22	32	40
$Ca_2[Fe(CN)_6]$	36	42	44
$K_3[Fe(CN)_6]$	31	39	45

$$Na_4[Fe(CN)_6] \longrightarrow 4\,NaCN + Fe + 2\,C + N_2$$

The alkali hexacyanoferrates are easily soluble in water; the solubility of the calcium salt is more than twice that of the sodium salt (Table 7). Double salts of low solubility may be formed in the presence of Ca^{2+} and K^+ or NH_4^+ ions.

2.3.1.2. Production

Historically, the preparation of hexacyanoferrates(II) was based on the fusion of potash with iron compounds and animal residues such as hide, horns, or dried blood, which led to the German name *gelbes Blutlaugensalz* (yellow salt of blood lye). Later on, the absorption of hydrogen cyanide from coal gas on iron hydroxide was used to produce hexacyanoferrates. Today, synthetic hydrogen cyanide, iron(II) chloride, and alkali or calcium hydroxide are the raw materials for the production of hexacyanoferrates(II) on a large scale.

Calcium hexacyanoferrate(II) is usually the primary product, which is subsequently converted to the potassium and sodium salts. For this purpose, liquid hydrogen cyanide and an aqueous solution of iron(II) chloride are mixed with calcium hydroxide solution in stoichiometric amounts in a stirred reactor:

$$3\,Ca(OH)_2 + FeCl_2 + 6\,HCN \longrightarrow Ca_2[Fe(CN)_6] + CaCl_2 + 6\,H_2O$$

After filtration, the solution is concentrated by evaporation of water under reduced pressure, and the calcium hexacyanoferrate(II) crystallizes with eleven molecules of water of crystallization. The relatively coarse-grained salt is then separated by filtration and generally used without drying. For the conversion of calcium hexacyanoferrate(II) to the potassium or sodium salt, two methods are used.

In the first, a stoichiometric amount of potassium chloride is added to the filtered calcium hexacyanoferrate(II) solution and the hexacyanoferrate(II) precipitates as a slightly soluble potassium calcium double salt [139]

$$Ca_2[Fe(CN)_6] + 2\,KCl \longrightarrow K_2Ca[Fe(CN)_6] + CaCl_2$$

The double salt is separated by filtration, redispersed in water, and then converted to the soluble potassium hexacyanoferrate(II) by the addition of potassium carbonate:

$$K_2Ca[Fe(CN)_6] + K_2CO_3 \longrightarrow K_4[Fe(CN)_6] + CaCO_3$$

After separation of the insoluble calcium carbonate, the potassium hexacyanoferrate(II) solution is concentrated by evaporation, and the potassium salt crystallizes with three molecules of water of crystallization. After filtration, the salt is carefully dried and packaged.

A similar method can be used for the production of sodium hexacyanoferrate(II): sodium carbonate is added to the calcium hexacyanoferrate(II) solution and the precipitated calcium carbonate is separated:

$$Ca_2[Fe(CN)_6] + 2\,Na_2CO_3 \longrightarrow Na_4[Fe(CN)_6] + 2\,CaCO_3$$

Usually, the sodium hexacyanoferrate (II) is synthesized directly from solutions of sodium cyanide and iron (II) chloride [140]:

$$6\,NaCN + FeCl_2 \longrightarrow Na_4[Fe(CN)_6] + 2\,NaCl$$

The crystalline $Na_4[Fe(CN)_6] \cdot 10\,H_2O$ can be obtained when the solution is concentrated; it must be dried very carefully to avoid loss of water of crystallization, which would influence its solubility properties.

The disadvantages of the double-salt process are the necessity of recycling the various mother liquors and the formation of large quantities of solid waste products.

In the second conversion method, the hexacyanoferrate(II) is precipitated from a solution of the calcium salt as Berlin white by the addition of iron(II) chloride solution:

$$Ca_2[Fe(CN)_6] + 2\,FeCl_2 \longrightarrow Fe_2[Fe(CN)_6] + 2\,CaCl_2$$

This iron(II) hexacyanoferrate(II) is separated by filtration, redispersed in water, and converted to sodium or potassium hexacyanoferrate(II) by the addition of stoichiometric amounts of sodium or potassium cyanide:

$$Fe_2[Fe(CN)_6] + 12\,KCN \longrightarrow 3\,K_4[Fe(CN)_6]$$

Potassium hexacyanoferrate(III) can be obtained by the oxidation of potassium hexacyanoferrate(II). Usually, anodic oxidation is applied by means of nickel electrodes [141]:

$$2\,K_4[Fe(CN)_6] + 2\,H_2O \longrightarrow 2\,K_3[Fe(CN)_6] + 2\,KOH + H_2$$

Alternatively, hydrogen peroxide can be used for the oxidation of hexacyanoferrate(II) when the reaction is carried out below pH 3 [142]:

$$3\,K_4[Fe(CN)_6] + H_4[Fe(CN)_6] + 2\,H_2O_2 \longrightarrow 4\,K_3[Fe(CN)_6] + 4\,H_2O$$

The dry, ruby-red potassium hexacyanoferrate(III) is obtained by concentration of the solution at reduced pressure and crystallization at ca. pH 7. The fact that H_2O_2 reduces hexacyanoferrate(III) to hexacyanoferrate(II) at higher pH must also be considered.

2.3.1.3. Commercial Forms, Specifications, and Packaging

The following commercial products are most frequently used:

Sodium hexacyanoferrate(II) [*69043-75-0*], sodium ferrocyanide, yellow prussiate of soda (YPS), $Na_4[Fe(CN)_6] \cdot 10 H_2O$, M_r 484.108, ϱ 1.46 g/cm^3, formed as light-yellow, monoclinic crystals, which are easily soluble in water (33.7 g in 100 g H_2O at 20 °C).

Potassium hexacyanoferrate(II) [*14459-95-1*], potassium ferrocyanide, yellow prussiate of potash (YPP), $K_4[Fe(CN)_6] \cdot 3 H_2O$, M_r 422.39, ϱ 1.85 g/cm^3, formed as lemon-yellow, monoclinic crystals, which are easily soluble in water (33.7 g in 100 g H_2O at 20 °C).

Calcium hexacyanoferrate(II) [*13821-08-4*], calcium ferrocyanide, yellow prussiate of calcium (YPC), $Ca_2[Fe(CN)_6] \cdot 11 H_2O$, M_r 490.296, ϱ 1.68 g/cm^3, formed as yellow monoclinic crystals, which are very soluble in water (148.4 g in 100 g H_2O at 20 °C).

Potassium hexacyanoferrate(III) [*13746-66-2*], potassium ferricyanide, $K_3[Fe(CN)_6]$, M_r 329.25, ϱ 1.858 g/cm^3, formed as orange to ruby-red, monoclinic crystals, which are easily soluble in water (46.4 g in 100 g H_2O at 20 °C).

The commercial products are fine, crystalline, free-flowing powders. Normally, they are packaged in polyethylene bags of 25-kg net weight, and must be stored under cool, dry conditions and kept away from acids. Hexacyanoferrates are not dangerous goods according to appendix C of the EVO, GGVS, RID, ADR, and IMCO Code. The specifications and average analyses for the commercially used hexacyanoferrates are listed in Table 8.

Uses and Economic Aspects. Sodium, potassium, and calcium hexacyanoferrates(II) are equally suitable for use in most applications, but sometimes one product may be preferred. For the production of blue pigments, the most important application of hexacyanoferrates, the potassium salt is used in Europe, whereas sodium is preferred in the United States. Generally, blue hexacyanoferrate(II) compounds are formed by the reaction of ferric salts with hexacyanoferrates(II):

$$FeCl_3 + K_4[Fe(CN)_6] \longrightarrow FeCl_2 + K_3[Fe(CN)_6] + KCl$$
$$\longrightarrow KFe(III)[Fe(II)(CN)_6] + 3 KCl$$

The production of blue pigments is more complicated, however, and is based primarily on the oxidation of iron(II) hexacyanoferrate(II) precipitates in the presence of alkali or ammonium salts [8]. The properties of blue pigments are not dependent only on the composition; the precipitation method and the finishing of the product are also very

Table 8. Specifications of hexacyanoferrates

	$Na_4[Fe(CN)_6] \cdot 10\ H_2O$ (anticaking of NaCl)	$K_4[Fe(CN)_6] \cdot 3\ H_2O$ (refining of wine)	$Ca_2[Fe(CN)_6] \cdot 11\ H_2O$ (anticaking of NaCl)	$K_3[Fe(CN)_6]$ (photographic industry*)
Assay, %	> 99.0	> 99.0	> 99.5	> 98.5
$[Fe(CN)_6]^{4-}$, %	> 43.3	> 49.7	> 43.0	< 0.5
H_2O, %	< 37.5*	< 12.0	< 41*	< 0.1
Ca + Mg, mval		< 60		
Cl^-, %	< 0.2	< 0.02	< 0.6*	< 0.25
SO_4^{2-}, %	< 0.07	< 0.005	< 0.01*	< 0.1
CO_3^{2-}, %	< 0.15*	< 0.1*	< 0.05*	
S^{2-}, ppm		< 100		
NH_3, ppm		< 100		
Heavy metals, as				
Pb, ppm	< 10	< 10	< 10	
Zn, ppm	< 25	< 10	< 25	
As, ppm	< 3	< 1	< 3	
HCN, ppm		< 100		
Insolubles, %	< 0.03	< 0.01	< 0.1*	< 0.05
pH (10% solution)	10–11*	10–11*	10–11*	6–8

* Average analysis.

important. The use of hexacyanoferrates(II) for the production of classified documents is also based on this reaction. Large amounts of all three types of hexacyanoferrates(II) are used as additives for rock salt to prevent caking [143]; hexacyanoferrates(II) act as crystal growth inhibitors in this application. The ability of hexacyanoferrates(II) to form insoluble or almost insoluble heavy metal compounds is used, e.g., in the production of citric acid by fermentation [144], in the refining of wine [145], in the electroplating of tin [146], and in analytical chemistry. Potassium hexacyanoferrate(III) and alkali hexacyanoferrate(II) combined with peroxydisulfates play an important role as bleaching agents in color photography [137]. Both hexacyanoferrates are also important depressants for the separation of molybdenum from copper by flotation [147].

2.3.2. Cyanides of Copper, Zinc, and Cadmium

Properties. *Copper(I) cyanide [544-92-3]*, CuCN, M_r 89.56, *mp* 473 °C, forms white monoclinic prismatic crystals, ϱ 2.92 g/cm³, which are insoluble in water, dilute mineral acids, and organic solvents; CuCN is soluble in ammonia and in alkali cyanide solution because of the formation of different equilibria of very soluble cyano complexes:

$$\text{CuCN} + \text{NaCN} \rightleftharpoons \text{Na[Cu(CN)}_2\text{]}$$
(slightly soluble)   (slightly soluble)

$$\text{Na[Cu(CN)}_2\text{]} + \text{NaCN} \rightleftharpoons \text{Na}_2\text{[Cu(CN)}_3\text{]} + \text{NaCN}$$
(slightly soluble) $\rightleftharpoons \text{Na}_3\text{[Cu(CN)}_4\text{]}$

These complex compounds can also be obtained as solids. In the absence of air and moisture, pure CuCN can be stored indefinitely without deterioration.

Zinc cyanide [557-21-1], Zn(CN)_2, M_r 117.42, mp 800 °C (decomp.), forms a white crystalline powder with a slight HCN odor, ϱ 1.852 g/cm^3 (25 °C), which is insoluble in water but soluble in aqueous solutions of alkali hydroxides or cyanides by formation of hydroxy ($[\text{Zn(OH)}_4]^{2-}$) or cyano complexes ($[\text{Zn(CN)}_4]^{2-}$). In the absence of air, carbon dioxide, and moisture, zinc cyanide can be stored indefinitely without deterioration.

Cadmium cyanide [542-83-6], Cd(CN)_2, M_r 164.45, mp 200 °C (decomp.), forms white irregular octahedral crystals, ϱ 2.23 g/cm^3 (25 °C), which are soluble in water (1.7 g in 100 g H_2O at 15 °C). Cyano complexes are formed when cadmium cyanide is dissolved in aqueous solutions of alkali cyanides: $[\text{Cd(CN)}_3]^-$, $[\text{Cd(CN)}_4]^{2-}$. In the absence of air and moisture, pure Cd(CN)_2 is stable and can be stored indefinitely; in moist air, it decomposes slowly.

Production. The preferred method for the production of copper cyanide is based on the reaction of copper(II) salts, metallic copper, and hydrogen cyanide in hydrochloric acid solution [148], [149]. The reaction solution can be recycled, which avoids environmental problems. The Cu(CN)_2 formed when copper(II) salts react directly with alkali cyanide or HCN is unstable and decomposes to CuCN and cyanogen, (CN)_2 (see Section 5.2).

Zinc cyanide can be obtained from the reaction of zinc salts with alkali cyanide in aqueous solution. However, the reaction of zinc oxide with hydrogen cyanide in acetic acid has the advantage that no additional salt is produced [150].

Cadmium cyanide can be produced by evaporation of a solution of cadmium hydroxide in aqueous hydrogen cyanide [151] or by precipitation from a cadmium salt solution with alkali cyanide [152].

Commercial Forms, Specifications, and Packaging. Commercially, CuCN, Zn(CN)_2, and Cd(CN)_2 are supplied in the form of white, fine crystalline powders, packaged in steel drums of 50 kg net weight. The specifications and the average analyses are shown in Table 9.

Water-soluble, complex copper and zinc cyanide, as well as mixtures of CuCN and Zn(CN)_2 with alkali cyanides (containing ca. 23 or 29% Cu or 23% Zn), are available for use in electroplating processes. For transportation, storage, and handling, the same precautions must be observed as for alkali cyanides.

Table 9. Specifications (DIN 50971) and average analysis of the cyanides of copper, zinc, and cadmium

	CuCN		Zn (CN)$_2$		Cd (CN)$_2$	
	DIN 50 971	Average analysis	DIN 50 971	Average analysis	DIN 50 971	Average analysis
Cu$^+$, %	> 70.0	> 70.0	< 0.001	< 0.001	< 0.001	< 0.001
Zn^{2+}, %	< 0.002	< 0.002	> 55.0	> 55	< 0.005	< 0.005
Cd^{2+}, %	< 0.001	< 0.001	< 0.005	< 0.005	> 67.0	> 67.0
CN$^-$, %	> 28.8	> 28.8	> 42.0	> 43	> 30.0	> 30.0
OH$^-$, %		< 0.05		< 0.05		< 0.05
CO$_3^{2-}$, %		< 0.1		< 0.5		
Cl$^-$, %		< 0.1		< 0.1		< 0.1
SO$_4^{2-}$, %		< 0.1		< 0.3		< 0.5
H$_2$O, %		< 0.1		< 0.5		< 0.5
Sb, ppm	< 10	< 10	< 10	< 10	< 100	< 100
As, ppm	< 10	< 10	< 10	< 10	< 100	< 100
Pb, ppm	< 20	< 20	< 50	< 50	< 50	< 50
Fe, ppm	< 100	< 100	< 100	< 100	< 500	< 500
Ni, ppm	< 20	< 20	< 50	< 50	< 50	< 50
Sn, ppm	< 10	< 10	< 10	< 10	< 100	< 100
Insoluble in NaCN solution, %	< 0.02	< 0.02	< 0.1	< 0.1	< 0.1	< 0.1

Uses and Economic Aspects. The simple cyanides of copper, zinc, and cadmium are used primarily in combination with alkali cyanides as raw materials in the preparation of electrolytes for the electroplating industry.

The cyanide complexes are preferred in some cases. Copper(I) cyanide is an important reagent for the introduction of cyano groups into aromatic rings to produce benzonitrile derivatives. It is also used as an insecticide and fungicide [153], in the preparation of special paints for ships, in the production of phthalocyanine dyes, and as a catalyst for the polymerization of organic compounds.

2.3.3. Cyanides of Mercury, Lead, Cobalt, and Nickel

The cyanides of mercury, lead, cobalt, and nickel do not have any major industrial importance. Water-soluble complex cyanides can be obtained by the reaction of the simple cyanides with alkali cyanides.

Properties, Preparation, and Uses. *Mercury(II) cyanide* [3021-39-4], Hg(CN)$_2$, M_r 252.65, ϱ 4.0 g/cm^3, forms white crystals, which are very soluble in water; it is highly toxic. The substance can be prepared by the reaction of HgO with aqueous HCN. When heated, it decomposes to Hg and (CN)$_2$, and may explode if detonated [151],

[154]. The complex salts Na[Hg(CN)$_3$], M_r 301.66, and K$_2$[Hg(CN)$_4$] [591-89-9], M_r 382.87, are white crystalline powders which are very soluble in water. Mercury(II) oxycyanide [2040-54-2], Hg(CN)$_2$ · HgO, M_r 469.26, is less stable and more sensitive to shock than picric acid [155].

Lead cyanide [13453-58-2], Pb(CN)$_2$, M_r 259.25, can be obtained by the reaction of aqueous solutions of lead salts and alkali cyanides in the presence of HCN [8]. In the absence of HCN, oxycyanides are formed, e.g., Pb(CN)$_2$ · 2 PbO.

Cobalt(II) cyanide [542-84-7], Co(CN)$_2$ · 2 H$_2$O, M_r 110.95, mp 280 °C (decomp.), ϱ 1.872 g/cm^3, forms red-brown needles which are insoluble in water. It can be obtained by the reaction of aqueous solutions of cobalt salts and alkali cyanides, and can be converted to soluble complex compounds with ammonia or excess alkali cyanide.

Complex cobalt(III) cyanides are formed when solutions of complex cobalt(II) cyanides are heated in the presence of oxygen:

$$2\ K_4[Co(CN)_6] + H_2O + 1/2\ O_2 \longrightarrow 2\ K_3[Co(CN)_6] + 2\ KOH$$

Cobalt cyanides are among the most stable complex compounds; therefore, cobalt compounds have been used as antidotes in cases of cyanide poisoning [156].

Nickel cyanide [557-19-7], Ni(CN)$_2$, M_r 110.74, mp > 200 °C (decomp.), ϱ 2.393 g/cm^3, forms brownish-yellow crystals which are insoluble in water. It can be obtained by the dehydration of Ni(CN)$_2$ · 4 H$_2$O, which is precipitated when alkali cyanide is added to a solution of nickel salt.

Potassium tetracyanonickelate(II) monohydrate [14323-41-2], K$_2$[Ni(CN)$_4$] · H$_2$O, M_r 258.97, forms orange-red crystals which are very soluble in water. The complex salt is formed when excess KCN is employed in the preparation; K$_2$[Ni(CN)$_4$] is used as an additive to electrolytes in the plating of gold, silver, and zinc, to improve the quality of the deposits.

2.3.4. Cyanides of Precious Metals

The precious metal cyanides can be regarded as the historical starting point of inorganic cyanide chemistry. Their importance for the recovery, refining, and electroplating of precious metals is as great today as it was 100 years ago.

Gold(I) cyanide [506-65-0], AuCN, M_r 222.98, ϱ 7.12 g/cm^3, forms yellow crystals which are insoluble in water and hot dilute acids, but soluble in ether, alkali hydroxide, and thiosulfate solutions. Water-soluble [Au(CN)$_2$]$^-$ complexes are formed when AuCN is dissolved in alkali cyanide solution or when metallic gold is attacked by alkali cyanide in the presence of oxygen or oxidizing agents.

Silver(I) cyanide [506-64-9], AgCN, M_r 133.89, mp 320 °C (decomp.), ϱ 3.95 g/cm^3, forms white hexagonal crystals which are insoluble in water but soluble in aqueous alkali cyanide solution, forming complex [Ag(CN)$_2$]$^-$ compounds. Like AgCl, AgCN becomes brown if exposed to light. The production of AgCN is based on the reaction of AgNO$_3$ with alkali cyanide in aqueous solution. Its main application is in electroplating.

Palladium(II) cyanide [2035-66-7], Pd(CN)$_2$, M_r 158.7, *mp* 210 °C (decomp.), is soluble in water.

Platinum(II) cyanide [592-06-3], Pt(CN)$_2$, M_r 247.27, is insoluble in water and dilute acids. Both palladium(II) cyanide and platinum(II) cyanide can be prepared from the corresponding chlorides by precipitation with Hg(CN)$_2$. Tetracyano complexes are formed with excess alkali cyanide: K$_2$[Pd(CN)$_4$] · 3 H$_2$O forms white rhombohedral fluorescent crystals, and K$_2$[Pt(CN)$_4$] · 3 H$_2$O forms rhombic yellow prisms with blue fluorescence. The barium salt Ba[PtCN)$_4$] · 4 H$_2$O is used to prepare the fluorescent coating of X-ray screens; the alkali tetracyanoplatinates are used in electroplating. An unstable Pd(CN)$_4$ is also known.

2.4. Cyanide Analysis

The cyanide content of solid alkali cyanides, alkaline-earth cyanides, and dissociating heavy metal cyano complexes such as [Zn(CN)$_4$]$^{2-}$ may be determined by the argentometric titration of a dilute aqueous solution of the cyanide, according to the method of Liebig–Denigés, in which potassium iodide is used as an indicator [157]:

$$2\,CN^- + Ag^+ \longrightarrow [Ag(CN)_2]^-$$

A very accurate indication of the end point of this reaction can be obtained potentiometrically by use of a silver–calomel or silver–Thalamide (Schott Glaswerke, Mainz) electrode pair.

For more stable cyano complexes or insoluble cyanide compounds [e.g., iron(II) hexacyanoferrate(II)] or in the presence of substances that disturb the argentometric titration, liberation and separation of the cyanide must be carried out before determination. This can be done by acidifying and boiling the cyanide solution or suspension, and absorbing the released hydrogen cyanide in sodium hydroxide solution [158]. Depending on the concentration, the cyanide content of the absorption solution may be analyzed argentometrically or colorimetrically, e.g., by using pyridine–barbituric acid reagents [159].

To evaluate the toxicity of wastewater, measurement of the "cyanide amenable to chlorination" is still in use [160]. A new and reliable analytical method to measure the "easily liberatable cyanide" has been developed, which includes all cyanides amenable to chlorination, especially hydrogen cyanide, alkali and alkaline-earth-metal cyanides, and cyanides of zinc, cadmium, silver, mercury, copper, and nickel. The complex cyanides of iron, cobalt, and gold and the nitriles [161] are not included. In this method, hydrogen cyanide is liberated at pH 4 in the presence of metallic zinc and ethylenediaminetetra-acetic acid. The HCN is transferred by an air stream into an absorption vessel containing an NaOH solution and is determined argentometrically or colorimetrically.

Quick tests to check the level of cyanides in wastewater are based on various colorimetric methods [159], [162]. Generally, the reactions permit detection of very small amounts of free cyanide, but they can be disturbed by various substances, e.g., reducing agents or thiocyanate, sulfide, and nitrite which all produce similar colors.

A new potentiostatic method is used for the continuous determination of cyanide in effluent streams, which also allows the detection of cyanide concentrations below 100 µg/L [163].

3. Detoxification of Cyanide-Containing Wastes

Whenever cyanides are manufactured or used, effluents and wastes containing various amounts of cyanide are produced. Because of the high toxicity of cyanide to all forms of life (see Chap. 6), the effluents and wastes must be treated to reduce the cyanide content to concentrations that are acceptable with regard to the particular environmental conditions [164]–[168]. Depending on the quantity and type of cyanidic waste, various detoxification methods are used [169]–[173]. In addition to effectiveness and cost of treatment, the formation of undesirable byproducts and additional salting of the wastewater are factors of growing importance in choosing an effluent treatment method.

3.1. Wastewater Treatment

Chlorination. Alkaline chlorination is the most frequently used process for the treatment of effluents containing < 1 g CN^- per liter [174]. In principle, this method allows the destruction of all the commercially used simple and complex cyanides, with the exception of complex iron cyanides which are only attacked at temperatures above 80 °C. The treatment can be carried out with chlorine and alkali [NaOH, Ca(OH)$_2$] or with ready-made hypochlorite solutions that contain about 12% NaOCl. At first, toxic cyanogen chloride is formed, which hydrolyzes quickly to cyanate and chloride at pH > 11 [175]:

$$CN^- + OCl^- + H_2O \longrightarrow ClCN + 2\,OH^-$$
$$ClCN + 2\,OH^- \longrightarrow OCN^- + Cl^- + H_2O$$

The cyanate may be oxidized further to nitrogen and carbonate by using excess hypochlorite

$$2\,OCN^- + 3\,OCl^- + 2\,OH^- \longrightarrow N_2 + 2\,CO_3^{2-} + 3\,Cl^- + H_2O$$

or hydrolyzed to ammonium carbonate at pH < 7 [176], [177]

$$OCN^- + 2\,H_2O \longrightarrow NH_4^+ + CO_3^{2-}$$

Heavy metals may be precipitated and separated as hydroxides, carbonates, or sulfides, when complex heavy metal cyanides are oxidized. The wastewater treatment can be carried out batchwise or continuously, and the process may be monitored and controlled automatically.

Additional salting of the water is the main disadvantage of alkaline chlorination: at least 5 kg of NaCl are produced per kilogram of cyanide for the oxidation of cyanate, and an additional 6.5 kg of NaCl per kilogram of cyanide results from total oxidation [178]. Sometimes, modified processes are used in electroplating works; a very fast, complete destruction of cyanide can be obtained when the plated parts are dipped directly into a hypochlorite detoxification bath after treatment in cyanide-containing electrolytes [179], [180]. Consumption of chemicals and additional salting of the water can be reduced when hypochlorite is produced electrolytically in the course of wastewater treatment [181]. However, this process and the anodic oxidation of cyanide [182], which form urea and oxalate in addition to cyanate, carbonate, and nitrogen, only work economically at higher cyanide concentrations. In the case of heavy metal cyanides, the heavy metals are deposited on the cathode.

Hydrogen Peroxide Oxidation. The use of hydrogen peroxide for cyanide detoxification has increased in recent years because, in this case, the oxidation of cyanide leads directly to cyanate without the formation of toxic intermediates and byproducts, and the oxidant does not cause additional salting [178], [183]:

$$CN^- + H_2O_2 \longrightarrow OCN^- + H_2O$$

The H_2O_2 process is used mainly for the batchwise treatment of effluents from organic nitrile synthesis and metal hardening plants, and can also be monitored and controlled automatically [184], [185]. One recently developed process allows delivery of very accurate amounts of H_2O_2 to continuous effluent streams, which is controlled by online measurement of the total oxygen demand [186]; this process is used, for example, in the treatment of tailings from a gold mine [187]. Sometimes the reaction of hydrogen peroxide with cyanide may be supported by the addition of formaldehyde [188] to form the cyanohydrin:

$$CN^- + H_2CO + H_2O \rightleftharpoons H_2C\!\!\begin{array}{c}\nearrow OH \\ \searrow CN\end{array}\!\! + OH^-$$

The final reaction products are cyanate, glycolic acid, and ammonia. This process is used for the treatment of effluents from electroplating works; a modified reaction is applied to the detoxification of wastewater from the wet-scrubbing systems of steel mills [189].

Ozone Oxidation. Very effective oxidation of cyanide to cyanate, carbonate, and nitrogen takes place with ozone [84], [190]; however, this requires high capital investment and causes problems with the adaptation to varying oxygen demand.

Other Methods of Treatment. In the case of highly concentrated cyanide solutions, the volatilization and catalytic oxidation of HCN after acidification of the solution may be applied [172]. Sometimes, the hydrogen cyanide can be reused by absorption in alkali hydroxide solution and conversion to alkali cyanide. The oxygen in air is occasionally used for the destruction of cyanide in weak solutions, in combination with activated carbon [82], [191] or microorganisms [192]. The natural degradation of cyanide can be used as a pretreatment to reduce chemical consumption, provided that the pond has an adequate retention time and short circuiting is avoided. Removal of cyanides by ion exchange and reverse osmosis does not solve the problem entirely because the concentrates require further treatment [193], [194].

The hydrolysis of cyanide at temperatures of 180–230 °C and high pressure may be used for the destruction of simple and complex cyanides, even in high concentrations [79]. The process does not require chemicals, but the investment for the plant is high.

The conversion of cyanide to thiocyanate by the addition of sulfur is a process that used to be employed but is now regarded more critically [169]. A more recent method is based on the treatment of effluents with a mixture of sulfur dioxide (2.5%) and air in the presence of small amounts of copper salts (> 50 mg Cu per liter) which act as a catalyst [195]:

$$CN^- + O_2 + SO_2 + H_2O \longrightarrow OCN^- + H_2SO_4$$

This method reduces the total cyanide content to below 1 mg/L.

The conversion of cyanide by reaction with iron(II) salts to hexacyanoferrate(II) in alkaline solution, followed by precipitation of hexacyanoferrate(II) as iron(II) hexacyanoferrate(II) at pH 3–4, is one of the oldest methods for cyanide removal [196]. However, this should be used only for the removal of complex iron cyanides, if at all, because the sludge must be disposed of in a manner that takes into consideration the protection of water resources. Furthermore, the filtrates have to be treated once more to destroy the remaining traces of cyanide.

In many cases, low concentrations of complex iron cyanides, e.g., < 20 mg of cyanide per liter, in effluents discharged to sewage systems are tolerated because they are precipitated in contact with heavy metal salts, e.g., the iron salts present in domestic wastewater, and are separated with the sludge of municipal sewage treatment plants.

3.2. Solid Wastes

For a long time, disposal of the solid wastes from metal hardening plants, which contain alkali cyanides, cyanates, nitrites, and nitrates in addition to other inorganic salts, was a special problem [69], [70]. Currently, new processes have been developed that, in many cases, allow regeneration of the salt baths without production of waste salts. In any case, solid wastes should not be treated or destroyed by a wet chemical process. The safest and most practical solution to the disposal problem at the present time is to deposit them in disused salt mines.

4. Cyanogen Halides

Halogen cyanides, usually known as cyanogen halides, have been the subject of considerable interest, and some aspects of their chemistry have been reviewed in detail [8], [197]–[202].

These compounds, except for cyanogen fluoride, have been known for a long time [203], [204]. Cyanogen fluoride was prepared only recently [205], [206]. Depending upon the electronegativity of the halogens compared to the cyanide group, the cyanogen halides can act as halogenating or cyanating agents. The cyanogen halides can be considered as pseudohalogen or interhalogen compounds with a definite nitrile structure (Hal–C≡N) [207]. The most important cyanogen halide is cyanogen chloride, which is used for the commercial production of cyanuric chloride, a starting material for pesticides, dyes, and drugs.

4.1. Properties

Physical Properties. The pure cyanogen halides are colorless, highly volatile, and poisonous compounds. Some physical properties are given in Table 10.

Chemical Properties. Pure cyanogen halides are stable at normal temperature. Impurities catalyze the exothermic trimerization to the corresponding cyanuric halides. Catalysts are proton acids, Lewis acids (e.g., heavy metal salts), and even bases. Relatively little is known about the preparative use of FCN [206], which explodes at −41 °C on ignition. Mixtures of FCN and air are more explosive than mixtures of acetylene and air [208]. In a catalyzed reaction, ClCN forms not only the trimer (cyanuric chloride [108-77-0]) but also a stable tetramer, 2,4-dichloro-6-isocyanodichloro-s-triazine [877-83-8] [209].

Table 10. Physical properties of the cyanogen halides [201], [202], [207]

	FCN [1495-50-7]	ClCN [506-77-4]	BrCN [506-68-3]	ICN [506-78-5]
M_r	45.02	61.47	105.92	152.92
mp, °C	−82	−6.5	52	148*
bp, °C	−46.2	12.5	61	subl. > 45*
Density ϱ, g/cm^3		1.24 (0 °C)	2.01 (18 °C)	2.84 (18 °C)
		1.19 (15 °C)	1.18 (54 °C)	
Bond length, pm				
C–Hal	126.0	162.9	179.0	199.5
C–N	116.5	116.3	115.8	115.9
Enthalpy of vaporization ΔH_v, kJ/mol	22.39	26.75	33.82	40.02
Enthalpy of fusion ΔH_f, kJ/mol		11.39		19.88
Enthalpy of sublimation ΔH_s, kJ/mol			45.21	59.90
Dipole moment μ, D	2.17	2.80	2.94	3.71

* Triple point, p = 132.1 kPa; subl. = sublimation.

Cyanogen bromide polymerizes more easily to cyanuric bromide [14921-00-7] than ClCN does to cyanuric chloride [210].

When heated to 130 °C in a sealed tube, cyanogen iodide does not polymerize [211]. Above 150 °C, ICN eliminates iodine:

2 ICN $\longrightarrow$ (CN)$_2$ + I$_2$

The reaction is not reversible [212].

Cyanogen chloride is an intermediate in the detoxification of cyanide-containing wastewater with sodium hypochlorite solutions; therefore, the cyanogen halide must be removed immediately by appropriate hydrolysis [213]. Alkaline hydrolysis of cyanogen halides proceeds in different ways:

XCN + 2 NaOH $\longrightarrow$ NaOCN + NaX + H$_2$O (X = Cl, Br)
ICN + 2 NaOH $\longrightarrow$ [NaOI] + NaCN + H$_2$O

Thus, ClCN and BrCN hydrolyze to form the cyanate and halide salts; ICN, however, forms hypoiodite, which disproportionates to iodide and iodate. This demonstrates the more electropositive character of iodine as compared with chlorine [202]. Therefore, ClCN can be used for the electrophilic introduction of a cyano group into a substrate molecule; use of ICN results in an electrophilic iodination. However, the reaction pathway of BrCN often depends on the reaction conditions and on the nature of the substrate [214]. Cyanogen halides add to olefins and acetylenes in the presence of acid catalysts [215], [216]. Aromatic ring systems can be cyanated by ClCN or BrCN under Friedel-Crafts conditions [207], [217]. The reactions of cyanogen halides — particularly ClCN and BrCN — with nitrogen, oxygen, and sulfur nucleophiles, which lead to cyanamides [218], [219], cyanates [220], [221], and thiocyanates [222], are well known.

In some cases, the newly formed compound undergoes further reactions at the C≡N triple bond to form guanidines [218], [219], ureas, or heterocyclic ring systems [223]. Cyanogen bromide reacts with tertiary alkylamines to form ammonium bromides, which decompose to give dialkyl cyanamides and alkyl bromides (von Braun reaction) [224], [225]. Reaction of sulfur trioxide with ClCN leads to the formation of the very reactive chlorosulfonyl isocyanate [226], [227]. Cyanogen halides undergo radical reactions at elevated temperatures by irradiation with UV light, or in the presence of peroxide promoters. Thus, cyanogen chloride and acetonitrile are converted to malonodinitrile [228]–[231]. In the gas phase, aromatic methyl compounds and cyanogen chloride react to give arylacetonitriles [232].

4.2. Production

Cyanogen fluoride is prepared by the reaction of tetracyanomethane and cesium fluoride [233]:

$$C(CN)_4 + CsF \longrightarrow FCN + CsC(CN)_3$$

The pyrolysis of cyanuric fluoride at 1300 °C and reduced pressure gives FCN in ca. 50% yield [205], [206], [208].

The preparation of the industrially important ClCN has been developed extensively; the synthetic pathways are as follows:

1) Electrolysis of an aqueous solution of HCN and NH_4Cl [234]
2) Reaction of complex cyanide salts (e.g., $Na_2[Zn(CN)_4]$) with chlorine below 20 °C [235]
3) Formation of ClCN from cyanide salts (mostly NaCN) and chlorine in an exothermic reaction; in a continuous process, sprayed aqueous NaCN solution is contacted with chlorine, and the reaction heat evaporates the ClCN [236].
4) Processes involving hydrogen cyanide and chlorine as the most convenient starting materials; the reaction is carried out in aqueous solution [237]–[242], in organic [243] and inorganic solvents [244], and in the gas phase [245], [246]. To avoid the byproduct HCl, attempts have been made to reoxidize HCl to chlorine with oxygen [241] and hydrogen peroxide [247], [248] in catalyzed reactions.
5) Chlorinolysis of cyanogen in the gas phase at 300–600 °C in the presence of a catalyst [249], [250]
6) Pyrolysis of cyanuric chloride at 600–900 °C in the presence of a charcoal catalyst [251]
7) High-temperature syntheses based on elemental chlorine, nitrogen, and carbon [252]

Usually, these processes lead to impure ClCN; therefore, in many cases the product must be purified before further use. Water can be removed by treatment with calcium salts or molecular sieves and by fractional distillation [253]. Chlorine is stripped with water (mixed with HCN and HCl) [254] or with solutions containing a soluble iron (II) salt and an insoluble carbonate [255]; chlorine may also be eliminated by treatment with γ-aluminum oxide [256]. Cyanogen chloride, free of HCl, is obtained by treatment with water [254] or with organic solvents [257], [258]. Hydrogen cyanide is removed by distillation with chlorine [259] or by washing with $FeSO_4$ solutions [260].

Cyanogen bromide can be prepared from bromine and cyanide salts or hydrogen cyanide [261], [262]. Cyanogen iodide is synthesized analogously [263], [264].

4.3. Storage and Transportation

In most cases, ClCN is used in the gas phase right after preparation. Relatively small amounts are condensed and stored in gas containers as liquids. Steel cylinders, which must meet specific requirements in each country, can be used for shipment. The condensed and bottled ClCN must be very pure; moreover, it must be mixed with a stabilizing agent (generally sodium pyrophosphate) to inhibit the exothermic polymerization caused by the impurities [265]–[267]. Impure BrCN can polymerize during storage [268]; the danger of explosion exists with closed BrCN bottles [269]. Cyanogen halides should be handled carefully because they are very toxic (similar to HCN) and have a strong lacrimatory effect.

4.4. Uses

Most of the cyanogen chloride manufactured is used for the production of cyanuric chloride. The reaction with amines leads to diphenylguanidines (vulcanization accelerators) [270]. Cyanogen chloride and bisphenols react to give cyanate esters, which can polymerize to polytriazine resins [220], [221].

Halogen cyanides and cyanamide react to form dicyanamide [504-66-5] [271], [272], which is used for the preparation of pharmaceutical bisbiguanides. The industrial production of malonodinitrile [109-77-3] [228]–[231], and chlorosulfonyl isocyanate [1189-71-5] [226], [227] based on cyanogen chloride has become increasingly important. Cyanogen chloride is the most economically significant of all cyanogen halides. Its worldwide production capacity is more than 100 000 t/a (mainly in Europe and the United States).

Cyanogen bromide and cyanogen iodide are commonly used in the laboratory, e.g., for the dealkylation of tertiary amines [224], [225], the selective cleavage of peptides [273], and the preparation of heterocyclic compounds.

5. Cyanogen

The first synthesis of cyanogen [460-19-5], dicyanogen, oxalonitrile, $(CN)_2$, M_r 52.05, was carried out by GAY-LUSSAC in 1815 via pyrolysis of silver cyanide [260]. The chemistry of this reactive compound has been investigated mainly in recent times [2], [274]–[279]. In many respects, cyanogen can be compared to the halogens. In addition, many of its properties are a result of the reactivity of the cyano groups. Although various procedures for the synthesis of cyanogen are known, commercial interest in cyanogen is limited.

5.1. Properties [2], [279]

Physical Properties. At room temperature, cyanogen is a colorless, flammable, and very poisonous gas with a pungent odor. Some important physical properties are listed below:

Melting point	−27.98 °C
Boiling point	−21.15 °C
Critical pressure, $p_{crit.}$	60.79 bar
Critical temperature, $t_{crit.}$	128.3 °C
Density	2.321 g/L at 0 °C
	1.25 g/cm^3 at −95 °C
	0.954 g/cm^3 at −21.17 °C

Vapor pressure

t, °C	−21.15	−4.88	20.88	44.43	72.40
p, bar	1	2	5	10	20

Solubility at 20 °C, water dissolves 4.5 times its own volume of cyanogen; diethyl ether, 5.0 times; and ethyl alcohol, 23.0 times.

Bond length:	
C–C	138 pm
C–N	113 pm
Bond angle: C–C–N	179° 38′
Enthalpy of fusion, ΔH_f	8.112 kJ/mol
Enthalpy of vaporization	23.341 kJ/mol
MAK	10 ppm

The values reported for the ignition limits of cyanogen–air mixtures vary tremendously, depending on experimental conditions, e.g., with humid air (1.7% H_2O); the reported values are 6.4–7.25 vol% for the lower limit and 26.2–30.6 vol% for the upper limit.

Chemical Properties. Cyanogen has a high thermal stability; nevertheless at high temperatures (and when irradiated with UV light), polymerization to paracyanogen occurs [279], [280], [281]. With stoichiometric amounts of air or oxygen, cyanogen burns with the hottest flame (ca. 5000 K) ever observed in chemical reactions [282], [283].

Cyanogen undergoes reactions which depend either on its properties as a pseudohalogen or on the reactivity of the C≡N triple bond [2], [277], [279]:

1) Hydrogenation of cyanogen at 550–675 °C gives hydrogen cyanide [284].
2) At 300–500 °C, chlorine and cyanogen react to give cyanogen chloride and, subsequently, cyanuric chloride in the presence of an active charcoal catalyst [285], [286].
3) The direct introduction of a cyano group into an aryl or alkyl compound proceeds via a radical- or heat-initiated reaction [287]–[290].
4) Two moles of hydrogen cyanide add to cyanogen to form diiminosuccinonitrile [291].
5) Cycloadditions (Diels-Alder) with organic dienes give cyanopyridines [292].
6) As for the halogens, alkaline hydrolysis leads to cyanide and cyanate [293].
7) The acid-catalyzed hydrolysis of cyanogen first gives cyanoformamide [294], [295] and then oxamide and oxalic acid. Oxamide [471-46-5] is a nitrogen fertilizer with good depot activity. Its industrial preparation has been published in several papers; most of the synthetic paths use the isolated $(CN)_2$ as an intermediate [296]–[301], but in situ procedures are also known [302].

Heterocyclic compounds are formed by the reaction of cyanogen and disulfur dichloride [303] or sulfur [304]:

$$(CN)_2 + S_2Cl_2 \longrightarrow \text{[heterocycle with Cl, Cl, N, S, N]}$$

$$2\,(CN)_2 + S \xrightarrow[\text{copper}]{\text{Dimethyl-formamide,}} \text{[heterocycle with S, N, NC, N, CN]}$$

5.2. Production

Numerous synthetic pathways to cyanogen have been devised [2], [275], [277], [279]. In the laboratory, the compound is normally prepared via cyanide salts; in this case, the reaction of alkali cyanides with copper(II) salts is more common than the thermal decomposition of $Hg(CN)_2$ or AgCN [279].

$$CuSO_4 + 2\,KCN \longrightarrow Cu(CN)_2 + K_2SO_4$$
$$2\,Cu(CN)_2 \longrightarrow (CN)_2 + 2\,CuCN$$

However, the oxidation of hydrogen cyanide in the presence of a catalyst has greater industrial interest. These reactions are based on the principle of JACQUEMIN [305] — the conversion of cyanides, by treatment with copper salts, to cyanogen in aqueous solution. In a continuous process, the copper(I) salt formed must be reoxidized rapidly to recycle the catalyst. Reoxidation of copper(I) to copper(II) can be carried out directly with air or oxygen [306], [307] or with NO_2 [308]; the NO formed in the second case can then be reoxidized with air or oxygen [309], [310]. The oxidation of hydrogen cyanide to cyanogen in nonaqueous solutions has also been achieved [311], [312]. Hydrogen peroxide is another reoxidation agent [313]–[316].

The continuous production of $(CN)_2$ from HCN and O_2 or H_2O_2 is utilized industrially on a small scale.

In addition to the liquid-phase oxidation of HCN, various gas-phase oxidations have been investigated [317]–[322].

Cyanogen iodide decomposes to cyanogen and iodine (Chap. 4). More recently, the reaction of cyanogen halides with trimethylsilyl cyanide in the presence of Lewis acids has been described [323].

Cyanogen prepared by the above procedures contains impurities such as O_2, N_2, NO, CO_2, or ClCN. Its purification can be achieved by fractional vaporization of the crude material [324] or by scrubbing the gas with aqueous H_2O_2 [325].

5.3. Storage, Transportation, and Uses

Storage and Transportation. The flammability and toxicity of cyanogen necessitate special storing and shipping conditions. Cyanogen should be stored in cool, well-ventilated locations, not near to flammable goods or open flames. Dry and pure $(CN)_2$ can be stored in steel cylinders under pressure and at normal temperatures [326]. Comprehensive studies have been made on the stability of pure cyanogen to heat, pressure, chemical additions, and severe mechanical shock [327]. Shipment must be in accordance with governmental regulations in each country.

Uses and Economic Value. Cyanogen is used mainly in the laboratory. In small quantities, cyanogen is used industrially for drug synthesis. The preparation of oxamide (a slow-release fertilizer) has been carried out on a small scale [286]. Large industrial-scale applications are not yet known.

6. Toxicology and Occupational Health

Toxicity and Cyano Compounds. Hydrogen cyanide and other cyano compounds that can form HCN or free cyanide ions are highly toxic to almost all forms of life. The lethal dose depends on a number of factors, and differs from species to species. A useful survey on this subject can be found in the Proceedings of a Workshop on Cyanide from Mineral Processing [167]. As shown in this report, the toxicity of metal

Table 11. Lethal concentrations of some cyanides (in mg/L)

Substance	Species		
	Fish	Daphnia magna	Escherichia coli
NaCN	0.05 (5 d) [329]	3.4 (48 h) [330]	0.0004–0.1 [331]
NaCN	0.3–0.7 (96 h) [332]		
NaCN	1.0 (20 months) [333], [334]		
$K_3[Cu(CN)_4]$ [335]	1.0	0.8	2500
CuCN	2.2 [165]		
$K_2[Zn(CN)_4]$ [335]	0.3	13.5	625
$Zn(CN)_2$	0.2–0.3		
$K_2[Cd(CN)_4]$ [335]	0.75	0.5	250
$Cd(CN)_2$	0.17 [165]		
$K_2[Ni(CN)_6]$ [335]	30	75	5000
$Na_4[Fe(CN)_6]$	500 [165]	> 600 [336]	> 1000 [337]

cyanide complexes is a function of the bond strength between the metal atoms and the cyanide ligands. Toxicity decreases with increasing bond strength, and complexes such as hexacyanocobalt(II) are nontoxic. Hexacyanoferrates(II) and (III) also have very low toxicities [328]. These differences in toxicity are particularly important when the acceptable concentrations of cyanides in wastewater are being considered. Some of the lethal concentrations are shown in Table 11. The LD_{50} values of complex hexacyanoferrates (rat, oral) are as follows:

$Ca_2[Fe(CN)_6] \cdot 11 H_2O$	> 5110 mg/kg
$Na_4[Fe(CN)_6] \cdot 10 H_2O$	> 5110 mg/kg
$K_4[Fe(CN)_6] \cdot 3 H_2O$	3613 mg/kg
$K_3[Fe(CN)_6]$	3503 mg/kg

Poisoning by Hydrogen Cyanide and Alkali Cyanides. Hydrogen cyanide is easily absorbed, even through intact skin and mucous membranes. Not even full respiratory protection can prevent hydrogen cyanide from penetrating into the organism. Symptoms [338] of the poisoning may vary somewhat, depending on the quantity of hydrogen cyanide absorbed or the quantity of cyanide ingested. Low concentrations of hydrogen cyanide irritate the nasopharyngeal cavity; occasional complaints involve headache, anxiety, and nausea. High cyanide concentrations in the blood, however, stimulate the respiratory center. Rapid inhalation of the lethal dose (ca. 50–100 mg) of hydrogen cyanide results in collapse of the individual. This apoplectiform syndrome, characterized by a short spasmodic stage with subsequent respiratory paralysis, results in immediate death. Poisoning symptoms following oral ingestion of cyanides, on the other hand, exhibit a slow progress, even with large doses (150–250 mg), because of the slow release of hydrogen cyanide. Case histories described in the literature involve poisoning durations up to several hours [339].

The following concentrations of hydrogen cyanide and other cyanides in air are considered acceptable at the workplace:

TLV–TWA	(hydrogen cyanide)	10 mg/m^3 [340]
MAK	(hydrogen cyanide)	11 mg/m^3 [341]
TLV–TWA	(cyanide dust, as CN)	5 mg/m^3 [340]
MAK	(cyanide dust, as CN)	5 mg/m^3 [341]

Therapy for Poisoning by Hydrogen Cyanide and other Cyanides. The progress and prognosis for hydrogen cyanide poisoning of accidental or deliberate origin depends primarily on the medical aid received at the place of emergency.

The first aid given to victims of hydrogen cyanide poisoning is not fundamentally different from the measures taken in other acute emergencies. The success of the treatment, however, depends on the immediate start of antidote therapy. Seconds may determine the success or failure of the treatment of cyanide poisoning. The therapeutic measures required depend on the particular poisoning syndrome encountered. The initial treatment is of supreme importance and must be aimed at safeguarding vital functions such as respiration and circulation. In cases of medium and severe poisoning accompanied by unconsciousness (reactive and a-reactive poisoning with respiratory paralysis), 5 mL of 4-dimethylaminophenol must be injected immediately i.v. [342], followed by a slow infusion of 50 – 100 mL of 10% sodium thiosulfate solution. The sodium thiosulfate accelerates the normal detoxification by enzymatic formation of thiocyanate. 4-Dimethylaminophenol splits the cyanide – iron bond in cytochromeoxidase.

Cyanogen Chloride, Cyanogen Bromide [338]. Because of their high volatility, cyanogen chloride and cyanogen bromide are strong irritants to mucous membranes (eyes, lungs). Their effects are essentially identical to those produced by hydrogen cyanide; however, they are rarely encountered because the irritating effects serve as a warning of potential danger. When handling liquid cyanogen chloride, attention must be paid to the potential for skin absorption. The exposure limit of cyanogen chloride (TLV – TWA) is 0.6 mg/m^3 [340]. The therapy is analogous to that for hydrogen cyanide poisoning.

Cyanogen [338]. Cyanogen irritates mucous membranes and is probably hydrolyzed in the organism to hydrogen cyanide and cyanic acid. The symptoms are similar to those produced by hydrogen cyanide poisoning.

TLV–TWA	20 mg/m^3 [340]
MAK	22 mg/m^3 [341]

The therapy is analogous to that for hydrogen cyanide poisoning.

7. References

[1] H. Bauer, *Naturwissenschaften* **67** (1980) 1–6.
[2] *Gmelin*, System no. 14, D 1, Kohlenstoff (1971).
[3] W. F. Giauque, R. A. Ruehrwein, *J. Am. Chem. Soc.* 1939, 2626–2633.
[4] G. E. Coates, J. E. Coates, *J. Chem. Soc.* 1944, 77–81.
[5] G. Bredig, L. Teichmann, *Z. Elektrochem.* **31** (1925) no. 8, 449–454.
[6] G. Kortüm, H. Reber, *Z. Elektrochem.* **65** (1961) no. 9, 809–812.
[7] B. Grüttner, M. F. A. Dove, A. F. Clifford: *Chemistry in Anhydrous, Prototropic Inorganic Solvents*, vol. **II**, part 1, Pergamon Press, Oxford, London, Edinburgh, New York, Toronto 1971.
[8] H. E. Williams: *Cyanogen Compounds*, 2nd ed., Arnold & Co., London 1948.
[9] M. H. Ford-Smith: *The Chemistry of Complex Cyanides*, Dep. of Scientific and Industrial Research, Her Majesty's Stationery Office, London 1964.
[10] T. Völker, *Angew. Chem.* **72** (1960) 379–384.
[11] Röhm & Haas, DE 1 020 320, 1956 (T. Völker, H. Zima).
[12] W. Ruske, *Wiss. Z. Humboldt Univ. Berlin. Math. Naturwiss. Reihe* **8** (1958) 557–572.
[13] J. P. Ferris, W. J. Hagan, *Tetrahedron* **40** (1984) no. 7, 1093–1120.
[14] *Ullmann*, 3rd ed., **5**, 629.
[15] H. Schaefer, *Chem.-Tech. Heidelberg* **7** (1978) 231–237.
[16] F. Götzen, *Hüttenwesen Symposium Int. Gas Union Karlsruhe* **12** (1972) 1–15; *Chem. Abstr.* **81** (1974) 65 808 m.
[17] *Ullmann*, 3rd ed., **5**, 631–633.
[18] Standard Oil, US 4 423 023, 1983 (L. J. Velenyi, H. F. Hardman, F. A. Pesa).
[19] Du Pont, US 4 164 552, 1979 (F. J. Weigert).
[20] V. D. Sokolovskii, Z. G. Osipova, S. Y. Burylin, *React. Kinet. Catal. Lett.* **15** (1980) no. 3, 311–314.
[21] Nitto Chem. Ind., EP-A 0 089 118, 1983 (Y. Sasaki, H. Utsumi).
[22] Monsanto, US 4 457 904, 1984 (J. W. Gambell, S. R. Anvie).
[23] Standard Oil, US 4 423 023, 1983 (J. L. Velenyi, H. F. Hardman, F. A. Pesa).
[24] Monsanto, US 4 425 260, 1984 (J. R. Ebner).
[25] BASF, DE 1 143 497, 1963 (G. Schulze, G. Weiß).
[26] Sumitomo, DE-OS 2 350 212, 1974 (T. Shiraishi, H. Ichihashi, F. Kato, E. Niihama).
[27] Distillers Ltd., GB 913 836, 1962.
[28] H. Bockhorn, R. Coy, F. Fetting, W. Prätorius, *Chem. Ing. Tech.* **49** (1977) no. 11, 883–895.
[29] L. Andrussow, *Angew. Chem.* **48** (1935) no. 37, 593–604.
[30] C. T. Kautter, W. Leitenberger, *Chem. Ing. Tech.* **25** (1953) no. 12, 697–701.
[31] L. Andrussow, *Chem. Ing. Tech.* **27** (1955) 469–474.
[32] Du Pont, US 4 128 622, 1978 (K. D. Loos, K. C. McCullough, D. Q. Whitworth).
[33] Johnson Matthey & Co., US 4 469 666, 1984 (D. J. Stephenson, A. E. Heywood, G. L. Selman).
[34] W. R. Rolingson, J. McPherson, P. D. Montgomery, B. L. Williams, *J. Chem. Eng. Data* **5** (1960) no. 3, 349–351.
[35] N. Kalkert, H.-G. Schecker, *Chem. Ing. Tech.* **51** (1979) no. 9, 895.
[36] Freeport Sulphur, US 2 590 146, 1952 (G. Barsky).
[37] DOW Chemical, US 2 899 274, 1959 (R. A. Smith, H. H. McClure, G. C. Bond).
[38] Toyo Koatsu Industries, US 3 112 177, 1968 (S. Fujise, N. Nagai, T. Numata).

[39] Du Pont, US 2 797 148, 1957 (H. C. Carlson).
[40] Du Pont, US 3 914 386, 1974 (N. R. Wilson).
[41] Du Pont, EP-A 0 030 141, 1983 (G. Levitt).
[42] F. Endter, *DECHEMA Monogr.* **33** (1959).
[43] F. Endter, *Chem. Ing. Tech.* **30** (1958) no. 5, 305–310.
[44] DEGUSSA, DE 959 364, 1954 (F. Endter).
[45] DEGUSSA, DE 882 985, 1953 (R. Wendlandt, G. Hoffmann, E. Kokert).
[46] DEGUSSA, DE 1 007 300, 1957 (F. Endter).
[47] DEGUSSA, DE 1 013 636, 1958 (F. Endter, M. Svendsen, L. Raum).
[48] DEGUSSA, US 4 471 712, 1984 (C. Voigt, P. Kleinschmit, R. Manner).
[49] E. Koberstein, *Ind. Eng. Chem. Process. Des. Dev.* **12** (1973) no. 4, 444–448.
[50] DEGUSSA, DE-OS 33 09 394, 1983 (R. Manner, H. Schaefer, C. Voigt, W. D. Pfeifer).
[51] DEGUSSA, DE 31 34 851, 1982 (H. Bruck).
[52] DEGUSSA, EP-A 0 072 416, 1981 (C. Voigt, P. Kleinschmit).
[53] DEGUSSA, US 4 289 741, 1981 (C. Voigt, P. Kleinschmit, E. Walter).
[54] DEGUSSA, US 4 387 081, 1983 (C. Voigt, P. Kleinschmit, G. Schreyer, G. Sperka).
[55] N. B. Shine, *Chem. Eng. Prog.* **67** (1971).
[56] Shawinigan Chem., CA 573 348, 1959 (H. S. Johnson, A. H. Andersen).
[57] Shawinigan Chem., US 3 097 921, 1963 (D. J. Kennedy, N. B. Shine).
[58] Shawinigan Chem., US 3 032 396, 1962 (D. J. Kennedy).
[59] *Hommel-Handbuch der gefährlichen Güter*, 4. Lfg., Merkblatt 42, Springer Verlag, Berlin, Heidelberg, New York 1980.
[60] R. Baxter, *Ind. Finish. Wheaton Ill.* **53** (1977) no. 12, 38–41.
[61] J. E. Johnson, E. R. Poor, US 2 753 248, 1956.
[62] Compur Electronic GmbH, Steinerstraße 15, 8000 München 70, FRG.
[63] A. G. Sharpe: *The Chemistry of Cyano Complexes of the Transition Metals*, Academic Press, London 1976.
[64] H. B. Northrup, *Trans. Am. Electrochem. Soc.* **60** (1931) 49–53.
[65] DEGUSSA, DE 1 149 035, 1963 (C. Albrecht, J. Müller).
[66] DEGUSSA, DE 1 178 668, 1964.
[67] L. Hackspill, R. Grandadam, *C.R. Acad. Sci.* **180** (1925) 931.
[68] L. Hackspill, R. Grandadam, *Ann. Chim. (Paris)* **5** (1926) no. 10, 237.
[69] W. Müller, L. Witzke, *Chem. Ing. Tech.* **45** (1973) no. 22, 1285–1289.
[70] H. Beyer, G. Dittrich, H. Kunst, F. Schmidt, *Chem. Anlagen + Verfahren* 1973, no. 11, 69–75.
[71] R. W. E. B. Harman, F. P. Worky, *Trans. Faraday. Soc.* **20** (1925) 502; *Chem. Abstr.* **19** (1925) 923.
[72] B. Ricca, G. D'Amore, *Gazz. Chim. Ital.* **79** (1949) 308–322; *Chem. Abstr.* **43** (1949) 8821.
[73] E. Deltombe, M. Pourbaix, *Proc. Meet. Int. Comm. Electrochem. Thermodyn. Kinet. 6th* 1955, 138–142; *Chem Abstr.* 1956, 6896.
[74] J. Zawidski, T. Witkowski, *Rocz. Chem.* **5** (1925) 515–528; *Chem. Abstr.* 1926, 3258.
[75] G. W. Wiegand, M. Tremelling, *J. Org. Chem.* **37** (1972) 914–916.
[76] Du Pont, US 2 773 752, 1953 (V. W. Kremer, C. H. Lemke).
[77] H. Sulzer, *Z. Angew. Chem.* **25** (1912) 1268–1273.
[78] G. W. Heisse, H. E. Foote, *Ind. Eng. Chem.* **12** (1920) 331–336.
[79] J. Hoerth, U. Schindewolf, W. Zbinden, *Chem. Ing. Tech.* **45** (1973) 641–646.
[80] E. v. Papp, J. Pogany, *Z. Angew. Chem.* **54** (1941) 55.
[81] W. Bucksteeg, *Purdue Univ. Eng. Bull. Ext. Ser.* **121** (1966) 688–695.

[82] D. C. Hoffmann, *Plating (East Orange, NJ)* **60** (1973) 157–161.
[83] M. R. Lowndes, *Chem. Ind.* **34** (1971) 951–956.
[84] C. Fabjan, *Galvanotechnik* **66** (1975) 100–107.
[85] DEGUSSA, DE 742 074, 1943 (H. Beier).
[86] J. P. Zumbrunn, *Inf. Chim.* **124** (1973) 189–194.
[87] H. Stamm, *Z. Angew. Chem.* **47** (1934) 579, 791–795.
[88] A. T. Kuhn, *J. Appl. Chem. Biotechnol.* **21** (1971) no. 2, 29–34.
[89] E. Schulek, P. Endröi, *Anal. Chim. Acta* **5** (1951) 252.
[90] P. D. Bartlett, R. E. Davis, *J. Am. Chem. Soc.* **80** (1958) 2513–2516.
[91] K. R. Leininger, J. Westley, *J. Biol. Chem.* **243** (1968) 1892–1899.
[92] L. G. Sillen, A. E. Mostell: *Stability Constants of Metal-Ion Complexes*, HMSO, London 1964.
[93] J. Bailar: *The Chemistry of the Coordination Compounds*, Reinhold Publ. Co., New York 1956.
[94] E. Rabald: *Dechema-Werkstoff-Tabelle*, 3rd ed., part 12, 1963.
[95] *Ullmann*, 4th ed., **8**, 253.
[96] V. Migrdichian: *The Chemistry of Organic Cyanogen Compounds*, Reinhold Publ. Co., New York 1947.
[97] *Houben-Weyl*, **8**, 247–358.
[98] F. and E. Rodgers, *Justus Liebigs Ann. Chem.* **41** (1942) 285.
[99] H. J. Castner, GB 12 219, 1894; GB 21 732, 1894; DE 90 999, 1894.
[100] DEGUSSA, DE 124 977, 1900; DE 126 241, 1900; DE 148 045, 1901; DE 148 046, 1901; DE 149 678, 1901; GB 265 639, 1926 (H. Freudenberg); DE 851 056, 1943 (W. Stephan).
[101] *Ullmann*, 3rd ed., **5**, 643–644.
[102] DEGUSSA, US 716 350, 1902 (F. Roessler).
[103] BASF, DE 1 064 934, 1959 (L. Zimmermann, H. G. Schwarz).
[104] Du Pont, US 2 993 754, 1961 (W. R. Jenks, J. S. Linder).
[105] Monsanto, US 2 708 151, 1955 (T. D. McMinn).
[106] Monsanto, US 2 726 139, 1955 (G. D. Oliver).
[107] Du Pont, US 2 993 754, 1961 (W. R. Jenks, J. S. Linder).
[108] Du Pont, DE 1 103 905, 1959 (W. R. Jenks, J. S. Linder).
[109] Du Pont, US 2 876 066, 1959 (B. N. Inman); US 2 949 341, 1960 (C. P. Green); US 3 015 539, 1962 (J. M. Snyder).
[110] Du Pont, DE 1 021 340, 1956; US 2 773 752, 1956 (V. W. Kremer, C. H. Lemke).
[111] Soc. Elec. Chimie, FR 1 338 413, 1962; FR 1 338 450, 1962.
[112] Du Pont, DE 1 150 367, 1958 (C. P. Green, W. R. Jenks).
[113] DEGUSSA, DE 1 089 738, 1960 (H. Klebe, K. Rinn).
[114] BASF, DE 1 144 246, 1960 (E. Hartert, H. G. Schwarz).
[115] Shawinigan, US 3 207 574, 1965 (F. B. Popper).
[116] Du Pont, DE 1 120 437, 1959 (C. P. Green); GB 842 077, 1959.
[117] I. G. Farben, DE 710 757, 1941 (K. Gabel).
[118] DEGUSSA, DE 111 154, 1898; DE 456 350, 1924 (C. Andrich).
[119] DIN 50 971, 30, Dec. 1984.
[120] Chemical Safety Data Sheet SD–30 (1967), Manufacturing Chemists Association, Safety and Fire Protection Committee, 1825 Connecticut Avenue, N.W. Washington, D.C. 20 009.
[121] Merkblatt M 002, Ausgabe 12/81. *Cyanwasserstoff*, Berufsgenossenschaft der chemischen Industrie, Verlag Chemie, Weinheim 1981.
[122] A. Bergmann, A. Guzman, G. Potter: *Regeneration of Cyanide*, A Case Study, Conference of Cyanide and the Environment, University of Arizona, Tucson, Dec. 1984.

[123] G. E. Sheriwan, G. G. Griswold, US 1 421 585, 1922.
[124] *Praktische Galvanotechnik*, 2nd ed., Leuze Verlag, Saulgau 1970, p. 291.
[125] L. Marshall, S. S. Meusel, *Engineering (London)* **181** (1956) 425–428.
[126] DEGUSSA, US 2 875 095, 1959 (J. Müller); DE 1 108 246, 1959 (J. Müller); DE 1 149 035, 1963 (C. Albrecht, J. Müller).
[127] Kolene Corp., US 3 317 357, 1967 (J. Müller).
[128] H. Kunst: *Aufkohlen in Salzschmelzen. Wärmebehandlung der Bau- und Werkzeugstähle*. 3rd ed., BAZ Buchverlag, Basel 1978.
[129] I. G. Farben, DE 529 506, 1928 (C. Schumann, R. Fick); DE 529 575, 1928 (R. Fick).
[130] G. H. Buchanan, *Trans. Am. Electrochem. Soc.* **60** (1931) 93.
[131] A. M. Golub et al.: *Chemie der Pseudohalogenide*, Hüthig Verlag, Heidelberg 1979.
[132] G. H. Buchanan, *Trans. Electrochem. Soc.* **60** (1931) 93–112.
[133] H. H. Franck, W. Burg, *Z. Electrochem. Angew. Phys. Chem.* **40** (1934) 686–692.
[134] Shin-inchi Ohno, *Bull. Chem. Soc. Jpn.* **40** (1969) 1765–1769.
[135] H. Bauer, *Münchener Beiträge z. Abwasser, Fisch. u. Flußbiol.* **28** (1979) 181–192.
[136] *The Chemistry of the Ferrocyanides*, vol. VII, American Cyanamid Co., New York 1953.
[137] *Ullmann*, 4th ed., **18**, 463–467. B. A. Hutchins, L. E. West, *J SMPTE*, **66** (1957) 764–768.
[138] J. Jostan, H. Mietz, DE 2 534 213, 1975 (J. Jostan, H. Miez).
[139] *Ullmann*, 3rd ed., **5**, 659.
[140] American Cyanamid, US 3 695 833, 1972 (O. F. Wiedeman, K. W. Saunders, M. N. O'Connor).
[141] G. Grube: *Die elektrolytische Darstellung des Ferricyankaliums*, Enke Verlag, Stuttgart 1913.
[142] DEGUSSA, DE 2 016 848, 1972 (H. Reinhardt, K. Treibinger, G. Kallrath).
[143] Ch. E. MacKinnon, *Symp. Salt, 2nd*, **1** (1965) 356–364.
[144] R. Noyes: "Citric Acid Production Processes," *Chem. Process. (Chicago)* **37** (1969) 70–120.
[145] A. Dinsmoor Webb: "Chemistry of Winemaking," *Adv. Chem. Ser.* **137** (1974) 278–305.
[146] Du Pont, US 2 512 719, 1950 (R. O. Hull); US 2 585 902, 1952 (A. G. Gray).
[147] R. D. Crozier, *Mining Magazine*, February 1979, p. 174.
[148] Du Pont, DE-OS 1 951 260, 1970 (W. T. Hess).
[149] H. J. Barber, *J. Chem. Soc.* 1943, 79.
[150] Du Pont, DE-OS 2 012 444, 1970 (J. D. Rushmere).
[151] W. Biltz, *Z. Anorg. Chem.* **170** (1928) 161–183.
[152] K. Masaki, *Bull. Chem. Soc. Jap.* **6** (1931) 143–147.
[153] BASF, FR 1 539 598, 1968.
[154] L. Wöhler, J. F. Roth, *Chem. Ztg.* **50** (1926) 761–763 and 781–782.
[155] H. Karst, *Z. Gesamte Schiess Sprengstoffwes.* **77** (1922) 116–117.
[156] G. Paulèt, *Presse Méd.* **66** (1958) 1435.
[157] G. Denigés, *Ann. Chim. Phys.* **6** (1895) 381.
[158] W. Leithe: *Die Analyse der org. Verunreinigungen in Trink-, Brauch- und Abwässern*, Wissenschaftl. Verlags GmbH, Stuttgart 1972.
[159] E. Asmus, H. Garschagen, *Z. Anal. Chem.* **138** (1953) 414–422.
[160] *Annual Book ASTM Standards*, American Society for Testing and Materials, Philadelphia 1983, D 2036–82.
[161] *German Standard Methods for the Analysis of Water, Waste Water and Sludge*, DIN 38 405, part 13, DIN Deutsches Institut für Normung e.V., D-1000 Berlin 30.
[162] D. J. Barkley, J. C. Ingles, Report 221, Canmet, February 1970.
[163] F. Dietz, H. D. Frank, G. Teske, WLB – Wasser, Luft und Betrieb 1/2 – 82, (1982) 17–21.

[164]　W. S. Spector: *Handbook of Toxicology*, vol. **I**, Saunders, Philadelphia, London.
[165]　P. Doudoroff: *Toxicity to Fish of Cyanides and Related Compounds, A Review*. Environmental Protection Agency Report No. EPA–600/3–76–038 (1976).
[166]　P. Doudoroff: *A Critical Review of Recent Literature on the Toxicity of Cyanides to Fish*, American Petroleum Institute, Washington, D.C., 1980.
[167]　J. L. Huiatt, J. E. Kerrigan, F. A. Olson, G. L. Potter: *Cyanide from Mineral Processing*, Proceedings of a Workshop, Utah Mining and Mineral Resources Research Institute, College of Mines and Mineral Industries, 209 W.C. Browning Building, Salt Lake City, Utah, 1983.
[168]　USEPA: *Ambient Water Quality Criteria for Cyanide*, United States Environmental Protection Agency. EPA-440/5-80-00 (1980).
[169]　B. F. Dodge, D. C. Reams: "A Critical Review of the Literature Pertaining to the Disposal of Waste Cyanide Solutions, part 2," *Plating (East Orange, N.J.)* **36** (1949) no. 571–577, 664.
[170]　A. K. Reed et al.: *An Investigation of Techniques for Removal of Cyanide from Electroplating Waste*, U.S. Environ. Protection Agency, Water Pollution Control Series, Program No. 120 10 EIE 11/71 (1971).
[171]　J. C. Ingles, J. S. Scott: *Overview of Cyanide Treatment Methods*, Cyanide in Gold Mining Seminar, Ottawa, Ontario, 1981.
[172]　J. Conrad, M. Jola: "Cyanidentgiftung heute," *Oberfläche Surf.* **13** (1972) no. 7, 143–148.
[173]　L. E. Lancy, W. Nohse, D. Wystrach: "Practical and Economical Comparison of the Most Common Metal Finishing Waste Treatment Systems," *Plating (East. Orange, N.J.)* **59** (1972) 126–130.
[174]　G. C. White: *Handbook of Chlorination*, Van Nostrand Reinhold Co., New York 1972.
[175]　W. Stumm, H. Woker, H. W. Fischer, *Z. Hydrol.* **16** (1954) 1–21.
[176]　R. Weiner, *Plating (East Orange, N.J.)* **54** (1964) 1354.
[177]　B. F. Dodge, W. Zabban, *Plating (East Orange, N.J.)* **38** (1951) 561–586.
[178]　H. Knorre, *Galvanotechnik* **66** (1975) 374–383.
[179]　Lancy Laboratories Inc. Corp., DE 1 270 504, 1963 (W. Götzelmann).
[180]　R. Weiner, *Galvanotechnik* **60** (1969) 605–615.
[181]　V. Rumel, M. Topinka, *Metalloberfläche* **23** (1969) no. 8, 225–231.
[182]　H. W. Lieber, *Metall (Berlin)* **18** (1964) 611–618.
[183]　J. P. Zumbrunn, *Inf. Chim.* **124** (1973) 189–194.
[184]　DEGUSSA, DE 2 352 856, 1973 (J. Fischer, H. Knorre, G. Pohl); US 3 970 554, 1976 (J. Fischer, H. Knorre, G. Pohl).
[185]　DEGUSSA, DE 2 917 714, 1983 (K. Eiermann, R. Goedecke, H. Knorre, R. Möller).
[186]　DEGUSSA, DE 3 125 452, 1981 (H. Knorre, J. Fischer, K. Stutzel); US 4 416 786, 1983 (H. Knorre, J. Fischer, K. Stutzel).
[187]　H. Knorre, A. Griffiths, *Conference on Cyanide and the Environment*, University of Arizona, Tucson, Dec. 1984.
[188]　Du Pont, US 3 617 582, 1971 (B. C. Lawes, O. Bertwell).
[189]　B. Fischer, H. Rüffer, W. Düppers, G. Nagels, H. Knorre, *Z. Wasser Abwasser Forsch.* **14** (1981) no. no. 5/6, 210–217.
[190]　C. A. Walker, W. Zabban, *Plating (East Orange, N.J.)* **40** (1953) 777–780.
[191]　G. Wysocki, B. Höke, *Wasser Luft Betr.* **18** (1974) 311–314.
[192]　R. S. Murphy, J. B. Nesbitt, *Eng. Res. Bull. P. State Univ. Coll. Eng.* **B 88** (1964); *Chem. Abstr.* **61** (1964) 11 740 d.
[193]　K. Marquardt, *Metalloberfläche* **23** (1969) no. 8, 231–236.
[194]　W. Götzelmann, *Galvanotechnik* **64** (1973) 588–600.

[195] E. A. Devuyst, V. A. Ettel, G. J. Barbely, J. Roy Gordon Research Laboratory, Sheridan Park, Mississauga, Ontario, presented at the 1982 AIME Annual Meeting, Dallas, Texas.
[196] W. Götzelmann, G. Spanier, *Galvanotechnik* **54** (1963) 265.
[197] V. Migrdichian: *The Chemistry of Organic Cyanogen Compounds*. Reinhold Publ. Co., New York 1947, pp. 97–124.
[198] *Houben-Weyl*, **8**, 90–93.
[199] G. Brauer: *Handbuch der präparativen anorganischen Chemie*, 2nd ed., vol. **1**, Enke Verlag, Stuttgart 1960.
[200] B. S. Thyagarajan: *The Chemistry of the Cyanogen Halides*, Intra-Science Chemistry Reports, vol. **2**, no. 1, Santa Monica, California 1968, pp. 1–89.
[201] J. MacCordick, *Bull. Soc. Chim. Fr.* 1973, 50–70.
[202] *Gmelin*, System no. 14, D3, Kohlenstoff (1976) 149–266
[203] C. L. Berthollet, *Ann. Chim. Phys.* **1** (1789) 30–35.
[204] L. J. Gay-Lussac, *Ann. Chim. Phys.* **95** (1815) 200–213.
[205] F. S. Fawcett, R. D. Lipscomb, *J. Am. Chem. Soc.* **82** (1960) 1509–1510.
[206] H. Schachner, W. Sundermeyer, *J. Fluorine Chem.* **18** (1981) 259–268.
[207] R. Ch. Paul, R. L. Chauhan, R. Parkash, *J. Sci. Ind. Res.* **33** (1974) 31–38.
[208] F. S. Fawcett, R. S. Lipscomb, *J. Am. Chem. Soc.* **86** (1964) 2576–2579.
[209] Geigy, GB 942 545, 1983.
[210] A. A. Woolf, *J. Chem. Soc.* 1954, 252–265.
[211] E. Mulder, *Recl. Trav. Chim. Pays-Bas* **5** (1880) 84–98.
[212] D. M. Yost, W. E. Stone, *J. Am. Chem. Soc.* **55** (1933) 1889–1895.
[213] P. L. Bailey, E. Bishop, *Analyst (London)* **97** (1972) 691–695.
[214] D. Martin, H.-J. Niclas, *Chem. Ber.* **100** (1967) 187–195.
[215] J. V. Bodrikov, B. V. Danova, *Zh. Org. Khim* **8** (1972) no. 12, 2462–2467.
[216] J. Iwai, T. Iwashige, Y. Yuma, N. Nakamune et al., *Chem. Pharm. Bull.* **12** (1964) 1446–1451.
[217] E. H. Bartlett, C. Eaborn, D. R. M. Walton, *J. Organomet. Chem.* **46** (1972), no. 2, 267–269.
[218] Nippon Carbide Industries Co., JP-Kokai 80 133 352, 1980.
[219] Bayer, DE 1 958 095, 1971 (E. Enders, K. Wedemeyer, D. Lesch).
[220] Bayer, DE 2 507 671, 1976 (G. Rottloff, R. Sundermann, E. Grigat, R. Putter).
[221] Bayer, DE 2 507 705, 1976 (G. Rottloff, R. Sundermann, E. Grigat, R. Putter).
[222] Bayer, DE 1 183 903, 1964 (K. Goliasch, E. Grigat, R. Putter).
[223] K. Lempert, G. Doleschall, *Tetrahedron Lett.* 1963, 781.
[224] J. v. Braun, *Chem. Ber.* **33** (1900) 1438–1452.
[225] G. Fodor, Sh. Abidi, *Tetrahedron Lett.* 1971, no. 18, 1369–1372.
[226] R. Graf, *Chem. Ber.* **89** (1956) 1071–1079.
[227] Hoechst, DE 940 351, 1956 (R. Graf).
[228] DEGUSSA, DE 1 911 174, 1979 (W. Weigert, T. Lussling, F. Theissen).
[229] DEGUSSA, DE 1 768 154, 1977 (T. Lussling, F. Theissen, W. Weigert).
[230] DEGUSSA, DE-OS 3 006 492, 1981 (A. Kleemann, P. Schalke).
[231] Lonza, DE-OS 2 449 013, 1975 (A. Egger, E. Widmer, A. Faucci, R. Gregorin).
[232] DEGUSSA, DE-OS 3 006 424, 1981 (A. Kleemann, P. Schalke)
[233] E. Mayer, *Angew. Chem.* **81** (1969) 627.
[234] Standard Oil, US 3 105 023, 1963 (R. W. Foreman, F. Veatch).
[235] H. Schröder, *Z. Anorg. Allg. Chem.* **297** (1958) 296–299.
[236] Nilok Chemicals Inc., GB 967 458, 1960.
[237] DEGUSSA, DE 833 490, 1949 (H. Huemer).

[238] DEGUSSA, DE 842 067, 1950 (H. Huemer, H. Schulz).
[239] DEGUSSA, DE 827 358, 1949 (1952) (H. Huemer, H. Schulz, W. Pohl).
[240] DEGUSSA, DE-OS 2 521 580, 1976 (L. Devlies, R. Hendricx, K. Henkel, N. Kriebitzsch, M. Petzold).
[241] Ciba-Geigy Corp., US 4 100 263, 1978 (R. Miller).
[242] Bayer, DE-OS 3 117 054, 1982 (H. Königshofen, D. Bruck, A. Nierth, M. Zlokarnik, H. J. Uhlmann).
[243] Agripat SS.A, DE-OS 1 801 311, 1969 (W. S. Durrell, R. J. Eckert, Jr.).
[244] SKW Trostberg, DE-OS 2 838 016, 1980 (G. Buchreiter, P. Kniep, K. Scheinost, H. R. Vollbrecht).
[245] DEGUSSA, DE-OS 2 154 721, 1973 (F. Geiger, W. Weigert).
[246] Bayer, DE-OS 2 157 973, 1973 (Z. Kricsfalussy, K. Blöcker, J. Pawlowski, B. Scherhag, R. Weiler).
[247] DEGUSSA, DE-AS 2 521 581, 1977 (W. Heimberger, G. Schreyer).
[248] DEGUSSA, DE-AS 2 521 582, 1979 (W. Heimberger, G. Schreyer).
[249] Röhm, DE-AS 2 442 161, 1975 (W. Gruber, F. Schnierle, G. Schröder).
[250] Asahi Chemical Ind., JP 5 020-224, 1980.
[251] SKW-Trostberg, DE-OS 3 103 963, 1982 (U. Kriele, H. Michand, A. Michlbauer).
[252] Du Pont, US 3 023 077, 1959 (S. V. R. Mastrangelo).
[253] DEGUSSA, DE 2 931 353, 1982 (F. Bittner, W. Heimberger, K. Henkel, N. Kriebitzsch, M. Petzold).
[254] Agripat, FR 1 555 981, 1969 (W. J. Evers).
[255] Ciba-Geigy Corp., US 4 255 167, 1981 (R. M. Babb, M. J. Guillory).
[256] DEGUSSA, DE-OS 2 363 867, 1975 (F. Geiger, W. Heimberger, H. Schmitt, G. Schreyer).
[257] DEGUSSA, DE-OS 1 809 607, 1970 (F. Geiger, W. Weigert, T. Lussling).
[258] DEGUSSA, DE-OS 2 154 721, 1973 (F. Geiger, W. Weigert).
[259] T. S. Price, S. J. Green, *Chem. News J. Phys. Sci.* **120** (1920) 101–102.
[260] American Cyanamid, US 3 226 182, 1965 (J. F. Martino).
[261] W. W. Hartmann, E. E. Dreger, *Org. Synth. Coll.* **2** (1943) 150–151.
[262] E. Schulek, V. Gervay, *Z. Anal. Chem.* **92** (1933) 406–417.
[263] B. Bak, H. Hillebert, *Org. Synth.* **32** (1952) 29–31.
[264] C. A. Goy, D. H. Shaw, H. O. Pritchard, *J. Phys. Chem.* **69** (1965) 1504–1507.
[265] M. S. Kharasch, A. R. Stiles, E. V. Jensen, A. W. Dewis, *Ind. Eng. Chem.* **41** (1949) 2840–2842.
[266] A. B. Van Cleave, R. L. Eager, *Can. J. Res. Sect. F* **25** (1947) 284–290.
[267] A. B. Van Cleave, H. E. Mitton, *Can. J. Res. Sect. B* **25** (1947) 430–439.
[268] A. B. Van Cleave, V. C. Haskell, J. H. Hudson, A. M. Kristjanson, *Can. J. Res. Sect. B* **27** (1949) 266–279.
[269] M. J. Grossman, *Chem. Eng. News* **58** (1980) no. 35, 43.
[270] American Cyanamid, US 1 727 093, 1929 (G. Barsky).
[271] W. Mandelung, E. Kern, *Justus Liebigs Ann. Chem.* **427** (1922) 1–34.
[272] Nilok Chemicals, US 3 429 658, 1969 (G. V. Vosseller).
[273] B. Witkop, *Adv. Protein. Chem.* **16** (1961) 221–321.
[274] *Houben-Weyl*, **8,** 256–265.
[275] T. K. Brotherton, J. W. Lynn, *Chem. Rev.* **59** (1959) 841–883.
[276] S. Nakamura, *Yuki Gosei Kagaku Kyokai shi* **26** (1968) 23–26.

[277] R. H. Fahnenstich, W. Th. Heimberger, F. M. Theissen, W. M. Weigert, *Chem. Ztg.* **96** (1972) 388–396.
[278] G. Brauer: *Handbuch der präparativen anorganischen Chemie*, 3rd ed., vol. **2**, Enke Verlag, Stuttgart 1978.
[279] H. W. Roesky, H. Hofmann, *Chem. Ztg.* **108** (1984) 231–238.
[280] H. J. Rodewald, *Chemiker-Ztg.* **89** (1965) 522–525.
[281] R. D. Dresdner, J. Merrit, J. P. Royal, *Inorg. Chem.* **4** (1965) 1228–1230.
[282] J. B. Conway, R. H. Wilson, A. V. Grosse, *J. Am. Chem. Soc.* **75** (1953) 499.
[283] J. Janin, A. Bouvier, *Spectrochim. Acta* **20** (1964) 1787–1798.
[284] N. C. Robertson, R. N. Pease, *J. Am. Chem. Soc.* **64** (1942) 1880–1886.
[285] Asahi Chemicals Ind., JP-Kokai 80 20 224, 1980.
[286] Asahi Chemicals Ind., JP-Kokai 80 422 242, 1980.
[287] N. B. H. Henis, Y.-H. So, L. L. Miller, *J. Am. Chem. Soc.* **103** (1981), 4632–4633.
[288] N. B. H. Henis, L. L. Miller, *J. Am. Chem. Soc.* **104** (1982) 2526–2529.
[289] Y.-H. So, L. L. Miller, *J. Am. Chem. Soc.* **103** (1981) 4202–4209.
[290] H. Bock, J. Wittmann, H. J. Arpe, *Chem. Ber.* **115** (1982) 2326–2337.
[291] Du Pont, US 3 564 039, 1971 (O. W. Webster).
[292] G. J. Janz, *1,4 [One,Four] Cycloaddition React.* 1967, 97.
[293] G. J. Janz, *Inorg. Synthesis* **5** (1957) 43–48.
[294] Hoechst, DE-OS 2 460 779, 1976 (W. Riemenschneider, P. Wegener).
[295] Asahi Chem. Ind., JP-Kokai 74 13 3326, 1970.
[296] DEGUSSA, FR 2 003 105, 1969.
[297] Asahi Chem. Ind., JP 74 13 110, 1974.
[298] Asahi Kasei Kogyo, DE 2 456 035, 1974 (S. Kamada, T. Yamashita, T. Ide, R. Ajiki, T. Kitamura).
[299] Asahi Chem. Ind., JP 76 29 429, 1976.
[300] W. Riemenschneider, *Chem. Ing. Tech.* **50** (1978) 55.
[301] V. D. Parkhomenko, V. K. Steba, A. A. Pivovarov, B. J. Mel'nikov, *Khim. Tekhnol.* **4** (1981) 3–4; *Chem. Abstr.* **95** (1981) 149 178 h.
[302] Hoechst, DE 2 427 269, 1982 (W. Riemenschneider, P. Wegener).
[303] L. M. Weinstock, P. Davis, B. Handelsman, R. Tull, *J. Org. Chem.* **32** (1967) 2823–2829.
[304] H. W. Roesky, K. Keller, J. W. Bats, *Angew. Chem.* **95** (1983) 904.
[305] G. Jaquemin, *Ann. Chim. Phys.* **6** (1885) 140.
[306] E. Pfeil, DE 1 163 302, 1965 (E. Pfeil, K. H. Schäfer, R. Nebeling).
[307] DEGUSSA, DE-AS 2 461 204, 1977 (W. Heimberger, G. Schreyer, H. Schmitt).
[308] Sagami Chemical Research Center, Tokio, DE-AS 1 297 589, 1969 (S. Nakamura).
[309] Asahi Kasei Kogyo K. K., DE-AS 2 355 040, 1977 (T. Yamashita, T. Ide, S. Kamada, T. Kitamura).
[310] Hoechst, DE-AS 2 427 268, 1979 (W. Riemenschneider, P. Wegener).
[311] Union Oil Comp. of California, US 3 630 671, 1971 (M. J. Block).
[312] Union Oil Comp. of California, US 3 769 388, 1973 (K. L. Olivier, M. J. Block).
[313] DEGUSSA, DE-OS 2 012 509, 1971 (W. Heimberger, T. Lussling, W. Weigert).
[314] DEGUSSA, DE-OS 2 022 454, 1971 (J. Heilos, W. Heimberger, T. Lussling, W. Weigert).
[315] DEGUSSA, DE-OS 2 022 455, 1971 (J. Heilos, W. Heimberger, T. Lussling, W. Weigert).
[316] DEGUSSA, DE-AS 2 118 819, 1974 (J. Heilos, W. Heimberger, T. Lussling, W. Weigert).
[317] Pure Oil Comp., US 2 884 308, 1956 (W. L. Fierce, W. J. Sandner).
[318] Röhm, DE-OS 2 341 370, 1973 (W. Gruber, G. Schröder).

[319] Mitsubishi, DE-OS 2 532 307, 1976 (T. Onoda, T. Yokohama, K. Shinpei).
[320] DEGUSSA, DE-OS 1 927 847, 1970 (F. Geiger, T. Lussling, W. Weigert).
[321] American Cyanamid, US 2 730 430, 1956 (J. K. Dixon, J. E. Longfield).
[322] DEGUSSA, DE 1 283 815, 1967 (R. Fahnenstich, H. Schulz).
[323] Bayer, DE-OS 3 229 415, 1984 (R. Fauß, K. H. Linker, K. Findeisen).
[324] W. Tsang, S. H. Bauer, M. Cowperthwaite, *J. Chem. Phys.* **36** (1962) 1768–1775.
[325] Asahi Chemical Industries, JP-Kokai 75 93 893, 1975.
[326] J. B. Conway, A. V. Grosse, *J. Am. Chem. Soc.* **80** (1958) 2972–2976.
[327] R. P. Welcher, D. J. Berets, L. E. Sentz, *Ind. Eng. Chem.* **49** (1957) 1755–1758.
[328] R. Dailey and A. Weissler: *Monograph on Ferrocyanide Salts,* Food and Drug Administration Report No. FDA/BF–79/28 Washington, D.C., 1978.
[329] Harsten, A., Blackhill, *Engineer (London)* **22** (1934) 145–174.
[330] Anderson, B. G., *Sewage Works J.* **16** (1944) 1156.
[331] Bringmann, G., Kühn, R., *Z. Wasser Abwasser Forsch.* **10** (1977) 87–98.
[332] C. Henderson, Q. H. Pickering, A. E. Lencke, *Eng. Bull. Purdue Univ. Ext. Ser.* 1960, no. 106, 120–130.
[333] W. P. Bridges: *Sodium Cyanide as a Fish Poison,* Special Scientific Report Fisheries No. 253, US Dept. Interior (1958).
[334] J. R. Hawley: *The Use, Characteristics and Toxicity of Mine-Mill Reagents in the Province of Ontario,* Ontario Ministry of Environment, 135 St. Clair Avenue West, Toronto, 1972.
[335] W. Bucksteeg, *Schweiz. Z. Hydrol.* **22** (1960) 407.
[336] B. G. Anderson: "The Toxicity Thresholds of Various Sodium Salt Determined by the Use of Daphnia magna," *Sewage Works J.* **18** (1946) 82.
[337] G. Bringmann, R. Kuhn: "The Toxic Effect of Waste Water on Aquatic Bacteria, Algae, and Small Crustaceans," *Gesund. Ing.* **80** (1959) 115–120.
[338] *Patty's Industrial Hygiene and Toxicology,* 3rd ed., J. Wiley & Sons, New York 1978, p. 4845.
[339] V. Schultz, Deutsches Ärzteblatt 1978.
[340] TLVs, *Threshold Limit Values for chemical substances in the work environment,* (Adopted by ACGiH for 1984–85) American Conference of Governmental Industrial Hygienists.
[341] MAK. Mitteilung XX der Senatskommission zur Prüfung gesundheitsschädlicher Arbeitsstoffe, 1984.
[342] Theml, Weger: "Behandlung der Blausäurevergiftung mit 4-DMAP – Treatment of Hydrocyanic Acid Poisoning with 4-DMAP," *Medizinische Klinik* 1969, 174.